THE McGRAW-HILL RECYCLING HANDBOOK

THE McGRAW-HILL RECYCLING HANDBOOK

Herbert F. Lund, Editor in Chief

Second Edition

McGRAW-HILL

New York San Francisco Washington, D.C. Auckland Bogotá Caracas Lisbon London Madrid Mexico City Milan Montreal New Delhi San Juan Singapore Sydney Tokyo Toronto

Library of Congress Cataloging-in-Publication Data

The McGraw-Hill recycling handbook / Herbert F. Lund, editor in chief. — 2nd ed.
p. cm.
Includes index
ISBN 0-07-039156-4
1. Recycling (Waste, etc.) 2. Recycling (Waste, etc.)—United States. I. Lund, Herbert F.
TD794.5 .M397 2000
363.72'82—dc 21 00-028382

McGraw-Hill

A Division of The ***McGraw·Hill*** *Companies*

1 2 3 4 5 7 8 9 0 DOC/DOC 0 6 5 4 3 2 1 0

ISBN 0-07-039156-4

The sponsoring editor for this book was Scott Grillo and the production supervisor was Pamela A. Pelton. It was set in Times Roman by Ampersand Graphics, Ltd.

Printed and bound by R. R. Donnelley and Sons, Co.

This book was printed on recycled, acid-free paper containing a minimum of 50% recycled de-inked fiber.

CONTENTS

SECTION II RECYCLING MATERIALS

SECTION III RECYCLING FACILITIES AND EQUIPMENT

SECTION IV RECYCLING OPERATIONS

SECTION V CASE HISTORIES

SECTION VI RECYCLING IN OTHER COUNTRIES

FOREWORD

Recycling is happening almost everywhere you turn. It is roaring through the halls of government and business like a freight train and there is no stopping it.

Cities and states across the country are passing recycling laws as fast as they can. As this is written, virtually every state in the country has some kind of recycling requirement on its books. Homeowners and businesses alike are now incorporating recycling into their day-to-day activities at an increasing rate.

Recycling is destined to become a permanent part of how we manage waste in this country, but it will take some patience on our part to get there. We will not turn around decades of past practices overnight; nor will we as a society embrace without qualification a practice that, in the end, does not deliver what we wanted.

Laws are now on the books, programs are being put in place, industry is making major investments in plants and equipment to process and reuse the materials collected for recycling. Markets are being developed and habits are slowly being changed. This is all for the good—but it will take some time and the expenditure of precious resources. But the basic system is coming together, and if we give it time, recycling will assume a permanent and major role in our lives.

Of course, this all presumes that we do it well, that the programs we put in place do what they are supposed to do. The recycling landscape has evolved so rapidly in the past few years, that keeping up with the changes is a major challenge. That is where this valuable new book comes in. In *The McGraw-Hill Recycling Handbook*, Herbert Lund has put together a collection of information from experts that covers all aspects of recycling in this country. While primarily aimed at all professionals responsible for planning and operating recycling programs, this book can be of enormous help to local government officials and to the business community as well.

From yard waste to TV sets, from setting realistic recycling goals to setting up reporting systems, this volume provides a comprehensive look at the state of recyling today and where it is likely to be tomorrow.

As a nation, we are moving to make recycling a permanent part of our lives. This Handbook can help make that transition as efficient as possible.

William D. Ruckelshaus
Former Chairman and CEO,
Browning-Ferris Industries, Inc.
and Former EPA Administrator

FOREWORD

PREFACE TO THE SECOND EDITION

The original *McGraw-Hill Recycling Handbook* was published in 1993. Both Bob Esposito, Executive Editor in the Professional Books Division, and I felt there have been numerous improvements and sufficient changes in the recycling field to warrant a new Second Edition. Not only have there been improvements but the whole field has grown up from immature infancy to a booming, sophisticated new industry.

In this Second Edition, we not only keep you up-to-date with current improvements and changes, but we look ahead for outstanding new developments. For instance, in the field of separating colored glass, we found a European–American new automatic technology that uses worthless mixed cullet and achieves a 70% separation. Also looking to the future, we present how to best apply computers to recycling operations. Also, although municipal solid wastes (MSW) combustible ash may or may not be economically recycled at present, John Booth, Director of Engineering, Solid Waste Authority of Palm Beach County, Florida reviews the economics and looks into the future.

Some of the new important handbook additions are recycling electronic devices, mercury-containing devices and lamps, textiles, and carpeting in the expanded materials section. In the recycling operations section, besides the role of computers, we offer an intense new chapter on recycling at large commercial facilities, with particular emphasis on theme parks and sports arenas. Our rules and regulations coverage has been updated, stressing recycling issues and followed by state-by-state information. Under collection operations, we have included the latest on weight-based systems.

There has been increasing interest in the Spanish Edition of the handbook. For this reason and due to more recycling developments abroad, we have added a significant new section on Recycling in Other Countries. EPA, The Office of Solid Waste, has updated Appendix A, "Recycling Information and Sources" for back-of-book references and contacts.

And as with the original 1993 handbook, my wife Belle and I have updated the Glossary and the comprehensive cross-referenced index to make sure this Second Edition continues to be your indispensible desktop reference to recycling.

Herb Lund

SECTION I

THE BASICS OF RECYCLING

CHAPTER 1
RECYCLING OVERVIEW AND GROWTH

JOSEPH A. RUIZ, JR.,
Former Vice President, Attwoods, Inc.
Coconut Grove, Florida

RECYCLING DEFINED

The Recycling Perplexity

Recycling? This is a seductive word to the environmentally aware among us. But what is it? What does it mean? What is it all about? Who does it? Why should I? These questions seem almost endless. Some have simple answers. Many answers are a part of a much more complex issue. Many more are yet unresolved and others are only now evolving. Just addressing the question of definition becomes an evermore complex issue as almost every governmental entity, industrial and commercial trade organizations, professional associations, academics, and practitioners attempt to define what it is. Because each has a different perspective and goal, each has a slightly different definition. Even a dispassionate search for a bias-free definition is difficult perhaps because of the circular nature or the subject.

When does the cycle begin or end? Does it ever? Does paper recycled for use as a raw material in making boxboard constitute a virgin material? What about the boxboard trimmings that were made from recycled materials? Are they waste again? If so, were they ever recycled? Perhaps a better understanding of how and why we got here can make any current definition more meaningful.

What It Is Today

Recycling today is, and must be understood as, a solid waste management strategy. A method of solid waste management equally useful as landfilling or incineration and environmentally more desirable. Today it is clearly the environmentally preferred method of solid waste management.

The Beginning

Early humans did not have a solid waste management strategy per se simply because the hunter-gatherer existence did not require one. Never staying in one place long enough to accumulate any significant amount of solid waste, as well as a need to utilize scarce resources to their highest degree, probably did not create any concern or action. However, as humans began to settle in permanent communities with higher concentrations of waste-producing individuals and activities, the need for waste management became evident. Although this occurred around 10,000 B.C. in some places, it occurred much later in others and remains much less a concern in the less populated and more rural areas of the planet even today.

By 500 B.C. Athens organized the first municipal dump in the western world, and scavengers were required to dispose of waste at least 1 mile from city walls. This imperative continued from place to place, going forward and backward relative to the desires and ability of governments. During the middle ages waste disposal continued to be an individual responsibility commensurate with the lack of enlightened authority by government.

In 1388 the English Parliament banned waste disposal in public waterways and ditches. A few short years later in 1400 garbage was piled so high outside the Paris gates that it interfered with the defense of the city. These examples are cited because they indicated a desire on the part of government to assume responsibility for this element of the health and safety of the community primarily when other responsibilities such as drainage and defense were involved. This growth in governmental concern for health and safety with regard to waste disposal lead to additional regulations and operations. By the 1840s the western world began to enter the "Age of Sanitation" as filthy conditions began to be seen as a nuisance that the public demanded government to resolve. Sanitarians employed by government primarily to deal with sewage disposal increasingly turned their attention to solid wastes.

Government's increasing assumption of solid waste management soon led to systematic approaches including the "destructor," an incineration system in Nottingham, England, in 1874. America's first municipal incinerator on Governor's Island in New York was built in 1885.

Government response continued to include a wide variety of innovative programs designed to address both specific elements of the solid waste stream as well as the broad brush approach of dumps and incinerators. Municipalities cleaned streets and sanitary engineers invented new technologies to reduce costs and volume. Fats and oils were recovered for reuse in manufacturing soap and candles. Incinerators generated steam for power and heat. Rags were increasingly recycled for use in making paper, and the inherent value of metals was always enhanced during war times to a sufficient level to promote public recycling programs. But environmental concerns were generally limited beyond the next hill, out at sea and out of sight. Ocean dumping and open space outside of the urban areas continued to be both environmentally acceptable and economical.

Recycling in Modern Times—Awakening to Solid Waste Responsibilities

Only after World War II did fast-growing populations, greatly enhanced scientific understanding of the environment, and later the concept of finite resources combine to truly afford an opportunity for a conscious examination of the detrimental nature of land or ocean disposal practices. A rapid expansion in understanding the long-term impacts of groundwater and air pollution began to demand even greater regulation of disposal practices. In many areas of the nation both open burning of solid waste at dumps and ocean disposal remained an acceptable practice well into the 1970s.

The inability of local governments to deal with these larger problems quickly led to a federal interest and assumption of responsibilities. The first federal solid waste manage-

FIGURE 1.1 Horse-drawn garbage collection carts at the turn of the century.

FIGURE 1.2 Horsepower replaces horses in this 1915 version of high-technology garbage pickup truck.

ment law was the Solid Waste Disposal Act (SWDA) of 1965, which authorized research and provided state grants. Three years later in 1968 President Johnson commissioned the National Survey of Community Solid Waste Practices. It provided the first comprehensive data on solid waste on a national basis. Two years later the Solid Waste Disposal Act was amended by the Resource Recovery Act, and the federal government was required to issue waste disposal guidelines.

The year 1970 also saw passage of the Clean Air Act, which established federal authority to combat smog and air pollution leading to the shutdown of many solid waste incinerators and the elimination of open burning of solid waste. Significantly the first Earth Day was celebrated that same year on April 22, 1970, indicating a worldwide heightened environmental awareness including that of the solid waste disposal dilemma. Within a year Oregon became the first state to pass a bottle bill, thereby creating a procedure for government regulation covering the reuse and recycling of designated portions of the waste stream during peacetime without the imperative of wartime economics. Although all 50 states had some kind of solid waste regulation by the mid-1970s, it was the Resource Conservation and Recovery Act of 1976 (RCRA) that created the first truly significant role for the federal government in solid waste management. The act emphasized conservation of resources, particularly energy conservation, and recycling as preferred solid waste management alternatives. It also provided for the national hazardous waste management program, recognizing the detrimental effect of hazardous waste on solid waste management alternatives as well as the environment in general.

The stimulus of the Arab oil embargo, the Public Utilities Regulatory Policies Act of 1978, which guarantees markets for small energy producers, and RCRA combined to encourage an explosive growth for waste-to-energy plants and to some extent the recovery of methane for fuel from landfills. The banning of open dumping of solid waste by the EPA in 1979 increased the attractiveness of waste-to-energy plants because of their volume reduction capabilities. In addition, waste-to-energy plants are generally perceived to be a form of recycling solid waste, as its use as a fuel to create energy does return a significant part of the waste stream to a useful product. This view was, and still is, enhanced by the continuing demand for electrical energy derived from nonfossil fuel sources. However, others, including the State of Florida, do not consider the burning of and recovery of energy from solid waste as recycling. Even today the debate continues with positions taken on both sides as to whether or not waste to energy is legitimately considered recycling.

Although previously preferred to landfills, waste-to-energy plants have now become almost as unpopular with communities unwilling to exchange potential groundwater pollution for potential air pollution. The disposal crisis created by the ever-diminishing lack of acceptable disposal capacity is exemplified by EPA's estimate that over 10,000 landfills (70 percent of the total) closed between 1978 and 1988.

All of those considerations have led to both a public and a legislated demand for recycling as the preferred solid waste management strategy today and in the future. The willingness of government to require and subsidize recycling when necessary has grown to enormous proportions. Significantly the dominant theme of Earth Day 1990 was recycling.

Defining Recycling

Recycling remains, however, one of those elusive concepts about which everyone thinks they have a clear understanding until they begin to practice it. Although most people understand the relatively simple tasks required by individuals in order for them to participate, the subtleties necessary for the interplay of both the public and private sectors needed to return those materials to industry as raw materials and the methods employed to do so require definitions other than common language and as a matter of law. In addition, the concept gives rise to other terms required to fully implement the concept. The terms recyclable materials, recovered materials, and recycled materials all are needed to define the concept of recycling and usually require definition in various state regulations. Therefore, only a dictionary definition of recycling can convey a general concept of a term that has been, and will continue to be, defined through committee discussions, contractual negotiations, and legislation designed to meet specific needs.

Public Perceptions

Although rapidly changing in response to local public awareness campaigns, the general public's perception of what recycling is remains largely limited to those visible elements including curbside programs, recycling centers, and so on, and a vague understanding that this is good for the environment because these materials do not go to a landfill or incinerator. This view also usually incorporates a demand for recycling a greater variety of materials than is practical or economically feasible at this time or a misunderstanding about what can or cannot be recycled.

Legislative definitions at this time in the evolution of legislation promoting and requiring recycling generally center on those materials in the waste stream that are selectively easy to separate and for which known and relatively stable markets exist. In addition, these definitions ignore previously established industrial recycling efforts based on purely economic needs of avoided cost of disposal and intrinsic value of industrial raw material-derived waste. This kind of legislative definition is directed at promoting additional recycling activities rather than accounting for existing economic considerations.

Current legislation focuses on providing for the promotion of recycling those materials that have not been recycled because the economic reasons to do so do not exist or at least are not readily apparent to the private sector of the economy.

FIGURE 1.3

THE WHY'S OF RECYCLING

Recycling occurs for three basic reasons: altruistic reasons, economic imperatives, and legal considerations. In the first instance, protecting the environment and conserving resources have become self-evident as being in everyone's general interest. Second, the avoided cost of environmentally acceptable disposal of waste has risen to a level where when combined with the other costs associated with recycling, it now makes economic sense to recycle many materials. Finally, in responding to both public demand and a growing lack of alternative waste disposal methods, government is requiring recycling and providing for a wide variety of economic and civil penalties and incentives in order to encourage recycling.

GROWTH

The support for recycling on both a state and federal basis continues to be explosive and generally responsive to widespread public support and demand. This is a demand that has in many instances outstripped both the public and private sectors of the economy's ability to meet the requirements and/or intent of legislation. In the rush to require recycling, the market for those materials has often been ignored or misunderstood. The entry of the public sector into a traditional and well-established private sector activity has created severe stresses and difficulties in the commodities marketplace for recycled materials. The commodities market is commerce and industry's source of raw materials. It is a traditionally volatile element of the economy that is very sensitive to the relationship between supply and demand for materials. The sudden growth in legislation promoting recycling created an external stimulus that increased the supply of newspaper in the northeastern United States to the point where a glut occurred. In 1989 the price fell not only to zero but was further depressed when communities that were prevented from landfilling recovered newspaper began paying to have it taken away. The phenomenon of negative prices for this commodity severely impacted public programs that were dependent in the past on revenues from this material and those portions of the private sector that were engaged in the recycling of these materials.

However, these kinds of lessons have been helpful and will contribute to better planned and thereby more effective programs in the future. Governments at all levels appear to be directing more and more of their legislation to ensuring markets by creating demand for recycled products through preferential procurement practices. In addition the concept of tax incentives to encourage both recycling and the use of products containing recycled materials continues to gain favor.

PROGRAM OPTIONS

Whether anyone wants to recycle for altruistic reasons or because "the law makes us do it," a wide variety of options for recycling are available. Although each option is discussed in detail in other chapters, a central issue that must be considered is that no single option yet available provides all the answers. It is most likely that there is no single option that is best for everyone. There is, however, a best option or combination of options for everyone when a careful evaluation is made to determine what is available to meet specific needs and circumstances.

Recycling debate has evolved into several broad configurations for which examples can be rapidly found in operation today. Innovation, creativity, and practicality provide many variations.

Both residential and commercial establishments can participate in recycling by separating materials before they are mixed with wastes. In these programs recyclable materials are kept separate in a variety of containers whether in the home or in the workplace. At appropriate intervals, they are placed for collection or transported to centralized collection and/or processing facilities.

Curbside Collection

Single-family residential units are often served by curbside programs. These programs may require residents to use one or more containers to separate and store recyclable materials that are diverted from the normal waste stream. The type and number of containers can vary dependending on the variety of materials collected and the degree of separation desired. The design, capacity, and construction of the containers can also vary. Some programs provide containers and others do not. Containers may be rigid, specialized plastics, paper or plastic bags; materials can even be bundled or contained at the participant's discretion.

Commercial Collection

Similar programs are also used for multidwelling residential units and commercial applications. However, of necessity these programs do not include curbside collection. These programs require recyclable materials to be placed in specialized containers of the type traditionally used in those applications. Therefore, if a multidwelling unit residence is normally served as a single-family unit for waste, then it can be served as a single family for recycling service. Similarly, if it is served as a commercial establishment for waste collection, then it will probably require recycling service in a similar manner.

Commingled or Source-Separated?

In both residential and commercial applications, the degree of separation may vary significantly. A great deal of commingling can be allowed or required in all instances if a centralized processing facility where commingled materials can be separated after collection is used. Even if a centralized processing facility is not available, single-family residential units can always be allowed to commingle materials by using a truckside sort collection method (Fig. 1.4). This method requires the collector to manually separate the material in the containers and keep them separate in the collection vehicles until delivered to markets or intermediate processing facilities.

Material Recovery Facilities

The use of material recovery facilities (MRFs) serving commingled residential programs is rapidly gaining popularity (Fig. 1.5). Commingled programs used in a multidwelling unit or commercial application can allow efficient collection methods where space available for placement of collection containers is limited.

FIGURE 1.4 Specialized curbside recycling materials: collection truck with compartments, dual sides, and hydraulic top loading.

FIGURE 1.5 Materials recovery facility (MFR) infeed conveyors for commingled materials at Community Recycling, an Attwoods Company in Dade County, Florida, that serves over 265,000 homes.

Drop-off/Buy-Back Centers

Voluntary participation in recycling programs is often related to the ease with which an individual can participate. Therefore drop-off centers that depend on altruistic motivations add a degree of inconvenience that can reduce participation. These centralized locations where recyclable materials are collected are easier and less expensive to implement than curbside programs. They are especially effective in areas where regular waste collection is not required or available.

Buy-back centers offer all of the benefits of drop-off centers and the increased incentives of monetary benefits to participants. They are, however, more expensive to operate because they must be staffed, secured, and handle cash.

Waste may also be segregated and collected by broader categories such as wet and dry, putrescible and nonputrescible, household waste or yard waste, etc., prior to recycling. It can still be recycled to greater or lesser extent even if it is mixed.

Recycling at Waste-to-Energy Facilities

Waste-to-energy facilities increasingly employ separation systems to recover nonorganic recyclable materials. The degree and type of material recovery depends largely on whether the materials are recovered prior to or after incineration. Recovery of materials after incineration, known as back-end systems, are frequently used with mass burn waste-to-energy facilities. They can recover high percentages of ferrous and nonferrous metals through the use of simple technology. Although the quality and quantity of the recovered materials may be diminished, back-end systems can offer a relatively inexpensive retrofit for existing facilities.

Front-end separation systems that remove recyclable materials prior to incineration are used in facilities that prepare a refuse-derived fuel by removing the inorganic fraction of the waste prior to incineration. These materials are then recycled or, when no markets exist, landfilled with other nonprocessible materials. Although many front-end designs and technologies are available, they are all more complex and therefore more expensive to construct and operate than back-end systems. However, they offer a greater opportunity to recover a wider variety and higher quality of recovered materials.

Composting

The methodologies and technologies used in separating materials from mixed waste can also be used in the preparation of compost. Long championed as the solution to the solid waste dilemma, the composting of mixed waste has met with limited success. Although easy to do on a small scale, its success has been hindered by a lack of markets and other applications in large quantities. Compost from mixed waste has recently fallen into disfavor with some environmental groups on the basis that it discourages other kinds of recycling. Its use is even being prohibited by several states because of the potential negative consequences of contaminants such as heavy metals that may be present in compost made from mixed waste.

Higher degrees of separation of metals and other contaminants will be necessary for successful composting of mixed waste in the future. Programs for composting clean yard waste and other homogeneous materials offer greater promise. Although still limited by a lack of markets, they are enjoying a much higher degree of success.

SUMMARY

All collection and processing methods are technologies that have their merits and limitations. There is no single answer or solution. Recycling, in whatever manner, is and must be part of integrated solid waste management strategies. When compared to the environmental risks associated with landfilling or incineration, recycling is the preferred solid waste management strategy.

CHAPTER 2
RECYCLING POLICIES AND EVALUATIONS

MICHELE RAYMOND
Raymond Communications, Inc.
College Park, Maryland

Since the famous Long Island "garbage barge" incident in 1987, all 50 U.S. states have enacted some sort of recycling legislation or policy. Even though there is no "landfill shortage" today, the laws designed to divert solid waste flourished between 1988 and 1994. State legislatures not only passed "comprehensive" recycling laws with ambitious goals, but they also tried to create new markets for collected materials. Thus, there were more than 500 recycling bills on the books as of 1999. While the number of bills passed each year declined in 1998–1999, the volume of recycling bills introduced in 1999 jumped by one third.

There is evidence that the recycling laws have made a difference in landfill diversion. The 1970 EPA figures from Franklin Associates (now a unit of McLaren-Hart) indicate that the United States had a 6% recycling rate for municipal solid waste (MSW). In 1997, the rate reported was 28%, up from 27% in 1996. In 1999, average figures compiled by *Biocycle Magazine* reported that recycling grew in 1998, with the recycling rate at about 31.5%.*

Waste generation continues to slowly rise. EPA reported that waste increased from 4.32 lbs per capita in 1996 to 4.44 lbs per capita in 1997. *Biocycle Magazine* reported that 27 states had an increase in waste between 1997 and 1998; Texas's trash grew 12 million tons and California's grew 11 million tons. Whereas EPA estimates municipal solid waste at about 210 million tons, *Biocycle* estimates waste from states to be 375 million tons; this is because EPA excludes construction and demolition debris, sludge, and industrial wastes, and many states include some of those in their waste generation figures.

There were 2400 landfills left as of 1996; however, the new landfills are much bigger. According to EPA figures, as of 1997, 35 states reported having 20 years of landfill capacity left; three had less than five. Landfills managed 55% of U.S. MSW; incineration managed 17% in 1996, according to EPA. The 1999 *Biocycle* figures indicated that only 7.5% of MSW was incinerated.

*The Franklin study is based on material flows analysis estimates and it excludes construction and demolition debris, industrial process waste, sludges, and other wastes that may find their way into MSW landfills. Biocycle's rate was arrived at by averaging state figures, which are based on inconsistent methodologies.

OUTLOOK

Despite 500 laws designed to push consumers to recycle and manufacturers to use more recycled material, local governments remained unhappy with recycling in 1999. The markets for paper and plastics have gyrated up and down, and most governments do not fund all recycling as a separate tax item; they depend on the materials income to support part of the programs. Thus, collecting glass is marginal (though few will drop it from programs) and plastics is expensive because of its high volume and difficulty to sort.

Industry, however, has maintained that it will only use recycled material when it is "economic" to do so. Critics complain that landfills are cheap and plentiful. Landfill experts, however, say that even with current liner requirements, today's landfills will leak.

With low landfill tip fees and cheap energy and virgin material prices, there is a need to "drive" more recycling beginning in 2000. The realization that some goals will not be met has spurred a number of states to plan more legislation in 2000.

DIFFERENT STATE APPROACHES

Each state has approached recycling policy in its own way. A few states take a mandatory approach, requiring local governments to achieve a certain recycling or landfill diversion rate by a certain date. The most popular rate was 50% by 2000. A total of 11 states (California, Connecticut, Iowa, Maine, Maryland, Massachusetts, North Carolina, New Jersey, Pennsylvania, Rhode Island, and Wisconsin) had mandatory recycling laws as of 1999. In the early years, a number of Eastern states mandated source separation by households. Most recycling is "voluntary" on the state level, though local governments vary as to whether there is legislation for mandatory recycling.

CURBSIDE PROGRAMS

Even though statistics later showed that the majority of waste comes from the commercial and industrial sectors, most state laws focused almost entirely on curbside consumer trash. *This has proved to be the most expensive approach.*

As of 1999, there were more than 9000 curbside recycling programs, according to Franklin Associates (9349, serving 139.5 million people, according to *Biocycle Magazine*). *Biocycle*'s annual survey found that 22 states had an increase in curbside programs, and only three reported decreases. A survey from Raymond Communications for *State Recycling Laws Update* (SRLU), which included responses from about 36 states, found that 11 reported increases in curbside programs; 12 said they stayed the same (Table 2.1).

Each state took a different approach to financing its ambitious curbside collection scheme. Wisconsin enacted a major corporate tax that generated nearly $40 million per year, though it was slated to end by 2000. The problem with the law was that any city could apply for money, so the state ended up with a cost-plus system for 1200+ entities instead of 50 counties.

Other states took money from oil overcharges and general revenues. A number of states used tip fee surcharges. Pennsylvania, for example, has a $2 tip fee surcharge that funds about $30 million in grants and loans each year.

TABLE 2.1 Top ten states recycling goals, rates, and funding sources

State	Recycling or reduction goal	Recycling rate 1997 (1996)	Mandatory recycling	Source of funds
Connecticut	40% 2000	25% (23%)	Yes, counties must mandate	Bonds
Illinois	N.A.	23.5% (26%)	Yes, counties must implement	Tipping fee surcharge, surcharge on tires or other items
Massachusetts	46%	34% (33%)	Yes, source separation	Bonds
Minnesota	50% 1996	45% (46%)	No	State General Fund, surcharge on tires or other items
Nebraska	50% 2002	25–30% (22%)	No	State Management Fund, tires surcharge, business tax
New Jersey	65% 2000	42% (42%)	Yes, source separation	Grants
Pennsylvania	35% 2002	26.2% (25.9%)	Yes, counties to develop source separation	Tipping fee surcharge
Tennessee	25% 2003	21% (37.7%)	No	State Solid Waste Management Fund
Washington	50%	32.9% (37%)	No	Tax on business or other entity
Wisconsin	N.A.	36% of MSW (1995)	Yes	Tax on business or other entity

*N.A. = not available.
Courtesy of Raymond Communications, Inc., College Park, MD.

FLOW CONTROL

A number of local governments charged extra tip fees at a government-owned landfills, incinerators, or transfer stations and then used the extra funds to finance curbside recycling. The catch was that these governments had to force haulers to use a specific facility. In the mid-1990's, the waste haulers sued and won at the Supreme Court level because interstate trash is considered "commerce" and they proved that forcing them to use a particular facility was in violation of the interstate commerce clause of the Constitution. These "flow control" decisions wreaked some havoc in certain states, especially Minnesota and New Jersey.

The "flow control" issue continued to heat up in 1999. Since New York City's Fresh Kills landfill was due to close by 2000, the City's Sanitation Department planned to offload 13,000 tons per day of trash in states such as Pennsylvania and Virginia. Five states

(Virginia, Maryland, New Jersey, Pennsylvania, and West Virginia) signed a letter stating that this plan was “an unacceptable policy.”

Egged on by an exposé in the *Washington Post,* several trash truck accidents, and consumer outrage, Governor James Gilmore of Virginia realized his state had become the nation’s second-largest trash importing state, pulling in about 4 million tons of extra waste per year. This clashed with the state’s pristine image—and its $11 billion tourism industry. Gilmore vowed to take action to stem the tide of imported trash, even though nearly every avenue of suggested legislation appears to be unconstitutional.

To crack down on haulers, inspectors from many states began an “East Coast garbage truck inspection blitz” on February 8, 1999. The inspection, conducted in eight states (Maryland, Virginia, Delaware, New York, Ohio, Pennsylvania, and the District of Columbia), discovered tired and poorly licensed drivers, heavy loads, flat tires, and bad brakes. Nearly 417 trucks were stopped in Maryland, D.C., and New Jersey on one day; 37 were ordered off the road because of potential hazards.

Houston-based Waste Management, Inc. completed a $15 million reconstruction of Port Tobacco in Charles City, one of Virginia’s oldest ports, on the James River with the intention of barging in as much as 4000 tons per day of New York City garbage. Waste Management runs five of the state’s seven biggest landfills, and under current agreements, Virginia receives approximately 3700 tons of Brooklyn’s trash per day. But New York’s plan has Virginia receiving 3900 tons a day by the 2001 closing of Fresh Kills. The state would also receive part of Manhattan’s daily 2600 tons, as well as the regular 4000 tons a day of commercial waste from apartment buildings and offices, according to officials. An average load of 4000 tons each day during the week would mean an additional 900,000 tons a year—a 30% increase in total imports.

The movement of trash across state lines is protected by the U.S. Constitution commerce clause, which permits only Congress to regulate imports of commodities such as waste. U.S. Representative Thomas Bliley, Jr., a Richmond Republican and chairman of the House Commerce Committee, got strong pressure from Virginia’s legislators to allow a vote on legislation to create exceptions to the clause. And U.S. Senator Charles Robb (Democrat from Virginia) filed legislation for a special exception for Virginia localities to have control over interstate trash.

Meanwhile, Gilmore did sign two bills in March 1999—one would cap landfill tonnage at 2000 tons per day, and the other would ban garbage barges on Virginia rivers. By fall, the haulers had already found one judge to stop enforcement of the garbage barge ban, based on the commerce clause.

Meanwhile, New Jersey and Minnesota’s local “flow control” ordinances were struck down as well. Thus, New Jersey governments faced some severe hardships in attempting to replace the money they had counted on for local facilities. Several billion dollars worth of public bonds hung in the balance, and as of late 1999, there was no long-term resolution, save increasing local taxes and cutting recycling.

Thus, the “flow control” issue will remain highly contentious in the future. Congress has failed to act on the issue, and observers doubt it will do so as long as Republicans control at least one chamber.

PACKAGING ISSUES

The “Stop Styro” campaign started in Berkeley in 1988, possibly affected by the famous “garbage barge” incident that led the media to conclude that there was a landfill shortage. Without looking at any numbers, environmentalists decided the MacDonalds’ polystyrene

(PS) "clamshells"* were clogging our landfills, and it became the symbol for a wasteful society. For various reasons, the campaign was successful.

Local governments started proposing and passing ordinances to ban PS foam cups and related items; media attention meant letters from children. In 1989, Minneapolis banned packaging not included in the local government's collection system. Even though the ban proved impossible to enforce, 40 local governments around the country did pass retail PS foam restrictions. Meanwhile, the packaging bug got to state lawmakers.

PLASTICS BASHING

Numerous recycled content mandates, bans, advance disposal fees, etc. on packaging were proposed by states between 1989 and 1991. For example, Massachusetts and Oregon got ballot issues that would have required all packaging to be 50% recycled or contain 50% recycled content. While defeated with plenty of industry money, 40% of the population of those states voted for the measures.

In 1989, amidst the packaging regulation hype, an event occurred involving plastic bags that had nothing to do with the environment. The then American Paper Council (now the American Forest and Paper Association) hired the Madison Group in New York to improve the image of paper grocery bags; they were losing market share to the resource-efficient plastic bags. The Madison Group did an efficient job of enlisting the General Federation of Women's Clubs, which promoted the degradability of paper sacks. So successful was this that to this day the average consumer thinks that the waste stream should be more degradable.

Not to be outdone, some plastic bag makers decided to start using degradable additives (Florida mandated this until 1995). However, there were so many green claims flying around at this point that the environmentalists, led by Barry Commoner, decided to "investigate." Most environmentalists do not like plastics in general, and they maintained that unless the plastic can degrade back to carbon dioxide and water, the claims were fraudulent.

Meanwhile, state attorneys general (AGs) were quietly egged on by large diaper makers, who were concerned about inroads from the new "degradable" diaper. The AGs did get into the act, and ultimately some of the degradable plastics users had to tone down their claims. Even Procter & Gamble was ordered to alter their new degradability claims. And the big players had to go underground with degradability research. (Degradable plastics have some very useful niche applications, such as agricultural mulch films, fishing nets, flushable sanitary products, medical applications, etc. Japan continued the research, with the goal of trying to make quality degradable polymers at an affordable price. Today there is an effort to introduce degradable yogurt cups in Germany, but since there already is an extensive recycling infrastructure, this is problematic.) But ultimately, this was a *market war;* it had little to do with the environment.

When the dust settled in 1993, the states had a dozen packaging laws, all different. Restrictions on plastic bags have been quietly watered down or repealed since then.

Florida had an advance disposal fee (ADF) on rigid containers (1993) until it ended in 1995. The law required a 2 cent ADF on containers that did not meet overall recycling rates. The law was a bit expensive to enforce, as the fee was collected like a tax, and small grocery stores did not comply because of the fee's complexity. However, state officials

*"Styrofoam" is not a word but a trade name of Dow Chemical; we use the real name of the resin—polystyrene.

feel it had a positive effect on new markets for recycled plastics, and they wished it had more time to work.

In 1994, the legislative interest in packaging mandates, besides that in a few states such a California, Oregon, and Wisconsin, began to falter. The issue was no longer fresh and most states backed down because of industry's success in preventing most stringent legislation.

California has a number of packaging mandates still on the books, including a recycled content mandate on glass, plastic trash bags (which was softened in 1998), and rigid plastic containers. With a strong economy, plastics waste is especially increasing as manufacturers use the material in more applications. As packaging technology allows more complex designs, creativity is interfering with the recycling of existing plastics streams. That is why the Miller Brewing Company's plastic beer bottle has caused such controversy, observers say.

1999 was a year of reborn interest in regulating plastic packaging. This is partially due to the controversy over the plastic beer bottle and the Democrats' return to power in California. The state of California was not only moving to enforce its recycled content mandate, but the Legislature was moving to strengthen the mandate. Moreover, lawmakers in Wisconsin were heartened by the news, and were attempting to expand their mandate as well.

California cities have a state mandate to reduce landfilled waste by 50% by 2000 or face fines. Plastics represent up 27% by volume of MSW, though it is about 7% by weight. Gyrations in the prices paid for recycled plastics have caused a number of recyclers to go out of business. Most large applications for recycled PET tend to be open-loop going into fiber applications. Consumer product companies claim they have no control over what consumers do with their bottles and cannot force their material into open-loop markets.

California's original 1991 law (SB 235) requires that for the years (beginning in 1995) in which industry does not meet an aggregate 25% recycling rate for rigid plastic containers, between eight ounces and five gallons either use 25% postconsumer recycled resin (PCR), reduce usage by 10% or make the containers reusable.

In 1996, the Integrated Waste Management Board (CIWMB) failed to reach a decision on whether the 25% rate was met. But when there was a new appointee added to the board, the tie was broken and they decided on a "range" of 23–25%, which let manufacturers off the hook. However, the decision was not made until January 1997. In addition, it should be noted that without redemption (deposit) containers, the rigid plastics recycling rate would be about 14%, according to CIWMB

Meanwhile, at the eleventh hour, the Republican-dominated Legislature amended the law, exempting food and cosmetic containers (1996) but not exempting the products from counting towards the recycling rate.

CIWMB ended up with seven compliance orders and one fine for companies. The board has requested compliance data from a total of 1500 firms that have failed to comply with SB 235.

In addition, the state Senate and the Assembly Natural Resources Committee passed SB 1110, which will expand the state's recycled content mandate to again include food and cosmetic plastic containers and increase the required recycling rate from 25% to 35% by 2001. If the rate is not met, all manufacturers would have to use 35% postconsumer recycled material in their rigid plastic containers.

Industry vigorously opposed expansion of SB 235, forming a new packaging alliance that includes more than 50 associations and companies. In addition, industry lobbyists have been arguing that even though there are 52 FDA letters of no objection on file for a number of PET and HDPE food applications, there are no viable content options for man-

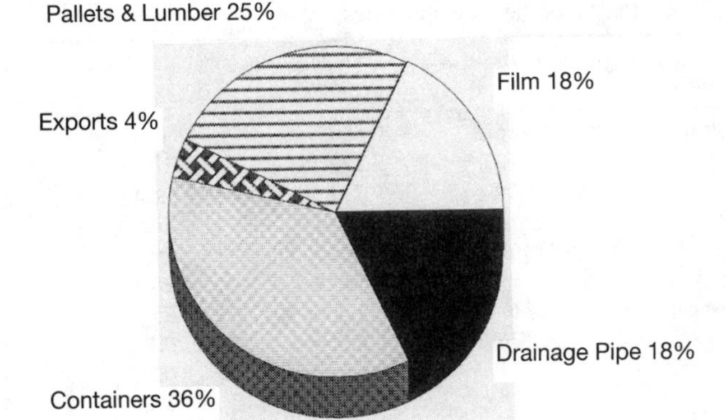

FIGURE 2.1 End user markets for HDPE by product category. (Source: *Modern Plastics.*)

ufacturers that use other resins, such as polypropylene (PP), polyvinyl chloride (PVC), and polystyrene (PS).

Consumer groups argue that 94% of all containers are made of PET or HDPE, and postconsumer material is available for those containers (Figure 2.1).

Oregon, too, ended up with mandates on rigid plastic containers (RPCs), requiring them to maintain a 25% recycling rate, or all containers would have to contain 25% postconsumer recycled material. Oregon has not had to enforce its mandate, since industry met the 25% overall recycling rate. Wisconsin also has a 10% recycled content mandate on its books for rigid plastic containers, but this includes industrial regrind. There were some small efforts to enforce the law, but there seemed to be little support in the mid-1990's.

PLASTICS RECYCLING DECLINES

The plastics recycling rate did decline in 1997, and slightly in 1998 as well. The bottle recycling rate was 23.5% in 1998, off from 23.7% in 1997. The rate for customer PET bottles was down at 10.1%; for PET soft drink bottles it was 35.6%. Recovery of HDPE bottles was up slightly to 25.2% from 24.7% in 1997 (Table 2.2).

Dennis Sabourin, VP of Wellman, told delegates at the Take it Back! '97 conference (November 18, 1997), that unless something was done to improve the situation for plastics recyclers, he feared his industry would crumble in the next few years. If that happened, manufacturers would end up facing another round of proposed state bans and content mandates they thought were gone two years ago.

Sabourin complained that plastics recyclers are unable to get enough feedstock (a few are importing from Canada and as far away as Europe, SRLU sources say) and that curbside recycling will only grow about 1% per year.

The American Plastics Council (a group of plastic resin producers) points out that there is double the capacity to process recycled plastics as there is feedstock. APC is urging communities to increase education for more collection of plastics bottles.

Recovery is barely keeping up with growth in use of plastics, which increased 424

TABLE 2.2 Plastic bottle recycling rates, 1998

Plastic Bottle type	Resin sales 1998	Pounds recycled 1998	1998 Recycle rate	1997 Recycle rate
PET soft drink	1628	580.3	35.6	35.8
PET customer	1278	129.7	10.1	10.2
Total PET	2906	710	24.4	25.4
Nat. HDPE	1415	433.7	30.7	30.6
HDPE Pig.	1497	300.2	20.1	19.3
Total HDPE	2912	733.9	25.2	24.7
HDPE base cups	6	1.5	25	30
PVC	152	0.2	0.1	0.1
LDPE/LLDPE	51	0.1	0.2	0.3
PP	145	5.3	3.7	4.1
PS	10	<0.1	<0.1	<0.1
Total bottles	6182	1451	23.5	23.7

Source: R. W. Beck for American Plastics Council 1999.

million pounds in 1998. The 1998 plastics recycling rate survey from the R.W. Beck shows that recovery of plastic bottles increased 7% between 1997 and 1998 to 1.45 million pounds. The survey mainly covers bottles and related packaging, not all plastic materials recovered. In addition, polypropylene and polystyrene recovery was down slightly; rates for all of the "other" resin bottles continue to be low, ranging from less than 0.1% for PS to 3.7% for PP.

In 1998, Jack Milgrom, of Walden Research, Inc., a New Jersey recycling consultant who has contacted numerous recycling sources, says the single-serve PET containers unfortunately end up in trash more often than curbside bins, thus depressing PET container recycling.

PLASTIC BEER BOTTLE CAUSES FUROR

In 1999, Miller Brewing Co. test-marketed a plastic beer bottle in cities across the United States. The bottle was composed largely of polyethylene terephthalate (PET). Both the Miller Lite and Icehouse bottles were amber-tinted with a five-layer barrier material (three PET, two nylon polyamide), and metal label and cap. Tinted green bottles are estimated to be approximately 25% of all PET soft drink bottles. Adding yet another color would further increase sortation costs. The metal caps and labels also present a problem, as they will need to be modified or eliminated to reduce or prevent additional processing costs for recyclers in recovery. The bottles currently bear a "#1-PETE" SPI code, whereas most state plastic container coding laws define multilayer bottles like Miller's as "#7, other."

Miller made the mistake of making Los Angeles one of its test markets. Heated debates over the bottle's recyclability ensued. The City of Los Angeles passed resolution 99-0 130 on February 3, 1999 by an 11-0 vote. The resolution stated that Miller would need to commit to developing markets for recycled PET by using at least 25% postconsumer recycled content in its plastic beer container. Miller would also have to address the issues without creating any additional cost to the city's recycling program and resolve the issues before extending or expanding its current test marketing. The City of West Hollywood passed a similar resolution, and Long Beach was considering one. As for the general recycling problem, city officials were rather blunt in their concerns. They were unsure what would happen to the new material or if it would cause them new sorting and cost headaches. They were also skeptical that new markets for amber PET could be developed in so short a time.

Miller had been working on this idea for two years, consulting closely with the Association of Postconsumer Plastics Recyclers, the National Association of PET Container Resources, and the American Plastics Council about a responsible, coordinated effort to address recycling of the bottle before it was even produced. Miller tried to assure skeptical city solid waste managers that its company was committed to using the recycled PET/nylon material with a premium buyback program.

Miller made a claim that Continental PET Technologies of Bedford, New Hampshire would be able to add up to 40% postconsumer recycled material to the package should they move beyond the test market phase. It offered to buy back amber PET bottles (all types) for one year, at a premium price. Public officials were skeptical that markets could be developed in one year.

The producers of the bottles said the two nylon barriers were less than 5% of the bottle's weight. Tests performed by Plastics Forming Systems, Inc. in New Hampshire indicate this material could be aspirated or floated off in the normal PET bottle processing. However, material recovery facility (MRF) operators say even with the small scale of Miller's test, the bottle may still cause problems for the unstable PET recovery industry.

In an effort to meet recycler's needs, the vendor (PPG) developed a coating technology to provide additional barrier properties that wash off in the reclaimer's bath. They assert that Miller can add the amber tint to its coating, which would also wash off. There has been no decision as to whether the technology will be used by beer makers.

With only 16% of recovered PET being remade into new containers, the growth in production of virgin plastic outstrips the increases in plastics recycling. As of September 1999, sources say no one has even shipped a load of plastic beer bottles. Continental PET claimed that even though there has been a major roll-out at stadiums and beaches, people won't put the new bottles in bins to get them recycled. Most material recovery facilities are discarding the bottles as a contaminant. Anhaeuser-Busch tested a plastic beer bottle with a different polymer mix, but quickly pulled out of the new market, although it re-entered the market in 2000.

Ironically, a 1999 consumer survey indicated most people will not buy beer in plastic except for use at stadiums and beaches because they look "cheap."

THE PROBLEM ISSUE

Even though the actual number of beer bottles was too small to wreck the PET recycling industry, the Miller issue apparently was the straw that broke the camel's back. It was clear that industry was introducing more and more complex multilayer glass-like bottles that were difficult to sort out in the recycling stream. For example, the packaging industry had already designed at least five different methods of making beer bottles, all using different layered polymers. In addition, dairies were testing small 8-ounce PET containers with full shrink sleeves for milk products. These containers needed a barrier layer as well, making them incompatible with most PET streams.

The new multilayer containers already on the market created additional costs for sorting facilities. These facilities largely relied upon manual sorting methods, even though many automatic sorting technologies were commercially available in 1999. Most MRF's could not afford the new sorting equipment or the expensive X-ray maintenance parts required. Moreover, the equipment was not yet designed for a multitude of PET bottle colors or multilayer containers with numerous different polymer layers.

The sorting and design conflicts threatened to carry over to high-density polyethylene (HDPE) as well. In 1998, several dairies were testing a white HDPE milk jug. Switching to pigmented milk jugs for a perceived improvement in flavor and nutrients would have cost the diary industry $56 million per year, and plastics recyclers would lose $14 million per year, according to a case study made in December 1997 by Recycle Worlds Consulting. According to several of the dairies' public statements, the reason for the switch to pigmented opaque bottles was to protect the milk's flavor from off tastes caused by light, and one dairy has cited vitamin degradation. However, if grocers added yellow shields on to their flourescent lighting in diary cases, it would cost just $95,000. Robert Byrne of the International Dairy Foods Association claimed that there was no strong evidence that light in a dairy case affected most vitamins in milk and concluded that pigmenting was largely a marketing tool.

DEPOSITS

Mandatory bottle deposits are the most popular recycling policy concept in America, but they are the most difficult to pass and they remained controversial in 1999. Deposits schemes—in which consumers generally pay an extra 5–10 cents per container and get the money back upon return to the grocer—are currently on the books in 10 U.S. states. In 1999, 17 states introduced 43 new container deposit bills; there were 36 bills introduced between January 1997 and June 1998.

Sales of "new age" drinks have increased dramatically since the deposits on traditional soft drinks, beer, and wine were passed. Environmentalists argue that the most efficient way to get back more bottles, especially plastic bottles - is the expansion of deposits. A referendum on the issue was defeated in Oregon in 1996, and industry has spent heavily to lobby against expanding existing deposit laws.

California has a complex "redemption" system. In California's system, the state takes all the money and consumers get a refund when they bring their bottles to separate redemption centers. Each material must pay for itself, however, so there is a glass processing fee that beverage makers pay the state, and plastics users must guarantee a floor price of about $1200 per ton for PET bottles. Instead of direct return, California requires establishment of "convenience zones" where people can return their containers.

In 1999, all eyes were on California again, as environmentalists were able to push a new bill (SB 332, Sher) that would expand the state's deposit redemption system from beer, soft drinks, wine coolers, and mineral waters to all carbonated water, noncarbonated soft drinks, sport drinks, fruit drinks, coffee, tea, and carbonated fruit drinks, sold in plastic, glass, bimetal, or aluminum containers in liquid, ready-to-drink form.

The soft drink industry traditionally has opposed any expansions in deposits, claiming that the costs would be extremely high because of the complexity of sorting by distributor. Currently, the soft drink industry has a federal anti-trust exemption allowing it to have exclusive territories. Once a deposit moves to fruit drinks, however, a container could be from any local distributor. Thus, in order to ensure that the right distributor gets credited with the deposit in a traditional state, the containers must be sorted by bar code. In California, expanding the deposit system will not require any sorting by distributor because all monies go to the state.

Maine is the only state with a traditional but expanded deposit law, but no one has financed an impartial study of the real costs of Maine's system. Maine did face overredemption problems from neighboring states, though no complaints about this were heard in 1999.

As for actual costs, the comptroller of the Alberta Bottle Depot Association told SRLU that the net cost in that Canadian Province for its expanded deposit is about 0.7 cents (Canadian) per container. Industry had estimated 10 cents.

One other problem facing bottlers is the small single-serve PET bottle. This segment has been growing, slowly replacing glass and some aluminum. The single-serve PET bottles are ten times more profitable for bottlers (when they are sold through convenience stores and vending machines) than cans.

However, consumers are not recycling the small PET bottles like they do the big ones, and this is reducing recovery rates, even in deposit states. Moreover, institutions are unhappy when bottlers replace aluminum in their vending machines with PET. They made a profit on the cans, but markets for small quantities of PET bottles are much more difficult to find.

The expansion of California's deposit system will become another bellwether test: the law adds 800 million more plastic containers to the redemption system. No on yet knows how this will play out or what it will cost, as many of the containers are not made of PET.

NEWSPRINT MANDATES

As of 1999, 13 states mandated recycled content in newsprint, while 15 other states have voluntary newsprint agreements. Arizona, California, Connecticut, and Missouri have goals that require 50% recycled content in newsprint by 2000. A few states have amended their laws since passing them in the early 1990's. Most laws have a number of loopholes that practically make use of the material a public relations move on the newspapers' part. For example, seven of the states have exemptions if the recycled material is not available, if it costs more than virgin newsprint, or if it does not meet quality standards. Moreover, most states do not specify that the recycled material be postconsumer.

Wisconsin's law did require use of postconsumer material, and in 1998 it became the only state that took enforcement action against several companies. The Legislature amended the law, scaling back targets to 33% by 2000, 37% 2002, and 40% by 2003.

The newsprint laws have served to increase use of recycled material in newsprint and expand investments in recycled-pulp mills, most observers acknowledge. A logistical problem arose because a large percentage of newsprint comes from Canada, and Canadian mills are located near their natural feedstock (trees), not near big cities with lots of old newsprint.

In 1998, the Northeast Recycling Council decided to move away from supporting newsprint mandates, choosing a voluntary reporting system instead. While demand for newsprint is expected to decline (due to improved newspaper publisher efficiencies and increased use of online services), existing newsprint mills are increasingly switching capacity to use recycled material. Recycled content in newsprint averaged 28–29%, in 1999, according to experts.

The combination of having to comply with the new cluster rules (air/water/sewer effluents), issues over chlorine bleaching, and some pressure from existing mandates, makes recycled newsprint more appealing. (It should be noted that, according to *Pulp and Paper Week,* the cluster rules are expected to cost industry about $2–3 billion to comply, not $10 billion as industry had predicted.)

The American Forest and Paper Association (AF&PA) estimated the 1998 recycling rate for newsprint to be 68.5%, up slightly from 1997. But as more capacity is switched, observers wonder where feedstock will come from if curbside recycling is not growing quickly enough. Experts suggest industry will have to improve the efficiencies of existing programs.

OTHER POPULAR IDEAS

Landfill Bans

Landfill bans remain a very effective means of reducing landfilled waste and pushing recovery of selected items, especially larger items with potential for recycling or reuse. Most states have banned something from their landfills, the majority passing laws between 1989 and 1995 (Table 2.3). As courts continued to rule in favor of haulers on control of waste flows, many local governments have been loathe to pass new landfill bans because they fear the haulers will simply take the trash to another county with no such restrictions.

Surveys from *State Recycling Laws Update* have shown some major examples of how certain landfill bans, if enforced, have been an effective policy. South Dakota, for example, has achieved a 42% diversion rate since 1993. State officials say this is because five of its major cities have decided to enforce a sweeping landfill ban law (1992), which requires them to restrict everything from used oil to appliances to packaging and paper.

North Carolina has about 32 local landfill bans that have been effective in reducing waste and increasing recycling, especially of corrugated in this manufacturing state. The bans also encourage small material recovery facilities to spring up at landfills when the market is good for certain items.

Wisconsin's SB 300 (1990) banned lead–acid batteries, tires, yard waste, white goods, motor oil, and all recyclable packaging by 1995, including fibers, old newsprint (ONP), magazines, corrugated, office paper, glass, aluminum, tin, bimetals, plastic containers, and PS foam from landfills. Technically, Wisconsin is not enforcing its sweeping bans of packaging and paper, because if local government meet complex requirements, they need not enforce them. (#3-7 plastics were exempted because of economics.)

The 1999 SRLU survey found that 22 states claim landfill bans are enforced; five said they were not. Of interest, the majority of respondents also felt landfill bans have been an effective tool for recycling; 19 reported that landfill bans have been effective in reducing landfilled waste, while 20 agreed they were useful in changing business habits.

Massachusetts enacted its regulation to ban cathode ray tubes (CRTs) from disposal, and one county in Florida has already banned commercially generated computers. If Massachusetts' move is effective, other states may also consider banning CRT's.

Mercury

U.S. states continue to consider legislation that would restrict mercury emissions; in 1999 there were 22 bills introduced, with Connecticut enacting a bill that requires labeling of mercury-containing products.

In late June 1999, EPA finally ruled on how to treat mercury-containing lamps: it decided on the option of treating them under the "Universal Waste" Rule (which now affects batteries and related items). This means that if the fluorescents are destined for recycling, they will be under the less stringent Universal Waste Rule, but if they are disposed of in commercial quantities, they must be treated as hazardous waste. Between 600 mil-

TABLE 2.3 Landfill bans—sample laws

State	Lead–acid batteries	Tires	Yard waste	Used oil	White goods	All house-hold-waste	House-hold batteries	Packaging	Scrap metal	Other	Local bans
California	X	X		X	X	NE	X		X	Merc, AF	Yes
Connecticut	X		X				X				Merc
Florida	X	X	X	X	X		X			C&D	Yes
Iowa	X	X	X	X				DEP		Liquid	No
Minnesota	X	X	X	X	X	X		X		PH, Merc, FT, AF OF	Few
Missouri	X	X	X	X	X	X					No
Nebraska	X	X	X	X	X	X					No
North Carolina	X	X	X	X	X			X (cans)		AF	Yes
South Dakota	X		X	X	X			X		All Pkg (1997) PP	No†
Vermont	X	X		X	X		X (ni–cd)			Paint (oil/ latex), merc	Yes
Wisconsin	X	X	X	X	X			X*		ONP	Yes

Acronyms: AF = antifreeze; C&D = construction and demolition debris; DEP = deposit; FT = flourescent tubes; merc = mercury, mercury products; OF = oil filters; PH = phone books; pkg = packaging; ONP = old newsprint; PP = paper products.

*Wisconsin passed a bill in 1996 allowing an exemption from the rigid plastics landfill bans.

†Counties can be exempted.

lion and 1 billion lamps are disposed of annually. Observers estimate that about 10% are recycled.

Under the Universal Waste designation, any hazardous material labeling would be applied only to transportation packaging, not to retail labeling.

Concern about mercury has already led certain states, such as Florida and Maryland, to ban mercury-filled batteries in landfills, and a few have gone further, regulating fluorescent lamps. Minnesota requires *all* mercury to be recovered before disposal, and it requires some labeling and commercial recovery requirements.

Mercury lamp makers acknowledge that without a technical breakthrough, it will not be possible to remove all the mercury from the lamps without destroying their energy efficiency, which provides many other environmental benefits.

In 1999, the National Electrical Manufacturers Association filed suit against Vermont for its mercury labeling law, claiming that the state cannot prove that mercury lamps are worse for the environment than standard light bulbs that burn out faster.

Purchasing Preference Laws

Nearly all 50 states have some sort of policy that favors government purchase of recycled products (Table 2.4). A total of 32 states have price preferences for recycled products, and 23 have set-asides. Whereas about 30 states have made efforts to buy recycled copy paper, and 13 are using recycled forms, only four reported even trying to purchase recycled plastic products, according to a 1997 SRLU survey.

President Clinton issued two Executive Orders (EOs) purchases of recycled products. Executive Order #13101, signed in October 1998, attempts to put some teeth in the federal government's recycled procurement policy. According to Federal Environment Executive Fran McPoland, federal auditors will be able to check federal agency compliance with recycled purchasing and issue fines as for other environmental laws. Under the EO, the federal government is supposed to purchase only recycled copy paper, and agencies are supposed to purchase "environmentally preferable" products. In October 1999, McPoland reported that the General Services Administration was in full compliance with the copy paper part of the Order.

A report from Raymond Communications, "Purchasing Preferences for Recycled Products:Primer and Guide to Laws," concludes that the purchasing laws have not accomplished the goal of mainstreaming recycled materials. The idea was that the purchasing preferences were to be temporary, not needed when use of recycled materials was common.

TABLE 2.4 Purchasing preference laws—samples

State	Year	Product	% Content*	Price preference	Set-asides
Arizona	1990	Paper	10% PC	5%	No
California	1977, 1986, 1989, 1993, 1994, 1998	All Engine coolant or antifreeze	Paper: 50% Sec., 10% PC; All PC must increase to 30% by 1/99 Grants variance from specs if chloride content less than 150 ppn	10%	Yes
Colorado	1989, 1990, 1993	Plastics and paper	Plastics: min. 10% RC Paper: 50% Sec., 50% PC min. 10% PC		Yes
Maryland	1988, 1990, 1993	All	EPA	5%	40%
Massachusetts	1989	All		10%	No
Oregon	1991, 1998	All	50% Sec. or 25% PC paper; 25% by 1997; 40% by 1999	5% = 12% on printing paper	Yes
Washington	1991, 1996	Paper, compost All		10%	Yes

*Sec = secondary; RC = recycled content; PC = postconsumer.

Tax Credits

Tax credits or incentives continue to be popular tools with lawmakers to sustain and promote recycling markets as the 17 tax-incentive-related legislative introductions in 1999 indicate, though the number of bills was down from 25 in the 1997–1998 period.

The types of credits vary widely, and may include a business income tax credit, sales tax credit, or even property tax credit (Table 2.5). Some states require creation of a certain number of jobs, most require investment in recycling equipment, and a few allow companies a credit for source-reducing their product or a switch to reusable crates.

A 1998 survey by *Resource Recycling Magazine* found a decline from 21 in 1993 to 10 in 1998 in the number of income tax credits offered, while the number of sales and property tax exemptions remained more or less the same. As for costs to state governments, the survey found that in some cases the tax incentive programs "ended up costing the state more than it had bargained for," and suggested that states are moving toward loans and grants. An official of the Maine State Planning Office was quoted as feeling that loans "seem a little more easier to manage."

Batteries

Just about time the U.S. states were introducing an avalanche of recycling and packaging legislation in the early 1990s, a number of states concurrently passed new laws that attempted to address toxicity in the waste stream. Industry formed the Portable Rechargeable Battery Association (PRBA) in 1991, but was unable to stop 13 states from enacting specific legislation relating to nickel–cadmium (ni–cd) batteries.

The laws usually banned various batteries from landfills, banned added mercury in

TABLE 2.5 Tax credits–samples

State	Type of credit	Purpose	Comments
Arizona	Sales tax	Recycling, pollution control equipment recycling	10% on the installed cost of equipment used for
Kentucky	Income tax/property tax/sales tax	Materials recovered	Exemptions for recycling equipment. 10% allowable credit, no sales tax in such cases
Nevada	Property tax exemption	Recycling process waste-to-energy plants: Tires, plastics, asphalt shingles, agricultural or municipal wastes	Property tax exemption and utility benefit for business with $50 million investment in a recycling process.
Oregon	Income tax	Pollution control recycling facility; reclaimed plastics; business energy recycling	Extended and expanded in 1995

batteries, and required ni–cds and small sealed lead–acid batteries to be labeled and easily removable from household electronics products. The battery industry had little trouble phasing out the mercury from consumer alkaline batteries; in fact, two major U.S. battery makers eliminated the mercury six months or more before regulatory deadlines.

Eight states also decided that manufacturers should take back old batteries. This requirement started what is the only example of modern "producer responsibility" legislation in the United States to date (unless one includes container deposits). Of the 13 states, eight specifically required industry to take back ni–cd batteries, five required recovery of small sealed lead–acid batteries, and one, Minnesota, technically required take-back of "other" rechargeable batteries.

These new laws placed industry in a very difficult situation: how can you get back small batteries, which become "hazardous waste" when they are accumulated anywhere, from consumers? Current hazardous waste laws have cradle-to-grave requirements on hazardous waste, recycling is very difficult, and landfilling is expensive. Consumers are allowed to trash hazardous batteries in most states (except New Jersey, Maryland, Minnesota, Iowa, and Florida), but businesses risked becoming "generators" under RCRA Subtitle C.

Industry planned to form a national collection scheme, but it needed to get the cooperation of retailers to make it work for the consumer. Observers say retailers did not want to be bothered with the issue, in addition to regulatory problems. Retailers were technically responsible under New Jersey law.

Off to Congress

The patchwork of requirements pushed industry to get Congress to pass the Mercury-Containing and Rechargeable Battery Management Act (Battery Act) in 1996. The federal act was designed to facilitate take-back by standardizing labeling requirements and codifying issues such as mercury, "easily removable." But to get a bill through Congress, there would have to be no take-back requirement, and membership in industry's new Rechargeable Battery Recycling Corporation (RBRC, formed in 1994) would be voluntary.

Meanwhile, industry lobbyists had to push through what is now known as the Universal Waste Rule at EPA, which would allow batteries and other items destined for recycling to be partially exempt from standard RCRA rules. It took five years to get the Universal Waste Rule passed; the industry had to get all 50 states to adopt the rule. Observers note that a few states still have not adopted the Universal Waste Rule. It was frustrating for organizers because the retail chains didn't want to educate employees in some states and not others. It was not until 1996 that California passed a law allowing ni–cd recovery, so that RBRC could fully implement its retail take-back scheme and expand it to 25,000 outlets in North America.

Although a change in battery chemistry was helpful, observers note that despite the take-back laws, it will take many years to get the mercury out of the battery waste stream because of consumer hoarding. Meanwhile, it appears that ni–cd makers are slowly being penalized for forming a take-back program. Manufacturers that can do so have been switching to lithium ion and nickel metal hyrdride batteries, which have no take-back requirement in the United States. Thus, the pool of ni–cd's to recycle has been shrinking. Moreover, with pressure to ban cadmium batteries in Europe, research to replace such batteries will accelerate. As of 1998, industry spent about $7.5 million to recover about 25% of ni–cd's in the United States. (This information courtesy of *Battery Recovery Laws Worldwide,* April 1999, Raymond Communications.)

FEDERAL RECYCLING POLICIES

Although EPA set a voluntary recycling goal of 25% in 1990 and increased in to 35% in 1997, there is little EPA or other federal agencies have been able to do to enforce these goals. Congress has never made any kind of recycling mandatory, and indeed, lawmakers have not reauthorized the Resource Conservation and Recovery Act. (If they did, there might be a few recycling mandates in it.) The federal government did move on batteries, only after industry begged to get 13 states off of their backs. It has made laws on marine dumping, plastic ring carriers, CFC's, and recycling of certain otherwise "hazardous" wastes (e.g., batteries and mercury lamps). President Clinton did sign two Executive Orders on government procurement of recycled products, though the orders had few teeth.

When it has funding, EPA will provide extra recycling staff positions in the states and conduct studies on different kinds of recycling programs and policies, and it will fund various conferences. In 1996–1999, it provided some funding for roundtable discussions on electronics recycling.

EPA's Office of Solid Waste maintains a small staff that is researching the concept of voluntary extended producer responsibility, and plans a web site in 1999 to help business. EPA's voluntary Waste Wise program now has 900 members. The members pledge to voluntarily reduce their waste and report back to EPA. In return, EPA provides technical support to help members. Sources say funding is so tight that the agency can only provide help for the Waste Wise members, not for local governments and other small businesses.

In terms of proposing any new policies, observers say EPA is put in a position of doing nothing because of politics. For example, it commissioned a study of container deposit policy in 1995. A member of Congress promptly introduced a rider on a funding bill that would ban EPA from studying deposits ever again!

In general, solid waste and recycling have been the purview of state and local governments.

SNAPSHOTS OF STATE RECYCLING POLICIES*

California

California has taken a mandatory approach to recycling. Local governments were required to divert 25% of waste by 1995 and 50% by 2000, with a reduction in per-capita waste. Incineration does not count as recycling. It also requires local governments to submit plans. As of 1998, the state claimed a 33% diversion rate; however, it should still be a difficult task to reach its goal for 2000. The law carries an unusual $10,000/day penalty for those governments not meeting the goal. Millions have been spent on planning documents.

Commercial recycling has become nearly a requirement because of the mandate to achieve the 50% reduction goal by 2000. There are local mandates in many cities and counties and some local governments require multifamily recycling as well.

*See Table 2.6 for a summary.

TABLE 2.6 State Disposal Capacity and Tipping Fees

State	Number of landfills	Landfill average tipping fee ($/ton)	Remaining capacity (years)	Number of incinerators	Incinerator average tipping fee ($/ton)	Daily capacity (tons/day)
Alabama	30	33	10	1	40	700
Alaska	322	50	N.A.*	4	80	210
Arizona	54	22	N.A.	2	N.A.	N.A.
Arkansas	23	27	20	1	N.A.	N.A.
California	188	39	28	3	34	6440
Colorado	68	33	50	1	N.A.	<20
Connecticut	3	N.A.	N.A.	6	64	6500
Delaware	3	58.50	20	0	—	—
District of Columbia	0	—	0	0	—	—
Florida	95	43	N.A.	13	55	18996
Georgia	76	28	20	1	N.A.	480
Hawaii	8	24	N.A.	1	N.A.	2000
Idaho	27	21	N.A.	0	—	—
Illinois	56	28	15	1	N.A.	1600
Indiana	45	30	N.A.	1	27.50	2175
Iowa	60	30.5	12	1	N.A.	100
Kansas	53	23	N.A.	1	N.A.	N.A.
Kentucky	26	25	19	0	—	—
Louisiana	25	23	N.A.	0	—	
Maine	8	N.A.	18	4	47	2850
Maryland	22	48	10+	3	51	3860
Massachusetts	47	N.A.	N.A.	8	N.A.	8621
Michigan	58	N.A.	15–20	5	N.A.	3700
Minnesota	26	50	9	9	50	4681
Mississippi	19	18	10	1	N.A.	150
Missouri	26	27	9	0	—	—
Montana	33	32	20	1	65	N.A.
Nebraska	23	25	N.A.	0	—	—
Nevada	25	23	75	0	—	—
New Hampshire	19	55	11	2	55	700
New Jersey	11	60	11	5	51	6491
New Mexico	55	23	20	0	—	—
New York	28	N.A.	N.A.	10	N.A.	10,350
North Carolina	35	31	5	1	N.A.	540
North Dakota	15	25	35+	0	—	—
Ohio	52	30	20	2	N.A.	N.A.
Oklahoma	41	18	N.A.	2	N.A.	1200
Oregon	33	25	40+	2	67	600
Pennsylvania	51	49	10–15	6	69	8925
Rhode Island	4	35	5+	0	—	—
South Carolina	19	29	16	1	N.A.	255
South Dakota	15	32	10+	0	—	
Tennessee	34	35	10	2	43	1250
Texas	181	N.A.	30	4	N.A.	N.A.

(continued)

TABLE 2.6 *(continued)*

State	Number of landfills	Landfill average tipping fee ($/ton)	Remaining capacity (years)	Number of incinerators	Incinerator average tipping fee ($/ton)	Daily capacity (tons/day)
Utah	45	N.A.	20+	2	N.A.	340
Vermont	5	65	5–10	0	—	—
Virginia	70	35	20	5	N.A.	N.A.
Washington	21	N.A.	37	5	N.A.	N.A.
West Virginia	19	48	20	0	—	—
Wisconsin	46	30	6	2	45.50	347
Wyoming	66	10	>100	0	—	—
Total	2314			119		

*N.A. = not available.
Reprinted with permission from *Biocycle,* April 1999.

The mandate has pushed local governments into an activist role on the policy side, as they do not have much choice but to push for more recycling.

Connecticut

The state of Connecticut requires source separation of nine mandated items for recycling. There is a state mandate for commercial recycling. The Department of Economical Development distributes $1 million in grants and loans towards recycling. Connecticut had a 40% recycling goal by the year 2000. The goal is a combination of recycling and reduction. The state claimed a 25% recycling rate in 1997, which met the 25% goal set for 1998.

Delaware

Delaware does not have curbside programs—it collects recyclables only through drop-off locations. The state's recycling budget is $3 million. In addition, Delaware claims to have a 60% diversion rate of material from landfills. The rate, driven mostly by markets, includes waste-to-energy and commercial recycling. Delaware Solid Waste Authority has developed regulations to abolish flow control. This will affect the state's ability to reduce exports and imports of waste.

Delaware has tax credits towards source-reduction recycling equipment. The amount is based on the percentage of source reduction or funds plus employees for use of recycled content. In 1995, a bill was passed that created an incentive program to clean brownfields at industrial sites for reuse. Delaware has a Green Industries Initiative, which is a program of tax incentives and loans for industry to reduce waste and use recycled content. The requirements on source reduction tax credits are a bit stiff: you must have reduced nonhazardous solid waste 50%, not for regulatory reasons, to receive a credit of $250 for each 10% increment. If you use 25% recycled materials removed from Delaware's solid waste stream, invest $200,000, and create five new jobs, then you can get $500 for each

$100,000 invested and each new job. Companies establishing new facilities in the state can get a credit of up to $750 on each. The Green Industries package also includes financial assistance for firms with 100 or fewer employees and expedited state permit applications.

Florida

Florida has a 38% recycling rate on a budget of $35 million annually. In 1988, the law set a 50% recycling goal of certain materials by 1992 by means of mandatory community recycling. It has met the goal. The state has given generously to local governments for recycling in the past, but each year the Legislature threatens to take most of the state money away—$12 million was diverted for weed control in 1999. The state recycling program suffers in part because of 3 million tourists that come though each year, and don't recycle.

Hawaii

There is a 1.5 cent advance disposal fee imposed on distributors for all types of glass containers in Hawaii. The law exempts containers holding under two fluid ounces. The funds are used for market development and recovery of glass. Hawaii is in an unusual situation because it can't melt the glass and cannot afford to ship it to distant markets. State officials believe the ADF has been effective, though they do not recommend it to other states.

Hawaii also has a statewide mandatory recycling for glass, newspaper, office paper, and corrugated. There are penalties for noncompliance. The state has a recycling goal of 50% by 2000.

Iowa

Iowa has diverted 34% of its waste from landfills, shy of its 2000 goal of 50%. There are landfill bans on lead–acid batteries, yard waste, used oil, whole tires, liquor bottles with a deposit, and container deposit bottles of all types.

The state requires local governments to collect a $4.25–4.75 per ton tip fee. Local governments that meet the diversion rate (25% in 1994) pay the government $2.80 per ton of waste landfilled. The local governments that do not meet the diversion rate have to pay the state $4.75 per ton. The law also requires those not meeting the goal to implement volume-based fees for trash.

The state does not count recycling rates, but officials note that poor paper markets have hurt many recycling programs. It has a 4% tax on amusements to fund environmental programs. The same bill allows grants to cities.

The state awards $3 million annually in grants and loans. There is financial assistance for recycling collection, processing, and market development. In addition, companies can apply for final assistance for containers and refurbishing processes. However, like many states, Iowa's laws are not being followed in rural areas, where landfills are not used and open burning is common.

Massachusetts

Massachusetts cannot pass a mandatory recycling law due to its type of government. Under an earlier Proposition, the state cannot make its towns do anything without funding.

However, with a $10.5 million recycling budget, the state is one of the nation's top recycling states, as it is recycling 34%, excluding C&D waste and industrial waste. Most of the states count some C&D and industrial wastes in their rates; we were able to confirm that Massachusetts does not.

The state has deregulated cathode ray tubes as hazardous waste when they are destined for recycling, and has banned CRT's from landfills by regulation in 1998. There is a six-point plan to encourage electronics recycling. These points include seed money, the procurement of recycled products, the establishment of 8–12 permanent collection centers, and the purchase of services through the state contract.

The state has given grants to the University of Massachusetts for research, and has a $4 million loan fund for businesses. A five-way partnership to recycle TVs and computers joins the DEP, the U.S. EPA, Goodwill Industries, the Salvation Army, and the University of Massachusetts. The first six months of the program have diverted 9000 CRTs from households for repair and recycling. On February 1, 2000 the state will be the first in the country to add CRTs to the list of items prohibited from disposal in solid waste facilities, which includes large home appliances, tires, and auto batteries. The state bans most packaging, paper, yard waste, whole tires, used oil, and white goods from landfills. Landfills are required to recycle 25%.

Minnesota

Minnesota is another of the nation's leading recycling states, with the Twin Cities Metro Area innovating in recovery of everything from electronics to mercury lamps to carpets. Much of the recycling was funded by a $12 per ton fee at the Hennepin County incinerator. Tip fees used to average about $95 per ton in Hennepin County. The Metro Area had a waste-designation ordinance that kept solid wastes from flowing to inexpensive rural landfills. However, because of federal court rulings on flow control of trash, many wastes now go to cheaper landfills in rural areas and other states and the incinerator tip fee has been reduced. The county also faced lawsuits over its old control ordinances.

Lead–acid batteries, yard waste, whole tires, used oil, brake fluid, antifreeze, power steering fluid, transmission fluid, oil filters, and white goods are banned from Minnesota landfills. Out-of-state waste is also banned unless it meets regulations of the generating state and does not contain commercially generated mercury or items banned from Minnesota landfills, including light bulbs, batteries, stretch wrap, and wood-transport packaging.

Minnesota requires all mercury to be recovered before disposal, and it requires some labeling and commercial recovery requirements. It continues to be the most serious state when it comes to toxics in the waste stream.

New Jersey

New Jersey claims to have a high recycling rate—61% overall, 42% of MSW. This higher figure includes C&D and industrial waste. Higher tip fees, mandatory recycling, mandatory commercial recycling, and recycling education programs have helped the state achieve its high rates. The state has mandatory curbside separation of three materials and leaves. The law requires a 60% (50% MSW) recycling rate and a 65% rate after achieving the 60% in 1995. Costs are recovered through rates at facilities. The state also mandates source separation and recycling in the residential, commercial, and industrial sectors as of

April 1987. According to state figures, New Jersey's in-state disposal dropped from 85% in 1985 to 25% in 1996.

One of New Jersey's biggest issues is flow control. Both New York City and Philadelphia export to or through the Garden State. New Jersey also exports a lot of its waste because so many landfills have closed. In the past, a state law allowed local governments the right to control the flow of trash and recyclables, so a number of incinerators and transfer facilities were built relying on this flow control, forcing haulers to use these facilities. When the courts struck down the law, the bonds were jeopardized. The state passed a bill in 1999 to allow cities to recoup the money lost due to flow-control being rejected. There was a bond issue that was authorized by the Legislature, but as of 2000, many communities were in a very difficult position financially. Moreover, this has affected funding for local recycling programs, and observers say some cities are just doing the minimum; recycling could actually decline in real terms. Governor Christine Whitman has joined Pennsylvania, Virginia, and Indiana in pushing for federal legislation.

New York

The state reports a 42% recovery rate for 1997, an increase from 1996 of 4%. New York City's Fresh Kills landfill is supposed to close by 2001, so the City is trying to offload about 13,000 tons per day of trash to other states, especially Pennsylvania and Virginia. The City continues to have local opposition to its transfer station proposals. Recent statistics show that commercial waste diversion in the City declined 27% between 1998 and 1999. Assemblyman Engelbright introduced the first electronics take-back bill in 1999. New York continues to introduce numerous recycling bills each year, but very few make it to law.

Oregon

Oregon has a 50% recovery goal from the waste stream by 2000. It mandates different recycling rates for different watersheds. The state also requires apartment owners to provide recycling containers, a waste composition study, and a statewide solid waste plan by 1994. Oregon's industry funds the Recycling Markets Development Council. Volume-based fees for trash pickup will be required for grants. "Opportunity to recycle" landfills must have drop-off sites, etc.

There are no mandates on commercial recycling. Portland, however, requires all businesses of all sizes to have a goal of 50% recycling of waste and report to their hauler. The hauler must pick up three of the biggest recyclables at no extra charge to customer (built into fees). The state may have to enforce its recycled content mandate on rigid plastic containers if plastics recycling continues to deline.

Vermont

Vermont has a 1998 mercury law that relates to the procedure for reporting and prosecuting illegal dumping and littering violations. In addition, it allows municipalities to establish requirements for the management of public and private recycling centers. The mercury labeling law was postponed until March 2000. The present law requires labeling and a landfill/incinerator ban of mercury-containing products. This applies to thermostats, switches, medical instruments, relays, lamps, and consumer batteries other than button batteries. It also deregulates such items on transportation for recovery as solid and hazardous waste. The National Manufacturers Electrical Association (NEMA) has sued Vermont's attorney general and secretary of the Agency of Natural Resources (ANR) over this law, claiming it interferes with interstate commerce and conflicts with the FTC Guides for the Use of Environmental Marketing Claims.

Wisconsin

Wisconsin enacted sweeping legislation in 1990–1991 that covered everything from landfill bans of packaging to a recycled content mandate on rigid plastic containers (10% only, no other options) to a big tax to fund recycling. The corporate tax raised about $40 million per year at its peak; the tax was slated to end in 1999.

The state bans lead–acid batteries, tires, yard waste, white goods, motor oil, and all recyclable packaging from landfills. This includes fibers, old newsprint, magazines, corrugated, office paper, glass, aluminum, tin, bimetal, plastic containers, and PS foam. Cities and counties with certified "effective" recycling programs will be exempt from attempting to enforce a sweeping landfill ban on all recyclable packaging.

Wisconsin requires its cities and counties to have a mandatory recycling ordinance and multifamily and commercial collection of a certain number of materials. Appliances must be recycled, staff must be hired, and volume-based fees must be put into place. Cities that are recycling 25% by volume of their waste are exempted.

The state also requires a "per-capita" minimum recycling of each material. For example, the town must recycle 47 pounds per person of newsprint, 7 pounds of corrugated, 1.8 pounds aluminum, 9 pounds steel/bimetal, 5 pounds plastic, 29 pounds glass, 0.4 pounds PS foam, and 9 pounds magazines in nonrural areas. Communities that export waste to Wisconsin have to have a certified program.The problem was that the law allowed any city or town to be a "responsible entity" and get recycling program reimbursement. Thus, instead of 50 counties applying for money, the state ended up with 1200 towns that did not cooperate with each other. This resulted in what has to be one of the most inefficient recycling systems in the nation.

In 1999, Governments were lobbying to get their recycling funding replaced; industry says it has done its part already. Observers say the lawmakers will try to develop a simple per-capita reimbursement system that will encourage local governments to be more cost-effective. However, apparently forcing smaller cities to join with other cities is not politically feasible.

The source for this chapter was *State Recycling Laws Update,* Year-End Edition, 1999, plus other updates of *SRLU.* For additional recycling information from states, refer to Appendix A: Recycling Information and Sources.

CHAPTER 3
CHARACTERIZATION OF WASTE STREAMS

DAVID S. CERRATO
Vice President, Malcolm Pirnie, Inc.
Tampa, Florida

OVERVIEW

Why Identify Garbage Contents?

Every man, woman, and child generates garbage. Our businesses, factories, and institutional establishments generate garbage. The question is not whether we will or will not generate garbage, but how much, what kinds, and whether there is any secondary use for solid waste before we decide to bury or burn it.

Figure 3.1 illustrates the steady increase in the solid waste stream in the United States over the past 30 years. As can be seen, annual solid waste generation in the United States has steadily increased from an estimated 82 million tons in 1960 to an estimated 155 million tons in 1990. This averages out to an increase of approximately 2 percent per year over the last 30 years. Our propensity to produce and discard more has put an ever-increasing burden on our society to effectively manage our solid waste stream.

In order to face up to this challenge we need to know what constitutes our solid waste stream. This is the only way we can plan environmentally sound disposal and, more importantly, efficient and effective resource management and recycling programs.

IMPACT OF CHARACTERIZATION RESULTS

Over the past, say, 20 years, a considerable amount of attention has been given to planning for the disposal of the municipal solid waste stream. Our efforts have focused primarily on the quantity of solid waste generated and to a lesser extent the compositional breakdown of the solid waste stream.

In the past, solid waste characterization was typically one component of a solid waste quantification study. The major concern focused on quantity and not quality. Through the years, however, an increasing awareness as to the composition of the solid waste has been

Anthony J. DeBenedetto, a research scientist at Malcolm Pirnie, assisted with the compilation and analysis of data contained in this chapter.

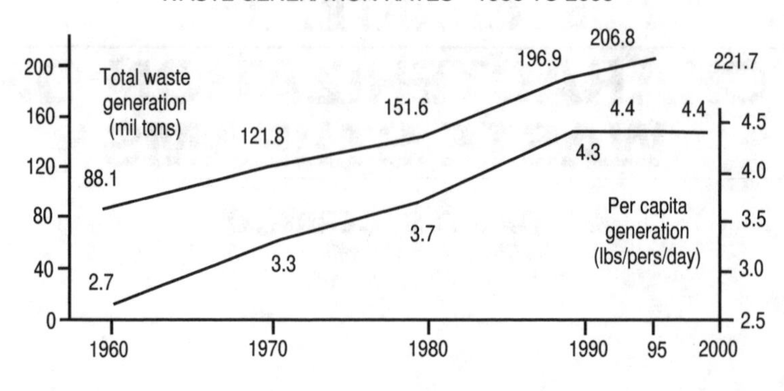

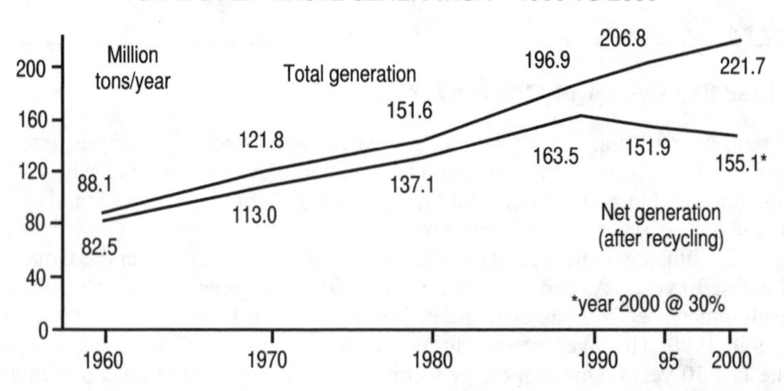

FIGURE 3.1 Estimated solid waste generation 1960–1990. (*Source: Franklin Associates, Ltd.*)

found to be essential for effective long-term solid waste management. As a result, solid waste characterization continues to be an essential element to adequately assess the feasibility of various disposal technologies.

The quantity and composition of the solid waste stream has a direct impact on the technologies selected for management and disposal. For example, the evaluation of waste-to-energy technologies requires a thorough understanding of the solid waste stream, including its value as a fuel source. The composition of the solid waste stream will ultimately determine its higher heating value as a fuel for generating power. The higher heating value (which is the measurement of the amount of energy released from a fuel, in this case solid waste, when burned) will have a direct effect on, first, the feasibility of a waste-to-energy technology: whether the solid waste will burn; and, second, on the sizing of the waste-to-energy facility: as solid waste burns hotter, its processing rate decreases. As a result, the rate of throughput at a waste-to-energy facility is adjusted based upon the higher heating value of the solid waste stream being processed.

The composition of the solid waste stream is also used to assess the potential environmental impacts associated with its disposal. The old adage is true: *You get out what you put in.* Again, using our waste-to-energy example, knowing the constituents of the solid waste stream to be processed in a waste-to-energy facility allows scientists to determine what chemical compounds and gases are likely to be formed during and after the combustion process. This knowledge then enables engineers to design state-of-the-art air pollution control systems capable of mitigating potential adverse environmental impacts.

In the instance of landfill disposal, the composition of the solid waste to be buried has an impact on the assumed in-place density, which in turn affects landfill capacity or landfill life expectancy. The solid waste characterization data are also used to determine what potential chemical compounds are likely to be released in the form of leachate as rain percolates through the landfill. This again enables scientists and engineers to design appropriate leachate collection and treatment systems to mitigate potential adverse environmental impacts.

In the past, solid waste characterization studies tended to be structured to address a limited number of solid waste management issues. The studies typically addressed an overall strategy for waste disposal. Let us call this the *macroapproach.* When conducting a solid waste quantity and characterization study, the macroapproach would typically identify the following solid waste constituents:

Paper and paperboard	Glass
Metals	Plastic
Rubber and leather	Textiles
Wood	Food wastes
Yard wastes	Other wastes
Miscellaneous inorganic wastes	

Although a solid waste composition study must characterize these solid waste constituents at a minimum, it has become essential that a *microapproach* be adopted to analyze each waste constituent by subcomponent. The microapproach identifies solid waste constituents by subcomponent. The microapproach provides information that enables the assessment of various recycling and materials marketing strategies as well as detailed information necessary to plan comprehensive waste management systems. The following is what the breakdown of solid waste subcomponents might look like:

- Paper: newsprint, corrugated, books, magazines, tissue-towels, commercial printed, office paper, packaging
- Glass: container glass (clear, green, amber), other glass
- Metals: aluminum cans, aluminum foil, ferrous, tin
- Plastics: polyethylene tetraphthalate (PET), polystyrene, clear high-density polyethylene (HDPE), colored high density polyethylene (HDPE), polyvinyl chloride (PVC)
- Food wastes
- Rubber
- Leather
- Textiles: fabrics, clothing
- Wood: stumps, pallets, furniture

- Yard wastes: leaves, grass, limbs
- Ceramics
- Construction and demolition debris
- Tires
- Waste oil
- Other wastes
- Miscellaneous inorganic wastes

Reliability of Characterization Study

Although a solid waste characterization study is based on statistical reasoning, the results are estimates. We can, however, develop a study protocol to achieve a level of confidence such that the data obtained would be within 5 to 10 percent of the true mean for all samples taken. The statistical reasoning utilized is explained later in the section on calculations (p. 3.7).

We can know within a reasonably tolerable margin of error the composition and characteristics of a particular solid waste stream. When disaggregated into its subcomponents, the solid waste stream under study can provide the solid waste planner with sufficient information to maximize the efficiency and effectiveness of solid waste recycling and disposal systems. Fluctuations in solid waste generation patterns, however, particularly on an annual basis, can still affect waste characterization study results. Thus, any solid waste composition study must be designed to allow data obtained during 1 year to be analyzed in later years to estimate predictable changes resulting from changing economical or socioeconomical conditions.

Characterization Helps Solid Waste and Recycling Planning

The underlying purpose of a solid waste characterization study is to provide useful information to enable the solid waste planner to assess feasible options for solid waste reuse, recycling, and disposal. If this were purely an economic analysis, the least costly alternative would always be the preferred option. Garbage, however, has become a major political issue, exacerbated by public perception, that goes far beyond pure economics; the least costly option is not always the preferred option. As a result, the characteristics of the waste stream become increasingly important as the solid waste planner develops reuse, recycling, and disposal options, seeking the most efficient and effective alternatives outside of a pure economic model.

The environmental idealist might say that all solid waste ("garbage") is recyclable. But a simple test of taking one day's generation of solid waste from your home and separating it into potential recycling categories reveals that this is not true. Although a significant portion of the waste stream can be recycled, ultimately a greater portion is likely to require disposal. For years, one perception held was that what one individual discarded as worthless was another individual's fortune. One of the best examples of recycling effectiveness can be found in developing countries where the "gold in garbage" myth may be dispelled once and for all.

In developing countries, recycling typically occurs at every stage of the solid waste generation, transfer, and disposal process. The generator carefully screens what is being disposed and removes materials that may have any secondary market value. The solid waste collection crews will then hand-pick through the waste before collecting it. Collection crews typically carry personal containers that are used to separate recyclables. Then, as part of their daily collection route, the crews will stop at various marketplaces to sell any materials of value that they have gathered. It is not uncommon for a collection crew to work all day at a particular collection location, say a marketplace, segregating marketable recyclables and bringing such materials to a known recycler for a fee.

When the collection crew has finally made it to the disposal location, for example, a municipal landfill, scavengers, sometimes numbering in the hundreds, await to pick through the garbage yet again. In the end, every piece of material that had any minimal value as a secondary marketable material has been removed from the waste stream. Regardless of the inefficiencies of such a recycling system, it is effective in removing more than 50 percent of the solid waste stream. In a previous discussion of the efficiency and effectiveness of this type of recycling system for *National Development Magazine,* November/December 1989, I wrote, "If there were gold in garbage, the people who scavenge city dumps for valuable, reusable or recyclable materials would park their Cadillacs at the base of the landfill every morning and could afford the equipment they need to protect their health as they work. If there were gold in garbage, the economics of the business would create a worldwide network of entrepreneurs, brokers, and markets capable of recycling the world's wastes. Recycling would not be a government- and environment-driven business."

The fact remains that although recycling is essential in any solid waste management system, garbage is ultimately just that—garbage. Its composition must be understood before the proper technology can be developed for its disposal. Understanding that our priorities are to maximize reuse and recycling with the knowledge that a significant amount of the solid waste generated will ultimately require disposal, we can develop solid waste planning processes, including solid waste characterization studies, that provide solid waste planners with essential information necessary to develop integrated solid waste management systems capable of providing the maximization of reuse and recycling, and environmentally sound disposal options.

METHODS AND PRACTICES

Developing Study Parameters

Understanding the ultimate objectives of the solid waste plan is essential to developing the solid waste characterization study methods. For example, if the study area desires to initiate waste reduction, the solid waste characterization study must include a microapproach to packaging. All types of packaging must be disaggregated and an analysis conducted as to what effect the reduction of various packaging components could have on the entire waste stream.

Another example would stem from a particular study area's concern over the effectiveness of existing recycling programs and/or "bottle bill" legislation. In this case, the solid waste characterization study must include an assessment of the amounts of recyclable materials remaining in the waste stream received at the disposal location after recycling was to already have taken place. The solid waste characterization study would therefore have to include an evaluation of solid waste generation and disposal patterns. This approach would also require an analysis of individual regions within the study area where recycling effectiveness can be studied on a case-by-case basis. This would allow data to be extrapolated across wider population strata.

As can be seen by just these few variations, the microapproach must be modified for different scenarios to gather information beyond basic solid waste stream characterization. However, prior to designing the various modifications to the microapproach, there must be a fundamental consistency to the basic approach. Consistency in the solid waste characterization study methods and procedures is an essential element if we are ever to have the capability of comparing and extrapolating solid waste characterization data between various regions and populations across the United States and abroad.

Industry Standards

As discussed earlier, the macroapproach to solid waste characterization is first modified to provide a more detailed understanding of solid waste composition. Understanding that the solid waste stream, once separated by component and subcomponent to the maximum extent possible, allows for more efficient and effective solid waste planning. Making the solid waste characterization study parameters consistent allows the solid waste planner to analyze data obtained from other studies to develop comparisons of regional solid waste characteristics.

The American Society for Testing and Materials (ASTM) considered procedures and methods submitted to them to allow solid waste characterization studies to be undertaken by standard protocol. From these submittals, one protocol in particular has attained some level of industry recognition. The particular protocol is known as ASTM Standards, Draft Number 2, October 21, 1988, "Method for Determination of the Composition of Unprocessed Municipal Solid Waste" (the "Standards"). This method provides proposed procedures for measuring the composition of unprocessed municipal solid waste by employing manual sorting. The proposed procedure allows the user to estimate the mean composition of solid waste based on the collection and manual sorting of solid waste samples over a period of time, usually 1 or 2 weeks. The 1- or 2-week procedures can then be repeated during various periods within a year to obtain seasonal solid waste characterization variations.

Summary of Methods. The number of samples to be taken and sorted to estimate solid waste composition is calculated based upon statistical criteria discussed later in this chapter. Vehicle loads of waste are designated for sampling or randomly selected, and sorting samples are collected from the various discharged vehicle loads. Each sorting sample is then manually separated into its individual waste components. Consequently, a weight fraction is calculated for each component, and the mean waste composition is determined using the results of the composition analyses from each of the sorting samples.

Described below are a group of terms used within the prepared Standards that will be used throughout this chapter. These definitions have been included in order to provide a better understanding of the language used when describing the methods involved with undertaking a waste characterization analysis.

The proposed Standards define *unprocessed municipal solid waste* as solid waste in its discarded form, or waste that has not undergone size-reduction or processing. Unprocessed municipal solid waste can be divided into *waste components,* consisting of materials of similar physical properties and chemical composition. These categories include ferrous metals, glass, newsprint, yard waste, aluminum, and so forth.

During the waste stream characterization, *sorting samples,* which are approximately 200- to 300-lb portions of a solid waste disposal vehicle load determined to be representative of the entire vehicle load, are taken and manually sorted into the various waste components. In addition, the waste characterization study may also include laboratory analyses of the waste stream. In this case, a sample such as a *composite item,* which could consist of multiple waste constituents contained in the original sorting sample would be analyzed in a laboratory to determine the higher heating value of the waste and chemical makeup of the waste stream.

Once all waste characterization activities are completed, all the data obtained during the course of the study are compiled and analyzed. The data regarding *solid waste composition* are then typically presented in terms of mass fraction or weight percentage.

Calculations. The Standards provide the formulas and calculations necessary to the number of sorting samples needed during a sampling session. The number of sorting samples required to achieve the level of confidence desired so that the samples taken are representative of the entire solid waste stream under consideration is a function of the solid waste constituency. The proposed ASTM equation for determining the number of samples n is

$$n = \left(\frac{t^* s}{e\bar{x}} \right)^2$$

where t^* is the Student t characteristic that corresponds to the desired level of confidence (see Table 3.1), s is the estimated standard deviation, and x is the estimated mean (see Table 3.2). When utilizing this formula, all values should be represented in decimal notation. For example, a precision value e of 20 percent is presented as 0.2. Table 3.2 provides the suggested values of s and x for waste components. Values of t^* are provided in Table 3.1 for 90 and 95 percent levels of confidence. Then, based upon the desired level of precision and confidence, estimate the number of samples n and components using the proposed ASTM equation provided.

Since the required number of samples varies by component, the sample size, that is, the number of samples to be sorted, is controlled by the component from which the total number of samples to be taken was derived. After determining the number of samples n to be taken, return to Table 3.1 and select the Student t statistic t^* that corresponds to n. Then recalculate using the same formula in order to determine the total number of samples to be taken during the solid waste characterization study.

IMPLEMENTING THE STUDY

Introduction

Waste characterization studies are typically undertaken during the first phase of planning for a comprehensive solid waste management plan that may include resource recovery, various recycling, and other processing and disposal systems. Before the actual field studies are undertaken, background information is required in order to adequately design the program methodology. To achieve a complete understanding of the waste stream, the implementing agency must determine how much waste is being generated, where it is coming from, and what it is made up of in general terms.

The solid waste characterization study must subdivide the solid waste stream into microcomponents, for example, plastics should be further separated into PET, HDPE, PVC, and mixed other plastics, if it is going to provide the information necessary for effective

TABLE 3.1 Values of *t* Statistics (t^*) as a Function of Number of Samples and Confidence Interval

Number of samples, *n*	90%	95%
2	6.314	12.706
3	2.920	4.303
4	2.353	3.182
5	2.132	2.776
6	2.015	2.571
7	1.943	2.447
8	1.895	2.365
9	1.860	2.306
10	1.833	2.262
11	1.812	2.228
12	1.796	2.201
13	1.782	2.179
14	1.771	2.160
15	1.761	2.145
16	1.753	2.131
17	1.746	2.120
18	1.740	2.110
19	1.734	2.101
20	1.729	2.093
21	1.725	2.086
22	1.721	2.080
23	1.717	2.074
24	1.714	2.069
25	1.711	2.064
26	1.708	2.060
27	1.706	2.056
28	1.703	2.052
29	1.701	2.048
30	1.699	2.045
31	1.697	2.042
36	1.690	2.030
41	1.684	2.021
46	1.697	2.014
51	1.676	2.009
61	1.671	2.000
71	1.667	1.994
81	1.664	1.990
91	1.662	1.987
101	1.660	1.984
121	1.658	1.980
141	1.656	1.977
161	1.654	1.976
189	1.653	1.973
201	1.653	1.972

Source: Proposed ASTM Standards, Draft Number 2, October 21, 1988.

TABLE 3.2 Values of Mean x and of Standard Deviation s for Within Week Sampling to Determine MSW Component Composition*

Component	Standard deviation, s	Mean, x
Mixed paper	0.05	0.22
Newsprint	0.07	0.10
Corrugated	0.06	0.14
Plastic	0.03	0.09
Yard waste	0.14	0.04
Food waste	0.03	0.10
Wood	0.06	0.06
Other organics	0.06	0.05
Ferrous	0.03	0.05
Aluminum	0.004	0.01
Glass	0.05	0.08
Other inorganics	0.03	0.06
		1.00

*The tabulated mean values and standard deviations are estimates based on field test data reported for municipal solid waste sampled during weekly sampling periods at several locations around the United States.

Source: Proposed ASTM Standards, Draft Number 2, October 21, 1988.

solid waste planning. In order to evaluate the potential for source reduction, the waste stream is broken down into specific and indirect waste products. As a means of reducing waste generation, particular attention is given to the waste stream, that is, packaging, advertisements, labels, and so on. Sampling programs must also provide for both seasonal and geographical fluctuations in the quantity and composition of waste types.

Additionally, waste composition programs must provide information on waste flow by specific generator types, enabling the agency undertaking the solid waste characterization study to target specific generators for recycling programs to increase the agency's potential for success. Developing the solid waste characterization study specific to program needs will save a significant amount of time and money and can help to assure an accurate accounting of the entire solid waste stream.

Presampling Activities

The importance of planning prior to initiating a solid waste characterization study cannot be overstressed. No matter how much time is spent planning from an office, without a full understanding of study objectives and knowledge of facility operations at the site where the field work is going to take place prior to initiating the solid waste characterization study, there is the potential risk of incorrectly designing the program methodology.

As previously discussed, understanding the full range of informational needs to meet study objectives is essential to developing the solid waste characterization study methodology. A solid waste characterization study that includes, for example, 4 weeks of sampling, one in each season of the year to capture seasonal variations, and complete laboratory testing is an expensive undertaking for any municipality. Such studies can range in cost from \$300,000 to \$700,000, depending on the level of detail required, number of sites for sampling, and so on. Since the majority of the study costs relate directly to the field work, it is critical that all informational needs be identified prior to initiating the solid waste characterization study. All study objectives must be clearly defined. Every effort must be made during the field sampling activities to acquire all essential data. Any data needs not adequately addressed could result in repeating field efforts. Repetition of any portion of the field effort will ultimately have a considerable impact on study costs.

To help eliminate the possibility of not addressing the full complement of study needs, the project team should at a minimum:

- Define the study area
- Review socioeconomic data within the study area
- Develop a list of all private and public waste haulers operating in the study area
- Contact all of the haulers identified and discuss the solid waste characterization study
- Visit each disposal location whether sampling is to take place there or not
- Review previous solid waste plans and recycling programs
- Analyze quantities and composition of recyclables removed from the waste stream prior to disposal
- Schedule the sampling periods to address possible seasonal and cyclical variations that may impact waste generation

Once the project team has clearly defined the study parameters, the solid waste characterization study, including these presampling activities, can begin.

Waste Categories. The next step following the presampling activities discussed in Section 3.2 is to define the waste categories to be sampled. A sample list of waste components for sorting is shown in Table 3.3. A description of some of the waste component categories provided in Table 3.1 is given in Table 3.4. Other waste components can also be defined and sorted as needed. Table 3.3 includes those components most commonly used to define and report the composition of solid waste. For consistency, it is recom-

TABLE 3.3 List of Waste Component Categories

Mixed paper	Other organics
High-grade paper:	Ferrous:
Computer printout	Cans
Other office paper	Other ferrous
Newsprint	Aluminum:
Corrugated	Cans
Plastic:	Foil
PET bottles*	Other aluminum
HDPE bottles†	Glass:
Film	Clear
Other plastic	Brown
Yard waste	Green
Food waste	Other inorganics
Wood	

*PET = polyethylene tetraphthalate.
†HDPE = high-density polyethylene.

TABLE 3.4 Description of Some Waste Component Categories

Category	Description
Mixed paper	Office paper, computer paper, magazines, glossy paper, waxed paper, other paper not fitting categories of "newsprint" and "corrugated."
Newsprint	Newspaper.
Corrugated	Corrugated medium, corrugated boxes or cartons, brown paper (i.e., corrugated) bags.
Plastic	All plastics.
Yard waste	Branches, twigs, leaves, grass, other plant material.
Food waste	All food waste except bones.
Wood	Lumber, wood products.
Other organics and combustibles	Textiles, rubber, leather, other primarily burnable materials not included in the above component categories.
Ferrous	Iron, steel, tin cans, bimetal cans.
Aluminum	Aluminum, aluminum cans, aluminum foil.
Glass	All Glass.
Other inorganics and noncombustibles	Rock, sand, dirt, ceramics, plaster, nonferrous, nonaluminum metals (copper, brass, etc.), bones.

mended that, at least, the left-justified categories in Table 3.3 be sorted. Similar breakdowns of solid waste composition would thus be available for comparison, if desired.

Inspecting the Site. To make sure that the solid waste characterization study will generate the information required and present a statistically reliable breakdown of the solid waste stream, those persons who will actually be involved with the field work should visit and tour each of the sites with the operators of the involved facilities. It should take no longer than 3 to 5 days to acquire the information necessary prior to starting the actual solid waste sampling program. During these site visits, staff should also examine the physical aspects of the study area and obtain information regarding the following:

- Identification and quantification of incoming waste
- Identification and quantification of all types of incoming waste routes, schedules and delivery information
- Private hauler information
- Facility operating procedure
- Utilities accessibility, e.g., electricity, water
- Identification of primary and secondary sampling areas

The information obtained will be used to refine and finalize the sampling procedures. It will enable easy access to vehicles to be sampled and allow for operations to take place as safely and efficiently as possible. This extra planning effort minimizes potential problems in the field, facilitates vehicle identification, familiarizes field staff with vehicle schedules, and involves facility operators and inspectors. The information obtained will be extremely valuable to field staff during the actual study.

The cooperation of the owner and operator of the solid waste management facility where the solid waste sampling will take place is critical to the success of the solid waste characterization study. All parties involved both directly and indirectly must understand the objectives, the procedure to be used, and data needs. In addition, private haulers utilizing the disposal location where the solid waste sampling is to take place must be contacted to obtain cooperation. It is recommended that private haulers and site operators be sent written confirmation of these discussions to ensure cooperation prior to initiating the field work. This information should include

- Waste sampling schedules
- Proposed methods and procedures that may affect operations
- Draft questionnaire to be used to obtain data from operators and facility users
- Weighing procedures
- Location of weighing and sorting areas
- Overall management of sorted solid waste

Crew Size. The size of the field sampling crew is governed by the scope of the solid waste characterization study, the amount of total incoming waste at the sampling location, and the recommended number of samples to be sorted. A composition study that involves only a waste characterization by component analysis usually will require approximately four to six sorters as well as one supervisor who functions as the scale operator and data recorder. If the solid waste characterization study includes visual inspections of incoming waste loads, commercial and industrial waste assessments, and hauler interviews, additional staff will be required. Each added activity may require up to an additional two staff persons. Actual staffing, however, is usually suggested by the agency proposing the study prior to actually staffing. The estimated staffing requirements presented here are provided as a guide to allow that sufficient personnel are available on site. It is recommended, however, that the study sponsor discuss personnel and staffing with the study performer prior to the commencement of the field work. This will enable the sponsor to modify or supplement staffing in a manner that may reduce the solid waste characterization study costs.

Field Safety Plan. The field safety plan is an important part of the field sampling program. Sorters manually picking through mixed municipal solid waste may encounter potentially hazardous conditions. In order to minimize the risks associated with waste sorting, a field sorter safety plan should be developed. This safety plan should include, at a minimum, the following:

- Standard operating procedures and precautions
- Identification of required protective equipment
- General first aid procedures
- Procedures in the event hazardous waste is encountered
- Emergency procedures

Prior to the initiation of site activities, the precautions and procedures for the study should be reviewed with operating and sorting personnel. It should be remembered that solid waste is likely to contain sharp objects such as nails, razor blades, and pieces of glass. It may also include hypodermic needles, poisonous sprays, punctured aerosol cans, and so forth. Personnel should be reminded of this danger and instructed to brush waste particles aside while sorting, as opposed to forcefully projecting their hands into the mixture. Personnel handling and sorting solid waste should wear appropriate protection, including heavy leather or rubber (puncture-proof) gloves, hardhats, safety glasses, and heavy duty footgear with steel toes.

During the process of collecting samples for sorting, large machinery will be involved.

This may include refuse collection vehicles, payloaders, dumptrucks, and so forth. The dumping process can scatter refuse, create dust, and throw waste products into the air. Such projectiles can include flying glass particles from breaking glass containers and metal lids from plastic and metal containers that burst under pressure when run over by heavy equipment. The problem is particularly acute when the waste handling surface is of high compressive strength, e.g., concrete. Field personnel should be made aware of the danger and wear eye and head protection. Additionally, personnel should be instructed to refrain from looking in the direction of unloading vehicles or of heavy equipment that is in the process of moving or placing solid waste. Field personnel must also be dressed appropriately in heavy-duty clothing and vests with iridescent markings. Safety lights should also be used around the sorting area.

Apparatus. Sufficient metal, plastic, or fiber containers for sorting and weighing each waste component, labeled accordingly, should be acquired for use throughout the solid waste sampling study. For components that will have a substantial moisture content (e.g., food waste), metal or plastic containers should be used to avoid absorption of moisture by the containers, and thus preclude numerous weighings to maintain an accurate tare weight for the container. Storage containers should be weighed at the beginning of each day, or more frequently if necessary, in order to maintain accurate tare weights. An electronic scale with a capacity of at least 200 lb, and a precision of at least 0.1 lb, should be used. These scales should be calibrated daily according to manufacturer's specifications. Tabletop scales should be positioned on a clean, flat, and level surface. Portable truck scales may also be employed when the sampling site is not equipped with a truck scale to determine arriving quantities of waste as well as a method to calculate waste density.

Sorting personnel should be equipped with heavy-duty tarps, shovels, rakes, push brooms, dust pans, hand brooms, magnets, a sorting table, first aid kit, miscellaneous small hand tools, traffic cones, traffic safety vests, puncture-proof gloves, hardhats, safety glasses, and heavy-duty steel-toed boots.

The area and location selected for performing designated manual sorting activities and weighing operations should be flat, level, and distinctly set away from the normal waste handling and processing areas at the study site. The surface should be swept clean or covered with a clean, durable tarp prior to discharge of the load. A sample layout of a typical sorting area is presented in Fig. 3.2.

Solid Waste Samples for Manual Sorting. As previously discussed, the number of samples to be sorted is calculated based upon a statistical formula. The methods and procedures described here summarize the ASTM proposed procedures.

Sorting samples are collected from normal refuse collection vehicles. A refuse collection vehicle could be carrying an 8- to 10-ton load. Therefore, a solid waste sample must be drawn from that load which is representative of the entire load. The sorting sample is then manually sorted into its waste components and weight fractions calculated for each component. The mean waste composition can be calculated later using the results of the composition of each of the sorting samples.

Vehicles for sampling can be selected at random during each day of the sampling period or can be preselected to provide a solid waste mix from varying socioeconomic strata within a regional jurisdiction. This decision is made during the presampling activities. For a weekly sampling period of x days, the number of vehicles sampled each day shall be approximately n/x, where n is the total number of vehicle loads to be selected for estimating waste composition. A weekly period can vary from 5 to 7 days.

Vehicles designated for sampling are directed to the area set aside for discharge of the load and collection of the sorting sample. The vehicle operator is instructed to discharge

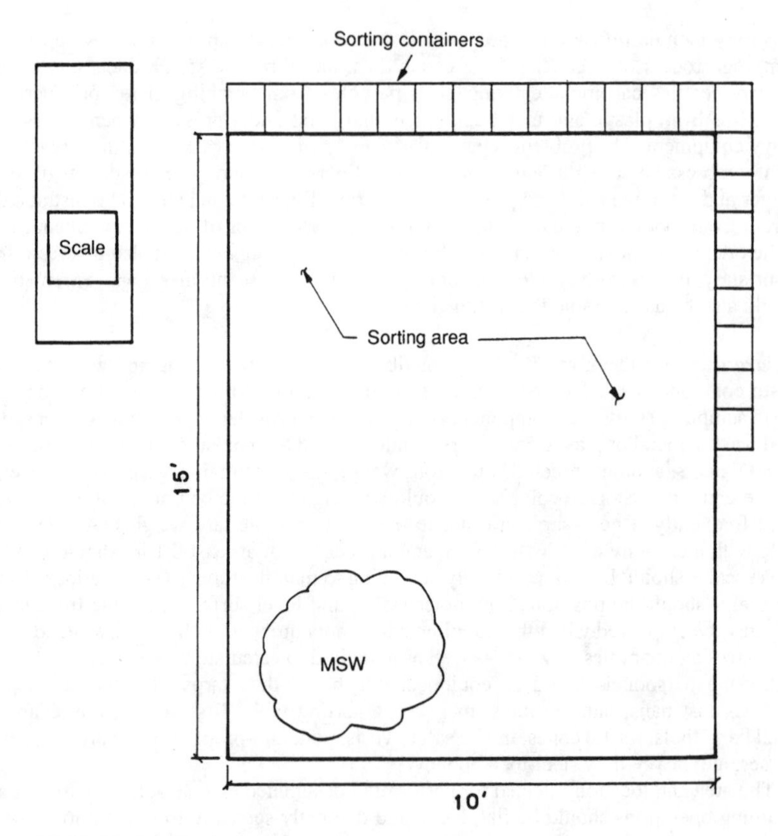

FIGURE 3.2 Sorting activities layout.

the load onto the clean surface in one continuous pile, that is, to avoid gaps in the discharged load. Information from the vehicle operator regarding the waste load is also collected at this time, prior to the vehicle leaving the discharge area. Once the vehicle has discharged its waste at the designated area, it is necessary to remove a manageable portion of the load that is representative of the entire vehicle load. There are basically two procedures that can be employed to accomplish this.

The first method requires using a payloader. The operator of the payloader uses the bucket of the payloader to thoroughly mix the entire vehicle waste load and levels it off into one continuous pile. Then the operator, once again using the bucket of the payloader, removes approximately one-fourth of the vehicle's load and places it away from the remainder of the discharged waste load. Again, the removed portion is mixed and one-quarter is separated. This procedure is repeated until the remaining quarter is approximately 200 to 300 lb. This final waste sample is then hand-sorted into its waste constituents.

The second method also requires a payloader and operator. In this instance, however, rather than mixing the entire vehicle's discharged waste load before removing a sample, the sample is removed longitudinally along one entire side of the discharged load. The sample should form a mass weighing approximately 1000 lb. The sample is then thoroughly mixed and one-quarter of it is selected to be manually sorted. This method is less time consuming and the results are consistent with those of the first method described.

Only one sorting sample can be selected from each designated collection vehicle load. All handling and manipulation should be conducted on clean surfaces. If necessary, the sorting sample should be removed to a secured manual sorting area. Waste material not selected for manual sorting should be removed from the area as soon as possible and disposed of.

Manual Sorting Procedures

To prepare a sort area, a tarp or a sheet of plastic is placed on a level area near the discharge location. The tarp or plastic sheet is surrounded (and held down) by the containers used to hold the sorted materials. The number of containers should be in approximate proportion to the expected waste composition. Corrugated cardboard may be too bulky to place in the containers, so a separate area should be identified for corrugated storage. The containers should be located around the sorting area so that, where possible, containers for the same component are located on opposite sides of the sample pile to provide easy access for the sorting crew. A sorting table can be used for the manual sorting process. A table makes the sorting process easier for the field crew members and enables the identification of fine materials that may be lost if sorting were to take place on the floor. Tables can also be designed to include cut-outs where screens of various sizes can be used to allow fine materials to fall through into containers for later evaluation.

The sort containers are weighed empty to obtain tare weights. The containers are labeled or numbered and the tare weights clearly marked on each container. The tare weight corresponding to each container label or number is recorded for future reference. The containers should be cleaned periodically to ensure a consistent and accurate tare weight.

To begin the sorting operation, the first portion of a sample is brought to the sorting area and dumped onto the main sorting table. The crew members begin sorting the sample by hand. The crew supervisor oversees the operation, checking each container for separation quality and assists in classifying questionable items. When about 90 percent of the first portion of the sample has been sorted, another portion of the same sample is dumped on top of the first portion and the sorting continues.

All components are sorted manually. After the entire sample has been sorted, the fines on the sort table are dumped into a container. The table is removed and the sheet is picked up at the corners and the contents are examined and placed into the appropriate containers.

The plastic sheet is then cleaned and placed back on the ground. The filled containers are weighed, and the gross weights recorded. The containers are emptied and placed in their appropriate locations for the next round of sorting.

Sample selection and sorting in this manner can be performed in approximately one hour. Time reductions can be achieved when

- The next sample to be sorted is being selected while the previous sample is being sorted.
- Multiple sorting areas are provided adjacent to one another and additional staff are provided.

When the sorting of one sample is completed and all data compiled, additional comments on the characteristics of the material may need to be recorded following visual analysis. For example, the approximate percentage of sheet stock and castings might be noted for the aluminum fraction. Similarly, a higher than normal moisture content in the newsprint or aluminum categories should be noted as well as weather conditions. The results of the survey, however, should be recorded with no adjustments for moisture characteristics, and so on.

Laboratory Analyses

If the solid waste characterization study includes an analytical assessment of the compounds found in the solid waste during sorting activities represented by the samples taken, the assistance of a certified laboratory will be required. The laboratory selected to perform the tests should be qualified to do so and should submit a full protocol prior to beginning any testing. Typically, the following tests are required:

- *Heating value:* Heating value is usually expressed in British thermal units per pound (Btu/lb) of waste and is a measure of the waste's energy content available through burning it.
- *Proximate analysis:* the determination of total moisture content, volatile matter, fixed carbon, and ash content.
- *Ultimate analysis:* includes the determination of ash, carbon, hydrogen, sulfur, oxygen, nitrogen, and chlorine in the waste.
- *Elemental analysis:* a broad category including the determination of parameters such as acidity, herbicides, asbestos, and dioxin.

The laboratory will be required to handle samples weighing between 10 and 20 lb, be able to grind and thoroughly mix the samples without contamination, and to conduct the analytical testing.

Samples are selected for laboratory analysis by either reconstructing those samples taken in the field in proportion (percent weight) to the original sample weight or by repeating a similar exercise to that of the quartering method where the 200- to 300-lb sample is placed in a conical pile and shaved until a 20- to 25-lb sample remains. Grab sampling is inappropriate because of the statistical unreliability of the data. Each sample taken in the field or a daily composite sample of all samples taken on a particular day for each generator segment (residential, commercial, etc.) is sent to the laboratory for testing. Proximate and ultimate analyses of the samples are conducted in accordance with ASME, ASTM Committee E-38 Standards and Procedures.

- *Proximate analysis:* total moisture, ash (including percent by volume), volatiles, fixed carbon, heating value (Btu/lb on an as-received and moisturefree basis)
- *Ultimate analysis:* ash, carbon, hydrogen, nitrogen, oxygen, sulfur

Incoming Solid Waste Survey

As part of the solid waste characterization study, it is important to quantify all incoming solid waste to the facility from which the sampling is taking place. Data should be collected related to all incoming waste quantities. Sufficient personnel should be made available to staff the solid waste facility scalehouse to collect information on waste quantity. Every incoming truck (to the extent possible) will be studied, weighed, and the driver surveyed. The drivers are queried as to the type of waste carried (commercial, residential, mixed, etc.), the area where the waste was collected, the type of establishment where the waste was generated, and percentage of the vehicle full. In addition, drivers of mixed loads are asked to estimate the percentage breakdown of the load by waste type.

The amount of waste entering the solid waste disposal facility is quantified according to its point of origin and generator segment (e.g., residential, commercial, or industrial). Additionally, every vehicle will be quantified as to its weight or volume and density. The quantity data for solid waste generated within the study area shall be defined within the following framework:

- Quantity of waste types by geographic points (origin).
- Specific route information, if available.
- Weight or volume and corresponding density characteristics expressed in terms of daily, average, peak, and minimum flow to the solid waste facility for the seasonal sampling in question and for the year as a whole.
- The total quantity of waste for the region, by type, shall be clearly depicted in the results. Weight, volume, and corresponding density characteristics expressed in terms of regional daily, average, peak, and minimum flow for seasonal periods and for the year as a whole.

Visual Inspection

Although not as reliable as the solid waste sorting process, a visual inspection of incoming solid waste is a recommended addition to any solid waste characterization study. The purpose of the visual inspection is to visually characterize by component all incoming solid waste. This effort allows for a complete understanding of the general characteristics of all solid waste being disposed of in the study area.

Sufficient personnel should be available to visually inspect all incoming waste loads. The waste inspectors are responsible for analyzing the waste to estimate its composition. This information will ultimately be cross-referenced with information obtained by the scalehouse data collectors to enhance overall data reliability.

During a visual waste inspection, two phases of study are undertaken. The first phase is a hauler interview and the second is a waste characterization. A hauler interview consists of a series of short-answer questions which are used to develop a "history" or background information on a particular delivery of waste. Some of the questions include inquiries as to where the waste was collected, from what types of establishments, the types of waste composing the load, and the capacity of the vehicle utilized. The waste characterization consists of disaggregating the waste by component. Following the hauler interview, the vehicle is observed during its normal discharging cycle. Once the vehicle discharges its load, staff estimates the composition of the waste. The compositional analysis here is not as detailed as the sorting analysis, but it can provide valuable information as to its general characteristics. The solid waste can be characterized by percent as residential, commercial, industrial, and so forth, and can provide insight as to the general character of the solid waste stream as a whole. The two phases serve to complement each other, where each is used to justify or qualify the other.

Samples of the types of survey forms used during a visual waste inspection for both a hauler/scalehouse interview and waste assessment are included as Figs. 3.3 through 3.7.

Seasonal Variations

Quantities of waste delivered to a disposal site can vary by the hour, day of the week, week of the month, month of the year, seasonally, and annually. As a result, fluctuations

INTERVIEW SURVEY FORM

Date:

MPI Personnel:

Vehicle I.D. (permit sticker):

Carter:

Vehicle Type (Circle One): PACKER ROLL-OFF TRANSFER TRAILER OTHER (Specify)

Vehicle Capacity (Cubic Yards):

Vehicle Capacity Utilized (%):

Net Weight (Pounds/Tons):

Waste Type (Circle Applicable: RESIDENTIAL COMMERCIAL/LIGHT INDUSTRIAL OTHER (Specify)

Percent of Total Load: RESIDENTIAL _____ COMMERCIAL/LIGHT INDUSTRIAL _____ OTHER _____

Waste Origin:	Municipality	Percent of Total Load	Percent Residential	Percent Commercial/Light Industrial
1.				
2.				
3.				
		100%	100%	100%

How much waste originates from apartment complexes?

Vehicle Destination (Circle One): LANDFILL SHREDDER

FIGURE 3.3 Survey form—equipment visual inspection.

SHREDDER/LANDFILL SURVEY FORM

Date:

MPI Personnel:

Vehicle I.D. (permit sticker):

Carter:

Vehicle Type (Circle One): PACKER ROLL-OFF TRANSFER TRAILER OTHER (specify)

Vehicle Capacity (Cubic Yards):

Waste Characterization:		Percent of Total Load
	RESIDENTIAL	
	COMMERCIAL/LIGHT INDUSTRIAL	
	OTHER (Specify)	______
		100%

Processable vs. Unprocessable:		Percent of Total Load
	PROCESSABLE	
	UNPROCESSABLE	______
		100%

FIGURE 3.4 Survey form—equipment visual inspection.

Residential waste analysis

MPI Personnel:

Hauler Name:

Vehicle Type (Circle One): Packer Roll-Off Transfer Trailer Other (Specify)

Waste Stream Analysis:

Component	Quantity, % (Circle One)																			
Paper	5	10	15	20	25	30	35	40	45	50	55	60	65	70	75	80	85	90	95	100
Bagged Household	5	10	15	20	25	30	35	40	45	50	55	60	65	70	75	80	85	90	95	100
Corrugated	5	10	15	20	25	30	35	40	45	50	55	60	65	70	75	80	85	90	95	100
Metals	5	10	15	20	25	30	35	40	45	50	55	60	65	70	75	80	85	90	95	100
Glass	5	10	15	20	25	30	35	40	45	50	55	60	65	70	75	80	85	90	95	100
Plastic	5	10	15	20	25	30	35	40	45	50	55	60	65	70	75	80	85	90	95	100
Wood	5	10	15	20	25	30	35	40	45	50	55	60	65	70	75	80	85	90	95	100
Yard waste	5	10	15	20	25	30	35	40	45	50	55	60	65	70	75	80	85	90	95	100
Tires/rubber	5	10	15	20	25	30	35	40	45	50	55	60	65	70	75	80	85	90	95	100
Brick/concrete	5	10	15	20	25	30	35	40	45	50	55	60	65	70	75	80	85	90	95	100
Dirt/fines	5	10	15	20	25	30	35	40	45	50	55	60	65	70	75	80	85	90	95	100
Textiles	5	10	15	20	25	30	35	40	45	50	55	60	65	70	75	80	85	90	95	100

FIGURE 3.5 Form—data collection, residential waste.

Landfill Survey Form

Date: ________ Hauler Name: ________ Vehicle I.D.: ________

Vehicle Type (Circle One): PACKER ROLL-OFF TRANSFER TRAILER OTHER (Specify): ________

Vehicle Capacity: ________ Vehicle Capacity Utilized (%): ________

Waste type (Circle Applicable): RESIDENTIAL COMMERCIAL INDUSTRIAL OTHER (Specify): ________

Percent of Total Load: RESIDENTIAL ________ COMMERCIAL ________ INDUSTRIAL ________ OTHER ________

Waste Origin:	Municipality	% of Total Load	Residential (%)	Commercial (%)	Industrial (%)	Other (%)
1.						
2.						
3.						
4.						
5.						

Waste Stream Analysis:

Component	Quantity, % (Circle One)																			
Paper	5	10	15	20	25	30	35	40	45	50	55	60	65	70	75	80	85	90	95	100
Bagged household	5	10	15	20	25	30	35	40	45	50	55	60	65	70	75	80	85	90	95	100
Corrugated	5	10	15	20	25	30	35	40	45	50	55	60	65	70	75	80	85	90	95	100
Metals	5	10	15	20	25	30	35	40	45	50	55	60	65	70	75	80	85	90	95	100
Glass	5	10	15	20	25	30	35	40	45	50	55	60	65	70	75	80	85	90	95	100
Plastic	5	10	15	20	25	30	35	40	45	50	55	60	65	70	75	80	85	90	95	100
Wood	5	10	15	20	25	30	35	40	45	50	55	60	65	70	75	80	85	90	95	100
Yard waste	5	10	15	20	25	30	35	40	45	50	55	60	65	70	75	80	85	90	95	100
Tires/rubber	5	10	15	20	25	30	35	40	45	50	55	60	65	70	75	80	85	90	95	100
Brick/concrete	5	10	15	20	25	30	35	40	45	50	55	60	65	70	75	80	85	90	95	100
Textiles	5	10	15	20	25	30	35	40	45	50	55	60	65	70	75	80	85	90	95	100
Carcasses	5	10	15	20	25	30	35	40	45	50	55	60	65	70	75	80	85	90	95	100
Coal mine waste	5	10	15	20	25	30	35	40	45	50	55	60	65	70	75	80	85	90	95	100
Sludge	5	10	15	20	25	30	35	40	45	50	55	60	65	70	75	80	85	90	95	100

FIGURE 3.6 Form—data collection, landfill survey.

Field Waste Study
MPI Personnel:
Name of Business:
Type of Business:
Waste Stream Analysis:

Component	Quantity, % (Circle One)																			
Paper	5	10	15	20	25	30	35	40	45	50	55	60	65	70	75	80	85	90	95	100
Corrugated	5	10	15	20	25	30	35	40	45	50	55	60	65	70	75	80	85	90	95	100
Metals	5	10	15	20	25	30	35	40	45	50	55	60	65	70	75	80	85	90	95	100
Glass	5	10	15	20	25	30	35	40	45	50	55	60	65	70	75	80	85	90	95	100
Plastic	5	10	15	20	25	30	35	40	45	50	55	60	65	70	75	80	85	90	95	100
Wood	5	10	15	20	25	30	35	40	45	50	55	60	65	70	75	80	85	90	95	100
Textiles	5	10	15	20	25	30	35	40	45	50	55	60	65	70	75	80	85	90	95	100
Others	5	10	15	20	25	30	35	40	45	50	55	60	65	70	75	80	85	90	95	100

FIGURE 3.7 Form—data collection, field waste study.

in the solid waste stream may occur over a period of time. Typically the solid waste characterization study analyzes variations on a seasonal basis. Seasonal estimates usually require four separate week-long programs (winter, spring, summer, fall) at each site. The week-long programs are considered a minimum and also provide estimates of variations with the day of the week.

Although a 4-week program (one sampling in each of the four seasons of the year) is the most preferred approach, it is usually possible to assess significant seasonal variations by conducting the solid waste characterization study in two of the four seasons, for example, winter and summer. Notwithstanding, it is essential that when scheduling the weeks of sampling that specific regional characteristics are considered. For example, certain industries may manufacture specific goods and services during certain times of the year. Their processes may have a considerable impact on the overall solid waste stream being generated during certain times of the year. Therefore, it is necessary that the solid waste characterization study coincide with their production schedules. Additionally, certain regions experience increased tourism during one or more seasons of the year that affects solid waste generation from both a quantity and quality standpoint. The solid waste characterization study must assess these associated impacts to solid waste generation.

With respect to solid waste quantities, seasonal variations in waste quantity are estimated from site records and waste stream assessment data. In order to better understand how seasonal variation has an effect on the waste stream, Table 3.5 has been provided and presents a summary of seasonal variation data obtained during several waste composition studies. As can be seen in Table 3.5, subtle changes can be noticed between seasons for certain materials; that is, newsprint, corrugated, plastics, and other wastes show greater variation than other paper, ferrous metals, aluminum, and glass between the seasons.

Geographical Variations

The composition of solid waste varies dramatically with respect to geography. Waste composition can vary from city to city, county to county, state to state, country to country, and even continent to continent. These differences could be due in whole or in part to any of the following factors: economics, population, proximity to collection, or political and social factors.

Table 3.6 provides a summary of seasonal variation data compiled from waste composition studies conducted in four regions of the United States.

TABLE 3.5 Seasonal Variation and Waste Composition

	Winter, %	Summer, %
Newsprint	5–10	8–12
Corrugated	5–10	8–12
Other paper	35–40	32–36
Plastics	8–12	10–15
Ferrous metals	3–6	2–6
Aluminum	<1	<1
Glass	5–10	8–12
Other wastes	25–30	18–30

TABLE 3.6 Geography and Waste Composition in the United States

	Northeast, %	Southeast, %	Midwest, %	West, %
Newsprint	8–10	8–12	5–8	2–12
Corrugated	9–12	8–12	5–12	5–15
Other paper	22–32	30–35	20–25	15–25
Plastics	7–10	5–10	5–10	6–10
Yard waste	10–17	3–7	9–13	2–12
Ferrous metals	2–6	1–5	3–6	5–8
Aluminum	1–3	1–5	<1	<1
Glass	6–12	5–8	2–6	2–8
Other wastes	18–23	20–25	5–15	5–10

Commercial versus Residential

In order to develop a comprehensive understanding of the total waste stream, the waste is often segmented according to residential and commercial-industrial generators. Therefore, it is necessary to sample a representative number of each of these waste generators. Since solid waste is delivered in open as well as closed refuse vehicles, a sampling, either physical or visual, of a representative number of both types of vehicles should be undertaken.

It should be understood that the composition of the residential and commercial waste streams can vary dramatically. In Table 3.7, a comparison between the commercial and residential waste streams of a typical suburban population is presented. As can be seen in Table 3.7, the amount of some waste constituents (e.g., corrugated) can differ significantly in composition between the residential and commercial waste streams.

In addition, the subcomponents of each of the waste types tend to differ between the residential and commercial sectors. For instance, the paper category in the residential sector consists of newsprint, tissue, paper towels, and other types of paper which are discarded with everyday household trash. But in the commercial sector, the paper category is mainly composed of office paper discards including computer paper. This is a prime example of what makes each of these generator sectors so unique. Often when solid waste

TABLE 3.7 Residential vs. Commercial Waste Composition

	Residential, %	Commercial, %
Paper	20–40	25–50
Corrugated	8–12	20–30
Plastics	6–8	10–15
Metals	4–8	2–5
Other wastes*,†	40–50	18–24

*Residentially generated "other wastes" are typically composed of the following and estimated at the following percentages by weight: yard wastes, 17%; aluminum, 1%; glass, 6%; textiles, 4%; food waste, 9%; wood, 5%; rubber, 2%; miscellaneous wastes, 3%. Total = 47%.

†Commercially generated "other wastes" are typically composed of the following and estimated at the following percentages by weight: glass, 3%; food waste, 9%; wood, 4%; miscellaneous wastes, 4%. Total = 20%.

characterization studies are in the planning phases, special and unique approaches need to be developed and tailored to the specific needs of each individual study. An example of one such unique procedure is discussed later in this chapter.

Waste Sampling and Composition

The sampling methodology represents the procedures for measuring the composition of solid waste generated within the study area. The procedure considers the mean composition of solid waste based on the collection and manual sorting of a number of samples of waste over the study period. This procedure is then duplicated for each additional sampling session. The procedure should identify

- A representative number of sorting samples to be representative of the total waste stream
- Manual sorting of the waste into individual components
- Data reduction
- Reporting of the results

Selected trucks are unloaded and at the sorting area. Samples of approximately 200 to 300 lb are to be selected from each load at random. Random identification can be performed either conceptually or physically but cannot be left to the discretion of the sort crew. A waste composition worksheet (Table 3.8) is needed to collect data of waste stream components.

ANALYZING THE RESULTS

Solid Waste Quantity and Composition

It is important to determine the level of detail required from the waste assessment program. The basic measures of solid waste quantity are tons per hour, tons per day, tons per week, and tons per year. These basic measures may be divided into components such as

- Tons per day of residential, commercial, industrial, and demolition wastes
- Tons per day collected by individual hauler or from specific routes in the study area
- Pounds per capita per day generated or collected, based on various socioeconomic characteristics

Quantity measurements may also include waste volumes expressed in cubic yards, but volume measurements must also be associated with weight information since variations in waste density can occur. These variations occur primarily because of differences in waste composition and compaction equipment.

The units most often used when referring to solid waste quantity are tons per day and pounds per capita per day. Tons per day is typically used in reference to the quantity of waste received at a facility, while pounds per capita per day refers to waste generation as a function of population.

Following the field survey, the quantity data for each day and week for each site should be analyzed. Load-density measurements should be computed by dividing net load weights by truck capacity (in cubic yards). It may be helpful to include a column on the data collection forms for this computation, as well as the computation of net weight.

TABLE 3.8 Waste Composition Worksheet

Sheet No. ____ of ____

Date:
Researcher:
Vehicle ID:
Sample Number:

Waste type: Residential Commercial Industrial Other (specify):

	Sample weight, lb.										
Waste stream components	A	B	C	D	E	F	G	H	I	J	Total
Newspaper											
Corrugated paper											
Mixed Paper											
Other paper											
Plastic beverage containers											
Plastic milk bottles											
Other plastics											
Plastic film											
Aluminum cans											
Other aluminum											
Ferrous scrap											
Tin cans											
Textiles/fabric											
Food waste											
Container glass											
Other glass											
Wood											
Dirt and debris											
Ceramics and fines											
Yard waste											
Rubber											
Leather											
										New weight:	

Local seasonality of waste generation is best estimated using quantity data kept by the site operator. Data on daily truck and/or vehicle arrivals, load weights if a scale is employed at the site, volume estimates, or billings records are useful for estimating seasonality of waste generation.

Solid waste composition data are typically compiled and analyzed for each of the individual samples taken. All samples are then combined to determine mean averages for each waste component. The means are then presented as representative of the solid waste as a whole for the study area.

Laboratory Results

The laboratory should be required to submit a separate report that presents the results of the tests undertaken. The test data should be correlated by sample number to each of the samples sorted in the field. This is extremely important if the laboratory tests the combustibility of only the combustible fraction of each of these samples taken. If this is done, the higher heating value of the solid waste stream is based upon only the combustible fraction. To determine the actual higher heating value of the solid waste on an as-received basis, the noncombustible fraction (metals, inerts, etc.) removed from each sample must be analyzed to reflect heat and energy loss. A new series of higher heating values is then developed for each of the samples. The series of values are reviewed and any clear outlines are removed and a range of higher heating values is presented. Typical ranges experienced in the United States are between 4200 and 5500 Btu/lb.

The chemical analyses conducted can estimate what chemical compounds are likely to be released during the disposal process in the form of gases from a waste-to-energy facility and leachate from a landfill. These estimates can be compared to parameters for emissions and leachate to determine whether the environmental discharges associated with the disposal alternative chosen will be within regulatory standards.

PRESENTING THE RESULTS

Once all of the data have been compiled and organized, they should be presented in an orderly and concise manner. Tables and graphics are valuable tools to illustrate solid waste characterization study results.

Pie charts, bar graphs, or other forms of visual representation all serve the purpose of creating a picture and personality for the solid waste stream. Some samples of graphics that could be used in presenting the results from a solid waste characterization study are provided in Figs. 3.8 through 3.12. Figure 3.9 describes the solid waste composition by component for a residential sector. Figure 3.10 describes the composition of an entire solid waste stream. Figure 3.11 presents residential tonnage deliveries to various waste desti-

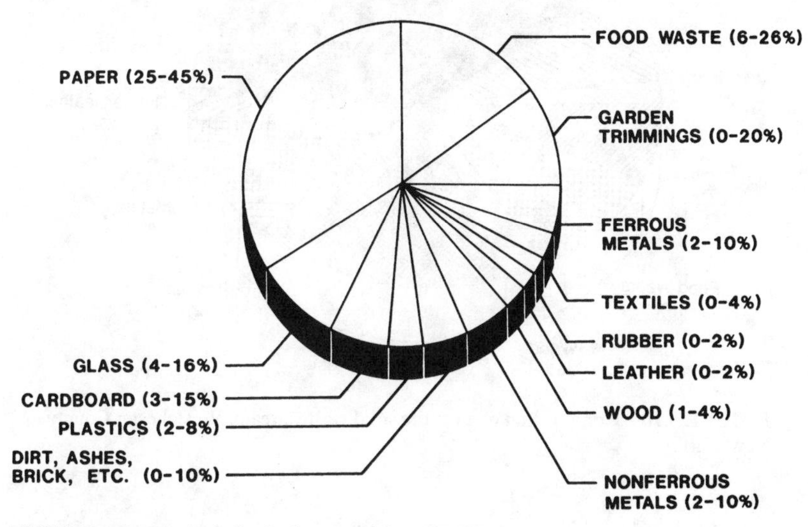

FIGURE 3.8 Typical physical composition of MSW.

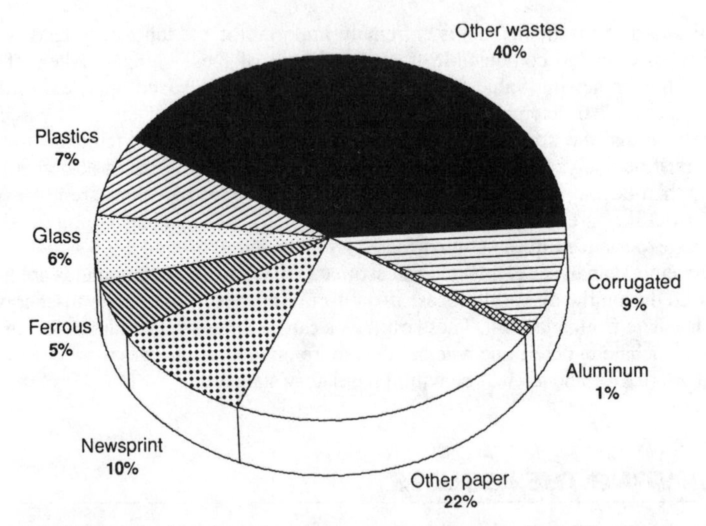

FIGURE 3.9 Residential sector waste stream composition, Westchester County, New York.

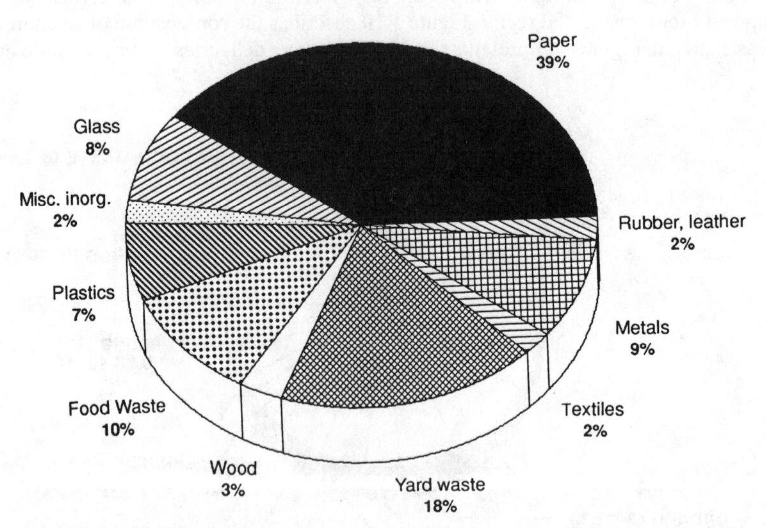

FIGURE 3.10 Composition of an entire solid waste stream, Westchester County, New York.

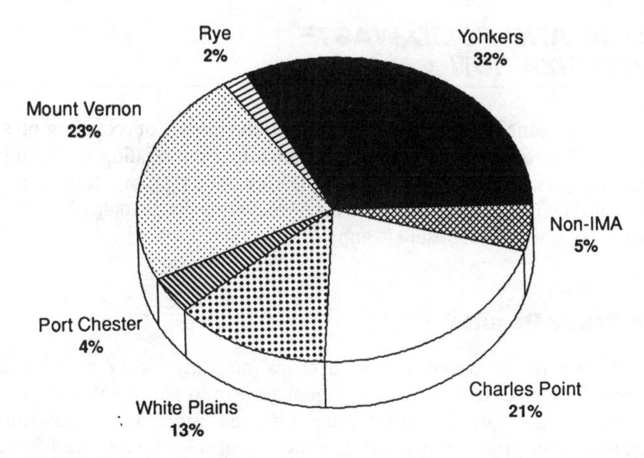

FIGURE 3.11 Residential tonnage deliveries, 1989, Westchester County, New York, by transfer station.

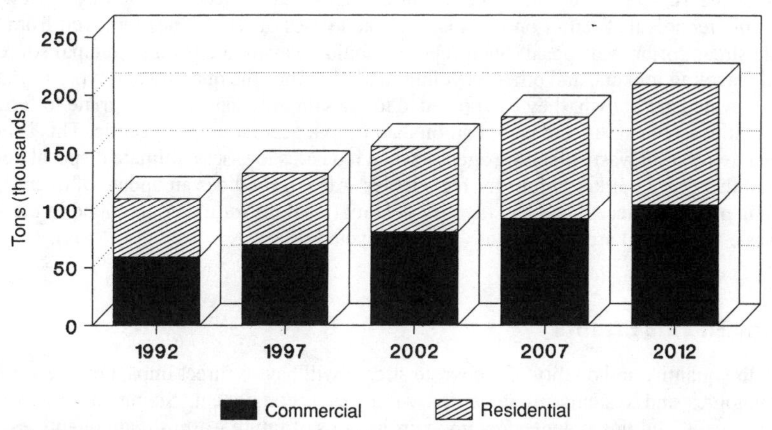

FIGURE 3.12 Commercial and residential recycling estimates and projections, Westchester County, New York.

nations in a study area. Figure 3.12 depicts commercial and residential recycling estimates and projections.

The final report format and the level of detail desired may be specified by the uses of the survey data. If not, the results should be reported in summary form with a brief review of the procedure used (assumptions, data sources, etc.). Two important items which shall be addressed are:

- Reporting the quantity and composition data by generator and region.
- Reporting the results of the survey in such a manner that others may use the data later and understand its limitations.

RECYCLING AND SOLID WASTE CHARACTERIZATION

There are two major considerations when assessing the impact of recycling on solid waste characterization. The first consideration is, what effect do existing recycling programs have on the data obtained during the solid waste characterization study and the second consideration deals with the long-term effects of recycling on the data obtained relative to the planning of solid waste disposal systems.

Effect on Study Results

If the study area where the solid waste characterization study takes place already has a recycling program in place, the solid waste characterization study will not provide a true picture of the generator patterns within the study area. Since all of the sampling takes place at the disposal or processing location, materials previously removed from the waste stream through recycling are not accounted for. Since it is the preferred approach to report solid waste from a generation perspective, it will be necessary to account for all recyclables removed from the waste stream prior to disposal.

Therefore, the solid waste characterization study must quantify recyclables in terms of overall quantity removed and its composition. This can be accomplished by reviewing available records that report on the quantities and types of recyclables removed from the waste stream prior to disposal. Such reports should be available from municipal sources, local recycling brokers, and other recycling markets. Once the quantity and types of materials recycled are known, they can be added to the sampling data as appropriate to formulate solid waste quantity and composition data from a generation perspective. This is beneficial to the solid waste planning agency when it desires to assess ultimate disposal needs should there be a market failure for recyclables. Also, it provides an update of overall recycling program success and can identify potential markets for materials found in the solid waste stream that are not currently being recycled.

Study Result Effects

Both the quantity and quality of the waste stream will have a direct impact on the various technologies and systems implemented within each component. Sizing of solid waste management facilities is dependent upon projections of future waste stream quantities and characteristics, and upon recycling capture rates that may not be readily defined before the recycling program is implemented. Undoubtedly, recycling will affect waste availability.

The amount and types of waste removed from the waste stream through recycling will have a direct effect on the energy content (higher heating value) of the waste stream and its in-place density at a landfill. Both of these effects will impact disposal capacity requirements. Typically, during the initiation phase of recycling programs, newspaper and other paper goods, including corrugated material, are removed from the waste at a far greater rate than cans, bottles, and other noncombustibles. When this is the case, there is a marginal decrease in the higher heating value of the overall waste stream. As a recycling program becomes more effective, larger quantities of recyclable materials are removed from the waste stream. These materials are typically noncombustible and cause the higher heating value of the waste stream to increase. At a 25 to 30 percent recycling level, there can be an increase in the higher heating value of up to 10 percent. This could reduce the waste-to-energy processing capacity by a like amount.

Combined with expected growth in commercial activity, potential and expected increases in materials such as packaging wastes will have a significant effect on the quantity as well as the processable composition of the refuse. As paper, plastics, and packaging increase, the higher heating value continues to increase. Similarly, as inert or low heating value materials such as glass and ferrous and noncombustible found in yard wastes are removed, the higher heating value of the remaining refuse also increases. Such increases have to be taken into consideration when planning for disposal capacity.

SPECIAL AND CREATIVE APPROACHES

In certain instances, a full solid waste characterization with waste sorting may not be warranted and unique and creative approaches may be acceptable.

A unique and creative approach was undertaken for Westchester County, New York, in an attempt to estimate the amount of recyclables in the commercial segment of its waste stream. In order to plan for commercial recycling, a comprehensive analysis of the quantity and composition of the commercial waste stream was required.

There are considerable advantages to having the commercial industry participate in municipal and private recycling programs. The advantages of initiating recycling programs in the commercial sector include

- The commercial waste stream typically consists of a large fraction of recyclable materials, such as paper, corrugated cardboard, and wood.
- Commercial recyclables are both readily identifiable and easily separated from the commercial waste stream.
- A significant amount of commercial recycling is usually already going on, and these efforts can be enhanced or used as examples for other programs.
- Due to increasing disposal and collection costs, commercial recycling becomes an economically attractive business decision.
- Commercial recycling activity will help municipalities reach mandated state recycling goals by increasing recycling rates and therefore reducing the overall commercial waste stream requiring disposal.

If presented clearly to the commercial community, such advantages can stimulate local business leaders to initiate recycling programs. This can be accomplished by advertising in local business journals, presenting the benefits of recycling at local business organizational meetings, and establishing a task force or committee of business leaders and public officials to develop the necessary incentives in the private sector to promote recycling.

Prior to initiating recycling programs in the commercial sector, the commercial solid waste stream should be assessed to determine its compositional breakdown. It is important to determine the components of the commercial waste stream so that the types and quantities of materials available for recycling can be identified. Waste quantification and characterization require the joint efforts of the municipality, commercial establishments, and private waste carting industry.

Usually to develop waste quantity and composition data to this level of detail, waste deliveries from commercial establishments are weighed, the weights recorded, and representative samples drawn for composition analyses as described earlier in this chapter. This method requires a minimum of several weeks of field samplings, is labor intensive and costly, and its results are affected by many variables, such as seasonality, daily disposal patterns, and business cycles.

However, there are other methods available to determine the type and quantity of waste generated by the commercial sector. For example, this can be done by analyzing scalehouse data and/or by conducting municipal surveys. A much more challenging task, however, is disaggregating the commercial waste stream by municipality and generator and assessing its content and quantity. This method allows for targeting specific commercial generators for recycling based upon their actual waste generation patterns.

During the development of Westchester County's Solid Waste Management Plan, an innovative method was proposed to determine the quantity and composition of the county's commercial generator types. The method proposed consisted of reviewing available information from waste composition studies conducted within other regions of the United States, conducting surveys of the commercial business community, disaggregating the commercial business sector based upon county planning data, meeting with and interviewing key members of the private carting industry within the county, and conducting limited waste sampling programs at specific generator locations (Fig. 3.13).

The first step in the study required the development of commercial waste generator segments. Based on information available through the county's planning department, the commercial sector was broken down into the six major business types or generator segments listed here:

FIGURE 3.13 Conducting field sampling.

Pounds per person per day

versus

Pounds per occupied square foot of floor space

FIGURE 3.14 Generation determination (slide).

- Office
- Industrial
- Transportation, communication, and utilities
- Retail
- Wholesale, warehouse, and distribution
- Public and institutional

Generation rates per square foot of occupied floor space were chosen as the variable to estimate waste production, since commercial waste production is a function of the type of business activity and not necessarily of the number of employees working at a particular business location (Fig. 3.14). Once the generator segments had been identified, various surveys and studies were conducted to develop estimates of the quantity and composition of the commercial waste stream.

The basic survey and study premise proceeded using the following guidelines: question, compile data, assess and analyze the results, and test. For example, large business complexes were contacted and asked to provide information relating to the size of their office space (square footage), number of employees, business type (office, retail, etc.), amount of waste generated, and waste storage capacity (number and size of containers). All of the information was *compiled* and *assessed* to determine specific generator patterns (pounds of waste generated per square foot of office space per week). These estimates were then *tested* for accuracy by field weighing and composition studies conducted at specific commercial waste generator locations. The following is a listing of the surveys and studies conducted:

- Municipal phone survey
- Commercial business survey
- Private hauler survey
- Specific generator studies
- Assessor's survey
- Field survey
- Major business survey

As previously mentioned, various sources of information and methods were utilized in an effort to develop an acceptable confidence level for determining the amount of waste currently being generated by the commercial sector. However, each method of investigation presented a varied degree of confidence with respect to each of the commercial waste generator segments studied (i.e., office; industrial; transportation, communications, and utilities; retail; wholesale, warehouse, and distribution; and public and institutional).

As a result, it was necessary to review the results from each method of analysis to estimate waste generation for each of the respective generator segments and to make an as-

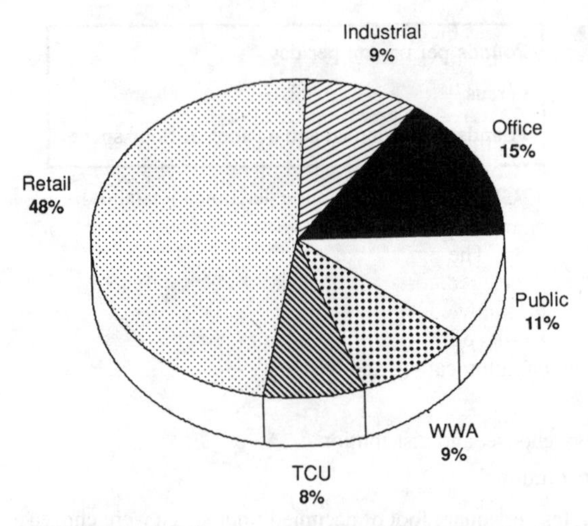

FIGURE 3.15 Classes of commercial waste generation, Westchester County, New York.

sumption as to which method presented the most reasonable results. The criteria used to establish reasonableness in the ultimate waste quantity determination was the selection of an estimate which could be supported by at least two of the other estimating methods conducted. Methods which resulted in generation rates clearly outside the range of the expected true mean for a specific generator segment were eliminated from further consideration.

Following the determination of the estimated commercial waste quantities, available commercial waste composition data was used together with the information obtained from the additional surveys and studies undertaken to provide a breakdown of the commercial waste stream into principal components. Each of the individual surveys was reviewed in terms of the generator's knowledge of the compositional breakdown of its

TABLE 3.9 Commercial Waste Quantity and Composition

Generator segment	Waste composition					
	Paper, %	Corrugated, %	Plastic, %	Metals, %	Others, %	Total tons
Office	65	15	6	2	12	54,290
Industrial	35	20	25	6	14	33,060
Retail	35	40	8	1	16	178,390
TCU*	20	15	15	5	45	31,500
WWA†	25	32	25	7	11	31,860
Public	45	10	5	6	34	42,000
Total tons	142,000	105,200	40,600	11,200	71,600	371,100

*TCU = Transportation, communications, and utilities.
†WWA = Wholesale, warehouse, and distribution.

waste stream. A matrix was then developed illustrating for each of the surveys undertaken the reported waste composition. The reported estimates were then analyzed during the field waste composition studies at each of the generator segment locations and revised according to actual field data. A graphic representation of this data is provided as in Fig. 3.15. Table 3.9 provides the estimated compositional breakdown of the commercial waste sector in Westchester County by generator segment.

The information gleaned on waste quantity and composition was then presented to the private haulers, who are responsible for collecting 90 percent of the commercial waste generated in the county. The private haulers found the waste quantity information to be accurate based upon their records and the waste composition data to be a reasonable estimate based upon their knowledge of waste composition throughout the county.

Based upon the results of this study, the county began to target large commercial generators of recyclable materials as participants in its recycling program. In summary, the design and implementation of a solid waste characterization study can include various methods and techniques other than those traditionally used such as manual sorting to achieve reliable results.

STUDY COSTS

As mentioned at the beginning of this chapter, the cost to conduct a comprehensive solid waste characterization study as set forth in this chapter could range from $250,000 to 700,000. There are many variables, however, that affect cost. They are

- The scope of the study
- Labor requirements
- The number of sites to be analyzed
- The number of weekly programs
- Sampling scheduling
- The amount of incoming solid waste
- The amount of samples to be taken to reach the desired level of precision and confidence
- Equipment requirements
- Analyses undertaken
- Laboratory testing

The majority of costs are found in the labor requirements of the study. For example, if a four-person sampling crew is needed for waste sorting, two additional persons for visual inspection and hauler interviews of the scale house, a total of 10 persons, including supervision, will be on the site. For the field effort alone, 10 persons working 40 h per week for 4 weeks equals 1,600 h of labor. Adding laboratory costs, data compilation, and review of data analyses, and program planning and reporting, it is easy to see how the labor costs alone reach the $250,000 range including administrative costs, overhead, and profit. Adding to this, expenses and special equipment needs and, in some instances, simultaneous study at several sites, one begins to realize that a comprehensive solid waste characterization study is an expensive undertaking.

Costs can be reduced, however, by lowering, for example, labor requirements and the number of seasonal samplings that take place. It is recommended, therefore, that planners review potential alternatives to labor and scheduling that could reduce study costs prior to initiating the solid waste characterization study. Additionally, alternative and creative approaches should be explored to potentially reduce cost. The creative approach developed for Westchester County, New York, resulted in savings in excess of $100,000 from manual sorting techniques and produced reliable and accurate data for planning purposes.

For planning purposes, however, the agency sponsoring a comprehensive solid waste characterization study should at for each week of sampling, the costs will range from approximately $65,000 to 175,000 depending on the level of effort required to meet the study's objectives and producing the information desired. Therefore, as discussed throughout this chapter, know your study objectives and informational needs, and discuss viable options.

REFERENCES

1. Franklin Associates, Ltd., *Characterization of Municipal Solid Waste in the United States, 1960 to 2000,* United States Environmental Protection Agency, July 11, 1986.
2. Franklin Associates, Ltd., *Characterization of Municipal Solid Waste in the United States, 1960 to 2000 (Update 1988),* United States Environmental Protection Agency, March 30, 1988.
3. United States Environmental Protection Agency, *Characterization of Municipal Solid Waste in the United States: 1990 Update,* United States Environmental Protection Agency, June 1990.
4. Cal Recovery Systems Incorporated, *Broward County Resource Recovery Project Waste Characterization Study,* Broward County, Florida Resource Recovery Office, February 1988.
5. William F. Cosulich Associates, P.C., *Quantitative and Qualitative Analysis of Broward County Solid Waste—Phase I,* Broward County, Florida, September 28, 1983.
6. William F. Cosulich Associates, P.C., *Quantitative and Qualitative Analysis of Broward County Solid Waste—Phase II,* Broward County, Florida, November 14, 1983.
7. SCS Engineers, *Solid Waste Assessment Guidebook,* Michigan Department of Natural Resources, Community Assistance Division, Resource Recovery Section, June 1986.
8. G. Savage, *Proposed ASTM Standards, Draft Number 2, Method for Determination of the Composition of Unprocessed Municipal Solid Waste,* ASTM Committee, October 21, 1988.
9. D. S. Cerrato, "Estimating Recyclables in the Commercial Waste Stream," *Resource Recovery Magazine,* August 1989.
10. D. S. Cerrato, "Is There Gold in Garbage?" *National Development Magazine,* November/December 1989.
11. D. S. Cerrato, *Recycling and Resource Recovery Facility Sizing,* Fifth Annual Winter Conference of the Resource Recovery Institute, Miami, Florida, February 22, 1990.
12. N. Starobin and J. Kornberg, "A Cleaner Bangkok," *National Development Magazine,* March 1989.
13. Malcolm Pirnie, Inc., *Solid Waste Management Plan—Phase II,* Westchester County Department of Public Works Solid Waste Management Division, May 1988.
14. Malcolm Pirnie, Inc., *ANSWERS Wasteshed Recycling Plan,* City of Albany, New York, July 1988.
15. Malcolm Pirnie, Inc., *Solid Waste Quantification and Characterization Phase I Sampling Program,* Ulster County Resource Recovery Agency, Ulster County, New York, March 30, 1988.
16. Franklin Associates, Ltd., Environmental Consulting and Technology, Inc., and Resource Integration Systems Ltd., *Dakota County Solid Waste Generation and Characterization Study,* Dakota County, Minnesota, February 1991.

CHAPTER 4
SETTING RECYCLING GOALS AND PRIORITIES

J. FRANK BERNHEISEL, QEP*
Vice President, Gershman, Brickner & Bratton, Inc.
Fairfax, Virginia

INTRODUCTION

Strategic planning is critical to the success of recycling programs. Numerous technical and institutional components need to be coordinated and managed to ensure smooth program start-up and eventual expansion. The central issue facing planners will be how to incorporate recycling into an integrated solid waste management system. As with any new venture, short- and long-term goals need to be developed so that the process can be guided and monitored along the way.

In general, short-term goals for a recycling program will be oriented toward planning and implementation. These will include developing a recycling plan; determining which recyclable materials will initially be targeted and how the residential, commercial, and institutional sectors of the community will be served; and securing market agreements and processing capacity. Long-term goals will usually pertain to the attainment of a mandated or self-imposed waste reduction-recycling target and optimizing the integrated solid waste management system. Typical short- and long-term goals are presented in Table 4.1.

The development of recycling programs across the United States has borne out one enduring principle—there is no one program that works for every community. Each community has its own geographic and demographic identity, a particular waste collection and disposal network, a unique set of legal and financial constraints, and specific market requirements. The most successful recycling programs, then, are planned with each of these local variables in mind. Priorities or goals are established according to the particular needs of a given community.

REVIEWING EXISTING PRACTICES

Recycling program planning for any community begins with a careful examination of the existing solid waste management picture. This includes understanding the types and amounts of solid waste generated. A waste stream analysis should serve as the basis for

*Qualified Environmental Professional

TABLE 4.1 Typical Short- and Long-Term Recycling Program Goals

Short-term goals	Long term goals
Draft and complete recycling plan	Achieve and surpass diversion goals
Determine recyclable materials to be collected	Secure long-term market arrangements
Secure marketing arrangements	Secure long-term processing capacity
Secure processing capacity	Explore methods to reduce costs
Design and initiate public education/promotional campaign	Expand list of recyclable materials
Plan and implement first phase of community	Identify additional markets
Plan and implement first phase of residential curbside program	Monitor program effectiveness drop-off program
Initiate outreach-technical assistance plan for commercial sector	Review need for mandatory legislation and implement if necessary
Select communities for expanded curbside program	
Determine need for and hire recycling staff	
Develop and institute record keeping system	

determining the sources, quantities, and characteristics of a community's solid waste. This type of detailed assessment yields valuable information on available recyclable materials and serves as a tool for planning the most appropriate collection, marketing, and processing options for these materials.

The current waste stream may be analyzed in one of two ways. The more accurate, yet costly and time-consuming, way is to physically sort composite samples of municipal solid waste into designated sort categories. Representative seasonal samples of waste may be taken at the landfill or transfer station, at another disposal or processing facility, or at the point of generation. Examples of sort categories are shown in Table 4.2. Alternately, when faced with cost or time constraints, a community may decide to use existing data (i.e., information from communities with similar demographics and waste sources, existing state or county planning data) combined with local knowledge (local haulers and facility operators) to develop a snapshot of the types and amounts of waste generated in the area.

Existing collection practices determine, to a large extent, how a recycling program is instituted. Since collection is frequently the most costly component of a local waste management system, recyclable material collection needs to be incorporated in as cost-effective a manner as possible. Program planners need to know the following:

- Is collection accomplished by the public sector, the private sector, or a combination of the two?
- What are personnel needs for collection? Is union labor being used?
- What types of vehicles are currently being used to collect solid waste (i.e., rear, side or front loaders; roll-offs and tilt frames; transfer trailers) and what is their availability for recycling?
- What types of containers are used for collection (i.e, wheeled carts, containers for mechanized collection; metal or plastic cans; paper or plastic bags; 55-gal drums)?
- Where are the residential and commercial points of collection (curbside; alley; backyard/on-property; drop-off centers)?

These and other elements of the local disposal system need to be well understood. In some rural communities, for example, residents take their garbage directly to a transfer station or landfill. In this case, recycling opportunities can be provided directly at the point of

TABLE 4.2 Waste Category Description List

Component	Description
Newsprint	Newspapers
Corrugated	Corrugated cardboard boxes
Office paper	Computer printout, white and colored ledger paper
Kraft	Brown paper bags and other kraft items
Magazines and glossy inserts	All magazines and glossy inserts from newspapers
Mixed Paper (Recyclable)	Junk mail, envelopes, shoe boxes, books
Other paper	Paper towels, wax-coated paper, milk and juice containers, tissues, food packaging, paper bags other than kraft
Plastic, PET	PET, containers with symbol 1
Natural HDPE†	Clear HDPE, milk, juice, water bottles
Plastic, colored HDPE†	Colored HDPE containers with symbol 2, detergent containers, shampoo, and cleaner bottles (non-milk and juice containers)
Other plastics	Clear food wrap, flexible food containers (trash bags, baggies, zip-lock, etc.), toothpaste containers, food containers such as margarine, plastic toys, formed parts (from appliances), disposable eating utensils, etc. Containers with symbol 3, 4, 5, 6, 7
Ferrous cans	Food and beverage cans that are magnetic
Other ferrous metals	All other magnetic metal items
Other aluminum	Foil, siding, cast aluminum products, other aluminum objects
Other nonferrous	All nonmagnetic metals other than aluminum
Glass containers	Beverage bottles, food jars, other containers
Noncontainer glass	Broken window glass, other glass non-containers, plate glass, light bulbs, etc.
Leaves	Leaves
Grass	Grass
Brush	Tree branches, shrubbery, etc.
Wood waste	Lumber, furniture, tool handles, wooden toys, wooden kitchen implements, etc. (not yard waste)
Food waste	Kitchen scraps or any other sortable and identifiable food items
Textiles	Clothing, shoes, rags, carpets, etc.
Tires	Tires
Household hazardous waste	Cleaners, solvents, paints, etc.
Miscellaneous organics	Sweepings and other items remaining after sorting; other items not fitting above
Miscellaneous inorganics	Rock, brick, stones, sheet rock, ceramics, flower pots, sand, and related items
Diapers	Diapers
Fines	Sweepings less than 1/4 in in diameter

*PET = polyethylene tetraphthalate.
†HDPE = high-density polyethylene.

disposal. On the other hand, residents accustomed to receiving municipal or private collection at the curb may desire a more convenient curbside program. Another issue that needs to be addressed is the presence of any municipal contracts with disposal facilities, such as waste-to-energy plants or landfills. These generally have "Put or pay" agreements, which obligate communities to deliver a specified amount of waste to the disposal facility.

The last component of the current solid waste management system that needs to be identified and quantified are any existing recycling programs or activities. Municipally sponsored recycling programs (pilot or full-scale curbside programs; drop-off or buy-back centers; government office paper recycling efforts, etc) should already be well documented. The identification of other types of activities, however, requires further investigation. For example, in many communities, metal recycling efforts, include segregation of old appliances "white goods" and other ferrous scrap delivered to the landfill. These materials are sent to scrap dealers for recycling. Aluminum beverage can recycling programs for charitable purposes have been widespread in the past. Large amounts of commercially generated corrugated cardboard may already be recovered through by baling at commercial generators or "dump and pick" operations at transfer stations or landfills or through contractual arrangements between the business or industry and private waste haulers. Commercial or hauler surveys can help to quantify these activities. The surge of interest in the environment has led to many school or college-based recycling programs. These programs commonly target paper (newspaper, high-grade and mixed papers) and aluminum. Finally, planners need to be aware of any special programs, such as those for used oil, antifreeze, batteries, or tires. In general, program planners should learn how a particular solid waste system works, who the players are, where the waste goes, and who controls the system.

IDENTIFY AND EVALUATE INCENTIVES TO PARTICIPATE

The ultimate success of recycling programs depends, in large part, on public participation. If sufficient quantities of high-quality recyclable materials are to be recovered and diverted from the landfill, every resident, employee, and business owner needs to incorporate recycling into his or her daily life. Incentives to participation must be created and maintained. For some citizens, participation will spring from a desire to "help the environment." For most, legal or economic incentives will be deciding factors.

In the 1980's there was a dramatic increase in state waste reduction legislation, designed to encourage communities to recycle, has served as a powerful impetus for the development of recycling programs. Most states have set recycling goals ranging between 20 and 50 percent to be met by the year 2000. Some of these are mandatory. These legislative initiatives differ in their approach. Some mandate that local governments (municipalities) pass ordinances requiring citizens and businesses within their jurisdictions to source-separate and recycle a specified number of materials. Others require local governments to provide citizens with recycling services but do not mandate that they adopt ordinances. Oregon's 1983 Opportunity to Recycle Act was an early example of this type of legislation. A third variety of legislation mandates only that local governments reach a certain waste reduction-recycling goal.

Several states have also enacted bans on the disposal of certain materials—usually yard waste, tires, used oil, white goods, and batteries. Wisconsin has focused and expanded on this approach to encourage recycling by banning the disposal of most recyclable materials from landfills and incinerators. Furthermore, the law requires municipalities to meet certain recycling program criteria by 1995 in order to have access to these disposal facilities.

Comprehensive state waste reduction laws frequently contain provisions that stimulate the development of markets for recyclable materials. For example, state agencies may be required to purchase products made from recyclable materials and some provide for a price premium of up to 10 percent for the products meeting recycled content specification. Companies making products from recyclable materials or seeking to purchase recycling equipment may receive tax credits or become eligible for grants or low-interest loans. Many state recycling laws also make grants or loans available to local governments for the development of recycling programs. Another financial incentive that some states offer is funding for market research and development.

Whereas some financial incentives stimulate recycling by encouraging the procurement of recycled products or the investment in recycling technologies, a few states have chosen to adopt mandatory deposit legislation aimed at containers and vehicle batteries. The intent of this type of legislation is to reduce litter and divert these materials from disposal by encouraging their redemption.

Another approach is to impose product fees or taxes on certain consumer goods, for example, tires. Such fees represent an attempt to incorporate the cost of disposal into the cost of production. Ultimately, manufacturers pass this cost on to the consumer. Although this approach is used throughout the Europe, it is not widely used in the United States. Florida's Solid Waste Management Act mandates that an advanced disposal fee (ADF) of 1¢ per container be assessed on all containers made from glass, plastic, plastic-coated paper, aluminum, and other metals if these containers were not recycled at a sustained rate of 50 percent by October 1992. The Florida legislature allowed the programs to end in October 1995.

Clearly, the passage of state legislation requiring local governments either to meet mandatory goals, provide recycling services, or pass mandatory recycling ordinances has boosted recycling efforts in many communities. Planners need to be fully aware of all legal and financial incentives at their disposal. Some communities, however, may choose to institute regulatory measures of their own.

In communities with an open collection system, where residents contract with a private hauler of their choice for garbage collection, the municipality may not have a strong economic incentive to recycle. In this case, the local government may pass an ordinance detailing the level of recycling services that each hauler must provide to its customers and require that hauler licensure be contingent on the provision of these services. The local government, in turn, can provide financial assistance by providing curbside collection containers, funding some or all of the promotional costs, or developing a materials recovery facility (MRF) for the processing and marketing of the recovered recyclable materials.

Increasingly, communities are beginning to consider volume-based waste disposal fees as a way to encourage recycling and source reduction. Under these "Pay as you throw" systems, residential waste generators pay according to the amount of garbage they generate. Residents can presumably be expected to attempt to minimize their bills, in this case by generating less garbage and availing themselves of recycling opportunities which are offered at reduced or no charge. Several options are available for implementing variable-rate collection systems, including charging customers based on the number of containers set out, the frequency of collection, or the weight collected.

CONDUCT RESEARCH ON COMMUNITY WILLINGNESS TO PARTICIPATE

Community participation is critical to the success of a recycling program. The efficient recovery of large volumes of high-quality recyclable materials depends on citizen involvement. Whereas public perception of the validity of recycling as a waste management strategy is growing, program planners need to anticipate some resistance to change. Not all residents and businesses can be expected to embrace recycling wholeheartedly at the outset. If community attitudes and/or objections to recycling behaviors can be identified and characterized, however, planners can design programs that achieve a maximum level of recovery with a minimum level of inconvenience.

It has been suggested that a portion of the public will participate in recycling programs no matter how well promoted or convenient the opportunities are. These individuals are probably motivated by a sincere environmental ethic. Therefore, there is no need to try to convince these people to participate (i.e., don't "preach to the choir"). On the other hand, another small segment of the population will be unlikely to participate regardless of the level or types of publicity. It is probably not worthwhile, therefore, to spend a significant amount of money or time trying to move the immovable. The majority of the populace, however, can be affected by responsible information and promotional programs, and it is this larger segment of the population that it is important to know more about and understand.

One way to gauge public interest in and support for recycling is to survey residents and businesses about their attitudes toward waste disposal and recycling issues. Written and phone surveys can be administered to a wide audience in the jurisdiction. Responses received are then tabulated, analyzed, and used to guide program design and public education efforts. Figure 4.1 presents a sample attitudinal survey form.

Another way to assess local attitudes toward recycling is to conduct one or several focus groups. Focus groups are a recognized, small group survey technique in which a moderator asks a series of open-ended questions to a small sample of individuals (generally 10 to 25 people), who are either chosen at random or demographically representative of a target population. This survey approach enables moderators to gather qualitative data that are not easily gathered through more traditional phone or mail survey techniques. Though the results are rarely statistically valid because the sample population is so small, focus groups do provide more detailed information about a wide range of issues than multiple choice or true-false questions. In addition, moderators can structure questions to assess knowledge levels at the start of the session and monitor how quickly the group's knowledge levels rise during the course of the meeting.

During the course of planning and implementing a recycling program, a municipality or a county may decide to develop a recycling facility. Siting solid waste processing facilities is typically met with opposition from some local citizens. This public outcry has come to be known as the not-in-my-back-yard (NIMBY) syndrome. Dealing with NIMBY opposition has generated much discussion among waste management professionals and public education experts. To what extent do you involve the public? How much technical information do you try to condense for a lay audience—often an audience armed with misinformation? Do public meetings present media opportunities for nay-sayers or promote consensus building? Does risk communication improve a community's understanding of solid waste projects or increase public anxiety that can lead to public opposition? Does an open communication policy lead to politically motivated solid waste management decision making at the expense of technical considerations? These questions are not easily answered, nor is one answer going to be appropriate for all communities. Ultimately, each jurisdiction will have to develop a communication strategy that is most effective for its particular citizenry.

An important component of a community outreach plan is the development of a citizen's advisory committee or task force. Such citizen groups can include local civic and business leaders (and others who have local influence) as well as local technical experts. An advisory committee or task force can serve as a valuable mechanism to build consen-

1. Which of the following best describes your home?

   (a) ( ) Single-family detached
   (b) ( ) Duplex or townhouse
   (c) ( ) Multifamily unit (up to 4 stories)
   (d) ( ) Multifamily unit (5 to 8 stories)
   (e) ( ) Highrise (9 or more stories)

2. How many people live in your house? ________________

3. Would you say you listen to/read about environmental issues?

   (a) ( ) Very closely
   (b) ( ) Somewhat closely
   (c) ( ) Not closely
   (d) ( ) Not at all

4. Have you heard or read anything in the news or in conversations among friends and associates about recycling in [name of municipality or community]?

   (1) ( ) Yes (2) ( ) No

   If yes, how have you heard or read about recycling (check as many as apply)?

   (a) ( ) Newspaper
   (b) ( ) Television
   (c) ( ) AM radio
   (d) ( ) FM radio
   (e) ( ) Magazines or newsletters
   (f) ( ) Billboards
   (g) ( ) Buses, subway stations
   (h) ( ) Other:________________________________

5. Do you currently recycle any of the following materials?

   (a) ( ) Newspapers
   (b) ( ) Other paper or cardboard
   (c) ( ) Glass
   (d) ( ) Cans (aluminum or tin)
   (e) ( ) Plastic
   (f) ( ) Leaf/yard waste
   (g) ( ) Other
   (h) ( ) Don't recycle now

6. If you do recycle, what is the principal reason?

   (a) ( ) Concern for the environment
   (b) ( ) Concern about availability of landfill space
   (c) ( ) My children encourage me to recycle
   (d) ( ) I get paid for my recyclable materials
   (e) ( ) Other:________________________________

7. If you do not recycle, what would you say is the principal reason you don't?

   (a) ( ) Inconvenience
   (b) ( ) Believe there are better ways to handle my garbage
   (c) ( ) Other:________________________________

FIGURE 4.1 Sample attitudinal survey form.

8. How long have you been recycling?

    (a) ( ) Less than 1 year
    (b) ( ) 1-2 years
    (c) ( ) 3-5 years
    (d) ( ) More than 5 years

NOTE: Questions 9 through 14 can be used in communities that plan to expand or modify drop-off program or assess the effectiveness of a drop-off program.

9. [*Name of municipality/community*] has placed bins or containers in several areas where residents may deposit [*list materials*]. Have you ever seen these recycling centers?

    (a) ( ) Yes
    (b) ( ) No
    (c) ( ) Don't know

10. Have you ever brought materials to one of these centers?

    (a) ( ) Yes
    (b) ( ) No

    Which one? ____________________

11. [*Name of community*] is likely to increase the number of places you can take recyclable materials soon. What can [*name of community*] do to make it more likely that you will take your recyclable materials to these recycling centers?

    (a) ( ) Pay for recyclable materials
    (b) ( ) Locate centers closer to my home
    (e) ( ) Make recycling mandatory
    (d) ( ) Provide more information
    (e) ( ) Not likely to go to recycling center

12. How much time would you be willing to spend [or do you now spend] driving *one way* to a recycling center?

    (a) ( ) Less than 10 min
    (b) ( ) 10 min
    (c) ( ) 11–15 min
    (d) ( ) 16–20 min
    (e) ( ) More than 20 min
    (f) ( ) Not likely to go to recycling center

13. What day of the week is more convenient for you to go to a recycling center?

    ____

14. What locations for recycling drop-off centers would be most convenient for you?

    (a) ( ) Fire or police station
    (b) ( ) Shopping area or grocery store
    (c) ( ) Park or recreation area
    (d) ( ) Centrally located special area dedicated to recycling
    (e) ( ) Other____________________

15. If [*name municipality or private haulers*] picked up cans, glass, plastic and other recyclable materials at your curb, would you be more likely to recycle than under the current system?

    (a) ( ) Yes, more likely
    (b) ( ) No, not more likely
    (c) ( ) No difference, I'd recycle anyway
    (d) ( ) No difference, I wouldn't recycle

FIGURE 4.1 (*Continued*) Sample attitudinal survey form.

16. If you answered "yes" to question 15, would you say you would:
    (a) ( ) Definitely recycle
    (b) ( ) Be very likely to recycle
    (e) ( ) Be somewhat likely to recycle
    (d) ( ) Would recycle reluctantly
17. How much extra would you be willing to pay on your monthly trash collection bill for curbside recycling?
    (a) ( ) $.50 or less
    (b) ( ) $.50–$1
    (c) ( ) $1–$2
    (d) ( ) $2–$5
    (e) ( ) Would not be willing to pay for curbside recycling
18. Do you think [*name of community*] residents should be required by law to recycle, or should it be voluntary?
    (a) ( ) By law
    (b) ( ) Voluntary
    (c) ( ) Unsure
19. Do you have any specific comments you'd like to make about [*name of municipality or community*] recycling program?

    ____

    ____

    ____

20. Age of respondent ____________________
21. Yearly salary of household (all residents combined).
    (a) ( ) Less than $9,000
    (b) ( ) $9000–$15,000
    (c) ( ) $15,001–$25,000
    (d) ( ) $25,001–$35,000
    (e) ( ) $35,001–$45,000
    (f) ( ) $45,001–$55,000
    (g) ( ) Over $55,000

FIGURE 4.1 (*Continued*) Sample attitudinal survey form.

sus, involve stakeholders in the decision-making process, solicit public input, harness local resources, educate possible opponents about the importance and value of the project, and shape public opinion through involving local opinion leaders.

MARKET RESEARCH AND IDENTIFICATION

In order for any community to achieve its recycling goals, markets must be available to absorb recovered recyclable materials. Over the years, many communities have experienced setbacks in their recycling programs due to the faltering of one or more markets. In some cases, separated materials were stored in anticipation of a short-term changing of market conditions and landfilled when markets did not improve. For some materials, markets will naturally grow as new supplies become available. For others, the public and private sectors are working together to promote growth in industries that can rely on secondary materials in their production processes. Planners should keep in mind that many markets (especially for lower grade materials) have had a cyclical history with fluctuating prices.

The community must concern itself with both existing and future markets for recyclable materials. The first step is to refer to any market research studies already completed. In states that have enacted comprehensive waste reduction and recycling legislation, it is likely that the state's environmental agency or department of commerce has conducted a recyclable materials market analysis. Additionally, regional studies performed for a group of counties can be consulted.

In the absence of existing studies, a local government may need to conduct its own research in order to identify and secure recyclable materials markets. Two types of markets should be investigated: intermediate and final. Intermediate markets include both processors and brokers of materials. In general, intermediate markets handle a variety of materials which they purchase from industrial and private sources as well as municipal recycling programs. They accumulate, process, store, and transport the recyclable materials to final markets. Final markets, generally manufacturing facilities, convert recyclable materials into new products. These markets usually handle only one material to produce one type of product (i.e., glass bottles, metal cans, newsprint, etc.).

Market identification can be systematically accomplished with the help of a variety of resources. The yellow pages in the local phone book contain a wealth of information. Following is a listing of some of the headings that can be referred to:

Recycling centers
Waste paper
Rubbish and garbage removal
Thrift shops
Junk dealers
Scrap metals
Landscapers
Nurseries
Lawn maintenance
Garden centers
Mulches
Sod and sodding services
Automobile wrecking

Trade associations can also be contacted for information on local markets. Following is a partial listing of some of these material-specific organizations:

The Steel Recycling Institute
Glass Packaging Institute
The Aluminum Association
American Forestry and Paper Association
Society for the Plastics Industry
National Association for Plastic Container Recovery (NAPCOR)
National Soft Drink Association
American Retreaders Association
Institute of Scrap Recycling Industries

In addition, recycling journals such as *American Recycling Market, American Metals Market, Fibre Market News,* and frequently publish useful market information.

Specialized local markets for certain recyclable materials should also be investigated. For example, some communities shred collected newspaper for use as animal bedding by local farmers, cooperative extensions, agricultural organizations, and agricultural service companies. Newspaper and mixed paper grades are also used in hydromulch applications by seeding contractors, land improvement contractors, or soil conservation services. Glass is used by some fiberglass manufacturers or as aggregate or glassphalt.

Once a list of potential markets has been compiled, these markets should be surveyed to determine the type and quantity of materials that they can accept. The survey should also establish price, material specifications, assistance that is available from the market for processing, transportation or storage of materials, and any other available information on market conditions (i.e., availability of short-term or long-term contracts). Figure 4.2 presents a sample survey form that can be used for interviewing prospective markets. In addition to serving as a useful tool for compiling necessary information, it can also be used either for mail or phone surveys.

Finally, program planners should be aware of any waste exchange systems that match industrial waste generators with potential users of these wastes. Waste exchanges typically deal with both hazardous and nonhazardous materials, such as acids, alkalis, and inorganic chemicals; solvents and other organic chemicals; oils and waxes; plastics and rubber; textiles and leather; wood and paper; and metals and metal sludges. Waste exchanges and other organizations have begun using the internet to match buyers and sellers.

If it becomes apparent that existing markets for recyclable materials are not sufficient to handle the materials collected in the region, a market development program can be initiated. The fluctuating nature of recyclable materials markets requires that program planners accept a certain level of uncertainty and design programs with flexibility in mind. Market analysis and development, then, will be an ongoing process, since recycling programs will need to continually respond to market changes. One of the most important roles that program officials can play in market development is to ensure that recyclable materials that enter the market meet industry specifications. They should also let potential markets know about the timing and availability of new supplies.

There are a variety of market development tools that have been used by public and private agencies to increase the markets for recyclable materials. Generally, these involve offering financial incentives and technical assistance to businesses and industries that use recycled materials. State, city, and county economic development agencies have attracted and retained these businesses by providing assistance with siting, zoning, financing, labor, real estate development, and environmental issues. The adoption of preferential procurement policies also helps develop markets by signaling local commitment to purchase products with recycled content.

Some smaller or more rural communities may not generate sufficient quantities of recyclable materials to interest potential markets. In this case, several counties or municipalities can consolidate their recovered materials to provide final market buyers with the quantities and quality of materials they consistently require. This consolidation has been termed regional or cooperative marketing and represents an effective market development initiative available to certain groups of communities. One of the earliest and best-known examples of such a cooperative is the New Hampshire Resource Recovery Association (NHRRA), created in 1981 to provide technical, educational, and marketing services to its members. NHRRA staff identifies market options and, in conjunction with a marketing committee, recommends buyers to municipal representatives on the board of directors. The board will then enter into a contractual agreement with a buyer for each specific recyclable material.

Company name:_______________________________ Date:__________
Contact name: ______________________________ Title:__________
Telephone number: ______________________________________
County in which facility is located:______________________
Address: ____
City: __________________ State: ______________ Zip:__________
Please complete the following chart, indicating what materials you accept, prices paid, and purchase requirements.

	Materials specifications (Baled-loose-crushed-shredded)					Will you pick up [P1 or must it be delivered [D]	Price/ton*	Minimum quantity, tons
	B	L	C	S	Comments			
Paper								
Newsprint (please specify grade #)								
Corrugated								
Computer printout								
White ledger								
Colored ledger								
Mixed Paper								
Glass								
Mixed cullet								
Green								
Clear								
Amber								
Metal								
Aluminum								
Tin cans								
Bimetal cans								
Heavy ferrous								
White goods								
Batteries (car)								
Plastic								
PET bottles								
HDPE bottles								
Other plastics								
Tires								
Other materials								

*We are aware that prices fluctuate frequently and do not expect you to be committed to these prices.

FIGURE 4.2 Sample marketing interview survey form.

1. Is your firm a Broker _____
   Dealer/processor
   Manufacturer _______
2. If you are a manufacturer, what products do you make from the recycled materials?
3. To enable us to gauge the current amount of recycling in your area, please provide an estimate of the amount of materials you purchase annually (please give a separate quantity for each material).
4. What kind of additional capacity do you feel you have available?
5. What is your firm's preferred method of transportation to receive materials?
   Rail _______
   Truck _____
6. Would you be willing to pick up trailer-load quantities of any material?
   Yes _______
   No ______
7. Would you be willing to provide a contract for the purchase of material?
   Yes ____ No ____
   If yes, for what duration? ______ Years
8. Would you be willing to provide storage and/or processing equipment or capability (Gaylord boxes, trailers, compactors, etc.)?
   Yes _____ No _____
   If yes, what type of equipment/capabilities?
9. Estimated cost to provide such equipment:
   $ ____________ per ton $ ____________ per load $ ____________ flat rate
10. Any additional requirements or comments:
   Return forms to:

FIGURE 4.2 (*Continued*) Sample marketing interview survey form.

Export Markets

Recyclable materials separated from the waste stream become commodities, similar to virgin materials. Program planners, then, should understand the many factors affecting the commodities' markets and, thus, the prices paid for these materials. Domestic production capacity, imports, consumption, energy and transportation costs, changing technology, new product opportunities, availability of substitute material, and other factors affect the markets for recyclable materials. Factors affecting international markets include these same forces plus foreign trade tariffs, currency exchange rates, trade policy and programs, and other political forces.

Export markets for recovered materials have provided a demand pull on the recycling industry in the United States. Paper is a prime example. Greatly expanded recycling programs on the East Coast have generated more paper than regional mills could absorb. At the same time, the growing economies of the Pacific Rim countries demanded fiber that could not be produced locally, causing them to import recycled paper from the United States and Europe. When the Asian economics had their downturn, the prices for recovered fiber in the United States dropped. Paper is still the largest single export (in annual tons) from the port of New York City.

The large number of factors involved in the export markets serve to increase the volatility of prices paid for recycled materials. It is important to understand that domestic users (markets) of recyclable materials compete with export markets for the same materials. All of these factors combined with the geographic location of the generating community should be considered when investigating export markets for recyclable materials.

INSTITUTIONAL ARRANGEMENTS

There are roles to be played by both the public and private sectors in solid waste management. The optimum balance of responsibilities for carrying out recycling programs depends on the specific program elements and on the objectives and philosophies of the local jurisdiction.

Municipal solid waste collection is usually accomplished through a combination of public and private sector efforts. Generally, the private sector collects from commercial and industrial establishments, and local governments often collect or arrange for collection from the residential sector.

One of the most important decisions that needs to be made is how to incorporate recycling into the existing system. Planners are cautioned against prematurely restructuring current collection services. In communities with open collection systems, for example, residents may be very loyal to their particular hauler and may not respond well to a change in service (i.e., a switch to municipal collection of recyclable materials). Planners should consult local haulers and citizen advisory groups on their interest in providing recycling services and involve them in the planning process. Successful recycling programs often depend on the solid support of haulers.

If a community decides to utilize existing haulers to provide recycling services, they must still monitor the program and retain some control (especially in states with strict diversion mandates). Reporting, verification, and inspection requirements are key components in a system that relies heavily on private-sector initiatives. These components are the responsibility of the local government. In these situations, the local government may wish to include recycling services and reporting in licensing requirements.

IDENTIFY POTENTIAL FOR WASTE REDUCTION AT THE SOURCE

Source reduction activities focus on preventing the generation of solid waste in the first place, generally by decreasing the volume and toxicity of materials produced and consumed. Methods of reducing waste include reducing the use of nonrecyclable materials; replacing disposable materials with durable-reusable materials; reducing packaging; minimizing yard waste generation; establishing volume-based garbage rate structures; and increasing efficient use of materials (including paper products, glass, metals, plastics, and other materials). An example of more efficient use is copying on two sides of the paper.

Some states have established reduction goals as part of recycling legislation. These goals generally range from 5 to 10 percent of the Municipal Solid Waste (MSW) stream.

However, these gains are difficult to measure. Legislation also exists to require or encourage waste reduction addressing the methods mentioned above.

Planners need to be aware of any incentives that are already in place as well as to identify existing consumer awareness campaigns and local reuse or salvage industries. These types of activities can have positive effects on a planned recycling program.

IDENTIFY MATERIALS TO BE RECYCLED

A common goal of recycling programs is to divert substantial quantities of material from the waste stream; an accompanying goal is to offset recycling system costs with material revenues to the maximum extent possible. Therefore, the materials a community selects for recycling depend, in part, on available markets. Most larger communities have well-developed markets for paper, metal, and glass. Yard waste, which averages 15 percent of the municipal waste stream can contribute significantly to the achievement of recycling goals; however, a community must commit to the development of yard waste processing facilities to handle projected amounts of these materials as well as markets and uses for the finished products. Plastic containers [high-density polyethylene (HDPE) and polyethylene tetraphthalate (PET)] are an easily targeted portion of the waste stream, but their high volume to weight ratio makes cost-effective collection difficult. With the advent of on-board compaction equipment and specialized recycling vehicles, however, plastics are being included in a growing number of recycling programs. Other types of plastics [i.e., polyvinyl chloride (PVC), mixed polymer containers, and film plastics] are more difficult to recover and market.

There are many other materials that are collected in recycling programs, including textiles, batteries, food waste, household ferrous scrap, and reusable items. A recent survey investigated the materials most commonly included in municipal recycling programs. The results are tabulated in Table 4.3.

Program planners should keep in mind that collecting a material without having secured a market can result in, unexpected storage or disposal costs. In addition, public opposition may arise if recyclable materials are dropped from the program due to lack of markets.

TABLE 4.3 Most Popular Materials Included in Municipal Recycling Programs by Percent

Material	%	Material	%
Newspaper	96.2	Waste oil	46.2
Glass	93.9	High-grade paper	41.3
Aluminum	88.3	Mixed paper	32.2
Plastic bottles	67.0	Other (batteries, tin)	15.2
Cardboard	60.6	Rigid plastics	11.0
Scrap metal	52.3	Chip board	6.4
Yard waste	47.3		

Source: Public Administration Review, May–June 1991, based on 264 recycling coordinators' responses to a survey conducted by David H. Folz, University of Tennessee, Knoxville.

EVALUATE COLLECTION METHODS

Many alternatives are used for collecting recyclable materials from the municipal solid waste stream. The collection system is usually the most expensive component of a recycling program. Therefore, careful consideration must be given to providing reliable and convenient collection services in a cost-effective manner.

Factors affecting the collection of recyclable materials from the generator are often quite similar—and in some cases the same—as those affecting collection of regular refuse. General factors affecting both recyclable materials and waste collection include crew size, vehicle size, and maintenance issues. Community-specific factors affecting residential collection of both recyclable materials and refuse include community size and housing density; quantities of waste and recyclable materials to be collected; available equipment; traffic patterns; weather; and institutional issues, such as wages to be paid to collectors, frequency of collection, and point of set-out (curbside, back door, etc.). In most cases, the collection of recyclable materials in a community will be superimposed upon an existing waste collection approach, either municipally or privately operated. The most economical approach is to integrate the two collection approaches, maximizing the benefits associated with regular refuse amounts decreasing as recyclable materials set-outs increase.

RESIDENTIAL CURBSIDE COLLECTION

Curbside collection of recyclable materials is the standard approach to the recovery of recyclable materials from the residential waste stream. Like regular curbside refuse collection, this method of collection provides participating residents with a convenient and consistent method of recovering recyclable materials for processing and marketing. Furthermore, properly operated and publicized curbside programs that provide residents with regularly scheduled pickup of recyclable materials (often on the same day as regular trash collection) have been demonstrated to be effective in capturing large amounts of recyclable materials.

The way in which recyclable materials are set out at the curb, and the containers used for home storage and placement at the curb, vary from program to program. In general, three main approaches are used to set-out and collection; commingled set-out with commingled collection; commingled set-out with curbside sorting of materials; and source-separated set-out with separated collection. The amount of sorting—either at the curb or at a centralized processing facility—is be determined at least in part by the conditions under which locally available markets will agree to purchase recovered materials. If, for example, no local market is available for mixed-color glass, color separation will need to take place in the home, at the curb, or at a processing facility.

The type of home storage container varies with the set-out approach chosen. Containers for commingled set-outs need to be sized appropriately to store expected quantities of target materials. These containers typically cost less than multiple bin systems. However, centralized processing to sort materials has been found to be the most cost effective. Commingled collection is considered to maximize participant convenience and minimize the presence of bulky containers in the household, thereby increasing the likelihood of participation.

Multiple bin systems, used for source-separated set-outs and collection, range from eight-can systems on a wheeled cart to 90-gal cans with removable baskets for collecting the separated waste streams. The most popular approach to multiple bin collection is the use of a stacked, three-bin system. The cost of these containers is greater than single bins, but their use may result in lower processing costs due to decreased sorting requirements. Other factors to consider when evaluating bin specifications include ease of storage in the home and cost of replacement. Bins wear out, are stolen, are appropriated for unintended purposes, and are treated less than gently by collectors. These issues need to be considered when determining the type of collection and set-out approach and the style of home storage containers employed.

In addition to reusable home storage container bins, some communities have opted to collect recyclable materials in plastic bags. Participating residents put their recyclable materials in some type of a "recycling bag" (usually polyethylene or woven polypropylene) and place it at the curb. In some programs that have used bags, recyclable materials and trash bags are picked up by the hauler at the same time in the same truck, and the recyclable materials are pulled from the regular waste stream at a processing facility (landfill, transfer station, materials recovery facility, or waste-to-energy facility). Once separated, the bags are opened, and recyclable materials are processed and marketed. Issues to keep in mind when considering bag co-collection with trash are breakage of glass, material contamination, material loss, and, an additional processing step.

The decision to use bags in some communities has been based on several factors. Bags are usually purchased by the system users (residents and businesses), whereas bins are often provided by the municipality. Residents might leave bins at the curb after collection, resulting in a higher incidence of theft and damage. Whether a community chooses to use plastic bags or bins affects other components of the recycling system, such as type of collection vehicle used, the amount of material separation required, and the types of processing equipment or capacity needed. There is no hard-and-fast rule for a community to apply when considering bags versus bins. Careful analysis and discussions with other communities that have faced similar decisions will be useful.

Frequency of collection is also a major consideration when designing a residential recycling program. Weekly collection is the standard for curbside collection. Some communities institute every-other-week or even once per month collection as a cost-saving measure. Once or twice monthly collections can reduce operational costs, but require more extensive promotional efforts to remind residents of their collection day. Furthermore, if residents miss a collection day, they may decide to dispose of recyclable materials that may have overflowed in the storage container(s) in order to avoid the nuisance.

MULTIFAMILY COLLECTION

The principle behind collecting recyclable materials from multifamily residential dwellings is no different from curbside collection from single-family residences—maximize ease of participation for the resident. Planning for the collection of recyclable materials from multifamily units must take into account the existing procedures for refuse disposal (residents may take their garbage to a trash room, may access trash chutes, or may be required to take their waste to an outside dumpster).

Home storage of the recyclable materials needs to be considered. Apartments typically have small kitchens and limited storage space. Commingled collection is generally advisable under such settings—one storage container takes up less space than three. Residents would then deliver the commingled recyclable materials to the set-out point, where they may be required to sort the materials into appropriate bins or place them in a single large container. Each multifamily situation will present unique challenges. Creative, flexible approaches to collection may be more appropriate than a strictly prescriptive approach.

RESIDENTIAL DROP-OFF COLLECTION

Drop-off recycling programs are the most common recycling collection systems currently in operation in the United States and are the preferred approach for small or rural communities. Drop-off centers may be publicly or privately operated or run by nonprofit community groups. Systems rarely are capital intensive or require high operation costs; they present a relatively low cost and flexible way for recycling programs to be designed to fit the specific needs of the community.

In communities where the majority of residents take their refuse to the landfill or a transfer station, the placement of drop-off centers at these sites may be the most appropriate way to recover recyclable materials. In other communities, however, planners need to weigh the benefits of a drop-off system (reduced capital expenditures and lower operating and maintenance costs) against potential drawbacks. Drop-off systems generally achieve lower diversion rates than more convenient curbside programs since they require that residents not only keep recyclable materials separate from other refuse but also deliver these materials to the drop-off facility during its hours of operation.

COMMERCIAL WASTE STREAM COLLECTION

When designing recyclable materials collection systems for the commercial sector, it is critical to understand what can be efficiently recovered from the waste stream of each establishment. Industry typically recovers and recycles a large amount of preconsumer scrap. Many other postconsumer commercial recycling activities exist; however, the most prevalent programs recover paper products, mostly corrugated cardboard. Office paper and commercially generated glass and cans also present a commercial sector recycling opportunity.

Designing a system for the collection of recyclable materials from a commercial establishment must take into consideration how the waste stream is generated and how it leaves the generator. Incentives for separation can encourage recovery rates. For example, in a corrugated recovery program, tipping fees could be reduced for corrugated-rich loads and/or the generator could be paid for the materials. Private paper dealers and/or haulers may put baling or storage equipment on site to encourage separation, collection, and storage of the target material.

The collection mechanism for recyclable materials in the commercial sector will probably vary little from the collection of garbage. Typically collection is mechanized. Sometimes adding a separate container for cardboard is all that is necessary. If cardboard makes up the largest component of the waste stream, a smaller container for noncardboard materials may be the only adjustment needed. Buyers of recovered office paper, on the other hand, may provide for separate collection from centrally located set-out points. Obviously, local markets, processing capabilities, and many other factors must be considered in starting a commercial collection system.

INSTITUTIONAL-GOVERNMENTAL COLLECTION SYSTEMS

Institutions are major solid waste producers, and significant quantities of recyclable materials can be removed from institutional and governmental waste. To successfully implement recycling programs in institutions, three issues must be addressed, either collectively or on an individual basis: (1) obtaining reliable markets and transportation for recyclable materials collected; (2) developing a well-organized and convenient internal recyclable materials' collection program accessible to employees and those being served by the institution; and (3) providing adequate educational and promotional support for the program. Many of the same principles involved in commercial recycling programs apply to institutional systems.

COLLECTION EQUIPMENT

A properly designed collection system with the most suitable collection vehicles forms the backbone of a successful recycling program. Selecting the most appropriate collection vehicles for a recycling program requires careful consideration and analysis of the entire program structure. The collection vehicle is vital to obtaining the best collection efficiency available, given the particulars of home storage systems, market requirements, transportation routes, and processing capabilities. In some communities, existing equipment can be used or modified, resulting in substantial cost savings. Other communities, however, may need to invest in the purchase of dedicated recycling collection vehicles. When selecting appropriate equipment, program planners should keep the following considerations in mind:

- Total system cost can be minimized if programs are designed with interchangeable equipment.
- Curbside collection vehicles for separate collection should be designed with a low materials loading height, sufficient capacity for full collection routes, readily accessible cabs, and quick off-loading of materials.
- Commercial collection vehicles should be able to maneuver well in tight areas near participating businesses.

PROCESSING ALTERNATIVES

A system (including one or more facilities) to receive and process recyclable materials generated by residential and commercial-institutional sectors is a critical element in an efficient, comprehensive recycling program. To ensure that the recycled materials are marketable, the system must have the capability to upgrade materials to a variety of specifications. It must also have the flexibility to adapt to new specifications should new markets be engaged. Meeting market specifications can be ensured to the greatest extent possible by processing the materials in a materials recovery facility (MRF). The MRF receives collected recyclable materials, removes contaminating materials, separates, and provides for storage and loadout of large quantities so that economies of transportation to markets are achieved. It is increasingly important to process and upgrade a large flow of material to meet a range of current and future market specifications particularly in periods of lowered demand and prices. Another strong argument for developing a MRF is that it allows haulers flexibility in the type of equipment they can use for collection. Large haulers with specialized recycling vehicles can be accommodated as easily as smaller haulers with trailers, stake-body trucks, or compactors.

Yard waste composting facilities may also be required to process increasing supplies of grass, brush, and/or leaves. Developing a yard waste processing system is attractive for several reasons. First, the market for the end product (wood chips or compost) is generally available at the local level (municipal or landscaping operations). Second, yard waste represents a significant portion of the waste stream in many communities and is easily identified and separated. Significant reductions in landfill disposal of waste can be achieved through yard waste recovery operations.

After determining current and anticipated processing needs for their program, planners must decide if public or private facilities will be used. Ownership, financing, and operation options range from full public to full private, and include a hybrid version comprising public and private elements. There is generally some correlation between privatization and the allocation of risks and rewards. The owner and operator of a processing facility will generally bear a greater proportion of the risks while expecting a greater share of the rewards or revenues.

A private company might be willing to assume certain risks (responsibility to ensure steady delivery and sale of materials) if it owns and operates the facility. These risks would necessarily rest with the owner-operator of a private "merchant" facility (one that is financed, owned, and operated by a private company without contractual guarantees from local government), but to a more limited degree might be borne by some nonmerchant owner-operator. A privately owned and operated facility holding recyclable materials delivery contracts with a municipality would have a guaranteed flow of materials, thereby incurring less risk. Planners should anticipate this in any tip fee or revenue sharing agreements with such a facility. The willingness of private firms that own and operate processing facilities to bear these risks can be determined more accurately through the Request for Proposal (RFP) process.

With a full-public facility, the public sector can be expected to bear the material delivery and marketing risks but would also garner the benefits of a strong market. A publicly owned, privately operated facility normally has some sharing of risk and reward. This often is accomplished through some kind of revenue sharing arrangement. Decision makers should be aware, however, that the vendor community has as yet shown only limited willingness to share recyclable materials delivery or market price risk. The general vendor view is that they should be protected against loss for events beyond their direct control. Delivery of recyclable materials will in many cases be a governmental responsibility. Although vendors should be held accountable for meeting market specifications, they are not generally willing to run much risk with respect to severe drops in market prices.

There are a number of important risks associated directly with the construction and operation of a processing facility. For these risks, ownership is of lesser importance than operation. With full-public option, the public entity operating the plant is completely responsible for most things that may go wrong. In theory, problems resulting from faulty design or construction should be the responsibility of the party that designed and built the facility. In practice, however, it is quite difficult to establish blame clearly or to secure a satisfactory remedy. While this may not be a very important consideration for many routine governmental functions, it should certainly be of concern when innovative approaches are implemented.

MEDIA RESOURCES TO PUBLICIZE PROGRAM

In order to encourage and sustain a commitment to recycling in the community, an aggressive educational program of instructional and motivational messages will need to be developed. Once the target populations have been identified, the messages likely to educate and motivate those populations can be developed. Selling the concept of recycling will be no less strategic than selling a product, and traditional marketing practices can be applied. Promotional and educational materials should be consistent, clear, colorful, and creative—representing an understanding of the technical issues and an appreciation of the needs of the target audiences. If illiteracy is identified as a barrier to participation, for example, the promotional materials developed to reach the target audience should be aural (radio and television public service announcements, for example) and/or pictorial in nature. If the target population is primarily composed of middle income families with two or more children living at home, a family-oriented theme may be more effective. In addition, if the target population is college students, media tactics to reach that population would be different from those chosen to reach the working sector or a senior citizen audience. The community should evaluate local media opportunities with regard to their typical audience and reach. Available media resources include newspapers, magazines, newsletters and other publications, radio stations, television stations, mass transit, billboards, and recreational facilities.

Newspapers, broadcast stations, and outdoor advertising firms typically maintain statistics regarding readership-viewership. For example, a local radio station should be able to identify the most prevalent age range of listeners during a variety of time slots during the day. If the target audience were senior citizens, a radio station whose midday programs achieve the highest local rating would be an excellent target for placing paid advertising. Morning and evening commuter time slots would be expected to reach a wider audience.

When possible, local media outlets should be recruited to cosponsor recycling programs as a community service. Public service announcements in newspapers, television, and on radio programs provide cost-effective promotional avenues. Other promotional tools include

- Paid media advertising
- Press conferences, news stories, media events
- Direct mail publicity to citizens and businesses
- Promotional signage at recycling sites, containers, and equipment
- Bumper stickers, buttons, t-shirts, doorhangers, magnets, etc.

MODEL RECYCLING PROGRAM CASE STUDIES

Rural County

Population: 50,000

Size: 250 mi^2

Total MSW: 48,000 tons per year (30 percent commercial, 60 percent residential, 10 percent institutional)

Legislative Background and Recycling Goals. The state recycling law requires that a plan be developed that describes how the county will achieve a 25 percent reduction in the solid waste stream through recycling and reuse by January 1, 1995. The issuance of building permits is contingent upon the achievement of this goal.

Current Solid Waste Management System

- The single county-owned and operated landfill has an expected capacity of 25 years.
- Seventy percent of the residential MSW is delivered directly to a county disposal facility (one of six staffed compactor sites or the county landfill) by residents. Approximately 50 percent of county waste is collected by private haulers.
- Most of the MSW generated in the county (from all sectors) is disposed of at the landfill. A smaller, unknown quantity is burned or buried on-site or illegally dumped.
- White goods are accepted at the compactor sites, but residents are encouraged to bring them to the landfill.
- The tip fee at the landfill for private haulers has recently been raised from $15 to $40 per ton. Residents may deliver garbage for free.
- Revenues for the waste management system are raised through tip fees and a special assessment on each improved lot in the county.

Existing Recycling

- White goods and scrap metal segregation at the landfill.
- Roll-off boxes placed at local military installation for scrap metal recovery.
- Old corrugated containers (OCC) separation at several local supermarkets.
- Office paper recycling programs in several county offices.
- Newspaper recycling through The Optimist Club (when markets are favorable).
- Local markets identified for newspaper, high grade office paper, white goods, and scrap metal.
- Regional markets identified for glass, OCC, metal cans, and plastic containers (HDPE and PET).

Setting Goals

Short-Term

- Analyze the county's waste stream and determine the potential quantities of recyclable materials available for recovery from residential, commercial, and institutional-governmental sources.
- Review existing recycling activities in the county and quantify amounts of materials currently recovered through these efforts.
- Identify markets available to the county for the targeted recyclable materials.
- Evaluate collection alternatives, including the legal basis for county involvement in recyclable materials collection activities.
- Analyze existing in-county processing options for handling targeted recyclable materials.
- Review regional processing alternatives, including the feasibility of establishing a regional materials recovery facility and/or sharing processing equipment with neighboring counties.
- Review possible ownership and operation structures for any needed facilities.
- Develop reporting requirements and data collection mechanisms necessary to accurately document the county's progress toward its goals.

- Review financing mechanisms available to the county to pay for the planned systems and programs.
- Take advantage of residential self-haul practices and pursue an aggressive residential drop-off system using existing compactor sites.
- Develop public education and promotion campaign.
- Document existing commercial recycling that can be credited toward goal.

Medium-Term

- Monitor progress of residential drop-off program. Modify existing sites as needed to accommodate increasing amounts of targeted recyclable materials or additional materials. Determine if curbside system is indicated in order to attain recycling goal.
- Take action to increase commercial sector recycling through a combination of technical assistance and financial incentives such as:

  Requiring or offering a recyclable materials collection service to commercial establishments.

  Facilitating cooperative marketing of commercial recyclable materials.

  Providing businesses with balers or separate dumpsters for OCC collection.

  Waiving tip fee or offering a reduced tip fee for loads of commercial recyclable materials.

  Prohibiting disposal of loads containing designated recyclable materials.

Long-Term

- Meet and maintain state-mandated recycling goal within the county's budget constraints.

Implementation Schedule

Short-Term

Residential Program

- Determine what materials will be collected and what types of containers will be used.
- Determine the role of the public and private sectors in operating the drop-off sites.
- Develop bid specifications for and procure drop-off equipment.
- Modify existing compactor sites to accommodate roll-off containers for recycling.
- Procure appropriate signage.
- Train compactor site attendants.
- Begin accepting targeted recyclable materials from county residents.
- Monitor collection and assess public attitudes toward the program.

Commercial-Institutional Recycling Program

- Conduct survey of selected commercial establishments to determine current level of recycling and to guide the development of a technical assistance outreach program.
- Meet with local haulers and processors to encourage private sector collection of commercial and institutional recyclable materials.
- Expand in-house recycling programs for county offices.

- Develop and institute reporting mechanism to quantify and document commercial and institutional recycling activities.

Marketing

- Work with state to identify and develop markets for targeted recyclable materials.
- Negotiate and secure sales agreements with preferred markets.

Processing

- Determine if private sector processing capacity will be utilized. If so, develop and issue procurement documents and draft contracts.
- If county owned and operated processing facility is chosen:

  Select and prepare site(s).

  Secure necessary permits.

  Develop and issue bid documents for building and equipment.

  Select preferred contractor(s), execute agreements, and monitor facility construction and testing.

Evaluate feasibility of private or regional yard waste processing facility. If county operated composting facility is chosen, secure necessary permits; select and prepare site; procure necessary equipment; and begin receiving yard waste.

Public Education

- Develop and institute aggressive promotional campaign aimed at both residential and commercial sectors.

Administration

- Hire appropriate staff.
- Determine reporting procedures and documentation needs for commercial and institutional generators and/or local haulers and processors.
- Develop and implement enforcement strategy.
- Develop procurement policy for recycled products.

Medium-Term

Residential Program

- Evaluate drop-off program and make changes as needed (i.e., need for additional sites).
- Evaluate potential sites for additional drop-off capacity at other county locations.
- Evaluate need for more aggressive residential collection measures (i.e., curbside collection of recyclable materials from those residents who use private haulers).

Commercial–Institutional Program

- Evaluate the contribution of commercial recycling activities toward recycling goal through reporting and documentation systems.
- Continue outreach-technical assistance program for this sector.
- Consider appropriate incentives and disincentives to encourage increased recycling.
- Expand recycling programs to include all county facilities; consider including multiple materials (scrap metal, yard waste, batteries, used oil, tires, etc.).

Marketing

- Monitor market conditions and existing market arrangements.
- Renegotiate market agreements as needed.

Processing

- Monitor process flow, receiving procedures, storage and handling methods, and adjust as needed.
- Procure additional equipment or services as needed.

Public Education

- Continue educational and promotional campaign for general recycling programs.
- Implement waste reduction campaign.

Administration

- Evaluate staffing and budget allocations and request adjustments as needed.
- Provide ongoing management of the recycling program.
- Document recycling recovery rates and contribution that each sector is making toward recycling goal; report to the state as needed.
- Assess progress toward goal and consider program adjustments if shortfall is encountered.
- Prepare timetable to enact or amend necessary legislation, ordinances, codes, and other governmental tools to achieve recycling goals.
- Monitor evolving federal and state legislative proposals and keep abreast of potential impacts on county program.
- Implement and expand procurement policy for recycled products.

Long-Term

Residential

- Monitor all existing collection systems and adjust as necessary.
- Implement curbside collection in indicated areas.

Commercial–Institutional Collection

- Monitor effectiveness of commercial-institutional collection programs and make adjustments as needed.

Marketing

- Monitor market conditions and existing market arrangements; renegotiate as needed.

Processing

- Monitor processing of recyclable materials and yard waste; adjust operating procedures as needed.
- Procure additional equipment as needed.

Public Education

- Continue ongoing educational and promotional campaigns.
- If mandatory ordinance is implemented, develop appropriate media relations strategy and public outreach efforts.

Administration

- Document recycling recovery rates and report to state.
- Implement mandatory recycling ordinance if necessary.
- Monitor legislative needs and initiatives.
- Evaluate staffing and budget allocations and request adjustments as needed.
- Provide ongoing management of the recycling program.

Large Municipality

Population: 3,500,000

Demographics: 1,000,000 households (single and multifamily)

Total MSW: 4,000,000 tons per year (55 percent commercial; 35 percent residential; 10 percent institutional)

Legislative Background and Recycling Goals. The state's comprehensive Waste Management Act requires that every county in the state reduce its waste streams by 50 percent by the year 2000. An interim goal of 25 percent reduction-recycling must be achieved by 1995. The act provides for the disbursement of grants and loans to help jurisdictions implement recycling programs. Recycling plans are required from each of the counties; however, municipalities with populations over 500,000 may choose to submit their own plans.

Current Solid Waste Management System

- Twice-per-week municipal collection from residential sector (single-family and multifamily up to four units) with city crews and vehicles.
- Currently, three-person collection crews manually empty garbage into compactor trucks. Under this system, over 700 city trucks are necessary to accomplish collection. City is considering a switch to semiautomated collection of wheeled carts in some residential areas and reducing collection frequency so that existing trucks can be diverted for recycling collection.
- Front-end loader collection trucks used to collect waste from dumpsters from city institutional buildings, housing projects, and a smaller number of condominiums.
- Schools use their own vehicles to collect their own waste.
- Commercial establishments receive private waste collection.
- All of the city's waste is delivered either to a privately owned and operated waste-to-energy facility or the city owned and operated landfill.
- Tip fee is $65/ton at the landfill and at the waste-to-energy facility.
- A tip fee surcharge imposed at the waste-to-energy facility and the landfill is credited toward a recycling fund, which can be used for program development and implementation.
- The city has retained the services of a consulting firm to conduct a waste stream analysis and to prepare its recycling plan in accordance with state regulations.

Existing Recycling

- White goods are segregated at the landfill.
- Ferrous metal is magnetically recovered from ash at the waste-to-energy facility.

- The city has recently launched a 2000 household pilot curbside program. City crews collect mixed paper, plastic containers (HDPE and PET), metal cans, glass containers, and leaves, from these residents on a weekly basis.
- There are 10 city-sponsored community drop-offs and 2 private buy-back centers.
- The city provides vacuum collection of leaves from city parks, recreation areas, and two upper-income suburban neighborhoods within the city. These leaves are taken to one of two composting facilities.
- Several city offices have instituted office paper recycling programs.
- Local markets have been identified for newspaper, old corrugated containers (OCC), office and mixed paper, plastic containers (HDPE and PET), and steel cans and scrap.
- The city is currently diverting 8 percent of its generated waste stream through existing recycling programs from residential and commercial–institutional sources.

Major Policy Considerations

- Achievement of a 50 percent reduction-recycling goal necessitates a full-scale recycling program serving all sectors of the population. Adequate financial, administrative, and legal resources must be committed if the program is to satisfy state requirements.
- Consultants have recommended that, in order to meet its recycling implementation deadlines, the city should couple its recycling program with a switch to a semiautomated collection system for refuse. As semiautomated collection of refuse is introduced in residential areas, manual collection vehicles will be diverted to provide recycling collection. The consultants have determined that the increased efficiency realized with semiautomated collection will offset the costs of implementing citywide residential recycling.
- The aggressive nature of the recycling program necessary to achieve the state mandated recycling goal will require the commitment of adequate staff. At a minimum, the city needs to hire qualified personnel to include division heads for recycling and waste reduction; coordinators or specialists to handle several distinct program areas (i.e., public education, marketing, commercial recycling initiatives, and residential recycling activities); and clerical or support personnel.
- Traffic congestion issues need to be considered when incorporating recycling into the current refuse collection system.

Setting Goals

Short-Term

- Apply for state funding for program implementation, facility development, and/ or equipment purchase through grant or loan program.
- Distribute wheeled carts for once-per-week refuse collection and implement once-per-week recycling collection in selected communities.
- Test yard waste collection alternatives and select most appropriate option(s).
- Allocate available collection resources (personnel and vehicles) to accommodate both refuse and recyclable materials collection.
- Continue expanding once-per-week refuse collection and once-per-week recycling collection to all city collected households.
- By working with private haulers, businesses, and property management firms, implement a commercial collection program.

- As curbside recycling programs are initiated in neighborhoods, develop program of education, warnings, and enforcement to encourage participation and compliance.
- Develop a record-keeping system that will be used to monitor all recycling activities (including drop-off and buy-back centers) and assess progress toward the recycling goal.
- Target recyclable materials that will be included in the city's program, based on the results of the waste stream analysis, potential contribution toward achievement of the recycling goal, and the presence of local or regional markets.
- Identify and secure markets for recovered materials.
- Identify and secure adequate processing capacity. If necessary, apply for funding to finance the development of a materials recovery facility and/or a yard waste composting facility.
- Develop an aggressive public education–promotional campaign for residential, institutional, and commercial sectors.
- Hire appropriate staff to manage comprehensive recycling program.
- Consider imposing landfill bans on designated materials.
- Ensure that all new public convenience centers or other waste-receiving facilities are designed with recyclable materials handling capabilities.

Medium-Term

- Monitor progress of programs and adjust as needed to assure attainment of 25 percent recycling goal.
- Expand program to include other materials that can contribute toward the 50 percent reduction goal.
- Consider mixed waste processing or composting to achieve 50 percent goal.
- Provide ongoing outreach and technical assistance to commercial and institutional sectors.
- Consider passing ordinances requiring private haulers to offer recycling services to their clients and making hauler licensure contingent upon the provision of these services.

Long-Term

- Achieve and maintain 50 percent reduction goal.

Implementation Schedule

Short-Term

Residential Program

- Expand pilot curbside program.
- Begin shift to semiautomated collection of refuse with once-per-week refuse and once-per-week recycling collection; begin training existing personnel in recyclable materials collection methods.
- Procure and distribute wheeled refuse collection carts and home storage recycling bins for household recyclable materials and yard wastes (as appropriate).
- Pass ordinance requiring homeowners in areas receiving curbside collection to keep recyclable materials separate from trash. Allow residents choice of participating in curbside program or taking recyclable materials to drop-off or buy-back center.
- Institute pilot multifamily collection programs.
- Monitor collection and assess public attitudes toward the programs.

Commercial-Institutional Recycling Program

- Identify selected commercial and institutional waste generators and issue survey to identify recycling potential and existing recycling practices.
- Implement technical assistance–outreach program to commercial sector.
- Develop and institute reporting mechanism to quantify and document commercial-institutional recycling activities.
- Establish a voluntary program whereby local haulers are encouraged to offer the collection of recyclable materials to their commercial-institutional accounts. If a good response is not received, pass ordinances requiring private haulers to provide their customers with recycling collection.
- Place special emphasis on programs to recover materials from schools to help foster the development of a recycling ethic in school-age children.
- Expand in-house recycling programs for city offices.

Marketing

- Investigate feasibility of collective marketing agreement with county.
- Work with state to identify and develop markets for targeted recyclable materials.
- Negotiate and secure sales agreements with preferred markets.

Processing

- Begin procurement of processing capacity through a Request for Proposals (RFP).
- Develop yard waste processing capacity through public or private sources.

Public Education

- Develop and institute an aggressive, general promotional campaign aimed at both residential and commercial–institutional sectors.
- Develop specific media relations and a public outreach strategy to offset potentially a negative response to mandatory recycling ordinances.
- Promote and initiate block leader programs in areas receiving curbside collection as a way of encouraging participation and generating community pride in and "ownership" of the program.

Administration

- Institute landfill disposal ban on lead-acid batteries and tires, concurrent with the development of recycling programs to handle these materials.
- Hire appropriate staff.
- Develop procurement policy for recycled products.
- Determine reporting procedures and documentation needs for commercial–institutional generators and/or local haulers and processors.
- Develop and implement enforcement strategy.

Medium-Term

Residential Program

- Evaluate ongoing curbside and drop-off programs, identify problems regarding participation, equipment, collection, and markets, and make changes as needed to ensure achievement of interim 25 percent recycling goal.

- Evaluate potential sites for additional drop-off centers.
- Continue expansion of curbside program.
- Expand multifamily programs working with building owners and managers and private haulers. Commercial-Institutional Program
- Evaluate the contribution of commercial-institutional recycling activities toward the recycling goal through reporting and documentation systems.
- Continue outreach–technical assistance program for this sector.
- Consider appropriate incentives and disincentives to encourage increased recycling.
- Expand recycling programs to include all city facilities; begin including additional materials, concurrent with identification of available markets.

Marketing

- Evaluate market conditions and existing market arrangements.
- Renegotiate market agreements as needed to accommodate increased quantities and additional materials in program.

Processing

- Evaluate process flow, receiving procedures, and storage and handling methods and adjust as needed for public or private MRFs.
- Procure additional equipment and/or capacity as needed.
- Begin project development work for mixed waste processing or composting if appropriate.

Public Education

- Continue general educational and promotional campaign for all sectors.
- Implement waste reduction campaign.

Administration

- Consider landfill ban on yard waste and other designated recyclable materials.
- Evaluate staffing and budget allocations and request adjustments as needed.
- Provide ongoing management of the recycling program.
- Document recycling recovery rates and contribution that each sector is making toward recycling goal; report to the state as needed.
- Assess progress toward goal and consider program adjustments if shortfall is encountered.
- Implement and expand procurement policy for recycled products.

Long-Term

Residential

- Monitor all existing collection systems and adjust as necessary.
- Expand curbside collection to include all city-collected residences.

Commercial–Institutional Collection

- Monitor effectiveness of commercial-institutional collection programs and make adjustments as needed.

Marketing

- Evaluate market conditions and existing market arrangements; renegotiate as needed.

Processing

- Monitor processing of recyclable materials and yard waste; adjust operating procedures as needed.
- Procure additional equipment and/or capacity (including mixed waste processing or composting) as needed.

Public Education

- Continue ongoing educational and promotional campaigns.

Administration

- Document recycling recovery rates and report to state.
- Monitor legislative needs and initiatives.
- Evaluate staffing and budget allocations and request adjustments as needed.
- Provide ongoing management of the recycling program.

Small Town

Population: Year round 36,000.

Demographics: Major state university is located in the town. Forty thousand students attend the university.

Total MSW: 44,000 tons per year (including 8000 tons per year generated by the university).

Legislative Background and Recycling Goals. The state's comprehensive recycling law sets a mandatory recycling goal of 30 percent to be achieved by 1997. The law states that each municipality must pass an ordinance mandating source separation and collection of recyclable materials from each resident, business, or institutional establishment within its borders. This provision requires all primary and secondary schools, colleges, and universities to implement recycling programs.

Current Solid Waste Management System

- The county in which the town is located is a member of a regional solid waste authority (the authority), which is the lead agency for the recycling program.
- The town's waste is taken to the authority's transfer station, where it is subsequently transported to a neighboring county's landfill.
- The authority has implemented a surcharge on each ton of waste disposed at its transfer station to fund the local portion of the capital costs of the recycling program.
- The authority's transfer station is being modified to accommodate an intermediate processing center (IPC) for recyclable materials and a scale house.

Existing Recycling

- Through a grant, the state environmental regulatory agency has provided funding to the authority for program planning, contract development, and equipment and facilities for the service area.

- The authority has prepared a recycling plan for its member municipalities describing markets for the sale of recyclable materials; an initial curbside recycling service area; locations for drop-off centers; volumes of materials that can be anticipated; and details on the proposed IPC.
- The proposed recycling program will be multimaterial, mandatory, regional, authority-sponsored, and will provide the weekly collection of recyclable materials from residences.
- The authority has hired a full-time recycling program manager to help implement the program and work closely with member municipalities.
- The town, as a member municipality, has entered into a recycling agreement with the authority. Under the terms of the agreement, the authority accepts responsibility for all program planning, financing, implementation, and administration; application for funding; adoption of regulations; and provision of public education. The municipality is obligated to adopt a mandatory recycling ordinance and to cooperate and assist the authority with the development of the recycling program.
- The state university located in the town has also entered into a recycling agreement with the authority. Under the terms of this agreement, the university purchases all equipment needed to collect recyclable materials and provides transport of the materials to the authority's IPC at the transfer station. In turn, the authority pursues state funding to cover part of the start-up costs for the recycling program.
- The town is reviewing a draft recycling ordinance developed by the authority.
- The authority is sponsoring recycling demonstration programs, including limited curbside collection, establishment of drop-offs, pilot commercial programs, and yard waste processing and tire recycling projects. It plans to expand these programs.

Setting Goals. By choosing to become a member municipality of a regional authority, the town has simplified its role in the development of a comprehensive recycling program for its residents. Essentially, the town's short-, medium-, and long-term goals can be combined as follows:

- Achieve the state's recycling mandate in the most economical manner by working through the authority and other municipalities.
- Adopt the mandatory source separation and collection ordinance.
- Through the town representative on the authority's board, monitor the recycling agreement with the authority to assure that the town stays in compliance with state law.
- Approve, through town council, any funds required or loans that need to be guaranteed in connection with the recycling program.
- Adopt any legislative measures necessary to comply with the state law or the achievement of the reduction goal.

CHAPTER 5
SEPARATION, COLLECTION, AND MONITORING SYSTEMS

ABBIE (PAGE) McMILLEN*

With a Special Section on Volume- and Weight-Based Collection Rates by

LISA A. SKUMATZ, Ph.D.
Principal
Skumatz Economic Research Associates, Inc.
Seattle Washington

INTRODUCTION

Since the mid-1980s, recycling has blossomed as an entrepreneurial activity within both government and the private sector. The public desire to recycle has created a demand for many new products and services. Products include specialized recycling containers, collection vehicles, and processing equipment/facilities; services include collection, separation, processing, transportation, marketing, and public information. Business and government have both been eager to fill these needs for products and services, in many cases competing for the opportunity to provide them.

This chapter will discuss the products and services at the "front end" of the recycling cycle. Once the decision has been made to recycle, there are literally hundreds of decisions that need to be made concerning how to go about it. Perhaps the first decisions that are faced (and certainly the decisions with the greatest public visibility) are how to separate the recyclables from the rest of the refuse stream and, once separated, how to collect them for processing and marketing. These decisions are crucial to the success of the recycling effort. They must be made with full consideration of the fact that postconsumer recycling is essentially a remanufacturing process which usually depends upon thousands of volunteers as the source of the raw materials. If these thousands of sources of raw material are reluctant or unreliable, the remanufacturing process will be expensive at best, or even unfeasible.

This chapter addresses both the selection process and the available options for a separation and collection system, as well as measuring the performance of the selected separation and collection system. A critical objective is "to move the material from the joint of generation to the market in the least costly and most efficient fashion."

*During the initial preparation of this chapter, the author was project director for Roy F. Weston, Inc. of Burlington, Massachusetts.

Separation

The most fundamental question concerning separation is whether and to what degree to rely on separation at the point of generation. *Mixed-refuse collection and processing* systems ask the least of the generator. They also require the least change in the established collection system. However, far more effort is required in processing, since the recyclable materials must be separated from the balance of the refuse by human and mechanical means. This is expensive. Source separation places far more reliance on the generator, and in addition requires that modifications be made in the established collection practice. This can also be expensive.

Collection

Completely interrelated with the question about whether and how much source separation to rely upon is the question about whether to institute a special collection of recyclables (commonly called *curbside collection* when applied to residential refuse), or whether to rely instead upon the generator to transport the recyclables to a *drop-off* or *buy-back* location. Curbside collections generally yield much more material per capita but are also much more expensive than drop-off or buy-back collection. Quantity and marketability of the collected recyclables are of paramount concern. If it can be shown that the collected recyclables are much greater in quantity or much more marketable if collected in an expensive way rather than in cheaper, alternative way, then the extra cost of the more expensive collection system might be justifiable.

Measuring Performance

As a practical matter, the total quantity of recyclables in the refuse stream cannot all be recovered for recycling. For example, newspaper that is used for wrapping putrescible refuse, and glass containers that are broken into pieces too small to sort, will not be available for recycling. The term capture rate is used to denote the weight percent of an eligible material in the total refuse stream that is actually separated out for recycling. This performance measure is of the greatest importance in measuring the success of a separation and collection program. Note that the capture rate is a term that usually applies to an individual recyclable material. For example, the capture rate of aluminum may be very different from the capture rate for newspaper.

For residential refuse, the individual unit of refuse generation is the household. For commercial and industrial sources of refuse, the individual unit is the business. *Participation rate* is the term used to denote the percent of households or businesses that regularly separate out recyclables; i.e., that separate some eligible items at least once during a given period of time. Although usually applied to participation in curbside collection, participation rate has also been applied as a measure of the effectiveness of drop-off or buy-back centers. In fact, participation in a curbside collection program could be lower than expected if there is an effective network of drop-off or buy-back centers, but the overall capture rate might be high. On a monthly basis, it is estimated that 80 percent of households will participate at least to some degree in a well-designed and properly publicized curbside collection program, but a high participation rate does not necessarily mean that the capture rate will meet expectations. Participation rate is a very loosely defined term with the potential to mislead people into believing that a recycling program is more effective in capturing recyclables than it actually is.

Other performance measures have been developed for specific program design purposes. For example, set-out rate indicates the participation on any given collection day. In a community with weekly curbside collection experiencing an 80 percent residential participation rate, the set-out rate may be in the range of 50 to 60 percent. This factor is important in determining collection vehicle requirements.

Recycling rate is sometimes used to denote the pounds of total recyclables that are collected per household per month. It has been expressed as per household served by the program, per household in the community, or per household participating. In other cases, it is used to denote the weight percent of total refuse that is recycled instead of being landfilled or incinerated. This term can be misleading unless the specific materials being recycled and the characteristics of the refuse stream are both specified. For example, if a town generates 20 tons per day (TPD) of total refuse (residential plus commercial) and sends to the remanufacturing industries a total of 2 TPD (net of processing residue and rejects), the recycling rate is 10 percent of total waste but probably about 20 percent of residential waste (assuming residential and commercial tonnages are roughly equivalent, as they are in many communities).

Another performance measure is the *net diversion rate*, which represents the weight percent of total refuse that is not landfilled (or, in some cases, not incinerated either). Thus, if the objective of the program is to minimize the weight of refuse (including processing residues and incinerator ash) sent to a landfill through a combination of strategies (such as source reduction, recycling, and incineration), the ultimate performance measure is the net diversion rate. It is also possible to estimate a *volumetric* net diversion rate by knowing the compacted densities and weights of the materials diverted. Although not commonly computed, this measure would be a useful one, since landfills do not become full because they have become too *heavy*, but rather because their available *volume* has been consumed. Thus, high-volume, low-weight materials (such as some plastics) can contribute to a more rapid depletion of landfill space than their refuse weight percentage might imply.

SEPARATION

As in other aspects of recycling, there is confusion concerning the definition of terms relating to separation, for example, *commingled, source-separated,* and *curbside-separated*. The confusion is enhanced with additional qualifiers such as "fully commingled." In this chapter, we shall adhere to the following guidelines for use of these terms:

Commingling is an attribute that can only be fully understood if the materials that are commingled are also listed (for example, "commingled food and beverage containers" or "all fiber and nonfiber recyclables commingled"). There does not seem to be an easy way out of this more lengthy description, if clarity of meaning is to be preserved.

Source-separated should be a term reserved to mean separated into any number of categories of refuse by the generator of the refuse. Thus, the act of separating all recyclables from all other refuse is an act of source separation, even though only two categories result (refuse, and all recyclables commingled). Refuse can also be source-separated into multiple categories of recyclables, plus a remainder that is still refuse.

Curbside-separated refers to the process by which the collector receives commingled recyclables and separates them into categories during the act of putting them into a compartmentalized collection vehicle. (For example, "residents source-separate newspapers and containers from the refuse, placing the newspapers and commingled containers at the curbside, where the collectors curbside-separate the commingled containers into four categories: three colors of glass, and all other.")

The "generator" of waste is considered to be the person discarding it. Ranking separation concepts from those that make the least demands on the generator to those that make the most demands on the generator results in a hierarchy somewhat like the one shown in Fig. 5.1.

In general, participation rates can be expected to be greatest for the separation concepts that place the least demand on the generator, although motivational factors are complex and it has been possible to achieve high participation with methods that place relatively large demands on the generator. This section describes separation concepts and their effectiveness in detail.

No Source Separation: Mixed-Refuse Collection/Processing

In addition to being the easiest for the refuse generator, mixed refuse is also the easiest for the collector. Participation in a no-sort system is by definition 100 percent. However, the challenges to the processor should not be underestimated. All refuse is discarded in a single container and it is up to a processor to separate out the usable components, then upgrade these components to meet marketing specifications.

Availability
There are few examples of unsorted residential refuse being processed to remove recyclables. "Although a few mixed refuse recycling facilities are operating in the United

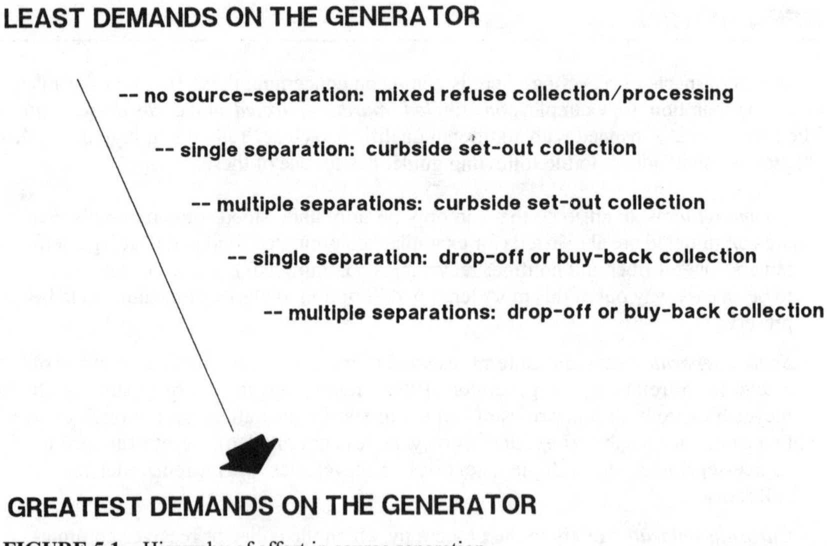

FIGURE 5.1 Hierarchy of effort in source separation.

States and Europe, common wisdom in residential recycling has always been that the process starts at the home. The responsibility for keeping recyclables out of the refuse bin and storing the materials rests with the resident." This assumption may be about to undergo a change, however, due to the increasing availability of innovative sorting and processing techniques, the generally higher cost of refuse disposal that makes such techniques cost-effective, and the inability of source-separation schemes to keep pace with refuse diversion goals set by policy makers.

In recent years, attempts have been made to engage the residents of inner-city areas in curbside collection. Where adequate data have been collected, it has been shown that participation and capture rates have been lower than experienced by curbside collection programs in the more affluent suburbs and in areas where the population is less transient. The reasons for this phenomenon are perhaps intuitively obvious, including the fact that economically stressed individuals may have more pressing priorities than source-separating their refuse, and transients are not in one place long enough to receive instructions. Communities are therefore taking an interest in mixed-refuse collection (followed by mixed-refuse processing) to recover recyclables. If the efficiency of the mixed-refuse processing system is high enough to recover a sufficient quantity of recyclables to meet program goals and mandates at an affordable price, the mixed-refuse processing might be a good solution for some cities.

In addition to the potential for recovering recyclables without effort on the part of the generator, there is the consideration of collection equipment cost. It is expensive to replace or supplement an existing fleet of refuse-collection vehicles with those designed to collect recyclables. Some cities' fiscal management policies require that they allocate the entire cost of publicly purchased recycling vehicles and in-home source-separation containers to the budget year in which they are purchased, rather than amortizing the cost over time. This results in the appearance of a large, immediate impact on the budget. To reduce this impact, a service contract with a mixed-waste processor could achieve diversion goals while spreading the cost out over time. (Another way, of course, is to phase-in the new vehicles over a period of years.)

Mixed-waste processing of the "dump and pick" variety is popular for reclaiming materials from the commercial waste stream. Mixed commercial loads tend to be more homogeneous and less noxious than residential refuse, so they are good candidates for mixed-refuse processing and in fact can be included to improve the economics of a mixed residential refuse collection and processing operation. For example, in Mann County, California, mixed loads of commercial refuse are hand-sorted, and cardboard, glass bottles, aluminum cans, scrap metal, paper, and wood are recovered. The nonrecoverable components (mostly nonrecyclable cardboard, food refuse, paper, and various plastics) go to a transfer station. It was found that plastics are so lightweight that they are difficult to pull out of the refuse, and their market prices were not high enough to cover the cost of recovering them.

Mixed-waste processing system designs differ substantially in the ratio of hand separation to mechanical separation. While it is not the purpose of this section to describe in detail the processing steps that take place in these facilities, in general they are quite similar to those used in the production of refuse-derived fuel. The mixed refuse is unloaded onto conveyors where workers hand-pick some easy-to-remove items (e.g., corrugated cardboard, newspapers, large pieces of brush, items that might damage the machinery). Bags are then broken open, and the refuse is separated into components using various combinations of shredding, screening, magnetic separation, eddy-current separation, air classification, and hand-sorting. New technology may soon be available that will distinguish among plastic resins.

A list of mixed-waste processing system suppliers is shown in Table 5.1, and these

TABLE 5.1 Mixed-waste processing system suppliers

Firm	Location	Telephone
American Recovery Corporation	Washington, D.C.	(202) 775-5150
Ashbrook-Simon-Hartiey	Birmingham, Alabama	(205) 823-5231
Bedminster Bioconversion Co.	Cherry Hill, New Jersey	(609) 795-5767
Catrel	Edison, New Jersey	(201) 225-4849
Enviro-Gro Technologies	Baltimore, Maryland	(301) 644-9600
Environmental Recovery Systems	Denver, Colorado	(303) 623.1011
Ebara International Corporation	Greenburg, Pennsylvania	(412) 832.1200
Fairfield Service Co.	Marion, Ohio	(614) 387-3335
Harbert Triga	Birmingham, Alabama	(205) 987-5500
K/R Biochem	Tulsa, Oklahoma	(918) 492-9060
Lundell Mfg. Co.	Cherokee, Iowa	(712) 722-3709
National Recovery Technology, Inc.	Nashville, Tennessee	(615) 329-9088
ORFA	Cherry Hill, New Jersey	(609) 662-6600
Omni Technical Services	Uniondale, New York	(516) 222-0709
Raytheon Service Company	Burlington, Massachusetts	(617) 272-9300
RECOMP	Denver, Colorado	(303) 753-0945
Refuse Resource Recovery Systems	Omaha, Nebraska	(402) 342-8446
Reuter, Inc.	Hopkins, Minnesota	(612) 935-6921
Riedel Environmental, Inc. (Dano)	Portland, Oregon	(503) 286-4656
Taulman Composting Systems	Atlanta, Georgia	(404) 261-2535
Trash Reduction Systems, Inc.	Overland Park, Kansas	(913) 661-9494
U.S. Waste Recovery Systems	Fort Worth, Texas	(817) 877-0147
Waste Management of North America Inc.	Oak Brook, Illinois	(708) 572-8800
Waste Processing Corp. (Dano)	Bloomington, Minnesota	(612) 854-8666
Waste Reduction Services, Inc.	Scottsdale, Arizona	(602) 483-8586
Wheelabrator/Buhier-Maig	Danvers, Massachusetts	(508) 777-2207
XL Disposal Corporation	Crestwood, Illinois	(703) 389-6312

Sources: Roy F. Weston, Inc., vendor files.
Note: Some of these systems may require some presorting of the refuse. Not all vendors have systems operating in the United States.

may be contacted for the details of their processes. A process flow diagram for one type of mixed waste processing system is shown in Fig. 5.2.

Effectiveness and Cost

The products most likely to be recovered by manual or mechanical separation from mixed municipal (primarily residential) refuse include aluminum cans, aluminum scrap, steel cans, glass containers, plastic containers, newspaper, corrugated cardboard, mixed paper, refuse-derived fuel, and compost substrate. The kind and amount of recyclables recovered depend on:

- The processing steps that are employed in the particular facility design
- The ability of those processing steps to remove the materials from the particular refuse stream presented to them as feedstock
- The existence of markets for the quality of processed recyclables that can be produced, given the facility design and the refuse feedstock

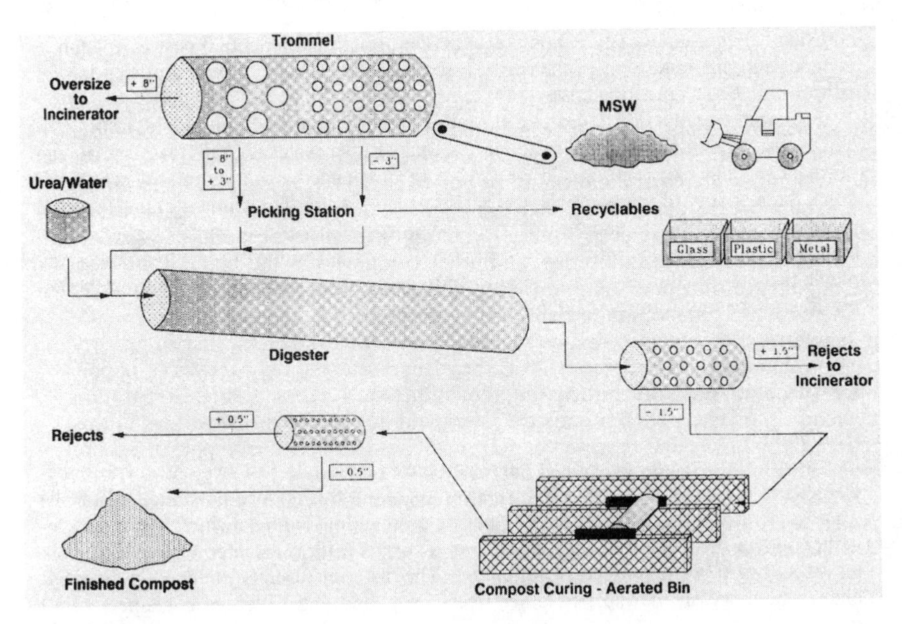

FIGURE 5.2 Process flow diagram.

Contamination of various degrees is bound to occur in a mixed-refuse collection. As market revenue rises, it becomes more worthwhile to employ ever more sophisticated and costly recovery and cleanup processes, whether manual or mechanical. It will be noted that one highly recyclable commodity, glass, is often not recovered in mixed-refuse collection and processing systems due to breakage both during collection and in the early stages of mixed-refuse processing, where refuse bags are being mechanically broken open and the waste tumbled around.

It is possible for mixed-waste processing systems to recover many recyclables (for example; steel cans) at capture rates that rival curbside collection. In addition, mixed-waste processing may have a very high landfill diversion rate through the production of fuel and/or compost. Some claim diversion rates in the neighborhood of 75 percent (not counting ash from combustion of the fuel fraction). Achieving this high a rate depends on being able to market the fuel and/or compost produced by the facility.

The cost of mixed-waste processing varies so widely that it would be misleading to generalize. However, it should be noted that all waste would typically be processed, so the facility would tend to be large and to have a relatively high capital cost (although much lower than a waste-to-energy incinerator of the same capacity). For example, if a curbside collection program were projected to capture enough recyclables to warrant construction of a 100-TPD intermediate processing center (IPC) or material recovery facility (MRF), then (assuming a typical curbside recycling rate of about 12 percent) a mixed-waste processing facility of about 830 TPD might be an alternative to the curbside-collection-plus-MRF scenario (since a total waste quantity of about 830 TPD would generate 12%—about 100 TPD—of recyclables). Whereas a 100-TPD MRF might be constructed for about $3 million, an 830-TPD mixed-waste processing facility might have a construction cost over $15 million, making it attractive for project financing. Associated with the curbside collection alternative would also be the capital and operating cost of collection vehicles and operating cost of the MRF: associated with the mixed-waste processing facility would be its operating cost.

A prominent recycling consulting firm, Resource Integration Systems Ltd. (RIS) concludes that "commingling of all materials (i.e., paper fiber materials with glass, metal, and plastic food and beverage containers) is not recommended. In programs where this is done, the recovery of all materials, especially paper, suffers. The amount of paper lost in processing is significant, and the quality of the processed paper is lowered by glass shards and other contaminants, limiting marketability. The trash residue of recycling processing plants that receive this material is significantly greater than for facilities that receive paper fibers separate from food and beverage containers."

Single Source Separation

Availability

In a single-sort system, generators segregate their refuse into just two categories: recyclable, and not recyclable. Next to no separation at all, this is the easiest method for the waste generator. Clearly, there have to be some limitations on what should be mixed together in the "recyclable" category. It would make little sense, for example, to mix leaves and yard waste with cans and bottles. The materials usually included in the recyclables category are newspaper, other fibers, and food and beverage containers of all types.

Usually, one-sort systems employ curbside collection. While it is theoretically possible to operate a drop-off or a buy-back program that accepts all recyclables (fiber and nonfiber) mixed together, there do not appear to be any examples of this approach. Some generators may choose to mix all fiber and nonfiber recyclables together in one container as a convenience in the home and separate them when they arrive at the drop-off or buy-back center. However, there are no data to support the supposition that this occurs. As a practical matter, it would seem to be neater, more convenient, and more efficient to separate newspapers and other fiber from cans, bottles, and other containers in the home if the resident is going to transport them to a drop-off or buy-back center.

There are, of course, limited drop-off centers, drop boxes, and reverse vending machines that accept only one material such as newspapers, plastic grocery bags, or aluminum beverage cans. If these represent the only recycling opportunity available in a town, then they constitute in a sense a single-separation "system." However, such a limited system will not be discussed further in this section. Drop boxes and reverse vending machines can be a useful part of a more comprehensive recycling system, however.

A "new" one-separation concept (which has actually been undergoing development in Europe for some time) involves separating refuse into "wet" and "dry" fractions. The dry (paper, plastic, metal, etc.) fraction is further separated at a processing facility into marketable recyclables and fuel, while the wet (organic putrescible) fraction is processed into compost. In parts of Germany, residents who separate out their wet garbage for recycling pay a lower refuse collection bill. In one town in the Netherlands, a two-sort, yellow and blue bag system was tried. The participation rate was 67%. The contents of the blue bags were composted with other organic waste. The contents of the yellow bags were further separated in a processing facility, but this proved to be technologically unfeasible. In another town, the dry fraction was more restrictively specified (cardboard and paper, tin cans, rags, and plastics). However, in this case, the separation of the dry fraction still proved unfeasible due to labor costs. The paper/plastic separation was found to be particularly difficult. Canada has also been experimenting with wet/dry systems, using green bags or covered buckets for food waste. However, most of the Canadian experiments also involve multiple source separations.

The most prominent example in the United States of a one-sort system is in Seattle, Washington. This city began a residential curbside collection experiment in 1988. In the southern half of the city, residents received a 60- or 90-gal container for the commingling of all recyclables. Recyclables in these containers were collected monthly. (For the northern half of the city, residents received three stackable bins for newspaper, mixed containers, and mixed paper. These containers were emptied weekly.) A single source separation has also been used in a pilot project serving apartments in Tukwila, Washington. The recycling dumpster is located outside of the building and accepts all paper, cardboard, glass, metals, and plastics. Detailed instructions are printed on the dumpster, which is emptied by the same contractor serving the Seattle collection. In this case, dumpsters were the preferred container because of their greater volume capacity to accept cardboard, mechanical unloading ability, and lower tendency to be stolen.

An alternative approach to providing separate collections for the two waste fractions (recyclables and nonrecyclables) is the provision of special bags in which residents place their recyclables. The bags are placed in the same container as the balance of the refuse and are collected in the same manner. At the processing facility, the bags are manually removed, opened, and the contents are processed. This method of collecting recyclables is termed *co-collection*. The chief advantage of co-collection is that high capture rates are theoretically possible without the cost of a special recyclables collection. There are several problems that remain to be fully resolved, however. To be durable enough to withstand the collection process without breaking, the special bags are relatively expensive (on the order of 30 to 50 cents apiece). People do not always put the right materials in them. Breakage of glass is a problem, especially if paper is not included in the bag to cushion the glass. And retrieval of the bags from the mixed refuse and opening them for processing is labor-intensive.

Effectiveness and Costs

Glass particles in the paper pose problems for recycled paper mills. Because of the glass breakage problem, the firm in Seattle, Washington, that has been collecting materials in the single-separation system is distributing plastic inserts that will fit into the 90-gal carts to hold glass containers, which will be separated by the collector. (See Fig. 5.3.) In addition, pickup will be more frequent than monthly in the future, since a cart of 90-gal capacity may not hold all the recyclables that a household generates in one month.

Multiple Source Separations

What Is the Optimal Degree of Source Separation?

By far the greatest number of recycling programs employ more than one separation on the part of the refuse generator. Theoretically dozens of categories of waste could be separated, leading to popular cartoons, such as the one shown in Fig. 5.4, which drives home the point that extra effort is required of the participants. The level of source separation required can be expected to have a direct impact on participation and capture rates. "As a general rule, the less residents have to do to participate in a recycling program, the more likely they are to participate. This convenience factor points out the need for multiple separations to be conducted somewhere other than at the household."

One thing seems to be clear: the responsibility of sorting into categories will not be the *overriding* consideration of residents in their decision to participate. For example, a sur-

FIGURE 5.3 Plastic insert for keeping glass containers out of the otherwise commingled recyclahles. (Photograph by A. McMillen, Mercer Island, Washington.)

FIGURE 5.4 Recycling cartoon. (Source: Don Addis in *The St. Petersburg Times*.)

vey conducted by the National Solid Waste Management Association (NSWMA) in 1986 showed a higher participation in Santa Rosa, California (70%), where residents were asked to sort into three categories (news, glass, and metals) than in Islip, New York (30%), where all containers could be commingled. (For factors leading to individual participation decisions, see Chap. 10 of this Handbook.)

The Center for the Biology of Natural Systems (CBNS) at Queens College in New York undertook an extensive recycling experiment in East Hampton, Long Island, New York. The researchers determined that four household containers would be needed for effective separation of about 70% of household refuse. As shown in Fig. 5.5, the four containers would receive:

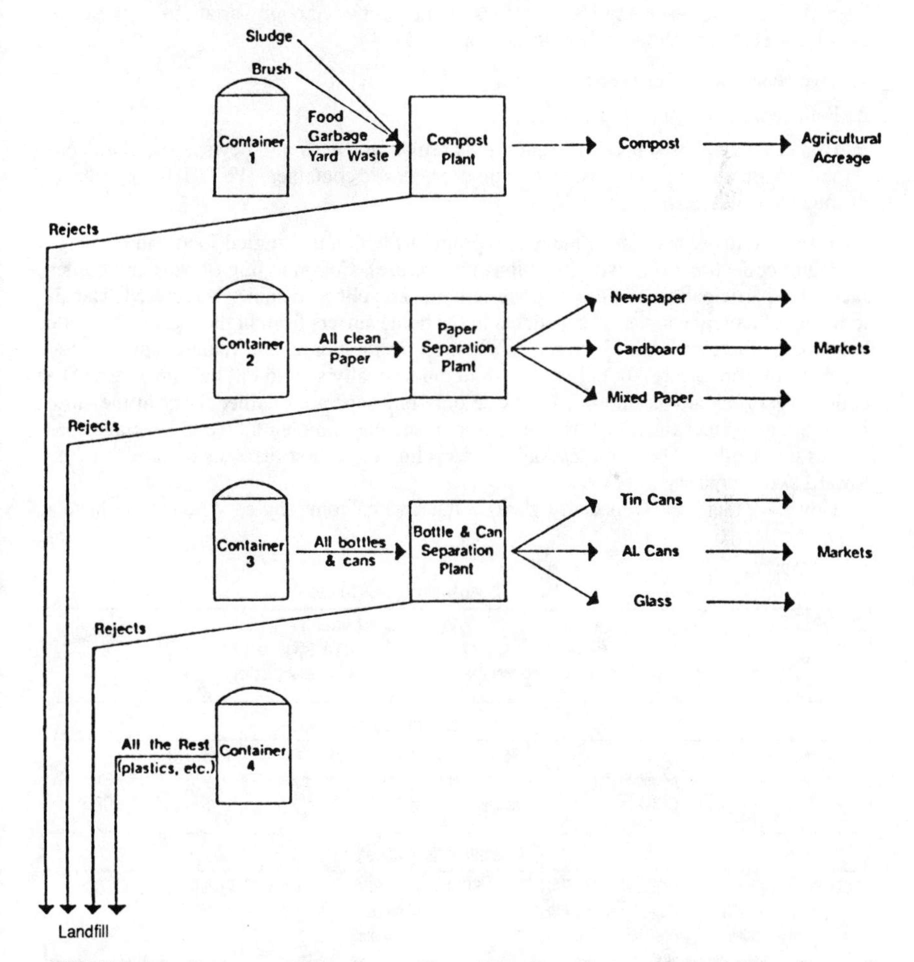

FIGURE 5.5 Basic separation scheme. (From "Non-Burn System for Total Waste Stream," *BioCycle,* pp. 30–31, April 1987.)

- Food waste (plus other organics such as disposable diapers and yard waste, to be mixed with sewage sludge and brush at a compost plant). This category functions like the "wet" category described earlier
- All clean paper (to be separated at a paper separation plant and sold)
- All bottles and cans (to be separated at a separation plant and sold)
- All other refuse (to be landfilled with the rejects from the other three plants).

Hazardous waste and bulky waste would require separate collection and disposal systems. However, this sorting scheme has not yet been widely implemented in its entirety, due to cost and current unavailability of mixed-paper separating facilities and mixed-organics composting facilities.

Three levels of multiple source separations (i.e., three degrees of commingling) have been attempted at various times in various municipalities (though variations of these are possible). These are shown schematically in Fig. 5.6:

1. Two separations plus refuse
2. Four separations plus refuse
3. Higher degrees of source separation plus refuse (for example, newspaper, glass containers, tin and bimetal cans, aluminum cans, plastic containers (PET/HDPE), and corrugated cardboard).

There are many successful municipal programs with commingled food and beverage container collection (i.e., two separations plus refuse). Commingling of cans and bottles facilitates participation because just one in-home recycling container is required. The alternative of having a series of containers in the home suffers from higher cost of in-home containers, containers taking up too much space in the home, and requirements for too much participant effort—both in sorting and in physically setting out the containers. The collector's efficiency is improved because only one separation is necessary at the curb; i.e., news and paper fibers in one compartment, and commingled food and beverage containers in the other. The major nationwide recycling collection firms seem to be trending toward two source separations.

However, materials (especially glass) collected in a commingled state tend to have a

2 separations, plus refuse

REFUSE	FIBER	CONTAINERS
	newspaper old corrugated containers	food/beverage containers (glass, metal, plastic)

4 separations, plus refuse

REFUSE	FIBER	GLASS	METAL	PLASTIC
	newspaper OCC	bottles jars	cans	PET HDPE

4 separations, plus refuse

REFUSE	BROWN GLASS	GREEN GLASS	CLEAR GLASS	NEWS	FERROUS	ALUMINUM	PLASTIC	OCC
	bottles, jars	bottles, jars	bottles, jars		tin cans bimetals	cans	PET HDPE	

FIGURE 5.6 Multiple recyclables sorting schemes.

lower market value, and research in New Jersey shows that smaller communities (especially those under 10,000 population) may find participation and capture to be entirely satisfactory in systems that require higher orders of source separation. In a pilot project, Fitchburg, Wisconsin's, 3,000 households are separating numerous recyclables into three bins plus an HDPE sack full of polystyrene. Participation in the polystyrene experiment has been 83% (but the weekly set-out rate for the sacks has been only about 11%, because it takes a long time for most homes to fill them).

Curbside Separation versus Source Separation

Many communities rely on two source separations plus further curbside separation by the collector. For example, Anne Arundel County, Maryland, asks the collector to separate recyclables into five truck compartments: three for the different colors of glass, one for mixed cans, and one for newspaper plus corrugated cardboard. In Palm Beach County, Florida, the collector sorts recyclables into four categories: glass (all three colors commingled), aluminum cans, plastics, and newspaper. Fitchburg, Wisconsin, which has three source separations, further separates at the collection truck into 10 bins.

Those who favor curbside sorting believe it "enables the community to sell quality separated materials to markets without installing an extensive processing system" with its attendant capital and operating costs. Curbside sorted material is delivered directly to a storage/aggregation facility rather than to an IPC or MRF. Less glass is wasted, since most of the breakage occurs after the glass is color-separated. Curbside sorting may be one answer for municipalities "whose operating budgets are more easily increased than their access to capital monies," and for smaller communities of about 10,000 or fewer stops.

In addition to the advantage of avoiding some of the costs of a processing facility, proponents of curbside sorting point out the educational advantage of curbside sorting: nonrecyclables are left in the container. Sorting at the curb allows the collector to correct a householder by leaving a "contamination note" attached to the reject in the bin. "Over time, the level of contamination in a commingled program actually gets worse . . . Our homeowner learns *the first time* he or she puts the wrong item in the box . . . so we get fewer and fewer contaminants."

Another advantage of curbside sorting is said to be its ability to accommodate change. "One of the advantages of initially beginning a curbside program with curbside sorting is the fact that the capabilities of the processing method can change as the program grows without changing the separation behavior of the residents."

On the other hand, curbside sorting does slow down the collection process. "In timing studies conducted in three communities comparing commingling with curbside sorting, the advantages of commingled collection are apparent. Two communities averaged slightly more than 30 seconds per stop with 5 separations at the curb. These same communities reduced time per stop between 7 and 10 seconds with commingled collection (one compartment for paper and another for mixed containers). This time savings resulted in extending route sizes substantially." (Note that other estimates for curbside separation have shown that it takes 50 to 60 seconds to serve one stop, and that 450 to 500 stops can be made in 9 hours of collection time.

Another problem with curbside sorting is that "When a collection truck is divided into four or five compartments it is difficult to size each compartment so that they all fill up at the same rate. . . . This improper ratio of compartment space results in the truck being forced off route with one or more compartments full (usually paper and/or plastic compartments). . . . Better utilization of truck capacity results in fewer trips to the processing/storage facility, therefore less nonproductive time and more efficient collection." For example, data from San Diego, California, shows that when six materials were collected, the two glass compartments (flint and colored) were only 35% full when the newspaper, mixed paper, mixed cans, and mixed plastic container compartments were 90 to 100% full.

Two of the major nationwide waste collection firms, Browning-Ferris Industries (BFI) and Waste Management, Inc. (WMI), are reported to hold opposing views on the value and cost of curbside sorting. BFI has taken the position that the improvement in material quality and reduced processing cost are worth the increased collection cost of curbside sorting, and WMI has taken the position that curbside sorting keeps the collectors on their routes too long.

In summary, commingled materials collection is cheaper than curbside separation, encourages participation, and less glass breakage occurs due to cushioning effects. But, contamination levels are probably higher. Separating at the curb may improve participation over requiring multiple source separations, but it slows down collection. Separation of recyclables that arrive at a processing facility in a commingled state requires relatively more extensive and expensive processing, but, in any source-separation system, some materials upgrading still must ordinarily be done by the collector or processor since residents cannot be relied on to make perfect sorts every time."

The availability of favorable markets should dictate choice of separation to a large degree; e.g., a high-paying market with a low tolerance for contamination may imply a need for either greater degrees of source separation or curbside separation, or a heavier investment in processing and cleaning.

Mandatory versus Voluntary Source Separation

Effect on Participation and Capture

The NSWMA survey conducted in the mid-1980s found that participation in mandatory programs averaged about 55% compared with 34% for voluntary programs. But the existence or nonexistence of mandates will not be the overriding consideration of residents in their decision to participate. Other factors are of equal if not greater importance, such as whether collection is curbside or through drop-off centers, whether in-home containers are provided, and the extensiveness of the public education effort. The survey showed a higher participation in Santa Rosa, California (70%), where participation was not mandatory than in Islip, New York (30%), where participation was mandatory.

The data in Tables 5.2 and 5.3 show that mandatory programs may generally achieve somewhat greater recovery and participation than voluntary programs.

Enforcement of Mandatory Source Separation

It would be very difficult to enforce source separation in programs where collection occurs by means of dropoff centers. In cases where residents must bring all refuse to a central drop-off point, it is conceivable that bags might be opened and violators of source-separation regulations could be fined. This confrontational approach is rarely used with residents, although Groton, Connecticut, uses indirect enforcement at the landfill. The tipping fee goes from $30 per ton to $100 per ton if recyclables are discovered in a collector's load. The private collectors therefore do not pick up refuse containing recyclables.

Enforcement is much more common where recyclables are curbside-collected. In Woodbury, New Jersey, and East Lyme, Connecticut, the inspector precedes the collection vehicles to see if the regular refuse cans and bags have recyclables in them. The State of Rhode Island suggests that a local recycling coordinator might ride on the recycling truck as it makes its pickups. The driver would point out nonparticipating households, and the recycling coordinator would write them a letter. In Woonsocket, Rhode Island (popu-

TABLE 5.2 Recycling rate for mandatory and voluntary programs

Municipality	Type of separation	Pounds per capita per year collected
Essex County, New Jersey	Mandatory	600
Madison, Wisconsin	Mandatory	50
Montclair, New Jersey	Mandatory	250
Portland, Oregon	Mandatory	75
Prairie du Sac, Wisconsin	Mandatory	250
Wilton, New Hampshire	Mandatory	600
Ann Arbor, Michigan	Voluntary	95
Austin, Texas	Voluntary	550
Champaign, Illinois	Voluntary	130
Kitchener, Ontario	Voluntary	100
Mann County, California	Voluntary	190
Mecklenburg County, North Carolina	Voluntary	115
Minneapolis, Minnesota	Voluntary	55
San Jose, California	Voluntary	115
Santa Monica, California	Voluntary	45
Urbana, Illinois	Voluntary	110

Source: Peters, Anne, and Pete Grogan, "Community Recycling," *BioCycle*.

TABLE 5.3 Curbside collection participation and recycling rates (1990 survey)

	Weekly programs	Biweekly or monthly programs	All programs
	Participation rates, %		
Mandatory programs			
High			90
Average	84		75
Low			49
Voluntary programs			
High			92
Average	72		70
Low			55
All programs			
High			92
Average	75	60	
Low			49
	Recycling rates (lb/household/month)		
All programs			
High	65.4	64.2	
Average	44.3	37	
Low	18.7	22.7	

Source: Glenn, Jim, "Curbside Recycling Reaches 40 Million," *BioCycle*, pp. 30–37, July 1990.

lation 50,000), the recycling coordinator reports that neighbors, landlords, and "busybodies" call to inform him that others are not recycling, and he calls the person being reported.

For apartments the problem is more complex. "When tenants are accused of failure to participate in recycling, they often say the landlord never told them about the law, or never provided recycling containers. When landlords are accused, they usually blame tenants for the lack of recycling."

Containers for Source Separation

Source separation requires home storage containers. The type and capacity of container (or multiple containers), and who provides them, depends to a large extent on the collection system.

Who Should Supply the Containers?

Containers are usually provided by the municipality (through a contract with a container supplier). If a private company is doing the collecting, that company may provide the containers. Sometimes, containers are supplied by the participants, particularly if collection is through drop-off centers.

The main argument for the public provision of containers (as opposed to having the private collector provide them) is that since a very large number of containers will be required (at least one per household), government procurement may be cheaper. There are several arguments for having containers provided by one entity (whether it be the municipality, a single collection company, or other supplier under contract to the municipality), as opposed to having the containers provided by the participants:

- Having one entity responsible for containers assures uniformity of appearance, capacity, design, and recognizability.
- Collection methods and equipment can be standardized if only one type of container is used.
- Uniform containers can be an excellent vehicle for publicizing the collection program at the point of startup. The distribution of a free permanent household collection container to program participants has a significant positive impact on participation rates and recovery levels; it provides an ongoing promotional reminder to residents about the program both in the home and at the curb.

Having containers provided to the potential participants seems unequivocally to increase participation. For example, Urbana, Illinois, supplied containers to three of four test routes. These routes had participation rates of 50, 75, and 85%. The route without containers had 25% participation. In Brampton, Ontario, the routes with containers had 50 to 60% participation and the routes without containers had 30 to 40% participation. In San Jose, California, providing containers resulted in 57% participation, while in routes without containers participation was 26%. Springfield Township, Pennsylvania, jumped from 40 to 60% participation when containers were provided. Kitchener, Ontario, went from 39 to 83%. Minneapolis, Minnesota, reported 54 to 97% increases.

Containers make it easy to collect materials in the home, remind people to do so, and when set out at curbside form a "not-so-subtle bit of peer pressure." For the collector, the uniform-appearing container is easy to spot, eliminates confusion about recyclables versus refuse, makes it easier to load the collection vehicle, and reduces litter, especially over having random boxes and bags set out.

Types of Containers

Four kinds of containers are employed for aggregating recyclables:

1. Single containers
2. Stackable multicontainer systems
3. Wheeled carts
4. Recyclables collection bags

Capacity, cost, uses, collection method, and other features of these containers are shown in Table 5.4.

What type of container is best? Clearly, if the collection program is of the one-sort, all-recyclables-commingled type, then a single container is the most efficient option. The choice between a wheeled cart, as used in Seattle, Washington, or a bag depends on the collection method (separate collection or co-collection). The alternative, a series of smaller containers, would have to be stored in the home until collection day, taking up valuable space.

For two or more source-separated recyclables, the following guidance is offered:

- Rectangular boxes are preferred over round buckets by both participants and collection crews; in addition to ease of collection, they are less susceptible to being blown away by high winds. They are somewhat more expensive, however.
- Containers should have holes in the bottom to release liquids.
- For a weekly collection schedule with all food and beverage containers commingled, recycling collection containers should have a capacity of 14 gal per household.
- Containers should not have lids (for ease of collection and cost).
- The municipality's recycling logo should be imprinted on each container.
- A flyer should be distributed along with each container, including basic information about acceptable materials, method of preparation and set-out, and collection schedule.
- A recycling "hot-line" telephone number can be printed on the container and on the promotional material.

There are circumstances that would result in exceptions to the above advice, of course. For example, in Mann County, California, a bank provided round buckets free of charge. From the bank's standpoint, the advertising printed on the recycling container was viewed as more valuable than advertising in the newspaper. Philadelphia, Pennsylvania, also favors buckets for their low cost, ease of carrying, small space requirements in the home, and lack of appeal for other uses such as storing laundry or records. La Porte, Indiana, also uses donated 5-gal pails that would otherwise have been sent to the landfill as refuse. It costs about 5 cents to stencil a logo onto each pail. This city claims to have diverted 20% of its refuse, including 7,500 pails. Palo Alto, California, supplies residents with two burlap sacks: one for aluminum and tin cans, the other for glass. In Albuquerque, New Mexico, residents commingle in one plastic bag or orange bag the following materials: aluminum cans, papers, cardboard boxes, and plastics. Glass containers are set out separately in a box. Some communities use clear plastic bags, so that recyclables can be easily distinguished from refuse. In the pilot program to collect polystyrene, Fitchburg, Wisconsin, uses plastic bags to contain the lightweight material and keep it from blowing around the neighborhood. East Lyme, Connecticut, began its program by passing out stickers to be affixed to the resident's own containers. In Newark, New Jersey, residents are supposed to provide their own sturdy reusable container for bottles and cans, and newspapers should be bundled and tied with string."

TABLE 5.4 Containers and their characteristics

	Capacity, ft^3	Cost, 1988 $	Uses	Collection method	Other features
Single containers	0.7 to 3	$5 to 20	Source-separated food and beverage containers	Curbside Drop-off	Round bucket Rectangular box
Stackable multicontainer systems	1.5 to 1.6 each	$30 to 45 per set of three	Source-separated newspapers, cans and bottle/jars	Curbside Drop-off	
Wheeled carts	32 to 96	$60 to 95	All recyclables commingled Yard waste, possibly all organic waste	Curbside Back-door	Most have lids Some have inner compartments for further separations Mechanical unloading possible All-weather, outdoor storage possible
Bags	Various	Less than $2 each	Source-separated food and beverage containers All recyclables commingled	Curbside Drop-off Co-collection	Higher replacement rates and handling costs when used in curbside collection

Sources: Glenn, Jim, "Containers at Curbside," *Biocycle*.

Lexington, Kentucky, uses a 45-gal wheeled cart with three inner compartments (but this was found to take more time unloading than desired). Plastic bags containing plastic containers are tied onto the outside.

In communities that have tested a variety of containers, the open-top box of approximately 14-gal capacity seems to be favored. Mecklenburg County, North Carolina, tested three different container designs (two kinds of rectangular boxes and one type of cylindrical pail) to determine that the householders and collectors preferred, and whether the estimated 1.5 ft^3 volume was sufficient. It was found that although residents had initially expressed concern that 1.5 ft^3 would be too small, it proved to be adequate. The householders requested boxes rather than pails. Drivers preferred a box shape rather than a pail, because the pail made it more difficult for them to sort recyclables and to identify nonrecyclable items. One problem was that the containers tended to fade in the sunlight, a problem that could be remedied by a warranty. Montgomery Country, Maryland, tested five kinds of container: the 14 gal open-top box, a 13 gal container with attached lid, a 16 gal container with detachable lid, an 18 gal container with attached lid, and a 24 gal wheeled cart with attached lid (see Fig. 5.7). Contrary to expectations, the largest container was set out the most frequently, but it was only 50% full on the average. The 14 gal lidless box, in addition to being the cheapest, was also the most full when set out (83% full), perhaps because it was not stored outside. The attractive 16 gal containers were not only stolen more often, but were not liked by the collectors because they had to remove and replace the lids. The collectors preferred the 14 gal open boxes.

Sources of Containers

Table 5.5 displays a list of home storage container suppliers. Location and telephone numbers are provided so that convenient suppliers may be contacted for the latest infor-

FIGURE 5.7 Different kinds of curbside containers tested in Montgomery County, Maryland. (Photograph courtesy *Resource Recycling* Magazine.)

TABLE 5.5 Home storage container suppliers

Firm	Location	Telephone
A-1 Products	Etobicoke, Ontario, Canada	(416) 626-6446
Advanced Recycling Systems	Washington, Iowa	(800) 255-5571
American Container	Plainwell, Michigan	(800) 525-1686
Ameri-Kart Corp.	Goddard, Kansas	(316) 794-2213
The Bag Connection	Newbeng, Oregon	(800) 228-2247 (503) 538.8180
Bonar Inc.	Montreal, Quebec, Canada	(514) 481-7987
Buckhorn Inc.	Milford, Ohio	(800) 543-4453
Busch-Coskery of Canada	Mississauga, Ontario, Canada	(416) 828-9898 (416) 897-8618
Champion Plastic Container	Ajax, Ontario, Canada	(416) 427-9756
Dover Parkersburg	Fall River, Massachusetts	(800) 225.8140
Grief Brothers Corp.	Hebron, Ohio	(614) 928.0070
Heil Co.	Chattanooga, Tennessee	(615) 899.9100
Household Recycling Products	Andover, Massachusetts	(508) 475.1776
IPL Products Ltd.	North Andover, Massachusetts	(508) 683-7668
Kirk Manufacturing	Houma, Louisiana	(800) 447-1687
Letica Corp.	Rochester, Michigan	(313) 652-0557
LewiSystems	Watertown, Wisconsin	(800) 558-9563
Management Sciences Applications	Upland, California	(714) 981-0894
Master Cart	Fresno, California	(209) 233.3270
Microphor, Inc.	Willits, California	(707) 459.5563
North American Rotomolding	Vancouver, Washington	(206) 693-6074
Otto Industries	Charlotte, North Carolina	(704) 588.9191
Plastican	Leominster, Massachusetts	(508) 537-4911
Philadelphia Can Co.	Philadelphia, Pennsylvania	(215) 223-3500
Piper Casepro	Dallas, Texas	(800) 238.3202
Refuse Removal Systems	Fair Oaks, California	(800) 231-2212
Rehrig Pacific Co.	Los Angeles, California	(231) 262-5154 (800) 421.6244
Reuter, Inc.	Hopkins, Minnesota	(612) 935-6921
RMI-C Division of Rotonics Molding Inc.	Itasca, Illinois	(708) 773-9510
Ropak Atlantic	Dayton, New Jersey	(201) 329-3020
Rotational Molding	Gardena, California	(213) 327-5401
Shamrock Industries Inc.	Minneapolis, Minnesota	(800) 822.2342 (612) 332-2100
Snyder Industries	Lincoln, Nebraska	(402) 467-5221
Spectrum International Inc.	Shrewsbury, New Jersey	(201) 747.1313
SSI Schaefer Systems	Charlotte, North Carolina Northbrook, Illinois	(704) 588-2150 (312) 498-4004
Toter Inc.	Statesville, North Carolina	(704) 872-8171 (800) 288-6837
Windsor Barrel Works	Kempton, Pennsylvania	(215) 756-4344
Zarn, Inc.	Riedsville, North Carolina	(919) 349-3323

mation on product specifications, prices, availability, lead times for ordering, and shipping information.

Container Volume Required
To calculate the volume of recyclables likely to be set out in containers and therefore the volume of the containers needed, one must have an idea of the weight of each material likely to be set out and the density of that material. To find the needed volume of the container, determine the weight of each recyclable that will be set out by the average household in the container each week (if the recyclables are to be collected weekly), and divide by the density of that recyclable, then add up the resulting volumes:

$$\text{Total volume required} = \Sigma_i \frac{\text{weight of recyclable } i}{\text{density of recyclable } i}$$

The weight of each recyclable that might be set out by the average household can be estimated by dividing the total weight of refuse generated in the community by the number of households, and multiplying the result by the weight percent of each component in the refuse as determined by composition studies:

$$\text{Per household weight of recyclable } i = \text{\% of recyclable } i \text{ in the total waste stream} \times \frac{\text{weight of total waste stream}}{\text{number of households}}$$

Table 5.6 gives approximate densities of common recyclables and a sample calculation of weekly volume. It will be noted that despite their light weight, plastics contribute a great deal to the volume requirements of in-home recycling containers (and collection vehicles, as well). Because households differ substantially in the number and size of plastic bottles used, a much larger container might be required for some households.

TABLE 5.6 Calculation of container volume required for 100% capture of selected recyclables (Milwaukee example)

Recyclables	Density, lb/ft^3 (1)	Weight, % of refuse (2)	Weight per household per week, lb (3)	Volume per household per week, ft^3 (4)
Newspaper	18	12.2	6.5	0.36
Glass bottles	15	8.1	4.3	0.29
Tin cans	8	2.5	1.3	0.16
Aluminum cans	2	0.6	0.3	0.15
PET bottles	0.9	0.5	0.3	0.33
Milk bottles (HDPE)	0.7	0.5	0.3	0.43
Total (excluding newspapers, that are set on top of container)				1.36 (allow 1.5 ft^3 or 12 gal)

Sources: (1) Glenn, Jim, "Containers at Curbside," *BioCycle*. (2) *City of Milwaukee Residential Waste Characterization* , Department of Public works, City of Milwaukee, Wisconsin. (3) Milwaukee's 241,000 households generate about 333,000 TPY or 33 lbs per household per week. The weight of each material per household per week is thus 53 lbs times the weight percent of refuse. (4) weight divided by density. (Gallon = 231 cubic inches.)

How Many Containers Should Be Provided?

The correct number of containers equals the number of categories into which participants are expected to source-separate their recyclables (with the possible exception of corrugated cardboard, which can be stacked alongside or placed under other containers, and newspapers, which can be tied or placed in paper bags). There are a number of arguments for the provision of just one container:

- A single container is more convenient than multiple containers for participants; it takes up less space in the home and is easier to move to the curb for collection.
- A single container can save collection crew labor if the nonfiber recyclables are placed into the collection truck in a commingled state. (On the other hand, if the collector has to sort the commingled containers at the curbside, collection labor is increased.
- A single container is obviously less expensive to provide than multiple containers.
- Placing food and beverage containers in a single collection container accommodates changes in the mixture of recyclables that may occur over time, especially if the program is expanded to include additional materials.

Container Replacement

A small percentage of containers will need to be replaced annually. Considering the demographic composition of the municipality, the replacement rate can be expected to be between approximately 2 and 10% per year. In some areas, losses as high as 5% per month have been experienced due to theft. One replacement bin is often given for free and after that they must be purchased.

Containers for Apartments and Condominiums

Collection of recyclables from apartments and condominiums presents special challenges, including transient populations, space constraints, lack of interest in recycling by building managers, and high cost of collection. Many experiments are being conducted to determine what works best.

Containers are not usually provided to the individual household, but larger containers are provided for the building, and they function much like drop-off centers. (Drop-off centers are discussed later in this section.) The types of containers that have been used include:

- Roll-out carts
- Sheds to hold paper materials and to conceal roll-out carts
- Roll-off boxes for large quantities of materials
- Sectionalized roll-off boxes or dumpsters
- "Igloos" or other containers for glass, aluminum, metal, and plastic

Collection vehicles are available that hydraulically unload many of these containers. Sheds containing 90 gal carts for mixed recyclables and a designated area for newspaper can be strategically placed in parking lots or areas adjacent to the building. The shed can be an attractive reminder of the recycling program and can encourage regular participation.

Mann Country, California, has a population of 90,000, of which 83% live in multifamily dwellings. There are 57 drop-off sites in the county. Each site has three differently colored, 2-yd^3 bins: green for mixed glass, brown for mixed cans, and blue for newspaper. These bins are replaced when half full. There were some scavenging problems, so the lids were modified so that materials could only be deposited into them and not retrieved from them without special equipment. (Scavenging is discussed later in this section.)

In York, Pennsylvania (population 45,000), where 45% of the total housing is rental, four types of containers are provided depending on the type of building.

1. For high-rise buildings (75 or more units), containers in "trash rooms" are located on each floor of the building. These are emptied into larger containers outdoors by building staff.
2. For smaller (15 to 75 unit) buildings, dumpsters or wheeled carts are provided outside that residents fill directly from in-home 6.5 gal buckets.
3. For 5 to 15 unit buildings, curbside collection buckets (6.5 gal capacity) are provided to each rental unit. Residents are responsible for labeling their own containers and placing them at the curb.
4. For buildings with less than 5 units, 14 gal rectangular curbside collection containers are provided to each rental unit.

Prince George's County, Maryland, is testing six different models, involving various combinations of 11 gal stackable bins, 90 gal wheeled carts, and recycling bags on racks. These are located in trash chute rooms, basement trash rooms, outdoor trash corrals, laundry rooms, or elsewhere, depending on the building architecture and space available. Preliminary results indicate that it is difficult to expect the collector to replace recycling bags on the racks, and that collectors as well as maintenance staff, residents, and building managers should definitely be involved in the design of the system from the start.

In St. Paul, Minnesota, 11,000 households in 309 buildings use 90 gal wheelable bins placed throughout the apartment complex, rather than in a few out-of-the-way areas. The most cost-effective pickups were found to occur when containers were within 25 ft of the truck. To improve participation, residents are given reminder flyers every 10 to 12 months; posters are hung every 3 months; and information flyers for new tenants are given to building managers. Contamination is the most important problem. "Unless residents feel well treated at their building, they feel they are doing the building owner a favor by recycling."

One experiment involving 896 units in Fitchburgh, Wisconsin, showed that providing apartment residents with plastic bags for recyclables without a strong educational program was not very effective (less than 5% diversion rate and 40% contamination by nonrecyclable items).

A firm in Miami Beach, Florida, is marketing a system in which a typical garbage chute in a high-rise apartment complex is "retrofitted" with a rotating turntable in the basement. Push-button controls on each floor allow residents to rotate the appropriate container to a position under the chute where it can accept one source-separated recyclable commodity at a time. Glass, plastic, aluminum, and newspapers have been collected from high-rise apartments using this device.

Containers for Source Separation at the Workplace

Recycling at offices and other workplaces is just beginning, even though in many cities the commercial sector accounts for at least 50% of total refuse generated. For example, in Palo Alto, California, the city's estimated population of 56,000 is doubled on normal work days by commuters. These workers and the companies that employ them discard a great deal of material. It is estimated that 20% of the total material landfilled is corrugated cardboard, with 75% of that coming from commerce and industry. Other materials that can be collected in addition to corrugated cardboard include aluminum cans, white office paper, computer paper, glass, and PET soft drink bottles, newspapers, and telephone directories.

Often, the recycler who picks up the material will provide some containers (such as large, wheeled hampers for paper). In general, it is easy in an office to separate high-grade paper (such as computer paper) from other paper if receiving containers are placed where the paper is normally discarded. Employees can also easily source-separate white paper in their offices, if supplied with a brightly colored bin. The bin should be large enough so that paper does not have to be stacked neatly in order to be inserted.

One of the pioneering companies in workplace recycling is Coca-Cola in Atlanta, Georgia, where 3000 employees generate over 1,000,000 aluminum cans each year. These are collected in large cardboard boxes lined with plastic bags. Recyclable paper is collected in small containers in each office and removed nightly by custodial staff members, who put the paper into 95 gal wheeled carts. Corrugated cardboard is compacted separately from other trash. In these respects, Coca-Cola's program resembles that of many other forward-looking companies, but Coca-Cola takes recycling a step further. Telephone books and glass containers from the laboratory are also collected. The company has also set up an area where employees can bring recyclables from home and place them in wheeled carts. The eight elements that make the program a success are considered to be:

1. Securing top management support
2. Appointing a coordinator
3. Knowing what wastes are generated by the company, and what it costs to dispose of them
4. Securing building managers' and custodial support
5. Using reliable recyclers to pick up the material
6. Providing employees with convenient drop-off locations
7. Communicating regularly with employees
8. Educating new employees.

It is also very important to most companies that sites devoted to recycling, both inside and outside the building, present a neat appearance at all times. Collection containers must be attractively designed and large enough so that there are no overflows.

Keys to Successful Source Separation

Publicity

If source separation is necessary, as it is for all but the mixed-waste collection/processing schemes, public willingness is essential. Publicity and education are the primary keys to successful source separation. Techniques are covered elsewhere in this Handbook, and include hiring of public relations firms, direct mail, news media events, a consistent graphic theme, billboards, brochures, public service announcements, public school information, adult education, and so on. Public education and information is a key to all successful recycling programs, and this is especially true with efforts aimed at apartment residents." Mailings must be addressed to the occupant and not just to the owner, even though apartment recycling typically involves a great deal of individual attention to building owners and managers. Once a few buildings are doing a great job of recycling, they can be used as examples when other landlords claim that recycling will not work in their buildings.

Convenience

Feelings toward recycling may be more important than convenience in encouraging a citizen to participate initially. Therefore, administrators should plan public information literature that stresses positive benefits of recycling, rather than convenience. However, continued participation depends on the quality of recycling service offered.

Containers
Another key factor in successful separation seems to be the provision of containers. A 1988 survey of over 20 different programs of various sizes and types concluded that providing containers and having citizens put them out on the same day as regular trash collection increases participatlon.

Lower Apparent Cost
A third incentive to participate in source separation is to make it more economical to the participant through refuse collection charges that reward source separation. From its inception, recyclables collection was subsidized by Seattle, Washington. In 1979, the variable-can rate was tested. Residents who set out less regular refuse paid less: recyclables were collected for free. The recycling rate was higher in the variable-can rate routes than in the others. Perkasie, Pennsylvania, has a pay-by-the-bag-system. When it was instituted, the town experienced a 40% decline in tonnage of residential refuse, with two-thirds of that attributable to recycling through curbside and drop-off programs. The economic incentive to participate will only occur if recycling services are offered to the public at a cheaper rate than regular refuse service. On a strictly computed cost-of-service basis, recycling is sometimes more expensive. ". . . no one, including the project directors, expects home sorting and collection of recyclables to break even or make money."

Deposit Legislation
Another incentive to source-separate is the "bottle bill." Although legislation has been enacted in nine states (some with many years of experience) and is being proposed for an additional half-dozen or so, container deposit laws remain controversial. Seven of the states with deposit legislation report high capture rates: 75 to 95% for aluminum, 88 to 95% for glass, and 70 to 90% for plastics. Container deposit legislation causes redemption centers (whether grocery stores or specialized centers) to function essentially as buy-back centers. If deposit legislation were extended beyond beverage containers, would the capture rates equal those experienced for beverage containers without the expense of curbside collection?

Lottery
In a 1982 study of 615 homes, Tallahassee, Florida, tried four approaches to increase participation:

1. "Prompting" (i.e., distributing leaflets about a week in advance of each scheduled pickup)
2. Payment for material (1 cent per pound of newspaper: news was fetching $40 per ton at the time)
3. A lottery prize of $5 drawn at random from the collection day's participants
4. Shifting the frequency of collection from biweekly to weekly.

The lottery demonstrated greatest improvement in participation (about 16%); the other methods increased participation about half as much. None of the methods was cost-effective, but long-term or residual effects of the methods were not analyzed.

Program Consistency
After separation habits have been established, it is very important that the program continue unchanged, except for the addition of new recyclables to the collection. "Once this

habit of separation is interrupted, tremendous efforts will be necessary to re-educate and re-interest the citizens."

COLLECTION SYSTEMS

Curbside Collection versus Drop-off and Buy-Back Centers

Most communities have found that the way to ensure high participation is to provide a convenient method of participating. Drop-off and buy-back centers are not as convenient as curbside collection, and will typically result in much lower participation and capture rates. Curbside collection can divert up to 20 to 25% of the refuse coming from the homes provided with collection service. However, there are examples of drop-off programs capturing as much material as curbside programs operating in the same community (for example, Bloomsburg, Pennsylvania; and Everett, Washington). Communities such as Wilton, New Hampshire, and Wellesley, Massachusetts, where residents have traditionally all residential refuse and recyclables to a central point, experienced an increase in *all* drop-off programs. Drop-off centers may be the only practical choice in areas where waste disposal is cheap and in rural areas where there is a long driving distance between residences.

Curbside collection may be more effective in recovering materials, but it is more expensive to establish and maintain. Drop-off centers offer the advantage that equipment, personnel, and maintenance needs are minimal. Another advantage of some drop-off centers is that a consistently clean supply of marketable materials is generated. Especially with centers that are staffed, greater control can be exercised over the quality of the material accepted. In this respect, staffed drop-off centers function somewhat like curbside separation, producing a product that does not have to be upgraded in an expensive processing facility.

Still another advantage of drop-off centers is that they can accept a greater variety of recyclables than are practical to collect at curbside. For example, used motor oil, automobile batteries, yard waste, construction debris, tires, furniture, white goods, latex paint, and pieces of scrap metal are all candidate recyclables for the "super self-help" stations being established in Milwaukee, Wisconsin. In Palo Alto, California, and other communities, yard debris is not collected; but gardeners and residents can bring it to a composting location at the landfill.

As in Milwaukee, Wisconsin, many locations offer both curbside collection and drop-off locations. As shown in Fig. 5.8, Seattle, Washington, expects that both will contribute significantly to material recovery in the future. A private collector with subsidiary companies operating in the San Francisco, California, area uses five methods for collecting recyclables: stationary buy-back centers, mobile buy-back centers, apartment house roll-out carts (originally for newspaper only, now for newspaper and glass), three-container curbside collection, and single-container curbside collection. Mecklenburg County, North Carolina (population 460,000), has curbside collection plus staffed and unstaffed drop-off centers. One of the staffed drop-off centers accepts household refuse and recyclables three days per week, taking household refuse only if the user brings in recyclables. In Palo Alto, California, half of what is recycled comes from curbside collection and half from the drop-off center. Some of the material dropped off comes from citizens of nearby cities where there is no curbside pickup.

Everett, Washington, decided to test whether curbside collection or drop-off centers would better serve its needs. Table 5.7 shows that weekly curbside collection would cost

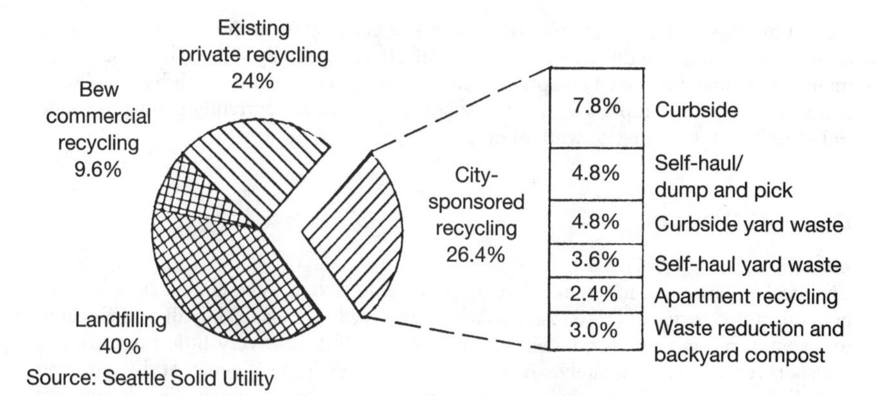

FIGURE 5.8 Contribution of programs to 60% waste reduction/recycling program. (*Source:* Tom Watson, "Seattle blazes new ground with diverse approach," *Resource Recycling,* pp. 28–31, 73–74, November 1989.)

the most but also would probably recover the most material. The low recovery from buy-back centers in this instance could be due in part to their less convenient locations.

Sometimes the quantity of material delivered to drop-off centers may actually rise upon the institution of curbside collection, as experienced in Durham, North Carolina, and Bloomsburg, Pennsylvania. However, typically, one would expect that capture at drop-off centers would decline when curbside collection is instituted. For example, quantities of materials have declined by 40 to 65% at Seattle, Washington, drop-off sites since the city began curbside collection in the spring of 1988. Charity newspaper collection drop boxes received about 33% less material after the first year of curbside collection, and 47% less after 18 months of curbside collection; noncharity drop boxes lost 47% after a year and 64% after 18 months. (Quantities delivered to buy-back centers also declined,

TABLE 5.7 Cost and recovery estimates for various collection alternatives Everett, Washington

	Recovery achieved in 6-month pilot, lb/households participating	Estimated costs for full program implementation (60,000 population), 1988 $
Curbside (weekly), 2 bins provided (newspaper/aluminum + glass)	131	599,000
Curbside (monthly), 2 bins provided (newspaper/aluminum + glass)	86	289,000
Drop-off igloos (in-home bags provided)	67	208,000
Buy-back	49	37,000

but not as dramatically). This phenomenon may help to explain the opposition of some established recycling industries to curbside collection, especially if there is a risk that the municipality may institute a competitive bid for the right to process its curbside collected material rather than simply giving it to the businesses that are currently getting the material at zero cost through a network of drop-off boxes.

Drop-off Centers

Location of Drop-off Centers

The ideal environment for an effective drop-off center seems to be in a community without collection services, where residents are accustomed to transporting all of their refuse to a central place, such as a transfer station or a landfill. It is very little extra effort to transport recyclables separately, and the presence of the drop-off center at the refuse aggregation point provides a weekly reminder to do so. In Peterborough, New Hampshire, incoming cars make stops at two open-air drop-off stations before proceeding to unload their mixed refuse. The first station is for glass, which residents separate into clear, brown, and green; the second is for plastics, aluminum, and paper.

Wellesley, Massachusetts, has never provided collection services, so people are accustomed to using the recycling and disposal facility. Located at a closed dump, the facility has a "parklike setting and layout" and is staffed with one full-time and six part-time employees. Refuse and recyclables are brought by residents along with materials to exchange, such as books and firewood. Of the town's 27,000 citizens, 75.2% participate voluntarily. High participation is attributed in part to demographics: the citizens are highly educated and in the upper economic strata. Since Massachusetts is a "bottle bill state," contributors are given receipts if they want to take a tax deduction for donating returnable bottles and cans. In 1985, when the town experienced a dramatic (over 42%) increase in refuse tonnage due to the return of privately collected refuse that formerly was taken to other disposal sites, the tipping fee for nonsegregated refuse was raised, but no tipping fee was charged for separated recyclables. This provided economic incentive to recycle.

In communities where refuse is collected, drop-off centers need to be in convenient locations, within 3 to 5 miles of the home. Candidate sites are shopping centers, grocery stores, schools, churches, and fire stations. For example, Woodridge, Illinois' drop-off center is between the library, the post office, the police station, and a soccer field. The drop-off center should be in a visible place, which helps to make patrons feel more secure and helps to control litter and illegal dumping. One rule of thumb for planning is to establish one drop-off center per 5,000 to 10,000 people, but it is more important to put the centers in frequently used locations and make them larger if necessary. Santa Monica, California, has 66 sites serving 70,000 residents, one within one-half mile of every resident. Participation is reported to be about 35%. In parts of Germany, there are drop-off centers for paper and glass with a density of 1 per 1,000 residents. Recycling at apartment complexes essentially involves the establishment of a "mini" drop-off center for residents of the complex.

The owner of the site must be fully cooperative. In Cobb County, Georgia, local businesses "adopt a site," providing space and controlling litter.

Larger-scale drop-off centers have been established at landfills. For example, in Deschutes County, Oregon, a recycling center was set up at one of the existing landfills through the cooperation of a refuse collection company, the county that owned the landfills, and a nonprofit corporation. One acre of land at the landfill is leased to the collector for $1 per year. The collector constructed a facility to reclaim materials such as cardboard, which is dumped and picked. The collector's investment is paid back by avoiding the landfill fees it would otherwise pay. The site is operated by the nonprofit corporation.

Drop boxes are widely used for specialized collections of single materials. For example, drop boxes for newspaper are used in the southeast, sponsored by recycled paper mills. Plastic grocery sacks are now widely collected in grocery stores, promoted by the makers of the sacks. The cost of the collection is largely borne by the grocery stores. Store customers' response has been "overwhelming," with returns at 15 to 20% of the weight of sacks purchased by the store.

Equipment for Drop-off Centers

The most noticeable pieces of equipment at drop-off centers are containers, which are increasingly evolving and becoming more specialized. Whatever containers are provided for recyclables, it is also important to provide a container for ordinary refuse to help control litter and the dumping of refuse in the same containers as the recyclables.

Three basic types of containers are found at drop-off centers:

1. Two- or three-cubic-yard bins, some compartmentalized. These cost $300 to $500 each.
2. Fifteen- to forty-cubic-yard roll-off containers, which are hauled to the processing facility as trailers. If uncompartmentalized, these containers cost $2,300 to $4,000 (in 1989). Compartmentalization can cost up to $2,500 more per container. Hoists to load the roll-offs cost from $10,000 to $23,000, depending on the size of the roll off.
3. Specialized containers that are emptied on site by special equipment; for example, a truck-mounted crane that costs from $18,000 to $30,000. These specialized containers can be:
   - Compartmentalized (for example, a box-shaped container with two 2.1-yd^3 compartments)
   - Separate for each recyclable (for example, the popular dome-shaped igloos that cost from $350 for 1.1 yd^3 to $650 for 4 yd^3

See the section on "Containers for Source Separation" for a description of containers for separation of recyclables at apartment and condominium complexes. Table 5.8 gives a list of drop-off equipment makers. Locations and telephone numbers are provided so that convenient suppliers may be contacted for the latest information on product specifications, prices, availability, lead-times for ordering, and shipping information.

Figure 5.9 shows one type of drop-off center designed for Hollywood, Florida. A state grant provided funds for 128 centers to be located throughout the city. The containers are emptied and maintained by a volunteer charitable organization at no cost to the city. Revenues go to the charity.

Improving Participation and Capture at Drop-Off Centers

Given their advantages (low cost and the fact that it is practical to accept many different kinds of recyclables at drop-off centers), it would be desirable to increase participation and capture at drop-off centers so that a level of refuse diversion acceptable to the community could occur without the expense of curbside collection and elaborate MRFs. It is possible that participation in drop-off centers could be improved if the municipality operating them would concentrate on two factors: (1) convenience to participants, and (2) public education. For example, home storage containers provided by the municipality would help a drop-off program in a similar manner as they do a curbside program. Direct purchase by the municipality of the recyclables collected would provide incentives to both volunteer and private drop-off centers. The quantities of recyclables received per

TABLE 5.8 Suppliers of drop-off center equipment

Firm	Location	Telephone
Specialized Containers:		
Environmental Container Corp.	Hemet, California	(714) 652-4339
Fibrex Plastics, Inc.	Hayward, California	(415) 887-0779
Igloo Recycling Systems	New York, New York	(212) 265-6426
SSI Schaefer	Northbrook, Illinois	(312) 498-4004
Specialized Container Cranes:		
Hiab Cranes and Loaders	New Castle, Delaware	(302) 328-5100
Omark Industries	Zebulon, North Carolina	(919) 269-7421
IMT Cranes	Gainer, Iowa	(800) 247-5959
Roll-off Containers:		
J. V. Manufacturing	Springdale, Arizona	(800) 858-8563
Parker IMP	Winamac, Indiana	(219) 946-6614
Quality Products/McClain Ind.	Kalamazoo, Michigan	(616) 381.2620
Rudco Products	Vineland, New Jersey	(609) 691-0800
Roll-off Hoists:		
Converto Mfg.	Cambridge City, Indiana	(317) 478-3201
Galbreath, Inc.	Winamac, Indiana	(219) 946-6631
G & H MFG	Mansfield, Texas	(817) 467-9883
Quality Products/McClain Ind.	Kalamazoo, Michigan	(616) 381-2620
Small Containers:		
Ameri-Kan	North Warsaw, Indiana	(219) 269-3035
Perkins Manufacturing	Chicago, Illinois	(312) 927-0200
Riblet Products	Elkhart, Indiana	(219) 264-9565
Small Container Collection Units:		
Mobile Equipment Company	Bakersfield, California	(805) 327-8476
Perkins Manufacturing	Chicago, Illinois	(312) 927-0200
Multipurpose Equipment:		
Hiab Cranes and Loaders	New Castle, Delaware	(302) 328-5100
Marrel Corporation	Hendersonville, Tennessee	(615) 822-3536
Multitek, Inc.	Prentice, Wisconsin	(715) 428-2000

resident and per drop-off center are shown for several municipalities in Table 5.9, and the figures show order-of-magnitude differences from one program to another.

Block-Corner Drop-Off

A variation on the drop-off center for densely populated areas is block-corner collection. Periodically (once every other week) residents carry recyclables to a designated location at the end of their blocks. This method is in use in Philadelphia, Pennsylvania, and is actually a hybrid drop-off/curbside program.

A municipal truck with three municipal employees collects the recyclables from each corner and takes them to a private purchaser of recyclables. Because the truck stops only once for the recyclables from 30 to 150 homes and because all three workers assist with the loading, collection efficiency is said to be high, with the crew handling 670 lb of recy-

FIGURE 5.9 A convenient corner drop-off center in Hollywood, Florida, accepts aluminum cans and three colors of glass. (Photograph courtesy of Haul-All Equipment Systems Ltd., Lethbridge, Canada.)

clables per labor-hour (including travel time to the buyer), compared with 180 lb in the city's curbside pilot. A three-person crew takes about 40 mm to pick up 1 ton of recyclables, and up to an hour to off-load 3.5 tons. Using these figures and estimates of driving time, the time to complete a route, and time to off-load could be estimated (it may take more than one trip to the buyer).

In Philadelphia, revenues from the sale of recyclables are returned to the volunteer neighborhood organization that organizes the collection and provides staff for the corners

TABLE 5.9 Quantity of recylables received at selected drop-off centers

Location	Population	No. of sites	Tons/year received	Lb/year (per person)	Tons/year (per site)
Champaign County, Illinois	171,000	15	1000	12	67
Columbia County, Pennsylvania	50,000	17	469	19	28
Cook & Lake County, Illinois	270,000	18	7140	53	397
Delaware Co., Pennsylvania	500,000	50	1800	7	36
Durham County, North Carolina	120,000	10	1200	20	120
Fairfax County, Virginia	75,000	8	1000	27	125
Kent/Ottawa County, Michigan	650,000	30	3200	10	107
Santa Monica, California	70,000	66	1398	40	21

on drop-off day, rather than to the city that subsidizes the collection and transportation cost of about $140 per ton. Success of a block-corner collection program depends on a strong neighborhood organizational structure and good cooperation between city officials and the neighborhood groups.

Table 5.10 lists and describes the eight steps to establishing a block corner collection program in urban areas.

To determine the amount of material that can realistically be recovered through a block-corner collection program, organizers can obtain (from local, state, or federal officials) the refuse generation rate per person of the materials to be solicited, then multiply by the number of people in the service area. The result should be discounted substantially, recognizing that not all residents in the area will participate and those that do will not contribute all their recyclables. A discount factor of 75% might not be unreasonable.

In deciding how materials should be prepared for block-corner collection, early experiments in Philadelphia, Pennsylvania, determined that newspapers are best tied or placed in large paper bags; aluminum cans can be handled in paper bags; glass should be separated at the corner into color; and, on rainy days, paper bags should not be used for glass.

Cost of Drop-Off Centers

In estimating the cost of drop-off centers (whether one center or a network), all of the factors listed in Table 5.11 must be taken into account. Drop-off centers vary as much in their costs as they do in their effectiveness: there is no such thing as a "typical" drop-off center.

Buy-Back Centers

Stationary Buy-Back Centers

Buy-back centers offer the advantage that collectors are paid for the recyclables they bring in. Popular in urban areas, these centers frequently grew up as an adjunct to long-established scrap metal and paper dealerships. Prices paid for recyclables brought in depend on the market, which for some recyclables is notoriously volatile, dropping to zero with little warning. Traditional buy-back centers have not proven to be a reliable means of capturing the desired amount of recyclable material from the refuse stream. If they were, there would have been little need to develop drop-off centers or curbside collection. However, particularly in inner-city urban areas, buy-back centers offer an incentive to the economically disadvantaged.

Several buy-back centers have attempted to increase interest among patrons by adopting a theme, such as a circus, with attendant storybook parklike atmosphere that appeals to small children. Originally, many of these theme centers were funded in part by beverage container interests to help thwart bottle bill initiatives. With the rise of curbside collection as an alternative, that support has diminished. However, some theme buy-back centers run by not-for-profit organizations are thriving, in part because they are less sensitive to market price fluctuations (especially the decline in paper prices) and in part because they receive several kinds of public subsidy such as free land, volunteer labor, and donations. Information on establishing a theme buy-back center is available from the Glass Packaging Institute in Washington, D.C.

A buy-back center was recently established in a rural area of North Carolina. Transporting and marketing the collected recyclables would have been a challenge, but an existing buy-back center in a nearby city agreed to pick up the recyclables for a small fee. Participation is clearly highest among residents who live closest to the center. The county

TABLE 5.10 Eight steps to establishing block-corner collection

Step 1. Form a well-organized and responsible neighborhood recycling committee, preferably an offshoot of a recognized neighborhood organization. Duties of the committee include:

- Coordinate with the city recycling officials
- Determine the geographic area to be served
- Estimate how much material can realistically be recovered, using the population of the area to be served
- Establish and clearly articulate the purposes of the project
- Make up a good publicity logo to be used on all promotional materials
- Serve as, and recruit other, block coordinators
- Estimate program expenses and revenues
- Determine where startup and ongoing funds will come from
- Determine how revenues (if any) will be distributed
- Open a bank account for the receipt of revenues
- Determine how the materials are to be prepared by residents: degree of separation required, etc.

Step 2. Find a buyer, using the estimate of the amount that will be collected:

- Look for buyers who are close to the neighborhood, (to minimize transportation costs) and visit the buyer
- Find out what the potential buyer's requirements and flexibility is concerning matters such as:

  What materials are purchased and for what price
  What the materials preparation specifications are
  How the weight of the materials is determined
  How payments are made
  Days and hours open to receive materials
  Availability of unloading machinery or labor assistance
  Whether the buyer also would be willing to pick up the materials

- Select a buyer who is reliable, preferably one who has dealt with other recycling groups that can be contacted as a reference.

Step 3. Arrange a collection and transportation method. This is the most expensive part of the operation.

- Find a truck to service the route. The options are:

  The city provides the service
  A private hauler provides the service (disadvantages: private hauler might want the revenue from the sale of recyclables and, if the revenues dropped, the private hauler might abandon the route)
  The neighborhood recycling committee operates a rented truck (disadvantages: might result in legal liabilities, reliance on volunteer labor that probably cannot be a permanent arrangement, and is expensive)

- Formalize the arrangement with the trucker in a written contract covering route, schedule, delivery to buyer, revenues, payments, etc.
- Map the exact route (include all pickup locations and the route to the buyer)
- Set a collection schedule (semimonthly pickup, not on a regular refuse day to avoid confusion: establish set-out time to minimize vandalism and scavenging)
- Equip trucks with compartments (e.g., pallets and barrels)
- Train truckers to recognize what materials are acceptable to the buyer and what materials must be left behind (but not as piles of litter).

(continued)

TABLE 5.10 Eight steps to establishing block-corner collection (*Continued*)

Step 4. Publicize the refuse crisis and the benefits of recycling, using the neighborhood association's newsletter or a weekly newspaper.

Step 5. Find block coordinators. Duties:

- Help select a corner
- Monitor the corner on pickup day
- Pass out leaflets and reminders
- Answer residents' questions

Step 6. Select a program startup date, taking into account the need for advance publicity.

Step 7. Several weeks before program startup, publicize the details of participation using all media means available (posters, flyers, articles, newsletters).

Step 8. Begin the program and measure progress:

- Continue sending reminder flyers for first several months
- Follow the trucks initially to ensure cleanup of broken glass and litter
- Assure that all goes smoothly with the buyer
- Measure program's success against its original goals
- Expand into materials, restaurant glass, office paper, and school projects

Source: Pierson, Robert W., Jr., "Eight Steps To Block Corner Success," *Waste Age.*

landfill diversion rate is estimated to be about 0.5% due to the buy-back center, although this figure is uncertain as waste is not weighed at the landfill.

Mobile Buy-Back Centers

If not carefully tended, stationary drop-off and buy-back centers can look like eyesores and collect debris. The use of a mobile buy-back center avoids this problem. It also costs less to equip and operate than an attended stationary center. Vallejo, California (population 100,000), a San Francisco suburb, has had a mobile buy-back service, making about 10 scheduled stops each weekday at convenient locations like schools and churches. The driver stays at each location for 15 to 45 mm. Materials are weighed and paid for on the spot. The truck used is a flatbed truck and trailer with 1 and 2 yd bins and a scale mounted on it.

Difficulties experienced by a mobile buy-back center include finding the right locations to park, considering the need for acceptable access and safety. Scheduling also probably presents a problem: if the vehicle is delayed in its scheduled route, people are unlikely to wait very long for it. In areas with severe winters, this collection method is likely to be impractical.

Reverse Vending Machines

A reverse vending machine functions as a mini-buy-back center for used beverage containers. There are some 7000 machines operating in the United States. The most successful are located in bottle-bill states. Depending on their design, the machines can be sited outdoors or indoors. They give cash or in-store credit. Some models dispense lottery tickets or operate as a slot machine. Costs range from $500 for small, hand-crank indoor units to $35,000 for giant outdoor versions. Many units accept aluminum cans; some accept steel cans, plastic bottles, and/or glass bottles. Some color-sort and shred or crush PET and glass bottles. Like any vending machine, they are subject to significant wear and tear. The leading suppliers operating in the United States are:

- Aries Aluminum Corp. (Hilliard, Ohio)
- CoinBak (New Rochelle, New York)
- Egapro Management AG (Zurich, Switzerland)
- Envipco (Fairfax, Virginia)
- Gadar Industries (Forest Lake, Minnesota)
- Kansmacker, Inc. (Lansing, Michigan)
- Tomra Systems Inc. (London, Ontario, Canada)

Curbside Collection

Diversity of Curbside Collection Programs

There are about 1,600 curbside recyclables collection programs in the United States in 1990 and the number is expected to increase, mostly east of the Mississippi. Table 5.12

TABLE 5.11 Factors influencing the cost of collecting recyclables at drop-off centers

Factor
Center design:
Site work required (roads, paving, grading, fencing, utilities, landscaping)
Type of structures (e.g., shed for workers, scales)
Receiving containers:
Number of categories of recyclables accepted
Density of materials being accepted
Number of containers needed
Capacity/type of containers
Frequency of replacement
Ease of unloading/trailering
Density of centers:
Number of centers serving geographic area
Driving distance/time between centers
Number of centers that can be serviced by transport vehicles
Workers:
Number of workers staffing each center
Number of staffed hours (including security, cleanup and maintenance)
Wages and fringes
Length of work day
Performance (attitude, efficiency, speed)
Transport vehicles:
Initial cost
Maintenance cost
Mileage efficiency
Cycle time for automatic lifting equipment
Unloading efficiency
Whether used for other uses (e.g., leaf pickup)
Weather
Distance/time to processing center from drop-off center
Amount of publicity required to achieve desired capture rates

TABLE 5.12 Examples of diversity of curbside collection programs (22 communities surveyed)

Program characteristics	Number of programs with this characteristic
Household separation	
Required?	
Voluntary	14
Mandatory	8
Containers provided?	
Yes	20
No	2
Source separation	
None	1
One	1
Two	6
Three	11
Four	1
Five	1
Materials separated from refuse	
Newspaper	22
Glass	22
Aluminum	22
Tin Cans	20
HDPE	11
PET	10
Mixed Paper	S
Cardboard	4
Number of materials collected	
Three	1
Four	6
Five	4
Six	8
Seven	2
Eight	1
Number of households served	
Less than 5,000	4
5,000 to 20,000	8
20,000 to 40,000	6
Over 40,000	4
Pounds collected per household served per year	
100 to 200	5
200 to 300	2
300 to 400	5
400 to 500	7
500 to 600	1
600 to 700	1

TABLE 5.12 Examples of diversity of curbside collection programs (*Continued*)

Program characteristics	Number of programs with this characteristic
Collection	
Day	
Same day as refuse	13
Different day	6
Some of each	3
Frequency	
Weekly	19
Bimonthly	1
Monthly	2
Private or municipal?	
Private	13
Municipal	9
Crew size	
One	11
One to Two	1
Two	4
Three	3
Three to Four	1
Route size range	
350 to 1,000	9
1,000 to 1,500	8
1,501 to 2,000	3
Unreported	2
Stops per route	
240 to 400	5
401 to 500	2
501 to 600	5
1,000	1
Unreported	9

shows the great diversity of curbside collection systems with respect to such program factors as:

- Who does the collection
- Size of crew
- Average route size
- Average stops per route
- Frequency of collection
- Degree of source separation

Planning for Curbside Collection

Many factors must be considered in planning for curbside collection of recyclables, among them the population density of the area to be served (possible stops per mile of route), degree of source separation the residents will be willing to tolerate and still participate, the desired capture levels, the amount of money available to pay for the program, and the source of funds. Planning for collection cannot be divorced from planning for processing and marketing. For example, Mecklenburg County, North Carolina, began planning for curbside collections in the fall of 1986. The pilot collection program began in February 1987. Countywide collections were originally scheduled to start in 1988, but were postponed to 1989 because of delays in the construction of the central processing facility.

Once a policy decision has been made to collect recyclables at curbside, adequate time must be allowed for municipal staff and collectors to undertake detailed program design and promotion. It takes 6 to 12 months after the decision to collect curbside to get trucks on the road. Some specific program design issues for curbside collection of residential recyclables include:

- Who will perform recyclables collection, the public sector or the private sector?
- If collection is to be performed by the private sector, will the municipality be divided into districts, or will multiple collectors be allowed to overlap their service territories (open collection)?
- Should recyclables collection be the same day as refuse collection? How many days per month? Will back-door collection service be permitted?
- Who will be responsible for education and promotion? For enforcement?
- How will collection be handled for apartments and condominium units?
- What will the program cost?

The issues concerning source separation and the provision of containers were previously discussed. In addition, there are questions concerning whether the municipality wishes to undertake additional collections for different kinds of recyclables, particularly yard wastes, which are readily compostable and make up a large fraction (up to around 20%) of some municipal refuse streams. For example, Seattle, Washington (population 500,000), mandated the source separation of yard waste. For $2 per month, residents of Seattle can set out as many as 20 cans, bags, or bundles of yard debris (branches up to 4 in diameter) to be collected at curbside (see Fig. 5.10). The collector composts the material at a site outside the city. Since yard waste is accepted in plastic bags, a challenge is to remove the bags from the compost. Self-haul of yard waste to the city's two transfer stations is another option for residents.

Public or Private Collection?

Four alternative approaches to the question of who should collect recyclables have been employed. Which of these approaches offers the best service and the lowest-cost service depends on local circumstances, and is typically a very "hot" political issue. The first three are private sector approaches while the fourth is the public sector approach.

1. Competitive-bid *contract collection*, with one or more private collectors providing services under contract to the municipality, which pays the contractor directly for the service.
2. *Franchise collection*, wherein the collector is granted a specific territory in which to collect recyclables, and the collection cost is billed directly to the collector's customers.

FIGURE 5.10 An overflowing container of yard debris awaiting pickup. (*Photograph by A. McMillen, Mercer Island, Washington.*)

3. Unrestricted private *open collection*, with multiple collectors contracting directly with individual residents and businesses for recycling services.
4. *Municipal collection* performed by municipal employees.

In the short term, the collection crews who are most familiar with the service area could be expected to perform the most effectively, and there seems to be a trend to favor the same arrangement as is used for collection of regular refuse. Table 5.13 indicates what might be expected to be the relative cost of these four options, whether the collection is of refuse or recyclables.

Collection workers are on the "front lines" of the recycling effort. They are the ones who create the public impression about the municipality's recycling program. They are frequently also charged with the responsibility of enforcement of a city mandatory ordinance, or at least are expected to notify the enforcement officer of violations. They are the ones who put the reminder notices in the recycling containers. Intelligence, diligence, and good public relations ability are required for this job if a recycling program is to make a good impression.

Municipal Collection

In some municipalities with public refuse collection, the collection workers face an ongoing challenge from advocates of "privatization" who are eager to see the service taken over by the private sector. Data from a 22-community survey seems to support the contention that private collection is generally more efficient (528 stops per day) than municipal collection (415 stops per day); however, without knowing the terrain of the route, this hypothesis cannot be conclusively proven. If for some reason the public sector employees cannot

TABLE 5.13 Relative costs of collection

Type of collection	Who collects?	Relative cost
Municipal	Employees of the local government	1.27 to 1.37
Contract	A private firm hired by the community provides service in a specific territory	1.00
Franchise	Same as contract, except the private firm bills the customers rather than being paid by the municipality	1.15
Open	Multiple private firms compete: no service territories specified by the municipality	1.27 to 1.37

Source: Stevens, Barbara J., President, EcoData, Inc., Westport, Connectictut.

compete with the private sector to provide efficient, low-cost recycling collection services (for example, due to union rules that require two, three, or four persons on a collection truck), then a municipality may need to consider the private collection of recyclables. With refuse quantities potentially decreasing significantly due to source reduction (for example, backyard composting and mulch mowing), one way for public collection workers to preserve their jobs would be to undertake efficient recycling collections. This was the approach of the sanitation workers' union in Milwaukee, Wisconsin, which agreed to having only one worker on a recycling truck (special recycling trucks were purchased).

If the municipality is already served by private refuse collection (whether contract, franchise, or open), the option to establish an entire "department" to carry out recycling collections would result in considerable debate and a relatively long startup period. A decision for public collection in this instance could also be portrayed as ignoring private business interests and not using collection resources that might be already available in the private sector.

However, early in the decision process over who is to do the collection, it is useful to identify services and quantify the associated costs as if the collection were to be done with public forces (employees and equipment). There are three reasons for this exercise:

1. It provides a benchmark to evaluate private sector bids. If the level of service and cost are not equivalent to, or better than, what the municipality would provide with its own forces, the argument for using private collection is considerably weakened.
2. Since private collectors' true costs are difficult to determine (most collectors consider the information confidential and privileged), the estimated cost of the collection program can be based on a public-sector model. Numbers of households to be included, expected capture rates, vehicle types, and other logistical aspects of collection can be used to develop and evaluate the cost of alternative program designs until the most desirable program is decided upon, and then put out for private sector bid.
3. If unsatisfactory bids are received for collection services, the municipalities could implement an already planned and costed public collection. This would be especially useful if a decision had been made to divide the municipality into districts and some of the districts were unsuccessfully bid.

Open Collection

This approach allows households to contract directly with collectors of their choice to provide trash and recycling collection service. This approach permits the greatest freedom to the private sector, particularly the very small collection firms. For example, citizens' intense loyalty to their trash haulers influenced the decision of Manheim Township, Pennsylvania, to revise its plans to have a bid system. Instead, it was decided that each hauler would be responsible for collecting both recyclables and trash. However, higher costs are often associated with this alternative because of inefficiency of service delivery. In addition, the multihauler approach is difficult and cumbersome for the municipality to monitor. Some collectors may simply dump the collected recyclables in the landfill. Recycling program promotion, and scheduling of the commencement of service, are more difficult under this alternative. Collections might begin sporadically in different parts of the municipality at different times and may never reach the entire municipality. Residents and businesses may be subject to payment for overlapping services. The ability to rely on recycling as a significant component of a comprehensive refuse management plan could be limited. For example, Portland, Oregon, has over 100 private collectors. But the recycling rate was only about 2% in Portland, compared with 10% and above for other curbside collection programs.

What would be the rationale for open collection of recyclables, besides customer loyalty? For one thing, the existing collectors are familiar with the territory, and they may exert considerable political leverage. To make open collection work, a municipality has to be willing to spend the time and money to put the program in place and monitor its progress, and take some responsibility for controlling the collectors' actions. Also, the collectors must support the program.

To remedy some of the problems with open collection of recyclables, the municipality could undertake some or all of the following responsibilities:

- Identify the specific services to be offered (collection frequency, day of collection, materials collected, etc.)
- Develop uniform education and publicity to support the program
- Provide household set-out containers to all households
- Provide technical assistance to collectors, including:

  Collection strategies for specific needs (e.g., single family, apartments, condominiums, etc.)

  Information regarding available equipment and latest collection techniques
- Provide a MRF or other processing facility for collected materials
- Assume market responsibility, either directly, or indirectly through MRF services (for example in Dakota County, Minnesota, the county's MRF pays the collectors for the recyclables delivered)
- Regulate the rates that the collectors charge their customers for recycling collections (for example, in Lenexa, Kansas, collectors were allowed a $1.50 per household per month additional charge as of November 1989)

To assure that services are performed in conformance with the minimal standards acceptable to the municipality, collectors will need to do some or all of the following:

- Procure the necessary equipment and designate necessary personnel
- Coordinate enforcement procedures between the municipality and households (materials, inappropriate set-out units, any participation requirements, etc.)
- Maintain a telephone service line for household inquiries and complaints
- Set appropriate standards for the physical appearance of crew and vehicles
- Conform to local safety regulations

- Apply the uniform municipal logo on set-out containers and recycling vehicles
- Use publicity and information that is consistent with the municipality's recycling program
- Deliver reports to the municipality on pickup routes, tonnages delivered to the MRF, participation rates, and other documentation needed by the municipality

Contract Collection

One way to help defray the costs of a recycling collection program may be to switch from open or municipal collection of refuse to contract collection for both refuse and recyclables. Advantages of the contract collection approach include the following:

- Offers the most control to the municipality of the private sector recycling collection options.
- Uses the existing refuse collection infrastructure to perform recycling services
- Ensures service to each household
- Allows rapid startup of the curbside collection program
- Ensures qualified collectors would be performing the service
- Promotes efficiency by controlling the number of collectors of recyclables operating within a service area
- Promotes orderly and routine collection
- Facilitates coordination of program promotion and the scheduling of pickups

The municipality may determine that some areas may not be appropriate for contracted collection; i.e., the competitive selection process might not identify collectors willing to perform collections at a price the municipality can afford. Alternative options for limited areas may involve municipal collection, or a decision to allow open collection. The jurisdiction can be divided into districts, and different services procured for the different districts (provided residents of a district do not subsidize a service that they do not receive). Newark, New Jersey (population 300,000; land area 24 mi^2), was divided into nine collection zones. In Newark, collection is divided between city forces and a local handicapped organization as an outside contractor. Districting also facilitates phasing in of the program.

Some companies offer processing and marketing services, in addition to collection of recyclables. Contracting for the recycling collection component separately would make sense under the following conditions:

- When the community has access to a preexisting public processing center (e.g., a regional or state-provided processing center such as in Connecticut or Rhode Island)
- When the community wants to encourage an existing private processing center (e.g., the municipality may want to retain and encourage existing scrap processors who may feel that their business would be threatened)
- If there is little competition in the area among full-service national firms (the larger firms are not typically interested if the population is under 50,000)
- If the community is procuring general refuse collection services and wants to add recycling collection into the same contract(s), perhaps at a community-specified percentage of the refuse collection bid price

A procurement document for recycling collection services should include the following elements:

- A list of what materials are to be collected, or at least the number of separate materials
- Provision for future change in the list of items
- The number of sorts that residents will set out and for which the collector must provide separate compartments in the collection vehicle
- Which households are to receive recycling collection service
- The correct number of households in each category to be served (e.g., single family, two-family, etc)
- When and how often recyclables will be picked up
- Where recyclables are to be picked up (curbside or otherwise)
- What kind of educational responsibilities the collector will have, (e.g., whether the municipality will retain approval rights over flyers, what kind of "reminder" notices are to be issued, etc.)
- Where the recyclables are to be taken for processing, and what the unloading configuration/limitations will be at the processing facility

To develop and issue the procurement document, the following tasks need to be accomplished:

- Assess municipal demographics and divide the municipality into recycling collection districts
- Develop education/publicity program and determine collector's role
- Secure markets for materials as-collected, and secure processing capabilities for collected materials not proceeding directly to markets
- Develop bid specification document describing all services requested and subject this procurement document to legal review
- Hold prebid conferences with collectors; potential topics:

  Division of the municipality into districts: are the divisions logical?

  Should the lines be drawn elsewhere?

  Services requested: can the collectors perform them with existing equipment?

  Are the requests reasonable?

  Compliance: What are the enforcement responsibilities of the collectors and of the municipality?

  Where will the collectors bring collected recyclables, and how far away is it?
- Release bid specifications for collection services
- Evaluate bid responses and award contracts according to:

  Ability to perform services with available equipment and personnel,

  Price bid for services,

  Other municipal procurement requirements (bid bond, legal certification, collection permits in place, etc.).

Franchise Collection

An arrangement wherein a municipality grants collectors an exclusive right to collect in a specific geographic area without a competitively bid contract is problematic. In some states, such arrangements have been viewed as unconstitutionally anticompetitive. In essence, the municipality is creating a monopoly. Unless a regulatory authority is established to oversee the collector's rate structure, residents are offered little protection from unnecessarily high costs. The problem is compounded when the collector is allowed to bill the customers in advance of the service being provided.

If, however, the collector has to compete for the collection district and the municipality regulates the rates charged, then the only real difference between franchise and contract collection is in who sends the bill to the customer.

In summary, "recycling collection should be as compatible as possible with existing refuse disposal practices and the needs of each area. For areas with private hauler infrastructure, we recommend that the municipality use the existing hauler infrastructure to collect recyclables, but in a manner whereby the municipality's needs are identified and met through contracted services."

Frequency and Timing of Collection

Recycling collections typically occur on a weekly, biweekly, or monthly basis. The more frequent the collection service the higher the operating costs. However, weekly collection offers two important advantages:

1. Participation levels are increased because residents do not have to remember which week is recycling week.
2. Capture rates are increased because residents do not have to store large quantities of materials in their homes between collections.

The timing of recycling collection is also a factor. Some programs collect recyclables on the same day as refuse, while others collect on a separate day. Same-day collection (i.e., same day as regular refuse service) does seem to improve participation. Among 13 required-participation programs, the average participation rate was 76.5% for same-day schedule and 41% for different-day schedule. But whether the recyclables collection occurs on the same day or on a different day, a regular schedule is essential. "As long as public education is continuous, and the pick-up schedule is highly reliable, the public seems willing to follow whatever schedule is established for collecting recyclables."

One way to compensate for the extra cost of collecting recyclables is to reduce the number of regular refuse collections per household per month. Reducing from twice per week collection to once per week typically reduces costs by 20%: the money saved can be applied toward the increased cost of a weekly or biweekly recyclables collection.

Vehicle Types, Capabilities, and Costs

It is expensive to replace or supplement an existing fleet of refuse collection vehicles with new vehicles specifically designed to collect recyclables. Therefore, many communities have tried at least initially to retrofit existing refuse collection equipment. Madison, Wisconsin, designed a newspaper rack welded to the frame of the refuse truck to collect bundled newspaper. But such retrofits usually are not well suited to the collection of many different source-separated recyclables.

Some communities use pickup trucks that follow the refuse truck. The recyclables are set out in clear plastic bags to make them easily distinguished from the refuse. A different approach involves using existing refuse compaction vehicles ("packer trucks") to collect newspaper or mixed fiber, with a separate vehicle (such as a pickup truck) for commingled containers. This approach would reduce capital expense at the startup of the program.

Packer trucks are used in some recycling programs to collect commingled containers, but more breakage of glass items occurs. In this case, more residue can be expected at the point of processing, especially from glass items. This reduces recovery levels and increases total net program cost.

Many municipalities have found it necessary to make the investment in a fleet of specialized recycling vehicles. There are three categories of specialized recycling collection vehicles: closed-body trucks, open-top trucks, and trailers. Although a stepvan is in the same cost range, it is less efficient because of its inability to off-load by lifting hydraulically, and the subsequent need to unload the vehicle manually. Factors to be weighed in the selection of a vehicle type include:

- Considerations to minimize worker fatigue, including:

  Ease of entering and exiting the truck

  Loading height

  Street-side driving/vehicle entry-exit

  Other drivers' safety and comfort features
- Overall net tonnage capacity (capacity should be large enough to allow the truck to be on the collection route for the entire day without having to make an intermediate stop at the processing facility, especially if it is far away)
- Ability to adjust material compartment sizes to allow for changes in collection quantities and to permit addition of new materials
- Material loading efficiency (e.g., ability to serve 90 gal carts from apartment buildings)
- Fuel efficiency
- Operating costs
- Flexibility of vehicle for other uses (e.g., in Milwaukee, Wisconsin, vehicles capable of carrying a snowplow were selected because the city's packer trucks have traditionally been used to plow snow)
- Safety and maneuvering considerations (e.g., attaching a trailer to a dump truck results in a very lengthy apparatus)
- Chassis and power requirements compatible with the terrain (more engine power for hilly rural areas; more maneuverability for more heavily populated areas)
- Design of the receiving area of the processing facility (or vice versa, if the truck fleet already exists, the processing facility receiving design should accommodate it)

In Palo Alto, California, a three-bin trailer system is used (glass, newspaper, and cans). Flattened cardboard and lightweight scrap metal are placed in the bed of the van. Used motor oil, packaged in unbreakable containers with tight-fitting lids, is placed in the space between the wheels of the trailer. The intent of this collection design is to recover both high-volume items such as glass and newspaper, and low-volume items with high pollution potential, such as batteries and oil. Three trailers are sent out each day, and each returns with full bins three times, unloading into roll-off containers.

Some collectors have designed their own trucks. For example, a collector operating in the Windsor Locks, Connecticut, area, has designed the SAC I and SAC II (for "separation at curbside") (see Fig. 5.11). SAC I holds 17 yd^3 SAC II holds 34 yd^3. Collectors sort recyclables at the truck into five categories: three colors of glass, newspapers, and cans. The company has also designed an onboard plastics densification machine. In another instance, a collector worked with a manufacturer and city officials to design a two-compartment collection truck (20 yd^3 for mixed paper, 11 yd^3 for commingled containers). The container side began filling up too quickly when residents began putting out mixed plastic containers.

FIGURE 5.11 SAC trucks in Vermont. Note the two on-board plastic densification receiving ports just behind the cab on the passenger side. (Photograph courtesy of-John Casella.)

It has been estimated that an 80-TPD container recycling plant will save landfill space equivalent to a building 20 ft high, 100 ft wide, and a quarter-mile long. This is also approximately the total volume of containers that will have to be hauled in a year by the collection vehicles serving the facility: 98,000 yd^3 or 3,900 truck load equivalents for a 25 yd^3 recycling collection vehicle (calculation ignores landfill cover material). Additional truck load equivalents would be required for any papers collected.

Table 5.14 gives a list of collection vehicles suppliers. As with the previous supplier lists, location and telephone numbers are provided so that convenient suppliers may be contacted for the latest information on product specifications, prices, availability, lead times for ordering, and shipping information.

Optimal Crew Size

For regular refuse collection, a two-person crew may be more efficient than a one-person crew, if households are quite close together and where low traffic permits service to both sides of the street simultaneously. The same would apply to the collection of recyclables, except that the effective distance between households is much longer because of set-out and participation rates, so one-person crews are more efficient in most cases. For example, in Mecklenburg County, North Carolina, trucks stopped at an average of about 32 of the 105 households passed per hour. "The personnel required for separate collection of recyclables is the largest operating cost component of residential recycling programs. It is therefore not surprising that the most cost-efficient programs use one-person collection crews servicing one side of the street at a time. The addition of extra crew members has been found to have a negligible impact on collection productivity, at considerable increased expense."

Collection Productivity

Typically, curbside recycling routes are comprised of about 1,000 households passed by in one day. However, there can be considerable variation between municipalities and be-

TABLE 5.14 Factors influencing the cost of collecting recyclables at curbside

Set-out containers:
Number per household or business
Capacity of containers
Frequency of replacement
Ease of unloading
Density of set-out locations and set-out rates
Driving distance/time between stops
Number of pickups per stop
Compartmentalization
Materials to be collected and their densities
Number of categories of recyclables set out
Number of compartments in collection truck
Volume of each compartment
Whether collectors sort at curbside or not
Drivers and workers:
Number on each collection vehicle
Wages and fringes
Length of work day
Performance (attitude, efficiency, speed)
Vehicles:
Initial cost
Maintenance cost
Mileage efficiency
Suitability for terrain (e.g., tight turns)
Cycle time for automatic lifting equipment
Unloading efficiency
Whether used for other uses (e.g., snowplowing)
Weather
Distance/time to processing center from end of collection route
Amount of publicity required to achieve desired participation, set-out, and capture rates

tween routes within an municipality. For example, in Mecklenburg County, North Carolina, passbys averaged about 105 households per hour, with 5.47 productive collection hours per day. East Lyme, Connecticut's, single truck and crew of two serves 4,100 to 4,200 homes each week, picking up newspaper, cardboard, and mixed containers.

The productivity of collection vehicles (the number of stops per day per vehicle) is affected by four major factors:

1. Demographic and geographic characteristics of the area including density of housing, flatness of the terrain, parked cars, and street configuration
2. The design and capacity of the vehicle used
3. The total nonproductive time per day (travel time to and from the processing facility; lunch and coffee breaks)
4. Materials collected, their densities and their capture rates, and whether they are source-separated or curbside-separated

The number of stops that can be attained in a day can be estimated by knowing the set-out rate, the average distance between the households that set out recyclables, the driving time between stops, the average time it takes to unload a household's set-out containers into the truck, and the daily hours spent in actual driving and collecting.

Number of Vehicles Required

The size of the recycling collection vehicle can be estimated by knowing the volume, in cubic yards, of the recyclables likely to be set out by the average household each collection day and the number of set-outs that a single vehicle can serve in a day, including the time it takes to drive to the route in the morning, to the processing facility, and back to the vehicle's overnight parking area. For example, if each household generates 0.0375 yd^3 of recyclables each week and the recycling vehicle can drive by 800 households stopping at 480 of them, then the vehicle should have a capacity of at least 480 times 0.0375 or 18 yd^3, assuming that the vehicles' compartments are the ideal volume for the materials collected and that the compartments can be completely filled. (In practice, conditions are not ideal, so a somewhat larger truck should be selected.)

The total number of such vehicles required is estimated by the number of collection days per week that collection is to occur, the total number of set-outs expected on each day, and the number that each truck can serve. In the above example, 10 vehicles would be required if collection occurred 5 days per week, each truck could serve 480 set-outs per collection day, and 24,000 households set out materials once each week.

Costs of Curbside Collection

Curbside collection is one of the most costly aspects of postconsumer recycling. In estimating the cost of curbside collection, all of the factors listed in Table 5.15 must be taken into account. There are numerous examples in the literature of attempts to justify recyclables collection cost on the basis of the "avoided cost of landfilling," but this approach can be very misleading.

Costs can sometimes be justified when set against the full costs of the system without recycling, including the costs of funds that will have to be used to close and monitor landfills. The cost for curbside collection of recyclables ranges from $1.50 to 3.25 per household served per month in. In Tallahassee, Florida, the collector is paid $1.32 per month per household for recyclables collection. The City of Rockford, Illinois, is paying the collector $1.78 per month per household for weekly collection of newspapers, magazines, glass, plastic bottles, tin, steel, and aluminum cans. These are curbside-separated into eight categories.

In any given municipality, recyclables collection is likely to have a cost similar to regular refuse collection because many of the cost elements are similar: labor earns about the same wages, recycling collection vehicles plus associated in-home containers cost about as much as garbage trucks, and vehicle operating costs are similar. The key to controlling costs is to establish an efficient collection and processing system. "It is extremely unlikely that inefficient collection and expensive recycling systems will be self-financing [*i.e., operate at zero net cost*] unless market prices for recycled goods are at historic highs and disposal fees near $200 per ton. Even the efficient collection and processing system will be self-financing only in areas of the country where the prevailing disposal fees are in the $25 per ton range" (parenthetical added).

To reduce the overall net cost of the entire solid waste management system, the whole picture needs to be examined to see if there are possible cost-savings measures. For example, considering the cost of collection, savings could be realized by eliminating back-door collection, switching to contract collection for refuse if municipal collection costs have gotten out of hand, or reducing the number of regular refuse collections from two per week to one per week. For example, a private firm operating in the middle-class suburbs of San Francisco, California, estimated the cost savings due to recycling 48 TPD [12,500 tons per year (TPY)] were the following:

- Savings through five fewer daily refuse collection routes (labor, capital, and operating expenses)—$787,000
- Savings through avoided transfer and disposal costs—$262,000

A comparison of the almost-twin towns of Longmeadow and East Longmeadow, Massachusetts, on a system-cost basis shows that the town with a recycling collection had lower net per-household cost in 1989. In this simple example, both towns employ the same collection firm and the same equipment (a 17 yd^3 right-hand drive, one-person, side-loading packer). The results of the analysis are shown in Table 5.16.

The reasons that East Longmeadow's system cost is less with recycling include the following:

TABLE 5.15 Comparison of system costs with and without recycling in two communities in Massachusetts

	Longmeadow	East Longmeadow
Demography:		
Population	16,380	12,910
Housing units	5,450	4,400
Road miles	94	83
Houses per road mile	58	53
Annual per capita income	$20,306	$12,693
Topography	Flat	Flat
Area, mi^2	9.0	13.2
Refuse tonnage:		
Frequency of refuse collection	Weekly	Biweekly
Annual refuse generation	6,605 tons	5,060 tons
Annual recycling:		
Newspaper	1,298	
Corrugated/paperboard	237	
Glass containers	70	
Aluminum	10	
Annual disposal	4,990	5,060
Percent waste reduction	24%	0%
Costs		
Cost of refuse collection	$181,000	$160,000
Recycling collection	68,000	0
Disposal @$34 per ton	169,660	172,040
Revenue, sale of recyclables	(30,460)	0
Total solid waste cost	$388,200	$322,040
Annual cost per residence	$71.23	$75.50

Source: Powell, Jerry, "Recycling is cheaper: the Massachusetts experience," *Resource Recycling*, pp. 37, 61, October 1989.

- Participation rate of 90%
- An inexpensive collection system: a cleaned-out packer truck is sent out one week to pick up old newspapers and the following week for corrugated cardboard and paperboard
- Adequate revenue from material sales (corrugated cardboard and paperboard received $30 per ton and newspaper revenue was $10 per ton when the costs were analyzed)

In addition to paying for the additional cost of recycling collections through surcharges and cost-savings measures adopted in other parts of the refuse management system, municipalities can sometimes obtain grants. For example, Coca-Cola donated $99,000 to Mecklenburg County for PET recycling support. It is also sometimes possible to rely to some extent, at least on a limited or temporary basis, on donated collection services by public-spirited organizations.

Collecting plastics for recycling presents a particularly difficult economic challenge. Because of the materials' low density, collection trucks carrying uncompacted plastic are carrying around a lot of dead airspace. In Rhode Island, it was estimated that adding plastics to a curbside collection program would result in an incremental cost of $54 to $108 per ton of plastic collected, while Milwaukee, Wisconsin, estimated that collection and processing of one ton of postconsumer plastic in its pilot program cost $975 compared with $115 for paper. Even with a maximum theoretical revenue on the order of $800 per ton, the curbside collection of undensified plastic may not be economical. (Actual average revenues for plastics are more in the range of $120 to $160 per ton, depending on resin type, and can be as low as $40 per ton for mixed plastics).

A common way to raise revenues to support recycling is to increase the tipping fees at the landfill, as is done in Mecklenburg Country, North Carolina.

MONITORING PERFORMANCE

The first step in monitoring performance of a separation and collection system is to have a clear idea of what the goals of the overall program are. Is the recycling program being undertaken to involve as many citizens as possible in recycling? Or to divert the maximum amount of material from the landfill? Or to minimize the cost of the solid waste management system in the long run? To employ as many people as possible? To foster the growth and development of local recycling industries? Or all of the above? The performance measurements chosen should reflect the goals of the program. For example, Mecklenburg County, North Carolina, collected data during its initial program phase that would allow it to compute and evaluate a number of performance measures including:

- Household participation and material set-out rates
- Material recovery rates
- Adequacy of the volume and design of the in-home container
- Impact of variables on collection production rates
- Collection vehicle design parameters, efficiencies, and costs
- Optimum size of collection routes
- Processing efficiencies and costs

Whatever measurements are taken, for all but the smallest of communities a computerized data base will be necessary. For example, the city of Philadelphia developed a dBASE III + program that incorporates census data so that participation can be predicted. Data on participation, set-out of unacceptable materials, and other parameters are recorded by block and entered daily, as are cost data.

Measuring Participation

The most common measurement taken of curbside collection programs is the participation rate. Participation is measured in various ways from pure guessing to actual sampling. Participation is commonly measured on a monthly basis since not all households set out recyclables each week.

Use of Bar Code Scanners to Record Participation

Bar coding of set-out containers is gaining in popularity. Advocates think that widespread use of bar coding might create some consistency in the current "hodgepodge" of methods for tabulation and estimation of participation rate. Bar coding offers the additional advantage that worker productivity is automatically registered, allowing program managers to adjust route sizes. In addition, nonparticipating households can be automatically identified, so that individualized attention and education can be given to them. In addition, if there are differential rates charged to participating households, the households that actually participate can be accurately rewarded for their efforts.

The technical challenge to successful bar code recording is to provide a suitable label that sticks to the bins despite their ultraviolet light-resistant coatings, and that does not become unreadable over time. Hand-held scanners seem to function without major problems (even in cold weather), and the procedures do not seem to slow the workers down inordinately. Different collection firms hold different views on the utility and difficulty of bar coding, however.

A pioneer in the use of bar coding is St. Louis Park, Minnesota (population 43,000) (Fig. 5.12.). This community uses a stackable, triple 10 gal bin system, and collection is bimonthly. The 2,500 TPY collected is said to be 16% of residential waste. It is important that participation be measured accurately, because a quarterly $6.60 recycling credit (from a basic garbage service charge of $11 per month) is granted to residents who set out recyclables at least three times within a 3-month period. (A credit is also given to citizens using buy-back and drop-off centers.) When the bar coding and credit system went into effect, participation rose from 75 to 87%. Bar code scanners, computers, documentation, and training costs were $25,000 for the 12,000 households covered in the collection. Each household was sent a set of bar code stickers that included the household's utility billing account number. Data from the hand-held scanners are transferred electronically to the utility's billing office. Bar coding is also used in Tonka Bay, Minnesota; Rolling Hills Estates, California; Munster, Indiana; Clinton County, Pennsylvania; and is soon to be used in Charlotte, North Carolina. Munster, Indiana, also uses a reduced-rate incentive to participate, and Rolling Hills Estates uses a lottery.

Measuring Participation by Sign-Up

In communities where residents must subscribe in order to receive recycling collection, participation is measured by the percent of residences that have signed up. In Seattle, Washington, sign-up for the three-bin collection on the northern side is 85.8%; sign-up for the one-cart system serving the south side is 62.4%, for a 72.7% sign-up rate overall. The difference in sign-up rate between the north and south side has been attributed to weekly pickup (therefore a weekly reminder) in the north, more brightly colored bins in

FIGURE 5.12 A worker scans a set-out container's bar code. (*Photograph courtesy of the City of St. Louis Park.*)

the north, and cultural influences (the south has more low-income residents and a wider range of ethnic backgrounds).

Other Methods of Estimating Participation

The traditional way of determining participation rate is to have the collection worker keep records. However, this slows down collection, places an additional burden on the collector, and results in a large data recording and analysis job for someone in an office.

Estimating participation rate from weekly set-out rates can be unreliable, unless sampling is done to calibrate the set-out rate. In Santa Cruz, California, 35 to 45% of residents with stackable recycling containers put them at the curb in any one week. A study route was chosen. Out of 640 residences, 417 distinct addresses set out materials during a 7-week period, for a participation rate of 65.2%. (Note that the rate would be expected to be smaller if the study period were the more typical 1-month duration)

Estimating participation rate for apartment complexes is very difficult unless each apartment has its own containers set out at the curb.

Lexington, Kentucky, asked participants in its 6-month pilot project to volunteer to keep weekly diaries of the number of containers (aluminum, glass, and plastic) and the number of newspapers they set out each week. Over 100 participants volunteered and completed at least one of the 6 one-page diary sheets mailed to them. This method is unlikely to be very useful over a long period of time, as it depends on self-reporting.

Measuring Capture

The quantity of refuse that could theoretically be recycled has been variously estimated. In one composition study, it was found to be as high as 50.3% on a weight basis and

63.3% on a volume basis. (Recyclables in this study included clean wood, leaves and twigs, textiles, and boxes.) There was a significant difference depending on the day of the week that the sample was taken, with refuse sampled early in the week being richer in recyclables. How much of this theoretically recyclable material is actually captured depends on what materials are accepted for collection, and how faithfully participants set out all that they can.

Not all participants set out for collection all the materials that are accepted for collection. For example, in Minneapolis, Minnesota, about half of all participants set out all materials, 25% set out only two items, and 25% set out only paper. Plymouth, Minnesota, claimed that 30% set out only paper. In Austin, Texas, less than 50% set out containers. In Santa Cruz, California, three-quarters of participants put out newspapers, two-thirds included glass, and one-half included cans. Many HDPE milk and water bottles were set out, even though they were not eligible in this program (although PET bottles were). In general, newspaper is the most frequent material set out. It usually makes up about 75% of the weight of material collected, with glass contributing 15 to 25%, and metal 5 to 10%.

Measuring capture rate requires two measurements: the amount of the recyclable correctly source-separated, and the amount that was not source-separated but instead placed in the refuse. The accurate way to do this is through periodic waste composition analysis from a statistically valid sample of randomly selected recycling routes, analyzing the composition of both the recyclables stream and the regular refuse stream. This is the approach used in Milwaukee, Wisconsin, and in several other cities. In Milwaukee, the waste composition analysis is conducted at the city's three transfer stations, and 50 separate materials are classified, including all materials currently thought to be recyclable by any means (even those that the city does not currently accept in its curbside and drop-off programs). In Mecklenburg County, North Carolina, capture rates were measured with a refuse sampling study. In 1987, the curbside collection program was estimated to recover about 8.75% of the residential refuse stream. This recycling rate is consistent with other similar curbside recycling programs.

Fitchburg, Wisconsin, has analyzed the capture rate for polystyrene in its experimental curbside collection of this material. The weight of polystyrene collected from a sample of residences was obtained from the collector. The weight of polystyrene left over in the refuse stream was determined by sorting over 9 tons of refuse from the same homes. Approximately 83% of the polystyrene was captured. Of the material captured, 8.5% was not polystyrene and another 8.5% was contaminated.

More information on waste composition analysis is found in Chap. 3 of this Handbook.

Measuring Costs of Separation and Collection

Measuring costs of the separation and collection program should be a straightforward matter. Why, then, does the literature on this subject tend to be somewhat obscure? Perhaps the answer lies in the fact that collection of recyclables is expensive. Looking only at the short-term economics, many recycling programs have a net positive cost to a community, and collection is largely responsible.

Recyclables collection cost is normally added onto the regular refuse collection cost. The factors determining the costs of recyclables collection are almost exactly the same as the factors determining the cost of regular refuse collection, and include crew wages, distance between houses, frequency of collection, point of collection (curbside or backyard), quantity of material to collect, terrain, weather, crew size, and vehicle efficiency (i.e., whether all compartments are sized so that they fill up at the same time). Without a serious community effort at source reduction, the total amount of refuse collected will not change appreciably whether it is picked up in a recyclables truck or a refuse collection truck.

It is possible that cost savings can occur in regular refuse collection as a result of diverting material to a recyclables collection. However, it takes several years to determine how much larger the regular collection routes can be as a result of the diversion of material into the recycling collection, and even longer to make the necessary cost-saving changes in the regular collection routes. In addition, lengthening the regular refuse collection routes does not result in a dollar-for-dollar cost saving . . . adding a recycling collection at the same frequency as regular refuse collection will cost about as much as would adding an extra weekly pickup to the regular refuse collection schedule." Increasing collection frequency from once per week to twice per week increases collection costs about 26%, and increasing frequency from twice a week to three times in a week increases total collection cost by about 18%. But, this assumes the same vehicles and crew can be used. The costs may actually be higher if specialized vehicles are needed.

If recyclables collection should happen to be cheaper than regular refuse collection for some reason (such as a different contractor collecting the material, cheaper collection trucks, shorter haul distance to unloading location, low degree of source separation, and no curbside separation required), then there should be a point at which the entire collection system cost is lower as a result of the recyclables collection. For example, Tukwila, Washington, found that since contract refuse collection service for apartments was more expensive per ton than contract recyclables collection service for the same apartments, there would be a point at which the public provision of dumpsters for recyclables would break even economically. For that program, the break-even point was estimated to occur when about 30% of the waste was recovered in the recycling collection. (In this instance, the recyclables were separated in a single source separation with all recyclables commingled.)

The concept of avoided disposal cost is frequently misused in discussing the cost of recyclables sorting and collection. The only accurate and rational basis to compare costs is on a system cost basis, where the full cost of refuse management is added up for the system with recycling and compared to the system without recycling (or with a different kind of recycling). Full system cost should include any quantifiable costs of landfill development and closure, as well as any cost savings in regular refuse collection and disposal. The subject of full system cost is exceedingly important, but it is beyond the scope of this section to provide a detailed method. Table 5.16 provides a list of many of the myriad cost and revenue elements that go into a full system cost calculation.

Public Support/Satisfaction

In trying to determine or predict public support and satisfaction with a program, all waste generators must be listened to if the program is to be truly successful. One cannot rely on "experts" to indicate what public reaction to a particular program feature will be. "It is argued that people will not cooperate on a voluntary basis. This argument is often stated by speakers who then indicate their own situation. In other cases, these people are influenced by objective arguments and calculations, but for sorting at the source they make decisions on the basis of their personal prejudices."

Information on public attitudes can be found in Chap. 30, *Public Awareness Programs*, of this Handbook.

TABLE 5.16 Elements of total solid waste system costs

System planning costs and source reduction costs
- Education and outreach
- Administration costs for volume-based rates
- Enforcement costs

Costs for collection from point of generation and delivery to initial drop-off location
- Equipment
  - Regular collection and special recycling collection vehicles, purchase cost of special bins or roll-off carts, if provided for either regular waste or recyclables, and replacement of same
- Labor
  - Salaries and wages for all collection workers and supervisors, regular waste and other special collection (e.g., recyclables, yard waste, white goods)
  - Fringe benefits for collectors
  - Other labor-related costs (e.g., overtime)
- Operating and maintenance costs of all vehicles, including:
  - Fuel usage
  - Tires
  - Truck maintenance
  - Insurance and licenses
  - Other
- [*Alternatively, the cost of collection may be the cost of contracts for collection.*]

Costs for transfer and processing facilities:
- Drop-off centers for recyclables, regular waste, yard waste
- Transfer stations
- Transfer stations for special materials (e.g., household hazardous waste)
- Intermediate processing facilities (IPFs)
- Material recovery facilities (MRFs)
- Composting facilities
- Special processing facilities (e.g., construction debris)
- Waste-to-energy facilities
- Mixed-waste processing facilities
- Site and equipment
  - Site investigation/selection/purchase costs
  - Permitting costs
  - Costs for construction labor
  - Site preparation costs (earthwork, utilities, fencing, etc.)
  - Costs for buildings, structures
  - Costs for on-site mobile equipment
  - Material processing equipment costs
  - Costs for controls, all other auxiliary equipment
  - Performance testing costs
  - Startup operations may also be included
- Operating labor
  - Salaries and wages for site workers and supervisors
  - Fringe benefits
  - Other labor-related costs

(continued)

TABLE 5.16 Elements of total solid waste system costs (*continued*)

Operating and maintenance costs for site and equipment
- Fuel
- Vehicle maintenance
- Equipment maintenance
- Site maintenance
- Equipment repair and replacement (including deposits to sinking funds)

Environmental and other permit compliance monitoring and testing costs

[*Alternatively, the cost of furnishing and equipping and/or operating, any of these facilities could be on a contract basis. In this case, the cost would be the contractually specified costs.*]

Costs for haul from transfer/processing facility to disposal sites and/or markets

Equipment
- Cost of mobile equipment used for hauling (tractors, trailers, spares, etc.)

Labor
- Salaries and wages for hauling workers and supervisors
- Fringe benefits for haulers
- Other labor-related costs (e.g., overtime)

Operating and maintenance cost of all vehicle, including:
- Fuel usage
- Tires
- Truck maintenance
- Insurance and licenses
- Other

[*Alternatively, the cost of hauling may be the cost of contracts for hauling. In some cases, markets may pick up the material for a somewhat lower revenue than if it were delivered.*]

Costs for disposal

Landfill development costs
- Site investigation/selection/purchase costs
- Permitting costs
- Costs for construction labor
- Site preparation costs (earthwork, utilities, fencing, etc.)
- Costs for buildings, structures, liners, leachate controls, methane controls
- Costs for on-site mobile equipment
- Costs for all other auxiliary equipment

Operating labor
- Salaries and wages for site workers and supervisors
- Fringe benefits
- Other labor-related costs

Operating and maintenance costs
- Fuel
- Vehicle maintenance
- Equipment maintenance
- Site maintenance
- Equipment repair and replacement (including deposits to sinking funds)
- Additional liners, cover material, etc., as cells are developed
- Landfill closure costs and post-closure monitoring costs (escrow fund deposits)

(*continued*)

TABLE 5.16 Elements of total solid waste system costs (*continued*)

Environmental and other permit compliance monitoring and testing costs
[*Alternatively, the cost of furnishing and equipping, and/or operating, a landfill could be on a contract basis. In this case, the cost would be the contractually specified costs.*]

General administrative costs
- Administrative labor and fringe benefits
- Education and public relations cost
- Costs associated with advisory committees
- Memberships, travel, and other professional costs
- Legal, accounting, and other professional services

Revenues
- Tipping fee revenues
- Contract revenues
- Franchise fee revenues
- General fund revenues
- Special assessment revenues
- User charges
- Revenues from the sale of recyclables and compost
- Revenues from the sale of energy (steam, electricity, and fuel products)
- Landfill reserve capacity fees
- Grants-in-aid (federal or state) and other subsidiaries
- Fines (where specifically dedicated)

Notes: All costs are likely to be financed by some mixture of debt and equity. The costs of financing, including letters of credit or other security, and underwriter's costs, would need to be added to the total construction and equipment cost to derive the capital cost for facilities and equipment.

Costs borne by individuals hauling material to drop-off or disposal or in "backyard composting" or other individual waste management activities are not commonly included in analysis of system costs.

Scavenging

No discussion of separation and collection would be complete without an acknowledgment that scavenging occurs. Scavenging was noted as a problem even in the early experiments with curbside collection.

In societies where reliance on highly mechanized refuse collection and disposal is not widespread, scavenging is the primary means of recovering valuable resources. For example, in Egypt, 15,000 people live in a refuse ghetto, paying for the privilege of collecting refuse in the city using donkey carts for collection vehicles. These people support themselves entirely by the hand-picking and sale of salvageable goods. Sorting is done in their front yards. Marketing involves a daily negotiated price with intermediaries, who sell the separated materials to conversion plants.

Although China has an extensive state-controlled waste recovery administration extending down to the neighborhood level, widespread informal scavenging takes place on the streets and at the dumps of other Asian cities like Cairo, Calcutta, Bali, Manila, and Bangkok. Intermediaries control the scavengers, and sometimes these operations function at the edge of legality. "Municipalities are now asserting their rights over wastes that formerly they were only too glad to have others deal with, and ownership issues are becoming contentious . . . public authorities see street picking as risky and degrading for the whole community; welfare groups wish to protect scavengers' traditional rights and access . . . health considerations raise some of the most difficult issues. Should groups be permitted to continue in unhealthy occupations if they so wish?"

Mexico City has a recycling infrastructure somewhat similar to Cairo's. In this city an estimated 5,000 to 20,000 scavengers make their living off the dump. These people have an average life expectancy of about 35 years and their infant mortality rate is almost 50%. Phasing out scavenging has been difficult for the government because scavenging is important to the fragile economy of the city and because of pressure from influential people who profit from it. There are many purchasers of reclaimed material, and the city's sanitation workers can almost double their income by sorting through the trash for salable items like cardboard. For the Mexico City metropolitan area, a study using linear programming analysis showed that the most cost-effective recycling would occur when businesses recycled their own waste, residents and street sweepers recycled residential waste; and recycling centers were sited at a density of one per 2 km^2.

In the United States, scavenging is a problem for some urban areas that undertake curbside collection. One solution is to adopt a strictly enforced antiscavenging ordinance. Some communities have chosen to overlook scavenging, as long as it does not create a public nuisance. In Milwaukee, Wisconsin, some citizens had been setting out their aluminum cans separately for years before the city had a curbside collection program, specifically so that scavengers could have easier access to them. A unique solution to a scavenging problem was adopted in Hollywood, Florida, where a scavenger (who would otherwise have been in violation of an antiscavenging law) was officially assigned the areas of the city that could not be served by city crews. In areas with many buy-back centers, there tends to be more scavenging. Buy-back centers have also had to deal with suspected thieves who have stolen valuable commodities like aluminum siding and copper wire from construction sites, guard rails, and, in some cases, from dwellings. If extensive, scavenging can undermine the economics of a curbside collection program since the scavengers typically take the most valuable materials (especially aluminum), the revenues from which are important for helping to offset recycling program costs.

Will scavenging become a greater problem for organized recycling in the United States as world resources become more scarce and costly? The answer to this question is currently unclear.

VOLUME- AND WEIGHT-BASED COLLECTION RATES

Background on Variable Rates

Introduction

Communities across the nation are interested in increasing recycling, reaching goals, and reducing costs. Program planners are considering a variety of programmatic strategies to improve the convenience and availability of recycling and composting opportunities, as well as considering augmenting the materials that can be recycled or composted. Variable rates (VR), a strategy with many advocates, is being adopted in thousands of communities as one possible strategy to induce additional recycling in the residential sector. Under a variable rates system, customers are provided an economic signal to reduce the waste thrown away, because garbage bills increase with the volume (or weight) of waste they put out for disposal.

Variable rates provide an incentive for residents to reduce waste and increase recycling. In most parts of the country, garbage is removed once or twice a week, with the revenues coming from one of two places:

- From a portion of the property tax
- From fixed bills for unlimited pickup—bills that do not vary with respect to the amount of garbage taken away.

Neither of these methods gives residents any incentive to reduce their waste. In fact, with the property tax method, residents never even see a bill, and generally have no idea how much it costs to remove their garbage every week. Areas with these methods of payment have sometimes implemented mandatory recycling programs to reduce the amount of garbage. Available information indicates that variable rates can significantly increase the performance of recycling programs, and lead to greater efficiency and equity in the solid waste management system (Fig. 5.13).

Variable Rates Features and Advantages

These systems are known by a variety of names—variable rates, pay by the bag, variable can rates, volume-based systems, pay as you throw, among others. However, the basic concept underlying all these terms is the same and is very straightforward: those households putting out more waste for collection pay more than those households putting out less. Other key features of variable rates include:

- Waste disposal/recycling behavior and choices affect the bill
- Nonmandatory—you can put out more waste for collection, but it just costs more

Variable rates programs provide a number of advantages for communities and residents:

- *Equity.* Variable rates are fair, with customers who use more service paying more. Customers receive a signal that now links their behavior and the bill they receive.
- *Reduction.* Variable rates reward all behaviors that reduce garbage thrown away for disposal—recycling, composting, and source reduction—unlike recycling programs, which only encourage recycling.
- *Efficient.* These programs can be very inexpensive to implement and, unlike recycling programs, do not require additional trucks down the street. They also help prevent overuse of solid waste services. Rather than fixed buffet-style charges, which encourage overuse of the service, volume-based rates encourage customers to only use the amount of service they need.

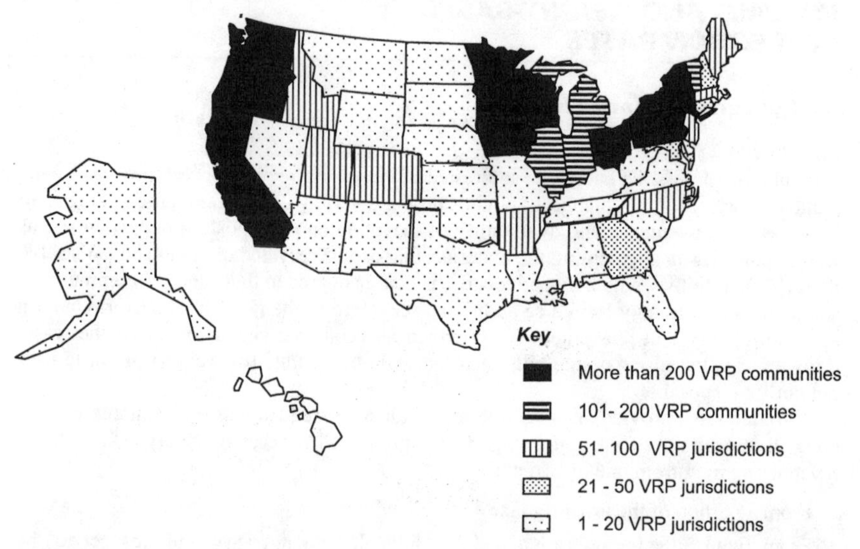

FIGURE 5.13 Variable rates program (VRP) distribution. SERA's 1999 survey has found more than 6000 VR communities covering about 20% of the population. Only 3 states lack programs. (*Source:* Skumatz Economic Research Associates Inc., Seattle, WA, 1999 survey © SERA. All rights reserved.)

- *Effective.* These programs have shown to be very effective in reducing disposal and increasing recycling. They provide a strong "shot in the arm" in reaching diversion goals.
- *Flexible.* These programs have been implemented in a variety of sizes and types of communities, with a broad range of collection arrangements.
- *Environmental and Other.* These programs can be very quickly implemented; one community installed a variable rates program in less than 3 months, although most take longer. In addition, the programs provide environmental benefits.

Types of Variable Rate Systems

The programs are very flexible, and adapt to a wide range of community types (Table 5.17). Variable rates programs can be broken into five major system types:

1. *Variable can or subscribed can.* Customers select a number or size of container as their "normal" weekly disposal amount. These are commonly one can, two cans, etc., or can be set as 30–35 gallons, 60–64 gallons, etc. Rates for customers signed up for two-can service are higher than rates for one-can customers. Some communities have also introduced minican (13–20 gallons) or microcan (10 gallons) service levels to provide incentives for aggressive recyclers.
2. *Bag program.* In this program, customers must purchase specially logoed bags, and any waste they want collected must be put in the special bag. Thirty to 35 gallon bags are most common, but some communities also sell smaller bags at a discounted price. Bags can be sold at city hall or community centers, but even more commonly, communities work with grocery stores or convenience store chains to sell the bags (some-

TABLE 5.17 Advantages and disadvantages of major variable rates system types

Hybrid System	Variable Can System	Bag/Sticker Systems	Weight-Based
Advantages: • Can often use existing containers (which can help limit "scatter") • Can be implemented quickly and inexpensively—easy transition from current collection • No capital investment for trucks, containers • No new billing system needed—continue to bill using current method, but now for more limited service • Can design "base" service amount to community needs—and can modify over time • High customer satisfaction because "out of pocket" can be limited (many will not exceed base units)—and easy transition from current system in customer minds • Can modify system later with little to no wasted expenditure • Stable revenues because "base" paid by all customers • Provides incentive at relatively low revenue risk to the system • Customers only need to buy extra bags/tags for waste beyond their can or base level—less inconvenient than programs for which they have to buy bags for all waste • Multiple haulers can be accommodated using different colored bags/stickers	Advantages: • Multiple can sizes can provide incentives/equity • Using relatively small first container limit can provide good incentives for reduction • Containers are sturdy, tend to reduce scatter • Revenues relatively stable • Possible to use existing containers if sizes are compatible • Experience in larger jurisdictions • Works with automated collection systems • Using standardized containers simplifies enforcement • Billing system can usually accommodate low income, other special services • Can develop rates with very flexible structures for incentives (can develop varying differentials) Disadvantages: • Customers must determine their "normal" service level for billing purposes • Customers must call to change service levels (some hassle) • System for handling occasional "extras" beyond subscribed service must be established (bag, sticker) • If standardized containers to be provided by community or hauler, purchase, distribution, and storage can be expensive	Advantages: • Smaller, more flexible increments of service available—easy to make multiple bag or sticker sizes—harder for cans • No billing system needed except invoicing retail sales outlets • Convenient outlets have been willing to sell bags/tags fairly readily in communities (sometimes without commission in exchange for foot traffic) • Easily handle multiple haulers by using colored bags/stickers • Pure bag/tag systems can be enhanced/modified with "base" customer charge (fixed), which can be easily billed, and can reduce revenue volatility • Bags and stickers are cheap; easily distributed (stickers even easily mailed). They are readily available from multiple firms. • Collection can be very fast—collection staff do not need to return to curb after collection • Collection is "clean"—nothing left on curb • Service is "prepaid" when the bag/tag is purchased. Revenues are received ahead of service delivery. Disadvantages: • Supply and distribution system needed (grocery/convenience stores, etc.)—need to order, distribute, and invoice distributors • Customers must buy bags/stickers for ALL waste (hybrid or can programs have reusable containers for some amount of waste)	Advantages: • More flexible—better recycling incentive for customers because they save for every bit removed from container • Fair and easily understood—customers used to paying for services by increments (water, electricity, etc.) • Flexible on a weekly basis—customers don't pay for can service they don't actually use • Equipment now available, certified—fully automated and semi-automated Disadvantages: • Some systems take additional time at the curb—others don't • No city-wide systems in operation in U.S. to date—many used overseas • Trucks need to be retrofitted with special scales and need to label containers with RF tags (or less efficiently, bar codes) • More complicated billing system needed • Billing procedures need to be established for equipment breakdowns

TABLE 5.17 Advantages and disadvantages of major variable rates system types (*Continued*)

Hybrid System	Variable Can System	Bag/Sticker Systems	Weight-Based
Disadvantages: • Customers don't have incentive to recycle below "base' service level • Need to set up bag/tag system for "extras" beyond base service level; customers need to learn/understand system and where to purchase bags • Customers may not see total cost of garbage system because billed in two portions	Disadvantages: (*cont.*) • Initial complications / administration when customers select initial service levels (billing, delivery of containers) • Coordination required (and expense) as customers want to change service levels • Slower collection—need to return to curb—and empty containers left on curb afterward • Multiple containers can be expensive to purchase, store, deliver/re-deliver, and estimating proportions customers will want up-front (for ordering) can be complicated • Small containers (especially ones suitable for automated / semi-automated col'n) difficult to find • No incentives for recycling below the smallest container	Disadvantages: (*cont.*) • Customers need to store/manage bags/tags and have bags on hand when they need them—need convenient distribution system with long hours • Does not work as easily with automated collection (unless bags are put in cans, which complicates enforcement)) • Revenue uncertainties relatively high—revenues depend SOLELY on number of bags/stickers sold (unless customer charge used in conjunction) • Need to explain to customers how system works and where to get bags/stickers (true for all systems, and for "extras" associated with hybrid and variable can programs also) • Stickers somewhat more complicated to explain to customers (size limits, etc.) • Bags may lead to scatter from animals (ammonia/vinegar in bag can reduce; bags can be put in cans, or stronger bags used) • Recycling not encouraged below smallest bag size (although customers may not put out waste each week) • Stickers are somewhat harder to enforce size limits—some hauler judgment required at curb • Structure of rate incentives is limited—a bag is a bag, so second bags can't be more or less expensive than first bags. Also, large bags cannot be priced with additional penalties—customers would just use multiple small bags	

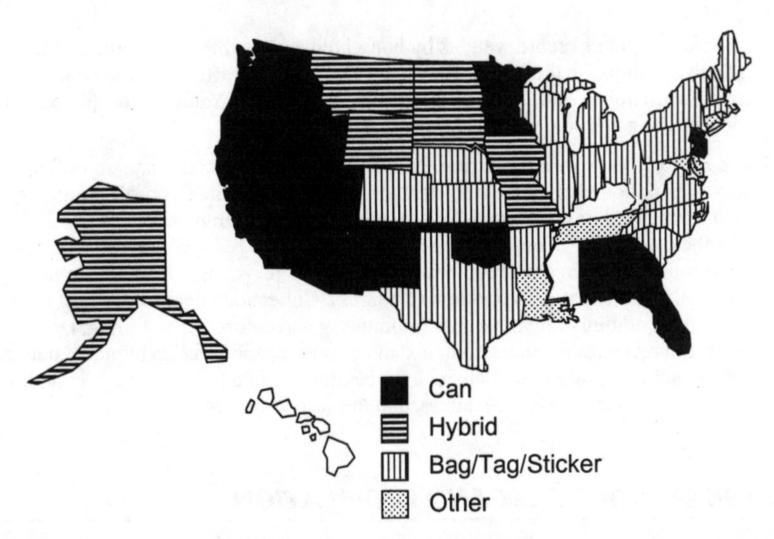

FIGURE 5.14 Most common variable rate (VR) program type by state. SERA's 1999 survey shows regional patterns in program type. (*Source:* 1999 Skumatz Economic Research Associates, Inc. survey. © SERA.)

times with a commission, and sometimes the foot traffic is enough reward to the retailer). The price of the bag incorporates the cost of the collection, transport, and disposal of the waste in the bag (as well as other programs, etc.). In some communities, they use the bag program in conjunction with a "customer charge," and in those cases, the bag price reflects only a portion of the cost of collection and disposal, with the remainder collected through the monthly charge.

3. *Tag or sticker programs.* These programs are almost identical to bag programs, except instead of a special bag, customers need to affix a special logoed sticker or tag to the waste they want disposed. The tags need to be visible to collection staff so they know whether the waste has been paid for. Like the bag program, tags are usually good for 30 gallon increments of service. Pricing and distribution options are identical to the bag program.
4. *Hybrid system.* This system makes a "hybrid" of the current collection system and a new incentive-based system. Instead of receiving unlimited collection for payment of the monthly fee or the tax bill, the system is changed so that the customer gets only a small limited volume of service for the fee. Typical limits for the base service in communities across the country are one can, two bags, or two cans as limits. Limits usually vary based on maturity of the program, disposal behavior, and availability and comprehensiveness of recycling options. Beyond the approved "base" service, customers are required to buy bags or stickers, as described above. Under this program, the "base" service level can be tailored to best suit the community or to achieve a variety of objectives. No new billing system is needed, and bags only need to be purchased for service above the base. Current collection and billing is retained with minimal changes, but an incentive is provided for those who are putting out higher levels of garbage.
5. *Weight-based.* This system uses a modified scale on trucks to weigh garbage containers and charge customers based on the actual pounds of garbage set out for disposal.

On-board computers record weights by household and customers are billed on this basis. Special "chips," called radio frequency (RF) tags are affixed to the containers to identify households, and these are read and recorded electronically on the on-board computer along with the weights for the household.

There may be other variations. Some communities (or especially haulers) offer variable rates as an option along with their standard unlimited system. Drop-off programs, using punch cards or other systems are also in place in other communities.

Using these systems, communities see savings through reduced landfill usage, efficiencies in routing, staffing, and equipment, and higher recycling. However, some negative aspects also arise that are important to address. Collection changes can lead to additional costs; new administrative burdens (monitoring and enforcement, billing, etc.) arise; rate setting and revenues are more complex and uncertain, and significant public outreach expenditures are necessary to implement a successful variable rates program. In addition, specific concerns arise. These are addressed in the section below.

PROGRAM ADOPTION AND LEGISLATION

These programs have become popular in the last decade, and they are in operation in thousands of communities across the United States. Furthermore, many states recommend these programs as strategies for increasing recycling and meeting diversion goals; a few even mandate the adoption of variable rates for communities in the state. Studies have found that:

- The program count and population coverage has increased dramatically in the 1990s. Increasing from about 100 to about 6000 currently, and the programs are available to more than 20% of the U.S. population. The research shows that programs exist in all but three states—Kentucky, Hawaii, and Alabama—and a community in Hawaii is currently considering VR.
- Nationally, can and bag programs are the most common, followed by hybrid programs. Sticker, optional and drop-off programs are somewhat less common. No weight-based programs are currently in full-scale operation in the United States. The frequency of types of programs is roughly ⅓ can, ¼ bag, and ⅙ each for hybrid and sticker programs. A few optional and drop-off variations of variable rates programs are also in place.
- Communities with variable rates programs range in size from about 50 to over a million in population.

Based on surveys with communities, this adoption has been driven by a number of factors, including increasing landfill costs, need to reach diversion goals, reports of successful programs, and legislative mandates, among other reasons. Note that four states mandate variable rates (Washington, Minnesota, Iowa, and Wisconsin), and a majority of the other states offer workshops, grants, and other strategies to encourage variable rates (Fig. 5.15). However, these programs are not suitable for all communities. Appropriateness for a community is affected by:

- Disposal issues and costs
- Availability of recycling and yard waste programs
- Markets for recycled materials
- Collection system type

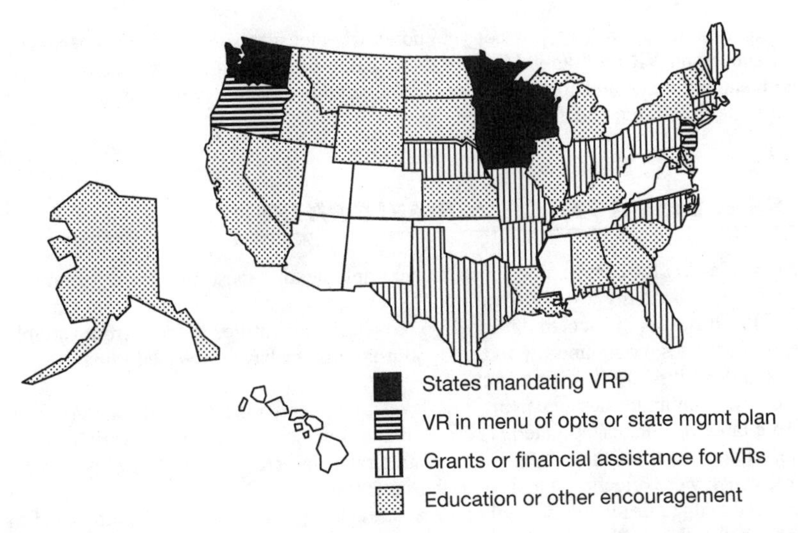

FIGURE 5.15 Variable rate (VR) legislation. Research found that four states mandate VR and more than 25 others offer incentives or encouragement. (*Source:* 1999 Skumatz Economic Research Associates, Inc. survey. © SERA.)

- Costs
- Acceptability and support
- Timing, economics, local factors, and other issues

Certainly, there are communities with many of these challenges that have successful programs, and others with similar conditions that have elected not to implement programs. Local priorities play an important role in these decisions. It is likely, however, that it makes sense to *examine* variable rates in all communities, to determine whether it is appropriate now or possibly later.

Communities often find it convenient to implement VR programs when they are implementing other changes; for example, changing collection system or contractors, modifying the recycling programs, or other changes. Key characteristics of successful variable rates programs include:

- *Rates that vary and provide incentives.* It is important that the rates for additional service levels be different enough to provide incentives, and that relatively small service levels be available to reward small disposers. As an example of rates that don't work, a community in California provides a minimum of 90 gallons of service for about $17, and customers may also select an additional 90 gallons of service for $1 more. This arrangement does not provide a significant incentive for recycling.
- *Recycling and diversion programs.* To make programs successful, customers must have access to moderately convenient recycling programs, composting options, source reduction information, and other alternatives to disposal.
- *Customer education.* Customer education about the program, alternatives, and how to make the program work is important to the success of VR programs.

- *Mandatory service or not.* It is not yet known whether mandatory service is necessary for successful VR programs. Many of the existing programs are in communities both with and without mandatory collection/fees. Certainly, revenue issues are simplified in communities with mandatory service.

POSSIBLE IMPACTS OF VARIABLE RATES

One key question for communities considering implementing these systems is "what will the advantages be to my town?" A variety of sources cite dramatic impacts from variable rates. EPA attributes impacts of 25% to 45% reduction in landfilled tonnage from variable rates. Other states, communities, and trade journal articles have published impacts from variable rates from 20% to over 60%.*

However, communities expecting this level of impacts on landfilled tonnage from variable rates will be disappointed. These estimates overstate the effects of variable rates alone, as they combine the effects from variable rates and recycling and yard waste programs, which were often implemented at the same time.

Specialized studies were undertaken to isolate the impacts that could be attributed to variable rates alone—that is, to identify the extra recycling and landfill diversion that would result from variable rates separate from additions or changes in recycling or yard waste programs. The key impacts communities have found from implementing variable rates fall into the following categories.

Tonnage shifts away from disposal—recycling, yard waste, and source reduction increases. The incentives from the program cause customers to try and reduce disposal and to increase recycling, composting, yard waste diversion, and source reduction. Although some report that variable rates lead to disposal decreases anywhere from 25% to 60%, this information is misleading because many of the communities made several changes at once, and variable rates was only one of them. Many implemented new recycling programs at the same time, so it would be a mistake to assume this large a decrease would be derived from implementing a variable rates program only. To provide more realistic expectations about the impacts of VR programs, several studies were conducted using data gathered from over 500 communities across the nation to clarify the impacts that could be attributed to the variable rates programs only. These studies found that variable rates decreases residential disposal by about 15% in weight, with 8–11% being diverted directly to recycling and yard programs. The reports also found that:

- 5–6% percentage points go to recycling (with similar increases for both curbside and drop-off programs)
- 4–5% go to yard waste programs, if any

*See for example, Solid Waste Program Development Unit, "A Guide to Volume-Based Fees for Garbage Collection," Minnesota Office of Waste Management, St. Paul, MN; Good, Linda, "Annual Report on the Borough of Perkasie: Per Bag Disposal Fee, Waste Reduction, and Recycling Program for the Year 1988," Borough of Perkasie, PA, 1989; Blume, Daniel, "Under What Conditions Should Cities Adopt Volume-Based Pricing for residential Solid waste Collection?," Master's Memo Study, Duke University, prepared for the Office of Management and Budget, Office of Information and Regulatory Affairs, May, 1992; EPA, "Pay As You Throw Manual," U.S. Environmental Protection Agency, Washington, DC, 1994; Skumatz, Lisa A., "Forecasting Solid Waste Tonnage: Techniques and Alternatives to Estimate Tonnage, Revenues, Source Reduction, and Program Performance," Skumatz Economic Research Associates, Inc. (SERA).

- About 4–7% is removed via source reduction efforts. This includes buying in bulk, buying items with less packaging, etc.
- The quantitative work also indicates that the impacts from variable rates were the single most effective change that could be made to a curbside (or drop-off) program. Implementing variable rates had a larger impact on recycling than adding additional materials, changing frequency of collection, or other changes and modifications to programs.
- These results are confirmed by other work. For instance, a survey in Iowa found that recycling increased by 30% to 100%, and averaged about 50%. When adjusted to the percent of the total waste stream instead of considering just increases in recycling, these results are very comparable. Recently completed work on California communities estimates the impact to be 3–4% for recycling and 3–4% for yard waste, for a total of 6–8% to programs. Surveys conducted by several universities and others also confirm the preliminary source reduction results—customers report taking the rates system into consideration when they are making decisions at the grocery store.

Reduced set-outs. The research finds that set-outs reduce dramatically—from 90 gallons to 30–45 gallons in many communities that have active recycling programs. Some of this is accomplished through actual tonnage reductions as the method above, but additional decreases are due solely to compaction—the "Seattle Stomp." Set-out decreases are important because they reflect the new unit of revenue, and are crucial to rate setting.

System cost impacts. This is difficult to measure, not only in general, but also even for specific communities. Communities rarely make only one change independently, and it is difficult even for them to attribute the cost impacts directly due to variable rates. As mentioned before, surveys in two states found that 60–65% of the communities reported that costs stayed the same or decreased after putting variable rates programs in place.

These results indicate that the potential additional diversion from variable rates may be on the order of 15% overall (Fig. 5.16), combining impacts from increased

- Recycling
- Yard waste diversion
- Source reduction

The diversion work conducted found that the impacts from variable rates were stronger than the impacts from "tweaks" to existing recycling programs—including adding new materials, collection frequency, and other changes. Further, SERA examined the costs from implementing these programs, and found that they were lower than most other recycling program changes, making it one of the most cost-effective changes available for increasing recycling.

Finally, the VR programs work well in a variety of situations, with successful programs currently working in conjunction with:

- Curbside and drop-off recycling and yard waste programs, commingled or separated programs
- Collection provided by municipal or hauler staff, under franchise, license, or purely private/competitive arrangements
- Large and small communities, in urban and rural settings
- Gargage pickup using manual, semiautomated, and fully automated trucks

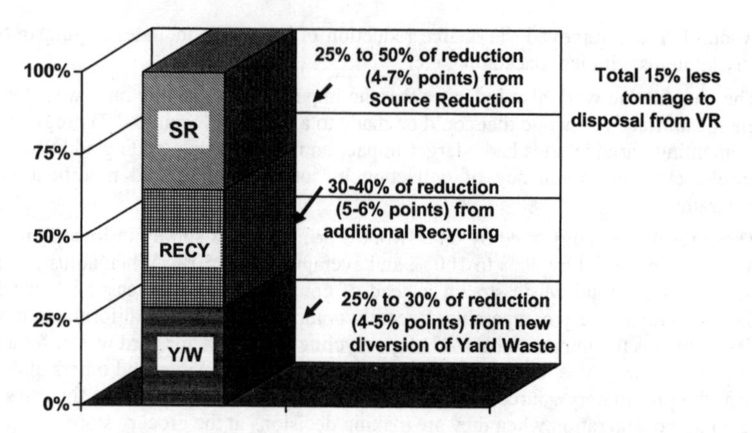

FIGURE 5.16 Impacts attributable to variable rates. Estimates of VR impacts show 15% less tonnage disposed—roughly ⅓ to recycling, ⅓ to yard waste, and ⅓ to source reduction. (*Source:* Skumatz, SERA. © 1998, all rights reserved.)

CHALLENGES OF IMPLEMENTING VARIABLE RATES

Although there has been a great deal of enthusiasm about these programs, they have drawbacks as well. Key concerns regarding variable rates focus on several major issues.

Illegal Dumping

Illegal dumping is one of the first worries when communities consider going to variable rates. However, in reality, dumping does not appear to be a serious problem. Illegal dumping exists in communities now; the question is whether illegal dumping will increase significantly in response to new variable rates systems. One issue complicating determining whether or not it is a problem is that very few communities have quantitative information on how big a problem illegal dumping is before they put in new rates. And since illegal dumping is almost always a fear, illegal dumping will be noticed, whether or not it actually increases over pre-variable rates levels.

Several studies have attempted to address the illegal dumping issue. In a survey of public officials in 10 Illinois communities with variable rate systems, respondents were asked to rank the dumping problem on a scale of 1–5 with "1" indicating the issue was not a problem. Illegal dumping along roadsides rated a 2.39; dumping into commercial and government dumpsters rated a 2.90. In a study of 14 cities, others found 42% reporting no problems, 29% reported minor problems, and another 29% reported notable problems. An analysis of contributing factors found that three of four communities with problems were rural, but not all rural communities in the sample had problems. Areas without easy methods for disposing of bulky items also had more difficulties. A Reason Foundation study of eight Massachusetts communities found no problem in five, and only minimal problems in two. One had most significant incidences of roadside dumping, but the community speculated it was actually from a neighboring community with high disposal fees.

A detailed report on illegal dumping and variable rates examined several kinds of data to identify whether illegal dumping has been found to be a problem. Surveys showed low incidence of illegal dumping problems. From interviews with over 1,000 communities that implemented variable rates programs, the report found that less than ¼ reported actual problems with variable rates, and all said that the problems were short-term and easily dealt with through fliers and education. A small percentage insisted that the focus and attention on illegal dumping actually helped them solve the problem and the situation improved. All of the communities felt that fears of illegal dumping should not be a deterrent to variable rates and that a variety of effective enforcement options were available to address the problem. Follow-up interviews with haulers showed that there was some initial increase in bags alongside commercial dumpsters, etc., but lockable dumpsters usually solved this problem. All communities recommended fines and visible enforcement.

Residential waste is not a large component of illegally dumped material. However, the most compelling information uncovered by the study was an examination of the composition of illegally dumped waste. Illegal dumping associated with VR programs was not significant. However, it was recognized that it is hard to find communities that tracked dumping before *and* after implementing VR. To try to determine if illegal dumping is a problem associated with VR. A study gathered information on the composition of illegal dumping and found over 75%–85% of it was nonresidential in origin (commercial waste). The largest components were C&D (over 25%), brush (almost 40%), and—the only important component with household origin—bulky items, sofas, and appliances (white goods). Therefore, communities recommend having a convenient bulky waste program to increase the success of the VR program and minimize the incentives for dumping these awkward materials.

Prompt cleanup, bulky waste programs, lockable dumpsters, burn bans, fines, and other strategies are recommended to reduce illegal dumping as a problem from the variable rates system. If a community is concerned that illegal dumping may be a problem, variable can or hybrid programs, which include some base level of service for all customers, may reduce the incentives for illegal dumping. The report includes a list of "tips" to help reduce illegal dumping concerns.

Other Problems and Solutions

Administrative workloads and costs. Concerns about costs are an issue for many communities. This is difficult to estimate because the costs depend on the current system in place and the type of system that the community is considering going to; that provides an impossible number of combinations for which to provide general cost guidance. However, two states, Wisconsin and Iowa, conducted surveys simply asking communities whether solid waste management and administrative costs increased, decreased, or stayed the same after putting in variable rates. Almost two-thirds of the communities indicated costs had decreased or stayed the same. This provides good evidence that implementing variable rates does not have to be expensive, and that appropriate system choices can be made to minimize the costs. In all cases, long-run costs were expected to be lower, but these surveys show that even short-run costs (even within a budget cycle) can be manageable.

Low-income customers and large families. Concerns are raised that the program is perceived as unfair to large families. It is important to separate concerns about large families from concerns about low-income households. Addressing just the large family issue, consider turning the argument around. Has it been fair all these years for small disposers to be subsidizing large disposers under fixed bill (or nearly fixed bill) systems? Opportunities to reduce waste are available to all households (recycling, etc.) and those who limit their waste can get control over a bill they previously could not reduce. Although there is some relationship between family size and amount disposed, all households have opportunities to reduce. In most communities, large households do not generally receive discounts on water service, groceries, or other services that might also vary by family size. However, subsidies for large families are not well justified. One area in which this concern may be more important is the combined impact on large, low-income families. SERA conducted a specialized study of low-income strategies. After analysis of rates and policies from hundreds of communities to identify programs with special considerations or rates for low-income customers. In a detailed report on this topic, it was found that low income or elderly discounts are provided in less than 10% of communities with variable rates. The communities provide discounts from 10% to over 60%. Eligibility is most commonly certified by mail and the assistance is provided for can, bag, sticker, tag, and hybrid systems. Low-income issues can be addressed through differential rates for "qualified" households, and through distribution of free or reduced-cost stickers or bags along with other assistance programs.

Revenue uncertainty. Variable rates programs, because they depend on customer behavior choices, will inherently lead to more volatile revenue streams than systems with fixed bills. This is very commonly a concern both for haulers and for municipalities. Revenues are no longer based on a stable number like households, but rather on the number of individual bags or cans of waste sold or disposed. The number of bags disposed can vary month-to-month and week-to-week, based on diversion program availability, seasonal factors, advertisements and promotions, and many other factors, and this can cause significant revenue headaches. However, a much greater source of concern is determining up front, before the program goes in place, the average amount of service that will be used by customers. This is vital for initial rate setting, and making sure that rates are established that will provide sufficient revenues to fund the solid waste management system. Appropriate rate setting is more complicated, but many firms have experience in this work. However, uncertainties associated with this process can be significantly reduced if data are available on current set-outs (volume of garbage and weight of waste set out for collection), remaining recycling potential in the sector, and other information. There are differences in the relative revenue volatility associated with different variable rates programs. If revenue uncertainty is a primary concern, systems with less volatility include variable can, hybrid programs, or bag/tag programs that include a customer change.

Multifamily buildings. Although the systems have not historically been available for large apartment buildings with shared "chutes," they are routinely implemented in garden apartments, townhouses, and apartments of about six or fewer units. Recall also, that larger multifamily buildings are already receiving a volume-based signal (although at the building and not tenant level) through dumpster charges, which are charged based on cubic yards of service. However, new hardware has become available that provides a workable variable rates system for large multifamily buildings with combined garbage chutes. Tenants push a button for garbage or recycling (up to six different streams). This makes recycling and garbage collection equally convenient; increases in recycling are 30%–300%, and payback is on the order of 3 years. More than 200 have been installed in new and retrofitted buildings to date, mostly in Florida and New York, and have led to significant increases in recycling and decreases in disposal. In addition, suggestions for variation on variable rates incentives that encourage recycling are being tried in communities across the nation. These recent developments show promise for removing a barrier to economic incentives for multifamily residents.

Acceptance and other concerns. Customers routinely view the programs as fair, and they end up being very popular with residents after the fact. Finally, any change always leads to confusion and resistance to change. Public education is strongly emphasized by all communities to improve success of the variable rates program. All systems establish weight limits for the cans and containers, to address both safety and equity concerns. Most importantly, even though there is generally resistance to change prior to implementation, numerous surveys have indicated that these programs are perceived as fair and are very popular after they have been implemented.

In summary, technical issues are seldom the problem in implementing variable rates. Variable rates programs have tremendous flexibility in their design and can usually be tailored to accommodate most concerns (Fig. 5.17). Instead, political will is usually the largest stumbling block to implementing variable rates programs.

WEIGHT-BASED RATES

Although the number of variable rates communities has been increasing dramatically, and the incentives are considerably improved over fixed fee systems, volume-based rates have several weaknesses. Many of the systems base charges on subscription rather than usage, with variable-can customers paying for a set number of cans on a weekly basis, whether or not the containers are filled. Other variable rate programs provide no incentives below the smallest can or bag size available. Weight-based systems offer customers stronger incentives, and provide fair, informative billing. They encourage all recycling and reduction efforts, without requiring a variety of different amounts for different materials and waste streams. Advances are promising for a number of systems, and various forms of the

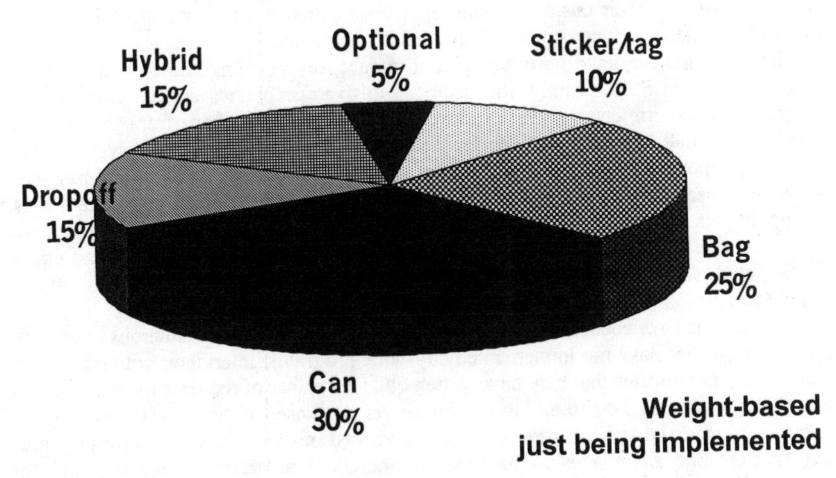

FIGURE 5.17 Frequency of variable rate (VR) program types. (*Source:* Skumatz, SERA survey, © 1997.)

equipment have been pilot tested in more than two dozen communities across the United States and full-scale programs in Australia, Denmark, Germany, and other countries. The first test, in Seattle, was nicknamed "Garbage by the Pound."

To make these systems work requires several basic components:

- *Weighing mechanism.* Scales have been retrofitted onto automated and semiautomated trucks.
- *Identification methods.* Generally, cans are labeled with radio frequency (RF) tags, although bar codes and coded route sheets have also been used.
- *Data storage and transfer.* On-board data storage is needed; the data is transferred to the billing computer via radio or direct download.
- *Billing system.* The billing program needs to be more complicated than traditional solid waste billing programs, but the needs are almost identical to those needed to bill for water service.

As of 1996, three companies had weight-based equipment certified as legal for trade and charging for variable rates. These include retrofitted semiautomated tippers with and without stops in the dumping cycle, fully automated tipping arms with hoppers, and commercial dumpster weighing systems. The systems are sold by Toter, Cardinal, and MCC. Depending on a variety of assumptions, residential weight-based systems may have paybacks of between 6 months to over 9 years, depending on landfill rates, and system types.

Pilot test communities in the United States and Canada include tests in Seattle (the first, in 1989). Columbia, SC, Lake Worth, FL, Durham, NC, Victoria, BC, Mendham Township, NJ, Milwaukee, WI, Farmington, MN, Minneapolis, MN, and others. Several haulers in Florida and Ohio are working with commercial weighing systems and are noting that the paybacks can be very rapid. They are finding that restaurant dumpsters are relatively heavy compared to office waste. Most are not strictly charging based on the week-to-week weights, but are using "averages" for the customer to determine more appropriate rates, including more appropriate dumping fee portions. One hauler met with their customers to renegotiate rates for the heavier customers. They were able to retain more than 90% of their customers, but significantly improved their bottom line, since they had been losing money on the "heavy-load" customers.

The programs can have fairly significant impacts on recycling and diversion. Based on data from the Seattle study, the decrease in average pounds set out from a pilot weight-based experiment was 15% (above and beyond the decrease from the volume-based system that had been in place for 7 years prior to the weight-based system). The customers reported their favorite features of the program were that they could pay only for what is used, see clearly what they were paying for, save money on garbage bills, and paying less than those overstuffing their cans. Least favorite features included: costs might increase, cans might be weighed incorrectly, the program is complex, and others might put waste in their can. Most of the concerns are parallel to those for volume-based systems.

The systems have been ready for some time, and there have been numerous pilot tests, but municipalities have not implemented city-wide programs. Interviews with equipment manufacturers indicates that they have gotten quite a number of requests for bids for systems by cities across the nation, but none have yet purchased systems for their cities. Interest in commercial systems seems to have advanced more rapidly, and since they provide bottom-line advantages to haulers, commercial systems may lead the way for municipalities. Given the significantly improved incentives, labor savings, flexibility in the systems, the move toward automation, and the programs in place overseas, residential systems may not be far behind.

EVALUATING AND IMPLEMENTING VARIABLE RATES

Evaluating Specific Variable Rates Systems

Certainly, each community needs to analyze whether variable rates makes sense for them, and if so, identifying which type of program provides the best fit. Certainly, each of the variable rates systems has pros and cons or, presumably, the systems would not be in place in communities around the county. The mix of pros and cons makes some more suitable for particular communities and their priorities than others, and the lists of important criteria often include:

- Increase recycling/decrease disposal/signals
- Equity/fairness
- Low implementation costs/easy transition
- Lower long-run costs for solid waste system
- Minimum disruption to operations/collection
- Revenue certainty/minimize volatility
- Flexible system/adaptable over time
- Customer acceptability issues/easy to explain
- Low incentives for illegal dumping
- Ongoing enforcement is low cost/easy
- Track record/well demonstrated success
- Easy to bill or no billing required

Relative Implementation and Administrative Costs

It is almost impossible to provide "rules of thumb" about how expensive it is to implement variable rates programs. The result hinges critically on the type of solid waste management and collection system in place in the community before, and the type of program they want in place afterward, and every community is different. Surveys asking about the changes in overall cost from implementing variable rates have been cited before. These studies found that for two-thirds of the communities implementing variable rates, they found costs stayed the same or decreased. From the Wisconsin and Iowa survey only one-third had an increase in costs. This finding indicates that: 1) these programs do not have to be expensive to implement, and 2) communities can find program types that fit well with their existing (or planned) solid waste management system.

The relative costs for implementing and operating the various variable rates systems are outlined in Table 5.18.

Local communities can assess the changes needed for their system, and help narrow the systems that are most suited for their needs. Using these same steps—analysis of key priorities and relative implementation burdens, communities may come to very different conclusions as to the types of programs that will work best for them. However, there are some patterns. For example, research finds that the percentage of variable can systems, for instance, is higher in urban areas, and bag programs are more common in suburban areas. This may relate to the greater prevalence of automated collection in urban areas (compatible with variable can programs) and concerns for low cost imple-

TABLE 5.18 Factors implementing costs for variable rates systems

- *Phone/customer service costs:* When changes occur, customers call with questions. Some communities handle the extra phone traffic as part of normal work, or are able to absorb it by deferring other administrative work that is not time sensitive. Some communities have been able to use staff from other municipal calling centers (water department, etc.). Others, with more elaborate changes, may require additional phone lines and new temporary staff for a month or two. This also includes costs related to on-going billing questions. Training of these staff may also be necessary.
- *Billing:* The systems differ in the need for new or enhanced billing systems to accommodate the new methods for billing. Variable can and weight-based systems need more complex billing systems than other program types. However, billing under bag/tag/sticker systems may be simpler to bill than current fixed bill systems because communities and haulers no longer need to bill individuals, but rather, they only need to invoice bag/tag distributors, like the grocery stores, etc.
- *Service level selection:* Under variable can programs, customers need to select a "basic" service level. This requires sending forms for customers to fill out, and entering the information into the billing system for each household. Other programs do not require this step.
- *Trucks and equipment:* Weight-based systems require changes to collection vehicles. However variable can programs can work with manual, automated, or semi-automated systems. Bag and sticker programs do not work very well with fully automated collection.
- *Containers:* Some systems include the purchase of new, uniform containers (variable can programs). Note that some communities, even with variable can programs, allow customers to use their own containers—but specify the size limits that are allowed. Under fully automated collection, it is still somewhat difficult to find stable, small-sized containers that will provide incentives for small disposers. Some communities with wind issues are using "inserts" to make larger containers smaller, but these increase costs.
- *Bag or tag purchases:* This includes designing, ordering, and storing bags or tags. Note that, on a smaller scale, bags or tags are needed for "extra" waste for several systems (including variable can and hybrid programs).
- *Bag or tag distribution:* This includes finding and negotiating agreements with grocery stores or convenience stores to sell bags or tags for the program. It is suggested that these outlets be used, and that communities not try to sell them through community centers or City Hall alone. Note that some towns have found that commissions for the sale of these items is not needed; in other communities, commissions on the order of 10% have been attached.
- *Advertising and outreach:* Costs of about $2–3 per household will be incurred in public education about the new program, including newsletters, bill inserts, or other media. Not one community that has been contacted on this issue wishes they had done less outreach. Suggestions on "message" are provided in a later section.
- *Service level enforcement:* Some programs are almost "self enforcing" (e.g., variable can systems with city-provided containers), and others may need more aggressive enforcement to assure that customers are not getting more service than they are paying for.
- *Illegal dumping enforcement:* Some of the programs lead to somewhat greater incentives to dump waste illegally (potentially bag and tag/sticker programs). For these systems, it may be appropriate to institute higher levels of enforcement of illegal dumping than would be needed with other programs.
- *Collection staff training:* A few of the systems may require modifications to the way in which the waste is collected, and others may require additional or new duties (for example, weight-based programs).
- *Rate study:* Rate studies will be needed to set rates. Variable can programs allow more flexibility in incentive structures, and therefore, have more complexities associated with the rate setting efforts. More information on the "distribution" of can sizes is needed to support a rate study for variable can rates than for a bag program, for instance.
- *Recycling and diversion program:* The recycling and diversion programs may need additional capacity to handle the increase in tonnage. This may mean more routes and staff and trucks; it may mean greater operating hours or additional capacity at processing facilities; or it may mean more frequent collection at drop-off sites for example.

mentation in rural areas, in combination with a variety of other community-specific factors.

RATE SETTING/DESIGN ISSUES AND TRADEOFFS

The final rate levels can have a major effect on the success of the VR system as well as the recycling and diversion programs. Each system type presents its own rate-setting opportunities and challenges but there are several rate setting issues that are common to all the systems.

Background

Rates accomplish two basic functions:

1. Recovering revenues
2. Creating incentives for customers to handle their solid waste as efficiently as possible

The amount of revenues the solid waste agency requires is determined locally based on operational costs, including salaries, equipment, facilities, disposal, and other costs. Traditionally, this amount of money is collected through a transfer from community's general fund or through billing customers a fixed rate. However a variable rate system provides flexibility in setting a rate that both pays the costs and acts as an incentive to reduce the amount of waste disposed. The range of incentives that a variable rate system will provide is determined by the final rate design.

Because of these dual functions of solid waste rates, it is critical that planners review their solid waste goals and priorities during the rate setting process. There is no inherently best way to design rates, and choices will need to be made based on an assessment of key priorities.

The process for setting rates requires several technical steps to be performed including population forecasting, waste generation forecasting, cost allocation, an economic analysis of the impact of the rate on waste generation, and other steps. While some communities may have in-house expertise, others may wish to hire outside expertise.

Key Steps and Policy Choices

There are three key questions crucial to setting appropriate rates. These include:

1. How much money is needed to cover cost
2. Number of paid garbage set outs (bags, cans, etc.) that will be set out for collection
3. What should the structure of the rates be to provide appropriate incentives?

The easiest way to estimate the costs for the new system is to look at current costs, and adjust for the types of changes that the variable rates program will bring. Guidance on some of the most important new costs has been provided elsewhere. The basic method for determining the number of bags or cans of waste is to examine current set-outs, and make two key adjustments: reduce the total tonnage by the amount that you expect will be diverted to recycling and yard waste programs and to source reduction. This adjustment may be something close to the 15% estimated in the section on "Impacts," or may be more than this figure if major changes are made to enhance the recycling or yard waste programs at the same time. The second adjustment is to estimate the amount of compaction that will occur. When a new volume-based system is implemented customers have an incentive to "stomp" the waste to put more waste into smaller (cheaper) containers. One additional factor that may be important is examining whether customers may have ways to avoid the system. Based on your weight limits per bag or can, you can then translate this waste volume into a number of bags or cans per household per week. It is prudent to adjust the expected amount of waste downward if customers can easily bring waste to a transfer station directly, etc. (for example in urban versus rural areas).

Setting rates for the bag program can be computed with the information available already. For example, the revised total revenue requirements on a per-household basis can be divided by the number of average bags estimated to be set out per month to determine the per-bag rate.

But the third major ratesetting question has to do with rate design, incentives, and acceptability. Setting appropriate rate levels requires balancing incentives against revenue security. Higher recycling incentives are provided through higher rate levels, bigger rate differentials, and smaller containers or increments in service levels. However, the structure of cost for providing service tends to work counter to stronger incentives. The greatest cost is the fixed cost of getting the truck to residents, regardless of how many cans or bags of waste are collected. To construct greater incentives requires shifting some portion of the fixed costs to higher can rates. If, however, the community is very successful in using rates to reduce disposal, they may find that they will have fewer garbage set-outs than they predicted and they will experience a shortfall in revenues. Fiscally conservative rate designs would have relatively small rate differentials. The low differentials is that they look very similar to flat rates, provide low incentives, and are not worth the extra administrative burden; flat rates might as well be maintained. The rate-setting key is to balance incentives with revenue risks.

Rate Levels, Steepness, and Program Fees

If a community finds the calculated bag rate to be unacceptable, it may make adjustments by introducing smaller bags as an option or introducing a "customer charge" to carry some of the burden of the fixed costs of the system. Other rate policy options and choices include:

- Steepness of rates for can programs
- Whether to incorporate recycling and program changes into the rates or to list them separately on the bill ("embedded" versus "line-itemed" program fees)

Rate Steepness

Variable solid waste rates provide economic signals to avoid using more service than needed, and reward those customers putting out less garbage. The signals come from two sources: the dollar level of the rates and the relative rate differentials. The percentage rate differentials represent the relative "extra" fee charged for extra containers of waste. Within limits, higher differentials tend to provide greater incentives to reduce waste and recycle—the user saves money. Higher differentials mean changing behavior and recycling more save more money.

Relatively high rate differentials provide incentives for recycling, but they have two other very important impacts:

TABLE 5.19 Summary of implementation and acceptability lessons from variable rates communities

Surveys of communities with VR programs provide suggestions for successful implementation:	A number of strategies can be used to help make rates more acceptable:
• *Plan for success:* Expect the recycling program to receive more tonnage throughput, and anticipate reductions of 15% for disposal. Expect that customers will downsize their containers because of the new incentives, and provide small-sized containers that will provide options for avid recyclers.	• *Provide options:* Continue to offer "premium" and optional services, but institute premium fees to help recover the costs of providing these more expensive services. This helps reduce complaints about reductions in service.
• *Pilot test:* Consider a pilot test to refine program up front and learn implementation lessons on a small, inexpensive scale, with time to modify the design for greater efficiency, lower cost, and greater success. Another option is to phase in implementation, gradually adding sections of the city.	• *Recall bills are paid, not rates:* Keep in mind that the actual amounts that customers will pay is not the rate levels per can or bag, but a contained impact from both the rate levels and their choices about behavior-rates times the number of set outs. They have control over their own set outs, and thus, have some control over their own bills.
• *Education is a* crucial component the program, both up-front and on an on-going basis. Research indicates that for communities in which implementation of VR "failed". It seemed that inadequate information/education was a crucial problem. Some communities spent very little per household and customers didn't understand how to make the systems work for them.	• *Keep "other" changes to a minimum:* If a community is trying to introduce a new variable rates program, the chances of success are improved if they do not implement expensive transfer station changes the same year. VR will be blamed for these extra costs, regardless of whether they are really responsible.
• *Political support is important:* Implementation timelines very from several months to several years, and program complexity is not generally the key factor in determining timelines. Instead, the level of effective political support seems to be the single most important factor in timely and effective implementation of VR.	• *Keep rates stable for* a period of time to increase confidence: It may be that the community will have mis-estimated its initial rate levels. It may be worth keeping rates stable for a year, drawing from reserves of the general fund, to build confidence in the system, and correct the rates in conjunction with other changes in the next year.
• *Involve many in decision-making:* Communities have found that involving a number of players in the decision-making process can increase acceptance and smooth implementation of the system. Stakeholders that have been included successfully include haulers, politicians, environmental groups, recyclers, and citizens (for example, through solid waste advisory committees).	• *Consider implementation changes* in the recycling programs to augment options: Enhancing the recycling programs with additional materials or other changes can help decrease the perception of less service and higher rates, and makes it easier for customers to work with the new system.
• *Program flexibility—offer choices:* Introducing a new system that appears to require lower service and higher fees with no options will be a tough sell. It is important to offer program alternatives (recycling, etc.), smaller service levels at reduced prices, and similar options to allow citizens to reduce their rate burden if they reduce the waste they set out for disposal.	
• *Convenient bulky collections* and other collections to reduce illegal dumping and make the program work for residents. These may be annual free cleanups, collections by appointment, collection with payments, or other options.	
• *Consider local conditions* make the program fit local collection systems and priorities, and try to retain characteristics of the current system that are popular with customers. "Cookie cutter" programs that work in some other community often need adjustments, information from several communities, and ingenuity to fit local conditions.	

1. *Revenue risks*. The vast majority of the costs of providing solid waste collection and disposal service are incurred in "getting the truck to the door," regardless of how much waste is collected. That is, there are high "fixed costs" in collecting waste, and it costs less than twice as much in labor, equipment, and disposal costs to get rid of twice the number of cans of garbage from a residence. However, the true cost relationship runs counter to the desire to structure significant price incentives for putting out less waste and recycling. Creating significant incentives requires shifting some of the fixed costs of waste collection to the variable portion of the charge. Higher incentives mean higher proportions of the fixed costs that have been allocated to "higher can levels", and the greater risk that the fixed costs of garbage collection will not be collected in rate revenues. This feature—revenue risk—is one of the pressures that argue for keeping rate differentials lower.
2. *Potential incentives for illegal dumping*. Higher rate differentials provide strong incentives for customers to reduce the amount of waste set out for collection, through recycling, source reduction, and littering, and, for some, illegal dumping, disposal in others' containers.

Therefore, a balance is needed between the incentives for greater reduction and the revenue and illegal dumping risks associated with aggressive rate structures.

A survey of rates across the nation finds that variable can rate differentials vary from just a few percent between "can" levels—10% to more than twice as much for additional cans. If the differentials are 100%, that is, two cans cost twice as much as one can, that is called "can is a can" pricing. Higher differential provide incentives, but increase revenue risk and incentives for illegal dumping. However, if certain "thresholds" in incentives—dollars or percentages—are not met, the switch to variable rates is probably not worth it because it will likely not modify customer disposal and recycling behavior.

Embedded versus Line-Itemed Program Fees

Some communities charge separately for their recycling programs (using a mandatory "line item" charge), and others embed the costs of the recycling program into a combined fee for garbage. This is almost purely a policy choice, with arguments on both sides of the question. Separate fees let customers know that recycling is not free, just cheaper than garbage collection (in many communities). It also diversifies the revenue source. Separate fees also provide a mechanism to keep garbage rates low. However, "embedded" fees can provide a way to increase the rate differentials and provide stronger incentives. Both strategies have advantages and disadvantages. The major impact of this policy choice is that under embedded fees, lower can users will pay a higher total bill than they would if the recycling rates were embedded in the garbage fees. Again, either option is acceptable, and can be justified (Table 5.18). It may be argued that low disposers are probably larger users of the recycling program and should pay more; others argue that they want higher penalties for large disposers. In that case, embedded fees help achieve that objective.

SUMMARY AND CONCLUSIONS

Variable rates have become very popular, and are currently in place in over 6000 communities and over 20% of the population in the United States (Fig. 15.18). Community experience shows that the programs provide a number of advantages, including:

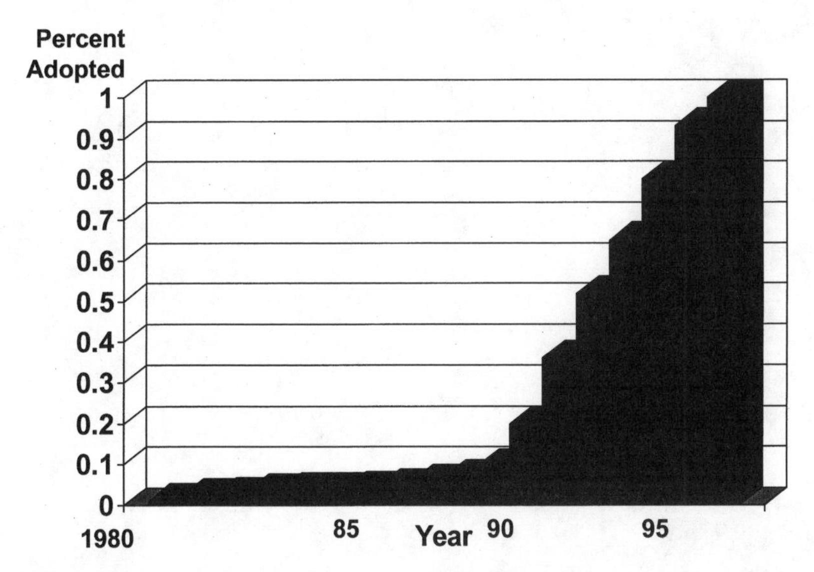

FIGURE 5.18 Community adoption of variable rate programs by year. (*Source:* Skumatz Economic Research Associates, Inc., SERA, Seattle, WA. © 1997.)

- VR programs are fair
- They provide customers with options and control over a bill they didn't have control over previously
- They are flexible, providing options and designs that fit a wide range of community and collection types
- VR programs provide consistent and recurring signals, reminding customers to reduce their disposal each time they pay a bill or buy a bag/tag/sticker

CHAPTER 6
PROCESSING FACILITIES FOR RECYCLABLE MATERIALS

THOMAS M. KACZMARSKI
*Former Manager, Recycling**
Waste Management, Inc.
Oakbrook, Illinois

WILLIAM P. MOORE
Vice President, Paper Recycling International, Inc.
Norcross, Georgia

JOHN D. BOOTH, P.E.
Director of Engineering, Solid Waste
Authority of Palm Beach County, Florida

INTRODUCTION–DIFFERENT PROCESSING ALTERNATIVES

Processing material from the municipal solid waste (MSW) stream is becoming more an art than an exact science. Factors such as demographics, collection practices, disposal costs, end-market uses, and most importantly, the desires of the residential or commercial customers being serviced, will determine the type of processing facility used to most effectively process the material. For example, at one end of the spectrum, there are small 10 ton per day (TPD) facilities accepting residential "source-separated" materials. These facilities may have nothing more than a hand sort conveyer, an aluminum can flattener, and a portable scale. At the other end are large-scale, mixed MSW facilities that can accept and process up to 2000 TPD of residential and commercial trash. This material typically has not been separated at its source.

The definition of source separation processing includes a collection system where residential and commercial customers are required to do some degree of "front-end" separation of the materials. This can range from implementing collection programs where homeowners are asked to place their recyclables [i.e., old newspaper (ONP), aluminum

*Coauthors from Waste Management include Tom Kaczmarski, Marty Felker, Russ Filtz, Dan Kemna, and Bill Moore.

cans, ferrous metals, three colors of glass, plastic soda bottles (PET), and milk jugs (HDPE), etc.] into separate containers. This technique is currently being practiced in many west coast communities. Another technique used is to have the homeowner do as little as separate "wet" material (i.e., organic food waste, yard waste, pet droppings, etc.) from the "dry" stream (i.e., all other noncompostable material) using two bins. This approach is used quite effectively in Europe and the provinces of Canada. These two techniques can also be applied to the commercial/industrial waste stream as well. A more detailed explanation of each program description is covered in the section on source-separated processes.

Mixed MSW processing, on the other hand, deals with processing the entire waste stream. Typically, the residential or residential and commercial/light industrial material is processed without any separation occurring by the generators. Although not as prominent as the source-separated approach, this processing technique is receiving more and more attention due to the simplification of the collection system. Various forms of mixed MSW processing will be discussed in more detail in the following section.

There are considerable differences both in terms of material balance, economics, and recovery rates between the three different systems described above. Table 6.1 presents a typical material balance of the various processing options. Table 6.2 presents a matrix of these options, including relative cost comparisons, appropriate materials recovered and projected diversion rates. *It is important to note that the information contained in these tables represents estimates only. They are not to be used as a baseline for establishing local program costs or budgets. A separate analysis factoring in local site conditions must be performed.*

MIXED MUNICIPAL SOLID WASTES PROCESSING APPROACHES

Introduction

As described above, this form of operation is generally used on large-scale waste streams running anywhere from 200 to 2,000 TPD. Four general approaches for MSW processing are typically used. They are refuse-derived fuel (RDF), composting, hybrid composting/RDF, and anaerobic digestion for methane recovery. Mixed MSW processing generally needs outlets for lower-quality recyclables than the source-separated approach. The outlet market must be able to accommodate large amounts of RDF or a low- to medium-quality compost. See Fig. 6.1 for a sample process flow diagram of a typical mixed MSW-RDF processing facility.

Europe vs. the United States

Beginning in the early 1970s, the United States was experimenting quite extensively on mixed MSW processing technology. During the latter half of the 1970s, a group of about a dozen MSW processing plants were built in the United States, many of which have since been shut down. The reasons that these facilities were closed vary. In some cases the processing technologies which were employed were borrowed from other industries and were ill equipped to handle a feedstock as variable as MSW. While the overall concept of processing was sound, problems such as shredder explosions and difficulties in material handling surfaced. In other instances, markets for the materials which were re-

TABLE 6.1 Residential Recycling Options: Material Balance Percent Recovered by Weight of RMSW*

	ONP	RMWP	Aluminum	Tin cans	HDPE/ PET	MRC	Glass	RDF	To compost	Residue % of material processed	RMSW % diverted from disposal
Curbside[†] recycling (TSS)	4	10	1	2	3	2	2	N/A	N/A	N/A	15 30 w/options
Curbside recycling (CSS)	4	10	1	2	3	2	2	N/A	N/A	N/A	15 30 w/options
Curbside recycling (MR)	4	10	1	2	3	2	2	N/A	N/A	N/A	30 w/options
Curbside recycling (FCM)	4	10	1	2	3	2	2	N/A	N/A	N/A	15 30 w/options
Yard waste source-separated	N/A	N/A	N/A	N/A	N/A	N/A	N/A	N/A	10	5	10
Wet/dry[‡]	4	30	1	3	3	4	2	N/A	25	35	65
RMSW processing RDF	0	0	1	3	1	0	0	60	20	15	75
RMSW processing compost	0	0	1	3	1	0	0	0	60	30	60

*RMSW—residential municipal solid waste; OCC—old corrugated cardboard; RMWP—residential mixed waste paper; MRC—mixed rigid containers; ONP—old newsprint; TSS—truckside sort; CSS—Curbside sort; MR—mixed recyclables; FCM—Fully commingled; RDF—refuse-derived fuels.

[†]Curbside recycling capture rates based on 70 percent participation.

[‡]Wet/dry, RMSW processing—RDF and compost, capture rates based on 100 percent participation.

TABLE 6.2 Residential Recycling Options

	Household container	Collection vehicle type	Collection cost	Processing methods	Processing capital cost	Processing operating cost	Material recovered	Recovered material quality	% of RMSW diverted from disposal	Significant issues
Curbside recycling, truckside sort (TSS)	Single, 18 gal	3-bin recycling truck	High	Conventional MRF	Med	Low–Med	ONP Aluminum HDPE/PET Tin cans Glass RMWP, OCC (opt.) MRC (opt.)	High High High High High Med–high Med–high	15 10 5	RMWP markets, PRA ability to take MRC
Curbside recycling, curbside sort (CSS)	3–11gal	3-bin recycling truck	Med–high	Conventional MRF	Med	Med	ONP Aluminum HDPE/PET Tin cans Glass RMWP, OCC (opt.) MRC (opt.)	High High High High High Med–high Med–high	15 10 5	RMWP markets, PRA ability to take MRC
Curbside recycling, mixed recyclables (MR)	Single, 18 gal	2-bin recycling truck	Med	Conventional MRF	Med	Med	ONP Aluminum HDPE/PET Tin cans Glass RMWP, 0CC (opt.) MRC (opt.)	High High High High Med–high Med–high Med–high	15 10 5	RMWP markets, PRA ability to take MRC

Curbside recycling, fully commingled (FCM)	Single, 18 gal	Residential packer	Low	BRINI MRF	Med	Med-high	ONP Aluminum HDPE/PET Tin cans Glass RMWP, OCC (opt.) MRC (opt.)	High High High High Med–high Med–high Med–high	15 10 5	Glass breakage
Yard waste, source-separated	Bags or cart	Residential packer	Low	Windrow composting	Low	Med	Compost	High	10	
Wet/dry	2-bin, 63 gal	2 residential packers or split packer	Low	Advanced BRINI MRF	Med–high	Med	ONP Aluminum HDPE/PET Tin cans Glass RMWP, OCC (opt.) Compost	Med–high Med–high High High Med Med High	50 15	Glass breakage, RMWP markets
Mixed MSW processing—RDF	Bags or 95-gal cart	Residential packer	Low	BRINI RDF processing	High	Med	OCC Aluminum HDPE/PET Tin cans RDF	Med High Med Med High	20 55	RDF markets
Mixed MSW processing—compost	Bags or 95-gal cart	Residential packer	Low	Buhler-type compost processing	High	Med	OCC Aluminum HDPE/PET Tin cans Compost	Med High Mcd Med Low–Med	5 55	MSW compost markets

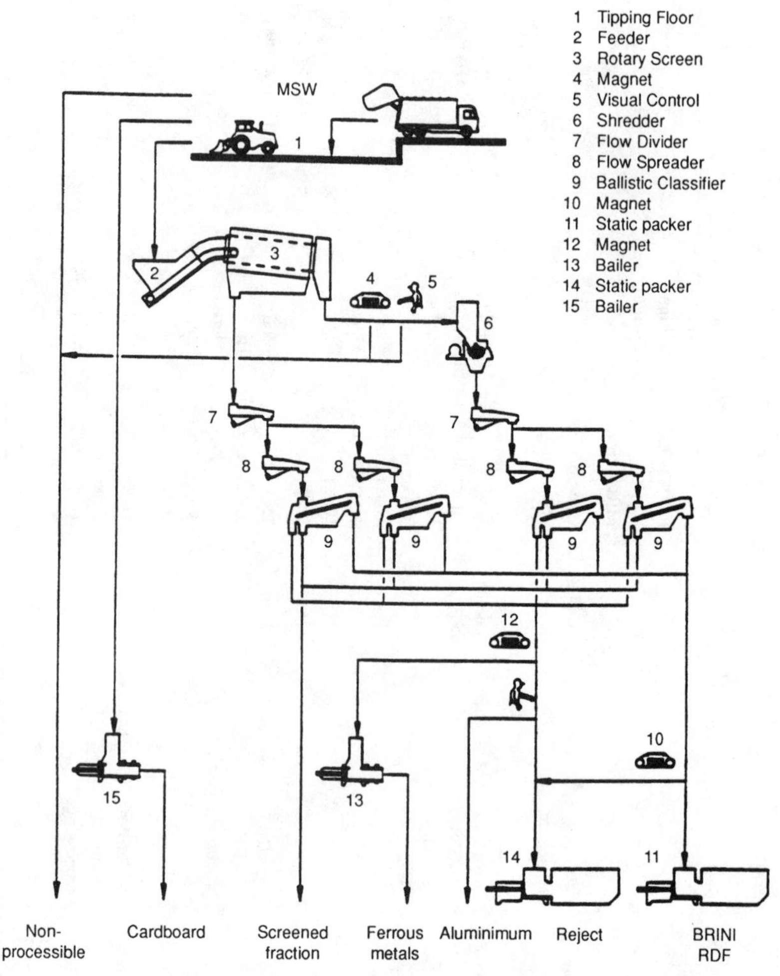

FIGURE 6.1 Diagram BRINI resource recovery facility. (*Courtesy Sellbergs Engineering.*)

covered or produced at these facilities were unstable. These factors combined to raise the overall cost of processing to levels which could not be justified in a time of relatively low landfill disposal costs.

In Europe, throughout the late seventies and eighties, there had been a fair amount of activity in mixed MSW processing facilities. This is due to some of the basic differences in the waste stream that existed in Europe (i.e., a higher vegetative content and lower packaging content). All of the industrialized nations saw a compositional difference which can be attributed to more convenience in terms of life-style. For example, frozen and packaged foods gained in popularity with consumers. In addition, the high cost of petroleum-based fuels led to more mixed MSW plants surfacing in Europe. RDFs were used to produce steam and hot water in numerous district heating systems.

Front-End Processing for Waste-to-Energy (WTE) Plants

There has been a small amount of front-end processing of mixed MSW prior to mass burn plants in Europe. We have yet to see any significant efforts in this area in the United States. The Environmental Protection Agency (EPA) is expected to promulgate regulations requiring a 25 percent recycling rate on the waste stream entering WTE plants. Once this legislation is in place, this could spark a renewed interest in the United States in front-end processing. The processing techniques used would be similar to those used in a mixed MSW plant. The goal of this frontend system, however, would be to pull out as much of the heavy metals and inert materials as possible. The fibrous material could conceivably be separated, however, but doing so would impede the Btu characteristics needed to fuel the burners.

RDF/Alternative Fuels

Several different types of technologies exist for making RDF from mixed MSW. Proprietary systems such as the BRINI, Buhler, Lundell, and several others are available in the marketplace. All of these approaches use various screening techniques, either trommels, disk screens, or other types of classifiers to perform the separation required. Exhibits 6.1 to 6.4 are photos of a typical RDF facility including incoming waste stream, ballistic separation equipment, and finished product. Many different grades of RDF can be produced from MSW. Generally speaking, the higher the fuel quality, the lower the fuel yield. For example, an RDF plant in Albany, New York, simply shreds the incoming waste and then passes the shredded material across a magnetic separator to remove the ferrous component. The fuel yield is roughly 95 percent, while the average Btu value of this fuel would

EXHIBIT 6.1 Typical mixed MSW waste stream at refuse-derived-fuel (RFD) facility.

EXHIBIT 6.2 Ballistic separator used to produce high-quality RFD material.

EXHIBIT 6.3 Shredded mixed MSW inside ballistic separator.

EXHIBIT 6.4 Pelletized RDF—finished product.

be similar to raw MSW. Conversely, in order to produce a pelletized fuel, much preprocessing must be done. Fuel yields in the neighborhood of 50 percent, when based on the total incoming waste, can be achieved, which would have a heating value which approximates 6,500 to 7,000 Btu/lb. The type of fuel that must be prepared in a mixed-waste processing facility must be dictated by the combustion equipment. Early RDF plants after shredding the waste stream, did a minimum amount of screening and separation and produced a rather inferior-quality fuel material.

Today's state-of-the-art facility shreds the MSW, screens it, and allows for the recovery of both an organic (yard and food waste) fraction for composting and also material recovery in the form of corrugated boxes, aluminum, plastic bottles, and steel. After the compost fraction is removed and the material recovered, secondary grinding forms the fuel fraction, which can either be delivered in a fluff form or a densified pellet or cube. Table 6.1 gives an estimated material balance for this type of facility.

In the United States today a handful of these types of plants are operating. One of the earlier generation plants still operating is the Ames, Iowa, facility. Buhler technology is in use today at the Eden Prairie, Minnesota, plant. Early large and very complex processing facilities in Milwaukee and Rochester have now been shut down. The success of the active operating facilities can be directly tied to the availability of combustion facilities for the fuels which are produced.

An example of a large-scale, complex processing facility (using Raytheon technology) is the northern Delaware facility. This facility is equipped to produce an RDF from the organic fraction while recovering such materials as aluminum and glass. Recently, an incinerator was added in order to burn the RDF on the site and produce electricity.

Composting—Windrow and In-Vessel

Three types of separation approaches are used for making compost from mixed MSW. Generally the greater the degree of front-end separation, the higher the cost of processing. Naturally this means that more equipment and/or sorters will be used in order to isolate the organic fraction of the MSW. After separation, there are two general approaches for making the compost. The first approach was described above in the RDF type plant, with an organic fraction coming off the MSW/RDF processing.

The second approach is to shred the entire waste stream and then compost it with minimal screening and material recovery while attempting to clean the material up after the composting process is complete. This system produces a product which tends to contain small pieces of glass and plastic and other inert contaminants. The end use for this material may be limited because of these physical contaminants, but market applications are currently being sought. One facility that employs this technology in Dade County, Florida, is Agripost. Several other operations in the United States employ similar technology.

The third approach to the separation focuses on a fair amount of front-end screening and material picking before finally grinding what's left of the waste stream for composting. This differs from the first approach by not making an RDF fraction. After the separation of the organics is complete, two general approaches are used for composting: windrow and in-vessel processing.

Windrow processing can be accomplished either out-of-doors or in an enclosed or semienclosed building. Buildings are used mainly to effect some control over the amount of moisture present in the compost heap and also to aid in odor control since the air which is captured in the building can be discharged through a biofilter or similar containment equipment. In either case, the windrows, which are triangular piles roughly 8 ft by 8 ft by 8 ft. are formed for the natural composting process to take place. Turning for aeration and some homogenization is accomplished by mechanical means as the material decomposes.

In the in-vessel approach, structures are used that contain either trenches or circular reactor type vessels which are filled with the organic feedstock. The organic feedstock is turned and mixed periodically to provide the proper environment to convert the material to compost. Many factors come into play when choosing a composting method. One must keep in mind, however, that the process of decomposition which takes place in either a windrow or a vessel is the same. How a technology is applied must be based on the local conditions. For instance, if ample space is available at a remote location, a windrow-type composting system may offer the best solution. However, where space is at a premium and odor control is critical, a system which would offer the greatest degree of process and odor control, such as a composting vessel, may be needed. The end-product market needs will also play a role in determining the type of processing to be employed. If it is desirable to produce a landfill cover, little up-front processing would be called for. However, if horticultural markets are targeted for the finished product, great care should be taken with the up-front collection and processing to ensure the highest quality of material possible.

Methane Recovery—Anaerobic Digestion

There have been some attempts to subject a full MSW waste stream to anaerobic digestion to produce methane. One of the most well-known in the United States was the RefCom project in Pompano, Florida. In this approach, the MSW was shredded and water added to a cylindrical reactor vessel, which is maintained in a mixed anaerobic state to produce methane.

Some systems attempt to remove contaminants and screen out undesirable materials. Many difficulties were encountered with handling and mixing the anaerobic reactors. The methane from the reactors is either upgraded for use in pipeline gas or used to fire turbines directly. At present, we know of no active U.S. methane recovery operations for MSW. In Europe, the Valorga technology is in use in several locations. This technology is similar to what was described in the above paragraph.

Many factors are responsible for impeding the development of anaerobic digestion facilities in the United States. These factors include:

- Heterogenic nature of the mixed-waste stream makes material handling particularly troublesome.
- Lack of available equipment specifically designed for methane recovery systems.
- Relative abundance of cheap fuels.

When compared with the European market, the economics of an anaerobic digestion process makes it difficult to justify. As fuel costs and tipping fees continue to rise, the economic viability of an anaerobic process is enhanced.

SOURCE-SEPARATED PROCESSING APPROACHES

Introduction

Recycling, a form of waste reduction, offers a variety of cost-competitive and environmentally safe alternatives to waste disposal by converting source-separated MSW into valuable resources. Source-separated processing as the name implies is characterized by producing very high quality materials for distribution to the end market. The processing techniques used are typically less complicated than those needed for a mixed MSW stream and therefore less costly to operate and maintain. Residential source-separated systems and techniques highlighted in this section include the following:

- Commingled approach
- Multiple-bin approach
- Truckside sort approach
- Fully commingled approach
- Wet/dry approach

Due to the similarities in handling techniques, a less detailed analysis of commercial/industrial source-separation programs will also be discussed.

Commingled Approach

Residential source-separation facilities are generally built to handle recyclable waste streams ranging from 10 to 250 TPD. In a commingled material recycling program only one container is provided to the homeowner. A commingled MRF's main objective is to receive, process, and ship to market, residential curbside-collected recyclables. These recyclables typically consist of glass, high-density polyethylene (HDPE) and polyethylene terephthalate (PET) plastics, and aluminum and ferrous metal containers mixed together (often referred to as "hard recyclables") and old newspapers (ONP). The ONP is always kept separate in this approach. The facility has a tipping floor designed to handle these two streams while keeping them separated (i.e., hard recyclables and ONP). A main processing line for the hard recyclables is necessary for sorting these materials by grade. The paper-processing line may or may not have a distinct sorting station, but the material should have at least a quality control inspection station before baling or loose filling into an open-top trailer. Since commercial loads accepting old corrugated containers (OCC) and high-grade paper are not much different to handle than ONP, most commingled MRFs should have adequate space to accept these materials and process them on the paper line.

At present, the commingled approach is a more cost-effective alternative than a multiple-bin program primarily due to the productivity gains achieved in the collection of the material. In other words, when looking at the total cost of providing the service, it is typically less expensive to collect and process commingled material than material collected in several containers. The rest of this section will present a typical design for an average 100-TPD commingled MRF. Process flow description and a diagram are included along with a plant layout. This design is not fixed and should be viewed as a general guideline to be incorporated into a specific project. Most MRFs will be unique designs, but there are still many common elements that can be used from this example.

Process Flow Description. A general arrangement for a typical 100-TPD commingled facility is presented on Fig. 6.2. The process flow diagram is shown on Fig. 6.3. Recyclable materials are delivered to this facility by specially designed two-compartment collection vehicles. The design of this facility is flexible enough to handle materials collected in the form of separated glass, plastics, and cans. This is often necessary when there is one facility servicing both residential and multifamily collection programs.

When designing an MRF, careful consideration must be given to a flexible equipment layout to allow for future expansion capability. The characteristics of the recyclable materials which are processed generally dictate two separate process lines. One line is configured to process the "light-fraction" recyclables including newspaper, cardboard, mixed paper, and plastic film. The second line is designed to handle various types and sizes of "heavy-fraction" materials (i.e., glass, aluminum, steel, wood, rubber, plastics, etc.)

For this facility the equipment has been selected with the capability to process additional materials as they are added to the collection program. Eventually this type of facility could also be converted to facilitate processing of the dry fraction of the municipal solid waste stream.

Newspaper is tipped directly onto the receiving floor adjacent to the in-floor feed conveyor and push wall (see Exhibit 6.5). The material is then pushed by a front-end loader onto the in-feed conveyor where it is conveyed up to the elevated sorting platform and picking stations. Manual sorting pulls out contaminants and drops them down into storage bunkers beneath sorting stations. It is critical that these material's storage bunkers be sized with adequate capacity to store enough material to make at least three bales. The concept of separating sorting from baling operations maximizes the availability and efficiency of the baler. The best approach is to follow a negative sort (i.e., remove contaminants and smaller-quantity recoverable materials and let the ONP accumulate in the last bunker of the sorting line). If less material is physically handled, it reduces labor costs and increases throughput capability of the processing line.

Commingled containers are tipped into a continuous-feed floor conveyor which is typically fitted with a rubber-lined skirt onto the receiving edge (see Exhibit 6.6). This skirt acts as a buffer to minimize the breakage of glass. Materials are conveyed up to the presort station by a slider belt conveyor. *Note:* The necessity to include a presort station depends on the level of contamination in the incoming feed stream. This would allow for the removal of contaminants such as steel paint cans, film plastics, and batteries typically not included in commingled recycling programs.

Ferrous materials (e.g., steel can) are the first items to be separated from the commin-

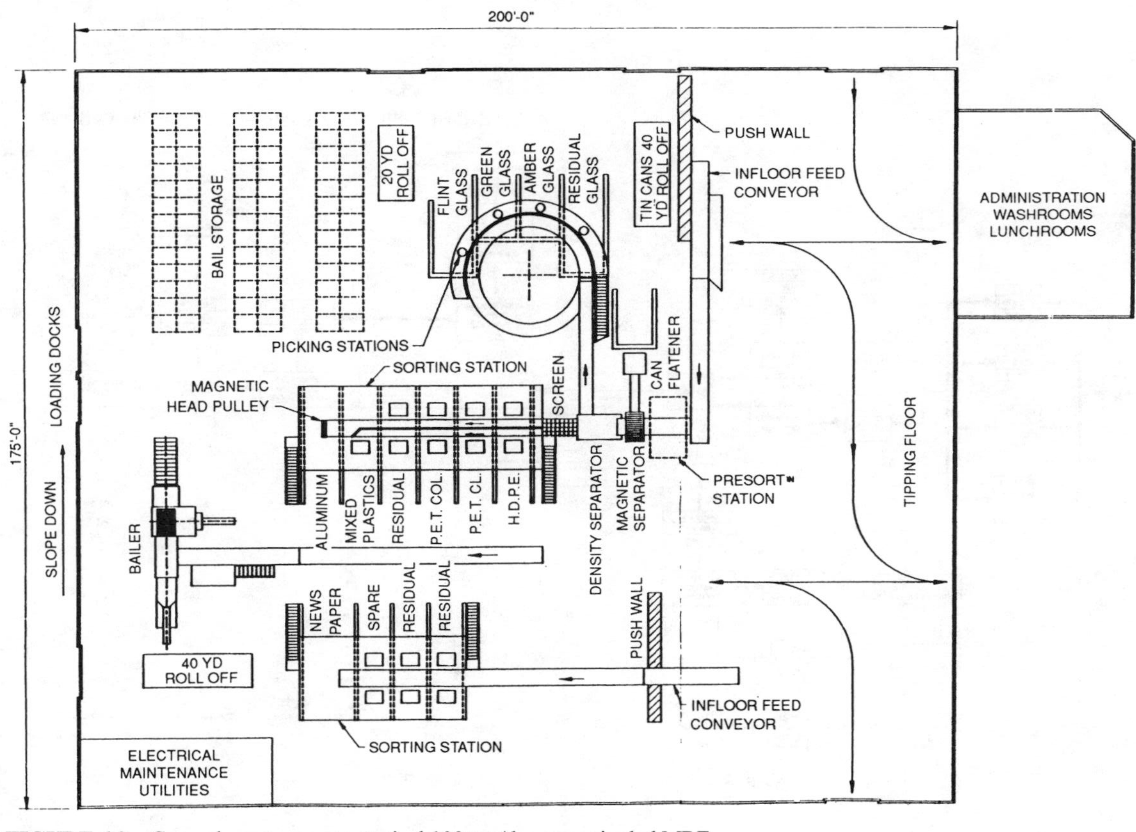

FIGURE 6.2 General arrangement—typical 100 ton/day commingled MRF.

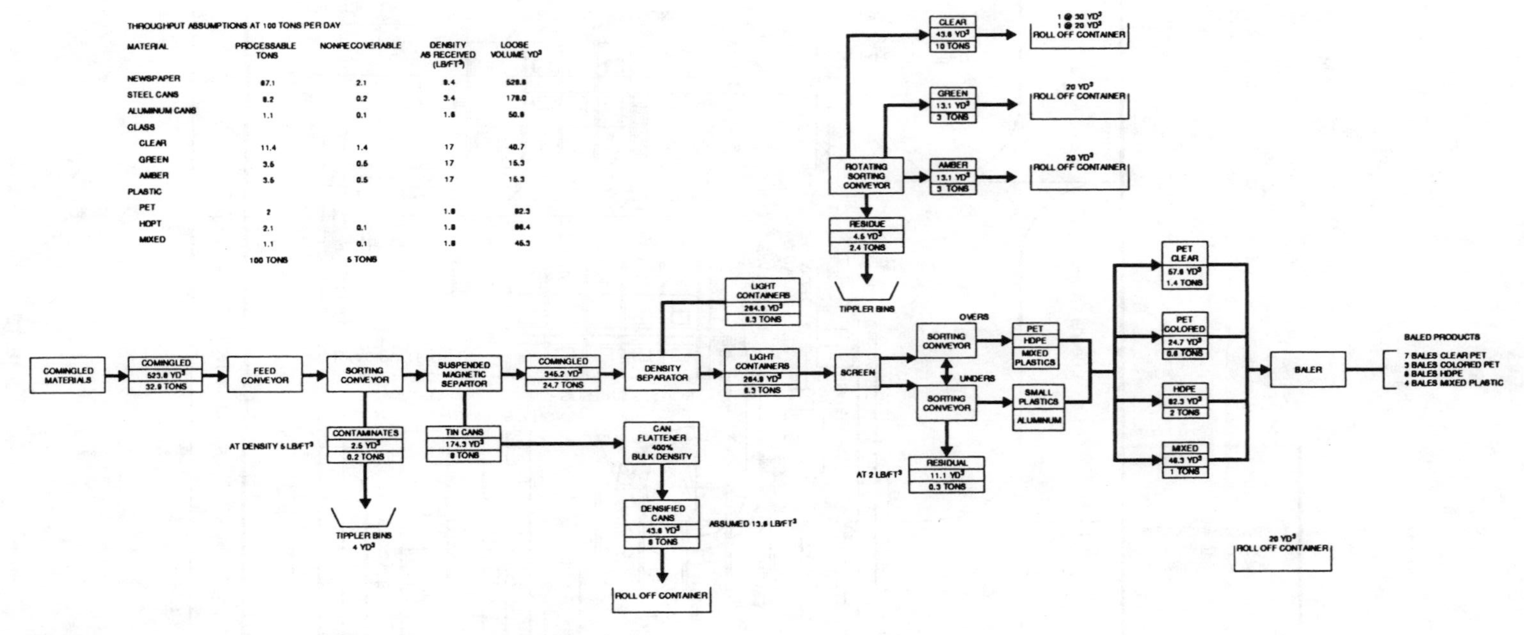

THROUGHPUT ASSUMPTIONS AT 100 TONS PER DAY

MATERIAL	PROCESSABLE TONS	NONRECOVERABLE	DENSITY AS RECEIVED (LB/FT^3)	LOOSE VOLUME YD^3
NEWSPAPER	67.1	2.1	9.4	528.8
STEEL CANS	8.2	0.2	3.4	178.0
ALUMINUM CANS	1.1	0.1	1.6	50.9
GLASS				
CLEAR	11.4	1.4	17	40.7
GREEN	3.6	0.5	17	15.3
AMBER	3.6	0.5	17	15.3
PLASTIC				
PET	2		1.8	82.3
HOPT	2.1	0.1	1.8	86.4
MIXED	1.1	0.1	1.8	45.3
	100 TONS	5 TONS		

FIGURE 6.3 Process flow diagram–100 ton/day commingled MRF.

EXHIBIT 6.5 Tipping floor and infeed newspaper conveyor.

EXHIBIT 6.6 Commingled plastics and cans infeed conveyor.

gled stream. This is accomplished through the use of a suspended magnetic separator. The separator will remove 90 to 95 percent of the steel present in the commingled stream. The separated cans are then passed through a can flattener to decrease volume, or alternatively, can be conveyed to a concrete bunker and flattened by front-end loader.

The remaining commingled materials are passed on to a density separator which splits the materials into a light and a heavy fraction (see Exhibit 6.7). Separators in place today

EXHIBIT 6.7 Density "air classification" system separates aluminum and plastic from glass. (*The Forge Recycling and Transfer Station, Philadelphia, Pennsylvania.*)

could be one of several devices (i.e., vibrating gravity table, air separation, density brush, etc.) A cost/benefit analysis to justify the addition of these components needs to be performed. Factors to be considered include labor rates, utility costs, equipment life-cycle cost, and projected volume.

The heavy fraction as it leaves the density separator is primarily glass containers. This glass may be conveyed either onto a straight sorting belt or collected onto a rotating sorting conveyor for subsequent manual color separation. The advantage of using a sorting ring conveyor is that temporary storage is provided. It also allows sorters to be allocated more effectively since they are not required on the sorting ring until it is at capacity.

The light fraction is passed through a trommel-type screen which separates according to size (see Exhibit 6.8). The resultant materials on the divided sorting conveyor are large plastic containers on one side and aluminum and small plastics on the other. The divided conveyor allows for negative sorting of the aluminum cans and the ability to transfer any aluminum cans which pass through to the large or oversize side of the sorting belt. Manual separation is required for all of the materials except for the aluminum, which is left on the sorting conveyor. The aluminum undergoes a second ferrous magnetic sort to assure product purity.

All materials are stored in the bunkers located beneath the picking stations on the elevated sorting platforms. When sufficient quantities of material have been accumulated, the bunker is emptied by front-end loader and the material pushed onto the baler feed conveyor. Glass containers are sorted into bunkers beneath the sorting ring, or alternatively, then can be dropped directly into roll-off containers for shipping or fed into crushers to increase density prior to shipping. The 100-TPD MRF represents the state of the art with respect to available technology and cost effectiveness. This facility primarily consists of two sorting systems; however, this general layout also allows flexibility for the future implementation of a ballistic separator capable of separating recyclables mixed together.

EXHIBIT 6.8 Size separation using Trommel screen. (*Note:* Aluminum and plastic separation.) (*The Forge, Philadelphia, Pennsylvania.*)

With this information, classified materials can be delivered to the facility "fully commingled" (i.e., newspaper, mixed paper, cardboard and containers can be mixed together during collection). Even greater collection productivity can be achieved with this approach.

Multiple-Bin Approach

The concept behind a multiple-bin MRF is to let the homeowner help out by doing most of the sorting. Any sort beyond the commingled approach of hard recyclables separated from paper is considered multiple-bin. The most familiar program is the three-bin system with glass/plastic, aluminum and ferrous cans, and ONP sorted separately by the homeowner. Plastics are a problem because the options are to mix and lose space, bundle ONP on the side, or add a fourth bin. All not very good choices, and what about truck design? The natural inclination is to believe that since the material is already at least partially sorted, that there will be savings in running a processing facility. Possibly, but there are other factors which outweigh the savings in sorting costs. These factors include:

- Increased collection costs
- Longer time at each stop to empty bins
- Possibility of one truck bin filling up first increases
- Longer to dump, once at the MRF
- Redundant processing lines to handle each partial stream

Given the above reasons it is reasonable to assume that in most instances the multiple-bin system should be avoided. However, successful programs still exist, especially in areas with small populations where constructing a large typical processing operation is described below.

Process Flow Description. A general arrangement for a typical 100-TPD multiplebin facility is presented in Fig. 6.4. The process flow diagram is shown in Fig. 6.5a to c. Recyclable materials are delivered to this facility by specially designed multicompartment collection vehicles.

Receiving bins are typically set below the floor grade next to the truck scales, which allow for weighed loading of material in the following mixes:

1. Glass and plastics.
2. Aluminum and tin cans.
3. ONP.

The trucks tip the newspaper on the tipping floor designed to hold typically 1 day's worth of material. Conveyors from the in-floor loading pits convey recyclables up to elevated sorting lines. These conveyors are typically slider-belt types with high-sided troughs. The conveyors run horizontally for the full length of the pit, inclining to their respective sorting conveyors.

Glass and Plastics. Glass and plastics sorting conveyor is also a slider belt type. Manual sorting pulls out contaminants and drops them down a chute onto a conveyor leading to a tippler bin. Manual sorting separates all plastics out of the material stream and drops down a chute to the plastic conveyor below.

Glass is color-sorted manually and dropped through chutes onto separate slider-belt conveyors that transport the material to exterior receptacles. A sliderbelt conveyor brings

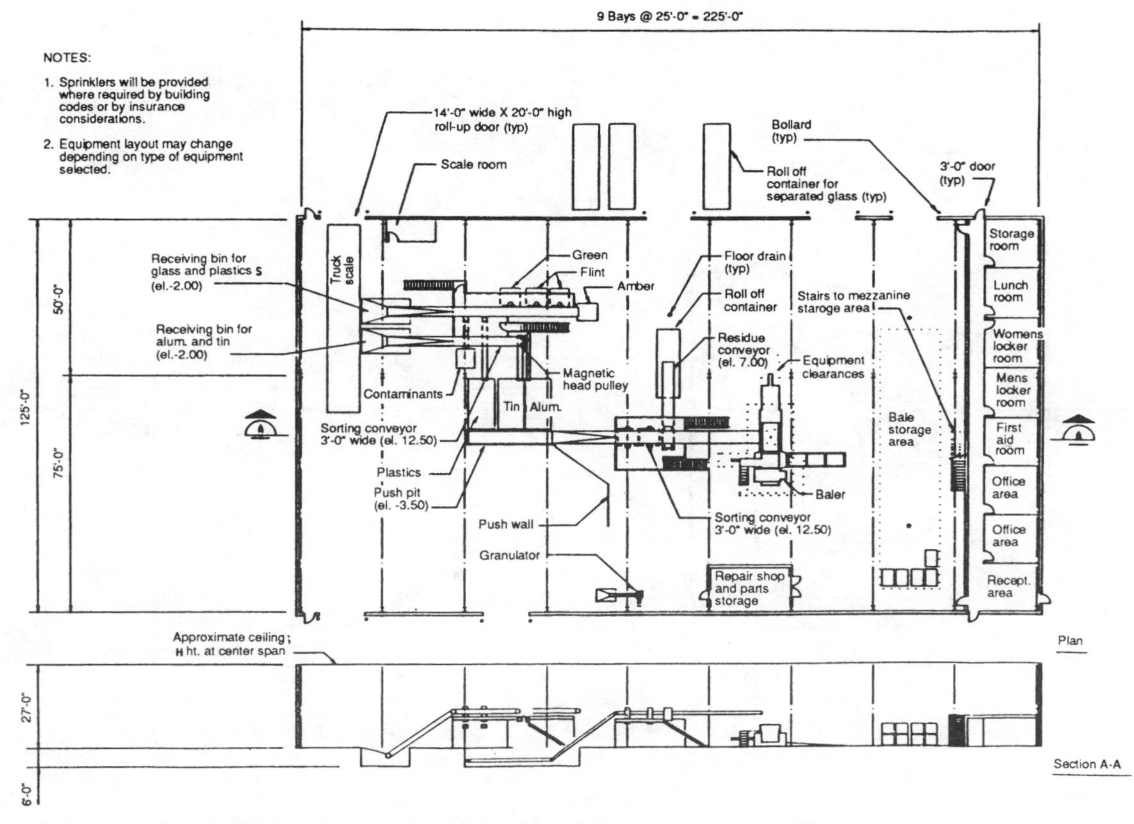

FIGURE 6.4 General arrangement—100 ton/day multiple-bin source-separated MRF.

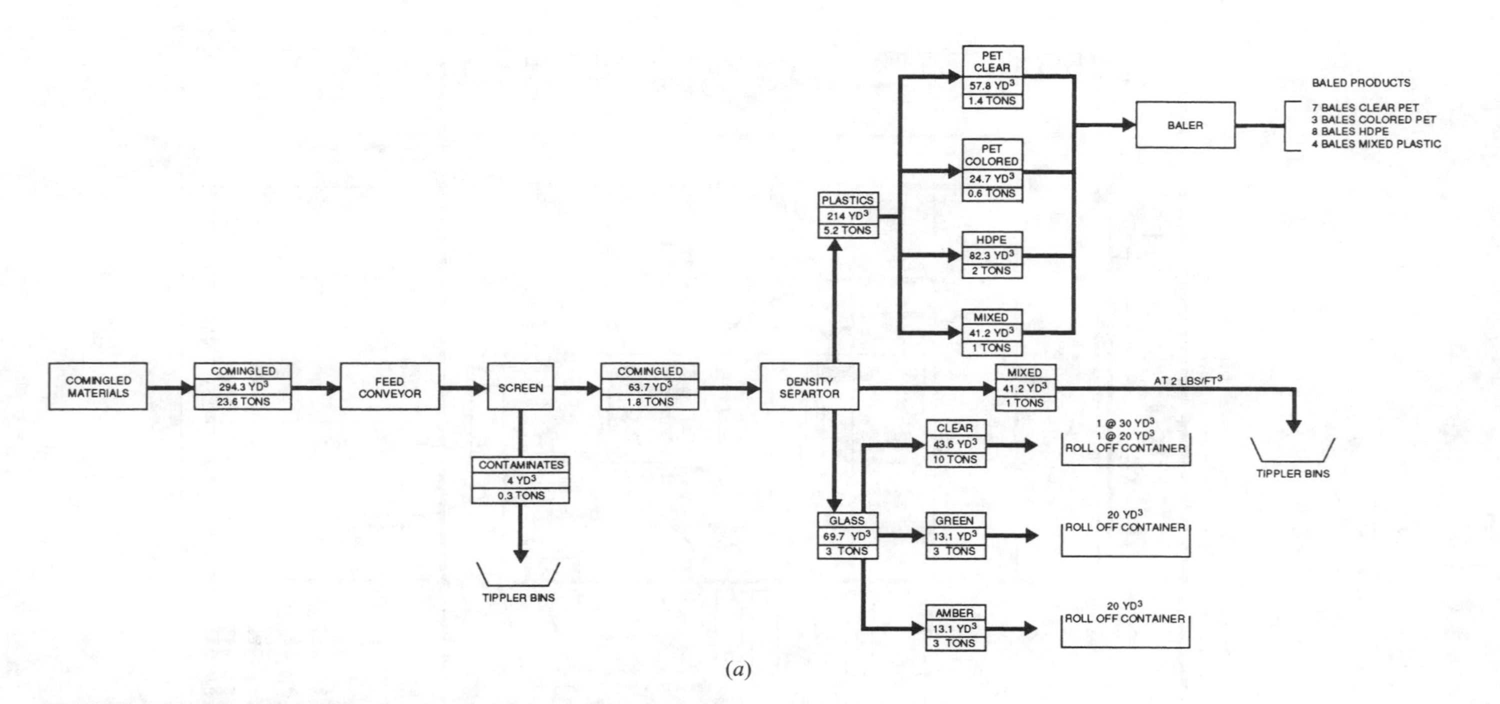

(*a*)

FIGURE 6.5 (*a*) Process flow diagram—source-separated glass and plastic containers.

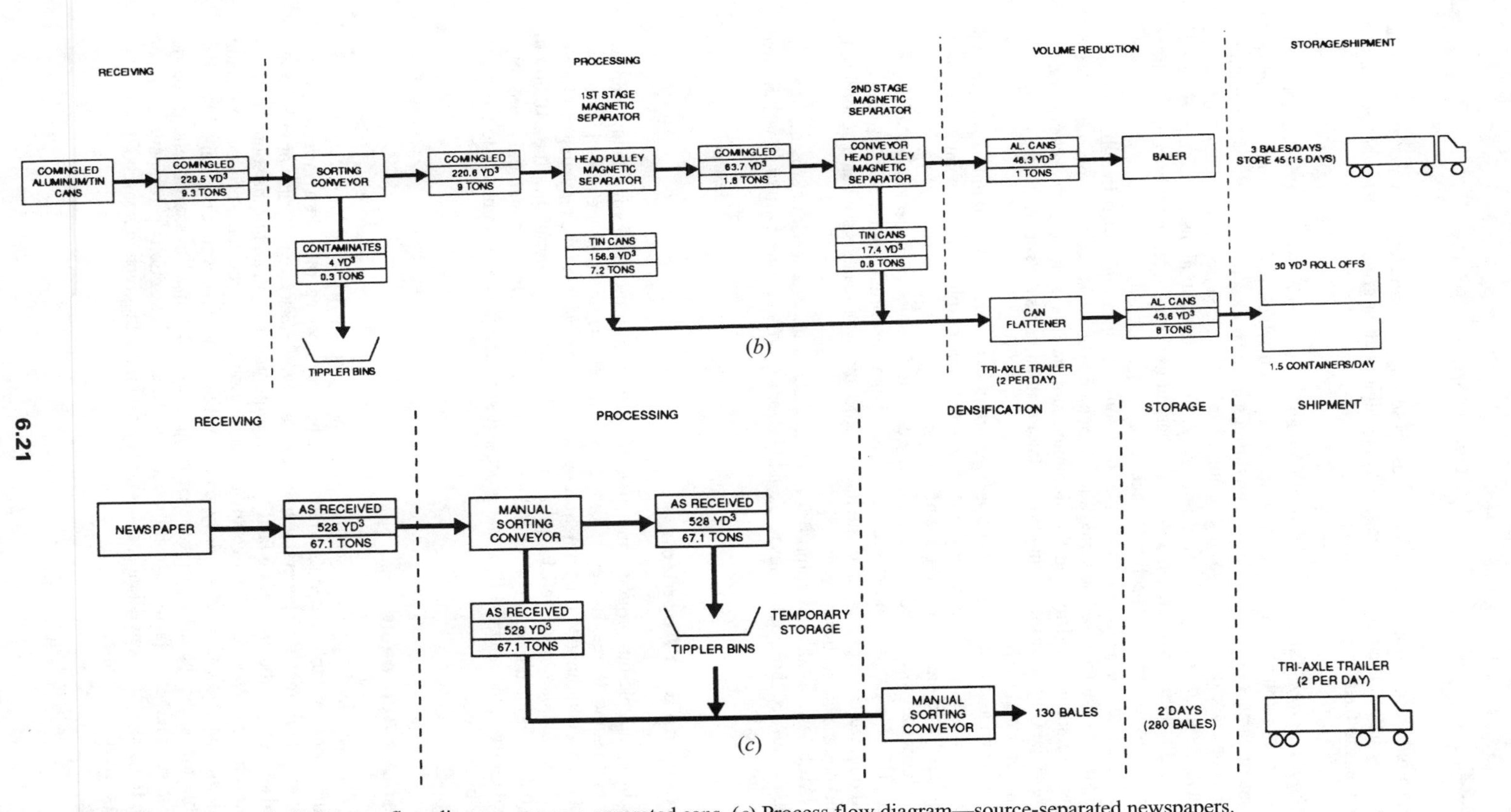

FIGURE 6.5 (*Continued*) (*b*) Process flow diagram—source-separated cans. (*c*) Process flow diagram—source-separated newspapers.

in sorted plastics from the glass and plastics sorting line and elevates material up into the intermediate self-emptying storage bin. The slider belt at ground level is horizontal, and inclines to feed the plastics storage bin.

Aluminum and Tin Cans. An aluminum and tin sorting conveyor at the elevated platform is horizontal and is a slider-belt conveyor on which manual sorting will pick out contaminants, (e.g., bimetal cans) and through a chute, direct them to the same contaminant conveyor described above. The aluminum and tin continue on the conveyor. At the discharge point a magnetic head pulley and chute network allow aluminum to be discharged to one conveyor, while tin is discharged to another conveyor, both of which are inclined to feed their respective storage bins.

The sorted tin conveyor is a slider-belt conveyor in a high-sided trough, inclined to feed the tin storage bin. The sorted aluminum conveyor is a slider belt conveyor in a high-sided trough inclined to feed the aluminum storage bin. A front-end loader is used to spread the ONP over the tipping floor. Major contaminants are sorted and the clean newspaper is stored for baling.

Tin and aluminum cans and newspaper are individually conveyed onto a sorting conveyor and onto an elevated conveyor leading to the baler inlet hopper opening. The material is baled to meet marketing specifications.

Plastics are processed in the same way as described above, except at the sorting station, HDPE milk jugs and mixed plastics are manually sorted from the PET soda bottles and directed through a metal chute into a tippler bin. The plastics are baled according to marketing specifications.

All bales are taken by a fork-lift truck to either a transfer truck or storage area. The glass deposited into tippler bins is taken to larger roll-off containers (30 to 40 yd^3) located outside the facility.

Truckside Sort Approach

As in a multiple-bin program, materials are delivered to the processing facility already sorted to some degree. However, for the homeowner, the materials are placed in one bin, just as in a commingled program. Sorting is done at curbside by the recycling truck driver. This method works well for pilot programs or smaller communities (i.e., 10,000 to 20,000 range) that have no or limited processing facilities. Once the material is presorted by the driver, the material flow and process description is similar in nature to the multiple-bin approach.

Wet/Dry Approach

As discussed in the previous section, MRFs are typically designed to process a selected number of recyclable materials which have been source-separated in a variety of ways. The recyclable materials generally arrive as one or more discrete materials or in commingled streams. At the processing facility the commingled stream and/or the source-separated materials are sorted to remove any contamination or nonrecyclable products collected with the recyclable products. This type of recycling system recovers a portion of the recyclable materials, the amount recovered ranging from 10 to 20 percent of the domestic waste stream. Many factors affect the recovery rate, such as public participation, collection frequency and convenience, collection system including type of household container, and collection vehicles.

There has been a trend to increase the number of materials being collected in such systems as markets for recovered materials are developed. This results in significant problems to the collection system and also at the MRF.

There are generally two existing methods for collecting residential separated recyclable materials. Both of these present systems separate into at least two categories: newspaper and containers. The options for collection after this level of separation can include curbside and subsequent vehicle separation into separate glass, plastics, and tin/aluminum can components. As a rule, a higher level of separation at the curbside translates to higher overall collection and processing costs. This is a factor of the increased time to sort materials at the curbside and the reduced capacity of the recycling collection vehicle.

In Europe over the past decade recycling programs have developed in many municipalities to recover significantly greater quantities of recyclable materials, in the order of 50 to 60 percent of the waste stream. This has been achieved by designing collection systems and MRF facilities which process the total dry fraction of the waste stream (see Exhibit 6.9). Rather than having a portion of the recyclable materials being available at the MRF, all these materials are delivered to the facility together with many other dry materials which may not be recyclable. The dry fraction has been defined as all the domestic waste except for food wastes, yard wastes, and wet or contaminated dry wastes. While this type of system has a significant effect on the residents' waste management practices, it has been found to be as convenient to the public as the system which involves a small recyclable materials container. These containers in themselves can limit the extent of recycling through their size. Public attitude surveys which have been carried out in Canada have indicated a strong support for the collection of the dry part of the waste stream. A number of such programs are being implemented in Ontario to meet a 50 percent reduction of the waste stream since this is a goal of municipalities, the provincial government, and the federal government. In Europe it has also been found that the remaining wet wastes can be processed readily and cost effectively into a contaminant-free high-quality

EXHIBIT 6.9 Dry fraction processing—wet/dry facility, Skara, Sweden.

EXHIBIT 6.10 Wet fraction processing—wet/dry facility, Skara, Sweden.

compost since the potentially toxic components in the waste stream are collected with the dry fraction and removed at the MRF (see Exhibit 6.10).

A general arrangement for a typical 200-TPD wet/dry facility is presented in Fig. 6.6. The process flow diagram is shown on Fig. 6.7.

Wet/Dry Program Profile.* As a result of the lack of landfill capacity in the County of Neunkirchen, Austria, during 1984, the county's waste management association considered many waste management options including: refuse-derived fuel, mass incineration, materials recovery processes, and export of wastes to Hungary. The materials recovery system was selected because it resulted in the greatest reduction in waste requiring disposal in a landfill, minimal environmental effects, and the lowest cost. Authority was given to commence the design of the facility in August 1985, with startup targeted for January 1986.

The project involved the separate collection of dry and wet wastes from approximately 100,000 residents and businesses in 43 municipalities within the County of Neunkirchen in Austria. The dry waste is processed to remove recyclable materials, the wet waste being composted. The technology presently of interest with respect to MRF design is the dry processing facility; however, the Neunkirchen plant is an integrated facility consisting of dry waste processing and composting.

In 1985 an operating company was formed as a joint venture between the County's Waste Management Association and the Hamburger Paper Mill company. Construction of the facility began in September and was completed by January 1986. The dry mixed waste is delivered to the facility, located on 3 acres, in collection vehicles serving the various municipalities and weighed on a truck scale. Records of the weights of materials from each municipality are recorded by computer for invoicing purposes. Waste is deposited on the tipping floor (10,000 ft^2) and pushed by small front-end loader onto a feed conveyor which discharges the material into a specially modified ballistic separator

*Extracted from technical paper written by Engineer Reinhard Goechl, Managing Director.

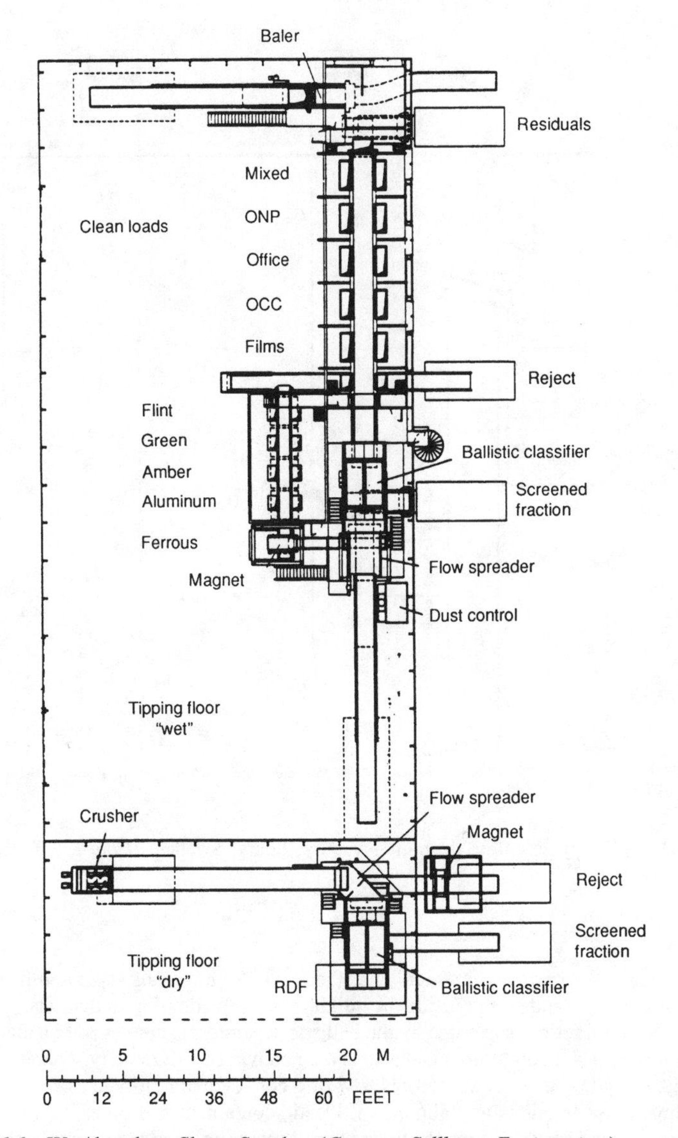

FIGURE 6.6 Wet/dry plant, Skara, Sweden. (*Courtesy Sellbergs Engineering.*)

(BRINI classifier). This separator, which has a capacity of 7.5 tons per hour, divides the mixed dry waste into light and heavy fractions and screens out particles less than 2 in.

The light fraction is discharged onto a sorting conveyor elevated 12 ft above the floor. Newspaper, cardboard, film plastic, and textiles are removed by five operators and deposited in live-bottom bunkers beneath the conveyor. The residuals, which consist of mixed paper, are also stored in a bunker. Materials in the bunkers are conveyed periodi-

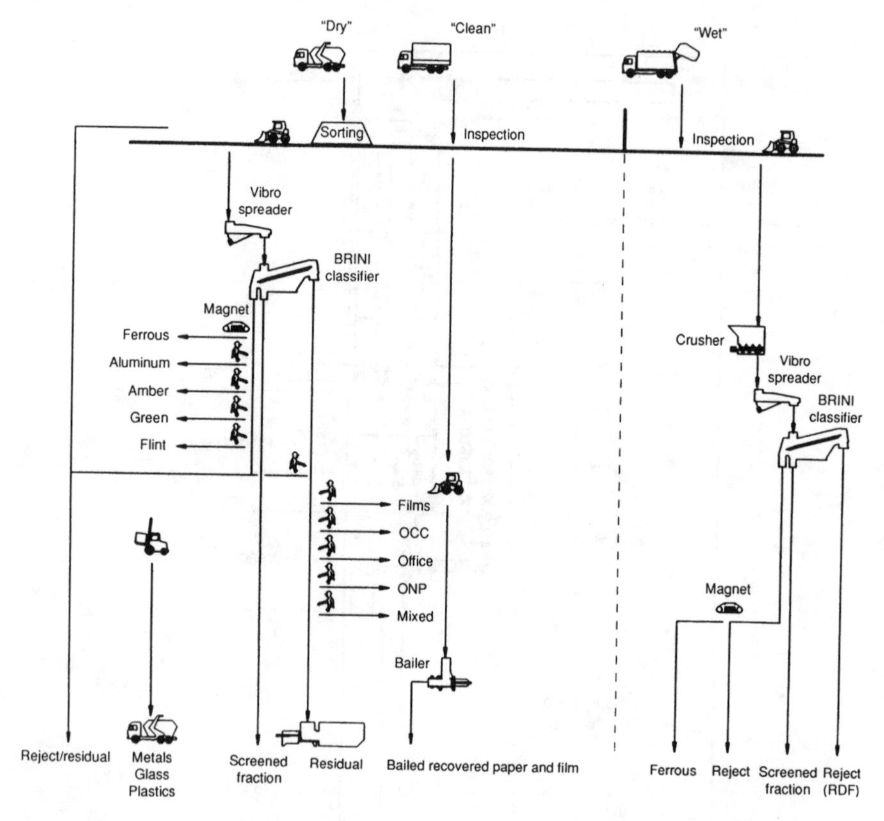

FIGURE 6.7 Process flow diagram—wet/dry, Skara, Sweden. (*Courtesy Sellbergs Engineering.*)

cally to an in-floor conveyor which feeds the baler. All materials separated from the light fraction are baled and stored inside the building prior to shipping to markets.

The heavy fraction, separated by the ballistic separator, is conveyed to a magnetic separator for ferrous metal removal. Waste is then conveyed to a density separator which diverts glass and other heavy materials to a ring conveyor for storage and manual sorting into three glass fractions and aluminum. Plastic containers and other light materials are conveyed to the sorting room for the manual separation of plastic containers. Once the materials are separated from the ring conveyor, hazardous materials are removed and the remaining is conveyed to storage bins for landfill disposal.

The facility is operated for two 8-h shifts, 5 days per week, 52 weeks per year. In the first 2 years of operation, the plant was only closed as a result of equipment maintenance for 1 day. Overall energy consumption of the process equipment in the facility is 175 kW (27 kW/ton) of which heating requirements are 30 kW.

Materials which are sold from this facility include newspaper, cardboard, mixed paper, plastic film, glass, ferrous metals, aluminum cans, and mixed nonferrous metals. The hazardous materials and wood recovered are disposed of separately from the plant residues. The materials meet specifications of markets in Austria and Germany consistently. In 1986 there were markets for textiles; however, currently no market exists.

The Hamburger Paper Mill purchases all the cardboard and mixed paper; newspaper is purchased by Lekykam, glass is purchased by Vetropack, and metals are purchased by a local scrap iron dealer.

The facility is operated by a staff of nine, which includes a clerk to operate the weigh scale, a plant manager, five sorters, one baler operator, and one mobile equipment operator.

The most significant aspects of this facility are that the facility averages 58 percent recovery of the dry wastes delivered, it is the first facility to utilize the ballistic separator on mixed dry waste, and that this facility was the first two-container wet/dry facility ever constructed.

Commercial/Industrial Approach

Commercial/industrial (C/I) MRFs are similar to many of the operations run by paper stock companies today. Most facilities are designed to bale clean loads of old corrugated containers (OCC), ONP, computer printout (CPO), and mixed paper. From a design basis, the plants can be compared to the paper portion of a typical residential MRF (see Exhibits 6.11 and 6.12). A similar system that handles strictly ONP can be adapted easily to sort other paper products.

Another type of C/I facility combines MRF activities with a transfer station operation. Many transfer stations today are recovering material, even if it's only select OCC loads. The setup is convenient since residue or contaminants don't have to be hauled twice. Having a transfer station integral to the C/I MRF will make it much easier to go after margin-

EXHIBIT 6.11 Typical commercial industrial dry fraction waste system.

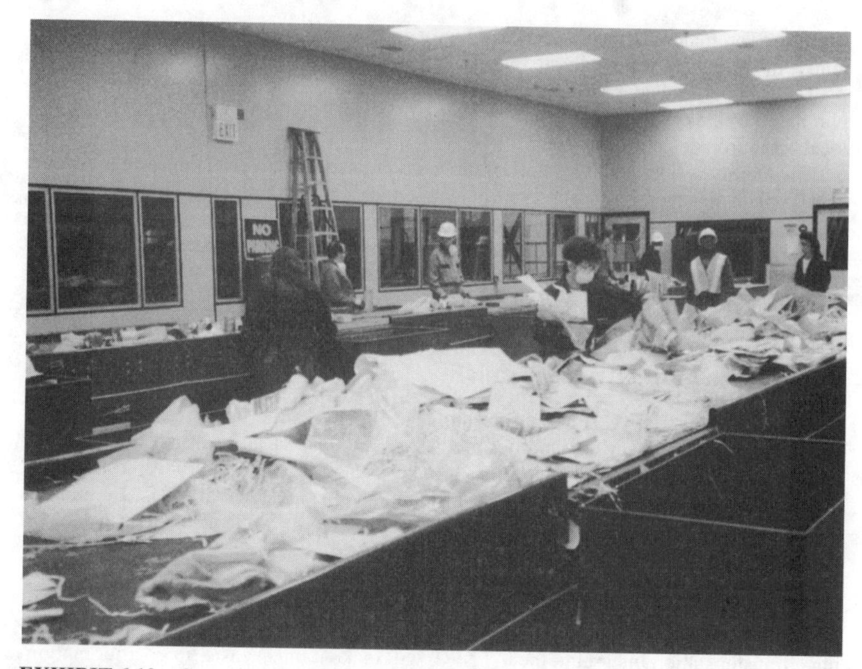

EXHIBIT 6.12 Commercial-industrial climate-controlled sorting section. (*Recycle America, Etobicoke, Canada.*)

ally clean loads. These loads normally wouldn't be handled by a standalone MRF and are destined for disposal. With a combined facility, the dirty loads can be sorted to attain maximum diversion and the residue is easily handled by the transfer portion. Siting considerations are easier since there's only one site instead of two. Zoning and permitting may be a little more complicated than with just one type of facility, but may be worth the effort. Figure 6.8 depicts a sample facility layout of a 300-TPD C/I MRF coupled with a 600-TPD transfer operation.

The typical method of separation used in these facilities is to dump the material onto a tipping floor and sort out contaminants by hand. If, however, an operation is bringing 50 percent corrugated loads to be separated, a cost/benefit analysis comparing the revenue potential versus the costs of processing the material is required. If the volumes of "dirty" commercial loads are great enough (more than 50 TPD), then the application of some sort of mechanical separation system should be investigated.

CONSTRUCTION AND DEMOLITION WASTE PROCESSING

Introduction

In addition to regular municipal solid waste and household garbage, a significant volume of mixed construction and demolition (C&D) waste is being hauled to landfills. This type

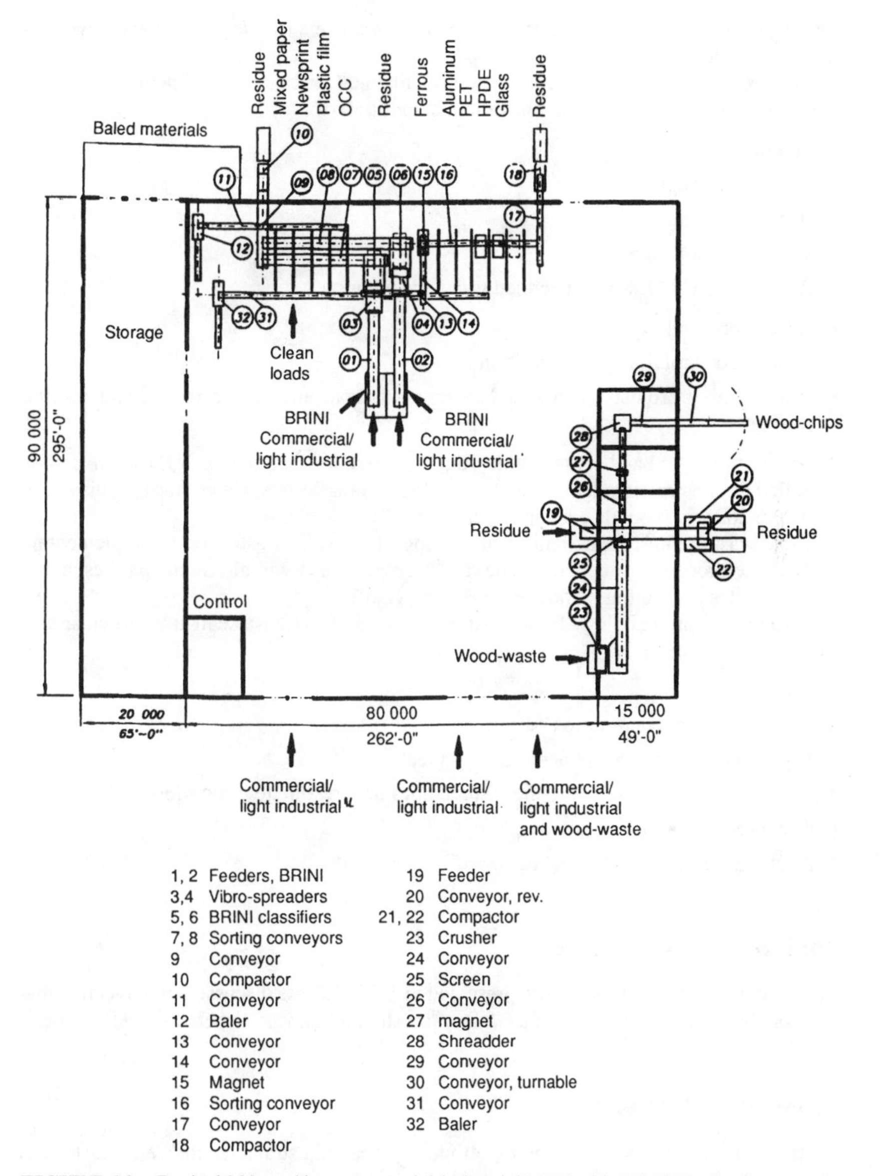

FIGURE 6.8 Typical 300-ton/day commercial-industrial MRF with 600 TPD transfer operator.

of waste, which is bulky and often not suitable for regular compaction, has a high index of recoverable and recyclable materials such as wood, aggregates, and metals.

The main objective for C&D processing facilities is *diversion* of the material from the landfill. The viability of any system depends on end uses for the materials. Local markets will determine the degree of "contamination" allowed for each type of product. For a system to be financially successful, it is essential to produce a *usable product* and have adequate outlets for the recovered materials.

The processing technology exists and is available. Construction and demolition waste is generated from construction, renovation, demolition of

- Buildings
- Roads and bridges
- Docks and piers
- Site conversion, etc.

Amounts of C&D waste generated is dependent upon

- Population level
- General construction and demolition activity
- Extraordinary projects such as urban renewal, road and bridge repair, and disaster cleanup

In recent studies, it has been estimated that *23 percent of municipal solid waste* can be classified as construction/demolition waste (1989 Franklin Associates study). This corresponds to 262 pcy (pounds per capita per year).

C&D waste represents a significant fraction of the solid waste stream. Implementing C&D waste processing, recycling, and diversion programs will aid municipalities, counties, and states in meeting mandatory recycling goals.

A suggested methodology for estimating volumes of C&D waste in any particular area includes analyzing

- Population trends
- Construction and demolition permits
- Types of construction and demolition projects
- Disposal estimates (be aware these sometimes underestimate generation)
- Past, present, and future trends
- Planning for "baseload" and "peakload" generation

Contents of C&D Waste

Typical contents of C&D waste are presented in Table 6.3. As discussed in the section that follows, the major components of a C&D waste stream will vary depending on its "type."

Types of C&D Waste

For the purposes of this discussion to introduce processing strategies for C&D waste, it is appropriate to categorize C&D waste into two main types.

Type I—Roadway and Site Conversion C&D Waste. C&D waste is typically classified as Type I if a large percentage of the waste stream (80 to 90 percent) consists of a small number of fairly "clean" fractions of material (i.e., rubble, wood). The rubble in Type I waste is typified by virgin soil/rock and concrete/asphalt bridges and pavements. The wood fraction often consists of stumps, brush, pallets, lumber, etc., and is typically generated by land clearing, landscaper activity, and residential home builders. Type I waste can

TABLE 6.3 Contents of C&D Waste

Waste type	Contents
Rubble	Soil, rock, concrete, asphalt, bricks
Tar-based materials	Shingles, tar paper
Ferrous metal	Steel rebar, pipes, roofing, flashing, structural members, ductwork
Nonferrous metal	Aluminum, copper, brass
Harvested wood	Stumps, brush, treetops and limbs
Untreated wood	Framing, scrap lumber, pallets
Treated wood	Plywood, pressure-treated, creosote-treated, laminates
Plaster	Drywall, sheetrock
Glass	Windows, doors
Plastic	Vinyl siding, doors, windows, blinds, material packaging
White goods/bulky items	Appliances, furniture, carpeting
Corrugated	Material packaging, cartons, paper
Contaminants	Lead paint, lead piping, asbestos, fiberglass, fuel tanks

be considered "clean" if large volumes are delivered for processing in a form that is easily separable into single waste types (material is dumped in separate pile or can be presorted with frontend loader, bobcat, etc.). Of course, Type I material is sometimes collected and hauled in mixed forms which make it difficult to separate.

A typical approximate composition of Type I C&D waste by *weight* might be:

Rubble	
Concrete, asphalt	40%
Soil, rock	20%
Wood	30%
Metals, plastic	10%

Type II—Construction and Interior Demolition Waste. C&D waste classified as Type II is generated from the construction and demolition of urban structures (office buildings, stores, etc.). Type II waste differs from Type I waste in that the material, as it exists in a building and in the way it is collected, is in "mixed" form (thoroughly mixed fractions of concrete, drywall, framing, ductwork, roofing, windows, corrugated, packaging, etc.) and thus difficult to separate. This material is typically collected and hauled in open-top trailers, roll-offs, etc.

A typical approximate composition of Type II C&D waste by *volume* might be:

Rubble	25%
Wood	33%
Metals	20%
Corrugated	12%
Other (carpet, residue, etc.)	10%

It should be realized that the *density* of C&D waste can vary dramatically. Material composed mainly of rubble can approach densities of 2000 lb/yd^3. Conversely, material containing large amounts of brush, insulation, or drywall can have densities as low as 250 lb/yd^3. If a single number is needed for an "approximate" density for *mixed* C&D waste, either Type I or Type II, 1000 lb/yd^3 is often a good estimate.

The throughput of a C&D processing facility will depend on the equipment used and more importantly on the type (rubble and wood versus building demolition) and nature ("clean" or mixed) of the material.

Primary reduction equipment used with Type I rubble and wood have large throughput capacities. Impactors and jaw crushers used to crush rubble can process anywhere from 50 to 400 tons/h depending on machine size and characteristics of the rubble. The throughput of hammermills and stump grinders used to shred wood typically range between 10 to 50 tons/h. Conversely, Type II C&D waste requires more hand sorting, which can lower system throughput dramatically. Facilities that receive urban building demolition waste and process 500 TPD or more would be considered a fairly large system.

C&D Waste-Processing Strategies

The waste stream composition and the end uses for the recovered materials determine the processing strategy and thus the equipment required for sorting and reduction. Table 6.4 presents end-use markets for recycled C&D waste.

In many cases, the rubble and wood received at facilities are bulky and vary in size. End-use markets typically require this material to be crushed or reduced into a smaller, consistent size. The primary reduction equipment that is characteristic to C&D facilities includes impactors or jaw crushers for rubble material and hammermills or stump grinders for wood waste. It should be noted that many of the processing system components are *modular* and can be added as necessary to address a certain fraction of the waste stream. For example, some C&D processors receive only rubble and therefore need only an impactor and operator, a front-end loader and operator, and possibly some screening equipment.

An expensive processing system is not needed for clean, separated material as the primary reduction equipment alone can provide quality end products. The sorting components of the system give access to the *mixed* material fraction of the waste stream. The type and nature of the mixed material determines the basic processing strategy:

1. Sort and separate, crush and reduce
2. Crush and reduce, sort and separate

It is important to note that with all mixed loads, the cost of separation versus disposal of the mixed fraction must be weighed. Certain loads may be so contaminated or mixed that separation may not be viable, with disposal being the only economic alternative.

Regardless of whether the material is Type I or Type II, all mixed material should be *presorted* as much as possible via judicious "tipping" and "picking" with bobcats, front-end loaders, etc. Generally, bulky items that are often presorted include major pieces of metal, large pieces of rubble or wood, white goods, furniture, and other undesirable materials such as carpet and tires.

Type I C&D Waste Processing Strategy. With Type I C&D material, the decision to sort first or crush first depends on the nature of the mixed material. Clean rubble and wood can be fed directly to an impactor and hammermill for reduction. In cases where mixed material contains any significant amount of plastics, paper, rags, or other contami-

TABLE 6.4 End-Use Markets for Recycled C&D

Waste type	End use
Dirt	Soil, soil conditioner, landscaping, landfill daily cover
Bricks	Masonry, landscaping, ornamental stone
Concrete, cinder blocks, rocks	Fill, roadbed, landfill haul roads
Asphalt	Road/bridge resurfacing, landfill haul roads
Tar-based materials	Mixed with used asphalt for resurfacing
Ferrous pipes, roofing, flashing	To scrap metal buyers
Aluminum	"The gold of C&D," remelted
Copper	"The other gold of C&D," reused
Steel, brass	To scrap metal buyers
Stumps, treetops, and limbs	Chipped for fuel, landscaping, compost bulking, animal bedding, manufactured building products, landfill haul roads
Framing, scraps	Chipped for fuel, landscaping, compost bulking, animal bedding, manufactured building products, landfill haul roads
Plywood, pressure-treated	May or may not be chipped for fuel, landscaping, compost bulking, animal bedding, manufactured building products
Creosote-treated, laminates	End use depends on local regulations concerning chemicals in material. If use is approved, uses are similar to those for other wood materials (listed above)
Used cardboard	Fuel pellets
Plaster, sheetrock	In place of sand in concrete/aggregate fill
Glass	In place of sand in concrete/aggregate
White goods/appliances	Scrap recyclers for crushing
Lead paint, asbestos, fiberglass, fuel tanks	None known

nants (paint, lead pipe, etc.), it makes sense to sort and separate, then crush and reduce. The combination of a disk screen and trommel can remove the fine soil and small rocks. Any contaminants, ferrous, nonferrous, and oversize rubble can be removed via magnets and picking, leaving "medium"size rock and wood for reduction.

When the remaining wood is mixed with rock, a flotation tank is sometimes used to separate the wood (which floats) from the rock (which sinks). With large amounts of rubble material the water tends to clean the product, which is beneficial. It should be noted that an air classifier could also be used to separate the lighter wood from the heavier rock. The air system costs more to operate because of the 50- to 75-hp blower that is required. However, depending on the type of material and the local environmental regulations, the wash water from a flotation tank may require treatment before discharge to a sewer or septic system. This could be costly and make the air system more attractive. It should be noted that a flotation tank *is not effective* for separating Type II material, which typically contains more fibrous contaminants which may become "soggy."

In cases where the Type I material is fairly clean with a large portion (80 to 90 percent) consisting of rubble (soil, rock, concrete, asphalt, etc.) and wood, and minimal contamination in the mixed fractions, it may be acceptable to crush and reduce, then sort and separate. This type of processing strategy is often used with roadway demolition projects and/or site conversion projects where there may be significant amounts of harvested and treated/untreated wood waste. A flotation tank is often effective with this Type I waste for separating the mixed wood from the rock, but environmental and wastewater treatment concerns must be addressed to determine if air classification is a better alternative.

The processing layout (Fig. 6.9) for this variety of Type I material displays a fundamental processing scheme that is being successfully implemented in various parts of the country. Note that this system is available in stationary or essentially self-contained, portable modes. The portable system can be desirable for maximum flexibility and reduction of hauling costs by processing "on site."

Type II C&D Waste Processing Strategy. In almost all cases, it is important that Type II building demolition waste be sorted and separated *before* being crushed. This waste stream may contain asbestos, paint, lead pipe, etc., that could become fragmentized if crushed, thus contaminating large amounts of material or causing environmental concerns.

After the bulky material has been removed via presorting, the mixed material is introduced to the system for separation and sorting. Building demolition processors have found that an effective first step for mixed material is to separate the soil and rocks *prior to hand picking* of the cleaned and uncrushed recyclables (sort and separate, then crush and reduce). This can be achieved by the combination of a specially designed trommel or disk screen system to separate two fractions of soil and rocks. Additional screening and air classification can be performed if needed. Hand pickers recover the various recyclables on a sorting platform that follows. This type of process has been shown, in some instances, to increase the efficiency of hand pickers and improve the recovery rate of recyclable materials such as wood, metals, and corrugated. In addition, the soil and rock fractions tend to be free of pieces of plastic, cardboard, paper, and other materials that are undesirable to fill or aggregate material. The recovered wood fraction can be shredded into a marketable form. Crushing, screening, and further classification of the cleaned rocks can be performed if required by local markets.

C&D Processing Equipment. In the area of C&D waste processing, equipment can essentially be grouped into three types:

- Conveying
- Crushing and reduction
- Screening and separation

Conveying Equipment. The majority of conveying equipment used is rubber-belted conveyors. Troughed idler conveyors are attractive for many systems as opposed to pan-type conveyors because of the grit associated with this waste stream. In certain ares of the system such as the main infeed point, heavy-duty steel-apron-type converters from the aggregate industry, for example, are used. Conveyors can be made portable rather easily for use in the portable systems that are available.

Crushing and Reduction Equipment. The actual types of equipment used will depend on what components of the C&D waste stream are to be reduced. For example, on rubble

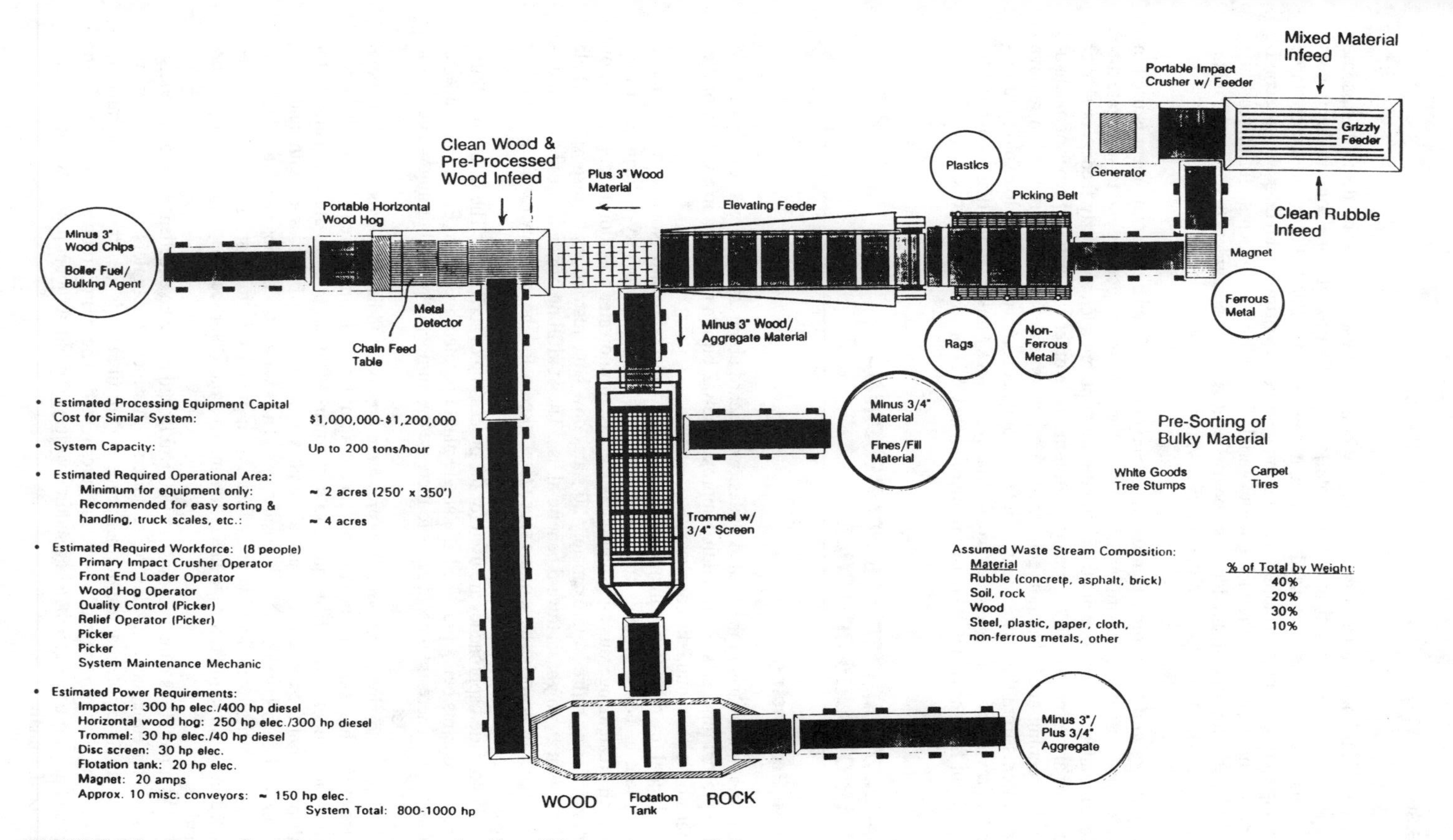

FIGURE 6.9 Process flow diagram—construction/demolition waste processing.

material, an impactor (rock crusher) or jaw crusher is used. For wood demolition, some type of hammermill (vertical or horizontal) is typically used at some point in the process because of the consistent-sized product it can provide. For bulky wood waste (large stumps, etc.), a stump-grinding machine can be used as a primary shredder with a secondary hammermill further reducing the material. For bulky waste such as furniture, white goods, etc., low-speed shredders can be used if it is more economical to reduce the material than to haul or dispose of the material whole.

Screening and Separation Equipment. Screening and separation equipment is used in C&D processing to split similar materials into various-size fractions and to segregate different materials from one another. Vibratory equipment such as grizzly feeders or shaker screens are common. Disk screens and trommels are other conventional types of mechanical screening and separation equipment. Various types of magnets are used to remove steel items from the waste stream. Flotation tanks are sometimes used to exploit the specific gravity difference between wood (floats) and rock (sinks). Separation of material by air (light from heavy) will be used more often as C&D processing evolves.

SPECIAL MATERIALS PROCESSING—VARIOUS APPROACHES

Tire Processing

On an annual basis in the United States, we discard about 250 million tires per year. At the present time, it is estimated that approximately 2 billion tires exist in a variety of illegal stockpiles around the country.

Illegal tire piles form a major safety and environmental hazard. Not only do tires make excellent breeding grounds for mosquitoes, but they are also very vulnerable to fires which create dangerous air pollution as well as groundwater pollution from potential oil runoff during a fire. Approximately 20 states now have restrictions on disposal of whole tires, with an emphasis toward beneficial use. The landfilling of whole tires has been discouraged for years because of the space they take up and the tendency for the tires to float to the top of the landfill.

Two varied approaches are used for the processing of scrap tires. The whole-tire-burning approach is being practiced at a single plant operated by Oxford Energy in Modesto, California. There are present plans to increase the number of plants burning whole tires in the United States.

The predominant form of processing scrap tires is shredding and disposal or beneficial use of the shreds that are produced. The output of these machines are tire chips that range anywhere in size from 1 in square to 6 in square. The larger chips are only suitable for disposal, but the reduction in size eliminates some of the difficult issues with landfilling tires, and does save valuable disposal space. The smaller-sized chips have value as either gravel substitutes in landfills or as fuels. As a gravel substitute, small tire chips can be used in constructing drainage layers in sanitary landfills. (See Chap. 18, "Tires," for details.)

As a fuel, tires contain almost 15,000 Btu/lb and make an excellent alternative energy source for either cement kilns, paper boilers, other industrial boilers, or utility operations.

As a future potential use, further processing of tire shreds (i.e., the use of granulators) to produce crumb materials is being investigated. The crumb rubber produced via the granulation process can then be used in a variety of molded rubber products and also rubberized asphalt.

Batteries

Lead-Acid Automobile Batteries. The largest amount of batteries by weight in the United States are discarded automobile batteries. The traditional lead-acid battery can pose an environmental threat due to its lead content when disposed with municipal solid waste. Many efforts are made to keep lead-acid batteries out of the waste stream.

The prime recycling method is for automotive shops, when replacing the batteries, to stockpile them for bulk shipment to battery recyclers. The battery recycling stream is called "battery breaking." This process, practiced by a handful of vendors around the country, involves the physical breaking of the case and removal of the lead components. In addition, the polypropylene plastic case is also recovered.

Household Batteries. Household batteries range from the traditional dry cell based on manganese and zinc, to mercury-containing button batteries, nickel-cadmium rechargeable cells, and alkaline cells. Some effort is being expended in an attempt to recycle household batteries. Various toxic materials such as mercury, cadmium, and zinc are found in individual batteries. The problem that exists today is that there are no good collection networks in the United States for the recovery of these materials. Dry cells and alkaline batteries have virtually no recycling processing capacity and are disposed of as hazardous waste. The so-called button batteries that are based on silver oxide and used in various small appliances, hearing aids, etc., and the nickel-cadmium batteries do have some recycling processing capability in the country. These processes, which shred and extract the valuable metals, hold some economic promise so that we may continue to recover more of these products from the waste stream. However, until a sufficient, steady quantity of these button batteries is available, so that a facility can be developed which could take advantage of the economies of scale, there will be little incentive to achieve an economically feasible recovery system. (See Chap. 19, "Batteries," for details.)

Waste Oil

As with batteries described above, waste oil is a problem in disposal facilities. In the category of materials known as household hazardous waste, waste oil is the largest single identifiable material in this stream. As with batteries, the collection infrastructure for household-produced waste oil is not good. Because of the toxic nature and classification of waste oils, more and more oil companies are beginning to collect and treat them. The processing of these oils consists of either direct reuse as a fuel or the dissolution and treatment of the waste oil to upgrade it for use again as a lubricant. These types of reprocessing facilities will continue to become more common so that this material can be reused rather than disposed of at a higher and higher cost. Like lead-acid batteries, the automotive and quick-change oil center collection infrastructure is in place.

Various companies perform the service of collecting commercially generated waste oils. Their processing consists of either direct reuse as a fuel, or the distillation and treatment of the waste oil to upgrade it for use again as a lubricating fluid.

MUNICIPAL SOLID WASTE (MSW) COMBUSTOR ASH

The recycling of municipal solid waste combustor ash (MSW combustor ash) has been slow to develop especially in the United States, for some very basic economic and environmental reasons.

First, from an economic point of view incinerator ash is a relatively dense material (in the range of 1250–1750 lb per cubic yard) and takes up little landfill space. The cost of landfilling ash is relatively low compared to other solid wastes. It is an easy material to handle, gives off no odor, and is very stable in wet,weather conditions. Ash either landfilled in a dedicated monofill or co-disposed with other solid waste therefore, is a beneficial material to the landfill operator for use in internal roads and for use as a daily cover material for MSW landfills as allowed by many states. Simply stated, it's cheap to landfill MSW ash.

From an environmental perspective, MSW ash is a difficult material to manage outside of a controlled environment such as a landfill. The disposal and/or use of MSW ash is highly regulated in the United States. A decision by the U.S. Supreme Court in the case *City of Chicago v. Environmental Defense Fund, Inc.* 1994 ruled that MSW ash was not exempt from regulation as a hazardous waste under Subtitle C of the Resource Conservation and Recovery Act (RCRA). Municipal solid waste, prior to incineration, is exempt. This places a cloud of uncertainty on the prospects of using MSW ash as a product rather than disposing of it in a permitted landfill. The concern for MSW ash is that certain heavy metals such as lead and cadmium, for example, would not meet the Toxic Characteristics Leaching Procedure (TCLP) requirements as established by RCRA Subtitle C and, therefore, require management as a hazardous waste.

This issue is further complicated by the fact that the fly ash portion of the MSW ash, which is removed from the air pollution control system, is generally the bad actor relating to heavy metal TCLP testing. The bottom ash material coming off the stoker grates is generally a benign material with good structural fill characteristics. It also contains valuable recyclable materials such as copper, aluminum, and ferrous metals.

The separate management of these materials, as is typical outside of the United States, where bottom ash is recycled separately, could, in effect, create a serious management problem for the fly ash component since it has little or no structural properties and could be regulated as a hazardous waste or at least require expensive health/risk assessments and monitoring.

Ash Recycling Processing

The low value of incinerator materials as a substitute for relatively cheap common fill or structural fill, its high density and ease of landfilling, and environmental issues associated with its use make recycling of MSW ash difficult.

The recycling of the ferrous and nonferrous metals and even coins contained within the ash, which can be extracted for sale, can and does often make it economical to treat and process the ash for reuse. This is especially the case for mass burn technology, where little or no front-end separation of materials is done. The ash clinkers coming out of the combustor can be sized and processed through magnetic and eddy current separators to remove valuable metals. The final product is usually a well-graded aggregate/sandy material that can be utilized as fill. Unfortunately, the requirements for controlled regulatory testing of the processed ash material and its low value often result in the ash being landfilled once it has been stripped of its valuable materials. In other words, it is not the ash that has value, but what is in it.

For refused-derived fuel (RDF) combustors, which remove a great deal of the valuable metals in processing MSW into a higher BTU fuel, very little value remains in the ash stream. Although the resultant ash is in a form that requires little or no additional processing for use as a fill material, the same environmental issues concerning its use remain.

There are numerous existing technologies, mostly European, in use today to process and screen MSW ash from mass burn facilities located around the world. American Ash Recycling, Inc. located in Jacksonville, Florida, for example, is currently processing MSW ash in several locations. In the United States, ash must typically be treated prior to or after processing before being utilized. The treatment of ash can be either a wet or dry process. In the dry process, ash conditioner is usually added to the fly-ash-only portion of the ash, which typically is about 25% of the total ash by weight. The conditioner is usually a portland cement type material that chemically reacts with the fly ash to stabilize or eliminate the metals of concern.

In the wet treatment system, liquid chemical is added to the combined ash to condition the ash to a more neutral PH and prevent the metals of concern from leaching out of the ash.

These systems are proprietary and can be purchased through companies such as Wheelabrator Technology, Inc. or Forrester Environmental Services, Inc., for example.

State permits for treated ash exist, but at present are the exception rather than the norm. These permits are granted for a specific use on a case-by-case basis usually for use as a road base fill material and cannot be used by any other MSW ash generator. Often the requirements for some type of controlled environment or monitoring simply take this low-value material out of the marketplace and it ultimately goes to the landfill.

Outlook for the Future

The outlook for a generally accepted beneficial use for ash on a state or nationwide basis is not very promising at this time. Some states such as Florida are attempting to provide guidelines for ash recycling, but again, permits are issued on a case-by-case basis and likely involve expensive health/risk assessments. Given the regulatory climate in the United States concerning combustor ash, uses of ash in which the material is covered by soil, liners, or other materials and is managed in a controlled environment will likely have the most success. These uses include, for example:

1. Daily and intermediate cover material for lined or Subtitle D approved landfills with leachate collection systems.
2. Structural fill for landfill-based construction above the water table and below the primary and secondary liner systems.
3. Fill material and gas transmission layers for landfill closures both below the liner (gas transmission) and above the liner (18″ sand layer), which is then covered by soil (6″ or more) and vegetation.

The use of this material in making concrete or concrete products, for example, was once considered a practical and very simple solution. This process tends to fix any metals and prevent future leaching. But environmental concerns over the runoff and dust from stockpiles at batch plants, not to mention transportation issues and monitoring, have all but taken this consideration out of economic reality. Nearly all recycling programs using ash as part of a concrete product has been abandoned for these reasons.

The future of MSW ash recycling is not certain unless supported by specific regulatory or legislative effort on either a state or national basis.

CHAPTER 7

MARKET DEVELOPMENT: PROBLEMS AND SOLUTIONS

GREGG D. SUTHERLAND
Former Eastern Regional Director
Resource Integration Systems, Inc.
Toronto, Canada

INTRODUCTION

Typical economic behavior tells us that price and quantity are directly related. As price increases, quantity supplied should increase to take advantage of the new price. This traditional behavior is supposed to work in the other direction as well, so that when prices fall, quantity supplied should decrease due to reduced price incentive.

However, recent recycling markets have not followed these traditional economic precepts, at least on the surface. In fact, for many grades of recyclable materials, the volume supplied has increased to record levels while prices have fallen to historic lows. How can such contrary behavior occur? The answer lies in the fact that the supply of recyclables has been artificially stimulated by government mandate, while the demand for recyclables has been largely unaffected. The result is growing supplies along with declining prices.

The development of end markets for recyclable materials is now advancing to the forefront of vital recycling issues. Without effective development of these markets, recycling cannot grow. Accordingly, market development is now receiving the attention from government policy makers, from recyclers, and from industry that it requires in order to "close the loop" on recycling.

This chapter will

- Define market development within the broader context of the overall recycling system.
- Delineate the appropriate goals of market development.
- Examine why market development is typically overlooked in the early stages of expanding recycling systems.
- Identify the major barriers to market development. This section will also explore actions and policies that can overcome each barrier in order to effectively promote market development.
- Discuss alternative markets as a way to help recycling programs cope with disruptions of traditional markets.

DEFINITION OF TERMS

When discussing market development, it is important to start with a clear understanding of the terminology. This is because the term *market* depends largely on perspective. For a generator of waste material, the "market" is the recycler who collects the material. That recycler then sorts, grades, and processes the material for shipment to his "market," a manufacturer who is an end user of that material as a raw feedstock for his product. The end user then sells the product to her "market," a consumer who purchases the finished product with recycled material content. Figure 7.1 illustrates this loop, showing the definitions as they are used in this chapter.

While recognizing that generally accepted terminology for many recycling terms is still widely debated, for the purposes of this chapter the following definitions will be used:

- A *consumer* is a person or other entity who purchases a finished product that may or may not include recycled content. After consuming the product, the consumer becomes a generator of any remaining waste material. Consumers and generators can be individuals, businesses, or government units. Examples include homeowners who produce old newspapers and scrap packaging or businesses that produce used corrugated boxes.
- A *recycler* is a person or other entity who collects scrap materials for sorting, processing, and shipping to a manufacturer. Manufacturers often make the point that, technically, the term "recycler" does not apply fully to the business that serves in this intermediate role, since these businesses usually do not complete the loop by turning the material into a finished product. Nonetheless, because they take the first step toward diverting material from the waste stream and they are widely dispersed and visible to generators, these businesses are commonly called recyclers anyway. In this chapter, that common definition will be used. Examples include scrap metal dealers, municipally mandated recycling facilities for residential recyclables, and paper stock dealers.

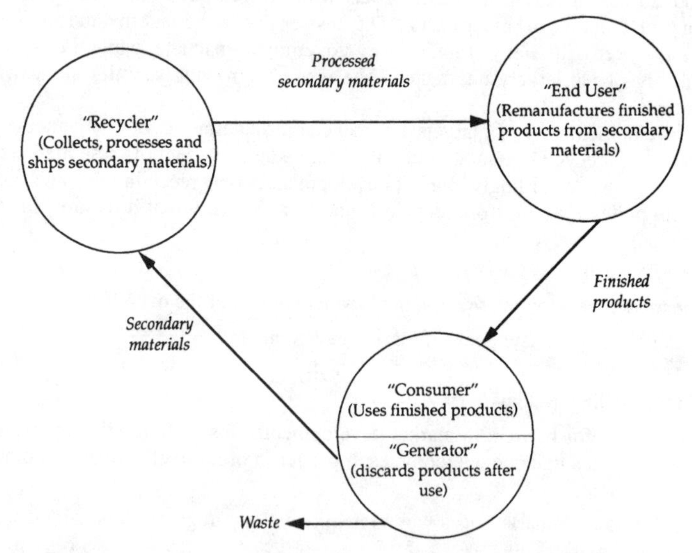

FIGURE 7.1 Definition of key recycling entities.

- An *end user* is a person or other entity who remanufactures a finished product from secondary, scrap material that has been diverted from the waste stream. This use of the term end user is based on the concept of secondary material as a raw material. Thus, the end user is the manufacturer who uses that raw material, as opposed to the consumer who then buys the finished product with recycled content. Examples include steel mills, paper mills, glass plants, and plastic bottle plants that use scrap as a feedstock. End users often combine secondary materials with virgin materials.

Given these definitions, *market development* means development of end users of secondary materials. This chapter addresses ways to promote the development of end users as the final market for recyclable materials. Because generators and recyclers must produce and separate these raw materials and because consumers must purchase the finished products with recycled content, these groups will be addressed as well, but only as they affect market development, the development of end users of secondary materials.

THE GOALS OF MARKET DEVELOPMENT

Before discussing ways to promote market development, it is important to understand the goals of this activity. There are several appropriate goals, summarized in Table 7.1, for market development.

Prevent Imbalances

The most obvious goal of market development is to match end-use infrastructure growth with collection infrastructure growth. Imbalances, where collection grows rapidly compared to end-use markets, have resulted in severe material gluts that threaten the operational and financial viability of many recycling programs.

Figure 7.2 shows graphically that collection programs and infrastructure can grow rapidly without a corresponding increase in diversion of materials from the waste stream. Recycling is like a pipeline, where the smallest-volume section determines the maximum flow through the entire system. When end markets do not grow with collection, end markets constrain the entire recycling system. Thus, the first goal of market development is to create a system balance between collection and end use.

Promote Economically Sustainable Recycling

While creating a balance between collection and end use is the most obvious goal of market development, it is not the only goal. In fact, from a long-term perspective, there may

TABLE 7.1 Goals of Market Development

- Prevent imbalances
- Promote economically sustainable recycling
- Minimize need for government intervention
- Promote economic development
- Conserve ancillary resources

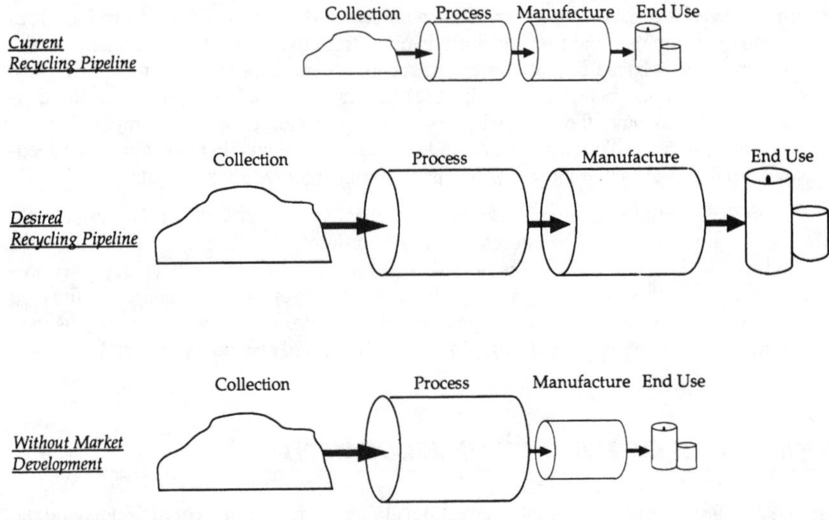

FIGURE 7.2 Consequences of ignoring market development.

be an even more important goal of market development: to promote economically sustainable recycling.

Much of the growth of recycling is shifting away from programs initiated and operated strictly for the sake of private sector profitability and toward government-mandated systems. Accordingly, the economics of government-mandated systems are coming under scrutiny. In many cases, the value of recyclable materials recovered from a municipal recycling program does not cover collection, processing, and shipping costs. Adding the avoided costs of disposal to that equation may or may not show a recycling program to be economically justified, depending largely on local disposal fees and the means by which those disposal fees are calculated.

Most government-mandated systems are not initiated strictly on the basis of economics, however. Municipal recycling is rapidly becoming a standard municipal service, just like sewage treatment, water supply, street maintenance, police and fire protection, and trash removal.

Nonetheless, as local governments assume more responsibility with less federal funding, they are encountering unprecedented budget problems. This means that expansion even of government-mandated systems is threatened by economics. For example, one large state had originally planned to build an entire network of materials recovery facilities (MRFs) to process recyclables from residential recycling programs. However, after funding the first MRF, the state discovered that it had inadequate funding for additional MRFs and, as of this writing, none have been built since. A key element of the problem is that the glut of household recyclable materials has driven revenue from these programs so low that they often require substantial government subsidies to operate. A particularly significant factor in the glut is old newspapers, which comprise the majority of household recycling collections.

The same economic problem occurs as governments attempt to promote recycling by commercial waste generators. Almost by definition, commercial waste generators already recycle most of the materials that are economically viable to recover. Government programs typically put the burden on commercial generators to recycle more by requiring recycling plans or by banning certain materials from disposal.

In some cases, this prompts businesses who were missing out on savings to realize their mistake and implement a recycling program. However, this approach often becomes a hidden tax to subsidize recycling, since mandatory recycling can force commercial generators to pay more than they normally would for recycling. This is because they no longer recycle only for economic reasons; they now recycle because it is illegal not to. As a result, many commercial generators are told to recycle beyond any directly measurable economic incentive to do so.

While some economic improvement can result from improved collection and operating technology, the variable with greatest impact on the viability of most recycling programs is the market value of materials collected.. This is where market development becomes critical, whether for residential or commercial recycling programs.

Because market development promotes the demand for recyclable materials, the price of those materials increases. Figure 7.3 illustrates a classic economic supply and demand curve, showing that price (as well as quantity) increases from point 1 to point 2 if market development can shift the demand curve out by causing a structural increase in demand.

Market development improves the economics of recycling. An improvement in the economic motives to recycle makes all varieties of recycling programs more viable and sustainable.

Minimize Need for Government Intervention

Another goal of market development is to minimize the need for governmental involvement in recycling programs. Currently, most public sector policy toward recycling is based on overcoming economic barriers to recycling by funding new collection and pro-

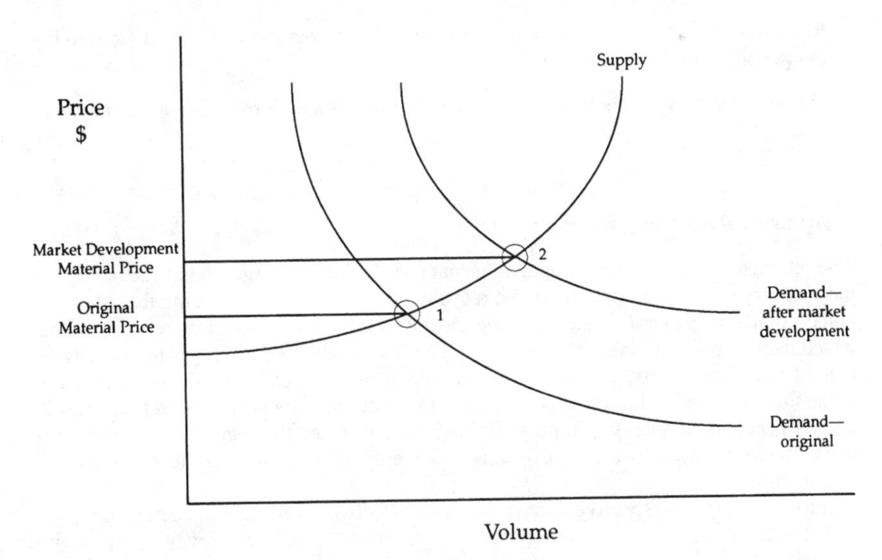

FIGURE 7.3 Graph: Market development effect on price.

cessing systems (in the case of residential recycling) or by mandating generators to bear the incremental costs of increased recycling that may or may not be directly economical (in the case of commercial recycling). To implement and enforce these policies can be expensive, intrusive, and bureaucratic.

Some public sector decisions are made without the realization that, before the advent of municipal involvement in recycling, millions of tons of scrap materials were already recycled each year by private industry for economic gain. While there is little dispute that increased government involvement is needed to achieve recycling targets, it is imperative to remember that the private sector can play an effective role in achieving those goals. The best way to harness the private sector is through economic incentives. Market development is the most effective way to provide those economic incentives.

To the degree that government can rely on private industry to achieve recycling goals, government can commit less of its own scarce resources. Government can also take a less intrusive and bureaucratic role in the process.

Promote Economic Development

Another goal for market development is that of general economic development. Most jurisdictions have committed resources toward economic development. As a matter of local public policy, these actions usually focus on providing information to prospective new businesses and on making special arrangements to attract specific, large-scale new businesses. Recycling end users are excellent targets for such market development efforts, because

- End users are often large-scale businesses that can be significant employers.
- End users engage in an activity, recycling, that holds widespread public appeal.
- Recycling is a sustainable use of resources and usually has significant energy savings over virgin material use.
- Recycling is perceived as a long-term growth industry that can lend economic stability to a jurisdiction.

Market development promotes recycling, which is often a legitimate economic development objective.

Conserve Ancillary Resources

One other goal for market development is that jurisdictions are eager to take advantage of the ancillary resource conservation that recycling can cause. Manufacturing that uses secondary materials generally takes less energy and water and produces less waste than that which uses virgin materials. Increasingly, proactive jurisdictions are eager to take advantage of this resource conservation. For example, some utilities have granted millions of dollars to paper mills that undergo retooling to install equipment that would utilize secondary paper rather than virgin pulp. Utilities have justified these grants on the grounds that using secondary paper is significantly less energy intensive than producing and using virgin pulp.

From a utility's perspective, grants are a cost-effective investment in energy conservation. From a market development perspective, grants are an effective way to initiate investment in remanufacturing capacity.

THE NEGLECT OF MARKET DEVELOPMENT

Because end markets are a downstream activity from the generation of waste, there has been a natural tendency for most people to pay less attention to market development than to collection. However, the reasons for market development neglect are often much more subtle than this.

Traditional steps in developing recycling systems have consistently led to market debacles. First, collection-based programs are launched, and a local collection and intermediate processing infrastructure is built. As a glut of materials builds, the jurisdiction implements procurement programs, only to discover that recycled-content products are in short supply. This history has been repeated in jurisdiction after jurisdiction for both residential and commercial recycling programs. The result of this approach, until the final step of market development is implemented, is to create both a glut of recovered materials for which there is an inadequate end market and procurement programs for which there are few suppliers.

The reasons for the neglect of market development are illustrated in Table 7.2 and described below:

- Whether targeted at the individual, collector, or processor, collection and procurement programs are generally inexpensive, especially on a unit-cost basis. For end users, investing in remanufacturing equipment for secondary materials is often, by contrast, expensive. For example, adding a drinking system to an existing moderately sized newsprint mill could cost over $50 million.
- The time frame for the expansion of collection and procurement programs is quite short, especially as measured in lead time for the key element of equipment. The time frame for market development is much longer, because typically it takes years to obtain an internal capital budget; apply for and receive zoning, land, water, air, waste, and operating permits for facility construction; and order, install, and debug the necessary equipment.
- Collection and procurement programs are typically implemented at the local, municipal level. Compared to federal or state levels, these jurisdictions can move relatively quickly to implement collection programs, especially in areas that enjoy widespread public support for recycling. However, promoting large-scale capacity expansion among end users is frequently beyond the limited scope of local government. It usually

TABLE 7.2 Reasons Market Development Is Neglected

Who recycles	How to increase	Unit cost	Lead time	Government level
Individual	Promote, provide collection container	$10–75	30 days	Local
Recycler/collector	Buy trucks	$125,000	60 days	Local
Recycler/processor	Buy balers	$300,000	90 days	Local
End User	Build cleaning systems	$75,000,000	3 to 5 years	State/regional
Consumer	Buy recycled content	$0	Immediate	Local

takes the authority and resources of a state, regional, or federal government to create the incentives and minimize the obstacles to develop significant markets. Gaining consensus and initiating action at these governmental levels is inherently more difficult and time-consuming than at a local level.

There are also perceptual reasons why market development is neglected. Collection is a highly visible activity that requires the participation of the individuals in the jurisdiction. As such, collection is an obvious magnet for activity by community, political, industry, and advocacy groups. The visibility and participation of collection programs are important to each of these groups.

By contrast, market development often involves complex negotiation with small corporate committees far outside the associated jurisdictions. This lack of visibility and grass-roots participation make it easy for many types of groups to overlook market development.

However, for all the reasons why ignoring market development is easy, there is one compelling reason why it cannot be ignored for long: *without market development, the overall recycling system cannot grow.* Without growth of the overall system, the visible and participatory collection and procurement programs start to fail in a highly visible way. As a result, well-deserved attention is now being focused on market development, particularly by state- and provincial-level governments and by industry.

BARRIERS AND OPPORTUNITIES

Market development is a complex process that calls for the careful integration of government, the public, and private industry across a well-developed set of policies and programs. This section will identify barriers to market development and opportunities to overcome those barriers.

There are several key areas for effective market development strategies:

- *Supply:* Market development depends upon a reliable, high-volume, high-quality stream of secondary raw materials.
- *Consumer demand:* Market development can only succeed if consumers purchase the finished goods that have recycled content.
- *Technology:* Effective market development will require investment in new technologies as industry moves away from using virgin raw materials.
- *Government role:* Government action can have a large impact—positive or negative—on market development.
- *Economics:* Market development is primarily economically driven.
- *Alternative end markets:* Developing nontraditional end uses can take the pressure off oversupplied markets.

Supply

Some government recycling authorities quietly endorse the concept of intentionally creating collection policies that produce a glut of secondary materials to "force" industry into expanding end markets. Others insist that collection programs should only be initiated after end markets are well established and the overall economics are favorable.

In fact, these opposing positions are a false dilemma. The expansion of markets and of collection should proceed in step with each other. Imbalances resulting from uncoordinated development only frustrate the overall system. In other words, collecting a supply and creating a demand are both necessary conditions, but neither alone is a sufficient condition.

In order to convince potential end users to expand their capacity, there must be an assurance of a reliable supply channel. End users typically invest millions of dollars in their manufacturing systems and must look to a long-term return on their investment. This has traditionally caused many manufacturers to integrate backward into enterprises that allow them to control their virgin raw material feedstocks. For example, paper mills often own large tracts of forest land. Plastics producers are often petrochemical producers, or they have established longterm agreements with such producers. The attitude of "controlling the raw material supply channel" is well entrenched in most large manufacturers.

There has long been concern among manufacturers about relying on a scrap collection network that is highly fragmented among thousands of small, regional recyclers whose programs are dependent on widely fluctuating and virtually unpredictable scrap material prices. Add to this a public and government that have had a checkered history of implementing sustainable recycling, and the result is that many large manufacturers (with some notable exceptions) have traditionally minimized the use of scrap as a significant raw material.

However, several factors (see Table 7.3) are changing to improve the quality and reliability of supply channels:

- Manufacturing firms are integrating backward into ownership of or joint ventures with large recycling companies. This technique, long practiced by traditional scrap end users, is now becoming common among new end users in order to secure their supply channels. In fact, one emerging model is that large trash hauling firms acquire traditional recyclers. These increasingly concentrated recyclers then form alliances with large end users. As long as they do not interfere with a competitive industry structure, such ventures can help to create markets through the development of secure, integrated supply lines that are integrated with end users.
- The collection of recyclable materials was once almost completely driven by spot markets and by mostly small, entrepreneurial recyclers. Such a system did achieve excellent economies for a given amount of recycling but produced sometimes wide fluctuations in price and quantity. Now that recycling is becoming a public policy goal and a public service expectation, collecting recyclables is an increasingly formalized, predictable process. This evolution serves to assure manufacturers that their investment in new systems will continue to have raw materials available to them, because the collection programs are government-mandated.
- One drawback associated with the growth of government-mandated recycling is that the quality of the secondary raw material has, in many cases, been reduced. This is

TABLE 7.3 Supply Opportunities

- Backward integration of manufacturers
- Government-mandated supply reliability
- Quality improvement of recyclable materials
- Development of robust cleaning systems
- Attention to market specifications
- Restrict contaminants

partly because the growth in recycling programs has sometimes put inexperienced municipal or private operators in control of operations, and that lack of experience has resulted in lower quality material. Also, the focus of recycling is now shifting toward maximizing diversion from disposal. This focus places a lower emphasis upon quality of material. For example, some office recycling programs that used to collect only white and computer paper have expanded into all office paper (often as a result of government mandates). This lowers the overall grade of paper collected. Solutions to this barrier include

The development of more robust cleaning systems by end users. Some tissue mills, for example, have achieved a significant competitive advantage by developing cleaning processes that allow them to use lower grade materials at lower prices.

More careful attention to market specifications by recyclers and by government-mandated collection programs. One large city added a wide diversity of plastic to its ongoing curbside recycling program, only to discover that the overseas end user could not actually use such a broad range of resin types. The city reverted to a limited range of plastic bottles for which secure markets were available. Another municipal program, located in an area infamous for disruptive gluts of low-grade paper, launched its collection program to include all household paper, not just newspaper. As a result, markets for this mixture of low-grade paper are often unavailable, prices are always low, and stockpiling the baled paper is common. While private and government recycling programs should challenge end users to take a broader range of materials, they do not promote market development by dumping scrap of unacceptable quality onto end users.

Materials that pose significant contamination hazards to end users are being restricted or banned. Manufacturing finished goods from recycled raw materials is a demanding process. Contaminants can disrupt an entire processing run, which works against market development. Increasingly, governments in the United States and abroad are restricting or banning materials that could cause such contamination problems at end users' facilities. Examples of some potentially problematic contaminants include polyvinyl chloride (PVC) bottles or components in polyethylene tetraphthalate (PET) bottle reclamation streams (some European countries have banned PVC bottles for this reason), hot melt glues in paper (some jurisdictions are considering prohibiting newspapers from printing advertising specialties that include such glues), and ceramics in glass bottle reclamation streams (some jurisdictions now prohibit specialty ceramic bottles or bottle components). Industry must ensure that its packages and products are compatible with the recycling infrastructure, with the assurance that governments will act to restrict them if they do not restrict themselves.

Effective market development tools relating to supply include developing secure supply channels and working to improve the quality of materials for recycling end use.

Consumer Demand

End users have traditionally shied away from recycled materials because of perceptions that these materials do not meet the quality standards of virgin materials. Even end users who do use secondary materials typically kept that fact quiet in order to avoid creating an impression that their product was second-rate. Without consumers willing to buy products with recycled content, market development will fail. Fortunately, consumers are convinc-

TABLE 7.4 Consumer Demand Opportunities

- Purchasing preference
- Recycled content mandates
- Labeling standards

ing end users that they often actually prefer products with recycled content. At the same time, end-user technology is maturing, which improves the quality of recycled-content products. Consumer demand methods for market development are listed in Table 7.4.

Purchasing Preference. An increasingly aware public is more interested than ever in buying products and packages with recycled content. This is true on an individual consumer level, as numerous market preference studies indicate. Some tissue mills have always had a very high percentage of recycled content, which was used initially to gain a cost advantage over their competitors. In marketing tests, some of these mills started labeling certain products to promote the recycled content. As a result of the labeling, products labeled with recycled content sell faster in certain markets than similar but unlabeled products with the same recycled content. Several large consumer products companies have also started promoting the recycled content of their plastic bottles and paperboard boxes, tapping into the consumer preference for such items.

Consumer demand is also growing among businesses and governments. Increasingly, these organizations demand recycled content in their packaging, paper, and other products. In fact, some organizations now demand similar procurement standards from their suppliers and vendors. This trend could expand the scope of procurement standards dramatically.

Recycled-Content Mandates. In addition to internally initiated procurement standards, there is an emerging trend toward externally imposed procurement standards. Newsprint is perhaps the best example, since the majority of the U.S. population now lives in states that have some form of recycled-content standards for newsprint. Several states and provinces are implementing similar recycled-content standards for a variety of products and packages. Plastic bottles and bags are favorite targets, but jurisdictions may set recycled-content rates for a wide variety of other materials as well. Some proposed legislation would set recycled-content targets for all packaging.

Labeling Standards. Ironically, the promotion of recycled content, which should facilitate market development, has created a need for standards for product labeling. If these labeling standards are addressed incorrectly, they could actually work against market development. In the absence of a uniform, national approach to labeling regarding recycled content, local jurisdictions will develop their own standards. With thousands of jurisdictions in the United States, the proliferation of such inconsistent and even contradictory labeling standards has already driven many manufacturers to eliminate labeling claims on their products and packages. Thus, whereas the label should give a competitive edge to the company that has made the investment and commitment to use secondary materials, that company may be forced to drop the recycled-content label entirely. This is because, as a national manufacturer, it cannot afford to set up differently labeled products to meet each individual jurisdiction's standards.

There are currently several competing approaches to labeling. Some labeling programs indicate a "seal of approval" approach. Examples include Germany's Blue Angel program, Canada's Environmental Choice program, and private initiatives in the United States. A seal of approval is consistent with many state-level rules, which allow the word "recycled" only if the product meets certain levels of recycled content (which may be different from other materials and from other jurisdictions).

Others are advocating a full disclosure type labeling that would describe a wider range of recycled content. With this approach, manufacturers who used recycled content can use the word recycled but must indicate the specifics about composition. One recycled paper dealer has even developed a sliding scale of recycled ratings, depending on the proportions of preconsumer, postconsumer, and virgin materials.

The subject of labeling issues and guidelines is currently being addressed by a wide variety of groups, including the Environmental Protection Agency, the Recycling Advisory Council of the National Recycling Coalition, the American Society for Testing and Materials, the National Association of State Purchasing Officials, several industry associations such as the Paper Recycling Coalition, state and local governments, and many others. While the resolution of labeling issues is beyond the scope of this chapter, it is valuable to establish market development goals for labeling:

- Labeling policies should reward those manufacturers which make a commitment to using recycling content. This reward should be proportional to the level of end-user recycled content. This means that variable, incremental labeling with full disclosure of the amount and type of recycled content is preferable to labeling that does not disclose those factors.
- Labeling should be nationally consistent. The expenses of small production runs, increased safety stocks, and distribution logistics, can prohibit end users from labeling products rather than attempt to meet a variety of standards.

Technology Development

To some degree, all end use of recycled materials is dependent upon effective technology. Technology for end users to utilize secondary materials is improving. Flotation systems for newsprint production can now utilize coated ground-wood paper; depolymerizing PET plastic allows that resin to be recycled back into soft drink bottles; detinners now routinely use old steel, tin-coated cans as part of their feedstock. However, other technologies are still struggling. Laser-printed paper is a problem for many paper and tissue mills; ceramics are an easily concealed contaminant for glass; and boxboard, although having a high recycled content, is not readily recycled.

Improving the technology of end users is an important market development tool. In addition, technology development is an area where government and industry have historically enjoyed a mutually productive partnership. Expanding technology development programs, through government research laboratories, universities, and technology transfer programs, can help promote market development.

Government Level

Collection of recyclables is usually a local government role. With residential programs, local governments typically establish curbside collection and drop-off programs. With commercial programs, local governments typically mandate that businesses separate specified materials and ban those materials from disposal. These types of activity are within the charter and authority of most local governments.

However, a local government is often severely limited in its potential impact on market development. This is because the local government usually does not have the authority or resources to significantly affect market development decisions. To significantly affect such decisions, large-scale procurement programs, product labeling standards, economic incentives, and technology support must be implemented. That level of authority and resources is usually placed at the state or federal level.

Unfortunately, obtaining a consensus and charting a course of action is often more difficult at these higher levels of government. On their own initiative as well as at the prompting of local government and industry, state governments are increasingly addressing market development issues in a meaningful way. Currently, most U.S. states have some type of market development program underway. In the absence of clear, definitive federal leadership, several states are banding together, formally or informally, to establish regional recycling initiatives, including market development programs. Such state-level and regional approaches can be very effective in implementing market development programs.

Economics

In the end, market development is driven by economics. In fact, as discussed under "Goals," one of the key goals of market development is to improve the economics of the entire recycling system. There are a variety of ways to improve the economics of end use of recyclable materials (see Table 7.5), but broadly, they fit into one of two categories:

- *Economic intervention:* One way to improve end-use economics is to intervene in the marketplace with direct economic intervention, typically with government funds. Examples include loans, grants, subsidies, tax incentives, and recycling credits.
- *Free market promotion:* In some ways, end markets are constrained due to restrictions on a free market that would normally favor recycling more than is currently the case. By promoting a free market, these variables would improve the economics of recycling markets compared to virgin materials. Examples include accurate disposal pricing and removal of virgin material subsidies.

Economic Intervention. Economic intervention can be a very effective market development tool, but it must be approached with caution. Inappropriate use of economic intervention can easily allocate scarce recycling resources into end uses that are ineffective and not sustainable. In general, policies that help to cover start-up or capital costs are preferable to policies that subsidize ongoing losses. End markets that require ongoing

TABLE 7.5 Economics

- Economic intervention
  - Loans
  - Grants
  - Subsidies
  - Tax incentives
  - Recycled content credits
- Free market promotion
  - Removal of virgin material subsidies
  - Disposal pricing
  - External cost accounting

subsidies are not, by definition, sustainable. In addition, committing public funds to ongoing losses subjects end markets to the uncertainty of the public sector budgeting process.

Loans and Loan-Guarantee Programs. Already, several state and federal programs facilitate loans or loan guarantees to new ventures in the name of business development. When applied to market development, such programs can be effective. This is because market development often requires a manufacturer in a capital-intensive industry to retool. Such a requirement can be a significant burden on the end user's balance sheet. A below-market or guaranteed loan can make the necessary funding available to the end user.

Grants. The same logic applies to grants. Grants often depend upon specific criteria established by the granting agency. Typically, such grants focus on feasibility studies and research projects. However, grants are also being used to fund capital and start-up costs as well. Some grants are targeted toward promotion of ancillary resource conservation, particularly energy conservation.

Subsidies. Subsidies usually address some ongoing portion of the end user's costs. Subsidies may be structured as a tax credit for the consumption of recycled materials. Some subsidies are structured to phase out over a set period of time. This minimizes the potential of funding an end user that is not sustainable.

Tax Incentives. Tax incentives can take many forms. However, many tax incentives are not effective because they have a relatively small financial impact. For example, some local jurisdictions offer credits on property, sales, income, and other taxes for specific market development investments and expenses. However, the tax burden imposed by these local jurisdictions is sometimes so small that it has little impact on investment decision making.

States may use tax incentives to better advantage, since their tax effect is usually larger. Several states now exempt property taxes and sales taxes for purchases of new recycling equipment. In the case of large manufacturing systems, these tax exemptions can be significant incentives to the economic decision to invest in market development.

In addition to tax exemptions, some states are implementing investment tax credits (ITCs). These credits can be very effective because they may be targeted to offset taxes that will significantly affect end-user decision making. For example, in several states, corporate income tax is a much larger dollar item than many other taxes. An ITC can be specifically structured to offset that significant tax. By focusing on a specific tax, the ITC can result in a significant dollar impact.

Recycled-Content Credits. One interesting approach toward using a market-based economic intervention tool is to create a market in recycled-content credits. This approach is patterned after federal air emission regulations, where power plants that emit air pollutants are each granted a limited number of "pollution credits." These credits could be bought and sold, which enables plants that cannot economically retrofit with pollution-control equipment to buy credits from other plants. Overall, air pollution is reduced, an economic incentive is created to install pollution-control equipment, and specific plants are spared the problems of trying to apply blanket regulations to a variety of cases.

This same approach could apply to recycled content for certain materials in specific applications. The federal government could issue a limited number of credits for the use of virgin materials. The amount of these credits would be limited and would decrease over time. Manufacturers who wish to continue using virgin materials would have to buy credits from manufacturers who use recycled, rather than virgin, materials. Bills along these lines have been introduced in the U.S. Congress for the use of recycled content in a variety of materials.

The benefits of such an approach are the same as for pollution control. The credits provide a direct economic incentive to utilize recycled materials, while specific manufacturers are not "forced" to switch technologies. The drawback is that a complex reporting and auditing mechanism would be required to implement and enforce this system of credits.

Free Market Promotion. Free market promotions often do not produce the quick, dramatic impacts of economic intervention steps, but their role can be much more significant over the long run. In our economy, there are several government-imposed market dislocation mechanisms that work against recycling. These mechanisms were established when national policy favored increased exploitation of natural resources for national security and economic development reasons. However, today those reasons are vague at best. The real impediment to removing these dislocations is the fact that many large industries have grown accustomed to taking advantage of them.

Recycling markets would be promoted if several specific market dislocations were removed:

- *Federal depletion allowances:* Federal depletion allowances give tax credits to the petroleum and mining industries based on the depletion of the resource in the mine or well. The effect is to subsidize the expense of the operation with tax dollars and promote the use of virgin ores and plastics over their recycled alternatives. This also keeps the cost of energy artificially low, which minimizes the energy savings that recycled materials usually have over virgin materials.
- *Forest Service policies:* Federal Forest Service policies promote logging by funding site preparation and reclamation without full compensation for these activities from the timber companies that harvest the timber. In effect these policies subsidize the expense of logging on federal lands with tax dollars. This also promotes the use of virgin pulp over recycled paper stock.
- *Timing of deductions:* The timing of deductions for exploration, development, and reforestation is accelerated for extractive and forestry activity. The effect of these deductions is to subsidize the investment of extraction and logging with tax dollars. This subsidy also promotes the use of virgin materials over recycled materials, which do not generally enjoy such accelerated deductions.

Calculating the impact of these subsidies for virgin materials is difficult. However, a recent Office of Technology Assessment estimate exceeds $1 billion per year in subsidies. On the other hand, other federal studies would indicate that the actual impact of these subsidies on recycling is slight.

In order to create a level playing field for recycled materials, there are three possible approaches:

- *Eliminate virgin materials subsidies:* While this is perhaps the most direct way to deal with the problem, it is not likely to happen quickly. Most of the subsidies are written into complex federal tax law, and it would require significant revisions to remove them. More significantly, large and powerful virgin-materials-based industries are not likely to let these subsidies go away without a battle.
- *Create subsidies for recycled materials equal to those that apply to virgin materials:* This is a complex task, since the subsidy structures are not directly transferable from virgin materials to recycled materials. Furthermore, economists point out that adding one market dislocation upon another is not the most efficient way to resolve a problem.

However, this may turn out to be the most politically acceptable compromise. This approach retains the existing subsidies for virgin materials, yet it has the potential to offset the financial advantages of those virgin material subsidies by creating comparable subsidies for recycled materials.

- *Create a tax for virgin materials roughly equal to the subsidies:* Known as a "virgin materials tax," this approach has already been introduced to the state and federal policy debate. It has the advantage of potentially offsetting the virgin material subsidies while adding a new source of revenue to the treasury. However, as a new tax, it is subject to attack from many sources and has not yet gained a solid constituency.

Disposal Costs. There is another type of subsidy that works against the development of recycling markets: artificially low disposal costs. While disposal costs are rising rapidly in many parts of the country, they are still free or nominal in many parts of the United States. In fact, even in areas with relatively high disposal costs, economists point out that a finite resource like disposal capacity should be priced on its replacement value, not just on its current operating costs. In other words, added into disposal costs should be the expenses of closure of the disposal facility, long-term monitoring and remediation of the facility, and development of new (and increasingly expensive) disposal facilities. Very few jurisdictions consciously include those long-term costs in their disposal fees.

By keeping the cost of disposal artificially low, more materials are diverted into the waste stream rather than into the recycling stream. This is because, economically, the effect of an increase in disposal costs is the same as an increase in market value of the material. Both improve the economics of recycling over disposal, which promotes market development.

The effect of artificially low disposal costs is that, since end markets must compete with disposal facilities, disposal facilities have a subsidized advantage over end markets.

External Costs. Many environmentalists and economists argue that the environmental impacts of virgin material extraction and use are not fully accounted for. Examples include loss of wildlife habitat, loss of recreational opportunity, process pollution, spills, and consumption of ancillary resources, such as power and water. By not internalizing those costs into the price of virgin materials, those materials enjoy an apparent economic advantage over recycled materials, which typically use fewer ancillary resources.

This topic is controversial because it raises all the difficult issues of external costs, such as how to account for them, how to measure them, how to weigh them against each other, and how to build those costs into the economic system.

ALTERNATIVE END MARKETS

Municipal recycling programs tend to collect steady streams of materials. However, the end markets for those materials are increasingly undependable. The result is that market fluctuations often threaten the viability of a collection program. Some recycling programs are able to protect themselves by developing long-term agreements with reliable end markets. Other programs are diversifying their end markets to include emerging, local, or revived end uses. For example, some programs that collect newspapers are shredding that paper into animal bedding rather than selling the paper to traditional end users like newsprint mills.

On the surface, developing alternative markets can be economically unattractive. Almost by definition, the reason the market is an "alternative" to traditional markets is that

TABLE 7.6 Alternative Markets

Material collected	Traditional markets	Alternative products and markets
Newspapers	Newsprint mills Paperboard mills	Shredded animal bedding; farmers Grocery bags and corrugating medium; kraft paper mills Compost; farmers, nurseries, public works
Corrugated boxes	Kraft paper mills Paperboard mills	Shredded animal bedding; farmers Compost; farmers, nurseries, public works
Glass bottles	Glass bottle plants	Asphalt additive ("glassphalt"); paving contractors, public works Drain bedding; public works "Sand" blasting and abrasive medium; industrial users, contractors, public works
Plastics	Multipurpose resin reclaimers Virgin resin replacement in products such as strapping and pipe	Plastic "lumber," concrete, and wood substitutes; public works, commercial and government establishments
Office paper	Tissue mills Writing and printing paper mills Paperboard mills	White linerboard; premium product packagers (fruit, office supplies, etc.)

the end value of the reclaimed product is lower or the processing costs are higher. However, such a limited analysis does not account for the political, disposal, environmental and community costs of disrupting a collection program during market downturns.

A more significant limitation of many alternative markets is that they tend to be small-scale. However, even if the size of an alternative market does not completely replace the traditional market, it can ease the pressure caused by market disruptions.

Table 7.6 shows a matrix of alternative end markets for a variety of commonly collected materials. Often, these markets can be developed locally, at least on a small-scale basis.

CONCLUSION

The current state of recycling is that a glut of many materials grows daily while prices for those same materials fall. This apparently noneconomic behavior is caused primarily by government-mandated collection programs that are out of sync with end markets. The solution to this problem is to promote the development of end markets.

Developing end markets has several goals:

- Prevent imbalances
- Promote economically sustainable recycling
- Minimize need for government intervention

- Promote economic development
- Conserve ancillary resources

Market development is not as visible as development of collection programs. Market development requires a comparatively high level of investment, a long lead time, and a high level of government involvement. Accordingly, it is all too easy to overlook market development, until that lack of markets begins to threaten the collection programs themselves.

Market development can be promoted in a variety of ways. Market development strategies fall into one of these categories:

- Supply
- Consumer demand
- Technology
- Government role
- Economics
- Alternative markets

Market development is a challenging process necessitating careful policy implementation at high governmental levels, but it is a challenge that must be met in order to advance recycling. Without market development for recyclable materials, there can be no growth in recycling.

CHAPTER 8

FINANCIAL PLANNING AND PROGRAM DEVELOPMENT

ROBERT HAUSER, JR.
Vice President, Camp Dresser & McKee, Inc.
Tampa, Florida

INTRODUCTION

Today, many communities are looking at recycling as an integral element of their solid waste management strategies. All of the benefits associated with recycling must be considered when evaluating a particular recycling program, or combination of programs, as a component of a total solid waste management system. However, implementing a recycling program in many communities often introduces a number of new issues not directly addressed by current solid waste systems. Of particular importance are issues associated with the financing or funding of the program and its management.

Recycling goals adopted by communities or mandated by legislation impose new responsibilities on a community and require the development of programs which the existing solid waste management structure may not be equipped to address. Therefore, it may be necessary to change or evolve a new management structure to address the new responsibilities.

The individual components of an overall recycling program, such as curbside collection, drop-off centers, processing facilities, and public education, all introduce new costs to a community. Depending upon the specific components, they may have high or low capital funding requirements and high or low operating and maintenance costs. Financial commitments may be short-term or long-term. Revenues from the sale of materials, state or other program grants, and/or avoided costs will offset the costs of the recycling program. However, these revenue sources are also highly erratic and introduce a high level of uncertainty in their realization. Finally, experience and financial planning show that recycling programs, particularly very aggressive programs, do not generate sufficient revenues to cover their costs. Thus, it becomes apparent that funding to support recycling programs is required.

The additional funding required should not be viewed as an unnecessary cost. All solid waste systems, including elements of collection, transportation, processing, and disposal, cost money. Recycling is part of an overall integrated program and it is as reasonable to expend funds for this purpose as it is to expend funds for other parts of the solid waste system. Moreover, recycling produces benefits not attained by other solid waste disposal systems.

This chapter addresses the financing, funding, and management alternatives available to implement recycling programs. These alternatives are basically the same ones available to implement any solid waste management system. However, the new responsibilities and opportunities available through recycling programs require some communities to look at these alternatives for the first time, while for others it requires a whole new look at their existing solid waste programs. No two communities are the same, and the selection of a funding and management program must be made considering that community's own unique set of goals and objectives as well as constraints.

RECYCLING PROGRAM IMPLEMENTATION RISKS

In any recycling program development, many decisions regarding financial planning and management of the program involve risks that must be recognized and properly allocated to the program participants. The risk of most concern to communities and private firms in recycling programs is monetary loss. Thus, the allocation of risk is the assignment of monetary loss, if it occurs, to a specific party prior to the actual occurrence of the loss. It is important to note that monetary loss does not refer to a net program loss but rather a loss exceeding that budgeted and funded using responsible assumptions.

The factors which lead to risk exposure for recycling programs can be grouped into five categories:

- Technology
- Waste stream
- Markets
- Legal and regulatory
- *Force majeure* (unforeseen circumstances)

The following paragraphs briefly review these five categories. It is important to note that these risk factors apply to all recycling programs, as well as individually to each element of the program whether it is a curbside collection program, drop-off centers, or a major processing facility such as a compost operation. The relative importance of the risks, however, may shift depending upon the specific element of the program.

Technology

Risks associated with technology include completing the construction of recycling and processing facilities on time and within a specified cost. This category also includes technical problems that might affect the ability to complete the facility as designed and unanticipated construction cost increases.

The procurement or bid documents for recycling and recovery facilities should require guarantees that the facility will be completed by a certain date, for a certain cost, and in compliance with specified performance and/or design criteria. These guarantees would be supported by a construction bond and/or insurance.

Technological causes of risk also include those associated with actual program operation—that the facility or operational program such as curbside collection will operate as planned and guaranteed, and that program requirements such as material quantity and quality are met. Most of those performance guarantees will be as specified in the bid documents or negotiated in the procurement process. Under a turnkey or full-service procurement, the contractor will be responsible for meeting the prescribed performance criteria, correcting any reasons for not meeting those criteria, or paying financial penalties for noncompliance.

It is advantageous if the risks associated with technological causes can be negotiated and allocated to the various project participants in an equitable and precise manner. These risks also demonstrate the critical nature of proper technology and specific program selection to meet the individual needs of a community.

Waste Stream

Another cause of risk to any recycling or materials recovery program is the assurance of an adequate and reliable supply of recovered materials and/or waste of the proper composition and quality. Waste stream control is a major cause of risk allocated to a city, county, or authority, whether or not that entity owns the facility. Waste stream control is particularly important in programs using tipping fees, surcharges, or other quantity-related revenue sources as the bases of support to program funding.

The degree of waste stream control is directly tied to the recycling program selected by a community and must be considered in the selection of that system. It further must be considered in terms of its applicability to residential wastes and commercial or industrial wastes and the various programs to recover materials from each of these waste streams.

A final consideration is the nature of any publicly mandated goals to achieve minimum levels of recycling or recovery. Many states have legislatively mandated recycling and recovery goals. In most cases, attaining these goals has been assigned as the responsibility of a city, county, or other "responsible" agency. The ability to achieve and exceed these goals must be met through a proper program which may require waste stream control at a level not initially envisioned.

Methods of waste stream control include:

- *Economics:* Program and/or facility tipping fees could be set or subsidized to be the least costly disposal option within the area, or economic incentives to recycle materials could otherwise be provided.
- *Legislation:* Responsible public agencies could submit special legislation to provide exemptions from regulatory antitrust statutes, empower the public entity to control the collection and disposal of recycled materials within a specified geographic area, and grant the entity waste flow control. This is particularly important in recycling programs where recycled materials are often excluded from the definitions of solid waste under regulatory programs. Waste generators within the area could also be required to deliver recycled materials to the location specified by the municipal jurisdiction. The implementation of this legislation, with appropriate enforcement and combined with one or more of the other methods of control, significantly reduces the risk of providing a reliable supply of waste and/or recovered materials.
- *Contract agreement:* The city, county, and perhaps an authority may designate the programs to be implemented and the facility to be used through agreements with contract collectors. Franchising of collectors for residential, commercial, and recyclable routes can be done.
- *Municipal collection:* The city, county, or authority could assure that the waste is delivered to the facility simply because its forces and equipment perform the operation of the program.

In addition, any of the above could be implemented in combination with the others.

Other causes of risk exist which are related to the quality of the waste stream rather than controlling waste delivery. For example, a central materials recovery system's guarantee may be contingent upon a specified quality of waste, including minimum percentages of paper, glass, and aluminum. Waste quality could change for numerous reasons (for example, enactment of beverage container deposit legislation). Public entities must often assume the responsibilities and costs of this risk.

Markets

In recycling and materials recovery programs, securing markets to purchase (or even to receive) recovered materials is critical. Historically, secondary material markets are very volatile with constantly changing price structures and quality requirements. For some materials, markets or outlets are still developing. It is difficult to secure long-term contracts to receive materials. Some legislatively mandated programs require recycling of materials irrespective of market conditions.

These factors affect not only program selection but, importantly, the ability to budget, fund, and operate the program. The risk associated with markets can be mitigated primarily by maintaining program flexibility with respect to the materials recovered and their quality, and sound budgeting that recognizes market factors. Establishing a good relationship with markets can help assure outlets for materials through market ups and downs. The budgeting process should allow, where necessary, the payment of a tipping fee to markets to accept the material, recognizing that this may still provide benefits and can be economical as compared to the cost of disposal associated with other methods.

Legal and Regulatory

This set of risk causes are not possible to anticipate. Changes do occur in state or federal legislation and associated rules and regulations which may affect recycling programs. In addition, it is difficult to anticipate how courts will, in the future, interpret present laws and regulations.

Significant legal and regulatory issues that could affect the project include

- *Tax laws:* Federal income taxation legislation has undergone major changes. Such changes may impact upon the financing options and revenues generated. Also, tax laws affecting the use of virgin materials as compared to recycled materials impact the markets.
- *Environmental protection regulations:* Environmental protection laws and regulations could change. For example, federal or state air or water (including groundwater) quality regulations could be altered and could result in significant constraints on the operational practices of an existing recovery facility. Changes in administrative practices could affect governmental procedures associated with various facilities (such as environmental renewals). Finally, regulatory changes may affect the goals of recycling programs, thus requiring additional program elements to meet goals or resulting in changes to the composition of the waste stream.
- *Antitrust challenges:* The program could be faced with federal antitrust challenges as a result of the need to direct material flow. Special state legislation can substantially reduce this risk.
- *Other regulatory issues:* Laws related to packaging legislation, bottle bills, government procurement policies, etc., can all affect recycling and materials recovery programs. Many can be positive.

Force Majeure

This category of risk causes is the term for all of the unanticipated occurrences that might affect the operation of the program. These causes include war, sabotage, and other occurrences beyond control, including acts of God, such as earthquakes and other natural disasters. Proper risk allocation for the *force majeure* causes requires a comprehensive listing of the potential causes and a program to allocate responsibility for such risks, particularly costs.

Summary

In determining a risk posture, consideration must be given to each of the above risk categories both in terms of each individual program element as well as the entirety of the recycling program. This will assist in fully defining the financial requirements of the program to assure proper funding of all capital, operating, and maintenance costs.

FINANCING AND OWNERSHIP ALTERNATIVES

A key element in the success of a recycling program is the development of a plan and structure for capital financing and ownership of any facilities required for the program. Decisions related to these issues extend into the selection of a procurement approach. These decisions must be made by each community on an individual basis, considering its overall program and its goals and objectives. The following describes the general framework of alternatives available to communities.

Financing Alternatives

A key element in the success of a solid waste management project is the development of a plan and structure for capital financing. The financing program must be structured to meet the overall objectives of the governmental sponsor. At the same time, it must also meet the needs of the financial community in order to attract adequate funds for project implementation, including the equitable allocation of risks among the project participants.

Public Ownership Financing Options. The major sources of capital funds for a publicly owned recycling facility include general obligation bonds, revenue bonds, and bonds and grants issued by state and private agencies.

1. *General obligation (G.O. bonds):* In order to finance many non-revenue-producing capital projects, local governments have typically issued long-term general obligation debt. While such debt could be repaid from project revenues, the bonds are secured by a pledge of the full faith and credit and taxing powers of the governmental sponsor. If project revenues fail to materialize, the debt service must be covered by tax revenues.

G.O. bonds and other tax-exempt debt instruments can be issued with lower interest rates than taxable bonds because the holders of G.O. bonds are not required to pay income taxes on the interest income they receive. Consequently, the public entity that issues G.O. debt pays less interest over the term of the bonds, which usually results in considerable savings. The spread between G.O. bonds and private financing alternatives depends upon the credit rating of the governmental sponsor and the specifics of the project.

2. *Project revenue bonds:* Typically, project revenue bonds are the preferred form of financing when a project is capable of producing revenues sufficient to support the project and entirely repay the bonded debt. Revenues include not only revenues from the sale of materials but also revenues from other pledged funding sources such as tipping fees. However, the rate of interest charged on such a bond issue will generally be higher than for G.O. bonds. It depends upon potential investors' perception of the overall economic viability of the project (the "coverage" of expenses and debt service by project revenues); the contractual requirements, if any, of the arrangements (including the financial strengths of the contracting parties); and the perceived value of any backup pledge revenues to further guarantee the project's fiscal integrity.

3. *Other bond and grant programs:* Many states provide opportunities for communities to finance recycling projects from state bond or loan programs. These have many of the characteristics associated with those previously financing methods described above. However, they represent an opportunity for a community to obtain an outside funding source.

Also, some states, federal agencies, and private companies or foundations offer specific grants which may be utilized by a community. The availability of such programs is generally limited, and further they are often restricted as to what purposes they may be applied. However, they can represent a significant source for financing projects or to defer their associated costs. These need to be investigated on a project-by-project basis.

Private Ownership Financing Options. The major source of capital funds for a privately owned solid waste facility would be private equity and industrial development bonds (IDBs). In addition, some private grant programs may be available to privately finance recycling facilities.

If the proposed facility is to be owned by a private party, then the capital financing would be obtained through either a combination of an equity capital contribution by a private party and the issuance of tax-exempt private activity bonds, or total private equity for less capital-intensive facilities. With a combination of private capital equity and private activity bonds, the equity capital contribution must be structured in order for the private party to be recognized by the U.S. Internal Revenue Service (IRS) as the facility owner, and thus be eligible for the tax benefits associated with facility ownership.

IDBs are tax-exempt bonds issued by a public agency on behalf of a private party proposing a project with a public benefit, such as economic revitalization or increased employment opportunities. Actions by the United States Congress in revising federal tax laws have put a cap on the annual issuance of IDBs in each state as well as other restrictions. The availability of funds and the methodology for allocating funds to projects varies in each state but represents a significant potential source of funding for recycling facilities.

Ownership Alternatives

In most cases, the financing and ownership choices must be made in tandem. Thus, with municipal G.O. bonds or municipal revenue bonds, the governmental sponsor will own the facility, and with the use of private equity capital and IDBs, a private entity will own the facility. This does not mean that the selection of the source of capital funds should precede the decision as to whether a public- or private-sector party should own the desired facility. In fact, both of these basic issues should be discussed separately but decided upon simultaneously.

Public Ownership. Public ownership of the facilities gives the governmental sponsor maximum control over providing solid waste disposal for its citizens. However, the governmental sponsor must also assume certain additional risks associated with ownership. Capital financing would be accomplished through either the sale of G.O. bonds or revenue bonds.

Advantages of Public Ownership

- Public-ownership financing may be less complex, less time consuming, and more assured of implementation than private-ownership financing because there are fewer parties involved.
- Ownership of the facility site will be retained at the end of the bond term. Even if the facility (for example, a materials processing facility) must be totally reconstructed at the end of the bond term, the facility site is a valuable resource.
- The governmental sponsor will benefit from the economic usefulness (residual value) of the facility at the end of the bond term.
- The governmental sponsor will enjoy more control over the operation of the facility and ongoing flexibility in implementing its recycling program.

Disadvantages of Public Ownership

- Certain additional financial risks accrue to the governmental sponsor, including equipment serviceability at the end of the bond term.
- Any potential cost savings associated with private-owner equity capital contributions cannot be realized.
- The public sector usually must assume a greater risk position.

Private Ownership. If the private ownership option is selected, the governmental sponsor will negotiate with a private party to obtain the services and facilities desired. The governmental sponsor will derive some financial benefit from the equity capital contribution associated with private ownership. In some instances, the private sector may provide all of the capital funding; in others, financing would consist of an equity capital contribution by the private party and the sale of IDBs issued by the governmental sponsor on behalf of the private party.

Under this approach, the governmental sponsor would negotiate with one or more private parties to finance, design, construct, own, and operate the facility. The allocation of risks and responsibilities would be documented in the service contracts governing construction, operation, and ownership. Disposition of the revenues of sales of materials must be included in the contract. The private participant is the owner of the facility and will always own that facility even if IDBs are used to finance it. The facility could be sold after the debt is retired at fair market value.

Advantages of Private Ownership

- The equity capital contributed by the owner will reduce the amount of borrowed capital, which should result in cost savings to the overall recycling program. (This assumes that the equity capital is less costly to rate payers than debt capital.)
- More of the financial and operating risks associated with the facility must be assumed by the private owner.
- There is a greater sense of security where the operator has at risk a sizable investment of its own equity capital (assuming the equity capital is contributed by the contractor).
- Property taxes (or payments in lieu of taxes) will accrue to the local governmental unit.

Disadvantages of Private Ownership

- If IDBs are used to finance the facility, rate payers have, in effect, retired the project debt in part through the tipping fees or other payments; however, unlike a publicly owned project, the rate payers do not own the facility when all debt is retired. To the extent the facility has remaining economic usefulness (residual value), that value may be lost to the community.
- Financing of a facility which is to be privately owned may be more complex, costly, and time consuming than financing a publicly owned facility.
- The financial rewards (net revenues after expenses and debt service) associated with the facility accrue to the private owner.

Procurement Alternatives

Procurement approaches indicate with whom responsibility for design, construction, and operation of the facility will rest. Three procurement approaches are available for publicly owned recycling and materials recovery facilities. These include

- *Classical architect and engineer (A&E) approach:* The public owner hires an engineer to design the facility. The facility is then bid for construction. Operations may be the responsibility of the public owner or another contractor.
- *Turnkey approach:* The public owner hires one firm to take responsibility for both the design and construction of the facility. That firm must meet performance specifications identified by the public owner. Upon meeting those specifications, the owner will take physical control of the facility. The owner or another contractor may operate the facility.
- *Full-service approach:* The public owner selects one firm to design, construct, and operate the facility. That firm must meet performance specifications upon completion of the project and during operation of the project.

In the case of privately owned facilities, the full-service approach is typically used, as all of the project functions will be under the control of the private owner of the facility.

Under the turnkey and full-service approaches, two procurement methodologies may be used. The first is sole-source. Under this approach, the governmental sponsor would simply select a firm to be the developer of the project.

The second approach is the RFQ/RFP approach. Under this approach, firms are invited to submit qualifications to finance, construct, and operate the desired facilities. Following an evaluation, firms deemed qualified are invited to submit technical and business proposals. One firm is then selected on the basis of predetermined qualitative and quantitative criteria.

Sole-source procurement is usually restricted to cases where time constraints are extremely critical and/or those projects which have such special circumstances that only one firm has the ability to meet the project's needs.

The RFQ/RFP approach offers several advantages over sole-source procurement. First, it allows an opportunity to structure the request for proposals to meet the specific project needs and to fully reflect local decisions. Second, through an open, competitive proposal process, the community has the best opportunity to receive proposals which meet its needs and which are priced competitively.

The A&E approach is often used for recycling and materials recovery facilities. This approach allows the public owner to actively participate in the design process, ensuring

TABLE 8.1 Design, Construction, and Operation Responsibilities for Various Procurement Approaches

	Procurement Approach		
Function	A&E	Turnkey	Full-service
Design	Consulting engineer or government staff	Vendor	System vendor
Construction	Best-bid contractor or government	Vendor	System vendor
Operation	Public or contractor	Public	System vendor
Ownership	Public	Public	System vendor or third party

that community needs and standards are incorporated in the facility. It also provides for price competition, through the bid process, for construction of the facility.

A summary of the responsibility of assignments under each procurement approach is provided in Table 8.1.

INSTITUTIONAL FRAMEWORK AND MANAGEMENT ALTERNATIVES

It is important that a recycling and materials recovery program adopted by a community be acceptable to its citizens, provide for efficient operation, and be established within the framework of legal constraints. The program should be backed by sufficient authority to provide adequate funding, site acquisition, and effective operation. It should have the adaptability and flexibility to meet changing conditions, and be structured so that it will have access to state and federal funds. The advantages of local control should be balanced against the economy of large-scale operation to reduce duplicated effort.

The following sections discuss these and other associated issues. It is important to recognize that the institutional and management issues discussed below relate to the program implementation responsibilities of public entities. Communities may, and often do, assign these responsibilities to the private sector. However, the ultimate responsibility for program implementation rests with a governmental entity.

Assignment of Responsibilities

As recycling programs have developed, they have often been undertaken apart from the remainder of a solid waste management system. Also, solid waste system planners have often neglected or underestimated the impacts of the system upon recycling programs. These problems have strongly demonstrated the need for an integrated approach to improving the total solid waste management system. The best approach to the problem is to establish clear and definite assignment of responsibility among involved governmental entities to meet particular local needs and preferences. It is then incumbent upon these governmental entities to assume the assigned responsibilities and to arrange the details of administration, financing, and operation.

The assignment of responsibility includes a determination of the various functions which need to be performed and the jurisdictional levels available.

A solid waste system includes several functional components, including:

- Collection
- Transfer and haul
- Source separation and recycling
- Processing
- Disposal

Solid waste management can be considered as a single function in all its aspects and dealt with accordingly; however, since each of the functional components has its own distinct activities with its own characteristics, it may be more suitable to treat these activities as separate management or jurisdictional functions.

Recycling programs, however, complicate this picture. Depending upon the type of recycling program, it may include elements crossing two or more of the functional components listed above. For example, curbside collection is, in effect, part of the collection system. Composting may be considered solid waste processing. Recycling programs will affect disposal programs and vice versa. Therefore, while each of the functional solid waste components may be managed and implemented by different jurisdictions, there must be coordination between the programs and common goals and objectives. This introduces a sixth component to the system, which is program management.

Another dimension to the problem of assignment of responsibilities is related to the geographic scope of existing governmental jurisdictions, which range from cities and counties at the most local level to multicounty jurisdictions. There are five potential geographic or jurisdictional configurations to which various functional responsibilities may be assigned:

- Cities
- Counties
- Service areas
- Multicounty areas
- Regional planning groups

Cities and counties require no explanation. They are the most local units of government, and citizen access to them is generally easiest. These entities also represent the configuration under which most solid waste responsibilities are currently assigned.

Service areas are defined by an existing solid waste system sharing facilities or other functions. They may comprise groups of counties (and associated cities) when they are sending waste to the same landfill sites.

Regional planning groups refer to existing regional agencies with planning responsibilities over a large regional area. The possibilities for assignment of primary functional responsibilities can be displayed in matrix form as shown in Table 8.2.

In theory, at least, each of the six solid waste management functions can be assigned as the primary responsibility of any of the five geographic levels, thus giving 30 alternative sets for evaluation. However, it is usually possible to reduce the number of alternatives to a handful by making a few straightforward assumptions based upon the existing system and other political considerations.

At this point, two additional matters require comment. While the assignment of functional responsibilities is a major component of the management system, it is not the only one. It helps assure multijurisdictional solutions by arranging the most desirable pattern

TABLE 8.2 Matrix for Assignment of Primary Functional Responsibilities

Functional area	Cities	Counties	Service areas	Multicounty area	Regional planning group
Program management					
Collection					
Transfer and haul					
Source separation and recycling					
Processing					
Disposal					

of intergovernmental responsibility and coordination, but other more detailed management components must be assessed. These other matters include financing, administration, and private versus public operation. The assignments made at this stage, however, determine what level of government bears responsibility for arranging these detailed matters.

Secondly, for each of the five jurisdictional frameworks outlined above, there is a distinct set of legal mechanisms that can be used to carry out the assigned responsibilities. Some legal mechanisms cannot be used at certain geographic levels or for certain functions. Since the availability of these mechanisms is a defining characteristic of each geographic level, they are presented in this section. The ease with which they may be employed is a factor in the selection of an assignment of responsibilities and is discussed in the next section.

Institutional Mechanisms

There are several kinds of institutional mechanisms which may be considered to manage recycling programs as well as solid waste management systems. The specific powers and legal parameters associated with these vary among the states. However, in general they may be grouped as follows:

- General governmental powers
- Interlocal service agreements
- Agreements for joint activities
- Special-purpose districts
- Public authorities

These mechanisms represent a continuum and, depending upon the specific arrangements associated with each mechanism, they may significantly overlap one another.

General Governmental Power. Each city and county has the authority to undertake solid waste handling and recycling activities as a municipal function within its borders. In some cases, specific charter provisions may place special conditions or limitations on these powers. In all cases, the authority extends to acquisition of property and financing with either general tax revenues or service charges. This mechanism is obviously available where primary functional responsibility is assigned at either the local level or at the county level.

Interlocal Service Agreements. Most states allow any governmental entity to enter into an agreement whereby it undertakes to supply solid waste management services to another unit. An interlocal agreement must be approved by each participating governmental unit. Interlocal service agreements are commonly used between counties and cities for the provision of services. They are flexible and allow a wide range of latitude between governmental units in negotiating them and assigning responsibilities. They can range from relatively simple arrangements to cooperate to detailed "contractual arrangements" between the governmental units.

Usually one governmental unit is designated as the lead agency and assigned primary administrative and control activities. They also may be negotiated to include oversight committees and requirements for review and approvals from all participating units before any action can be taken by the lead agency. Interlocal service agreements provide a good mechanism where it is necessary to administer and manage activities at a level higher than each individual county. However, the total powers under such an agreement cannot exceed the powers of any governmental unit participating in the project. Moreover, the assignment of such powers by an entity only extends as far as included in the agreement.

Agreements for Joint Activities. These agreements are similar to interlocal service assignments. The major difference is that activities are carried out jointly and a joint governing body may be created comprising designated elected officials of participating governmental entities or their appointees or others as specified. Agreements for joint activities may also be used to carry out responsibilities at a service-area level or multicounty level or wherever appropriate. These agreements include the same considerations as described for interlocal service agreements.

Special-Purpose Districts. Counties and cities may create special solid waste collection and disposal districts. Such district operations may be financed through special benefit assessments or *ad valorem* assessments against the properties in the district. The special service district may be an appropriate legal mechanism for carrying out functional responsibilities at a service-area level. Districts would probably have a small role to play unless countywide operation of a solid waste function was impossible to organize for political reasons. The primary benefits of this option are related to funding mechanisms.

Public Authorities. A public authority (sometimes called a public corporation) is generally a corporate instrument of the state, usually created by the state legislature or referendum for the furtherance of self-liquidating public improvements. As creations of the state, such authorities can be formed for a multiplicity of purposes and with a wide range of powers. The ultimate structure of an authority will be determined by the cities and counties and the state legislature. The powers invested in the authority may range only from those necessary to coordinate activities to very strong powers overruling those of individual counties and cities (including the power to tax and to acquire facility sites).

Evaluation Considerations

A number of factors are relevant to selecting a jurisdictional arrangement to handle recycling program functions. Some of these factors are derived from economic and technological characteristics of the physical system. Others have their origin in legal or political aspects of available organizational mechanisms. All of the factors are pertinent to each recycling program function, but they are not of equal importance. Nine factors which must be considered are discussed below.

Economy of Scale. One of the most important considerations in organizing the management system is the realization of any efficiencies of scale that may be inherent in various techniques of recycling program management. Here, where investigation, dissemination of information, organization of markets, and influencing procurement practices are the major activities, a single entity can be more efficient than parallel activity by several smaller entities. Also, curbside collection and the construction of materials recovery facilities may also be more cost efficient at larger scales.

Duplication of Technical and Administrative Services. Certain technical and administrative services are always required but, once available, can easily be applied to many activities. Thus, this factor represents a special case of economies of scale.

Local Control. It is important that the parties who obtain the benefits and incur the costs of various recycling functions be able to influence the decision-making process. Generally speaking, people have more influence in a smaller group, which suggests that more local control can be achieved when functions are assigned at the local level. The need for local control, however, is not the same for all functions. Curbside collection or drop-off centers are highly visible and of keen local concern, whereas remote activities such as a composting operation are more removed and of lesser local concern for most people.

Financing Considerations. There are two ways in which matters related to financing can affect assignment of functional responsibilities. First, certain financing techniques are not available under some legal mechanisms. These must be evaluated very carefully under any strategy. The second financing consideration relates to constitutional limits on local debt and taxing powers and to municipal credit ratings. Counties and cities are subject to constitutional real estate tax limits. This limits the financial ability of any entity to undertake expensive projects on a regional basis.

Ease of Establishing Appropriate Mechanisms. One of the most significant factors to consider is whether it will be necessary to create new legal mechanisms in order to implement a given arrangement. Other things being equal, it is more desirable to employ existing governmental units rather than add layers of local jurisdiction to solve the solid waste management problem. By this reasoning, there is an advantage to assigning functions at the county or city level.

In order to place primary responsibility at the service-area level or multicounty level, it is necessary to create new legal mechanisms—either interlocal agreements or an authority.

Management Flexibility. This factor concerns the extent to which the jurisdictional framework constrains responsible agencies to focus on a narrow range of alternatives. The greater the opportunity to analyze and implement a wide range of choices in operating the system, the greater the management flexibility.

Access to Federal and State Programs. It is important that any jurisdictional arrangement have access to all potential forms of funding and program assistance. Given the very limited availability of such programs and funding from such sources, any mechanism which can provide access to these services would be favorable. In fact, given the need to maximize the effectiveness of funding to provide real solutions, the regional programs and organizations have a much better opportunity to obtain funding and program assistance.

Power to Obtain Necessary Facility Sites. Any jurisdictional organization must be able to obtain sites, including the power to condemn sites. On a local basis, cities and counties have such powers. However, their political ability to exercise such powers for a regional solution is questionable. Therefore, this power should be performed by the highest level of governmental power.

Ability to Adapt. This factor is closely related to management flexibility and becomes particularly important where consideration is given to assigning functions to a service-area mechanism. If, at some point in the future, developments indicate a change in service areas is appropriate, it is easier for a regionwide agency to adapt to these changes than for a service area or county to make the adjustment.

FUNDING

As a community proceeds with the implementation of a recycling program, a number of important issues must be addressed. One significant issue is the development of a funding mechanism to pay the costs of the program. The financing alternatives discussed previously only address capital financing. Communities will be responsible for paying all system costs, including any debt service charges and annual operating and maintenance (O&M) costs net of revenues from the sale of materials. Even with a private system, this cost will be paid by the community through a service fee.

The method of funding recycling programs directly impacts the recycling program selected for implementation. Program elements such as drop-off centers have much lower costs than elements such as curbside collection or various processing facilities. However, the lower-cost programs generally result in lower recycling rates of a limited number of materials. As program goals to achieve higher recycling and reuse rates increase either by local decision or legislative mandates, the program must become more extensive and the associated costs rise. This in turn requires more funding and greater flexibility in the use of funds.

The remainder of this section discusses the issues associated with funding and the alternative funding mechanisms.

Key Issues

Any community evaluating alternative funding mechanisms must consider a number of issues. For the most part, these issues are interrelated. A decision with respect to one will affect the direction of the decisions on the others. The final selection of a funding alternative, or alternatives, will involve elements of these key issues and compromises which must be made. The key issues which must be considered in the evaluation are as follows:

1. *Funding of existing solid waste systems:* An important consideration is the funding system currently, used by a community for collection, disposal, etc. Minimizing changes to a funding system is always easier than changing the system. Conversely,

some communities may use the recycling program as a vehicle to change the entire solid waste funding mechanism. Many of the alternative funding mechanisms will affect the existing solid waste system, including the mix of private and public activities. The impacts will include the manner in which private haulers operate, and the rates and level of service to both residential and commercial users. Finally, the community's position with respect to encouraging, promoting, and funding commercial and industrial recycling activities must be considered as these mechanisms may be quite different from those used to fund residential programs.

2. *Avoided costs:* Costs savings to the overall solid waste system are likely to occur as avoided costs of collection, processing, and disposal because of the wastes are diverted by a recycling program. The ability of a management system to accurately measure these costs and credit them to the recycling program can be an important mechanism to provide funding to a recycling program from funding services used to support the solid waste system.
3. *Waste flow control:* The degree to which waste flow control and the level of enforcement associated with it is exercised can be directly impacted by the funding method used. This issue affects the sizing of any recycling facilities, the degree of risk that must be accepted in procurement, operation, and the financing strategy for the overall program.
4. *Flow of funds:* For almost all of the various financing and ownership options available, as well as for many of the operating contract strategies, there will be a requirement to assume the payment of funds on a regular basis. The degree to which the flow of funds to the program can be controlled to assure the availability of funds on a regular, reliable basis will impact the financeability and cost structure of the project and the level of risk.
5. Ad valorem *taxes versus user fees:* Each community's position with respect to user fees, *ad valorem* taxes, other taxing mechanisms, or a mix of these must be considered.
6. *Availability of infrastructure:* Several of the funding alternatives available require development of an extensive data base and the commitment of resources to compute and issue bills, receive and account for funds, and provide for enforcement. The extent to which such services can be performed within the existing organizational structures and the cost of implementing such a system must be considered when evaluating funding mechanisms.

Evaluation of Funding Mechanisms

The preceding discussion presented a number of key issues which a community should consider in identifying alternative funding mechanisms. These may limit the number and feasibility of the alternative funding mechanisms available to that community and impact the selection of its recycling program. For a community to further evaluate alternative funding mechanisms, several evaluation criteria are presented as the basis that a community may use for evaluation within the context of its own specific goals and objectives. These criteria are discussed below.

Equity. Equity is a concept basic to all rate making, whether it be for recycling programs or other types of utilities. Perfect equity means that each user of a system pays exactly the full cost of the service provided to that user. Because every customer is somewhat different in terms of participation and contribution to the system, the rate structure should ideally be able to reflect the differences of each customer to provide perfect equity. Perfect equity is never achieved in practice. However, to be considered acceptable, a funding mechanism and rate structure must be generally perceived as equitable by rate makers, the decision-making body, and users.

Effectiveness. An effective funding mechanism or rate structure is one that encourages generators to recycle and/or dispose of their wastes in the manner directed by the governing body. For example, one measure of effectiveness is the extent to which participation in the recycling program is encouraged. The funding mechanism must encourage the use of recycling facilities and not provide incentives for users to bypass the system.

Adequacy. Another criterion is the ability of a given funding mechanism or rate structure to generate adequate revenues to meet all of the financial requirements of the recycling program.

Legality. The extent to which an alternative complies with statutory and case law is an important consideration. Legality is intended to indicate an alternative's ability to successfully withstand legal review and possible challenges.

Administrative Simplicity. This criterion relates to the time, cost, and effort required to implement, collect, and update the rates based upon the selected funding mechanism.

Alternative Funding Mechanisms

Tipping Fees. Tipping fees or gate fees are charges collected from each vehicle passing through the gate of the disposal facility. These fees are revenues collected at the time service is provided based upon the amount of service required in terms of tonnage disposal.

Where tipping fees are currently in use to fund part or all of a solid waste system, many communities and some states have imposed surcharges on the tipping fees or receive a percentage of the receipts with the money specifically earmarked to support recycling programs. This may be done in most areas, whether the facility is privately or publicly owned and operated. The use of differential tipping fees may be used to encourage and promote recycling. An example would be the imposition of a higher tipping fee on waste loads at a landfill containing recyclable materials which have not been recovered.

Tipping fees are usually equitable in funding solid waste system costs as each user pays according to use of the disposal system. As a funding method for recycling programs, however, waste generators are paying for the program although they may not be directly involved. The use of tipping fees is generally easy to implement, but if fees rise too high or if enforcement is low, illegal dumping could result.

User Charges. User charges differ from tipping fees in that they are charged by sending a bill directly to the service customers or to those receiving the benefits from the program. Customers are required to pay for the service in direct proportion to their actual or potential system use.

User charges may be collected in several different ways. A line item for solid waste service may be included on another utility bill or the user charge may be billed separately to each customer. A user charge could also be imposed by a separate line item on the annual tax bill identified specifically as a solid waste fee. Finally, user charges could be collected through private haulers who could directly bill their customers, then remit the receipts to the community. This system would, of course, require all customers to receive such service.

User charges can be a good funding mechanism. The major potential problem with user charges is that rates are based upon averages for a class of generators. Acceptable rates can be readily established for residential customers, but opposition by commercial customers can be a problem. Thus, the process used to establish commercial rates must be perceived as fair and equitable, and effective commercial recycling programs must be in place and/or not interfere with existing commercial recycling.

User charges can also encourage recycling in that the fee will have already been paid and there is little incentive not to participate. However, tipping fees must still be used to some extent in order to regulate disposal and to identify system users. If properly developed, user charges should provide adequate revenues.

User charges are becoming more widespread for all types of utilities including recycling programs due to the desirability of matching costs with benefits received. However, user charges can be difficult to administer, particularly during implementation. Constant supervision and enforcement are required.

Ad Valorem *Taxes.* Recovery of recycling program costs can be accomplished through *ad valorem* taxes. This mechanism is currently used by many communities around the country. A rate could be calculated annually and levied upon properties as part of a community's general revenues or as a separate line item on the tax bill. A significant restriction to their use are the legal limits on *ad valorem* rates.

This method of solid waste system cost recovery is often considered poor on an equity basis. This assertion is made because property values or other bases upon which the taxes may be calculated bear no relationship to the level of recycling activity. *Ad valorem* taxes can be an effective mechanism for solid waste cost recovery. The reason, similar to that for user charges, is that no financial incentive exists to avoid participation in the program.

Special Assessment Districts. Properties receiving a special benefit from some particular improvement or service may be levied a special assessment. Generally such a fee for a recycling program would be a user fee or *ad valorem* tax. The implementation of such a district and its exact form varies greatly from state to state, and there may be very restrictive legal constraints. However, in some areas it provides a mechanism to provide funding and to avoid other restrictions associated with the use of general *ad valorem* taxes. It also allows the fees or taxes to be applied to a limited area rather than, for example, an entire county.

Other Funding Mechanisms

State and Federal Funding Alternatives. A number of states provide funding to local communities for the implementation of recycling programs, including grants for the construction of facilities and/or funds to defray operating costs. Also, as discussed earlier, loan or state bond programs may be available. The state funds used are generated from a number of sources, including general tax funds, sales taxes, special trust accounts, tipping fee surcharges, fees or taxes on specific products and/or materials, and occasionally federal funds. State funds may have significant restrictions that limit their use to specific purposes.

Communities should be monitoring and locating sources of state funding. Such programs are always being developed and modified. Moreover, the level of funding can significantly change from year to year depending upon revenues generated and the general fiscal condition of the state. States providing such funding usually have a lead agency and contact person who can assist communities in identifying the availability of such funds. Local state legislators also can assist.

With respect to federal funding, there are essentially no programs currently being funded to provide monies for recycling program implementation. Occasionally, grants from the United States Environmental Protection Agency (EPA) and the Department of Energy (DOE) are available, usually to demonstrate new technologies. However, as in the past, the federal government may consider instituting a grant program for funding recycling programs in the future.

Private Grants. In the past and currently, some private companies and industry associations representing manufacturers in glass, paper, plastics, steel, and various consumer products make grants available to communities to assist in implementing recycling programs. These funds are generally used to demonstrate a specific program or technology. Many of them also provide assistance to communities in setting up recycling programs.

Product Taxes and Fees. Some states have implemented programs to tax or levy fees on the production, use, and sale of various products such as batteries, newsprint, and certain container packaging. The funds generated are used to a for programs related to these materials and/or become available to communities through state programs. Generally, such funding mechanisms are only available at a state level and extremely rarely as a local community source of funding. Similar programs may, in the future, be imposed by the federal government as a funding source.

Avoided Costs. As discussed earlier, avoided costs of collection, processing, and disposal are likely to be realized as a result of a recycling program. Often the problem is quantifying these savings. Recycling programs funded and managed apart from the overall solid waste management program may be able to receive reimbursement from the overall management program through its funding sources for documentable savings.

Other Fees or Charges. Some communities have utilized creative variations on the user charge concept. One category that has been successfully used by a number of communities to fund recycling programs as well as to encourage recycling is based upon the collection system. A community or company will sell special bags to users in which to place their recyclable materials. These bags are then collected as recyclables and the charges for the bags are used to fund the program. They have been particularly successful for the collection of commingled materials and yard waste. Other programs have used differential fees for the collection of recyclable materials versus nonrecycled materials. However, many of these programs are used more for enforcement than as a source of funds.

Impact fees have been used by some localities to recover the cost of providing capacity in utility systems for new customers. They potentially could be used to assist in funding capital facilities. There are many legal restrictions associated with the use of impact fees and they are difficult to implement.

Summary

As discussed above, many methods are available to a community to fund its recycling program. The correct funding method in any particular community is dependent upon the specific goals and objectives of that community, its recycling program, and its political environment. The existing solid waste system and funding mechanisms will significantly impact the evaluation of funding mechanisms for recycling. However, each community has a number of options to choose from.

PROGRAM DEVELOPMENT

The development of a recycling program demands that a number of issues be identified, evaluated, and resolved. One of the major reasons for projects failing to achieve significant recycling rates has been that the community is often unable to identify or resolve critical issues. While any issue may not have seemed critical earlier in the program, it later became of sufficient importance to stop or substantially delay the project.

The selection of an organizational strategy to implement the recycling program and the evaluation of financing and funding options are an integral part of the program development. In fact, these issues may directly impact the feasibility of various recycling components in a specific community. The discussion below contains an outline for the development of a recycling program. The specific requirements will vary in each community. However, an important consideration is to retain as much flexibility as possible throughout the implementation program. Changes in goals and objectives and the experience gained as a result of initial programs will impact the direction at later stages.

Design of the Program

The first step in any recycling program is the design of the program. This step includes the planning, evaluation, and decision making necessary to develop a detailed implementation program. The elements of this step include the following:

- *Governmental requirements:* Determine local and/or state requirements for recycling program that the community must meet.
- *Existing solid waste system:* The existing system, residential and commercial, will impact the direction of any recycling programs and the feasibility and ease of implementation of various programs.
- *Solid waste quantities:* The total solid waste quantities, existing and projected, as well as the composition must be determined leading to a determination of the quantity of various potentially recyclable materials.
- *Materials markets:* A detailed evaluation of material markets must be performed, including local markets such as end users and brokers as well as available national markets. Material quantity and quality constraints should be determined. The price structure for various materials, including past changes and projected future costs, must be evaluated. This should consider transportation costs and other requirements, as necessary. The willingness of a buyer to enter into a contract should be established. This evaluation should look beyond existing markets and consider the potential development of new markets, locally and nationally, for materials such as compost.
- *Alternative recycling methods and programs:* A detailed evaluation of alternative recycling methods and programs must be made specific to a community's needs. The evaluation should include likely recovery rates, material quality and program costs, including all costs and revenues. This analysis should consider commercial programs as well as residential programs.
- *Organization:* Appropriate organizational and management alternatives should be evaluated to identify and assign responsibility for implementing the entire program as

well as each element of the program. At this time, the potential for developing regional or multijurisdictional programs should be considered.

- *Financing and funding:* The financing requirements for each alternative should be identified, both short- and long-term, and funding needs determined. Alternative financing and funding programs should be evaluated.

In performing the above evaluations, total programs should be evaluated, considering the following factors:

Level of service
Impact upon existing solid waste system
Siting considerations
Environmental considerations
Flexibility
Comparative costs
Funding
Ability to achieve goals and objectives
Implementability

Operations Plan Development

A detailed operations plan should be developed prior to recycling program startup. This plan should detail personnel, schedule, pricing, daily operating procedures for each program component, transportation, administration, and public education program.

Steps which should be included are

- *Public education and information:* The public education program should start well before the actual recycling program. Public education is the most important means of ensuring public acceptance and the participation necessary to generate high recovery rates. Public education programs should continue throughout the recycling program operation. A substantial portion of the overall project budget may need to be earmarked for this program element. During this stage, a public education program strategy should be developed
- *Economic analysis:* The financing analysis and funding requirements conducted previously must be developed in detail to reflect the overall program. The financing and funding program must be implemented.
- *Personnel and equipment:* Based upon the organizational structure selected and facilities required and the procurement/operational strategy (i.e., public or private), the appropriate personnel should be hired and put in place. At a minimum, a day-to-day operations program coordinator and appropriate support staff should be hired. At this time, procurement or purchase of necessary equipment should be undertaken.
- *Site selection:* Sites required for any and all facilities required by the plan must be undertaken. Specific criteria that need to be considered include convenience to population centers, accessibility for residential and truck traffic, security, available space, zoning requirements, suitable access roads, and building permits.
- *Evaluate:* It is very important that an ongoing evaluation of the recycling program be included as part of the operations. This should include review not only by staff but by

public officials and the general public. As required, changes to the plan should be made based upon this input.

- *Recordkeeping and reporting system:* In order to justify any recycling program from the point of view of solid waste reduction and materials reuse, it is essential to establish a practical, accurate recordkeeping and reporting system of total generated solid wastes and tonnages of various recyclables collected, processed, and reused. In many states where mandatory percentages of reduction are legally established, reporting percent recycled compared to total solid waste generated has become the responsibility of municipalities and county governments.

CHAPTER 9
THE PSYCHOLOGY OF RECYCLING

PENNY McCORNACK
President
The EarthResource Company
Portland, Oregon

BACKGROUND OF RECYCLING BEHAVIOR

Speaking about a plan to reduce the amount of packaging going to landfills by 50%, William Rathje, head of the Garbage Project at the University of Arizona in Tucson, said, "I don't think it will happen. We're not dealing with garbage; we're dealing with lifestyle."[1]

Everywhere across the nation, solid waste officials and involved citizens are asking, "How can we increase recycling?" Are they not really asking, "How can we encourage Americans to change their behavior?"

The process of recycling necessarily begins with the individual. Therefore, the individual must have sufficient education and motivation to participate in solving America's waste disposal crisis. It sounds like a simple equation, yet the methods by which to modify attitudes to collectively achieve the desired result remain shrouded in the psychology of human behavior.

The roots of psychology are ancient, yet as a science, psychology emerged just over a century ago. It is generally recognized that psychology often lacks the accuracy and precision of other sciences such as chemistry, physics, or biology. Furthermore, the subject matter of psychology is variable. The behavior and mental processes of subjects change all the time. Available scientific research on the subject of recycling and conservation behavior is scant. Few theories have been comprehensively applied to this aspect of environmental involvement. However, it has been perceived by many that such research could potentially have a profound effect on what is statistically a crisis in the making. How the environment affects behavior has been the focus of most studies. The bulk of research within the emerging discipline of environmental psychology has failed to consider the reciprocal relationship between environment and behavior.

Not only do recycling rates suffer from the lack of useful research, the industry is plagued by underfunding, which in turn impedes technological progress. Increased rates require efficient and cost-effective technology balanced with the participation of household and commercial recyclers, to create a clean and well-sorted waste stream. As with other issues on the global priority list, there are no simple answers or large-scale proven applications that will move our nation forward at a pace of repair which exceeds our current rate of depletion. As with many social trends that emerge over time, it is likely that recycling will increase collectively as a result of community-based influence. Overall national policy and public opinion will be operative in establishing need, while local influence will induce citizens to act.

Although the 1960s inaugurated what is now thought of by many as the environmental movement, the 1970s and 1980s were decades in which vigorous efforts were made to create a more environmentally educated American public. Has this resulted in a significantly more informed and motivated public? An opinion poll published in the August 6, 1990, issue of *Chemical and Engineering News* and performed by Roper Organization of New York, found that the worst solid waste problem is perceived to be disposable diapers; however, these throwaways account for less than 2% of the waste stream.[2] Apparently, the citizens who responded to this survey did not know that one of the most recycle-friendly materials, paper (including paperboard), comprises the largest portion of municipal solid waste at 41%, followed by yard wastes at 19.9%.[3]

In an interview with environmental specialist Jessica Tuchman Mathews, Ph.D., Bill Moyers spoke of polls which show that Americans do care deeply about the environment. Eighty-one percent are in favor of not allowing toxic waste to be dumped around the country; 74% strongly agree that government should be doing more to clean up the environment.[4] While statistics such as these are encouraging, Moyers shares the caution of many in expressing uncertainty about whether such optimism is justified.

Psychological literature supports the outcome that what people say about their attitudes toward the environment is not consistent with their actions. One study that exemplifies this involves monitoring whether persons drop or retain a handbill given to them by an experimenter. Of those observed having littered, only 50% admitted to having dropped the handbill.[7] Another study involved the observation of college students walking past trash that had been intentionally put in their path. While 94% agreed that it should be everyone's responsibility to pick up litter, only 1.4% actually picked up the litter.[8]

Introducing the need for recycling and other conservation behaviors has not captured society's interest as did the introduction of advances such as the polio vaccine, the automobile, the television, or the flush toilet. These were perceived as enhancements to the American lifestyle. Although inherently vital to environmental safety and productivity, why are recycling, energy issues, and other conservation behaviors not also perceived as influential factors in augmenting the quality of life in the United States?

The adoption of new ideas in American society is mandatory for increasing recycling and other conservation behaviors. Recycling and solid waste management are relatively new public concerns. Historically, these have not been issues of widespread social importance. Even atrocities such as Hanford have only recently gained considerable attention. In the decades preceding the 1970s, the attention paid to the threats of environmental hazards was minimal. With regard to solid waste, the public simply had their garbage hauled away, burned it, or buried it. Recycling did not exist as a municipal philosophy or practice, and little public attention was paid to local waste management practices. For most, it was a simple matter to dispose of the garbage, utilize the plumbing, and leave anything extraneous to the soil.

Applying a framework for examining behavioral solutions to environmental problems, B. F. Skinner concluded that many environmental problems arise out of conflicts between the positive consequences of short-term behavior and the negative outcomes of long-term behavior.[9] An example of this is the well-known "tragedy of the commons" in which a grassy square was set aside for common use in the center of towns and hamlets, altering the private use of the land once available to inhabitants. Because this encouraged individuals to increase their herds, the cows eventually all died, having exhausted the grazing.[10]

In the long run, this was collectively punitive, although the original intent was individually reinforcing.

Closer to home and current issues, one might view the outcome of the oil-dependent lifestyle of Americans as a prime example of conflict resulting from a lack of long-term objectives. Some environmental psychologists argue that without the guidance of a unified policy organized at a system level (e.g., federal and state programs), individuals will continue to practice ineffective conservation behaviors consistent with current policy.

Although the United States has no comprehensive federal-level recycling policy, the implementation of effective community programs is on the rise, and advances are being made at the state level. Chapter 2 in this Handbook speaks to these initiatives.

Globally, there is evidence that world leaders are uniting as a team to collectively address environmental issues that are no longer viewed as nationally segregated concerns. A new sense of shared destiny is emerging. The signing of the international treaty to protect the stratospheric ozone is an example of progressive effort to reduce worldwide environmental degradation.

At a meeting of the American Psychological Association, *New York Times* science news editor Daniel J. Goleman, Ph.D., presented a lecture entitled "The Psychology of Planetary Concern: Self-Deception and the World Crisis." In explaining why citizens do not alter their lifestyles to preserve rather than destroy their ecology, he cited denial, avoidance, and global indifference to the potential destruction of our planet as commonplace attitudes. At the end of the address, E. Scott Geller, Ph.D., professor of psychology at Virginia Polytechnic Institute and State University, asked Dr. Goleman "Whether it might be wise to define behaviors and contingencies that need to be changed in order to protect and preserve the environment and then set out to intervene for such a change, instead of contemplating reasons for human denial of environmental problems."[11]

In agreement with Dr. Geller's call for action, this chapter is intended as an investigation of various bodies of recycling and conservation research, to explore possibilities that may enhance new approaches in the search to psychologically motivate more people to recycle. If a common goal is assumed—that recycling rates must increase and that we must first understand how to contribute to this through greater motivation and commitment on behalf of American citizens—then it is necessary to approach the psychology of recycling.

MOTIVES FOR RECYCLING

A nationwide public opinion poll by Maritz AmeriPoll revealed why some people do not participate in recycling:

- Thirty percent say it takes too much time.
- Nineteen percent ask, "Why should I?"
- Twelve percent say they don't know how to do it.
- Eight percent say it's too messy.
- Another 8% say they have no curbside collection provided.
- Twenty-three percent cite other reasons.[12]

A news brief in an issue of *Recycling Today* reports that, "If curbside collection was easier, 91% of nonrecyclers in Great Britain would separate recyclables from their trash."

Such summaries are typical of the kind of information frequently printed in the press and reported in recycling industry publications, yet in what way can this information be translated to yield an effective result? What does this information reveal? That recycling needs to be more convenient and cleaner? That curbside needs to be easier and more widely available? That citizens need to be taught how to recycle? The answer to these questions may be "yes"; however, a more important question involves examining what might motivate the public to relinquish these considerations which are used as reasons for not recycling. Surveys indicate that recyclers report experiencing the same inconveniences that nonrecyclers view as deterrents, yet they recycle anyway.

Motivation, as it relates to recycling and conservation behavior, has been determined to be influenced by numerous components and combinations of components. The range is broad. Listed below are several factors that influence motivation, all of which have an effect on outcome:

- The credibility of a source of information or request
- The context in which information is delivered
- The frequency with which information is delivered
- The relativity of a request for action
- The degree to which an incentive is social or monetary in nature
- The extent to which an individual is already attitudinally disposed to a desired behavior.

Justification

A predominant theme in motivational behavior involves the role of justification. People are motivated to arrive at the conclusions they want to; however, doing so is constrained by their ability to construct seemingly reasonable justifications for their conclusions.[13] Much research in conservation behavior suggests that one of the most important factors underlying conservation behavior is an individual's ability to identify a reasonable justification for such behavior. This is certainly not unique to conservation behavior. The inclusion of justification is pervasive in nearly every act of human behavior; yet, despite the obvious nature of this fundamental supposition, it is often inadequately considered in approaches to influence recycling and conservation behavior.

Universality and Gratification

There is no evidence to indicate that recyclers represent a population segment with unique characteristics. A University of Michigan study which focused on conservation behavior and the structure of satisfaction themes conclude that conservation behavior is potentially satisfying to a broad cross section of a population.[14] Although the study included a small number of participants, all of whom had a known interest in conservation, findings revealed universal themes which suggested that those who conserve do not have a special or unique outlook.

Research supports the notion that recyclers possess attitudes and motivations, which when combined with other "favorable conditions," create a difference in behavior. These other conditions include elements such as situational incentives, social norms, accessibility of attitude-appropriate behaviors, and other contingencies. Without these favorable conditions, even people with proconservation attitudes may not engage in conservation behavior.

This is encouraging, for it proposes numerous alternatives within varying clusters of individuals and lifestyles. With regard to approaches based on existing motivation, one cluster may be motivated to conserve resources to save money, while another cluster may be motivated by self-image and the desire to participate in reducing waste. With regard to approaches to modify behaviors, one cluster may be influenced by a particular persuasive communication, while another cluster may be influenced by a considerably altered message and method of delivery.

Consider the different messages that might be designed to encourage recycling for such diverse audiences as MTV viewers and members of a community senior citizen network. These audiences generally respond to significantly different approaches. A musical message delivered by fashionably dressed youth, or a musical icon of the time, encouraging recycling for environmental reasons may appeal to MTV viewers yet have little impact on senior citizens. An appeal to seniors, who often live on limited incomes, might include a message delivered by a senior who is known and respected in the community, encouraging recycling as a means to curb future waste disposal bills.

In his environmental broadcast production "After the Warming," writer and producer James Burke provides a frightening portrayal in a futuristically proposed commercial. Two parents wake their sleeping children, load them gently into a Mercedes, and drive down a dark street devastated with the scars of human disregard and destruction. They arrive at a steaming mountain of waste, a landfill. The children are led out of the car and left standing as the parents drive away. The tag reads, "Don't leave this to the children." While this evocative message would likely have an impact on viewers from a variety of social clusters, it might have the greatest effect on the segment of the population who are parents.

A 1983 study describing a postcampaign analysis of California's "War on Waste" indicates that there may be an advantage in presenting the benefits (of recycling) to the individual rather than emphasizing the benefits to society.[15] This is based on the premise that the extent to which one perceives gratification through a message, to justify time and effort, will determine how much attention is paid to the message.

A look at the $110 billion advertising industry supports this premise. Most ads are built around a conveyance designed to impart one primary perceived outcome: gratification. A car with a high miles-per-gallon (MPG) rating is not touted as contributing to the solution of the global energy crisis. It is advertised with the appeal of saving the consumer money.

Although messages that direct appeal toward the individual are commonly used, and may be thought to provide a more immediate and accessible link to justification, the preference of this notion is not consistently supported by conservation behavior research. Studies suggest that an individual's sense of community involvement can play an important role in motivation. Messages that depict a dual appeal may therefore be extremely effective.

Intrinsic and Extrinsic Motivation

The role of intrinsic and extrinsic motivation has been the basis of numerous environmental behavior studies. Intrinsic motivation might best be described as self-generated desire or reasoning based on any number of personally held justifications. An individual who is intrinsically motivated to recycle may partake in the activity for any number of reasons that do not involve specific external reward conditions such as receiving something tangible in exchange for recycling. Today, many people recycle because they consider it an environmental responsibility.

Extrinsic motivation involves externally induced incentives, such as monetary reward, prizes, "payoffs," and other rewards (which was once thought to have promising potential). The greatest criticism of the use of external incentives, as it relates specifically to recycling, is that once the external incentive is withdrawn, the desired behavior often diminishes significantly or ceases entirely. The use of extrinsic motivation to promote recycling is also not a cost-effective alternative in today's recycled commodities marketplace. If incentives that cost money are to be offered, it is expected that the funding should, at least eventually, be sustainable through the operation of the recycling program.

A 1979 paper recycling project at a Florida trailer park involved a newspaper recycling reward program whereby children delivering collected newspapers were awarded their choice of a toy within a particular price range, based on the amount of paper they gathered.[16] The average weight of newspaper collected per week increased dramatically; however, removal of the "prompt and prize" condition caused the recycling activity to return to previous levels.

A 1986 University of Michigan study investigated the role that intrinsic motivation and satisfaction play in the relatively ordinary conservation behaviors of household recycling and reusing.[17] The study suggests that a factor often overlooked in approaching environmental behavior studies involves satisfaction in terms of goals and rewards derived simply from participation in an ongoing activity.

Questionnaires were distributed that were designed to measure conservation behavior, satisfaction, and motivation. Of 959 surveys distributed, the study's data analysis was based on 263 responses. Findings support the notion that "involvement with a conservation activity can be seen as satisfying in its own right, suggesting that ecologically responsible behavior might be encouraged by helping people to discover that there are intrinsic payoffs (such as satisfactions derived from living by an ecological ethic, saving energy, having a chance to participate, being a member of an affluent society, etc.) associated with such activities. The satisfaction scales that emerged from the survey included frugality, participation, and prosperity. The motivation scales which were developed revealed that monetary reward was not a dominant motive for recycling.

A University of Colorado study points out that some applications of material incentives have been shown to evoke conservation actions that have no inherent relationship to an attitudinal disposition to favor such actions.[18] Energy tax credits demonstrate this point. There is no evidence that participation in such a program is motivated by the intent to make a contribution to resource conservation. It is expected that monetary incentive has been the basis for participation in energy tax credit programs. It is interesting to note here that access to energy-efficient systems does not necessarily result in energy savings. Lifestyle and motivation have been found to override the potential effectiveness of available energy-saving systems.

Seattle, Washington, enjoys a successful recycling program of national acclaim that involves 55% of the city's households. A rate incentive is the factor commonly attributed to the program's success. However, the $12 per year difference between collecting one can compared to two cans of refuse did not appear to represent substantial savings when considered in the context of an overall household budget. Does this savings alone constitute the reason for high motivation to recycle, or are other behavioral components contributing to the success of this program? Answers to these questions require further study.

Public Commitment

Public commitment has been found to be a promising behavioral change technique. This approach is based on the premise that when attitudes are publicly stated, they remain relatively stable and are likely to increase the performance of the behavior consistent with them. The use of public commitment techniques generally has included the employment of persuasive communication followed by a commitment request.

A study introduced an alternative social psychological approach derived from formulations of the minimal justification principle.[20] Under conditions of modest external pressure, three groups of a newspaper recycling study were (1) asked to sign a statement saying their household would participate in the project, (2) asked to make a verbal commitment to recycling, or (3) informed about the project through a leaflet.

Findings revealed favorable outcomes using the commitment technique. Among the three groups, those who signed the agreement recycled significantly more than the other two groups. There was only a marginal difference between those who made a verbal commitment and those who were left a leaflet. (This latter group had no personal contact and recycled minimally.)

In a follow-up period, when they were no longer bound to their commitment, those who had signed the agreement continued to recycle, while the performance of the other two groups became almost identical: the recycling rate declined.

In a 1986 study to evaluate the incidence of increased recycling in a citywide program, a public commitment technique was applied, with additional emphasis on employing persuasive communication.[21] The persuasive communication incorporated variables that were considered to have demonstrated value in influencing attitude and/or behavior change. These included (1) the use of a credible source of information, (2) the use of recommendations closely related to an individual's accepted beliefs and practices, (3) the inclusion of information from reference groups relevant to the individual, (4) the use of a moderate fear arousal component, and (5) supplying specific recommendations.

These components were integrated into three different treatments, delivered by Boy Scouts (a credible source), that included (1) the presentation of oral information, (2) a request for public commitment, or (3) both. Included were a total of 201 nonrecycling households that were selected from a 6-week baseline study and 132 homes that served as a control group and received no treatment.

The findings of this study further support the effectivenss of a public commitment. Of those who received both treatments, 42% began recycling. Of those who received the public commitment condition, 42% began recycling while 39% of those who received the oral information condition began recycling. Only 11% of the control group began recycling.

INFLUENCING VARIABLES

In concluding with factors that have been found to affect recycling and conservation behavior, reviewed here are other components that appear independently and in combination in numerous studies and recycling programs.

Source Credibility

Consider the source is a popularized adage and a widely accepted condition applied to the process of evaluating the credibility of information. In the study that utilized the Boy Scouts as a credible source of information, a different outcome might have been anticipated had the study employed work-release convicts or other socially perceived offenders.

A credibility factor is also built into a strategy referred to as modeling, which includes the demonstration of specific behaviors and the desired response, often followed by the presentation of pleasant or unpleasant outcomes. The use of opinion leaders and credible information on sources can greatly enhance messages designed to promote recycling and conservation behaviors.

Relativity

Noted earlier was a discussion on how perceived gratification can determine to what degree a receiver pays attention to a message. It is further suggested that persuasive communication which deviates too far from existing beliefs and practices will have a significantly diminished impact, if any at all. As this might relate to recycling, there is an implied learning curve. Introducing the various processes of recycling may best be accomplished incrementally and include messages within a reasonable range of the receivers beliefs and experience relative to such increments.

Reference Groups

The concept of providing information from *reference groups* which are relevant to the individual is aligned with studies that point to the importance of a perceived community link, and with the premise that individuals are more likely to respond when they believe others are participating in the same or similar activity. An example of this might include the presentation of information which depicts recycling activities being undertaken by local organizations, and the positive results of the programs. This is thought to diminish an individual's feeling that his or her actions alone will have little impact.

Perceived Evidence

Humans are more likely to be influenced when *perceived evidence* is available. In the 1976–1977 drought experienced in California, there was a high correlation between the direct observation of empty reservoirs and the information disseminated about the need to conserve. This correlation led to increased general acceptance.

By contrast, an implied condition which is unobservable or which is introduced under circumstances lacking factual clarity is not likely to lead to an attitude of general acceptance. Consider conflicting information on the issue of a fuel shortage which includes publicized negations such as excess profits in the oil industry. The consequence of this scenario can result in the assumption that a shortage is not actual, but implied.

Fear Arousal

There is evidence of a relationship between *fear arousal* and persuasion. It has been suggested that fear appeals should include recommended actions and emphasize how such actions might diminish a proposed threat. The appeal should also have personal relevance.[22]

Punishment

As a disincentive, *punishment* is found to be largely unpredictable; however, research points to several possible outcomes: (1) through avoidance of the source of punishment, its effect may be minimized; and (2) punishment may only suppress, but not eliminate undesired behavior. It has been suggested that punishment is only effective when positive reinforcement for an alternate behavior is offered.

Demands

Demands, rather than polite requests, are thought to potentially elicit behaviors other than those desired. Of two antilitter handbills which were distributed at a public swimming pool, those that said "Don't You Dare Litter!"were littered significantly more than those that said "Help Keep Your Pool Clean."[23] Many individuals might not only feel threatened by demands, but they may also elicit contrary behavior to assert their freedom.

Mandatory Recycling

Mandatory recycling has so far been an alternative recourse that has been minimized. Recycling has been mandated in 10 states and a significant number of communities. Los Angeles, a world leader in waste per capita (producing waste on the average of 6 lb per person per day), is expected to introduce mandatory recycling.

In many cases it has been implemented when voluntary programs have produced poor results. Low recycling rates led the town of Hamburg, New York, to mandatory recycling. The participation rate is 100%; landfilled trash has been reduced by 34%; and disposal costs were saved during the first year. Those who fail to separate recyclables from their trash face baneful consequences. After four steps in a process of attempts, the city ultimately stops picking up their garbage.[24]

Until mandatory recycling is studied from a scientific perspective of human behavior, the psychological mechanics of its effect will not be known. Research warns that because attitudes which follow behavior change are most significant, attempts to mandate behavior change through disincentives and punishments are not favored because they frequently elicit negative attitudes.[25] It is interesting to note that the use of disincentives becomes more frequent as a resource crisis deepens.[26]

Careful study will answer questions that will aid in the future design of optimum recycling programs. Is compliance with mandatory recycling generally progressive or immediate? How do attitudes change? What program components influence behavior changes the most? Does contrary behavior result?

The Media

More statistical data about the state of our environment are available than ever before. When utilized effectively, the *mass media* can be powerful as an educational tool in encouraging proenvironmental behavior. It can also serve as a source of information leading to public confusion and mistrust.

One of four dominant attitudes revealed in the postcampaign analysis of California's "War on Waste" was that "Conflicting messages regarding the national energy crisis have confused many people."[27]

Research on public information campaigns points out that reliance on the mass media as a sole source of information as a basis for behavioral change is ineffective without supportive interpersonal communication and social reinforcement. This implies an advantage in strengthening specific publicity with door-to-door campaigns, newsletters, etc., in conjunction with media coverage about recycling, rather than relying solely on the mass media as a source of motivating information.

It is also suggested that the greatest impact on public awareness of solid waste issues is derived at the local level, followed by national publicity, and that for both, the degree of awareness achieved is directly correlated to the amount of coverage and publicity.

Commercial use of the mass media as an advertising-promotional avenue to reflect ecological accountability has increased. While much of this communication is educative, some is harmful. Potlach Forests, Inc., was dumping 40 tons of suspended organic waste into the Clearwater River and employees were washing their cars before leaving work each day to protect them from the pulp plant's corrosive sodium sulfate emissions, yet the company's national ad campaign conveyed purity and diligent environmental responsibility. The headline under a picture (taken 50 mi upstream) of natural beauty read, "It Cost Us a Bundle but the Clearwater River Still Runs Clear."[28]

Although this discrepancy was exposed, such practices are not extinct. This type of media use is indeed contraindicated in the development of a more ecologically motivated society. It is hoped that members of the press and those who use the media to spread their message will take a strong lead in presenting information that is motivating and that encourages citizens toward more responsible recycling and conservation behavior.

For instance, in reporting on the seriousness and expense of the closure of a community landfill, a newscaster might take the opportunity to mention that the average waste generated per person per day is 3.5 lb. If each citizen begins recycling today to save half of their daily waste from the new landfill, the life of the landfill could be extended by (insert accurate calculation for the number of years), saving (insert accurate calculation for the thousands, possibly millions of dollars saved). In communities where recycling data are collected, it may also be practical to report daily quantitative totals of recycling activity to demonstrate community participation. There are numerous ways in which the mass media can actively encourage recycling.

CONCLUSION

This chapter has been a discussion of recycling and conservation behavior with a focus on the role of motivation, which depends largely upon communication. Several critical factors specifically relevant to communication should be considered future application.

A greater link is needed between those who develop policy or implement recycling programs and those who are engaged in the study of conservation behavior. While it may not be possible to scientifically evaluate the components of the many types of recycling programs before they become widespread practices, such analysis should be pursued. To enhance communication that will lead to increased recycling rates, intervention strategies that encourage the development of intrinsic motivation, and that provide a rationale for prorecycling behavior, are needed.

As information about recycling is disseminated, careful attention should be given to the design and content of the message, with the intent to motivate and provide balanced perspectives on issues that may lead to stronger conservation and recycling behavior. Information such as that which simply states "more people would recycle if it was more convenient," has little impact on the prevailing problem. Consistent with several studies, an alternate approach might offer a discussion on the benefits of recycling rather than attempt to diminish perceived inconveniences.

YOU CAN HELP SAVE OUR FORESTS. ONE TON OF RECYCLED NEWSPRINT WILL SAVE 17 TREES!

News reporting should provide more than simple statistics and might seek to include information that may assist in providing a basis for citizens to gain a greater understanding of the scope of America's waste problem. For example, instead of reporting simply that plastics are not being recycled in a given community, information on how residents could make efforts toward establishing a successful program might be included. Perhaps the reason that the Roper Organization survey reflected the mistaken belief that diapers are the leading solid waste problem is because disposables have been an attractive news feature due to the controversial nature of the topic. While such emphasis may seem interesting, a more comprehensive approach could be more effective in influencing behavioral changes. For instance, news reporting on diapers could include a reminder that paper comprises the largest part of the waste stream; that paper is easy to recycle; and where more information may be obtained for those who do not currently recyle paper.

Where motivation does not exist, access to recycling alone may not be a sufficient incentive by which to achieve higher recycling rates. Simply offering containers or curbside collection may not be enough. Although no recipe exists for motivating everyone to recycle, there is promising evidence that effective use of innovative approaches to communication can have a dramatic effect on future recycling rates.

REFERENCES

1. "Canada Package Cutback Plan Criticized as Overly Ambitious," *Plastics News,* June 4, 1990, p. 6.
2. Robert J. Samuelson, "Diapers: The Sequel," *Newsweek,* April 16, 1990, p. 65.
3. *Decision-Makers Guide to Solid Waste Management,* U.S. Environmental Protection Agency, November 1989, p. 25.
4. Bill Moyers, *A World of Ideas,* Doubleday, New York, 1989, p. 301.
5. John D. Cone, *Environmental Problems/Behavioral Solutions,* Brooks/Cole Publishing, California, 1980, p. 13.
6. Cone, loc. cit.
7. B. F. Skinner, *Science and Human Behavior,* Macmillan, New York, 1953.
8. G. Hardin, "The Tragedy of the Commons," *Science,* Vol. 162, 1968, pp. 1243–1248.
9. E. Scott Geller, "Applied Behavior Analysis and Social Marketing: An Integration for Environmental Preservation," *Journal of Social Issues,* Vol. 45, No. 1, 1989, pp. 17–36.
10. *Resource Recycling's Plastics Recycling Update,* July 1990, p. 3.
11. Ziva Kunda, "The Case for Motivated Reasoning," *Psychological Bulletin,* Vol. 108, No. 3, November 1990, p. 480.
12. Raymond De Young and Stephen Kaplan, "Conservation Behavior and the Structure of Satisfactions," *Journal of Environmental Systems,* Vol. 15, No. 3, 1985–1986, pp. 233–242.
13. Mark A. Larson and Karen L. Massetti-Miller, "Measuring Change after a Public Education Campaign," *Public Relations Review,* Vol. 10, No. 4, Winter 1984, pp. 23–32.
14. P. D. Luyben and J. S. Bailey, "Newspaper Recycling: The Effects of Rewards and Proximity of Containers," *Environment and Behavior,* Vol. 11, 1979, pp. 539–557.
15. Raymond De Young, "Encouraging Environmentally Appropriate Behavior: The Role of Intrinsic Motivation," *Journal of Environmental Systems,* Vol. 15, No. 4, 1985–1986, pp. 281–292.
16. Stuart W. Cook and Joy L. Berrenberg, "Approaches to Encouraging Conservation Behavior: A Review and Conceptual Framework," *Journal of Social Issues,* Vol. 37, No. 2, 1981, pp. 73–107.
17. Ken Stump and Kathy Doiron, "The System Works: Seattle's Recycling Success," *Greenpeace Magazine,* Jan./Feb. 1989, pp. 16–17.
18. Anton U. Pardini and Richard D. Katzev, "The Effect of Strength of Commitment on Newspaper Recycling," *Journal of Environmental Systems,* Vol. 13, No. 3, 1983–1984, pp. 245–254.

19. Shawn M. Burn and Stuart Oskamp, "Increasing Community Recycling with Persuasive Communication and Public Commitment," *Journal of Applied Social Psychology,* Vol. 16, No. 1, 1986, pp. 29–41.
20. Burn and Oskamp, op. cit.
21. J. W. Reich and J. L. Robertson, "Reactance and Norm Appeal in Antilittering Messages," *Journal of Applied Social Psychology,* Vol. 9, 1979, pp. 91-101.
22. Jon Naar, *Design for a Livable Planet,* Harper & Row, New York, 1990, p. 17.
23. Geller, op. cit.
24. Cook and Berrenberg, op. cit.
25. Larson and Massetti-Miller, op. cit.
26. "Pollution: Puffery or Progress?" Newsweek; December 28, 1970, pp. 49–51.

SECTION II
RECYCLING MATERIALS

CHAPTER 10
ELECTRONIC DEVICES

JOSEPH C. YOB, JR.
Vice President
Creative Recycling Systems, Inc.

INTRODUCTION

In the years ahead new molecular integrated circuits capable of producing computers 100 billion times as fast as today's most powerful personal computers could become so pervasive and inexpensive that they would literally become an integral part of every human-made object. The basic technology for these integrated circuits was recently developed by a team from Hewlitt-Packard and UCLA This technology could spell rapid obsolecence for every electronic device in the world in the not too distant future.[1]

The Electronic Product Recovery and Recycling Baseline Report (EPR2 Baseline Report) published by the National Safety Council's Environmental Health Center estimates that 275 million pounds of electronic equipment was recycled in the United States in 1998 alone.[2] Additionally, this report estimates that more than 75% of end-of-life electronics products received by electronics recyclers and third-party organizations come from electronics OEMs (original equipment manufacturers) and large-scale users of electronic equipment (those with more than 500 employees).[3] Where is the electronic equipment from small-scale users going? Why isn't more of it recycled? What options are available now to recycle electronic devices? What future partnerships will be required to build the necessary recycling infrastructure? This chapter will attempt to answer these questions.

Electronic Recycling

For the purposes of this chapter, electronic recycling is defined as the diversion of end-of-life/surplus electronic devices from the waste stream by the re-use and/or recycling of electronic devices. It is essentially a front-end waste prevention process. A new model is needed for this process: Wastes are materials that do not have a cost-effective use based on a cost–benefit analysis. As shall latter be demonstrated, an appropriate cost–benefit analysis reveals that in fact few electronic devices should be classified as waste.

Electronic Recycling System

An electronic recycling system is a cost–benefit controlled management process that utilizes end-of-life/surplus electronic devices as a resource in an attempt to recover the greatest return from the materials at the lowest possible cost while complying with applicable environmental laws and regulations. Materials are reused/recycled by utilizing decision tree flowcharts that include multiple decision steps and processes (see Figure 10.1).

Electronic Device

An electronic device is a product, a piece of equipment, and/or a mechanism utilizing electric circuits in which electron flow is controlled in a manner designed to serve a special purpose or perform a special function by the methods or principles of electronics. Examples are computers, data processing equipment, communications equipment, integrated circuits, semiconductors, display devices, and other like products and equipment.

Cost Item

When taken as a whole, an electronic device that has a net cost to reuse/recycle is referred to as a cost item. An example would be a defective cathode ray tube in a television or computer. In many cases, the net value return from these devices is less than the cost incurred to refurbish and/or recycle them. Cost items are the greatest challenge in electronic device recycling.

Margin Item

When taken as a whole, an electronic device that yields a net margin when the device is reused/recycled is referred to as a margin item. For instance a personal computer "box" containing a motherboard, CPU (central processor), memory, and input/output devices other than video displays is an example of a margin item. Presently, in most cases the value realized after they are refurbished/recycled is greater than the cost incurred to refurbish and/or recycle them. Margin items can offset cost items, and in many instances yield a return to the customer who recycles combinations of them.

Closed Loop System

This is a process that ensures that a significant portion of the electronic devices that are managed under an electronic recycling system will be diverted from the waste stream. The objective of such a system is zero waste.

Flow Control

Flow control is a government sanctioned control of reusable/recyclable electronic devices by declaring them to be waste and/or directing the manufacturer to take back its electronic devices at the end of the device's useful life. It is similar to a closed loop system but does not utilize the free market to make an appropriate cost–benefit analysis. Typically, flow control is an after the fact approach that prevents a total cost accounting for electronic devices during their life cycle. Historically, such approaches yield much litigation and little recycling of products because partnerships between regulating authorities and stakeholders (people and organizations that are affected) are not established prior to the issuing of the regulations implementing flow control.

Scope

This chapter is an overview of electronic recycling from a business perspective. Electronic recycling is a relatively new field and its development will take the efforts of a broad-

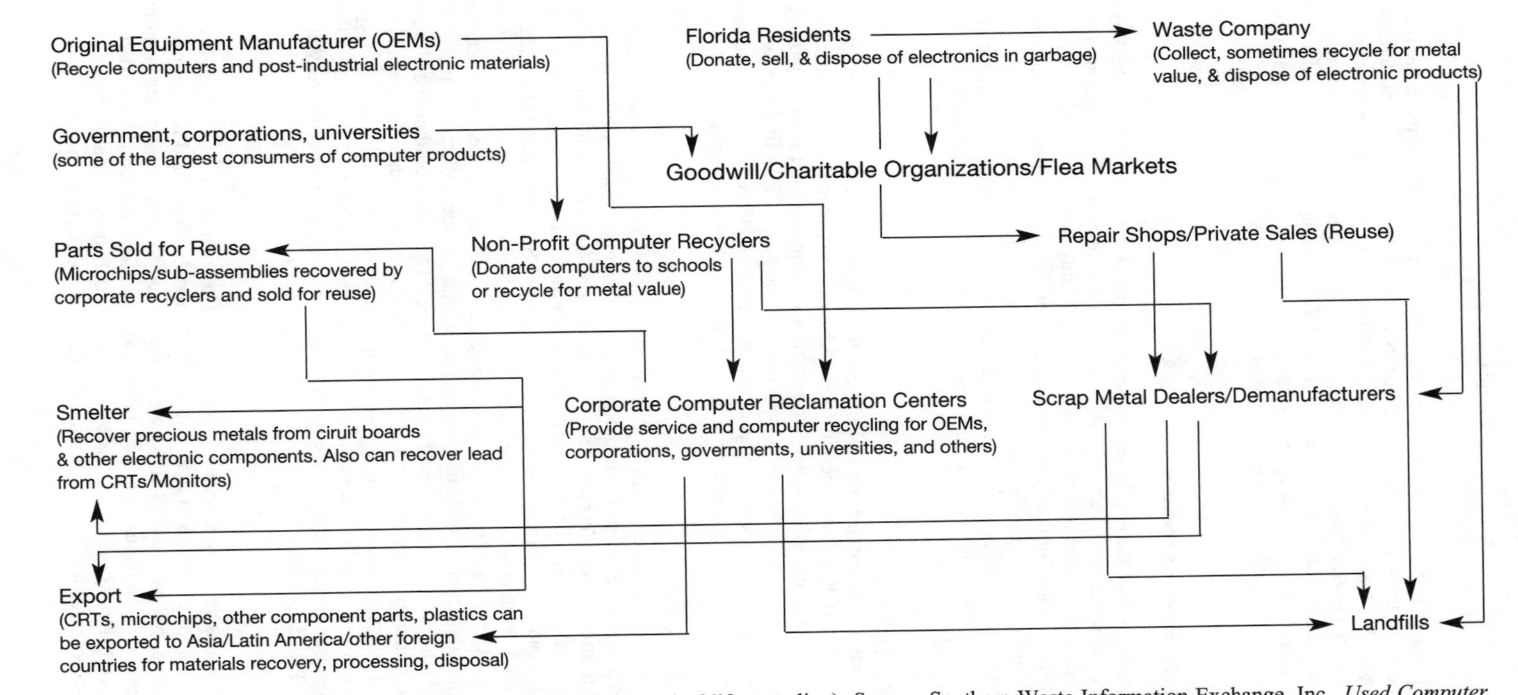

FIGURE 10.1 Typical electronic management systems flow (end-of-life recycling). *Source:* Southern Waste Information Exchange, Inc., *Used Computer Recycling and Management in Florida: A Resource Guide.*

based coalition in order to meet the challenges of the future. To increase the recycling rates of these products the collaborative efforts of not only local, state, and federal government agencies concerned with recycling, but also electronic device owners and manufacturers is imperative. Organizations and persons that are interested in the recycling of electronic devices will find this information useful in developing electronic device recycling programs. Governmental agencies charged with regulating recycling activities as well as OEM's role as perceived by the author are also included.

Regulatory Environment

Congress, in the Hazardous Solid Waste Amendments (HSWA) of 1984 to the Resource Conservation and Recovery Act (RCRA), established a timetable for restricting land disposal of hazardous wastes unless properly treated.[4] Only consumers and small-quantity generators are exempt from the provisions of this act. Several exclusions from hazardous waste rules are provided for the recycling of electronic devices, most notably recycled whole and shredded electronic circuit boards, under certain conditions.[5] All the specific regulations set forth in this act can be found in Parts 260 through 279 of Title 40 of the Code of Federal Regulations (CFR).

Cathode Ray Tubes

Recent preliminary research from researchers at the University of Florida indicates that cathode ray tubes (CRTs) contained in some video displays (computer monitors and terminals), televisions, and other electronic products contain lead and may be hazardous.[6] This research would tend to support the contention that some CRTs fit the seven factors (40 CFR 273.81) for inclusion as a waste regulated under 40 CFR 273, and subsequently, states could include this waste stream in the state list of universal wastes.[7]

Universal Wastes Rule

In the event a determination was made that certain CRTs do in fact meet the requirements to be listed as universal wastes and state(s) added them to their list of universal wastes, certain implications arise. According to the provisions of 40 CFR 273.60, universal waste destination facilities must be permitted RCRA (Resource Conservation and Recovery Act) Part B Treatment, Storage, and Disposal Facilities (TSDFs) or recycle their materials without storing. Neither glass recyclers nor secondary lead smelters are typically permitted as RCRA Part B TSDFs. Neither of these types of facilities is likely to process their incoming material streams without storage, i.e., within 24 hours of receipt per 40 CFR 261.6§(2). Thus, under universal waste regulations, CRTs likely would not be allowed to go to either of their primary current recycling destinations. As discussed in the comments on the U.S. Environmental Protection Agency universal waste rule, this remains a significant disincentive to recycling universal wastes because obtaining a RCRA Part B Permit "is time consuming, cost prohibitive, and in most cases unprofitable for the recycling facilities" (59 FR 25534, May 11, 1995).[8]

Superfund

The Comprehensive Environmental Response, Compensation, and Liability Act (CERCLA), commonly referred to as "Superfund," and the Superfund Amendments and Reauthorization Act of 1986 (SARA) were established to clean up leaking hazardous waste disposal sites. In the case of the United States v. Monsanto Co., 858 F.2d 160 (4th Cir. 1988), cert. Denied 490 U.S. 1106 (1989), the court held that a generator of hazardous substances was liable under section 107(a)(3) if the government could prove that:

- The generator's hazardous substances were, at some point in the past, shipped to the facility
- The generator's hazardous substances, or hazardous substances like those of the generator were present at the site
- There was a release or threatened release of any hazardous substance at the site
- The release or threatened release caused the incurrence of response costs[9]

Material Sources

As previously discussed, 75% of end-of-life electronics products received by electronics recyclers and third-party organizations come from electronics OEMs (original equipment manufacturers) and large-scale users of electronic equipment (those with more than 500 employees. It would be difficult to measure electronic disposal for households and small businesses, and little reliable data is available as to how and where these materials are disposed of. Electronic devices vary as to their life cycle. Anecdotal evidence would tend to support the conclusion that household and small-scale organizations are a significant portion of the electronic device end-of-life stream.

The EPR2 Baseline Report indicates that as many as 499.8 million personal computers will have become obsolete between 1997–2007.[10] By the year 2003, CRT monitor obsolescence will reach 26.1 million per year.[11]

The public schools installed computer base was estimated at 6 million units in the 1996–1997 school year.[12] U.S. Households that have PCs were estimated to be 44% in 1998.[13] In the United States today, virtually all households have televisions and some have VCRs.[14] As flat screen displays replace CRT displays in video displays and televisions, these obsolescence rates are likely to increase dramatically. As DVD players become more affordable, VCRs will rapidly become obsolete.

Historical Perspective of Electronics Recycling

Electronic devices have been recycled since the dawn of the electronic age. Principally, these devices were recycled for their precious and base metals. Mainframe computers in years past contained a significant amount of gold and other precious metals in the form of plating, as these metals were much less expensive than they are today. The high gold prices of the late 1970s caused a major economizing in the use of gold, so people in the electronic recycling business in the years passed have referred to the pre-1970s as the "golden age." Additionally, devices consisted of copper, aluminum, and steel, which were demanufactured by scrap processors and sold to smelters, refineries, and mills.

Although there is little statistical information about electronic device recycling from years past, there is an abundance of anecdotal information to suggest that mainframe computers and telecommunication scrap recycling rates were quite high. There is also anecdotal evidence to support the contention that many parts from televisions were in fact cannibalized from defective units to be used as repair parts in other units. PCs were also recycled from the beginning. In fact, there were companies recycling microcomputers as far back as 1979.

The household waste exemption to the hazardous waste rules has enabled the landfilling/incineration "cost" of end-of-life electronic devices. Small-scale users are undoubtedly disposing of these items. There is a lack of statistical information on exactly what quantities of electronic devices are disposed of in this manner. From the author's observation of waste operations, the quantity is significant.

Original Equipment Manufacturers (OEM) Initiatives

Several OEMs have started design for recycling initiatives.) Dell Computer's chassis designs are now fully recyclable. Dell was one of first computer manufacturers to concentrate on lowering the total cost of ownership. Dell's leasing program relieves the computer user of the end-of-life disposition problem for their computer equipment. This program appears to be a model for the future. These solutions were implemented not just in European countries where they were required by law, but all over the world.[15]

A growing number of HP (Hewlitt-Packard) products—including recent models of HP Vectra and HP Brio PCs, HP Kayak PC Workstations, HP OfficeJet All-In-Ones, and HP LaserJet and HP DeskJet printers—are designed to be easier to take apart and recycle. Many components simply snap apart, making it easier to separate metal from plastic. HP has also reduced the number of different types of plastic and metal parts used, and put material identification codes on plastic parts.

Government Initiatives

Presently, there are initiatives at local, state, and federal government levels concerned with the recycling of electronic devices.

Common Sense Initiative

The Common Sense Initiative (CSI) represents a new approach for EPA in creating policies and environmental management solutions that relate to industry sectors. CSI examines the environmental requirements impacting six industries: automobile manufacturing, iron and steel, metal finishing, computers and electronics, printing, and petroleum refining. For each industry, EPA convenes a team of stakeholders that look for opportunities to change complicated and inconsistent environmental policies into comprehensive sector environmental strategies for the future. The process is producing better, more applicable environmental protection strategies that are developed by those who have to live with them, avoiding costly and time consuming adversarial processes later. The web site for the common sense initiative is http://www.epa.gov/commonsense/computer/index.htm.

Electronic Product Recovery and Recycling Residential Collection Pilots

The CSI Computers and Electronics sector has embarked on a project designed to 1) determine the composition of the waste stream and the types and volume of equipment to be collected; 2) assess the economic viability of a residential postconsumer collection/demanufacturing program for end-of-life electronic equipment; 3) determine residents' willingness to pay for this disposal option; and 4) evaluate any available data on other residential postconsumer pilot collection programs.

Two pilot communities have been identified—Somerville, Massachusetts, and Binghamton/Broome County, New York. One collection was completed in the fall of 1996 and an analysis of the types of products recovered is underway. A second collection day took place in 1997.

(CSI) RCRA Regulatory Barriers to Cathode Ray Tube (CRT) Recycling

Computers and televisions use a cathode ray tube (CRT) for viewing. The CRT contains lead to shield users from the radioactivity required to produce the image. Improper disposal of CRTs can place lead in the waste stream; this represents not only a health hazard, but also the loss of a recyclable natural resource. Lead recovered from used CRTs can be safely and practically reused to produce new CRTs.

A work group is developing a strategy for removing perceived federal regulatory barriers to recycling CRTs. Its goal is to apply common sense to hazardous waste requirements for this waste stream while maintaining high standards for health concerns. The strategy will take into account potential economic benefits, as well as potential risks to the environment, the community, and workers. The project team will document the basics of CRT recycling; existing federal regulatory issues relating to CRT recycling; environmental and worker safety risks posed by CRT recycling methods; and the economic and environmental benefit and risk issues of CRT recycling. Upon completion of this work, options for improving the current system will be presented to the Computers and Electronics Subcommittee and the CSI Council.

CSI Electronic Product Recovery and Recycling (EPR2) Project

This project of the National Safety Council's Environmental Health Center provides education, information exchange, and the building of productive relationships among diverse stakeholders concerned with managing outmoded computer equipment. The workgroup cosponsored a very successful conference on electronic product recovery and recycling held in February 1997 and attended by over 200 people. The conference was the kick-off for an independent roundtable being established to facilitate sound management of end-of-life electronic equipment over the long term. The Roundtable will identify and promote resolution of emerging issues related to better management of unwanted computer equipment. CSI's partner for carrying out the initial phases of this project is the Environmental Health Center, a division of the National Safety Council. The EPR2 Project promotes environmentally safe, responsible, and cost-effective management of electronic equipment that has reached the end of its useful life or no longer meets the needs of its original owners. The project will help identify and prioritize ways to overcome market, economic, regulatory, administrative, and institutional barriers to effective management of electronic equipment throughout its life cycle. Their web site is http://www.nsc.org/ehc/epr2.htm.

Industry Trends

The electronic recycling industry is undergoing unprecedented growth. End markets for recycled electronic products will drive this industry. Commodity prices have taken wide swings in the last few years and some commodities sell for a price lower than the cost of recycling. If commodity prices drop to a point where disposal is more cost effective than recycling, a system of charges will have to be implemented or a much lower level of recycling will be realized. New government regulations that do not provide for cost–benefit analysis will either result in subsidization of the electronic recycling industry or a significant reduction in electronic recycling.

Incentives to Recycle Electronic Devices

In cases where the recycling of margin items exceeds cost items, the incentive will be payment to the customer for their materials. The savings plus the cost avoidance of disposal would be an incentive to recycle. In the case of organizations that are not exempt from hazardous waste regulations, the cost to recycle electronic devices that may be classified as hazardous waste regulations in most cases will be less expensive than disposing of electronic devices as hazardous waste. These costs can be directly compared.

In cases where pollution prevention via recycling of electronic devices is cheaper than disposing of hazardous waste created by incineration and/or disposal of exempt entities (e.g., household hazardous waste) the benefits of reduced costs at the "end of the pipe" provides the incentive of lower overall costs. The present household waste exemption is allowing electronic devices to enter the waste stream. If there were no cost savings between the methods, there would be the incentive of risk reduction from future regulatory requirements. This benefit is immeasurable, as without knowing what those regulation costs are, one cannot adequately determine the value of the benefit.

If solid waste authorities lose their exemption for household hazardous waste by recycling but maintain the exemption for disposal, this condition would provide an extreme disincentive to recycle electronic devices. The effects of such regulations would stifle recycling.

If electronic devices were banned from landfills/incinerators and adequate recycling infrastructure was not in place prior to the ban, it is very likely that illegal disposal of electronic devices would occur. A cost–benefit analysis in terms of law enforcement costs and property damage would need to be made in advance of any such bans as a matter of sound public policy.

RECYCLING ELECTRONIC DEVICES—A SYSTEMS APPROACH

Regulatory/Legal Concerns

As was previously discussed, proper management of end-of-life electronic devices is essential to avoid liability for noncompliance with environmental laws and regulations. At a minimum, an organization should contact its local waste authorities and the applicable state regulatory agencies for assistance in setting up an electronic device recycling program. Written requests for information on the recycling of electronic devices and electronic device recyclers demonstrating compliance with applicable laws and regulation should be requested. Analyzing recycling solutions will also assist organizations in compliance with these rules and regulations.

Identifying Electronic Devices to Be Recycled

The first step in the electronic device recycling system is to identify the electronic devices to be recycled some examples are:

- Computers and data processing equipment
- Communications equipment
- Integrated circuits and semiconductors
- Display devices (e.g., video displays in computer monitors, terminals, televisions, and other CRT-containing devices)
- Other scrap electronic devices

Quantifying

After determining what type of electronic devices are to be recycled, the next step is to quantify the electronic devices to be recycled. The following questions are essential to this process:

- What is the quantity of electronic devices specifically, by type?
- Where are the electronic devices located by quantity and type?
- What shipping facilities are available at each location?
- What available modes of transportation are available for each location?
- Who are the necessary personnel required to coordinate with the shipment of the electronic devices?

Triage (What Goes Where)

In order to lower costs and increase return for electronic devices, a triage process must take place prior to shipment. The process includes the following:

- Whenever possible, package and stage like items together.
- Prepare electronic devices for shipping. Packaging should ensure that CRTs contained in video display devices such as TVs, computer monitors, and other electronic devices will not be broken in shipment.
- Document specifically by type, model, and quantity each electronic device to be recycled.
- Obtain material safety data sheets as necessary for all electronic devices to be shipped.

Analyzing Recycling Solutions—Cost–Benefit Approach

It may be more cost effective to have an electronic device recycler handle many of the above functions. In many cases, costs can be reduced by consulting with professionals to determine which steps in the process will be done internally and which should be contracted out. A total cost analysis should examine the direct and indirect costs of each step and select a service provider for steps that can be performed by external contractors at a lower cost. When determining total costs, the opportunity costs of having internal personnel and assets utilized should not be discounted. In many instances, a well-thought-out program of recycling electronic devices will yield a net return to the organization for their recycled assets.

Selecting a Recycler

The following due diligence steps should be performed as a minimum prior to selecting an electronic device recycler:

- Visit the facility where the processing is to be done. Inspect the general housekeeping to ensure that there is an orderly system for processing and accounting for your materials. If you require proprietary information to be removed from you equipment, ensure that this is plainly stated in any contracts you enter into. Confirm with your management information systems person that this will meet your requirements or plan to have this done in-house.
- Obtain a copy of all applicable permits, registrations, and licenses from the recycler. If permits are not required for the recycler's operation, ensure that you receive documentation from the recycler that states that no permits are required.

- Ensure that the recycler and its vendors have an environmentally sound method for handling printed circuit boards, CRTs, mercury-containing devices, and ni–cad batteries.
- Ask for a list of all vendors used by the recycler to outlet processed materials; this should include contact numbers and names. Contact these vendors in writing to ensure that they are in compliance with all applicable environmental laws and regulations.
- Contact (preferably in writing) your local and state governmental environmental agencies and request any relevant information they have on the prospective recycler and its vendors. Check the documentation obtained from all vendors and contractors with the applicable agencies to ensure that the response from the recycler and its vendor are in compliance with all applicable environmental laws and regulations. If the responses are not forthcoming, contact the EPA Ombudsman in your state for assistance.
- Require detailed written accountability for the material the recycler has processed for your organization. At a minimum, the ultimate destination for all recycled electronic devices should be reported from each electronic recycler.
- Follow up with, at a minimum, an annual audit of the recycler and its vendors to insure your material is being handled, processed, and accounted for in compliance with the applicable environmental laws and regulations.

Purchasing with Recycling in Mind

Whenever possible, buy electronic devices that have been engineered for recycling. Ask the sales representative about the recyclability of the products they are attempting to sell you. If they cannot answer these questions, request that they query the companies that manufacture the electronic devices for this information and provide it to you. Demand for recyclable products will increase the supply of them.

ELECTRONIC RECYCLING PROCESSES

The following section summarizes the recycling process and flow of products through the electronics recycling industry, first by presenting a generic flow model and then by summarizing data collected about the sources and destinations of end-of-life electronic equipment.

Figure 10.2 is a generic flow diagram for the electronics recycling process. It is most applicable to PC recycling but can also be used to illustrate the recycling process for other electronic equipment. Not all the steps shown are taken by all recyclers, nor are the steps necessarily taken in the order shown. In general, however, the four stages described below represent the major decision/action points in the process.

Determination of Potential for System Reuse

The first stage consists of testing the computer to determine its usability; this can be carried out by primary recyclers, secondary recyclers, or third-party organizations. Two primary factors determine usability.

The first determinant of usability is age. For PCs in particular, the age at which a computer becomes "old" has decreased over time. However, particular developments in

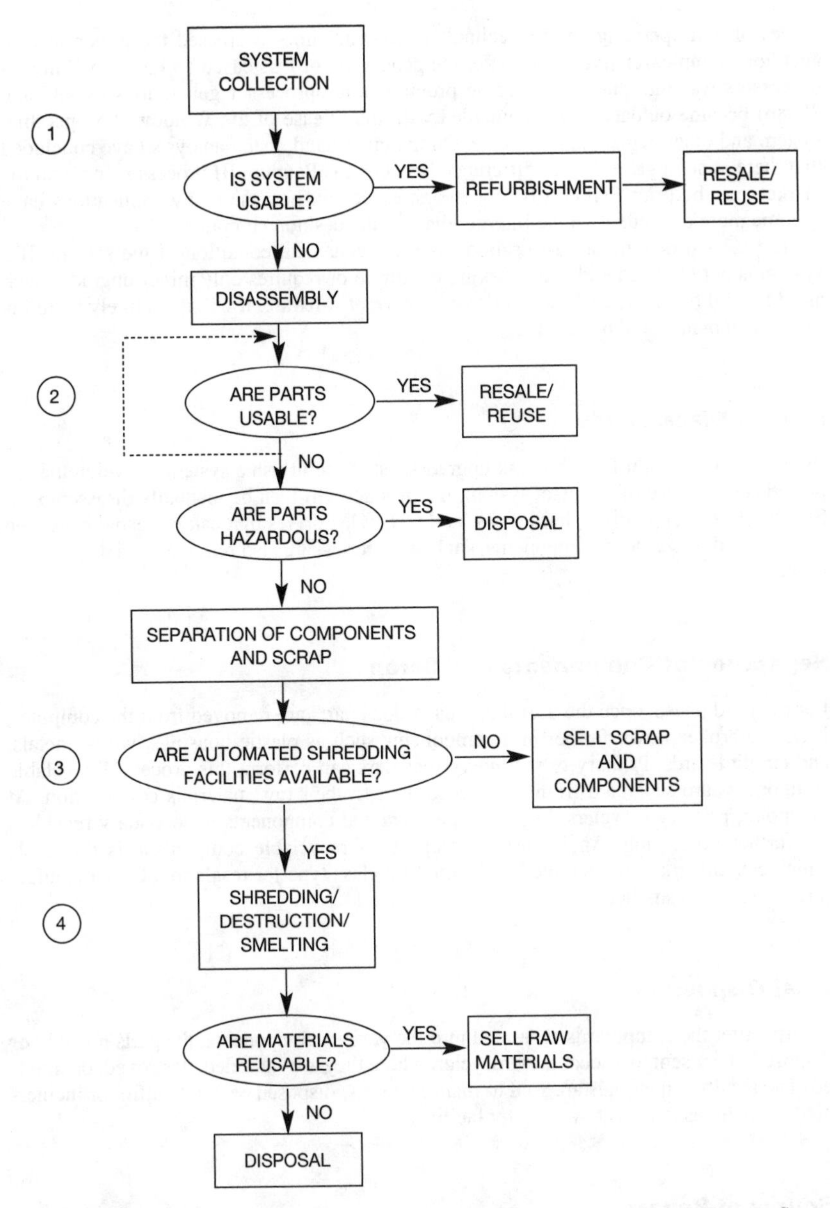

FIGURE 10.2 Generic flow diagram for the electronics recycling process. (*Source:* Stanford Resources, Inc., 1999 and *EPR2 Baseline Report,* approved by the National Safety Council.)

processor and operating system technology have at times decreased the importance of age. For example, relatively soon after the generation of PCs based on Pentium™ microprocessors was introduced, PCs using previous microprocessor generations (80486 and 80386) became outdated. This coincided with the release of the Windows 95 operating system and other demanding software. On the other hand, many analysts have concluded that there is not a great deal of difference between the Pentium III processors now on the market and the prior Pentium II processors, especially because many applications have become more dependent on the Internet than on the desktop computer.

The second determinant of usability is the mechanical condition of the system. If a system is not outdated and is in working condition or requires only minor upgrades such as additional memory, a CD-ROM drive, or current software, it will most likely be refurbished and then resold or donated.

Manual Disassembly

In the second stage, if the labor and upgrade costs to refurbish a system outweigh the expected selling price of an intact system, the product will then be manually disassembled. Such parts as floppy disks, hard disks, and CD-ROM drives that can be resold or reused are removed. Hazardous components, such as batteries, are also removed and disposed of properly.

Separation of Components and Scrap

For the third phase, once the useful and hazardous parts are removed from the computer, it can be broken down further into components such as plastic housings, wires, metals, and circuit boards. Primary or secondary recyclers can perform this process. Recyclable components are separated from scrap according to their raw materials composition. At this point, primary recyclers distribute the separated components to secondary recyclers for further processing. An important category of recyclable components is the CRT, which can either be disassembled and sorted by glass type for resale to a CRT manufacturer or sent to a smelter.

Final Disposition

Finally, after the computer is separated into the various components, the parts not sold or disposed of are sent to a secondary recycler, where they are shredded, destroyed, or smelted. The resulting materials are sold to manufacturers, disposed of in a landfill, or incinerated (sometimes in a waste-to-energy facility).

Source of Report

It is recommended that all organizations interested or concerned with the recycling of electronic devices obtain a copy of the Electronic Product Recovery and Recycling Baseline Report from the National Safety Council's Environmental Health Center, 1025 Connecticut Avenue, NW, Suite 1200, Washington, DC 20036. They may be reached by telephone (202) 293-2270. Their internet website is http://www.nsc.org/ehc/epr2.htm.

THE FUTURE OF ELECTRONICS RECYCLING

Governments at all levels should actively encourage the environmentally sound recycling of end-of-life electronic devices. This encouragement should include funding of sound recycling efforts. A partnership at all levels of government could create a certification system that could relieve end-of-life electronic device owners of environmental liability by recycling these devices with certified recyclers. Governments would in turn need funding to then monitor the certified electronic recyclers. This would be a proactive approach creating a partnership between industry and government.

Marketing Incentives

Incentives to recycle could include support of electronic device recycling infrastructure growth by assistance in the form of grants, loans, and aid in the development of end markets for recycled materials. Pollution prevention and recycling commodity futures and options could be used as part of a market system to increase demand for these products and services. Continued support of solid waste exchanges such as the Southern Waste Exchange, Inc. in Tallahassee, Florida so they may continue working as clearinghouses for pollution prevention and waste diversion from the resource stream. Many of these organizations have yielded benefits of up to 30 times cost in the minimization of waste and are model organizations in resource conservation.

The Role of Solid Waste Authorities

Solid waste authorities should be funded for front-end solutions that would facilitate greater recycling rates of electronic devices. Substitute funding mechanisms should be implemented to allow the authorities to obtain offsets for recycling and not be forced to lose funding by diversion of electronic devices from the waste stream, thus forfeiting tipping fees without a replacement funding mechanism for recycling.

The Role of Industry

Industry should be given incentives to recycle electronic devices instead of merely being penalized for not recycling. Incentives could be in the form of liability risk avoidance and/or tax credits for recycling electronic devices. In return for industry support of such a program, mandated take-back initiatives of end-of-life electronic devices would not be instituted. International Association of Electronics Recyclers, Inc. (IAER) is the first and only trade association for the electronics recycling industry. The IAER was formed to represent and serve its interests as a key element in the development of an effective and efficient infrastructure for managing the life cycle of electronics products. One area, in particular, that it is important to the future of the industry is the support and promotion of high standards of environmental quality and regulatory compliance. Additionally, the IAER is promoting Standard Practices and Certification, which is a process to certify business practices and compliance, industry standards, guidelines, and best practices. Their website is http://www.iaer.org. IAER can be reached at P.O. Box 16222, Albany, NY 12212-6222. Toll-free phone: 888-989-IAER (4237); Fax: 877-989-IAER (4237); Email: info@IAER.org.

REFERENCES

1. *New York Times*, July 16, 1999, pp. Al and C17, Tiniest Circuits Hold Prospect of Explosive Computer Speeds.
2. *Electronic Product Recovery and Recycling Baseline Report*, National Safety Council, Environmental Health Center, May 1999, p. vii (hereafter referred to as *EPR2 Baseline Report*.
3. Ibid., p. viii.
4. EPA Small Business Ombudsman, January 1999, p. 26.
5. Ibid, p. 26.
6. 1999 EPR2 Conference, State Electronics Policy Summaries.
7. Ibid.
8. Ibid.
9. *Environmental Law In a Nutshell*, Third Edition, West Publishing Company, 1992, p. 246.
10. *EPR2 Baseline Report*, p. 29, Table 6.
11. *EPR2 Baseline Report*, p. 39, Figure 18: Forecast Shipments, Obsolescence, and Recycling.
12. *EPR2 Baseline Report*, p. 35, Table 16: Quality Education Data, Inc., data from www.qualityeducationdata.com.
13. CEMA Research Center, as sited in *USA Today*, June 22, 1999.
14. *EPR2 Baseline Report,* p. 39, Figure 18: Forecast of U.S. CRT Monitor Shipments, Obsolescence, and Recycling.
15. Direct from Dell, by Michael Dell, *HarperBusiness*, 1999.

CHAPTER 11
PAPER

TOM FRIBERG, PhD*
Project Manager
Weyerhaeuser Company
Tacoma, Washington

LISA MAX†
President
Better World, Inc.
Miami, Florida

LYNN M. THOMPSON‡
Manager, Environmental Affairs and Communications
Bell South Advertising and Publishing
Jacksonville, Florida

BACKGROUND

History of Paper Recycling

Papermaking. Paper was invented in China around 2000 years ago and was made from vegetable fibers. All paper was made by hand until 1799, when the first papermaking machine was invented. Around 1850, wood fiber started to replace the agricultural fibers. [1] Today in the U.S., although some paper is made from wood specifically grown for pulp, a large amount of the paper and board is made from the residuals of lumber production—the trimmings, shavings and even sawdust.

Paper Recycling. The first recorded use of recovered paper to make new paper was in 1031 AD in Japan. Bleaching had not been invented, so the recycled paper was grayish. It was not until the major effort to recycle during World War I that de-inking occurred in commercial quantities. [1] The use of recovered paper again increased during World War II and the use of recycled material in printing and writing paper peaked in 1950 at 20%. Today, much larger quantities of recovered paper go toward corrugated containers and paperboard packaging.

*Processing of recycled paper.
†Office papers.
‡Telephone directories.

Paper Recycling—Present Status. In 1996, Americans recovered 42.3 million tons of paper. That is 45% of all of the paper used in the U.S. Approximately 75% of the 500 paper and board mills in the U.S. use recovered paper. Of these 200 depend entirely on recovered paper. How much paper is that?—a train of box cars 20 miles long every day (Figures 11.1 and 11.2) [3].

Worldwide Paper Recycling. Recovered paper use in North America and in Europe is relatively stable with only moderate growth. The real interest is in Asia. Some of the needs of the Asian market will be satisfied by internal collection, but a large amount will come from the U.S. and Europe (Tables 11.1 and 11.2).

RAW MATERIALS SUPPLY (PAPER RECOVERY)

Sources

Preconsumer (postindustrial) or postconsumer paper comes from:

- Commercial/industrial
- Offices

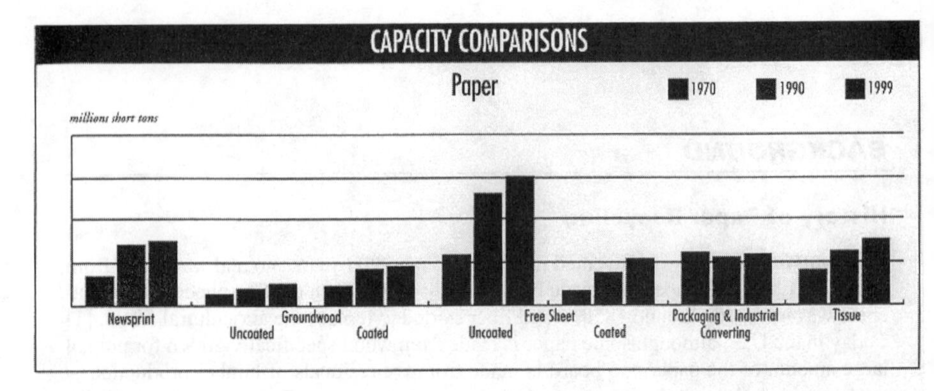

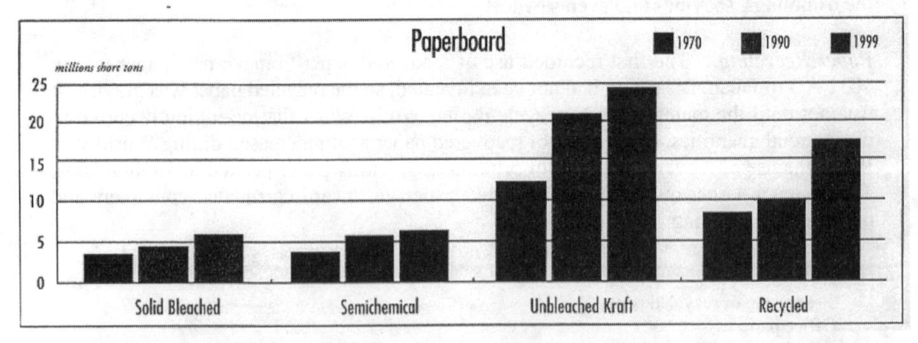

FIGURE 11.1 Total U.S. paper and paperboard capacity [2].

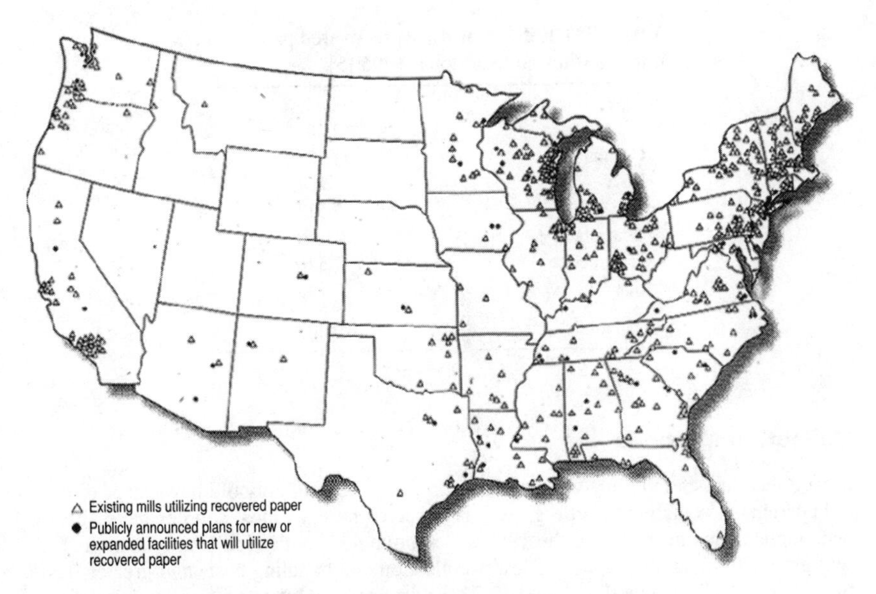

FIGURE 11.2 U.S. map of mills using recovered paper [4].

- Institutions
- Small generators
  - Small offices
  - Multitenant buildings
  - Residential

The traditional sources of recovered paper are the easiest ones to access. In recent years, it has been necessary to "go deeper into the waste stream." This has meant accessing the small generators. This is a large source, as one half of the printing and writing paper is used in the home, yet it is difficult to effectively collect.

TABLE 11.1 Worldwide paper recovery (thousands of tons) 1995 [5]

Europe	32,237
North America	41,999
Canada	2,694
USA	39,305
Asia	34,276
Australasia	127
Latin America	4,711
Africa	933
World total:	114,283

TABLE 11.2 Worldwide recovered paper imports (thousands of tons) 1995 [5]

Europe	7,629
North America	2,387
Canada	1,923
USA	464
Asia	7,513
Australasia	0
Latin America	1,546
Africa	133
World total:	19,207

Collection Methods

Most collection started with "source separation"—collecting only old newspaper or only old corrugated containers. Little growth is expected here, as most sources have already been tapped. However, during the 1990's, "commingled" collection" really started. This will grow significantly because of lower collection and handling costs and greater flexibility in responding to market demands. These drivers have led to greater numbers of material recovery facilities (MRFs).

Sort Plants

Today's sort plants handle not only paper but also glass, aluminum, ferrous metals, and plastic (Figure 11.3). While they are very labor intensive, a variety of mechanical equipment is used. The one critical device used is the baler, since the common unit for all papers is the 1000–2000 pound bale. Other equipment include trommel and screens, sorting belts, ferrous magnets, eddy current magnets, and pneumatic transfer pipes.

Upon arrival at the sort plant, the incoming paper is weighed and moved around the plant by carts, Bob Cats, or front-end loaders. If source separated, the loose paper may be directly baled. When paper sorting is required, it may be cleaned by pulling out contaminants as it is pushed around on the plant floor or it may travel by two to 20 people standing next to a sort belt. These people positively sort for high-grade paper or pull out contaminants (Figure 11.5). Automated sorting is just starting to be used. Once sorted, the paper is baled and shipped as soon as possible because plant inventory space is usually very small.

Specifications and Quality

The Institute of Scrap Recycling Industries, Inc. (1325 G Street, N.W., Washington DC 20005) sets the standard specifications for a number of residual materials including paper stock, both domestic and export. There are 51 standard specifications and 33 specialty grades. The grade is based on a general description keyed to the origin of the paper and the level of contamination. The most common grade descriptions can be grouped as pulp substitutes, computer printout, white ledger, colored ledger, news, old corrugated and mixed. The level of contamination is defined by both "outthrows," those materials that

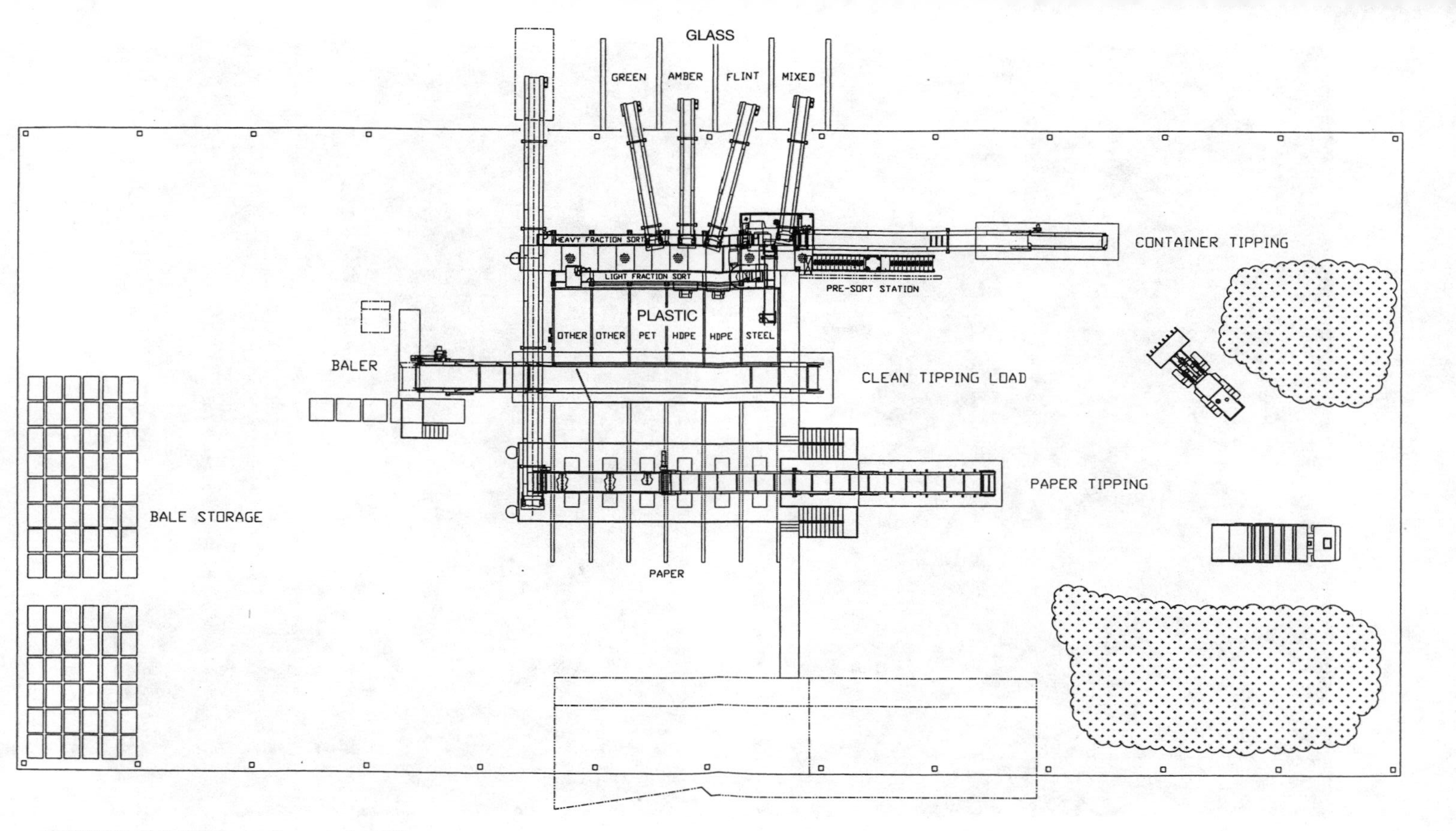

FIGURE 11.3 Diagram of a sort plant [6].

FIGURE 11.4 Sort plant [6].

FIGURE 11.5 Manual sort belt [6].

are paper but not the paper described in the specification, and "prohibitives," those materials that are not paper. These two criteria may range from zero to 5% depending upon the grade. (See Appendix for Paper Grade Definitions.) In addition to the published grades, there are custom grades that meet the specific needs of one or a similar group of mills.

Because of the size and density of the paper bales, quality is difficult to measure and track. Four and six sided inspections are common but they only address bale integrity and surface contamination. As time goes on, more emphasis will be put on "fitness-for-use," which is based on fiber type—fiber mechanical pulp (groundwood), unbleached chemical pulp, or bleached chemical pulp.

A major problem for mills are "stickies." These are polymeric materials that originate with the wood (wood resins) and synthetics that are added in paper converting and during use (adhesives, sizings, coatings, etc.). When present in small amounts, these materials agglomerate during repulping and papermaking, creating significant runability problems. As little as a tenth of an ounce in a ton of paper can cause trouble.

Recovered Volumes (Major Grades)

See Table 11.3 for a summary of recovered volumes by grade.

Costs (to Collect)

Recycling, hauling and handling is a very competitive industry, so getting cost numbers is problematic. It seems that every time the paper is touched it is going to cost between $5 and $10/ton. The various activities include: collection (more $$), receiving, sorting (more $$), baling, inventory, and shipping. As for shipping, local adds $5/ton, regional adds $15/ton, national adds $30/ton, and international adds $50/ton.

Prices (to Mill, Including Ranges and Trends)

The supply/demand curve for recovered paper is extremely inelastic (Figure 11.6). Even though millions of tons of paper move annually, small changes in available volumes

TABLE 11.3 Total recovered paper in U.S.—1999 estimate (thousands of tons) [8]*

Material	Total U.S.	Percent of total fiber	New England	Middle Atlantic	East North Central	West North Central	South Atlantic	East South Central	West South Central	Mountain & Pacific
Total recovered paper†	37,691	36.1	2,313	4,076	8,282	1,671	7,225	3,755	4,071	6,298
Mixed papers	5,107	4.9	357	479	1,492	268	1,055	534	404	518
Newspapers	5,436	5.2	381	679	978	114	1,138	287	508	1,351
Corrugated	21,146	20.2	949	2,289	3,646	1,152	4,158	2,351	2,611	3,989
Pulp substitutes	2,470	2.4	471	343	778	80	289	169	164	175
High-grade de-inking	3,532	3.4	156	285	1,387	57	584	413	383	266

*Totals are not adjusted for differences in rounding.
†Includes construction grades and molded pulp products.
Source: American Forest and Paper Association

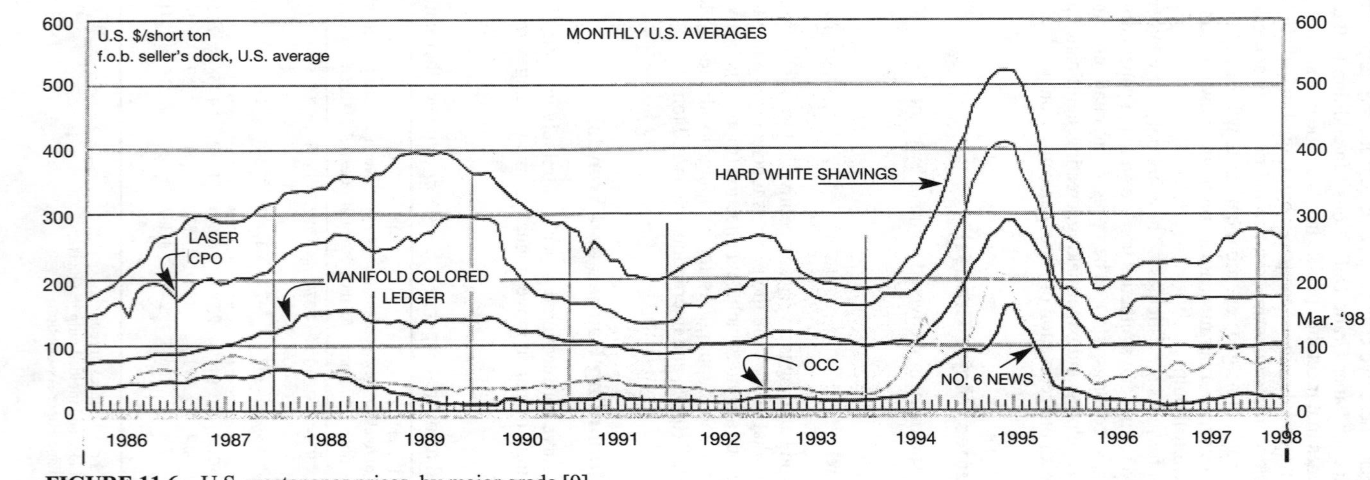

FIGURE 11.6 U.S. wastepaper prices, by major grade [9].

cause significant price fluctuations. This is because if a highly capital-intensive mill needs paper to continue to run, it will pay what it needs to get more paper. Conversely, if a recycler with collection contracts and high inventory has paper building up inventory, it will take whatever it can get to move paper.

Legislation and Regulations

Most states have recycling and waste reduction goals that have been called "rates and dates." These generally have a percentage reduction of the waste stream or recovery of specific grades of paper to be accomplished by set dates. On rare occasions, jurisdictions have banned specific papers from going to landfill, like OCC. For specifics, refer to the chapter on legislation and regulation.

PRODUCTION CAPACITY (MILL USE)

Technology

Fiber Quality (Fitness for Use). The traditional grading of recovered paper has been based on the origin of the fiber, such as old newspaper or old corrugated. With today's processing technology and wide variety of end-uses for paper and board, this is no longer adequate. The specific end-product performance requirements dictate the "fitness-of-use" of recovered fiber. Important considerations are the type of pulping that the original fiber had: 1) mechanical (groundwood), 2) unbleached chemical, and 3) bleached chemical. Figures 11.7 and 11.8 show the change in properties on multiple recyclings for bleached Kraft (chemical) and mechanical fibers [10].

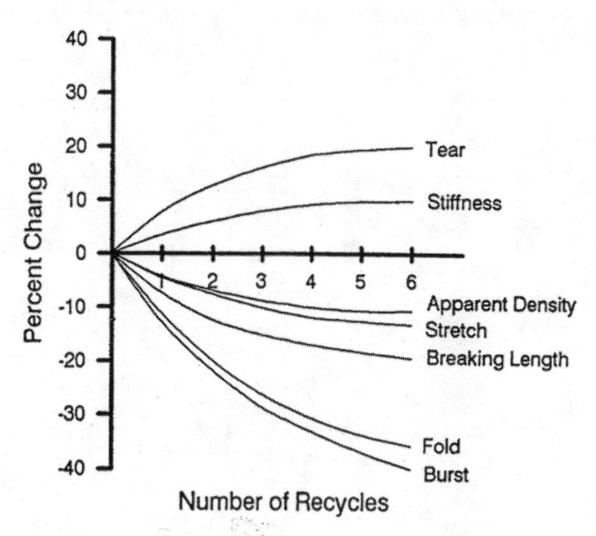

FIGURE 11.7 Property changes with multiple recyclings of unbleached and bleached pulp.

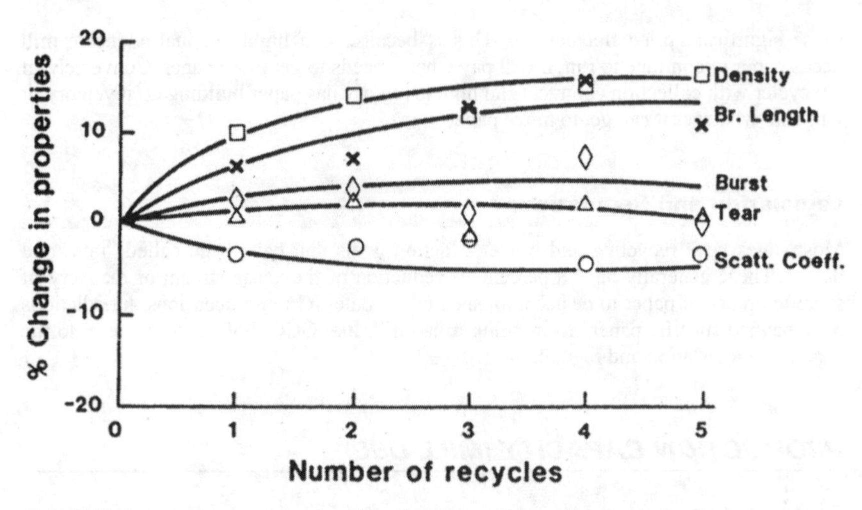

FIGURE 11.8 Property changes with multiple recyclings of mechanical pulp.

Fundamentals of Decontamination

Equipment (Pulping, Screening, Cleaning, De-inking–Washing, and Flotation). Whether used for making liner for tissue, corrugated containers, or printing paper, the recovered paper goes through some or all of the steps shown in Figure 11.9 [11] The type of equipment and the order in which it is used is dictated by the type of recovered paper use, what contaminants are present, and the type of end product.

Once the large contaminants are removed, each piece of equipment is supposed to remove certain sizes of the remaining small contaminants. The traditional idea for the

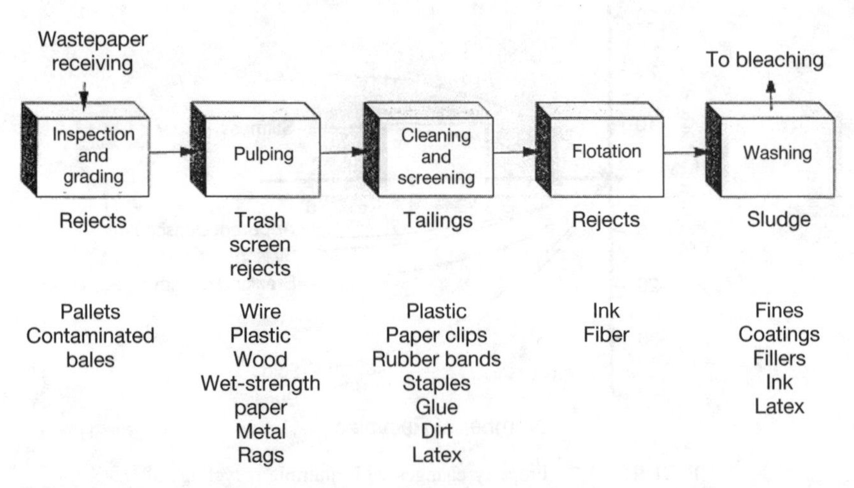

FIGURE 11.9 General steps in paper repulping.

regimes of size separation are shown in Figure 11.10 [12]. Recent research (Figure 11.11) indicates that the removal rates are not discretely distributed [13]. All of this come together in Figure 11.12, which shows the major process steps for each of the major fiber categories [12].

After de-inking, the fibers are usually bleached. Oxidative chemicals like sodium hypochlorite, hydrogen peroxide, oxygen or ozone are used to solubilize or whiten the remaining lignin or other contaminants. Inks must be removed by the de-inking stages because they generally do not bleach out. Reductive chemicals like sodium hydrosulfite and formamidine sulfinic acid (FAS) are used to modify paper dyes; this is also called color-stripping [17]. Enzymes are increasingly used to help bleach, de-ink, and/or refine recycled paper. This is because they can be the basis of a system that is not highly capital-intensive.

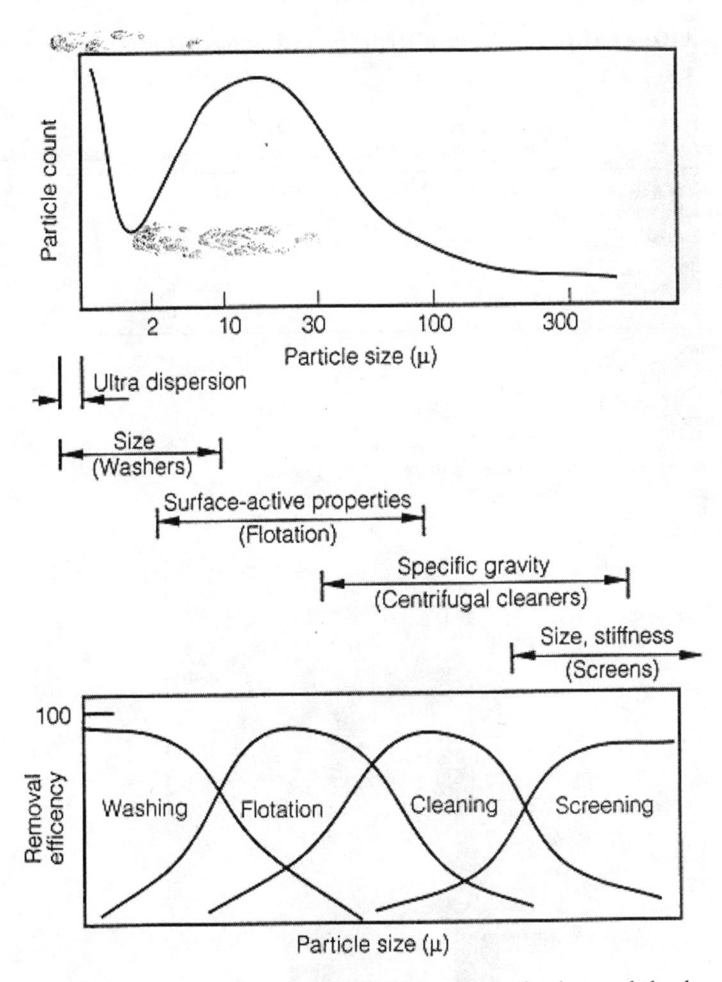

FIGURE 11.10 Top—Fiber and particle size distribution in recycled pulp. Bottom—Theoretical removal efficiencies for various separation equipment.

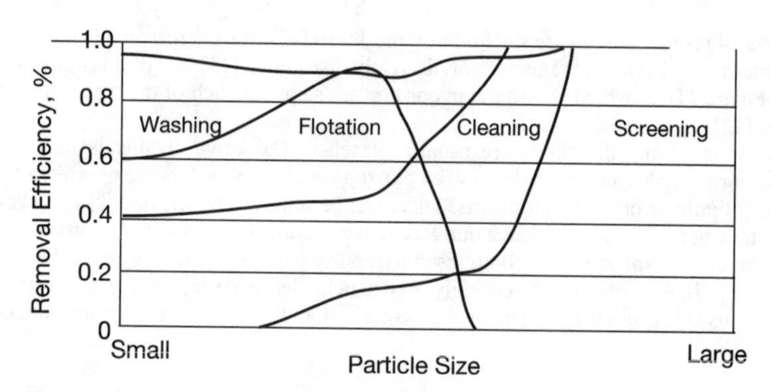

FIGURE 11.11 Actual removal efficiencies for various separation equipment.

Process	Pulping	Screening	Cleaning	Asphalt dispersion	Washing	De-inking flotation	Dispersion
Wastepaper category							
Pulp substitutes	1						
Mixed paper	1	1	1				
Corrugated	1	1	1	2			
Newspaper	1	1	1		1	3	4
De-inking	1	1	1		1	3	4

1 = Always.
2 = Many in use.
3 = Most new facilities.
4 = Future.

FIGURE 11.12 Major processing steps for fiber grades.

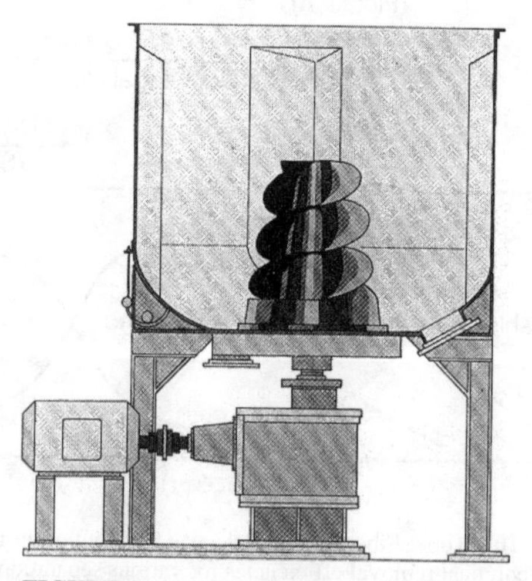

FIGURE 11.13 High-consistency pulper [14].

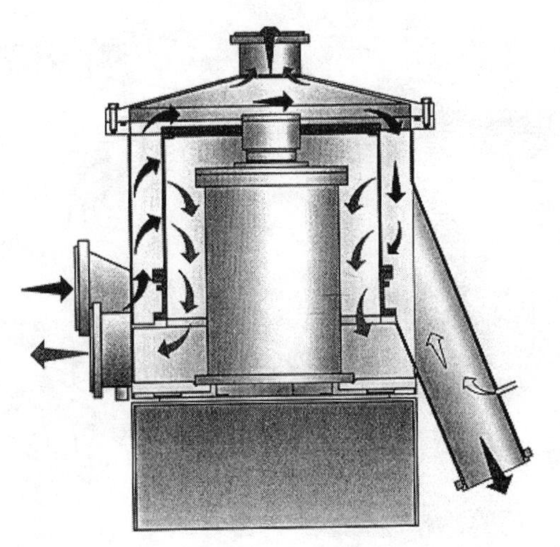

FIGURE 11.14 Pressure screen [15].

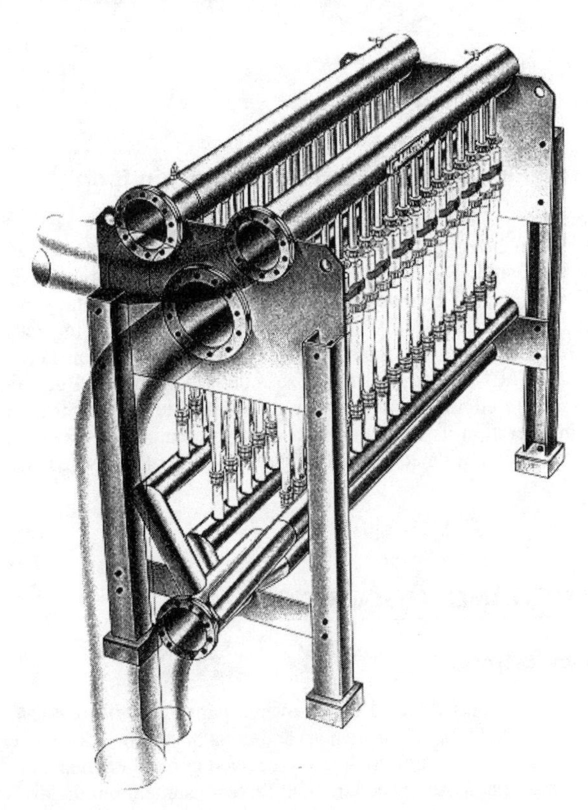

FIGURE 11.15 Cleaners [15].

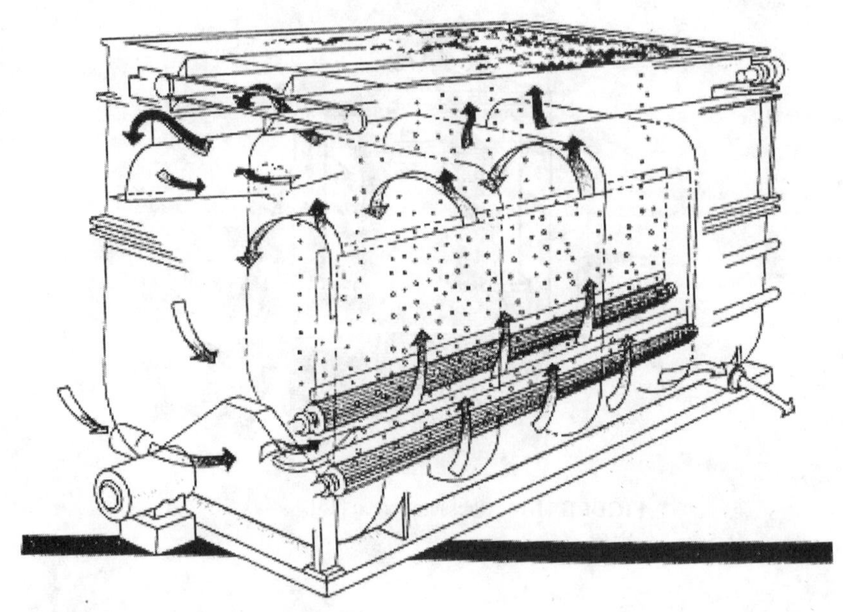

FIGURE 11.16 Flotation unit [16].

Environmental Impacts (Recycle versus Virgin Pulp)

Besides the numerous life cycle interconnections shown in Figure 11.17, there are complex relationships in energy use and air, water, and solid emissions when virgin and recycled pulp are compared (Figures 11.18–11.20). The most comprehensive evaluation to date is the Environmental Defense Fund's study on environmentally preferable paper [18]. This and other studies identify the tradeoffs in environmental impacts. For example, many virgin pulp and paper operations burn their solid waste and generate their own power. So, a "stand-alone" recycle mill may use more fossil fuel and generate more solid waste than an integrated virgin facility. But, to complicate matters, if the recycle mill is part of a virgin operation its environmental impacts change considerably. Examples of this are tissue mills or liner/medium mills that combine both recycled and virgin operations on the same site.

MARKET DEMAND (UTILIZATION)

Market Expectations

Until the 1980s, the supply/demand for recovered paper was discrete (old news to new news and old corrugated to new corrugated) and in balance. The waste diversion goals of the 1980s upset the balances and the lure of recycled content created further disruption. As the artificial driversion diminished, market forces resumed, but the dislocations led to

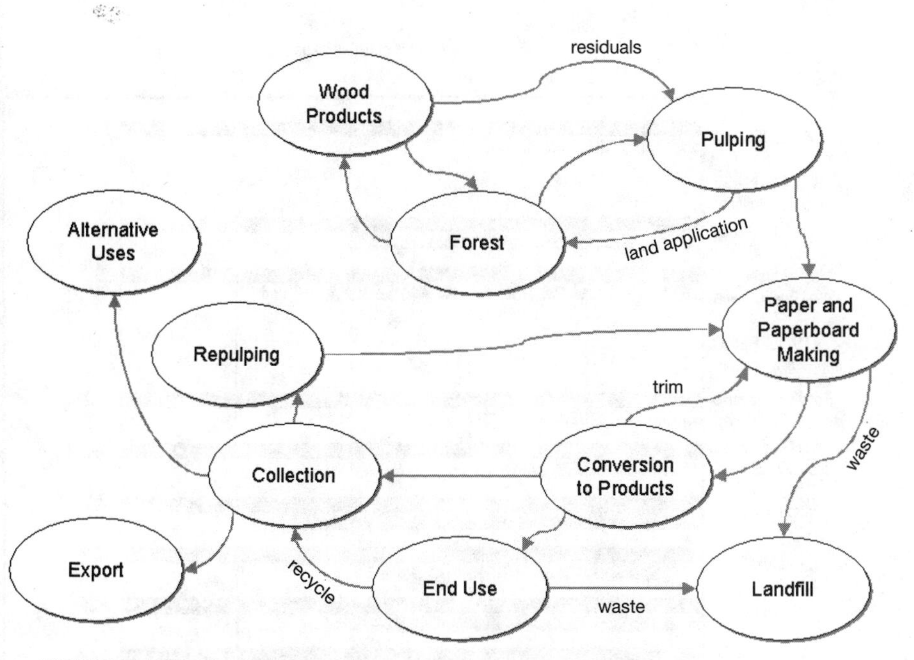

FIGURE 11.17 Cycle change and interconnections.

the price swings of the 1990s. These swings were very dramatic because of exceptional inelastiticy in this market, as described under the section on prices.

Legislation and Regulations/Guidelines

Recycled content requirements vary for materials and markets. California has newsprint content requirements of 40% and there are Federal purchasing guidelines for printing and writing paper. For more specific information, refer to Chapter 2.

Utilization (What Grades Go to What End Uses)

As mentioned before, the utilization loops used to be well defined. Old news went to new news, etc. That is no longer the case as technology has increased the ability to cross-utilize the paper streams (Table 11.4). For example, it is possible to lightly cook OCC and then bleach the fibers to make white printing, writing, and tissue papers. Magazines can be blended with ONP to make newsprint. So overall numbers are much harder to track. Figure 11.21 shows some of the flows [19].

Export

The United States is a supplier to the world of high-quality recovered fiber; see Table 11.5 [21].

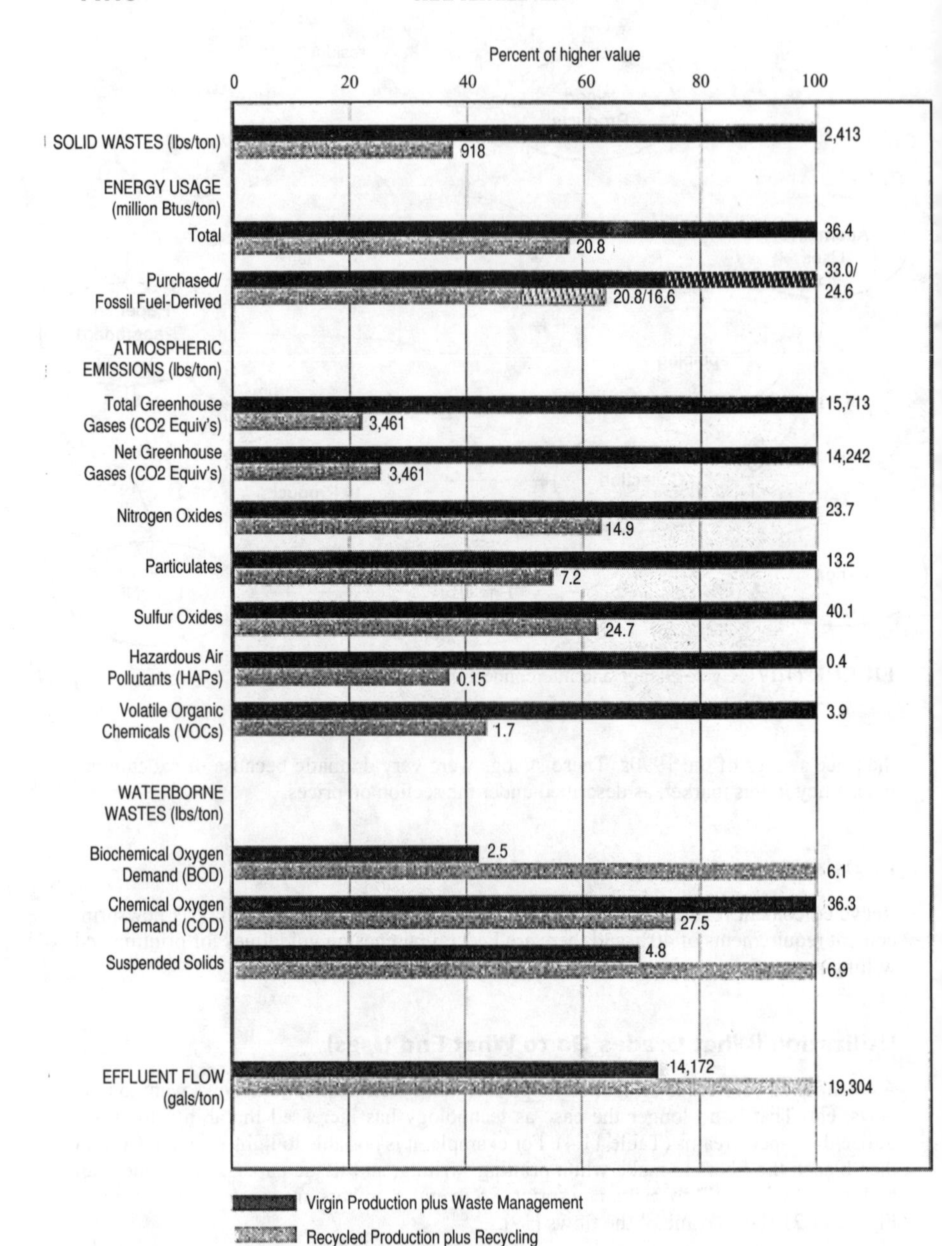

FIGURE 11.18 Recycled versus virgin newsprint [18].

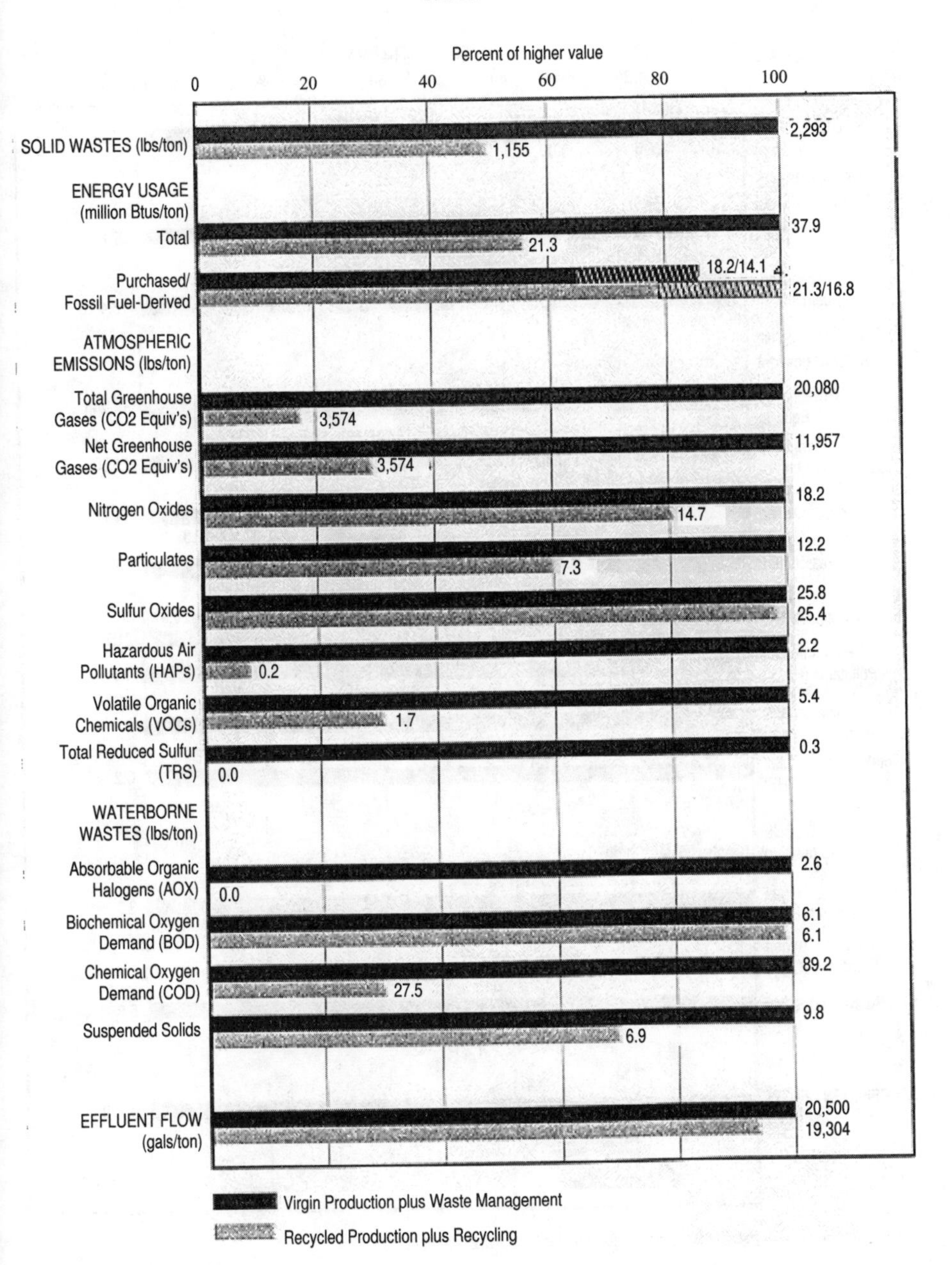

FIGURE 11.19 Recycled versus virgin office paper [18].

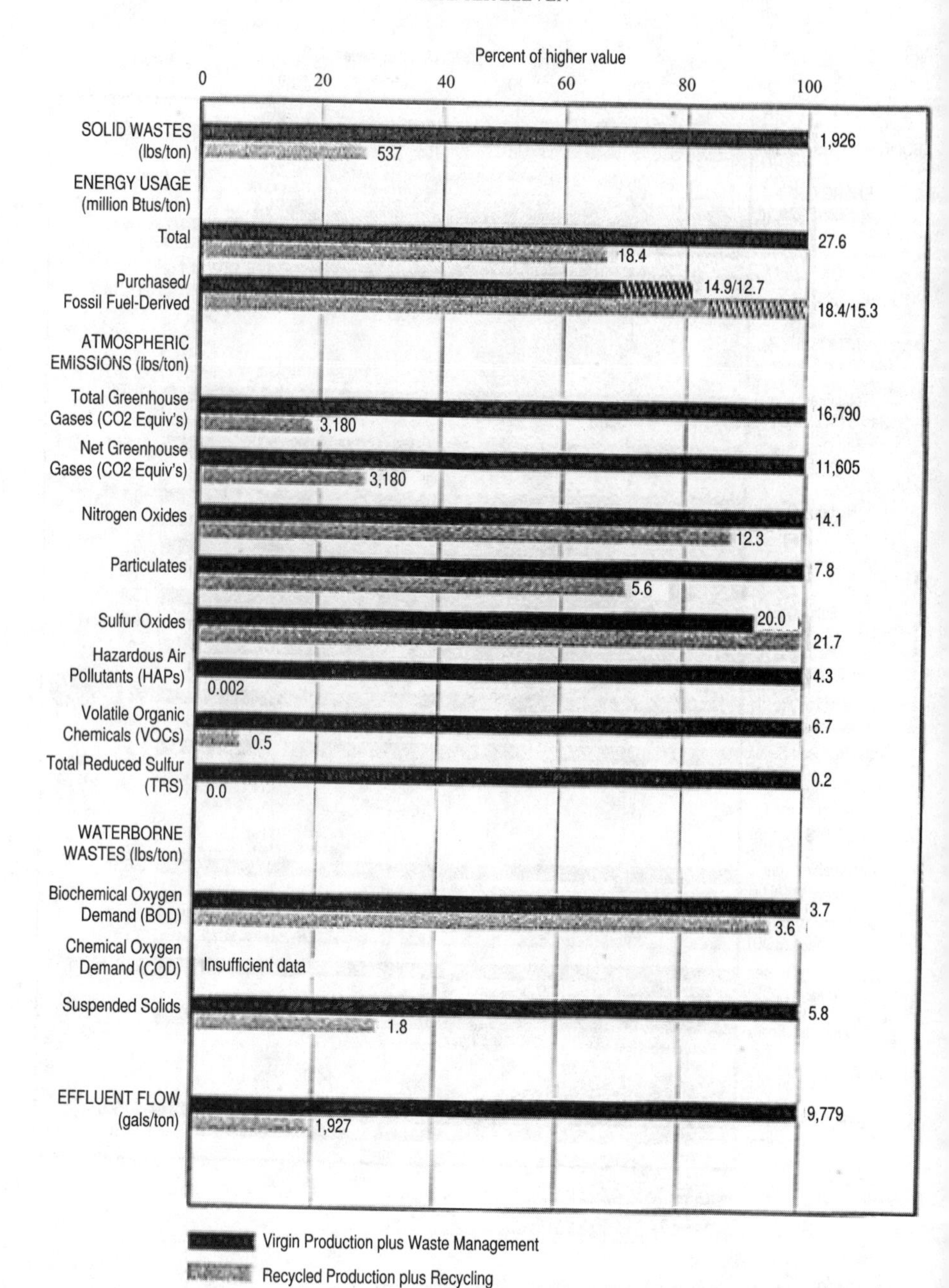

FIGURE 11.20 Recycled versus virgin corrugated boxes [18].

TABLE 11.4 Utilization of recovered paper—1999 estimate (thousands of tons) [20]

Material	Total	Mixed papers	Newspapers	Corrugated	Pulp substitutes	High-grade de-inking
Total all grades	37,691	5,107	5,436	21,146	2,470	3,532
Total paper	11,083	1,753	3,408	1,124	1,718	3,080
Newsprint	3,144	180	2,711	—	13	240
Printing, writing, and related	2,949	749	178	4	905	1,113
Packaging and independent converting	1,229	62	15	887	245	20
Tissue	3,761	762	504	233	555	1,707
Total paperboard	25,513	2,962	1,733	19,718	657	443
Kraft linerboard	4,638	248	19	4,225	10	136
Other Kraft, bleached, and unbleached	435	131	—	248	48	8
Semichemical	2,223	170	8	2,043	2	—
Recycled containerboard	10,278	603	243	9,287	76	69
Other recycled	7,939	1,820	1,463	3,915	521	230
Construction paper and board	1,095	392	295	304	95	9

Source: American Forest and Paper Association

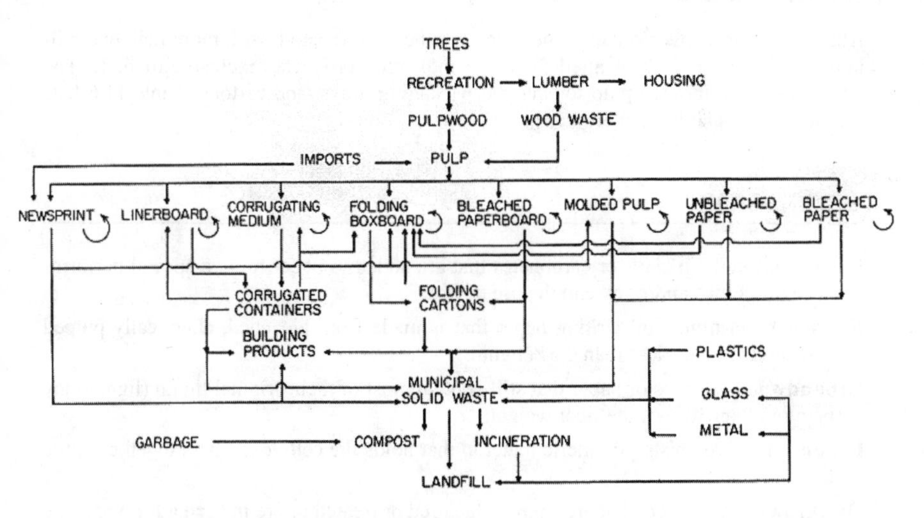

FIGURE 11.21 Paper life cycles by grade.

TABLE 11.5 Export of high-quality recovered fiber, 1996 (thousands of tons)

	Total Recovered	Exported
Old corrugated	21,585	2,618
Printing/writing	9,810	1,560
Newspapers	7,386	1,048
Other	3,492	1,933
Total	42,273	7,159

TABLE 11.6 Alternative uses for recovered paper, 1990 and 2000 (tons per year)

	1990	2000 (estimate)
Molded pulp	500,000	900,000
Insulation	300,000	600,000
Fillers and fibers	260,000	260,000
Animal bedding	125,000	300,000
Internal packaging	100,000	100,000
Hydromulch	100,000	200,000
Wallboard	90,000	90,000
Medium density board	80,000	80,000

Alternative Uses

Although most recovered paper goes back into new paper and board, more than one million tons per year in the United States go into other products (exclusive of fuel). The largest use is in molded pulp for internal packaging, trays, and cartons. Table 11.6 lists the various uses and volumes. (22)

GLOSSARY

Fillers Ground up clays or carbonates that are added to paper to give special performance properties and/or extend the use of fiber.

Freesheet Printing and writing paper that is made from bleached, chemically pulped fiber and contains less than 0.5% lignin.

Groundwood Fibers or paper that still contain most of their original lignin (lignin may be more than 30% of the fiber weight.

Lignin The brownish, polymeric material that holds the cellulose fibers together in the tree.

Outthrows All papers that are so manufactured or treated or are in such a form as to be unsuitable for consumption as the grade specified.

Prohibitives Any materials, which by their presence in a packing of paper stock, in excess of the amount allowed, will make the packaging unusable as the grade specified. Any materials that may be damaging to equipment.

Rejects All materials that do not slurry into individual fibers during reprocessing.

Repulpable Paper and board that slurries into individual fibers with the use of standard equipment

Wet Strength Resins Polymeric additives that give paper and board good performance in high humidity or in water. They tend to make the paper difficult to repulp.

REFERENCES

1. Thompson, C. *Recycled Papers—The Essential Guide*. MIT Press, Cambridge, MA, 1992, pp. 21–33
2. *Capacity and Fiber Consumption, 37th Annual Survey*. American Forest and Paper Association, Washington, DC, December 1996, p. 9.
3. *Recovered Paper Statistical Highlights—1997 Edition*. American Forest and Paper Association, Washington, DC, 1997, p. 2.
4. *Paper Matcher—A Directory of Paper Recycling Mills*, 4th ed. American Forest and Paper Association, Washington DC, February 1996, p. 53.
5. Matussek, H. and Stefan, V. *International Fact and Price Book 1997*. Pulp and Paper International, 1997, pp. 274–281.
6. CP Manufacturing, National City CA, 1998.
7. Institute of Scrap Recycling Industries. *ISRI Scrap Specifications Circular 1997*. Washington, DC, 1997, pp. 27–38.
8. Reference 2, p. 23.
9. *Paper Recycler*. Miller Freeman, San Francisco, CA, January 1998, p. 3.
10. Spangenberg, R. J., *Secondary Fiber Recycling*. TAPPI Press, Atlanta GA, 1993, Ch. 2, p. 9.
11. Patrick, K. *Paper Recycling—Strategies, Economics, and Technologies*. Miller Freeman, San Francisco, CA, 1991, p. 122.
12. Reference 2, p. 68.
13. Moss, C. "The Contaminant Removal Curve—Theory vs. Reality." 1997 TAPPI Recycling Symposium, April 1997, pp. 1–6.
14. Thermo Black-Clawson
15. Ahlstrom
16. Doshi, M. and Dyer, J. *Paper Recycling Challenge, Volume 2—De-inking and Bleaching*. Doshi & Associates Inc., Appleton WA, 1997, p. 76.
17. Reference 15, pp. 197–204.
18. The Paper Task Force. *Paper Task Force Recommendations for Purchasing and Using Environmentally Preferable Paper*. Environmental Defense Fund, New York, NY, 1995, pp. 81–96.
19. Hamilton, F., Leopold, B., and Kocurek, M. *Pulp and Paper Manufacture, Volume 3—Secondary Fiber and Non-Wood Pulping*. The Joint Textbook Committee of the Paper Industry—TAPPI / CPPA, Atlanta, GA, 1987, p. 145.
20. Reference 2, p. 24.
21. Reference 3, pp. 6–24.
22. Friberg, T. "Alternative Uses for Recovered Paper." *Resource Recycling*, January 1993, pp. 26–33.

APPENDIX PAPER GRADE DEFINITIONS*

The definitions which follow describe grades as they should be sorted and packed. *Consideration should be given to the fact that paper stock as such is a secondary material produced manually and may not be technically perfect.*

*Courtesy of Paper Stock Institute, Institute of Scrap Recycling Industries, Inc., "Guidelines for Paper Stock."

Outthrows

The term "Outthrows" as used throughout this section is defined as "all papers that are so manufactured or treated or are in such a form as to be unsuitable for consumption as the grade specified."

The term "Prohibitive Materials" as used throughout this section is defined as:

a. Any materials which by their presence in a packing of paper stock, in excess of the amount allowed, will make the packaging unusable as the grade specified.
b. Any materials that may be damaging to equipment.

Note: The maximum quantity of "Outthrows" indicated in connection with the following grade definitions is understood to be the *total* of "Outthrows" and "Prohibitive Materials." A material can be classified as an "Outthrow" in one grade and as a "Prohibitive Material" in another grade. Carbon paper, for instance, is "Unsuitable" in Mixed Paper and is, therefore, classified as an "Outthrow"; whereas it is "Unusable" in White Ledger and in this case classified as a "Prohibitive Material."

(1) Mixed Paper. Consists of a mixture of various qualities of paper not limited as to type of packing or fiber content.

Prohibitive materials may not exceed	2%
Total Outthrows may not exceed	10%

(2) (Grade not currently in use)

(3) Super Mixed Paper. Consists of a baled clean, sorted mixture of various qualities of papers containing less than 10% of groundwood stock, coated or uncoated.

Prohibitive materials may not exceed	½ of 1%
Total Outthrows may not exceed	3%

(4) Boxboard Cuttings. Consists of baled new cuttings of paperboard such as are used in the manufacture of folding paper cartons, set-up boxes and similar boxboard products.

Prohibitive materials may not exceed	½ of 1%
Total Outthrows may not exceed	1%

(5) Mill Wrappers. Consists of baled wrappers used as outside wrappers for rolls, bundles or skids of finished paper.

Prohibitive materials may not exceed	½ of 1%
Total Outthrows may not exceed	3%

(6) News. Consists of baled newspapers containing less than 5% of other papers.

Prohibitive materials may not exceed	½ of 1%
Total Outthrows may not exceed	2%

(7) Special News. Consists of baled sorted, fresh dry newspapers, not sunburned, free from paper other than news, containing not more than the normal percentage of rotogravure and colored sections.

Prohibitive materials	None permitted
Total Outthrows may not exceed	2%

(8) Special News De-ink Quality. Consists of baled sorted, fresh dry newspapers, not sunburned, free from magazines, white blank, pressroom overissues, and paper other than news, containing not more than the normal percentage of rotogravure and colored sections. This packing must be free from tare.

Prohibitive materials	None permitted
Total Outthrows may not exceed	¼ of 1%

(9) Over-Issue News. Consists of unused, overrun regular newspapers printed on newsprint, baled or securely tied in bundles, containing not more than the normal percentage of rotogravure and colored sections.

Prohibitive, materials	None permitted
Total Outthrows	None permitted

(10) (Grade not currently in use—See Specialty Grade 29-S)

(11) Corrugated Containers. Consists of baled corrugated containers having liners of either test liner, jute or kraft.

Prohibitive materials may not exceed	1%
Total Outthrows may not exceed	5%

(12) (Grade not currently in use)

(13) New Double-Lined Kraft Corrugated Cuttings. Consists of baled corrugated cuttings having liners of either kraft, jute or test liner. Non-soluble adhesives, butt rolls, slabbed or hogged medium, and treated medium or liners are not acceptable in this grade.

Prohibitive materials	None permitted
Total Outthrows may not exceed	2%

(14) (Grade not currently in use)

(15) Used Brown Kraft. Consists of baled brown kraft bags free of objectionable liners or contents.

Prohibitive materials	None permitted
Total Outthrows may not exceed	½ of 1%

(16) Mixed Kraft Cuttings. Consists of baled new brown kraft cuttings, sheets and bag waste free of sewed and stitched paper.

Prohibitive materials	None permitted
Total Outthrows may not exceed	2%

(17) Carrier Stock. Consists of new unbleached kraft cuttings and sheets, wet strength treated, with printed or unprinted clay coating.

Prohibitive materials	None permitted
Total Outthrows may not exceed	2%

(18) New Colored Kraft. Consists of baled new colored kraft cuttings, sheets and bag waste, free of sewed or stitched papers.

Prohibitive materials	None permitted
Total Outthrows may not exceed	1%

(19) Grocery Bag Waste. Consists of baled, new brown kraft bag cuttings, sheets and misprinted bags.

Prohibitive materials	None permitted
Total Outthrows may not exceed	1%

(20) Kraft Multi-Wall Bag Waste. Consists of new brown kraft multi-wall bag waste and sheets, including misprint bags. Stitched or sewed papers are not acceptable in this grade.

Prohibitive materials	None permitted
Total Outthrows may not exceed	1%

(21) New Brown Kraft Envelope Cuttings. Consists of baled new unprinted brown kraft envelope cuttings or sheets.

Prohibitive materials	None permitted
Total Outthrows may not exceed	1%

(22) Mixed Groundwood Shavings. Consists of baled trim of magazines, catalogs and similar printed matter, not limited with respect to groundwood or coated stock, and may contain the bleed of cover and insert stock as well as beater-dyed papers and solid color printing.

Prohibitive materials	None permitted
Total Outthrows may not exceed	2%

(23) (Grade not currently in use)

(24) White Blank News. Consists of baled unprinted cuttings and sheets of white newsprint paper or other papers of white groundwood quality, free of coated stock.

Prohibitive materials	None permitted
Total Outthrows may not exceed	1%

(25) Groundwood Computer Printout. Consists of papers which are used in forms manufactured for use in data processing machines. This grade may contain a reasonable amount of treated papers.

Prohibitive materials	None permitted
Total Outthrows may not exceed	2%

(26) Publication Blanks. Consists of baled unprinted cuttings or sheets of white coated or filled white groundwood content paper.

Prohibitive materials	None permitted
Total Outthrows may not exceed	1%

(27) Flyleaf Shavings. Consists of baled trim of magazines, catalogs and similar printed matter. It may contain the bleed of cover and insert stock to a maximum of 10% of dark colors, and must be made from predominantly bleached chemical fiber. Beater-dyed papers may not exceed 2%. Shavings of novel news or newsprint grades may not be included in this packing.

Prohibitive materials	None permitted
Total Outthrows may not exceed	1%

(28) Coated Soft White Shavings. Consists of baled coated and uncoated shavings and sheets of all white sulphite and sulphate printing papers, free from printing. May contain a small percentage of groundwood.

Prohibitive materials	None permitted
Total Outthrows may not exceed	1%

(29) (Grade not currently in use)

(30) Hard White Shavings. Consists of baled shavings or sheets of all untreated white bond ledger of writing papers. Must be free from printing and groundwood.

Prohibitive materials	None permitted
Total Outthrows may not exceed	½ of 1%

(31) Hard White Envelope Cuttings. Consists of baled envelope cuttings or sheets of untreated hard white papers free from printing and groundwood.

Prohibitive materials	None permitted
Total Outthrows may not exceed	½ of 1%

(32) (Grade not currently in use)

(33) New Colored Envelope Cuttings. Consists of baled untreated colored envelope cuttings, shavings or sheets of bleached colored papers, predominantly sulphite or sulphate.

Prohibitive materials	None permitted
Total Outthrows may not exceed	2%

(34) (Grade not currently in use)

(35) Semi Bleached Cuttings. Consists of baled sheets and cuttings of untreated sulphite or sulphate papers such as file folder stock, manila tabulating card trim, untreated milk carton stock, manila tag; and should be free from any printing, wax, greaseproof lamination, adhesives or coatings that are non-soluble.

Prohibitive materials	None permitted
Total Outthrows may not exceed	2%

(36) Colored Tabulating Cards. Consists of printed colored or manila cards, predominantly sulphite or sulphate which have been manufactured for use in tabulating machines. Unbleached kraft cards are not acceptable.

Prohibitive materials	None permitted
Total Outthrows may not exceed	1%

(37) Manila Tabulating Cards. Consists of manila-colored cards, predominantly sulphite or sulphate, which have been manufactured for use in tabulating machines. This grade may contain manila-colored tabulating cards with tinted margins.

Prohibitive materials	None permitted
Total Outthrows may not exceed	1%

(38) Sorted Colored Ledger (postconsumer). Consists of printed or unprinted sheets, shavings, and cuttings of colored or white sulphite or sulphate ledger, bond, writing, and other papers which have a similar fiber and filler content. This grade must be free of treated, coated, padded or heavily printed stock.

Prohibitive materials	½ of 1%
Total Outthrows may not exceed	2%

(39) Manifold Colored Ledger (preconsumer). Sheets and trim of new (unused by consumer) printed or unprinted colored or white sulphite or sulphate paper used in the manufacturing of manifold forms, continuous forms, data forms, and other printed pieces such as sales literature and catalogs. All stock must be uncoated and free of laser and office paper waste. A percentage of carbonless paper is allowable.

Prohibitive materials	½ of 1%
Total Outthrows may not exceed	2%

(40) Sorted White Ledger (postconsumer). Consists of printed or unprinted sheets, shavings, guillotined books, quire waste, and cuttings of white sulphite or sulphate ledger bond, writing paper, and all other papers which have a similar fiber and filler content. This grade must be free of treated, coated, padded, or heavily printed stock.

Prohibitive materials	½ of 1%
Total Outthrows may not exceed	2%

(41) Manifold White Ledger (preconsumer). Sheets and trim of new (unused by consumer) printed or unprinted white sulphite or sulphate paper used in the manufacturing of manifold forms, continuous forms, data forms, and other printed pieces such as sales literature and catalogs. All stock must be uncoated and free of laser and office paper waste. A percentage of carbonless paper is allowable.

Prohibitive materials	½ of 1%
Total Outthrows may not exceed	2%

(42) Computer Printout. Consists of white sulphite or sulphate papers in forms manufactured for use in data processing machines. This grade may contain colored stripes and/or impact or non-impact (e.g., laser) computer printing and may contain not more than 5% of groundwood in the packing. All stock must be untreated and uncoated.

Prohibitive materials	None permitted
Total Outthrows may not exceed	2%

(43) Coated Book Stock. Consists of coated bleached sulphite or sulphate papers, printed or unprinted in sheets, shavings, guillotined books or quire waste. A reasonable percentage of papers containing fine groundwood may be included.

Prohibitive materials	None permitted
Total Outthrows may not exceed	2%

(44) Coated Groundwood Sections. Consists of new printed, coated groundwood papers in sheets, sections, shavings or guillotined books. This grade shall not include news quality groundwood papers.

Prohibitive materials	None permitted
Total Outthrows may not exceed	2%

(45) Printed Bleached Sulphate Coatings. Consists of printed bleached sulphate cuttings, free from misprint sheets, printed cartons, wax, greaseproof lamination, gilt, and inks, adhesives or coatings that are non-soluble.

Prohibitive materials	½ of 1%
Total Outthrows may not exceed	2%

(46) Misprint Bleached Sulphate. Consists of misprint sheets and printed cartons of bleached sulphate, free from wax, greaseproof lamination, gilt, and inks, adhesives or coatings that are non-soluble.

Prohibitive materials	1%
Total Outthrows may not exceed	2%

(47) Unprinted Bleached Sulphate. Consists of unprinted bleached sulphate cuttings, sheets or rolls, free from any printing, wax, greaseproof lamination or adhesives or coatings that are non-soluble.

Prohibitive materials	None permitted
Total Outthrows may not exceed	1%

(48) #1 Bleached Cup Stock. Consists of baled, untreated cup cuttings or sheets of coated or uncoated cup base stock. Cuttings with slight bleed may be included. Must be free of wax, poly, and other non-soluble coatings.

Prohibitive materials	None permitted
Total Outthrows may not exceed	½ of 1%

(49) #2 Printed Bleached Cup Stock. Consists of baled printed formed cups, cup die cuts, and misprint sheets of untreated coated or uncoated cup base stock. Glues must be water soluble. Must be free of wax, poly, and other nonsoluble coatings.

Prohibitive materials	None permitted
Total Outthrows may not exceed	1%

(50) Unprinted Bleached Sulphate Plate Stock. Consists of baled bleached untreated and unprinted plate cuttings and sheets. May contain clay coated and uncoated bleached board.

Prohibitive materials	None permitted
Total Outthrows may not exceed	½ of 1%

(51) Printed Bleached Sulphate Plate Stock. Consists of baled bleached untreated printed plates and sheets. May contain clay coated and uncoated bleached board. Must be free of nonsoluble ink or coatings.

Prohibitive materials	None permitted
Total Outthrows may not exceed	1%

Specialty Grades

The grades listed below are produced and traded in carload and truckload quantities throughout the United States, and because of certain characteristics (i.e., the presence of wet strength, polycoatings, plastic, foil, carbon paper, hot melt glue), are not included in the regular grades of paper stock. However, it is recognized that many mills have special equipment and are able to utilize large quantities of these grades. Since many paper mills around the world do use these specialty grades, they are being listed with appropriate grade numbers for easy reference.

The Paper Stock Industries Chapter of Institute of Scrap Recycling Industries (ISRI) not establishing specific specifications, which would refer to such factors as the type of wet strength agent use, the percentage of wax, the amount of polycoating, whether it is on top of or under the printing, etc. The specification for each grade should be determined between buyer and seller, and it is recommended that purchase be made based on sample.

These specialty grades are as follows:

1-S White Waxed Cup Cuttings
2-S Printed Waxed Cup Cuttings

3-S Plastic Coated Cups
4-S Polycoated Bleached Kraft-Unprinted
5-S Polycoated Bleached Kraft-Printed
6-S Polycoated Milk Carton Stock
7-S Polycoated Diaper Stock
8-S Polycoated Boxboard Cuttings
9-S Waxed Boxboard Cuttings
10-S Printed and/or Unprinted Bleached Sulphate Containing Foil
1l-S Waxed Corrugated Cuttings
12-S Wet Strength Corrugated Cuttings
13-S Asphalt Laminated Corrugated Cuttings
14-S Beer Carton Scrap
15-S Contaminated Bag Scrap
16-S Insoluble Glued Free Sheet Paper and/or Board
17-S White Wet Strength Scrap
18-S Brown Wet Strength Scrap
19-S Printed and/or Colored Wet Strength Scrap
20-S File Stock
21-S New Computer Print Out (C.P.0.)
22-S Ruled White
23-S Flyleaf Shavings Containing Hot Melt Glue
24-S Carbon Mix
25-S Books with Covers
26-S Unsorted Tabulating Cards
27-S Colored Tabulating Cards
28-S Carbonless Treated Ledger (N.C.R.)
29-S (Not currently in use)
30-S Plastic Windowed Envelopes
31-S Textile Boxes
32-S Printed TMP
33-S Unprinted TMP

OFFICE PAPER

Collection Systems and Marketing

Office paper recycling programs are relatively simple to devise and operate. Many local governments offer instruction and assistance to businesses and other local governments including literature outlining entire sample programs and marketing advice. Some will provide supplies such as desktop containers, posters, and other promotional materials. Where local government assistance is not available, the same items are offered commercially and by nonprofit associations and industry representatives (Fig. 11.22).

Many recycling firms offer collection of office paper. Most, however, find it unprofitable to collect less than 2400 lb/week from any one account. Most will offer to evaluate office paper recovery potential at a particular business if a cursory examination indicates more than the minimum could be recovered (Fig. 11.23).

There are three basic types of office paper separation systems in wide use today. They are the desktop or deskside system with internal collection containers, the internal collection container system, and the external collection container system fed by internal containers.

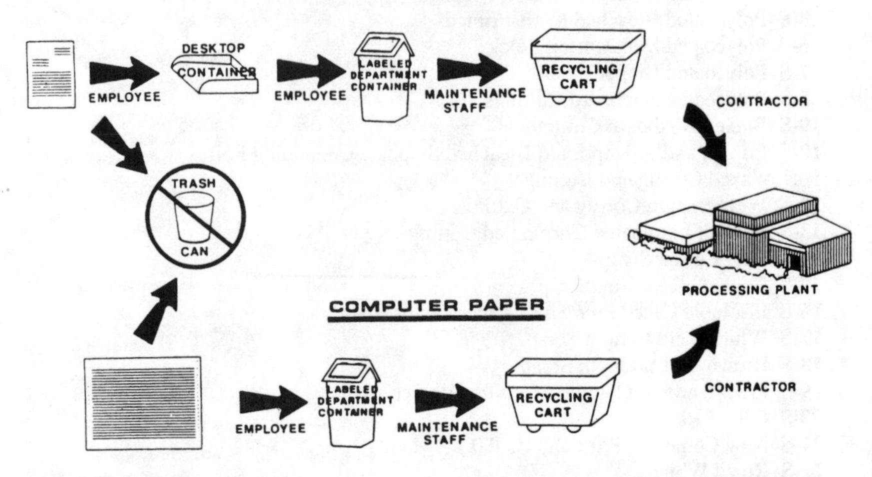

FIGURE 11.22 Office and computer paper recycling flow sheet. (*Source: Florida Business and Industry Recycling Program.*)

The desktop or deskside system with internal collection containers is used most frequently where paper is generated primarily by desk-oriented employees and departments. A desktop tray or box is used to contain paper designated for recycling, while all other desk-generated waste goes into the usual waste receptacle. A box may be placed at the side of the desk instead of on top. Several central containers are placed in high-traffic areas and each is clearly marked to indicate which type of paper is to be placed in that con-

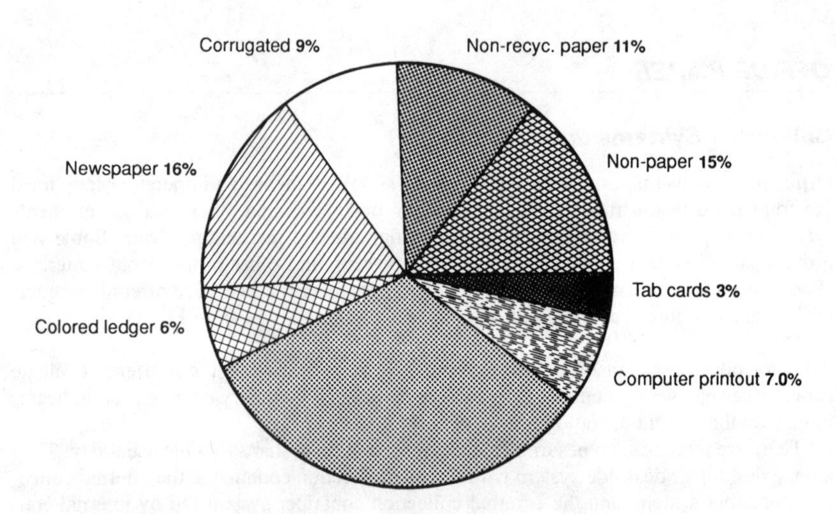

FIGURE 11.23 Composition of solid waste-general office. (*Source: SCS Engineers.*)

tainer. The central containers are serviced by the recycling company. Employees are carefully trained to recognize the paper to be recycled and to separate it properly from all other desk waste. They are instructed to empty the desktop or deskside containers into the appropriate central containers as needed or when their duties will take them near the central containers.

The system using only central containers is implemented where paper generation is not desk-oriented, such as in print shops and data processing departments.

External containers fed by internal central containers are used where large volumes of one paper grade or mixed grades will be recovered.

With any of these systems continuing employee education must be a well-planned part of the program. The systems may be combined to meet the needs of very large generators and where paper generation is both desk-oriented and not desk-oriented.

Impact of Office Paper Recycling

The American Paper Institute estimates that as much as 85 percent of an office building's waste stream by weight is high-grade recyclable paper. This means, in essence, that we are throwing away tremendous savings and monies each year. Recovered office paper primarily feeds mills that manufacture tissue, paper towels, and toilet paper; however, the manufacture of recycled bond and printing paper is on the increase. Many domestic paper mills are under construction and/or retrofitting current equipment to handle the influx of secondary materials. In addition, much of the office paper recovered in this country is exported to nations that lack the resources required to make paper. Recovered secondary paper is the second largest item exported out of New York City harbor, and in many western ports ships loaded with waste paper can be found headed for Pacific Rim nations.

Step-by-Step Office Paper Recycling Program

Instituting an office paper recycling program can be a relatively simple task to undertake. The steps involved are easy and flexible enough to allow for many variations. Perhaps the most economical and least-complicated system is one where the materials being separated for recycling are done so at the source. This system is advocated primarily because it involves a change in habit and in mindset. By having to directly handle what we generate as discards, it makes us think. It is this thought process that is key to a successful recycling program. As behavioral changes occur, the spread of office paper recycling is quick and painless.

Step 1: Conduct Office Waste Audit. The first step initially involves conducting a waste audit to determine precisely what materials should be collected in the program (Table 11.7). Usually orchestrated by the chosen waste paper dealer or municipal recycling representative, it requires some degree of expertise. If the building being assessed is multitenant, then utilization of a questionnaire is oftentimes helpful. The information obtained from these tests assists in determining which recycling system is optimal.

Step 2: Create a Practical Collection System. The next step is to create a workable collection system. Since the success of any given office paper recycling program is dependent on employee participation, it is crucial to develop a convenient and simple program. While many people may be enthusiastic about implementing a recycling program at their

TABLE 11.7 Office Paper Audit

BUSINESS USAGE SURVEY

How much of the following is used in your company's office per month?

1. *White* paper (color of ink is not important), letter paper, envelopes, copy machine paper (No glossy finish)
   ____________ pounds
2. Bond quality computer paper (Please attach a small sample of computer paper to this sheet.)
   ____________ pounds
3. Computer print out cards
   ____________ pounds
4. Brown or manila envelopes or folders
   ____________ pounds
5. Newspaper (black & white only, no comics or advertising inserts)
   ____________ pounds
6. Corrugated boxes
   ____________ pounds
7. Other (Please list different kinds and amounts separately.)
   a. ________________ ________ pounds
   b. ________________ ________ pounds
   c. ________________ ________ pounds

For any questions that you might have just call:

Florida Business and Industry Recycling Program

(407) 678-4200 or 1-800-FLA-BIRP

place of work, this eagerness will soon dissipate if the system introduced is too complicated. Convenience is the lifeblood of a healthy office paper recycling program.

Step 3: Keeping Informed of Improvements. Keeping informed of updates and changes, particularly for the designated recycling coordinators, is equally important for the prosperity of a recycling program. Thus, attendance and participation in scheduled educational and informational seminars is critical. Taking the time necessary to learn about the recycling program from the onset helps avoid potential problems and pitfalls later on.

Step 4: Management Support. Soliciting upper management support for office paper recycling is required to properly implement any corporate program. This is particularly relevant when it comes to educating employees about recycling. Oftentimes overlooked, lack of proper training can kill an office paper recycling program later on. High levels of contamination along with a dropoff in enthusiasm results when training is ignored. Thus, be committed to properly educating participants about the when, where, why, and how of office paper recycling. Waste paper dealers and/or municipalities can often assist in this area. Obviously, the best way to encourage and inspire your staff to recycle is by setting a good example yourself. Show management support and provide detailed information on the advantages of recycling and the procedures of the collection method to ensure a successful program.

Step 5: Monitor and Maintain Interest. Finally, promote, monitor, and maintain the program on a constant basis. Be flexible enough to alter the program if necessary, and provide updates through newsletters, memos, meetings, and/or events. Promote and advertise the program's progress to motivate employees and reward those that excel.

Problems and Solutions

Problem: The annual disposal fee for a large 20-story office building in the heart of a bustling U.S. metropolis averages approximately $60,000. This fee is based on a 40-ton refuse container being pulled about two times a week, even with a compactor maximizing space. As a result of new disposal facilities recently incorporated into the region, tipping fees at the local landfill are rising more than 30 percent in the upcoming quarter. The building's management company has allowed for only the previous amount for waste services in its budgetary framework. Low occupancy rates combined with a recessionary economy prohibit an increase in expenditures across the board. Orders have come down from the top, reduce this potential expense or face a possible salary cut throughout your department to cover the additional costs.

Solution: Implementation of a building-wide office paper recycling program would immediately reduce the amount entering the dumpster, thus necessitating fewer pulls, which translates into less monies required for disposal. Additionally, revenues would be generated from the sale of the recovered materials, which could offset the initial costs required to initiate the program and subsidize any future promotional campaigns. Ultimately, a smaller refuse container would be more than sufficient to handle the reduced solid waste stream, thus corporate objectives will have been achieved and even surpassed with the advent of a less costly method of handling the building's trash flow.

The above scenario depicts only one advantage of establishing an office paper recycling program. Environmental integrity is to many, the true motivation behind initiating office paper recycling; however, a multitude of other positive benefits can be derived.

Financial consequences are often a big impetus in establishing office paper recycling programs. The economics associated with the startup of a recycling program can be offset through savings from circumvention of escalating waste disposal costs and from the revenues from the sale of recycled materials. Most programs realize a complete recapture of startup costs within the first 6 to 12 months of operations.

Computer Printout

Computer printout is the most valuable of the high-grade office papers. It may have colored bars and may be impact or nonimpact (laser) type. Many markets prefer to purchase colored-bar paper. Usually it is a green bar. Many markets prefer impact type or not to mix impact and laser. It is important in planning an office paper recycling program to determine precisely, what the market requires.

Computer printout is generally considered too valuable to mix with other high-grade papers and should be kept separate and free from contamination. Some manifold white ledger paper may appear to be similar to computer printout. Where there is not absolute certainty, a market expert should be consulted.

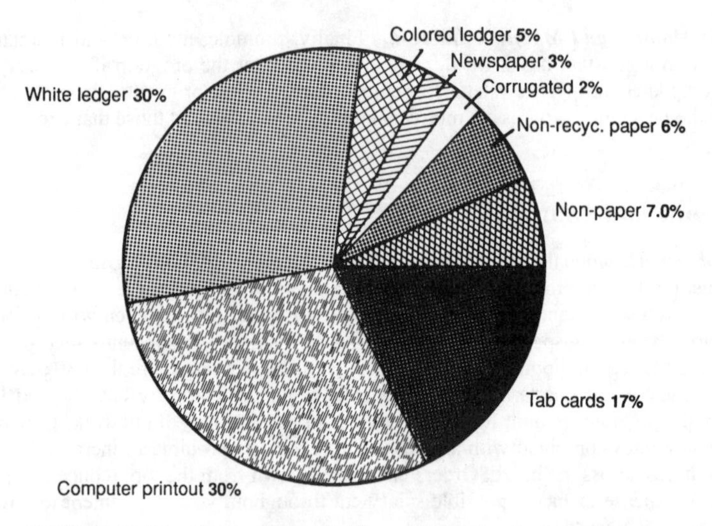

FIGURE 11.24 Composition of office solid waste—bank and insurance companies. (*Source: SCS Engineers.*)

White Ledger

White ledger should generally be considered any fine white writing, printing, typing, or copy paper containing no color other than black. The market must always be consulted for precise specifications. Certain qualities of book pages, completely separated from covers and bindings, might be allowed in this grade by certain markets.

Other Office Paper

Colored ledger should generally be considered any fine writing or printing paper containing any color other than black. The ledger grades exclude paper that is treated, coated, padded, or heavily printed.

In banks and insurance companies, computer printout and white ledger average about 30 percent each and colored ledger averages about 5 percent of the total waste (Fig. 11.24). In the general office, computer printout averages 7 percent; white ledger, 33 percent; and colored ledger, 6 percent of total waste. Seventeen percent of bank waste is tabulating cards.*

Computer printout, white ledger, and colored ledger are high-grade de-inking papers which are used to make tissue and fine writing and printing papers. High-grade office paper is used to produce tissue, napkins, and paper towels more often than it is used to produce more office paper.

Other office grades that are recyclable, depending on market demand, are tabulating cards, file stock, and mixed grades. Banks and similar institutions produce significant quantities of tabulating cards, both manila and colored. File stock should be considered

*Source: SCS Engineers.

for recycling whenever files are purged. Mixed-grade recycling can be convenient for companies desiring not to incur sorting expense.

TELEPHONE DIRECTORIES

History of Phone Book Recycling

Interest in the recycling of phone books began in the late 1980s, when America was facing a landfill crisis. Space in existing landfills was being depleted, with land suitable for siting new landfills rapidly dwindling. Although old telephone directories (OTD) made up only 0.5% of all material going to landfills, the public's perception, fueled by the media, was that OTD was a major problem.

At about the same time, several states were proposing legislation mandating that directory publishers collect all the old directories they had distributed. As of 1998, only one state in the U.S. had passed such legislation. Most realized that OTD represents only a small portion of the larger solid waste picture, and that to impose such restrictions on only one type of waste producer was arbitrary and discriminatory.

Major directory publishers across America began their involvement in recycling programs at about this time. A campaign in Jacksonville, Florida yielded 111 tons of OTD in one year. In 1991, the Yellow Pages Publishers' Association (YPPA) implemented its first Environmental Council Action Plan, with annual updates following. The plan specified publishing guidelines for environmentally responsible directories and measures that could be taken toward source reduction. It also called for member publishers to ". . . work with local communities, governments, businesses and agencies to develop on-going programs for the collection and recycling of old directories."[1]

Impact on the Environment

Communicating the importance of recycling to the public is easier if they are aware of the impact of their actions. Commonly accepted figures for the resources saved by recycling paper may be used to show this. Each ton of paper recycled saves approximately:

- 3.3 cubic yards of landfill space[2]
- 17 trees[2]
- 3700 pounds of lumber[3]
- 24000 gallons of water[3]
- 3 barrels of oil[2]

Multiplying the tons recycled in a given campaign or state by these figures often gives impressive totals of resources saved.

Large figures are more meaningful, however, if they can be related to "real world" things that people can picture in their minds. For example, saying that "enough lumber to build 200 houses was saved" is more meaningful in publicity than saying, "1000 tons of lumber was saved" by recycling phone books. Using houses, animals, football fields, or

[1]Yellow Pages Publishers Association Environmental Council Action Plan, June 1996 edition.
[2]*Resource Recycling* magazine, May/June 1984
[3]University of Southern Mississippi study, Appendix B, "Solid Waste Facts."

other visual objects allows people to picture the amount in their minds. This is a more powerful message and will be remembered longer than mere numbers.

The Public-Private Partnership

Phone book recycling programs work best when they involve a mix of private firms and public-sector representatives. In most instances, cities and counties already have collection infrastructure for recyclables in place, as well as public education programs to promote these methods. Waste haulers and processors have a vested interest in the recycling industry, and may be willing to provide some services at cost or gratis in exchange for publicity. Major directory publishers are interested in community involvement and publicity, as well, and will usually assist with promotion or costs of the campaign.

COMPOSITION OF PHONE BOOKS

As with any recyclable, knowing exactly what phone books are made of is important. The books' composition can affect the end user selected, the collection methods, or even whether the books are recycled at all.

Needs of Paper Mills

The components a particular mill can handle will depend on the product into which the OTD is being made and upon that mill's technology for handling inks, glues, etc. A basic understanding of the papermaking process is useful in matching the optimal end user to OTD from a specific community.

Most mills begin the recycling of OTD by shredding, or "hogging," the books. This may have been done earlier, prior to baling, depending on the mill's specifications and who processed the OTD for shipment to them. The baling strips are removed, and the shredded books are processed with chemicals and water to remove ink, coatings, and glue. De-inking technologies vary by mill, and are closely guarded trade secrets.

This cleaned fiber may be mixed with some virgin pulp fiber, depending on the recycled content of the end product. The wet paper mixture is thinned and pressed into sheets, dried by hot air blowers as it runs through a series of rollers, then is fed onto long spools. Finally, the paper rolls are cut into standard sizes and wrapped for shipment to customers.

Grades and Prices of Recycled Paper

OTD is considered to be one of the lower grades of recycled papers. This, coupled with its lower volume in comparison to other grades, may be why OTD's price is not quoted in common sources, such as *Waste News* or *Recycling Times*. Prices paid by the same mill may differ if the OTD is baled or loose. Other mills may pay a lower price, but include transportation. Still others may have agreements with directory publishers for a standard price. All of these factors make it difficult to pinpoint a prevailing market price for OTD.

Since end users are limited, the most reliable method is to call the intended end user and negotiate a price directly. Prices for all paper grades are somewhat interrelated, so monitoring of the markets and the factors that affect them is advisable. One grade of special interest is Number 8 Newspaper (ONP), as many of the mills who accept OTD also use this grade of ONP. This is discussed in more detail in the "Effect of Market Factors" section of this chapter.

In soft markets, some processors move OTD by mixing it into loads of ONP or residential mixed paper (RMP). This is an option for smaller amounts, but there are usually restrictions on how much OTD can be in a load of RMP, depending on the end use for the paper. This is also something that must be discussed individually with the end user, as some may consider phone books to be contamination in loads of ONP or RMP.

Other Materials in Phone Books

Besides paper, phone books contain glue, inks, dyes, and coatings. Mills differ in their capabilities to process some of these. Depending on the de-inking and cleaning processes, some may prefer water soluble glue, while others prefer hot-melt. Different types of ink (e.g., flexographic) may also present a problem to some mills. Awareness of both the mill's needs and the composition of all directories in circulation in the community will decrease the likelihood of rejected loads.

Printing Changes To Accommodate Recycling

Most major directory publishers in the United States now manufacture their books to be recyclable. This mandated a change for many in the materials used to print the books. Industry standards are set by the Yellow Pages Publishers' Association Environmental Council, including representatives from YPPA member publishers.

Since many paper mills prefer the use of water-soluble glue over hot-melt, most major directory publishers now specify water-soluble as their preferred glue. Soybean-based inks have replaced petroleum-based as the industry standard, along with vegetable-based dyes and environmentally friendly coatings for the directory covers. YPPA member publishers also complete the recycling loop by buying recycled paper, with 40% recycled content becoming standard as of January, 1998.

Some publishers may not adhere to these industry standards. Knowing the environmental policy of all publishers with directories in circulation within a given community is essential to a successful phone book recycling program.

CHALLENGES OF PHONE BOOK RECYCLING

Several obstacles to a successful phone book recycling program can be overcome by careful planning. Awareness of these obstacles is the first step toward conquering them.

Nature of Old Telephone Directories (OTD)

OTD, as a commodity, is difficult for paper mills to process. The paper used in phone books is thinner and stronger than newsprint, to keep book size manageable and to withstand a year of frequent use. The covers and tabs are a different type of paper than the pages. OTD's fibers are shorter, sometimes becoming lost in the longer cleaning process that is required to remove the inks and dyes. They are also more difficult to process by nature—they are a "glob" of paper, bound together, as opposed to loose newsprint.

Changes in the yellow pages publishing industry have also affected the nature of OTD. The prevalence of "white knock-out" ads in today's phone books (ads with a white background appearing in the yellow pages) mean that publishers are *printing* the yellow color onto white paper with ink. Yellow pages used to be printed on paper that was *dyed* yellow. Since dyes are removed differently than inks, this affects the cleaning process for OTD and the treatment of the mill's wastewater.

Some end users don't need to de-ink the paper. These include manufacturers of insulation, hydromulch, or any other product consisting chiefly of shredded paper. Others don't need a high brightness for their product (e.g., egg cartons or packing material). Some actually *want* a lower brightness (e.g., wallboard manufacturers) and will use OTD to get it. Depending on their location in relation to the source and available transportation, one of these end users may be a better option than manufacturers of newsprint, writing, or directory paper.

Economics

Many costs must be factored into the equation for recycling phone books. These begin with collection (including placement and servicing of containers), publicity and advertising of the campaign, and transportation to the processor.

Separation costs for the processor cannot be ignored. These companies are in the recycling business, and must remain profitable. Since phone books are not a year-round commodity, no automated separation systems exist for them. Most processors hire a staff specifically to pull phone books from the mix of paper at the materials recovery facility (MRF). Storage of the collected books may also be a cost factor, as they must be kept dry for recycling. If the processor has no covered area for storage, the books must be kept in a trailer or warehoused elsewhere. The use of prison labor or subcontracting with an organization for the disabled are possible ways to reduce processing costs.

If the end user requires that the OTD be hogged (shredded) and baled prior to shipment, there will be a cost involved. Some processors have their own machinery for accomplishing this task, but others subcontract it out. If the OTD must be transported for hogging and baling, this is an additional cost. This could be a determining factor in which end user to select; some will accept OTD loose-loaded.

Transportation to the end user must also be accounted for. Depending on the MRF's location, and that of the end user, options include rail, truck, or containerized shipping. Most processors are aware of the various options, since they ship other recycled materials year-round, and can select the optimal transportation mode. Trailers usually hold 20-ton loads of OTD; transportation chargebacks may result for less than a full load. In cases where the campaign has yielded less than 20 tons, comarketing the OTD with a nearby community may be an option.

If a load of OTD arrives too wet for the selected mill to use, it can be rejected. This results in both freight chargebacks and a "homeless" load of OTD. Careful storage of the collected material prior to shipping can avert this problem. While some dampness is to be expected, be aware of the mill's maximum allowable amount.

If an export end market is used, there may be tariffs or other charges involved. It is best to allow an exporter or broker to handle such matters, as they are aware of markets and customs regulations in the countries with which they regularly do business. Also ask the exporter about their fees; this may be an additional expense.

Once accounting for all the above costs, the possibility of a profit on the OTD seems unlikely. However, depending on the end user selected, their location in relation to the OTD source, and the price they're paying for the material, a profit may be possible.

Reducing some of the costs through partnership with the parties involved offers some help. The value of publicity for their company's involvement must be greater than the revenue they'd otherwise gain for providing the service. Community service has value to companies, mainly for the publicity they gain from it. This is not because companies are inherently evil or greedy, but they must make a profit or they cease to exist. Supporting partners through the publicity they need is essential to success of the partnership.

Effect of Market Factors

As with any other commodity, the price of OTD is affected by many external factors. Even economists can't fully predict commodities prices, but an awareness of the factors affecting them can help in planning for phone book recycling campaigns with some assurance of having an end market for the books collected.

The economy in general is the first major factor affecting OTD prices. This includes the U.S. economy, that of the region in which one is collecting and marketing OTD, and that of the world markets.

The availability and price of labor will affect the cost of doing business for the mills and recyclers in the region. Other economic trends to watch include:

- Inflation
- Personal consumption expenditures in relation to income
- Willingness of financial institutions to lend money
- Investment spending on equipment
- Real estate investment on both the residential and commercial sides
- Government spending (national, state, and local)
- International trade

All of the above affect the demand for paper and other products made from OTD, as well as the cost of manufacturing those products and moving them to market. And don't forget about noneconomic factors, such as weather, that can affect collection and transportation efforts.

Other factors, such as the amount of OTD available by month, also have an impact. Since directories are being issued *somewhere* throughout the year, and the size of these communities varies greatly, the flow of OTD is not constant. Even the largest consumers of OTD have a finite capacity. Demand for the products made from OTD is another important factor.

Within the paper market, related grades to watch include old newspaper (ONP—especially #8 news) and residential mixed paper (RMP). While prices for both of these are generally higher than for OTD, they are similar grades of paper that are often consumed by the same end users, and the prices may rise or fall similarly.

Public Apathy/Laziness

When phone book recycling began in the late 1980s, many communities saw it as a way to turn the expense of waste disposal into a revenue-producing operation. Once the reality—that recycling is merely an alternative form of waste disposal that still costs money—was realized, enthusiasm began to wane. Keeping the public excited about recycling is not an easy task when those in positions of leadership are blasé about it.

Public education is the best tool for overcoming this apathy. Appeals to recycle because "it's the right thing to do" have proven to be only mildly effective. Realize that recycling something takes extra effort on the part of the consumer; what *benefits* do they receive in exchange for this effort? They must be made aware of the savings in resources and landfill space, if not in cost. Appeal to their dislike of bigger, more intrusive government; if they recycle voluntarily, they won't be forced to by law.

Educating children is also quite effective in this effort. Children are more receptive to messages about "saving the planet," and are great motivators of families. They love projects that allow them to be creative, and can incorporate basic information about recycling into such projects.

Multiple Publishers In A Community

Another challenge is when more than one publisher issues directories in the same community. While most major publishers manufacture their books to be recyclable, some publishers do not. If the optimal end user cannot accommodate hot-melt glue, for instance, and a book bound with hot-melt has been distributed in the area, you must choose between selecting a different end-user and separating the hot-melt books from those to be recycled.

Various publishers may also differ in their willingness to support the recycling campaign. While many large publishers offer assistance to communities, smaller ones may not be in a position to do so. Making them aware of program needs and *asking* for their help is a first step to ensuring future participation. Involvement of nearby competitors in the campaign may help to motivate them!

Tracking Results Accurately

Reporting your results from a recycling campaign can be one of the most difficult aspects. If there is only one processor and one end user involved, freight bills or reports from the end user can be used to determine tonnage. If the books are marketed to different end users, however, it will take several phone calls to get tonnage from each. If some of the OTD has also been commingled with other paper grades, make sure to factor this into the totals.

Some large entities may collect, process and market their OTD separately from those collected in curbside and drop-off locations. Examples are military bases, colleges, and universities. If some of these are in the community, call their recycling coordinators to get tonnage.

In the absence of reliable figures from *any* sources, you may use surveys to estimate the percentage of phone books that were recycled in the community. Publishers can provide figures for the previous year's delivered tonnage, and applying the survey percentages to this total will yield a recycled tonnage figure.

If surveys are used to estimate, make sure that the sample size is large enough to be statistically projectable to the general population, and that the sample is randomly selected across all demographics. There is also a tendency among respondents to give the answer they think you want, so a portion of the "yes, I recycled my old phone book" answers should be subtracted to account for this bias.

SPECIFICS OF COMMUNITY PHONE BOOK RECYCLING PROGRAMS

There are several elements that must be in place to have a successful phone book recycling program. The absence (or loss) of any of these elements can result in the collapse of an otherwise promising program.

Finding End Markets

Although their role in the program is at the end, a market for collected OTD is the *first* piece that must be in place for the program to occur. Without an end user, there is no reason to collect or process the books for recycling. There are several ways to identify and cultivate potential end markets.

Networking at paper and recycling industry events is essential to keep abreast of market trends and uncover potential end users. Membership in groups such as SWANA, National Recycling Coalition, and Keep America Beautiful affiliates also reflects your support for recycling, while attendance at their meetings give you access to those in the industry. All of these groups are also looking for officers, which makes you more well known to the membership. Local Chambers of Commerce and groups such as the American Forestry and Paper Association are other networking avenues.

Serving as a speaker at some of these events is a good way to get some visibility and become well known to industry leaders. This can result in even more leads on potential markets. Exhibiting at industry expositions is another means of gaining exposure to end users. Strolling through such expositions may also introduce you to manufacturers who need your OTD to make their products.

Subscriptions to recycling publications will provide you with important industry news that affects end markets, as well as leads on mills of which you may not have otherwise been aware. *Waste News, Recycling Today, Waste Age, World Wastes,* and *Recycling Times* are a few to try. Don't neglect community business publications: chamber newsletters, statewide business magazines, and even the daily newspaper are additional sources of leads.

Many of the above-listed groups and publications also have websites on the Internet. Use a search engine to find them, then bookmark them and check back often. They can be a wealth of useful information.

If local markets dry up, the export market may be an alternative. Although transportation costs are higher, many manufacturing operations are based overseas. Most countries of the world generate only a fraction of the waste paper generated in the United States, so these countries cannot supply the mills located there with enough recycled fiber to meet their needs. Egg cartons, packaging material, and school paper are among the items manufactured overseas.

It's important to support end markets by encouraging the purchase of products made from OTD. Use any publicity opportunity to inform the public about these products. Displays and flyers at local chamber trade shows, Earth Day, America Recycles Day, and other environmental events should feature samples of the products and information about the brands and where they can be purchased. This helps to ensure that the end market will remain viable in the future.

Partners' Support

The participation of several partners is essential to a successful phone book recycling program. Each provides a unique service that is difficult to find without their support. The planning committee for a phone book recycling program should include representatives of each of these partners.

City and county governments usually have collection infrastructure for a number of recyclables. This may be public, or contracted with private haulers and processors. They are experts on recycling programs and public education, and many have staff who handle recycling full-time.

In privatized operations, private waste haulers collect phone books in the program, whether from drop-off sites or in curbside programs. As with public collection methods, the books are heavy and are out of the ordinary for them to handle. Private haulers will often provide and service some containers for the phone book program at no charge.

The operators of the materials recycling facilities (MRFs) that process phone books once they're collected, whether public or private, are important partners. They sort and store the OTD, sell it to the end market, and arrange for transportation there. MRFs with an adjacent rail site expand the transportation options, which increases the end markets available and helps control transportation costs. The recycler also arranges for any necessary shredding and baling before shipment to market.

Directory publishers provide services ranging from end market cultivation to public education, and may include financial support. The costs they cover may include advertising, containers, special collection events, hauling, processing, transportation, or contest sponsorship. They also support recycling campaigns by printing their directories on recycled-content paper. The Yellow Pages Publishers Association (YPPA) set a 1998 industry standard of 40% recycled content in directory paper for its member publishers. Most of the members meet or exceed this standard, purchasing over 600,000 tons of recycled-content directory paper in 1997.

Schools are excellent partners because children are a receptive audience to the recycling message. If the campaign occurs at a time when school is in session, a collection contest provides a great way to recover OTD. Many school boards have a full-time recycling coordinator who can be the liaison to individual schools, channeling information and drumming up support for the contest.

Local businesses can be enlisted to provide prizes for contests or incentives for recyclers. These can include restaurants, retailers, recreational venues, shopping centers, or sports teams. Some may volunteer to serve as drop-off sites for phone books during the campaign. Offers may include discounted admission, a free soft drink, a free round of miniature golf, or a free promotional item in exchange for the books.

Media partners can be an invaluable asset. They may provide free airtime or advertising space, or offer a discount on purchased advertisements. If a broadcast partner has a news department, they may provide coverage of collection events to boost attendance. Ask about appearance on a morning talk show for additional educational value.

A local end user is an excellent partner. Having a local market for the OTD eliminates major transportation costs and supports local industry. If they pay a good price for the books, a profit may even be possible.

Most of these partners provide whatever services they donate in exchange for publicity. Giving them credit for their contributions to the campaign's success cannot be overlooked. Logos should be included in all advertising and publicity. Good public relations for the partners helps to ensure future participation.

Collection Methods

There are many ways phone books can be collected during the campaign. Each has pros and cons, some of which can be avoided with proper planning and education.

Curbside bins—*Pros:* Easiest for consumers. Usually significantly boosts results for the campaign. Once the public is used to recycling old phone books in their curbside bins, it will be difficult to convince them to do otherwise. *Cons:* Some haulers and processors may not want to handle phone books in a curbside program, as they must be commingled on the trucks and require extra personnel at the MRF to be sorted from other recyclables. There is currently no automated way to separate the books, so costs go up for the recycler.

Tip: Publicity should include instructions to hold the books until a sunny day, as they can't be recycled if they get wet.

Drop-off sites—*Pros:* A good way to collect "pure" loads of OTD. *Cons:* Contamination can occur at unattended sites. It can be difficult to convince some people to go to this much effort to recycle their old phone books when they can more easily just throw them in the trash. *Tips:* Covered containers must be used to keep the OTD dry. In smaller towns, you may be able to persuade some local businesses, such as fast-food restaurants, to contribute some type of incentive to phone book recyclers at their locations. An incentive seems to increase the recycling rate.

Collection contests—*Pros:* Schools, churches and other volunteer groups are always looking for funds. Involving them is an excellent way to educate them on recycling. Competition can boost results. Loads of OTD are clean. Directory publishers are usually happy to sponsor these contests, and other local businesses may also be persuaded to offer prizes. *Cons:* If not carefully constructed, payoffs to participants can be higher than budgeted. Uninsured groups may cause liability risks. *Tips:* Provide collection safety tips in the contest package. Use tally sheets for individuals, listing all directories in circulation within the area, to provide good recordkeeping for determination of award winners. Awarding a special prize for the oldest or farthest-away phone book can add interest.

Collection events—*Pros:* A good way to recover books from those who don't have a curbside or convenient drop-off option. A directory publisher or other business partner may handle arrangements. *Cons:* requires the cooperation of a local sports team, recreational venue, or festival organizing committee to get permission to set up the collection site and perhaps offer a discounted admission or other incentive to recyclers. *Tips:* Containers should be placed immediately prior to the event and picked up just after it to avoid contamination. Advance publicity is a must to ensure a good turnout of recyclers. Provide staffing volunteers with a tent in case of rain or strong sun, a table and chairs, and sufficient water to drink in hot weather. Free attendance after they have worked their shifts, and/or special promotional shirts or badges are also good motivators.

Collection by volunteer groups—*Pros:* Good way to collect phone books from area businesses. Badges can be provided for scouts, or contributions to other groups, in exchange for their efforts. *Tips:* Scouts, woman's clubs, churches, garden clubs, and environmental groups are some possibilities. Ensure that any group enlisted has liability insurance to cover them for any injuries or property damage that may occur during the collection.

Promotion of Campaigns

There are two aspects of public awareness of a phone book recycling campaign: first, that people know it's going on and why it's important to recycle phone books, and second, that they know how to recycle the books.

Advertising can be effective at communicating the "how-to's" of the program. Tell people to keep the books dry, and include partners' and sponsors' logos. Often, broadcast media will offer a discounted price advertisements, or may run additional free spots, as a community service. The key to getting such deals is to *ask* for them. Publicity is *not* the same as advertising; it's free! Press releases should always be used to notify the media about the upcoming recycling campaign with "gee whiz" information. Appearance on local morning TV talk shows is another means of communicating your message at no charge. Take samples of products into which the books will be made, photos from past years' campaigns, or anything else visual that will be memorable to the audience. Lastly, don't neglect internal promotion! Make sure all employees are aware of the whys and hows of recycling. This will allow them to answer questions from others outside the organization.

Awards Programs

Everyone loves to get awards! Payoffs of collection contests should be swift, with local media present if possible. School contests can be paid off at school board meetings or awards assemblies.

Recognition should also be given via press releases, including a photo of the winners receiving their prizes and program partners who provided valuable services. These stories stand a better chance of running in smaller newspapers, so target them with your releases.

OTHER RECYCLABLE PAPER

Book Stock

Hard-bound books are occasionally sought by markets. Prices paid for book stock are relatively low because of the processing expense. Pages of hard-bound books are recovered as ledger, provided the pages meet the specifications for ledger. Soft-bound books may be recycled in certain low-grade mixes.

CHAPTER 12
ALUMINUM CANS

DURENE M. AYER
Associate
Malcolm Pirnie, Inc.
White Plains, New York

GENERAL INTRODUCTION

Sources, Amounts, and Types of Aluminum Products

To determine the potential sources, amounts and types of aluminum products and to confirm that aluminum cans are one of the most feasible items to recover for recycling purposes, two factors must first be considered: aluminum consumption by market sector and product durability. Figure 12.1 illustrates domestic consumption percentages for the major aluminum market sectors: transportation, containers and packaging, construction, exports, electrical, consumer durables, machinery and equipment, and other, minor, sectors. The transportation industry utilizes approximately 28% of total aluminum production, making it the largest market for raw aluminum. The containers and packaging industry follows, with a consumption of approximately 23%. The construction industry consumes an estimated 14%, 13% is exported, and 7% of total production is estimated to be used for electrical purposes. The consumer durables and machinery and equipment markets utilize approximately 6% each, and the remaining 3% is utilized by other, minor, markets [1]. The transportation, packaging, and construction sectors together account for a total of approximately 65% of aluminum consumption.

Materials most suited for recycling are those intended for popular short-term consumer usage; they are discarded quickly and are present in large quantities in the solid waste stream. Aluminum products from the transportation and construction market sectors generally have long-term uses, measured in years, and are therefore less likely than packaging to be present in the municipal solid waste stream in sufficient quantities for efficient recycling. Packaging materials, however, constitute approximately 30% of the products discarded in the municipal solid waste stream (Figure 12.2) and have a consumer usage span of only a few days or months. Therefore, a large amount of aluminum that can be efficiently recovered from the municipal solid waste stream for recycling is considered to come from the packaging sector.

Figure 12.3 shows the major types of packaging materials found in the municipal solid waste stream. Of the approximately 30% of municipal solid waste discards that are attributed to packaging materials, aluminum accounts for less than 1% by weight, which does not appear to be significant, unless volumetric comparisons are made. More than 80% of aluminum in the municipal solid waste stream is from used beverage containers (UBC) [2]. The remaining 20% consists of items such as aluminum foil, flexible packaging, ap-

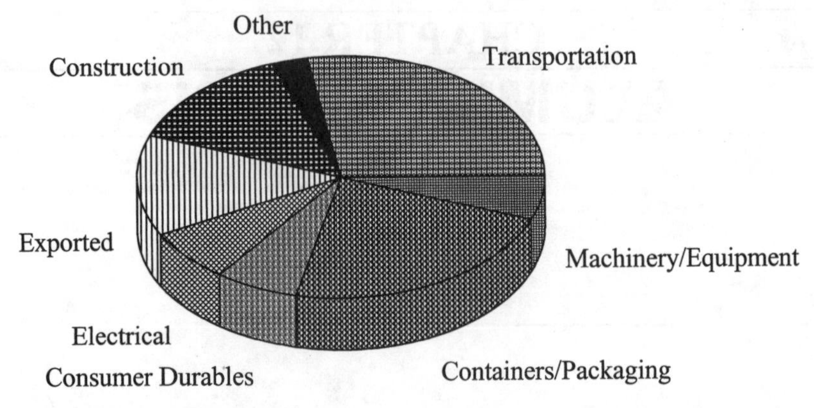

FIGURE 12.1 Aluminum market sector consumption. (*Source:* The Aluminum Association, *United States Industry—A Quick Review*, 1997.)

pliances, furniture, etc. Because the majority of aluminum found in the municipal solid waste stream is in the form of UBCs, aluminum recycling efforts should focus on the recovery of this form of packaging.

Recycling of Aluminum Beverage Cans

Aluminum beverage cans are typically included in recycling programs. Increased concerns by the public regarding the environment, including concerns over decreasing landfill capacity, littering, and increasing energy prices, have caused aluminum can recycling to steadily increase over the past two decades. As illustrated by Figure 12.4, the number of aluminum cans collected through recycling programs has been steadily increasing on an annual basis since 1972. Figures for 1996 indicate an increase of approximately 1.3%

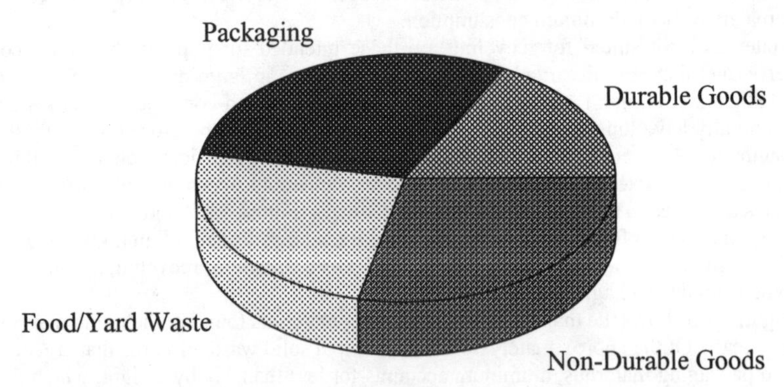

FIGURE 12.2 Types of products discarded in the MSW stream. (*Source:* The United States Environmental Protection Agency, *Characterization of Municipal Solid Waste in the United States: 1996 Update*, June 1997.)

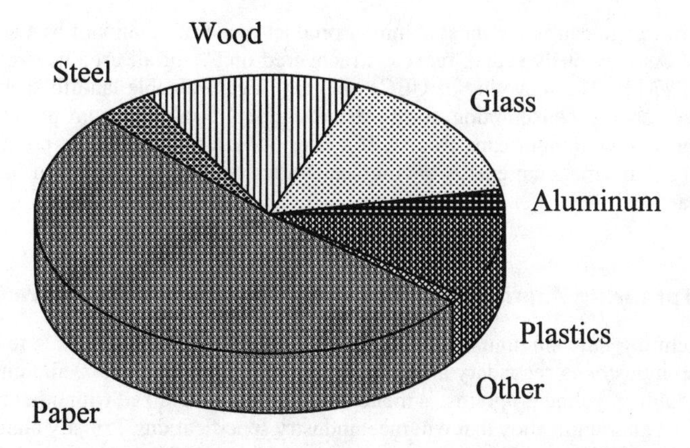

FIGURE 12.3 Packaging materials generated in the MSW stream. (*Source:* The United States Environmental Protection Agency, *Characterization of Municipal Solid Waste in the United States: 1996 Update*, June 1997.)

over the 62.2% aluminum can recycling rate attained in 1995. Data for 1997 indicate a larger increase—approximately 3% over the 1996 rate of 63.5% (66.5%).

Aluminum cans are one of the most common items recovered through municipal and commercial recycling programs because they are easily identifiable by residents and employees. They also provide higher revenues than other recyclable materials. Also, as landfill space continues to become scarce in many areas throughout the United States, the widespread use and recycling of aluminum cans will help to mitigate the depletion of remaining landfill capacity.

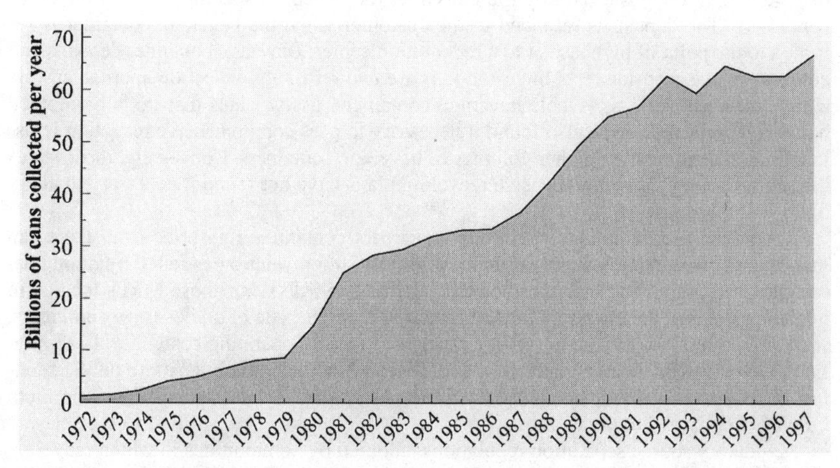

FIGURE 12.4 U.S. aluminum can collection. (*Source:* The Aluminum Association, Can Manufacturers Institute, Institute of Scrap Recycling Industries.)

The beverage can is the most common product made of aluminum by the packaging industry. As previously stated, recyclers recovered 66.5% of all used beverage containers in 1997 [3]. The recycling of UBCs not only saves valuable landfill space, but also minimizes energy consumption during the manufacture of aluminum products. Manufacturing new aluminum cans from UBCs uses 95% less energy than producing them from virgin materials, an energy savings equivalent to tens of millions of barrels of oil each year [4].

Manufacturing Aluminum from Used Beverage Containers (UBC)

Manufacturing new aluminum products from used aluminum materials is referred to in the scrap industry as secondary aluminum production. In this process, aluminum, recovered through recycling programs, is melted in a furnace and mixed with other materials to produce an aluminum alloy that will meet industry specifications. Primary aluminum (virgin aluminum) is also added to ensure proper material specifications required for the final end-use product. After heating, the molten mixture is then cast into ingots, sheets, or aluminum products.

Approximately 95% of the UBCs collected nationwide are melted down and formed into aluminum sheets to be utilized in the manufacture of new aluminum cans. The remaining 5% is utilized by foundries in the production of ingots for other uses, and a small percentage is exported. The aluminum from used beverage cans will often be found in the form of new beverage containers on supermarket shelves in as few as 60 days, thereby completing the recycling loop [5].

Legislation

Existing or proposed legislative actions may guide the development of recycling programs. Comprehensive recycling laws and "bottle bill" legislation are both examples of laws that can impact aluminum can recycling activities. Bottle bill legislation typically mandates that a deposit, paid by the consumer, is placed on specific types of beverage containers. The deposit is refunded to the customer when the beverage container is returned to the point of purchase or to a redemption center. Beverage container deposits and comprehensive curbside recycling programs are two recycling collection approaches that compete for valuable recyclable beverage containers. Many states that initially adopted bottle bill legislation have also found it necessary to pass comprehensive recycling legislation in order to recover higher volumes of beverage containers. Conversely, those states that initially adopted comprehensive recycling plans have not found it necessary to adopt bottle bill legislation [6].

Recovered used aluminum beverage containers demand a high price from the scrap market and consequently supply a major portion of the revenues generated by municipal recycling programs and material recovery facilities (MRFs). For those MRFs located in nondeposit states, the operators rely on revenues from the sale of all beverage containers, including aluminum UBCs, to offset a portion of facility operating costs. For those publicly owned MRFs located in states with a deposit law (or bottle bill), more public funding is required to offset operating costs because fewer revenue-generating aluminum UBCs are delivered to the facility.

Mandatory source separation regulations, adopted by communities in which recycling programs are being developed, may also affect recycling efforts. Legislation can significantly impact program operation, especially if source separation and market preparation

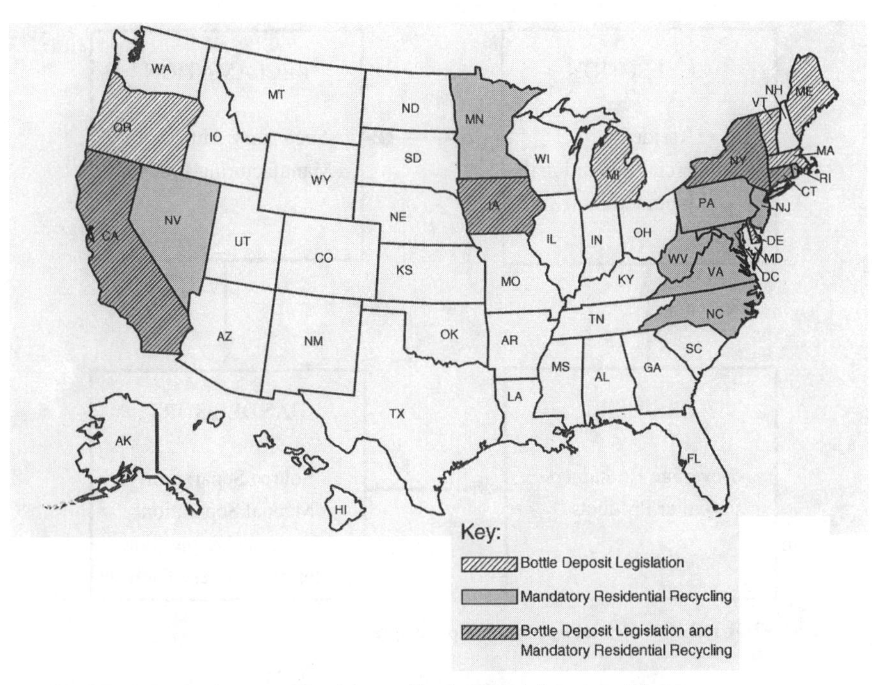

FIGURE 12.5 Mandatory residential recycling laws and deposit legislation.

of recovered materials is mandated. For example, some communities have entered into intergovernmental agreements whereby member municipalities are required to bring their recyclables to a certain market, preprocessed per market specifications. In this circumstance, participating communities benefit from economies of scale in marketing their recyclables, since consistent, large quantities of materials generally demand a higher price from markets. In other locations, municipalities are required to include aluminum cans in their recycling program, but are permitted to market the materials on their own. Currently, although only 13 states have mandatory source separation recycling laws, almost all states have either established recycling goals or landfill bans on recyclable materials (Figure 12.5). Ten states have adopted deposit legislation for the recovery of aluminum cans [7]. As shown on Figure 12.5, four states have enacted both mandatory recycling and deposit legislation.

ALUMINUM CAN RECOVERY PROCESS

Implementing An Aluminum Can Recycling Program

As outlined by Figure 12.6, a successful aluminum recycling program must have interaction between various entities, including those involved with collection, sorting and processing, reclamation, and reuse. There are three generator sectors from which aluminum beverage containers can be recovered: residential households, commercial institutions,

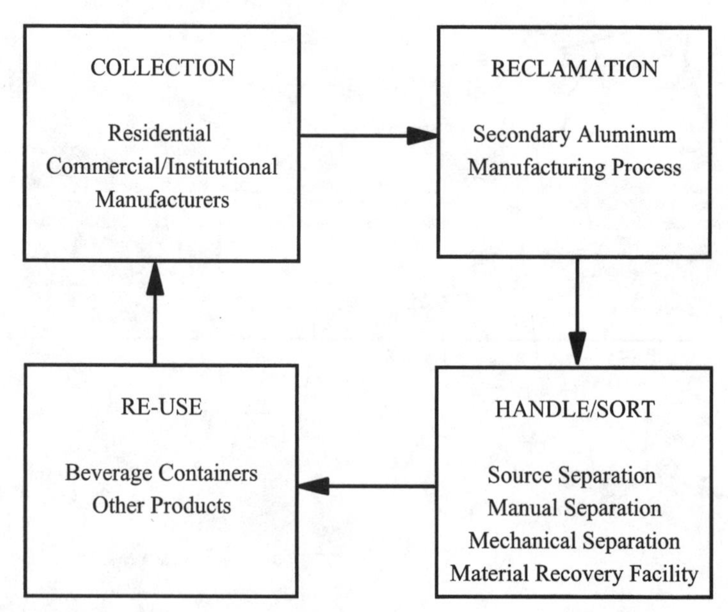

FIGURE 12.6 Recycling process flow diagram.

and manufacturing entities (other than those producing aluminum products). Collection practices in each of these sectors are outlined below.

Residential Collection. Communities have three basic options available for the collection of aluminum cans: drop-off (depot) centers, buy-back centers, and curbside collection programs. Depot centers and buy-back centers require residents to bring aluminum cans and other source separated materials to a specific location. Depending on the size of a particular community, multiple depot or buy-back sites may be necessary to make collection of recyclables convenient for all residents. Depot and buy-back centers differ only in that buy-back centers pay for the recyclables brought in by the residents. An evaluation of programs throughout the nation indicates that these two types of collection programs, although effective in certain circumstances, are generally associated with overall low recovery rates [8]. Curbside collection programs, considered the most convenient collection method because the resident places recyclables at the curb, capture relatively large quantities of recyclables. Aluminum UBCs can be separated as an individual commodity or commingled with other recyclables for collection, as seen by the incidence of curbside collection programs implemented throughout the nation. For those communities providing curbside programs for recyclables separated by material type, collection is generally performed using compartmentalized vehicles. Recyclables are taken directly to processors or brought back to a central location for disposition into containers for transport to a processor. Commingled recycling programs, on the other hand, enable residents to mix various recyclable materials in one container. Compartmentalized vehicles are not needed in this scenario because the commingled material is typically brought to a material recovery facility (MRF) for separation and processing prior to sale to a secondary materials market. MRFs sort and densify the individual components of a recyclable materials mix, such as aluminum cans, glass bottles and jars, and plastic beverage containers. In most cases, depot programs and buy-back centers do not offer commingling as an option and require residents to separate their recyclables by individual material type.

Reverse vending machines are also utilized to recover aluminum cans from residential sources and basically function as "unstaffed" buy-back centers. Rather than accepting coins and dispensing items for sale, reverse vending machines accept cans and return cash or an equivalent store credit coupon. Reverse vending machines are usually found in supermarkets and other retail trade establishments, and may be utilized in conjunction with other recovery methods described in this section. Reverse vending machines are also used in bottle bill states.

Commercial/Institutional Collection. The commercial sector can be a large generator of recyclable materials, depending on the number of commercial establishments within the program area and the types and volume of business conducted. In designing a commercial recycling program to recover aluminum cans, larger commercial establishments should consider designating a recycling program coordinator who would be responsible for program design, implementation, and oversight of operations. Determining what type of recyclable materials are generated by the business is the initial step in developing a program. The next step would be to determine the approximate amount of used aluminum that is generated in order to ascertain the resulting volume or quantity of material. Markets should then be contacted to determine how the aluminum cans must be prepared. The market may provide a pickup service or processing equipment for the recovered aluminum cans. Market services should be evaluated early in the planning stages as they may affect municipal collection procedures currently in place for the commercial establishment. The marketing questions listed later in this chapter should be reviewed when selecting a market for the recovered aluminum cans. Based on the materials specification standards of the market, as well as the approximate volume of aluminum generated and the procedure by which the material is going to be collected (i.e., in cafeterias, at central locations throughout the business, etc.), the type and number of collection containers needed can be determined. Furthermore, the type of processing equipment needed can be selected if required by the market and if volumes warrant. It is also important to determine storage space requirements, whether indoor or outdoor, to sufficiently and safely store the aluminum cans for the scheduled pick-up. Resolution of transportation issues, such as whether delivery would be better handled by the market, a private hauler, the municipality, or by the business establishment itself, is critical. Under any scenario, records should be kept of the volume or quantity of aluminum cans collected through the program in order to facilitate accounting procedures and determine program success.

One major component of a successful aluminum can recycling program is the implementation of an effective publicity and education program for the employees. The program should be promoted periodically, utilizing initiatives that would provide maximum motivation.

Manufacturing Entities Collection. Manufacturing entities that have an in-house smelting process would most likely recover aluminum scrap or waste generated by their in-house manufacturing process. Scrap cast-offs or products not meeting specifications generated during the production of their specific commodity may, in certain circumstances, be returned directly to the manufacturing process and thus never enter the solid waste stream.

In general, a limited quantity of UBCs would be generated from this type of manufacturing process. Typical generation would be the result of an on-site food service, cafeteria, beverage vending machines, or brown-bag lunches. However, manufacturing entities collecting UBCs, regardless of their source, would generally collect the UBCs in a manner similar to the recovery process previously described for the commercial and institutional sectors.

Methods to Remove Aluminum from Other Recyclables

There are several methods for removing aluminum from other recyclables when collected in a commingled state. Manual separation is a labor-intensive option, but is utilized in some MRFs. The method entails employees, located along conveyor belt picking lines, performing the physical separation of the commingled recyclable waste stream into its various components. However, mixed recyclable processing systems incorporate a magnetic separator within the processing line for the purpose of removing ferrous materials, thereby making it easier for employees to identify aluminum cans.

Manual separation of UBCs is normally deemed to be too labor-intensive for larger operations. In such cases, mechanical separation methods are often used. One mechanical separation system widely utilized as part of MRF processing systems is the nonferrous separator or eddy current magnet (see Figure 12.7). The most common eddy current systems incorporate the use of opposing magnetic fields as a primary method to separate or divert aluminum from mixed plastic food and beverage containers. When the commingled recyclables traversing the conveyor belt reach the position of the magnetic field, the aluminum, due to its ability to hold an electric charge, is thrown into a catch hopper or conveyor by the magnetic field. The concept is similar to that of holding the positive and negative poles of two magnets within a short distance of each other, thereby creating a noticeable force as the magnets repel each other.

FIGURE 12.7 Eddy current separator (*Source:* Photo Courtesy of Dings Magnetic Group.)

MARKETING

Market Specifications

When UBCs are manufactured into new products, the closed-loop recycling cycle can be considered complete. The most important component of an aluminum can recycling program is the identification, selection, and securing of markets for the recovered UBC material. The method of collection instituted for a recycling program and the form in which the material is sold will depend on market specifications. There are three major types of markets for aluminum cans: brokers, processors, and end-users.

Aluminum scrap brokers are business entities that buy and sell recovered recyclable materials in processed or unprocessed form. In general, brokers do not process materials but merely serve as a middleman between the generator and a processor, or between the processor and an end-user. Thus, brokers purchase, consolidate and resell materials, providing a viable market outlet for many recycling programs.

Processors accept aluminum cans from municipal programs, postindustrial or postconsumer entities, and brokers. Aluminum is also accepted from some MRFs that separate but do not bale or densify the material. In the case of aluminum, the processor may buy loose aluminum cans from a municipality and bale them for sale to an end-user.

End-users are those manufacturers that clean and melt the aluminum into aluminum sheets, ingots, or blocks for reuse in the manufacturing of new cans or other items such as airplane or truck bodies. Many aluminum end-users purchase processed and unprocessed materials directly from municipal programs in close proximity to their facilities.

Prices paid for recyclables can vary between markets, making it important to obtain as much information on markets as possible in order to secure the best deal for each individual recycling program. Other factors to take into consideration when selecting a market include material preparation requirements and market location. The quality of material being sold will be the most significant factor in determining the price paid by potential markets. Since recovered aluminum will become the raw material for the manufacture of new products, clean, uncontaminated material will be the most valuable.

Aluminum markets have material specifications that regulate the extent of contamination allowed in each delivery, as well as the method by which materials will be prepared (Figure 12.8). For example, some markets prohibit aluminum foil and pie pans from being commingled with aluminum cans. Also, where some markets require aluminum to be baled, others may accept flattened, loose material. Material specifications for the markets should be evaluated prior to initiating collection activities in order to ensure that the recycling program and operating scenario selected for implementation will prepare materials in a manner acceptable to the intended market. As previously mentioned, market location should also be taken into consideration when deciding which markets to utilize. A market which is in close proximity and requires minimal processing is usually of greatest interest to small-scale municipal programs. Costs usually associated with the marketing of aluminum cans include labor and energy for material preparation as well as vehicle operation and maintenance costs for material transportation.

Marketing services, offered by some aluminum end-users, should also be considered, and may include the provision of storage containers, processing equipment, and/or pickup services. A market may also be willing to assist in fostering an aluminum can recycling program by providing public relations services and assistance.

In many instances, municipalities have made arrangements with the aluminum recycling industry for the use of can flatteners and blowers to assist in reducing storage and shipping costs. Furthermore, some municipalities have established agreements whereby the market assumes responsibility for the transport of materials from the municipality to

LOOSE FLATTENED UBC*

UBC must be flattened using commercial flatteners and not compressed by other means.
Flatteners must be equipped with magnetic separators.

BALED UBC

12 to 17 lbs. per cu. foot for unflattened UBC scrap.
12 to 20 lbs. per cu. foot for flattened UBC scrap.
Bale must be dense enough to permit movement by fork lift.
Bales should be of uniform size.
Bale Size:
- Minimum of 30 cu. ft. with minimum dimension of 24″ in one direction and a maximum of 72″ dimension in another.
- Preferred bale size is 3' × 4' × 5' or 60 cu. ft.
- Bales of two or more individual bales bonded together to meet preferred bale size specifications are not acceptable.

Banding:
- 2.5 lbs. per bale deductor.
- Four to six 5/8″ × .020 steel or aluminum bands.
- Six to fifteen #13 gauge steel or aluminum wired.
- Not Acceptable: bands or wire of other material; and use of support sheets of any material.

SPECIFICATIONS FOR LOOSE FLATTENED OR BALED UBC

Moisture not to exceed 1%.
Material to be stored indoors.
Any non-UBC material will be subject to deduction.
Shipment received meeting permissable moisture level and not otherwise contaminated, will be accepted.
Receiving facility has the option of accepting or rejecting a load.
If material does not meet the specifications as detailed above, the vendor will contact the deliverer prior to processing and review all deductions to be applied.
The prevailing weight is determined by the receiving facility.
Materials not covered in these specifications are subject to special arrangements between the buyer and seller.

BRIQUETTES

Density: 35 to 45 lbs. per cu. foot
Size:
- 10 ¾″ × 10 ¾″ × 7 ¾″
- 13 ¼″ × 20 ¼″ × 7″
- 13 ½″ × 13 ½″ × 6 ½″
- 14″ × 10 ¾″ × 7″

Banding:
- 7 lb. per bundle deductor will be taken for banding.
- Banding slots in both directions to facilitate bundle handling.
- 5/8″ × .020 steel straps minimum
- One band per row.
- Minimum two horizontal bands per bundle.

Bundle Specifications:
- All briquettes comprising a bundle must be of uniform size.
- Bundle sizes: 41″ to 44″ × 51″ to 54″ × 54″ to 56″ (L×W×H).

Quality Specifications: See bale specifications.
General: Items not covered in this specification are subject to special arrangement between buyer and seller.
*Used beverage container.

FIGURE 12.8 Aluminum can market specifications.

the processing location. A good marketing plan will include markets that require minimum transport and labor costs. However, a higher market price may compensate for increased transport costs, especially when utilizing a regional marketing strategy.

The willingness of the market to provide a municipality with processing equipment, storage containers, and/or pickup service will usually depend on the ability of the program to recover large quantities of clean, uncontaminated aluminum cans on a regular basis. A consistent supply of UBCs that continually meet market specifications will minimize marketing problems.

Potential revenue is understandably a major factor when selecting a market. Prices paid for UBCs vary depending on many of the previously mentioned factors. However, revenues generally increase in accordance with increased levels of material processing, returning to the concept of more money paid for a better product. Prices paid for aluminum cans are based upon preparation levels and can be quoted in a variety of ways. For example, UBCs may be whole and loose, flattened, densified, or shredded (uncommon).

When contacting markets, answers to the following questions should be obtained:

- What types of aluminum are currently purchased (e.g., UBCs, scrap aluminum siding, food trays, etc.)?
- How should the recovered aluminum cans be prepared for sale (material specifications)?
- What minimum (or maximum) quantities will the market accept?
- Is the market willing to provide storage containers, processing equipment, and/or a pick-up service?
- What are the hours that the market site is open for delivery of materials?
- What is the current price being paid for each specified grade?
- How is payment made?
- Is a material purchase contract optional or required?

Material Preparation for Marketing

As previously mentioned, aluminum cans recovered through municipal recycling programs can be flattened and blown into a trailer or collected loose in bulk and delivered to a processor. In turn, the processor will flatten and/or bale the aluminum cans for sale to an end-user who smelts the material into aluminum sheets or ingots for use in the manufacture of new aluminum cans or other aluminum products. Aluminum cans collected for recycling by commercial and industrial establishments are often delivered to processors in baled form. In some instances, processors and brokers provide storage containers and pick-up services to recover aluminum cans from the commercial sector. However, commercial programs are similar to residential programs in that if either service is provided by the processor, a minimum tonnage must generally be guaranteed by the commercial establishment in order for the processor to provide special services. Sample material specifications are presented in Figure 12.8.

End-Use Market

UBCs are commonly utilized by sheet manufacturers because the aluminum alloy used to make beverage cans is consistent among most can manufacturers and can be smelted into the proper concentration required for the manufacture of new aluminum sheet. Approximately 95% of the UBCs recovered from the municipal solid waste stream are utilized directly by sheet manufacturers who produce new aluminum sheet for cans. A small amount of all UBCs are purchased by secondary smelters for the production of ingots. Secondary smelters sell the majority of their aluminum to foundries, which can only tolerate a minimal amount of magnesium, an element used in the manufacture of aluminum cans. Therefore, the demand for UBC by secondary smelters is minimal.

A small percentage of UBCs not utilized by secondary smelters or sheet manufacturers is exported. Exportation is minimal because of the overwhelming demand for the material by U.S. manufacturing entities. Manufacture of aluminum sheet is also the dominant UBC market overseas.

CASE STUDIES

The purpose of this section is to provide a description of different types of recycling programs that recover aluminum UBCs. Program issues of relevance include the amount of aluminum recovered, revenue generated by the municipalities from the sale of the aluminum cans, avoided tipping fee benefits, and program operating costs. It is important to note that although the avoided tipping fee for municipalities recovering only aluminum UBCs may not amount to a substantial savings when compared to avoided tipping fees associated with newspaper, yard waste, or glass recovery programs, there are benefits that often outweigh the savings from avoided tipping fees alone. These include: 1) the energy conservation experienced in the manufacturing process, 2) the volume of landfill space conservation encountered through eliminating UBCs from the waste stream, and 3) the generation of revenue from the sale of aluminum.

In cases where municipalities collect multiple categories of recyclable materials through the same collection program, it is difficult to ascertain the costs associated with the collection, processing, and marketing of the aluminum can fraction only. However, an estimate of the program costs for aluminum can recycling programs has been estimated for the recycling programs described next. Please keep in mind that these estimates are based on the total program budget, the total amount of recyclables collected, the revenue generated through the sale of the aluminum cans, and the avoided tipping fee costs for each program. Therefore, the operating costs as described in the case study summary are to be utilized for comparative and informational purposes only.

Curb-Sorted Collection Programs

Township of Mahwah, New Jersey. The Township of Mahwah has a residential population of approximately 22,000 and a daytime population of almost 40,000. The Township has had a mandatory recycling program in effect since September 1987. Biweekly curbside collection of recyclables is provided to residents, augmented by a depot facility, located at the Department of Public Works, which also provides service to commercial establishments. The program requires that the material to be recycled be source separated by the residents; commingling of recyclables is not permitted. The Township provides its own transportation of the material to the markets by means of a roll-off vehicle. The Township's Recycling Division consists of seven full-time employees, assigned daily to three crews, and one roll-off operator. Completely separated material has made it possible for the program to receive a steady stream of revenue in each of the years since 1987.

Materials collected at the curb include newspaper, magazines, cardboard, glass (separated by color), aluminum cans, tin cans, and plastic containers. These materials are also accepted at the depot, along with used motor oil, car batteries, car and truck tires, scrap steel, and mixed paper.

The Division has several vehicles that are specialized for recycling programs (Figure 12.9). For curbside collection there are two Ford 600 trucks that pull two Eager Beaver compartmentalized trailers, and one Ford 350 truck for newspaper collection. An Autocar

FIGURE 12.9 Township of Mahwah, New Jersey: recycling vehicle.

roll-off and containers are used to transport material as well as for curbside collection of leaves. Also for leaf collection are two grip claw attachments for the front end loaders. There is a tubgrinder that is used for all aspects of the waste vegetation program. At the depot facility there is a can crusher for processing aluminum cans as well as magnetically separating any tin and ferrous metal cans.

In 1997, the program collected approximately 20 tons of aluminum cans, generating $15,000 of revenue, compared to 57.5 tons of tin cans, generating approximately $2,000 in revenue.

Commingled Collection Program

Borough of Park Ridge, New Jersey. The Borough of Park Ridge, New Jersey has a resident population of approximately 8100 persons. Park Ridge is composed of approximately 2500 single family residences, approximately 300 condominium/apartment units, and approximately 75 commercial establishments within 2.6 square miles. During 1997, Park Ridge generated approximately 6500 tons of solid waste. Of the total solid waste generated within the municipality, approximately 61% was attributed to the residential sector, and the remaining 39% (2566 tons) was generated by the commercial sector.

In March 1988, when it initiated its recycling program, the Borough hired three men and bought two used garbage trucks for the curbside collections of mixed paper, magazines, and cardboard, and commingled containers, which included glass, aluminum and steel cans, and #1 and #2 plastics. The Borough alternated biweekly collections. In 1991, the Borough privatized its collection of household garbage, which included the collection of recyclables. As a result, a private hauler currently provides collection of all recyclables on a biweekly basis and weekly grass pickup between April and September. The program has basically remained the same since 1991, except that in 1997, aseptic containers, styrofoam, aluminum foil and types 3–7 plastics were added to the commingled mix. The Borough also operates a recycling center that accepts all the materials collected curbside, in addition to tires, waste oil, oil filters, textiles, scrap metal, appliances, brush, concrete, asphalt, automotive batteries, and household batteries (Figure 12.10).

The Borough was able to process 30.9 tons of aluminum cans for 1997. The tipping fees cost the Borough $101/ton for household garbage and $117/ton for trash before the fees were struck down by the State Supreme Court. The current rate is $54/ton. This current cost translates to a savings of $2865 in tipping fees for the months of January to November and $140 for December, due to the recycled aluminum cans. It should also be noted that in order to pay the debt for its transfer station, the county is charging $30/ton to each municipality for all waste going through the facility. By reducing waste and expanding Park Ridge's recycling program to include more materials, it is limiting waste going through the transfer station. Based on the 30.9 tons of aluminum cans recovered during 1997 and the county's $30/ton fee, the Borough saved an additional $927 by recycling aluminum cans. In total, for 1997, the cost savings for the Borough of Park Ridge of recycling aluminum cans was approximately $3932.

Buy-Back Center Program

City of Sunnyvale, California. The City of Sunnyvale (population 120,000) utilizes a buy-back center to divert its recyclables. This technique is a supplement to the main diversion techniques, which are curbside pickup of recyclables and yard waste, processing

FIGURE 12.10 Borough of Park Ridge, New Jersey: depot center.

mixed waste to recover recyclables from refuse, and facilitating private recyclers' collection of recyclables from commercial and industrial generators. The City recycles old newspaper (ONP), old corrugated cardboard (OCC), mixed paper (MXD), white paper, scrap steel, aluminum, glass, tin, PET and HDPE plastics, and phone books. Residents who bring materials to the buy-back center receive money for glass, PET plastic, and aluminum. For the fiscal year 1997/1998, 126.47 tons of aluminum cans were recycled at the buy-back center. Since the tipping fee to dispose of municipal solid waste is $42.44 per ton, the amount the City saved by diverting this waste from a landfill is approximately $5367. The City also saved approximately $26,975 by diverting glass and PET from the landfill. Since the amount reimbursed to residents was only $10,397, the City saved approximately $21,945 through the buy-back center in the fiscal year 1997/1998.

Depot Collection Programs

Village of Ridgewood, New Jersey. The Village of Ridgewood, New Jersey (population 24,152), generated approximately 9900 tons of municipal solid waste during 1997, excluding recyclables. Of this amount, approximately 83% was generated in the residential sector and 17% in the commercial sector. The Village encompasses approximately 5.797 square miles, with 8619 households and 307 commercial establishments. As far back as the early 1970s, the Village's recycling activities consisted of a drop-off depot located in a convenient part of the municipality. The facility grew to accept aluminum, glass, plastic, steel cans, corrugated cardboard, high-grade paper, newspaper, textiles, and light iron, in multiple 12-by-18-foot concrete bins and roll off and boxed containers at the depot. The total cost for the original five concrete bins was $12,000. An additional bin was added at a cost of $9000. The depot is open Monday to Saturday from 8:00 a.m. to 3:00 p.m. and Tuesday from 8:00 a.m. to 8:00 p.m.; it is closed Sundays and holidays. The depot is staffed by at least one person at all times. In January 1991, the Village implemented a curbside recycling program to enhance its recycling efforts. Presently, the Recycling Department consists of one recycling coordinator/enforcement officer and eight laborers.

In 1997, the Village received a range of approximately –$15 to +$1000 per ton for the nearly 5664 tons of recyclables collected, generating approximately $92,000 in revenues. The price paid for the material varied slightly each month. The Village saved approximately $540,022 in tipping fees alone for the diverted recyclables, making the total cost savings of the recycling program approximately $632,022. In addition, a recycling grant of $148,221.67 from the Bergen County Utilities Authority was received by the Village. The total cost of the program for 1997 was approximately $431,000. Ninety-six percent of the costs were attributed to salaries; operation and maintenance of the depot and supplies for curbside collections constituted approximately 2.4% of the total cost, and 1.6% was due to the tip fee charged by various vendors.

As a supplement to curbside collection, the depot now accepts commingled aluminum and steel cans and glass and plastic bottles, corrugated cardboard, mixed-grade papers, newspapers, textiles, light iron, household and lead–acid batteries, tires, and there is a pilot program for concrete.

City of Hollywood, Florida. The City of Hollywood, Florida covers approximately 55 square miles and consists of 30,000 single family and duplex homes. Until 1996, the City had a recycling program that enabled residents to visit any of the mini-depot centers and other drop-off centers located throughout the community in order to deposit their glass and aluminum containers. In 1996, the Environmental Services Division implemented a citywide curbside, commingled recycling program. City crews service the western portion of Hollywood and a private company collects recyclables from the remaining areas of the City. Due to the implementation of curbside recycling, the drop-off program was reevaluated and most stations were deactivated. More than 100 drop-off centers were removed; some were relocated to the Orangebrook Golf Course and some were used as replacements on the beach. Some drop-off stations were recycled to other communities. In addition, approximately eight drop-off sites remain open in parks throughout the City.

Since the implementation of the program, resident participation dramatically increased, with recyclable tonnage up 60% for 1996. In fiscal year 1997, Hollywood, Florida crews picked up more than 80 million pounds of garbage from customers and recycled almost 16.5 million pounds of household materials and 24 million pounds of yard waste. The Environmental Services Division recycles newspapers; food and beverage containers made from plastics, metal, and glass; brush; scrap wood and metal; old appliances; furniture; junk mail; catalogs; cardboard, and magazines. The Division also promotes waste reduction techniques such as backyard composting of leaves, grass clippings and fruit and vegetable scraps. Residents are also encouraged to leave grass clippings on the lawn rather than bagging and disposing of them. The Division mails out its newsletter, *Reusable News*, on a quarterly basis to keep residents informed of collection procedures and recycling tips.

With the development of this new program, the City was able to reduce the amount of waste going to the county incinerator by 1600 tons and reach more customers with better service. The curbside recycling program was effective, with the quantity of recyclables increasing greatly.

The City is able to process 3000 tpy of commingled mixed recyclables and 6000 tpy of newspaper. The county pays for the tipping fees, which are $24/ton. This cost translates to $72,000 for commingled mix and $144,000 for newspaper. The county has a contract with BFI to process the recyclables. The out-of-pocket costs for the City are $80,000 for the area for which the City provides collection and $300,000 for the privatized collection areas. The City received $350,000 in revenue from the sale of the recycled material [9].

Case Study Summary

Table 12.1 provides a cost comparison of the various case study program scenarios described in the previous sections. As previously stated, annual UBC program costs are estimated. Annual costs were determined based on the total tons of recyclables collected and the total program budget for that year, resulting in an approximate cost per ton for all recyclables collected through the municipality's recycling program. The cost per ton figure was then applied to the total tons of UBC collected to determine the estimated UBC recycling program cost.

When making program comparisons, please keep in mind the variables that effect program costs in each municipality. These variables include:

TABLE 12.1 Case study summary

Project costs					
Municipality	Total program budget	Tons of recyclables collected	Estimated total recycling program operating cost/ton	Tons of UBC collected	Estimated UBC recycling program cost
Mahwah	$338,820	2696	$126	20.0	$2514
Park Ridge	$90,000	4645	$19	30.9	$599
Sunnyvale	Not available	525	—	126.5	—
Ridgewood*	$431,000	5664	$76	39.6	$3017
Hollywood*	$380,000	9000	$42	63.0	$2660

Project savings/revenue			
Municipality	UBC avoided tipping fee savings	UBC material revenue	Total UBC program savings & revenue
Mahwah	$1080	$15,000	$16,080
Park Ridge	$3932	$0	$3932
Sunnyvale	$5367	Not available	—
Ridgewood	$3780	$644	$4424
Hollywood	$1512	$2450	$3962

*Tons of UBC collected for Ridgewood and Hollywood were calculated using 0.7% by weight as the average percent UBC collected of the total recycled for the other municipalities. UBC savings and revenues for Ridgewood and Hollywood were calculated as 0.7% of each municipalities total savings and revenues.

- Type of collection program implemented
- Population of a given area
- Publicity and education efforts
- Program enforcement efforts
- Age of program and number of materials collected
- Market utilized (revenue)
- Availability of grants/loans
- Utilization of nonprofit groups for certain program tasks

TRENDS AND FORECASTS

Demand

The United States is the primary producer of aluminum and is also the world's largest market for aluminum products. The United States consumes approximately one-quarter of the world's primary aluminum alloy. In the United States, the aluminum can and packaging markets are the second largest consumers of aluminum production. The demand for aluminum has been increasing as new uses for the metal are developed. This trend is expected to continue. For example, the automobile industry is now the largest market for aluminum, due to its expanded use in the construction of body frames and engines to increase the fuel efficiency of vehicles. In addition, it is anticipated that the demand for aluminum beverage containers will remain strong, as aluminum is viewed as a cost-effective packaging material. However, plastics continue to compete with aluminum in the container market. The aluminum industry is also concerned with competition from aluminum producers overseas. New primary aluminum production capacity continues to be devel-

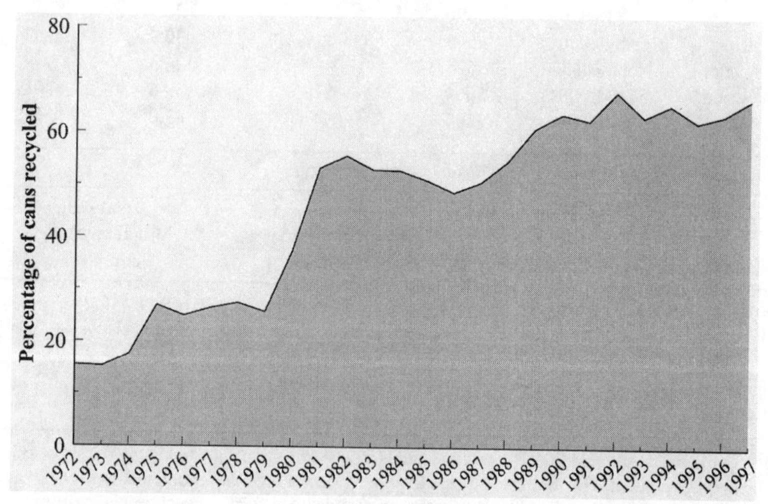

FIGURE 12.11 U.S. aluminum can recycling. (*Source:* The Aluminum Association, Can Manufacturers Institute, Institute of Scrap Recycling Industries.)

oped in countries that possess lower electrical costs. As a result, it is expected that U.S. aluminum producers will be placing an increased emphasis on aluminum can recycling. This trend will continue to expand UBC recycling efforts, which have been on the rise since 1972 (Figure 12.11).

The aluminum industry's initiatives, however, do not end with recycling. The aluminum industry has also fostered source reduction activities to assist in minimizing the impact of UBCs on diminishing landfill space and to decrease manufacturing costs. As indicated in Figure 12.12, the industry has developed methods to produce more cans per pound of aluminum. In 1975, aluminum manufacturers produced 23 cans per pound of aluminum, compared to 33 cans per pound in 1997 [10].

Price

As indicated by the trend for most recyclable materials in the 1990s, the price paid for aluminum cans appears to be declining. The street price paid for aluminum cans dropped from approximately 59¢/lb in July 1997 to 46¢/lb in July 1998 [11]. Although there has been a downward shift in the price paid for aluminum UBCs, resulting in less revenue generated by municipalities for the sale of recovered UBCs, it is still a high-revenue-generating material when compared to other types of recyclables.

Figure 12.13 graphically presents the average revenue generated by material type. In parts of the U.S. where solid waste disposal costs have skyrocketed, the avoided tipping fee costs for recovered recyclable materials has more of a financial impact on communities than does the revenue generated from the sale of those materials. Therefore, a fluctuation in the price paid for UBCs in those areas would not effect recycling programs to the same degree as it would in communities where tipping fee costs are still minimal. Programs more apt to be effected by fluctuations in price include not-for-profit organizations that provide recycling program services in exchange for the revenue generated from the

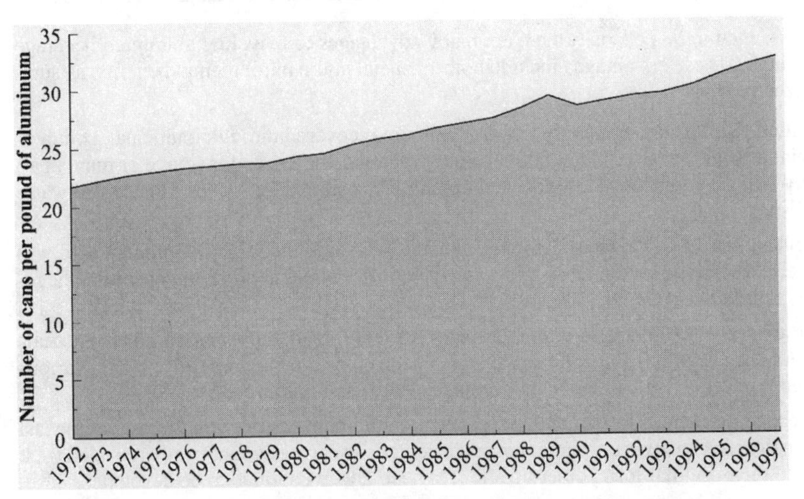

FIGURE 12.12 Aluminum can weight and source reduction. (*Source:* The Aluminum Association, Can Manufacturers Institute, Institute of Scrap Recycling Industries.)

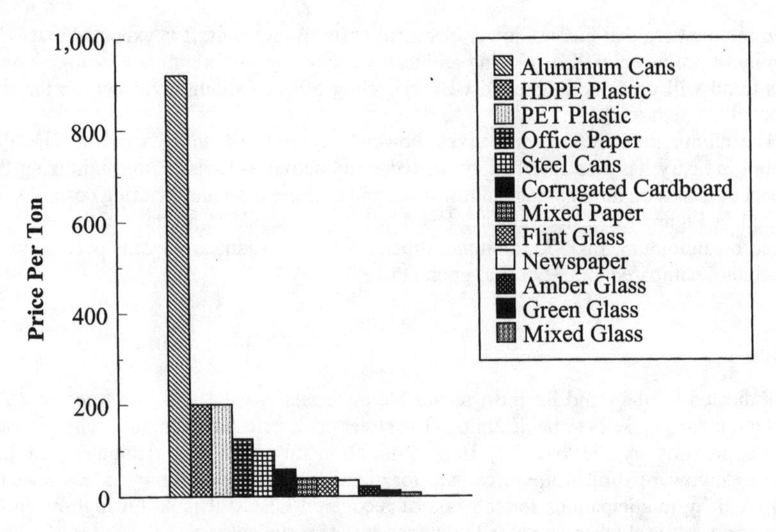

FIGURE 12.13 Average revenue generated per material. (*Source:* On-Line Market Prices, *Recycler's World*, August 18, 1998 and *Recycling Manager Market Summaries*, Chilton Media Inc., July 27, 1998.)

sale of the recyclable materials, and those programs that donate material revenues to charity or nonprofit groups.

CONCLUSION

As presented in this chapter, there are many advantages to recycling aluminum beverage containers. These advantages include both financial and environmental benefits, as summarized below.

Avoided tipping fee. For every ton of aluminum recovered through municipal, commercial, institutional, or industrial recycling programs, the generator source or municipality will save or avoid the associated tipping fee costs charged at their local solid waste disposal facility.

Potential source of revenue. Revenue generated through the sale of aluminum cans will help to offset a portion of the costs associated with implementing and operating the recycling program.

Ease of recovery. Because aluminum cans are easy to identify among other materials found in the municipal solid waste stream, and since they are compatible with various types of recycling collection programs, they are easily recovered.

Increased demand. The demand for secondary aluminum has been on the rise for the last ten years. Although aluminum beverage containers can be economically recycled back into their original form, other uses for recycled UBCs continue to be developed.

Energy savings. An energy savings of 95% is associated with the production of new aluminum beverage containers from UBCs rather than from virgin materials. Thus, the recovery of UBCs makes economic sense for aluminum manufacturers because reducing energy consumption lowers operating costs and conserves petrochemical resources.

Public, private, or nonprofit organizations instituting a new recycling program should consider the recovery of aluminum cans. Aluminum has traditionally been one of the most stable materials to recycle in terms of markets, revenues, and program participation. The added advantages of energy savings and avoided disposal fees combine to make aluminum cans one of the most attractive components of any recycling program.

ADDITIONAL INFORMATION SOURCES

Alcan Aluminum Corporation, 6060 Parkland Blvd., Cleveland, Ohio 44124-4185. (440) 423-6600

Alcoa Recycling, 10 County Line Rd., Somerville, NJ 08876-6007. (908) 722-4440

Aluminum Association, 900 19th St., NW # 300. Washington, DC 20006-2168. (202) 862-510

Anheuser Busch Companies Inc., One Busch Pl., St. Louis, MO 63118-1852. (314) 577-2000

Can Manufacturers Institute, 1625 Massachusetts Ave., NW #500, Washington, DC 20036-2245. (202) 232-4677

The Council for Solid Waste Solutions, 1275 K Street, NW, Suite 400, Washington, DC 20005. (202) 371-5319

Institute of Scrap Recycling Industries, 1325 G St. NW, # 1000, Washington, DC 20005-3104. (202) 737-1770

Keep America Beautiful Inc., 1010 Washington Blvd., Stamford, CT 06901-2202. (203) 323-8987

The National Recycling Coalition Inc., 1727 King Street, Suite 105, Alexandria, VA 22314-2720. 683-9025

National Solid Waste Management Association, 4301 Connecticut Ave. N.W., # 300, Washington, DC 20008. 244-4700

Reynolds Aluminum Recycling Co., 404 Stagecoach Rd, Bristol, VA 24201-8359. 669-5109

Solid Waste Association of North America, P.O. Box 7219, Silver Spring, MD 20910-7219. (301) 585-2898

United States Environmental Protection Agency, Office of Solid Waste, 401 M. Street, SW, Washington, DC 20460. (800) 424-9346

ACKNOWLEDGMENTS

Kathleen Scully, an engineer at Malcolm Pirnie, contributed greatly to the compilation of updated information and analysis of data contained in this chapter.

REFERENCES

1. *United States Industry—A Quick Review.* The Aluminum Association. Industry Profile. 1997.
2. United States Environmental Protection Agency. *Characterization of Municipal Solid Waste in the United States: 1996 Update.* 1997.
3. *Aluminum Can Reclamation.* The Aluminum Association. April 1998.
4. *Recycling: The State of the Art.* Malcolm Pirnie Technical Publication. 1988.
5. *Aluminum Recycling: America's Environmental Success Story.* Aluminum Association. 1989.
6. *Why Comprehensive Recycling is More Effective than Beverage Container Deposits.* Glass Packaging Institute.
7. "The State of Garbage in America." *Biocycle.* April 1998.
8. *Beyond 25 Percent: Material Recovery Comes of Age.* Institute for Local Self-Reliance. April 1989.
9. *Environmental Services Division.* Hollywood Department of Public Works Web page. July 1998.
10. *The Aluminum Can.* The Aluminum Association. February 1998.
11. "U.S. Recyclable Commodity Prices." *Recycling Manager.* July 1998.

CHAPTER 13
GLASS BEVERAGE BOTTLES

TAMMY L. HAYES
Senior Business Development Specialist
Camp Dresser and McKee Inc.
Tampa, Florida

ANTHONY LAME*
President
GR Technology, Inc.
Haverford, Pennsylvania

RICHARD LEHMAN*
Technical Director
GR Technology, Inc.
Haverford, Pennsylvania

INTRODUCTION

Less than a generation ago, jars and bottles were made only of glass. Over the last 20 years, high-density polyethylene (HDPE) and polyethylene terephthalate (PET) plastics have been used to make food and beverage containers. During the past 10 years, laminated paper materials and foil have also been combined to containerize foods and beverages. Still, the glass industry estimates that every person in the United States throws away approximately 85 lb of glass each year, and 7 billion glass containers are recovered and returned for remanufacture annually (Fig. 13.1).

Container glass is the glass that is used to make jars and bottles. It is the glass in soft drink bottles, beer bottles, mayonnaise and pickle jars, baby food jars, wine and liquor bottles, and many other containerized foods and beverages.

Container glass is the only glass that is being recycled in large quantities at the present time. Window panes, light bulbs, mirrors, ceramic dishes and pots, glassware, crystal, ovenware, and fiberglass are not recyclable with container glass and are considered contaminants in container glass recycling.

The common glass jar or bottle is unique in the recyclables manufacturing industry. One 12-ounce glass bottle, melted down and reformed, yields one 12-ounce bottle with-

*Color separation section.

FIGURE 13.1 Variety of food and beverage containers.

out any loss of quality. No waste or by-products are generated in the remanufacturing process, and the same glass can repeatedly make and remake one 12-ounce bottle. This trait makes glass one of the few manufactured goods that is 100% recyclable (Fig. 13.2).

Container glass is common in everyday use, yet it has unique properties that make it a special recyclable. For example, glass is made from common inert raw materials including white silica sand, soda ash, and limestone. Slag, salt cake, feldspar, aragonite, and cullet (crushed glass) are other ingredients typically used to manufacture glass containers. These raw and secondary materials are not in short supply; they are plentiful and easily obtainable.

The most unique or special consideration in marketing container glass is the need for color separation. Permanent dyes are used to make different-colored glass containers. The most common colorings are green, brown, and clear (or colorless). In the industry, green glass is called emerald, brown glass is amber, and clear glass is called flint. In order for bottles and jars to meet strict manufacturing specifications, only emerald or amber cullet can be used to make green and brown bottles, respectively.

Glass itself is not a threat to the environment because it is inert; it is not biodegradable. If exposed to weathering forces, glass breaks down into small particles of silica, basic beach sand, which is one of the most common elements on earth.

While only container glass is used to remake glass containers, glass cullet can be used in other manufacturing processes and industrial applications. For example, crushed and broken glass can be part of the aggregate used in bituminous road paving—we know this product as *Glasphalt*. Other uses of cullet as an aggregate substitute are discussed later in this chapter. Examples of glass reuse range from glass wool insulation and fiberglass to telephone poles and fence posts made from glass cullet and plastic polymer mixtures. These represent only a few of the newer markets that have been developed for cullet in recent years.

FIGURE 13.2 Remanufactured glass containers.

Using recycled container glass as cullet to make new glass container products conserves energy and reduces glass manufacturing costs. Energy is conserved because cullet melts down at a lower temperature than that required to combine the raw materials that go into making glass. This not only reduces energy costs, but increases furnace life as well. Depending on the amount of cullet being used, furnace life can be extended by as much as 15 to 20%. The conservation of energy, in turn, conserves natural resources such as our depleting supply of fossil fuels. For every 1% increase in the use of cullet, 0.25% of the energy needed is saved. In more practical terms, 9 gal of fuel oil is saved for each ton of glass that is recycled. Energy reduction and the extension of furnace life enable glass manufacturing plants to run more efficiently, thus reducing overall costs.

The recycling of glass containers has more impact on enhancing a solid waste recycling program than it does on reducing waste collection and disposal requirements. Glass containers represent approximately 2% of the solid waste volume. While every little bit counts, other wastes such as paper and yard waste comprise greater portions of the total waste volume. On the other hand, glass containers represent 7 to 8% of the weight of total solid wastes. Thus, the reduction of glass from the waste disposal system can be a significant contributor toward meeting recycling and landfill avoidance goals that are typically measured as a percentage of total weight.

Some communities use waste-to-energy or resource recovery facilities (plants that produce energy from the combustion of wastes) to reduce the volume of solid waste prior to its final disposal. The removal of glass from these waste streams is beneficial because plant maintenance is reduced, and the overall efficiency of the plant's operation is increased due to the removal of the noncombustible glass containers.

In general, glass recovery processes based on hand-picking or screening are effective in removing container glass from the disposable waste stream. Once recovered, glass containers are storable, transportable, and processible as a future feedstock to glass remanufacturing and other industrial processes.

It is estimated that all glass containers are manufactured using some amount of glass cullet. The percentage of cullet being used varies among manufacturers, but it is generally considered to be increasing. Overall industry averages indicate that approximately 25 to 35% of raw material needs are currently being supplied by cullet. The glass manufacturing industry expects to increase this cullet usage to 50%.

Most importantly, the continued recovery and recycling of glass containers is evidence of the stability of one industry to produce a desirable consumer product—the glass container—in a form that is totally recyclable as a remanufactured glass container. Thus, the glass container can be removed from the postconsumer waste stream and returned as usable feedstock to the glass remanufacturing process.

GLASS CONTAINER RECOVERY

For years, the glass container was a reusable product that was returned to the bottler or food packer for washing and refilling. Familiar examples of this recovery process are the returnable glass milk bottle, returnable soft drink and beer bottles, and prepared jars of food stuffs such as "canned" vegetables, fruits, and jams. Foods and beverages that were not packed in jars and bottles were packed in tin cans.

Traditionally, cullet was the glass recovered from breakage or rejects in the manufacturing process or in the washing and bottling processes. The age of "no deposit—no return" glass containers and other forms of "new and improved" food and beverage packing (e.g., aluminum and plastic containers) sent the majority of glass containers into the disposable waste stream.

Changes in postconsumer glass disposal have come about along with changes in solid waste collection practices. In general, glass containers that are recovered and returned for remanufacturing (Fig. 13.3) are the result of materials recovery practices that

- Recover glass containers in response to local bottle bills that prohibit landfill disposal and provide for the payment of container deposit money.
- Recover glass containers at decentralized collection depots for separated recyclables.
- Recover glass containers that have been separated from curbside refuse.
- Recover glass containers from commercial sources of food and beverage products (e.g., bars and restaurants).
- Recover glass containers from loads of mixed recyclables, typically including paper, glass, aluminum, and plastic materials.
- Recover glass containers from solid waste processing plants.
- Recover in-plant breakage and rejects in the glass container manufacturing process and in the food and beverage packaging industry.

Bottle Bill Glass Recovery

Beverage container deposit legislation, commonly referred to as a "bottle bill," is usually a state or local government law enacted to impose monetary deposits on all beverage containers (not only the glass ones). The imposed deposits are refunded to persons returning beverage containers. Accompanying these laws are usually restrictions on the disposal of beverage containers. (More discussion about the ongoing bottle bill debate is provided later in this chapter.)

In most systems, beverage retailers are required to act as a container drop-off depot

FIGURE 13.3 Recovered green glass bottles.

because a monetary fund is paid for each returned beverage container including glass, aluminum, bimetal, and plastic beverage containers. However, only beverage containers and not other food packages (e.g., glass jars, plastic jars, tin cans, etc.) are subject to the monetary deposit and refund.

Various systems exist to return the beverage containers to recycling markets such as the bottling industry, glass container manufacturers, and plastic bottle users and manufacturers. The local retailer or beverage wholesaler acts as the receiver of the postconsumer beverage containers and the refunder of the deposit money.

As discussed later in this chapter, bottle bill legislation has been in effect for more than 20 years (Oregon was the first state to enact a bottle bill in 1971). Currently, only nine states use bottle bill legislation to recover beverage containers.

Drop-off Centers

Glass containers are frequently recovered from drop-off centers that collect a variety of source-separated recyclables (e.g., paper, aluminum, plastic, and glass). Users of these facilities are mainly individuals participating in voluntary programs, but such facilities are incorporated into many types of voluntary and mandatory recycling programs. Glass containers either arrive already separated, or they are easily separated upon receipt. Glass containers are usually stored in bunkers according to color (green, brown, clear, and mixed). The glass may or may not be processed on-site for shipment to market outlets. When processing is involved, it typically consists of

- Volume reduction by breaking or crushing
- Cleaning by screening to remove metal neck rings, paper labels, and foreign debris
- Containerizing by color in gaylord boxes, drums, or truck beds for bulk delivery

Curbside Separation and Collection

Glass containers are collected from the residential solid waste stream on a large-scale basis through curbside collection systems. Residents are typically asked to separate specific recyclables from the rest of the refuse set out. Thus, recyclables are typically segregated by type or in mixtures that can be further sorted at the curb by the collector or later at a separation and processing facility. These types of source-separation curbside collection systems appeal to citizenry due to their relative convenience. However, public education programs must be intensive and specific to encourage voluntary participation and to educate residents about cleaning and color-separation requirements prior to curbside setout.

After the curbside collection, the mixed or separated recyclables are stored for processing and bulk shipment to prearranged markets.

Recovery from Primary Commercial Sources and Multifamily Residences

Primary commercial sources of glass containers are restaurants, taverns, and other select public places (e.g., schools, recreation areas, and hotels) where considerable quantities of food and beverages are consumed, leaving empty glass containers for recycling or disposal. Experience shows that greater quantities of glass containers are recoverable from commercial sources than from the residential sector.

Since commercial sources were first approached as possible participants in large-scale recycling programs, various systems have been developed to increase the convenience, efficiency, and sanitation aspects of storing large quantities of empty, rinsed, or nonrinsed glass containers. For example, behind-the-bar glass bottle crushers have been installed to keep glass bottles separate from kitchen refuse. Services that provide daily or near-daily pickup of glass recyclables are the most effective in minimizing potential sanitation nuisances and reducing storage needs for empty containers.

Another segment of the commercial waste stream that produces glass containers is the multifamily dwelling (apartments and condominiums). While these are technically residential dwelling units, municipalities frequently defer refuse and recyclable collections to the commercial haulers.

Multistory apartment complexes having interior refuse chutes and low-rise complexes having parking lot refuse dumpsters require special collection considerations. In these areas, collection practices for recyclable materials have been developing much slower.

In communities where multifamily housing styles represent significant sources of recyclables, innovative concepts are beginning to be implemented to recover select recyclable materials. For example, user-based systems that rely on residents to containerize recyclables in special bags are being demonstrated. Bags of recyclables are deposited on each floor or gathered in the basements of each complex. These methods have been reported as cumbersome, inefficient, and resulting in low participation and recovery rates.

For collection systems using parking lot dumpsters, compartmentalized dumpsters have achieved some success in recovering recyclables. Like refuse collection, these systems require considerable equipment maintenance and supervisory control.

An example of a new equipment line has been implemented in Miami, Florida, to serve high-rise complexes using interior refuse chutes. The new equipment consists of an electric carousel bin arrangement located in the basement of each complex. On each floor, residents use a remote selection device to indicate which type of recyclable or refuse will be entering the chute. The rotary bin system responds to an electronic message and rotates the correct receiving bin into the proper position before signaling for the chute delivery. Staff labor services the collection bins.

Recovery from Mixed Recyclables

In response to user claims of inefficiency, inconvenience, and costliness, service vendors have begun to look at collection services for recyclables and refuse from commercial sources as a wet-dry issue. One result is a collection system that mixes dry recyclables (e.g., paper, aluminum cans, glass containers, and plastic) for collection and later separation. Regular "wet" refuse (e.g., food, soiled paper, diapers, fruit peelings, etc.) is collected in separate bags and handled as garbage.

Glass containers are usually recovered whole or slightly broken at the commercial collection facility. Hand-picking or screening is performed to recover glass from the mixed recyclables.

Recovery from Solid Waste Processing Systems

Solid waste processing systems have taken on several configurations since the early 1970s. Most often the glass component is handled according to its physical properties such as density and particle size. Glass usually becomes part of a "grit" fraction that is either marketed as an aggregate or disposed of in landfills. Processing systems that manufacture refuse-derived fuel (RDF) usually remove glass from the fuel product.

Newer solid waste processing systems are based on producing an organic product that can be marketed as commercial compost. In lieu of marketing arrangements, the compost product is often applied to marginal lands for land restoration. In these facilities, glass containers are broken or crushed in the separation and volume-reduction process and become part of the compost mixture. As a sand-size particle, glass is silica which is a useful component of compost posing no deleterious effects in the land restoration operation.

In solid waste processing systems, recyclables such as ferrous, aluminum, and plastic are removed prior to entering the process line. Recovery usually consists of hand-picking. If desirable, glass containers could be removed at the front end (the picking and separation stage) as whole containers for color sorting, processing, and marketing to glass container remanufacturers.

Recovery as In-Plant Cullet

The recovery of breakage and rejects in glass manufacturing plants and in bottling and packaging plants is still performed using traditional recovery methods. In a glass container manufacturing plant, a process line will result in a small percentage of breakage and imperfect containers. These are simply recovered as a by-product and reintroduced as cullet into the mixture at the appropriate time.

It is necessary to remember that glass container manufacturers are separate and distinct from bottlers and packagers of beverages and food products. In these plants, empty or partially filled containers that break or are rejected can be recovered, washed, and returned to the glass manufacturing process.

In summary, there are various methods of separating and recovering glass from the waste stream as there are for other recyclable materials. These methods are generally classified into two categories: source separation and processing. Source-separation methods include curbside collection, drop-off sites, and buy-back centers. Of these three, curbside seems to produce the best results, but this is highly dependent upon the characteristics of the community. It is one of the most convenient methods of residential recycling, and the generator performs the initial separation of materials. It also makes people think about what they're throwing away and where they're throwing it. However, curbside recycling programs can be expensive to set up and implement.

Overall, source separation is generally the preferred method of recycling by the industry and by the participants. It seems to achieve the highest participation and recovery rates and reduces contamination problems. However, because glass must be color-separated, three separate bins are necessary for storage purposes. This has sometimes been considered problematic due to space limitations. However, flexible bin arrangements, frequent shipments, and reliable market outlets can minimize these problems.

Usable container glass can also be recovered from commercial and industrial sources such as restaurants, bars, and glass manufacturers. Experience has shown that greater quantities of glass can be recovered from commercial sources than from the residential waste stream. Commercial glass recovery should be strongly considered as part of a community's initial and long-term glass recycling efforts for waste reduction purposes and to meet recycling goals.

PROCESSING GLASS CONTAINERS

Processing glass containers is directly related to the types of products that will be manufactured and the types of materials that will be replaced using postconsumer cullet. In the glass manufacturing industry, in-plant cullet has always been reintroduced into the production batch because it was a reliable, contaminant-free secondary material. However, the reuse of recovered postconsumer container glass took many years to become a bona fide segment of the recycling industry.

The basic requirement for using recycled glass containers to make new glass containers has not changed since in-plant cullet was first introduced as a secondary material ingredient. Glass must be clean, free of metal caps and neck rings, and most importantly, color-sorted. Due to these standard manufacturing criteria, glass processing has evolved over time to include a number of steps that assure a usable secondary material.

One of the most common elements of a grassroots volunteer recycling program was the time-consuming processing of glass containers. Countless hours were devoted by dedicated workers crushing green, brown, and clear bottles and jars. The most typical method involved a worker standing over a 55-gal metal drum using a hand-held tamper or mall to break a few bottles, and crush the glass pieces. The glass containers were usually cleaned before crushing and metal neck rings and caps could be removed first or screened out later. Paper labels were often removed before crushing. Crushing was usually required prior to marketing so that potential contaminants could be removed. Crushed glass was also more economical to ship because it had a higher density than whole bottles and jars.

The glass recovered and processed by small volunteer recycling centers was often an inconsistent material; therefore, other uses were gradually developed. For example, Glasphalt became a popular outlet for grit and glass that contained some foreign matter. However, the early uses of Glasphalt for road paving often resulted in substandard street surfaces. Bits of paper and metal were exposed to weathering, and wear was excessive. Consequently, glass processing had to be improved to produce a reliable, contaminant-free, and, when necessary, color-uniform material in order for recovered glass to be marketed in large quantities.

The basic container glass processing steps are:

1. Initial rinsing, cap, and lid removal
2. Color separation
3. Volume reduction by breaking or crushing
4. Packaging for market shipping
5. In-plant beneficiation

These steps are performed at various stages after postconsumer recovery and the intended marketing of the processed glass (Fig. 13.4).

Initial Cleaning and Color Separation

In residential recycling programs, the trend has been to require glass containers to be rinsed and have the caps and lids removed before placing them at the curb or taking them to a drop-off center. Some programs require color-sorting by the resident recycler, but curbside separating or hand-sorting at a recycling center is becoming more common. This tends to increase the convenience of setting out recyclables by the resident, and some amount of separation occurs at the collection or transfer center as a control measure anyway.

Recovery programs for mixed recyclables may be designed to include glass containers. When glass bottles and jars are recovered, it is usually by hand-picking conveyor operations. Glass containers can be systematically sorted as they are picked from the process conveyor line. Some processing conveyors are designed so that hand-picking simply sorts the glass containers into individual conveyors that direct color-sorted containers to the breaking, screening, and bulk storage processes.

Glass Breaking and Crushing

Glass breaking is not desirable if it occurs before color separation. Broken glass is not readily separated from the mixed waste stream and becomes part of a mixed glass material that is of no real value to cullet users. In certain mixed-waste stream processing systems, the glass fraction of the waste stream simply becomes part of the grit residue which

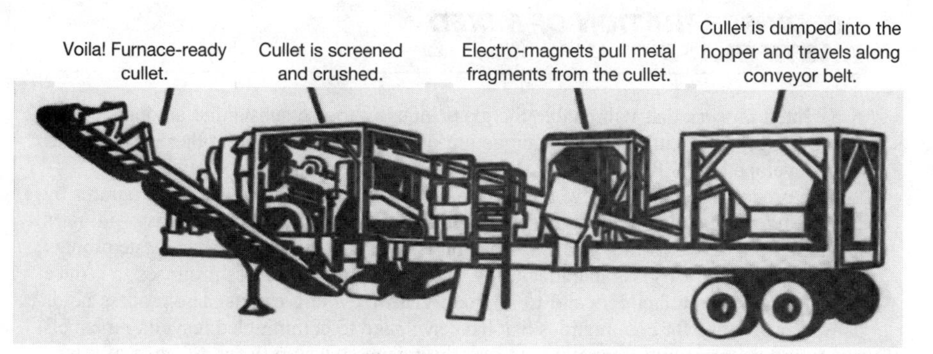
Voila! Furnace-ready cullet.
Cullet is screened and crushed.
Electro-magnets pull metal fragments from the cullet.
Cullet is dumped into the hopper and travels along conveyor belt.

FIGURE 13.4 Glass processing equipment.

FIGURE 13.5 Storage of bulk cleaned, processed cullet.

is landfilled or a component of a composted waste product. In compost, glass particles are beneficial because they have the same physical properties as sand.

If glass containers are to be recovered for marketing to glass container manufacturers or to other users of clean, contaminant-free glass cullet, then color-sorting needs to occur before breakage; metal neck rings, paper labels and food debris can be cleaned and screened from the glass after initial breakage and/or crushing; and storage of the processed cullet must assure that the bulk material is kept clean until it is packed for market (Fig. 13.5).

COLOR SEPARATION OF MIXED WASTE GLASS

A patented process that will enable the glass industry to use substantial amounts of unsorted mixed color cullet in the manufacture of glass containers and other products has been developed by G R Technology, Inc. (GRT) of Haverford, Pennsylvania.

The process promises to have a major positive impact on glass recycling programs by facilitating the use of mixed cullet—containing amber, green and clear glass—in glass manufacturing. By enabling glass manufacturers to use what is currently a waste product as a raw material for glass manufacturing, the CulChrome® process promises to reduce costs for glass manufacturers and to improve glass recycling efforts. The process holds particular promise for communities that have switched to commingled recycling in an effort to reduce costs since commingled recycling generates significant amounts of mixed color cullet.

Heretofore, glass manufacturers have generally insisted that recyclers provide them with color-sorted cullet for use in container manufacturing, i.e., amber cullet for amber containers and green cullet for green containers. Mixed color cullet has been considered a troublesome by-product of recycling efforts. Although alternative uses, such as use as a paving material, have been developed, a large amount of mixed color cullet ends up being relegated to landfills at considerable expense in tipping fees.

Utilizing mixed color cullet in glass container manufacture has been difficult because of the adverse effects of amber cullet in green glass batches and green cullet in amber glass batches. Prior attempts to solve this colorant problem have met with limited success and principally involved the use of "batch compensation." In batch compensation, lower percentages of mixed color cullet are used in the glass batches (as a percentage of the overall glass batch) when off-color cullet levels are high and threaten to cause the resultant glass batch to fail to meet the container glass transmission specifications. At the same time, higher levels of virgin batch are used in the overall glass batch and higher levels of virgin colorizing compounds are added to colorize the virgin glass batch and the flint components of the mixed color cullet to the target glass color. Thus, in batch compensation, the green cullet in amber glass batches and the amber cullet in green glass batches are essentially ignored. However, the presence of such off-color colorant levels, particularly the presence of a significant amount of chrome in the recovered mixed color cullet, has remained a significant "colorant barrier" to the use of recovered mixed color cullet in glass manufacturing (Table 13.1).

Batch compensation, of course, only permits the use of a very limited amount of mixed color cullet in glass production. By contrast, this new process will permit the use of 70% or more of mixed color cullet in glass production (Table 13.2).

The patented process overcomes the "colorant barrier" by providing techniques for (1) selectively decolorizing green glass and for colorizing the flint and decolorized green glass in mixed color cullet to amber color, (2) selectively decolorizing amber glass and for colorizing the flint and decolorized amber glass in mixed color cullet to green color, and (3) selectively decolorizing amber and green glass in mixed color cullet to get flint (clear) glass.

To cite one example, the process solves the colorant problem posed by chrome through the selective combination of certain decolorizing agents and colorizing agents in a glass batch to offset the iron and chrome colorants.

In June of 2000 the U.S. Patent Office issued a Notice of Allowability to the company

TABLE 13.1 Commingled cullet compositions

	Clear	Amber	Green
Approximate USA production	58.0%	32.0%	10.0%
East/West Coast mix, (USA plus 20% green import)	48.3%	26.7%	25.0%
One third clear removed from USA production	47.9%	39.7%	12.4%
Two-thirds clear removed from USA production	31.5%	52.2%	16.3%
Trend to Amber (USA with 30% more amber)	52.9%	38.0%	9.1%
Beer belt blend (heavy amber, few imports)	55.0%	40.0%	5.0%
Special mix (clear glass removed)	0%	50.0%	50.0%
Min	0	26.7%	5.0%
Max	58.0%	52.2%	50.0%

Source: GT Technology, Inc., Haverford, PA (1999).

TABLE 13.2 Process potential percentage recovery of separated colored glass from waste mixed glass

Source of Waste Mixed Glass	Color of Production glass		
	Clear	Amber	Green
East/West Coast Mix (USA plus 20% green import)	10%	65%	>70%
Beer Belt Blend (heavy amber, few imports)	25%	>70%	>70%
Special Mix (clear glass removed)	N/A	60%	>70%

Courtesy of GR Technology, Haverford, PA (1999).

for a sophisticated process methodology that permits the flexible and dynamic use of the process to produce uniform melted glass color in the presence of variable cullet color ratios and other changing batch conditions. This process methodology allows the glass manufacturer to specify the virgin glass raw materials, the desired target glass properties, the composition of a batch of mixed color cullet, and the quantity of the cullet to be used in the glass melt. A computer program then determines the proper amounts of raw materials to add to the batch of mixed colored cullet so that glass is produced having the desired coloring oxides, redox agents, and glass structural oxides in the proper proportion.

Although the CulChrome® process has not yet been applied in a production melt, a series of test melts at Corning Laboratories in New York has demonstrated its potential.

Test melts conducted at Corning Laboratories confirmed production of amber or green glass which meets industry standards using as much as 70% of mixed color cullet in the glass batch. Greatest potential for the process is in the manufacture of green and amber containers such as beer bottles.

Packaging and Shipping

Container glass is a low-density material until it is broken and/or crushed. It then becomes a high-density material. Glass storage is usually required until enough of one color has been accumulated for cost-effective shipment to market. One example of the range between the amounts of the different glass colors is shown in the following breakdown from a recycling facility proposal made to Mercer County Improvement Authority, New Jersey, in 1988.

Flint (clear)	38.5 percent
Amber (brown)	26.1 percent
Emerald (green)	15.4 percent
Mixed glass	20.0 percent

The above example ratio of colored- and mixed-glass cullet was an estimate for a proposed 180-ton/day processing facility of mixed recyclables.

Large amounts of glass cullet are frequently shipped as a bulk material in roll-off containers. Occasionally, gaylord boxes have been used to ship small amounts of exceptionally clean, uniformly colored glass to high-quality crushed glass users.

In-Plant Final Processing

Industry surveys indicate that the container glass used to make remanufactured glass bottles and jars is processed by intermediaries to meet the requirements of the manufacturer. In practice, color-sorted glass containers are shipped whole, broken, or crushed to the end users. Final cleaning is performed at the manufacturing plant by specialized beneficiation equipment to remove residual metals, plastics, and paper labels.

The cullet is then mixed with the raw materials used in the production of glass. After mixing, the batch is melted in a furnace at temperatures ranging from 2,600 to 2,800 °F, depending on the percentage of cullet contained in the batch. The mix can burn at lower temperatures if more cullet is used. The melted glass is dropped into a forming machine where it is blown or pressed into shape. The newly formed glass containers are slowly cooled in an annealing lehr. They are inspected for defects, packed, and shipped to the bottling company (Fig. 13.7).

In summary, the most important rule of thumb for recovering and marketing glass containers is to clean and color-sort them in order to produce a high-quality recyclable product. It is not necessary to thoroughly wash glass containers in order to recycle them; a quick rinsing is usually sufficient, and paper labels do not need to be removed. Generally speaking, if the containers are clean enough to store in the home for a week, they are clean enough to be recycled. Many metals, stones, ceramics, and other foreign materials do not melt in the furnace with the materials that form glass and create stones or bubbles in the bottles. These bubbles or stones not only cause aesthetic problems, but weaken the bottle's wall as well. In the glass furnace, iron and lead contaminants settle to the bottom of the furnace tank and corrode its brick lining. Larger materials (e.g.. steel lids and ceramics) often block feed lines from the furnace causing temporary production shutdowns. At the present time, there are no mechanical means of color sorting. Research and devel-

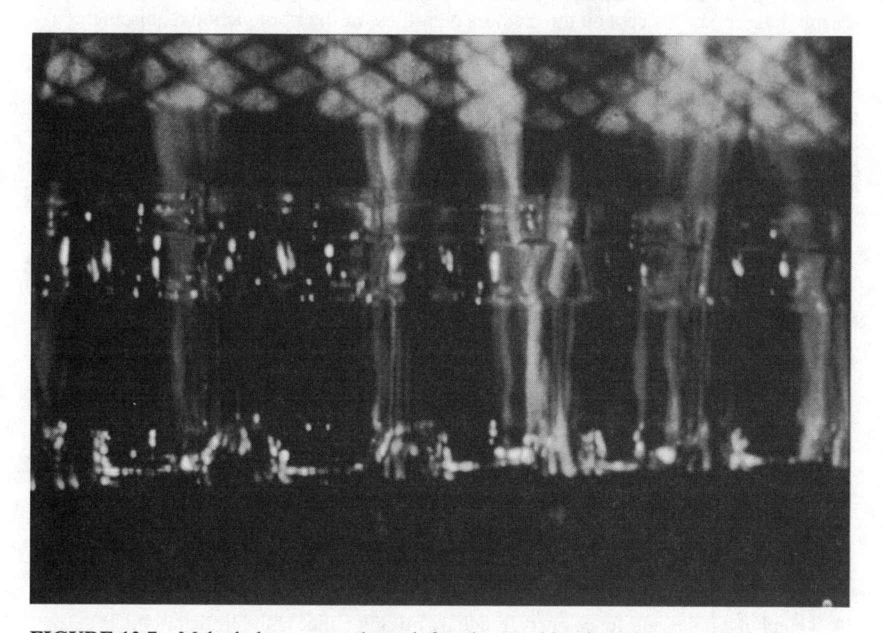

FIGURE 13.7 Melted glass passes through forming machine for new glass shape.

opment in this area and in ceramic detection are promising; however, these functions are currently performed manually. Industry representatives indicate that meeting these quality requirements through consistent processing is the most difficult challenge in establishing and implementing successful glass recycling programs.

CONTAINER GLASS MARKETING

The successful recycling of recovered container glass depends on marketing a color-sorted and contaminant-free secondary material. Thus far, the biggest market has been the glass container manufacturing industry. When recovered glass does not meet manufacturing specifications, it can be used as an aggregate in Glasphalt or as a beneficial component of a soil conditioner product. These types of outlets usually depend on the local and/or regional availability of industries that would incorporate processed container glass into their manufacturing operations on a regular basis.

The primary end market for glass cullet is glass bottling manufacturing plants—there are currently over 80 of them throughout the United States. A vast majority of the glass recovered from the waste stream is used to make more glass containers (some studies cite up to 90%). These markets are generally available throughout the country. Glass manufacturing plants indicate a desire to increase their use of recovered glass. Bottle manufacturers are capable of using 80 to 90% cullet and most would like to use at least 50 to 60%. In fact, one Anchor Glass plant in Pennsylvania ran successfully on 100% used glass for seven weeks during the winter when frigid weather in Wyoming caused a disruption in the supply of soda ash to manufacturers around the country.

Markets for the three different colors of glass may vary by geographic location because some glass manufacturing plants only produce bottles of one or two colors. Large manufacturers may accept all three colors regardless of their production requirements and transport the color(s) they can't use to a "sister" plant in another location. However, hauling distances can affect the prices paid for different colors of cullet. As there is a greater demand for clear glass containers, prices are generally somewhat lower for brown and green cullet. If any gluts in the market do occur, it will be in the market for green glass because most of the green bottles used in this country are from imported beers, yet very few domestic products are contained in green bottles.

It is important to the glass container manufacturing process that the percentage of cullet used remain consistent over periods of time. It cannot vary on a daily basis. Therefore, glass manufacturers are usually conservative in determining the percentage of cullet used in the batch in order to ensure the availability of an adequate, steady supply.

Other markets for cullet have been identified through continuing research. Glass is or can be used in the manufacture of:

- Glasphalt, an asphalt made using a percentage of crushed glass for roadway applications
- Building and construction materials such as clay brick and tiles, masonry block, and Glascrete; as a lightweight aggregate in concrete and plastics; in glass polymer composites and extrusions; and FoamGlas for construction board and insulation
- Reflective paint for road signs (made from small glass beads)
- Glass wool insulation
- Telephone poles and fence posts (made by mixing cullet with plastic polymers)

- Agricultural soil conditioners to improve drainage and moisture distribution
- Artificial sand for beach restoration
- Fiberglass
- Abrasives
- Many other materials associated with the construction and textile industries

Most of these applications have been proven. A few are currently being used more frequently. The demand for glass by bottle manufacturers dropped in 1991 because of oversupplies from local recycling programs.

GLASS RECYCLING PROGRAM CONSIDERATIONS

In comprehensive solid waste management systems, recycling has become a typical program element along with refuse collection and waste disposal (Fig. 13.8). Many states mandate recycling goals in terms of waste disposal avoidance. Waste minimization programs are beginning to target the packaging industry as a primary place to reduce the generation of waste materials.

There are several issues that need to be addressed when considering glass recycling. In general, it is unlikely that glass containers would be the only targeted recyclable; it is typically one of the "big four" recyclable materials that also include aluminum, paper, and plastic.

First, it may be necessary to justify why glass recycling is important and if glass recycling can be cost-effective. After all, glass is made from relatively abundant and inexpensive virgin materials. Recycling glass does not save trees, for example, and therefore becomes a less emotional issue. However, the conservation of our energy supply (through the use of recycled glass cullet) is just as easily understood and accepted. Reducing the waste stream is also a major accomplishment when considering today's waste disposal problems. Although glass comprises a relatively small portion of the waste stream by volume (approximately 2%), glass recovery and recycling can have a significant impact on waste reduction by weight (7 to 8% in comparison) because glass is one of the heavier materials found in municipal solid waste.

In conjunction with other evaluation tasks, and as part of a comprehensive analysis of a community's solid waste generation characteristics, the amount of glass and sources of glass containers should be determined. The potential for marketing recovered glass to local and regional users of reclaimed glass should also be determined.

It is well known that glass containers are common in the residential waste stream. However, restaurants, taverns, recreational facilities, and institutions (e.g., schools) need to be considered as major generators of postconsumer glass as well. Restaurants and taverns in particular can derive many benefits from a glass recycling program. Although additional storage space is required, these businesses usually contract with private haulers for waste disposal on a per ton price basis and can therefore save a substantial amount of money on disposal costs due to the weight of glass. Many also generate a steady stream and large enough quantities of glass to negotiate good market prices. There are organizations throughout the country to assist businesses with setting up and implementing successful glass recycling programs. Some of these organizations are listed at the end of this chapter.

FIGURE 13.8 Collecting separated glass in refuse truck.

Other aspects of program implementation that need to be considered in the evaluation phase are legislative issues, cost factors, and program flexibility.

Legislative Issues

The Bottle Bill

The most infamous legislation affecting the use and recycling of glass beverage containers is known as *beverage container deposit legislation* (BCDL). Commonly referred to as "bottle bills," these laws have been enacted to establish monetary deposits on beverage containers. Nine of the 50 states currently have bottle bills in effect: Connecticut, Delaware, Iowa, Maine, Massachusetts, Michigan, New York, Oregon, and Vermont. Legislation to enact bottle bills is pending in many other states as well, and efforts have been made to enact a national bottle bill on and off for almost 20 years (although they have not yet been successful). This has been very controversial among state governments, citizens, the recycling industry, and bottle manufacturers.

Bottle bill advocates believe that bottle bills provide a number of advantages:

- They reduce litter and litter-related costs such as pickup and disposal.
- They help to reduce the waste stream by diverting these materials from the landfill.
- They encourage reuse and recycling because of the economic incentives.
- They create markets to recycle the aluminum, glass, and plastic that is returned for deposit.
- They provide all of these benefits without cost to government.

Bottle bill advocates are also quick to point out that states with bottle bills are achieving higher recovery rates for beverage container materials than those without, and believe that industry is against such legislation in order to increase profits through the production or disposal of throwaways.

Those who oppose bottle bills include manufacturers and recyclers and their representative organizations. Their arguments against bottle bills include:

- They decrease the use of glass because retailers tend to shy away from selling glass containers due to the amount of space returned containers take up in the store, and consumers tend to use more aluminum and plastic containers because they are lightweight and easier to return.
- They force higher beverage prices.
- They are discriminatory because they only apply to beverage bottles and do not affect food containers and other packaging.
- The waste stream reduction achieved is much less than could be accomplished through successful recycling programs.
- They actually hurt existing recycling efforts by diverting the flow of beverage containers to retailers instead of recyclers.

An argument also exists between refillable versus nonrefillable glass. Nonrefillable bottles that are recovered from the waste stream are recycled, while refillable bottles are returned to the bottler to be used again. Both methods help to reduce the waste stream. However, fewer and fewer refillable bottles are being purchased and returned due to the inconvenience to the consumer. It is estimated that the average number of trips made by these bottles—as high as 50 in 1950—is now about 8.5.

Whatever one's standpoint on this issue, the effect and status of bottle bill legislation on the state and national levels should be considered when setting up a recycling program.

State Recycling Goals

Other legislation related to glass-recycling programs includes state-regulated waste reduction goals and mandatory recycling programs. Many states have adopted legislation that sets forth waste reduction goals by specified dates. Some of this legislation also mandates that a certain percentage of some materials be recycled. Glass is usually one of the "big four" materials specified in such legislation, along with aluminum, plastics, and newspaper. The status of state legislation should also be considered when setting up a recycling program. For example, if legislation in one s state mandates a 50% reduction in the amount of glass (and other specified materials) that is currently being disposed of to achieve say a 25% reduction in the overall waste stream, a local glass-recycling program will probably have to include the recovery of glass from commercial as well as residential sources in order to achieve the state-mandated waste reduction goals.

Mandatory Recycling Ordinances

Although most communities prefer to initiate recycling efforts through voluntary programs, many have already turned to mandatory programs in order to meet state or local recycling and waste reduction goals. Both voluntary and mandatory recycling programs can be effective. The success of voluntary programs depends largely on education, while mandatory programs rely on the various enforcement measures. This must also be considered when setting up specific program elements such as determining the amount of money necessary for public education programs versus the amount of money and effort required for enforcement.

Cost Factors

There are a number of cost factors that must be taken into account when setting up any recycling program. However, costs specific to glass recycling programs are mainly those associated with the color separation requirement. Additional containers are required for storage purposes, and additional labor costs are incurred because glass must be color-separated by hand.

Expenses associated with glass-recycling programs generally include:

- Collection costs
- Sorting costs (if collected mixed)
- Storage costs
- Transportation costs
- Other costs (e.g., public education and training and containers)

In order to determine these costs, several factors must be considered, such as the population served, characteristics of the geographical area (residential versus industrial, urban versus rural, etc.), the estimated weight of the glass expected to be recovered, the refuse collection service used (public or private), the method(s) of recovery (curbside collection, buy-back centers or drop-off sites), the type and location of the market(s), and the mode of transportation used to deliver the glass to the market(s).

Material sales revenues can be used to offset some of these program costs. A breakdown of material sales contributions and their associated values for typical programs are shown in Tables 13.3 and 13.4. Since less than half of the costs associated with typical multimaterial curbside programs are covered by sales revenues, remaining program costs must be met by other revenue sources (e.g., contract payments, grants, tax and surcharge revenues, and waste diversion credits).

End-use markets currently pay $40 to $75/ton for uncontaminated color-separated glass. The prices paid by independent recyclers vary in accordance with the prices they receive from the end-use markets, transportation costs, the quantities of glass they receive and many other factors. There will be differences in the prices paid for recovered glass depending on whether it is sold to a recycling company or directly to a glass manufacturer. It is important to determine where the potential market is and what prices are being offered prior to implementing any glass-recycling program.

Avoided disposal costs should also be accounted for when determining the costs and revenues associated with glass recycling programs. For example, one Florida restaurant estimates that it saves approximately $28,000 in disposal costs by separating its glass for recycling. The restaurant does not receive any revenues for the glass but rather donates it

TABLE 13.3 Sales revenue contribution by material

Material	Percent of tonnage	Percent of revenue
Newspaper	56–70	57–65
Glass	15–22	10–20
Aluminum	1–2	15–20
Other	5–15	5–16

Source: "Comprehensive Curbside Recycling, Collection Costs and How to Control Them," Glass Packaging Institute, Washington, D.C.

TABLE 13.4 Materials sales values*

Materials	Value, $/ton
Newspaper	25–35
OCC (corrugated)	35–90
Glass, clear (flint)	15–25
Glass, brown	12–25
Glass, green	–20–0
Plastics, PET	120–150
Natural Colored	125–175
Plastics, HDEP	80–120
Aluminum	960–1050
Steel cans	15–50

Prices effective 1998–1999 and based on clean, sorted materials delivered to end users. Low-range prices generally reflect prices in east and central United States. High-range prices generally reflect prices in western United States.

to a Goodwill Industries recycling center because the avoided disposal costs alone are worth the effort. A New Jersey recycler estimates that the restaurants and bars involved in his program save $600 to $1200 monthly in refuse collection costs.

Program Flexibility

Flexibility is a key factor in any successful recycling program. Programs must be flexible enough to switch in and out of materials in order to accommodate changing market conditions. They must also be able to absorb fluctuating market prices. For glass, this can mean finding alternative markets for one or more colors and/or pursuing emerging markets for mixed cullet as the need arises.

Recycling programs should be assessed on a regular basis (at least annually) in order to determine what changes, if any, should be made to maximize participation, recovery rates, and sales revenues. These changes may be related to collection practices, education efforts, processing technologies, and/or materials marketing. Program flexibility will allow for such needed improvements without impacting other successful program elements.

SUMMARY

Potential Problems and Solutions

Representatives from various factions of the recycling industry cite contamination as the most common problem associated with glass recycling programs (mixed colors and/or foreign materials). Other problems include lack of communication, storage space, serviceability, and theft. Most of these problems are not specific to glass but also to other re-

Glass Recycling Made Easy

Acceptable

Glass food and beverage containers can be easily recycled by glass container plants. Generally speaking, metal caps and lids should be removed but labels can remain.

Not Acceptable

The following materials are not recycled by glass container plants and should not be mixed in with container glass.

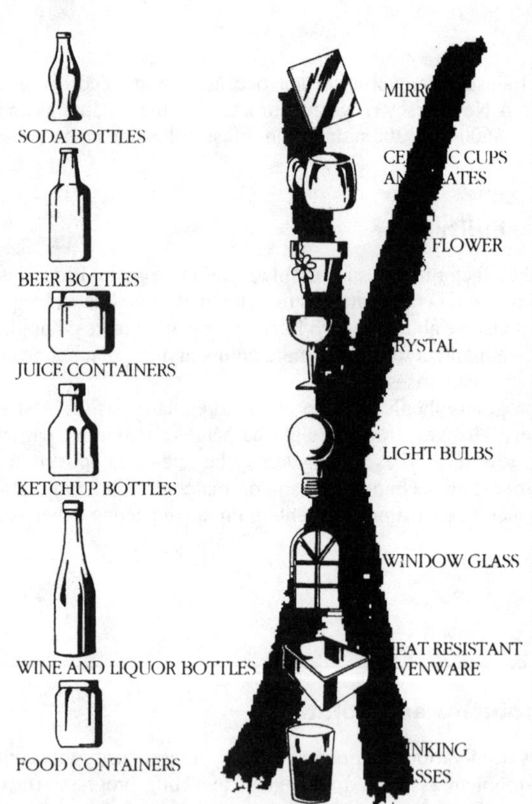

FIGURE 13.9 Glass Recycling Made Easy (*Source:* Southeast Glass Recycling Program, Clearwater, Florida.)

FIGURE 13.10 Igloo containers from a drop-off center.

cyclable materials. Color contamination is specific to glass recycling, and storage space can be somewhat more of a problem for glass because of the color separation requirement.

The solution to most of these problems can be summed up in one word: *Education* (Fig. 13.9). The importance of early, continuous, long-lasting public education programs cannot be overemphasized in the field of recycling. In glass recycling, public education is needed for maintaining quality (i.e., eliminating contamination). Large producers of used glass can also overcome the storage problem by using a glass crusher. Just be sure that the potential market will accept the glass in crushed form. Many independent recyclers prefer to collect the glass whole to ensure that colors have not been mixed (Fig. 13.10) and that other contaminants have been removed. However, glass manufacturing plants will generally accept recovered glass in either form.

Program Assistance

Due to the heightened public awareness of environmental issues in recent years and a strong commitment and effort by the glass industry in general, glass recycling has experienced the most rapid growth of any recyclable material. The industry itself has established ambitious goals and has assisted many communities in setting up successful glass-recycling programs.

CHAPTER 14
PLASTICS

TIM BUWALDA
Consultant to the American Plastics Council
R. W. Beck, Inc.
Orlando, Florida

INTRODUCTION

Collecting and disposing of municipal solid waste (MSW) is a non-value-adding service; MSW is collected and hauled to a site for disposal. In the United States, it is the responsibility of local governments to ensure that MSW is frequently collected to protect the health and safety of their citizens. In most cases, it is the responsibility of counties to ensure that disposal sites are available that protect the health and safety of their citizens, and the environment.

In the late 1980s, there was a widely held belief that the United States was running out of landfill space, and as a result the cost of disposal would increase tremendously. Because of this perception, communities started collection programs for recyclables, not in an attempt to turn a profit, but as an additional disposal service option that could divert materials that otherwise would go to a landfill, thereby preserving landfill life while also benefiting the environment. These community-sponsored recycling programs made plastics recycling personal for most Americans, even though there had been a private plastics recycling industry quietly operating behind the scenes for years.

Contrary to the fears of the late 1980s that jump-started these community recycling programs, the United States has not run out of landfill space. Although many smaller community landfills have closed, they have been replaced by regional mega-landfills with just as much total capacity. The national average cost of landfill disposal has also remained fairly level over the last several years, largely as a result of competition among landfills for revenues from waste disposal. For most jurisdictions the current cost (on a per-ton basis) of waste collection and disposal is less than the current cost of recycling collection and processing. Certainly, more can be done to improve the economics of recycling. The plastics industry, through the American Plastics Council (APC), has sponsored research aimed at driving down the cost of collecting, processing, sorting, and reclaiming plastics.

Despite its cost, recycling continues to be highly valued by the American public and others who take a broader viewpoint, and communities continue to offer collection service for recyclables. Recycling is valued for:

- Conserving natural resources
- Creating jobs and additional economic activity
- Reducing the environmental impacts of virgin material extraction

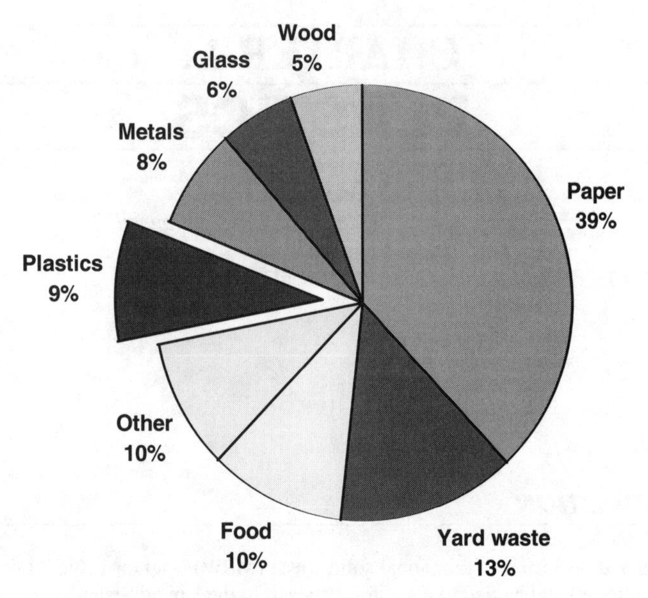

FIGURE 14.1 Materials generated in MSW by weight.

About 40 billion pounds of plastics are generated as MSW in the United States, according to the United States Environmental Protection Agency (USEPA, 1996 statistics), or 9.4% by weight of MSW produced in this country (Fig. 14.1). Although packaging of all types comprises only 33% of MSW, it receives the most national attention, and communities have focused the bulk of their attention on collecting various types of packaging (including plastics packaging) for recycling. Plastics packaging comprises 12% of all packaging, although it only makes up 4% (16 billion pounds) of all MSW by weight (Table 14.1).

Most plastics that are recycled come from the thermoplastic family of plastics, which represents about 90% of all plastics sold. Thermoplastics melt when heated to high temperatures and can be readily recycled. Thermosetting plastics do not melt when heated; instead, they harden, much like an egg hardens when heated, and thus are not as readily recycled. The entire spectrum of packaging plastics is thermoplastic.

The plastics industry annually recycles approximately 4 billion pounds of postconsumer and preconsumer scrap plastics. This does not include the recycling of several bil-

TABLE 14.1 Packaging composition by material

Paper	56%
Glass	16%
Plastics	12%
Wood	9%
Metal	7%

lion pounds of process scrap (regrind) that is incorporated straight back into the manufacturing process. The 4 billion pounds of plastics recycled can be split into approximately 1.4 billion pounds of preconsumer scrap plastics and 2.6 billion pounds of postconsumer plastics. Discards of plastics in MSW after recovery for recycling are 36 billion pounds, or 12.3% of total MSW discards (Fig. 14.2). The rest of this chapter will focus on the recycling of postconsumer plastics.

The successful plastics recycling that is occurring has five necessary parts in place:

- **Collection** programs
- **Processors** who locally prepare collected plastics for market by sorting and densifying them into a form for economical long-distance shipment
- **Reclaimers** who transform the recovered product into feedstock materials
- **End users** (manufacturers) who convert the reclaimed material into recycled-content products
- **Customers** who purchase the recycled-content products

The infrastructure for collecting, processing and reclaiming postconsumer plastics grew tremendously from the late 1980s through the late 1990s (Figs. 14.3 and 14.4). Before 1998, postconsumer plastics recycling consisted primarily of soft drink bottles collected from a few states requiring deposits, and the plastic casing from returned lead–acid vehicle batteries. Today, community collection programs offer curbside pick-up or drop-off site collection of select types of plastic products to nearly 80% of the United States population. Recovery of postconsumer plastics through community collection programs, however, has not kept pace with reclamation capacity and end-user demand for postconsumer plastics. As a result, most U.S. reclamation plants operate well below their capacity.

End-use markets for recycled plastics are diverse; a myriad of products are produced, such as plastic bottles, carpet, strapping, pipe, and plastic lumber. Some states have

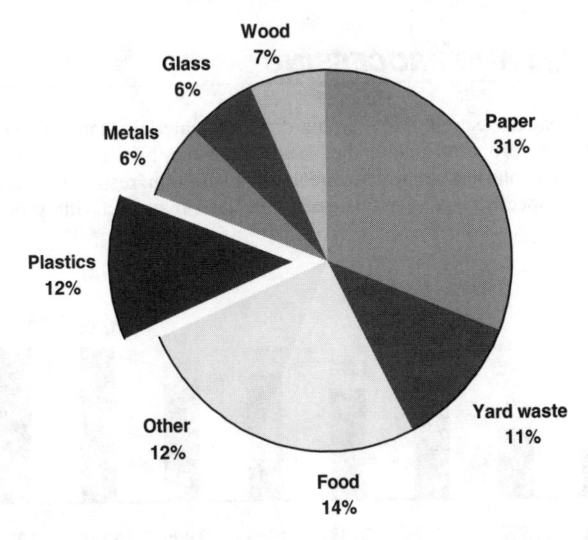

FIGURE 14.2 Materials discarded (after recovery) in MSW by weight.

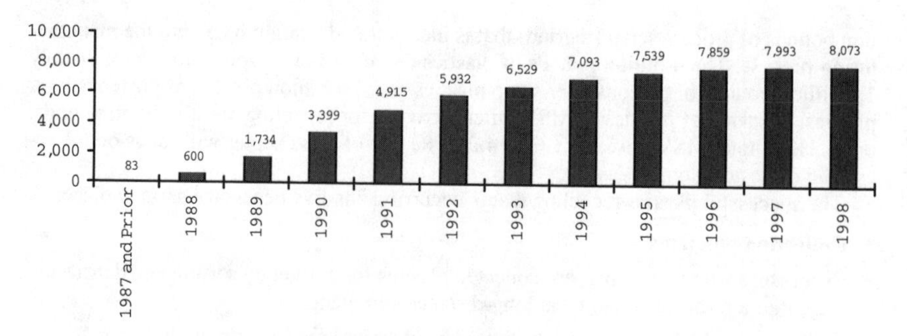

FIGURE 14.3 Growth of communities with curbside collection of postconsumer plastics.

sought to increase plastics recycling rates by passing demand-side legislation that mandates that manufacturers use certain recycled content levels for some products. While these actions have certainly increased the demand for some types of recycled plastics, they have done little to divert more plastics from disposal. Most individuals who participate in recycling programs do so because of an environmental ethic and because they have been provided with a convenient collection service. The bottom line for plastics is that the level of market demand or scrap prices paid has little impact on postconsumer plastics recycling rates. Other, more effective, mechanisms to improve diversion include increasing community recycling education and awareness efforts, expanding recycling collection programs, and implementing a "pay-as-you-throw" (PAYT) program wherein homes pay different amounts based on the amount of waste they dispose of, as opposed to the flat-rate system that is common today. PAYT fee systems provide a direct financial incentive for individuals to recycle more.

COLLECTION AND PROCESSING

Plastics are a broad collection of similar but different materials that can be manufactured to have almost any property imaginable: clear or opaque, hard or soft, stiff or flexible, brightly colored or colorless. In order to create such variation, resin producers tailor-make plastic resins for specific products and specific product-manufacturing processes. Some-

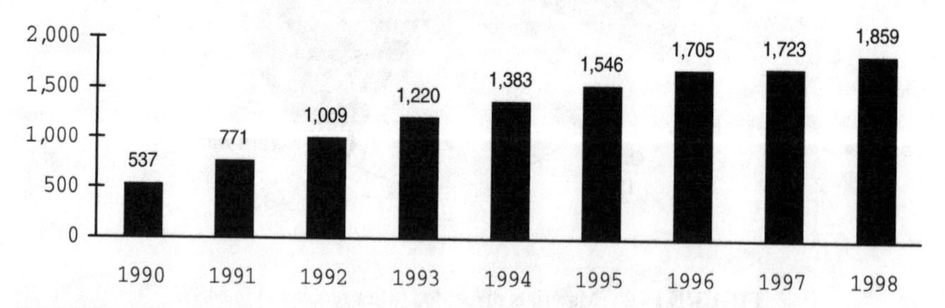

FIGURE 14.4 Growth of companies processing and reclaiming postconsumer plastics.

times, a single type of resin is made to be runny when it is melted, other times it is made to be viscous when melted. Different types of plastics resins are also made from different precursors, and are different on the molecular level. These molecular differences cause most plastics to be incompatible with other plastic types when melted together; they tend to separate much like oil and water separate when mixed together. For these reasons, different types and grades of plastics must be separated from each other at some point in the recycling process if the recycled resin and final recycled product are to be of high quality and have predictable properties. This separation is often problematic because many types of plastics look alike and behave alike because they are chemically identical or have similar physical properties (such as density).

Most states require rigid plastic containers from 8 ounces to 5 gallons capacity to be labeled with a symbol and numerical code to identify the generic family of plastic resin the container is made from. This coding system was developed by the Society of the Plastics Industry to provide a uniform national system for coding that meets the needs of the recycling industry, as defined by the recyclers and collectors themselves, and assists recyclers in manually sorting plastic containers by resin type.

The code is a three-sided triangular chasing arrows symbol with a number in the center and letters underneath. The three-sided chasing arrows symbol was selected to distinguish the code from other markings and to identify it as an aid in sorting for recycling. The number inside and the letters indicate the generic resin from which the container is made; containers with labels or caps of a different material may, if appropriate, be coded by their primary, base resin:

1 = PET (polyethylene terephthalate)
2 = HDPE (high-density polyethylene)
3 = V (vinyl)
4 = LDPE (low-density polyethylene)
5 = PP (polypropylene)
6 = PS (polystyrene)
7 = Other

Plastic products most commonly targeted for recovery through community recycling programs include PET and HDPE bottles. There is strong market demand for these types of bottles and they compose 94% of all plastic bottles sold and 78% of all rigid containers (from 8 ounces to 5 gallons) sold (Fig. 14.5). Research conducted by APC, however, has shown that most consumers are easily confused by the resin coding system. Communities that use it in their education and awareness programs to target only PET and HDPE bottles end up collecting tubs and bottles of other resin types as well because of the public's confusion. The levels of nontargeted items in those communities is close to the levels experienced by communities that simply ask for all plastic bottles. In fact, research has demonstrated that communities actually divert more PET and HDPE bottles from the waste stream and achieve higher recycling rates by asking for all plastic bottles because consumers are less confused and participate more fully in recycling. For this reason, APC promotes the collection of all plastic bottles.

The most common mechanism of recyclables collection for single-family homes is a system wherein householders place their recyclables in one or more bins and set those bins out periodically (usually once per week or every other week) for collection. This system is often referred to as a curbside collection system or a blue-box collection system. Recyclables from multifamily homes are frequently collected from large carts. Collection trucks then circulate through the neighborhoods and the bins and carts are emptied into

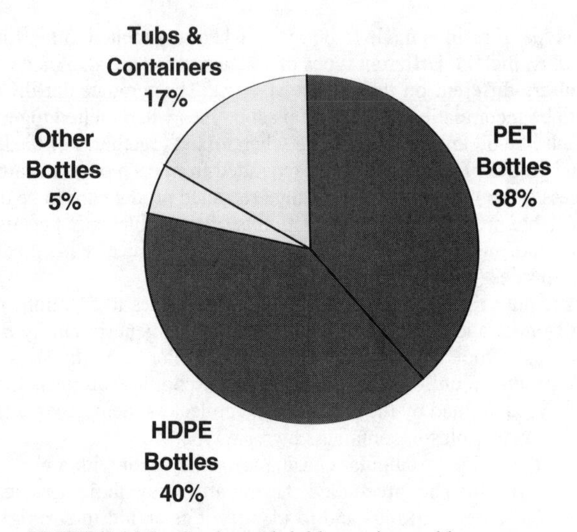

FIGURE 14.5 Bottles and rigid container sold.

the trucks. Depending on the local processing infrastructure for collected recyclables, the contents of the bins may or may not be sorted into different compartments of the truck at the curb. In the cases where there is truckside sorting, plastics of various types are typically separated from other nonplastic recyclables and loaded as mixed plastics into a compartment.

The high volume and low density nature of plastics makes compaction an attractive option when collecting this material. Although the body of evidence demonstrates that light compaction helps to reduce recyclables collection costs and does not lead to significant increases in residue, in practice, few collection programs currently use compaction. The reasons for not doing so are numerous and varied, including:

- **Useful life of existing fleets**. Haulers or communities are typically less likely to invest in new vehicle technology when the useful life of the existing fleet has not been exhausted.
- **Existing contracts**. For communities that contract for recycling services, contract length often affects vehicle selection. For example, some haulers may not be interested in investing in more capital-intensive compacting vehicles for shorter-term contracts.
- **Material quality concerns**. In spite of evidence that suggests that glass breakage is minimal in light-compaction vehicles, program operators may not want to cause even minor amounts of additional breakage that could lead to marginally higher amounts of unrecycled residue.

Even if compaction is not available in the collection vehicle, communities should educate program participants to flatten plastic bottles prior to placing them in the recycling bin. This step not only saves space in the collection truck, but also provides space for more recyclables in the collection bin.

Once collection is complete, the collection trucks off-load their materials at either a recyclable materials processing facility (called a recyclable materials wholesaler by the U.S. government), or at a materials recovery facility (MRF) for local processing. MRFs are distinguished from recyclable materials processing facilities because they primarily accept commingled recyclables, whereas recyclable materials processing facilities are primarily equipped for receiving separated materials. Both types of facilities remove contaminants and densify plastics by baling them for economical shipment to downstream markets. Because most downstream markets specialize in recycling only one broad family of plastics (e.g., PET or HDPE) or manufacture only one type of recycled product, most MRFs and many recyclable materials processing facilities further sort plastics into three categories prior to baling: PET bottles, natural (unpigmented) HDPE bottles, and pigmented HDPE bottles. All of this further sorting at the local level is done manually.

For those local processing establishments that choose not to sort by plastic type, there are several intermediate plastics recycling facilities located in the United States that specialize in sorting large amounts of baled mixed plastic bottles with automated sorting systems. These automated systems use X-ray and infrared spectroscopy to identify the color and type of plastic the bottle is made from, and sort them using jets of air at a cost that is less expensive than manual sorting. The equipment also meets stringent purity specifications, which manual sorting alone cannot do. Plastics intermediate recycling facilities typically grind the sorted bottles to produce small flakes (called "dirty flake") that are ready to be washed. These streams of dirty flake are then shipped to downstream reclamation and manufacturing markets.

Baled plastic bottles have a high per-ton scrap value compared to other postconsumer recyclables targeted in municipal programs. Only aluminum cans are more valuable. Because plastics are source reduced and light in weight, however, they typically make up only 5% or less of the total weight of recyclables collected through community recycling programs, and thus contribute about 10% of the total revenues received by the programs.

In addition to the residential recyclables collection and processing infrastructure described above, plastic products are recovered through other mechanisms as well. PET soft drink bottles are also recovered through deposit–return systems in ten U.S. states, including California's redemption system. These deposit systems—often called "bottle bills"—target on average only about one quarter of the rigid plastic containers sold within their states. Originally designed to reduce litter, bottle bills are not a substitute for (and are somewhat redundant to) more comprehensive recycling collection programs. Recovery of PET bottles through these systems accounts for 14% of all postconsumer plastics recovered in the United States.

Other postconsumer plastic products recycled in large quantities include retail bags, which are most often collected by supermarket chains; polypropylene vehicle battery casings, which are most often collected by auto repair or auto parts shops; and low-density polyethylene stretch wrap and protective films, which are most often recovered by warehouse and distribution centers. There are still many other opportunities for entrepreneurs to recover other plastic products for recycling, including plastics from durable goods. For example, collection programs and reclamation plants are emerging for used carpet, televisions, and computers. The APC is also sponsoring research in the durable goods recycling area to develop technologies to identify, sort, and remove coatings from durable goods plastics.

RECLAMATION

Once recovered plastics have been collected and processed on the local level, they are shipped to a reclamation facility where industrial systems clean and purify the materials and return them to a feedstock form ready for remanufacture. These facilities annually reclaim from tens to hundreds of millions of pounds of plastics per year, and the larger facilities source nationally and internationally to fulfill their market demand. Many reclamation facilities are integrated facilities where resin is reclaimed and then converted into intermediate or finished recycled-content products under one roof. For example, at one facility, baled PET bottles enter the processing system and finished polyester carpet made from the bottles comes out the other end.

The reclamation process starts with equipment that breaks apart the bales of plastics. At certain reclaimers (primarily PET reclamation facilities) plastic bottles go through automated whole-bottle separation systems that sort by color and resin. The sorted bottles are then typically ground into small flakes of about one quarter inch (Fig. 14.6). These flakes pass through an air classifier that uses opposing gravitational and air current forces to achieve separation of lighter fines and labels from the heavier flakes. Next, the flakes are washed using water and mild detergents to remove additional contaminants. The clean flakes pass through a hydrocyclone, which uses centrifugal force to effectively separate one resin type from another based on differences in density. The flakes are then dried. At many reclamation facilities, the clean flakes are passed through a high-speed automated flake-level sorter that separates out differently colored cap material or other plastic contaminants, so that the final clean flake is very nearly pure (i.e., contamination is only a couple of parts per million). The final step for some reclaimers is to melt and convert the clean flake to small pellets (similar to virgin plastic resin, Fig. 14.6).

There are now more than 140 automated sorting systems in place, mostly at reclamation facilities. As the number of units sold has increased, the price has decreased, which means that these sorting systems are now becoming affordable to establishments processing smaller volumes of material. Although some smaller units are being installed in MRFs, it is uncertain whether automated plastic bottle sorters will become commonplace in local recyclables processing facilities.

As previously mentioned, a single type of resin may commonly be produced in differ-

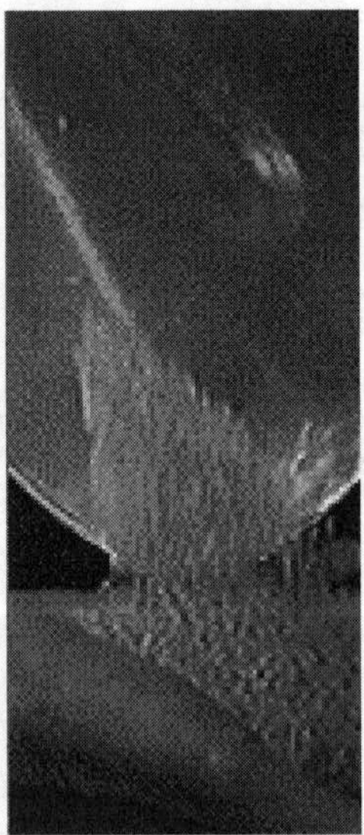

FIGURE 14.6 Plastic flakes and pellets.

ent grades that: (1) have radically different properties when melted; and (2) fit into significantly different market segments. For example, the low-viscosity injection molding grade of HDPE used to make margarine tubs is a significantly different material from the high-viscosity blow molding grade of HDPE used to make milk jugs. Despite the technological progress in automated sorting systems, there are not currently any high-speed automated sorting systems that can distinguish between grades of plastics within the same resin type—manual sorting is still required to separate these different products from each other.

Every time plastics are melted, they degrade to some extent. Some types of postconsumer plastics, such as PET, can be rejuvenated through a "solid-stating" process to restore the recycled resin to virgin-equivalent properties. Other resins, such as HDPE, can be blended with higher-quality virgin resins or other additives to produce a recycled resin blend with virgin-equivalent properties.

Some types of plastics, such as PET and nylon, can be recycled through a depolymerization process in which the plastic is broken down to its basic building blocks, called monomers. These monomers can be refined to a very high purity, including removing colorants, and the plastic can be repolymerized so that it is in all respects equal to virgin plastic. This process offers the greatest promise for unlimited recycling of plastics back into virgin-equivalent resins. The greatest obstacle to this technology is achieving large enough economies of scale so that the recycled resin is competitive with preconsumer and virgin resins.

END USE

The final step for closing the recycling loop is for recycled plastics to be incorporated into recycled-content end products, which are in turn purchased by consumers. Research has shown that consumers are generally not willing to pay more for a product just because it contains recycled material, even if the price difference is nominal. Recycled postconsumer plastic resins, therefore, must be competitively priced compared to virgin and recycled preconsumer resins. "Green" (i.e., environment-friendly) marketing tactics can, however, be effective if the recycled content product can be offered at an equivalent price.

Most recycled resins do not achieve the level of consistency and purity of the initial grade of virgin resin, which is a concern of product manufacturers. Recycled resins can be produced with virgin-equivalent properties; however, the cost to achieve this quality can sometimes price the recycled resin out of competitive range. In addition, it is nearly impossible to remove colorants from recovered plastic products unless a chemical recycling process is used. Moreover, it is very difficult and costly to perform additional sorting of recovered plastics by color. Even attempts to hide or mask colors in recycled resins by overcoloring the resin to a darker color or hiding it in the interior layer of a multi-layer product adds cost.

One of the saving graces of recycling of high-quality plastics, such as plastic bottles, is that they were manufactured as premium resins in the first place, often for use in food-contact applications. When recycled, they normally exceed the quality and performance standards for many product applications and so can preferentially compete against lower-grade virgin materials in performance and price.

The APC has identified over 1400 products made with recycled plastics. Tables 14.2, 14.3, and 14.4 show the dominant recycled-content products and markets that consume most recycled U.S. plastic resins.

The use of recycled plastic in manufacturing plastic lumber and related products is

TABLE 14.2 Consumption of recycled PET

Fiber	51%
Strapping	14%
Sheet products	10%
Bottles	7%
Other (molding)	5%
Export	13%

Source: R. W. Beck, Inc.

TABLE 14.3 Consumption of recycled HDPE

Bottles	29%
Pipe	13%
Lawn and garden	10%
Pallets, crates and buckets	9%
Film and sheet products	7%
Plastic lumber	5%
Other	17%
Export	10%

Source. R. W. Beck, Inc.

TABLE 14.4 Consumption of other recycled plastics

Vehicle battery casings	30%
Plastic lumber	15%
Bags	9%
Other	31%
Export	15%

Source. R. W. Beck, Inc.

growing at a rate of approximately 40% per year. Supporting this growth is the development of American Society for Testing and Materials (ASTM) test methods, standards, and specifications, which has provided engineers the data they need to use these lumber-like products for engineered applications that include decks, docks, and railroad ties. Although most plastic lumber produced is made from sorted plastics and is of high quality, some manufacturers fill local niches where they will take all mixed plastics collected by a community for recycling, including tubs and some film plastics, and convert the mix to plastic lumber.

It is important to note that export markets compete against U.S. reclaimers for high-quality recovered plastics and play a significant role in the overall market. Export markets also source mixed or lower grades of plastics for which there are limited domestic markets. Reclamation markets are truly national and international in scope, and there is no need for every state to have plastics reclamation facilities. In fact, many U.S. reclaimers are having a difficult time fully utilizing the capacity of their facilities.

Health and safety concerns limit the products that plastics are recycled into, and little recycled plastic is returned to food or beverage container use. Glass, steel, and aluminum packaging containers are recycled in a closed-loop with little worry because of the high processing temperatures involved and the impermeability of those materials. Plastics, however, are processed at much lower temperatures and are permeable to some chemicals. The concern is two-fold:

1. Homeowners occasionally use plastic containers to store hazardous materials, such as pesticides, gasoline, or other chemicals. These contaminated containers may be introduced into the recycling system.
2. Some virgin plastics (including certain grades of PET and HDPE) are not refined to high purities because they are not intended to come into contact with food. Containers made from these plastics are often included in recycling systems.

The ultimate concern is that the public could potentially be exposed to hazardous materials that could migrate from the recycled-content container into the food or beverages they consume.

The use of recycled plastics in food and beverage containers is not expressly forbidden. The U.S. Food and Drug Administration (FDA) evaluates individual company requests on a case-by-case basis. Each company must submit an application and describe in detail its sourcing controls, its reclamation system, testing protocols, and the results of test runs of plastics that have been intentionally contaminated with chemicals. The hopeful end result is that the FDA will issue a "no objection letter" to the company that says the company may recycle the plastic specified in the application into food or beverage containers, if: (1) the plastic is recycled through the company's recycling system; and (2) the recycled plastic is restricted in use (e.g., used only at low temperatures or only used in the middle layer of a multilayer container). Since 1990, the FDA has only issued 57 letters of "no objection."

OTHER DIVERSION OPTIONS

It is technically possible to recycle or reuse virtually all types of discarded plastics; however, the cost to do so would be prohibitive. The key to a healthy plastics recycling industry is for plastics to be recycled when and where it makes sense. In other words, the recycling of a particular type of plastic product should be environmentally and economically responsible and sustainable.

Recycled postconsumer plastics must compete with preconsumer recycled plastics and virgin plastics in terms of both price and performance. Recycled preconsumer plastics are generally clean and reasonably uniform as to resin type and color, and are generated in large quantities at a limited number of plastic product manufacturing facilities. Consequently, the cost to collect and process preconsumer plastics into a useful raw material is low. This economic fact suggests that clean preconsumer material will be consumed first and preferentially to postconsumer material. Virgin plastics prices ultimately set a price cap that recycled resins must remain below in order to be competitive.

For many postconsumer plastic products, the full cost to collect, process, and reclaim (less avoided disposal costs) exceeds the value of the recycled resin when compared to competing preconsumer and virgin resins. This is particularly the case for certain plastic products that are source-reduced and very resource-efficient in the first place, such as film plastics (bags) and foamed plastics (polystyrene), where the amount of plastic used to make the product is very small.

Because plastics are essentially energy (natural gas and oil) embodied in solid form, with more BTU value than coal, there are diversion opportunities for recovering energy from them rather than putting them in a landfill. Most of the MSW combustion currently practiced in the United States incorporates recovery of an energy product (generally steam or electricity). The resulting energy produced reduces the amount needed from other sources such as natural gas, oil, coal, or nuclear power, and the sale of the energy helps to offset the cost of operating the facility. Today, there are 103 waste-to-energy plants operating in the United States with a design capacity close to 100,000 tons per day, generating enough electricity to meet the power needs of 1.2 million homes and businesses. These plants operate in 32 states, although they are concentrated in the Northeast and the South.

Over 20% of discarded plastics (9 billion pounds annually) is burned for fuel in the United States as part of a mixed waste stream in waste-to-energy plants, or as part of a specially prepared fuel called refuse derived fuel or process engineered fuel (RDF or PEF), which is burned in industrial boilers. Approximately 5400 tons per day of refuse-derived fuel is prepared from plastics, paper, or other combustible materials.

An additional recovery technology, called pyrolysis, has also been attempted by some parties. In this process, plastics are heated to high temperatures in the absence of oxygen, so that the plastics decompose to oil, gas, and carbon black. To date, this technology has not proven to be economically feasible.

PLASTICS AND RESOURCE CONSERVATION

When the topic of resource conservation and plastics arises, many focus solely on recycling, but recycling is only part of the story. Conserving resources means using less raw materials and energy throughout a product's entire life—from its development and manufacture to its use, possible reuse or recovery (including energy recovery), and disposal.

The production, distribution, and consumption of products—regardless of the material from which they are made—uses energy resources in some way (fossil fuels, hydroelectric or nuclear power). Only about 4% of the United States' total energy consumption is used in the production of all plastic products. The use of plastics in products often conserves resources during a product's life cycle when compared to the alternatives. For example, plastics' unique characteristic—light weight, durability, formability—enable manufacturers to minimize the material used, energy consumed, and waste generated in manufacturing products ranging from coffee cups to automobiles.

Plastics often help product manufacturers do more with less material, which is known as resource efficiency or source reduction. Source reduction not only increases manufacturers' profitability and keeps consumer prices down, it also yields significant resource conservation benefits. For example, just 2 pounds of plastic can deliver 1,000 ounces—roughly 8 gallons—of a beverage. You would need 3 pounds of aluminum, 8 pounds of steel, or 27 pounds of glass to bring home the same beverage quantity. In addition to consuming less packaging resources, it takes fewer trucks and less fuel to transport plastic products or the goods they contain to the consumer. For example, it takes seven trucks to carry the same number of paper grocery bags that fit in one truckload of plastic grocery bags.

Reuse provides another significant way to conserve resources, reduce trash disposal costs, and extend landfill capacity. Plastics' durability allows many products and packaging to be reused over and over again. Not surprisingly, in a 1997 survey, Wirthlin Worldwide found that more than 80% of Americans reuse plastic products and packaging in their homes. Many U.S. businesses have also made the decision to receive their supplies and ship their products in reusable plastic shipping containers rather than single-use corrugated boxes. Over a two-year period, the Ford Motor Company eliminated more than 150 million pounds of wood and cardboard packaging by asking its suppliers to use returnable plastic shipping containers and durable plastic pallets. Returnable containers are also making major inroads in the produce and meat packaging industries.

Plastics also help conserve energy in homes and businesses because of their superior insulation properties. Vinyl windows, for instance, can save the average homeowner between $150 and $450 each year on heating and cooling costs compared to other types of windows. Similarly, 53 billion kilowatt hours of electricity are saved every year by improvements in major appliance energy efficiency made possible to a great extent by plastic polyurethane foam insulation.

GLOSSARY OF PLASTICS TERMS

Blow molding A manufacturing process in which molten plastics with the consistency of chewed-up bubble gum are blown like a bubble inside of an empty mold, which is in the shape of the object to be produced.

Coding system for plastic containers A coding system developed by the Society of the Plastics Industry designed to identify by resin type the most common plastic packaging resins used in the manufacture of rigid plastic containers.

Depolymerization A special recycling process by which certain types of plastics (such as PET and Nylon) can be broken down to their basic building blocks, allowing them to be refined to high purities and remanufactured into virgin-equivalent resin.

Injection molding A manufacturing process in which plastics that are runny when melted are injected into a mold in the shape of the product to be produced.

Monomer A "one-unit" building block material that when connected to other units forms long repeating chains called polymers ("many units") or plastics.

Packaging Materials, containers, and components used in the containment, protection, movement, and display of a product or commodity. This includes pallets, cardboard boxes, and other types of transport packaging.

Plastic lumber A lumber-like product made wholly or in part from recycled plastics that doesn't rot and is not attacked by termites or other insects.

Postconsumer A product that has entered the stream of commerce and served its intended purpose. This includes used products from commercial and industrial as well as residential sources.

Pre-consumer A product or scrap material that has not entered the stream of commerce or served its intended purpose.

Reclamation facility A facility that takes recovered products and materials and size-reduces, cleans, purifies, and returns them to a feedstock form ready for manufacture.

Recyclable materials processing or wholesaling facility A facility primarily equipped for receiving separated recyclables that removes contaminants, may do some limited sorting, and then densifies those materials for economical long-distance shipment to markets.

Resin A term that refers to the base plastic, usually in the raw material feedstock form, from which plastic products are made.

Reuse A source reduction activity involving the recovery or reapplication of a package, used product, or material in a manner that retains its original form or identity.

Recovery of materials Removing MSW from the waste stream for the purpose of recycling (including composting).

Regrind Clean preconsumer manufacturing scrap plastics, which normally is incorporated straight back into the manufacturing process by plastic product manufacturers.

Source reduction Activities that reduce the amount or toxicity of wastes before they enter the municipal solid waste management system.

Thermoplastic A family of plastics that can be melted and reformed when reheated.

Thermoset A family of plastics that cannot be melted and reformed after initial manufacture.

CHAPTER 15
STEEL RECYCLING

GREGORY L. CRAWFORD
Vice President, Operations
Steel Recycling Institute
Pittsburgh, Pennsylvania

INTRODUCTION

The North American steel industry has been recycling for decades. The industry recognized that the products it had already produced were a valuable resource for creating more steel, making recycling an economic decision. Today, economics and environmental considerations work hand-in-hand in creating North America's most recycled material: steel.

The properties of steel as a metal are what make it recyclable, but recycling occurs only through a well-developed infrastructure, and consumer, public-, and private-sector participation that moves used products from homes and businesses into that infrastructure.

As a result of the economic and environmental benefits, old steel has become a necessary raw material for making new steel. The steel industry currently recycles about 65% of all steel produced; it is referred to by the industry as steel "scrap" (Fig. 15.1), and it comes from a range of sources, including steel mills, fabrication processes, and used steel products. The steel products recycled include cars, cans, appliances, steel from construction and demolition projects, and other developing sources to be discussed later in this chapter.

Most existing steel products represent a "living" inventory of resources for new steel production. Today's food can may become tomorrow's car, and that car may ultimately be recycled to become the steel used in a new home. There is no limit to the number of times steel can be recycled.

END MARKETS

Overview

Steel mills are the major end market for most out-of-service steel products. With more than 120 operating steel mills in the United States, the steel industry has recycled nearly 340 million tons of steel scrap of all types in the past five years.

The two types of steelmaking processes are the basic oxygen furnace (BOF) process, which typically consumes a minimum of 25% scrap steel to make new steel, and the elec-

FIGURE 15.1 Steel scrap is the primary steelmaking ingredient today, and is increasingly collected from postconsumer use.

tric arc furnace (EAF) process, which consumes virtually 100% scrap steel. The BOF currently produces about 60% of the steel in the United States and the EAF produces about 40%. Foundries, which produce iron and some steels, generally use a 30–40% purchased scrap mix, in addition to the same proportion of self-generated scrap, to achieve their final product. These facilities use steel cans as a scrap resource for making ductile and gray iron products as well as cast steel.

Basic Oxygen Furnace Steelmaking

Integrated steel mills use a blast furnace to reduce iron ore to molten iron (Fig. 15.2). Then the BOF consumes scrap steel together with molten iron to make new steel. The BOF derives its name from the high-speed blast of oxygen used in its steelmaking process.

During the process, prepared steel scrap is magnetically hoisted into a scrap charging hopper (a large open container resembling an open-top railcar) and dumped into the furnace. A ladle of molten iron is then poured into the furnace, and a lance is lowered through a funnel-shaped opening at the top of the furnace to blow oxygen onto the metal bath. The oxygen reacts with the carbon in the molten iron in a sustained chemical reaction, generating enormous heat and creating steel.

Each "heat" of steel takes approximately 45 minutes to produce, with the actual conversion of iron to steel taking about 20 minutes. The final product contains a minimum of 25% steel scrap (or in today's environmental terminology, recycled content). Some products made with steel from the basic oxygen furnace include appliance shells; automotive

FIGURE 15.2 In the basic oxygen furnace, molten iron is mixed with the scrap steel and blasted by oxygen.

parts; food, paint, and aerosol cans; 55-gallon drums; and studs (steel "2×4s") used in residential steel construction. The major required characteristic of steel for these products is ductility.

Electric Arc Furnace Steelmaking

Whereas the BOF combines raw material and steel scrap to produce new steel, EAFs are charged with virtually 100% steel scrap, although some are using increasing proportions of direct reduced iron (DRI) or other scrap substitutes (Fig. 15.3). Specific grades of steel scrap are charged with small amounts of fluxing material into the furnace. Charging typically takes place through the furnace roof, which is lifted or swung aside. During the steelmaking process, three large cylindrical electrodes are lowered through openings in

FIGURE 15.3 The electric arc furnace consumes virtually 100% scrap steel to make new steel.

the roof to melt the steel scrap. Some of the products made with steel from the EAF include structural beams, steel plates, and reinforcement bar. The major required characteristic of steel for these products is strength.

Ferrous Scrap Processors

By definition, ferrous scrap dealers process iron and steel scrap (Fig. 15.4). In some cases, that is exclusive; in other cases, they handle other materials. Ferrous scrap dealers process a wide array of materials for recycling, including iron and steel scrap. With about 1500 such operations located across the country, many scrap dealers now handle retail and wholesale quantities of steel products received from the public, curbside programs, drop-off programs, industry, and resource recovery plants. The dealers prepare the steel scrap according to end-market specifications and ship truckload or railcar quantities to the steel mills and foundries. Detinners and other broker/processors also prepare steel cans for use as raw material feedstock for the copper precipitation and ferro alloy industries.

SOURCES OF SCRAP STEEL

The steel industry's steady, increasing demand for steel scrap has had notable consequences. First, the United States has developed the most efficient steel recycling infrastructure in the world. The United States is also the world's largest exporter of steel scrap,

FIGURE 15.4 The ferrous scrap processing industry has been a long-term steel industry partner in helping to obtain and prepare steel scrap for recycling.

exporting nearly ten million tons of steel scrap in 1997 alone.

Second, like any other raw material, steel scrap has true economic value. It is more economical to make new steel from old steel than to produce virgin steel. In fact, today's modern BOF requires scrap steel as part of the technological process. As a result, steel scrap is collected and prepared for recycling in the BOF or EAF for its market value as well as for the energy savings and natural resource conservation it provides to the steel industry.

Types of Steel Scrap

The steel industry has traditionally obtained steel scrap for recycling from three resources: mills, manufacturers, and postconsumer sources.

Steel mills themselves were once their own best suppliers of steel scrap. Mill scrap—often referred to as "home" scrap, which is leftover pieces of steel from steelmaking and defective or rejected products—was collected and recycled into new steel right at the mill. However, breakthroughs in technology, such as the BOF and continuous casting, have increased the quality and efficiency of steelmaking and significantly reduced the amount of home scrap generated in the mill. Today, steel mills are dependent upon outside sources of steel scrap.

Outside sources of steel scrap include manufacturing scrap and postconsumer scrap. Manufacturing scrap (also called "prompt" scrap) is steel scrap created during the manufacture of steel products. This scrap includes, for example, trimmed or punched-out pieces of steel sheet from an appliance manufacturer or a can-making operation. This scrap is typically recovered from the manufacturer by a ferrous scrap processor for preparation for recycling.

Postconsumer scrap includes steel products that have fulfilled their useful purpose and are collected for recycling. Steel mills depend on collected appliances, used steel cans, retired automobiles, and old bridges and other construction materials to meet their scrap needs (Fig. 15.5).

Inherent Recycled Content of Steel

The pre- and postconsumer content of steel products in the United States can be statistically determined for the calendar year 1997 using a variety of information from the American Iron and Steel Institute (AISI), the Institute of Scrap Recycling Industries (ISRI), and the U.S. Geological Survey. Additionally, a study prepared for the AISI by William T. Hogan, S.A., and Frank T. Koelble of Fordham University is used to establish pre- and postconsumer fractions of purchased scrap in these calculations.

Individual company statistics are not applicable or instructive because of the open loop recycling capability that the steel and iron industries enjoy, with available scrap typically going to the closest melting furnace. This open loop recycling allows, for example, an old car to be melted down and used to produce a new soup can, and then, as the new soup can is recycled, it is melted down to produce a new car, appliance, or perhaps a structural beam used to repair some portion of a bridge.

Recycled Content Usage Calculations

The following text describes a method of calculating recycled content usage, both for the

FIGURE 15.5 Steel scrap found at a ferrous scrap processing facility includes postconsumer goods such as cans, cars, appliances, and construction materials.

basic oxygen and the electric arc furnace processes.

Basic Oxygen Furnace. The basic oxygen furnace facilities consumed a total of 19,551,500 tons of ferrous scrap in the production of 64,500,000 tons of liquid steel during 1997. Based on U.S. Geological Survey statistics, 7,628,600 of these ferrous scrap tons had been generated as unsalable steel product within the confines of these steelmaking sites. In the steel industry, these tons are classified as "home scrap," and are therefore by definition, preconsumer. Additionally, these operations reported that they consumed 132,200 tons of obsolete scrap (buildings and warehouses dismantled on-site at the mill) during this time frame. This volume is classified as postconsumer scrap.

As a result of the above, based on the total scrap consumed, outside purchases of scrap equate to 11,790,700 tons [19,551,500 – (7,628,600 + 132,200)]. According to the findings of the Fordham University study, the post-consumer fraction of the purchased ferrous scrap would be 83.4%, and 16.6% of these purchases would be preconsumer. This equates to 1,957,300 tons of preconsumer scrap (11,790,700 × 16.6%). This "prompt scrap" is mainly scrap generated by manufacturing processes for products made with steel.

Therefore, based on the above information, it is determined that the total recycled content to produce the 64,500,000 tons of liquid steel in the BOF is as follows:

$$19{,}551{,}500 \div 64{,}500{,}000 = 30.3\%$$

(Total Tons of Ferrous Scrap ÷ Total Tons of Liquid Steel =
Total Recycled Content Percentage)

Also, based on the above, it is determined that the post-consumer recycled content of BOF steels is as follows:

$$(11{,}790{,}700 - 1{,}957{,}300) + 132{,}200 = 9{,}965{,}600$$

and

$$9{,}965{,}600 \div 64{,}500{,}000 = 15.4\%$$

(Post-Consumer Scrap Consumed ÷ Total Tons of Liquid Steel =
Post-Consumer Recycled Content Percentage)

Electric Arc Furnace. The electric arc furnace (EAF) facilities consumed a total of 42,128,000 tons of ferrous scrap in the production of 43,000,000 tons (reported figure) of liquid steel during 1997 (with the addition of scrap substitutes). Based on U.S. Geological Survey statistics, 7,615,200 of these ferrous scrap tons had been generated as unsalable steel product within the confines of these facilities (steelmaking sites). Again, in the steel industry, these tons are classified as "home scrap," and are therefore by definition, preconsumer. Additionally, these operations reported that they consumed 39,200 tons of obsolete scrap (buildings and warehouses dismantled on-site at the mill) during this time frame. This volume is classified as postconsumer scrap.

As a result of the above, based on the total scrap consumed, outside purchases of scrap equate to 34,473,600 tons [42,128,000 – (7,615,200 + 39,200)]. According to the findings of the Fordham University study, the postconsumer fraction of the purchased ferrous scrap would be 83.4%, and 16.6% of these purchases would be preconsumer. This equates to 5,722,600 tons of preconsumer scrap (34,473,600 × 16.6%). This "prompt scrap" is mainly scrap generated by manufacturing processes for products made with steel.

Based on the above information, it is determined that the total recycled content to produce the 43,000,000 tons of liquid steel in the EAF is as follows:

$$42{,}128{,}000 \div 43{,}000{,}000 = 98\%$$

(Total Tons of Ferrous Scrap ÷ Total Tons of Liquid Steel = Total Recycled Content Percentage)

Also, based on the above, it is determined that the post-consumer recycled content of EAF steels is as follows:

$$(34{,}473{,}600 - 5{,}722{,}600) + 39{,}200 = 28{,}790{,}200$$

and

$$28{,}790{,}200 \div 43{,}000{,}000 = 67\%$$

(Post Consumer Scrap Consumed ÷ Total Tons of Liquid Steel = Post-Consumer Recycled Content Percentage)

The above discussion and calculations demonstrate conclusively the inherent recycled content of today's steel in North America.

As the recycled content of BOF and EAF steels is understood, one should not attempt to select one steel producer over another on the basis of a simplistic comparison of relative scrap usage or the recycled content percentage. Rather than providing an enhanced environmental benefit, such a selection could prove more costly in terms of total life cycle assessment energy consumption or other considerations. Steel does not rely on "recycled content" purchasing to incorporate or drive scrap use. It already happens because of the economics. Recycled content for steel is a function of the steelmaking process itself. After its useful product life, regardless of its BOF or EAF origin, steel is recycled back into another steel product. Thus, steel with almost 100% recycled content cannot be described as environmentally superior to steel with 25% recycled content. This is not contradictory because they are both complementary parts of the total interlocking infrastructure of steelmaking, product manufacture, scrap generation, and recycling. The recycled content of EAF relies on the embodied energy savings of the steel created in the BOF. And the BOF infuses a greater supply of new steel into service to provide for continued economic development.

AUTOMOTIVE RECYCLING

For decades, steel recycling has been the driving force behind automobile recycling efforts (Fig. 15.6). The roots of automobile recycling lie in the steel industry's need for ferrous scrap. For reasons of strength, economy, and durability, steel has been the material of choice for automobile manufacturing since the beginnings of automobile manufacturing. Over time, manufacturers and scrap processors alike have come to realize the additional benefits of steel's infinite recyclability. In 1997 alone, the steel industry recycled enough steel from old automobiles to produce almost 13 million new automobiles. Because of the economic and environmental benefits of recycling automotive steel, automobiles are too valuable a resource to simply bury in a landfill when they are no longer in service.

The steel industry, together with the scrap processing industry, is responsible for laying the groundwork for the highly efficient recycling infrastructure for automobiles that

FIGURE 15.6 Steel is the material that "drives" automotive recycling due to the volume of steel that is recovered from a scrapped automobile.

exists today. To provide more steel scrap to the growing steel industry and to reduce the automobile's impact on the environment, the two industries partnered in the early 1960's to develop the first automobile shredders. Today, a network of dismantlers and shredders effectively processes the millions of vehicles taken off the road each year in a cost- and resource-effective manner.

Automobile Dismantlers

Automobiles begin their recycling trip with a brief but essential stop at one of the estimated 12,000 automobile dismantlers in North America. Auto dismantlers remove potentially hazardous materials and salvage selected components from an automobile.

These car "surgeons" remove tires, batteries, fluids, and any reusable parts. Even a car in the very worst shape may still contain valuable working parts that can be used to repair other vehicles (Fig. 15.7). Selected items such as engines and transmissions, as well as other auto parts in relatively good condition, are resold to the public or auto repair garages and body shops.

After removing reusable components, auto hulks are flattened and shipped to a ferrous scrap processor, where they are weighed for payment and unloaded to await the next step.

The Shredder

At a major ferrous scrap yard, the shredder is the primary piece of equipment for prepar-

FIGURE 15.7 Before a car is recycled into a new product, its reusable parts are removed for resale.

ing automobiles for recycling. Shredding a car breaks it down into small pieces so its basic materials may be separated for recycling. In addition, steel mills often prefer shredded steel scrap because it can be handled and recycled in the furnaces more easily. While cars are the commodity most often fed to a shredder, appliances, bicycles, and other steel products are also shredded for recycling (Fig. 15.8).

There are more than 200 scrap yards in North America equipped with automobile shredders, with most of these yards found in the United States. Generally, an automobile shredder consists of a sprawling network of conveyors and a large, rectangular central unit that houses the actual shredding equipment.

Automobiles are fed into the shredder, and in about a minute, they are reduced to fist-sized chunks of steel and other material. Steel components, which comprise the majority of the automobile, are magnetically separated and discharged from the conveyor system to accumulate in large piles of shredded scrap in preparation for shipment by rail or water to the consuming steel mill or foundry. Nonferrous metals are sorted manually or mechanically from a conveyor belt and shipped to their appropriate end markets. About 800,000 tons of nonferrous metals are recovered and recycled each year. The remaining materials are referred to as nonmetallic fluff. Fluff consists of bits of plastic, rubber, fabric, and glass. This material is currently landfilled.

APPLIANCE RECYCLING

Appliances are made with recycled steel and are recyclable at the end of an abundantly long service life. Appliances including refrigerators, freezers, washers, dryers, mi-

FIGURE 15.8 The shredder helps to tear down automotive steel into fist-sized chunks that are magnetically separated from the remainder of the material for recycling.

FIGURE 15.9 The steel and iron in appliances ensure that they will go to the scrap yard instead of the landfill.

crowaves, and many other household items are recycled because of the economic and environmental value of their steel and iron components (Fig. 15.9).

Appliances, sometimes called "White Goods," are a significant recyclable resource that can easily be processed for recycling and will conserve natural resources for future generations. On average, about 75% of an appliance, by weight, is steel. Appliances also yield other valuable materials, such as copper wiring and other metals, which can be recycled. Additionally, the appliances themselves are often quite bulky, and landfilling them would take up large volumes of space in our nation's landfills.

Despite continuing design changes over the years, steel is still the primary material used in appliance manufacturing. And as long as this is true, a recycling infrastructure will economically divert tons of appliances from landfills. However, in recent years, an increasing volume of plastics and other materials have been incorporated into appliance manufacture. During the recycling process, these other materials must be separated manually or mechanically, which increases the volume of materials being sent to the landfill, since nonmetallic fluff has not, to date, proven to be marketable for any economic reuse.

How Appliances Are Prepared for Recycling

Many communities have established periodic or permanent collection programs to ensure that appliances are not landfilled. These appliances are purchased or accepted by ferrous scrap processors who prepare them for recycling by the steel industry. In some cases, specialty recycling companies perform initial preparation.

To prepare the appliances for recycling, processors typically remove electric motors, capacitors, switches, and other mechanical parts. Some appliances have special processing considerations before a ferrous scrap processor can accept the appliance for recycling, unless they perform this work themselves.

Legislation enacted more than 20 years ago has significantly altered the way appliances are processed for recycling today. This legislation resulted in a U.S. Environmental Protection Agency (EPA) mandate that, prior to recycling or disposal, any refrigerant gases deemed as ozone-depleting and used in appliances must be captured for recycling. The chlorofluorocarbons (CFCs) and hydrochlorofluorocarbons (HCFCs) that are reclaimed are cleaned by a refrigerant manufacturer and then reused in the maintenance and repair of other units. In some areas, scrap dealers have the CFC removal equipment and certified technicians to capture the gases (Fig. 15.10). In other areas, specialty recycling companies provide this service, either independently or in association with appliance dealers or the local government. In any case, appliance recycling processors are responsible for ensuring that the refrigerant gases have been captured for reclamation.

After initial preparation, some appliances are crushed and bundled for subsequent shipment directly to a steel mill or to an automobile shredder. Appliances are normally fed into an automobile shredder by a large crane with an electromagnet, which loads the appliance onto a steel conveyor belt. The appliance is crushed and ripped apart as it enters the shredder (Fig. 15.11). Inside, free-swinging hammers shred the hulk into fist-sized chunks. The material then exits the shredding unit and continues down a conveyor belt for mechanical sorting. The majority of the shred stream is the steel components which are first magnetically separated and stockpiled for shipment to a steel mill for recycling. Nonferrous metals are manually or mechanically sorted for recycling, and the non-metallic fluff remaining on the belt is ultimately discarded in a landfill (Fig. 15.12).

FIGURE 15.10 Refrigerant gases may be removed by a qualified technician either at the scrap yard or at a specialty recycling company.

FIGURE 15.11 The versatile shredded does double duty with appliances as well as automobiles.

FIGURE 15.12 Steel's magnetic attraction is a big benefit to recycling, since steel can be so easily separated from other materials.

CONSTRUCTION AND DEMOLITION RECYCLING

For decades, the steel removed from demolition of commercial structures such as buildings, bridges and stadiums has been recycled. This vast resource of recyclable steel continues to be tapped for new steelmaking. The estimated recycling rate for steel from demolition sites is over 50%, and for miscellaneous steel—including reinforcement bar—the rate is estimated at more than 40%. Additionally, any steel left over at a construction job site, though far less abundant than that from demolition, is also recycled.

Residential Construction

Residential steel framing and roofing are relatively new markets for steel. Builders and home owners alike benefit from the advantages of using a "new" material—steel—to build their residences. Using steel to build homes also means turning old appliances, automobiles, and other scrap into recycled products.

Design and Construction

Almost everyone has heard of a "two-by-four," which refers to a 1½-inch by 3½-inch rectangular piece of wood. Naturally, builders and architects are thoroughly familiar with the concept of "two-by-four" construction. In residential light-gauge steel framing, the two-by-four wood studs are replaced by steel studs with similar or identical physical dimensions (Fig. 15.13). Steel framing does not twist, warp or split, so there is no need to sort out damaged or poor quality product. This translates into truer walls and more nearly perfect corners during construction and the life of the structure. Among steel framing's other construction advantages for the builder is its light weight. Steel studs weigh one-third as much as wood. Steel also has the highest strength-to-weight ratio of any construction material. Steel framing typically arrives at the job site precut to the builder's specified dimensions. This saves time and reduces waste. The minimal leftover steel scrap that may remain at the job site can be collected and delivered to a ferrous scrap processor for recycling.

Steel roofing comprises about 80% of the metal roofing market. A prepainted steel roof can easily last 30 years or more. It is gaining popularity in heavy snowfall areas due to its snow-shedding ability; on the West Coast it is used in seismic areas because of its light weight and noncombustability. Steel roofing is, of course, recyclable when it has completed its useful life.

Recycled Content/Recyclability

Have you ever heard of the saying "Waste not; want not?" Residential steel framing is the epitome of this concept in home construction. Steel framing and roofing are recyclable, and environmentally sensitive architects are encouraged to incorporate job site recycling requirements into their builder specifications.

In addition to being recyclable, all steel framing and roofing contains a minimum of 25% recycled steel. In this fashion, the refrigerators and steel food cans that were in the kitchens of yesterday's homes are helping to build the new homes of tomorrow.

FIGURE 15.13 Steel studs are similar in dimension to wood studs, but they are C-shaped instead of solid as a wood stud is.

Even if they last for more than 100 years, all buildings and homes eventually come to the end of their useful lives. Steel framing and construction materials can be dismantled and returned to the steel industry to be recycled into new steel products. As with all other steel products, the steel is only "borrowed" until it is returned for recycling.

STEEL CONTAINERS

What Is a Steel Can?

Steel cans are food, beverage, paint, aerosol, and general-purpose containers made from very thin gauge steel sheet. In order to protect the contents from corrosion, the steel for cans usually has an extremely thin coating of tin, about 30 millionths of an inch. Steel food cans, which make up more than 90% of all metal food containers, are often called "tin" cans because of this coating of tin. Some steel cans (such as tuna cans) and lids, are made with tin-free steel. Empty steel cans are completely recyclable by the steel industry and should be included in any recycling program (Fig. 15.14).

From the can production standpoint, there are two types of steel cans: three-piece cans and two-piece cans. The three-piece can typically contains a wide assortment of products. To produce these cans, the lids and bottoms of the cans are punched or cut out of rolls of tin plate that are gradually unwound. Once filled with holes, the original tin plate is called a "skeleton" and sent back to a detinning company or steel mill as prompt scrap, to produce new steel. Other rolls of tin plate are cut into the size of cans and rolled. The side seams are then welded at speeds higher than 500 can bodies per minute. The majority of welded cans have tin plate bodies, but have either tin plate or tin-free steel ends, which are mechanically crimped to the body.

The other type of steel can, the two-piece can, is usually referred to as draw–redraw (DRD) because of the production process. At present, more than one-third of all steel cans are two-piece cans, which tend to be smaller cans, such as tuna and soup or vegetable cans. Production of a two-piece can eliminates both a side seam and a separate bottom end piece. The entire body and bottom are drawn from one piece of steel, so that the only seam is between the single-unit body and the lid. And this seam is simplified because there is no side seam overlap. Most steel two-piece cans are made of tin-free steel. Tin plate may be used for the lids if a shiny appearance is desired, and for bodies of cans that contain foods likely to cause corrosion. Steel coils, coated on both sides, are usually the starting material for producing these cans, although flat sheet stock can also be used.

In the DRD production process, a shallow cup is produced from a flat, circular blank that was punched or cut from the can sheet. The diameter of the cup is reduced as the can is deepened. The can-making presses usually allow for two or more cans to be formed with each press stroke. Beading is performed to add strength to the can. Beads are grooves or ribs formed on the side of the container to stiffen the can body and to improve label retention. Lids are formed and crimped to the body as in the three-piece method.

Collection Sources and Processing of Steel Cans

Steel cans are recycled through various collection programs, processing methods, and end markets. Collection methods include curbside collection, drop-off collection sites, multicommodity buy-back centers, resource recovery plants, and waste processing facil-

FIGURE 15.14 Steel cans include food, beverage paint and aerosol cans; when empty, they are all recyclable.

ities. Because steel cans are magnetic, they are typically separated from other recyclables or from municipal solid waste through overhead magnetic separation, making them the easiest recyclable to sort and handle in any recycling program.

Curbside Recycling Programs

Communities have several options to choose from when implementing curbside collection programs. In the "commingled" curbside collection system, residents separate recyclables from their household trash and place them into one recycling box or bag that is later placed at the curb (Fig. 15.15). The recyclables usually include steel and aluminum cans, glass jars and bottles, and plastic containers. Newspapers may be bundled for collection as well. In many programs, the mix of steel cans may include empty paint and aerosol cans in addition to food and beverage cans. The steel can's magnetic property plays a valuable role in this recycling option, since magnetic separation is used in the ma-

FIGURE 15.15 Curbside recycling allows consumers to simply place a bin of mixed recyclables at the curb for pickup.

terial recovery facility (MRF) that receives, separates, and processes the recyclables to the specifications of the appropriate end markets for each material.

In the case of a commingled curbside program, the collection vehicles enter the MRF and tip their loads into a receiving pit or onto a tipping floor. The mixed recyclables are loaded onto a conveyor system from the pit or floor, and magnetic separation is used to remove steel cans. An overhead transverse magnetic separator conveyor is preferred over the slightly lower cost, end-pulley design because it sorts and handles higher volume easily, with almost no entrapment of other materials.

Some MRFs are labor-intensive and use minimal equipment. In these facilities, after magnetic removal of steel, the other recyclables are hand sorted from a conveyor line by workers. Other MRFs are more capital intensive and use less labor. Their systems are normally proprietary, so equipment design varies. As in any other system, magnetic separation for steel cans is the first and easiest part of the process.

A second and more rarely used form of curbside collection requires more participation from the residents but minimal sorting at the MRF. In this case, each type of recyclable material is source-separated into its individual bin. Steel and aluminum cans may be placed together in one bin. The truck used by the hauler has different storage compartments to keep the materials separate from one another. Steel cans are placed in the same compartment as aluminum cans because the cans are later magnetically separated at the MRF. No extra truck compartment is needed for the two metals.

A variation of this type of curbside collection uses a single commingled box at the curb, but the hauler hand-sorts the materials into the different compartments on the truck when the material is picked up at the curb.

The MRF's processing for source-separated programs, such as those described above, differs somewhat from commingled programs. In this case, the processing requires relatively little equipment, since metals, glass, and plastic have already been sorted, but careful inspection is necessary to avoid accidental mixing and contamination of these materials. Glass may need to be sorted by color, and plastic by type, depending on the details of the program. As noted, magnetic separation allows steel cans to be separated easily from aluminum cans. Each material is then processed for its respective end market.

Drop-Off Recycling Programs

A drop-off program enables the public to deliver packaging and other selected recyclables to a designated collection site, which may have collection boxes for commingled steel and aluminum cans as well as for other materials. Drop-off collection sites provide a recycling option for communities where curbside programs are impractical because of low population density (Fig. 15.16). Also, some communities develop drop-off collection sites as an interim program before beginning a curbside program.

Multicommodity Buy-Back Centers

Another option for communities is the multicommodity buy-back center. Originally, these centers purchased only aluminum beverage cans for recycling; however, many cen-

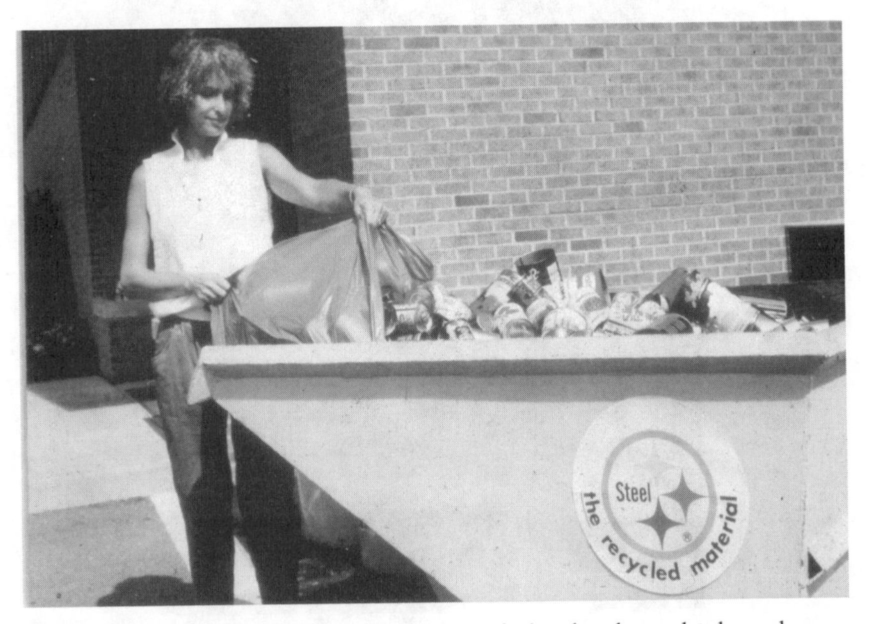

FIGURE 15.16 Drop-off sites are especially handy for less densely populated areas because the residents take recyclables to the site, rather than having a truck traverse a sparsely populated geographic area.

ters now accept recyclable materials such as steel, glass, plastic, or old newspaper. Even though recycling centers were set up to handle beverage containers, some multicommodity centers also accept or buy clean steel food cans.

Dockside Recycling for Commercial and Institutional Food Service

While initially limited to residential programs, recycling has spread to the commercial area to include office buildings, restaurants, and hotels, and to the public sector to include institutions such as hospitals, schools, and correctional facilities. The programs vary greatly depending on the nature of the establishment. Food service operations use the large one gallon steel food containers extensively (Fig. 15.17).

The first step in dockside steel can recycling is to rinse the steel cans to remove most

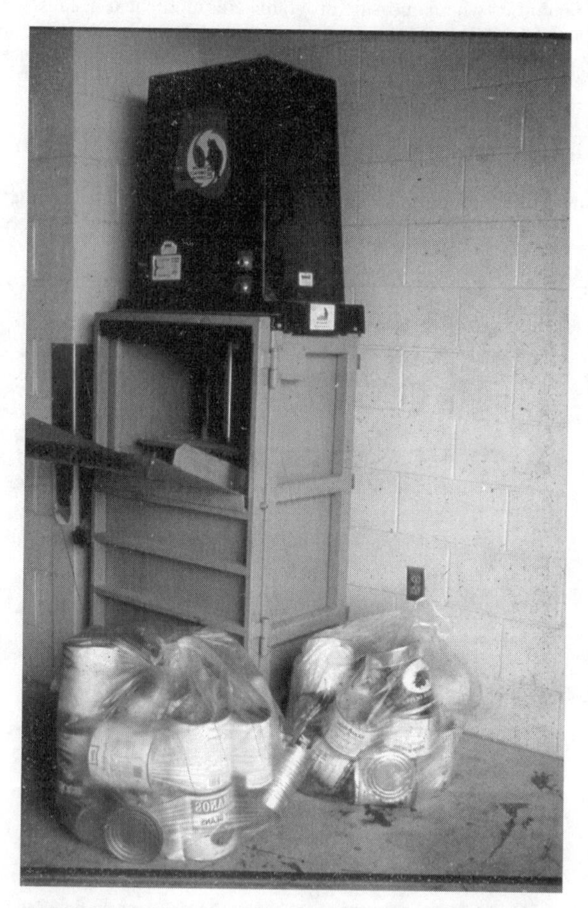

FIGURE 15.17 Cans used in food service preparation are generally the one gallon size, which allows the food service facility to purchase in bulk.

FIGURE 15.18 When cans have been rinsed, they are ready for flattening and storage to await pickup.

food particles for sanitation reasons (Fig. 15.18). But cans may be rinsed without wasting water by running them through empty spaces in the automatic dishwasher. They may also be rinsed in the leftover dish water that was used to wash large pots, pans, and utensils.

Once the cans are rinsed, they should be flattened to accommodate compact storage. Flattening steel cans may be accomplished mechanically or manually. If the establishment does not have a mechanical can flattener, the process can be completed manually by stepping on the side of the can. This can be done more easily if the bottom of the can is removed with a can opener. The lid and bottom of the can should also be recycled.

The third step in dockside recycling is to ship the steel cans and other recyclables from the establishment. Often, arrangements can be made with the waste hauler, who will provide for recycling steel cans and other materials. This normally means that the hauler provides and maintains a container for the recyclables. Another possibility for the business or institution is to work with a ferrous scrap processor or independent recycler. In this case, the steel cans will either be picked up by or delivered to the intermediate processor.

Resource Recovery Plants

Resource recovery plants require no separation of household or commercial/institutional trash. Instead, trash is simply collected by garbage trucks and hauled to the plant. Waste-to-energy plants burn the trash directly to produce energy in the form of steam or electricity. Refuse-derived fuel plants process the trash to remove unburnable components and produce organic combustible material. This material is sold to other facilities for power generation in lieu of or as a supplement to other fuels.

At most resource recovery plants, steel cans and other postconsumer steel products are magnetically separated from the solid waste stream, postburn or preburn, for recycling. The recovered steel normally consists of about 50% steel cans and 50% other postconsumer ferrous. Since magnetic separation at these plants results in steel recycling rates of well over 90%, this might be referred to as "automatic recycling." There are about 118 resource recovery plants in operation across the country; almost all the larger operators magnetically separate steel cans and other postconsumer steel products for recycling.

Mixed Waste Processing Facilities

A mixed waste processing facility is somewhat like a MRF; however, all trash, rather than just the recyclables, is hauled to the facility for processing. Some communities may refer to this facility as a "dirty MRF" or even as a resource recovery facility. After the trash stream is sized or shredded, perhaps with a flail mill, steel cans and other iron and steel scrap are magnetically separated or physically pulled from the municipal waste. Once separated, the recyclables are processed and shipped to end markets. The unrecoverable garbage is hauled to a landfill.

Variations of waste processing facilities may operate at metro area transfer stations. During the emptying and transloading process, transferring material from smaller neighborhood trucks into larger 40 foot trailers, magnetic separation and manual sorting can be used to remove steel cans and other iron and steel items for recycling.

Some landfills have also set up waste processing facilities on-site to remove steel cans and other iron and steel materials by magnetic or physical separation for recycling. This has helped to minimize landfill use.

End-Market Role

The locations and specifications of end markets must be taken into account before beginning any recycling program. Collected steel cans are sold to a variety of intermediate processors or, in some cases, directly to end markets. Intermediate processors, such as ferrous scrap dealers, provide a convenient service for smaller recyclers because they have the capacity to process truckload or railcar quantities of steel cans and ship them to the end market. Steel mills, iron and steel foundries, and detinning companies share the end-market role for recycling steel cans, as previously noted.

CASE STUDIES: EXAMPLES OF THE VARIOUS TYPES OF PROGRAMS

A Curbside Recycling Program

For over a decade residents of Milwaukee, Wisconsin have participated in a commingled curbside recycling program that includes the following recyclable materials: steel and aluminum cans, glass bottles and jars, PET and HDPE plastic, newspapers, magazines, and corrugated cardboard.

Utilizing semiautomated, 95 gallon split carts, recycling is collected from nearly 200,000 backyards by city crews on a monthly basis. These materials are hauled to a USA CRINC MRF for processing. USA CRINC then sells the steel cans to local end markets.

Since the first year of the program, when 423,000 pounds of steel cans were collected, steel recycling has increased to a level of about 2.5 million pounds per year, totaling nearly 16 million pounds over the last 10 years.

A Drop-Off Recycling Program

Conroe Recycling Company, located in Conroe, Texas, provides drop-off recycling sites that accept steel cans for residents of Kingwood, the Woodlands, Walden on Lake Conroe, and Huntsville. The recyclables from these drop-off centers, including steel and aluminum cans, glass containers, cardboard, and high-grade paper, are processed at the Conroe Recycling Center, to which this community's residents may also bring materials for recycling.

The company collects approximately 30,000 pounds of steel cans per month, and with anticipated company expansion and greater consumer participation, the number of steel cans recycled will increase still more.

A Multicommodity Buy-Back Center

The Adams/Brown Recycling Station, located in Georgetown, Ohio, is a multicommodity buy-back center that serves Adams and Brown County recyclers. Opened in 1981, the center has grown from an aluminum, glass, and paper recycling center to one that accepts a large variety of materials, including aluminum cans and scrap, glass containers, various types of paper, steel food cans, empty aerosol, paint, and beverage cans, plastic containers, used motor oil, car batteries, red metals, zinc, and lead.

In addition to accepting such a variety of materials, the Recycling Station has made recycling more convenient to consumers by providing a mobile buy-back center that operates in any one of five locations in Adams and Brown Counties. Officials from the Recycling Station recognize that convenience is a major concern to consumers. In fact, the organization's director suggested that eventually consumers will realize that making a trip to the recycling center is similar in concept to taking clothes to the dry cleaners. Both are necessary errands, and both can easily be incorporated into one's schedule.

As part of a community action program called Adams/Brown Counties Economic Opportunities, Inc., the Recycling Station receives funding from the Department of Natural Resources, Ohio's solid waste authorities, various private companies, and other sources.

A Resource Recovery Plant

The Metro–Dade County Resources Recovery Facility (Florida), one of the largest in the country, is also a leader in front-end magnetic separation of postconsumer steel products. When the plant opened nearly twenty years ago front-end separation has facilitated the recycling of postconsumer steel products. As of 1998, the plant recovers over 600 tons of this mixed ferrous material per week. An estimated 50% of the ferrous material recovered from resource recovery facilities is steel cans. The magnetically separated iron and steel is processed on site and sold to end markets.

Ferrous materials are processed in two sections of the plant. Household garbage waste is sized and crushed, and postconsumer steel cans, along with other small steel items, are magnetically separated with the use of belt magnets. Oversized household waste and bulky municipal trash are shredded in another area of the plant, and steel is separated through a system of drum magnets.

Current capacity is 4200 tons of municipal solid waste per day; 919,000 tons were processed in 1996, with 318,728 mWh of electricity exported and 26,176 tons of steel and 133 tons of aluminum recovered. In the first quarter of 1998, 8281 tons of steel and 160 tons of aluminum were recovered.

Montenay Power Company has operated the Metro–Dade facility since June 1985.

Advantages of Steel Can Recycling

Steel cans should be part of every community's recycling program because they are household items that are easy to collect and process, and have stable, well-established, long-term markets. In addition to providing the steel industry with a much-needed resource, steel can collection helps to avoid landfill usage. In 1997 alone, the steel industry recycled about 1.7 million short tons of steel cans. This equates to a little over four million cubic yards of landfill space saved.

Other Environmental Benefits

Recycling steel cans does a great deal more than divert usable materials from landfill. Although steel cans are made with very common and inexpensive natural elements—iron ore, coal, and limestone—using steel cans in the manufacturing process preserves domestic natural resources for future generations. Every time a ton of steel cans is recycled, 2500 pounds of iron ore, 1400 pounds of coal, and 120 pounds of limestone are preserved. It only takes twenty households annually to yield one ton of steel cans.

Steel can recycling saves energy. It is about 75% less energy intensive to make new steel from recycled steel than to start with iron ore. When 2 pounds of steel cans are recycled (the average weekly consumption in a household), enough energy is saved to illuminate a 60-watt light bulb for more than 2 days.

RECYCLING PROGRAMS FOR OTHER STEEL PRODUCTS

Oil Filters

Not so long ago, used oil filters were routinely disposed of in a landfill. Today, however, oil filters are being recycled into new products, saving energy, natural resources, and landfill space.

Used oil filters are recyclable because they are made of steel (Fig. 15.19). The Filter Manufacturers Council estimates that in 1997, about 30% of used oil filters were recycled.

Some states have banned used oil filters from the landfill, while others have placed restrictions on how they can be discarded. The U.S. Environmental Protection Agency requires used oil filters to be drained of all free-flowing oil before they are discarded or recycled.

FIGURE 15.19 Empty used oil filters should be recycled from the shop or from the home.

Tire Wire

Ten tires, when left intact, can occupy more than a cubic yard of space in a landfill. For years, that has been the fate of out-of-service tires. Today, however, tires are taking a new spin on life, one that extends their use after their days as tires are gone.

Tires were initially recycled for their rubber content, which is chipped, ground and/or melted into products such as asphalt and playground padding, as well as tire-derived fuel. But thanks to advances in technology, recycling steel tire wire is also an environmentally responsible means of collecting a high-quality source of steel scrap and conserving landfill space. The average passenger tire contains approximately 10% steel wire by weight, which helps make the tire stronger and more rigid. By chipping tires and recovering the steel wire, up to 99% of the average passenger car tire can now be captured for recycling.

In 1995 alone, 10,000 tons of steel tire wire from nearly eight million tires was shipped to be recycled into products such as the steel soup cans and appliances in your kitchen, as well as the car you drive and the tires that move it.

LIFE CYCLE ANALYSIS

In September 1996, the Steel Recycling Institute, along with the American Iron and Steel Institute (AISI), commissioned Scientific Certifications Systems to conduct a study to develop a credible life-cycle analysis for steel from "cradle" through "the gate," or actual shipping from the steel mill, and on to the "grave," or the final disposition of the steel product when it is recycled at the end of its useful life..

The study is intended to provide information about steel's environmental performance that reaches beyond the data typically achieved through Life-Cycle Inventory studies. This study, the Life Cycle Stressor Effects Assessment (LCSEA), was developed to evaluate the environmental performance of industrial systems while integrating both Environmental Impact Assessment and Risk Assessment factors.

The LCSEA study focused on galvanized sheet steel, used to produce items such as residential framing materials, residential siding and roofing materials, automotive panels, heating and air conditioning ducts, and many other products, including farm equipment, buckets, and mailboxes.

Based on this research, the steel industry will be able to make certified statements of environmental achievement in the form of certified eco-profiles. They will reflect the large reductions achieved in overall environmental impacts from mining through all aspects of steelmaking.

Certified eco-profiles of steel as an industrial material will be available for a wide range of applications. The study will also produce an eco-profile for steel framing used in residential construction to serve as a basis for comparison with wood. The Executive Summary, completed in 1999, follows.

EXECUTIVE SUMMARY

Certified Life Cycle Impact Profile of North American Steel Production. Life Cycle Impact Assessment (LCIA) Study Conducted by Scientific Certification Systems Oakland, California

Study Overview. The North American steel industry contacted with Scientific Certification Systems a recognized neutral third-party certifier of industrial environmental claims, to evaluate the science of life-cycle assessment (LCA), and to examine its potential applications in assessing the environmental performance of steel production. Drivers for this work included customer demand, competitive considerations, and the industry's internal need to assess improvements.

While past environmental assessments have portrayed the steel industry as energy intensive and highly polluting (DOE 1996), the steel industry believed that a more robust, comprehensive scientific methodology such as advanced LCA might more accurately reflect its performance and achievements.

The study, now completed and peer reviewed, demonstrates that advanced LCA—specifically, life-cycle impact assessment (LCIA) conducted in accordance with international standards—does satisfy the need for a rigorous and comprehensive scientific assessment method capable of presenting an accurate environmental impact profile of steel production. Moreover, the study findings show the significant environmental progress that has made by the North American Steel Industry in its steel production operations.

It is anticipated that the results of this study will be of assistance to the steel industry for market-based claims, to support public policy positions, and to help guide regulatory reforms.

Methodology. During the course of research, it became evident that historic LCA techniques focused on life cycle inventory (LCI) alone could not produce environmentally relevant results, and as such, could not supply information suitable for environmental decision making to the industry, its customers or stakeholders.

To overcome the shortcomings of this traditional approach, advanced LCIA techniques were utilized for this study. By integrating environmental data from other techniques such as traditional environmental impact assessment (EIA) and risk assessment (RA) into the overall LCA calculation framework, the methodology converts raw LCI data into environmentally relevant impact indicators. The methodology addresses all relevant environmental issues of resource depletion and emission and waste loadings, providing a quantitative basis by which to establish an overall LCIA impact profile for products and materials.

Goal of the Study. The primary goal of the study was to conduct a site-specific LCIA of steel production, including all significant upstream processes—in particular, coal mining and iron ore mining. The results of the study were then to be used to develop a Certified Life-Cycle Impact Profile for steel production, and a Certified LCIA profile for steel framing for residential construction to be used as the basis of comparison with wood framing.

Key Findings

- **Steel Resource Depletion.** Under LCIA, the depletion of resources is assessed rather than merely analyzing the amount of resources used. Depletion calculations take into account the rate of use, the size of reserve bases, recycling rates, and natural accretion, providing a more accurate measure of the impacts an industrial system on future availability of the resource. This study represents the first formal integration of the available steel currently in use (i.e., standing stock) as part of the overall reserve base of overall iron resources. The integration of the standing stock reserve base with ongoing recycling rates allowed for the accurate calculation of iron resource depletion rates. The study confirms that the depletion of iron resources is approaching zero, indicative of a "sustainable' resource. As such, this is the first study to put recyclable resources such as steel on a comparable footing with renewable resources such as wood. The result should help move the debate about resources from renewability versus nonrenewability to the more fundamental issue of sustainability.
- **Energy Resource Depletion and Embodied Energy of BOF Steel Production.** The depletion of energy resources was assessed rather than merely analyzing the amount of energy resources used. In this study, the most significant depletion at energy resources was found to come from the use of natural gas for electricity supplied from the grid, and not from the usage of coal. This study, furthermore, separated out energy use as heat from energy resources used to reduce ore into iron. This separation allowed for the calculation of the total residual embodied energy inherently bound into reduced iron. These results, in turn, can be used to more equitably allocate overall energy resources depleted between BOF and EAF production.
- **Physical Disruption.** Direct impacts to terrestrial and aquatic habitats from physical operations are accounted for under the impact indicator, Physical Disruption. Most of the physical disruption from steel production is associated with mining activities. Significant differences were noted between the physical disruption associated with abandoned mines as compared to reclaimed mines, demonstrating the value of reclamation. This physical disruption was quantifiable. When put into the context of the amount of steel produced annually (2.3 million metric tons), the area of physical disruption attributable to steel production was less than one percent of the "best case" physical disruption from sustainability managed forestry operations producing an equivalent number of wood framing materials.
- **Greenhouse Gas Loadings.** The most significant greenhouse gas loadings were attributed to the CO_2 emissions from steel making and coking, and surprisingly, to methane released during mining. If current efforts at recovering the methane liberated from the

mining operations are successful, the LCIA profile would show a significant reduction in total greenhouse gas loadings

- **Acidification Emission Loadings.** Most of the acidification loadings documented in this study were associated with SO_X and NO_X emissions from the combustion of coal for both coking and electricity unit operations. The calculations showed that only a small fraction of the emissions result in measurable effects on the environment These calculations demonstrate how misleading it can be to report "worst case" LCI data to customers, government and stakeholders.
- **Criteria and Hazardous Air Pollutant Emission Loadings.** Air pollution control is an area in which the steel industry is heavily regulated and which receives the most significant attention from public interest groups. Over the years, the industry has made significant process changes to reduce air pollution, such as eliminating the sintering process, resulting in significant reduction in environmental loadings from emissions.

The criteria and hazardous air pollutant indicator results were surprisingly low for all steel making processes. These indicator results reflect the extensive nature of both equipment, air pollution controls and administrative controls to ensure that operations stay below threshold levels for PM-10 and other air contaminants. In addition, risk assessment of emissions of benzene soluble organic compounds (BSO's) were also addressed in the study and found to be below deminimus. These indicator results have significant regulatory implications for the steel industry.

- **Water Emission Loadings.** Similarly, the blast furnaces and coke batteries studied demonstrated sophisticated wastewater treatment, which after extensive assessment was shown to have virtually eliminated any toxic aquatic emission loadings from their wastewater effluent.
- **Waste Management.** The study reconfirmed that much of the waste generated from individual unit processes is used as feedstock for other unit processes or is recycled for use in roadbeds and other beneficial uses.

WHITE GOODS*

Definition

Generally, when reference is made to "white goods" in the scrap industry, one is talking about large appliances such as refrigerators, freezers, washers, dryers, stoves, furnaces, and water heaters. These items contain significant amounts of steel (mostly sheet). Many refrigerators, for example, contain at least 80 lb of steel, most of which is located in the doors and cabinets.

Background

During the past decade (1980 to 1990), U.S. steel mills have shipped an annual average of about 3 billion lb of steel to appliance manufacturers. Shipments of major appliances from appliance manufacturers have averaged about 50 million units per year over the past 5 years–up from just over 35 million units per year during the previous 5 years.

According to published statistics, large appliances are replaced after about 15 years'

*This section was prepared by R. Jordan, former Vice President, The David J. Joseph Company, Cincinnati, Ohio use by the original buyer. An estimated 29 million units were replaced in 1991. Obviously not all of the units being replaced are ready for retirement. Other second-hand appliances are coming out of service at the same time. These figures, however, illustrate the huge quantity of discarded white goods showing up in scrap piles.

What does this mean for the environment? What happens to retired appliances if they are not recycled? Let us assume for a moment that all these large appliances were buried—with no processing. How much space would this take up? Using the appliance industry's estimates of expected replacements during 1991 and the approximate average size of these units, a pit 5 ft deep by 18 ft wide (the approximate width of a two-lane road) would be required to bury this scrap, which would stretch nearly 500 mi, or long enough to reach from Cleveland, Ohio, to New York City. That is just in 1 year!

The environmental contribution from recycling these products is indeed as great as the benefits of recovering the metals. Landfill space grows more precious yearly. Eight states already ban the disposal of large appliances in landfills: Florida, Louisiana, Massachusetts, Minnesota, Missouri, North Carolina, Vermont, and Wisconsin. As more states pass similar regulations, and as landfill disposal costs rise, the benefits for recycling will become more pronounced.

As growing concerns about the availability of landfill space reach the public's eye, more attention has been devoted to recycling. But recycling metals from appliances is certainly nothing new. The scrap industry has been doing it for over a century. Fortunately for municipalities—and anyone else interested in keeping appliances out of landfills—the demand for scrap metals over the years has supported the creation of a sophisticated recycling infrastructure. Development of markets for recycled metals is not something that suddenly needs to be encouraged just for appliances or to address concerns about landfill space.

Appliances are actually just one item in the recovery process of iron and steel from obsolete products. In most processing facilities, appliances are mixed with other items containing ferrous material—most notably automobiles. We can view recycling in basically three steps: recovery or collection, processing, and marketing.

Recovery (Collection)

Impact of Demand. How do "retired' appliances make their way into the recycling stream? What incentives are there for recycling? The heating contractor who installs a replacement furnace, the plumber who installs a new hot water tank, and the appliance dealer who delivers a new washer and dryer—all generally take old unusable units with them. They can either drop these at a landfill site (if their state allows) or they can sell them to a scrap processor. What does the individual home owner do with appliances if a contractor or delivery service is not involved—take them to a dump site or scrap processor? Just abandon them in a field or alongside the road? The answer to each of these questions is based on economics.

There are labor and transportation costs associated with getting scrap to a processor. Whether obsolete appliances are collected or not is largely determined by whether the scrap processor is offering a price high enough to compensate these costs. If the price is not high enough to keep the home owner from dumping the old washer along the roadside, the price may be high enough to give a peddler incentive to pick it up and deliver it to the scrap yard. Also, it may depend on the weather. It takes more money to get someone out to collect scrap in the winter than it does in the spring. Obviously, it is an individual decision, but the point is that the higher the price offered by the processor, the more scrap will be delivered to the yard. And the more scrap the processor gets, the less he or she must pay for more. It is all simple supply and demand, and the market is quite elastic.

Economic Factors. The scrap processor, therefore, controls the flow of obsolete appliances and other scrap items into the yard by adjusting the price offered for material. To increase the flow, the processor must increase the price and vice versa. Factors that can affect the price include the proximity of populated areas (where scrap is generated) to the processor, ease of access to the yard (major highways, etc.), and even local restrictions on landfills. Also, the closer a processor is to another scrap buyer, the more he or she will have to compete for suppliers, and the more of a price increase it will take to attract additional supply.

Generally, scrap supplied in larger quantities will receive a higher price. The processor must compete more aggressively for a large scrap collector's business than for an appliance from an individual. The collector, whose livelihood is scrap, is willing to travel farther to collect and to deliver scrap in order to get the best price than is an individual carrying a single appliance. The scrap collector's time and transportation costs become proportionately less as the load size increases.

Appliances are delivered to scrap yards by all different types of haulers: installation contractors, scrap peddlers, individual home owners, and even municipal waste haulers. In a typical yard operation, the scale house (where scrap is weighed) is open to the public-anyone is free to bring in scrap, whether it is in a tractor-trailer, pickup truck, station wagon, or even a push cart (Fig. 15.20).

FIGURE 15.20 Peddler delivering appliances to a recycling scrap processor.

Processing

Densification. What processing is performed on scrapped appliances depends on several factors such as volume, equipment costs, what final product local consumers want or need, and even space constraints. Processing options range from expensive shredding operations down to cutting sheet steel with a torch. Some processing may even be done by municipalities or scrap collectors before hauling the appliances to a processing yard. For example, to save transportation costs, a municipality may employ a baler or logger to densify or bundle appliances, which increases load weight per shipment and reduces per unit freight cost. This initial processing may also increase the scrap's value to the processing yard since densification increases handling efficiency. Once these bundles or logs reach the scrap yard, they may then be shredded. The two most common methods of processing appliances are baling and shredding.

Environmental Concerns: PCBs and CFCs. Before an appliance can be purchased by a scrap processor, certain environmental concerns must be addressed. Two harmful substances found in appliances are gaining more attention: polychlorinated biphenyls (PCBs) and chlorofluorocarbons (CFCs).

Removal and Disposal of PCBs and CFCs. PCBs are suspected carcinogens that do not break down. Since the substance may end up in rivers, it poses a threat to fish and other wildlife. PCBs may be found in capacitors manufactured before 1979 (federal law prohibited the use of PCBs in capacitor production beginning in 1979). Potential capacitor-

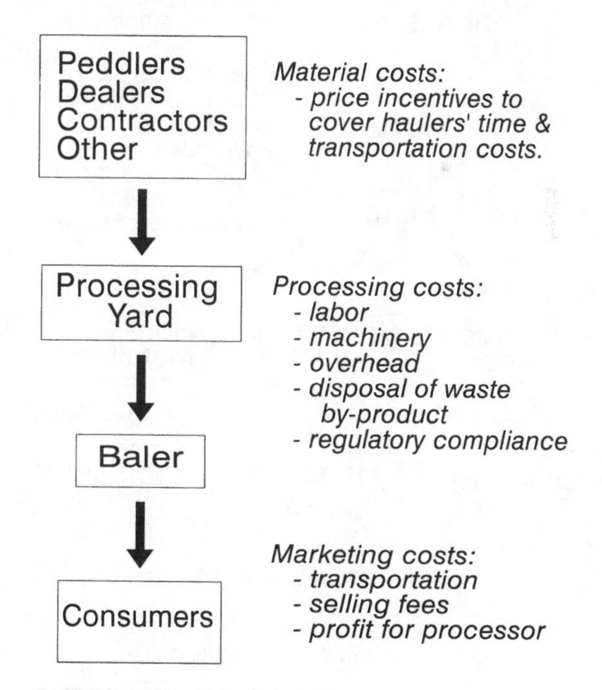

FIGURE 15.21 Baler flowchart.

containing equipment includes refrigerators, freezers, washing machines, microwave ovens, televisions, fluorescent light fixtures, heating and cooling equipment, and electronic equipment. Shredding of scrap products containing PCB capacitors and fluorescent light ballasts could produce PCB-contaminated waste by-products at a shredder.

The Toxic Substances Control Act (TSCA) addresses PCBs. Of particular concern are "running capacitors" (as opposed to starting capacitors), which are designed to increase a motor's efficiency. The capacitors are filled with oil- which may contain PCBs—to help dissipate heat. Running capacitors are often identified by their rectangular metal casings. Capacitors containing PCBs must be removed and disposed of properly prior to recycling; otherwise, most scrap processors will not accept the appliance.

Emissions of CFCs are linked to depletion of the earth's ozone layer. CFCs may be found in wall panel foam in refrigerators and freezers, and in refrigerants for refrigerators, freezers, and air conditioners.

Programs including recycling of CFC refrigerants should specify that reclamation of the refrigerant gas must take place prior to disposal and recycling of the unit in which the refrigerant is contained. Technology exists for recovering CFCs from refrigeration units. The cost ranges from approximately $1000 to $8000 per reclamation machine.

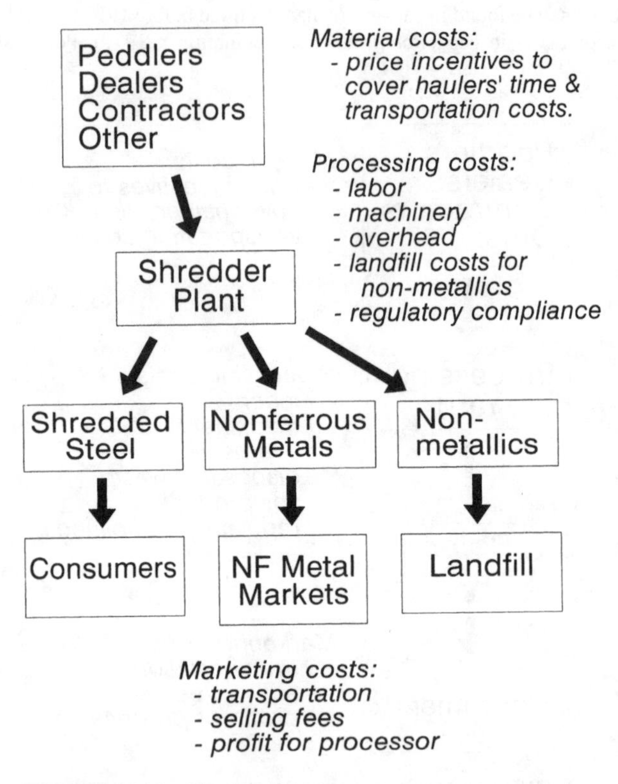

FIGURE 15.22 Shredder flowchart.

FIGURE 15.23 Shredded steel processed from appliances.

Processing White Goods. The flowcharts in Figs. 15.21 and 15.22 illustrate the movement of appliances through a processing facility. Figure 15.21 shows a typical baling operation, Fig. 15.22 a shredding facility. These charts depict the basic costs or value involved at each stage. The shredder operation is addressed in more detail.

Shredding. Appliances represent only part of shredder feed. Automobile shredders are expensive investments; machines can cost $3 million to $6 million per installation. A typical shredder can process 1 to 2 tons of appliances per minute, producing finished material averaging from as small as a fingernail to as large as a fist (Fig. 15.23). In an hour, a shredder can typically produce 65 to 100 tons of shredded steel.

The shredding process produces three output streams: shredded steel, nonferrous metals, and nonmetallics such as plastics, rubber, etc. (referred to as shredder residue). The ferrous and nonferrous metals are saleable. Shredder residue is landfilled.

Marketing

Consumers of processed ferrous scrap include domestic steel mills and foundries, and export markets. A marketing infrastructure is well-established to handle prompt industrial and obsolete scrap generated in the United States. Old appliances fit neatly into this structure all across the country.

Over the past 10 years, the domestic market for ferrous scrap has been about 35 to 45 million net tons annually, exports have ranged from 7 to 13 million net tons annually. Even though U.S. steelmaking capacity has declined 34 percent since its peak in 1973, dependence on scrap as feedstock has grown due to technological changes in steelmaking, resulting in a proportional increase in scrap as a raw material.

CHAPTER 16
CARPETING

LYNN PRESTON
Technical Environmental Manager
Collins & Aikman Floorcoverings, Inc.
Dalton, Georgia

DONALD FREEDLAND
Director of Public Works
City of Deerfield Beach, Florida

CHERYL MILLER
Recycling Specialist
City of Deerfield Beach, Florida

THE PROBLEM OF RECYCLING CARPETING

An estimated four billion pounds of carpeting are disposed of annually in the United States. This equates to roughly 200,000 trailer loads of waste. Not only is carpet bulky, it is made up of materials and components that vary widely among manufacturers. This diversity of carpet components has made industry's advancement as a unified force in recycling carpet almost nonexistent. Recycling programs to date have been made up of individual company efforts. Although options for recycling residential carpet are limited, considerable progress in recycling has been made in the commercial sector.

Fiber manufacturers mainly recover carpet to reclaim fiber, whereas carpet manufacturers tend to focus on recycling the entire carpet. In the Evergreen Nylon Recycling program, a joint venture between Honeywell International and DSM Chemicals, fiber from broadloom carpet is "depolymerized" through a chemical process to recover caprolactam, a raw material used by the company to produce new nylon 6 fiber. BASF also has a similar program. DuPont collects all types of waste carpet and processes it for the production of automotive parts, utility flooring, sod reinforcements, wood-like products, padding, and soundproofing material. They are also developing a "depolymerization" process for nylon 6,6 fiber.

Most commercial carpet manufacturers incorporate recycled materials into the face yarn and/or backing of their products. Image Industries, Inc., for example, uses recycled beverage bottles to produce polyester yarn for its products and Collins & Aikman recycles old carpet into new carpet backing. Shaw recycles carpet in the Evergreen Nylon Recycling program and incorporates recovered nylon 6 from the program in its yarns. Milliken & Company refurbishes carpet for reuse by cleaning, texturing, and printing old carpet tiles. These companies as well as many others are developing the "next generation" of recycled content floor coverings as well as the processes needed to recycle them.

A Case Study

Collins & Aikman Floorcoverings, based in Dalton, Georgia, achieved some of the industry's early successes in carpet recycling. The company created a "closed loop," carpet-in, carpet-out system, which resulted in the industry's first product line of modular carpet tile with 100% reclaimed backing.*

Product Development History

In the early 90s, the company set out to develop the industry's first "green" competitive product, one that could match the quality, price, design and long-term performance of products made from virgin material. Researchers explored ways to use 100% reclaimed material to make new backing for a commercially viable floor covering product. The new product design would recreate the backing again and again from carpet waste rather than virgin material.

The greatest technical challenge was developing a process to blend rather than separate nylon 6,6 and vinyl, two distinctly different polymers. Prior carpet recycling research involved separation of components—a costly step without a recycling option for all components. However, after several years of research and a significant financial investment, the company developed the extrusion technique that made 100% recycling of carpet waste possible. Extruding above the melt temperature of vinyl allowed the vinyl portion of the waste carpet to melt, leaving the nylon fiber intact. This technique results in a nylon fiber-reinforced composite.

Development began by extruding the reclaimed carpet into thick "profiles." Although many product applications, such as bird houses, park benches, and picnic tables were evaluated, an industrial flooring application proved most successful. The block flooring product is produced from reclaimed carpet and postconsumer polyethylene and is itself closed-loop recyclable. Its durability and water resistance are considered advantages over the creosote-treated lumber flooring it was designed to replace.

Once the technique of carpet recycling was mastered, the company began to research methods to produce a sheet of uniform thickness. After several successful experiments, the company purchased a calender for use in further development of carpet backing. Tests showed that carpet could be recycled again and again in the process without degradation. Once the process was optimized, the company successfully produced a fully reclaimed content carpet backing. The success of the product spurred the company on to incorporate the recycled backing to its entire line of carpet tile.

The company also developed floor covering products utilizing recycled content nylon 6,6 yarn. Universal Fiber Systems LLC uses special equipment and process technology to recycle postindustrial nylon waste into a fiber containing a minimum of 82% recycled content. The yarn is used by Collins & Aikman in a new line of carpet products. Similarly, Collins & Aikman incorporates DuPont Antron(r) Lumena nylon 6,6 yarn in a variety of carpeting. Many Lumena colors contain from 6 to 10% recycled content.

The Recycling Process

After its useful life, postconsumer vinyl-backed carpet is sent to Collins & Aikman manufacturing facilities in Dalton for recycling. The entire carpet is recycled without separat-

*U.S. Patent Numbers 5,728,741 and 5,914,343; other patents pending.

FIGURE 16.1 Guillotine process.

ing any components. Postconsumer carpet, along with the company's manufacturing waste and vinyl waste from the automotive industry are utilized in the process. Reclaimed carpet is chopped into strips by a guillotine process and the strips are ground by a granulator. The granulated material is then densified into pellet form.

Carpet pellets and ground automotive vinyl are vacuum transported to feeders equipped with holding bins. A computer-controlled feeding system allows the materials to be blended by weight. The material is fed into a twin screw extruder where it is extruded into a cylindrical, continuous rope. A conveyor feeds the rope to a calender where it is then pressed between large steel rolls to form a six-foot-wide sheet. In a subsequent process, the recycled content backing sheet is bonded to the carpet face. The finished carpet is either rolled into six-foot-wide rolls or cut into 18 × 18 inch carpet tiles. The entire process is illustrated in Figures 16.1 to 16.6.

Results

Closing the carpet-to-carpet loop substantially reduced the quantity of material landfilled by the company. Since virtually all manufacturing waste is recycled, landfill waste de-

FIGURE 16.2 Pelletized carpet.

FIGURE 16.3 Extruded rope.

FIGURE 16.4 Extruded rope fed to calendar.

FIGURE 16.5 Backing sheet exiting calendar process.

FIGURE 16.6 Finished recycled carpet product.

creased by 90% per square yard since 1990 while carpet production increased by 110%. Waste dropped from 13.6 ounces per square yard in 1990 to fewer than 1.3 ounces per square yard in 1999. By recycling carpet, companies that participate in Collins & Aikman's program also help reduce the burden on community landfills. For example, in 1999 one company kept more than one million pounds of carpet out of its local landfill by choosing to recycle more than one million-square-feet of carpet.

MUNICIPALITY STARTS CARPET RECYCLING*

The City of Deerfield Beach Solid Waste/Recycling Division initiated its carpet and carpet padding recycling program in the Spring of 1999. Knowing that 200 million pounds of nylon #6 waste carpet was going into the waste stream annually, the City recognized the large diversion and cost savings potential.

*This section contributed by Donald Freedland, Director of Public Works, and Cheryl Miller, Solid Waste/Recycling Specialist, City of Deerfield Beach, Florida.

PROGRAM DEVELOPMENT

During the previous year, The Solid Waste/Recycling Division and American Carpet Recyclers had established an agreement whereby carpet from specific locations would be collected and transported to the nearby American Carpet Recycling facility. Since all residential carpet looks the same and because a limited range of materials make some carpet recyclable for the manufacturing process, American Carpet Recyclers agreed to accept all residential carpet collected. However, the only carpet processable is #6 nylon residential. Commercial carpeting tends to be #66 nylon and was not part of the agreement.

American Carpet Recyclers is a supplier for Honeywell International Inc., which has the world's first nylon #6 closed-loop carpet recycling operation. Number 6 nylon is easily recyclable because it can be broken down to its base raw material and recolored to become like virgin fiber again. Honeywell can convert 100 million pounds of postconsumer carpet per year. The strength and stability of Honeywell Inc. and of American Carpet Recycling led to a sure partnership with the City.

STEPS TOWARD SUCCESS

One of the first questions which faced the City was how to collect the largest amount of carpet in the most efficient way. A second was how to keep the material dry (this being Florida!). The answer was simply to purchase forty yard roll-offs, modified to be completely covered. This covered roll-off ensured that a large dry load of carpet was delivered, which equated to a quality load (Figure 16.7)

FIGURE 16.7 Recyclable carpeting loaded onto special oversized container to avoid rain damage. (Courtesy of Deerfied Beach, FL, Department of Public Works)

The next question to resolve was where to collect the materials. After researching all the carpet suppliers in the City, the Deerfield Beach Solid Waste/Recycling staff visited each location to determine how to effectively coordinate and accommodate the recycling program for each of them. The staff found five carpet locations with large lots and facilities to accommodate the carpet recycling roll-offs. Other locations with lots too small for roll-offs will be provided at a future date with a specially designed dumpster with special water-resistant lids to deflect rain.

DOCUMENTING CARPET TONNAGE

One of the most exciting parts of the program was the steady increase and growth of carpet being recycled and diverted from the incinerator. In order to maintain accurate figures, tare weights of the empty special containers were established to determine the actual weight of carpeting collected.

EDUCATION

Of course, the next major step was to educate the carpet suppliers and installers about the proper material to be collected and what materials to avoid to eliminate contamination. A flyer that showed the acceptable and nonacceptable carpet and carpet padding materials was produced (see Table 16.1). The flyer was distributed to each participating carpet dealer. Along with the flyer, an example of each acceptable and nonacceptable material was made available to each dealer, who in turn shared a copy of the flyer and examples with each of their own installers. A couple of months into the program, a meeting including refreshments and special thank you gifts was scheduled to review the program and to discuss questions participants may have had in regards to the program.

ISSUES

Although the roll-offs were generally placed behind the places of business, the covered roll-off seemed to be a convenient sleeping arrangement for some people. Locks were im-

TABLE 16.1 Acceptable/not acceptable carpeting

Acceptable	Not acceptable
Residential carpet	No commercial
Rolled or folded	No razors, metal, nails, screws, tack strips
Four (4′) foot widths	*Padding not accepted*—waffle or rubber
Carpet pads accepted—rebond and prime	
NO DEBRIS INSIDE of carpet	

mediately placed on the roll-offs and keys were distributed to the carpet dealers. Drivers have matching keys for ease of opening the containers when delivery to American Carpet Recyclers is made. Occasionally, garbage would find its way into the carpet roll-offs from night travelers and people wondering in inappropriate places. The locks and keys provided a means to help alleviate that problem as well.

DETERMINING NOT ACCEPTABLE CARPETING

Contamination arose due to the legitimate inability of carpet dealers to determine #6 nylon from other types of carpets. Once the carpet roll-off arrives at the American Carpet Recyclers facility, it is unloaded by hand and separated into three categories: acceptable, not acceptable, and garbage. The first category is the acceptable recyclable #6 nylon carpet. The #6 nylon can only be distinguished from #66 nylon and other types of nonrecyclable fibers by a specialized highly technical laser "gun"-style apparatus. The gun is waved over the carpet and the type of fiber is revealed from on a small screen on top of the gun.

The second category, nonrecyclable or not acceptable carpet, is tossed aside to be placed in the garbage roll-off. The third category is regular garbage, which consists of the bags of scrap nails and screws that mistakenly get tossed into the container periodically. A simple document was established to determine the weight of acceptable vs. not acceptable carpeting. The document describes the source of the roll off container and the approximate amount of carpet in each category from that location.

CLOSED LOOP PROCESS

The amazing closed loop process begins and end with the consumer. By reusing the fibers in the carpet to make new fibers that consumers can purchase again and again, we are closing the recycling loop. Therefore, American Carpet Recyclers, Honeywell Inc., and the City of Deerfield Beach are only part of the integral flow of the circle toward the relife of carpet. The process flows like this: used residential carpet is removed, returned to the carpet supplier, then transported to a processing facility; baled postconsumer carpet is moved to the Evergreen nylon recycling facility; caprolactam is recovered from the fiber and is separated from the backing; backing is utilized in a waste-to-energy conversion process and caprolactam is converted into nylon; the recovered nylon is used to make new nylon 6 fiber for carpeting. The consumer can now purchase carpet made from postconsumer nylon.

ECONOMICS AND SAVINGS

During the last half of 1999, the gross tonnage of carpeting diverted from the Broward County Incinerator ranged from 40 to 60 tons per month. Of that, the City was able to accept 20 to 35% as recyclable material. With a tipping fee of $80.90 at this time, the savings averaged $4,050 per month or, projected annually at this rate, $50,000 per year.

CHAPTER 17
TEXTILES

BERNARD BRILL
Executive Vice President
SMART (Secondary Materials and Recycled Textiles Association)
Bethesda, Maryland

EARLY HISTORY

In the early 1990s, the United States Air Force unveiled a new fighter plane that was undetectable by radar. Due to its unusual shape and materials, the new A-117 Stealth Fighter could secretly fly into enemy territory and not be seen by radar. Not unlike this new stealth technology, textile recycling has been going on undetected by the general public for hundreds of years with little or no recognition. Back in the days of ancient Egypt, when great monuments were being created to commemorate dynasties, linen wiping cloths were being used by royal decorators to clean up the edges of painted friezes, murals, column capitals, and for a multitude of polishing and burnishing tasks. Linen, the fabric of the Egyptians and all the peoples throughout early recorded time, was a durable, hand woven fabric made from flax. It was the first recorded wiping material. Later Tutankhamun, Ramses, and many others benefitted from its usefulness. This is the first documented evidence of textile recycling.

The recycling of textiles, unlike other recycled commodities, has always been market driven. In recent years, local governments and communities have mandated collection programs for paper, plastics, glass, *and* textiles. Textiles were added later to collections because of their value. Post-consumer textile wastes account for 4 to 6% of residential wastes according to the U.S. Environmental Protection Agency (EPA).

CLASSIFICATIONS OF TEXTILE RECYCLING

Textiles are defined as items that are made from woven or knitted cloth, such as wool and cotton fibers, vinyl and other artificial fabrics, and items made from fur or animal skins.[1] The industry is divided into companies dealing with "preconsumer" and "postconsumer" textiles waste. While there are no two companies that do exactly the same thing exactly the same way, there are some commonalities.

The textile recycling industry, with approximately 3000 companies, is made up of mostly small, family owned businesses. Many of these companies were founded during the early 1900s and today are in the third, and even fourth, generations of ownership. Because the majority are small, closely held businesses, the public is generally unaware of their existence and their contributions to society. However, what is particularly fascinating is that these firms are capable of recycling 93% of what they receive—which is already considered a waste product!

PRECONSUMER MATERIALS

Preconsumer textile materials are those items considered factory waste generated by apparel producers, textile manufacturers, knitting operations, nonwoven paper producers, needle punch producers, and dye houses. Examples are clippings, cuttings, mill-ends, remnants, thread waste, or goods damaged during production. Goods may also be attained through insurance sales, business closures, and damaged materials. These products are gathered by the manufacturer and shipped to a textile materials recovery facility (MRF) for sorting, grading, inspection, cutting, packaging, and shipping.

This segment of the industry began during the industrial revolution, when textile manufacturing moved from the home to the factory.

INDUSTRIAL WIPERS

One of the end products derived from this process is institutional and industrial wipers. Industrial wipers are manufactured from raw materials supplied by both pre- and postconsumer sources.

Background

It was actually during the last decade of the 19th century that "wipers" took on their own identity as a necessity of the Industrial Revolution. With the rapid growth of machinery manufactured to meet the needs of the expanding industries, fitting tolerances were kept loose, as it took more time to machine close fits (tolerances) on moving parts. The demand for rapid production was too great. Consequently, machines such as locomotives, lathes, presses, engines, etc., with loose-fitting parts, leaked oil and had to be wiped down constantly.[2]

As the mighty machines grew in both number and size there was one constant factor—without wiping materials the plant could not operate. Machines became more sophisticated and complex while dust and grit became an ever-growing problem threatening their successful operation. The demand for wiping materials grew exponentially.

In 1919, new mill ends began to surface, and items such as gauze remnants, unbleached sheeting, and misprints from various printing mills were offered to the wiping materials industry in sizable quantities. The clean mill ends were purchased by wiping cloth manufacturers for processing. The volume expanded year after year until many old rags were supplanted by mill ends in the wiping materials industry.

It is interesting to note that from World War I onward, a large segment of the industry washed various cotton fabrics to produce wiping cloths. Examples of this were window shades, which had crinoline and sizing that was washed out. Also, sugar bags and flour bags, which were made of cotton, were cut open and washed.

Today, each dealer has their own standards for grading the material. The number of grades available will vary depending on the sophistication of the wiper dealer. Both pre- and postconsumer textiles are sorted in a similar manner. This cloth is then separated by cotton content or by fabric type, such as knits, wovens, and toweling. The material may be further separated within more specific categories, such as tee shirts, polo jerseys, fleece, sheeting, flannel, oxford, towels, linens, etc. These textiles can be sorted again into white and color (color versus white tee shirts), dark and light colors (darker flannel shirts versus lighter flannel pajamas), or heavy and light weight material (shirts and blouses versus pants and denim). The possibilities are almost infinite and are determined by both the particular market area and by the end user's requirements.

After separation and grading, the material is cut. There are very few options for cutting. Materials are cut into wipers by hand; using a one or two arm "rag cutter," which has a smooth enclosed rotating blade. The operator either stands or sits while guiding the clothing through the blade (Fig. 17.1). The arms and the legs of clothing are cut open and buttons and pockets are removed. Several preconsumer textiles, such as mill ends or roll goods, may be cut on tables or on advanced automated machinery. There are wiping materials that are cut in clean room environments and are highly specialized for high-tech and specific manufacturing operations. These materials are uniform and are certified to contain no more than a specific number of lint particles. Other uses include the automotive industry and circuit manufacturing. Once cut, the wipers are either boxed or baled and prepared for shipment to the end users directly or through wholesale suppliers.

Large items such as sheeting or mill remnants are cut to a specific size. The size of the wiper is determined by the weight of the fabric. The wiper should cover one's hand when "scrunched" in the palm of the hand. Since lighter-weight fabric can be compressed easily, the wipers made from them are usually cut into larger pieces. As a result, wipers can often vary in size but are generally cut into squares of 15 by 15 inches. Both pre- and

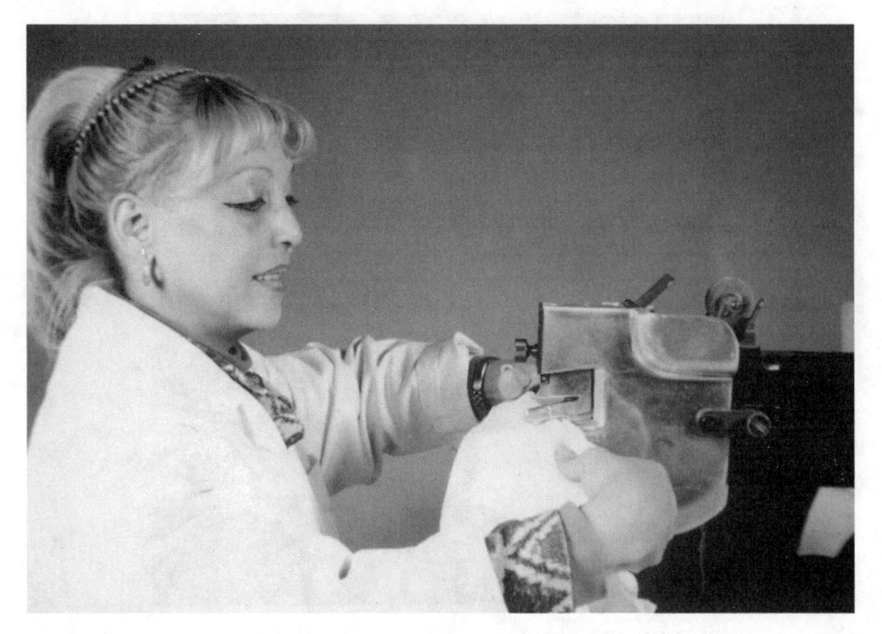

FIGURE 17.1 Wipers are cut to size before packing.

postconsumer textiles are used for cleaning surfaces in all types of manufacturing, automotive, janitorial, and food service applications.

The type of wiper (size, color, cotton content) used by particular business or industry is specifically determined by its application within the operation. For instance, white, pastel, or unbleached wipers are used in industries where chemicals might cause the release of color. For example, cheesecloth and gauze are used predominantly for staining or polishing. Furniture manufacturers and automotive painting and body shops would require low linting and soft-textured materials (Fig. 17.2).

Colored wipers are used mainly to absorb or clean grease and oil from industrial surfaces. The product is sold to both government and private industry. The United States General Services Administration uses many thousands of pounds of wipers per year. Delivery of these materials is done via UPS, common carriers, or company owned trucks.

Pricing

Prices for both new textile materials and manufactured wiping cloths are determined by supply and demand. Preconsumer textile remnants could be low-priced, whereas the postconsumer equivalent could be high. For example, corduroy made a recent fashion comeback, so the price on new corduroy remnants are low-priced. If corduroy has not been in style for a number of years, there would be very little in the waste stream, resulting in a price rise. When cloth diapers had a brief comeback, diaper wiping cloth prices fell. Now that disposable diapers dominate the market, used cloth diapers are virtually nonexistent, and therefore very costly. Like other recyclables, supply and price are cyclical.

FIGURE 17.2 Institutional and commercial wipers come in various colors and fabrics.

CLIPPINGS AND CUTTINGS

Clippings and cuttings are scraps of materials left over from the sewing room floors of apparel manufacturers and other industrial textile users. They arrive at the textile MRF from their facility of origin by way of truck or van, packed in boxes or bags, or baled in various sizes and weights.

The material is taken to the work area, inspected, and sorted by hand; it is then tested for cotton content. Wool and synthetic fiber are checked in a similar manner. Single-component fibers are most likely to be recycled and would have the highest value in the market. The materials are then sorted by size, material content, color, and waste. All unusable items are removed, such as markers, paper, cones, and assorted debris that would not be deemed acceptable in the shipment to the customer. Keeping waste clean and free of contaminants can enhance the chances it can be recycled as well as give it more value, but even that won't guarantee it can be used in the limited markets available for these kinds of waste.[3]

The various packings of clippings and cuttings are sorted within the categories of fiber, content, and color. Separating color is the easiest way to begin. White material is separated from the colored material and generally commands the highest value. The next step, for both colored and white material, is the cotton content stage. One hundred percent cotton, polyester–cotton, and 100% polyester represent the general packing of clippings. The contents of these grades vary and are tailored to the end-users needs. Other fiber grades (such as nylon, acrylic, polyester, and polypropylene) are organized in a similar fashion.

At the final stage of handling, the clippings and cuttings are sorted into approximately 1000 pound bales. These bales are placed in inventory until ready for shipment to their destination in full truck or container lots that eclipse the 40,000 pound mark (Fig. 17.3).

Often, these clippings and cuttings are sent to a reprocessing operation where the materials are chopped and torn apart by large machines. The finished product, known in the

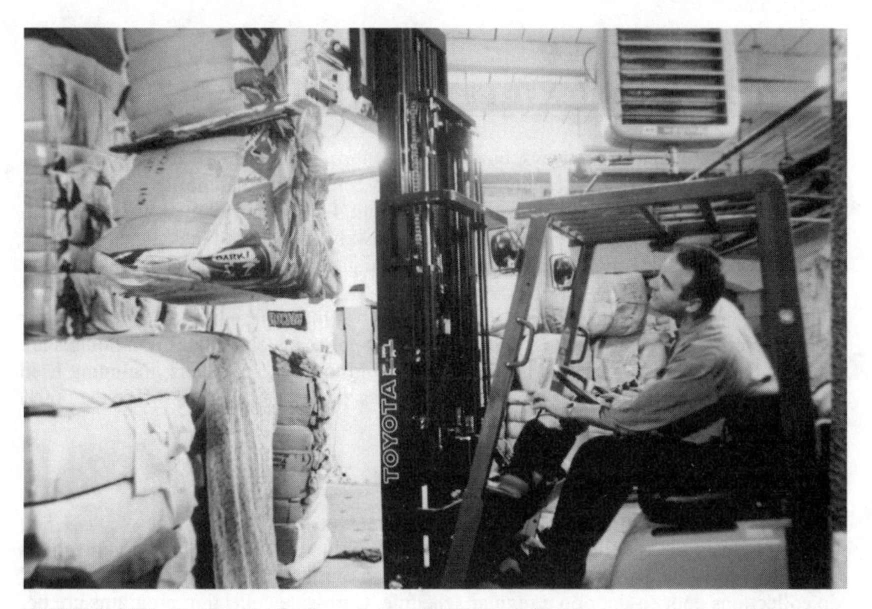

FIGURE 17.3 Recyclable textiles are stored until needed.

trade as shoddy, is both sold domestically and exported. The primary uses for this product are stuffing for cushions and pillows in the furniture trades, and for the manufacture of padding in the form of carpet underlay, mattress padding, and molded padding used in the automotive industry. It can also be used as stuffing for spill containment booms and in caskets. One newer application for this material is open-end spinning. Here, the shoddy is sorted by color and is spun into yarn. This process alleviates the need for dying and creates a lower-end yarn product suitable for opening-price socks, sweaters, and gloves. With the finished product, the recycling process is brought full circle.

Examples of high-end products are cotton knit rags or flat-knit tee shirts; because they are clean, they can be processed back into the fiber stage, and can be blended into "bleached shoddy," which has applications in hygiene products such as cotton swabs and cotton found in bottles containing prescriptions and other medicines. Colored thread wastes, on the other hand, are a lower grade that is frequently used for export, as are textiles composed largely of synthetic fibers such as nylon, polyester, and acrylic.

New Yarn

Textile recyclers in Italy and South Asia are big users of this material, particularly wool, and polyester. A product known as "tri-blend" wools contains a large percentage of reprocessed fibers. Companies in this segment of the industry buy presorted fiber and respin it into yarn.

Paper Manufacturing

The paper manufacturing industry is and has been traditionally the greatest user of 100% cotton bleached rags. In Japan, 100% cotton bleached rags are used in the production of mushrooms.

POSTCONSUMER TEXTILES

Recycling is defined in many ways, but the highest form of recycling is reuse, where minimal amounts of energy and resources are used to convert the material to another form. Postconsumer textile product waste comprises about 4–6% of residential waste.[4] Examples of postconsumer waste are clothing, drapes, towels, sheets and blankets, table cloths, belts, handbags, paired shoes and socks, and clean rags. It is estimated that there are today 2000 companies involved with diverting 2.6 billion pounds of postconsumer textile materials from the waste stream. This 2.6 billion pounds of material represents 10 pounds for every person in the United States. Of this amount, approximately 500 million pounds are used by the collecting agency, with the balance sold to textile recyclers, including used clothing dealers and exporters, wiper manufacturers, and fiber recyclers. More than 60% of these materials are for the export market.

Collections

Charitable institutions collect the majority of used textiles in the U.S. Some organizations utilize a drop-off center, drop-off box, or telephone routing system in which a truck will do collections door-to-door on a regular schedule. Curb-side collection programs are becoming more popular across the country as municipalities include textiles in their recycling programs. Today, more municipalities are deciding to add textiles to the existing collection programs to meet their goals and quotas. At curb-side, textiles are typically placed in a separate compartment on the recycling truck or picked up with the paper. In some cases, the textile MRF will supply a special plastic bag for textile collections. Bags are either dropped off at each home or distributed through schools and grocery stores, or by other methods. Ideally, the textiles should be brought indoors for sorting, baling, and loading into trailers.

Some of the larger nonprofit collection agencies are Goodwill Industries International, Inc., The Salvation Army, and St. Judes. These three organizations are the major collection forces within North America and have established drop-off centers. Proceeds from these operations are used for their charitable and rehabilitation efforts to help the disadvantaged. Because they collect far more than they can sell in their stores, the excess materials are sold by weight to individual textile MRFs.

Other organizations most frequently use telephone routing in suburban and urban areas. Telephone banks are set up and operators call residents in specific neighborhoods and ask them to set their items on the front porch for pick-up on a specific date. Porch pick ups help discourage scavengers from stealing or damaging the materials.

Drop-off boxes are common sights in supermarket and shopping center parking lots. People bring their items to the box at their convenience. These boxes are sponsored by a charitable organization or can be placed there by a private business. Over the years, the problems with collection boxes has been pilferage, people placing garbage in the boxes, and the failure of the sponsoring organization to make regular pick-ups. It is for these reasons that many organizations have abandoned this collection method in favor of manned collection centers.

Economies of Collection

Municipal recycling officials are concerned with the costs of adding new items to their existing collection costs. Therefore, before textiles are added, there is a considerable amount of planning that is needed before a program is put in place. First, estimates must be made to determine frequency of collection, whether the vehicle's compartments can handle the additional volume, and what the variable costs are in handling an additional item.

Second, how are the items going to be processed? Will the municipality be responsible for separating the textiles from the paper? Or will textiles be collected in plastic bags? Will the items be baled? If so, can the current machinery be used or will an additional baling machine be needed? (It is recommended that vertical balers be used, as horizontal machines often shear the materials.)

As cities and counties are forced to meet mandated recycling goals, textile recycling is becoming more attractive. One reason for this is that there is constant demand for used clothing, and the revenue received for these goods helps offset the expenses for collecting other recyclables. In some communities, textiles have helped to offset collection costs by 10 to 20%.

Tips for starting a municipal textile collection program:

1. Educate local recycling organizers as to the importance of including textiles in their collection programs.
2. If possible, work with local charitable programs already in place.
3. Notify state recycling officials and encourage them to include textiles in their recy-

cling goals.

4. Work with haulers at the local level to fit textiles into their collections.
5. Contract with a textile recycler to buy goods.
6. Educate residents on how to handle their discarded textile items.
7. Conduct regularly scheduled programs and publicize success stories.
8. Remember that education and promotion are key to the success of every program.

Carroll County, Iowa; St. Paul, Minnesota; San Jose, California; and Somerset County, New Jersey are examples of municipalities that have curbside collection programs in place. Aberdeen, Maryland collects textiles at curbside once a year as does the Solid Waste Authority of Palm Beach County, Florida. Calvert and Montgomery Counties, Maryland and Cobb County, Georgia, have added textiles to a long list of materials accepted at drop-off sites.[5] Some of these municipal programs have partnered with local charities and nonprofit organizations. The City of Los Angeles is working with the Salvation Army in select neighborhoods to collect textiles.

Unlike other recycled materials that are collected at curbside, (bottles, glass, plastics, cans, and newspapers), textiles must be kept dry at all times during the collection process. Natural fiber textiles will decompose or become moldy if wet. Although rare, such decomposition can generate heat which could lead to spontaneous combustion and cause facility fires when stored in baled form. That is why clothing must be kept clean and dry during the collection process.

Textile recyclers pay from $80 to $150 per ton for the materials. This used clothing is received at the MRF in large bales of approximately 1000 pounds each (Fig. 17.4). Because the clothing is mixed, it must first be sorted and inspected. For example, it may go

FIGURE 17.4 Textiles must be kept clean and dry and not commingled with other recyclables.

FIGURE 17.5 Used clothing is inspected and sorted for various markets. Materials such as used clothing, shoes, drapes, towels, bed linens, and table cloths are recyclable.

through a primary sort in which the graders are looking for certain products that may be sold domestically at vintage clothing stores and thrift shops (Fig. 17.5). The clothing can be sorted in literally hundreds of different ways, depending on the used clothing dealer's markets. The typical used clothing MRF will sort men's shirts, suits, pants, children clothing, tee shirts, polo shirts, women's blouses, pants, coats, paired shoes etc.

Vintage Clothing

Popular clothing from past generations is still popular today! Jeans, prom dresses, evening gowns, shoes, leather jackets, and bowling shirts are just a few examples of the items sold in today's vintage clothing stores. Styles from the 50s, 60s, and 70s are often made popular again by college students and are often seen in the movies and Broadway plays. Vintage clothing shops are found everywhere today, even in the trendiest shopping areas.

Once sorted and inspected, these textile materials are then repackaged for shipment to international markets. Currently, these markets exist mostly in developing countries in Africa, South and Central America, and the Far East.

Pricing

Pricing of used clothing is extremely elastic. For the purposes of this discussion, used clothing should be viewed as a commodity such as wheat or apples, as price is determined based upon supply and demand, world economic conditions, the strength of the U.S. dollar, shipping conditions, freight rates, and tariff schedules. Another situation that dictates trade is when war or civil strife breaks out in a particular export country. When this happens, trade comes to a halt because transportation within that country is dangerous, and there is very little hard currency to pay for the goods. There are also issues, such as the availability of insurance under these circumstances, that makes exporting a very dangerous and risky endeavor.

Depending upon the markets, prices also vary according to the types and condition of the clothing.

Because not all the clothing sorted at a textile MRF is suitable for reuse, ripped or stained items may be sold to produce shoddy that will be used as auto insulation, industrial wipers, roofing materials, and in blankets. It is estimated that less than 40% of every bale contains items suitable for export as used clothing. As a result, most of the bale is often converted to materials other than clothing and sold below the cost of the bale. In order to cover costs, the used clothing must be efficiently handled and processed and sold at a price that covers those materials that are sold off at losses.

As mentioned earlier, most recycling firms are small, family owned businesses, with fewer than 500 employees. The majority of these companies have between 35 and 100 employees and industry sales are estimated at $700 million annually. These recycling firms are usually inner city employers and hire people from nearby communities who might otherwise be unemployable. Many of the workers are unskilled, semiskilled, or physically or mentally challenged. At the same time, these businesses, through their taxes, contribute to the revenue bases of federal, state, and local governments.

Used Clothing

The used clothing market is very abstract in terms of market and pricing and at best difficult to estimate. According to the U.S. Department of Commerce, used clothing is this country's eighth largest export item behind automotive parts and wheat.

ITEMS UNSUITABLE FOR CLOTHING

Wipers

Clothing that is graded out to be unsuitable for wear because it is too worn, stained or torn, can be cut into wipe cloths. Fabrics used most often in this process are corduroy, denim, flannel, and knits. These individual pieces are often cut into wiper size after the buttons, zippers, collars, etc. are removed. These materials are then boxed or baled and shipped elsewhere for use in other various industries.

Shoddy

Materials unsuitable for clothing or wipers are sent to a fiber converter. Here the clothing is machined and garnetted. Garnetting is the process of chopping, ripping, and tearing the material so as to return it to a fibrous state. From this blend of fibers comes high-quality carpet underlay for commercial and residential use, mattress filler, stuffing for pillows and cushions, insulation for housing, deck panels, and sound deadening materials for the automotive industry.

Another use of shoddy is the reintroduction of the reprocessed fiber into the manufacturing process by the manufacturer who generated the waste originally. This is the process that is referred to as "commission picking."

Of the two types of textile waste, preconsumer and postconsumer, preconsumer waste will logically begin to diminish as a result of increased efficiencies, the production moving off-shore, and through overall source reduction. At the same time, many manufacturers today are finding their own ways to reprocess the textiles. Some companies have installed their own fiber picking/processing machinery.

DISTRIBUTION

As long as textile manufacturing plants generate large amounts of waste, there will be a large marketplace for dealers and brokers. Today's manufacturers want to be able to make their waste disappear. The manufacturer may or may not expect revenue in return. The generator may be satisfied to avoid disposal costs. It is not unusual for brokers and dealers to earn 5–10% by moving the preconsumer materials.

On the selling side, many of the users of shoddy are small to medium sized companies who do not bid their raw material purchases. One of the larger users of shoddy is the automotive industry, which uses this material to deaden sound. There is nearly 80 pounds of this material in every automobile. It can be found in the door panels, roof liner, under the hood, and in the trunk.

Once a product is found that works, there is significant reluctance to change suppliers. These factors reinforce the strength of a broker or dealer network. Market intelligence on new manufacturers, new products, and changing raw materials needs, which are the "bread and butter" of a successful dealer/broker, also reinforce the value and thereby the existence of a broker/dealer network. Many companies have found the use of brokers/dealers in addition to direct sales to be a very economical way to gain market information from various parts of the United States and around the world.

CONCLUSION

Textile recycling is just now being discovered by municipal recycling and government officials as a method of further reducing the amount of waste being deposited into our nation's landfills. It is estimated that only 15% of textile materials are being diverted from the waste stream for recycling purposes..

Unlike other recyclables whose collections are mandated by government, recycled textile materials have been market driven. For example, demand for high-quality, low-cost industrial wipers, used clothing, fiber, and related materials have made this a profitable industry for decades. More than 90% of the "waste" that is received by this industry is recyclable.

Today, communities of all sizes across North America and Europe are aggressively exploring new ways to economically collect these materials. The key to their success will depend on their ability to educate their citizenry about the importance of properly disposing of their textile materials and buying products made with recycled textile content.

Each community should carefully review its collection programs and consider adding textiles to the mix. By either working directly with a local charity or through a used clothing recycler, recycling coordinators can develop programs to remove textiles from the waste stream.

TEXTILE ASSOCIATIONS AND INFORMATION SOURCES

Secondary Materials and Recycled Textiles Association
7910 Woodmont Avenue, Suite 1130
Bethesda, MD 20814
Phone:301/656-1077
Fax:301/656-1079
www.smartasn.org
email: smartasn@erols.com

Council for Textile Recycling
7910 Woodmont Avenue, Suite 1130
Bethesda, MD 20814
Phone:301/718-0671
Fax:301/656-1079
www.smartasn.org

Institute for Local Self Reliance
2425 18th St., N.W.
Washington, DC 20009-2096
Phone:202/232-4108
Fax:202/332-0463
www.ilsr.org
email: ilsr@igc.apc.org

ACKNOWLEDGMENTS

Special thanks to members of the Secondary Materials and Recycled Textiles Association and the Council for Textile Recycling for their contributions. Their time, effort, and patience in gathering this information was greatly appreciated.

REFERENCES

1. E. Jablonowski and J. Carlton. "Textile Recycling." *Waste Age Magazine*, January 1995, p. 83.
2. *The Wiping Materials Story*. International Association of Wiping Cloth Manufacturers, 1982, p. 17.
3. P. Kron. "Recycle—If you Can!" *Apparel Industry Magazine*, September 1992, p. 74.
4. U.S. Environmental Protection Agency.
5. *Weaving Textile Reuse into Waste Reduction*. Institute for Local Self Reliance, Washington, D.C. 1997.

CHAPTER 18
TIRES

MICHAEL BLUMENTHAL
Executive Director
Scrap Tire Management Council
Washington, D.C.

INTRODUCTION

The information provided in this chapter updates the information provided in the first edition of this handbook. An attempt has been made to provide a realistic description of the scrap tire industry. The majority of the information in this chapter is the result of research by the Scrap Tire Management Council (STMC). The STMC is part of the trade association representing the United States tire manufacturers. One of the first observations that can be made is that there has been a considerable amount of change since the original information was presented. This chapter will attempt, where possible, to illustrate some of the more significant changes that have occurred since publication of the first edition in 1993.

Scrap tires represent slightly less than 2% of the total solid waste stream in the United States, yet they still present a special disposal/reuse challenge because of their size, shape, and physico-chemical nature. Scrap tires are still not collected with household waste by municipal authorities. Thus, scrap tires remain classified as a "special waste" or as a "durable product." This chapter will present information concerning the manner in which scrap tires are currently managed, as well as some of the problems commonly associated with scrap tires. This section will:

- Identify and evaluate the level of progress made with the various market applications
- Identify new markets and technologies that have emerged since 1991
- Identify market, technical, institutional, and other barriers to the expanded use of these markets and technologies

TIRE MAINTENANCE, RETREADING, AND REPAIR

Reduction is the first order of importance in any solid waste hierarchy. Proper tire maintenance is necessary to prolong the tire's useful life, enabling tire owners to derive full value and benefit from their purchase. A simple maintenance program consists of rotating tires every three to four thousand miles, proper balancing of tires, aligning the vehicle's front end, and properly maintaining a tire's air inflation. Not only will this result in even and longer wear of the tire, but it will increase gas mileage and reduce auto emissions. Proper driving techniques will contribute to safety and increase tire life. Finally, proper vehicle maintenance, especially alignment, will add to tire and vehicle life.

In 1998, approximately 33 million tires were retreaded, enabling them to be returned to useful applications as tires and keeping them from entering the waste stream. Retreading processes have been in use almost as long as tires have existed. Retread and repair technologies continue to increase the useful life of tires, thereby ensuring fuller utilization of the resources built into the tire.

OVERVIEW OF FINDINGS

The information provided in this chapter is based on the market survey conducted by the Scrap Tire Management Council (STMC) at the end of 1998 and the first quarter of 1999. The survey findings indicated that, as compared to 1996 data, there was a decrease in the use of tire-derived fuel (TDF) and an increase in the number of tires being used in civil engineering applications and that the size-reduced (ground) rubber market increased slightly (Table 18.1). In 1998, 178 of the 270 million scrap tires generated went to a market. This compares with the 202 of the 266 million scrap tires generated in 1996 that had markets (Table 18.2). Each of these general market areas will be detailed in their respective sections of this chapter.

The changes that have occurred demonstrate that there remain a series of nontire issues that impact on the markets for scrap tires. Whereas the reduction in TDF use can be attributed to a series of factors, it must be stated that there were no direct environmental reasons for the decrease. Furthermore, research suggests that the number of scrap tires going to markets should increase over the next two to three years, although at a lower rate than the previous years.

From the total number of tires going to a market, it can be seen that fuel continues to be the most significant market for scrap tires. The other markets for scrap tires (i.e., export, stamped products, etc.) remained at approximately the same level for several years. One potential market, pyrolysis, although the subject of continuing interest, has not demonstrated an ability to consume many, if any, scrap tires. It would appear there are no commercially operating pyrolysis facilities in the United States, although there are a series of pilot projects, demonstration facilitates, prototypes, and announcements of proposed commercial-scale facilities.

TABLE 18.1 Scrap tire market changes (millions of tires)

Application	1996	1998
Tire-derived fuel	152	118
Civil engineering	10	20
Ground rubber	14	17.5
Punched products	8	8
Export	15	15
Agricultural	1.5	1.5
Miscellaneous	1	3

TABLE 18.2 Scrap tires going to market

Year	Percentage of tires going to market
1990	11
1992	38
1994	55
1996	76
1998	66

LAND DISPOSAL ISSUES

In addition to the markets that have been mentioned, several states allow scrap tires to be landfilled or monofilled. In general, landfilling of scrap tires is allowed when there are no viable markets for scrap tires within an economically accessible distance. Many factors, including transportation costs and limited scrap tire volumes, may make it virtually impossible to have substantial scrap tire markets in some locations. Where this is the case, it is understandable that landfilling may be the most reasonable and cost-efficient management option. Nevertheless, landfill disposal of scrap tires should be considered a last resort solution, and not considered a market.

Since 1994, the use of scrap tire monofills (a landfill, or portion of one, that is dedicated to one type of material) has become more prominent in some locations as a means to manage scrap tires. Currently, some 10–12% (27–32 million) of the scrap tires generated in the United States are landfilled. In certain cases, monofills are being used because there are no other economically viable markets available and mixed landfills are not accepting, or are not allowed to accept, tires. In other cases, monofills are being portrayed as a management system that will allow for the long-term storage of scrap tires without the problems associated with their above-ground storage. In theory, monofilled processed scrap tires can be harvested when markets for scrap tire material improve.

Placing scrap tires into monofills is preferable to above-ground storage, especially if a pile is not well managed. Available data indicate that there are no negative environmental impacts from placing tires into monofills. Although monofilling scrap tires is environmentally sound, the belief that these tires will be readily available for the marketplace at some time the future is questionable. One question that goes unanswered is what happens to the dirt used to cover the tires in the monofill. In general, dirt will contaminate the processed tires, and render them inappropriate as feedstock for ground rubber production (also see the section on stockpiled scrap tires). It is also likely that economic factors would limit retrieval of monofilled tires for either civil engineering or fuel uses.

In addition, monofilled processed scrap tires will have to compete with both the annually generated scrap tires plus any tires that are made available to the market from tire-pile abatement programs. Typically, tires from abatement programs are subsidized by state scrap tire project funds, giving them an economic advantage over monofilled tires, since monofills are usually operated as private ventures, and are not recipients of state funds.

State solid waste regulators must carefully evaluate permitting of tire monofills. The presence of monofills, and landfills that still accept whole or processed scrap tires, tends to impact on the regional scrap tire flow. Typically, landfills and monofills should be the least-cost, legal, disposal option available to the marketplace. Consequently, a significant quantity of scrap tires will be directed toward these facilities, and away from other market applications. The existence of monofills and landfills will restrict the development of other markets for scrap tires, and in the end become the disposal option in that region.

GENERATION RATES FOR SCRAP TIRES

To remain consistent with the previous market surveys, data concerning the weight of tires remains the same as described in the first edition of this book; that is, one "scrap tire" is equal to 20 pounds of tire-derived fuel or 12 pounds of size-reduced rubber. This method of accounting for scrap tires also maintains consistency with the vast majority of other reports on scrap tires, in particular the market analysis done by the United States Environmental Protection Agency (USEPA).

As reported in earlier editions, this manner of tracking the number of scrap tires is limited in that it fails to indicate which of the various types of scrap tires generated—passenger, light truck, heavy truck, off-road tires, etc.—are being used. The rationale for maintaining this approach is threefold: first, it is consistent with all earlier studies; second, the use of a tire-weight-based reporting system (i.e., tire equivalents) would distort the size of the scrap tire situation; and third, regardless of how anyone calculates the total number or weight of scrap tires, the main issue is the creation of markets for all the scrap tires generated.

As it so happens, the majority of the scrap tires that have markets are the passenger, light truck, and bias ply truck tires. The majority of the heavy truck tires, off-road tires, and agricultural tires are only slowly coming into market applications. Consequently, while we are certainly pleased with the dramatic increases in scrap tire markets, we know there are still significant challenges that need to be addressed before the more difficult classes of scrap tires find their way into the marketplace.

In order to understand the differential between the percentage of the total number of scrap tires generated versus the weight of those tires; refer to the Table 18.3.

SCRAP TIRE GENERATION

One basic set of data needed to understand the scrap tire market is the measurement of the annual volume of scrap tires. How many tires are discarded annually and how is that volume calculated? A basic assumption in making a scrap tire estimate is that there is one scrap tire discarded for each replacement tire sold. In addition, it is assumed that tires are also discarded when the vehicle on which they are mounted is discarded. Thus, national estimates can be made based on the total volume of replacement tires sales and the total

TABLE 18.3 Tire generation/weight differential

Type of tire	Percent of units in the market	Percentage of total weight
Passenger/light truck	84	65
Medium/heavy truck	15	20
All other tires	1	15

TABLE 18.4 Scrap tire generation

Year	Number of scrap tires generated
1990	240
1992	246
1994	253
1996	266
1998	270

volume of vehicles scrapped each year. Over the past eight years, the number of scrap tires generated has increased (Table 18.4).

SCRAP TIRE STOCKPILES

All parties involved in scrap tire management understand that there are actually two separate but interrelated aspects to sound scrap tire management. The first aspect is dealing with the newly generated scrap tires, the 270 million new scrap tires created annually by the normal process of use of tires. The second problem is dealing with the legal and illegal stockpiles of tires that are the residue of past (and some current) methods of handling scrap tires. One of the major issues in dealing with tire stockpiles is the sheer size of the problem: how many scrap tires do we have stockpiled across the country?

Stockpiles were created as a means of disposing of tires outside the normal landfill destination for most solid waste. In some locations, many tires went to landfills, and some states still allow the practice, at least for shredded or cut tires. Stockpiles were an alternate disposal option. Also, some stockpile operators thought they were collecting "black gold," that the stockpiles they controlled contained highly valuable energy that would someday be of great value. With the threat to oil availability and rising oil prices in the 1970s and 1980s, many operators thought they would eventually be wealthy. In the meantime, the tip fees they collected to take tires provided an income. Stockpiles also resulted from cost avoidance: where landfills sought to exclude tires, the "tire jockeys" found other, illegal and cost-free sites in which to deposit tires. Out of the way ravines and woods became the sites of illegal dumping, often without the property owner's knowledge. In time, these illegal dumps could contain upwards of several thousand tires each.

Due to state programs that have focused on stockpile abatement, there are considerably fewer stockpiled scrap tires than were once reported. In 1996, it was estimated that the total national stockpile was 700 to 800 million-scrap tires, based on counts provided by a substantial majority of states. Since 1994, many states have actively worked to reduce tire stockpiles and the vast majority of states have undertaken efforts to quantify the number of stockpiled scrap tires

The information used to determine the number of stockpiled scrap tires was obtained from state agencies. Not all states had or reported their stockpile information. From the 40 states that did respond, it has been determined that the number of scrap tires in stockpiles was reduced by approximately 500 million. This reduction can be understood based on three factors: better estimates of tires in stockpiles, aggressive state-sponsored abatement programs, and tires lost in tire fires.

MARKET APPLICATIONS FOR STOCKPILED SCRAP TIRES

Tires in stockpiles have the same characteristics as those scrap tires generated in the present year. One of the first misconceptions concerning these inventoried tires is what actually happens to them while in storage. There are concerns that the tires lose heating value, that they begin to decompose, that toxic elements are leached from the pile, or that the tires emit potentially dangerous fumes. All of these allegations, theories, or beliefs are unfounded. Tires are extremely stable products, and are not subject to any of these processes.

What does occur in these piles is that water, dirt, and nontire wastes collect in and on the pile. These "contaminants" do not change the chemical composition of the tire. What they do cause is a limit to the market applications for these tires. To the best of our knowledge, no scrap tires removed from long-standing, outdoor stockpiles have ever been used as a feed source for ground rubber. The reason, as indicated, is that these tires can be quite dirty. This dirt has two negative impacts: it increases the wear on processing equipment and will "contaminate" any ground rubber generated. A ground rubber high in nonrubber material is typically not a valuable commodity. Washing the tires is impractical, since it would add to the cost of processing and create a secondary waste disposal problem—the dirty wastewater.

IMPACT OF STOCKPILE ABATEMENT PROGRAMS ON THE EXISTING MARKET INFRASTRUCTURE

One approach to effective scrap tire management is to address the flow of scrap tires generated annually. Once there is sufficient market demand for these tires, then attention should be placed on stockpile-abatement programs.

The rationale for this recommended approach is simple. If scrap tires are removed from stockpiles and sent into the existing market with no increase in demand, these tires will supplant the newly generated scrap tire. Consequently there is a zero net gain in the total of scrap tires going to markets. Furthermore, the new scrap tires will end up going to a nonmarket disposal option; in certain cases that would be to a stockpile.

It should be recognized that this approach of developing markets for newly generated scrap tires has not always be accepted. This is especially true where there has been a history of fires in a stockpile, or where there is the threat of a fire in a stockpile located in an environmentally sensitive area.

Some states have made substantial progress in abating scrap tire stockpiles, including Illinois, Wisconsin, Minnesota, Maryland, Florida, Virginia, and Oregon among others. Some of these states, including Illinois, Maryland, and Florida, have had success in feeding these stockpiled tires into the market flow without adversely affecting markets for current generation tires. Illinois, for example, has created market capacity for around 120% of annual generation. The development of market capacity is also the key to the success Maryland and Florida have had with abatement programs.

Should any state begin massive clean up of stockpiles before substantial markets have been developed, the result will likely be that tires from stockpiles will displace the current flow tires. In turn, those tires will be going to landfills (if allowed) or to stockpiles.

MARKETS FOR SCRAP TIRES

This section provides a status report on the various market applications for scrap tires. Overall, the information obtained during our market survey suggests that there continue to be three major markets for scrap tires—tire-derived fuel (TDF), products that contain recycled rubber, and civil engineering applications. There are three lesser uses for scrap tires—export, agricultural, and other miscellaneous uses—that do not fall into the preceding market areas. Finally, this chapter offers an analysis of pyrolysis, which does not appear to be making a contribution to the markets for scrap tires.

Tire-Derived Fuel

Since 1985, TDF has been the largest single market segment for scrap tires in the United States. From 1994 to 1996, TDF continued its upward trend, increasing from 101 million units annually to a permitted level of 152.5 million units annually. From 1996 to 1998, there was a decrease from the permitted capacity of 152 million to a permitted capacity of 135 million scrap tires. The actual number of tires used for TDF in 1998 was 118 million.

There are nine components to the TDF market—cement kilns, pulp and paper mill boilers, utility boilers, industrial boilers, dedicated scrap-tire-to-energy facilities, resource recovery facilities, copper smelters, iron cupola foundries, and lime kilns. Each one of these combustion technologies has its own set of engineering considerations and fuel requirements. This report will not review the technical issues for each of the combustion technologies. Information on this topic was reported in the 1990 and 1992 market surveys, which can be, obtained from the Scrap Tire Management Council.

In general, there are two methods of using tires as fuel: whole tires and processed tire-derived fuel. Whole tires are used in two combustion technologies—cement kilns and dedicated scrap-tire-to-energy facilities. Approximately 60% of the tires going to cement kilns are whole tires. This occurs for three reasons: first, the configuration of the combustion technology can accept whole tires; second, whole tires are easier to handle than processed TDF; and third, the facility that accepts whole tires also receives a tip fee. There are two dedicated scrap-tire-to-energy facilities; one only accepts whole tires, and the other can accept either whole or processed tires. The rationale for accepting whole tires in these facilities is virtually the same as for kilns. Another combustion technology that can accept whole tires is wet-bottom boilers; however, none are currently using TDF.

The trend in processed tire derived fuel (normally referred to as TDF; however, for the balance of this report TDF will mean all tire fuel use) is toward a more refined fuel chip. With advances in processing technology, a 1½ inch chip, with over 95% of all steel removed, is becoming the industry standard. While no official standards exist yet, this size fuel chip appears to be at the edge of the cost–revenue spectrum. This is to suggest that while the production of this size fuel chip is relatively more expensive than larger fuel chips, the return on investment is greater for the smaller-sized fuel chip than the larger chip. Moreover, certain types of combustion technologies are only capable of using the 1½ inch chip (i.e., cyclone boilers).

The use of TDF can be attributed to three factors. The first is that TDF has been used successfully in a wide array of end use markets that are both environmentally sound and cost efficient. This means that the use of TDF has not caused a facility to exceed its permitted air or other emissions limits. Furthermore, the use of TDF has been demonstrated to be economically competitive, offering a clear incentive to the end user.

The second factor, based in part on the first, is the development of an efficient private sector supporting the scrap tire industry. More established firms are operating in this industry, and advances in processing technology are taking place. These technical improvements are employed to produce a more refined fuel chip, allowing for easier handling and more complete combustion.

Although the private sector has grown, it is not risk free. There is considerable turmoil in the scrap tire industry. There is still a significant failure rate among start-up companies, and even established companies have experienced serious financial difficulties. Anyone interested in entering into this marketplace should proceed with the utmost care and research prior to making any type of financial investment.

The third factor that has contributed to the growth of the TDF markets has been the efforts of certain state scrap tire program managers. There are some states that have been actively developing markets (Maryland, Virginia, and Illinois, to mention a few of the more successful ones). The efforts of dedicated state employees can have a significant impact on the number of facilities using tire-derived materials.

Regulatory Issues

Before discussing the specifics of the various TDF markets, a review of state and Federal regulations on the use of TDF is in order. Many states have established permitting requirements specifically for facilities using TDF. In addition, any facility using tires as fuel must operate within established air emission standards. All combustion facilities using TDF have gone through this process.

Currently, TDF is considered a fuel by the USEPA (see Section 129 of the Clean Air Act). As part of its continuing implementation of the Clean Air Act Amendments (CAAA) of 1991, the USEPA will be promulgating at least two new standards that may impact on the use of TDF.

It is understood that USEPA will be surveying a wide range of combustion facilities to develop information for this process. It is to be hoped that the agency will also make use of the substantial body of data that has been developed in many states as a result of the growing use of TDF.

The second issue that could potentially impact the use TDF could be the proposed USEPA emission standards on particulate material emission limits, known as PM 2.5. These proposed standards are not specifically designed for TDF users, but for all combustion processes.

Scrap Tire Fuel in Cement Kilns

The use of scrap tires as a supplemental fuel in cement kilns continues to be a viable market, although the numbers of tires used has decreased from 1996 to 1998. Presently, 30 cement kilns are using TDF, with fuel placement ranging from three to 30%.

Kilns currently using scrap tires have individual permitted volume capacities ranging from 250,000 to three million scrap tires per year. The driving forces behind the current and anticipated use of TDF have remained constant. These factors are:

- Improved emissions
- Operational considerations
- Decreased fuel costs

The impediments to the further use of tires as a supplemental fuel in cement kilns have remained constant, although there are several new issues as well. The factors that impacting on TDF use are:

- Delays due to difficulty in obtaining a permit or modified permit from the state regulatory agency
- Kilns operating at full capacity
- Reliability of tire/TDF supplies in isolated areas
- Local opposition to tire combustion
- The use of competing supplemental fuels

Given the factors that currently impact this industry, it appears that the demand for TDF will be a function of production level as compared to fuel costs. Where and when TDF does not adversely impact on the rate of production for cement and where and when the economics are favorable, TDF will be used. With several cement kilns currently pursuing the use of TDF, it appears reasonable that the number of scrap tires consumed in this market niche could return to the 1996 rate within the next two years.

Scrap Tire Fuel in Lime Kilns

The use of scrap tires as a supplemental fuel in lime kilns was a short-lived market niche for TDF. In 1994, no lime kiln was using TDF. Lime kilns, like their cousins, cement kilns, are energy intensive technologies. The high level of heating value in TDF is clearly of interest to this industry. The differences between lime kilns and cement kilns are that lime kilns are shorter in length and often produces a different color end product. Some commercial lime must be white, whereas cement is usually gray. The use of TDF in lime kilns, under certain circumstances, can darken the product. If the color of the lime is a critical element for its sale, the use of TDF will be not be possible

In the 1995–1998 period, one lime kiln used TDF on a production basis, and consumed around 10,000 tons of TDF annually, or around 1 million tires. Although there continues to be some modest interest in this fuel market, at present no lime kiln is using or testing TDF. This lack of use appears to be based on the impact TDF has on the color of lime. The likelihood that these markets niche will provide demand for TDF is very low.

Scrap Tire Fuel in Pulp and Paper Mill Boilers

Tire-derived fuel can be used as supplemental fuel in pulp and paper mill boilers. The technology is proven and has been in continuous use in the United States since the early 1980s. Consumption of TDF in U.S. pulp and paper mills has almost doubled since the mid-1980s to approximately 23 million tires per year. There are currently 14 mills known to be using TDF on a continuous basis.

Environmental constraints associated with the use of TDF vary widely, depending on facility characteristics (e.g., type, age, pollution control equipment, etc.) and the local regulatory climate. In general, interviews with mill managers and environmental professionals indicate that permitting processes have become less complicated as access to air emissions data has increased; however, the permitting process remains an impediment.

The main reasons given for the use of TDF in pulp and paper mill boilers continue to be:

- Decreasing the cost of fuel
- Improving emissions
- Improved combustion efficiency

The principal impediments to the further use of TDF in the U.S. pulp and paper industry continue to be:

- Marginal cost advantage of TDF over typical mill fuels
- Permit modification requirements; inconsistent regulatory guidance
- Remote location of many mills (higher transportation costs)
- Reluctance of state officials to accept out of state emission data
- Reliability of TDF supplies in remote locations
- Variable quality in some fuel chips
- Elevated zinc levels in effluent collection systems
- Changes in state scrap tire programs

The likelihood of continued TDF use in this market sector appears to be high; mills using TDF will continue to do so. There are several mills that have concluded their testing programs or that are considering the use of TDF. Consequently, there is an expectation that this market sector should gradually increase its consumption of TDF over the next two years.

Scrap Tire Fuel in Electricity Generating Facilities

This section will discuss the use of TDF in various types of electrical generating facilities and industrial boilers, except dedicated scrap-tire-to-energy facilities. Overall, there are three types of facilities covered: large-scale utility boilers, industrial boilers, and resource recovery facilities. Overall, the use of TDF in the combined market segment typically is as a supplemental fuel, not exceeding 10% of the total fuel mix.

Utility Boilers. In the utility market, the use of TDF in wet-bottom, cyclone, stoker, and fluidized-bed boilers is proven. Our survey indicates that 12 utility power plants use TDF. The use of TDF in this market niche has decreased over the past two years. The main issues that have impacted on this market are the deregulation of the utility market, companies preparing for the implementation of the Clean Air Act (CAA), and the termination of state scrap tire programs.

Overall, the use of TDF in large-scale utility boilers is relatively small, ranging between 1 to 3% of the fuel. At these rates of substitution, there is little to no impact on the emission or operations of these facilities. The use of TDF is often as one of several alternate fuels.

The decrease in the number of utility boilers using TDF can be traced to the deregulation of the industry. In this case, TDF would have been used in an older boiler, one which was shut down due to company policies (i.e., looking to sell the plant or company). The CAA has also impacted this industry. There was one case where a utility using TDF stopped using solid fuel and switched to natural gas as their way of dealing with the CAA regulations. Finally, in Wisconsin, the state scrap tire program ended its subsidies for the use of Wisconsin-generated scrap tires. This change caused several utilities to stop using TDF, since their TDF suppliers were unable to competitively price TDF. Whether or not TDF use expands in this market sector will be function of the impacts of the CAA and deregulation.

Industrial Boilers. Nine industrial power plants are currently using TDF. While TDF has not demonstrated any adverse affects on these facilities, this market sector has seen a dramatic reduction of permitted facilities using or that could have used TDF. Since 1996, 10 facilities permitted to use TDF have either shut down or did not begin using TDF.

Although there is no one reason for this development, TDF supplies a relatively small percentage of a boiler's fuel. When competing fuels offer price advantages or when TDF supplies are inconsistent, the operations manager is quick to stop the use of TDF and use a fuel with fewer drawbacks. In general, however, TDF continues to be an attractive alternative fuel and should continue to be used on a regular basis. Over the course of the next two years this market niche should hold steady, if not expand gradually.

Resource Recovery Facilities. In 1998, the STMC completed a survey of municipal resource recovery facilities (RRFs). The results indicated that there are nine RRFs currently using TDF on a continuous basis, generally in the range of 1–2% of their total fuel intake.

The use of TDF in RRFs can be attributed to two factors: the demise of flow control and increased participation in recycling programs. Although the use of TDF in RRFs is still limited to a range of 2–5% of their fuel mix, more RRFs are allowing scrap tires to be fed into their systems. The benefits derived from the use of TDF in these combined markets continue to be decreased fuel costs, a source of fuel to complete a facility's mass balance, improved emissions, and improved combustion efficiencies.

The principal barriers to the further use of TDF in electricity generating facilities are varied. There is the marginal cost advantage of scrap tire material over competing fuels, environmental permit modification requirements, inconsistent regulatory guidance in some states, the conservative/risk adverse nature of a particular industry, the inability to blend fuel (TDF becomes economical unviable when pulverized coal is the main fuel source), limits of ash handling systems, and the variable quality of some TDF. Still, the use of TDF in this market sector is expected to continue, and should increase over the next two years.

Scrap Tire Fuel in Dedicated Tire-To-Energy Facilities

There remain two dedicated scrap-tire-to-energy facilities in the United States, one in California and one in Connecticut. These two plants consume 16 million tires annually. While the technology is sound, the economics impacting these facilities may change due to the deregulation of the utility industry, especially for the California facility.

In the period of 1996 through 1998, there was another dedicated tire-to-energy facility that was built, but never opened. The Chewton Glen facility, built in Ford Heights, IL (outside Chicago), filed for bankruptcy due to changes in a State program. This project did not fail because of the failure of the combustion technology, the availability of scrap tires, or failure to comply with emission standards. Rather, the facility was closed due to a change in the Illinois Retail Rate Law that provided subsidized power rates for certain facilities, including Chewton Glen, using solid waste as fuel to generate electricity. In April 1996, the Governor of Illinois signed a bill eliminating the retail rate provision. This made the operation of the Chewton Glen facility economically unviable. The result was Chapter 11 filing.

The development of any other large-scale, dedicated scrap-tire-to-energy facility appears to be unlikely due to the pending deregulation of the utility industry. Deregulation will result in downward pressure on the rates utilities pay for electricity, and will be especially hard on facilities using alternative fuel sources that have traditionally enjoyed preferential rate treatment.

Given the current market conditions, the potential usage of scrap tires in dedicated tire-to-energy facilities will likely continue to remain at its present level of 16 million a year. How long this rate of tires is consumed will depend on market conditions, especially in California.

USE OF SIZE-REDUCED RUBBER

The market for size-reduced rubber continues to grow, although the last two years have seen both positive and negative developments. On the negative side, the Federal mandate to use rubber-modified asphalt in federally aided highway construction was repealed, significantly reducing a major potential ground rubber market. On the positive side, a number of new applications for ground rubber have been developed and several other uses have been expanded. In addition, the automotive industry, one of the largest markets for new rubber products, has expressed its desire to see recycled rubber used as an ingredient in the new rubber parts it purchases. While it may be a few years before many parts containing a significant amount of recycled rubber are actually installed in new cars, the rubber parts industry is currently developing and testing new products using ground rubber. This market has also seen an influx of new producers, greatly increasing the potential material supply. Unfortunately, the supply capacity continues to exceed the market demand, placing some pressure on prices.

The overall market demand for size-reduced rubber in the United States was approximately 460 million pounds at the end of 1998. This compares with the 1992 market of 160 million pounds of size-reduced scrap tire rubber, the 1994 market volume of 240 million pounds, and the market volume of 400 million pounds in 1996. Although the increase is substantial, this market segment is still has a number of impediments.

There are two sources for tire-derived size-reduced rubber: tire buffings and processed whole scrap tires. Of the total market volume of 450 million pounds generated, about 210 million pounds, or 46.5% is obtained from tire buffings. The balance of 240 million pounds, or 54.5%, was obtained from whole scrap tires (17.5 million scrap tires). We estimate that 12.5 million scrap tires are being reduced to ground rubber. No attempt is made to differentiate between buffing dust and scrap tire rubber neither in identifying markets, nor to differentiate between cryogenically produced or ambiently ground scrap tire rubber.

The need to rely on scrap tires as the main source of increased capacity in this market is due to the finite supply of tire buffings. Buffings are a by-product from the retreading industry, and are created when a used tire is being prepared to accept new tread. The existing tread, and sometimes the shoulder and sidewall of a tire, are removed by a high-speed buffer. The buffings, relatively long, tubular shaped particles, are collected, packed and sold to the producers of size-reduced rubber.

The estimated quantity of buffings available in the United States is 250 million pounds. It appears that these quantities will not increase, since the number of tires retreaded annually has leveled off at approximately 30 to 33 million units. According to industry experts, the only likely growth potential for retreaded tires is in the utility and recreational tire markets.

Standards for Ground Rubber

One of the most important recent developments for this market segment was the publication in late 1996 of two revised standards for ground rubber. Both standards were developed by Committee D11.26 of the American Society for Testing and Materials. The first standard, D5603-96, Standard Classification for Rubber Compounding Materials—Recycled Vulcanizate Particulate Rubber, provides a set of size standards for all types of particulate ground rubber. The second standard, D5644-96, Standard Test Method for Rubber Compounding Materials—Determination of Particle Size Distribution of Recycled Vulcanizate Particulate Rubber, provides a standard method for measuring particle size. The combined effect of these two standards, developed jointly by ground rubber producers and users, should be to greatly enhance the possibility that ground rubber can became a commodity material and be freely bought and sold in the marketplace. (The standards are available from ASTM, 100 Bar Harbor Dr., West Conshohocken, PA 19428)

Market Applications for Ground Scrap Tire Rubber

There are seven general categories of markets for ground rubber. Market availability is a function of cost, product availability, product characteristics, and substitute material availability. While all these factors deserve explanation, this survey will only give a general description of the markets.

Bound Rubber Products. Ground scrap tire rubber is formed into a set shape, usually held together by an adhesive material (typically urethane or epoxy). The rubber can also be mixed with another polymer before the adhesive is added. Bound rubber products include, but are not limited to carpet underlay, flooring material, dock bumpers, patio floor material, railroad crossing blocks, and roof walkway pads.

This market segment currently consumes an estimated 134 million pounds of recycled rubber, and increased by 10% in 1997. The major limiting factors in this market segment are: (1) competition from other recycled or virgin materials, and (2) the fact that tire rubber is vulcanized (bonding of the carbon atoms from rubber with sulfur). Since this bond cannot be broken, no other polymer can chemically bond with the rubber.

New Tire Manufacturing. Powdered scrap tire rubber can be used as a low-volume filler material in two components of a tire: the tread and sidewall. In addition, powdered rubber produced from inner tubes is used in the inner liner. In general, scrap tire rubber is limited to a maximum of 1½% of the tire (by weight). Powered rubber can be used in off-road tires, intermodal tires, bias ply truck tires, and solid tires.

Currently, this market segment consumes 48 million pounds of recycled rubber annually. While there is no expected dramatic increase in the use of recycled scrap tire rubber in new tire manufacturing in the next few years, there is the possibility of a significant increase by 1999 or later.

One tire manufacturer, Michelin North America, has reported that it is developing a tire that will incorporate up to 10% recycled scrap tire rubber. Depending on the outcome of developmental testing to ensure that the tire meets performance and durability requirements, this time might be available, according to the company, for installation on model year 2000 vehicles.

The problems that must be overcome in reusing scrap tire rubber in new tire construction, and what any tire manufacturer seeking to increase recycled content must deal with, are relatively well known. In the past, the use of recyclable ground scrap tire rubber in new tire construction resulted in higher hysteresis loss, heat build up, and increased rolling resistance. High hysteresis loss returns less energy during stretching, flexing, and compression. The unreturned energy is converted primarily into heat, which has a negative impact on the rolling resistance of a tire. Increased rolling resistance, in turn, causes increased friction between the tire and the driving surface, which increases traction and decreases fuel economy. With today's emphasis on longer-lasting, more fuel efficient tires, adding recycled scrap tire rubber would normally be counterproductive. Abrasion resistance has also been a reported problem, with resulting loss of tread ware.

Rubber Modified Asphalt (RMA). Ground rubber can be blended with asphalt to modify the properties of the asphalt in highway construction. Size-reduced scrap tire rubber can be used either as part of the asphalt rubber binder, seal coat, cape seal spray, or joint and crack sealant (generally referred to as asphalt–rubber), or as an aggregate substitution (rubber-modified asphalt concrete, or RUMAC).

Repeal of ISTEA Mandate. One of the major hopes for the development of a major ground rubber market was the enactment of Section 1038(d) of the Intermodal Surface Transportation Efficiency Act of 1991 (ISTEA). This provision would have required all states to begin using rubber-modified asphalt for a specified portion of their asphalt paving beginning in 1994, with the requirement increasing each year until 1997, when 20% of all federally aided asphalt paving would have had to contain a specified volume of rubber. When originally enacted, it was calculated that by 1997 this requirement would have utilized the equivalent of 80 million tires in paving applications, or nearly 1 billion pounds of ground rubber. However, the provision generated major opposition from the asphalt paving community and from state highway administrators. In 1994 and 1995, riders to federal transportation appropriations legislation prevented funds from being spent to enforce the Section 1038(d) mandate. Ultimately the mandate was repealed as part of the Federal Highway Systems Act passed in late 1995. In its place was left an unfunded demonstration grant program.

During 1996, this market segment consumed around 168 million pounds of recycled rubber, mainly in states that have regularly utilized rubber-modified asphalt. Since the repeal, the number of projects outside of these states has declined dramatically. Currently, the only states with any substantial asphalt rubber activity are Arizona, California, and Florida, with lesser activity in Kansas and Texas. In general, asphalt–rubber, or the "wet process," has proven to be the most successful product, representing approximately 95% of the RMA market.

Asphalt Rubber Research. The technical, environmental, and economic constraints associated with the use of scrap tire material in asphalt paving applications have only moderately changed since 1992. The major issues with RMA are the lack of standard mix designs, continuing questions about its ability to be recycled, worker health and safety, and its cost. To address the worker safety issue, The Federal Highway Administration (FHWA) has contracted with the National Institute of Occupational Safety and Health (NIOSH) to conduct research into the worker safety issues. NIOSH has conducted a series of field studies in several states and so far has found no major adverse health issues. As for the recycling issue, several tests were undertaken which revealed no difficulty recycling pavements containing rubber.

A multistate cooperative research project was organized by the FHWA to conduct a comprehensive study of RMA performance and develop engineering guidelines and standardized application procedures. This is a 60 month study involving several engineering schools and research institutions, with a goal of providing detailed engineering criteria for use by all states. A separate major study of RMA is being conducted with federal funding at Texas A&M University.

To a large extent, the acceptance of RMA by any state's department of transportation (DOT) will depend upon the results of these studies and their own field trials. Any large-scale increase in the use of RMA is dependent upon the state's DOT acceptance of these test results and the willingness to begin their own state and local level programs. Even with some degree of acceptance by a DOT, the demand for size-reduced rubber will not explode overnight. Rather, a more gradual increase over the next five years is expected.

Athletic and Recreational Applications. This market segment has been one of the fastest growing markets for ground rubber over the last two years. In 1996, some 24 million pounds of ground rubber was used in the wide array of products in this category. While the total poundage only equals 6% of the total ground rubber market, this total represents a 40% increase in its usage.

Examples of this market segment include, but are not limited to, the use of rubber in running track material, in grass-surfaced playing areas, or for playground surfaces. Particulate rubber generally makes the playing surface and the running tracks more resilient and less rigid, while allowing the surface material to maintain traction and shape. One case in point is the running track at the White House, which contains particulate rubber. Another example of this application is the use of particulate rubber as a soil additive combined with organic material in playing fields that are subject to soil compaction. The addition of this rubber–organic material compound has demonstrated the ability to allow grass to become better rooted, resulting in improved drainage and reduced hardening of the playing surface.

A new use that has emerged in the last few years is the use of ground rubber as a turf top dressing. Developed at Michigan State University, this process is now patented as Crown Turf III. Ground rubber in a range of 4 to 10 mesh is applied to turf as a top dressing. The rubber granules fall through the grass leaves and remain on the soil surface. They will mound around the crown of the turf grass and will protect it from physical harm. In addition, it will insulate the grass plant and allow faster recovery in the spring. Principal applications to date have been in heavy wear areas of golf courses and on athletic fields.

The use of chipped tires as a loose laid playground cover has gained wide acceptance in the last two years. The use of chipped tires provides the highest level of shock attenuation of any competing material. The issues typically raised about the use as playground cover concern the toxicology of tire chips (what if a child swallows a piece?), whether the chip will scuff clothes (black coloring), and whether there are any harmful emissions. To date, all test results have indicated that there are no adverse affects from the use of scrap tires in any of these applications. Tire chips have also been used in equestrian arenas where they provide a superior, noncompacting riding surface.

This market sector continues to have the potential to be a large growth area in the next several years. One of the more critical factors in this market's growth potential will be school system budgets, since schools represent one of the largest potential markets.

Friction Material. Friction brake material in brake pads and brake shoes contains particulate rubber. This is a mature industry with little to no growth expected.

Molded and Extruded Plastics/Rubber. Particulate rubber can be added to other polymers (rubber or plastic) to extend or modify properties of thermoplastic polymeric materials. Examples of this application are injection molded products and extruded goods. There appears to be a significant market potential for this application, due to the continuing research and development of products using a surface-modified rubber.

Currently, 18 million pounds of particulate rubber is used in this market segment. This market segment increased considerably in 1997–1998. Although the market information suggests a significant potential market for molded and extruded co-polymeric products, there is not enough practical experience to suggest whether the inability to devulcanize rubber or other technical issues will impact on the compounding and/or bonding ability of recycled rubber.

Automotive Parts. As noted in the introduction to this section, the automotive industry has expressed strong interest in seeing that rubber parts they purchase contain recycled content. Nontire automotive rubber parts are the largest single rubber products market after tires. Every automobile contains literally hundreds of rubber-based parts, from hose and belts to weather-stripping, body and motor mounts, and tiny O-rings in fuel injectors. Already, there are some parts containing recycled rubber that are being used as original equipment on new cars. With the automobile manufacturers asking for more recycled content, the rubber parts manufacturers are doing the testing and developmental work needed to incorporate recycled rubber without affecting the performance, durability, or price of the parts. Given the long lead time on qualifying new parts for installation as original equipment on new cars, it is unlikely that this application will result in any substantial ground rubber consumption much before the 2000 model year, if not later. Current consumption is limited to those few parts already qualified and being used, and to rubber parts manufacturers engaged in testing and development.

Producer Issues

Of all the market segments for scrap tires, the interest in entering the ground-rubber business continues at a high level. While potentially a lucrative business, the current ground-rubber market is facing some very significant issues. First, there are some 200 companies across the United States that are either producing, or claim to have the capacity to produce, ground rubber. Despite this high total number of processors, less than a dozen companies generate and sell some 80–85% of ground rubber. Hence, the majority of companies must struggle to capture the relatively small remaining market share.

Even if the market demand for ground rubber would double, the impact on the industry would be minimal. Currently, many companies are operating at less than half their production capacity, which drives up costs. These increased production costs cannot be passed along because: (1) the market for recyclable rubber is not elastic; (2) ground-rubber supply greatly exceeds demand; and (3) the market forces are placing a downward pressure on prices. The result of all these factors is that many of the marginal producers have been or will soon be forced out of business.

While the longer-term outlook for the ground-rubber segment suggests a more stable market, there does not appear to be any overwhelming market or technological breakthrough that will radically alter the market dynamics in the short-term. The industry forecast is for continued and sustained growth, perhaps 10–15% annually. Furthermore, the larger, more established producers will continue to expand and the marginal producers will likely be forced out of business. Even with this situation, there will probably continue to be new actors seeking to penetrate the production market. Finally, with the beginnings of a true marketplace for ground rubber the emphasis will be placed on the quality of the ground rubber. This development should cause prices to stabilize, giving the ground-rubber market a more secure basis, albeit with a smaller number of producers.

One of the more discussed issues in this market segment is the difference between processing technologies: ambient versus cryogenic processing. Suffice it to say that both technologies can be performed in an efficient manner. The question asked is, which processing system is the more desirable? The answer is, it depends. What it depends on is the market the processor is attempting to sell its material in. Put in its simplest terms, the type of processing system needed will depend upon the characteristics wanted by the end user. If a rough surface particle is asked for, then the ambient system will be the logical choice. If a smooth particle surface is asked for, then the right move is to cryogenics. If it makes no difference, then the deciding factor will be the cost of processing.

Ground-Rubber Market Limits

The United States produces around 3.3 million metric tons of scrap tires annually. If all these scrap tires were recycled, it would produce around 2.3 million metric tons of rubber annually. The market for all ground-rubber applications in 1996 consumed about 400 million pounds or about 7% of the total potential volume. In order for the ground-rubber markets to consume all of the potential scrap-tire-derived ground rubber, the markets would have to increase more than twenty times.

This can be examined from a different perspective. The United States consumes about 3.119 million metric tons of virgin rubber, natural and synthetic combined each year (1994 numbers) for both tire and nontire applications. If ground, recycled rubber could replace an average of 10% of all virgin rubber in all new rubber products, this would be a total market of 312 thousand metric tons, nearly three times the current market. However, this would still leave about 2 million metric tons of ground rubber needing markets. In time, those markets might develop–in asphalt paving or as soil additives or in the manufacture of products not normally manufactured from rubber. But for the foreseeable future, all markets will be needed just to fill the demand for all recycled tires.

Volume Characteristics

Although the overall consumption of size-reduced rubber in 1998 was 450 million pounds, the STMC is only tracking the use of size-reduced rubber from scrap tires. From that estimated total, the use of size-reduced rubber from scrap tires is approximately 17.5 million scrap tires. Within two years, the expectation is for another 10% increase, barring any technological breakthrough.

The overall increase in the use of scrap-tire rubber will be a function of the quantity of tire buffings and increased market demand for RMA, athletic/recreational applications, and molded/extruded products. The factors that affect this increased demand are:

- Ability to enhance properties and characteristics of polymeric compounds (molded–extruded and RMA)
- Lower-cost additive (molded/extruded)
- Provides enhanced safety and performance (athletic/recreational)

The factors that currently impede the increased use of size-reduced rubber include:

- The generally negative attitudes of most state highway departments toward rubber-modified asphalt
- Competition from other recycled materials (molded/extruded products)
- Budgetary limits (athletic/recreational applications)
- Environmental concerns (i.e., leachate relating to athletic/recreational applications)

New Technologies

There are two new technologies that may have an impact on the ground-rubber industry: surface modification and devulcanization. Both attempt to deal with a basic problem encountered when trying to incorporate ground rubber into new rubber compounds: ground rubber is not a chemically active material. The process of vulcanization makes rubber a temperature-stable material by making largely irreversible chemical and physical changes to it. When vulcanized rubber is ground up, it retains these properties. Therefore, ground rubber is not bound into a new compound by the same chemical and physical bonds that affect the virgin materials. Rather, it will be bound into the new material largely by mechanical adhesion, which is a relatively weak bond, and will affect the physical properties of the final product.

Devulcanization. Devulcanization is the reversal of the sulfur–carbon bonding that makes rubber a stable material. There are several companies that report having technologies that can reverse, at least in part, the carbon–sulfur bond, and there are several researchers working on this issue Some of the more imaginative research involves the use of ultrasonics to break the carbon–sulfur bonds. Another researcher advocates a shear-extrusion process. Of all these companies, there is one that is close to, if not already, commercially generating and selling a devulcanized scrap-tire rubber. This technology uses a proprietary chemical agent mixed in an open mill. Although this technology could assist the growth of this market segment and would indeed be a great technological advance, there remain several critical questions.

Perhaps the most significant question is the quantity of scrap rubber that will be consumed through this process. It appears that, currently, the majority of rubber that will be processed through this devulcanization process will be factory, nontire scrap (preconsumer material). While the importance of this application should not be diminished, the issue is the impact this technology will have on scrap-tire rubber. It appears that this technology will likely only have a minimal impact on the scrap-tire rubber market, although there could be some trickle-down effect.

This conclusion as to the limited effect on scrap-tire rubber also is based on the composition of the tire itself. Today's tire is a combination of several types of polymers (rubbers). Even if devulcanization were possible, the resulting material would be a combination of the three to four polymers used in a tire. Add in several other types of tires, all of which contain various polymers, and you have series of devulcanized polymers, none of which have the characteristics or qualities of a virgin material, and hence no applicability for new tire manufacturing.

In order for this devulcanized material to be successful, it will have to be engineered into a rubber formula and tested. At this point, it appears that there have been some successes with devulcanized, nontire rubber. Actual applications will determine the ultimate value.

Surface Modification. Surface modification is a process in which size-reduced rubber is treated to make it more reactive in new compounds. As its name implies, there is no attempt to alter the vulcanized material. There are two basic surface-modification technologies. The first is to coat the rubber with a bonding agent that will make it more chemically active in a new compound. The second is to treat the rubber particle with a caustic gas to "activate" the surface of the rubber, allowing the material to bond with other polymers, usually urethane.

With respect to the coating process, there are several companies marketing either pretreated particles or the coating material to treat one's own ground rubber. There is one company marketing gas-treated rubber. In either case, the resultant materials are used together with various virgin polymers to manufacture new products. The major benefit of surface modification is that it can reduce the final cost of a finished product, as the treated ground rubber is used with high-value virgin materials to make a finished product at a lower material cost without affecting, or perhaps even improving, product performance.

Cut, Stamped and Punched Rubber Products

There has been no change in this market segment over the past several years. The process of cutting, punching or stamping products from scrap-tire carcasses is one of the oldest methods of reuse of old tires. This market encompasses several dozen, if not hundreds, of products, all of which take advantage of the toughness and durability of tire carcass material. The basic process uses the tire carcass as a raw material. Small parts are then die cut or stamped, or strips or other shapes are cut from the tires.

Examples of small die cut parts include muffler hangers and snow blower blades. Strips cut from tires have been used to make door mats as well as other types of mats for years; cut strips bonded together are used to make floor tiles for heavy traffic areas. Stamped shapes are also used to make wheel chocks, dock bumpers, support pads for back hoes and other construction equipment, commercial fishing equipment, blasting mats, solid tires for use on bush hogging equipment, wearing pads for front-end loader buckets used in resource recovery facilities, and a multitude of other uses.

One limitation of this market is that it generally uses only bias ply tires or fabric-bodied radial tires. The steel belts in most radial tires and the steel body plies in an increasing percentage of medium truck radials are not desirable in these applications. As the number of bias ply and fabric bodied medium truck tires being produced declines, it appears unlikely that this market segment will increase substantially. In the meantime, the value of scrap tires that can be used for cutting or stamping will likely increase slightly.

Larger, bias ply tires are another possible raw material for this market, and thus it may provide a reuse opportunity for some of the large off-the-road tires that are otherwise difficult to handle. Another real challenge to this market segment is state laws that subsidize tire shredding without regard to other markets. The highest value for the scrap tire may be the money the state is willing to provide the tire shredder, even if the shreds have no markets.

Because of the demand in this market, virtually all of the scrap bias ply medium truck tires that are collected by major truck casing dealers are finding their way to a cutting or stamping operation.

The estimate of the size of this market segment is eight million tires. This segment is expected to remain stagnant, since there is a (current) finite number of bias ply tires. If no new supply of bias ply tires can be secured, it is likely that this market segment will decrease slightly over the next five years as the supply of bias ply tires diminishes.

SCRAP TIRES IN CIVIL ENGINEERING APPLICATIONS

The civil engineering market encompasses a wide range of potential uses for scrap tires and scrap-tire-derived material. In virtually all civil engineering applications, the scrap-tire-derived material will be used to replace some other material currently used in construction (e.g., dirt, clean fill, gravel, sand, etc.). It is potentially a major market for scrap tires and one of the best fits for tires from stockpile clean-ups, as the presence of dirt is not usually a problem but other nontoxic contaminants preclude other, higher-value-added uses.

ASTM Guidelines

One of the major difficulties in expanding the civil engineering use of tires is a general lack of published information about the physical characteristics of scrap tires as an engineering material, or the lack of any general guidelines for these uses. In order to rectify this situation, the Scrap Tire Management Council, working in conjunction with Professor Dana Humphrey of the University of Maine, one of the foremost researchers into the civil engineering uses of tires, approached ASTM (American Society for Testing Materials Committee D 34.15, and proposed a set of guidelines to fill this need. After nearly three years work, the Guidelines for the Use of Scrap Tires in Civil Engineering Applications were approved by ASTM.

Heating Incidents

The growth of the civil engineering markets was substantially curtailed in early 1996 when three heating incidents involving deep scrap tire fills were detected. Two of the incidents were in Washington State and involved road embankment fills; the other occurred in Colorado and involved a fill over a rockslide area on a canyon wall. Although the cause of the heating was not initially known, the mere fact that they all involved tires used as fill material was enough to cause most planned projects to be canceled or postponed. All three of the projects were dug up and replaced with conventional fill materials.

Investigations undertaken by Washington State and by the Federal Highway Administration did not reveal a definitive cause for the incidents, although several factors are suspected. One of the first factors noted was that all three incident involved very deep fills: one Washington site had nearly 50 feet of tire fill, while the other two sites had more than 20 feet of tire fill. Other conditions that may have aided in the heating were the presence of water at some critical time, access to oxygen, presence of organic fill material, presence of rubber "fines," or small particulate matter, and the presence of large amounts of loose or exposed steel wire.

In order to seek answers to the issues raised by these incidents, the Scrap Tire Management Council convened a meeting of interested parties that became the ad hoc Committee on Civil Engineering Applications. As a result of committee investigations, it was determined that there were more than 70 tire fill projects at depths up to 15 feet of tires across the country. Furthermore, and equally as important, none of these projects had ever experienced a similar heating incident. Accordingly, the committee drafted a set of guidelines for the use of shredded tires in thin fill applications based on the successful uses of tire shreds. The premise for developing these construction guidelines is that techniques for using tires in such applications can be identified that will eliminate the possibility of such a heating event taking place.

Still, with all the negative reporting and misinformation concerning civil engineering applications, some 10 million scrap tires were used in 1996. This amount is only 20% less than the quantity used 1995, which is far less of a decrease than originally predicted. The strength of the civil engineering market was concentrated in three states; Maine, Texas, and Virginia. There were clearly other civil engineering applications in other states in 1996, but these three were the biggest single market areas. In general, the largest applications were landfill cover, leachate collection systems, and road insulation material. By the end of 1998, 20 million scrap tires were used in an array of applications. The combination of ASTM specifications, a FHWA approved construction guideline, and the availability of information on the various projects has allowed this market sector to expand.

Description of Civil Engineering Applications

Subgrade Fill and Embankments. Work has been done in several states (i.e., North Carolina, Virginia, Vermont) using shredded scrap tires as a subgrade fill in the construction of highway shoulders or other fill projects. The principal engineering advantage that scrap tire material brings to these projects is lighter weight. Scrap tire material can provide significant benefits where the use of a lightweight fill is indicated. Projects using this feature of scrap tires include the construction of an interstate ramp across a closed landfill in Colorado, construction of mine access roads across bogs in Minnesota, and the reconstruction of a highway shoulder in a slide-prone area in Oregon. Scrap tire material has also been used to retain forest roads, protect coastal roads from erosion, enhance the stability of steep slopes along highways and reinforce shoulder areas.

Research is also underway into the use of shredded tire material as backfill for retaining walls in Maine. Initial results suggest that the lower density of the scrap tire material significantly reduces the pressure on the retaining structure and permits substantial savings in the material used to construct it. This project is also of significant importance because it is the first major project to do field sampling of scrap tire leachate. While still in its preliminary stages, the reports have, to date, indicated that the leachate from scrap tires poses no health or environmental risks.

Landfill Construction and Operation. Although the aim of most scrap tire market development efforts is to finds uses for scrap tires that keep them out of landfills, many landfill operators have found that scrap tires can be used beneficially in the construction and operation of the landfill, and can replace other materials that would have to be purchased.

Several operators have used whole or chopped scrap tires to assist in constructing leachate collection systems in new landfill cells (Texas, New York, Pennsylvania, Florida, and Oregon). In addition, several states permit the use of shredded scrap tires as a partial daily cover material. The processed scrap tires are mixed with clean fill. This combination offers two advantages. First, processed scrap tires replace fill dirt that might otherwise have to be purchased. Second, the weight of the processed tires is effective in keeping the compacted material in the landfill. In landfills where the management wants to allow water to infiltrate the landfill, using a greater percentage of tires in the landfill allow a greater amount of water to enter the working face of the landfill.

Breakwaters and Artificial Reef Construction. Breakwaters are used to prevent coastal erosion and to provide boats protection from wave action. Scrap tire breakwaters are constructed by tying tires together with nylon bolts and rubber strips. Georgia and New Jersey have constructed scrap tire breakwaters and have reported no technical difficulties.

Artificial reefs are designed to prevent scouring, protect coastal roads, and provide habitat for aquatic life. Tire reefs are made by bundling punctured tires weighted down with concrete and anchoring them to the ocean floor. Major artificial reefs have been constructed off the coasts of Florida, Maryland, and New Jersey.

Septic System Drain Fields. Chopped or shredded scrap tire material can be used in small-scale, homeowner-level civil engineering applications. Shredded tire material has been used in some areas as a drainage medium around house foundations. Shredded tire material is also used in several states to construct leaching fields for septic systems. The lower density of the shredded tire material greatly reduces the expense and the labor to construct these leaching fields, and the material provides performance equal to that of traditional stone backfill material. South Carolina has issued specifications for the construction of tire-shred leaching fields.

The outlook for this market sector is very positive, although certainly not without issues that must continuously be addressed. Overall, the indications are that the use or scrap tires in civil engineering applications should increase 10–20% over the next two years. This forecast is based on the assumption that there will be a continued effort by states and the scrap tire industry to make use of the characteristics of scrap tires in civil engineering applications.

PRODUCT RECOVERY VIA PYROLYSIS

There is a continuing interest in thermal distillation or pyrolysis of scrap tires as a strategy to manage scrap tires, even though there has been no success in the marketing of pyrolysis by-products. Pyrolysis is the use of heat in the absence of oxygen to decompose a material. As a basic chemical technology, it has existed since the time of the ancient Greeks. Broad interest in tire pyrolysis began as a result of the world-wide concerns for the availability of petroleum. Pyrolysis was thought to be a method to liberate the liquid hydrocarbons in the tire. A return to easier world access to petroleum at reasonable cost led most investigators to abandon their research on pyrolysis.

The fact is that pyrolysis technology does work, in the sense that tires can be rendered and converted to three by-products. The key to whether pyrolysis can be more than a technological curiosity is the potential market for the by-products. The gas generated by pyrolysis can be used to provide the heat necessary to run the operation. There is, however, not normally enough produced to make it economically feasible to sell, so any excess gas is flared off. The solid fraction is often incorrectly referred to as carbon black. The proper term is carbon char. While it does have a high carbon content, that is the only thing it has in common with carbon black, a highly engineered form of carbon used to reinforce rubber. Pyrolytic carbon char, after extensive refining, has found limited markets as a filler in some materials and as a coloring agent for some plastics. In these markets, it faces strong competition from, among other things, off-specification carbon black, which cannot be sold to primary carbon black markets.

The liquid fraction is a hydrocarbon material variously presented as being comparable to a home heating oil or a diesel fuel. While synthetic rubber is manufactured from petroleum, contaminants resulting from the processing of the material and its pyrolysis render the liquid generally unfit to use directly, except as a waste fuel or as a feedstock for further refining. Its acceptability in either application depends on the receiving facility's ability to handle the material. To date, even experimental pyrolysis facilities have had limited success in identifying end-users for the liquid fraction.

There has been some research conducted both at universities and at private facilities to seek uses for, or to upgrade the pyrolysis by-products. At the current time, none of this recent work appears to be commercially successful. While the possibility always exists that a significant, economically viable use can be found for the pyrolytic by-products, it must be assumed that based on current information and technology, pyrolysis has no substantial role to play in long-term scrap-tire management. The principal barriers to the use of tires in pyrolysis applications are that this technology has yet to be proven on a short- or long-term basis. Questionable economics, no existing demand for pyrolitic products, and poor perception of product quality continue to plague this technology.

It is our contention that virtually no scrap tires were consumed by U.S. pyrolysis facilities in 1998. Extensive investigation indicated that there were no commercially viable facilities in the United States. The majority of the activities associated with this technology have been focused around the sale of pyrolytic equipment or the acquisition of companies. This has given rise to the perception that progress has been made in this market segment. Given the highly questionable economics and speculative nature of pyrolysis, it is difficult to estimate any future growth in the use of this technology as a means to address the scrap-tire situation.

EXPORTS OF TIRES

Export of sound used tires constitutes a major market for tires removed from initial service. As discussed earlier, many tires when initially removed from a vehicle are still sound and have adequate tread depth to be used as tires. In addition, many tires without proper tread are usable as candidate casings for retreading. Both categories of tires have ready markets both within the United States and North America and in many other parts of the world and are regularly sold into these markets. Virtually all tire-producing countries also participate in this world trade in used tires. Used tires and casings exported from the United States are assumed not to return to the U.S. for ultimate disposal.

In 1996, this market faced a major international challenge when an effort was made in the Technical Working Group of the Basel Convention to have used tires listed as a hazardous waste. The Basel Convention is a major world-wide treaty regulating the flow of hazardous waste materials between developed and developing nations, and was formed to limit the dumping of hazardous and toxic waste from developed countries in less-well-developed countries. A few South and Central American countries sought to use the Basel Convention to limit the importation of used tires into their countries, in part because of internal market competition for tires. The mechanism for this action under the Basel Convention is a listing on the so-called "A" list of banned materials. A concerted effort by the major tire manufacturing countries of the world has so far been successful in having tires instead placed on the "B" list, which identifies materials that are safe for international trade.

The export market routinely ships slightly more than one million tires per month, or more than 15 million tires per year, based on the estimates of participants in those markets. This constitutes about 5% of the annual volume of discarded tires.

AGRICULTURAL USES

Scrap tires are regularly used in agriculture in a variety of ways. Used tires not legally fit for highways may be used on low-speed farm equipment. Tires are also used to weigh down covers on hay stacks and silage, or for other purposes where an easily handled weight is needed. Tires can be used as feeding stations, to construct stock feeders, or to protect fence posts and other structures from wear and damage by livestock. Tires may also be used in erosion control and for other land-retention purposes. It is estimated that about two and one-half million tires are used in agricultural applications each year, or about 1% of the total volume of scrap tires in the U.S.

MISCELLANEOUS USES

There are a wide variety of uses for scrap tires that do not fit neatly into any of the preceding categories, which ranges from one of the most popular uses as a scrap tire swing, to more exotic uses as an art material.

One of the most interesting of these may be the use of tires to make houses. In this use, tires are stacked up like bricks and used to contain compacted earth or other recyclable materials. The odd spaces between the tires are filled with cans and bottles and the entire structure is covered with stucco or adobe. The developer of this construction technique intends to use it to construct highly ecological buildings, using recycled materials and taking maximum advantage of the mass of the structure both for insulation and for passive heating.

In terms of volume, however, these uses do not consume any substantial quantity of scrap tires. We estimate there are about three million tires used in all of these innovative applications.

CONCLUSIONS

This chapter has presented information on the various markets and management options for scrap tires. In general, there continue to be three major markets that form the basis of the demand for scrap tires. These markets are tire-derived fuel, products, and civil engineering applications. The market survey also reported on several additional markets for scrap tires, that include export, agricultural uses, pyrolysis, and miscellaneous uses, accounting for less significant, long-term growth potential.

It is clear that TDF will continue to be the most significant market in the near- and long-term. The use of TDF has proven to be both environmentally sound and cost effective. With the implementation of the Clean Air Act Amendment of 1991, and with utilities becoming more competitive, the indications are that TDF will be considered in a more positive manner.

The uses of scrap tires in civil engineering applications regained lost market share and increased over the last two years. The development of industry guidelines and construction practices should assist greatly in continuing the expansion of this market.

The use of size-reduced rubber from scrap tires remains on the verge of becoming a larger market segment. The reasoning is two fold. First, there is a finite limit to the quantity of tire buffings that are available, and that limit is rapidly being approached. Second, the markets for the various uses for size-reduced rubber are likely to increase over the next two to five years.

Of the other markets for scrap tires, the indications are that export, cut, stamped, and punched products, agricultural and other uses will not increase to any great extent over the next two years. The data evaluated also indicates that the pyrolysis of scrap tires does not appear to be a technology that will have any impact on the market for scrap tires.

Finally, it is evident that the scrap tire industry is entering into a new era, one where the processed products generated from scrap tires will have specifications. The level of processing efficiency has improved significantly, with improvement expected to boost productivity even further. While there are many positive indicators in the field, there are also some indications that there will continue to be consolidation and attrition in the scrap tire industry.

ABOUT THE SCRAP TIRE MANAGEMENT COUNCIL

The Scrap Tire Management Council is an advocacy organization created by the North American tire industry. The Council's primary goal is the creation of adequate markets and management to properly address 100% of the annually generated scrap tires in the United States. The Council serves as the focal point of industry concerns regarding the disposal of scrap tires. The mission of the Council is to assist in developing and promoting the utilization of scrap tires as a valuable resource. To this end, the Council supports all technologies and uses for scrap tires that are both environmentally sound and cost efficient.

The Council does not represent nor have any vested interest in any product derived from scrap tires or used in the processing of scrap tires. The Scrap Tire Management Council promotes the concept that scrap tires can be a resource that can be used in a wide array of applications.

CHAPTER 19
BATTERIES

ANN PATCHAK ADAMS
Former Project Leader, Roy F. Weston, Inc. Detroit, Michigan

INTRODUCTION

The disposal of batteries has become an ever-increasing topic of discussion in the last few years because they contain heavy metals such as mercury, lead, and cadmium. In response to these concerns, there has been an increase in battery collection programs and legislation controlling the production and disposal of batteries. This brings forward some key questions, What is a battery and why has its disposal become such an issue? What is the impact of the disposal of batteries on the environment? What are the mechanisms for removing batteries from municipal solid waste (MSW), and what do we do with the batteries once they have been removed?

To answer these questions it is best to distinguish between lead–acid automotive-type batteries and typical household-type batteries that are used in consumer items such as flashlights, radios, and watches, because the means by which these two types of batteries are handled and disposed of are quite different. This chapter attempts to answer these questions and, in turn, discuss the major issues concerning the collection of batteries.

DEFINITION AND COMPONENTS OF BATTERIES

A battery is an electrochemical device that has the ability to convert chemical energy to electrical energy. The basic battery consists of an anode (negative electrode), a cathode (positive electrode), and an electrolyte (a liquid solution through which an electric current can travel (Fig. 19.1). The potentially hazardous components of batteries include mercury, lead, copper, zinc, cadmium, manganese, nickel, and lithium. These components serve various functions; mercury, for example, is most commonly used to coat the zinc electrodes to reduce corrosion and thereby enhance battery performance.[34]

Lead-Acid Batteries

The lead–acid battery, also known as a wet battery, is typically used in automobiles and other motor vehicles. Most automotive lead–acid batteries contain sulfuric acid and approximately 18 lb of lead.[34] The smaller lead–acid batteries are used in items such as video camcorders and power tools.

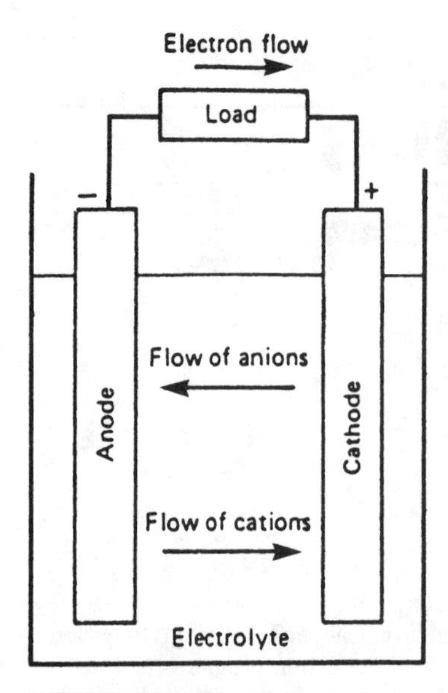

FIGURE 19.1 Electrochemical operation of a battery cell. (*Linden, 1984*)

Household-Type Batteries

The household battery industry is estimated to be a \$2.5 billion industry[27] with annual sales of nearly 3 billion batteries.[10] These batteries, also known as dry cells, are used in over 900 million battery-operated devices. The average family owns about 10 such devices and purchases approximately 32 batteries per year.[27] There are two basic types of household batteries: single-use primary cells and rechargeable secondary cells.[27]

There are five common primary-cell batteries: alkaline–manganese, carbon–zinc, mercuric oxide, zinc–air, and silver oxide (Fig. 19.2 and Table 19.1). The alkaline–manganese battery (Fig. 19.3) is the most common and is used in items such as flashlights, toys, radios, cameras, and some appliances. Typically these batteries come in sizes AAA, AA, C, D, and 9-V. Until 1989, the typical alkaline battery contained up to 1 percent mercury, by weight of each battery. During 1990, at least three large domestic battery manufacturers began manufacturing and marketing alkaline batteries with less than 0.025 percent mercury, by weight of each battery.[22] These newly marketed alkaline batteries contain approximately one-tenth of the amount of mercury contained in the typical alkaline batteries.[8] The National Electrical Manufacturers Association (NEMA) estimates that 4.25 alkaline–manganese batteries are sold per capita per year.[22]

The carbon–zinczinc battery (Fig. 19.4), similar to the alkaline battery, is available in the same sizes as the alkaline–manganese battery. These batteries contain up to 0.01 percent mercury, by weight of each battery.[8] NEMA estimates that 3.25 carbon–zinczinc batteries are sold per capita per year.[22]

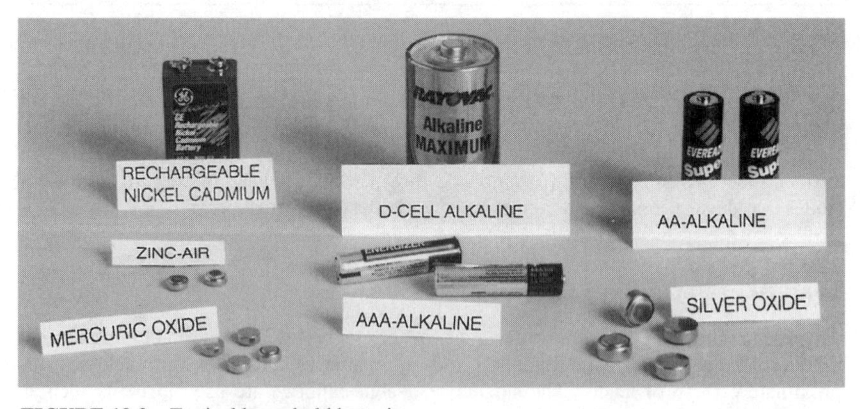

FIGURE 19.2 Typical household batteries.

TABLE 19.1 Common Household Batteries: Types, Components, Sizes, and Uses

Popular types and sizes	Cathode (negative electrode)	Anode (positive electrode)	Electrolyte	Common uses
Alkaline: 9-V. D, C, AA, AAA, button	Manganese dioxide	Zinc	Alkaline solution	Cassettes, radio, etc.
Carbon-zinc: 9-V, D, C, AA, AAA; Heavy-duty carbon-zinc: 9-V. D. C, AA, AAA	Manganese dioxide	Zinc	Ammonium and/or chloride; Zinc chloride	Flashlights, toys, etc.
Lithium: 9-V, C, AA, coin and button	Various metal oxides	Lithium	Organic solvent or salt solution	Cameras, calculators, watches
Mercury: D, C, AA, AAA, button, some cylindrical	Mercuric-oxide	Zinc	Alkaline solution	Hearing aids, pacemakers, photography
Nickel-cadmium: 9-V. D, C, AA, AAA	Nickel-oxide	Cadmium	Alkaline solution	Photography, power tools
Silver: button	Silver-oxide	Zinc	Alkaline solution	Hearing aids, watches, photography
Zinc: button	Oxygen	Zinc	Alkaline solution	Hearing aids, pagers

Source: Information reprinted from Minnesota Pollution Control Agency, "Household Batteries in Minnesota: Interim Report of the Household Battery Recycling and Disposal Study."

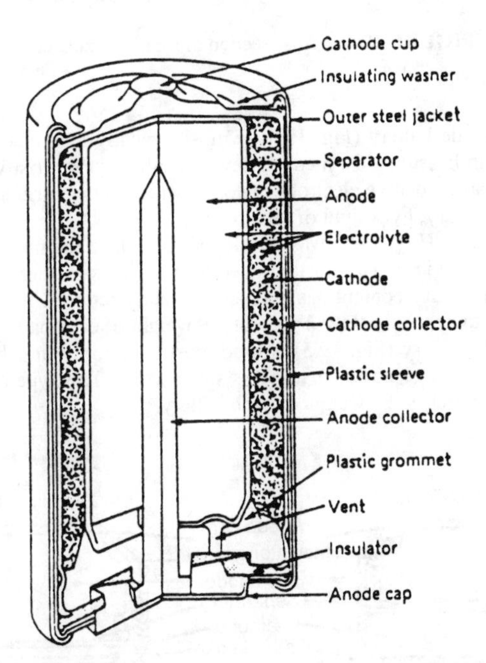

FIGURE 19.3 Cross section of cylindrical alkaline–manganese dioxide cell.

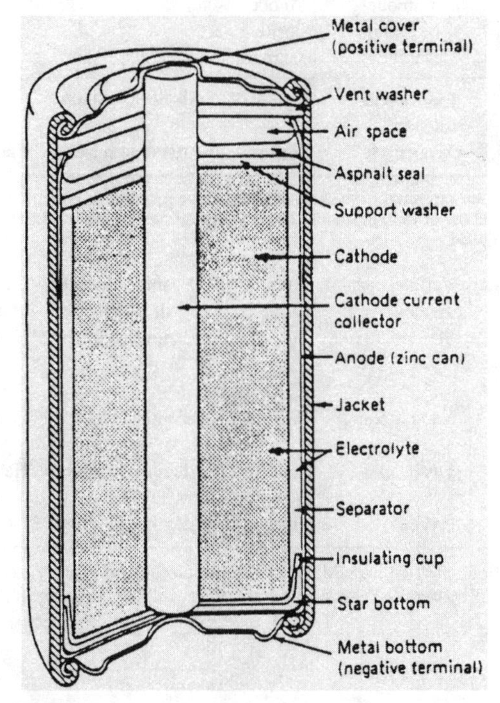

FIGURE 19.4 Cross section of a carbon–zinc cell.

The mercuric oxide battery (Fig. 19.5) is most typically marketed in a small button shape and is used in hearing aids, medical devices, calculators, wrist watches, and cameras. This battery has a positive electrode material, which is mercuric oxide, and contains 35 to 50 percent mercury, by weight of each battery.[24]

The zinc-air battery (Fig. 19.5) was developed to replace the mercuric oxide battery used in hearing-aid devices. The positive electrode for these batteries is oxygen taken from the air. The mercury content has been significantly reduced from that of the mercuric oxide battery to approximately 2 percent, by weight of each battery.[8]

The silver oxide battery (Fig. 19.5) is used most commonly in calculators, watches, and cameras. Its positive electrode material is silver oxide. This type of battery contains less than 1 percent mercury, by weight of each battery.[8]

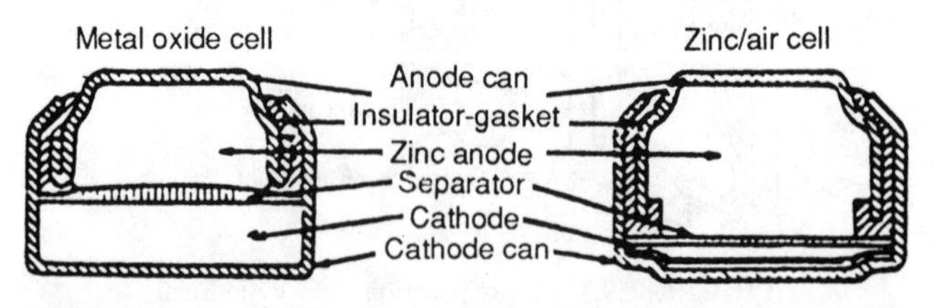

FIGURE 19.5 Cross section of metal–oxide and zinc–air button cells.

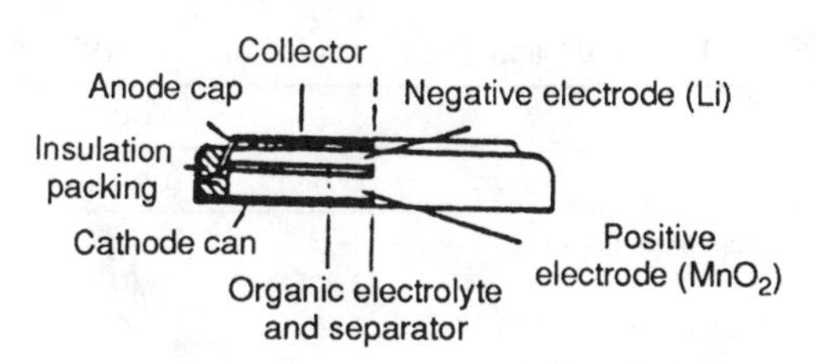

FIGURE 19.6 Cross section of lithium manganese dioxide flat cell.

The lithium battery comes in many shapes and sizes but most are marketed in a small button or cylindrical shape (Fig. 19.6) and are used most commonly in cameras, watches, memory back-up systems, and in many industrial and military applications. The positive electrode is lithium and a variety of materials are used for the negative electrode: sulfur dioxide, manganese dioxide, and carbon monofluoride to name a few.[15] The manganese dioxide is the most common type of consumer battery. Lithium batteries are known for their long shelf life, which is up to 10 years. Two issues currently limit the increased marketing of lithium batteries. First is the development of a cost-effective consumer battery that could compete with alkaline batteries. Second is the fact that lithium is a combustible material in water, which raises concerns about consumer safety.

The most common secondary-cell or rechargeable battery is the nickelcadmium battery (Fig. 19.7 and Table 19.1), which is commonly found in rechargeable appliances. The negative and positive electrodes are cadmium and nickel-oxide, respectively. These batteries are 17 percent cadmium, by weight of each battery.[22]

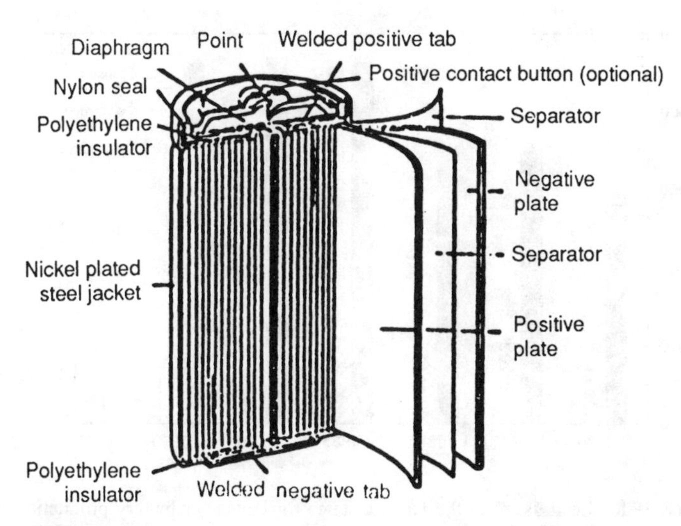

FIGURE 19.7 Cross section of a sealed, coiled-type, sintered nickel-cadmium cell.

TABLE 19.2 Lead and Cadmium Discarded in MSW 1970 to 2000

	Year		
	1970	1986	2000
Lead from lead–acid batteries, in short tons	83,825	138,043	181,546
Percent of total lead discards	50.9	64.6	64.4
Cadmium from household batteries, in short tons	53	930	2,035
Percent of total cadmium discard	4.4	52	75.8

BATTERIES AND HEAVY METALS IN THE ENVIRONMENT

In a report prepared for the U.S. Environmental Protection Agency (EPA), Franklin Associates[10] estimated that in 1986, approximately 941,000 tons of lead were used in the production of batteries. It is also estimated that 78 percent of the 941,000 tons (700,000 tons) of lead, from approximately 78 million batteries, was discarded. Of the total amount of lead discarded, 562,000 tons were reclaimed and 138,000 tons from approximately 15 million batteries were discarded in the MSW stream (Table 19.2 and Fig. 19.8). The disposal of lead–acid batteries produces approximately 65 percent of the lead in the MSW stream.

The amount of lead consumed in the production of batteries is expected to increase through the year 2000,[10] by approximately 28 percent over 1986 production levels, to 1.2 million tons in the year 2000 (Fig. 19.8 and Table 19.3).

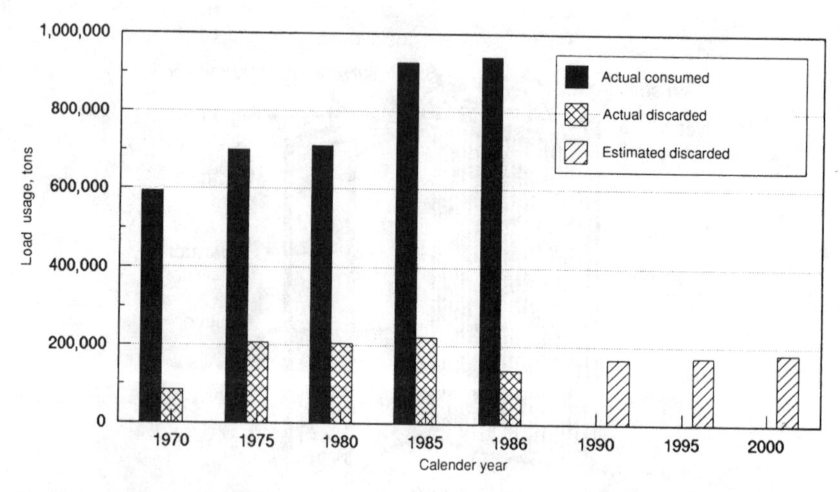

FIGURE 19.8 Lead usage in the United States for consumer battery production. (*From "Characterization of Products Containing Lead and Cadmium in Municipal Solid Waste," Franklin Associates*)

TABLE 19.3 Lead, Cadmium, and Mercury Consumed in the Production of Batteries (in short tons)

	Year		
	1970	1986	2000
Lead used in lead–acid batteries*	593,453	941,155	1,240,000†
Cadmium used in nickel–cadmium batteries*	167	1268	2285
Mercury used in batteries‡	753§	695	62¶

*Information provided from "Characterization of Products Containing Lead and Cadmium in Municipal Solid waste in the United States, 1970–2000," Franklin Associates, Ltd.

‡Data extrapolated from 1986 data. *Information provided by the National Electrical Manufacturing Association (NEMA).

In 1986, the U.S. consumption of cadmium was 4800 tons.[10] Of the cadmium consumed in 1986, 26 percent of the 4800 tons (1268 tons) of cadmium was used in the production of batteries. It is also estimated that 73 percent of the 1268 tons (930 tons) of cadmium entered the MSW stream (Tables 19.2 and Fig. 19.9). The disposal of nickel-cadmium batteries into the MSW stream accounts for 52 percent of total cadmium entering the MSW stream each year.[10]

The amount of cadmium consumed in the production of batteries is also expected to increase through the year 2000[10] by approximately 80 percent over 1986 production levels, to 2285 tons in year 2000 (Figs. 19.8 and 19.9).

In 1988, the U.S. consumption of mercury was 1755 tons. Of the mercury consumed

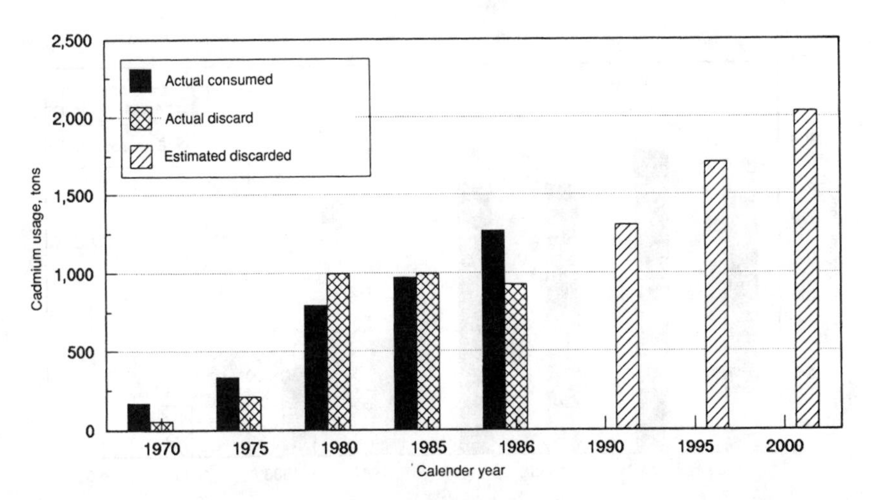

FIGURE 19.9 Cadmium usage in the United States for consumer battery production. (*From "Characterization of Products Containing Lead and Cadmium in Municipal Solid Waste, Franklin Associates*)

in 1988, 13 percent of the 1755 tons (225 tons) was used in the production of batteries. Of that 225 tons, approximately 73 percent (173 tons) is estimated to have been used in the production of mercuric oxide batteries. Of the 173 tons used in mercuric oxide batteries, approximately 126 tons were used in nonconsumer batteries (i.e., medical, military, and other industrial applications). Therefore, at least 56 percent of the mercury used in the production of batteries is used in nonhousehold-type batteries. One would assume that essentially all of the mercury from household batteries enters the MSW stream, and an unknown amount of mercury from nonconsumer batteries enters the MSW stream.

Unlike lead and cadmium, the amount of mercury consumed in the production of batteries is expected to continue to decrease. In 1983, the mercury consumed in the production of batteries was 753 tons; in 1988 mercury consumption decreased to 225 tons and is estimated to decrease to 62 tons in 1990 (Table 19.3 and Fig. 19.10).[22] No further projections on future mercury consumption are available. However, Franklin Associates, Ltd., is preparing a report for the EPA which will characterize mercury-containing products in MSW. This report is expected to be released in 1991.

Most household batteries are disposed of in MSW and are sent to either a sanitary type II landfill or to a municipal waste combustion (MWC) facility.[26] A smaller number of batteries are sent to a composting facility. The metals' contribution to the environment from the incineration, landfilling, or composting of household batteries will pose different concerns, depending upon the method of disposal.

Understanding the environmental fate of metals contributed by batteries in a landfill is a function of the conditions of the batteries when landfilled and the conditions of the landfill itself. The casings of household batteries are most commonly made of paper, plastic, or metal. The various conditions that can develop in a landfill affect the rate at which the casings will degrade or decompose. A 1978 study conducted in England[14] indicated that the following conditions affect the rate of degradation: the nature of the casing; the degree of electrical charge left in the battery; the extent of exposure to landfill leachate; and the oxygen content of the landfill.[18] The mobility of the metals in a landfill and the potential

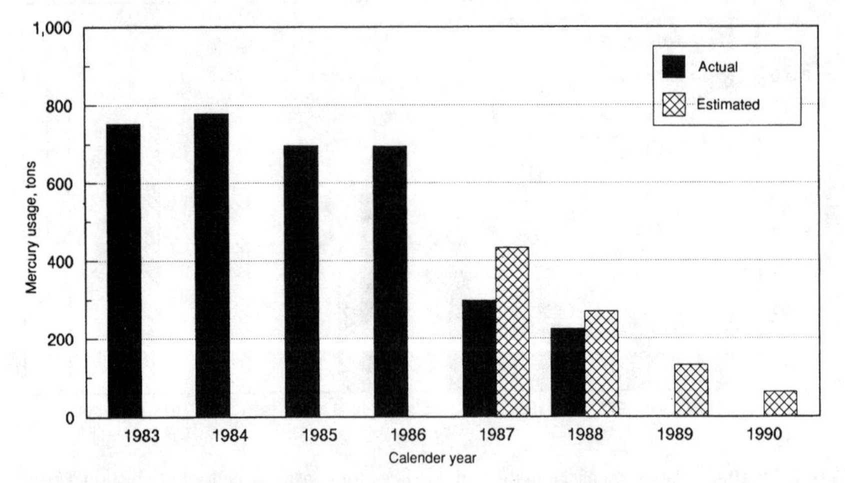

FIGURE 19.10 Mercury usage in the United States for consumer battery production. (*From the National Electrical Manufacturing Association (NEMA)*)

for groundwater contamination are also controlled by numerous conditions. These conditions include the design, construction, operation, and maintenance of the landfill (e.g., the liner, soil characteristics, leachate collection-and-detection systems, daily cover, final cover, etc.).

The release of metals from a battery in a landfill may not however, in and of itself, be problematic. The principal issue is the potential for those metals to contaminate groundwater, which is a function of the landfill's construction, its soil characteristics, and its proximity to groundwater.

The incineration of batteries also poses two major potential environmental concerns. The first is the release of metals into the ambient air, and the second is the concentration of metals in the ash that must be landfilled. Generally, mercury is more likely to be emitted in the stack gas and cadmium and lead will concentrate in the ash.[18] The fate of metals released from batteries during incineration is mainly a function of the boiler combustion temperature, the metals' volatilization temperature, and the presence of other nonmetallic compounds.[18]

The fate of metals in the ash, once landfilled, will be the same as those previously stated for landfills. However, the ash from many incinerators is disposed of in monocells (also known as monofills or landfills that accept only MWC ash). According to an EPA study, it appears that leachate from these facilities and even from codisposal facilities (i.e., where MSW and ash are landfilled together) is nonhazardous and predominantly less potentially harmful than the MSW-only landfill leachate.

Composting is the process of changing MSW into a humuslike (i.e., soil-like) product. Of the three disposal methods, composting is the least used method of disposal, although its use is increasing. The environmental concerns associated with composting are the quality of the humuslike product and the limitation on the uses of the humus. Also, potentially objectionable odors are produced. According to the Minnesota Pollution Control Agency (MPCA) report, *Household Batteries in Minnesota*,[18] most composting facilities sort the MSW prior to composting to minimize the amount of metals in the compost. The MPCA noted that batteries do not appear to pose a composting problem for facilities that manually and mechanically sort the MSW. However, for those facilities which only manually sort the MSW, there is a higher probability that because of their size, the small button batteries will not be removed from the compost material. The MPCA estimated that two mercury button batteries in a kilogram of compost contain enough mercury to limit the end use of the humus.

One difficulty in determining the impact of the metals contained in the batteries on the environment is that the requirements for pollution control at landfills and MWCs are becoming increasingly more stringent. Also, pollutant measurement and risk assessment analyses are being refined. Consequently, landfill leachate data may be from landfills that were constructed years earlier and, therefore, may not reflect the more recent stringent design-and-construction regulations now imposed upon the owners and/or operators of landfills. In addition, it is often difficult to establish that the leachate was contaminated with heavy metals specifically from batteries. Similarly, air emissions data from MWCs may not necessarily reflect either the state-of-the-art air pollution control technologies that are still being researched or completely accurate and precise emissions data.

LEAD–ACID BATTERIES

Attitudes toward recycling batteries vary significantly among the participants. The consensus among interested parties, industry, legislators, municipalities, independent researchers, consultants, and nonprofit organizations, seems to be that it is beneficial to reclaim lead–acid batteries. The Battery Council International (BCI) has developed model legislation to be used by states in developing a lead-battery collection program (see Appendix A). NEMA and the Battery Products Alliance (BPA) have also developed model legislation to assist individual states in their recycling efforts for small, nonvehicular lead–acid and nickel–cadmium batteries.[17]

The main provisions of the BCI model legislation are as follows: (1) Any person is prohibited from disposing of a lead–acid battery into mixed MSW. (2) Each battery improperly disposed of constitutes a violation subject to a fine. (3) And there is also a take-back provision, stipulating that any person selling lead–acid batteries must accept used lead–acid batteries and post written notices indicating that batteries can be returned.

As of June 1990, in the United States, 28 states have enacted disposal prohibitions for lead–acid batteries (Fig. 19.11) and 25 states have mandatory take-back provisions for lead–acid batteries (Fig. 19.12).

Various states have enacted legislation controlling the disposal of lead–acid batteries. In March 1990, the state of Michigan enacted Public Act Number 20, to govern the disposal of lead–acid batteries and to establish a legislative committee to review the effectiveness of the lead–acid battery regulations. The main provisions of Michigan Act Number 20 pertaining to the disposal of lead–acid batteries are the following:

1. A person shall dispose of a lead–acid battery only at a retailer, distributor, manufacturer, or collection center.
2. Retailers of lead–acid batteries must post a written notice indicating that spent batteries are accepted for recycling.
3. Beginning January 1, 1993, a purchaser of a lead–acid battery must exchange a used lead–acid battery for the battery purchased or pay the retailer a $6.00 deposit. The legislative committee submitted a report in December 1990, evaluating the lead–acid battery disposal programs and concluded that no changes are currently necessary to the existing management of lead–acid batteries and that the effectiveness of this program should be reviewed again before December 3, 1993.

Most lead–acid batteries are collected at local automotive service or repair garages. Some of these batteries are collected through local household hazardous-waste collection programs operated by local governments. Overall, the collection and recycling efforts for lead–acid batteries appear to have been successful.[25] The lead–acid battery collection and recycling programs tend to be successful because the automotive garages and repair centers serve as centralized collection points. Additionally, there is no or very little inconvenience caused to the consumer. Ultimately, the primary motivation for the recovery of automobile batteries is profit from the sale of lead.[34]

According to EPA estimates, approximately 80 percent of lead–acid batteries are being reclaimed. The batteries are collected by battery distributors and are then transported to reclamation centers where lead and polypropylene are recovered, generating a toxic residue that requires special disposal. Recycling technologies are developing in which the lead, polypropylene, and sulfuric acid will be recovered with a resulting nontoxic residue.

There currently are no formal domestic programs for collecting small lead–acid batteries. Consequently, it is difficult to determine the number of these batteries that are being recycled and their contribution of lead to the environment. It is reasonable to assume that the disposal and recycling rates for small lead–acid batteries purchased for home use parallels the disposal and recycling patterns of other household batteries.

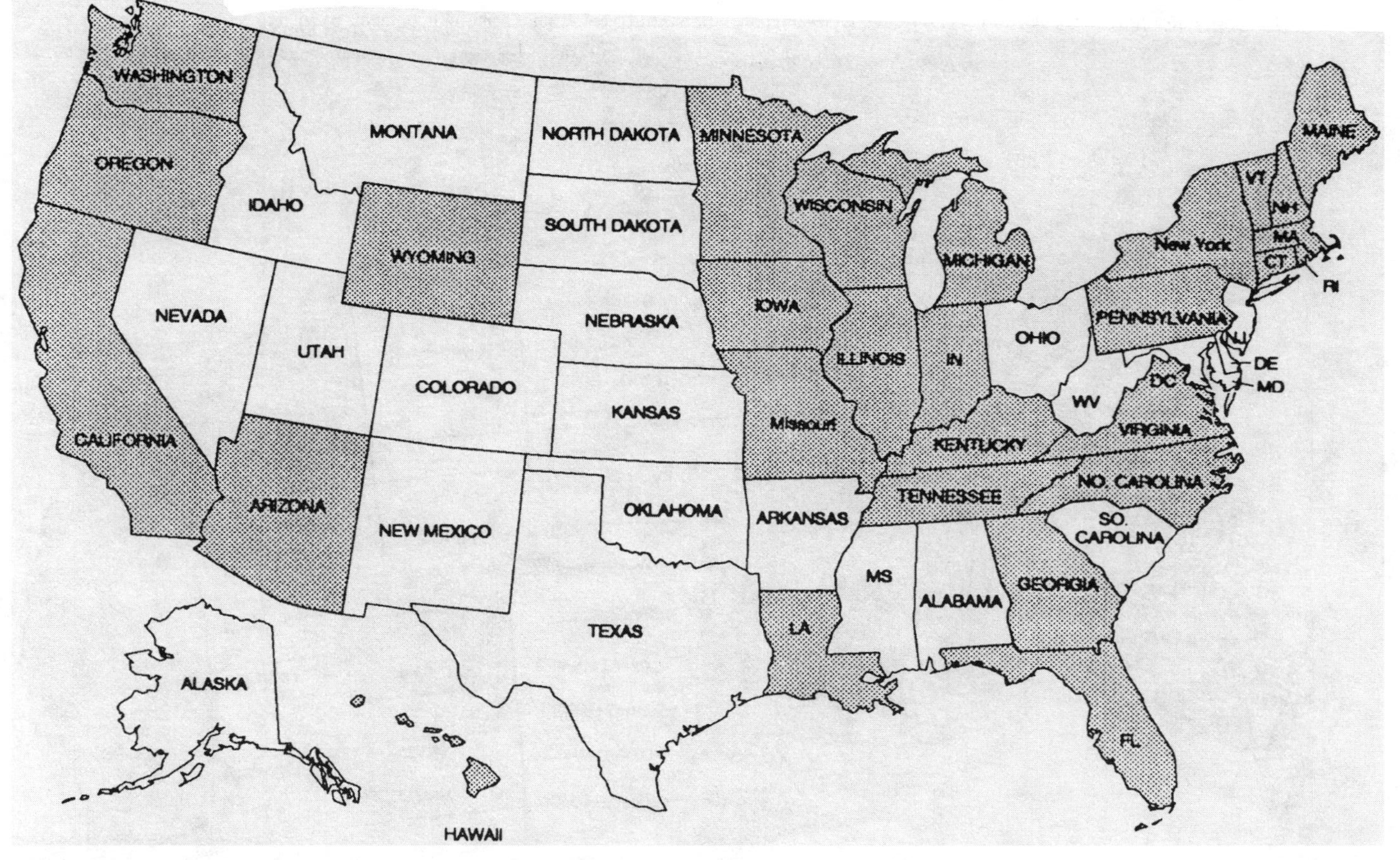

FIGURE 19.11 States that have enacted disposal prohibitions for lead–acid batteries. (*From the Municipal Solid Waste Program, OSWER, EPA*)

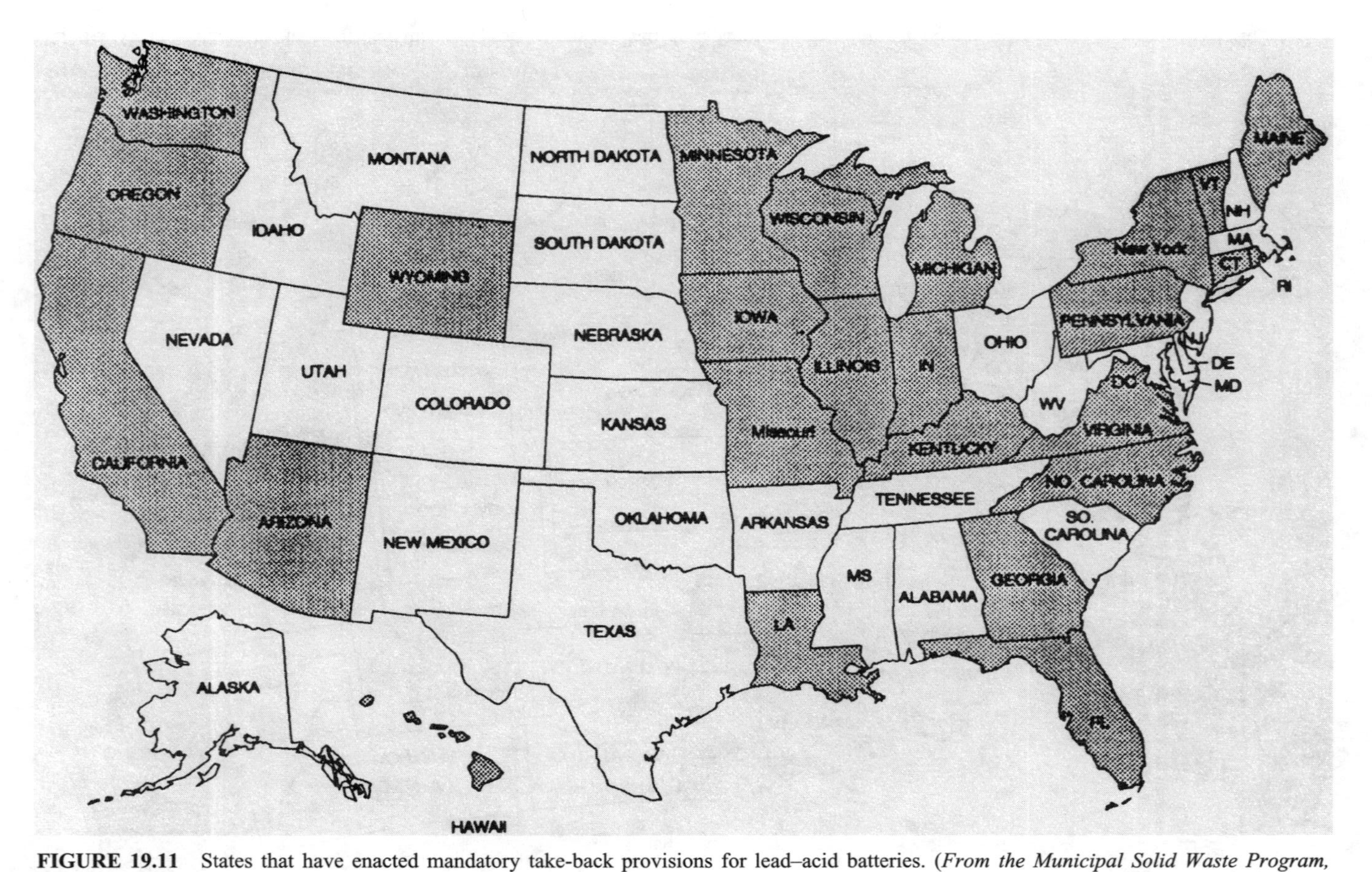

FIGURE 19.11 States that have enacted mandatory take-back provisions for lead–acid batteries. (*From the Municipal Solid Waste Program, OSWER, EPA*)

HOUSEHOLD BATTERIES

Attitudes Toward Recycling Household Batteries

A consensus on the collection of household batteries has not developed as quickly as that for lead–acid batteries. Attitudes toward recycling household batteries vary from no recycling to recycling a limited number of batteries (specifically mercuric oxide and nickel–cadmium) to recycling all types of household batteries.

For years, the battery industry has maintained that consumer disposal of batteries into municipal landfills and MWCs does not pose a hazard to human health or the environment. Regardless of the battery industry's position, the industry has responded to public concerns regarding the disposal of batteries. Low-level mercury-alkaline batteries and hearing-aid (zinc-air) batteries have been developed. Additionally, NEMA and BPA are developing model legislation which would require that nickel–cadmium batteries be removed from appliances and that these batteries then be collected.[1] The battery industry's response to concerns regarding the disposal of batteries eventually will redefine the issues associated with battery collection programs.

Program managers for the Hennepin County, Minnesota, Battery Collectionand-Recycling Program and New Hampshire/Vermont Solid Waste Project have outlined an approach that reflects the NEMA-proposed legislation and the policies being adopted in Europe by the European Economic Community (EEC).[13] They concluded that, instead of collecting all types of household batteries, mercury and cadmium can be more effectively removed from the MSW stream through the following:

1. Legislation that would limit the amount of mercury in alkaline batteries to 0.025 percent, by weight of each battery.
2. Efforts focused on recycling button batteries and larger mercuric oxide batteries, rather than all household batteries.
3. Legislation which would require that rechargeable batteries be removed from appliances.

The EEC-NEMA approach indicates that battery collection programs will need to be established to collect the nickel–cadmium and button-size batteries. Concern has been expressed that the button batteries are not well labeled. Consequently, programs that target button batteries will have difficulty separating reclaimable mercury-oxide and silver oxide batteries from lithium and other batteries that cannot be reclaimed.

Most battery programs in the United States, however, are designed to collect all types of batteries (Table 19.4). Switzerland similarly has recently developed legislation that requires the separate collection and recycling of all household batteries regardless of the toxicity. Sumitomo, a Japanese company that has developed a pilot facility for reclaiming metals from all household batteries, has recently been awarded a contract to develop a 2000-ton/day recycling facility in Switzerland.[32]

Another prevalent attitude is that more effort should be placed on and directed toward collecting batteries utilized by industry, hospitals, the utilities, the military, and the communications market. Examples of these types of batteries include power supplies for mainframe computers, emergency lighting, battery-operated medical tools, medical monitoring devices such as the Holter heart monitor, back-up battery systems for radio and television broadcasts, and police radios. Under the federal Resource Conservation and Recovery Act (RCRA) batteries collected from these sources would be classified as a hazardous waste and, consequently, are subject to transport, storage, and disposal require-

TABLE 19.4 1990 Household Battery Collection Programs

Battery collection program	Contact	Type of collection program	Type of batteries collected	Management of batteries
AL-Huntsville	Karen Schoening 205-880-6054	Curbside, with recyclables	All types	Will be landfilled in a HW landfill
FL-Gainesville	James Abbott 904-495-9215	Drop off rural SW collection centers and at HHW collection days	All types	Seal in concrete and store up to one year in 55 gallon drums before sending to municipal solid waste landfill
KS-Overland Park	Ron Tubb 913-381-5252	Drop off at 6 retail stores	Button batteries	MERECO
KY-Louisville	Ray Hilbrand 502-625-2788	Pilot drop off at recycling center	All types	Sent by waste Management of Kentucky to a HW incinerator
MI-Detroit	Phil Brown 313-876-0449	Drop off at recycling centers and neighborhood City Halls	All types	Combination of landfilling and reclamation
MI-SOCRRA	Tom Waffen 313-288-5150	Libraries, schools, recycling centers, and curbside pickup	All types	Combination of landfilling and reclamation
MN-Minneapolis	Mark Oyass 612-348-6157	Curbside with recyclables, and drop off at retail stores	All types	Sorted by type, placed in 55 gallon containers, then hauled to HW landfill
MO-23 counties	Marie Steinwachs 417-836-5777	Drop off at retail stores	Button batteries	MERECO
NH/VT-28 towns	Carl Hirth 603-543-1201	Drop off at transfer stations, municipal offices and retail stores	All types	Sent to HW landfill in South Carolina
NJ-Warren County	Mary Briggs 201-453-2174	Drop off at municipal offices, recycling centers in 14 communities, curbside in 9 communities	All types	Buttons, lithium, and nicads go to MERECO and all others go HW landfill
NJ-Somerset County	Mike Elks 201-231-7031	Curbside with recyclables—4 times per year	All types	Batteries are stored in plastic lined barrels for 90 days then sent by HW contractor to HW landfill

TABLE 19.4 1990 Household Battery Collection Programs (*Continued*)

Battery collection program	Contact	Type of collection program	Type of batteries collected	Management of batteries
NY-Little Valley	Richard Preston 716-938-9121	Drop off at transfer stations, curbside, looking to expand stores	All types	MERECO
NY-New York City	Sean Hecht 212-677-1601	Drop off at retail stores	Buttons and Nicads	Only storing until check on MERECO's status
NY-Poestenkill	Lois Fisher 518-283-5100	Drop off at landfill/recycling center	All types	MERECO
NY-Rochester	Alice Young 716-244-5824	Drop off in municipal buildings and retail stores	Button batteries	MERECO
NY-Rye	Frank Culross 914-967-7604	Drop off at DPW recycling center	All types	MERECO
NY-Scarsdale	Jim Rice 914-723-3300	Drop off at recycling center and incinerator	All types	MERECO
NY-Slingerlands	Mike Hotaling 518-765-2681	Drop off at highway garage	All types	MERECO
NY-Southold	Jim Bunchuck 516-734-7685	Drop off at retail stores	All types	MERECO
NY-Woodstock	Bill Reich 914-679-6570	Drop off at recycling center and retail stores	All types	MERECO
PA-Pottstown	Jim Crater 215-323-8545	Drop off at community recycling center	Button and Alkaline	MERECO
VA-Chesapeake	Jennifer Ladd 804-420-4700	Curbside	All types	Tidewater Fiber is storing, plan to send to MERECO
WA-Bellingham	Lisa Schnebele 206-384-1057	Drop off at recycling center and retail stores	All types	Alkaline and carbon-zinc sent to hazardous waste landfill in Oregon, all others to MERECO

Source: Dana Duxbury & Associates.

ments (Subtitle C requirements). Consequently, few household battery collection programs have targeted these batteries. The contribution of heavy metals to the environment from these sectors is unknown.

As an alternative to voluntary battery collection programs, advocates of battery recycling have suggested placing a refundable deposit on batteries, similar to that required for aluminum and glass beverage containers. Requiring a refundable deposit appears to be becoming a common practice for lead–acid batteries and is being reflected in much of the recently drafted and enacted legislation. A refundable deposit on household batteries, however, is more complex than a deposit on lead–acid batteries. A deposit on household batteries requires that the following questions be addressed:

1. Who will pay for collection containers, collection, and disposal of the batteries?
2. At which point, if any, in the collection process will the batteries become classified as a hazardous waste?
3. What types of precautions, if any, will the collectors (i.e., retail stores) have to consider to accommodate the volume of returned batteries?

Legislation Affecting the Collection of Household Batteries

Currently, there are no federal regulations governing the collection, recycling, or disposal of household batteries. However, a few states have adopted legislation controlling the disposal of batteries.

In April 1990, the Minnesota State Legislature enacted a law (Minnesota Statute 115 A.9155) regulating used dry-cell batteries (see Appendix B). The provisions of the law include the following:

1. Individuals are prohibited from placing into the MSW stream mercuric oxide, silver oxide, nickel–cadmium, or sealed lead–acid batteries purchased for use by a government agency, or an industrial, communications, or medical facility.
2. The manufacturer of a button-cell battery has to ensure that each battery is clearly identifiable as to the type of electrode (i.e., mercuric oxide, zinc oxide, or silver oxide) used in the battery.
3. After February 1, 1992, alkaline batteries with a mercury concentration greater than 0.025 percent, by weight of battery, were not sold.
4. After January 1, 1992, button-cell batteries with a mercury concentration greater than 25 mg were not sold.
5. After July 1, 1993, a manufacturer (excluding exemptions) can not sell, distribute, or offer for sale a rechargeable consumer product unless the battery can be removed.

The state of Michigan through Public Act Number 20, in addition to the lead–acid requirements, required the joint legislative committee to study safe use and disposal of nickel–cadmium and mercury batteries. In a report, the legislative committee developed some recommendations that paralleled the requirements of the Minnesota legislation, namely:

1. All batteries must be removable from the products or devices in which they are used.
2. The mercury content of alkaline and carbon–zinc household batteries must be less than 0.025 percent, by weight of each battery.

Through the EEC, some states in Europe have issued a proposal that became law in January 1992, to control the manufacturing and disposal of household batteries. Some of the key provisions include a ban on batteries containing more than 30 percent mercury, by weight of each battery, and a ban on alkaline batteries containing more than 0.025 percent mercury, by weight of each battery. The EEC proposal would also require that batteries be removed from products and that EEC states establish separate collection, disposal, and recycling programs.

It is reasonable to assume that as more states and countries adopt legislation requiring batteries with lower levels of mercury and easy removal of rechargeable batteries, that lower-level mercury batteries and removable nickel–cadmium batteries will become the most widely available, if not the only, household batteries available on the market.

Establishing a Battery Recycling Program

Although the removal of batteries from the MSW stream prior to disposal by landfilling or incineration may be desirable, there are many issues to consider:

1. Do all batteries need to be removed? If not, which ones?
2. Who should remove the batteries?
3. What is the best approach for removing the batteries?
4. Who should pay for the battery removal program?
5. What is the most cost-effective means for removing the batteries from the MSW stream?
6. What should be done with the batteries?
7. Will the ultimate disposal of the batteries be environmentally sound?

These questions must be answered prior to establishing a successful battery collection program.

The recycling efforts for household batteries are not as well developed as those for lead–acid batteries. The following discussion describes approaches to establishing a household-type battery collection program.

Prior to developing a recycling program, the first step is to determine the types of batteries to be collected. The types of batteries collected will affect public relations strategies, locations of collection centers, types of collection containers, and final disposal. Three of the more common collection strategies include (1) only button batteries; (2) button batteries and nickel–cadmium batteries; and (3) mixed batteries (i.e., button, nickel–cadmium, carbon–zinczinc, and alkaline batteries). Another fundamental step in developing a recycling program is thorough knowledge of local, state, and federal regulations for collection, transport, storage, and disposal of batteries. This preliminary research must precede the design of the battery collection program.

These regulations are often different in each state and can be ambiguous; however, at the federal level, it is clear that wastes generated in a household are nonhazardous (i.e., Subtitle D wastes) and/or exempt from RCRA regulations for transport, storage, and disposal (i.e., the Subtitle C requirements).

Prior to starting a battery collection program, it is essential to know the state regulations controlling the collection of batteries. In the state of Michigan, household batteries are classified as nonhazardous from collection to disposal. However, in the state of Minnesota, batteries are regulated as a hazardous waste, at the last point of consolidation prior to disposal.[30] Classification of household batteries as a hazardous waste will control the manner in which the batteries are transported, stored, and disposed.

After completing the preliminary research, the next step in developing a battery collection program is to develop a public relations program for education and awareness. The following is a list of potential promotional strategies:

1. Develop a theme and logo.
2. Establish frequency of promotions.
3. Develop newsletters, posters, and notices.
4. Utilize public service announcements (PSAs), TV, radio, print, school-education programs, retail advertisements, and community groups.

The third step is to establish battery collection centers, which could occur simultaneously with the development of the public awareness program. The various options that exist appear to be a function of the type of batteries to be collected. The New Hampshire/Vermont Solid Waste Project, which is responsible for a rural area of 38 towns, with a total population of 75,000, targets all types of batteries. As collection centers, this program uses retail stores that sell dry-cell batteries, town offices, and transfer station-recycling drop-off centers. The city of Detroit, Michigan, program, which also targets all types of batteries, uses 12 neighborhood city halls, fire stations, police stations, schools, and neighborhood recycling centers. The Hennepin County, Minnesota, pilot program uses both curbside pick-up and retail stores as collection centers for all types of discarded, spent batteries. The Environmental Action Coalition (EAC) program in New York City focuses on button batteries and uses physicians' offices, hearing-aid centers, and retail stores selling watches and cameras as the collection centers. The EAC program has recently been expanded to include nickel–cadmium batteries and uses retail stores as collection centers. Experience gained in the Hennepin County program indicated that curbside recycling was six times more successful for total number of batteries collected than the retail store drop-off method. Conversely, in the collection of button batteries, EAC indicates that retail stores and physicians' offices appear to have been more successful as collection centers.[7]

Determining who will collect the batteries is often a function of the types of collection centers used. For example, those programs which use retail stores as collection centers have often utilized volunteer groups, neighborhood block clubs, League of Women Voters, senior citizens, Boy Scouts, Girl Scouts, etc., to collect and sort batteries. The city of Detroit has used the city mail trucks to perform two services: deliver-collect mail and pick up batteries. In Hennepin County, officials negotiated with the organization collecting the other curbside recyclables to also collect the batteries. EAC tried two different approaches: retail stores were provided with mailback containers that could be mailed to EAC when filled; or EAC staff or volunteers collected the batteries. EAC indicated that there was more response when EAC collected the batteries because less effort was required by the retail stores.

There are various containers that are acceptable for the collection of batteries. For individuals to use at home for storage of spent batteries, most programs have small plastic bags printed with a recycling logo, mailed to households. The bags are also placed on display at local drop-off centers. Curbside collection programs use a similar type of resealable plastic bag. Retail store collection centers have also been provided with different sized buckets to store the spent batteries; 2-gal plastic minnow buckets appear to have been particularly successful because they are not too heavy to lift when filled and they have a spring-loaded lid which ensures that the batteries are stored securely (Fig. 19.13). The New Hampshire/Vermont program uses 5-qt silver-colored buckets. The EAC program uses small cardboard boxes (provided by the Mercury Refining Company in Latham, New York), which have an opening in the lid for disposal of button batteries and a second opening that can accept up to D cell batteries. Citizens are asked to dispose of only rechargeable batteries in this part of the box.

In the development of any recycling program, the frequency and timing of collection must be established and well publicized. The collection schedule may differ at each collection point, and is a function of the type of collection center used and the type of batteries being collected. For programs that utilize retail stores and collect all types of batteries, determining pick-up frequency will require experimentation because some stores will collect more batteries more quickly than others. For programs that utilize retail stores, but collect only button and/or nickel–cadmium batteries, the pickup frequency may be less. In the event more frequent pickups are necessary, all of the collection centers should have a contact with a telephone number who can be called for pick-ups. Pickup in curbside programs will most likely coincide with the collection schedule for other recyclables.

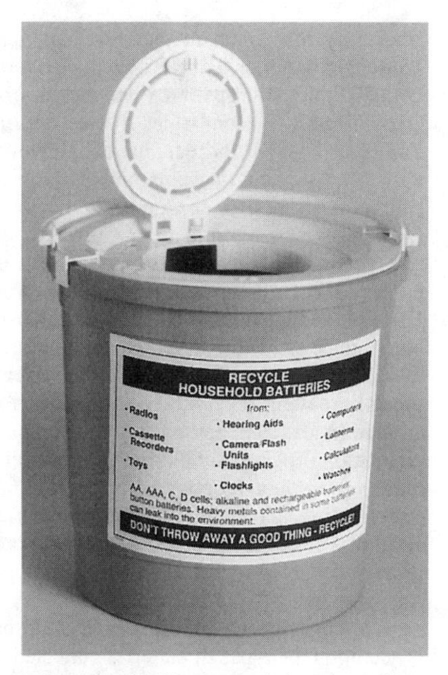

FIGURE 19.13 Typical battery collection container.

It must be determined how the batteries, once collected, will be stored until a sufficient number of batteries have been collected for economical and efficient disposal. Depending on whether it classifies household batteries as a hazardous or nonhazardous waste, each state will indicate how batteries should or must be stored. For states that classify the batteries as a hazardous waste, the collected batteries may have to be stored in polyethylene 55-gal plastic drums that are corrosion-resistant. A full 55-gal drum can weigh up to 800 lb. In states in which batteries are classified as nonhazardous, although a polypropylene plastic drum may be prudent, the use of any type of container is permitted. The more frequently a collection program disposes of its batteries, the less concern there is over the type of container. Regardless of how batteries are classified, all collected batteries should be stored in an adequately ventilated area with appropriate safety and fire-prevention measures.

When a sufficient number of batteries have been collected, they may be disposed of or recycled. Those batteries being sent for recycling or disposal (i.e., to a landfill) must be transported with U.S. Department of Transportation (DOT) shipping papers. A hazardous waste manifest is required in those states which classify batteries as hazardous waste. As mentioned earlier, each state determines how collected batteries must be handled. Also, note that regardless of how a specific state may classify batteries (i.e., hazardous or nonhazardous waste), the disposal or recycling facility (particularly if it is located in another state) may require the batteries to be packaged and manifested as though they were a hazardous waste.

Disposal and recycling opportunities should be located and contractually established in the initial stages of developing a recycling program. Battery recycling programs are frequently misnamed in that few batteries are actually recycled. At present, the only household batteries that can be recycled (i.e., actually reclaimed or remanufactured) in the United States are silver oxide, mercuric oxide, and nickel–cadmium batteries (Table 19.5). Most U.S. companies that accept nickel–cadmium batteries send the batteries to France or Sweden for reclamation. However, one company, Inmetco, located in Ellwood City, Pennsylvania, accepts nickel–cadmium batteries. They extract the nickel and the cadmium is sent to Zinc Corporation of America in Palmerton, Pennsylvania, who manufactures a zinc-cadmium alloy. Currently, there are no companies in the United States that can reclaim the components of alkaline or carbon–zinczinc batteries. Consequently, most collected batteries currently are deposited in a hazardous waste landfill. Interestingly, batteries in the typical MSW stream are deposited in type II sanitary landfills. The regulatory compliance status of all of the companies listed in Table 19.5 should be investigated prior to use as a disposal facility, in order to ensure that there are no restrictions or environmental problems that could prevent batteries from being sent to these facilities.

The final step in any battery recycling effort is to develop an evaluation program. This evaluation program should include a comparative statistical analysis to other battery collection programs, assessment of the number of households served, percentage of public participation, percentage of types of batteries collected, and strategies to maximize participation. The development of a recycling program can take up to 3 months, and such programs may change frequently based upon periodic evaluation. A timeline of typical activities is shown in Fig. 19.14.

The cost for developing a household battery collection program varies significantly based upon the types of batteries collected, the method of collection, and whether the batteries are classified as a hazardous or nonhazardous waste. It may be difficult for municipalities to utilize volunteer labor or to rely upon volunteers for a long-term program, and, consequently, most battery collection programs will have to incur the cost of a program coordinator and staff. Most nonprofit groups, however, cannot afford the costly land disposal fees for mixed batteries and, consequently, may limit their collection to specific types of batteries (e.g., EAC, which collects only button and nickel–cadmium batteries). The most significant expenses will be for transportation, disposal, and salaries. Printing, advertisements, collection containers, storage containers, and storage are often only a small portion of the total cost. The Southeastern Oakland County Resource Recovery Authority (SOCRRA) in Michigan has established a battery collection program.[29] Table 19.6 shows the breakdown of costs for the SOCRRA battery collection program. Because batteries in Michigan are classified as nonhazardous, operating a collection program is less expensive than if the batteries were classified as hazardous. Table 19.7 shows the cost of operating the same program assuming the batteries are classified as a hazardous waste. A reasonable current assumption for the transportation fees for hazardous waste is approximately $6 per loaded mile. Hazardous waste and land-disposal fees for the New Hampshire/Vermont Solid Waste Project are approximately $425 per drum. This rate will likely vary between disposal facilities and from region to region. As can be noted from Tables 19.6 and 19.7, the operation of a battery collection program where the batteries are classified as a hazardous waste is approximately 39 percent more expensive than a collection program where the batteries are classified as a nonhazardous waste.

Household Battery Collection Program Effectiveness

The collection of household batteries requires that the consumer play an active role in the process. The success of any battery collection program will depend, in large part, upon the ability of the public to change its MSW disposal habits, and change is dependent upon education (i.e., public awareness programs). The collection of household batteries has not

TABLE 19.5 Firms that Accept Waste Batteries for Disposal or Reclamation

Mercury Refining Co., Inc.
(MERECO)
790 Watervliet-Shaker Road
Latham, NY 12110518) 785-1703, (800) 833-3505

Accepts all household batteries, will pay for mercury–silver. Nickel–cadmium batteries accepted for free. Charge for alkaline and lithium batteries. Mercury and silver oxide batteries refined on-site, other cells marketed for disposal or reclamation.

Quicksilver Products, Inc.
200 Valley Drive, Suite 1
Brisbane, CA 94005
(415) 468-2000

Accepts mercury batteries. Charge for batteries, depending upon type and volume. Require specific packaging. Currently accept only commercial and industrial batteries. Anticipate accepting household batteries in 1992.

Environmental Pacific Corp.
PO Box 2116
Lake Oswego, OR 97055
(503) 226-7331

Accepts all batteries.

Inmetco
PO Box 720, Rt. 488
Ellwood City, PA 16117
(412) 758-5515
Accepts nickel–cadmium batteries only. Charge for batteries. Extract nickel and send cadmium to Zinc Corporation of America.

Bethlehem Apparatus Co.
890 Front Street, P0 Box Y
Hellertown, PA 18055
(215) 838-7034

Accepts mercury batteries.

NIFE
Industrial Boulevard
PO Box 7366
Greenville, NC 27835
(919) 830-1600

Accepts nickel–cadmium batteries only. Charge 0.70¢ per pound. Will accept batteries from household hazardous waste programs as long as there is a guarantee that the batteries are nickel–cadmium. Generator receives certificate of disposal. Batteries are broken up on-site and sent to a plant in Sweden for reclamation of cadmium.

Universal Metals and Ores
Mt. Vernon, NY
(914) 664-0200

Accepts nickel–cadmium batteries only. Pays 0.15¢ per pound for loads over 2000 pounds. Will accept batteries from household hazardous waste programs as long as there is a guarantee that the batteries are nickel–cadmium. Batteries are exported to Asia or Europe for reclamation as long as batteries are dry cells. Batteries can be shipped as a nonhazardous material.

F.W. Hempel & Co., Inc.
1370 Avenue of the Americas
New York, NY 10019
(212) 586-8055

Accepts nickel–cadmium batteries only. Charge for batteries. Will accept batteries from household hazardous waste programs as long as there is a guarantee that the batteries are nickel–cadmium. No processing performed on-site. Batteries are sent to France for processing.

Kinsbursky Brother Supply
1314 N. Lemon Street
Anaheim, CA 92801
(714) 738-8516

Accepts lead–acid and nickel–cadmium batteries. Charge 0.40¢ per pound for nickel–cadmium batteries. Pay for lead–acid batteries depending upon type and quantity of batteries. Will accept batteries from household hazardous waste programs as long as there is a guarantee that the batteries are nickel–cadmium. Lead plate and nickel sent to smelter. Cadmium sent to France. Batteries must be received as a hazardous waste.

BDT
4255 Research Parkway
Clarence, NY 14031
(716) 634-6794

Accepts alkaline and lithium batteries. Charge per pound. Crush, neutralize, and dispose of as hazardous waste.

Source: Information provided by the Environmental Action Coalition, and personal communication with company representatives.

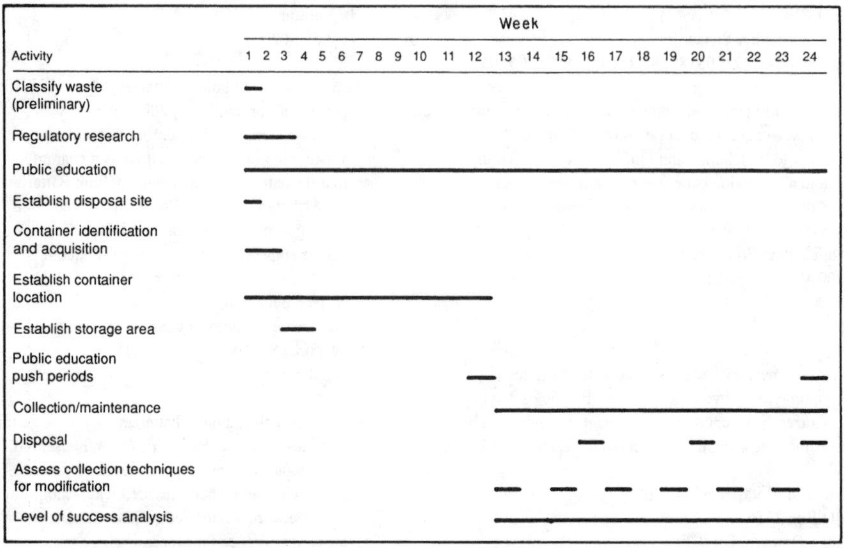

Week

Activity 1 2 3 4 5 6 7 8 9 10 11 12 13 14 15 16 17 18 19 20 21 22 23 24

Classify waste (preliminary)

Regulatory research

Public education

Establish disposal site

Container identification and acquisition

Establish container location

Establish storage area

Public education push periods

Collection/maintenance

Disposal

Assess collection techniques for modification

Level of success analysis

FIGURE 19.14 Example of a timeline for establishing a battery collection program.

been successful because of the large volume of batteries used, the lack of public education, the lack of centralized collection centers, and the inability to remove most rechargeable batteries from the appliances in which they are used. The access to rechargeable batteries is a significant issue, because approximately 80 percent of the nickel–cadmium batteries are permanently affixed to an appliance for consumer safety and cannot be removed.[13]

The New Hampshire/Vermont Solid Waste Project began collecting batteries in May 1987. Between May 1987 and October 1990, the New Hampshire/Vermont Solid Waste Project collected more than 13 tons of household batteries.[23] The capture rate of total battery volumes for the project area is estimated to be 18 percent. The Greater Detroit, Michigan, Resource Recovery Authority initiated a battery collection program. Within six months approximately 10 tons of batteries were collected.[11] The SOCRRA program collects approximately 3 tons a month.[29]

Various countries have been collecting batteries for a number of years. At least 11 European countries, including Sweden and Austria, have initiated battery collection programs. Japan, which has an established recycling program to collect alkaline and carbon–zinczinc batteries, has experienced less than a 10 percent recovery rate for cylinder-shaped batteries and 27 percent of button-shaped batteries.[34]

TABLE 19.6 Estimated Annual Cost of Battery Collection Program in States That Classify Batteries as Nonhazardous*

Buckets and decals	$700
Flyers	1100
Plastic bags	2300
Disposal	2830
Labor	4500
Miscellaneous	800
Total	$12,230

*Program for 6000 lb/month servicing area population of 325,000.

Source: Information provided by Tom Waffen, general manager of Southeastem Oakland County Resources Recovery Authority.

TABLE 19.7 Estimated Annual Cost of Battery Collection Program in States that Classify Batteries as Hazardous

Buckets and decals	$700
Fliers	1100
Plastic bags	2300
Labor	4500
Miscellaneous	800
90 polyethylene drums	6700
Transportation (assume 50 mi @ $6.00/loaded mile)	1200
Disposal	2830
Total	$20,130

Trends in Household Battery Production and Reclamation

Carbon-zinc and alkaline–manganese primary-cell batteries have the largest share of the market. A Duracell study indicated that 75 percent of all the batteries purchased in the United States by 1990 will be alkaline. The sales of carbon–zinczinc batteries are approximately 20 percent that of alkaline batteries and are declining.[18] The sale of alkaline batteries is expected to continue to grow.

According to the Office of Technology Assessment of the U.S. Congress, the market share for mercuric oxide batteries within six years decreased from 72 to 68 percent, while the demand for zinc–air batteries increased from 14 to 40 percent. The sales of silver oxide batteries have increased slightly. The sales of nickel–cadmium batteries are expected to significantly increase through the year 2000.[10]

There are alternatives to recycling, other than incineration and landfilling. These alternatives include finding substitute materials to replace the heavy metals currently used in batteries. Three domestic battery companies have recently begun marketing low-mercury-level batteries in the United States, as have already been marketed in Europe. Other battery manufacturers will likely develop and market similar batteries. There also is ongoing research by the battery industry to replace lead, mercury, and cadmium in batteries with nickel hydride, rechargeable lithium, and lower amounts of mercury.

As the battery industry seeks substitute materials for the hazardous constituents currently used, there is research by at least two foreign companies (i.e., Sumitomo in Japan, and Recytech, S.A., in Switzerland) to develop recycling technologies for all types of primary-cell household batteries.[9] A configuration of a treatment process is shown in Fig. 19.15. Currently, however, the technology to recycle all types of household batteries has not been developed or used in the United States. As landfill costs continue to increase and additional environmental regulations are enacted, the economic incentives to reclaim battery components are likely to increase.

SUMMARY AND CONCLUSIONS

1. According to EPA studies, approximately 80 percent of the discarded lead–acid batteries are being collected and recycled.

2. The unreclaimed 20 percent of lead-automotive batteries is estimated to contribute approximately 65 percent of the lead found in MSW. Consequently, the federal govern-

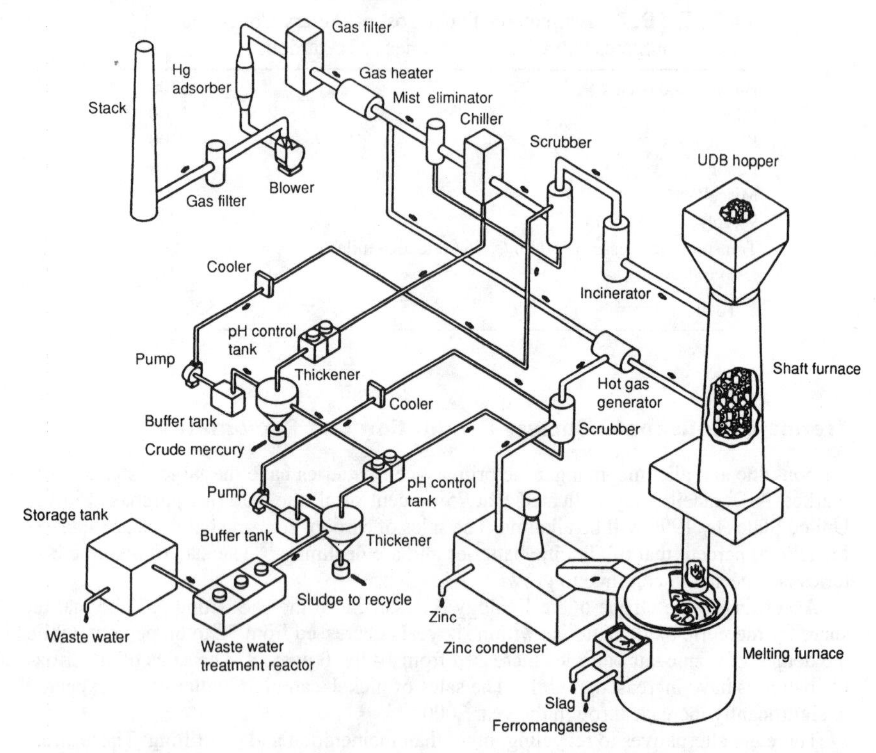

FIGURE 19.15 Configuration of the Sumitomo Treatment Process. (*Fiala-Goldiger et al., "The Status of Battery Recycling in Switzerland," Presented to the Second International Seminar on Battery Waste Management*)

ment and many state governments, with support from the battery industry and trade associations, have proposed legislation mandating the recycling of lead–acid batteries and requiring that the retailers of these batteries collect used batteries.

3. Through enactment of new legislation to ensure that an even higher percentage of lead–acid batteries will be recycled and removed from the MSW stream, the amount of lead found in MSW will decrease, thus decreasing the amount of lead released to the environment.

4. Although most participants in the legislative process agree that the components of lead–acid batteries should be reclaimed, controversy remains about the reclamation of the components of household-type batteries.

5. While the battery industry is researching alternative components to the heavy metals currently used in batteries, the date is unknown as to when these alternative components will be used. Low-level mercuric batteries are already being marketed in the United States.

6. Currently, most batteries collected through household battery collection programs are disposed of in hazardous waste landfills.

7. Currently, there are no recycling facilities in the Unites States that can practically and cost-effectively reclaim all types of household batteries, although domestic facilities that can reclaim button batteries do exist; most nickel–cadmium batteries are being reclaimed in Europe.

8. Although some battery collection programs are directed toward the collection of button, nickel–cadmium, and nonconsumer batteries, most programs collect all consumer batteries and no nonconsumer batteries.

9. There are insufficient incentives to encourage extensive recycling of household batteries.

10. It is likely that batteries which contain less mercury or other nonhazardous substitute materials will become widely marketed in the United States before large-scale reclamation of all types of batteries occurs in the United States.

ACRONYMS

BCI	Battery Council International
BPA	Battery Products Alliance
EAC	Environmental Action Coalition
EEC	European Economic Community
EPA	Environmental Protection Agency
GDRRA	Greater Detroit Resource Recovery Authority
MDNR	Michigan Department of Natural Resources
MPCA	Minnesota Pollution Control Agency
MSW	Municipal solid waste
MWC	Municipal waste combustor; also known as a solid waste incinerator.
NEMA	National Electrical Manufacturing Association
RCRA	Resource Conservation and Recovery Act
SOCRRA	Southeastern Oakland County Resource Recovery Authority

APPENDIX A. LEAD–ACID BATTERY RECYCLING LEGISLATION

Model Legislation*

Be it enacted by the legislature of the State of __________

Section 1. Lead–Acid Batteries; Land Disposal Prohibited

a. No person may place a used lead–acid battery in mixed municipal solid waste, discard or otherwise dispose of a lead–acid battery except by delivery to an automotive battery

*Courtesy of Battery Council International, Washington, D.C.

retailer or wholesaler, to a collection or recycling facility authorized under the law of (state), or to a secondary lead smelter permitted by the Environmental Protection Agency.

b. No automotive battery retailer shall dispose of a used lead–acid battery except by delivery to the agent of a battery wholesaler, to a battery manufacturer for delivery to a secondary lead, smelter permitted by the Environmental Protection Agency, or to a collection or recycling facility authorized under the law of (state), or to a secondary lead smelter permitted by the Environmental Protection Agency.

c. Each battery improperly disposed of shall constitute a separate violation.

d. For each violation of this section a violator shall be subject to a fine not to exceed $____ and/or a prison term not to exceed ____ days (as appropriate under state code).

Section 2. Lead Acid Batteries; Collection for Recycling

a. A person selling lead–acid batteries at retail or offering lead–acid batteries for retail sale in the state shall:

(1) accept, at the point of transfer, in a quantity at least equal to the number of new batteries purchased, used lead–acid batteries from customers, if offered by customers; and

(2) post written notice which must be at least 8-½ inches by 11 inches in size and must contain the universal recycling symbol and the following language:

(i) "It is illegal to discard a motor vehicle battery or other lead–acid battery."

(ii) "Recycle your used batteries."; and

(iii) "State law requires us to accept used motor vehicle batteries or other lead–acid batteries for recycling, in exchange for new batteries purchased."

Section 3. Inspection of Automotive Battery Retailers. The (appropriate state agency) shall produce, print, and distribute the notices required by Section 2 to all places where lead–acid batteries are offered for sale at retail. In performing its duties under this section the division may inspect any place, building, or premise governed by Section 2. Authorized employees of the agency may issue warnings and citations to persons who fail to comply with the requirement of those sections. Failure to post the required notice following warning shall subject the establishment to a fine of $_______ per day (as appropriate under state code).

Section 4. Lead Acid Battery Wholesalers. Any person selling new lead–acid batteries at wholesale shall accept, at the point of transfer, in a quantity at least equal to the number of new batteries purchased, used lead–acid batteries from customers, if offered by customers. A person accepting batteries in transfer from an automotive battery retailer shall be allowed a period not to exceed 90 days to remove batteries from the retail point of collection.

Section 5. Enforcement. The (appropriate state agency) shall enforce Sections 2 and 4. Violations shall be a misdemeanor under (applicable state code).

Section 6. Severability. If any clause, sentence, paragraph, or part of this chapter or the application thereof to any person or circumstance shall, for any reason, be adjudged by a court of competent jurisdiction to be invalid, such judgment shall not affect, impair, or invalidate the remainder of this chapter or its application to other persons or circumstances.

Analysis of Proposed Lead-Acid Battery Recycling Model Legislation*

Section 1. Lead Acid Batteries; Land Disposal Prohibited

a. The legislation would prohibit individuals from disposing of used lead–acid batteries except by delivery to the following:
- Battery retailers or wholesalers
- State-authorized collection or recycling facilities, or
- A secondary lead smelter permitted by the US Environmental Protection Agency

b. Battery retailers would be required to deliver used batteries to:
- The agent of a battery wholesaler
- A battery manufacturer for delivery to a secondary lead smelter permitted by EPA
- A state-authorized collection or recycling facility, or
- An EPA-permitted secondary lead smelter

c. Each battery improperly disposed of would constitute a separate violation, with such violation subject to the penalties deemed appropriate by each state.

Section 2. Lead Acid Batteries; Collection for Recycling

a. Battery retailers would be required to:
- Accept at least as many used batteries (if offered by customers) as new batteries purchased
- Post a written notice of the size and content required by the statute

Section 3. Inspection of Automotive Battery Retailers. The model legislation would direct the appropriate state agency to distribute the notices required by Section 2 to all battery retailers, and would grant that agency the authority to enter for inspection any covered place, building or premise. Failure to comply with Section 2 may result in warnings or citations as well as monetary penalties.

Section 4. Lead Acid Battery Wholesalers. Battery wholesalers must accept at least as many used batteries (if offered by customers) as new batteries purchased. Wholesalers accepting used batteries from retailers must remove the batteries from the retail point of collection within 90 days.

Section 5. Enforcement. Violations of Sections 2 and 4 would be a misdemeanor under the model legislation.

Section 6. Severability. This section contains standard severability language.

APPENDIX B. STATE OF MINNESOTA DRY-CELL BATTERY LEGISLATION†

[An Act] relating to waste; prohibiting the placement of certain dry cell batteries in mixed municipal solid waste; requiring labeling of certain batteries by electrode content; estab-

*Courtesy of Battery Council International, Washington, D.C.
†Chapter No. 409, H.F. No. 1921.

lishing maximum content levels of mercury in batteries; requiring that batteries in certain consumer products be easily removable; providing penalties; proposing coding for new law in Minnesota Statutes, chapters 115A and 325E.

Be it enacted by the legislature of the state of Minnesota:

Section 1. [115A.9155] [Disposal of Certain Dry Cell Batteries.]

Subdivision 1. [Prohibition.] A person may not place in mixed municipal solid waste a dry cell battery containing mercuric oxide electrode, silver oxide electrode, nickel–cadmium, or sealed lead–acid that was purchased for use or used by a government agency, or an industrial, communications, or medical facility.

Subdivision 2. [Manufacturer Responsibility.)

(*a*). A manufacturer of batteries subject to subdivision 1 shall:

(**1**). Ensure that a system for the proper collection, transportation, and processing of waste batteries exists for purchasers in Minnesota; and

(**2**). Clearly inform each purchaser of the prohibition on disposal of waste batteries and of the system or systems for proper collection, transportation, and processing of waste batteries available to the purchaser.

(*b*). To ensure that a system for the proper collection, transportation, and processing of waste batteries exists, a manufacturer shall:

(**1**). Identify collectors, transporters, and processors for the waste batteries and contract or otherwise expressly agree with a person or persons for the proper collection, transportation, and processing of the waste batteries; or

(**2**). Accept waste batteries returned to its manufacturing facility.

(*c*). A manufacturer shall ensure that the cost of proper collection, transportation, and processing of the waste batteries is included in the sales transaction or agreement between the manufacturer and any purchaser.

(*d*). A manufacturer that has complied with this subdivision is not liable under subdivision 1 for improper disposal by a person other than the manufacturer of waste batteries.

Section 2. (325E.125J [General and Special Purpose Battery Requirements.]

Subdivision 1. [Identification.] The manufacturer of a button cell battery that is to be sold in this state shall ensure that each battery is clearly identifiable as to the type of electrode used in the battery.

Subdivision 2. [Mercury Content]

(*a*). A manufacturer may not sell, distribute, or offer for sale in this state an alkaline manganese battery that contains more than .30 percent mercury by weight, or after February 1, 1992, 0.025 percent mercury by weight.

(*b*). On application by a manufacturer, the commissioner of the pollution control agency may exempt a specific type of battery from the requirements of paragraph (a) if there is no battery meeting the requirements that can be reasonably substituted for the battery for which the exemption is sought. The manufacturer of a battery exempted by the commissioner under this paragraph is subject to the requirements of section 1, subdivision 2.

(*c*). Notwithstanding paragraph (a), a manufacturer may not sell, distribute, or offer for sale in this state after January 1, 1992, a button cell alkaline manganese battery that contains more than 25 milligrams of mercury.

Subdivision 3. [Rechargeable Tools and Appliances.)

(*a*). A manufacturer may not sell, distribute, or offer for sale in this state a rechargeable consumer product unless:

(**1**). The battery can be easily removed by the consumer or is contained in a battery pack that is separate from the product and can be easily removed; and

(**2**). The product and the battery are both labeled in a manner that is clearly visible to the consumer indicating that the battery must be recycled or disposed of properly and the battery must be clearly identifiable as to the type of electrode used in the battery.

(*b*). "Rechargeable consumer product" as used in this subdivision means any product that contains a rechargeable battery and is primarily used or purchased to be used for personal, family, or household purposes.

(*c*). On application by a manufacturer, the commissioner of the pollution control agency may exempt a rechargeable consumer product from the requirements of paragraph (a) if:

(**1**). The product cannot be reasonably redesigned and manufactured to comply with the requirements prior to the effective date of this section;

(**2**). The redesign of the product to comply with the requirements would result in significant danger to public health and safety; or

(**3**). The type of electrode used in the battery poses no unreasonable hazards when placed in and processed or disposed of as part of mixed municipal solid waste.

(*d*). An exemption granted by the commissioner of the pollution control agency under paragraph (c), clause (1), must be limited to a maximum of two years and may be renewed.

Section 3. [325E.1251] [Penalty.] Violation of sections 1 and 2 is a misdemeanor. A manufacturer who violates section 1 or 2 is also subject to a minimum fine of $100 per violation.

Section 4. [Application; Effective Dates.] Section 1 is effective August 1, 1990. Section 2, subdivisions 1 and 2, are effective January 1, 1991, and apply to batteries manufactured on or after that date. Section 2, subdivision 3, is effective July 1, 1993, and applies to rechargeable consumer products manufactured on or after that date. Notwithstanding section 2, a retailer may sell alkaline manganese batteries from the retailer's stock existing on the effective dates for the two levels of mercury in section 2, subdivision 2, and rechargeable consumer products from the retailer's stock existing on the effective date of section 2, subdivision 3.

REFERENCES

1. Baum, Barry G., "Model Legislation for Nickel Cadmium and Small Lead Battery Recycling," presented at the Second International Seminar on Battery Waste Management, November 5–7, 1990.
2. Bureau of Mines, Minerals Yearbook, "Mercury," 1987.
3. Bureau of Mines, Minerals Yearbook, "Lead," 1988.
4. Bureau of Mines, Minerals Yearbook, "Cadmium," 1988.
5. Council of the European Communities, Draft Common Position on Batteries and Accumulators Containing Certain Dangerous Substances, September 1990.

6. Dickinson, Paul, Ultra Technologies of Kodak, Personal communication, March 1991.
7. Environmental Action Coalition, Personal communication with Mr. Sean Hecht, program coordinator, November 1990 to January 1991.
8. Eveready. Personal communication with Mr. Dave Dibell, product quality manager, November 1989.
9. Fiala-Goldiger, J., and M. A. Rollor, "The Status of Battery Recycling in Switzerland," presented at the Second International Seminar on Battery Waste Management, November 5–9, 1990.
10. Franklin Associates, Ltd., *Characterization of Products Containing Lead and Cadmium in Municipal Solid Waste in the United States, 1970 to 2000,* U.S. EPA, January 1989.
11. Greater Detroit Resource Recovery Authority, Personal communication with Mr. Phil Brown, program coordinator, November 1990 to January 1991.
12. Hinchey, Maurice D., "Household Batteries, Management or Neglect?" A Staff Report to the chairman, New York State Legislative Commission on Solid Waste Management, 1988.
13. Johnson, R., and C. Hirth, "Collecting Household Batteries," *Waste Age,* pp. 48–52, 1990.
14. Jones, C. J., P. J. McGugam, and P. F. Lawrence, "An Investigation of the Degradation of Some Dry Cell Batteries," Journal of Hazardous Materials, vol. 2, pp. 259–289, 1978.
15. Kodak Ultra Technologies, personal communication with Mr. Paul Dickinson, March 1991.
16. Linden, D. (ed.), *Handbook of Batteries and Fuel Cells,* McGraw-Hill, New York, 1984.
17. Michigan Legislative, "Report of the Joint Legislative Committee on Batteries," State of Michigan, December 1990.
18. Minnesota Pollution Control Agency, *Household Batteries in Minnesota: Interim Report of the Household Battery Recycling and Disposal Study,* March 1990.
19. National Electrical Manufacturers Association, "Written Statement of the NEMA and the Battery Products Alliance Concerning Household Battery Disposal," 1989.
20. National Electrical Manufacturers Association, personal communication with Mr. Fred Nicholson, section staff executive, November 1989.
21. National Electrical Manufacturers Association, "Mercury Usage in the U.S. Consumer Battery Production," from the NEMA public information document, 1990.
22. National Electrical Manufacturers Association, personal communication with Mr. Fred Nicholson, section staff executive, February 1991.
23. New Hampshire/Vermont Solid Waste Project, personal communication with Mr. Carl Hirth, program coordinator, December 1990.
24. New York State Legislative Commission on Solid Waste Management, September 1988.
25. Pillsbury, Hope, "Battery Recycling and Disposal in the United States: A Federal Perspective," presented at the Second International Seminar on Battery Waste Management. November 5–7, 1990.
26. Roos, C. E., J. Kearley, R. Quarles, and E. J. Summer, Jr., "Reducing Incineration Ash Toxicity due to Cadmium and Lead in Batteries," presented at the Second International Seminar on Battery Waste Management, November 5–7, 1990.
27. Seeberger, Donald, "A Study of Two Collection Methods for Removing Household Dry Cell Batteries from a Residential Waste Stream," Division of Environment and Energy, Hennepin County, Minnesota, 1989.
28. Serracane, Claudio, Electrolytic Process for the Recovery of Lead from Spent Storage Batteries," presented at the Second International Seminar on Battery Waste Management, November 5–7, 1990.
29. Southeastern Oakland County Resource Recovery Authority, personal communication with Mr. Tom Waffen, general manager, March 1991.
30. State of Minnesota, personal communication with Ms. Karen Arnold, February through March 1991.
31. Taylor, Kevin, David J. Hurd, and Brian Rohan, "Recycling in the 1980s: Batteries Not Included," *Resource Recycling*, May/June 1988, pp. 26–59.
32. Toshio, Matsuoka, "Sumitomo Used Dry Battery Recycling Process," presented at the Second International Seminar on Battery Waste Management, November 5–9, 1990.
33. U.S. Environmental Protection Agency, "Characterization of MWC Ashes and Leachates from MSW Landfills, Monofills, and Co-disposal Sites," Office of Solid Waste and Emergency Response, Washington, D.C., EPA 530-SW-87-008A, vol. 1, October 1987.

34. U.S. Congress Office of Technology Assessment, "Facing America's Trash: What's Next for Municipal Solid Waste," OTA-0-424, U.S. Government Printing Office, Washington, D.C., October 1989.
35. Wallis, George, and S. P. Wolsky, "Options for Household Battery Waste Management," presented at the Second International Seminar on Battery Waste Management, November 5–7, 1990.

CHAPTER 20
CONSTRUCTION AND DEMOLITION DEBRIS

EDWARD L. VON STEIN
Solid Waste Consultant
Stamford, Connecticut

INTRODUCTION

Historical Perspective—Economics Spur Recycling and Recovery

Construction and demolition debris waste or remaining building materials was typically co-disposed with other solid wastes until the mid-twentieth century. Recycling of construction and demolition debris was first conceived of as a response to the scarcity of building materials and the cost of disposal. In the United States, it was not until the introduction of incineration in the early 1900s, and later of resource recovery in the 1970s, that the separate disposal of largely incombustible rubble justified the added expense. In these systems, high maintenance costs and machine wear caused by heterogeneous waste streams favored separation of the construction and demolition debris fraction. In Europe, after the destruction of World War II, millions of tons of building rubble remained to be handled. Since rebuilding the transportation system was a priority, Germany, France, and other nations developed the recycling of rubble into new highway construction products. By 1987, some 100 million tons of rubble had been processed into aggregate and other products in Berlin alone [2], primarily using manual sorting methods.

Since the 1980s, separate disposal in the United States has usually meant separate landfills. In addition to separate construction and demolition (C&D) or bulky waste landfills, stump dumps to receive tree stumps resulting from land clearing, and pits to receive the residue from asphalt paving commonly served larger jurisdictions.

As interest in recycling to control the burgeoning municipal solid waste stream grew in Europe and the United States during the 1970s and 1980s, attention was focused on separately collected waste streams, including construction and demolition debris. With the cost of landfilling waste reaching $100 per ton or more in many municipalities at the end of the 1980s, the economics of diverting construction and demolition debris from disposal has become more attractive to the public sector. Once the incombustible material was separated, consideration could be given to reclamation.

The objectives of recycling C&D, as in recycling any other waste, are generally to:

- Realize an economic benefit
- Conserve material resources, (i.e., not waste the once-used materials)

- Conserve valuable landfill space
- Protect the environment by limiting the potential for discharge to ground water of leached constituents from landfills, or release to the atmosphere of air-borne pollutants during combustion
- Comply with regulations that prohibit co-disposal with municipal solid waste (MSW) or contribute diversion toward a recycling goal

The construction industry is the key market for recycled C&D. One significant impediment to recycling, however, is the lack of any established market for products, or acceptance of products created from C&D by the usual marketplace, the construction industry.

Characteristics of Construction and Demolition Debris Bulky Waste

Construction and demolition debris is generated by construction activity. It can result from the construction of buildings or other structures, or the demolition of old structures. Broadly categorized as C&D are waste streams generated by highway repaving, bridge demolition and construction, and remodeling and renovation.

The following regulatory definitions illustrate the variety of items specifically included in the definition.

> "Bulky waste" means landclearing debris and waste resulting directly from demolition activities other than clean fill [2].

> "Construction and demolition debris" means uncontaminated solid waste resulting from construction, remodeling, repair and demolition of structures and roads; and uncontaminated solid waste consisting of vegetation resulting from land clearing and grubbing, utility line maintenance and seasonal and storm related cleanup. Such waste includes, but is not limited to, bricks, concrete and other masonry materials, soil, rock, wood, wall coverings, plaster, drywall, plumbing fixtures, non-asbestos insulation, roofing shingles, asphaltic pavement, electrical wiring and components containing no hazardous liquids, and metals that are incidental to any of the above. Solid waste that is not construction and demolition debris (even if resulting from the construction, remodeling, repair and demolition of structures and roads and land clearing) includes, but is not limited to, asbestos waste, garbage, corrugated container board, electrical fixtures containing hazardous liquids such as fluorescent light ballasts or transformers, carpeting, furniture, appliances, tires, drums and containers, and fuel tanks. Specifically excluded from the definition of construction and demolition debris is solid waste (including what otherwise would be construction and demolition debris) resulting from any processing technique, other than that employed at a construction and demolition debris facility, that renders individual waste components unrecognizable, such as pulverizing or shredding [3].

> "Construction/demolition waste"—Solid waste resulting from the construction or demolition of buildings and other structures, including, but not limited to, wood, plaster, metals, asphaltic substances, bricks, block and unsegregated concrete. The term also includes dredging waste. The term does not include the following if they are separate from other waste and are used as clean fill:

Uncontaminated soil, rock, stone, gravel, unused brick and block and concrete.

Waste from land clearing, grubbing and excavation, including trees, brush, stumps and vegetative material [4].

Construction and demolition (C&D) waste is "solid waste resulting from the construction, demolition or razing of buildings, roads and other structures" (s. NR500.03(50), Wis. Adm. Code) [5].

Detailed analyses of the composition of C&D based on field sampling programs are rarely performed. Estimators have typically relied on extrapolations of data obtained from other similar localities, or on national averages. However, C&D composition has been researched by various professional investigators, including Davidson and Wilson [6], and others. Composition has often been found to vary by location. Tables 20.1 and 20.2 summarize typical composition of domestic buildings and of building roofing waste as reported by various published sources. Included in Table 20.1 are the apparent densities of materials for which both volume and weight are available.

TABLE 20.1 Typical building content in the United States* [7]

Material	Mass content in buildings in US, %	C&D, %†	Volume (yd^3)	Density (pounds/yd^3)
Steel and iron	1.57	2.73	0.05	1090
Copper	0.05	0.02	neg.	na
Lead	0.06	0.06	neg.	na
Aluminum	0.01	neg.	na	na
Concrete	63.33	53.75	0.90	1190
Brick and clay	15.01			
Brick	na	21.21	0.35	1210
Wood	19.64	22.01	1.10	400
Glass	0.33	0.22	neg.	na
Plastic	<0.01	neg.	na	na
Total	100.00	100.00	2.4	830

*"na" = not available; "neg" = negligible.
†C&D composition based on 1 ton of waste.

TABLE 20.2 Typical roofing waste composition [8]

Material	Percent
Asphalt cement	36.0
Hard rock granules (–No. 10 to +No. 60 sieve size)	22.0
Filler (–No. 100 sieve size)	8.0

Note: Minor amounts of coarse aggregate of about 1 inch in size, and of cellulose fiber felt, glass fiber felt, asbestos felt, and polyester films may also be present.

Quantities

C&D may be a significant component of local or regional waste streams, often accounting for as much as 10–30% by weight of the total waste stream. Table 20.3 indicates the quantity of C&D requiring disposal as part of the waste stream in several domestic locations.

Table 20.4 lists estimates of the bulk densities of various raw materials that commonly become C&D materials at the end of their useful life. These densities do not reflect the voids that exist in a pile of C&D, however. Overall bulk density of such a pile is considerably lower.

The quantity of C&D being generated will continue to increase in the near future. Modernization or replacement of aging urban infrastructure (bridges, highways, roads and sidewalks, sewerage and drainage) will generate C&D. As industrial occupants of factory space give way to commercial, residential, or other tenants, the attendant renovations will create more C&D. However, C&D generation for any locality many increase, stabilize, or perhaps decrease in the future. Local conditions, including principally economic conditions, will govern generation.

Certain variables discussed below may influence the quantity of C&D generated. Davidson and Wilson [6] presented their suggested approach to estimating the volume and material content of demolition debris, based on the material content of an "average" building. They determined the quantity of demolition debris disposed in 15 cities, estimated the probable materials contained in buildings demolished, and extended the estimates to the United States by using the national census of the number of housing units au-

TABLE 20.3 Quantities of C&D wastes [9, 10, 11]

	Locations		
Material	South Central Connecticut region	Town of North Hempstead, NY	Monroe County, NY
Paper products	42.9	36.2	33.7
Yard waste	15.5	15.3	14.8
Glass	8.0*	8.1	6.0
Metals	7.8*	4.3	4.9
Plastics	9.0*	6.8	8.5
Other combustibles	5.8*	19.2	23.1
Other noncombustibles	10.9*	N/A	N/A
Wood	above†	above	above
Lumber	above	above	above
Rock, brick, concrete	5.0	above	0.23
Asphalt	above	above	above
Masonry	above	above	above
Dirt, sand	above	4.1	5.89
Dirt	above	above	2.90
Other rubble	above	above	2.99
Pallets	above	above	0.62
Total	100%	100%	100.1%

*C&D waste is contained within these material categories.

†"above" indicates that the entry for the item is contained in another category.

TABLE 20.4 Densities of typical C&D materials [12]

Material	Density, pounds per ft^3	Specific gravity
Steel, cold drawn	489	7.83
Glass, common	162	2.4–2.8
Timber	38–42	0.61–0.72
Rubble masonry	137–156	1.9–2.7
Dry rubble masonry	110–130	1.8–2.3
Brick masonry	103–128	1.4–2.3
Earth excavated	63–126	1.0–2.1
Asphaltum	81	1.1–1.5

thorized for demolition. The result of this approach was as shown in Table 20.1. Davidson and Wilson then attempted to correlate factors that they expected might influence the flow of demolition wastes, as listed in Table 20.5.

DISPOSAL PRACTICES—CO-DISPOSAL VERSUS DEDICATED DISPOSAL

Much shredded or crushed C&D appears soil- or stone-like. Its disposal in the past has often been only minimally regulated. C&D landfills are similar to MSW landfills in several ways. Current regulations generally require that such landfills include leachate collection, storage or treatment, and monitoring systems; periodic and final cover; and other elements familiar to MSW landfill operators.

In many states, the construction cost of C&D landfills parallels that of MSW landfills, on a per-acre basis. Landfill liners, leachate collection systems, and gas control facilities may be required. However, since local regulations may not require the application of daily cover, operating costs can be limited and 5–10% of the volume reserved for MSW cover can provide useable C&D disposal volume.

Other methods of disposal of C&D wastes (i.e., the production of wood chips from wood wastes) are common in the northeastern United States. Wood chips from C&D are often used as a boiler fuel in the large-scale generation of industrial power.

TABLE 20.5 Factors influencing the generation of C&D wastes [5]

Public sector demolitions
Vacancy rate of homes
Age of homes
Private sector demolitions
Economic obsolescence

SOURCE SEPARATION—COLLECTING MATERIAL FOR USE

Construction Site Source Separation

At the construction site, laborers are typically assigned to clean up the site periodically. Workers dump materials into 20–50 yd^3 open-top roll-off containers provided by disposal firms. The items listed in Table 20.6 can typically be segregated for reuse or resale before being placed in the container. These materials may be destined for incorporation into new work at the same construction site, if allowed by contract, or for use on another project. Alternatively, resourceful contractors may have ready external markets for materials. Larger demolition projects may warrant a separate roll-off box for materials set aside. Projects involving construction several stories above grade may include job-site fabricated plywood chutes to drop materials directly into roll-off boxes at grade, or wheelbarrows may be wheeled to elevators for the trip to street level.

Little is published regarding the actual economics of source separation at construction sites. The advantages to a contractor include the ability to contain some material and operating costs, as well as the ability to offer C&D recycling as a "green" add-on to the basic construction service.

Collection

Aberdeen, MD, set up a demonstration project for curbside collection of scrap wood. Tree limbs, yard trimmings, and large quantities of C&D were excluded. About one in eight households participated, and the set-out averaged 174 pounds. A permanent drop-off center later was established [13].

MARKETS AND MARKETING

On-Site Uses

The most accessible market for source-separated demolition debris is the construction project underway at the site itself. Minimal or no marketing overhead is related to such use, and transportation, processing and storage costs are minimized. Although some dem-

TABLE 20.6 Typical construction site target materials

Material	Raw material
Construction-grade lumber	Wood
Ornamental trim	Wood
Metals	Steel, nonferrous
Tiles	Clay
Bricks	Clay
Electrical hardware and wire	Copper, aluminum
Plumbing hardware and copper pipe	Copper, bronze, brass

TABLE 20.7 Target construction site recyclables

Material On-site uses	On-site function
Wall studs, other construction grade lumber and timber	Temporary or permanent framing and general construction
Plywood	Concrete forms, floor protection, as a replacement for new plywood
Used brick and tile	Decorative facades
Electrical hardware	Electrical hardware

olition contracts may allow or stipulate reuse (i.e., historic renovation), the use of any but virgin materials in the final product may be prohibited in other cases (i.e., governmental contracts). Only recently has the routine practice of biasing publicly funded construction specifications toward the use of only new products made from virgin materials come under scrutiny. Purchase specifications of many state governments, e.g., Connecticut, New York, California, Minnesota, and Wisconsin, permit the limited use of recycled products. Typical products that may be reused on-site are listed in Table 20.7.

Homeowners' residential do-it-yourself projects typically absorb a portion of the wood, plywood, and other materials recovered from renovations underway at the same property. Waste lumber from one demolished wall of a home may comprise the structural framing of an added or renovated space.

Markets

Table 20.8 lists typical markets for C&D materials. Markets are discussed in this section. Historically, reuse in place of new wood and burning as fuel have been important outlets for wood waste. Most other materials do not have long market histories.

In third-world countries, wood scavenged from demolition sites may be used in home heating. Treated wood, including railroad ties, telephone poles, and pilings, is unsuitable for use as fuel, however. When not burned, wood waste may be sold to wood mills as a composition board component, or chipped for use as ground cover. Demolition wood may be contaminated with metals fasteners or paints, but may be salable.

As discussed in a later section, pavement recycling is becoming widely accepted. Recycled asphalt pavement competes not only with the application of new bituminous pavement, but also with glasphalt that contains crushed recycled glass as a portion of its aggregate and with asphaltic material with recycled rubber tire content. Therefore, the asphalt pavement industry is a preferred market for at least three competing recycled materials and one or more products from virgin materials.

Crushed concrete from which reinforcing steel has been separated can be used as a replacement for natural aggregate in foundation subgrades for building construction, road construction, or other applications. Coastal areas have used concrete rubble to construct artificial reefs. Pinellas County, Florida operates one of the largest such programs in the United States [14].

Scrap metal, including ferrous and non-ferrous metals such as aluminum (from window demolition), brass and copper (from old roofs, roof flashing, electrical and plumbing fixtures, and decorative uses), and others often find ready markets in urban areas. Scrap yards act as market intermediaries between contractors who separate the materials, or

TABLE 20.8 Typical markets for recyclables from C&D

Material	Typical markets
Wood waste	Fuelwood
	New Construction
	Remodeling
	Mulch, landscape material
	Animal bedding
	Particle board
	Construction forms
Asphalt pavement	Asphalt pavement
Concrete	Foundation stone
	Road Construction
	Aggregate for low-strength concrete
Masonry	Foundation stone
	Road construction
Brick	Decorative facades
Structural steel	Reinforcing steel
	Structural steel
	Other steel
Aluminum	Aluminum fabrications

waste disposal firms that provide the on-site waste containers, and the end users such as wiremills, aluminum extruders, and the like.

Waste Exchanges

Regional governments and others have established waste exchanges as an approach to assisting industry with waste disposal. The waste exchange concept provides for the creation of an information clearing house that publishes periodic, anonymous listings of available industrial waste materials and of raw material needs.

Of the seventeen waste exchanges operating in North America in 1990, most are able to deal with C&D. Exchanges have been created nationwide, including:

Seattle Exchange, King County, Washington

Renew Exchange, Texas Water Commission

Northeast Industrial Exchange, New York

International Marketing and C&D Recycling Overseas

International markets have historically developed at a faster rate than domestic markets. In Germany, for example, reserves of naturally occurring construction materials such as gravel have already been depleted in the vicinity of Berlin and in North Germany, and are projected to be exhausted by the year 2000 near Hanover [15]. Recognizing this situation, in 1980 in the GDR (West Germany, until 1990), some 47% of excavated material, building rubble, and road debris (140 million tonnes) was reused as fill or was processed into new raw material.

CONSTRUCTION DEBRIS RECYCLING METHODS

Table 20.9 summarizes C&D recycling methods. These practices are discussed in this section.

Structural Steel

Construction contractors may be responsible for disposition of excess structural steel upon termination or completion of a building project. Such steel can be sold "as is" through advertisements in public or trade media, or can be collected at the site by market intermediaries in quantities economic for shipping and returned to the steel fabricator for credit. It may be sold to local scrap dealers. It may ultimately be remanufactured, or made into reinforcing steel or other shapes.

Contaminants may include field welds, surface finishes such as mill primer, or rust. Such contaminants may be acceptable if the steel is to be reused as is for other construction, or they may need to be removed prior to remanufacture. Processing is usually limited to manual separation with or without mechanical assistance from mobile equipment at the site.

Wood Waste

Extra construction-grade dimensional lumber, timber, or plywood is often collected by the contractor and incorporated into other construction. Contaminants typically include nails, water or insect damage, rot, surface finishes, or chemical preservatives. Construction workers typically cut away visible contamination prior to incorporation into new

TABLE 20.9 C&D recycling methods

Material	Practice	Equipment
Concrete	Manual separation	Hand tools only
	Crushing	Bulldozer
		Rock crusher
		Jaw crusher
Structural steel	Manual separation	Hand tools only
Wood waste	Manual separation	Hand tools only
	Shredding	Wood hog
	Chipping	Chipper
Nonferrous metals	Manual separation	Hand tools only
Roofing material	Manual separation	Hand tools only
Steel	Magnetic separation	Magnetic separator
	Manual separation	Hand tools only

work, since contamination could render the construction unacceptable to the building owners or inspectors, or lead to structural failure.

Aluminum

Unused but unneeded extruded aluminum is typically sold to an aluminum ingot producer, a fabricator, or a metals broker. Aluminum scrap from fabricators has historically been acceptable for reuse internally at the fabricator's shop, or as raw material for another industry member. It is economical to extrude new aluminum from a mixture containing as much as 40% aluminum scrap.

Contaminants include surface finishes and corrosion products. Typical problems caused by surface finishes are exemplified by the experiments of aluminum extruders who tried to use postconsumer beer cans in production runs, but found that the lacquer surface finishes cause problems in the furnaces. Aluminum alloyed with other products likewise can cause an entire batch of material to exceed specified tolerances.

Other Nonferrous Metals

Extruded, rolled, forged or manufactured metal items, drawn copper wire, and other types of nonferrous metals are recovered manually at the construction site. Typical contaminants include plastic such as wire coatings or faucet handle inserts, fibers or rubber such as faucet washers or fiber wire insulation, and metallic sheathing of electrical conduit. The processes available to recover these nonferrous metals are visual inspection and manual separation.

DEMOLITION DEBRIS RECYCLING PRACTICES

Structure Demolition Practices

General. Recycling practices in building demolition projects are similar to those of the construction industry. The principal difference is the increased heterogeneity of materials. Practices vary with building size, demolition permit conditions, the demolition contractor's resources, and the economics of available disposal alternatives .

Concrete. Concrete requiring disposal results from demolishing building foundations, floors, and occasionally, roofs and structural elements. Removing or repairing sidewalks, storm or sanitary sewer appurtenances, and the like also generate waste concrete. Markets for concrete were discussed in a previous section.

Crushed concrete >5 mm in diameter can replace virgin coarse aggregate in new concrete. Fine aggregate (up to 5 mm) may also be used [16]. Quality requirements are difficult to meet because the concrete industry is reluctant to accept the material, and because pollution remains a concern [17]. A major issue in using recycled aggregate concrete is the increase of shrinkage [18]. Contaminants include reinforcing steel, either bar shapes or wire mesh; fasteners; adherent surface finishes, including ceramic, asphaltic, or other tiles; and adherent bricks. Mortar may be considered a contaminant if strength requirements are strict in the reuse application.

Crushing concrete is often performed by running a tracked bulldozer over the material several times, although this practice may generate noise and dust. At a Japanese demolition project, a jaw crusher produced a noise level of 75 dB(A) at a distance of 20 m [19]. Other processing includes manual sorting, the use of mobile jaw crushers, and magnetic separation of metals.

Roofing Materials. Table 20.2 provides an estimate of the composition of roofing material. Laboratory test results by Paulsen et al. [8], indicate that "acceptable paving mixtures that include roofing waste can be made." The paving industry, however, has not widely adopted the practice of incorporating such material. This condition may be due in part to the relative difficulty of securing adequate quantities of source-separated roofing waste. Equally important may be the costs of disposing of unused portions of the old roofing. Contaminants may include the original roofing substrate, which may be wood or metal.

Steel and Nonferrous Metals. Contaminants found in white goods may include capacitors; plastics in the form of wire sheaths and controls; insulation, such as fiberglass batts in refrigerators or dishwashers; and off-spec metals. Electrical equipment may generate electric motor "fluff,"* plastics, and paper.

Plumbing systems may include off-spec metals, such as tin/lead solder used in joints of copper pipes; components of valves such as rubber washers and stem packings; and petroleum-based lubricants.

Wood. The lack of appropriate grading rules and engineering design values limits the efficient use of old timber [20].

Facilities in New Hampshire and California experimented with wood recovery. The key issues were the ability to charge a competitive tipping fee and to make a value-added end product that would contribute to revenues [21]. Contaminants may include surface treatments, as discussed in the "Wood Wastes" section, gypsum or other wallboard, plaster, lath, electrical components, floor or wall covering, fasteners, and plumbing pieces. Manual separation and shredding is the most common waste wood processing method.

Restoration and Reuse. The simplest C&D recycling strategy is often to renovate all or part of a structure and rededicate it to a new use. This alternative should always be considered as part of the demolition planning process. Successful examples abound in most cities. Examples include reuse of factory space as offices; reuse of schools as public-use structures, including governmental offices and senior centers; and the reuse of railroad stations as commercial space. Typical planning concerns include those listed in Table 20.10.

TABLE 20.10 Typical concerns in planning for reuse of a structure

Adaptability to future use
Life-cycle cost of rennovations
Remaining useful life of structure
Environmental/public health liability
Structural integrity
Local zoning/land use requirements and preferences
Cost of demolition

*Shredded insulation.

Technology. The earliest C&D recycling projects in (West) Germany were designed to produce material that could compete with virgin supplies. This approach proved to be too costly. Second-generation projects employed simpler technology (i.e., jaw crushers alone or in conjunction with impact crushers and magnetic separators) to process material [22]. The most prevalent approach to recycling demolition debris at present involves visual inspection and manual separation.

Heavy-duty, preengineered, or custom-designed equipment is available in several configurations to facilitate these processes. Systems include mobile, semimobile, and fixed arrangements.

Mobile plants use a flatbed trailer as a platform for prescreening, crushing, magnetic separation, and final screening equipment, together with conveyors, chutes, and controls. Systems can typically be set up in less than a day by deploying hydraulically jacked legs and raising and aligning the equipment for proper materials flow. Semimobile plants, although also delivered to a site by truck, are larger than mobile units and may require up to 3 days to make ready for operation at a site.

Ocean County Recycling, a C&D recycling center located in Toms River, NJ, accepts three types of C&D and makes sixteen products. The center is vertically integrated in producing products and includes a contracting company, equipment, labor, as well as other product and business ventures [23].

In Europe, fixed or portable plants have been used to process excavation or building rubble into fresh raw materials. Fixed plants that can process in the range of 300 to 400 Mtph of infeed (12,000 × 1000 mm, or 4 ft × 3 ft) typically involve the following processes:

- Infeed
- Screening system, either single- or multiple-stage
- Crushing system, usually two-stage
- Magnetic belt separator

Mobile plants can operate up to 100 tph, assuming feed material of the same size, and using magnetic separation and a screening system. One class of mobile plants, mobile sorting systems, provide stations for laborers, a sorting conveyor belt, and often a magnet to ease the removal of ferrous metals. Belts are typically mounted at waist height for the most safe and efficient picking.

The sorting receptacle is typically at either side of the sorter. Up to six stations can be accommodated on a trailer that contains an in-feed hopper and an inclined belt section as well as a horizontal picking section.

To provide an overview of the range of equipment available from the manufacturers producing equipment to serve this field, Table 20.11 lists selected equipment and manufacturers. Detailed information can be obtained from the manufacturers.

Fixed plants are permanently mounted and provide the greatest range of capacity. Equipment in fixed plants is simple, and generally includes tracked bulldozers and hydraulically operated mobile jaw crushers.

Ravensburg, Germany Facility. A large fixed, highly mechanized C&D recycling facility in Ravensburg, Germany, has been processing C&D, "trade" or commercial waste, and bulky items into ferrous, wood, cardboard, and aggregate fractions since it began operating in November 1988. This plant was originally built with a partial construction grant from the German federal government to develop two pieces of equipment. Site specifications are as summarized in Table 20.12, and a block flow diagram is presented in Figure 20.1.

Material arriving at the facility is weighed and dumped in a tipping hall. Unacceptable

TABLE 20.11 Selected equipment manufacturers

Equipment	Manufacturer	Website
Knuckle Boom	Aurora Crane Corporation	www.auroracrane.com
	Ran-Paige Co., Inc.	Not available
	Samson Hoist Co.	Not available
Crushers, Asphalt and Other	American Pulverizer Co.	Not available
	Buffalo Hammer Mill Corp.	Not available
	Jeffrey, subsidiary of Penn Crusher	Not available
	Williams Patent Crusher & Pulverizer Co.	Not available
Crushers, Concrete	Cruendler Crusher	Not available
	Stedman Machine Co.	www.stedman-machine.com
Crushers, Jaw	Bayliss Machine & Welding Co.	www.baylissmachine.com
	Pennsylvania Crusher Corp.	www.penncrusher.com
	Stedman Machine Co.	Not available
	TM Engineering, Ltd.	Not available
Hogs, Wood	American Pulverizer Co.	Not available
	Jacobson LLC	www.jacobsonmn.com
	Jeffrey, subsidiary of Penn Crusher	Not available
	Williams	Not available
Chippers, Brush and Bark	Nicholson Industries	www.nicholsonindustries.com
Magnetic Separators	Dings Co., Magnetics Group	www.dingsco.com/magnetics/
	Eriez Magnetics	Not available
	Industrial Magnetics, Inc.	www.magnetics.com
	Shred-Tech	Not available
	Mayfran International	www.mayfran.com
Screens, Vibrating and/or Shaking	Andritz Sprout-Bauer	www.thomasregister.com/asb
	Derrick Corp.	www.derrickcorp.com
	FMC Corp., Material Handling Systems Division	www.fmc.com
	Midwestern Industries, Inc.	www.midwesternind.com
	TM Engineering	Not available
	Triple/S Dynamics, Inc.	Not available
	The Witte Company, Inc.	www.witte.com
Screens, Drum and/or Trommel	McNichols Co.	Not available
	Triple/S Dynamics, Inc.	Not available
	Shred-Tech	Not available

material and homogeneous loads are diverted into containers without processing. Material to be processed is visually inspected and elevated from an infeed pit to a bucket screen which diverts "larger pieces such as pallets, milled lumber . . . furniture and gutters" [24] to an enclosed manual picking room for trade waste and bulky materials where cardboard, wood, and ferrous metals are removed prior to loadout as reject.

The small-sized fraction remaining after the bucket screening (up to about 1 ft) is screened, sifted and mechanically sorted into "three debris fractions, two ferrous metal fractions, and one miscellaneous" portion [24]. The two smaller-sized debris fractions are

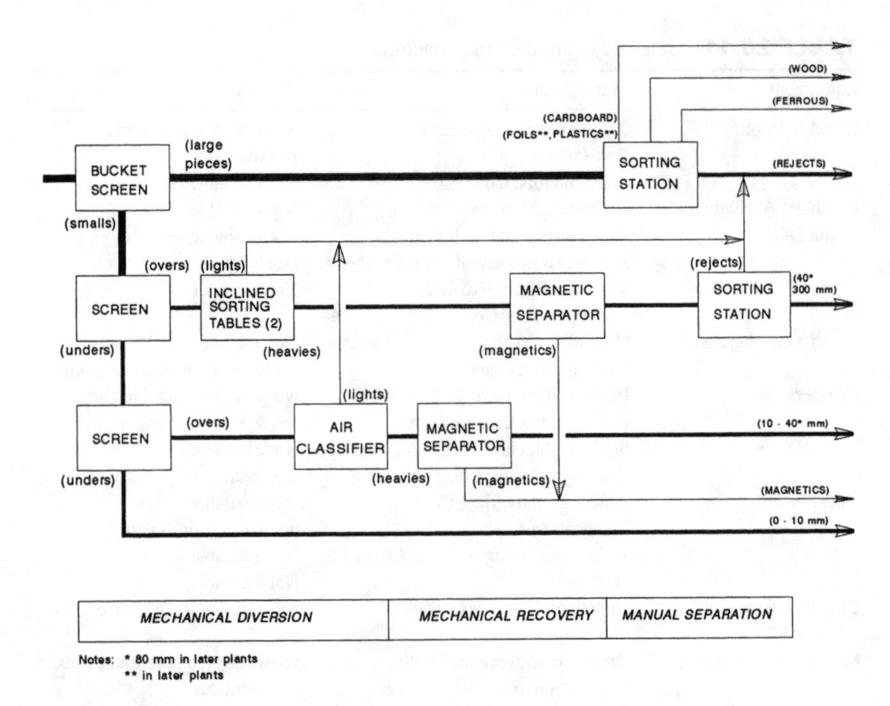

FIGURE 20.1 Flow diagram of Ravensburg, Germany, C&D recycling facility.

TABLE 20.12 Site description: Ravensburg, Germany C&D recycling facility [24]

Design parameter	Value
Area of Site	12,000 m^2
Area of building:	950 m^2
Design capacity, C&D	80 cubic m^3h 40 ton/h
Trade & bulky	150 m^3/h
Installed power	155 kW
Screening machines	2 (Type BSM)
Inclined sorting machines	2 (Type SSM)
Pneumosifter (air classifier)	1
Magnetic separators	2
Compactor	28 m^3
Air pollution control	34,000 m^3/h 38.5 kW 4 intakes

Source: Ref. 14.

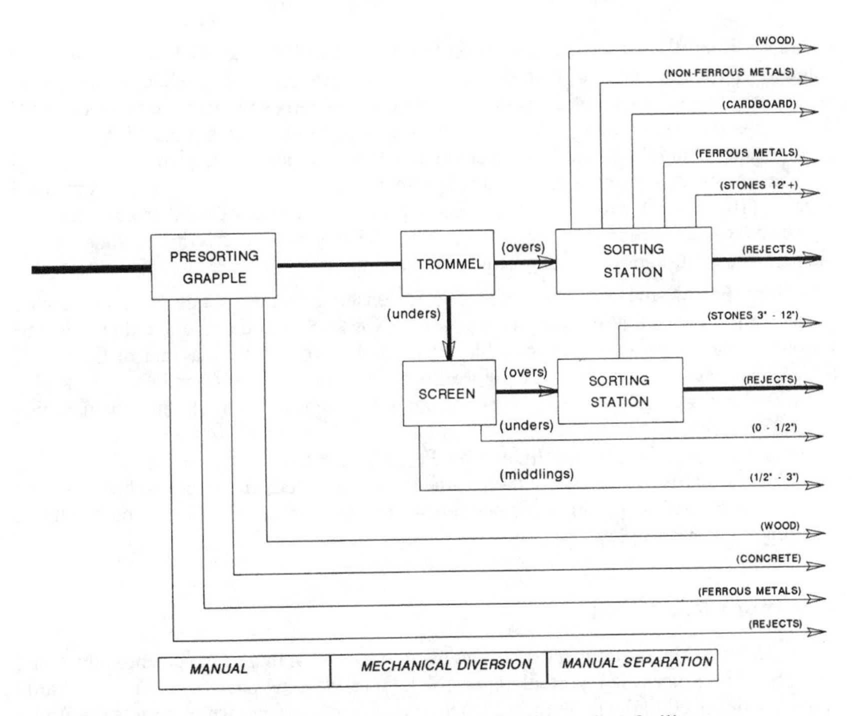

FIGURE 20.2 Flow diagram of Basel, Switzerland, C&D recycling facility.

FIGURE 20.3 C&D processing facility—Basel, Switzerland. (*Photograph courtesy of Lindemann Recycling Equipment, Inc.*)

used as fill and the largest-sized fraction has been stored on-site. The finest material is screened into three fractions, with the two larger-sized streams being delivered to inclined sorting machines for separation by density into usable debris and miscellaneous fractions. The debris fraction is conveyed past a magnetic separation stage and through a manual picking station and the miscellaneous fraction is treated as a reject stream.

The finest-sized fraction (0–40 mm) is screened into two fractions. The larger-sized fraction (10–40 mm) is air classified to remove light waste and the smaller-sized fraction (0–10 mm) is conveyed directly to a bunker. The larger-sized fraction is subjected to magnetic separation prior to being stored in a bunker.

Modern Mechanized Plants. The Ravensburg facility is one of several plants operating in Europe that recovery products from C&D. Selected additional domestic and European plants are identified in Table 20.13, and a block flow diagram of the Basel, Switzerland plant is provided in Figure 20.2. The Basel, Switzerland facility is also illustrated in Figure 20.3. This list is not meant to be a census, but rather merely identifies several representative projects.

At a recycling facility constructed by Fuji Electric Co., Ltd. in Fussa City, Japan, bulky waste is crushed, shredded, and magnetically separated; aluminum is then removed and it is air classified. Products include ferrous metals, aluminum, plastic, combustibles, and noncombustibles [25].

Pavement Recycling

Jones [26] presented a summary of the status of in-situ recycling of pavement in the United States. The process is reportedly appropriate in resurfacing pavements in which hardening has occurred in the uppermost ¾ inch of pavement surface, which results in surface failures, including cracks. Decreasing availability of new raw materials and consequent increases in cost have fostered interest in the process. Steps in recycling generally include pavement removal, pavement crushing and heating, adding aggregate and asphalt and/or an asphaltic modifier.

Mix-In-Place Bituminous Pavement Recycling. Existing bituminous surface is removed to a depth of 3–4 in using a tractor-mounted mechanical planer. The loose material is crushed and sized, then consolidated with a roller. An additive such as hot bitumen emulsion is sprayed onto the surface. Clean aggregate may be added. Finally, a conventionally wearing course of 1–3 inches of hot mix asphalt is applied. Costs of this process are variable, and may range from a savings of 20–40% to no savings over conventional pavement replacement.

Hot, In-Place Recycling. The pavement is softened by heating, then scarified or hot milled to a depth of ¾–1½ in and mixed. New hot mix material may be added. The sur-

TABLE 20.13 Selected European C&D facilities

Location	Status	Capacity
Ravensburg, Germany	Operating	600 yd^3/shift
Bad Reuzen	Operating	100 tons/h
Basel, Switzerland	Operating	1000 yd^3/shift
Zurich, Switzerland	Operating	1000 yd^3/shift

face may also be compacted, then a new wearing course added. Recycling in place involves heater scarifying, adding a modifier, screeding, and rolling.

To facilitate procurement and control of pavement recycling projects, standard specifications have been established by a trade association, the Asphalt Recycling and Reclaiming Association [26].

Issues

Constituents of C&D waste that could be harmful to health or the environment include thermostats, light switches, and lamps containing mercury; batteries from exit signs, emergency lights, and smoke alarms; lighting ballasts containing PCBs; lead pipes, and roof flashing [27].

Radon. The health affects of exposure to radon have long been known; recently, however, is apparent that sources of radon production may include typical construction substances such as bricks, concrete, and foundation stone. Therefore, the reuse of these materials is increasingly reconsidered in light of their potential adverse health impacts.

Research by Lettner and Steinhauser [28] indicates that the release of radon (^{222}Rn) from gypsum products may be related to and may increase with the moisture content of the material. Experiments indicated that increasing moisture content of concrete, soil, and shale caused an increase in the release of ^{222}Rn compared to dry materials. Also, some gypsum products may release higher levels of radon than most other common construction materials, according to Lettner and Steinhauser.

Asbestos. Building demolition debris from older structures typically contains some quantity of asbestos, which is specifically regulated by federal and state agencies. Because of the costs involved in separating asbestos contamination, which must often be performed by specialized firms, the presence of asbestos may render such C&D uneconomical to recycle.

Other Contaminants. Wood waste contamination may include lead-based paint, or preservatives such as creosote, pentachlorophenol, any of several water-born preservatives, and fire retardants. Arsenic may also be present.

REGULATORY AND LEGAL CONSTRAINTS

Federal Regulations

In the United States, national solid waste disposal legislation is contained in the Resource Conservation and Recovery Act (RCRA) of 1976, as amended. Congress has directed the Administrator of the Environmental Protection Agency to establish regulations and guidelines for waste disposal in a range of areas. However, no federal-level national recycling legislation mandates the recycling of C&D.

State Requirements

Many states in the United States have enacted recycling requirements, targets, and goals, and some states, including Connecticut, mention specific materials in their legislation.

Current legislation does not mandate the recycling of C&D in any state, however. Since states are establishing high recycling diversion targets, however, consideration of the inclusion of C&D in local programs is often necessary to meet goals.

State Permits

The Connecticut state statutes discuss solid waste management in Chapter 446d, Sections 22a-207 through 22a-255. A C&D recycling facility with a throughput capacity greater than 2000 pounds per hour (1 tph) and that may include named equipment such as pulverizers, compactors, shredders, and balers must operate under permits. The Commissioner of the Connecticut Department of Environmental Protection (CDEP) is authorized to issue permits to construct, once the need for the facility has been demonstrated, and permits to operate. Key provisions of the regulations are as follows:

- Recyclable materials may be removed from the waste stream at the point of generation, or before disposal, and may be transported to a facility for recycling.
- The CDEP may use a revolving fund for financial, legal, and technical planning by municipalities.
- The Commissioner may designate recyclable materials, and those items are prohibited from disposal with the waste stream.
- Construction materials may be regulated as litter if disposed outside a permitted site; a fine of up to $10,000 per day may be levied.

Application requirements for a permit to construct a C&D recycling facility include the preparation of an engineering description of the site and the facility, investigation of site hydrology and geology, preparation of operating plans and manuals, site life, closure, and postclosure. Specific technical requirements include incoming material storage capacity equal to 24 hours if facility capacity is greater than 100 tpd. But a facility may not store incoming material on site for more than 48 hours.

ECONOMICS

Little information is available on the economics of recycling C&D. This situation arises in part because of the relatively short track record of the practice. Projects such as the Ravensburg, Germany plant, discussed above, may reflect higher than expected operating costs if the throughput of C&D has not reached design capacity. Further, many recycling projects have been undertaken by the private sector; commercial recyclers consider their particular economics to be part of their competitive advantage, and are unwilling to reveal details. Equipment vendors report, however, that operating costs for centralized facilities may be as low as $12 to $17 per ton, although these estimates are unconfirmed. Since C&D is of relatively low density, the cost of its hauling and disposal will make on-site recycling particularly attractive.

Table 20.14 provides a list of selected equipment costs, based on second-quarter 1991 information. Table 20.15 provides the cost categories for an estimate of capital costs for a centralized C&D recycling facility, and may be used as a basic checklist. Table 20.16 lists the operating cost categories for a centralized facility.

Equipment vendors report that construction of a 35,000 to 40,000 square foot building to enclose a sorting system was completed for $2,650,000, and sorting equipment cost

TABLE 20.14 Capital costs of selected C&D recycling equipment

Equipment name	Cost range
Tub grinder, mobile	
with knuckleboom	$200,000
without knuckleboom	$100,000
Pedestal crane systems	$35,000 to $150,000

TABLE 20.16 Categories of annual operating costs for a centralized C&D recycling facility

Operation
Labor
General Manager
Weighmaster
Shift Supervisor
Mobile Equipment Operator
Sorters
Fixed Equipment Operators
Maintenance
Laborers
Utilities
Consumables
Fuel
Spares and lubricants
Baling wire and other supplies
Administration and overhead
(Varies with each installation)
Residual Disposal
Tip fee times tons tipped
Transportation of residuals

TABLE 20.15 Capital cost categories for a centralized C&D recycling plant

Site costs (minimum 3.0 acre site)
Acquisition
Permitting
Development
Buildings (typically, enclosed process building, administrative space, and scalehouse)
Design
Construction
Fixed process equipment (Typically, sorting, conveying, baling, and densifying)
Mobile equipment (Typically, front end loader, high lift, and yard tractor)

$1,825,000. After the inclusion of all project costs, however, and assuming that the site can be developed and a building can be built for the relatively low figure of $50.00 per square foot, it is probable that at least $5,000,000 will be required for construction.

Organizations Provide Contacts and Technology Transfer

Table 20.17 lists several key trade groups and professional organizations engaged in C&D recycling.

TABLE 20.17 List of organizations related to C&D recycling

ARRA	Asphalt Recycling and Reclaiming Association #3 Church Circle, Suite 250 Annapolis, MD 21401 Phone: 401 267 0023, FAX: 401 267 7546 Home Page: www.arra.org
ICRI	International Concrete Repair Institute 1323 Shepard Drive, Suite D Sterling, VA 20164 4428 Phone: 703 450 0016, FAX: 703 450 0019 Home Page: Not available
ISRI	Institute of Scrap Recycling Industries, Inc. 1325 G Street NW, Suite 1000 Washington, D. C. 20005 3104 Phone: 202 737 1770, FAX: 202 626 0900 Home Page: www.isri.org
NADC	National Association of Demolition Contractors (1972) 16 North Franklin Street, Suite 200B Doylestown, PA 18901 3536 Phone: 215 348 4949, 800 541 2412 Home Page: Not available
NARI	National Association of the Remodeling Industry 4900 Seminary Road, Suite 320 Alexandria, VA 22311 Phone: 703 575 1100, FAX: 703 575 1121 Home Page: www.nari.org/
NFFS	Non-Ferrous Founders Society 1480 Renaissance Drive, Suite 310 Park Ridge, IL 60068 Phone: Not available Home Page: www.nffs.org
NSWMA	National Solid Waste Management Association 4301 Connecticut Avenue, NW, Suite 300 Washington, D. C. 20008 Phone: 202 244 4700, FAX: 202 966 4841 Home Page: www.envasns.org/nswma

(continued)

TABLE 20.17 *(continued)*

SRI	Steel Recycling Institute 680 Crawford Drive Pittsburgh, PA 15220 2700 Phone: 800 876 7274 Home Page: www.recycle_steel.org
SWANA	Solid Waste Association of North America 1100 Wayne Avenue P. O. Box 7219 Silver Spring, MD 20907 7219 Phone: 800 467 9262, FAX: 301 589 7068 Home Page: www.swana.org/ 1 6:53 PM 06/19/00

REFERENCES

1. Heckoetter, Dr.-Ing. Ch. Recycling of Building Rubble, Aufbereitungs Technik, Nr. 8, 1987, p.443.
2. Section 22a-209 of the General Statutes of the State of Connecticut, Regulations of the Department of Environmental Protection, Concerning Standards for Solid Waste Landfill.
3. 6NYCRR Part 360, Solid Waste Management Facilities, Effective December 31, 1988, NYS DEC, Division of Solid Waste, Albany, NY.
4. 25PaCode Ch. 271, Municipal Waste Management, General Provisions, effective April 9, 1988, EQB, Harrisburg, PA.
5. www.dnr.state.wi.us/org/aw/air/reg/asbestos/asbes6.htm
6. Davidson, Thomas A. and David Gordon Wilson, U.S. Building-Demolition Wastes: Quantities and Potential for Resource Recovery, *Conservation and Recycling,* Vol. 5, No. 2/3, pp 113–132, 1982.
7. Wilson, David Gordon, P. Foley, R. Weisman, S. Frondistou-Yannas, Demolition Debris: Quantities, Composition and Possibilities for Recycling, a paper in the Proceedings of the Fifth Mineral Waste Utilization Symposium, Chicago, IL., April 13–14, 1976.
8. Paulsen, Greg, Mary Stroup-Gardiner and Jon Epps, Recycling Waste Roofing Material in Asphalt Paving Mixtures, Transportation Research Record 1115.
9. City of New Haven, A Request for Qualifications for a Front End Solid Waste System, December 20, 1990.
10. Town of North Hempstead, Nassau County, New York Final Request for Proposals—Recycling System Design,Equipment Installation and Operations.
11. Monroe County Solid Waste Management Plan/Draft Generic Environmental Impact Statement.
12. Baumeister, Theodore, Ed., *Standard Handbook for Mechanical Engineers,* 7th Ed., McGraw-Hill, Inc., New York, 1967.
13. Litke, Jim, Recovering Scrap Lumber at Curbside, *Biocycle,* Vol. 37, No. 3, pp 33–34, March 1996.
14. Brochure, Pinnellas County Artificial Reefs, Pinellas County Department of Solid Waste Management, undated.
15. Oldengott, Dr.-Ing. M., Recycling of Building Rubble—Importance and New Crusher Technology, *Aufbereitungs-Technik,* Nr. 6/1985.
16. Sri Ravindrarajah, R., Utilization of Waste Concrete for New Construction, *Conservation Recycling,* Vol. 10, No. 2–3, 1987; Recycling of Material, *Selected Papers from the Fifth Internatinal Recycling Congress,* Berlin, West Germany, pp. 69–74, October 29–31, 1986.

17. Hendriks, Ch. F., *Certification System for Aggregates Produced from Building Waste and Demolished Buildings,* pp. 821–834, Elsevier Science BV, Amsterdam, 1994.
18. Tavakoli, Mostafa and Parviz Soroushian, Drying Shrinkage Behavior of Recycled Aggregate Concrete, *Concrete International,* Vol. 18, No. 11, pp. 58–61, Nov. 1996.
19. Kagawa, A., *Recycling of Concrete Waste Material from Residential Apartment Buildings,* Konkurito Kogaku (Japan), Vol. 32, No. 4, pp. 5–14, April , 1994.
20. Falk, Robert H., David Green, Scott C. Lantz, and Michael R. Fix, Recycled Lumber and Timber, *Restructuring: America and Beyond Structures Congress—Proceedings,* Vol. 1 1995, pp. 1065–1068, ASCE, New York, 1995.
21. Anon., Sorting C&D into Wood Products, *Biocycle,* Vol. 37, No. 11, pp. 39–40, 42, November 1996.
22. Stein, V., Recycling of Demolition Waste and Its Influence on the Market of Natural Mineral Building Materials, *Conservation and Recycling,* Vol. 10, No. 2/3, pp. 53–57, 1987.
23. Sanzaro, Lou, Cashing In on C&D: Diverse and Conquer, *World Wastes,* Vol. 40, No. 8, pp. 38–43, August 1997.
24. Brochure, Sorting Construction-Demolition Waste Ravensburg Project, Maschinenfabrik Bezner GmbH & Co. KG, undated.
25. Matsumoto, S., Y. Maruko, M. Harada, Recycle Systems, *Fuji Jiho* (Fuji Electric Journal), Vol. 70, No. 3, pp. 21–24, March 10, 1997.
26. Brochure, Jones, George M., PE, In Situ Recycling of Bituminous Pavements, undated.
27. www.enveng.ufl.edu/homepp/townsend/Research/DemoHW/Guide/Dmgdintr.htm
28. Lettner, H. and F. Steinhausler, Radon Exhalation of Waste Gypsum Recycled as Building Material, *Radiation Protection Dosimetry,* Vol. 24, No. 1/4, pp.415–417 (1988).

CHAPTER 21
HOUSEHOLD HAZARDOUS WASTES

BUFF WINN, P.E.
Project Manager
The IT Group
Portland, Oregon
and
Emcon Associates
San Jose, California

INTRODUCTION

Household hazardous waste (HHW) arose as an issue in waste management in the 1980s. During that decade, various communities across the United States addressed the issue by developing plans, programs, and facilities to keep HHW out of the solid waste stream and to provide options for properly managing these wastes. In a relatively short period of time, alternative HHW programs have become a mainstream waste management practice.

Concern over HHW arose as the presence of hazardous waste in the municipal solid waste (MSW) stream became a well-recognized matter. Recognition of the matter has been brought about by a variety of factors, including (1) the results of environmental monitoring at solid waste landfills (some of which did not receive commercial or industrial wastes), (2) the continued lowering of detection limits of laboratory instrumentation, (3) the results of solid waste characterization studies, (4) observations made in load-checking programs at solid waste landfills and transfer stations, and (5) hazardous incidents involving refuse collection workers and equipment. The increased sorting and handling of wastes in the growing number of recycling operations is further placing the MSW stream under observation.

Without HHW management alternatives, most HHW is improperly disposed of in the refuse, down the drain, or in the soil. Furthermore, many HHWs are simply stored for long periods of time. These storage and disposal practices can result in various health, safety, environmental, and legal problems. This chapter discusses HHW issues and programs for properly managing HHW.

HAZARDOUS WASTE IN MUNICIPAL SOLID WASTE

Definition of Household Hazardous Waste

Hazardous wastes are defined under federal law as discarded materials that are not specifically excluded from regulation as a hazardous waste and that either (1) exhibit the characteristics of ignitability, corrosivity, reactivity, or toxicity or (2) are specifically listed as hazardous wastes. [Details of the federal definition are given in the Code of Federal Regulations (CFR), Title 40, Part 261.[1]] Of particular note here is that household wastes are one category of materials that is specifically excluded from regulation as hazardous waste under federal law, i.e., the Resource Conservation and Recovery Act (RCRA).

States also establish their own definition of hazardous wastes; state regulations can be no less stringent than federal regulations. Some states follow federal regulations exactly, while others typically follow federal regulations closely in organization and content with some additional restrictions (e.g., additional listed wastes, more rigorous extraction procedures for determining toxicity, reduced exclusions from regulation as a hazardous waste).

HHWs are discarded materials from residences that meet the criteria of hazardous waste. Although exempted from regulation by federal law and most state laws, HHWs contain the same chemicals as may be found in industrial hazardous wastes. This regulatory ambiguity has led to much of the confusion and inaction in managing HHW. However, there is widespread agreement in the waste management field that HHW should be kept out of the solid waste stream even where legally it is not a hazardous waste. It should also be noted that some states do not exempt HHW from regulation as hazardous waste.

Typical Wastes

Examples of HHW are given in Table 21.1. The table is organized alphabetically according to common material name and gives the typical hazard class of the material. In some cases, more than one common name for the same material is listed to aid in using the table.

It should be noted that product constituents may vary among manufacturers of similar products. For example, some drain cleaners are bases, some are acids, and some are noncorrosive, chlorinated solvents. Product constituents may also vary with time for the same manufacturer. Some pesticides have changed ingredients several times while retaining the same product name. Therefore, a qualified chemist should be responsible for actual determination of a material's hazard class.

Examples of hazardous materials in commercial products are given in Table 21.2. In some cases, a chemical is present as an ingredient in the product; that is, the chemical is intentionally used in the product for specific purposes. In other cases, the chemical may be present as a "contaminant" because it is. used as a precursor in the synthesis of another chemical in the product or because it may be a by-product in the synthesis of another chemical in the product. It is often very difficult to obtain a pure chemical compound because of the reaction kinetics in its formation.

For example, methylene chloride is commonly used as a key ingredient in paint removers, whereas benzene may be present in trace amounts (above the detection limit of laboratory instrumentation) because of the reaction kinetics in the formation of another product ingredient such as toluene.

Some chemicals are now under various degrees of restricted use in commercial products, for example, benzene and trichloroethylene. However, it is very common for some household products to be stored in garages and basements for over 20 years. Such storage practices have been repeatedly observed in HHW collection programs. Therefore, materials entering the waste stream may contain chemical constituents that were in use 20 or more years ago.

TABLE 21.1 Common Household Hazardous Wastes

Material	Typical hazard class*	Material	Typical hazard class
Acetone	Flammable liquid	Methyl ethyl ketone	Flammable liquid
Aerosols	Flammable gas, nonflammable gas	Mineral spirits	Flammable liquid
Alcohols	Flammable liquid	Moth balls	ORM-A
Ammonia (NH4OH; 12% < NH4 < 44%)	Corrosive (base)	Muriatic acid (hydrochloric acid)	Corrosive (acid)
Ammonia (NH4OH; NH4 < 12%)	ORM-A	Nail polish	Flammable liquid
Ammunition (small arms)	Explosive C	Nail polish remover	Flammable liquid
Antifreeze	Poison B†	Naphtha	Flammable liquid
Batteries—automotive	Corrosive (acid)	Naphthalene	ORM-A
Bleach (sodium hypochlorite: Cl < 7%)	ORM-B	Naval jelly	Corrosive (acid)
Bleach (sodium hypochlorite: Cl > 7%)	Corrosive (acid)	Nitric acid (<40%)	Corrosive (acid)
Brake fluid	Flammable liquid	Nitric acid (>40%)	Oxidizer, corrosive (acid)
Butane	Flammable gas	Oil—lubricating	Combustible liquid
Camphor oil	Combustible liquid	Oil—motor	Combustible liquid
Carbon tetrachloride	ORM-A	Oven cleaner	Corrosive (base)
Chlorine (pool)	Oxidizer	Paint—oil-based	Flammable liquid
Chloroform	ORM-A	Paint—water-based	‡
Contact cement	Flammable liquid	Paint remover	ORM-A
Degreasers	ORM-A	Paint thinner	Flammable liquid
Diesel fuel	Combustible liquid	Pesticide	Poison B, flam. liquid
Drain cleaner	Corrosive (base)	Phosphoric acid	Corrosive (acid)
	Corrosive (acid)	Polyurethane coatings	Flammable liquid
	ORM-A	Pool acid	Corrosive (acid)
Fireworks	Explosive C	Propane	Flammable gas
Flare	Explosive C	Rubber cement	Flammable liquid
Floor polish	Flammable liquid	Rug cleaner	ORM-A
Fuel oil	Combustible liquid	Shellac	Flammable liquid
Fungicides	Poison B	Shoe wax	Flammable solid
Furniture polish	Flammable liquid	Silver nitrate	Oxidizer
Gasoline	Flammable liquid	Spot remover	ORM-A
Glue—epoxy	Flammable liquid	Sterno	Flammable solid
Glue—model airplane	Flammable liquid	Strychnine	Poison B
Hydrochloric acid	Corrosive (acid)	Sulfuric acid	Corrosive (acid)
Hydrogen peroxide	Oxidizer	Toilet bowl cleaner	Corrosive (acid)
Ink	Flammable liquid	Transmission fluid	Flammable liquid
Insecticides	Poison B, flammable liquid	TSP (trisodium phosphate)	Corrosive (base)
Kerosene	Flammable liquid	Turpentine	Flammable liquid
Lacquer	Flammable liquid	Upholstery cleaner	ORM-A
Lighter fluid (charcoal lighter)	Flammable liquid	Varnish	Flammable liquid
Lime (calcium hydroxide)	ORM-B	Warfarin	Poison B
Linseed oil	Flammable liquid	Weed killer	Poison B, flam. liquid
Lye (sodium hydroxide)	Corrosive (base)	White gas ,	Flammable liquid
Mercury (metallic)	ORM-B	Wood preservative	Poison B
Methylene chloride	ORM-A	Wood stain	Flammable liquid

*Typical hazard class is given; however, product constituents may vary. Hazard class for a particular material should be determined by a qualified chemist. In some cases, more than one hazard class may apply.

†Not a hazardous waste under federal law; however, some states regulate as a hazardous waste.

‡Water-based paint disposal regulations vary by state. Water-based paints are generally nonhazardous. However, water-based paints may be hazardous if they contain elevated concentrations of heavy metals, particularly mercury.

TABLE 21.2 Hazardous Materials in Commercial Products

Chemical	Potential products containing chemical	Chemical	Potential products Containing chemical
Benzene	Dry cleaning fluids, fumigants, gasoline, insecticides, motor oil, paint brush cleaner, paint remover, rubber cement, solvents (various), spot remover	Tetrachloroethylene	Degreasers, dry cleaning fluids,drying agents, heat transfer medium, paint remover, spot removers, vermifuges
Carbon Tetrachloride	Degreasers, dry cleaning fluids, drying agents, fire extinguishers, fumigants, laquers, propellants, refrigerants, solvents (various)	Toluene	Adhesives, dry cleaning fluids, dyes, gasoline, motor oil, paint, paint remover, perfumes, pharmaceuticals, solvents (various), spot removers, wood putty
Chloroform	Anaesthetics, fluorocarbon regfrigerants, fumigants, insecticides, laquers, pharmaceuticals, solvents(various)	1,1,l-Trichloroethane	Aerosol propellant, degreasers, drain opener, furniture polish, oven cleaner, paint remover, pesticides, rug cleaner, septic tank cleaner, shoe dye, shoe polish, solvents (various), spot removers, upholstery cleaner
1,2-Dichloroethane	Degreasers, finish removers, fumigants, gasoline, paint remover, penetrating agents, scouring compounds, soaps, solvents (various), wetting agents	Trichloroethylene	Adhesives, degreasers, dry cleaning fluids, dyes, fumigants, fur cleaner, paint, pharmaceuticals, shoe cleaner, shoe polish, solvents (various)
Ethylene Dibromide	Fire extinguishers, fumigants, gasoline, solvents (various), waterproofing preparations	Vinyl Chloride	Adhesives for plastics, intermediate in polymer production
Methylene Chloride	Aerosol propellant, degreasers, dewaxers, fumigants, furniture refinishers, hair spray, oven cleaner, paint, paint brush cleaner, paint remover, septic tank cleaner, shoe cleaner, shoe polish, solvents (various), spot removers, wood putty	Xylene	Caulking compounds, dyes, gasoline, insecticides, motor oil, paint, paint remover, rubber cement, shoe dye, solvents (various)

TABLE 21.3 Quantity of Hazardous Waste in Municipal Solid Waste*

Location	Date	Percent hazardous waste				Reference
		Residential	Com./ind.†	Self-haul	Total	
Mission Canyon Landfill (Los Angeles County, Calif.)	1979	—	—	—	0.13	2
Puente Hills Landfill (Los Angeles County, Calif.)	1981	0.0045	0.24	—	0.15	2
Marin County, Calif.	1986	0.40	—	—	—	3
San Mateo County, Calif.	1987	0.29	—	0.59	—	4
Portland, Oreg.	1987	0.01	0.21	0.05	0.09	5
Santa Cruz, Calif.	1988	—	—	0.39	0.36	6
Berkeley, Calif.	1988–89	0.20	0.60	—	0.40	7
Sacramento County, Calif.	1989	0.27	0.20	0.29	0.25	8
San Antonio, Tex.	1989–90	0.40	0.50	0.30	0.34	9
Burbank, Calif.	1990	0.40	1.09	—	—	10
Sunnyvale, Calif.	1990	0.09	0.45	3.09	0.83	11
Palo Alto, Calif.	1990–91	0.37	0.08	—	—	12
Tulare County, Calif.	1991	0.47	0.83	0.04	0.38	13
Del Norte County; Calif.	1991	0.83	0.09	—	—	14
Stockton, Calif.	1991	0.28	0.01	1.22	0.30	15

*Disposed waste stream.
†Commercial/industrial.

Quantities in the Waste Stream

A number of waste characterization studies have provided estimates of the quantity of hazardous waste in MSW. Table 21.3 summarizes the results of various studies. The table shows a breakdown according to source (i.e., residential collection vehicles, commercial collection vehicles, and self-haul) as well as for the total waste stream.

The results vary considerably. Measurements of the amount of hazardous waste in the residential waste stream range from 0.0045 to about 1 percent. Estimates for hazardous waste in the total waste stream range from approximately 0.1 to 1 percent.

Composition of HHW

A breakdown of the composition of HHW is presented in Table 21.4. The based on operating experience at periodic collection programs and per collection facilities. Unfortunately, different operators report such data use different component categories. An effort was made to standardize the categories as much as possible in the table. By far the most common category of paint.

PROBLEMS ENCOUNTERED

Worker Injuries

Improper management of HHW can result in injuries to workers in waste management operations. Such workers include refuse collectors, material recovery facility sorters, and

TABLE 21.4 Composition of Household Hazardous Waste[16–20]

Component	Alameda County, Calif. (1987)	San Francisco, Calif. (1988–89)	Ontario, Canada (1989)	Milpitas, Calif. (1990)	Palm Beach County, Fla. (1990–91)
Latex paint	30.8	8.4	30.0	12.9	18.5
Oil-based paint	32.8	25.0	26.0	17.0	24.2
Waste oils	12.3	26.6	10.0	20.2	22.7
Misc. flammables*	3.1	20.9	14.0	10.8	14.4
Poisons	5.0	4.2	7.0	3.1	2.9
Corrosives	3.3	7.2	10.0	4.2	3.4
Oxidizers	0.5	1.3	1.0	0.3	0.3
Aerosols	2.3	3.4	1.0	2.7	1.4
Batteries	0.0	1.0	0.0	11.6	9.6
Other†	9.9	2.0	1.0	17.2	2.6
Total	100.0	100.0	100.0	100.0	100.0

*Includes miscellaneous flammable liquids and solids.
†Reported material categories vary among programs.

equipment operators at material recovery facilities, transfer stations, waste-to-energy facilities, and landfills. As recycling operations expand in solid waste management, there will be increased sorting and handling of waste materials; correspondingly, workers will have increased exposure to hazardous wastes if present in the waste stream.

The National Solid Waste Management Association has documented a variety of worker injuries due to disposal of hazardous waste in the solid waste stream.[21] Exposure to hazardous waste has resulted from spills, spraying (e.g., from packer trucks during compaction), touching, fumes, fires, and explosions. Injuries have included burns (acid, caustic, and thermal), blinding, eye irritation, respiratory problems, rashes, nausea, and unknown chronic problems.

Equipment and Property Damage

Hazardous wastes disposed of in refuse also can cause equipment and property damage. The most common incidents involve fires in refuse collection trucks or transfer trailers. Virtually every company or municipal agency that collects refuse has experienced a vehicle fire due to improperly disposed hazardous wastes. Fires are typically the consequence of either flammables coming into contact with an ignition source or incompatible materials mixing and reacting. Sometimes the materials come from different sources (e.g., brake fluid from one home and pool chlorine from another). Compaction vehicles tend to liberate materials from containers, thereby contributing to hazardous incidents, but hazardous incidents also occur in loose loads. In addition, landfill equipment such as dozers and compactors have been damaged as the result of improperly disposed hazardous wastes.

Waste processing facilities have also experienced damage due to hazardous wastes disposed of in the waste stream. The most serious incidents reported have been shredder explosions. Most of the problems to date have occurred in refuse-derived fuel (RDF) processing plants, but the same hazard exists for wood and yard waste processing operations and MSW composting operations.

Environmental Contamination

The potential impacts of hazardous wastes in MSW landfills can be evaluated by examining the leachate and landfill gas generated at these sites. For purposes of comparison, 10 organic constituents are examined here: (1) benzene, (2) carbon tetrachloride, (3) chloroform, (4) 1,2-dichloroethane, (5) ethylene dibromide, (6) methylene chloride, (7) tetrachloroethylene, (8) 1,1,1-trichloroethane, (9) trichloroethylene, and (10) vinyl chloride. The 10 compounds were selected because of the extensive data available on them. Each of these compounds is also examined in Table 21.2.

The U.S. Environmental Protection Agency (EPA) has correlated leachate data from 53 landfills.[22] The concentrations of various hazardous constituents are shown in Table 21.5. All 10 chemicals were detected in MSW leachate. Methylene chloride was found at the highest concentrations (220,000 parts per billion) and at the greatest number of sites (60 percent).

Analysis of landfill gas also provides an indication of constituents in landfills. The results of extensive testing of landfill gas by the California Air Resources Board are presented in Table 21.6.[23] The table summarizes results from 288 sites—271 nonhazardous waste sites and 17 hazardous waste sites. The terms "hazardous" and "nonhazardous" were used in the study to distinguish between sites that are known to have accepted hazardous waste and sites that are not known to have accepted hazardous waste.

The analyses were conducted on samples drawn from wells within the landfill. All of the 10 specified chemicals were detected in landfill gas. The lowest value observed is approximately equal to the detection limit for that compound. Benzene, methylene chloride, tetrachloroethylene, and trichloroethylene were found in more than half of the nonhazardous waste landfills; 1,1,1-trichloroethane and vinyl chloride were found in just under half of the nonhazardous waste sites.

Some interesting comparisons may be noted between the leachate and landfill gas analyses. Methylene chloride was found at the highest concentrations in leachate as well as in landfill gas at nonhazardous waste sites. Ethylene dibromide and carbon tetrachloride were found at the lowest percentages of sites in both the leachate and landfill gas analyses.

The California Air Resources Board concluded that the "overall composition of landfill gases from hazardous and nonhazardous sites appear to be similar, with no major distinguishing characteristics which would indicate from what type of landfill the sample

TABLE 21.5 Concentrations of Organic Constituents in Leachate from MSW Landfills[22]

Chemical	Concentration range, ppb	Percent of sites where detected
Benzene	4–1,080	34
Carbon tetrachloride	6–398	4
Chloroform	27–31	15
1,2-Dichloroethane	1–11,000	11
Ethylene dibromide	5–5	2
Methylene chloride	2–220,000	60
Tetrachloroethylene	2–620	21
1,1,1-Trichloroethane	1–13,000	25
Trichloroethylene	1–1,300	32
Vinyl chloride	8–61	11

TABLE 21.6 Concentrations of Organic Constituents in Gas from MSW Landfills[23]

	Nonhazardous waste sites*		Hazardous waste sites†	
Chemical	Concentration range, ppbv‡	Percent of sites where detected	Concentration range, ppbv	Percent of sites where detected
Benzene	500–29,000	51	500–791,000	82
Carbon tetrachloride	5–2,100	8	< 5	0
Chloroform	2–171,000	27	2–200	29
1,2-Dichloroethane	20–34,100	18	20–12,000	24
Ethylene dibromide	1–2,000	7	1–55	12
Methylene chloride	60–260,000	56	60–42,000	59
Tetrachloroethylene	10–62,000	72	10–10,000	88
1,1,1-Trichloroethane	10–21,000	49	10–14,000	59
Trichloroethylene	10–20,000	68	10–5,700	82
Vinyl chloride	500–120,000	47	500–60,000	53

*271 sites.
†17 sites.
‡ppbv = parts per billion by volume

was obtained." The Board goes on to state, "The data show that, based on landfill gas testing, there is hazardous waste in 86 percent of the landfills tested, regardless of what type of waste the site is known to have accepted. The landfill gas testing results show that nonhazardous landfills may contain concentrations of toxic gases equal to or exceeding those of hazardous landfills."

HHW MANAGEMENT PRACTICES

Source Control

The most important aspect of HHW management is source control. The objectives of source control are to reduce the amount of HHW generated and to prevent improper disposal of those wastes that are generated. Source control is thus aimed at preventing problems before they happen. Two key elements of source control are public education and prohibited waste control programs at waste management facilities (i.e., transfer stations, material recovery facilities, waste-to-energy facilities, landfills).

Public education is vital in providing the community with information about HHW management alternatives. The public must be informed about what types of household materials are hazardous, why they are hazardous, how to use nonhazardous materials in place of products that are hazardous, and how to properly dispose of hazardous wastes that are generated. A public education flier is illustrated in Fig. 21.1. As the public becomes more aware of HHW issues, progress can be expected in reducing the generation of these wastes. Public education is discussed further later in this chapter.

Prohibited waste control programs are aimed at preventing prohibited wastes, such as hazardous and other specific prohibited wastes, from entering a waste management facility. Other prohibited wastes depend on facility permit restrictions and may include latex paint, liquid wastes, sludges, ash, asbestos, dead animals, infectious wastes, and various additional materials. Prohibited waste control programs are sometimes referred to as load-checking programs or as hazardous waste exclusion programs, although they usually include control of certain nonhazardous wastes as well.

Prohibited waste control programs have been instrumental in forcing the issue of providing disposal options for HHWs. These programs result in the rejection of HHW at waste management facilities and also at the curb. At the same time, they educate residents about the hazardous nature of certain wastes they produce.

There are six major components of prohibited waste control programs:

- Customer notification
- Personnel training
- Waste characterization
- Waste inspection
- Record keeping
- Management of wastes identified

Customer notification consists of (1) informing refuse haulers, residents, businesses, and local agencies that the facility does not accept the prohibited wastes specified and that a load-checking program is in place at the facility (e.g., periodic mailed notices); (2) posting signs at prominent locations around the facility, including the site entrance, gate-house, and tipping areas (Fig. 21.2); (3) placing decals on waste containers; (4) responding to customer inquiries concerning waste acceptance policies; and (5) providing public education about proper waste management alternatives for the prohibited wastes.

The effectiveness of the prohibited waste control program depends in large part on the capabilities of the facility staff. Therefore, it is very important to thoroughly train facility personnel, including management, gate-house attendants, equipment operators, traffic coordinators (spotters), recycling sorters, and load-checking staff. In addition, refuse collection workers play an important role in identifying and preventing problems at the source of waste generation. Many refuse collection companies instruct workers to remove HHW from garbage cans and leave it at the residence or to simply leave the entire can as is. Personnel should be trained in (1) identifying prohibited wastes, (2) understanding the effects of these wastes on human health and the environment, and (3) proper handling and response procedures for their job category.

To judge whether a waste can be accepted at the facility, procedures must be established for characterizing the waste. The first level of information is often provided by the customers themselves; they generally know what their wastes consist of or at least what they were used for. The next level involves physical assessment by facility personnel, e.g., examination of product labels, detection of odors, and observation of unusual materials or containers. If the waste cannot be identified by the above procedures, field chemistry methods may be employed, e.g., pH measurement or flammability tests. In some cases, additional assessment may be needed, such as laboratory analysis.

The key component of prohibited waste control programs is often a random load-checking program. The load-checking program serves to detect and deter the disposal of prohibited wastes at the facility. A typical load-checking program consists of randomly diverting a specified number of loads per week for detailed inspection. The loads are spread and an inspector carefully examines the contents for prohibited wastes. Residential self-haul loads are a common source of HHWs. In addition to the load-checking inspections, all site personnel have the responsibility in the conduct of their work for observing prohibited wastes delivered to the facility.

Record keeping provides documentation of the prohibited waste control program. Records should be maintained for (1) load-checking reports, (2) event reports (i.e., emer-

Learn to Use These Safe Substitutes as Alternatives to Toxic Household Products

Product	Alternatives and preventative methods
HOUSEHOLD CLEANERS	
AEROSOL SPRAYS	Choose non-containers, such as pump spray, roll-on, or squeeze types.
ALL-PURPOSE CLEANERS	Mix 1 quart warm water with 1 tsp. borax, TSP, or liquid soap. Add squeeze of lemon or splash of vinegar. Never mix ammonia with chlorine bleach.
AR FRESHENERS	Open windows and doors and use fans to ventilate. Place box of baking soda in closets and refrigerator. Simmer cloves and cinnamon in boiling water. Houseplants help clean the air and herb sachets provide a pleasant smell.
CHLORINE BLEACH	Use borax or baking soda to whiten. Borax is a good grease-cutter and disinfectant. If you use bleach, choose the non-chlorine, dry bleach. Never mix chlorine bleach with ammonia or acid-type cleaners.
DEODORIZERS	For carpets, mix 1 part borax to 2 parts cornmeal, sprinkle on liberally, and vacuum up after 1 hour, For kitty litter, sprinkle baking soda in bottom of box before adding litter.
DISINFECTANTS	Use ½ cup borax in 1 gallon of hot water. To inhibit mold or mildew, do not rinse off the borax solution.
DRAIN OPENER	To prevent clogging, use drain strainer on every drain, Pour boiling water down the drain once a week. To unclog, use rubber plunger or metal snake.
FLOOR CLEANER	For vinyl floors, mix ½ cup white vinegar or ¼ cup TSP with 1 gallon of warm water, Polish with club soda. For wood floors, mix ¼ cup oil soap with 1 gallon of warm water.
FURNITURE POLISH	Dissolve 1 tsp, lemon oil in 2 pints mineral oil, Or use oil soap to clean and a soft cloth to polish, Rub toothpaste on wood furniture to remove water stains.
GLASS CLEANER	Mix ¼ cup white vinegar in 1 quart warm waler, apply to glass, and rub dry with newspaper.
MILDEW CLEANER	Scrub mildew spots with baking soda or sponge with white vinegar. For shower curtain, wash with ½ cup soup and ½ cup baking soda, adding 1 cup white vinegar to rinse cycle.
OVEN CLEANER	Mix 3 tbsp, of washing soda with one quart warm water, Spray on, wait 20 minutes, then clean, For tough stains, scrub with very fine steel wool pads (0000) and baking soda.
RUG AND UPHOLSTERY CLEANER	Use non-aerosol, soap-based cleaner.
SCOURING POWDER	Use brand that does not contain chlorine, or better yet, use baking soda.
SPOT REMOVER	Dissolve ¼ cup borax in 3 cups of cold water. Sponge it on and let dry, or soak fabric in the solution prior to washing it in soap and cold water, Use professional dry cleaner for stubborn stains.
TUB AND TILE CLEANER	Use scouring powder or baking soda.
PAINT PRODUCTS	
PAINT AND STAINS	Latex or other water-based paints are the best choice. Enamel paint, stain, and varnish are available in water-based forms. Clean-up does not require paint thinner.
PAINTS FOR ARTISTS	Use with good ventilation. Never put brush in mouth. Powdered paint is hazardous if inhaled, so wear protective gear or use pre-mixed paints.
PAINT REMOVER	Use heat gun and scraper to remove paint, wearing proper protective gear Strong alkali-type paint removers are available. A strong TSP solution (1

	pound to 1 gallon hot water) may do the job. Brush on, wait 30 minutes, then scrape off.
PAINT THINNER AND SOLVENTS	Hold brush cleaner in closed jar until paint particles settle to bottom. Pour off clean liquid and reuse. Save paint sludge for collection.
WOOD FINISHES	Shellac, tung oil, and linseed oil are finishes derived from natural sources. Shellac is diluted with an alcohol solvent, whereas the oils are diluted with turpentine for better application, Use with proper protection and ventilation.
WOOD PRESERVATIVES	Avoid using ones that contain pentachlorophenol, creosote, or arsenic. When possible, use decay-resistant wood, e.g., cedar or redwood.
PESTICIDES & FERTILIZERS	
CHEMICAL FERTILIZERS	Compost, which can be made in your own backyard from grass clippings, food scraps, and manure, is the best soil amendment. Other organic soil amendments include manure, seaweed, peat moss, and blood, fish, and bone meal.
FUNGICIDES	Remove dead or diseased leaves and branches. Sulfur dust, sulfur spray, and dormant oil spray (which dues not contain copper) are the least-toxic products to treat plant diseases.
HERBICIDES (weed killers)	Pull or hoe weeds prior to weeds going to seed. Use mulching (alfalfa hay is a good mulch) to keep weeds down in garden area.
INSECTICIDES (indoor)	Good sanitation in food prep and eating areas will prevent pests, while weather-stripping and caulking will seal them out. For crawling insects, use boric acid or silica aerogel in cracks and crevices. To keep out flies, keep door and window screens in good repair. Use fly swatter and sticky flypaper.
INSECTICIDES (garden)	Hose off plants with water using a jet spray nozzle. Use beneficial insects, e.g., lady beetle and praying mantis. When only a few bugs are found, spot treat with rubbing alcohol. A bacteria, B.T., is effective against caterpillars. Less-toxic sprays include insecticidal soap, pyrethrum, or a homemade garlic/red pepper spray.
INSECTICIDES (pets)	Vacuum frequently and dispose of bag afterwards. Use good flea comb and flick fleas into soapy water Dietary. supplements my be helpful, e.g., brewers yeast. Use herbal or d-limonene shampoos or dips.
MOTHBALLS	Place cedar chips, dried lavender or herb sachets in drawers or closets to discourage moths.
SNAIL AND SLUG KILLERS	Fill shallow pan with stale beer and position at ground level. Or, to capture snails during the day, overturn clay pots, leaving enough room for snails to crawl underneath. Snails also like to attach to boards. Collect and destroy.
AUTOMOTIVE PRODUCTS	
ANTIFREEZE	Small amounts may be diluted and put down a drain connected to the sewer system (not a septic tank).
BATTERIES	Old auto batteries can be exchanged when purchasing new battery or recycled at battery recyclers.
GASOLINE	For cleaning off grease, use non-toxic degreasers.
MOTOR OIL	Synthetic motor oil lasts longer than regular motor oil, thus reducing amount of oil used. Motor oil can he recycled at participating service stations.
TRANSMISSION FLUID AND BRAKE FLUID	May be mixed with waste oil and recycled at participating service stations.

FIGURE 21.1 Public education flier. (*Courtesy of San Diego Environmental Health Coalition.*)

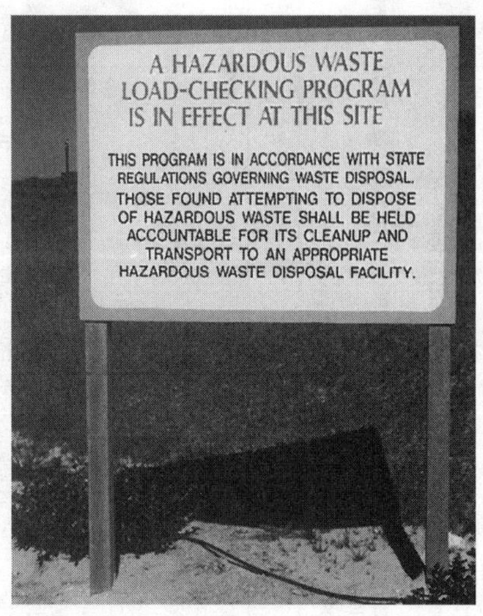

FIGURE 21.2 Prohibited waste control signs.

gencies, injuries, or special incidents), (3) training records and (4) wastes shipped from the site.

If prohibited wastes are identified, they must be properly handled, stored, and disposed of. If the source of the waste is known, the waste may be returned to the source or shipped to a hazardous waste treatment, storage, or disposal facility (TSDF) at the source's expense. If the source is not known, the facility must have established procedures for disposing of the waste and for assumption of generator liabilities. It is common for cities or counties to assume generator liabilities in the case of HHWs from unknown generators.

While source control measures help reduce the amount of HHW generated and prevent its improper disposal, the public must also be provided with options for taking HHWs that are generated. The following sections discuss these types of HHW management programs.

One-Day Collection Programs

The most common type of HHW management program to date has been the 1day collection program. Such programs have been referred to by a variety of names in different communities, including Toxic Away Days, Toxic Disposal Days, Amnesty Days, and Toxic Round-Ups. In these programs, HHWs are brought to a specified site on a specified day. The wastes are then taken from the public, usually free of charge, and properly packaged for recycling, treatment, or disposal. Participation rates in 1-day collection programs have typically been in the range of 1 to 5 percent of households in the community.

A typical layout for a 1-day collection program is illustrated in Fig. 21.3. The major functional areas of a one-day collection program are

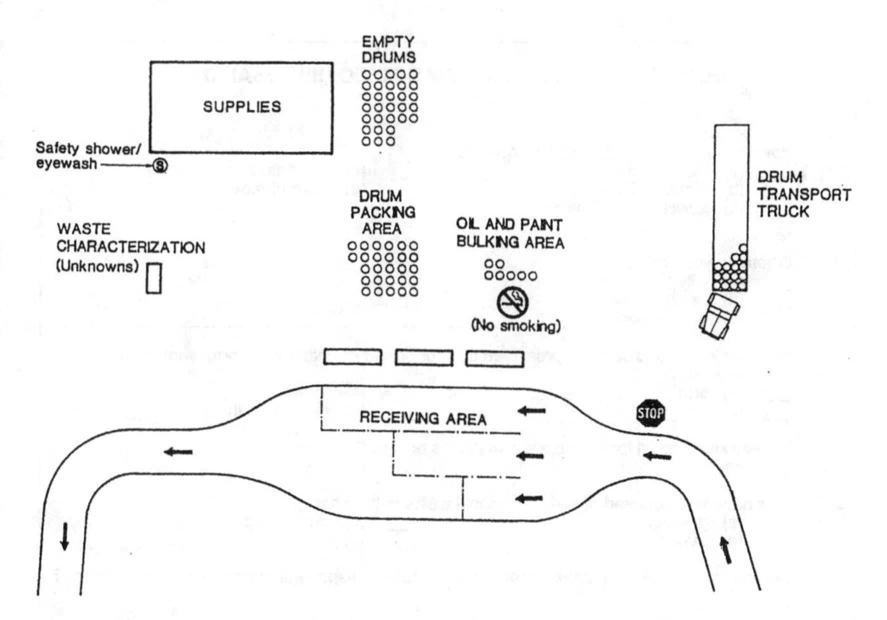

FIGURE 21.3 Layout for 1-day household hazardous waste collection program.

- Entrance area
- Receiving area
- Sorting area
- Packing area
- Storage area
- Loading area

Signs at the site entrance as well as along roads in the site vicinity direct traffic to the program. Upon arrival at the program, the public is typically given questionnaires and fliers regarding the program. A sample questionnaire is shown in Fig. 21.4. An example of a flier is given in Fig. 21.1. The questionnaires collect data about the program users, wastes delivered to the site, previous disposal practices, and program needs of the community. The fliers provide public information about proper household hazardous waste management, including source reduction. The questionnaires are either collected just before entering the receiving area or at the receiving area itself.

The receiving area is preferably divided into multiple lanes to facilitate traffic flow and waste removal. Traffic cones and signs are used to divide the lanes. At the receiving area, people in the vehicles are met by site personnel. Only qualified site personnel should be allowed to handle the hazardous waste once vehicles reach the site. The site personnel inspect the wastes delivered and ask the users about the contents of any wastes that are not in their original containers or are not readily identifiable. (If the resident does not know what the substance is, they may at least know what it was used for, e.g., to kill weeds.) This information is conveyed to the sorting personnel to assist in minimizing the amount of unidentified wastes received. The site personnel then remove the wastes from

QUESTIONNAIRE FOR SALINAS TOXIC DISPOSAL DAY

1. How did you hear about Toxic Disposal Day?
   ___ (a) flyer at work ___ (d) TV/radio announcement
   ___ (b) direct mail piece ___ (e) word of mouth
   ___ (c) newspaper ad/article

2. Where do you live?
   ___ (a) City of Salinas
   ___ (b) North County
   ___ (c) other (specify) ________________________________.

3. How many households are represented by your delivery of hazardous household waste materials?
   ___ (a) one ___ (c) three
   ___ (b) two ___ (d) more than three _______.

4. Did you know about toxic household materials previously?
   ___ yes ___ no

5. How have you disposed of toxic household waste in the past?
   ___ (a) garbage ___ (c) dump on ground
   ___ (b) sewer ___ (d) other ________________.

6. Did you know before this project that it was not safe to dispose of toxic waste in the garbage?
   ___ yes ___ no

7. Do you feel this is a worthwhile service for the community?
   ___ yes ___ no

   Why? ________________________________.

8. How often do you think this service should be available?
   ___ (a) weekly ___ (c) twice a year
   ___ (b) monthly ___ (d) once a year

9. Would you be willing to pay a fee (50 cents or $1) for this service?
   ___ yes ___ no

10. Comments:
   ____
   ____
   ____
   ____

FIGURE 21.4 Questionnaire for household hazardous waste collection program.

the vehicles and take them to the sorting area. Rollable carts are helpful for moving the wastes around the site. After the wastes are removed from the vehicles, the vehicles exit by the designated route. Special parking areas should be available for vehicles requiring extra attention (e.g., nonhousehold wastes, unidentified wastes, spills).

At the sorting area, trained personnel sort the wastes according to Department of Transportation (DOT) hazard class (and specific disposal site categories if necessary). The sorted wastes are either lab-packed or bulked into drums for transporting to recycling, treatment, or disposal facilities. The drums are appropriately labeled and manifested.

The drums are then loaded onto a hazardous waste transport truck and removed from the site within a relatively short period of time—ranging from one to several days in accordance with prevailing regulations.

In some communities, 1-day collection programs have evolved into periodic collection programs, in which the procedures described above are repeated on a regular basis one or more times per year. Some of the shortcomings encountered in 1-day collection programs include (1) long lines and waiting times for users of the program, (2) lack of scheduling convenience, (3) high costs per day of operation, and (4) limited recycling of wastes. However, these programs have been very successful in diverting hazardous wastes from the solid waste stream and in educating the public. One-day programs allow communities to offer an HHW management program without large capital investments, while retaining the flexibility to develop a more comprehensive program in the future.

Permanent Facilities

In efforts to develop long-term solutions to the management of HHW, some communities have developed permanent HHW facilities. Such facilities typically consist of a building constructed for the processing and storage of HHWs delivered to the site by residents. Wastes are shipped from the facility to recycling, treatment, or disposal facilities when full truck loads are accumulated or when storage time limitations are reached. The facility essentially functions as a hazardous waste transfer station (restricted to HHW unless otherwise permitted).

Permanent facilities offer residents the convenience of year-round disposal services. Residents do not have to store up wastes for the next 1-day collection event or sit in long lines to deliver their wastes. One of the most common occasions for HHW disposal is upon moving, and seldom does a resident's move coincide with a 1-day collection event. From the operator's perspective, the distributed waste flow (as compared to 1-day programs) results in a more controlled and manageable operation.

Another major benefit of permanent HHW facilities is increased recycling and treatment of hazardous wastes with a corresponding decrease in land disposal. Some permanent facilities report less than 10 percent of wastes being sent for landfilling. In contrast, 1-day collection programs often involve lab-packing the majority of wastes received and sending such wastes to landfills. In comparison to 1-day programs, recycling is enhanced at permanent facilities as a result of (1) facility features for bulking and treating wastes, (2) protection provided by the building against weather, (3) more evenly distributed waste flow rates, (4) increased processing time available, and (5) increased testing capabilities to verify a waste container's constituents.

A layout for a permanent HHW facility is presented in Fig. 21.5. The major functional areas of permanent HHW facilities are

- Entrance area
- Receiving area
- Sorting area
- Waste characterization area (laboratory)
- Treatment and/or bulking area
- Packing area
- Waste storage area
- Supplies storage area
- Loading area
- Office area
- Waste exchange area
- PPE changing area
- Restrooms
- Lunchroom

The flow of vehicles and wastes at a permanent facility generally follow that described for the 1-day collection programs. The major differences include lower daily traffic and

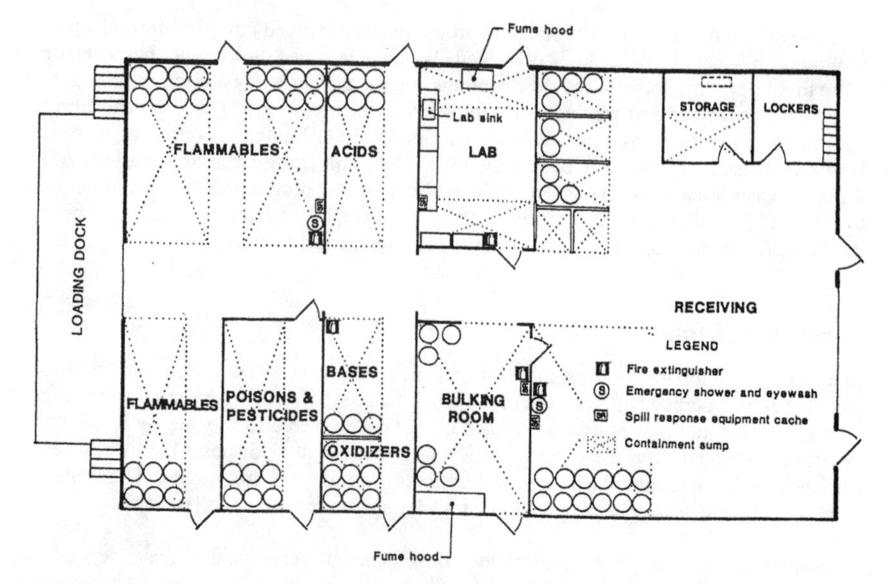

FIGURE 21.5 Layout for permanent household hazardous waste facility.

waste volumes, increased waste characterization prior to packing, and additional bulking and treatment operations.

The design and construction of hazardous waste storage facilities is governed by a set of overlapping regulations, including building codes, fire codes, Subtitle C of RCRA, state hazardous waste regulations, and local ordinances on the storage of hazardous materials. Some of the key building requirements pertain to secondary containment, separation of incompatible materials, fire suppression, ventilation, electrical systems, and building materials. These requirements result in the costs of HHW facility design and construction being relatively high.

A permanent HHW facility can function in an HHW management system in a variety of roles, including (1) a single facility serving the community, (2) a core facility with satellite stations located around the community to provide greater service and convenience, and (3) a core facility for a mobile collection program. HHW facilities can also provide storage for hazardous wastes found in load-checking programs at solid waste management facilities. In addition, HHW facilities play a valuable role in educating the public about proper HHW management.

Portable Facilities

A lower capital-cost alternative to a permanent HHW facility is a portable storage facility. Portable storage facilities usually consist of one or more prefabricated 'storage units. An example of a portable storage facility is given in Fig. 21.6. These units are available with a variety of options. The base units typically include secondary containment sumps, chemical-resistant coatings, passive ventilation, static grounding connections, and locking doors. Options include fire suppression systems (water, dry chemical, or foam), ex-

FIGURE 21.6 Portable hazardous waste storage facility.

plosion-proof lighting an foam electrical systems, heating and/or air conditioning, forced-ventilation, explosion-relief panels, floor grates, customized interiors (walls, shelving, cylinder racks), and alarm systems.

The storage capacity of the largest units typically does not exceed sixty 55-gal drums. The smallest units may have capacities of less than 10 drums. In general, portable storage facilities have less storage capacity than permanent facilities; however, the modular nature of the storage units could allow siting multiple units to expand overall capacity.

The units can usually be moved with a forklift. Some are designed to be moved by a roll-off truck or trailer. This mobility offers advantages for certain applications. For example, portable storage facilities can serve as satellite stations for a permanent facility or for storage of hazardous wastes found in load-checking programs, at solid waste management facilities.

Portable storage facilities offer flexibility in implementing an HHW management system. Portable facilities can be installed during an interim period with a permanent facility implemented at a later date. The portable storage facility can then still be used in conjunction with the permanent facility in one of the various roles discussed above.

Limitations of portable facilities include restricted bulking capabilities and storage space. Restricted bulking capabilities reduce the amount of wastes that can be recycled. The importance of these limitations depends upon the size and characteristics of the community.

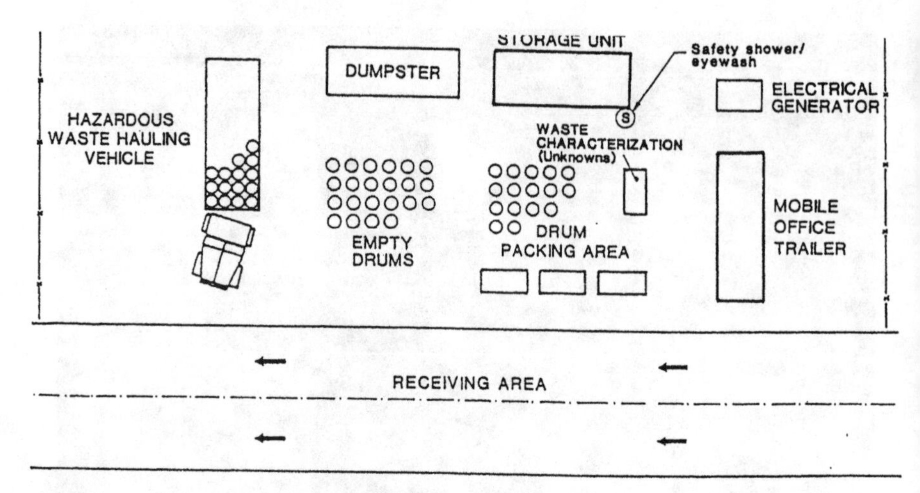

FIGURE 21.7 Layout for mobile household hazardous waste collection program.

Mobile Programs

Mobile HHW collection programs have been developed to provide more convenient service to residents. Mobile programs can be broken down into two categories: (1) door-to-door collection programs and (2) mobile collection facilities.

Door-to-door collection programs involve picking up HHWs at individual homes. The wastes must be categorized and secured for transport to a hazardous waste facility, e.g., a hazardous waste transfer station (fully permitted TSDF or HHW facility). Hazardous waste hauling permits are thus required for operation of door-to-door collection programs. These programs are typically run on an appointment basis. Such programs are not widespread at this time but can be offered as an extension of the services provided by a core HHW facility.

Mobile collection facilities combine some features of 1-day collection programs and some features of portable storage facilities. A layout for a mobile collection facility is shown in Fig. 21.7. Collection sites are set up at different locations throughout the year. The collection site may be open at a given location for a period ranging from several days to several weeks. The collection sites may make use of a portable storage facility, or wastes may be hauled to a storage facility elsewhere in the area.

TECHNICAL REQUIREMENTS FOR MANAGING HHW

Personnel Training

Well-trained personnel are crucial to the success of an HHW program. Personnel training requirements will depend on the type of HHW program implemented and on the job category of each employee—operational personnel may vary from technicians to chemists. Examples of topics to be addressed in training programs are given in Table 21.7. The op-

TABLE 21.7 Sample Training Program Topics

Topic	Items covered
Health and safety	Toxicology; health effects of hazardous materials; first aid, CPR; accident prevention-safety procedures; medical surveillance program
Use of protective equipment	Levels of protection; use of respirators (air-purifying, supplied-air, SCBA); glove alternatives and compatibilities; coverall alternatives and compatibilities
Hazardous waste regulations	40 CER (federal hazardous waste regulations); 49 CFR (federal transportation regulations); state regulations; local regulations
Chemistry of hazardous materials	Overview of chemistry basics; incompatibilities of chemicals; uses of hazardous materials in consumer products; hazardous material classification (e.g., DOT)
Waste identification procedures	Product labels; customer information; physical assessment; field chemistry; material safety data sheets (MSDS)
Facility operation and maintenance procedures	Waste receiving; waste sorting; waste packing (lab-packing, hulking, manifesting, labeling); vehicle operation; routine clean-up procedures; routine inspection procedures (daily, weekly, monthly); site security
Emergency response	Personnel responsibilities; spill containment; fire control; emergency support services; evacuation procedures; decontamination; notification requirements
Record-keeping procedures	DOT manifest; drum inventories; incident logs; facility inspection records; training records; medical records; internal reports (monthly, annual); reports to regulatory agencies

eration's safety officer has ultimate responsibility for determining the specific training needs of each individual. All applicable OSHA requirements must be complied with.

All employees should receive initial as well as refresher training. Much of the training required involves information that is not taught in academic programs, but rather is learned "on the job." Furthermore, regulations change frequently, thereby necessitating training updates.

Records should be maintained for all employee training. These records provide a good check on training needs and also document training for review by regulatory agencies.

Equipment and Materials

A list of typical equipment and materials used in HHW operations is presented in Table 21.8. The list is presented as a guideline. Specific requirements for equipment and materials should be determined for each program.

TABLE 21.8 Equipment and Materials used in HHW Operations

Category	Items
Safety equipment	Protective clothing (e.g., Tyvek, saran-coated Tyvek, polycoated splash suits) Protective gloves (e.g., neoprene, nitrile, viton, latex, polyethylene) Safety eyewear (glasses, goggles, face shields) Dust masks Boots Hard hats
Emergency response equipment	Fire extinguishers (ABC-rated) Air-purifying respirators (half- or full-facepiece) Respirator cartridges (organic vapor-acid gas cartridges, pesticide prefilters) Supplied air emergency masks (5- to 30-in) Self-contained breathing apparatus (SCBA) First aid and burn kits Emergency blankets Emergency shower and eyewash units Hose (connected to running water) Spill containment and cleanup kits (e.g., spill pillows, dikes) Noncombustible absorbent material (including dry sand) Decontamination solutions Plastic buckets Hazardous material recovery containers with lids Overpack drums Barricade tape Plastic tarps Brooms Dust pans (nonsparking) Shovels (nonsparking)
Packing materials	Drums Spare drum lids and rings Absorbent Funnel Drum liners Drum pump Drum wrenches (drum bolt, bung) DOT labels
Material handling equipment	Hand trucks Rollable carts Plastic tubs
Communication equipment	Telephone Two-way radios Alarm system
General supplies	Tables Markers, pencils, clipboards Record-keeping forms (including manifests, drum inventory forms) Reference documents Field chemistry kit Tool box Signs Traffic cones
Mobile equipment	Fork lift (with drum grabbers) Emergency response van Hazardous waste transport truck (provided by hauler)

Safety

Intrinsic to the definition of hazardous wastes is the potentially harmful nature of the materials so classified. Therefore, safety is of paramount importance in the management of these materials. To help ensure a safe operation, workers should be properly trained, wear protective equipment, and follow carefully established operating procedures.

During routine operations, workers should wear protective clothing, protective gloves, safety eyewear, and footwear appropriate to the task being performed. Table 21.8 lists various types of safety equipment. The different materials used in protective clothing and gloves exhibit different chemical compatibilities. The program's safety officer is responsible for making determinations of proper safety equipment.

Emergency Response

An emergency response plan should be developed for every HHW program. Periodic drills should be conducted at permanent HHW facilities. The emergency response plan should identify the responsible roles of site personnel (including designation of an emergency coordinator), describe response procedures for different types of emergencies, and identify emergency services to be called if necessary. Emergency telephone numbers should also be listed at the site, including fire, police, ambulance, hospitals, spill reporting agencies, and CHEMTREC (emergency information service of the chemical industry).

The emergency coordinator is responsible for evaluating the emergency situation and for directing the response or seeking assistance from emergency services. Following are general procedures for qualified personnel to respond to emergencies.

The area of the spill, fire, or explosion should be cleared of all persons not wearing protective equipment or not trained in emergency response techniques. Except for minor spills, all waste receiving should be halted and the general public kept a safe distance upwind. Emergency response personnel should don protective equipment appropriate to the incident, with the nearest qualified personnel to the incident responding first. Other site personnel should provide support from a safe distance as needed, e.g., assist in communications, deliver supplies, and monitor the situation.

In the event of a hazardous liquid spill, clean-up operations begin with spreading a noncombustible absorbent material around the perimeter of the spill. The absorbent material is then carefully placed on the liquid until absorption is complete. The materials are cleaned up by sweeping into an appropriate container.

In the event of a hazardous solid spill, the material is cleaned up by directly sweeping it into an appropriate container.

In the event of a fire or explosion, the responsibility of site personnel is that of fire control (if safe) rather than fire extinguishing, unless the incident is minor. The fire department should be notified immediately, and only trained site personnel should assist in fire control efforts while awaiting the fire department's arrival. Dry chemical fire extinguishers and/or inert materials (e.g., dry sand) may be useful in fire control. To the extent possible, other potential ignition sources should be removed from the area.

After mitigation of the emergency incident, it may be necessary to decontaminate equipment and surfaces that came into contact with the hazardous materials, depending on the material. The waste generated by the decontamination process and the protective clothing worn by emergency response personnel should be disposed of as a hazardous waste.

Finally, appropriate agencies should be notified of the emergency incident and applicable recordkeeping should be completed.

Characterization of Unknowns

One of the greatest operational difficulties in HHW programs is dealing with unknowns. At a minimum, the hazard class of each material must be determined to allow for safe and legal storage, transport, and disposal. The amount of unknowns can be reduced by collecting information about any wastes received that are not in their original containers or are not readily identifiable as described in "One-Day Collection Programs," earlier in this chapter. However, some capabilities for identifying unknowns are still required.

Field chemistry techniques, similar to those used in spill response, are well suited to characterizing unknown HHWs. An example of this approach is the HazCat method developed by the California Department of Industrial Relations. Field chemistry utilizes tests such as pH, flammability, solubility, oxidation, and reactivity to categorize unknowns into DOT hazard classes. Field chemistry kits are assembled to provide the supplies necessary to conduct the categorization tests.

Storage Requirements

Storage of hazardous wastes is subject to a variety of requirements, including

- Secondary containment
- Separation of incompatibles
- Distance to property line
- Distance to structures on site (depending on fire-wall ratings)
- Fire suppression
- Ventilation
- Electrical systems
- Building materials

Hazardous material storage criteria are established in building codes, fire codes, RCRA Subtitle C, state hazardous waste regulations, and local ordinances. Specific requirements are determined by the local building official and fire marshall; the requirements will depend upon the occupancy classification given and other discretionary rulings. The storage of flammables usually invokes the most stringent requirements with regard to HHWs.

Record Keeping

Some record keeping is regulatory mandated, while other record keeping is motivated by good operational control. An example of required records are hazardous waste manifests, which must be completed for all hazardous wastes shipped from a site. Manifest records must be maintained for 3 years. Many disposal sites also require waste profiles and drum inventories for all lab-packed wastes (i.e., an itemization of each container in the lab pack with a description of its contents). Permit conditions may also stipulate certain record-keeping and reporting requirements. Record-keeping needs are greater for storage facilities than for 1-day collection programs. Examples of records and reports to be maintained for an HHW operation include

- Hazardous waste manifests
- Drum inventories
- Employee training records

- Employee medical records
- Permits
- Operation plan
- Facility inspection records
- Incident log
- Internal reports
- Reports to regulatory agencies

RECYCLING AND TREATMENT ALTERNATIVES

Paint

As the largest component of most household hazardous waste streams, paint is a major target for recycling and/or treatment. Potential benefits include cost savings, material recovery, and diversion from land disposal. Paint is classified into two basic categories: (1) latex (water-based) and (2) oil-based. Recycling-and treatment alternatives for these categories of paint vary accordingly.

Latex paint recycling is a fairly common practice. There are various types of paint recycling' programs, including waste exchange, low-grade recycling, and high-grade recycling. Paint exchange typically involves either giving away the paint in its original container (i.e., paint in good condition and in full or nearly-full containers) or bulking (consolidating) the paint into drums for giving away or sale at a relatively low price. Low-grade recycling generally consists of bulking the paint into 55-gal drums, reprocessing the paint by a paint manufacturer into a low-grade paint, and returning the paint to the program for a fee. Because the recycled product is a low-grade paint, it is typically given away to community groups or local agencies; uses of the paint include graffiti control and general painting where specific requirements are not important. Outside use is often recommended. High-grade recycling involves bulking the paint for reprocessing by a paint manufacturer into a salable product.

Some latex paint is not recyclable because either it is in poor condition or it contains hazardous constituents. Potential hazardous constituents include mercury, lead, and other heavy metals. Mercury has been used as a biocide in paint. It was banned from interior latex paint in 1990 but is still allowed for exterior latex paint. Lead was used as a pigment in paint until it was banned in 1973. HHW programs still receive paint containing lead and can expect to receive paint containing mercury for many years. Paint containing high levels of heavy metals could be screened out of the recycling program by means of checking container labels and testing suspect containers when labeling information is not available.

Recycling oil-based paint is not commonly practiced; however, some programs (particularly permanent facilities) bulk oil-based paint for use as a supplementary fuel. Waste exchange programs serve to reclaim additional oil-based paint. Recycling oil-based paint has been demonstrated to be technically feasible and may provide greater alternatives in the future.

Oil

Recycling used motor oil is probably the oldest and most common HHW recycling practice. The rates of oil recycling have varied over the years as a result of several factors, including oil prices, oil purchasing patterns, and regulatory controls on used oil. Ambiguities over the designation of used oil as a hazardous waste resulted in a reduction in oil recycling in the 1980s. The importance of used oil collection and recycling is increasingly recognized at federal, state, and local levels.

In addition to HHW programs, used oil is collected at automotive service stations, recycling centers, transfer stations, and landfills. Oil is -also collected in some curbside recycling programs. The EPA estimates that less than half of the oil from do-it-yourselfers who change their own oil is collected; the remainder is put in the garbage, poured on the ground, poured down sewers, etc.[24] So more complete collection of used oil is still a big challenge for communities.

Used oil collected in HHW programs can be readily recycled. The oil should be bulked into either an oil storage tank or 55-gal drums. Used oil will normally contain some contaminants from its use in an engine (e.g., water, gasoline, sediments, and heavy metals), but it is important to guard against unusual contaminants in the oil received, particularly PCBs and chlorinated solvents. The presence of such major contaminants will likely result in the load being rejected by the oil processor. Disposal of used oil as a hazardous waste rather than recycling it will be substantially more expensive.

Used oil can be rerefined into lubricating oil many times with only minor can losses. Rerefining used oil takes only about one-third the energy of refining crude oil into lubricating oil. Used oil is also commonly processed into fuel oil.

Solvents

Solvent recycling is a common practice for industrial hazardous wastes and can also be employed for HHW. The solvents must first be bulked into drums for storage and shipment. Depending on the solvent recovery services available to the program, it may be necessary (or at least advantageous) to separately bulk chlorinated and nonchlorinated solvents as well as polar and nonpolar solvents. In some cases, these different categories of solvents may be shipped to different facilities. Solvents may also be bulked for use as a supplementary fuel.

Permanent HHW facilities offer an advantage for bulking and recycling solvents by providing protected space and fume hoods.

Antifreeze

Recycling antifreeze has become a common HHW management practice. The main ingredient in automotive antifreeze is ethylene glycol, which can be recovered for a variety of uses. Recycling antifreeze in HHW programs typically requires bulking the fluid into a drum or tank before shipping to a processing facility.

Ethylene glycol is not regulated as a hazardous waste by the EPA, but it is by several states. Used antifreeze may contain heavy metals (e.g., copper, lead, zinc) or organic contaminants from gasoline or oil that render it a hazardous waste under federal standards. Such contaminants are an important consideration in the processing method used for recycling as well as in the use of the recycled product.

The most common processing method to recover ethylene glycol is distillation. Distillation generally produces the highest-quality recycled antifreeze. However, some less expensive methods such as filtration can also be used. Recycled antifreeze can be used as automotive antifreeze, deicing solution for aircraft, or an agent in mineral and cement processing.

Batteries

Lead-acid batteries are used in motor vehicles, marine vehicles, and industrial applications (see also Chap. 19). Thus, they represent a potentially large contribution to the waste stream. The hazardous constituents of these batteries are the lead (approximately 15 to 20 lb per battery) and sulfuric acid (approximately 1 to 2 gal per battery).

Fortunately, the majority of lead-acid batteries are recycled. The sulfuric acid is drained and either recycled or neutralized. The lead is removed from the plastic housing of the battery and smelted to produce secondary lead, most of which is used for new batteries. Some battery processors also reclaim the plastic from the battery.

Household batteries are produced in a variety of forms, including alkaline, mercury oxide, silver oxide, zinc-air, lithium, and nickel-cadmium (see also Chap. 19). Recycling programs for household batteries are not widespread. These programs have been hampered by the limited number of processing facilities available. Efforts continue in trying to expand household battery recycling.

Waste Exchange

From a materials management perspective, waste exchange offers the simplest solution to dealing with waste from a given source. Such waste actually becomes an input material for another user. Many containers received in HHW programs have never been opened and others are in good condition and nearly full. Similar to the paint exchange previously described, a variety of materials in good condition can be directed to potential users rather than disposed of.

Numerous HHW programs have employed some form of waste exchange. One-day collection programs and mobile programs often set up a special table for usable products. Permanent and portable storage facilities sometimes designate a special storage area for usable products. Some materials (e.g., paints, solvents, cleaners, automotive products) may also be used in the operation and maintenance activities of government agencies or solid waste facility operators.

From a legal perspective, any programs involving hazardous waste should be managed very carefully. Some HHW programs that offer waste exchange require anyone taking materials to sign a liability waiver. Each program should obtain its own legal counsel on the issue of waste exchange. If legal hurdles can be overcome, waste exchange can provide an efficient means of managing certain wastes generated by households.

PUBLIC EDUCATION

Content

One of the most important objectives of HHW management programs is public education. There are several key messages to communicate, including

- What constitutes a household hazardous waste
- Problems resulting from improper use, storage, or disposal
- Alternatives to hazardous products
- HHW management alternatives in the community (e.g., collection programs, facilities), including date, time, location, types of wastes accepted and excluded, and quantities of wastes allowed

- Appropriate methods to transport HHWs to the HHW program
- Costs of the community's HHW management program
- Phone numbers and contacts for further information

The success of the HHW management program depends in turn on successful public education.

Approaches

A variety of approaches have been used in carrying out public education programs about HHW. Indeed, it is best to use more than one approach to increase the number of people reached with the information. Methods available include

- Prepared fliers, pamphlets, brochures, and posters
- Utility bill inserts
- Direct mail
- City newsletters
- Homeowners Association newsletters
- Door hangers or trash can hangers
- Press releases to newspapers, radio stations, and television stations
- Public service announcements
- Purchased advertising on newspapers and on radio
- Presentations
- Educational programs for schools

An example of a public education flier is given in Fig. 21.1.

Some of the methods are available at only the cost of preparing the information to be communicated, (e.g., public service announcements and city newsletters). Other methods can be relatively costly (e.g., direct mailing and purchased advertising). The public education program should employ cost-effective approaches, balancing available budget and audience reach.

As the HHW program moves from start-up to ongoing operation, the public education content and approach should adjust accordingly.

ECONOMICS

Program Costs

Cost information for HHW programs is presented here as a guideline. Costs will vary as a function of numerous factors, including quantity of materials received; types of materials received; amount of materials bulked versus lab-packed; amount of materials recycled or treated; local labor and materials costs; distance to recycling, treatment, and disposal facilities; and costs charged at the recycling, treatment, and disposal facilities used. Moreover, hazardous waste disposal costs continue to increase with time. A more accurate assessment of costs would require an analysis of capital and operating costs for a specific program. Such an analysis can be made when information about program details and local costs is developed.

TABLE 21.9 Capital Cost Guidelines for HHW Facilities

Facility	Typical range of values
Permanent facility	$100–250/ft^2
Portable facility	$10,000–40,000

*Total cost.

Capital cost guidelines for HHW facilities are presented in Table 21.9. Costs for permanent facilities depend on site development costs, facility size, facility features (e.g., bulking, treatment, laboratory), code requirements, building materials, and the fire-suppression system used. Costs for portable facilities depend on storage capacity and options selected; portable facility options are discussed earlier in this chapter.

In lieu of detailed capital and operating costs, overall program costs can be estimated using the following equation:

$$C_t = P \times H \times G \times C_u \tag{21.1}$$

where C_t = total program cost, $

P = participation rate, %/100

H = number of households in program service area

G = waste generation rate, lb/household

C_u = unit waste management cost, $/lb

Typical parameter values are given in Table 21.10. Units of pounds have been used above in the waste generation rate and unit waste management cost parameters. Alternatively, units of drums can be used in these parameters if preferred. Table 21.10 presents values in both units. Units of pounds are based on weight received; units of drums are based on volume produced (i.e., shipped off site). Because of the generalized nature of the ranges given, the two sets of units are not interconvertible.

The use of Eq. (21.1) is illustrated in the following example. An HHW program has a participation rate of 2 percent out of a total of 25,000 households in the community. The waste delivered by each participant averages 80 lb. and the cost of managing the wastes averages $L50/lb. The total program cost would then equal $60,000.

Equation (21.1) can be used to approximate costs for any type of HHW program—1-day collection programs, permanent facilities, portable facilities, or mobile programs—

TABLE 21.10 HHW Program Cost Parameters

Parameter	Typical range of values
Participation rates	1–2%
Waste generation rate:	
Pounds/household	50–100
Drums/household	0.2–0.4
Unit waste management cost:	
$/pound	1.00–2.00
$/drum	300–600

given the appropriate parameter values. The values given in Table 21.10 cover the majority of these programs.

Funding Alternatives

A variety of methods have been used for funding HHW programs. These include

- Surcharges on refuse collection bills
- Surcharges on tipping fees at solid waste disposal sites
- Solid waste enterprise funds
- General funds
- User fees
- Grant programs

Adding the costs of HHW management programs to refuse collection fees is a common means of funding such programs. This approach allows the costs to be spread out over the entire rate base. Although HHW costs are relatively high on a user basis, the costs become more reasonable when distributed throughout the community. The additional HHW management charge on the refuse collection bill is given by Eq. (21.2):

$$C_r = C_t/H/12 \tag{21.2}$$

where C_r equals HHW cost on monthly refuse collection bill ($ per household per month).

In the previous example, where the total program cost ran $60,000 and had a cost of $120 per participant (80 lb per participant × $1.50/lb), the additional charge on the refuse collection bill would be $0.20 per household per month.

Surcharges on tipping fees also spread out HHW management costs over the community. However, this approach imposes these costs on the commercial and industrial sectors as well as the residential sector. Consequently, the effective cost to each household is less than if only placed on residential refuse collection bills.

Solid waste enterprise funds have been established in some communities for purposes of financing solid waste management. These are generally self-sufficient funds that receive their revenues from a variety of sources, including refuse collection charges, solid waste facility tipping fees, land use fees, interest income, landfill gas recovery, and waste importation fees. HHW management costs can be covered through the solid waste enterprise fund.

General funds are obviously a possibility for funding any governmental expenditure. The suitability of paying HHW management costs with the general fund depends on the local government's fiscal condition.

The relatively high costs of HHW programs per user make it difficult to fund program costs through user fees. In many cases, the costs for HHW disposal (or even recycling) are higher than the purchase price of the product. High user fees would be a disincentive to participate in the HHW collection program. An alternative is to partially fund the HHW program with a modest user fee. Several A communities are employing user fees in this manner.

Grants have been used to fund many HHW programs around the country. Grants are best suited to helping establish new programs; they typically are not a reliable means of providing long-term funding. Nevertheless, in a developing field such as HHW management, grants are likely to continue to play an important role. Furthermore, some states have established funds (e.g., using statewide tipping fee surcharges or waste-management-related taxes) that will provide grant monies at least into the medium term.

Other funding alternatives may be available in a particular state or local community and should be explored. The best approach or combination of approaches must be determined locally to suit the specific conditions in a given community.

REFERENCES

1. Code of Federal Regulations, Title 40, Part 261, U.S. Government Printing Office, Washington, D.C., 1999.
2. Association of Bay Area Governments (ABAG), *Solid Waste Technical Memorandum No. 11,* ABAG, Oakland, California, 1984.
3. Rathje, W. L., Wilson, D. C., and Hughes, W. W., *A Characterization of Hazardous Household Wastes in Marin County, California,* Bureau of Applied Research in Anthropology, University of Arizona, Tucson, Arizona, March 1987.
4. SRI International, *Waste Characterization Study: Assessment of Recyclable and Hazardous Components,* Menlo Park, California, June 1988.
5. Metropolitan Service District, *Metro 1987 Waste Stream Characterization Study,* Portland, Oregon, 1987.
6. R. W. Beck and Associates, *Waste Stream Composition Study Final Report for the City of Santa Cruz,* April 1989.
7. Cal Recovery Systems, Inc., *Waste Characterization Study for Berkeley, California,* 1989.
8. R. W. Beck and Associates, *Waste Stream Composition Study Final Report for Sacramento County,* August 1989.
9. Cal Recovery Systems, Inc., *Waste Characterization for San Antonio,* Texas, 1990.
10. EMCON Associates, *City of Burbank Source Reduction and Recycling Element,* 1991.
11. Cal Recovery Systems, Inc., *City of Sunnyvale Source Reduction and Recycling Element,* 1991.
12. EMCON Associates, *City of Palo Alto Source Reduction and Recycling Element,* 1991.
13. EMCON Associates, *County of Tulare Source Reduction and Recycling Element,* 1991.
14. EMCON Associates, *County of Del Norte Source Reduction and Recycling Element,* 1991.
15. EMCON Associates, *City of Stockton Source Reduction and Recycling Element,* 1991.
16. Association of Bay Area Governments (ABAG), *Toxics Away! The Alameda County Pilot Collection Program for Small Quantity Generators of Hazardous Wastes,* ABAG, Oakland, California, 1988.
17. Johnston, K., and Kehoe, C., "San Francisco's Household Hazardous Waste Collection Facility: A Two Year Summary," *Proceedings of Hazmacon 90,* Anaheim, California, April 1990.
18. Laidlaw Environmental Services, Ltd., *Household Waste Paint Reuse,* St. Catherines, Ontario, 1990.
19. EMCON Associates, *City of Milpitas Household Hazardous Waste Element,* 1991.
20. Gregory, D., "Palm Beach: A Model Facility," *Household Hazardous Waste Management News,* May 1991.
21. National Solid Waste Management Association (NSWMA), "Examples of Small Quantities of Hazardous Waste in Trash," NSWMA, Washington, D.C., 1984.
22. United States Environmental Protection Agency, *Report to Congress on Solid Waste Disposal in the United States,* EPA/530-SW-80-011B, Washington, D.C., October 1988.
23. State of California Air Resources Board, *The Landfill Gas Testing Program: A Second Report to the California Legislature,* Sacramento, California, June 1989.
24. United States Environmental Protection Agency, *How to Set Up a Local Program to Recycle Used Oil,* EPA/530-SW-89-039A, Washington, D.C., May 1989.

CHAPTER 22

MERCURY-CONTAINING DEVICES AND LAMPS

RAYMOND P. JACKMAN
Engineer

JOHN L. PRICE
Environmental Manager
Hazardous Waste Management Section
Bureau of Solid and Hazardous Waste
Florida Department of Environmental Protection
Tallahassee, Florida

MERCURY IN THE ENVIRONMENT AND THE CONNECTION TO MUNICIPAL SOLID WASTE

Since 1989, a series of health advisories have been issued by the Florida Department of Health (DOH) to warn people about eating freshwater fish contaminated with mercury. The fish affected include largemouth bass, bowfin, and gar, among others. DOH issues such an advisory when fish in a freshwater body have more than 0.5 parts per million (ppm) of mercury in their tissue. About 80 freshwater streams, rivers and lakes making up more than 2 million acres of freshwater areas, including the Everglades, have been identified as contaminated with mercury. Recently, these health advisories have expanded to include saltwater king mackerel caught in some areas of the Gulf of Mexico off Florida's west coast.

Mercury is unique among the metals in its significant vapor pressure, especially in its metallic form, covalent bonding to organics, and its ready conversion between three oxidation states under typical environmental conditions. Background levels of mercury in the environment vary by media and geographical area. The concentrations of mercury in the air is significantly less over the ocean then over rural or urban areas. The ready conversion between oxidation states and its inorganic and organic forms create a complex cycling process in the environment. Mercury that ends up in the solid waste stream exists in three general forms: elemental mercury, and amalgams, organic mercury and mercury salts.

Total mercury emissions have doubled or tripled since 1900, but the sensitivity of the total mercury burden to the atmosphere is uncertain at present. However, people have caused changes in the atmosphere that may be increasing mercury deposition rates. Recent studies show that air deposition is a primary source of mercury in lakes in areas without industrial or municipal mercury sources. Both increasing atmospheric mercury levels and increases in the ozone and acid pollutants are likely to be responsible for recent increases in freshwater fish tissue mercury concentrations. The presence of mercury in the solid waste stream and the use of solid waste and medical waste incineration is a contributing factor in this scenario. Waste that is incinerated produces mercury vapor, which is either captured by pollution control equipment or emitted into the atmosphere. The rate of capture varies from 10 to 90% of the mercury, depending on the control equipment utilized. Other significant contributors of mercury emissions include electrical power plants, especially those using coal as a fuel. The use of more efficient lighting would reduce overall electrical consumption and likewise reduce mercury emissions from these sources.

Mercury in surface water is primarily from wet deposition in precipitation. This can lead to elevated mercury concentrations in surface water, even in remote lakes. The primary human pathway for mercury is through the consumption of contaminated fish. The concentration levels in fish are strongly magnified due to bioaccumulation factors in the fish tissue. Like any trace metal, the subsurface fate and transport of mercury is primarily controlled by geochemical processes. Mercury can readily change oxidation states under typical subsurface conditions. Unlike all other metals, mercury is a liquid at room temperature and has significant vapor pressure and can thus move in the subsurface and can volatilize from the soil environment.

Mercury exposure has had profound effects on human health. The primary source of human exposure to mercury is from the consumption of fish, which may contain as much as 1.2 parts per million (ppm) of methyl mercury. Long-term exposure to mercury can permanently damage the brain and kidneys and can cause severe defects in developing fetuses. Mercury is eliminated from the body through the kidney and intestines. Damage to the nervous system or to a developing fetus occurs when the amount of mercury entering the body exceeds that which can be cleared. When the amount of mercury in the blood crosses a threshold for toxicity, the resulting damage may be permanent. Symptoms can subside but may recur later in life. Routes of absorption can be oral, dermatological, and by inhalation.

Recycling rather than discarding mercury-bearing wastes effectively eliminates the potential for the mercury contained in products like boat bilge pump float switches, thermostats, thermometers, and fluorescent lamps from reaching and polluting the environment. By the year 2000 mercury-containing devices and lamps are predicted to be the first and second largest sources of mercury in municipal solid waste (MSW), respectively, accounting for just over 2 tons of mercury, or more than 90% of the total mercury predicted to enter Florida's MSW stream. Mercury-containing devices consist of products commonly known to contain mercury, like thermometers, thermostats, and blood pressure manometers, and some surprises such as bilge pump float switches in pleasure boats and a variety of industrial switches and relays. Fluorescent lamps are the most common mercury-containing lamp but industrial and outdoor applications often use high-intensity discharge lamps such as mercury vapor, metal halide, and high-pressure sodium lamps, which also contain mercury. Some neon lamps also contain mercury.

The intent of this chapter is to provide a brief overview of information relating to the manufacture, content and effects of mercury-containing devices and lamps and the methods and considerations regarding their recycling. Fact sheets prepared by the Florida Department of Environmental Protection are included as working summaries of current Florida regulations and guidelines.

MERCURY-CONTAINING DEVICES

Some of the many mercury-containing devices are thermometers, thermostats, silent electrical switches, manometers, and devices that require mercury for their use. They are widely used in homes, medical facilities, office buildings, and many commercial or industrial facilities. These devices employ mercury for their operation, used in its liquid form, and contained in an enclosed glass ampoule or tube. They may contain varying amounts of mercury but usually have from 1 to 3 grams of liquid mercury. Occasionally, these devices may contain a pound or more of mercury.

FLUORESCENT LAMPS

A typical fluorescent lamp is composed of a sealed glass tube filled with argon gas at a low pressure, as well as a low partial pressure of mercury vapor. The tube is under a partial vacuum. The inside of the tube is coated with a powder composed of various phosphor compounds. Tungsten coils, coated with an electron-emitting substance, form electrodes at either end of the tube. When a voltage is applied, electrons pass from one electrode to the other. These electrons pass through the tube, striking argon atoms, which in turn emit more electrons. The electrons strike mercury vapor atoms and energize the mercury vapor, causing it to emit ultraviolet radiation. As the ultraviolet light strikes the phosphor coating on the tube, it causes the phosphor to fluoresce, thereby producing visible light. The life of the lamp is determined by the life of the electron-producing coating on the cathode, which diminishes as the lamp is operated. The most commonly used fluorescent lamp is the 40 watt, 4 foot long tube, although smaller, larger, and differently shaped lamps are also used.

The amount of mercury in fluorescent lamps varies considerably with manufacturer, and even possibly within manufacturers. The National Electric Manufacturers Association (NEMA) estimates that in 1990, the average fluorescent lamp contained 41 milligrams (mg) of mercury per lamp; this decreased to 23 mg by 1995. NEMA predicts that this will decrease to below 10 mg by the year 2000.

Fluorescent lamps are widely used in business, as they provide an energy efficient source of lighting. The commercial and industrial sectors dominate usage of fluorescent lamps, accounting for over 90% of total usage. Approximately five hundred million lamps were manufactured in 1991. It is possible that this number will increase, as the EPA and the State of Florida promote the use of fluorescent lighting as part of its Green Lights Program, which is designed to reduce energy consumption. Each lamp has a lifetime of four to six years under normal use. The lamps are designed so that approximately half of them will operate after 20,000 hours of operation. Where these lamps are being used on a small scale, they are generally replaced as they burn out, one at a time. However, in large companies and industries, this method is not practicable, and, therefore, group relamping is done on a regular basis. Typically, group relamping is performed at 15,000 hours, or 75% of the lamp's rated life. This translates to replacement every two years for continuous operations, and every five to six years for noncontinuous operations, which is much more common. Approximately 20% of all lamps are currently replaced annually. The process of group relamping operations generate large quantities of lamps to be disposed of at a single time.

HIGH-INTENSITY DISCHARGE (HID) LAMPS

Mercury vapor and metal halide lamps consist of an inner quartz arc tube enclosed in an outer envelope of heat-resistant glass containing 5–10% lead oxides. Depending on the lamp type, the envelope is either clear-coated with a simple diffusing coating, or coated with one or two different phosphor materials. The quartz arc tube contains a small amount of mercury, ranging from 20 mg in a 75 watt lamp to 250 mg in a 1000 watt lamp. The arc tube contains a small amount of the inert gas argon. It also contains trace amounts of other materials used as an emission mix on the electrode, similar to those used in fluorescent lamps. The outer jacket of HID lamps are filled to a pressure slightly less than atmospheric, and implosion may occur if the lamps are broken. Unlike fluorescent lamps, elemental mercury in liquid form is contained in the arc tube (ampoule) and is separate from the outer element.

High-pressure sodium lamps, the third type of HID lamps, consist of an inner, high purity alumina ceramic tube enclosed in an outer envelope of heat-resistant glass that contains 5 to 10% lead or lead oxide. Depending on the lamp type, the envelope is either clear or coated with a diffusing material. The ceramic tube contains a small amount of sodium/mercury amalgam, ranging from 8.3 mg mercury in a 50 watt lamp up to 25 mg in a 1000 watt lamp. The outer bulb is fully evacuated and may implode when broken.

HID lamps are manufactured in a variety of physical configurations. Applications are primarily parking lot and street lighting and high-ceiling indoor work or warehouse areas.

THE RECYCLING PROCESS

Figure 22.1 depicts the general process and materials flow of lamp recycling operations commonly used in the United States.

Step 1: Component Separation

The first step in current lamp and device recycling operations is to process lamps or devices so as to produce separated components such as glass, phosphor powder, metal end caps, or separated glass ampoules and send the separated mercury-containing components for further processing to reclaim the mercury. Examples of separation technologies include crush and sieve, wet process separation, and specialized or manual processes for the disassembly of HID lamps or devices such as thermostats.

The most common of these methods used for separating fluorescent lamp components appears to be the crushing and sieve, or sizing process. This process entails crushing the lamps by mechanical means using a grinding or shearing machine so that the contents are broken into small pieces that can be readily separated. An implosion method can also be used in place of the crushing device. Separation takes place through the use of screens to sort and size the material. The phosphor powder, which has the smallest physical size, passes through one screen and is transported within the system to a collection container, which is usually sealed to the atmosphere. The glass is cleaned by shaking and glass-on-glass contact in the screening process, and is also separated by a larger screen and then transported to an individual collection container. This container can be a 55 gallon drum or a hopper. The glass can also be conveyed to a roll-off container or truck trailer. The metal (aluminum) end caps fail the screening size condition and are also transported internally to their own collection container. This process is conducted under vacuum or negative air pressure, with the resultant air being processed by several mercury and particulate removal devices before discharge or recirculation. Glass is sold or shipped to glass recycling companies, the metal end caps are sold or

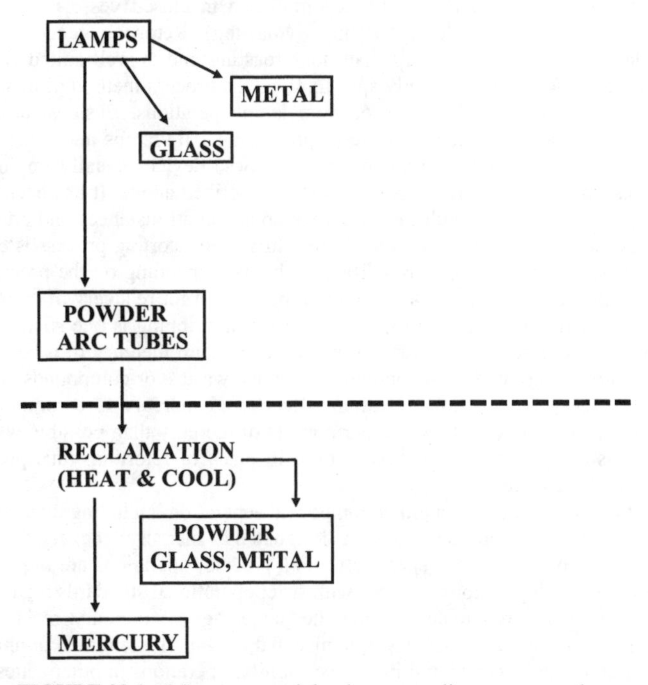

FIGURE 22.1 Mercury-containing lamp recycling process and materials flow.

shipped to metals recycling companies, and the phosphor powder is sent to retort or mercury reclamation operations for further processing and recovery of the mercury, the second step in the recycling process.

Manual processing of devices and HID lamps involves removing and separating the intact mercury-containing ampoule or arc tube from a device or lamp. The mercury ampoules or arc tubes are then sent to other facilities for mercury recovery. This separation is usually done by hand or direct labor means. The ampoules and arc tubes are stored separately for retorting and the device or lamp remains (metal, glass, and plastic) may be recycled or disposed of. This process is similar to the lamp separation process but is conducted manually due to the complexity and diversity of the devices.

Step 2: Mercury Reclamation Operations

Retort operations are used because of their ability to heat and, through vacuum and cooling, recapture the mercury. Lamp recyclers occasionally retort their own mercury on site if the technology and equipment is available to them. The objective of retort operations is to extract and recover elemental mercury from miscellaneous mercury-containing materials. The process consists of heating the mercury-containing material, either under vacuum or not, and collecting and cooling the off-gas stream to condense the liquid elemental mercury. This can be accomplished in closed vessels (retorts) or in open hearth furnaces, ovens, or rotary kilns (roasting). Retorting generally provides higher reclamation rates of mercury than does roasting and is well suited for volatile forms of mercury. Retorting currently appears to be the process method of choice for recovery of mercury in phosphor powder from lamp operations. Besides the phosphor powder, retort operations appear capable of processing HID lamps and other mercury-containing ampoule devices. The ampoules from these devices usually implode under heat and pressure and the mercury is collected as described above. It has been observed that the implosion of the ampoules does not take place in all instances and crushing operations may have to be performed prior to retorting. The retorting process is conducted in cycles, or batches that require from 10 to 36 hours, depending on the process device and material introduced. Other processes might be utilized for recovery of mercury from miscellaneous mercury-containing materials. Chemical stripping is one such process. In the context of this review, processes that extract or reclaim mercury of a commercially marketable purity from mercury-containing materials, wastes, or compounds are considered to be mercury reclamation operations. Recovered mercury is commonly further processed by distillation to achieve higher purities of mercury than possible with the retorting process. Sometimes distillation is used in place of retorting, thus producing a higher purity of recovered mercury.

Florida facilities that conduct either component separation, including manual processing, or mercury reclamation operations are designated as mercury recovery facilities and mercury reclamation facilities, respectively. Both types of operations are required to obtain a permit prior to operations, comply with strict operational standards regarding mercury content of recovered materials and the percentage of mercury reclaimed from processed materials, provide financial assurance in the event of intentional or unintentional closure, and maintain substantial insurance. Similar operations in other states are generally not as closely regulated as those in Florida.

Lamp Crushers **(Drum Top)**

This method is primarily used by lamp generators for size and volume reduction. No separation of component products occur and no recovery of the mercury takes place. The simplest of units consists of a lamp crusher mounted on top of a container, usually a 55 gallon drum, but this process might be conducted on a larger or smaller scale. Lamps feed into the device, are crushed and deposited in the drum or container for disposal or recycling. Devices of this type may or may not be operated under negative air pressure and do not necessarily have air pollution control devices. Further, material from this process may not be suitable for introduction into the crush and sieve devices described earlier, so recovery of the mercury would then take place through retort or other reclamation processes. The amount of time that the metal end caps remain in contact with the phosphor powder increases the mercury fixation to the metal and makes mercury recovery more difficult. The crushed materials resulting form volume reduction operations may or may not be acceptable at recycling facilities and if accepted usually results in additional costs for processing. Finally, if this equipment is not properly designed or maintained, emissions of mercury may occur and result in employee exposure and environmental contamination.

FACT SHEETS ON RECYCLING

The fact sheets presented below summarize the Florida laws, regulations, and guidelines for recycling mercury-containing devices and lamps.

Managing Discarded Mercury-Containing Devices (MCDs) in Florida*

Mercury-containing devices are electrical products or other devices, excluding batteries and lamps, that contain mercury as a necessary component for their operation. Some examples include mercury thermostats, thermometers, electric switches and relays, marine float switches and manometers. Due to the decline of mercury use in batteries and lamps and the larger quantities of this toxic heavy metal found in these products, mercury-containing devices are expected to be the largest source of mercury in municipal solid waste by the year 2000.

Mercury-containing devices do present special disposal considerations due to the quantity of mercury they contain and since they are usually considered to be hazardous wastes when disposed of. The amount of mercury in a device is relatively large. For example, a thermostat can contain as much mercury as 75–100 fluorescent or other mercury-containing lamps. Mercury is a toxic metal that in its various forms can accumulate in living tissue and cause adverse health effects. When a device is broken and is disposed of in a solid waste landfill or incinerator, the mercury can contaminate the air, surface water, and ground water. Mercury contamination in Florida is most evident from the Department of Health's warnings of high mercury levels in fish in a number of our lakes and in the Everglades.

Because of this, these types of devices, including those from households, have been banned from disposal at solid waste facilities, including landfills and incinerators, since January 1, 1996, in any quantity.

Florida businesses and other generators discarding mercury-containing devices ("Generators") have two options for managing them: either recycling or hazardous waste disposal. For either of these options, if a pound of more of mercury is contained in a shipment container (e.g., a little more that 100 mercury thermostats would contain this much mercury), the generator will need to ship the devices in accordance with the US DOT Hazardous Material Regulations.

1. You are encouraged to recycle mercury-containing devices by following the Chapter 62-737, Florida Administrative Code regulations outlined on the back of this fact sheet. Devices destined for recycling and managed in accordance with these regulations are considered to be universal wastes in Florida and do not count toward your facility's hazardous waste generator status. In addition to the following requirements and guidelines, check with the receiving storage or recycling facility for its guidelines on packaging and transportation. A list of recycling facilities in Florida can be obtained by calling 1-800-741-4337. *RECYCLING IS THE RECOMMENDED MANAGEMENT OPTION FOR THESE DEVICES!*
2. Mercury-containing devices may be managed at permitted hazardous waste treatment and disposal facilities and would count toward your facility's hazardous waste genera-

*A fact sheet for Florida businesses and government agencies.

tor status. Before they can be disposed of at a hazardous waste landfill, however, they will need to be shipped and treated at a permitted hazardous waste facility (e.g., a mercury recovery/reclamation facility) to remove most of the mercury in accordance with the EPA's land disposal restriction regulations.

Mercury-Containing Device Recycling Requirements and Guidelines

Generator Requirements

- Does not place used devices from business, industry, or institutions in the regular trash.
- Stores devices in an area and in a manner that will prevent them from breaking. *Does not stuff too many or too few devices into the shipping container and use adequate cushioning material for packing.*
- Labels the devices or each container as *"Spent Mercury-Containing Devices for Recycling" or "Universal Waste Mercury Devices", or "Waste (or Used) Mercury Devices". For thermostats, substitutes "Thermostats" for "Devices" in the last two labeling categories.*
- If devices are accidentally broken, immediately contain the breakage and store them in a tightly sealed container. It is recommended that you mark the container as "Spent Broken Mercury-Containing Devices For Recycling".
- Trains employees in the proper device handling, packaging and emergency cleanup and containment procedures. Non-device residues containing mercury and that are generated as a result of a device cleanup are to be managed as hazardous waste.
- Does not intentionally break, treat or dispose of devices.
- If on-site storage is not feasible, devices may be transported to a central accumulation point at one of your own facilities, to a registered handler facility, or directly to a permitted recycling facility. If you transport your own devices, you also need to comply with the Department's transporter regulations. *See the Transportation Requirements and Transporter Requirements below.*

Handler Facility (Nongenerator) Requirements

- Annually registers with the Department as a small or large quantity handler and receives or renews a DEP ID number.
- *Small quantity handler accumulates* up to 100 kilograms (220 pounds) of devices indoors at any one time for no longer than one year.
- A *large quantity handler facility* accumulating more than 220 pounds of devices at any one time must also register as such and submit to the Department: *a one-time $1,000 registration fee, an operational plan, and a closure plan including financial assurance.*
- Follows other requirements listed above for *Generators.*

Record Keeping Guidelines for Generators and Handlers

- Keep receipts for shipments of devices off-site to show DEP and local inspectors that devices were properly handled. Receipts should have the following information: the quantity of devices shipped or received, the date of shipment or receipt, and the name and address of the handler or recycling facility receiving the shipped devices.
- Records of receipts and shipments of devices are required for large quantity handler facilities (including generators) and shall be kept for 3 years from the date of shipment or receipt

Reverse Distribution Program Requirements

- Sponsored by a device manufacturer or distributor (which may include a business distributing devices to its facilities).
- Sponsor assumes responsibility for collection and recycling of discarded devices.
- Annual registration with the Department, receipt or renewal of a DEP ID Number, and submission of a program description including all participating transporters, handlers and recycling facilities.

Transportation Requirements

- When shipping devices within Florida, a hazardous waste manifest and a licensed hazardous waste transporter are not required for shipments to a handler or recycling facility within Florida in accordance with these requirements.
- When shipping out of Florida, follow the intermediate and receiving states' requirements.
- When shipping into Florida, you may use a shipping paper *unless* your state or an intermediate state requires a hazardous waste manifest; then you must follow those states' requirements.

Transporter/Transfer Facility Requirements

- Annually registers with the Department and receives or renews its DEP ID number.
- Uses only totally enclosed trucks in good condition.
- May store properly packaged devices on the truck or at a another area of a registered transfer facility for up to 10 days.
- Trains drivers in proper handling, packaging and emergency cleanup and containment procedures & keeps these procedures on the trucks.
- Complies with any applicable Department of Transportation (DOT) regulations, including the Hazardous Material Regulations.

Managing Spent Fluorescent and High Intensity Discharge (HID) Lamps*

Fluorescent or High Intensity Discharge (HID) lighting is a good business choice. Compared to incandescent lighting, fluorescent and HID lighting use less energy and produce less heat. Less energy and heat not only result in lower lighting and cooling costs, but they also result in utility power plants emitting less air pollutants such as mercury, lead, nitrogen oxides, and sulfur dioxides. If you are considering switching to high-efficiency fluorescent or HID lighting, don't hesitate to make the change.

Although fluorescent and HID lighting save energy and money, they do present special disposal considerations. Fluorescent and HID lamps (as well as some types of neon lamps) contain mercury and in most cases are considered to be hazardous wastes when disposed. Mercury is a toxic metal that in certain forms can accumulate in living tissue and cause adverse health effects. Although the amount of mercury in each lamp is small, several million lamps are discarded by Florida businesses each year, making these lamps one of the largest sources of mercury in our garbage. When a lamp is broken or placed in a landfill or incinerator, the mercury can contaminate the air, surface

*A fact sheet for Florida businesses and government facilities.

water, and ground water. Mercury contamination in Florida is most evident from the Department of Health's warnings of high mercury levels in fish in a number of our lakes and in the Everglades.

Because of this, these types of spent lamps, excluding those from households, containing any amount of mercury have been banned from solid waste incineration since July 1, 1994, in any quantity. Since most of these types of lamps contain hazardous levels of mercury, they should not be disposed of at solid waste landfills in Florida if more than 10 lamps per month are generated by a business from any one location. Local solid waste departments are the final authority for landfill disposal and may decide to refuse to accept any spent lamps from generators, regardless of the amount of mercury contained in the lamps, especially in those counties or municipalities that also operate solid waste incinerators.

Florida businesses and governmental facilities generating spent fluorescent and HID lamps ("Generators") have two options for managing them: either recycling or landfill disposal.

1. You are encouraged to recycle fluorescent and HID lamps, even those with lower mercury content, by following the Chapter 62-737, Florida Administrative Code regulations outlined in this fact sheet. Hazardous waste lamps destined for recycling and managed in accordance with these regulations are considered to be universal wastes in Florida and do not count toward your facility's hazardous waste generator status. Check with the receiving storage or recycling facility for its guidelines on packaging and transportation. A list of recycling facilities in Florida can be obtained by calling 1-800-741-4337. *RECYCLING IS THE RECOMMENDED MANAGEMENT OPTION FOR THESE LAMPS!*
2. (a) Generators of 10 or less spent lamps per month per location may dispose of these lamps with the regular trash going to a permitted, lined solid waste landfill. Low mercury, non-hazardous waste spent lamps may also be disposed of at permitted, lined solid waste landfills in any quantities. However, contact your local solid waste management department for any final guidance or restrictions on the landfill disposal of these lamps.

   (b) If more than 10 spent hazardous waste lamps are generated per month, they may be disposed of at a permitted hazardous waste landfill and would count toward your facility's hazardous waste generator status.

Recycling Requirements and Guidelines

Generator Requirements (Continued on Back)

- Does not place used lamps from business, industry, or institutions in the regular trash.
- Stores lamps in an area and in a manner that will prevent them from breaking. *Does not stuff too many or too few lamps into the shipping container. Recycling facilities request that you do not tape lamps together for storage or shipment and may not accept lamps that are taped together.*
- Labels the lamps or each container as *"Spent Mercury-Containing Lamps for Recycling" or "Universal Waste Mercury Lamps", or "Waste (or Used) Mercury Lamps".*
- A business or institutional generator location may accumulate and store up to 5,000 kilograms of lamps (20,000 lamps) at any one time and for up to one year, *if the lamps are destined for recycling*, without being subject to notification requirements (EPA Form 8700-12).

- If lamps are accidentally broken, immediately contain the broken lamps and store them in a tightly sealed container. It is recommended that you mark the container as *"Broken Spent Mercury-Containing Lamps For Recycling"*.
- Trains employees in proper lamp handling, packaging and emergency cleanup and containment procedures. Non-lamp residues containing mercury and that are generated as a result of a lamp cleanup are to be managed as hazardous waste.
- Do not intentionally break or crush lamps unless you are complying with the *"Drum-top Crushers" requirements below.*
- If on-site storage is not feasible, lamps may be transported to a central accumulation point at one of your own facilities, to a registered handler facility, or directly to a permitted recycling facility. If you transport your own lamps, you also need to comply with the Department's transporter regulations. *See the Transportation Requirements and Transporter Requirements below.*

Drum-top Crusher Requirements (For *Generators* Only). Most recycling facilities prefer unbroken lamps, and they may charge more to accept crushed lamps. Mercury may adhere to the drum, the container, or the metal end caps causing mercury contamination and increased costs for recycling or disposal especially under humid conditions or longer storage times. However, use of this equipment is allowed by a generator only per paragraph 62-737.400(6)(b), F.A.C., as long as the crushed lamps immediately enter the final accumulation container from the drum-top crusher equipment and crushing is done under the following conditions:

- Crushing poses employee health and environmental risks if mercury vapors are released. Releases of mercury vapors or other contaminants shall be prevented, and the user shall comply with all applicable OSHA standards.
- The crushing unit shall be properly maintained (e.g., adequate filter changes), operated per the manufacturer's written procedures, and the employees using this equipment shall be thoroughly familiar with these procedures.

Handler Facility (Nongenerator Collection) Requirements

- Annually registers with the Department as a small or large quantity handler and receives or renews a DEP ID number.
- *A small quantity handler facility* accumulates up to 2,000 kilograms (8,000) of lamps indoors at any one time for no longer than one year.
- A *large quantity handler facility* accumulating 8,000 or more lamps at any one time must also register as such and submit to the Department: *a one-time $1,000 registration fee, an operational plan, and a closure plan including financial assurance.*
- Follow other requirements listed above for *Generators* except that crushing of lamps as described above is only allowed by generators without a permit.

Record Keeping Guidelines for Generators and Handlers

- Obtain and keep receipts for shipments of lamps off-site to show DEP and local inspectors that lamps were properly handled. Receipts should have the following information: the quantity of lamps shipped or received, the date of shipment or receipt, and the name and address of the handler or recycling facility receiving any shipped lamps.
- Records of receipts and shipments of lamps are required for large quantity handler facilities (including generators) and shall be kept for 3 years from the date of shipment or receipt

Reverse Distribution Program Requirements

- Sponsored by a lamp manufacturer or distributor (which may include a business distributing lamps to its facilities).
- Sponsor assumes responsibility for collection and recycling of spent lamps.
- Annually registers with the Department, receives/renews a DEP ID Number, and provides a program description including all participating transporters, handlers and recycling facilities.

Transportation Requirements

- When shipping lamps within Florida, a hazardous waste manifest and a licensed hazardous waste transporter are not required for shipments to a handler or recycling facility within Florida.
- When shipping out of Florida, follow the intermediate and receiving states' requirements.
- When shipping into Florida, you may use a shipping paper *unless* your state or an intermediate state requires a hazardous waste manifest; then you must follow those states' requirements.

Transporter/Transfer Facility Requirements

- Annually registers with the Department and receives or renews its DEP ID number as a transporter and/or transfer facility.
- Uses only totally enclosed trucks in good condition.
- If registered as a transfer facility, may store properly packaged lamps on a truck used in the actual transportation of lamps or at an indoor location for up to 10 days.
- Trains drivers in proper handling, packaging and emergency cleanup and containment procedures and keeps these procedures on the trucks.
- Complies with any applicable Department of Transportation (DOT) regulations, including the Hazardous Material Regulations.

Note: Transporters and handlers collecting lamps from generators of 10 or less lamps per month and who do not accumulate more than 100 kilograms (400 lamps) at one time are exempt from the annual registration requirements outlined above.

PCB and Other Light Ballasts

- Ballasts containing PCBs (polychlorinated biphenyls) cannot be disposed in Florida. Send to a processor for removal of PCB components and disposal at approved facilities outside of Florida. Non-PCB components may be managed and recycled in Florida.
- About 25% of non-PCB ballasts contain DEHP (di (2-ethylhexyl) phthalate) which is classified by EPA as a hazardous substance. Disposal of about 1600 of these ballasts would trigger the reportable quantity requirement under the federal Superfund laws. The Department recommends that ballasts of this type not be disposed of at solid waste landfills.
- The Department recommends the recycling of all discarded light ballasts.

SECTION III

RECYCLING FACILITIES AND EQUIPMENT

CHAPTER 23
TRANSFER STATIONS

R. CHRISTIAN BROCKWAY, P.E.
Project Manager, Black & Veatch Corp.
Kansas City, Missouri

INTRODUCTION

Historically, solid waste was collected in "packer"-type collection vehicles which delivered the waste directly to landfills. As landfills closed, haul distances became greater, giving rise to the use of transfer stations in which the waste is transferred to large-capacity transfer trailers. The trailers are then hauled to the landfill.

In recent years, transfer stations have also been used for diverting, collecting, and transporting recyclables as well as incorporating material processing systems into the same transfer facility. The discussions in this chapter deal primarily with the transfer of solid waste, however the principles are also applicable to the processing of recyclable material.

Transfer stations currently being designed are typically enclosed in a building to reduce problems associated with noise, odor, and blowing litter and to provide an aesthetically pleasing facility. Advantages associated with transfer stations have resulted in a rapid growth in the number constructed in the past three decades. The principal benefits derived from a transfer station are:

1. *Economy of haul*. Transfer truck legal payloads of 18 to 25 tons can be obtained, compared to the 4 to 10 ton legal payload of most collection trucks. This results in fewer trips to the disposal or processing site, allowing the collection fleet more time on the route to perform collection service. An overall reduction in capital and operating cost for the collection fleet can result.
2. *Labor savings*. Many route trucks operate with two- or three-person crews. The additional travel time of the truck to the disposal site keeps these workers from their collection duties. Since transfer trucks require only a one-person crew, a reduction in nonproductive time can be achieved.
3. *Energy savings*. Over-the-road fuel use of collection equipment and transfer tractors are similar. Significant fuel savings will be experienced as a result of the fewer trips required to the disposal or processing site.
4. *Reduced wear and tear*. A total mileage savings will result from the fewer trips. However, just as important is the reduction in the number of flat tires and damage to power trains and suspension systems that results from operation on muddy and irregular landfill surfaces.

5. *Versatility*. The flexibility of a transfer system allows the solid waste manager the freedom to shift the waste destination with minimal impact on collection operations.
6. *Reduction of landfill face*. Since the length of the landfill dumping face is generally determined by the number and type of vehicles using the site, a reduction in the number of vehicles will result in a smaller working area, less daily cover requirements, and safer conditions at the landfill due to reduced traffic. A landfill that receives only waste hauled in transfer trailers may require a working face less than half that required for a landfill receiving a similar quantity of waste hauled in packer-type vehicles.

The concept of multiple transfer stations serving a disposal site is common today. Historically, several disposal sites served a large city or metropolitan area. As land use becomes more urbanized, public resistance to new disposal sites increases. The current trend is to use a network of transfer stations from which waste is transported to a remote sanitary landfill, processing facility, or energy-recovery facility. Transfer stations can be located on relatively small parcels of land and are perceived by the public as more compatible with urban development than sanitary landfills. The remainder of this chapter deals with various aspects of transfer stations including site considerations, station types, transfer trailers, transfer economics, recycling, and other considerations.

STATION TYPES

Transfer station types include:

- Direct dump—no floor storage
- Direct dump—floor storage
- Compactor
- Pit
- Combination

All types are well established and many successful examples of each type of transfer station are in operation. There are a large number of compactor stations, in part because this concept has been promoted by sales representatives of equipment manufacturers. The pit concept has traditionally been popular on the west coast. The direct-dump concept has gained popularity as improved self-unloading, open-top trailers have been developed.

Direct Dump—No Floor Storage

The direct-dump station is a two-level facility in which collection vehicles on the upper floor discharge waste through hoppers directly into open-top transfer trailers on the lower floor. A typical direct-dump transfer station is shown in Figs. 23.1 and 23.2. The concept is inherently efficient because equipment and labor necessary to load the trailers are minimized. A significant feature of this concept is that the trailer-loading operation must be capable of handling wastes as they are received.

The concept dates back at least to the 1950s, when it was used by the Los Angeles County Sanitation District. It is appropriate for small or large stations with the capacity determined by the number of direct-dump hoppers. Several variables control the capacity of a hopper, including the payload of collection vehicles, average unloading time, number

FIGURE 23.1 Typical direct-dump transfer station.

FIGURE 23.2 Trailer receiving waste from above.

unloading simultaneously at a hopper, and capacity of the transfer trailer. Representative values for these variables are:

Average collection vehicle payload	7 tons
Average unloading time per truck	6 min
Collection trucks unloading simultaneously in each hopper	2
Capacity of the transfer trailer	20 tons
Time to level load and change transfer trailers	7 min

Based on these conditions, the 20-ton transfer trailer would be loaded in 12 min and 7 min would be required to level the load, move it out, and replace it with an empty trailer. Three transfer trailers would be loaded each hour, providing a peak capacity of 60 tons per hopper per hour. Achieving this loading rate depends on collection trucks being available to unload and empty transfer trailers also being available. This situation only occurs a few hours per day, so that a single hopper with peak collection vehicle deliveries during 4 hours of the day and no deliveries at other times would have a capacity of approximately 240 tons per day.

Because transfer trailers usually can be loaded at a faster rate than the round-trip time to the disposal site, extra transfer trailers are needed to store the waste during peak hours. Peak storage for this type of station is in transfer trailers rather than in a pit or on the receiving floor. Although it is possible to provide receiving-floor space for storage, that would negate the simplicity and efficiency of the direct-dump concept. The extra trailers used for peak storage are temporarily stored on the transfer station site until tractors are available for the trip to the disposal site. A yard tractor can be used to move the loaded trailers from the transfer station to the yard storage area and to move empty trailers to the transfer station.

Direct-dump transfer stations typically use a stationary clamshell device, shown in Fig. 23.3, to distribute the solid waste in the transfer trailer. The clamshell will also provide a degree of compaction of the solid waste in the trailer. The clamshells are typically provided with a specially designed grapple, as shown in Fig. 23.4. The grapple can be opened and closed to move solid waste around in the trailer and it can be closed to allow it to be used in compacting the solid waste. A disadvantage of this type of station is lowered efficiency if the clamshell is out of service for maintenance. It cannot be replaced with another unit. Some stations keep a backhoe available for temporary use when this happens.

Rubber-tired mobile excavators, shown in Fig. 23.5, can be used as an alternative to the stationary units. In general, these units are two to three times as expensive ($150,000 to $175,000 each) as stationary units. However, they can provide more flexibility than stationary units. One mobile unit can serve several hoppers that are not continuously being loaded. If a mobile clamshell requires servicing, it can be removed from the area and another unit can take its place.

This particular type of transfer station is not conductive to recycling because the waste is dumped directly into the trailer with no opportunity to recover recyclable materials. However, this type of facility can be used to transfer recyclables where large amounts of materials are involved and hoppers and trailers are available.

Direct Dump—Tipping-Floor Storage

Many stations that have the capability for direct dumping of waste from the collection vehicles to the transfer trailer also utilize floor storage to increase station capacity during

FIGURE 23.3 Stationary clamshell distributes solid waste into trailer.

peak hours. A station with floor storage is shown in Fig. 23.6. This concept results in substantial changes to both station construction and operation.

The station construction cost is significantly increased due to the larger tipping floor area required to accommodate the stored waste. This larger tipping floor area results in the need for larger building space and therefore significantly increases cost.

Operation methods and costs are also increased. Storage of wastes on the tipping floor

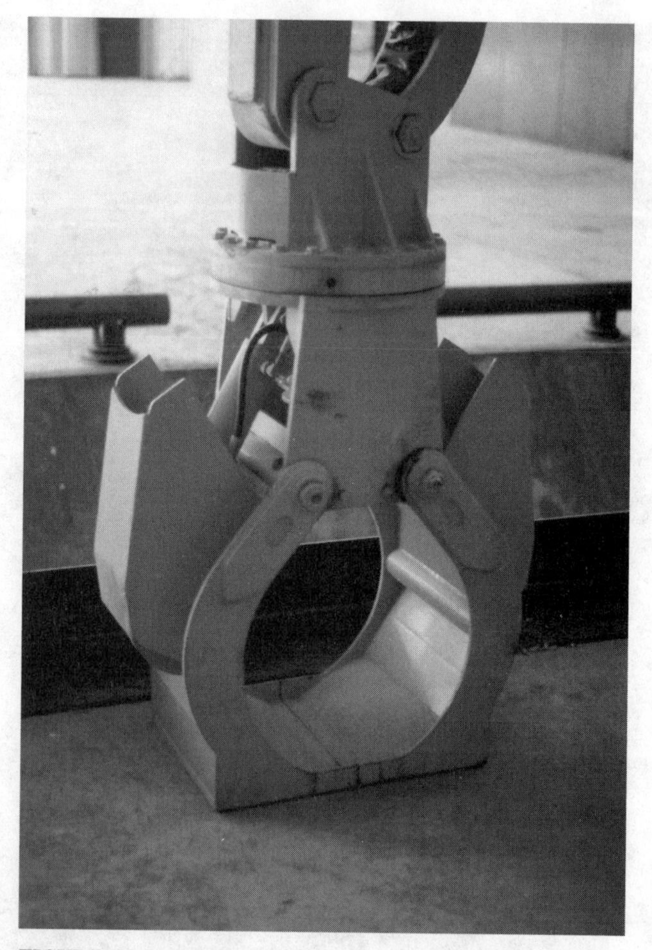

FIGURE 23.4 Specially designed grapple for clamshell. (Photo courtesy of Crane Equipment Corp.)

results in the need for a large wheel loader (Fig. 23.7) to load waste into the open-top trailer. There are equipment and labor costs associated with the wheel loader. The wear of the wheel loader's bucket on the tipping floor also increases maintenance cost. Stations of this type typically utilize hardened concrete toppings to protect the underlying structural concrete. Iron aggregate toppings are the most common type used to protect the concrete surfaces. These toppings are typically 1 to 2 inches in thickness and cost from $10 to $15 per square foot installed. Many operators of transfer stations consider the high cost of toppings to be worthwhile compared to the cost and problems associated with having the station out of service for long periods while concrete repairs are made. It should be noted that although the hardened concrete toppings will wear better than ordinary concrete, they too will eventually wear away and need to be replaced. Many operators now use replaceable rubber wear blades on the bottom of the wheel loader bucket to reduce wear on the floor.

FIGURE 23.5 Mobile excavator.

FIGURE 23.6 Transfer station floor storage (tipping floor).

FIGURE 23.7 Large-wheel loader operates on tipping floor.

The advantage of this concept is that during peak periods, the station capacity is not limited by the rate at which transfer trailers can be loaded. There are economic trade-offs between station costs and transfer trailer costs that must be evaluated by the design engineer.

The direct-dump type of facility with tipping-floor storage is also more suitable for a combined transfer station and materials-recovery facility. The tipping-floor storage provides the ability to separate materials that can be recycled before the waste is loaded into the trailer. Provided there is enough space, barriers can be erected on the tipping floor to provide storage bunkers for various materials such as yard waste, old corrugated cardboard (OCC), and other recyclable materials.

Compactor

Many compactor transfer stations have been constructed in the United States in the past few decades. Typically, they are a two-level operation with collection vehicles unloading onto a receiving floor or into a hopper at the upper level. The solid wastes are then moved into the compactor and compacted into a transfer trailer at the lower level. Figure 23.8 shows a typical compactor.

When compactor stations use the receiving floor as a waste storage area, a wheel loader is used to pick up the waste and load it into the hopper. Depending on the loader equipment capability and the operators' skill, it may be practical for one operator to load more than one compaction hopper.

Another compactor station concept is to provide a large bunker to receive wastes directly from the collection vehicle. The receiving bunker is at a right angle to the compaction hopper and the transfer trailer. The waste is pushed by a large hydraulically operated blade from the receiving bunker into the compaction hopper and then compacted into the transfer trailer. The transfer trailers are closed-top mechanical or hydraulic system types sized to handle the maximum legal payload. The concept has proven successful and

FIGURE 23.8 Compactor at transfer station.

generally has high reliability. The hydraulically operated equipment requires reasonable maintenance.

Historically, compactors have compacted the waste within the trailer. This requires the use of heavy-duty trailers. Within the last few years, compactors that compact the waste within their own chamber (see Fig. 23.9) and then discharge (or push) the waste into the transfer trailer have become widely used. Compactors such as this allow the use of lighter-weight trailers, since no compaction takes place within the trailer.

Pit

The pit concept has been in use for many years in some of the country's largest transfer stations. The principal advantage is the large storage capacity provided by the pit. It provides storage for peak deliveries and allows transfer haul to be operated on as much as a 24-h basis, if desired. The pit concept is shown in Fig. 23.10.

Whether waste is stored overnight in the pit depends on regulatory requirements and operator preference. It is common to leave some waste in the bottom of the pit to reduce damage to the pit floor from the track-type dozer equipment used to crush and handle the waste. This equipment damages the floor at a rapid rate unless some protection is provided.

Pit stations have proven to be capable of handling bulky waste. Equipment working in the pit can crush and break up the bulky waste in preparation for loading it into transfer trailers.

The pit-type transfer station is not conducive to recycling efforts. The emphasis with this type of transfer station is to unload the material as quickly as possible into the pit. Once in the pit, recyclable material is difficult to remove.

FIGURE 23.9 Compacting waste within compactor chamber.

FIGURE 23.10 Pit concept (tipping floor) operates 24 h/day.

Combination

Larger transfer stations may include more than one transfer method. Several stations include both pit and direct-dump. The pit portion provides storage for waste received during peak periods. The direct-dump portion reduces handling of the waste and provides economical operation. The collection vehicle maneuvering area is generally in the center of the station and serves both the pit and direct dump. The Montgomery County, Maryland, Transfer Station, constructed in the mid-1980s, is an example of this combination concept. Figure 23.11 shows a plan view of the Montgomery County Transfer Station.

STATION SITE CONSIDERATIONS

Obtaining a suitable site is a critical step in establishing a transfer station. It merits careful planning and consideration of several interrelated factors. Site considerations include:

- Required land area
- Site layout
- Access and traffic impact
- Location
- Zoning
- Public acceptance
- Land cost

Required Land Area

The transfer station site must have adequate area for on-site roads, utilities, surface-water drainage, and auxiliary facilities. In some parts of Florida, the land requirements to meet drainage regulations can be as such as 25% of the total site. Auxiliary requirements can include offices for staff, a scale house and scales, transfer vehicle storage, maintenance structures, vehicle wash bay, fuel storage and dispensing equipment, and employee parking.

Environmental awareness has led designers of transfer stations to place greater emphasis on minimizing adverse impacts such as noise, dust, and odor. Considerable additional area may be required for landscaping and screening berms to minimize adverse environmental impacts.

Land needs vary, but generally fall in the following ranges for stations involving only transfer and not processing of the solid waste:

Station capacity, tons per 8 hour shift	Site area, acres
less than 100	2–5
500	4–8
1000	8–16
1500	12–20

Large sites have less visual impact and reduced noise levels at the property line. Large sites also provide adequate roads on-site to accommodate queuing of collection vehicles during peak periods and eliminate backup onto public streets.

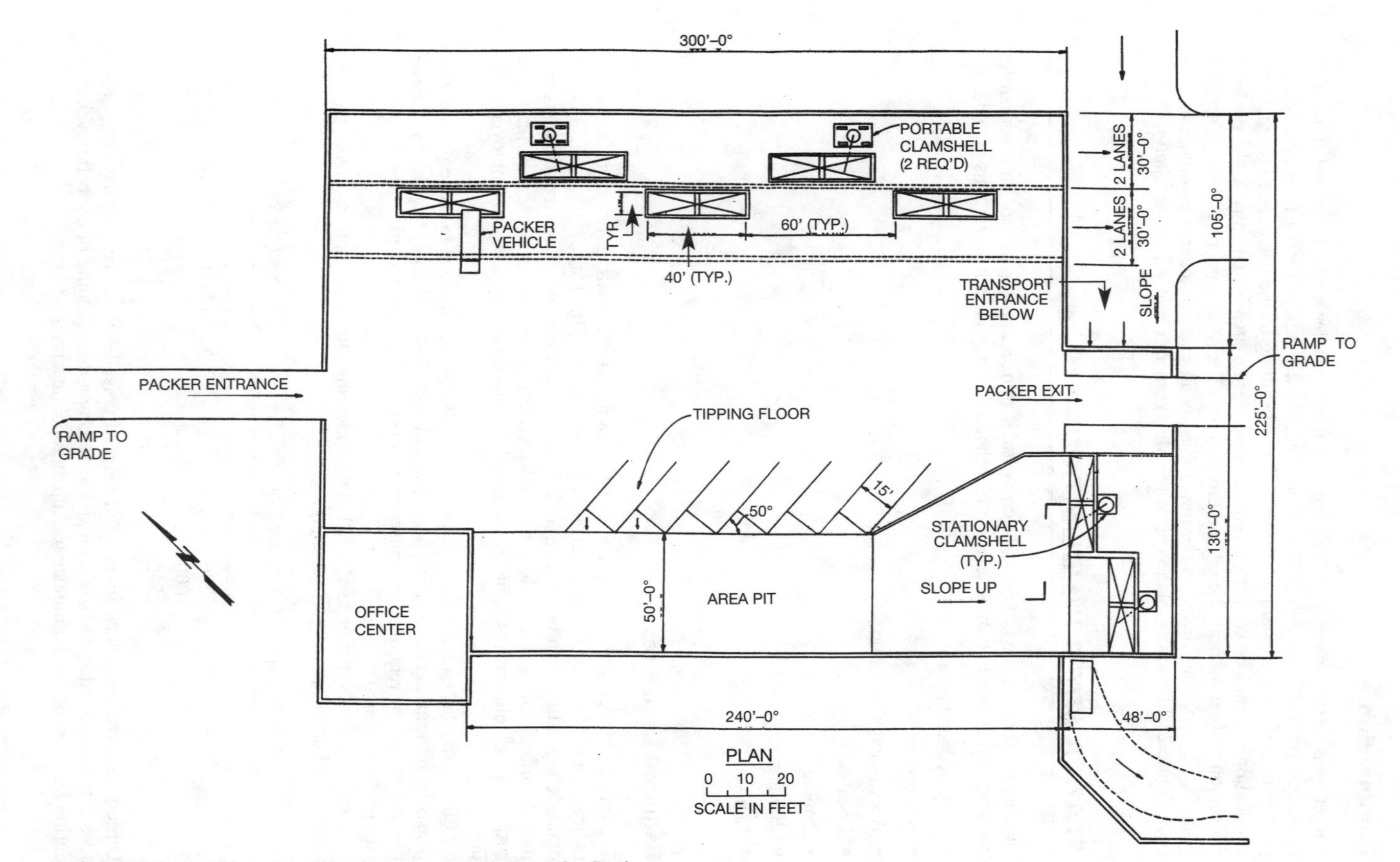

FIGURE 23.11 Plan view, Montgomery County Transfer Station.

Site Layout

The layout of buildings and roadways on a site is one of the most important factors in designing a successful transfer station. A typical site layout of a transfer station is shown in Fig. 23.12. A counterclockwise flow of traffic will result in fewer crossing traffic patterns. Left turns are also easier to make, since the driver is on the left side of the vehicle and has a better view. These rules may not apply when a recycling facility is also incorporated in the site, as traffic patterns become more complex. The scale house should be located where suitable queuing distance can be provided, such that vehicles will not back up onto access roads.

Public users of the facility should be kept separate, if possible, from the private haulers and transfer trailers. This is true not only within the transfer building itself, but also on the site roadways.

Ramps are required for most transfer stations. There will be either a ramp up to the tipping floor, a ramp down to the transfer level, or a combination of the two. The height difference between the tipping floor and the trailer loading level is typically 15 to 19 feet. Splitting the difference between elevating the tipping floor above grade and lowering the trailer loading level below grade will minimize the slope and length of the ramps. Locating the trailer level below grade may not be practical due to factors such as groundwater and potential flooding. Locating the tipping floor at grade will lower the overall height of the transfer station. The height of a station can be of great concern to local residents. However, locating the tipping floor at grade will increase the cost of the transfer level, due to its greater depth below grade.

Adequate area should be provided for trailer storage, employee parking, and visitor parking. Employees and visitors should not be required to walk across lanes of traffic unless absolutely necessary. Maintenance, wash, and fuel facilities should be located such that they do not interfere with the flow of traffic around the station site.

Sites should be selected and laid out to allow room for future expansion of the transfer station. Additional processing such as a material recovery facility, wood processing, and waste-to-energy should be considered in the layout, providing that sufficient room is available at the site.

Adequate signs should be provided along the access roads to the site, at the site entrance, and along the roadways on the transfer station site. It is especially important that sufficient signs be provided for the general public that uses the facility. These people will not use the site often and should be carefully directed to the appropriate locations around the site.

Access and Traffic Impact

Transfer stations are best served by major traffic arteries. Two chief advantages are easier access and reduced impact on adjacent development. Easy access can reduce travel time and result in low collection and transfer haul costs.

To reduce impact on existing traffic, improvements to service roads may be needed to accommodate transfer station traffic. Consideration should be given to the need for signals, turnoff lanes, and additional traffic lanes.

Location

Station location is important to secure the best system economy. Transporting solid waste and recyclables in collection vehicles is expensive, especially if a large collection crew

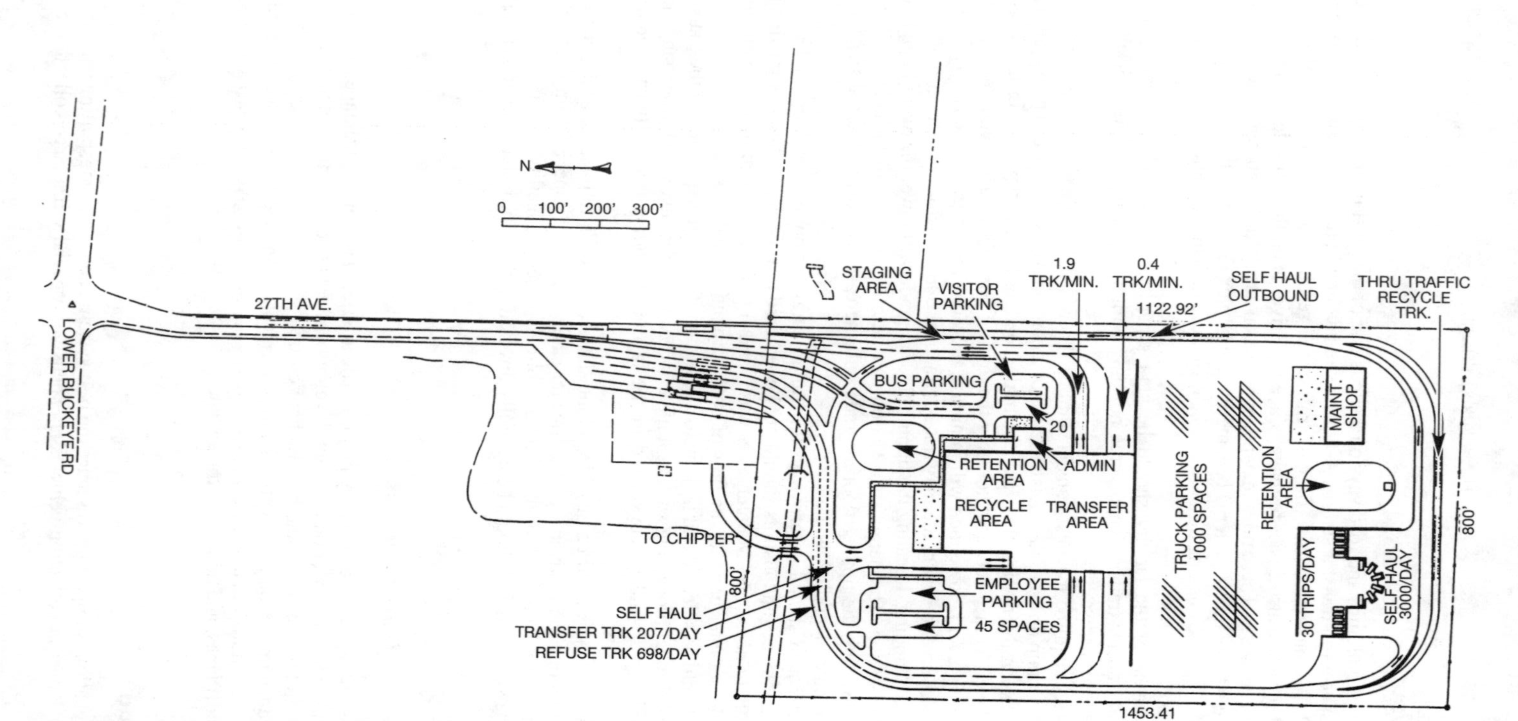

FIGURE 23.12 Site Plan, typical transfer station.

travels with the vehicle. To minimize collection costs, the transfer station should be located within the area of the collection routes served. The site having the lowest total hauling cost is typically located between the center of the collection routes served and the disposal site.

Zoning

Local laws and ordinances determine zoning requirements for a transfer station. Land for a transfer station usually must be zoned to be commercial or industrial and often a special use permit is required. Obtaining a change in zoning can require extensive effort. Zoning hearings require expertise from engineers, environmentalists, and lawyers. Personal contacts with the public perceived to be affected by the proposed facility are essential to success and effective visual aids to communicate the concept to the public and to zoning officials will be needed.

Zoning requirements can significantly increase the costs of developing a transfer station.

Public Acceptance

Obtaining a transfer station site can be a major challenge, due to public reluctance to accept a solid waste facility. Public acceptance is enhanced if the factors discussed above are properly handled. For example, the site should be properly zoned, large enough to provide screening, and served by an adequate road system.

It will probably be important to involve citizens and community leaders in the transfer station site selection. This can be accomplished by forming a task force to assist with siting and design of the transfer station. If representatives of community organizations are convinced the site is reasonable, organized public opposition can be minimized.

Public acceptance can sometimes be more easily obtained by locating a transfer station on public land. It may be practical to locate the station at a site previously used for solid waste disposal. If this is possible, precedence has been established for similar land use and opposition may be less.

Architectural treatment and landscaping are tools designers can use to enhance the acceptability of a transfer station. The architectural treatment of the station should be compatible with adjacent development. If there are commercial buildings nearby, the transfer station can be designed to have the appearance of a commercial structure similar to that shown in Fig. 23.13. Exterior treatment of transfer stations can include metal siding, tilt-up concrete panels, and precast concrete panels. Preengineered metal buildings placed on a concrete foundation are often used for transfer stations. Figure 23.14 shows a transfer station using a combination of metal siding and precast concrete panels for architectural treatment.

Landscaping and screening should be used where appropriate to reduce visual and noise impacts of the transfer operation and associated traffic. Screening can be provided by trees, shrubs, fencing, or walls. Earth berms can minimize visual impact and reduce noise levels.

The impact of a transfer station on its surroundings is due to several factors. Even though the installation is attractive, landscaped, and screened, it must be operated conscientiously to be a good neighbor. A common complaint is trash along roads near the station. Consistent policing of roads near the site is essential to maintain a reputation as a good neighbor.

FIGURE 23.13 Exterior design of transfer station.

FIGURE 23.14 Exterior of the station showing use of precast concrete panels and metal siding.

Land Cost

Land is one of several items making up the total cost. It is usually a relatively small part of the total capital and operating cost. Land cost would typically be a very few percent of the station construction cost. It may be wise to pay the amount necessary for a site that best meets other criteria. If a higher land price also means reduced public opposition or savings in site development or access road costs, it can be a good investment. Transfer station costs, including land costs, are covered later in this chapter.

TRANSFER TRAILERS AND UNLOADING EQUIPMENT

Transfer trailers are an integral part of the transfer system and the type required is directly related to the station type. For example, a direct-dump or pit transfer station requires an open-top trailer, as shown in Fig. 23.15, whereas a compactor-type station requires an enclosed trailer, as shown on Fig. 23.16.

Open-Top Trailer with Push-Out Blade

Hydraulically operated push-out blades can be used to unload open-top trailers. The trailer must be constructed to withstand forces from the push-out blade without cross-bracing that would interfere with top loading. Such a trailer is heavier than other open-top trailers and therefore, due to highway weight limits, provides a smaller legal payload.

FIGURE 23.15 Open-top trailer.

FIGURE 23.16 Enclosed trailer.

Moving Floor in Open-Top Trailer

The moving floor concept comprises a series of undulating floor slats, which move the wastes out of the trailer. Figure 23.17 shows a moving floor bottom. The moving floor systems, such as those made by Keith Manufacturing Co., are made of aluminum or steel floor slats that extend the full length of the floor bed.

The slats are hydraulically powered and move in a four-step process, as follows, for unloading the trailer:

1. Every third slat moves forward 6 to 10 inches, sliding under the load. The load does not move, since the majority of the load is supported on the two-thirds of the slats that do not move.
2. A second group of one-third of the slats moves forward underneath the load. Again, the load does not move, since two-thirds of the slats remain stationary.
3. The third group of slats moves under the load as described in steps 1 and 2. Again, the load does not move.
4. All floor slats move together (6 to 10 in), in one motion, toward the rear of the trailer. Since all slats are moving at the same time, the entire load moves toward the rear of the trailer.

The open-top trailers vary in capacity, with 110 yd^3 capacity being quite common.

Open-Top Trailer with Landfill Tipper

A landfill tipper, shown in Fig. 23.18, may be used to unload open-top trailers at a disposal site. It is a semistationary piece of equipment that is periodically moved to unload the

FIGURE 23.17 Trailer with moving floor. (Photo courtesy of Brothers Industries, Inc.)

solid waste reasonably near the working face. A principal advantage of such equipment is that it requires no unloading mechanism built into the trailer, so that payloads are greater and maintenance is less. A disadvantage is that the unloading point is not always at the working face of the landfill, making it necessary to move the solid waste from the unloading point to the active face. This may increase the exposure of the trash to wind and birds. The concept has been used at major sites for several years and has proven successful.

Enclosed Trailer with Push-Out Blade

The enclosed compactor-compatible transfer trailer with push-out blade is a reliable and proven design. The higher compaction achieved in this trailer results in a legal payload in

FIGURE 23.18 Trailer at landfill tipper unit.

a smaller volume. Usually, a legal payload can be achieved with a trailer volume in the range of 65 to 75 yd^3. This compares with 100 to 110 yd^3 for open-top, noncompaction trailers. Maintenance requirements are difficult to document, but the sturdier construction inherent in these trailers tends to reduce maintenance.

TRANSFER ECONOMICS

Anticipated lower system costs are typically the major factor in the decision to construct a transfer station. A cost model can be used to predict transfer haul costs. Although the input to the model is unique to each locality, the approach is generally applicable. A transfer haul model includes the following elements:

- Transfer station capital cost
- Transfer station operating cost
- Transfer haul equipment capital cost
- Transfer haul equipment operating cost

The total transfer haul cost is the sum of the four elements. Costs are expressed in terms of dollars per ton of waste handled.

Station Capital Cost

Station capital cost is influenced by a number of variables, including station capacity, land area and cost, site drainage requirements, subsurface conditions, screening and landscaping, utility needs, type of architecture, extent of building enclosure, and ancillary facilities such as employee amenities, maintenance facilities, and scales. A conceptual plan is needed as a basis for estimates of capital costs. Since many of the costs are site-specific, it is desirable to develop a site plan for the cost estimate.

Engineering economics are used to determine an annual cost premised on an estimated useful life, salvage value, and interest rate. The annual cost can be converted to a per-ton cost by dividing the total annual cost by the annual tonnage.

Station Operating Cost

Elements of operating cost include labor, utilities, supplies, and maintenance. Labor, utilities, and supplies are relatively easy to estimate. Labor cost should include any extra cost for overtime or staff for extra shifts as well as all applicable overhead and fringe benefit costs. Maintenance costs can be estimated by using a small percentage of the construction cost.

Operating costs should be calculated on an annual basis and converted to dollars per ton by dividing the annual operating costs by the tons to be processed per year.

Transfer Equipment Capital Cost

Transfer haul equipment includes the tractors and trailers for over-the-road hauling as well as yard tractors for maneuvering trailers at the transfer station. The capital cost must also include extra equipment needed to handle peak quantities and spare equipment to replace tractors and trailers out of service for maintenance. The amount of equipment needed for peaks varies with the type of station. For example, the direct-dump station with no floor storage requires trailers for storage of waste received during peak periods.

Spare equipment needed for service during repair of regular equipment depends on original quality, age, condition, and operator care. An allowance of 10 to 20% is considered adequate by many operating agencies. Transfer equipment costs should be amortized over the anticipated useful life to obtain an annual cost. The annual equipment cost can be converted to dollars per ton by dividing by the tons to be processed each year.

Transfer Equipment Operating Cost

Two categories of transfer haul operating costs are as follows:

- Variable costs, which are directly related to the miles driven
- Fixed costs, which are a function of time rather than miles

Variable costs include items such as repairs, tires, fuel, and lubrication. These costs can be calculated as the product of cost per mile times the miles per round-trip, divided by the average number of tons per load.

Fixed costs include labor, insurance, and taxes. Overtime, fringe benefits, and applicable overhead costs should be included.

The total transfer equipment operating cost is the sum of the fixed and variable costs. As before, these costs should be expressed in terms of dollars per ton of waste transferred.

Transfer System Cost

The transfer system cost is the sum of the four components. This total transfer haul cost may then be compared with the cost of direct haul in collection vehicles. An example of an economic analysis of transfer versus direct haul costs follows (see Fig. 23.19).

Detailed Economic Analysis

Making a cost comparison of direct haul in a packer-type vehicle versus transfer haul requires analyses of three components. These components are direct haul cost, transfer haul cost, and transfer station cost. The development of examples of these costs is based on a government-owned and operated system. Therefore, no costs are included for either taxes or profit. To utilize the cost model for private operations, allowances for these costs should be included.

TABLE 23.1 Direct-haul cost data for a 25 yd^3 rear-loading packer*

Annual fixed costs	
Truck	
Capital investment, $/unit	75,000
Estimated service life, years	5
Salvage value, $	14,000
Cost of debt, %	9.0%
Body	
Capital investment, $/unit	33,000
Estimated service life, years	10
Salvage value, $	1,600
Cost of debt, %	9.0%
Truck and Body	
Licenses, personal property tax, and insurance	12,000
Spare equipment for downtime, %	10.0%
Labor cost including supervision:	
1-person crew	20,000
2-person crew	40,000
Overhead, %	15.0%
Average payload, tons/load	8
Average speed of truck, mph	40
Variable costs per mile	
Fuel: 4 mpg	$0.24
Oil, lube, service, etc.	$0.01
Tires: 20,000 mi/set	$0.10
Truck repairs	$0.11
Subtotal	$0.46
Overhead	$0.07
Variable cost per mile	$0.53

*Representative travel variables: Miles per trip (one way): 10, 20, 30, 40, 50. Average unloading time: 15 min.

TABLE 23.2 Fixed direct-haul cost for a 25 yd^3 rear-loading packer

Annual fixed costs	1-Person crew	2-Person crew
Capital costs:		
Truck	$15,700	$15,700
Body	4,900	4,900
Licenses, personal property tax, and insurance	12,000	12,000
Subtotal	$32,600	$32,600
Spare equipment for downtime	3,300	3,300
Labor cost (including supervision)	20,000	40,000
Subtotal	$55,900	$75,900
Overhead	8,400	11,400
Total annual fixed cost	$64,300	$87,300
Fixed cost per hour (based on 1800 effective hours per year)	$35.72	$48.50

Direct-Haul Cost

The costs for direct haul in a 25 yd^3 rear-loading type vehicle are developed in Tables 23.1 through 23.3. Equipment costs used in the tables were obtained from equipment manufacturers. Table 23.1 shows the data on which the cost model is based. Table 23.2 develops the fixed costs, which are not closely related to mileage driven. The fixed cost shown in Table 23.3 equals the round trip travel time plus the unloading time multiplied by the fixed cost per hour (Table 23.2) and divided by the average payload in tons. Table 23.3 also develops the variable costs, which are dependent on miles driven. The variable-haul cost equals the variable cost per mile shown in Table 23.1 multiplied by the round-trip mileage and divided by the tons hauled. Table 23.3 shows the total direct haul cost expressed as both cost per ton and cost per ton-mile. Costs are developed for both a one- and two-person crew. The model shows that direct haul in a 25 yd^3 collection vehicle with a one-person crew costs at least $0.38 per ton-mile.

TABLE 23.3 Direct-haul cost summary for a 25 yd^3 rear-loading packer

	Miles per trip (one-way)				
	10	20	30	40	50
One-way travel time, min	15	30	45	60	75
One-person crew					
Fixed cost, $/ton	3.35	5.58	7.81	10.05	12.28
Variable cost, $/ton	1.33	2.65	3.98	5.30	6.63
Total cost, $/ton	4.68	8.23	11.79	15.35	18.91
Total cost, $/ton-mile	0.47	0.41	0.39	0.38	0.38
Two-person crew					
Fixed cost, $/ton	4.55	7.58	10.61	13.64	16.67
Variable cost, $/ton	1.33	2.65	3.98	5.30	6.63
Total cost, $/ton	5.88	10.23	14.59	18.94	23.30
Total cost, $/ton-mile	0.59	0.51	0.49	0.47	0.47

Transfer Haul Cost

Table 23.4 shows the assumptions used in developing capital, operation, and maintenance costs of transfer haul for trailers capable of hauling an average of 18 tons per trip. An allowance for additional trailers or tractors has been made. During peak periods, trailers may be filled faster than haulers can transport them to the landfill, thereby necessitating extra trailers. Tractor downtime can account for up to 25% of the fleet, but trailers have little downtime.

The haul distances shown in Table 23.4 are converted to trips per day by assuming a 40 mph travel rate, 20 min loading, and 15 min unloading times throughout an 8 hour workday. Time should also be allocated to break time and refueling.

TABLE 23.4 Transfer haul cost data

Annual fixed costs, $	
Tractor	
Capital investment, $/unit	65,000
Estimated service life, years	5
Salvage value, $	22,000
Cost of debt, %	9.0%
Trailer	
Capital investment, $/unit	50,000
Estimated service life, years	7
Salvage value, $	6,500
Cost of debt, %	9.0%
Licenses, personal property tax, and insurance	13,000
Allowance for spare tractors, %	15%
Allowance for spare trailers, %	15%
Labor per unit, $	45,000
Base trailer repairs, % of capital cost	15%
Additional trailer repairs per trip, % of capital cost	0.5%
Overhead, %	15%
Average payload, tons/load	18
Average speed of truck, mph	40
Variable costs per mile	
Fuel: 4 mpg	$0.25
Oil, lube, service, etc.	$0.01
Tires: 20,000 miles/set	$0.35
Tractor repairs	$0.20
Subtotal	$0.81
Overhead	$0.12
Variable cost per mile	$0.93
Representative travel variables	

Miles per one-way trip	Resulting trips per day
10	8
20	5
30	4
40	3
50	3

TABLE 23.5 Fixed transfer haul cost summary

	Miles per trip (one-way)				
	10	20	30	40	50
Trips per day	8	5	4	3	3
Capital cost per tractor, $/year	11,100	11,100	11,100	11,100	11,100
Licenses, personal property tax, and insurance, $/year	13,000	13,000	13,000	13,000	13,000
Allowance for spare tractors, $/year	3,615	3,615	3,615	3,615	3,615
Capital cost per trailer, $/year	8,600	8,600	8,600	8,600	8,600
Allowance for spare trailers, $/year	1,290	1,290	1,290	1,290	1,290
Labor per unit, $/year	45,000	45,000	45,000	45,000	45,000
Trailer repairs, $/year	9,500	8,750	8,500	8,250	8,250
Subtotal, $/year	92,105	91,355	91,105	90,855	90,855
Overhead, $/year	13,800	13,700	13,700	13,600	13,600
Total annual fixed cost, $	105,905	105,055	104,805	104,455	104,455
Cost per day, $ (260 day/year)	407	404	403	402	402
Fixed cost, $/ton	$2.83	$4.49	$5.60	$7.44	$7.44

In Table 23.4, trailer repair costs are considered to vary depending on the number of trips traveled. Other costs are summarized in Tables 23.5 and 23.6. Table 23.7 shows that the cost per ton for transfer haul ranges from $3.86 for a 10 mi haul to $12.61 for a 50 mile haul. Costs on a ton-mile basis are highest ($0.39/ton-mile) for the 10 mi haul and lowest ($0.25/ton-mile) for the longest haul.

Transfer Station Cost

The calculation of transfer station costs must be adapted to site-specific information. Transfer station costs will vary substantially depending on the type of station and the features included. The input data in Table 23.8 show capital cost for a 1500 ton per day transfer station as approximately $7,000,000. This cost is representative of recently constructed transfer stations. For simplicity, it is assumed that all costs at the transfer station are fixed costs, independent of the throughput experienced each operating day.

Table 23.9 shows annual costs totaling $1,543,300. When this cost is spread over the expected annual throughput of 390,000 tons, the transfer facility cost is about $3.86 per ton, as shown in Table 23.10.

TABLE 23.6 Variable-transfer haul cost

	Miles per trip (one-way)				
	10	20	30	40	50
Trips per day	8	5	4	3	3
Variable cost per round trip, $	18.60	37.20	55.80	74.40	93.00
Variable cost per ton, $	1.03	2.07	3.10	4.13	5.17

TABLE 23.7 Transfer haul cost summary

	Miles per trip (one-way)				
	10	20	30	40	50
Trips per day	8	5	4	3	3
Fixed cost, $/ton	2.83	4.49	5.60	7.44	7.44
Variable cost, $/ton	1.03	2.07	3.10	4.13	5.17
Total cost per ton, $	3.86	6.56	8.70	11.57	12.61
Total cost per ton-mile (one-way), $	0.39	0.33	0.29	0.29	0.25

TABLE 23.8 Input data for transfer station cost

Facility size, tons/day	1,500
Days of operation each year	260
Building and site:	
Development cost, $	7,000,000
Service life, years	20
Cost of debt, %	9.0%
Maintenance costs (% of initial cost)	2.0%
Land:	
Cost (10 acres/1,000 tons per day @ $25,000/acre)	375,000
Amortization period, years	30
Cost of debt, %	9.0%
Transfer station mobile equipment:	
Cost, $	400,000
Service life, years	5
Salvage value (% of initial value)	10.0%
Cost of debt, %	9.0%
Maintenance cost (% of initial cost)	5.0%
Station yard tractors:	
Cost, $	65,000
Service life, years	5
Salvage value (% of initial value)	10.0%
Cost of debt, %	9.0%
Maintenance cost (% of initial cost)	3.0%
Annual utilities allowance, $	25,000
Annual insurance allowance, $	15,000

Annual costs for labor

Position	Number	Annual salary, each
Supervisor	1	30,000
Equipment operators	6	27,000
Scale operators	2	22,000
Laborers	4	17,000
Fringe benefits allowance (% of annual salary)	35.0%	

TABLE 23.9 Annual Fixed Transfer Station Costs

Station labor position	Number	Annual salary (each)	Salary ($)
Supervisor	1	30,000	30,000
Equipment operators	6	27,000	162,000
Scale operators	2	22,000	44,000
Laborers	4	17,000	68,000
Subtotal	13		
Fringe benefits allowance (35% of salary)			106,400
Total labor cost			410,400
Land			36,500
Buildings and site development			766,800
Station mobile equipment			112,600
Station yard tractors			15,000
Station utilities			25,000
Building and site maintenance			140,000
Station mobile equipment maintenance			20,000
Station yard tractor maintenance			2,000
Insurance allowance			15,000
Total			1,543,300

The effect on the local economy of the addition of such a facility would be the addition of numerous construction jobs over a one- to two- year construction phase and the addition of several full-time employees once the facility became operational. Significantly more employees could be needed if the transfer station were to encompass aspects of recycling, composting, or other solid-waste-processing practices. There would be some workers associated with direct-haul operations who would be displaced as the result of the introduction of a collection system incorporating the transfer station.

Break-Even Analysis

Figure 23.19 depicts the cost of direct haul and transfer haul. Direct-haul costs are on a per-ton basis and are linearly related to transfer-haul distance. Transfer-haul cost consists of a flat cost per ton to operate the transfer station and a nearly linear relationship of transfer haul to haul distance.

The point where these two lines intersect is the break-even point, i.e., the distance at which it costs the same to direct-haul waste as to transfer and haul it in trailers. For shorter distances, direct haul is more economical. For greater distances, transfer haul is more economical. Figure 23.19 illustrates that transfer haul would be more cost effective than direct haul in a 25 yd^3 packer vehicle with one-person crew if the distance to the landfill is more than approximately 20 mi.

TABLE 23.10 Transfer station cost

Facility throughput, tons/year	390,000
Transfer station total cost, $/ton	3.96

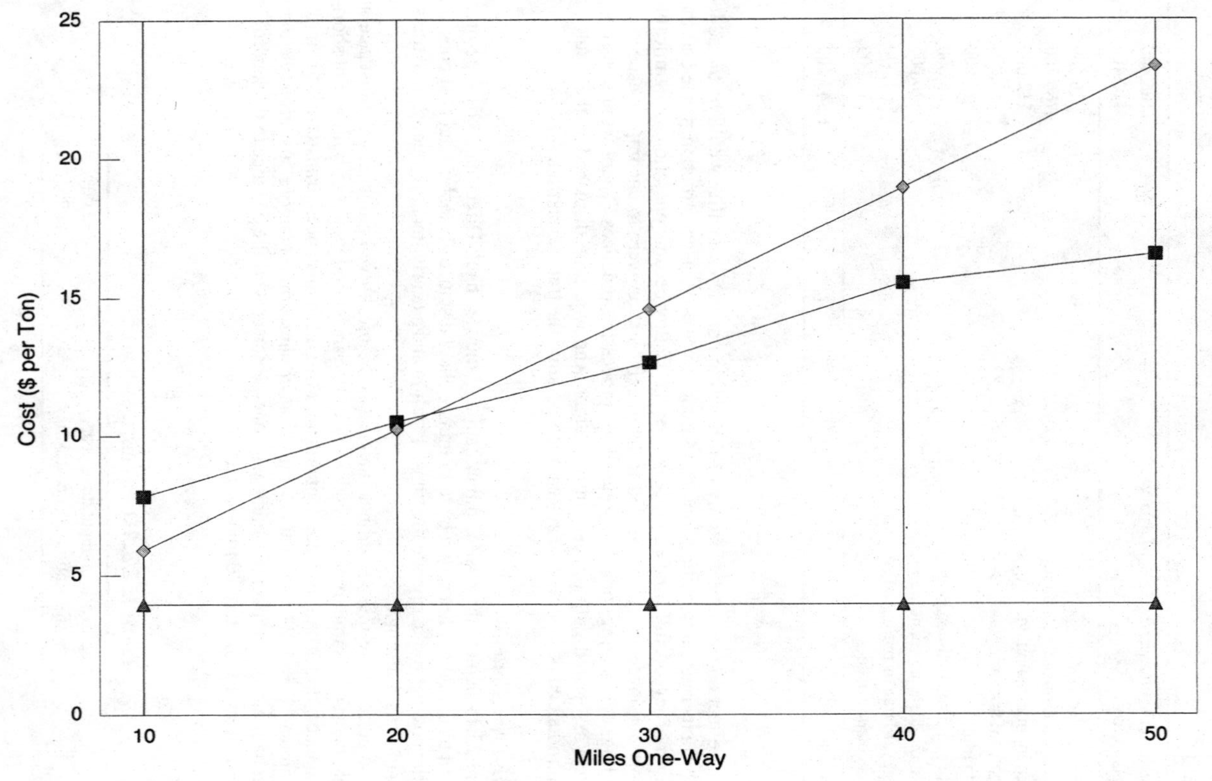

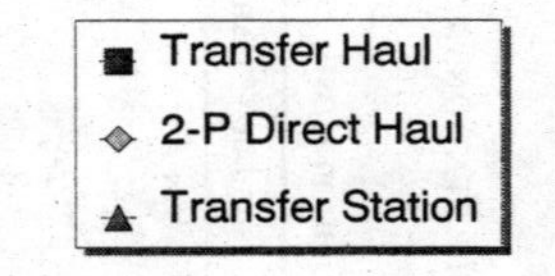

FIGURE 23.19 Direct haul versus transfer cost graph.

OTHER CONSIDERATIONS

Transfer stations can be adapted to the unique needs of the system. Auxiliary facilities may include office space, restrooms, scales and scale house, employee showers and dressing room, lunch room, maintenance facilities, fuel storage and dispensing area, and truck storage area. These facilities can have a substantial impact on site and structure requirements. Their availability may complement the basic station and enhance its usefulness.

Recycling and Energy Recovery Compatibility

Material recovery considerations have become an important element in the overall transfer station plan. Almost all new transfer stations being designed allow for some type of material recovery. Many existing transfer stations are being retrofitted to handle material recovery operations. These operations may include separation and handling of materials for recycling or processing of waste into fuel for energy-recovery facilities. New transfer stations can be designed to initially include material-recovery facilities (MRFS) or can be designed to add a MRF at a later date. The MRF and transfer station can be housed in the same building or the MRF can be in a separate building. Provisions should be made for rejects (unrecoverable materials) to be conveyed from the MRF to the transfer station for disposal.

Recycling
Materials recovery at transfer stations can include manual separation of materials on the tipping floor or along processing lines, mechanical methods of separating materials, or a combination of both. Many transfer stations (see Fig. 23.20) provide roll-off containers for individuals to drop off recyclables. Floor separation of materials most often includes

FIGURE 23.20 Roll-off containers at transfer station.

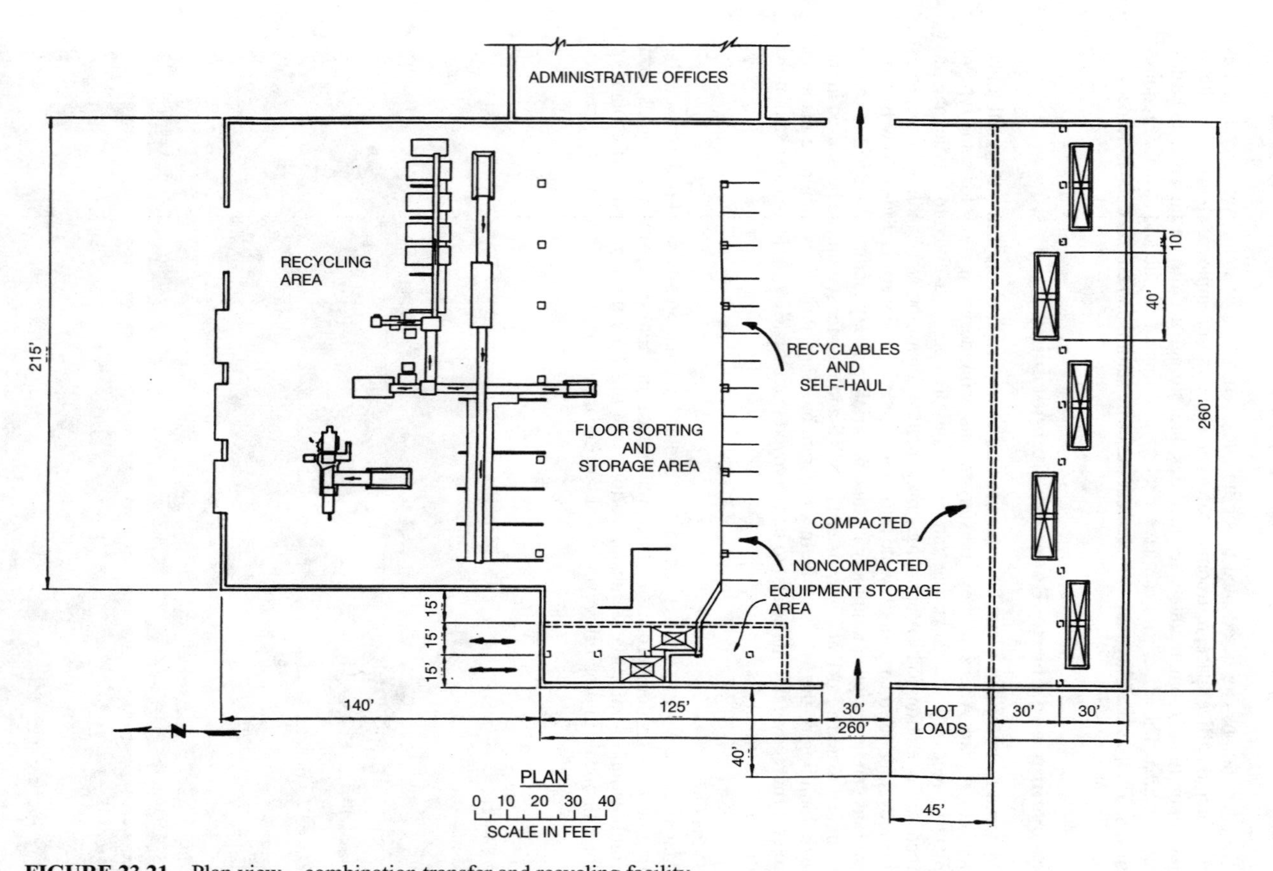

FIGURE 23.21 Plan view—combination transfer and recycling facility.

the recovery of corrugated and wood materials. Commercial collection vehicles often contain loads with a high percentage of these materials, making them economical to recover. Conveyors with manual picking stations can also be used to recover materials from the waste stream.

Separation and handling equipment such as magnetic separators, air classifiers, trommel screens, disk screens, glass crushers, and balers can be added to increase capacity and efficiency of the recycling operations. Figure 23.21 shows a plan view of a combination transfer and recycling facility. Compacted solid waste is processed on one side of the building and recyclable material is processed on the other side.

These material recovery operations can use considerable floor space in an existing transfer station. It is important to evaluate the space requirements carefully to make certain that there is enough room for both material recovery and transfer operations in the same building.

Energy Recovery

Transfer stations have been used as processing facilities for refuse-derived fuel. Transfer stations can also be planned to service a future energy-recovery facility. The waste receiving area can serve as a receiving area for an energy-recovery plant. Stations can be constructed with a removable wall to allow a future energy-recovery facility to be constructed adjacent to the transfer station.

FURTHER READING

1. U.S. Environmental Protection Agency. "Collection and Transfer." In *Decision-Maker's Guide to Solid Waste Management,* Second Edition. Washington, D.C., EPA/530-R-95-023, August 1995.
2. Brown, M. D., T. D. Vance, and T. C. Reilly. *Solid Waste Transfer Fundamentals,* Ann Arbor, MI, Ann Arbor Science Publishers, 1981.
3. Schaper, L. T. "Transfer of Municipal Solid Waste." In *The Solid Waste Handbook,* New York, Wiley, 1986.

CHAPTER 24

MATERIALS RECOVERY FACILITIES

THOMAS D. KNOX, P.E., D.E.E.
Engineering Manager, Black & Veatch Corp.
Kansas City, Missouri

R. CHRISTIAN BROCKWAY, P.E.
Project Manager, Black & Veatch Corp.
Kansas City, Missouri

INTRODUCTION

History

Materials recovery and processing facilities were initially developed in response to the need to handle the growing quantities of recyclables. They have become a "self-fulfilling prophecy" in that their success has increased the popularity of recycling, requiring the construction of still more processing capacity. Refuse and recyclables initially came from numerous uncontrolled sources and the quality varied. Processing facilities served as central collection locations where materials were processed to market specifications.

Processing facilities serve as brokers, collecting from several haulers and distributing to several markets. Alone, small haulers could not meet market specifications or profitably deal with buyers for a wide range of materials. Through the processing facility, materials from several sources, collected through several haulers, could be matched with several markets.

Many processing facilities have been combined with solid waste transfer operations. This is a logical combination, because both have similar site requirements (access to transportation routes, industrial zoning, and a central location) and both use similar facilities (tipping floor for receiving, areas for processing material, and shipping facilities). Also, some materials initially directed to a transfer facility, such as commercially generated old corrugated cardboard (OCC), are often taken to a processing facility instead. Combining the two facilities into one location eliminates a transfer step in the operations.

Definitions

Several types of processing facilities have evolved to accommodate the variety of recycling systems. Unfortunately, the recycling community has not established a consistent set of labels for these various facilities. Several facilities are defined in this chapter in order of increasing complexity of processing. These definitions will be used throughout this chapter; however, they are not universally accepted by the industry.

Some facilities operate as receiving and shipping centers. They simply accept various materials, ensure that materials are separated correctly, and ship them to other processing or manufacturing facilities. Such centers, called "drop-off facilities," include no processing equipment. If a drop-off facility pays patrons for recyclables, it is called a "buy-back center." Depending on their locations, drop-off and buy-back operations may be enclosed, but they are different from the facilities discussed in this chapter, because drop-off and buy-back centers deal directly with the public and do not typically contain significant processing equipment.

If recyclables are source-separated, the facility does not need sorting equipment. Source-separated recyclables that are collected through a curbside program are taken to one or more places for storage and subsequent shipment. The facility would need only hoppers for receiving and storage of the source-separated materials until they are shipped. The receiving hoppers may be located outdoors if they are protected from pilferage and vandalism. Such a facility may include provisions for the removal of contaminants, but generally, the materials received are already separated. Such a facility could be called a "load-out" facility because it serves as a central receiving and shipping center but does not process any materials.

Although equipment is not needed for separation at a load-out facility, equipment may be added for processing to increase the value of the recyclables. Processing is usually limited to preparing the material for markets by flattening, crushing, or baling. If the facility contains processing equipment, it is called an "intermediate processing center" (IPC). However, because these facilities do not require equipment for separation, they differ from the processing facilities described below.

Recycling programs that collect various recyclables together produce a stream of intermixed, or commingled, recyclables. If a facility receives commingled recyclables, separation equipment is necessary as well as equipment for preparing the materials for market. Such a facility is commonly called a materials recovery facility (MRF). This term has been used to describe virtually every type of facility, from drop-off center to mixed waste processing facility. In this chapter, MRF means a facility that receives commingled recyclables. The term "commingled recyclables" has different meanings in different parts of the country. Most commingled processing systems separate only the containers (glass, plastic, and aluminum) and do not attempt to separate fibers, from either the containers or from other types of fiber. In this chapter, commingled refers to only containers, unless otherwise noted.

Refuse that has not been separated to concentrate the amount of recyclables would generally go to the type of recycling facility with the most complex processing system. Unprocessed waste is called raw garbage or mixed waste. A facility that receives unprocessed waste is called a "mixed waste processing facility" or simply a "waste processing facility" (WPF). WPFs generally recover less than 20% of the waste stream they receive. The remainder includes organic material that can be composted or processed into refuse-derived fuel (RDF) for waste-to-energy facilities. The type of processing equipment used at these facilities differs depending on whether the final product is compost or RDF.

An attempt to establish categories of systems is usually confounded by overlapping categories. An example is a type of collection–recycling system that combines source separation with mixed waste processing. In this approach, recyclables are bagged separately from the mixed waste, often in blue bags, but collected in the same vehicles as the mixed waste. This approach is called "co-collection" or "blue bag" collection. At the processing facility, the bags of recyclables are removed from the other waste and processed as commingled recyclables. The remaining waste may be processed as mixed waste or not processed and landfilled.

Another type of facility defies classification into a single category because it accepts both mixed waste and commingled recyclables. Such facilities process both waste streams efficiently because they are designed to process mixed waste, but they allow source-separated recyclables to be introduced at a later stage in the processing line. In this way, commingled recyclables, which have already been separated, bypass early stages of separation in the process line.

In the United States, MRF designs vary widely depending on the incoming materials and the technology and labor used to sort the materials. In 1995, there were 310 MRFs in the United States; 196 low-technology (predominantly manual sorting) and 114 high-technology (mechanized sorting) systems. Approximately 74% of the MRFs are privately owned and approximately 85% are privately operated.[1]

SYSTEM ECONOMICS

The cost-effectiveness of a processing facility is more than a function of the facility costs. The costs of developing, implementing, and operating a processing facility should include all the elements of the processing system, including collection, transportation, processing, and shipping. For example, an IPC will have a lower initial cost than a WPF because the WPF has more processing equipment. However, the collection costs for the IPC system, which requires separate collection of source-separated recyclables, will be greater than the collection costs for the WPF. During the planning and evaluation of alternative collection and processing systems, the system should be evaluated as a whole. It can be misleading to consider the costs of only the physical facilities.

Collection costs can be a significant portion of the total cost of recycling systems, and collection systems for source-separation programs are more complex than those for simple trash collection. Commingled and source-separated systems typically require the collection route to be covered twice. The collection route would be covered once for trash and once for recyclables. Commingled recyclables can be collected with existing collection vehicles, but source-separated recyclables require the additional cost of special compartmentalized recycling vehicles. The compaction feature of existing packer collection vehicles, however, may not be usable on commingled recyclables because it may cause excessive breakage, depending on the collected recyclables. However, co-collected or blue bag waste systems offer separation at the source without additional collection vehicles.

Separation of recyclables involves a cost, whether it is done within the facility or at the source. Recyclables that are source-separated by the generators include additional costs of time and inconvenience that are paid by the generators. Commingled recyclables that are sorted as they are emptied into the collection vehicle require additional time during collection, although the need for additional separation at the facility is eliminated.

The choice of recycling system also influences the degree of participation, which can control the cost-effectiveness of the recycling program. Mixed waste processing does not require additional effort on the part of the generators and therefore, has 100% participation. Source-separated systems require more work on the part of the generators and thus have lower participation rates.

The cost of containers for collection also influences the cost of the program. Commingled, source-separated, and co-collected recyclables require buckets, bins, or bags. Mixed waste recycling does not require additional containers. However, some containers also offer a method of distributing the costs of the collection system to the users. For example, approved bags that must be purchased by generators establish a "pay by the bag" system.

CONCEPTUAL DESIGN

Objectives

A well-conceived design should optimize the many variables involved in producing an efficient facility. Among the variables to be considered are the following:

- Capital cost
- Collection cost
- Operation and maintenance cost (processing)
- Shipping cost of the processed material
- Material storage space
- Public and employee safety
- Public education
- Product quality

Some of these variables may compete with each other. The optimum solution varies from site to site, depending upon local and regional characteristics and priorities. With these variables in mind, this chapter will present some of the design considerations that the owner or designer should be aware of when developing the design concept for a MRF.

Market Specifications

The designer should know the market the system will be supplying and understand its specifications and dynamics. Material specifications change with supply and demand. For example, if a supply of color-separated, high-density polyethylene appears, a prior specification calling for clean, mixed plastics may become obsolete. The ways the supply and demand for various materials affect the facility design are dependent on the materials.

Paper
The paper market is perhaps the most dynamic of the recyclables markets. Privately operated paper sorting lines should be adjusted to optimize the mix of sorting labor to meet various market specifications. When foreign markets are available, paper that undergoes little sorting can have a strong market. The low cost of labor overseas (the Pacific Rim and Mexico) makes this a strong option. Specifications normally deal with the quality of the product and the form in which it is shipped. Haul distances to recycled paper plants usually make baling advantageous. Specifications for paper typically list a baled density, a maximum moisture content, and a maximum percentage of contamination by nonspecified paper.

Glass
The glass market requires that glass be separated by color, and it may or may not call for delivery of cullet. Some processors claim that crushed glass is easier to handle. The end market probably has a glass crusher, so crushing is done by the processing facility to facilitate transportation and handling. Bottle caps and other contaminants are usually characterized with a maximum allowable limit. Color separation is essential. Manufacturers of glass containers generally accept flint (clear) glass in batches of amber or green glass, but green and amber glass are considered contaminants in batches of any color but their own. Plate glass is also considered a contaminant in container glass because it has a lower melting temperature.

Plastics
High-density polyethylene (HDPE) is generally the most readily marketable plastic. Polyethylene terephthalate (PET) markets are less predictable but appear to be growing. Color separation of PET enhances its marketability. Clear and natural-colored PETs are generally more marketable than green. Consequently, if bales are not color-separated, there may be an upper limit on the allowable percentage of green PET. Since bottle caps are a contaminant to both HDPE and PET, they may be limited in the specifications.

Metals
Specifications for aluminum cans usually call for densification. A smaller brick of cans can bring a small premium over larger, less dense bales on the market. Because the detinning process to which tin cans are subjected involves dipping in chemical baths, the specification for their delivery may limit the density of the bales. Bimetal cans (cans with tin sides and aluminum ends) are generally not acceptable to either aluminum or tin processors.

Sizing

MRFs can generally be classified into two types: municipal MRFs and merchant MRFs. This classification has to do with the relationship between the facility and its supply of recyclables. Municipal MRFs are typically guaranteed a flow of recyclables from a community. Merchant MRFs, on the other hand, must compete in the market for their supply. Municipal MRFs may be operated by local government or, more often, by a partnership between the public and a private enterprise. Such an arrangement allows the project to have the public's access to capital markets for the construction of the facility. It also allows for the operational flexibility that an entrepreneur with performance-based incentives brings to the facility.

Merchant MRFs offer their services in a competitive environment and are often privately owned and operated. Mandatory recycling in east coast communities has fostered many merchant MRFs. Sizing is critical to MRFs, whether municipal or private, but it may be approached differently. The waste stream to a municipal or merchant MRF is a function of the fluctuations in material deliveries caused by competition, markets, and the following variables.

- Households in the service area
- Sign-up rate
- Capture rate
- Level of setout

Before embarking on the design of a large MRF, it is advisable to run a pilot project in neighborhoods that typify the entire community. If a pilot program is not practical, the characteristics of neighboring or similar communities could be substituted. Large service areas result in economies of scale, but smaller service areas have lower collection costs. In practice, the size of the service area is often set by geographical or political boundaries.

In a voluntary program, experience has shown that only 40 to 60% of those who signed up can be expected to participate. However, in communities with effective public education programs or where landfilling of recyclables is banned, the sign-up rate can approach 100%. Signing up should not be mistaken for participation in either voluntary or mandatory recycling. The argument has been advanced that capture rates are higher for programs requiring less effort on the part of the participant. Recent findings, however, indicate that capture rates may depend more on public education efforts than on the type of program.

The magnitude of setout is dependent upon the materials covered by the program. Programs that include the collection of plastics, aluminum, glass, and tin cans can anticipate typical monthly contributions of 20 to 30 pounds per household. If newspapers are included, this weight can double.

Using these general guidelines, an estimate for the daily throughput for a MRF processing commingled recyclables in a community of 100,000 households is as follows.

Assumptions:

100,000 households
70% sign-up rate
50% capture rate
20 pounds per month per household of recyclables

Calculation:

$$100{,}000 \text{ households} \times 70/100 \times 50/100 \times 20 \text{ lbs/mo/household} = 700{,}000 \text{ lbs/mo}$$

Assuming the collection is distributed evenly over 22 working days per month, daily tonnage received at the MRF from this community of 100,000 households would be the following.

$$700{,}000 \text{ lbs/mo} \times 1 \text{ mo}/22 \text{ days} \times 1 \text{ ton}/2000 \text{ lbs} = 15.9 \text{ tons/day}$$

Before proceeding with sizing the facility, the operating assumptions of the MRF should be well defined. Questions relating to operation to be addressed include the following.

- What are the normal operating hours?
- Can hours be expanded on an as-needed basis?
- Is operating double shifts an option?
- How much system redundancy must be provided?

With the answers to these questions, the designer can begin to develop a concept based on vendor systems with known throughput capacities or to conceptualize a new system. Figure 24.1 presents a material flow diagram for an industrial waste MRF processing system. Such a material balance flowchart is a useful tool for developing an understanding of the system processing requirements. Throughput tonnage should be separated into constituent materials. The flow of material should be expressed in both weight and volume, because both are used in material handling evaluations.

Site Layout

An important consideration in the site layout for a MRF is a sound traffic plan prepared by a competent traffic engineer. Factors that the traffic engineer should consider include the following variables.

- Incoming materials (number and type of collection vehicles)
- Outgoing materials (number and type of vehicles)
- Material weighing requirements
- Estimated peak hour traffic
- Public access
- Auxiliary functions (transfer station, public drop-off, household hazardous waste drop-off)

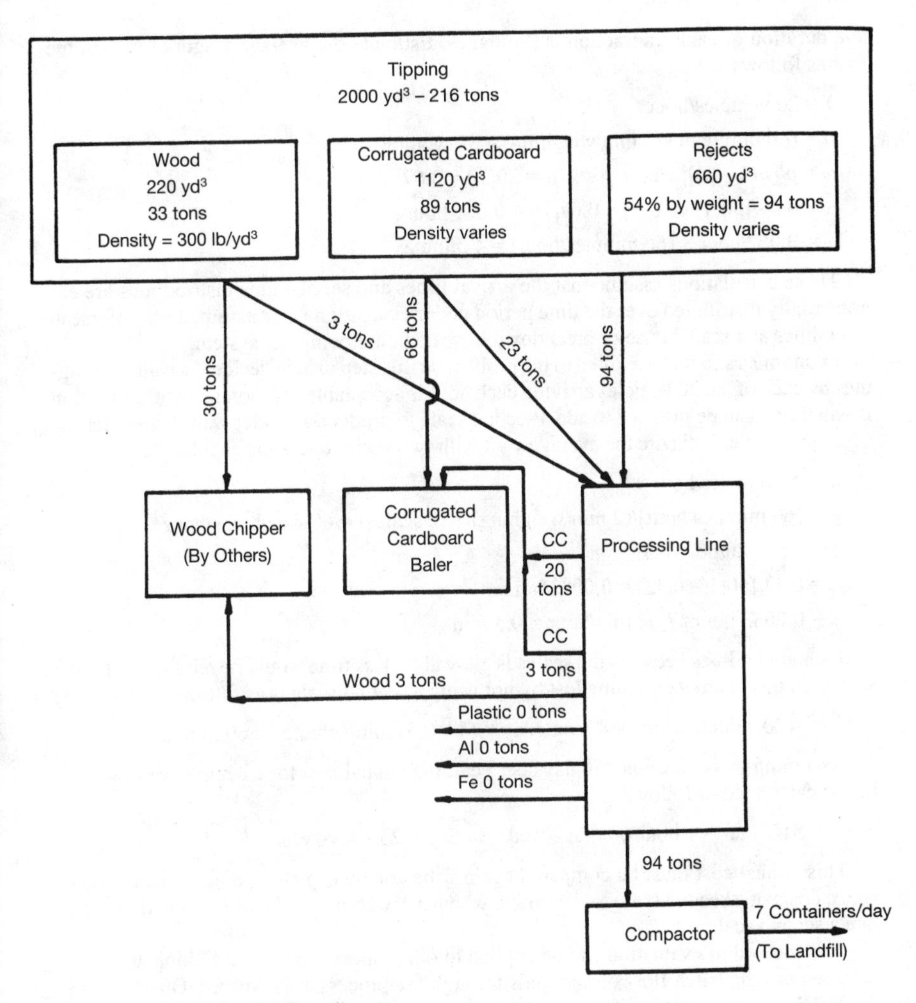

FIGURE 24.1 Material flow diagram.

The traffic information should be incorporated into a site plan that minimizes the number of traffic intersections, lane changes, and merges. In particular, the public's opportunities to interface with other traffic should be minimized.

Because of the many steps involved in the normal use of the facility, it is important to have a basic understanding of queuing theory for the site and facility layout. MRFs are labor-intensive, and whenever the personnel in either the MRF or the collection vehicles are idle, the efficiency of the operation is reduced. Ideally, material would arrive and be processed at a uniform rate throughout the facility's working hours, and the facilities would be sized accordingly. Determining the number of scales at a scale house is an example of how a facility should be sized, taking into account the results of queuing analysis.

Consider that a scale house receives an average of 20 vehicles each hour, and the average duration of each transaction is 2 minutes. Estimate the average length of stay in the line as follows:

$\lambda = 20$ vehicles/hour

$\mu = (60 \text{ min/hour})/(2 \text{ min/weighing}) = 30$ weighings/hour

$\omega = \rho/[\mu(1 - \rho)]$; where $\rho = \lambda/\mu = 20/30 = 0.67$

$\omega = 0.67/[30(1 - 0.67)] = 0.67/10 = 0.067$ hour

$\omega = 0.067 \text{ hour} \times (60 \text{ minutes/hour}) = 4$ minutes

These calculations assume that the arrival times and service time distributions are exponentially distributed over the time period under evaluation.[2] Not assumed are unforseen truck lines at a scale house or breakdown of scale house computer system.

Economics should be applied to the problem at this step to help decide if a wait of 4 minutes by each of the 20 vehicles arriving each hour is acceptable. Economically, the question is whether it can be justified to add a second scale to produce a shorter waiting time for the average vehicle. Estimate the average wait with two scales operating as follows.

$\lambda = 20$ vehicles/hour

$\mu = (60 \text{ minutes/hour})/(2 \text{ min/weighing}) \times (2 \text{ scales}) = 60$ weighings/hour

$\rho = \lambda/\mu = 20/60 = 0.33$

$\omega = 0.33/[60(1 - 0.33) = 0.0083$ hours

$\omega = 0.0083 \text{ hours} \times 60 \text{ min/hour} = 0.5$ min

If changing lines between the scales is allowable, less time would be required with two scales. In this example the time lost by not using a second scale is as follows.

$$20 \text{ vehicles/hour} \times 8 \text{ hours/day} \times (4 - 0.5) \text{ min/vehicle} = 560 \text{ min/day}$$

Assuming this condition 250 days per year, the annual cost for an employee to wait in line is calculated as follows.

$$\$12\text{/hour} \times 1 \text{ hour/60 min} \times 560 \text{ min/day} \times 250 \text{ days/year} = \$28{,}000\text{/year}$$

This annual cost must be compared against the annual capital, operating, and maintenance costs of a second scale to determine whether the layout of the MRF should incorporate a second scale.

This method of evaluation can be applied to other operations where waiting and interference with the batch flow of materials through the process are involved. Other areas at the MRF that may benefit from such an evaluation may include the following:

- Drop-off of recyclables into designated slots
- Loading recyclables onto conveyors
- Loading balers

Facility Layout

The principal guideline in the layout of a recycling system is to limit the handling of materials while still meeting the market specifications. Most existing systems require a high degree of manual handling in order to meet the needs of the material markets. Any effort to limit this handling must begin with the layout of the processing lines. An example for limiting the handling of materials is to arrange the hoppers and containers so that the system delivers directly to the container or vehicle in which the material will exit the facility.

Another important guideline is to arrange the sorting lines so that "negative sorts" of the material with the largest volume will occur. Material that is negatively sorted is not physically removed from the material flow. It simply falls off the end of the belt. On each sort line the product that goes off the end of the line should be the material with the largest volume, because that material is not handled by the sorters. The process should also be designed so that negatively sorted material meets market specifications.

Adequate storage area should be provided in the layout. Discussions with operators of existing facilities reveal that few facilities have floor space that is underutilized. Storage is required for incoming material and for material awaiting shipment to market. Space should be provided for at least one day's delivery of recyclables. In this way, if the processing equipment is not functioning for one shift, the facility can still accept and store recyclables for processing during an emergency second shift. Additional storage may be required if the normal operating hours of the facility exceed the period during which collection is carried out. For example, a facility that operates two 8 hour shifts daily, while collection is done during only one shift, will need storage for 8 hours worth of material as standard operating practice. If the system is to be designed to allow for occasional downtime, additional space may be required.

Turning radii and operating heights of collection vehicles should be taken into consideration when laying out a new facility. Most new MRFs should be designed with maximum clear spans and operating heights of 25 to 30 feet in areas where mobile equipment will be operating. MRFs are noisy and busy. Therefore, the designer must be aware of layout considerations that might affect worker safety and comfort. Washing facilities should be located away from heavy traffic lanes. Handling of large quantities of glass can be particularly noisy. For this reason in some MRFs, glass bunkers are located outside of the processing building.

It is important to design redundancy into the system. Equipment failure in one area of the MRF should not shut down the entire system. Incorporating multiple sorting lines and alternative handling methods will help to keep the system functional when parts of the system shut down. Equipment should be located so that maintenance can be performed without disruption to other system functions.[3]

PROCESSING SYSTEMS

Single Vendor Systems

Some processing systems are preengineered to serve a wide variety of applications. The systems often include proprietary equipment and configurations. Such systems are commercially available from companies that will design, construct, operate, and provide financing for the facility. Preengineering is no guarantee that the components or the system will work well, but some of the uncertainty can be eliminated by evaluating the system's performance in existing applications.

Figure 24.2 shows a preengineered system. This system is designed to process unsorted, mixed municipal solid waste. It has a capacity of up to 200 tons per day, according to the manufacturer. This system can handle higher capacities by adding processing lines.

Component Systems

A preengineered system may not be appropriate if the processing needs are nonstandard. The dynamic status of current recyclable markets combined with the many different col-

FIGURE 24.2 Buhler/Reuter mixed waste processing system, Eden Prairie, Minnesota. (*Courtesy of Buhler/Reuter*)

lection methods in use has not allowed for the development of a generic processing system that functions in multiple locations. If a preengineered solution is not appropriate, the alternative is to design a processing system to meet the site-specific requirements. Such a system may include customized equipment, but more often, such recycling systems consist of standard components (screens, conveyors, etc.). The challenge in designing a processing system for a particular application is to select the correct equipment, size it appropriately and cost-effectively, and arrange it efficiently. Thus, the major difference between systems is the arrangement and combination of the components.

Types of Systems

The configurations of both vendor systems and custom systems depend upon the type of waste to be processed. Systems can be designed to process any of the following materials:

- Presorted streams of a single recyclable material
- Presorted streams of mixed recyclables
- Mixed streams that are high in recyclables (for example, commercial waste that is high in corrugated cardboard)

Processing the first type of waste stream is relatively simple; the process consists of size reduction (shredding or baling) and screening. Processing the second type of waste stream is frequently the goal of a processing facility. Figure 24.3 shows a layout for one such facility. This layout for commingled recyclables also provides locations where materials can be introduced into the line midway to bypass unnecessary processing steps. A system for processing the mixed waste stream is shown in Figure 24.4. This system also

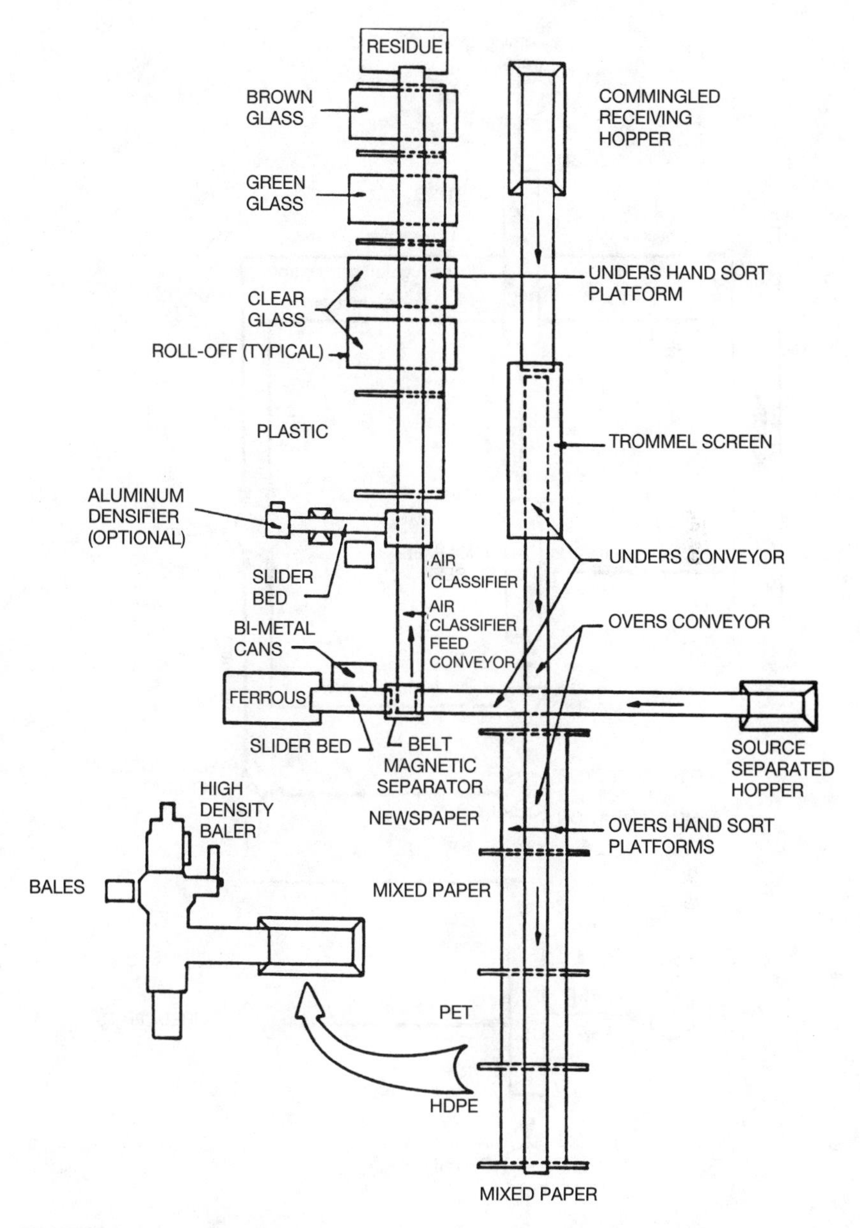

FIGURE 24.3 MRF processing equipment layout.

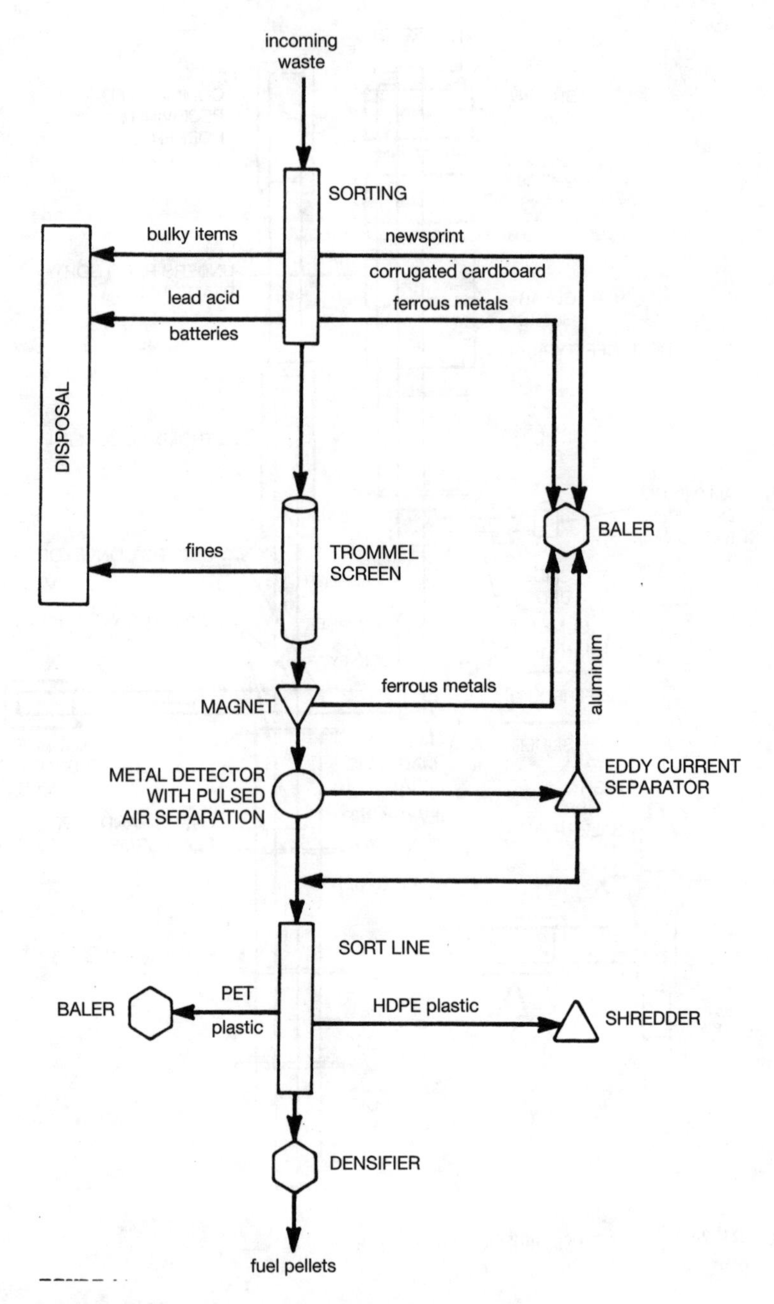

FIGURE 24.4 Mixed waste processing flow diagram.

provides for removal of some hazardous and nonprocessible materials, such as lead–acid batteries and automobile tires.

PROCESSING EQUIPMENT

Preparation

Some processing lines reduce the size of the material for ease in further processing and conveyance. A hammer mill is commonly used for reducing the size of solid waste. It consists of one or more spinning shafts to which hammers are attached. The hammers may swing freely or be rigidly attached; the swing hammer is most commonly used in solid waste processing. Hammers may be sharp for chopping or blunt for beating. The tearing and beating motion of the hammer makes the hammer mill particularly effective on paper fiber and brittle material such as glass. The particle sizes of the solid waste are reduced through impact with the hammers and the grate across the output opening. Output particle size is controlled primarily by the size of the grate openings.

Size reduction equipment for municipal solid waste requires a motor in the range of 200 to 1500 hp. A minimum of 10 hp is generally required for each ton per hour of capacity. Power requirements increase exponentially as particle size decreases.

Shredders have exploded while processing municipal solid waste. For this reason, they are often housed in an explosion-protected enclosure and their use is preceded by visual inspection. Because they homogenize the waste, they are seldom used at the front of the processing line if significant materials recovery is planned. Ferrous metal is an exception, because it is easily recovered from shredded waste. More commonly, shredding is used after initial screening, if at all. Shredding is also used at the end of the process to prepare materials for markets or to process the "rejects" for beneficial use as RDF or compost.

Separation

The essence of processing is the separating of materials. Several methods have been developed, each with its strengths and preferred applications. Some combination of methods is usually required to process a mixed waste stream.

Hand Sorting

Hand sorting is a part of most processing systems. Hand sorting may take place on the tipping floor, on a sorting line, or both. Large materials such as corrugated cardboard, hazardous materials, and materials that could damage the processing equipment are removed by hand on the tipping floor. On a sorting line, a belt conveyor moves the waste stream past workers, each of whom is responsible for removing one or more types of material. Figure 24.5 shows a sorting line.

Sorting conveyors must be designed around productivity levels of workers. They are often designed to travel at variable speeds for optimum performance and flexibility under various loadings. Belt speeds generally vary between 10 and 60 feet per minute.

Screens

The purpose of the screen is to separate items according to particle size. Different types of screens are commonly installed in a processing line and typically oscillate, vibrate, or rotate. Screens not only separate the waste stream, thus improving the manual sorting effi-

FIGURE 24.5 Materials recovery sorting lines.

ciency, but also remove loose dirt and residue, thereby improving the quality of the recovered materials.

Trommel Screens

A trommel screen is a cylindrical, rotating screen that sorts material by size. Figure 24.6 shows a trommel screen. The trommel is set at a slight angle, and the refuse tumbles through the cylinder as it rotates. Oversized objects (overs), such as large plastic bottles, move through the cylinder while undersized objects (unders), such as cans, bottles, and other small items, fall through holes in the cylinder.

The trommel may also be designed to open bags of waste and reduce the size of breakable items, especially glass. Spikes or knives fastened to the inside of the cylinder can help rip bags open. Trommels are designed with paddles, called lifters, fastened to the inside of the rotating trommel, which lift and drop the waste. The impact helps to open the bags and break glass. Trommels can be designed to limit glass breakage by controlling the rotational speed and designing the paddles to keep material in the lower portion of the trommel. Spikes and lifters can be bolted, rather than welded, to the trommel to facilitate replacement.

Trommel screens have been used for decades to sort industrial and manufacturing materials and, more recently, to sort solid waste. The lifting, dropping, and tumbling in the trommels also serve to untangle and separate materials; this is important for effective screening.

Trommel sizes range from approximately 5 to 12 ft in diameter and from 20 to 70 ft in length. Trommels larger than this cannot be shipped because of the size and weight limits for most trucks and even for rail hauling. Large trommels can be shipped in segments and erected in the field, but field construction may not be as good as shop fabrication. Trom-

FIGURE 24.6 Trommel screen. (*Courtesy of Triple/S Dynamics, Inc.*)

mels may be enclosed in shrouds for dust control. Screens may be bolted or welded to the frame. Bolting is preferred so that the screen can be removed for maintenance or to change the size of openings.

A series of trommels with different-sized openings can be used to sort materials into several size categories. Alternatively, a single trommel can sort for multiple categories by varying the size of openings between different sections of the trommel. For example, the first half of the trommel might contain 2 in openings and the last half 6 in openings. This will result in sorting the waste stream into groups of less than 2 in, 2 to 6 in, and larger than 6 in.

Chain and trunnion are two types of drive systems used for trommel screens. With the chain drive, the trommel is connected to a motor by a chain. With the trunnion system, the trommel cylinder is supported at several points by drive wheels. The wheels drive the trommel through friction, which results in less direct transfer of power than with a chain drive. This can create drive problems, especially if the trommel must be started under load.

Table 24.1 presents the characteristics of some standard-sized trommels. Many trommels, however, are designed for site-specific requirements. The costs of trommels range from approximately $60,000 to $500,000, excluding structural support, shipping, and erection. The costs vary depending on sizes, dust enclosures, bolted versus welded screens, and other options.

Disk Screens

The disk screen consists of successive rows of vertical rotating disks. The tops of the rotating disks form a moving surface that tumbles and conveys the solid waste, and smaller material falls through the openings between the disks as shown in Figure 24.7. Disk screens occupy less space than trommels, have lower capital cost, produce less dust, and require less operating power. However, disk screens may not separate materials as well as trommels because they do not agitate the waste as vigorously. For example, a heavy, wet load of waste may travel over the disks as a solid mass.

The opening size on disk screens can be varied by changing the size and spacing of the disks. Disk screens are most effective when they are sized so that the majority of the material falls through the disks. Screens separate by size only, so it is possible that light and dense materials, such as paper and metals, will be sorted out together if they have the same size. Therefore, screening is often followed by air classification.

Air Classification

Air classifiers sort by density and aerodynamic properties of the materials. The materials to be separated are introduced into an air stream. Based on size, shape, and density, some ma-

TABLE 24.1 Trommel screen characteristics

Nominal length, ft	Nominal diameter, ft	Rated capacity, tons/hour*	Screen length, feet	Screen characteristics	revolutions/ min	Weight, lbs†	Angle, degrees
40	8.5	55	30	As required	Varies	40,000	5, others available
50	10.5	75	40	5 in diameter holes, 55% open area, other sizes available	13	80,000	5, others available
60	12.5	120	48	5 in diameter holes, 53% open area, other sizes available	11-12	120,000	3–8, adjustable

Source: Based on trommel designs offered by the Heil Co.
*Based on density of 15 lb/ft^3 for unprocessed refuse.
†Including trommel, hopper, dust cover, and bases, excluding concrete piers.

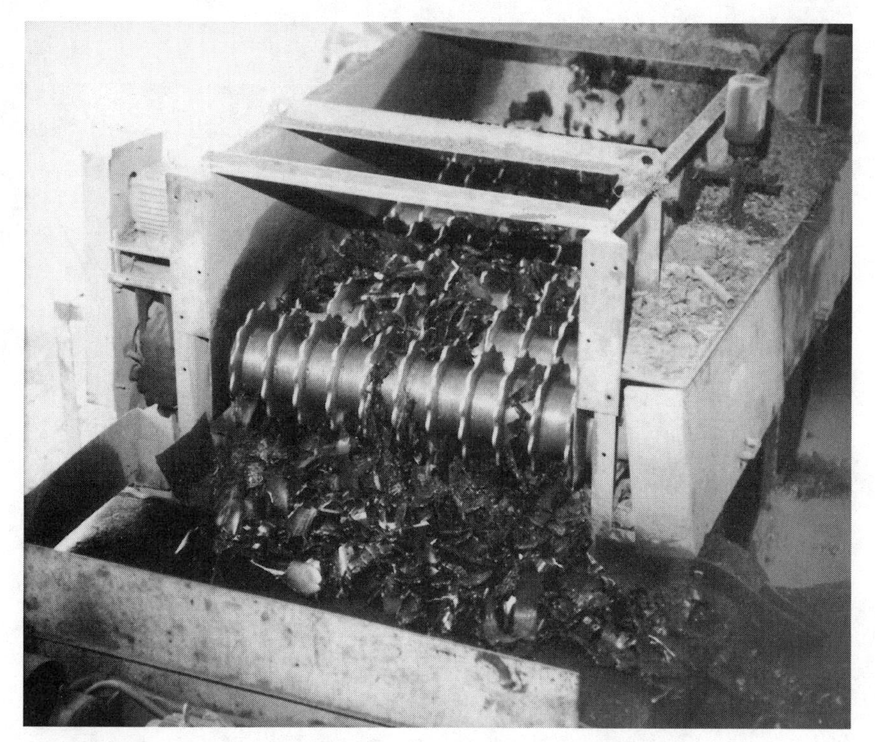

FIGURE 24.7 Disk screen. (*Courtesy of Rader Resource Recovery, Inc.*)

terials are entrained in the air stream and carried off while others drop out. Unfortunately, air classifiers may treat particles of the same material differently, depending upon their size and shape. For example, an uncrushed aluminum can or a flat piece of foil may be sorted with the light fraction, whereas a crushed can or a balled piece of foil may be sorted with the heavy fraction. The air supply to an air classifier should be adjustable to accommodate local conditions and changes in the composition and moisture content of the waste.

Most air classifiers are followed by a cyclone separator that removes the lightest materials, including dust, before the air is exhausted through a filter to the atmosphere. The two major categories of air classifiers commonly used in solid waste processing are vertical classifiers and air knives.

Vertical Air Classifiers

In vertical air classifiers, the material is introduced from the top and air is introduced from the bottom, as shown in the schematic diagram in Figure 24.8. Several variations of the design have been produced by incorporating zigzags, baffles, and other obstacles that increase turbulence. The added turbulence and longer residence time help break up aggregated particles for better sorting. A feature common to many vertical air classifier designs is a sharp bend in the path of the air stream. Presumably, heavier waste particles with higher inertia will not be able to make the bend and will thus be sorted into the heavy fraction.

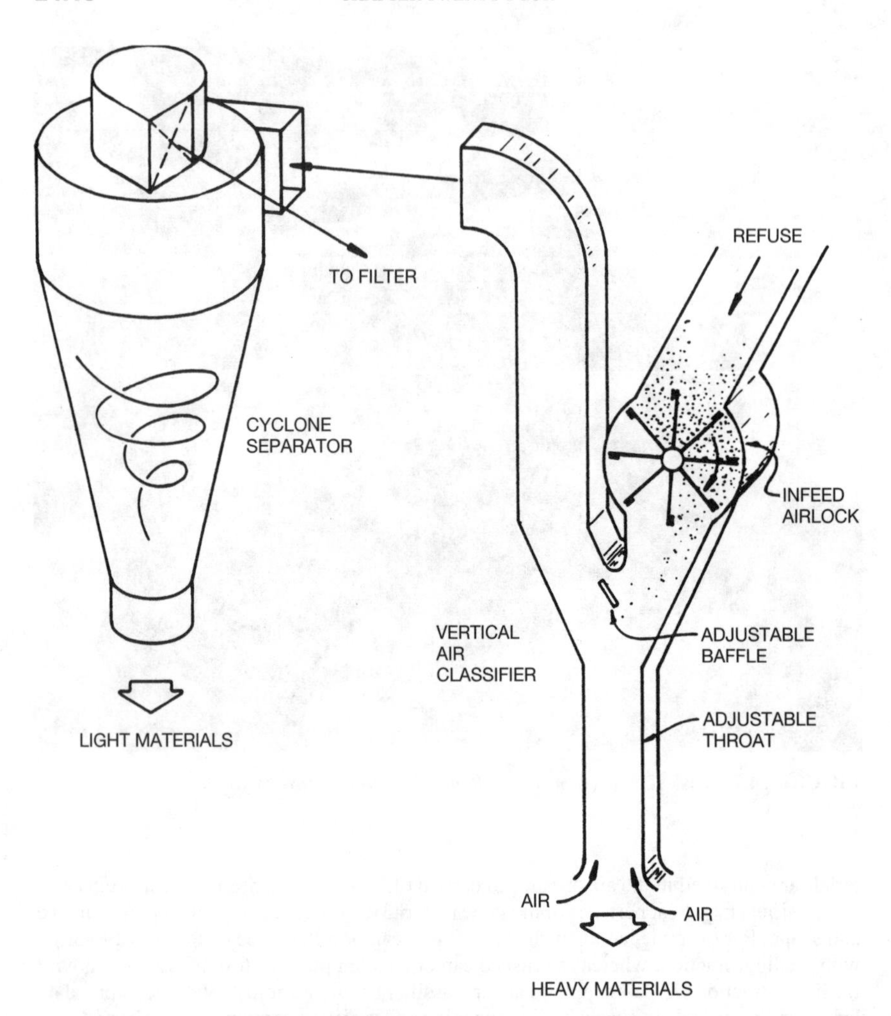

FIGURE 24.8 Vertical air classifier.

Air Knives

The air knife does not attempt to entrain light materials. Rather, it introduces a sharp curtain of air, through which the material passes. Light materials are propelled out of the flow and sorted with the light fraction, as shown in the schematic diagram in Figure 24.9. The air knives have capacities as high as 125 tons per hour and require 40 to 50 horsepower motors. They are designed for a finer separation of materials of different densities. For this reason, air knives may be used to clean up mixed nonferrous metals, such as those generated by eddy current processing. The cleaning consists of separating the heavy fraction such as castings, forgings, and rolled stock from the light fraction such as cans and other light-gauge metals.

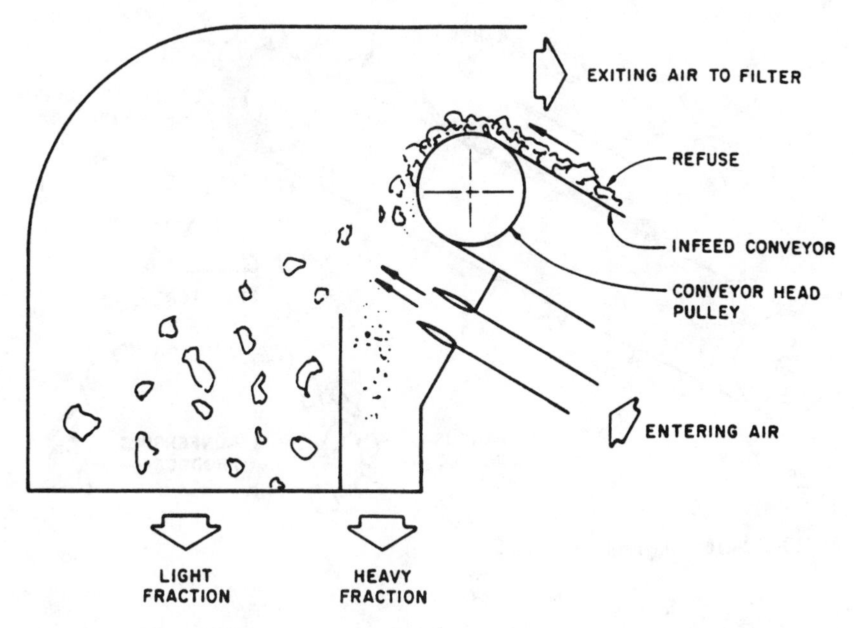

FIGURE 24.9 Air knife.

Eddy Currents

Nonferrous metals can be removed from waste streams by inducing repulsive magnetic fields that propel the nonmagnetic materials out of the stream. The repulsive forces are the result of eddy currents developed by alternating magnetic fields. This process is known as eddy current separation. Eddy current separation is most often used to remove aluminum after initial screening. It can remove up to approximately 98% of nonferrous metals with a similar level of product purity. Its energy consumption is generally low compared to other separation methods.

Magnetics

Magnetism is the property that allows ferrous metals to be separated from other materials. Although magnetic separation cannot separate tangled ferrous and nonferrous materials, it can often produce a marketable steel product for a relatively small investment. In addition, the removal of ferrous metals enhances the processes that follow. In magnetic separation, the materials to be separated pass near a magnet that removes the ferrous materials, transports or deflects them, and deposits them into a bin or onto another conveyor. Magnets are either permanent or electromagnets. The three major types of equipment used for magnetic separation are the head pulley, the suspended drum, and the suspended belt magnet.

The *magnetic head pulley* is presented in Figure 24.10. The forward pulley in the conveyor is a magnet. Nonmagnetic materials, unaffected by the magnet, fall onto the next conveying device. Magnetic materials travel through the free-fall zone and are dropped on the other side of the splitter. The conveyor belt should be cleated for positive discharge of the ferrous materials. Some nonmagnetic materials such as paper, plastics, and textiles

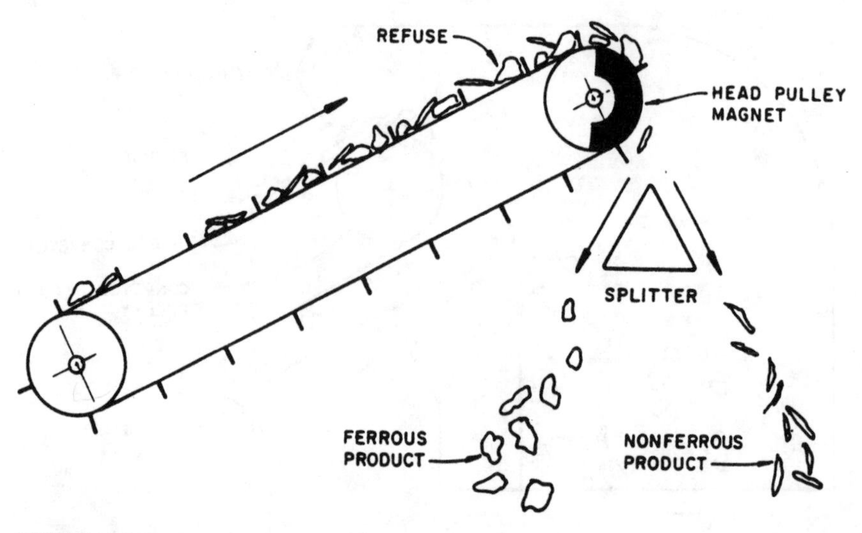

FIGURE 24.10 Magnetic head pulley.[4]

may be trapped with the magnetic material, so an additional separation step may be necessary.

The *suspended drum magnet* consists of a stationary magnet inside a rotating drum. Ferrous metals are pulled to the face of the drum, held against the drum as it travels, and then discharged free of the nonmagnetic materials. Figure 24.11 presents an arrangement of two drum magnets. The first is installed for underfeed, and the second for overfeed. By reducing the size of the magnet or increasing the distance to the magnet, the magnets can be made to discriminate between large and small ferrous particles. This can result in cleaner separation even among ferrous materials of different sizes.

The *suspended belt magnet* consists of a fixed magnet located between the pulleys of a conveyor. The magnetic conveyor is suspended above the flow of materials, either in-line (parallel) or transverse to the flow as shown in Figure 24.12. The in-line arrangement is preferred by some for cleaner separation and a more efficient operation. Unlike the head pulley and drum magnet, the belt magnet has belts that are subject to wear. The belts may be protected by installing stainless steel wear plates on the belts. Suspended drum and belt magnets are less likely than head pulley magnets to entrap nonmagnetic materials because these materials are released as the magnetic materials travel through the air.

Multiple magnets may be used in belt magnets, as shown in Figure 24.13. The pickup magnet must be the strongest to draw ferrous materials from a greater distance. Transfer magnets do not need to be as large as the pickup magnet. An additional separation step can be incorporated into the belt magnet by placing a second pickup magnet following a discharge location, as shown in Figure 24.13. This allows entrapped materials to be released.

Conveyors

From the time materials arrive at the processing facility, they must be conveyed through the various processing stages. Several means of conveyance are typically used. On the tipping

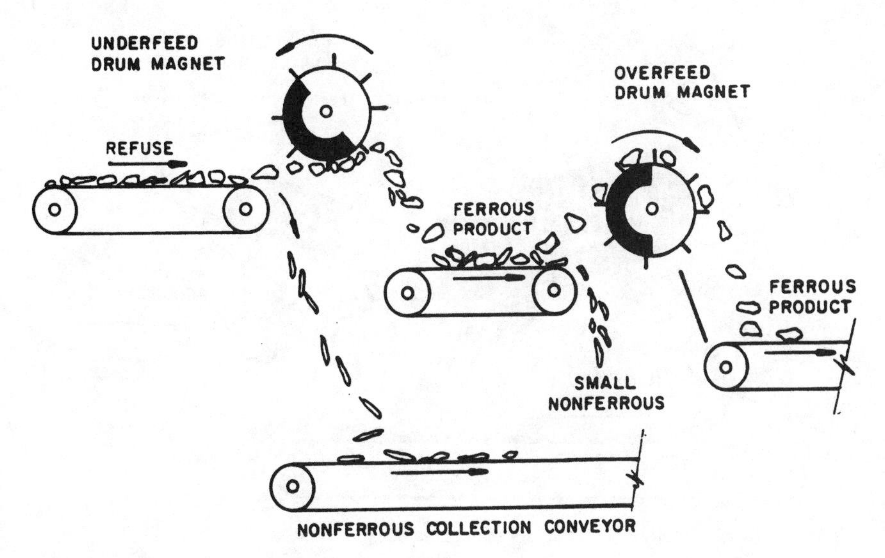

FIGURE 24.11 Magnetic drum arrangement.[4]

floor, materials that are not suited for processing and large recyclable items may be conveyed manually or with mechanical equipment. The remaining materials are likely to be pushed by a loader into a receiving pit. Conveyors are often used to move the materials horizontally and vertically. Inclined conveyors usually have risers or paddles attached to the belt for positive lifting. Conveyors can incline up to approximately 30 ft, and up to 45 ft if the height of the risers is increased. Materials may also be conveyed by gravity down chutes and slides, and light materials may be conveyed through ductwork by air.

The depth of the material on the conveyor, the conveyor width, and its travel speed determine the flow rate of the material. The depth of material can be increased or decreased by transferring the material to a slower or faster conveyor. Waste streams can also be divided or combined from several belts. The proper flow rates of the various conveyors are critical to synchronizing the elements of the processing system. The speed of individual

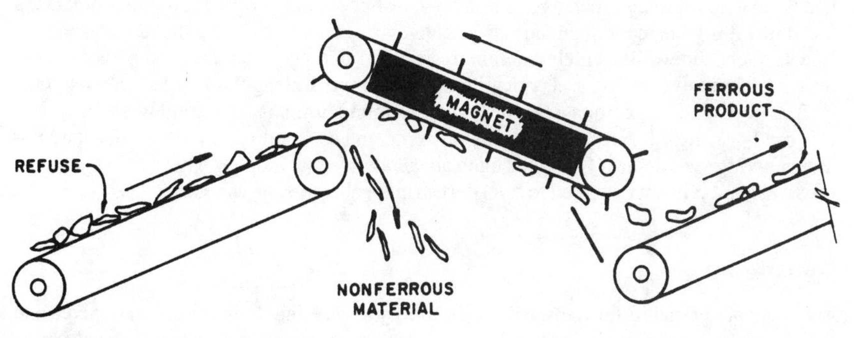

FIGURE 24.12 Belt magnet.[4]

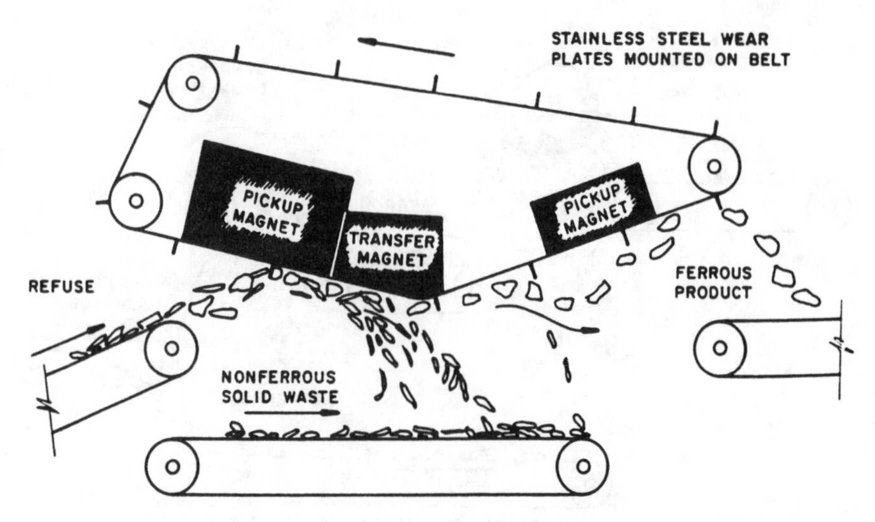

FIGURE 24.13 Multiple belt magnet arrangement.

conveyors or of the entire system may be controlled automatically for improved efficiency. For example, the speed of the conveyor feeding a shredder may be controlled by the loading on the shredder so that the feed rate will drop when more difficult material is processed. Conveyors and other parts of the processing system can usually be shut down by tripping an emergency shutoff switch.

There are two major drive systems for belt conveyors. The belt may be attached to a chain drive system, in which case it is not part of the drive system, but merely "goes along for the ride." An advantage of chain-driven conveyors is that they can change gradient. For example, a single chain-driven conveyor can travel horizontally for a given distance, then incline upward, and then travel horizontally again. A system of conventional belt-driven conveyors would require three separate belts to achieve this same effect. Alternatively, the belt may be driven by motors attached to its pulleys. Although these belts do not last as long as chain-driven belts, they may cost only half as much.

Rubber belts are the most popular types of conveyors. They are relatively low in cost, quiet, and do not interfere with magnetic separation. When conveyors are subjected to heavy, damaging loads, such as sharp, heavy, or hot objects, or great dropping heights, a metal pan design is recommended. These conveyors consist of hinged metal plates, usually with metal sidewalls, which are more resistant to damage. Metal pan conveyors, however, are expensive, noisy, and typically require more maintenance than belt conveyors.

The horizontal auger, or screw conveyor, has found limited use in solid waste handling. It is used primarily for dry, free-flowing material and has application for bin discharge or as a metering feed device. One of its main advantages is that it opens bags as it conveys the materials, but it is not designed to transport stringy, abrasive, or wet materials.

Densifying

Materials are prepared for transport to markets by decreasing their volume and increasing their density by using compactors, balers, densifiers, or glass crushers. The decision whether to densify material before shipping should be based primarily on market specifications and transportation costs. Operators must decide on the form in which material is delivered to the markets and, in most cases, densification of recyclables is preferred. Table 24.2 presents material density estimates that can be used in evaluating transportation costs.

Newer balers come equipped with sophisticated controls that allow a single baler to bale multiple products, each to specified dimensions and weights. Some balers allow for unattended baler stations. Since most markets are equipped to receive baled materials, selecting a compactor over a baler could limit the markets available to the MRF operator. This factor must be evaluated against the lower capital costs and higher transportation costs associated with a compactor.

Recycled fiber mills are generally equipped to handle standard-sized bales, and baled material generally brings a higher price. A compactor costs less than a baler but cannot consolidate the material to the same density as a baler. Compactors can be used where an intermediate market is involved, because the intermediate market will rebale the material to meet the specifications set by the mill.

Baling affects other recyclable materials in similar ways. The network for plastics recycling is not as well established, but baling is the preferred method of material handling. Aluminum is typically baled in smaller densifiers. The selection of densifying equipment for aluminum should focus on the appropriate size of equipment for the throughput quantity. Some materials are difficult to bale because they have a "memory" of their original shape. Such materials can be handled easier if a fluffer is installed above the baler charge hopper. The decision to include a glass crusher in a system depends upon the intended means of transportation. Glass plants are invariably equipped with their own glass crushers, capable of converting whole bottles into furnace-ready cullet or intermediate cullet. Because of its high density, the amount of crushed glass transported by highway is often limited by weight restrictions.

Processing

After materials are shipped from the MRF or WPF, they are further processed and purified before use as feedstock in manufacturing new products. Such processing may include washing, rinsing, drying, and further size reduction or separation and removal of contaminants. Although these processes and the related equipment are integral to recycling, they are not discussed here because they do not take place in the MRF or WPF. However, these processes and the associated equipment could be incorporated into the design of a recycling facility to increase the value of the recyclables.

TABLE 24.2 Relative densities of recyclable materials

Recyclable material	Loose, lbs/yd^3	Crushed, lbs/yd^3	Baled, lbs
Aluminum	50–74	250	730–2,230
Glass	600–1,000	800–2,000	—
HDPE	25	65	400–800
Mixed paper	600	—	900
ONP	360–800	—	720–1,000
OCC	300–350	—	1,000–1,200
PET	30–40	—	400–500
Solid waste	400	—	1,200
Steel	150	—	850

FINAL DESIGN

Architectural Considerations

The architectural design of a MRF depends largely on its location. MRFs have been successfully constructed in existing buildings or warehouses. In such retrofit installations, the modifications to the building are typically not of an aesthetic nature, as the structure is modified only to accommodate the processing equipment.

At the other end of the spectrum is an attractive building designed to be accepted by the surrounding immediate community as part of a larger solid waste transfer operation. The architectural effort to make a larger transfer or MRF operation acceptable to its community can be significant. Such a facility can be designed to resemble a commercial office building, to make it compatible with surrounding structures. Another architectural approach is to make the facility visible in an attractive manner, to draw the community's attention to the efforts required to manage solid waste effectively. It is the design architect's responsibility to recognize the aims and needs of the owner and to utilize his or her professional skills in designing an attractive facility that offers a safe and pleasant working environment. An example facility is shown in Figure 24.14.

Mechanical Considerations

Designing a conveying system to link the different sorting technologies is critical for an effective system. Layout and design of conveyor systems requires a thorough knowledge

FIGURE 24.14 Phoenix, Arizona, 27th Avenue Solid Waste Management Facility, which includes a materials recovery facility.

of mechanical engineering principles and extensive operating experience. Some of the guidelines to consider in laying out a system include the following:

- Limit changes in direction
- Allow drops of approximately 2 ft between conveyors
- Use metal conveyors whenever heavy impacts are anticipated
- Generally increase the speed of travel as material progresses through the system

The first step in designing a conveyor system is to select the dimensions of the belts. The following equation can be used to select the width and travel speed for a flat belt:

$$C = V \times A_n$$

where C = belt capacity (ft^3/min)
V = belt speed (ft/min)
A_n = nominal cross-sectional area (ft^2) = $[h_n \times (b - 2s)]/144$
h_n = nominal material heights (in)
b = belt width (in)
s = standard edge distance = $0.055 \times b + 0.9$ (in)

A reasonable belt speed for a conveyor charging the system is approximately 10 to 20 ft/min. Hand sorting can be efficiently done with variable belt speeds in the range of 20 to 50 ft/min. In the layout of such hand-sorting operations, when selecting the belt width, it is important to consider the reaching and twisting motions of the pickers. Belts should be 24 to 30 inches wide if sorting from one side and not more than 5 feet wide if sorting from both sides of the conveyor. The belt height should also be comfortable for the pickers and is typically near 3 ft above the sorting platform about counter height.

Drive requirements can be calculated according to the following formula:

$$P = (T_e \times V)/(33{,}000 \times E)$$

where P = power (hp)
T_e = effective tension (lbs)
V = belt speed (ft/min)
E = motor efficiency

T_e is the effective tension required to drive the belt at the drive pulley. The main forces to be overcome as material is moved on a conveyor include the following:

- Gravitational load to lift material
- Frictional forces at design speed
- Force to accelerate the material as it is fed onto the conveyor

Calculating the effective tension of the conveyor by identifying these forces requires an understanding of the conveyor equipment proposed. The Conveyor Equipment Manufacturers Association (CEMA) handbook presents the basic formula for calculating effective tension for rubber belts as

$$T_e = LK_t(K_x + K_yW_b + 0.015W_b) + W_m(LK_y \pm H) + T_p + T_{am} + T_{ac}$$

where L = length of conveyor (ft)
K_t = ambient temperature correction factor
K_x = friction factor of idlers (lbs/ft)
K_y = factor for resistance of the belt and load to flexure (lbs/ft)
W_b = weight of belt (lbs/ft)
W_m = weight of material (lbs/ft)

H = vertical distance (ft)
T_p = tension resulting from pulleys (lbs)
T_{am} = tension resulting from force to accelerate (lbs)
T_{ac} = tension from conveyor accessories (skidboards, plows, trippers, and belt scrapers)

Many of the values of these coefficients and associated forces can be estimated from the CEMA handbook. The CEMA handbook also provides guidelines for graphically estimating power requirements for rubber belt conveyors.[5] However, because solid waste and recyclables are generally at the lighter end of the material spectrum, graphical solutions with a high degree of accuracy are difficult, and analytical solutions are preferable after initial estimates by graphical methods. Considerations to keep in mind for designing recyclables conveying systems, some of which are reflected in the above equation, include the following:

- Frictional load is significantly increased by the inclusion of plows
- Plowing of nonhomogeneous material is difficult
- Belt skirts significantly affect frictional load to the point that wider belts may be more economical than belt skirts

Electrical Considerations

Allowances for additional floor storage will enable the facility to continue receiving material in the event of a power outage. The storage requirements resulting from power outages are normally considerably shorter than the full day of storage recommended for these facilities, so with flexible operating conditions, the inconvenience resulting from power outages will be relatively insignificant and provisions for backup power are generally not made.

The control system for a MRF should provide for the following capabilities:

- Sequenced startup of conveyors and other equipment
- Local control of conveyor belt speed
- Remote control of system (optional)
- Local emergency shutdown controls

Maintaining accurate records of total plant input and output is an important part of system monitoring and can be accomplished without sophisticated instrumentation. The nominal height of material on the infeed belt conveyor is an additional parameter that my be monitored with instrumentation. This measurement allows control of the conveyor's belt speed, so that a constant feed rate can be maintained.

OPERATIONS

Personnel

As mentioned previously, flexibility is crucial to the success of a MRF. After equipment is selected and installed, the adaptability of the MRF to changing market demands is largely dependent upon the flexibility of its management and operation.

Many private MRFs are constantly adjusting their systems to optimize their output.

Their managers utilize techniques that reward the line workers for surpassing established goals. Another common practice is to rotate personnel through the plant so that their tasks are varied.

Several municipally owned MRFs, on the other hand, have shown initiative by employing lower-cost work forces such as institutionalized laborers. The decision to reduce operating costs in a labor-intensive facility in this manner should be evaluated in light of the reduced efficiency or lack of flexibility that may be associated with such a labor force.

Equipment

Regularly scheduled maintenance of the equipment as suggested by suppliers is essential. If equipment failures result in frequent downtime, not only do operations become difficult but the facility may also lose public confidence. The redirecting of recyclables as a result of extended downtime in a MRF could reduce public interest in recycling.

Several high-volume facilities (over 100 tons per day) schedule a shift each day for maintenance to clean conveyor belts and other equipment. The abrasion caused by glass can be particularly damaging to the system unless regular cleaning is performed. Other maintenance tasks, such as lubricating idlers, can be performed less frequently at intervals recommended by the supplier.

In general, a MRF is not as maintenance intensive as other solid waste handling systems such as RDFs or composting facilities. The waste stream received at a MRF is generally more homogeneous and therefore less abrasive to the equipment. In addition, the processing equipment is intended to separate the stream of recyclables into its constituents, not to pulverize the waste.

COST

Capital Cost

Processing facilities labeled MRFs come in a variety of types and sizes. Recycling continues to be an emerging market, and materials handling equipment manufacturers, as well as specialty equipment manufacturers, are continuing to gain experience. Processing capacities and materials recovered vary from one system to the next, so it is difficult to develop a basis of comparison of costs. Costs per ton are higher when only containers are recovered and when more mechanized systems are used. When fibers are included, the unit costs are lower, even though the total capital outlay may be considerably higher. From historical data, however, the capital cost per ton of daily processing capacity for a MRF can range from a low end of $10,000 to $30,000 as compared to a high end of $60,000 to $120,000.

Operating and Maintenance Costs

Equipment operating and maintenance costs vary with the equipment selected. For general estimating purposes, annual equipment maintenance costs can be assumed to be approximately 5% of equipment capital costs and site maintenance costs can be assumed to be approximately 1% of the structure and site costs.

Manual labor often constitutes a large portion of the operating costs. In determining the number of manual pickers needed, the following average ranges can be applied to the projected throughput stream that is to be hand-sorted:

Glass sorter: 500–800 lbs/hour
Plastic sorter: 300–500 lbs/hour
Corrugated and other paper sorter: 800–1,500 lbs/hour

In addition, personnel will be needed to operate rolling stock and to operate and maintain other major pieces of equipment. These personnel can add costs of $10 to $20 per ton recovered. Obviously, systems with higher capital costs should have a trade-off in lower operating and maintenance (O&M) costs as a result of reduced staffing requirements. It is important to develop a conceptual design and to estimate the associated staffing requirements for a proposed MRF. Because there are many concepts for MRFs, only after developing cash flow projections and income statements for specific facilities can the capital and O&M costs be estimated with accuracy.

Additional Guidelines

Experience has shown that under today's changeable market conditions, MRFs that recover material at a total system cost less than the revenues they receive for their recyclables are few. The cost to process material received at a MRF is about equal to the revenues from the processed materials. This does not take into account the additional costs assignable to a MRF outside of processing costs, such as the collection cost and the disposal cost of unrecyclable material.

In addition to these financial costs assignable to a proposed MRF, a developer or municipality should consider the frequently significant economic system costs. Avoided costs for collection and disposal must be considered by the owner in order to make informed decisions.

ACKNOWLEDGMENTS

The authors would like to acknowledge Keith R. Connor and David A. Dorav for their research and preparation of the first edition of this chapter.

REFERENCES

1. U.S. Environmental Protection Agency. *Characterization of Municipal Solid Waste in The United States: 1996 Update*. EPA/530-R-97-015. May 1997.
2. Ruiz-Palá, E., C. Ávila-Beloso, & W. W. Hines. *Waiting-Line Models, An Introduction to their Theory and Application*. New York: Reinhold Publishing Corporation, 1967.
3. U.S. Environmental Protection Agency. *Decision-Maker's Guide to Solid Waste Management*, Second Edition. EPA/530-R-95-023. August 1995.
4. U.S. Environmental Protection Agency. *Material Recovery Facilities for Municipal Solid Waste*. EPA/625-91/031. September 1991.
5. Conveyor Equipment Manufacturers Association (CEMA). *Belt Conveyors for Bulk Materials*. 3rd Edition. Rockville, MD: CEMA, 1988.

CHAPTER 25
INTEGRATING RECYCLING WITH LANDFILLS AND INCINERATORS

ERNEST H. RUCKERT, III, P.E.
Senior Engineer
EMCON/OWT Solid Waste Services
Mahwah, New Jersey

INTRODUCTION

Integrated waste management includes landfilling and incineration (waste-to-energy) as components of the total integrated system. The basic role of landfills and incinerators in an integrated system is to manage nonrecyclable wastes. Accordingly, they are at the bottom of the integrated waste management hierarchy. Landfills and incinerators can further contribute to integrated waste management by serving as sites for recycling operations.

Recycling operations at landfills and incinerators can be instrumental in helping communities meet their recycling goals. National, state, and local recycling percentages have been adopted, which has increased the attention placed on the role that landfills and incinerators can play. This chapter examines the types of material recovery operations in practice at landfills and incinerators, and discusses key issues associated with such operations. In the case of landfills, the beneficial use of certain materials (e.g., low-grade recyclables, materials otherwise disposed of in landfills) in landfill construction and operation is discussed.

ADVANTAGES TO IMPLEMENTING RECYCLING AT LANDFILLS AND INCINERATORS

Siting Advantages

Existing landfills and incinerators are generally well-suited places to conduct recycling operations. They are the sites in our communities to which wastes are delivered and accordingly provide access to the materials in the waste stream. They are permitted waste management facilities, and the public is usually accepting of these locations being used for site management operations (albeit with a resigned acceptance in many cases). Therefore, the difficulties involved in siting and permitting a new solid waste facility can often be avoided by locating recycling operations at existing landfills and incinerators.

Some recycling operations require a large area, and some are most economically conducted outdoors. Appropriate sites may therefore be difficult to locate in urban areas. Landfills often provide good sites for these operations. Many incinerators also have substantial areas available around them, although others are more tightly constrained.

Operational Advantages

Landfills and incinerators also offer operational advantages for many types of recycling activities. Site features that offer operational advantages to recycling at landfills and incineration facilities are compared in Table 25.1.

Landfills typically have large areas of land available, which is advantageous for composting and stockpiling for periodic processing (e.g., concrete and asphalt recycling). The relative isolation of most landfills results in less impact of the recycling operations on surrounding areas. Also, landfills are in some ways an ongoing construction project that uses materials. This creates an internal market for compost, soil, concrete, asphalt, etc.

Waste-to-energy facilities can often provide a covered building for material recovery operations, which offers protection from weather and correspondingly enhances paper recovery. Concrete floors in these facilities provide better working conditions for employees than dirt or mud (i.e., in comparison to landfills), cleaner materials for recovery, and structural support and utilities for processing equipment. Waste-to-energy facilities are also more developed sites, providing paved roads, parking, etc., thereby facilitating use by the public for drop-off centers, buy-back centers, and household hazardous waste collection facilities.

The operational links between recycling and landfilling–incineration also offer a variety of advantages, as discussed later in this chapter. These include shared equipment and back-up capacity. In addition, the close proximity between the recycling and landfilling–incineration operations results in efficient and inexpensive hauling of residues.

TABLE 25.1 Comparison of site features at landfills and incinerators with respect to recycling operations

Facility	Typical site feature	Advantage to recycling
Landfill	Large land area	Composting Stockpiling for periodic processing
	Isolated	Less impact of operations on surrounding land uses
	Use of materials on site	Creates internal market for compost, soil., concrete, asphalt
Incinerators	Covered building	Protection from weather Paper recovery
	Concrete floor	Better working conditions for employees Cleaner materials for recovery Structural support for processing equipment
	Paved roads	Facilities use by public for drop-off and buy-back centers

Recycling, in turn, offers many advantages to landfilling and incineration operations. To landfilling, recycling offers saved air space—an ever more valuable asset to landfills. Saved air space translates to longer site life and greater total revenue flows to the landfill. Recycling can be beneficial to incineration operations by improving fuel efficiency, reducing ash disposal costs, downsizing boiler requirements, and removing large items that can cause material handling problems. Impacts on fuel efficiency are discussed in the next section and impacts of specific material recovery operations are discussed in the section entitled Problems Encountered.

Regulatory Advantages

The growing volume of recycling regulations and policies makes it advantageous for landfill and incinerator operators to conduct their own recycling operations. Such regulatory and policy issues include percent diversion requirements or goals, taxes on landfilled quantities, and permit conditions requiring recycling. Running their own programs gives landfill and incinerator operators internal control over meeting recycling goals. It also provides for optimizing the integrated waste management system by linking the components together. Finally, where tipping fee taxes are imposed on disposed quantities, it gives the operators the opportunity to avoid the tax on materials they divert from disposal.

OPERATIONAL LINKS BETWEEN RECYCLING AND LANDFILLING–INCINERATION

Traffic

When recycling operations are integrated with landfilling or incineration operations, traffic coordination is extremely important. This applies to: 1) vehicles using recycling versus disposal areas; 2) vehicles carrying processible versus nonprocessible wastes; 3) public versus commercial collection vehicles; and 4) vehicles delivering wastes to the site versus site vehicles. These various categories of vehicles should be kept separate to the extent possible. In addition, cross-traffic (i.e., vehicles crossing in front of other vehicles) should be minimized.

A well-planned entrance facility can serve a valuable role in traffic coordination. Some recycling operations (e.g., drop-off and buy-back centers) are best located as part of the entrance facility. There are trade-offs between locating them before or after the gatehouse, as discussed in a later section. In either case, traffic flow should be carefully considered. The smooth flow of traffic leaving the gatehouse for various tipping areas should also be incorporated into the site design. Figure 25.1 illustrates different approaches to locating recycling operations at entrance facilities.

Traffic coordination can be achieved by good facility design, signage, use of traffic coordinators (also referred to as spotters), and directions from the gatehouse attendant.

Materials Flow

Efficient materials flow is also important when integrating recycling with landfilling or incineration operations. Processible materials should be kept separate from nonprocessible materials; clean loads should be kept separate from mixed loads. The flow of materials toward recovered products versus toward landfilling–incineration should be distinct

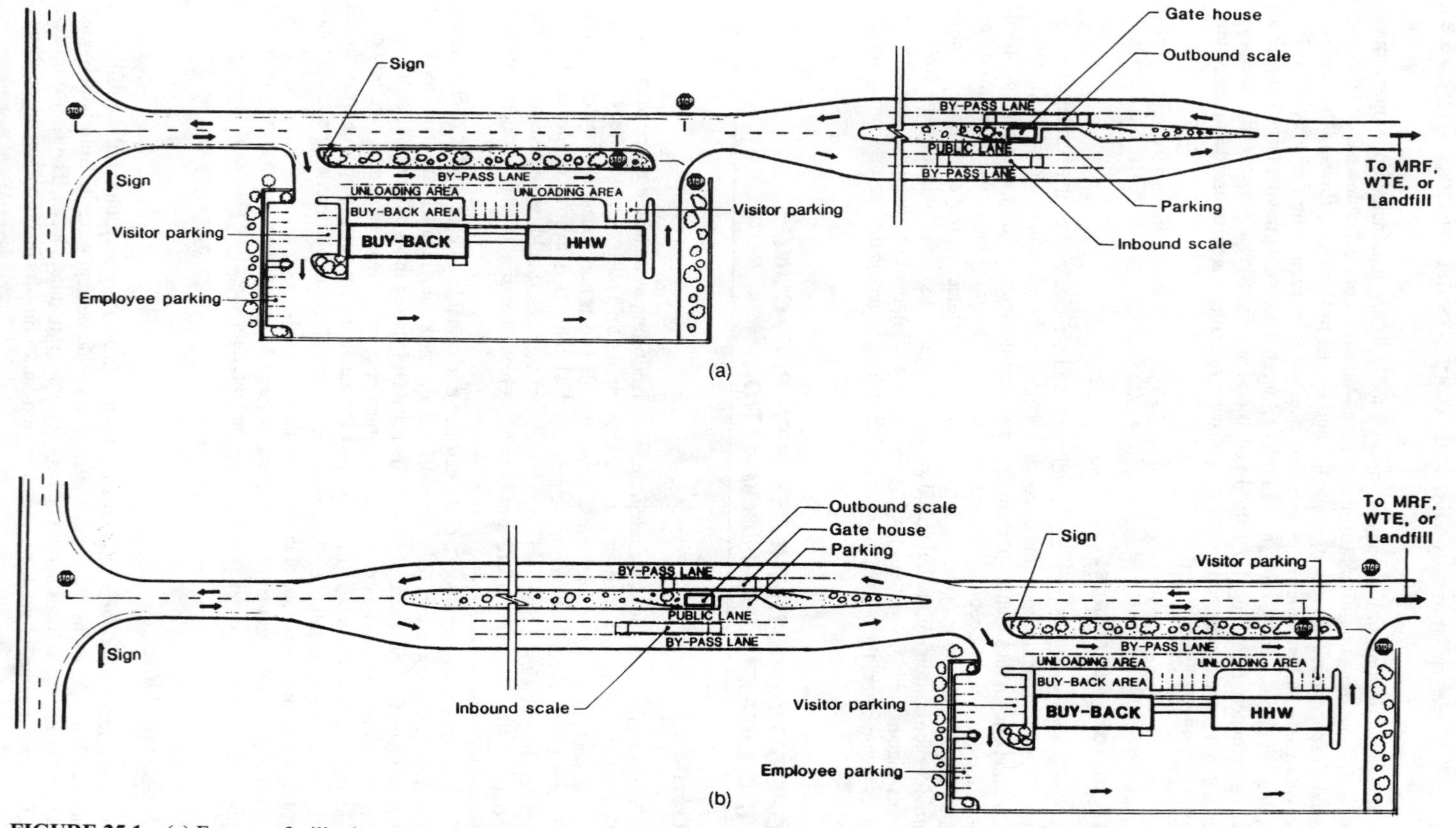

FIGURE 25.1 (a) Entrance facility layout with recycling before gatehouse; (b) entrance facility layout with recycling after gatehouse.

and efficient. Residue from recovery operations should end up in a location convenient for handling in the landfilling or incineration operation. Finally, the capability to readily load out recovered materials to markets should also be provided. A materials flow diagram for an integrated waste management facility is shown in Figure 25.2.

Material Properties

By removing materials in recycling locations, the material properties of the waste stream entering the landfilling or incineration operations are changed. These changes include both physical and chemical properties. The resulting impacts on incineration operations offer many benefits. Recycling can reduce ash content, lower the toxicity of the ash, raise the ash fusion temperature, provide a more consistent heating value, raise the heating value (in many cases), reduce air emissions, and reduce chloride content. As a consequence, ash disposal costs are reduced, and boiler efficiency is increased. Slagging is reduced due to the increase in ash fusion temperature. Decreased chloride content reduces boiler wall and tube corrosion.

Other material properties affected include bulk density and moisture content. These properties may either increase or decrease in value, depending on the recycling operation.

Equipment Requirements

Mobile equipment can often be shared between recycling and landfilling operations or between recycling and waste-to-energy operations. Potentially shared equipment includes wheel loaders, dozers, and roll-off trucks. Although this equipment is used for other operations at the facilities, it often is not being used full-time. By coordination and scheduling, this equipment can serve the needs of the recycling operations, especially for small-scale operations. Large-scale operations require backup equipment, which can be used in the recycling operation rather than sitting idle. Substantial recycling operations will clearly require dedicated equipment, but again there is a potential for shared backup equipment rather than separate backup equipment for each operation. Shared equipment thus results in improved overall economic efficiency of the recycling–landfilling operations or the recycling–incineration operations.

Backup Capacity

Landfilling and incineration operations offer an on-site backup to recycling operations if the recycling operations incur downtime or if peak waste flows exceed the throughput capacity of the recycling operation. These periods should be minimal in a well-run operation; nevertheless, backup capacity is another functional link between recycling and landfilling–incineration operations.

TARGET MATERIALS

Materials

Materials commonly targeted for recovery in recycling operations at landfills and incinerators are shown in Figure 25.3. The figure indicates the type of vehicle in which these ma-

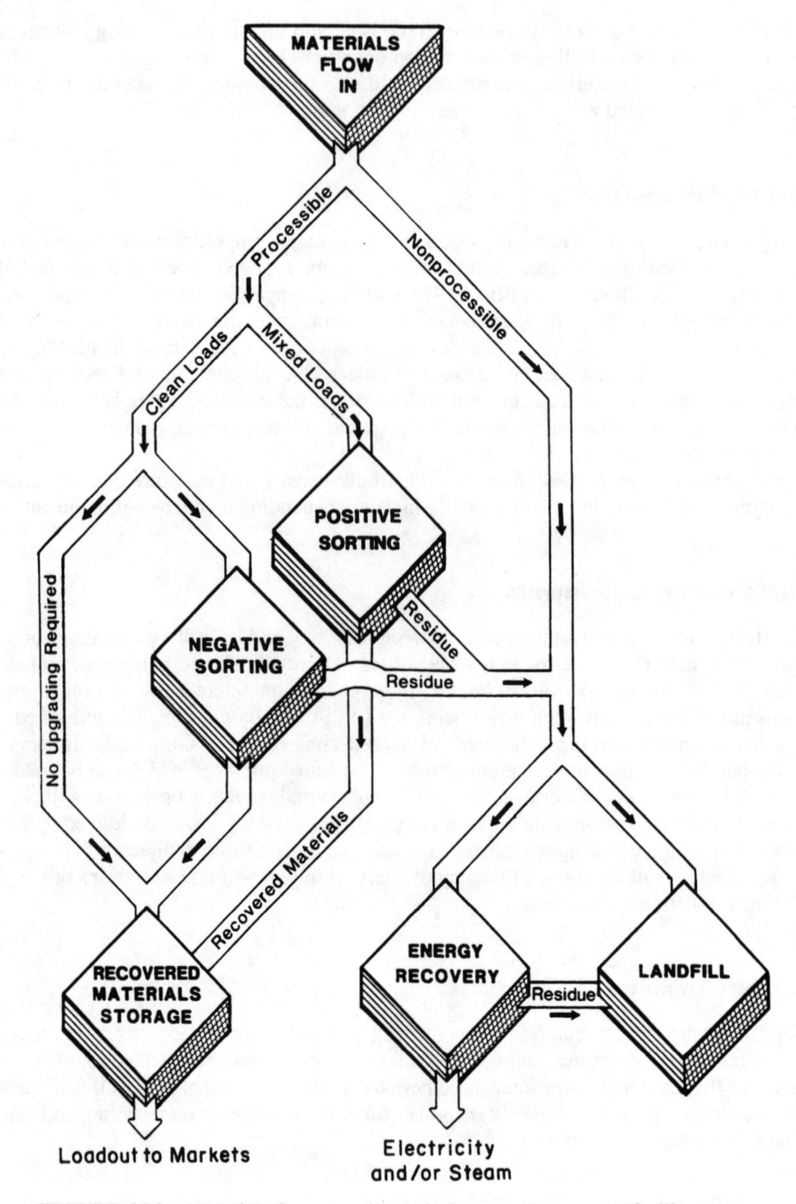

FIGURE 25.2 Materials flow at an integrated waste management facility.

MAJOR CATEGORY	RECOVERABILITY BY VEHICLE TYPE				
	RESIDENTIAL PACKER TRUCKS	COMMERCIAL PACKER TRUCKS	LOOSE ROLL-OFFS	COMPACTED ROLL-OFFS	SELF-HAUL
Asphalt	○□	○□	●■	○□	●■
Concrete	○□	○□	●■	○□	●■
Glass	◑◪	◑◪	◑◪	◑◪	○◪
Metals	◑◪	◑◪	●■	○◪	◑■
Paper	◑◪	●■	●■	●■	◑◪
Plastic	◑◪	●■	●■	●■	◑◪
Soil	○□	○□	●■	○□	●■
Tires	○◪	○◪	◑◪	○◪	●◪
Wood Waste	◑◪	◑◪	●■	◑◪	●■
Yard Waste	●□	○□	◑◪	○□	●■

KEY: SOURCE RATING ● Major ◑ Moderate ○ Minor RECOVERABILITY RATING ■ High ◪ Moderate □ Low

FIGURE 25.3 Recoverability of target materials.

terials are typically delivered to the facility, and the relative recoverability of the materials from the various sources. The ratings shown in the figure may vary by community and can be refined from this general rating to a more site-specific analysis and/or to include subcategories of materials. Figure 25.3 is intended to present a guide to targeting materials in the waste stream for recovery.

The materials targeted in various recycling operations differ as a function of waste stream characteristics, the distribution of waste-generating sources (i.e., residential, commercial, and industrial), site constraints, local climate, and local market conditions.

Various reusable goods are also often recovered at landfills and incinerators. Such goods include appliances, automotive parts, bicycles, electric motors, flower pots, furniture, lawn mowers, and mattresses. In some cases, repairs are required, but in other cases, the goods are in suitable working condition. Salvaged goods are either sold directly to the public or to another business for reuse or resale.

Sources

The key to recovery is economically segregating the materials relatively free of contamination from other types of materials, for example, recovering corrugated paper free of other paper grades, plastic, glass, metals, etc. The degree of difficulty in doing this depends on the condition of the materials delivered to the facility.

Source-separated materials are, of course, the easiest to recover. The next easiest are high-concentration loads (i.e., loads consisting primarily of one material). Typically, high-concentration loads include corrugated paper, office paper, yard waste, wood waste, concrete, asphalt, and soil. Mixed waste loads are the most difficult to process and consequently influence the operator's selection of target materials. Nevertheless, a variety of materials are successfully being recovered from mixed waste loads at landfills and incinerators. Commercial, industrial, and self-haul loads are generally more recoverable than residential packer-truck loads.

MATERIAL RECOVERY OPERATIONS

The various types of material recovery operations in practice at landfills and incinerators are shown in Table 25.2. The operations are broken down according to their function in waste receiving or waste processing. These operations are discussed in greater detail in other chapters of this handbook. Each operation is summarized in the following section, with particular consideration given to its integration with landfilling and incineration facilities.

Drop-off Centers

Drop-off centers are the simplest means to receive source-separated materials from the public. Such facilities are also the most common recycling facilities at landfills and incinerators. Drop-off centers can be implemented with relatively minor capital expenditures, although site development, paving, and retaining wall costs at the better-constructed facilities can cost several hundred thousand dollars. Operational costs are low; the required staffing may be either one person or unattended. Site supervision is generally recommended to control the types of wastes left at the facility and to oversee the unloading of materials into the properly designated locations. Since the materials are received free and facility costs are low, the net revenues to the facility operator are about the same as or close to the gross revenues received from the sale of materials.

Drop-off centers can be located either before or after the gatehouse at the landfill or incineration facility; each location has its advantages and disadvantages (see Figure 25.1). Drop-off centers are often located before the gatehouse to provide an incentive to recycling before entering the disposal site. Materials left at the recycling facility are not charged for at the gate. In addition, residents may readily use the drop-off center even if they do not have any refuse to discard at the landfill or incinerator. Locating the drop-off center after the gatehouse provides for better site surveillance and security. It is still possible to not charge users for recyclables if the drop-off center is after the gatehouse, but it may be operationally awkward.

TABLE 25.2 Material recovery operations at landfills and incinerators

Function	Operations
Landfill	Drop-off center
	Buy-back center
	Diversion of high-concentration loads
Processing	Dump-and-pick (salvaging)
	Mixed waste processing lines
	Composting
	Wood waste processing
	Concrete and asphalt recycling
	Soil recovery
	Tire processing
	Curbside recycling processing
	Postincineration recycling
	Household hazardous waste facilities

Buy-back Centers

Buy-back centers typically achieve greater recovery rates than drop-off centers. They can be readily located at either landfills or incinerators. Capital costs are only slightly greater than for drop-off centers, with additional costs typically required for scales and a minor amount of equipment (e.g., conveyors, magnetic separator). Operating costs are slightly higher, due to the equipment and staffing. Because operators control the price differential between that paid by the brokers or mills and that paid to the public, buy-back facilities are usually profitable undertakings for operators. Difficulties can occur at times of depressed market prices, particularly in communities with multiple buy-back centers.

Buy-back centers are usually located before the gatehouse at landfills and in incinerators. In addition to serving residents, many buy-back centers do substantial business with small commercial customers who make a living collecting recyclable materials for sale.

Diversion of High-Concentration Loads

Many loads arriving at landfills and incinerators contain high concentrations of a single material. As noted earlier, common types of high-concentration loads include corrugated paper, office paper, yard waste, and wood waste. In communities experiencing a high level of construction activity, high-concentration loads of soil, concrete, and asphalt are also received. Diversion of high concentration loads can be coupled with most of the processing operations listed in Table 25.2 (as appropriate to the specific material) and correspondingly contribute to increased material recovery rates. At landfills, high-concentration loads can be diverted to designated processing or storage areas. At incinerators, high-concentration loads can be diverted either to a separate tipping area within the building or to an outside tipping area (e.g., for asphalt, concrete, wood waste, yard waste).

Dump and Pick or Salvaging

Salvaging from mixed waste piles, often referred to as "dump and pick," is the simplest method for recovering materials from a mixed waste stream. It is a very old practice that is receiving renewed use at both landfills and incinerators around the country. Some dump and pick operations at disposal sites have been ongoing for decades.

At certain facilities, dump and pick is practiced on the entire waste stream (i.e., whatever loads are available at the site). Because dump and pick operations at landfills are strongly affected by weather, their focus is often limited to metals and reusable goods. In addition to these materials, dump and pick operations in covered buildings (such as at incinerators) may target paper, particularly corrugated cardboard. Dump and pick operations are best suited for low- to medium-volume landfills or incinerators.

Mixed Waste Processing Lines

Some of the more aggressive material recovery operations at landfills and waste-to-energy facilities use mixed waste processing lines consisting of mechanical and/or manual separation processes. Indeed, the refuse-derived fuel (RDF) approach to waste-to-energy is based on processing the waste prior to combustion and separating a fuel fraction from the noncombustible fraction(s), often including recovery of materials, such as ferrous metals, aluminum, and glass. The use of mixed waste processing lines is growing at landfills in efforts to increase material recovery rates. Some landfills utilize portable conveyor systems unprotected from weather; other landfills have constructed enclosed processing buildings (i.e., MRFs) as part of the entrance facility to the landfill.

Operations focused on material recovery generally place a greater reliance on manual separation, whereas operations focused on energy recovery (e.g., RDF facilities) use mechanical separation to a greater degree. Manual separation is often required to meet the demanding specifications of secondary materials markets; contamination levels must be much lower than are typically allowed in fuel markets. Most of the mechanical waste processing experience in the United States to date is based on RDF processing, dating back to the St. Louis resource recovery plant in 1972. However, mechanical waste processing technology is being increasingly applied to material recovery operations.

Unit processes commonly employed in mechanical processing include screening, air classification, magnetic separation, eddy current separation, ballistic separation, flotation, and shredding. Manual separation equipment consists of conveyors and sorting platforms. A sorting platform provides a safer and more comfortable working environment for the sorters and helps achieve a productive and efficient recovery operation (e.g., as compared to a dump and pick operation). Manual separation can be conducted either in: 1) a positive sort mode, where target recoverable materials are removed from the mixed waste stream; or 2) a negative sort mode, where contaminants are removed and the remaining material becomes the recovered product.

At landfills using mixed waste processing lines, it is common to process only specific types of loads and to divert others directly to the landfill face. A similar approach can be employed at waste-to-energy facilities. As illustrated in Figure 25.3, highly recoverable loads include commercial packer trucks, roll-offs, and self-haul waste.

Composting

Composting continues to grow as a solid waste management practice at landfills and incinerators. The most common material composted is yard waste. Much attention has been turned to yard waste because it typically constitutes 20 to 30% of the residential waste stream; composting programs can therefore contribute substantially to meeting a community's percentage recovery goals. (The amount of yard waste may be higher or lower for a given community depending on geography, season, and landscaping characteristics.) Yard waste composting results in a high-quality product. Additional materials being composted include food processing wastes and sewage sludge. Composting raw municipal solid waste (MSW) or a processed MSW fraction (e.g., a middlings or undersize fraction from a trommel screen) is also receiving increased attention.

It is usually advantageous to keep yard waste out of waste-to-energy facilities because its high moisture content tends to reduce the heating value of the fuel and its high nitrogen content can contribute to elevated NO_x concentrations.

Landfills generally have more available space than incinerators; however, successful composting operations are in place at several incinerators. The turned windrow system is by far the most commonly used composting technique at landfills. In-vessel systems are, however, being used increasingly for composting yard waste or other solid waste materials. In-vessel systems could offer advantages for incinerators and other solid waste facilities with space constraints.

Wood Waste Processing

Where markets are available, wood wastes are being diverted from the waste stream and processed for recovery. Such operations process either only wood waste (e.g., lumber, pallets) or combined wood wastes and yard wastes. Wood waste processing operations are in place at many landfills.

Wood waste processing can also provide several benefits to incineration operations. Large wood items, such as logs and thick lumber, typically do not undergo complete combustion (particularly in mass-burn facilities). In RDF facilities, much wood and yard waste ends up in reject streams (e.g., air classifier heavies fraction, trommel screen unders). Lumber and tree trunks and limbs can also present material handling problems in waste-to-energy facilities, such as bridging in feed hoppers in mass-burn plants or in mechanical processing equipment before the primary shredder in an RDF plant. Finally, as mentioned in the previous section, yard waste tends to adversely affect fuel properties.

The major markets for recovered wood waste materials are for wood chip fuel and soil amendment. Some wood chips are being sold for ornamental landscaping and some for use in sewage sludge composting operations. Another potential use is in particleboard manufacturing.

Typical equipment used in wood waste processing comprises a grinder, a screen, and related conveyors. A variety of grinders are in use, including tub grinders, hammer mills, and shear shredders. Likewise, several different types of screens are in use, including trommels, disk screens, and flatbed screens. The wood waste is first ground, then screened, with the oversized material being sold as wood chips and the undersized material (often referred to as "fines") as a soil amendment. The fines may be composted or sold directly to buyers (such as wholesale soil dealers) who blend them with other soil materials.

Wood waste processing operations at some facilities use a low-technology approach of diverting, crushing, and hauling the wood wastes to another wood waste processing site. In this type of operation, a track dozer or track loader is typically used to crush the wood waste.

Moisture content is an important parameter affecting the marketability and market price of wood chip fuel. Wood chip fuel is often purchased on a dry ton basis. Furthermore, the dry ton price itself may vary with different moisture content ranges. For example, the dry ton price for a load of wood chips with a moisture content between 30 and 40% may be lower than for a load with a moisture content between 20 and 30%. Drying yard wastes before they are processed for wood chip fuel is therefore beneficial.

Landfills typically have an advantage over incineration for drying yard waste because of the greater land area available. The material is stockpiled after it enters the site and the oldest material is processed first.

Concrete and Asphalt Processing

In terms of tonnage or recovery rate percentage, concrete and asphalt recycling can offer one of the greatest reductions in the amount of waste entering disposal sites. Obviously, concrete and asphalt are also good materials to keep out of an incinerator. The two general approaches to concrete and asphalt recycling are: 1) establishing a processing facility at a site; and 2) stockpiling materials for periodic processing by mobile equipment brought to the site. Concrete and asphalt processing equipment is expensive, and therefore substantial quantities of wastes are required to economically justify the equipment. Accordingly, concrete and asphalt recycling is usually best practiced on a regional basis using one of the two approaches described above.

Processing equipment consists of screens, crushers, magnetic separators, and conveyors. Size reduction may be achieved using a multistage approach (i.e., multiple crushers), with each unit designed for a specific range of size reduction.

Crushing equipment used includes hammer mills, impactors, and jaw crushers. Grizzly screens are often used in the early stages of processing, and single-deck or multi-deck flatbed vibratory screens are commonly used for grading the final products. Other single-stage size reduction units, which are transported to a job site as a single trailer, incorporate recirculation of oversize material back into the crusher in order to control maximum product size. These single trailer units are readily transported and include all conveyors, a crusher, screen, and magnetic separator for a complete size reduction system capable of producing multiple ranges of product size. The products are sold as a variety of construction materials, such as road base and aggregate, or can be used on site.

Area requirements are large, in the range of 5 to 10 acres. The large area is consumed by stockpiles of the incoming waste materials, the processing system, and stockpiles of the product materials. Again, landfills tend to be better suited than incinerators for this type of an operation, although some incinerators do have sufficient surrounding area.

Also worth noting is that recovered concrete and asphalt materials are commonly used at landfills in place of virgin construction materials (e.g., rock and gravel) for road base, winter tipping pads, drainage construction, erosion control, ditch lining, and leachate collection systems. In these applications, the materials are often worked over by dozers to break them up. The benefits of using the concrete and asphalt materials are thus twofold: 1) substituting for natural construction materials; and 2) saving landfill space by not being disposed of in the landfill burial cell.

Soil Recovery

Loads of clean or relatively clean soil are sometimes delivered to disposal sites. These materials would generally require little, if any, processing prior to being provided to end users, including utility companies, road maintenance crews, and occasionally, homeowners. Most landfills make good use of these materials for cover and on-site construction, and some have even resold them for off-site use. Soil recovery is more readily carried out at landfills than at incinerators.

Tire Processing

Tire processing can provide a variety of benefits for landfills and incinerators. The problem of tires rising to the surface of landfills is well known. Tire processing prevents this problem as well as saves landfill space. Since whole tires often do not combust completely in MSW incinerators, tire processing can be helpful by recycling the tires or at least shredding them. Tire shredding results in more complete combustion of tires in MSW incinerators.

Shear shredders are the most common type of processing equipment used. However, shredding is not always necessary for tire recovery to off-site markets. Tires are often shredded for efficient storage and transportation and for mitigation of vector problems, especially mosquito breeding.

Curbside Recycling Processing

As residential curbside recycling programs are implemented in more and more communities, facilities for processing the collected materials are often located at landfills or incinerators because of the ease of siting. Materials from both commingled and multicontainer recycling programs are being processed at landfill and incineration facilities. The processing operations may be enclosed, partially enclosed, or open. The approach taken is usually driven by climate.

Processing equipment requirements for multicontainer systems are, of course, less than those of commingled systems. Equipment used includes conveyors, magnetic separators, can flatteners, glass crushers, balers, and scales. Equipment varies among facilities and depends upon market requirements, available funding for the facility, and the size of the operation.

Curbside recycling programs help keep glass and metals out of incineration facilities. As a consequence, heating value is increased and ash content is decreased.

Beneficial Use of Materials at Landfills

As pointed out in previous discussions, certain low-grade recyclables and materials otherwise disposed of at a landfill can be used in the construction and operation of a landfill, subject to operational considerations, cost, and acceptance by permitting agencies. The benefit of using mixed glass, chipped tires and/or processed construction and demolition debris, could include a reduction in the cost of constructing certain components of landfills (compared to the use of natural materials), minimizing the use of available on-site materials, and receiving revenues for use of these materials in the landfill construction. The use of these materials is generally allowed by regulatory agencies on a case-by-case basis. Several examples of the beneficial use of materials follow.

Mixed broken glass, resulting from material recovery facility operations, has been used as a horizontal gas venting layer in landfill caps beneath a heavy geotextile and flexible membrane liner. The mixed broken glass can be minimally processed, if necessary, to meet state regulatory requirements for permeability and particle size for the gas venting layer. Permeability and particle size requirements can vary based on the state where the landfill is situated.

Shredded car and truck tires have been used in landfill construction. Tire chips have been effectively used for roadways, leachate collection blankets, and gas venting layers. Tire chip sizes of approximately 2 inches have been used in these applications.

Processed construction and demolition debris from urban areas has been used as daily cover at landfills. Typically, the acceptable material is trommel unders from sorting and processing systems. Rigorous testing protocols are established by regulatory agencies for use of processed construction and demolition debris. Long-term use of this material has demonstrated that it is an acceptable substitute for daily cover soil.

Postincineration Recycling

The most common type of postincineration recycling is recovery of ferrous metals from the ash by means of a magnetic separator. Screens (e.g., trommels) are often used to process the ash prior to magnetic separation. Some facilities also recover nonferrous metals from ash. In addition, ash can be processed for other uses, including aggregate, road base, and concrete construction blocks.

Concern over heavy metals concentrated in the ash has limited ash recovery. The application of solidification processes on incinerator ash and the implementation of alternative recycling programs for such materials as batteries, can mitigate the heavy metal problem and allow more beneficial uses of the ash. Ash from tire incineration facilities is sometimes processed for zinc recovery.

Household Hazardous Waste Facilities

Household hazardous waste facilities also can be integrated into landfill and incinerator operations. These facilities serve several roles in an integrated system, including: 1) preventing the improper disposal of hazardous wastes in solid waste facilities; 2) providing proper handling and storage of hazardous waste identified in other operations (e.g., recycling, transfer); and 3) recycling of various hazardous materials (e.g., paints, solvents, batteries). Household hazardous waste facilities are discussed in greater detail in Chapter 21.

Household hazardous waste facilities can be very beneficial to an incineration facility by reducing heavy metals in the ash and flue gas. Collection of such items as batteries can help the incinerator meet emission limits and potentially decrease the cost of ash disposal.

The entrance facility to the landfill or incinerator is a good location for the household waste facility—either before or after the gatehouse, depending on available space, traffic flow, and security considerations. Locating the hazardous waste facility before the gatehouse generally provides for easiest use by the public, but there may be overriding considerations or simply operator preferences that lead to its placement after the gatehouse.

Integration of Components

The alternative material recovery operations can be combined with landfilling and incineration operations in a variety of ways as is best suited to a particular community's waste stream and site conditions. An example of integrating material recovery operations at a landfill is presented in Figure 25.4. A similar example for integrating material recovery operations at an incineration facility is shown in Figure 25.5.

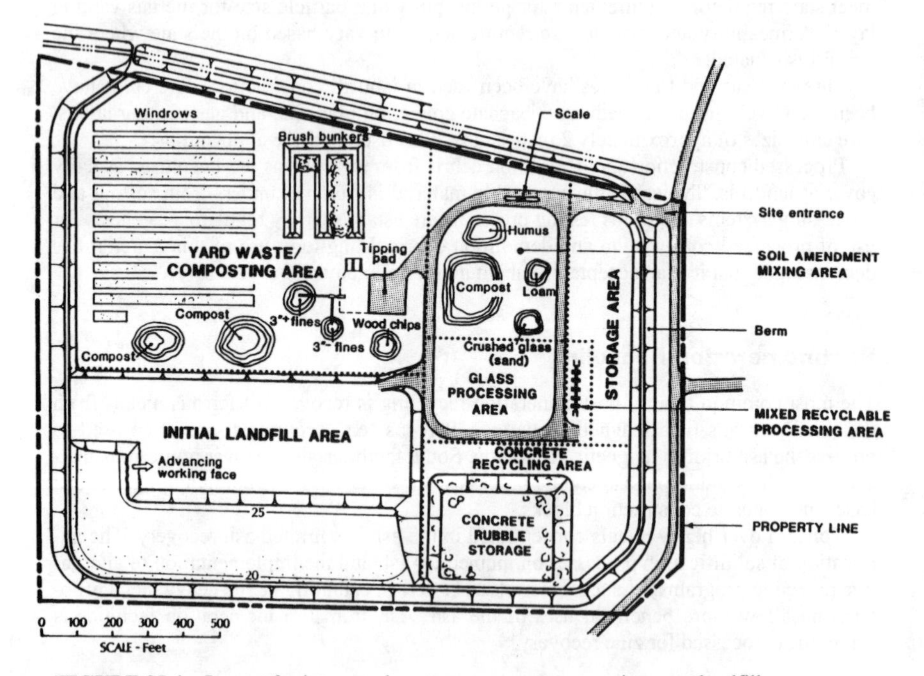

FIGURE 25.4 Layout for integrated waste management operations at a landfill.

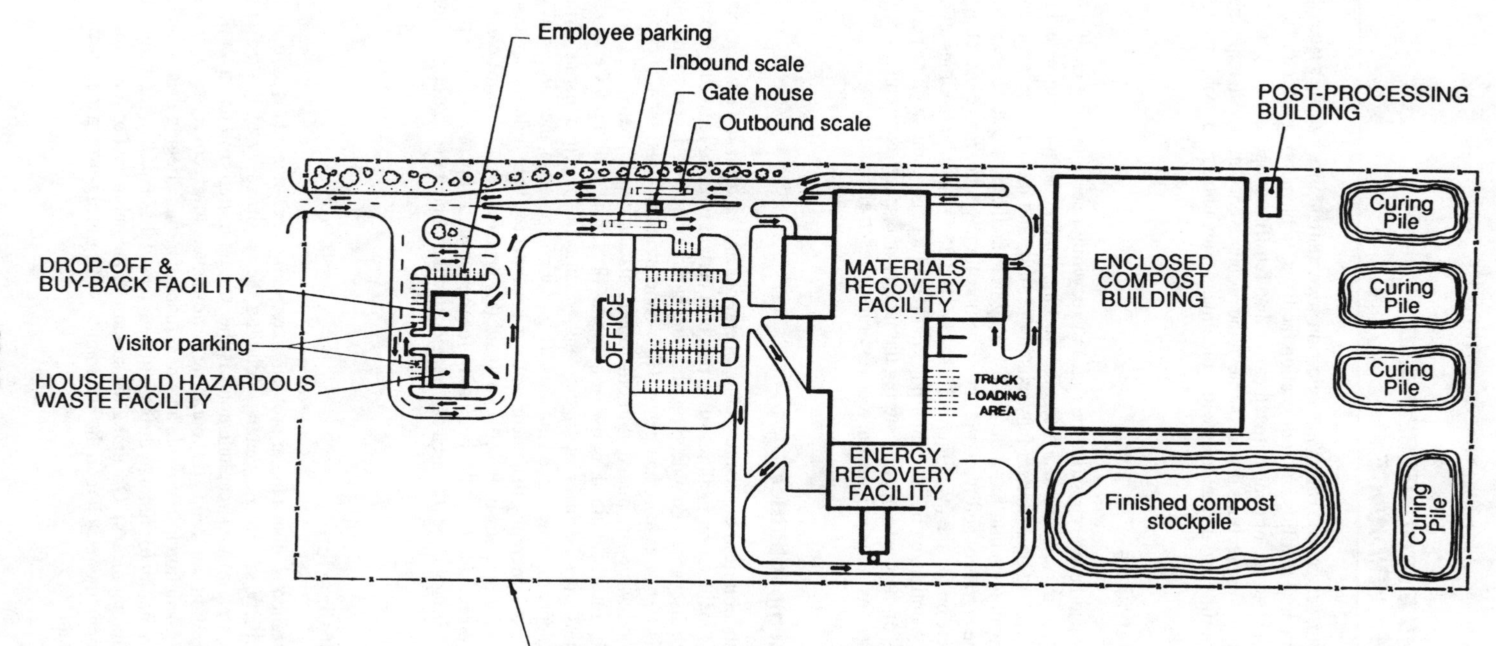

FIGURE 25.5 Layout for integrated waste management operations at an incineration facility.

PROBLEMS ENCOUNTERED AND PROBLEM AVOIDANCE

Proper planning, design, and equipment selection are important in achieving successful recycling operations at landfills and incineration facilities. Implementing recycling at a new landfill or incinerator is, of course, much easier than retrofitting operations at an existing facility. The following sections discuss some of the problems encountered in implementing recycling at landfills and incinerators with guidelines for avoiding such problems.

Space Constraints

Lack of space can severely restrict the types of recycling operations that can be implemented. Space constraints can also affect the working conditions, appearance, and recovery rates of recycling operations. Many existing landfills and incineration facilities were not designed with recycling in mind. In designing new facilities, ample space should be provided in the entrance facility, to the site as well as for any other major recycling operations elsewhere on the site (see Figures 25.1, 25.4, and 25.5). In older facilities, available areas should be optimized by analyzing potential recycling operations with respect to the characteristics of the waste stream entering the facility, the space requirements of the alternative operations, and the potential recovery rate for each alternative.

Design Incompatibilities

Design incompatibilities between recycling operations and landfills or between recycling operations and incineration facilities often preclude certain types of recycling. For example, if vehicles discharge into a deep pit at an incineration facility, access to the material for recovery is lost or at least hampered. High-concentration loads, such as paper, yard waste, and wood waste, are mixed with other materials and contaminated. This type of a problem can be mostly avoided by use of a tipping floor instead of a pit. Certain design incompatibilities may be inherent in the facility's functional requirements. Although landfills may appear to have large land areas (e.g., for composting), they are usually configured to have a topography that sheds water. Consequently, very little flat area is available on completed portions of the fill. To avoid this problem, either: 1) sufficient native land should be set aside from the beginning for the recycling operation; or 2) recycling operations requiring a large, flat area can be relocated throughout the phased development of the fill.

Undercapitalization

Undercapitalization is often a major problem in recycling operations. This problem can be based on a variety of specific factors, including the harsh reality of limited available funds. An example of undercapitalization is the misapplication of a down-stroke baler. Undercapitalization can lead to discouraging performance, low recovery rates, and messy operations. The solution is to provide adequate funding through tipping fees at the landfill or incineration facility (or funding through tipping fees at the landfill or incineration facility or other funding means). Obtaining the support of the public and of local government officials for including recycling in the community's integrated waste management system is also important.

Safety

Greater attention is being paid to safety issues, as recycling becomes a mainstream waste management practice. There are many potential safety hazards in material recovery operations at landfills and incineration facilities. These include: 1) working around heavy moving equipment, such as loaders, dozers, and collection vehicles; 2) picking and handling of waste materials with the associated risk of injury from such items as sharp-edged metals, hypodermic needles, and broken glass; and 3) possible exposure to hazardous wastes. To protect workers, a safety program for the facility must be developed, including safety training, regular safety meetings, refresher training, and personal protective equipment and clothing standards for all employees. A program to detect and deter hazardous wastes from entering the site is also a valuable tool in protecting site workers (see Chapter 21). Finally, good facility design is essential in providing safe working conditions.

Weather

Weather has a big impact on outdoor recycling operations, which are typical of landfill recycling operations. Working conditions are difficult in rain, snow, or mud. Wind and wetness make paper recovery difficult. The extent of weather problems depends on geographic location. The solution to such problems is to provide shelter from the elements in an enclosed or partially enclosed structure.

Markets

For any recycling operation to be successful, it must have stable markets for its recovered materials. This is true for recycling operations at landfills and incineration facilities as well as at any other locations. Markets have been addressed in detail in other sections of this handbook. The key point noted here is that in planning operations at landfills or incineration facilities, a thorough market analysis should be undertaken.

Permitting

Permit conditions can actually restrict recycling operations at some landfills and incineration facilities. This often comes as a surprise to facility operators when trying to implement a program. Restrictions commonly arise in land use permit conditions. The problem is becoming worse as other land uses surround waste management facilities, particularly in urban areas. Another type of problem occurs when established landfill end-use plans (i.e., after the landfill closes) 30 conflict with new plans for ongoing recycling activities at the site. This occurs, for example, if the site is designated to become a park. The best approach to avoid permitting restrictions is to include recycling operations in the facility's permits as early as possible.

ECONOMICS

Economic Incentives

There are a variety of economic incentives for recycling at landfills and incineration facilities. These are broken down according to facility operators and users in Table 25.3.

TABLE 25.3 Economic incentives for recycling at landfills and incineration facilities

Party	Economic incentive
Landfill operator	• Saved air space • Revenues from sales of materials • Tax on landfill quantities
Incineration facility operator	• Reduced ash disposal costs • Reduced incinerator capital costs • Reduced incinerator O&M costs
Facility user	• Free drop-off of recyclables • Buy-back of recyclables • Preferential tipping fee for recyclable loads • Reduced base tipping fee resulting from recycling savings

More and more landfill operators are recognizing the value of saving airspace for the economic potential it represents. For every cubic yard of material recycled, the landfill operator can resell that airspace. Recycling also brings revenues to the landfill operator from the sale of recovered materials. Another economic incentive is the cost avoided from taxes on landfill quantities; such taxes are not assessed on recycled quantities.

A more economic incentive for incineration facility operators to implement recycling is to reduce ash disposal costs. They also receive revenues from the sale of recovered materials. Because recycling results in a lower throughput capacity for the incinerator, it can be downsized and thereby reduce capital costs of the facility. Additional savings may be realized from lower incinerator operation and maintenance costs.

Facility users are very important in helping make recycling work. Various economic incentives can be given to users to encourage their participation and assistance in achieving good recovery rates in the different recycling operations. It is common to offer free drop-off of recyclable materials at landfills and incineration facilities; therefore, facility users are only charged for wastes disposed of.

Buy-back centers offer further economic incentives by paying for the recyclable materials. To encourage users to deliver loads in a manner that facilitates recycling, a preferential tipping fee may be set for relatively clean, high-concentration loads. Finally, users may benefit (at least in the long term) from a reduced base tipping fee that results from including recycling in an integrated waste management system

Capital and Operating Costs

The capital and operating costs of the alternative recycling operations are presented in other chapters of this handbook. The reader is referred to the corresponding chapters for details. The focus of the discussion on costs of recycling at landfills and incineration facilities is on cost avoidance, as described in the following section.

Cost Avoidance

When looking at the big picture of integrating recycling with landfill and incineration operations, one of the key economic issues that arises is cost avoidance. This section presents a methodology to account for cost avoidance in analyzing the economics of a recycling operation at a landfill or incineration facility. Cost avoidance can be realized as the result of various factors, including avoided disposal costs, avoided hauling costs, avoided taxes, reduced operation and maintenance costs, and saved air space (at a landfill).

To be economically profitable, tipping fees must cover the total costs incurred in running an operation (i.e., O&M, closure, and postclosure costs plus amortized capital costs), offset by revenues from the sale of recovered materials and avoided costs. At break-even,

$$T_0 = C - R - A \tag{25.1}$$

where T_0 = break-even tipping fees
C = total costs (O&M, closure, and postclosure costs plus amortized capital costs)
R = revenues from sale of materials
A = avoid costs

It is important that these parameters are all expressed on the same basis. The analysis presented here is based on unit costs of dollars per gate ton, where gate ton refers to tons entering the facility. Some cost parameters may need to be converted from dollars per recovered ton to dollars per gate ton. For example, unit revenues from the sale of materials are typically given in terms of the weighted average market price (i.e., dollars per ton sold). Unit conversion is given by:

$$(\$/\text{gate ton}) = (\$/\text{recovered ton}) \times (\text{recovery rate}) \tag{25.2}$$

If the weighted average market price for materials recovered is $100 per ton and the recovery rate for the operations is 75%, then this is equivalent to $75 per ton of material in the gate (i.e., $75 per gate ton).

Figure 25.6 illustrates the use of the recycling cost analysis methodology with accounting for avoided disposal costs. The figure depicts the break-even cost of a recycling operation as a function of tipping fees, revenues from the sale of recovered materials, and avoided costs. To serve as a general cost analysis tool, the cost and revenue terms have been normalized by the total unit cost of the operation. Tipping fees are given on the horizontal scale. Positive values represent a fee charged for accepting the materials at the facility; negative values represent a price paid for the materials. The vertical scale indicates revenues from the sale of recovered materials. The band of lines cutting across the figure account for avoided costs. Several examples of using the figure follow.

Assume the total unit cost to run a recycling operation is $30 per ton, the operation achieves a 60% recovery rate, the market price for recovered materials is $40 per ton, and there are no avoided costs. The normalized revenues would then be equal to 0.8 (i.e., 0.6 × $40/$30). Moving across the figure to the zero avoided cost line and vertically down to the tipping fee axis yields a normalized tipping fee of 0.2, which is equivalent to $6 per ton (i.e., 0.2 × $30).

If the above example were a landfill operation subject to a $6 per ton tax on landfilled quantities (A/C = $6/$30 = 0.2), then the break-even tipping fee would be $0 per ton.

Assume another operation for target materials with a relatively high market price of $120 per ton, a 75% recovery rate, and a total unit cost of $50 per ton to run the operation. Even with no avoided costs, the break-even tipping fee would be minus $40 per ton. In other words, the facility operator could pay $40 per ton for this material and still break even. The figure can also be used in a variety of other recycling cost analyses, such as determining the market price needed to break even for operations with no tipping fees, or determining tipping fees if recovered materials had to be given away while still accounting for avoided costs.

Privately run facilities usually would like to make a profit on their operations (although a particular break-even operation within the overall facility operations may be

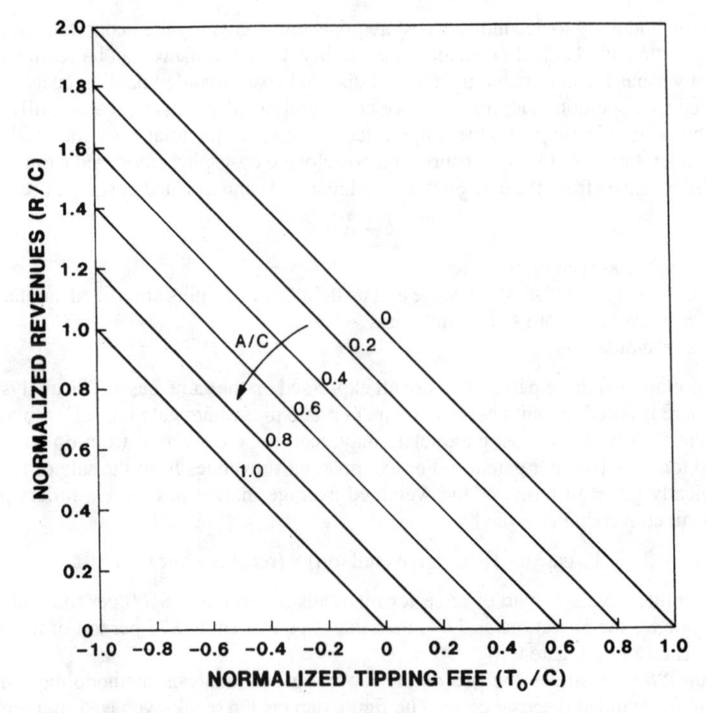

FIGURE 25.6 Break-even cost analysis of recycling operations.

suitable to private operator in certain cases). The cost analysis methodology presented here still applies, with the following adjustment:

$$T_p = T_0 + PC \tag{25.3}$$

where T_p = tipping fee with profit
P = profit margin

If in the first example above, which yielded a break-even tipping fee of \$6 per ton, a profit margin of 15% was applied, then the tipping fee including profit would be \$10.50 per ton (i.e., \$6 + 0.15 × \$30).

CHAPTER 26
PROCESSING YARD WASTE

RICHARD J. HLAVKA
President and Consultant
Green Solutions
South Prairie, Washington

INTRODUCTION

Yard waste is defined as the leaves or leaf fall, grass clippings or grass trimmings, and woody wastes—branches, trimmings, stalks, and roots—found in the municipal solid waste stream. The United States Environmental Protection Agency estimates that yard waste constitutes about 18% of the national municipal solid waste flow. Recovery of yard waste by processing into compost or mulch is an increasingly popular method of recycling solid waste. Further, yard waste recovery has the advantages of ease of separate collection, since the materials are generated outside of the home, low costs of processing, and products familiar to many users.

States use different methods to encourage yard waste recovery. Some states, for example, Illinois, North Carolina, and Ohio, have legislation that bans landfilling of all yard wastes. Pennsylvania and New Jersey ban the landfilling of leaf waste. Florida bans the landfilling of yard waste in lined landfills and effectively requires recovery of yard waste by adopting a high recycling goal. Many states encourage market development by requiring or giving preferential status to the use of compost and mulch made from yard waste.

The expanding experience in yard waste programs has resulted in methods of estimating quantities of yard waste, efficiently collecting yard waste, processing the materials into marketable products, and distributing the products to constructive uses.

GENERATION OF YARD WASTE

The two major sources of yard waste found in the municipal solid waste stream are households and commercial activities. Household yard waste consists of the vegetative wastes resulting from the maintenance of lawns, trees, planting areas, and gardens by residents. Commercial activities of landscaping and grounds maintenance such as found at institutional establishments, golf courses, and cemeteries generate large quantities of yard

The material on pages 26.1 to 26.9 was written by Ron Albrecht and is reprinted from Chapter 16 of the first edition of this book.

waste. A difference between the two sources of waste is that residential yard wastes are generally collected with other wastes as part of municipal solid waste services, whereas commercial firms are responsible for transportation and disposal.

Although the EPA estimate of 18% can provide overall guidance for some situations, a better approach is to determine the amount of yard waste in a specific waste stream. It works best to conduct an intensive multiseasonal waste stream characterization or to estimate the yard waste content on the basis of data developed by a nearby similar community. Yard waste generation rates vary widely with locations, climate, and type of development, for example, suburban versus urban or rural, maturity of the area (whether newly built with small trees and immature lawns or large trees and established lawns), and local weather (whether a time of excess rainfall or drought). These all have definite impact on the waste stream. Examples of wide differences in yard waste production are Florida, where a long growing season and plentiful rainfall cause yard waste to exceed 30% of the total waste stream, and Pennsylvania where yard waste collected in rural counties is estimated as low as 5 percent because of other disposal options readily available to residents.

The blend of yard waste components, that is, the relative amounts of grass, leaves, and wood wastes, is variable. Relative amounts are specific to a particular study area. For purposes of preliminary planning the data given in Table 26.1 are accepted estimates.

Yearly generation patterns vary with the location, climate, and weather. A typical yearly pattern is shown in Fig. 26.1.

Generation patterns in the temperate areas follow a seasonal pattern. January and February typically produce the minimum yard waste because growth is in the dormant stage. March and April's mild weather brings spring cleanup and grass starts to grow. Grass trimmings then become the major yard waste component until mid-October when leaf fall starts to dominate. A decrease in grass production because of drought conditions during July and August can be expected. Figure 26.1 does not show the decrease during these months and may illustrate the effects of local weather in a specific year.

The density of yard waste is an important factor in developing a management plan. Almost all solid waste management planning is done by mass, that is, on the basis of tonnage. Yard waste is frequently measured by volume because most yard waste sites do not have vehicle scales. Therefore an understanding of yard waste densities is important in keeping accurate records of yard waste activities. Table 26.2 presents a range of yard waste densities. It is advisable to randomly sample and weigh yard waste components to ensure the accuracy of records.

The causes of wide density variations in yard waste components are the moisture content and method of collection and transportation. Wet leaves compacted in rear-end-loader compactor trucks are more dense than leaves collected by vacuum trucks, as shown in Fig. 26.2. Similarly wet grass is heavier than grass cut during dry weather.

Many communities have established programs to reduce yard waste generation by en-

TABLE 26.1 Types of yard waste as percentage of total yard waste (by weight)*

Leaves	19–28%
Grass	54–64%
Woody waste	17–18%

*Based on Mid-Atlantic data.

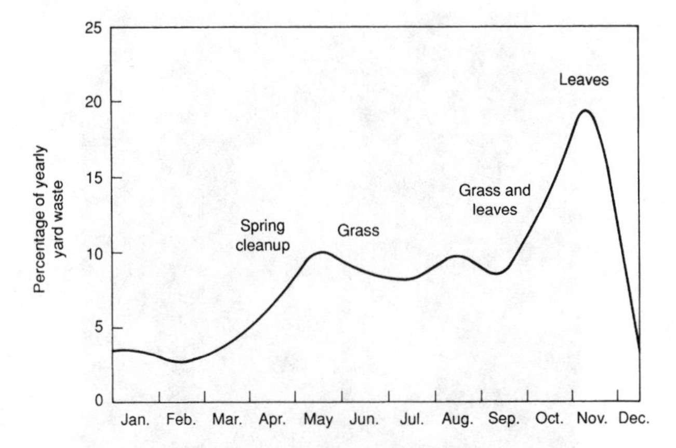

FIGURE 26.1 Seasonal distribution of yard waste.

couraging backyard composting, changes in landscape practices, and changes in turf management. Backyard composting programs encourage homeowners to compost yard waste on their properties. This encouragement ranges from the simple programs of giving out "how to" literature through the distribution of backyard composting units. Changes in landscape practices focus on the reduction of vegetative wastes by encouraging the use of mulches or decorative stone cover in place of turf. In some areas the concept of xeriscaping, creating a desertlike landscape with minimal water requirements and foliage production, is accepted. Changes in turf management involve leaving the grass trimmings on the lawn to recover the nitrogen content and reduce the quantity of waste. All of these practices can serve to reduce the amount of yard waste. Many of these programs conflict with traditional landscape practices and are not well received by homeowners. Further, it is difficult to measure the effectiveness of these source- reduction programs. Figure 26.3 shows a sample pamphlet on home composting.

TABLE 26.2 Density of yard wastes

Material	Condition	Typical density, lb/yd^3
Leaves	Loose and dry	100–260
Leaves	Shredded and dry	250–350
Leaves	Compacted and moist	400–500
Green grass	Loose	300–400
Green grass	Compacted	500–800
Yard waste	As collected	350–930
Yard waste	Shredded	450–600
Brush and dry leaves	Loose and dry	100–300
Compost	Finished, screened	700–1200

Source: "Yard Waste Management—A Planning Guide for New York State."

FIGURE 26.2 Vacuum truck collecting leaves.

COLLECTION OF YARD WASTE

Yard wastes are generally collected by separate curbside collection or at drop-off sites. Curbside collection generally involves additional expenses because of the labor and equipment involved in additional pickup. Drop-off sites are generally voluntary and less expensive than separate collection, but are not as effective as curbside pickup because they are not as convenient to the residents. Collection strategies differ according to the type of yard waste.

Leaf collection is a seasonal operation beginning in October and usually ending in mid-December. Some communities limit the collection to street leaves while others allow residents to rake leaves to curbside for collection. The frequency of collection varies from once a year to weekly during periods of leaf fall. Leaves are collected in various ways, either loose or bagged, by different types of collection vehicles.

Bagged leaves can be collected in three types of containers: nondegradable plastic bags, biodegradable plastic bags, or biodegradable paper bags. Experience shows that the biodegradable paper bags simplify processing the leaves. However, biodegradable bags are more expensive than alternative plastic bags. Nondegradable plastic bags are the least expensive but do complicate processing by littering the compost site and blinding the compost screens. In addition, pieces of the bags detract from the appearance of the compost. Some communities debag the leaves at time of collection or at the processing site. Debagging leaves involves a considerable amount of labor and is not considered cost-effective. The experience with biodegradable plastic bags is that the bags will deteriorate into plastic fragments which do not degrade during composting. Therefore the nondegrad-

Steps For Making Compost:

1) Gather your materials and pile them at least 3 feet wide and 3 feet tall. Avoid packing leaves as this slows down decomposition. Your compost pile can be placed anywhere convenient in your yard.
2) Water as you combine all ingredients. The pile should be as moist as a squeezed out sponge.
3) Indent the top to collect ralnwater.
4) Turn the pile monthly and water each time. Turning provides needed air and moisture. In winter it is best to leave it alone.

It can take several months or longer before your compost will be ready for use. How quickly your compost is ready depends on:

- pile size,
- aeration,
- moisture, and
- how often you turn the pile.

Benefits of Using Compost:

Well decomposed compost is perfect for most home gardening because it is usually near neutral pH (between 6.5-7.0).

Compost can be:

- Tilled directly into your soil to improve water and fertilizer holding capacity.
- Used as a mulch, 2 inches to 3 inches deep, to prevent water loss through evaporation.
- Used in a soil mix for houseplants. Adding 1/4 compost to any soil mix adds organic matter and small amounts of nutrients.

Compost is not a fertilizer, but improves soil structure. Both sand and clay soils can be helped with the addition of compost.

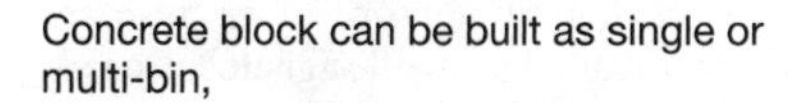

Concrete block can be built as single or multi-bin,

Recycled pallets make a great compost bin.

FIGURE 26.3 Pamphlet on home composting.

ed plastic fragments interfere with processing and detract from final product quality, similar to the nondegradable plastic bags. Both types of plastic bags are available in several different colors. The problem of identifying yard waste set out for separate collection is simplified by requiring the use of a specific-color bag.

Biodegradable paper bags are expensive, but have the advantage of degrading and disappearing into the compost. Many communities distribute preprinted paper bags through their government offices and supermarket chains. Residents are charged for each bag. The community then limits leaf collection to leaves bagged in the designated paper bags (Fig. 26.4).

A variety of methods are used to collect loose leaves. Residents are required to rake the leaves to curbside where different types of equipment are used to load the leaves onto collection vehicles. Vacuum trucks and hoses are used to suck the leaves into trucks (Fig. 26.5). Some vacuum collection systems are equipped with shredders to size reduce the leaves and increase their density. Front-end loaders are used to scrape the leaves from the roadway and load the leaves into dump trucks or rear-end-compactor trucks. Tractors with attachments and street sweepers are also used to collect leaves.

Brush and woody wastes are generated throughout the year. The spring and fall are the periods of greatest generation. Most communities require that woody wastes be restricted to items 2 in and under in diameter, cut to lengths no greater than 4 ft, and tied in bundles. The bundles facilitate loading and do not interfere with processing. Some communities offer special pickup of large quantities of brush. Collection vehicles include rear-end compactors, dump trucks with low sides, and scow trucks equipped with booms for pickup of large quantities.

The collection of grass clippings presents special problems in a yard waste system because the grass is readily degradable and can cause odors. If sealed bags are used, anaerobic conditions can develop in a very short period of time. In addition, bags of grass can be heavy because of the high moisture content. Perforated plastic or paper bags are made to try to control the development of anaerobic conditions. Plastic containers specially designed to promote the circulation of air are used by some communities for grass collection. The containers should be sized so that lifting them will not unnecessarily burden the collection crew.

Drop-off areas are inexpensive methods of collecting yard waste. The costs of labor and equipment needed to operate a drop-off yard waste site are below the costs of curbside pickup. Many of the drop-off center programs issue identification and restrict deliveries to community residents. Since the program is voluntary and participation requires effort, the effectiveness of drop-off programs is well below the effectiveness of curbside collection programs. Drop-off centers can be complementary to developing curbside programs and can be used to offer services in outlying areas where curbside collection is not cost-effective (Fig. 26.6).

Appropriate locations for permanent drop-off centers include transfer stations, recycling centers, landfills, and other government-owned properties where space is available and the traffic will not create a public nuisance or hazard. Collection containers are often used as scheduled rotating drop-off points in areas convenient to residents. Most permanent drop-off centers have supervision in order to prevent dumping of other trash and garbage. The collected yard waste should be promptly hauled to the designated processing site in order to prevent overloading the drop-off center and the possible development of odor and disease-vector problems.

Yard waste transfer stations are designed to receive materials from collection trucks. The primary purpose is to reduce the cost of hauling to processing sites; however, they can be used to provide interim storage if the rate of collection exceeds processing capacity at citizen drop-off centers, and to receive yard waste from commercial generators. In

FIGURE 26.4 Paper yard waste bag.

FIGURE 26.5 Rear-end loader discharging leaves. (*Courtesy of Brian Golob, DPRA, St. Paul, Minnesotta.*)

FIGURE 26.6 Citizen yard waste drop-off. (*Courtesy of Brian Golob, DPRA, St. Paul, Minnesotta.*)

addition, many yard waste transfer stations are constructed adjacent to municipal solid waste (MSW) transfer stations and can use the truck scales conventionally provided for MSW transfer. All yard waste transfer stations should be attended during the hours of operations. Records should be kept of the types and sources of the waste. Facilities should operate on a scheduled basis and be fenced to prevent vandalism and illegal dumping. Since yard waste processing facilities are generally located in remote areas, transfer stations can reduce the overall cost of the system.

INTRODUCTION TO COMPOSTING YARD WASTES

The following sections provide information on equipment and processing systems that can be used for composting yard waste. Many options are possible for both equipment and approaches, and the best choice of yard waste processing system for a given situation will depend on many factors, including:

- The volumes to be processed.
- The types of yard waste to be composted. For instance, if brush and other woody materials are to be composted, shredding or chipping prior to composting will be necessary or extremely long composting periods will be needed. If the incoming material is bagged, one or more steps in the process will have to be directed at removing the bags.
- The markets that are available for the finished products and the quality demanded by those markets. Processing can be minimal if a large market exists for low-quality material, but in general, the higher the quality is, then the easier it will be to find viable markets for the finished product.
- The budget that is available for purchase, operation, and maintenance of the site and equipment.
- Site conditions, especially if an existing site is being used or if land use is a constraint in the area.
- Desired turnaround time to produce a final product.
- Local and state regulations, which may affect the need for runoff controls and other operational parameters.
- The choice of collection method, which will affect the quality and condition of the incoming material.
- Existing equipment that may be available.

BACKGROUND FOR COMPOSTING

Handling Large Branches

Large branches and small amounts of soil are also frequently included in yard wastes. The degree to which larger branches can be handled effectively depends upon the processing equipment that is employed and the end markets. Successful composting of woody materials such as large branches, land clearing debris, scrap lumber, waste from logyards (timber storage yards) and similar materials, within a typical time frame of 6 to 18 months, generally requires that this material be chipped or shredded to provide greater surface area for microbial action. Wood waste can also be processed to produce landscape mulch or a fuel.

Basic Microbial Activity

Composting is accomplished by microorganisms that use organic materials as a food source. For composting, the most important microorganisms are bacteria and fungi. The specific microorganisms that are active in a compost pile depend on the temperature, raw materials placed in the compost pile, and the stage of the composting process. Oxygen is required by many of the types of bacteria and by most fungi. These types of bacteria and fungi are classified as aerobic microorganisms.

When oxygen is depleted, as can occur in the interior of compost piles, the composting process becomes anaerobic (meaning "without oxygen"). Under anaerobic conditions, microorganisms cannot break down organic materials as quickly or as completely. This causes the composting process to slow down and contributes to odor problems due to the formation of partially oxidized compounds. The partially oxidized compounds generated by anaerobic microorganisms can also be toxic to plants.

The Impact of Carbon–Nitrogen Ratios

The carbon–nitrogen (C/N) ratio has a significant impact on the ability of microorganisms to break down yard wastes. To be efficiently composted, a raw material must provide these elements in the proportions required for the respiration and reproduction of the microorganisms. A C/N ratio in the range of 20 to 35 is best. The C/N ratios of many common materials can be found in resource books or can be determined by testing. The C/N ratio of mixtures can also be calculated through the use of a weighted average of the C/N ratios for each component of the mixture.

The C/N ratio for grass clippings is about 20, which places it at the low end of the acceptable range (i.e., too much nitrogen). More importantly, however, grass clippings alone have a tendency to become too compact and proper aeration is difficult to maintain. With excess nitrogen and poor aeration, some of the nitrogen will be converted to ammonia and subsequently lost through volatilization and leaching. These factors are the primary cause of the odors for which grass clippings are notorious.

The C/N ratio for leaves is too high (too much carbon), which causes the composting process to proceed more slowly. The actual C/N ratio for leaves varies depending upon the type of tree. To the extent that grass clippings and leaves can be mixed, near-ideal composting conditions can be created. The strategy employed in some locales is to mix leaves from the previous fall with grass clippings generated in the spring and summer.

A Word about Windrows

Yard waste is often placed into windrows (long piles) for composting (see Figure 26.7). Generally, the only circumstances cases in which windrows are not used are very small scale composting (at the household level) or larger-scale operations where an in-vessel system is employed. Windrows provide beneficial composting conditions, make efficient use of space, and allow access for turning and watering. The shape of windrows in cross

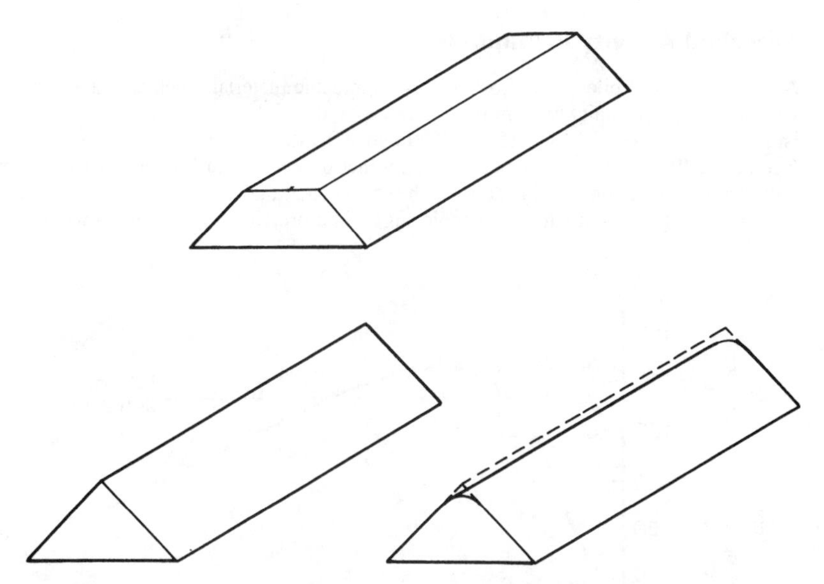

FIGURE 26.7 Three compost windrows, including one with a trapezoidal shape (at the top of this figure) and two with triangular shapes (at the bottom on the right side is shown a pile with a slightly-rounded top that should still be considered a triangular shape).

section (height and width) can be roughly triangular (coming to a point on top) or trapezoidal (flat on top), with the length varied as desired. Whether the shape of the windrow is triangular or trapezoidal will depend on the materials and their condition, the width of the pile versus the equipment used to create it, and other factors. Some degree of control over the windrow shape is desirable, however, the trapezoidal shape can be used to increase absorption of precipitation whereas the triangular shape sheds water and loses heat more rapidly.

The overall size of the windrow can also be used to control temperatures and moisture levels. For instance, the larger the cross section of the pile, the greater the internal temperature achieved. Smaller piles lose water more rapidly through evaporation. Thus, pile size and shape can be used to regulate temperature and moisture content. On the other hand, the larger piles also become anaerobic much more quickly, and even with an active aeration system, oxygen may not reach the interior of the pile. The dimensions and shape of the windrows need to be adapted to local and seasonal conditions.

As the composting process proceeds, compost piles must be "turned" occasionally. Turning includes in-place mixing, rolling the piles over, and mixing as the pile is moved over. Turning of compost piles provides a number of benefits, including:

- Mixing to produce a more uniform product
- Aerating to provide oxygen and remove carbon dioxide
- Breaking up clumps of materials that may be present, to provide mixing and improved composting
- Moving materials on the outside of the pile to the interior of the pile and exposing them to higher temperatures to destroy pathogens and weed seeds

Microbial Activity in Windrows

A typical compost pile, made from readily degradable materials such as grass clippings, will warm up very quickly from heat released by the biological activity of microorganisms. Temperatures of 140 to 150 °F (60 to 66 °C) will be achieved, usually within days (see Figure 26.8a). At this temperature, only thermophilic (heat-loving) microorganisms will survive and continue to break down the organic material. As oxygen is consumed and the readily degradable materials (simple sugars and proteins) are broken down, the bio-

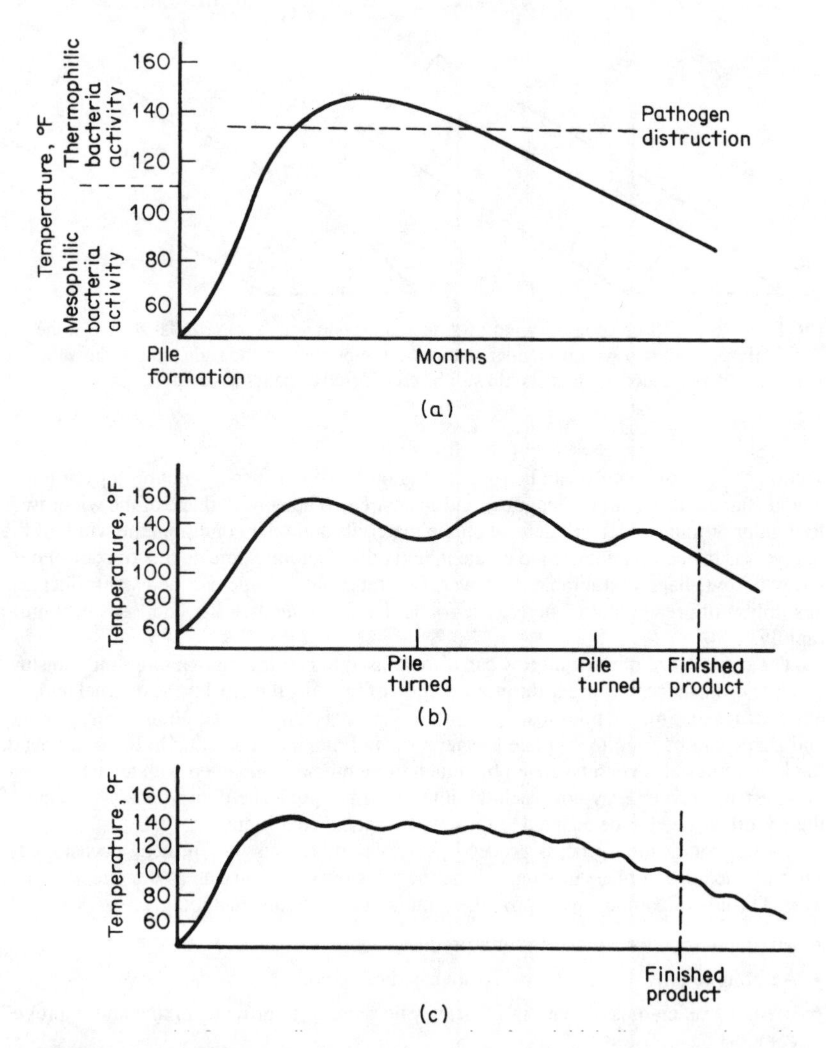

FIGURE 26.8 (a) Compost pile temperature versus time from initial pile construction, with no turning of the pile. (b) Compost pile temperature with occasional turning. (c) Compost pile temperature with frequent turning.

logical activity slows down. If compost piles are left undisturbed, the temperatures will eventually drop below 110 °F (45 °C) and mesophilic (medium-temperature) microorganisms will take over.

Composting yard waste in a pile without turning will cause some materials to be incompletely decomposed during the composting period. Even when pile temperatures drop due to a decrease in biological activity (normally signaling the end of the composting period), materials on the inside of the pile will have composted anaerobically and will not have been fully stabilized. Materials on the outside of the pile may not have decomposed much at all due to dry and cool conditions. Frequent turning of the piles will alleviate these problems and will sustain high pile temperatures until the raw materials are broken down into the finished compost. Less-frequent turning will achieve the same result but more slowly, as biological activity is alternately slowed by the lack of oxygen and then increased by turning and aeration (see Figure 26.8b). Whatever frequency of turning is conducted, pile temperatures will eventually cool and will not be increased much by turning or other additions of oxygen. Barring moisture problems or excessive loss of heat due to cold ambient temperatures, a lower pile temperature generally signals that the composting process is nearing completion. At this point, the original organic material has not been completely broken down, but has been degraded to a point where it is relatively stable and continued degradation will occur much more slowly.

Reductions in Mass and Volume

The biological activity that takes place in a compost pile causes substantial reductions in mass and volume. The microbial activity that takes place in the composting process causes up to one-half of the mass to be lost as carbon dioxide and water. Along with this, the volume of the yard waste is greatly reduced because composting results in smaller particle sizes and increased densities. With or without shredding, typical densities change from a range of 50 to 100 pounds per cubic yard for incoming materials to 1500 to 1800 pounds per cubic yard for the finished compost. Hence, the volume of the finished compost may be only 25% or less of the volume of incoming materials. As the volume of materials is reduced, windrows may have to be combined or shortened to maintain adequate pile (cross section) size.

PROCESSING EQUIPMENT

A variety of equipment can be employed in the composting process, depending upon the desired level of processing and types of material handled at a specific facility. An overview of the primary types of equipment is provided below.

Shredding/Grinding/Chipping Equipment

Shredding equipment is often used at yard waste processing facilities to reduce the size of the incoming materials. Shredding will substantially improve the results of the composting process, especially for woody materials, biodegradable bags and bulky items. Shredding helps to mix materials, creates larger amounts of surface area for microbial activity, and helps create denser piles, which facilitate microbial activity. If possible, removal of contaminants should be done prior to shredding to avoid spreading undesirable materials throughout the compost mix (plastic bags) and damage to equipment (metals and other foreign objects).

The chips resulting from shredding woody materials can be mixed with yard waste for composting, sold as a mulch, or sold as a fuel ("hog fuel"). If the chips are to be composted, they should be reduced to two inches in size or less. In the case of biodegradable bags and bulky items,

Shredders can be stationary or they can be combined with trailers to be used as mobile units. A shredder can also be combined with a crane to allow it to be self-feeding. Screens can be attached to most types of shredders to allow the product to be sized accurately, but matching the proper screen to a shredder can be difficult and has caused serious start-up problems for some facilities. Screening after grinding is often a better approach.

A variety of shredders for yard wastes are available, including hammermills, tub grinders, screw-type grinders, and chippers. These are described in further detail below.

Hammermills, also called "hogs," break up materials through the pounding action of "hammers." They can be vertically or horizontally fed. Vertical hammermills may be gravity-fed from conveyors (see Figure 26.9). Horizontal models employ feed conveyors and may have other power feeding elements, such as a spiked roller to feed material into the hammers. Power feed mechanisms such as this can also be designed to control the amount of material that is fed into the hammers to avoid overloading the grinder. Throughputs up to several hundred tons per hour are possible. The cost of this equipment ranges from $25,000 to $400,000.

Tub grinders can handle a wide variety of material, including leaves, brush, logs up to one foot in diameter, construction and demolition wood waste, and other materials (see

FIGURE 26.9 Vertical hammermill. (*Photo courtesy of Cedar Grove Compost Co., Seattle, WA.*)

FIGURE 26.10 Tub grinder. (*Photo courtesy of Cedar Grove Compost Co., Seattle, WA.*)

Figure 26.10). Throughputs as high as 50 tons per hour are possible. The cost of this equipment ranges from $70,000 to $400,000.

Screw-type and rotary shear grinders have performed well for wood waste. The screw-type shredder consists of a set of slowly rotating (10–30 rpm) screws moving in opposite directions. Material falls between the threads and is slowly crushed and sheared by them. This type of grinder is typically gravity-fed. The design of this type of shredder allows it to be self-feeding because material is pulled in as it is caught by the screws. The angle of the screw threads, the depth of the threads, and their rotational speed can be varied to accommodate different materials. Rotary shear shredders employ a somewhat different design. Throughputs as high as 120 tons per hour are possible. The cost of this equipment ranges from $50,000 to $600,000.

Chippers employ a rotating disk with blades on the surface to cut up the material. The wood is fed against the face of the disk and the blades chip off pieces of the wood. Chippers are rated by the size (diameter) of material that they can accommodate, and the size of the chipper should match the size of the material to be processed. Chippers are available in sizes sufficient to handle trees with diameters up to 20 inches. The larger models may have difficulty handling smaller branches and brush effectively, and it would be expensive to operate them for this purpose. Chippers may produce a stringy product when used for smaller, fresh branches, especially for equipment with low horsepower and/or dull blades. To avoid this problem, it is necessary to maintain the chipper in good condition and/or allow brush to dry before chipping. Processing rates up to 40 tons per hour are possible. The cost of this equipment ranges from $10,000 to $100,000.

Windrow Turners

Windrow turners provide effective mixing and turning of the windrows. Some success has been reported with the use of windrow turners for bag ripping and removal, but it is generally more effective to remove the bags before composting. Windrow turners are not essential to the operation of many composting facilities. For small facilities, a dedicated piece of equipment such as a windrow turner would not be cost-effective, and equipment such as front end loaders (that may already be available and used for other activities at the site) can be used instead. For medium to large facilities, and where high quality is necessary for the finished product, windrow turners would be more essential.

A variety of windrow turners are available. Models are available that can handle windrows up to 22 feet wide (in two passes) and 11 feet high. The operation of the windrow turners vary, with some models turning and mixing the compost piles in place (see Figure 26.11) and some models mixing as they pick up and move the compost piles over. Either method has its advantages depending on the site conditions and mode of operation. Windrow turners are available as self-propelled units or as attachments to tractors and loaders. Turning rates as high as 4000 tons per hour can be achieved. The cost of this equipment ranges from \$20,000 (for tractor attachments) up to \$400,000 (for self-propelled units).

Screening Equipment

Screening is an essential step for producing high-quality compost. A variety of different types of screens have proven useful in processing yard waste, including trommel screens, disc screens, and vibrating screens. When screening is done prior to composting, it removes some contaminants, diverts larger materials (that need to be shredded), and helps the composting process by breaking up clumps and mixing the materials. When done after composting, screening removes noncomposted materials (contaminants and raw materials that need further composting) and produces a more uniform, fine-grained material that has a higher level of acceptance by potential consumers.

Disc screens are often used to screen incoming yard waste to separate woody and bagged materials from grass clippings, leaves and other small materials (see Figure 26.12). A disc screen consists of several rotating shafts with discs on each shaft that are spaced to allow only a specific size of material to fall between the discs. The small-sized material can be sent directly to the composting area; the oversized material is sent to a shredder. With disc screens, screening rates as high as several hundred tons per hour can be achieved. The cost of this equipment ranges from \$5,000 to \$75,000.

Trommel screens have been used for many years to screen soils and peat moss. This

FIGURE 26.11 Windrow turner. (*Photo courtesy of Cedar Grove Compost Co., Seattle, WA.*)

FIGURE 26.12 Disc screen. (*Photo courtesy of Cedar Grove Compost Co., Seattle, WA.*)

screen consists of a rotating drum with holes in the sides of the drum. The drum is set at a declining angle (downward slope) so oversized material that is fed into the high end will move to the other end for removal. Trommel screens are relatively versatile and resistant to clogging, if operated correctly. To prevent clogging, rotating brushes are often placed outside of the trommel (see Figure 26.13). Trommel screens can be designed with more than one size of opening along the length of the screen so that the material can be separated into several different fractions, but this feature is generally not necessary for yard waste or compost. Trommel screens are generally best used to screen the finished compost or as a secondary screening step. With trommel screens, screening rates as high as 150 tons per hour can be achieved. The cost of this equipment ranges from $15,000 to $700,000.

Vibrating or shaking screens are simply screens that shake. These screens are set at a slight angle, with the incoming material fed in at the upper end and the oversized material removed from the lower end. Material that goes through the screen is collected below. With vibrating screens, screening rates as high as 300 tons per hour can be achieved. The cost of this equipment ranges from $10,000 to $125,000.

Miscellaneous Other Equipment

Several types of nonspecialized heavy equipment can be used in the composting process. Front end loaders are a key piece of equipment for moving material around on-site, pushing it into piles for storage or forming windrows. Watering trucks are useful in dry climates, or during dry periods, so that water can be applied to maintain adequate moisture levels for composting. Watering trucks can also be used to control dust at facilities if this is a prob-

FIGURE 26.13 Trommel screen. (*Photo courtesy of Pacific Topsoils, Inc., Seattle, WA.*)

lem. A variety of other trucks are also useful, especially dump trucks, which are used to transport compost to markets or, at some facilities, to move material around the site.

The use of aeration equipment is an option that can speed up the composting process while also allowing for more controlled conditions. This equipment typically consists of tubing with numerous holes laid beneath the compost windrows. Air is blown or sucked through the tubing. This system can have high maintenance costs, however, and short-circuiting of the air flow (i.e., the air flow moving through large channels instead of diffusing throughout the entire pile) can greatly reduce its effectiveness. In general, aeration is not cost-effective where only yard wastes are being composted, the exception being where aeration is a critical element of a required odor control system.

Various types of enclosures can be used for composting. Composting operations can be covered with a simple pole barn or similar structure in rainy climates to reduce runoff, and some facilities are using troughs or other container systems for composting. Recent developments have helped make composting vessels or troughs more appropriate for yard waste composting, although generally it is still not cost-effective to use enclosed systems for composting only yard wastes. Enclosed systems are more appropriate and cost-effective for locations where there is a high degree of sensitivity to odors or other impacts, or where the incoming materials require a higher degree of containment for vector and/or odor control (such as when food waste or biosolids are included in the compost mix).

PROCESSING AT THE SITE OF GENERATION

Processing at the site of generation primarily falls into two categories:

- Backyard composting, which is generally performed by a residential generator using little or no processing equipment.

- The use of mobile equipment, such as the mobile shredders used by commercial generators (landscapers, tree cutters, and others) and public-sector agencies (public works and parks departments).

Backyard Composting of Yard Wastes

The act of composting materials generated on-site at residential properties is typically called "backyard composting" because it is often performed in the backyard. This term has come to encompass a broader variety of activities, including composting on-site by businesses, institutions, and apartment buildings, and possibly even including the use of "worm bins" for handling food wastes. Since the focus of this chapter is on larger-scale processing, only the most important points for backyard composting are summarized below. Backyard composting has a number of advantages, including:

- *Cost*. Backyard composting can be done with little or no direct expenditures.
- *Efficiency*. The raw materials and the finished product do not have to leave the site of generation.
- *Benefits to soil quality*. The finished product (compost) is very beneficial to almost any type of soil as a soil amendment or mulch.

Some potential problems with backyard composting are:

- *Aesthetics*. Aesthetic concerns include appearance and odors. The pile may look messy unless an investment is made in some type of enclosure. Odors may be a short-term problem if large quantities of grass clippings or animal manure have been added to the pile.
- *Vermin*. Experience has shown that well-managed compost piles usually do not provide food or habitat for rodents and other pests. In some urban areas, however, this possibility has caused a great deal of concern and ordinances have been enacted regulating the materials that can be composted or even requiring rodent-proof enclosures.

Management of backyard composting piles consists primarily of occasional turning to provide aeration and mixing. If a variety of raw materials are available at the same time, such as leaves and grass clippings, these should be added to the piles in layers. If only one material is being added, as is often the case, this material should be mixed with finished or partially finished compost, or with small amounts of soil.

The length of time until a finished product is ready depends upon the raw materials, the intensity of management and composting conditions (moisture content, ambient temperatures, etc.). The following practices will decrease the amount of time required to produce a finished compost:

- The compost pile must be large enough to retain heat and moisture, but not so large as to prevent adequate aeration of the interior of the pile. Ideal dimensions will vary depending upon the climate and season, but for free-standing piles this generally translates to a pile four to five feet high and five to six feet at the base. For smaller enclosed systems, such as a bin or drum, sunlight can help bring the pile up to the proper temperature.
- Frequent turning will promote aeration and provide mixing.
- In drier climates and seasons (or for piles in sunny locations), water should be added to keep the pile moist. Watering should be sufficient to make the pile as damp as a wrung-out sponge.

- In very wet climates, covering the pile (but not so tightly as to prevent aeration) will prevent it from becoming too wet.
- The addition of nitrogen will speed composting of leaves, corn stalks, and other high-carbon materials.
- For grass clippings and other materials that have a tendency to compact, premixing or layering with a coarser material or with partially finished compost will help provide aeration.
- For coarse or woody materials, shredding to a size less than two inches is necessary.
- For all materials, finished or partially finished compost can be added to provide a source of microorganisms.

Backyard composting can be accomplished using:

- Free-standing piles, the least expensive method that can be used (see Figure 26.14)
- Bins made from wood, concrete blocks, or other materials (see Figure 26.15)
- Pits or depressions, which may be useful in dry climates to avoid moisture losses
- Barrels or drums, which are set up so that they can be rotated to promote mixing
- Wire fencing, which can be used to provide a temporary or adjustable enclosure
- Inside plastic bags (anaerobically), with the addition of water (for dry materials), nitrogen (for materials with high C/N ratios), and lime (to offset the greater amount of organic acids produced by the anaerobic process)

Backyard composting can be accomplished using no equipment beyond lawn mowers

FIGURE 26.14 Free-standing backyard compost pile.

FIGURE 26.15 Backyard compost bin from King County, Washington distribution program.

and rakes, but additional equipment can be used to improve the results. Besides the variety of enclosures that are sold for backyard composting, there are available a number of chippers and shredders that are specifically designed for residential use. Chippers and shredders can be used to break down leaves, brush, and other materials to speed up the composting process. Brush can be difficult for the homeowner to handle, but if chipped or shredded, it can be used as a mulch or added to the compost pile.

Shredding equipment designed for household use varies from 1.2 to 16 horsepower, with gasoline or electric motors, and can handle brush up to three inches in diameter. Some of this equipment can be used to produce a mulch (wood chips) or compostable material (smaller particle sizes) with the use of removable screens that control the size of the finished material. The cost of these units ranges from $250 up to $2000.

The Use of Mobile Equipment for Processing Yard Wastes

Mobile equipment includes:

- Portable shredders that are used by private companies and public agencies for brush generated by tree-cutting operations
- Large shredders brought to central facilities to handle stumps and bulky wood waste that has been collected or stockpiled over a period of time
- Shredders set up temporarily at work sites to handle land clearing waste

The use of shredders to process branches and other wood waste from tree and brush removal is widely practiced as a method of waste reduction. These chips can be mixed with yard waste for composting or used as a coarse mulch. As a mulch, the chips can be applied around trees in parks and along streets, used on trails, offered to homeowners, or left on-site.

Some facilities have found it more economical to hire a shredder to come to their site to eliminate a backlog of stumps or logs, rather than for the facility to purchase the shredding equipment required for this job. These facilities include landfills that encourage separation of yard material and bulk topsoil dealers that accept yard waste and land-clearing debris. Such facilities typically already have the equipment to handle other yard wastes, but may not be able to justify the significant expense for a larger shredder to handle a small quantity of stumps.

Chippers and shredders have also been set up temporarily at sites where land is being cleared for development or, less frequently, where timber is being harvested.

PROCESSING AT CENTRAL SITES

To divert substantial amounts of yard waste from the municipal solid waste stream, many areas choose a combination of waste reduction methods (backyard composting and mulching of grass clippings) and the use of central processing facilities. The central facilities are designed to handle the yard waste that homeowners and commercial generators (lawn services, landscapers) are unable, or unwilling, to handle on their own property. The design of these central facilities ranges from very simple sites with hardly more than an access road and space for composting windrows to highly capitalized sites with paved roads, structures, and many pieces of specialized equipment.

Processing Methods

The costs and benefits of different designs for central facilities vary depending on the area and the needs. The wide range of processing options and combinations is illustrated by the following examples of processing methods:

- *No processing.* An option used by some companies and agencies involves the use of "static piles," where the yard waste is simply placed in one large pile of no particular shape and allowed to sit undisturbed until composting is finished. This approach may be the least expensive, but requires additional time (three to four years) and space. This approach generates an end product of fairly low quality, due to the dependence on anaerobic decomposition methods, the lack of screening, and other factors.
- *"Low-tech" processing method.* This method employs windrows that are turned occasionally using front-end loaders or similar nonspecialized equipment. Very little other equipment is needed for this approach and it generates a low-quality end product (but better quality than the static pile approach described above) at a low cost. This approach is appropriate for areas with lower quantities of yard waste to be composted and where land is available at low cost. The low-tech processing method is described in greater detail in the following sections of this chapter.
- *"High-tech" processing method.* This method also employs windrows, but these are turned frequently. Additional processing at the beginning and end of the process produces higher-quality compost. The cost of this approach is moderate as long as large quantities are being processed. Depending on the equipment used at this type of facility and available markets for end products, this approach generally makes it possible to

handle additional material such as brush and other wood waste. The high-tech processing method is described in greater detail in the following sections of this chapter.

- *Aerated static piles*. With this method, piles are constructed over perforated tubing and left undisturbed during the composting period. Air is blown or drawn through the tubing to provide aeration. Although some areas have reported success with this approach, it is not widely used. This approach requires the use of special equipment such as blowers and tubing. Problems with the aerated static pile approach include the nondegradation of materials on the exterior of the pile, uneven aeration and moisture conditions, lack of mixing, uneven quality of the finished product, and the need to operate and maintain the aeration equipment.
- *Forced aeration*. Aeration is used more effectively at facilities that also provide some mixing, and mixing at these facilities may be done as frequently as daily. The aeration can be automatically or manually controlled to provide temperature control for the piles. Subsurface trenches are sometimes used with this approach. As with static piles, there are generally better, more cost-effective approaches for yard wastes, but this method may be employed where very quick turnaround times are desired, in highly urban areas where total containment of the composting process is necessary, or where sewage sludge and/or solid wastes are included in the incoming materials.
- *Enclosed vessels*. Examples of enclosed vessels include large horizontal cylinders that rotate or stationary circular tanks with mixers. As with forced aeration, above, this option is generally not cost-effective for yard wastes alone. Such systems are typically only necessary if there is a significant need to decrease the composting period or control odors because of site conditions or location. In addition, if wastes are being added that present a risk of odors or pathogens, such as sewage sludge or municipal solid wastes, the greater odor and temperature control provided by an enclosed system may be necessary.

Two of these approaches are discussed in greater detail in the remainder of this chapter; the low-tech processing method and the high-tech processing method. Both of these methods employ windrows for the active composting period. The low-tech operation employs no specialized equipment. The high-tech operation requires more equipment and personnel, including specialized equipment such as windrow turners. The capital and operating costs of the high-tech option are much greater than the low-tech operation, but the quality of the end product and the potential throughput for a given facility size is also much higher with the high-tech approach.

For areas with plenty of available land and low quantities of yard wastes, a "low-tech" approach may be best. The low-tech approach requires longer composting periods to produce a lower-quality material, requires more acreage per annual ton of material, and is dependent upon easy markets for the finished product. For areas where larger quantities of yard wastes provide economies of scale and/or land is relatively expensive, the "high-tech" approach may be best. Although this approach requires more space for associated operations (shredding, screening, runoff ponds, etc.), a high-tech facility can handle more tons per acre due to a quicker throughput rate. The high-tech approach also produces higher-quality compost that is more easily marketed.

The choice of approach may be dictated in part by state or local regulations. Some states require permits and additional controls for facilities that exceed a given size (generally measured in tons per year). Regulations in some states also address capacity (cubic yards per acre), windrow size, buffer distances, distance to groundwater, environmental controls, and other site and operational parameters. In some areas, the low-tech approach may be prevented by state or local regulations that require a greater level of control over the composting process than this approach allows.

Siting Central Facilities

Siting requirements for low-tech facilities are made easier by the fact that they are typically smaller, and so require less environmental controls (i.e., runoff collection and treatment) in addition to requiring less land. In addition, because the cost to construct a low-tech facility is minimal, a decentralized system of several low-tech facilities can be set up to serve a given area (such as a county). The larger capital investment required for a high-tech facility generally necessitates the construction of fewer facilities that serve a large area or handle larger quantities of material.

Siting factors for most central facilities include:

- *Distance to source(s) and markets*. The distance to the source of the yard waste to be processed should not be so great as to entail excessive transportation costs. Likewise, the distance to markets for the finished compost should also not be too large.
- *Housing*. Proximity to housing can be a problem due to the potential for the generation of odors and other impacts. Even small facilities using the low-tech approach present the potential for odors because the piles are turned infrequently and will probably become anaerobic between turnings. High-tech facilities will create more noise and dust due to the increased level of activity and equipment operations. Buffer zones of 50 to 500 feet are required by some states.
- *Traffic*. The ability of access roads to handle truck traffic must be considered. Facilities will have truck traffic bringing in raw material and removing finished compost. Nonresidential streets are preferable in most cases. Also, facilities must be easily accessible to the general public if a drop-off program is to be used.
- *Surface water*. Facilities should not be located immediately adjacent to surface water, such as lakes and streams, to avoid water quality impacts. Flood zones should also be avoided. Separation distances up to 1000 feet between composting operations and surface water are required in some states.
- *Groundwater*. Processing facilities should not be placed in areas upgradient from shallow wells that are used for drinking water. There is a possibility that nitrates and other compounds may leach from the composting materials and thus contaminate drinking water supplies. Ideally, composting should be conducted on impermeable soils or composting "pads" made from concrete or asphalt to prevent or minimize the potential for groundwater contamination.
- *Utilities*. Low-tech facilities can usually do without utilities, such as electricity and water, but high-tech facilities will need these services.
- *Co-location*. Locating compost facilities near solid waste or public works facilities has both advantages and disadvantages. For instance, co-location with landfills provides some economies for equipment usage, allows poor-quality compost or wood chips to be used for cover material or temporary roads, and siting can often be accomplished without a problem. However, co-location with landfills may be a problem due to the potential for either operation to be disrupted by the odor or water contamination problems of the other. Additional co-location possibilities include mining operations, topsoil companies, public works or parks departments facilities, and other locations where heavy equipment may already be in use or where similar operations are being conducted.
- *Permits and ordinances*. Siting may be affected by various permit requirements, local ordinances, zoning codes, land use regulations, and solid waste management plans for states, counties, and other levels of government. All of these may affect the location of compost facilities, buffer zones, and operational parameters such as the need for runoff control.

The Low-Tech Approach to Composting of Yard Wastes

The low-tech approach to composting involves piling yard waste into windrows and occasionally turning these windrows to provide mixing and aeration. This type of composting typically requires 1–2 years to produce a finished product. A longer period may be required if woody material is included or if the end use requires a highly finished and stabilized product.

This approach requires more space per ton of capacity than high-tech processing methods due to the lengthy residence time of the yard wastes. Since incoming materials do not leave the facility for 1–2 years, and there may be a lag time in marketing the finished product, at a minimum the site must be sufficiently large to contain the amount of yard waste received in a two-year period, plus buffer areas and access roads. Yard waste can be dropped off at the facility by the public and/or brought in by larger generators (such as lawn services, landscapers, and separate collections by waste haulers). Yard waste is generally brought to the facility in bulk (i.e., no bags) or debagged by the generator as it is dropped off. The low-tech approach assumes no screening equipment is available. The facility should be staffed to ensure that contaminants are kept to a minimum, or the contaminants will have to be removed manually. Manual removal can be expensive and ineffective for many types of contaminants. The site monitor can also help to promote the program by assisting people who are dropping off yard waste or picking up compost, answering questions on the use of compost, and providing information on related topics.

The incoming yard waste is typically deposited initially in a specific receiving area. As space is needed in the receiving area, the yard waste is pushed into a windrow for composting. This windrow can be moved away from the receiving side as it is turned during the next 12 to 18 months. The other side of the site can be used for removal of the finished product. Other modes of operation are also possible and the actual approach should be tailored to the conditions at a specific facility.

The approximate dimensions of the windrows should be six to twelve feet high, twelve to thirty feet wide at the base, up to six feet wide at the top, and as long as necessary (or as the site will allow). A flat top will allow rainfall and other precipitation to be captured by the windrow more effectively, which may be necessary to maintain moist conditions in the pile for the long composting period required by this approach. If the incoming materials are sufficiently moist, as is typically the case with grass clippings, a triangular shape (for the cross section of the pile) can be used to shed water more effectively. If the incoming materials are dry, watering should be done to provide sufficient moisture initially, after which it is then optional in all but the driest of climates.

If proper conditions are attained, microbial activity will cause the height of the windrow to diminish rapidly over the first few months, and the pile should be turned and reshaped as necessary to maintain a minimum height of five to six feet. During cold weather, piles with smaller dimensions will cool off rapidly and microbial activity will slow down or stop. Larger pile sizes are often used in very cold climates for the initial piling of fall leaves to allow the composting process to proceed over the winter months. In warm weather, piles can be lower in height to maintain better aeration without losing excess heat. Smaller pile sizes should also be used for potentially odorous materials, such as grass clippings, to prevent anaerobic conditions and thus increased odors.

This option requires the least investment in new equipment. The operation of a low-tech facility generally requires only occasional use of a front-end loader. Other pieces of nonspecialized equipment may be necessary for site preparation (graders and other road-building equipment) and maintenance.

Preparation of a site for low-tech composting includes clearing trees and brush (some should be left in buffer zones to visually screen the facility from neighboring areas), and grading to provide a slope of 1–3%. The slope is to provide adequate drainage of the site. Care must be taken to avoid directing the runoff where it will impact surface water bodies or other sensitive areas. Grading and windrow orientation should be designed to allow runoff water to move between the windrows.

The site should be prepared with clay or other impermeable material where the composting will take place, all-weather access roads, and access control using gates and fences. Depending upon the location, the entire site may not need to be fenced. The roads can be surfaced in some areas using gravel or wood chips, but mixing of these materials with the yard waste should be avoided.

The quality of the final product will be relatively low. The presence of sticks and the small amounts of contaminants (bottles, cans, small scraps of plastic) that inevitably show up will detract from the marketability of the compost. With publicity and education of potential users, however, most programs using this approach have not had any difficulty in marketing the compost if it is offered free or at a low cost. Primary markets have included homeowners, public works and parks departments, and landscapers. A possible layout is shown in Figure 26.16. This layout shows the receiving area close to the main entrance because there will be a greater number of vehicles bringing in materials than picking up finished compost, since there are significant volume reductions during the composting process. Three windrows are shown. Windrow One has been formed from incoming material. Windrow Two is material that has been at the site for about one year, and which began in the same position as Windrow One but has been moved over as it was turned and mixed. Windrow Three is finished material that is available for pick up by the general public and small contractors. This arrangement allows yard waste to be moved from one side of the site to the other as the compost piles are turned.

The costs of developing and operating a typical facility are shown in Table 26.3. This cost estimate assumes a site capable of accepting 500 tons per year. This translates to about 1000 tons of on-site capacity to allow for a composting period of 18 months and short-term storage of finished product. Some of the costs are shown as wide ranges because actual expenses will vary depending on site conditions and local needs. Ideally, a municipality would be able to use land and equipment currently owned, and would not incur any additional capital cost for these items. Co-location of the compost facility with another facility will also reduce expenses by allowing shared use of utilities, equipment, buffer zones, access roads, and other space.

The High-Tech Approach to Composting of Yard Wastes

The high-tech approach is more involved, but it allows the handling of a greater amount of materials and produces a higher-quality end product. This type of facility may be open to the public, but typically it primarily receives materials from yard waste collection programs and private companies (landscapers, land developers, etc.). The increased degree of processing at this type of facility may allow a greater variety of materials to be accepted. Wood waste can be accepted and shredded to produce chips that can be sold as a mulch material or added to the compost pile.

This approach can result in finished compost within 90 days or less. To accomplish this requires maintaining control over the temperature, moisture, and oxygen content of the windrow, as well as controlling the size of the incoming material and the dimensions of the windrow. It is also necessary to turn and mix the piles frequently with one or two screening steps at the end of the process. In addition to the use of general equipment such as front-end loaders, this approach requires specialized equipment for chipping, grinding or shredding, turning windrows, and screening.

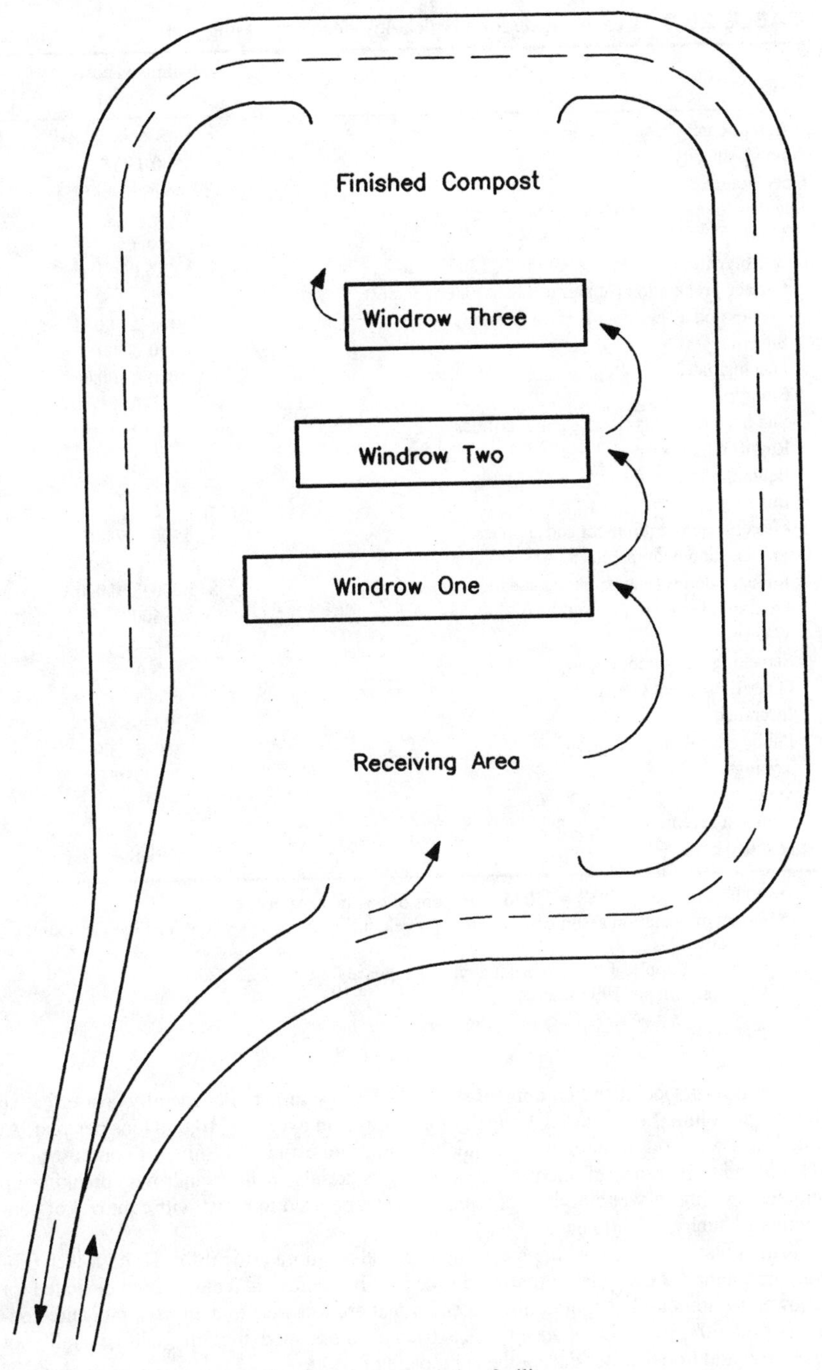

FIGURE 26.16 Site layout for a low-tech processing facility.

TABLE 26.3 Cost estimate for a typical low-tech processing facility

Item	Estimated cost or amount
Assumptions:	
Annual capacity	500 TPY*
Open hours	32 hours/week**
Site development and other capital costs:	
Land	5 acres
Site preparation (clearing, grading), $3000/acre	$15,000
Surface preparation (impermeable base for compost piles and all-weather surface for main roads)	$5000–$40,000
Signs	$150–$500
Fencing/gates	$1,000–$100,000
Siting/permitting	NA***
Site improvements (landscaping, utilities)	NA
Runoff controls	NA
Equipment	NA
Buildings	NA
Miscellaneous equipment and supplies	$500–$5000
Operations and maintenance:	
Annual salaries (one part-time site monitor)	$15,000–$30,000
Front-end loader rental, $600/day	$6000
Watering	NA
Screening and shredding	NA
Other Equipment O&M	NA
Insurance	$300–$3000
Public education/promotion	0–$5000
Testing	0–$1000
Revenues:	
Compost revenues	0
Avoided costs	Variable

*500 tons per year (TPY) = 750 to 1,000 tons of on-site capacity.

**Hours of operation assumes 40 hours per week during the busy season(s) and fewer hours during the slow season(s).

***NA = not applicable, item is not a typical expense for this type of site.

All costs shown are 1997 figures.

The use of specialized equipment such as windrow turners is generally cost-effective for larger volumes of material, typically for amounts in excess of 10,000 tons per year. At this level, windrow turners provide quicker and more efficient turning of compost piles. The turners are more efficient at mixing the materials in the windrows, breaking up clumps, mixing of wet and dry materials, and can be used to assist with removal of contaminants such as plastic bags.

The moisture content of the incoming materials is more critical for the high-tech processing method if the quick turnaround time is to be achieved. Water should be added (or allowed to drain and evaporate for materials that are too wet) to achieve a moisture content of 40–60% by weight. Once this has been accomplished, the piles will generally not need any additional water during the composting period.

Land requirements per ton of annual capacity is less than the low-tech approach due to the shorter composting time, even though additional space is needed for related functions. Additional space is typically needed for offices and garages, equipment and processing area for incoming materials, a stabilization pile, screening and storage of finished materials, and runoff retention ponds.

Figure 26.17 shows a possible layout for a high-tech processing facility. Figure 26.18 shows the flow of materials through this type of facility. As shown, the process begins with a coarse screening to separate finer materials such as grass clippings and leaves from the larger material such as brush. The brush is diverted to a shredding operation prior to being mixed with the finer fraction for composting. After mixing, the materials are placed

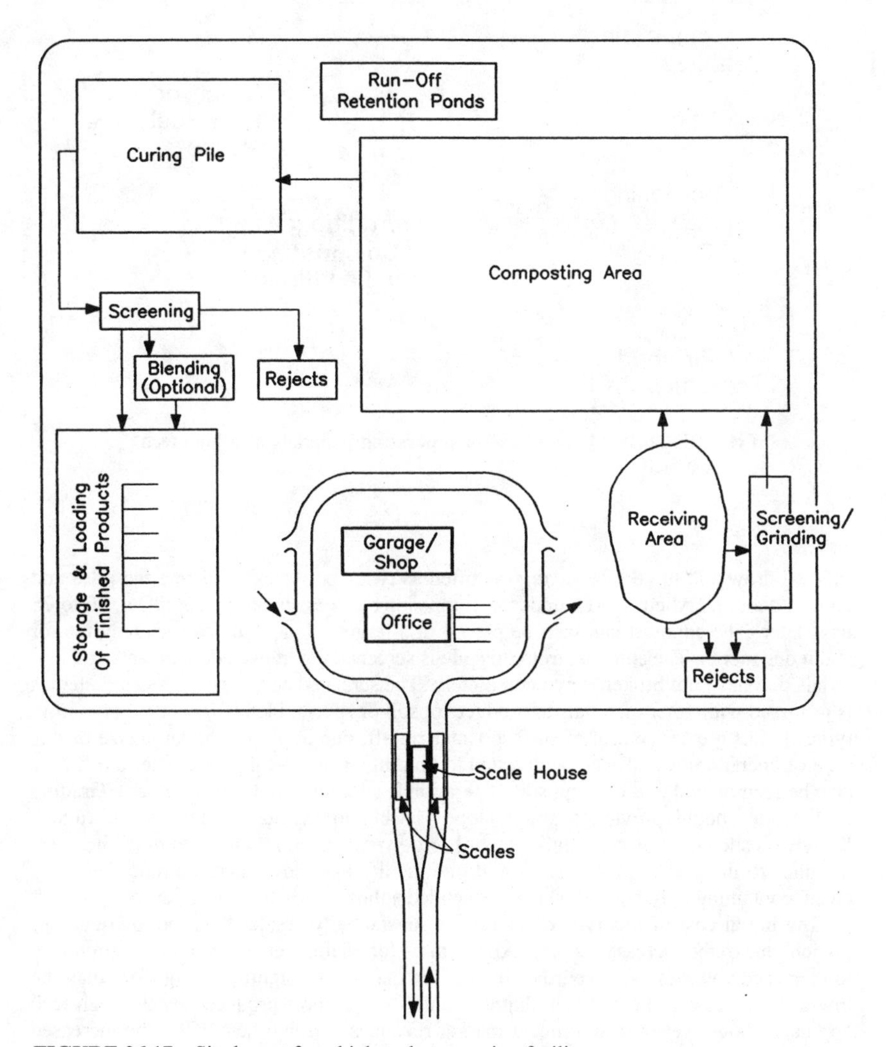

FIGURE 26.17 Site layout for a high-tech processing facility.

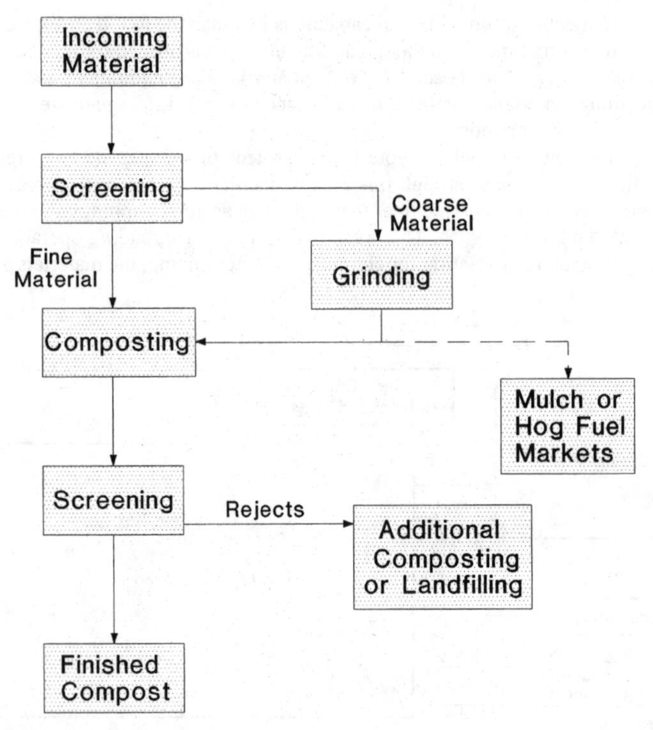

FIGURE 26.18 Flow chart for processing materials at a high-tech compost facility.

in a windrow and turned frequently (as often as twice per week) to aerate the piles and control odors (J. Allen, 1991), until the composting process is nearly complete, 60 to 75 days later. The compost can then be placed in a larger pile and allowed to stabilize for about one month. The compost from this pile is screened to remove contaminants and materials that have not broken down completely. The screened compost can be marketed as is or mixed with other materials to produce topsoil or special blends. This type of facility typically has greater control of surface water runoff, due in part to the large size of this type of operation. Runoff can be diverted to holding or treatment ponds. The runoff may also be recirculated to the composting operation if additional water is necessary. Grading of the site should provide a slight slope (1–3%) to facilitate collection of runoff. Windrows should be placed running uphill or downhill to allow runoff to move between the piles to the ponding area. Testing at one facility has shown that the runoff is fairly clean, containing only high BOD and suspended solids (J. Allen, 1991).

The initial cost of this type of facility is substantially greater than the low-tech approach, due to the increased capital expenditures for equipment and site preparation. Due to increased volumes, however, the cost per ton may not be significantly greater than the low-tech processing method. The higher quality of compost produced by the high-tech approach should generate significant market revenues that will help offset the increased capital and operating costs.

Table 26.4 shows typical costs for a composting facility that employs the high-tech processing approach. This table assumes a site capable of handling about 50,000 tons of yard and wood waste per year. The hours that this site is open reflect extended hours to accept deliveries, but only one operating shift is assumed. As with the costs for the low-tech facility, some of these costs are difficult to accurately project, due to the impact of local conditions, and so are simply shown as "variable." This is especially the case for land costs, siting and permitting, runoff controls, and salaries. In Table 26.4, site preparation expenses includes clearing, grading, installing an impermeable composting pad, and paving of the main access roads on the facility grounds. The expense for buildings

TABLE 26.4 Cost estimate for a typical high-tech processing facility

Item	Estimated cost or amount
Assumptions:	
Annual capacity	50,000 TPY*
Open hours	10 hours/day, 6 days/week
Site Development and other capital costs:	
Land	45 acres**
Site preparation, $3000/acre	$135,000
Surface preparation (compacted compost pads, paving)	$100,000–$250,000
Signs	$500–$10,000
Fencing/gates	$20,000–$100,000
Siting/permitting	Variable
Shredders (hammermill and tub grinder)	$400,000
Windrow turner	$250,000
Screens (trommel and disc screen)	$200,000
Front end loader	$150,000
Trucks (2)	$150,000
Conveyors	$32,000
Buildings	$200,000
Site improvements (landscaping, utilities)	Variable
Runoff controls	Variable
Engineering	Variable
Miscellaneous equipment and supplies	$500–$10,000
Operations and Maintenance:	
Salaries (15–22 full-time employees)	Variable
Equipment O&M, including fuel	$50,000–$150,000
Testing	$500–$10,000
Insurance	$10,000–$100,000
Public education/promotion	$1000–$50,000
Disposal of rejects from screening	1500–3000 TPY
Revenues:	
Compost revenues	$4–$10/ton***
Avoided costs	Variable

*TPY = tons per year.
**Site size provides sufficient space for peak flow and possible future expansions.
***Usually sold by the cubic yard.
All costs shown are 1997 figures.

assumes a very simple structure for a maintenance shop and a trailer for the office space.

ISSUES AND ANSWERS

This section addresses select issues concerning processing and composting of yard wastes.

Odor and Odor Control

Odors are often a significant problem for composting facilities. Few composting facilities are located far enough away from homes and businesses to avoid at least occasional complaints about the odors they generate, and several facilities have been forced to shut down because these complaints could not be resolved.

Odor control is only partially an equipment issue. The creation of odors is often caused by factors out of the control of equipment, such as site design, weather conditions, amount and condition of feedstock, and proximity of homes. There are two areas where equipment can be used to help control odors:

- Use of typical processing equipment to maintain proper composting conditions
- Specific odor control equipment and technologies

The use of equipment to maintain proper composting conditions includes appropriate turning schedules, maintaining proper aeration and moisture levels, and other pile management activities. Equipment can also assist with odor control to the extent that it can help maintain the proper pH and C/N ratios, although these are more of a function of the incoming feedstocks. For the best odor control, a combination of activities must be tailored to local conditions and incoming feedstocks. For instance, maximizing aeration is generally the best approach to avoid odors, by maintaining adequate oxygen levels to avoid the production of malodorous compounds, but aeration may also create a situation where odors are vented instead of being contained within the windrows. In another example, grinding over-sized pieces is generally best for proper composting, but one operator using the static pile method claims to reduce odors by not grinding (the larger pieces apparently act as a bulking agent and increase aeration).

Odor control systems require that the odors be collected and then treated. Collection for odor control purposes requires either an enclosed facility or an effective system for drawing air through the piles and into piping, which then directs it to a treatment process. Many facilities are simply not designed as an enclosed facility or to easily accommodate in-ground piping, and so are not easily amenable to effective odor control. These facilities must rely on odor minimization techniques, perhaps the use masking agents. Masking agents are generally adequate only for short-range or localized applications, however, and are often ill-advised because they simply add to the total amount of odor.

Once collected, odors can be treated using a variety of systems:

- Biofilters
- Bioscrubbers
- Multistage chemical scrubbers
- Carbon adsorption
- Chemical counteractants

Biofilters are one of the most commonly used treatment methods, and some of these even use finished compost in the filter bed.

In summary, the best odor control strategy at a specific facility generally begins with odor minimization and containment techniques at the operations level, and any active odor collection and treatment must be tailored to local conditions and feedstocks (or more precisely, the specific odor-causing compounds released by the feedstocks).

Biodegradable Bags

Most attempts to incorporate biodegradable bags into the composting process have produced poor results. Some pilot efforts have observed that any type of bag initially retards composting and acts as barrier to mixing, aeration, and wetting of its contents until it begins to break down. Others have concluded that the bags may be degradable, but they do not break down as quickly as the yard waste does and so detract from the appearance of the finished compost. Biodegradable bags made of paper have broken down more completely than degradable plastic bags in some cases, and the larger pieces of nondegraded paper can always be screened out and returned to composting piles to finish composting.

Improvements have been made recently in the formulation of plastic degradable bags. Agencies or companies interested in using biodegradable bags of any sort should conduct pilot efforts first to test the performance of the bags in their system.

Compost Maturity versus Nitrogen Availability

In the composting process, bacteria that are instrumental in breaking down the organic compounds may temporarily tie up all of the available nitrogen in their cell structures. As the composting process nears completion, there is a die-off of the bacteria, which then release the nitrogen so that it is available to plants again. With any compost, however, there is the risk of a surge in bacterial activity when the compost is mixed into garden soil or blended with other materials to produce a topsoil or other product. This surge can again tie up available nitrogen temporarily and hinder plant growth.

To avoid nitrogen availability problems, compost should be mixed into gardens at least two to four weeks before planting, depending upon soil temperatures and the amount of compost added. If this is not possible or if the compost is of questionable quality, nitrogen fertilizers should be added with the compost. Topsoil mixtures containing compost should be monitored for one to two weeks after blending to check for the generation of heat as an indication of bacterial activity. If heat is detected, it would be best to hold the mixture for a short time before distribution to avoid consumer problems with plant stunting.

Additives for Composting

A number of materials can be used to enhance the composting process, although some of these additives are of questionable value. Potential additives include:

- *Fertilizers*. To compost high-carbon materials (materials with high C/N ratios), it is beneficial to add a fertilizer that contains nitrogen or to add another raw material that is high in nitrogen. Nitrogen is generally the only nutrient of concern; rarely are other nutrients present in such low quantities that they limit the growth of microorganisms. If the addition of fertilizers is being conducted solely to improve the nutrient content of the finished product, the additional nutrients should blended into the finished compost to avoid losses during composting.

- *Lime*. Lime is considered by some to cause an increase in the rate of decomposition and a reduction of odors. An increase in decomposition rate is suggested because the lime may offset organic acids produced in the early stages of decomposition. These acids may decrease the pH of the compost pile to a point where microbial activity is hindered. Although lime has been shown to increase the rate of decomposition in this manner, the lime may also convert ammonium nitrogen to ammonia (Rosen et al., 1988). The subsequent off-gassing of ammonia will cause nitrogen loss and increased odors. Unless raw materials other than yard wastes are being composted, and these materials are inherently very acidic, the addition of lime is generally not necessary or advisable.
- *Inoculants and enzymes*. Some products are being sold on the basis that they will improve composting rates and results. These products contain inoculants ("starter" bacteria and fungi) and/or enzymes that are supposed to help break down yard wastes. While these products may help get the composting process off to a quicker start, it appears that such products are not vital to the composting process. Except in special circumstance or unusual feedstocks, these products are generally not cost-effective.

Measuring Amounts of Yard Wastes Composted

Measuring the amount of yard waste that is being composted will be necessary in some cases. It may be necessary not only in areas that are striving to meet established recycling and composting goals, but it will also be helpful as a public information tool to encourage participation. Installation of truck scales at all composting facilities would allow for easy measurement of incoming and/or outgoing quantities, but this cost cannot be justified for smaller facilities. Simply weighing the finished product may be helpful but would not determine the weight of material diverted from the waste stream due to the significant losses of weight (up to 50%) that occur as a result of the composting process. One method that could be used is to conduct a survey by occasionally weighing individual deliveries by different types of vehicles to derive an average figure for the weight of a load by type of vehicle. By then counting the number of each type of vehicle, the total weight of incoming material can be estimated.

In the absence of other methods, the weight of material received at a facility can be estimated based on volume and density measurements. This should be done a few weeks after the initial formation of the windrows. This amount of time will avoid substantial losses caused by biological activity, but will allow time for the piles to settle and for moisture and density to even out. The volume of a windrow can be determined by first deciding if the shape of a cross section is closest to a trapezoid or triangle, then taking measurements and making calculations using formulas appropriate to the shape.

If the windrow is closest to a triangular shape, the volume can be determined by:

$$\tfrac{1}{2}\,WHL$$

where W = the average width at the bottom, H = the average height of the pile, and L = the length of the windrow (see Figure 26.7).

For a trapezoidal shape, volume is determined by:

$$\tfrac{1}{2}\,(W_1 + W_2)HL$$

where W_1 = the average width at the bottom, W_2 = the average width at the top, and H and L are as defined above (see Figure 26.7).

In either case, the density is determined by extracting a sample of the pile sufficient to fill a container of known volume and then weighing the sample. This must be done with care so as to avoid fluffing or compacting the sample. This procedure must be repeated a number of times from a number of locations, and the results averaged to yield a figure that can be applied to the entire pile. A minimum of four to six measurements should be taken, depending on the size of the pile and the variance encountered with the results of the first few samples. This process should be conducted separately for each windrow at a site. Once the total volume has been determined and an average density figure has been derived, determining the total weight of a windrow is a matter of simple math and checking to ensure that consistent units are being used.

Marketing Information

Markets are briefly discussed here because they have a strong influence on the choice of processing system. A number of materials can be produced from yard waste, wood wastes, and other compostable wastes, and each of the end products can be used by a variety of groups. Generally, the more intensive processing methods will yield higher-quality composts that can compete for a greater variety of markets.

To some extent, the type of waste material predetermines the end product. For example, it is difficult to produce a mulch material from yard wastes such as grass clippings. Other materials may lend themselves to greater flexibility for the end product. Wood wastes and brush, for instance, can be chipped and added to a composting system, sold as mulch, or sold as hog fuel. Yard waste can be composted to varying degrees and then directly land-applied, sold as a soil amendment, or mixed with soils to produce a topsoil mixture. End products should be designed based on the capacity of available markets and the specifications of those markets.

The following products can be derived from yard and wood wastes:

- *Compost.* Low-quality compost usually cannot be sold, but can be given away for use in gardens, agricultural purposes, erosion control, and applications where aesthetics are not a major concern. Composted yard wastes of high and medium quality can be sold in bulk or bagged as a soil amendment. Bagging operations require special expertise and equipment. It is generally best to initially subcontract the bagging of compost so that the market for this material can be tested before making a substantial investment in the equipment and training necessary for bagging (J. Allen, 1991).
- *Mulch material.* Various grades of wood chips may be marketed as a mulch material in bulk quantities or bagged for retail sales. These chips can replace bark traditionally used for landscaping and other uses. Other uses include park trails, temporary roads, farm yards, and other areas where stabilization of the surface soil is desired.
- *Topsoil mixtures.* Blending compost with soil to produce topsoil (bulk) or potting soil (bagged) can be done. For markets that intend to use topsoil mixtures or compost for growing plants, the compost must be highly stabilized before use or a nitrogen-containing fertilizer must be added in sufficient quantities to ensure that some free nitrogen is available for plant growth. Also, mixtures should be monitored for one to two weeks after blending to check for the generation of heat as an indication of bacterial activity.
- *Hog fuel.* Wood wastes and other woody materials from land clearing debris can be ground or shredded to produce hog fuel. This requires the removal of soil and the production of medium to large chips that can be burned for heat in industrial boilers.
- *Specialty products.* Specialty products include animal bedding, coarse mulch for erosion control, landfill cover, and organic material for land reclamation and remedial ac-

tion at contaminated sites. These applications may require significant marketing efforts unless there is an existing demand for the product.

The following groups may act as markets:

- *Public agencies and government contractors*. Procurement policies and practices for public agencies and their contractors should encourage the use of compost and related products. In doing this, as with other market development efforts, it is important to avoid displacing products that are currently in use and that are derived from waste materials.
- *Nurseries, orchards*. High quality compost could be used by nurseries and landscapers for some applications, such as top dressings to conserve moisture and reduce weeds, and as part of a mix to be used for potting trees and plants.
- *Garden centers*. Garden centers and related retail outlets (grocery and hardware stores) can sell bulk and bagged wood chips, compost, and topsoil mixtures. These outlets typically serve the general public, and so demand high-quality products.
- *Soil dealers and distributors*. Soil (and bark) dealers can handle a variety of products. As dealers of bulk materials, they may be able to handle low-grade products.
- *Farms*. Farmers can use low-quality composts to improve their soil, but this group will object to visual contaminants such as plastics. Chemical contaminants can also be of concern to them, especially regulated metals that may be a limiting factor for land application.
- *National parks and forests*. Forested areas and national parks can act as markets for compost where soil preparation or top dressing is needed. National parks and other recreational settings can also use wood chips as a substitute for bark on trails or as a mulch.
- *Local residents*. Local residents will pick up compost or wood chips if it is offered for free or at a low cost.
- *Landscapers*. Landscapers can use products similarly to residential users, but may be able and willing to use a wider variety of materials once they are familiar with the possible applications for different grades of products. For instance, landscapers could use large quantities of low-quality material for things such as in-place production of topsoil and building berms.
- *Industry*. Industrial markets include the use of wood chips as hog fuel and some of the specialty applications mentioned above; in addition, they can consume compost and mulch materials for use on their property.

GLOSSARY

aerobic: with oxygen; typically used in reference to a biological decomposition process that occurs in the presence of and through the use of oxygen

anaerobic: without oxygen; typically used in reference to a biological decomposition process that cannot occur in the presence of oxygen

C/N Ratio: the ratio of carbon to nitrogen in the raw materials or the finished product

compost: the relatively stable end product of a process employing biological decomposition

composting: the controlled biological degradation of an organic material

hog fuel: chipped or shredded wood wastes (including bark) that are used as a fuel, typically in industrial boilers

humus: an organic material consisting of a mixture of organic compounds such as humic acid, fulvic acid, and humin; "humus" is often used interchangeably with "compost"

inoculant: a source of microorganisms (bacteria and fungi) for the composting process

maturity: the degree to which the compost is stable and the process of rapid degradation is finished

microorganisms: includes bacteria, fungi, and other microscopic plants and animals

mulch: a material that is applied to the surface of soil to reduce weed growth, conserve moisture, protect dormant plants from freezing, and/or enrich the soil

soil amendment: a material, such as compost, that is mixed into soil to improve beneficial properties such as the ability to retain water and nutrients

windrows: long piles typically used for composting yard wastes

yard wastes: generally defined as grass clippings, leaves, small branches, garden wastes, and related materials. In some areas, wood wastes such as larger branches and stumps may be included

REFERENCES

J. Allen. 1991. Personal interview, February 28, 1991, J. Allen, P.E., Cedar Grove Compost Co., Seattle, WA.

C. Rosen, N. Schumacher, R. Magaas, & S. Proudfoot. 1988. *Composting and Mulching: A Guide to Managing Organic Yard Waste*, AG-FO-3296. Revised. Minnesota Extension Service, Department of Soil Science, University of Minnesota.

CHAPTER 27
COLLECTION EQUIPMENT AND VEHICLES

BOB GRAHAM, P.E.
Director Engineering
Enviros RIS, Ltd
Toronto, Canada

INTRODUCTION

Collection of recyclables, trash and organics in North America has seen considerable innovation over the past 10 years. With the focus on reducing the quantity of material sent to final disposal, municipalities, private haulers, and vehicle manufacturers have designed and pilot tested new collection vehicles to serve the specific needs of the wide variety of collection programs.

Early attempts at collecting curbside recyclables usually involved recovering only one material, such as newspaper, using a pickup truck or van. This type of collection from residential units was not complicated, but even then, operators recognized the limitations of using a small-capacity vehicle that was not specifically designed to provide efficient, safe material collection. As the number of materials added to a program grew, these problems became more apparent.

As more and more municipalities became serious about collecting not only a wide variety of curbside recyclables, but also organics generated in the kitchen and/or yard, truck manufacturers struggled to respond to the potential market opportunities and design challenges created by these demands.

This chapter explores recent developments in vehicles used for collection of trash and recyclables from residential, commercial, business, and multifamily generators. Although vehicles used to collect residential and commercial trash have become more sophisticated in their design, the greatest change in collection technology over the past decade has occurred in collecting recyclables and in co-collecting multiple waste streams in the same collection vehicle. This chapter will primarily focus on development of vehicles for this purpose.

CURBSIDE COLLECTION

Program Design

Curbside collection programs, by their nature, are best suited to single family residential dwellings or those multifamily or townhouse units that have convenient ground floor access to the curb. The selection of collection equipment for a multimaterial curbside program is contingent not only on the details of the equipment itself, but on numerous other program design considerations related to the separation and storage activities at the household, curbside setout and collection details, and the proposed processing and marketing arrangements. It is important to understand how these factors interact and how they impact on the selection of collection equipment.

The fundamental concept behind a multimaterial curbside collection program is to rely on the householder to properly separate recyclable materials from what is normally thrown out in the trash and to routinely place these materials at the curb for collection in a manner prescribed by the program designer. Some of the major issues that must be resolved in the collection program design include:

- The materials to be included in the program
- How materials are to be set out at the curb and the type and specific design details of the household container (or containers) to be used
- The collection frequency
- The degree of material sorting that will occur at the vehicle as materials are loaded into the truck

Obviously, many of these issues are interrelated and will have a direct bearing on the type of collection vehicle required. For instance, materials to be included in the program are usually governed by market availability and material specifications. The efficiency of the collection operation and the complexity of the processing operation are governed by the collection frequency, the number of items included in the program, the size and type of container used, and the degree of mixing materials together (commingling).

The variety of system components related to the operations that occur in the household, at the collection point, and at the processing facility are shown schematically in Figure 27.1. The program designer's role is to select the components that are most appropriate for the jurisdiction in question and to make sure that they are properly integrated to ensure an efficient, cost-effective program with maximum recovery of recyclables.

Materials

The majority of curbside programs in North America began by collecting a core group of materials that typically consisted of newspaper, glass jars and bottles, ferrous and nonferrous beverage and food cans, and PET (polyethylene terephthalate) plastic beverage containers. Most programs now include a much broader list of materials as a result of efforts to maximize diversion from conventional disposal alternatives such as incineration and landfill. Many programs now include a full range of fibers such as magazines, telephone books, household corrugated, boxboard, mixed paper, as well as other items such as textiles, sheet plastics, and often a broader range of other plastic containers such as milk and water jugs, detergent containers, margarine tubs, etc.

The materials selected for program implementation as well as the materials that might be considered at some future date for incorporation into the program will obviously impact the choice of collection equipment. Individual material recovery rates and associated volumes are important considerations in choosing the type of household container that might be used as well as the type and design of collection vehicle. It is important that when selecting this equipment, as much consideration as possible be given to the flexibility of the system to accommodate either new materials that might be added to the program and/or increases in the recovery levels of each material.

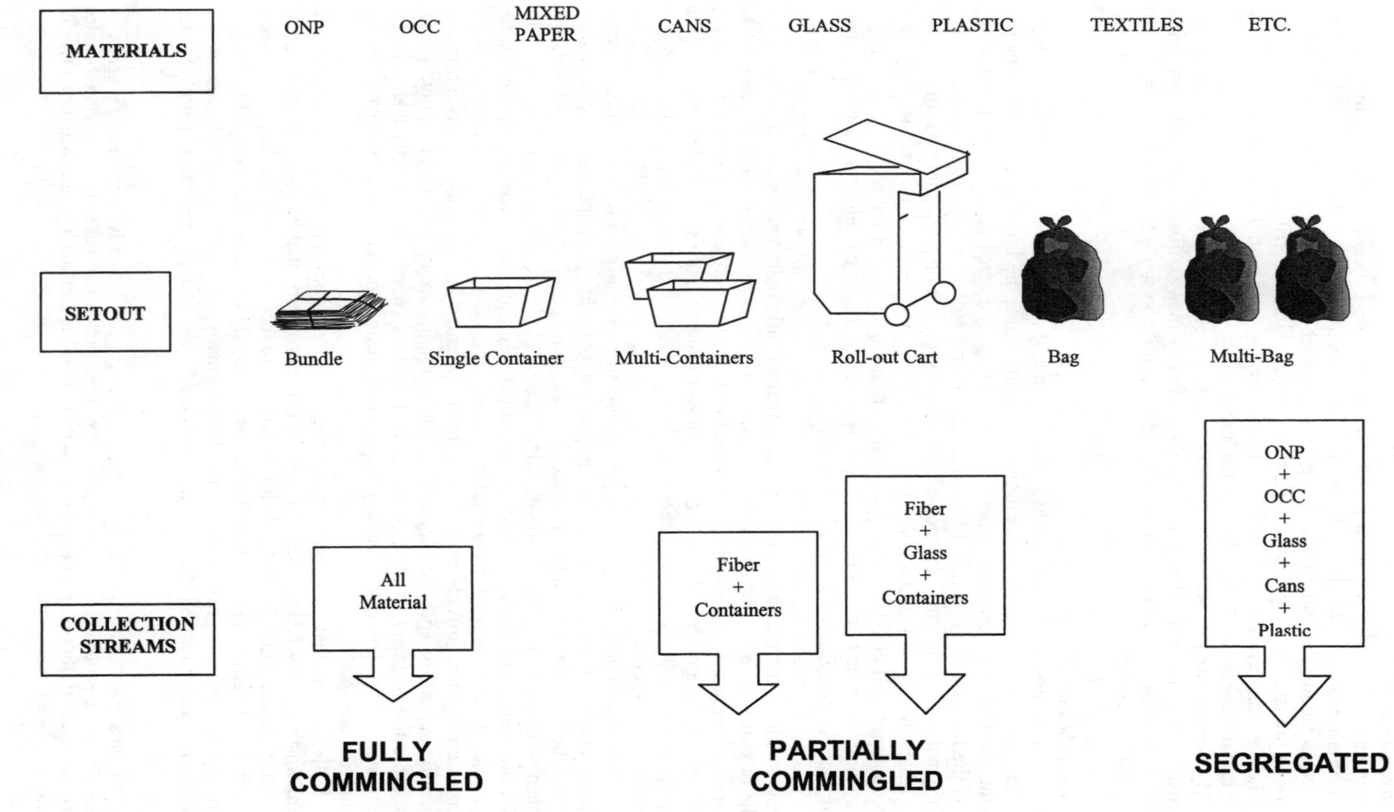

FIGURE 27.1 Schematic of curbside collection of recyclables—integration of system decisions.

Convenience

Experience has shown that for a multimaterial curbside collection program, material recovery is maximized when the program is made convenient for the householder. Most programs require the householder to separate recyclables from the trash, undertake a few steps to prepare the recyclables (e.g., rinse bottles, bundle corrugated, etc.), store them until collection day, and carry them out to the curb for collection. Each of these steps poses a certain level of inconvenience to the householder.

By complicating or adding to any of these steps, convenience drops, as do participation and recovery rates. This is a primary reason why many curbside collection programs have moved away from more onerous multisegregation systems to more commingling of recyclable streams.

Household Containers

The use of a household container has been shown to dramatically increase material recovery rates for curbside collection programs and illustrates the importance of convenience to the resident. The container, whether a box, cart, bag or some combination of these, assists the resident in minimizing material separations, storage of recyclables, and trips to place materials at the curb. The container also helps collection personnel to distinguish recyclables from trash that may be set out at the same time, may result in higher collection productivity, and also has value as a visual promotional tool to remind the public about recycling. There are generally four basic types of household containers in use today: rectangular boxes, cylindrical buckets, roll-out carts, and bags.

The most popular household container for the majority of North American curbside programs has been the single, rectangular container (originally known as the "blue box"). The box may be used either inside or outside the house (wherever most convenient) to store recyclables. Residents are encouraged to place full containers of recyclables at the curb on the designated collection day. In most programs that use a single container, the residents set out all loose containers in the box and place bundled or bagged fibers on top of or beside the box.

Convenience plays an important part in container selection, since the number of containers and their size may have an impact on program participation. The container should be of sufficient capacity to contain the anticipated volume of recyclables between scheduled collections and yet not so large that the weight of a full container poses a problem either for set-out or for collection. Multiple containers for designated recyclables have been used successfully in many curbside programs. For instance, some curbside programs provide several household containers to residents for separation and curbside set-out of specific recyclables. These require more household storage space, are more inconvenient to carry to the curb and are obviously more costly than a single container. They may also adversely impact collection productivity, as discussed in the following section.

Roll-out carts that were previously used only for automated collection of trash and yard waste are now being used in many curbside programs that collect a broad range of recyclables, since they have the ability to hold more recyclables and thereby encourage a less frequent collection schedule. These carts typically range in size from 30 to 90 gallons, are convenient to roll to the curb and are normally lifted and unloaded by means of a hydraulic mechanism on the collection vehicle.* Cart storage may pose a problem for

*In 1998, the City of Los Angeles, after completing a comprehensive pilot testing program on three alternative collection systems to replace its manual "yellow bin" curbside service, implemented a single stream (fully commingled) system using a 90 gallon cart and a regular automated type of collection vehicle that was already in its fleet.

some residents if space is limited or if multiple carts are required. In addition, the carts are relatively expensive, and by their nature dictate that virtually no sorting of recyclables or of contaminants will be done at the collection vehicle.

A growing number of communities are now using large, translucent color-coded trash bags for the collection of recyclables (and other materials such as kitchen organics). The primary advantages of bag systems are that the bags can be collected efficiently at the curb and can often be collected in the same truck as trash, potentially reducing system collection costs. Disadvantages include the potential of bags ripping, the cost of removing recyclables from the bag for further processing, finding an efficient, cost-effective, sustainable and convenient means of bag distribution for the householder, and identifying a market for the bags once recyclables have been removed.

Curbside Collection Operations

Once recyclables are set out at the curb, there are several tasks that the collector typically may be required to perform. These include:

- Sorting recyclables into individual truck compartments
- Segregating nonrecyclables or contaminants from the materials loaded
- Returning empty containers or carts to the curb after loading

Whether to collect recyclable materials commingled or segregated is a key issue because of its implications to overall program design and especially collection efficiency. Selection of one or the other modes of set-out and collection influences materials to be included in the program, collection vehicles utilized, and the design of the processing facility.

The degree to which recyclables are mixed in the collection vehicle can be defined as follows:

- *Full commingled.* Recyclables are set out together and loaded into the collection vehicle unsorted (fiber and containers together)
- *Partial commingled.* At a minimum, the collection vehicle is divided into at least two compartments, usually to separate fiber materials from the remaining container materials. A truck with more than two compartments may have further segregation of materials (e.g., plastics combined with glass, or cans combined with plastics, etc.)
- *Segregated.* Each material (i.e., glass, cans, newspaper, etc.) is placed in its own compartment in the collection vehicle. All plastics or all colors of glass may be commingled, but different materials are not mixed together

Partial commingling or full segregation of materials typically results in slower collection times per stop than commingled collection. Obviously, the more sorting done at the truck, the more time spent per stop. Therefore, as collection staff sort recyclables into more and more categories, the number of stops achieved per hour typically decline. In addition, it is increasingly difficult to add new materials to a curbside program with multiple curbside sorts. This will be discussed in more detail in the following section.

The householder provides the first step in recyclables separation (separating recyclables from trash and storing these in defined material streams) but there is often a second opportunity to further segregate recyclables as they are loaded into the collection vehicle. Even if the recyclables are all set out in one container, or recyclable streams are mixed in multiple containers, the collector may be required to segregate the recyclables in the collection vehicles to integrate with the processing operation at the processing facility. As already mentioned, the opportunity for segregating recyclables from roll-out carts or bags is limited.

An additional task for collection staff might be to sort out items which are not part of the recycling program. In many jurisdictions, these are left behind in the household container (sometimes with an accompanying note to the householder) so that the householder becomes educated as to the materials that are and are not acceptable. In programs where collection staff unload bagged recyclables at the truck, there is an operational issue of what to do with the bag after the contents are unloaded. Assuming that the bag cannot be left at the curb, an additional compartment may be necessary in the truck to store these materials. Again, for commingled collection programs that utilize large bags or roll-out carts, sorting of contaminants and nonrecyclables at the collection vehicle is not practical.

The final step for the collector before returning to the cab is usually to return the household container(s) or roll-out cart to the curb when empty.

Selection of Curbside Collection Vehicles

The foregoing illustrates how many of the program design details dictate many vehicle design features. Key criteria to be considered in selecting the appropriate collection vehicle are:

- Weight and capacity
- Vehicle dimensions
- Flexibility
- Design features
- Cost

The following sections will examine each of these items and discuss changes that have occurred over the past 10 years and future trends in collections vehicle designs.

Capacity and Weight

The desired capacity of a vehicle is a function of the quantity and volume of materials to be collected in a given period and size or weight regulations that exist in a given state or local municipality. Early recycling collection vehicles varied in theoretical capacity from 15 to 31 cubic yards (usable capacity is typically less). As programs expanded both in terms of materials collected and households served, greater focus was placed on maximizing collection efficiency and minimizing costs. Many programs are now commonly served by collection vehicles with 40–46 cubic yard capacities.

When capacities and payloads increase, so do concerns about being overweight. Canada and Mexico have higher weight limits than are allowed in the United States (34,479 lbs on a tandem axle in Canada and 42,990 lbs in Mexico compared to 34,000 lbs in the United States). Most refuse trucks with high compaction capabilities already exceed payload weight, and the trend in both refuse and recycling vehicles is for lighter truck bodies to increase payloads. One way to do this is by using higher tensile steel that is stronger and lighter than that currently used.

Vehicle Dimensions

The physical shape of the vehicle, its length, width, and height, may be important in some jurisdictions. Narrow streets or lanes, limited clearance at underpasses or bridges, headroom restrictions at processing facilities, turning radii in cul de sacs, etc. may influence selection decisions. Some vehicles feature a shorter wheelbase to improve turning capabilities.

Flexibility

One aspect that is often overlooked is how flexible the collection vehicle is to adapt to changes in material types collected or variations in material volumes brought about by seasonal fluctuations during the course of the curbside program. One sure fact is that things change. More and more programs are looking to increase diversion of materials from traditional disposal options such as landfills by adding more materials to the curbside program. The addition of these materials and/or program design changes that the operator would like to put in place as a result of these additions are often hindered by the limitations of the vehicles already in place.

The vehicle that was originally purchased to serve a specific program design may now not have the flexibility to accommodate changes in the program. For instance, is the current collection vehicle suitable if including one more material is added to the collection program? Does this material require further segregation in another truck compartment and can one more compartment be added? If not, what is the associated impact on the processing operation of adding this material to the current commingled stream? Many curbside operators deal with these and similar questions on an ongoing basis.

Design Features

The primary consideration in selecting a collection vehicle for use in a multimaterial curbside collection program should be collection efficiency. Although this is a function of many of the program design elements previously discussed, the truck design itself greatly impacts the collection efficiency. It is therefore important to understand what comprises the daily collection tasks.

The normal workday for a curbside collection crew will involve a certain portion of nonproductive or noncollection time. This is the amount of time during a workday that is devoted to start-up, vehicle safety checks, travel time to the collection route, travel time to the MRF to unload, unloading time, lunch and breaks, etc.—all of the time in a normal workday except the time spent actually collecting recyclables.

The actual time on the route is comprised of two parts: picking up material at each stop and driving time between stops. The driving time between stops is not influenced by the type of collection vehicle. It is a function of a number of factors including the set-out rate (the number of households on a route that have recyclables at the curb for pickup), the density of the households on the route and the amount of traffic delays encountered (stop signs, traffic lights, congestion). The minimum travel time on a given route would be the time for the truck to travel the route under these conditions without making any stops.

The type of truck utilized directly impacts the time spent loading material, the collection efficiency, and ultimately, the cost of the collection operation. Vehicles designed specifically for curbside collection are more efficient than regular trucks (like cube vans or pickup trucks). Some of the specific factors that affect collection efficiency are itemized below.

Right-hand drive, step-out cab. If the driver also loads material, a right-hand drive cab with a low profile, curbside step-out design minimizes the time it takes for the driver to exit and enter the vehicle during the collection. This design also minimizes the amount of wear and tear on the driver, a factor that reduces driver efficiency as the day progresses. Most curbside collection vehicles now provide this feature and many have both left and right hand cab configurations to allow loading from either side.

Length of vehicle. With earlier side-loader designs, loading a longer, multicompartment vehicle with a single operator would take more time than a regular-length truck.

Many side-loading vehicles now have loading hoppers directly behind the cab with provision for up to three compartments.

Loading height. Vehicles with a low loading height are more efficient to load. Loading heights in manual, side-loading trucks range from 40″ to 100″, with most vehicles exhibiting a loading height of about 45″. Generally, the higher the loading height, the more inefficient the loading operation.

Hydraulic loading. Many collection vehicles now have hydraulic lifting devices on the side, front, or rear, to assist in loading recyclables into the truck. Low-level hoppers are used on some vehicles to store the recyclables (segregated or commingled). This provides easy, accessible loading of recyclables into the hopper, and when the hopper fills, the operator engages the lifting mechanism to empty the recyclables into the truck. This hydraulic loading system has replaced older, manual side-loading vehicles that were popular in many curbside collection programs 10 years ago. The lifting mechanisms are usually also equipped to handle one or more handle roll-out carts at a time. Most vehicles require 15–20 seconds for the lift and load cycle. Nevertheless, hydraulic loading mechanisms typically increase collection efficiency by 15%.

Vehicle capacity. Capacity indirectly impacts collection efficiency, in that fewer trips are required to unload material and thus less "nonproductive" time is spent in the entire collection operation. Generally, the larger the capacity, the more efficient the collection. However, in many jurisdictions, vehicle size is limited either by physical constraints (height, length, etc.) or by weight restrictions.

Degree of sorting. As mentioned earlier, stop time per household increases with each additional material sort that the driver does at the truck. Commingled collection is faster than segregated collection using the same vehicle. Depending on the physical features of the truck and the type of household container(s) used, each additional sort at the curb will add to the vehicle stop time. Furthermore, the more sorts that are required, the more chance that any one compartment or bin will fill up or "cube out" before another. Once one of the compartments fills, the truck must return to the MRF to unload, even if the other compartments are not full. This is inefficient use of truck capacity.

Unloading. Some vehicles are easier to unload than others. Again, the more time spent off-loading recyclables, the less time available in a given day to spend on the collection route.

In addition to the foregoing, innovations in vehicle controls are constantly being developed by vehicle manufacturers. For instance, many vehicles now feature antilock brakes, electronic transmission controls, hydrostatic drive technology, and on-board weighing devices to make the vehicles both safer and more efficient.

Collection Vehicle Types

While almost any type of truck can be and has been used to collect curbside recyclables, dedicated, closed-body vehicles have proven to be the most popular. These vary according to how materials are loaded into the vehicle, the degree of material separations possible, and whether or not the materials can be compacted.

No attempt has been made here to document details of all of the manufacturers and models of curbside collection vehicles currently available in North America. This list is constantly changing and expanding. The reader is referred to waste management and recycling trade journals that periodically publish this information.

Manual side-loader. This vehicle consists of an enclosed, compartmentalized body that is typically loaded from the side(s) and unloaded from the rear. Movable, lockable, top-hinged interior dividers create compartments for segregated materials. By unlocking successive dividers and tipping the body, segmented materials are unloaded. This type of vehicle is commonly used in municipal collection programs that rely upon multiple material separations at the curb. The vehicle capacity is typically about 31 cubic yards, and capital cost is about $95,000 (Figure 27.2).

Hydraulic side-loader. Many manufacturers now offer a closed-body truck that hydraulically loads recyclables through the use of a side trough. When sufficient recyclables accumulate in any compartment, they are lifted and loaded into the top of the vehicle. These trucks are typically larger in rated capacity (40–46 cubic yards) than the manual versions. Prices average about $120,000 (Figure 27.3).

Automated side-loader. A variation of the side-loader is a vehicle that has a hydraulic arm to permit collection using roll-out carts containing either refuse or recyclables. These allow either fully automated collection (where the operator does not leave the cab) or semiautomated collection (where the operator moves the cart to the lifting device). Prices range from $150,000 to $180,000 (Figure 27.4).

Hydraulic front-loader. This vehicle (Figure 27.5), which was previously almost exclusively used to collect individual bins of commercial/industrial waste, has been adapted to collect curbside recyclables and refuse. A permanent front bucket permits convenient curbside loading of recyclables. In many communities, a divided bucket permits segregation of up to three materials (see co-collection). Capacities typically range from 38–49 cubic yards and prices range from $110,000 to $140,000.

Costs

Consideration should be given to both the capital and operating costs of a collection vehicle. Dedicated curbside collection vehicles generally range in price from about $80,000* (basic manual side-loader) to $140,000–$160,000 (hydraulic side-loader), depending on the generic type of vehicle and its design features.

Whereas capital costs of the vehicle is perhaps the most obvious cost to consider in selecting a truck, variable operating costs are a larger portion of the annual collection cost and should be carefully examined. At least 50% of the annual operating cost of a one-person-operated vehicle is typically comprised of salary and benefits. Obviously, as the crew size increases, this percentage grows. This is why much of the focus of recycling vehicle designers has been on trying to provide efficient material collection utilizing only a single driver/loader.

Many of the design features previously reviewed impact not only on the vehicle capital costs, but also on the operating cost. Although the capital costs of incorporating right-hand drive capabilities with convenient step-out to the curb may be high, the reduction in operating costs resulting from the increase in operator efficiency will more than offset this. Similarly, other physical features that impact the collection efficiency, such as loading height and the cycle time of hydraulic mechanisms, will also directly impact collection efficiency and thus operating costs.

Other obvious contributors to operating costs are fuel economy, maintenance and repair costs, and insurance and license fees. When all other considerations are equal, some

*U.S. dollars

Advantages	Disadvantages
• one-person operation	• high capital cost
• dual drive	• no auto loading capability for 90 gallon carts
• easy exit and entry	• not useful for O.C.C. collection
• low loading height	
• compartment flexibility	
• hydraulic unloading	
• larger operational capacity than standard closed body truck	
• folding, hinged side panels	

Advantages	Disadvantages
• one-person operation	• high loading height as compartments fill up —maximum volume is not achieved
• dual drive	• high capital cost
• easy exit and entry	• not useful for O.C.C. collection
• compartment flexibility to handle changing material mix	
• hydraulic unloading	
• adaptable to multi-family with auto-loading feature	
• distinct, specialized recycling truck contributes to promotion effect	

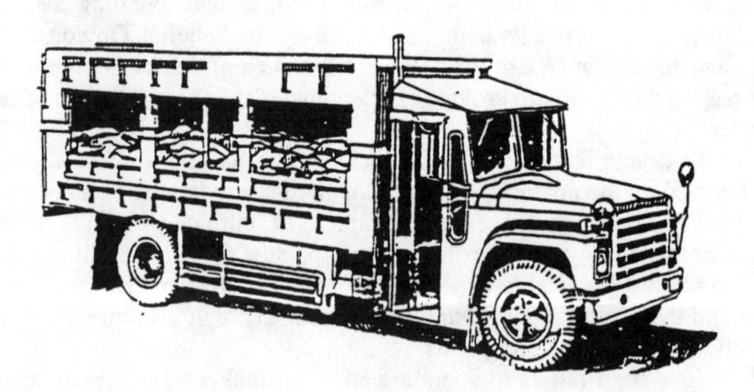

FIGURE 27.2 Manual side-loaders. Top: low-profile closed-body truck. Bottom: closed-body truck.

Advantages

- one-person operation
- dual drive
- easy exit and entry
- compartment flexibility to handle changing material mix
- hydraulic unloading
- distinct, specialized recycling truck contributes to promotion effect
- hydraulic side buckets result in constant low loading height for curbside collection
- capable of automatically servicing 75 gallon collection carts from apartments and commercial establishments
- full volume capacity can be utilized (31 cubic yards)

Disadvantages

- high capital cost
- not useful for O.C.C. collection
- roof height when loading (some models)

FIGURE 27.3 Hydraulic side-loading truck.

FIGURE 27.4 Automated side-loading truck.

FIGURE 27.5 Hydraulic front-end loading truck.

operators may select collection vehicles based on the availability of a distributor to supply parts, the maintenance and repair records of a given vehicle model, or the individual preference of the vehicle operator or mechanic.

MULTIFAMILY COLLECTION

In many respects, designing collection systems to service the wide variety of multifamily housing types is more difficult than for single-family residences, especially if an attempt is made to integrate collection equipment for both services.

Recycling system options within existing multifamily buildings are typically governed by design, space, and operational constraints, making it difficult to plan and implement effective recycling systems. Very often, space is limited to even handle the garbage generated within these structures, let alone add a new requirement to provide a separate system for recyclables. When residents are required to carry recyclables to the ground floor or to the basement, they are often supplied a recycling bag to make this effort more convenient.

Many townhouse and apartment complexes utilize various sizes of roll-out carts, igloos, or front-end containers to store source segregated recyclables. In many cases, the system to be used parallels the system used for garbage collection. Roll-out carts are the most popular means of recyclables storage and they can be collected with conventional compartmentalized curbside collection vehicles that are equipped with hydraulic lifting mechanisms, or alternatively, a rear-packer truck (Figure 27.6) with a similar lifting device can be used if the recyclables are fully commingled. These carts can be located either inside or outside of the building, depending on available space, and when full, can be wheeled to the curb or to some convenient location on the property for collection.

Igloos (bell-shaped containers that are lifted using a crane truck) are attractive, distinctive containers that have become popular in many noncurbside recycling applications, such as at drop-off depots (Figure 27.7). These containers are usually lifted over a compartmentalized, open-top crane truck and the contents unloaded through the bottom of the igloo into the respective compartment. Crane trucks are relatively expensive and are limited both in total capacity and in the number of compartments that can be used.

Advantages

- available and familiar in most communities
- well suited for garbage collection —may integrate recycling with garbage collection
- easy to load and unload
- high cargo weight
- suitable for collection of O.C.C.
- low loading height
- loader can ride on back
- if clean, suitable for "news only" collection

Disadvantages

- difficult to adapt to multi-material collection
- normally two- or three-person operation
- high capital and operating costs related to hydraulic system
- only one compartment available for one material
- contamination problems when also used for garbage collection, could render collected material useless
- confusion with regular garbage collection is disincentive for residents (not if good signage)

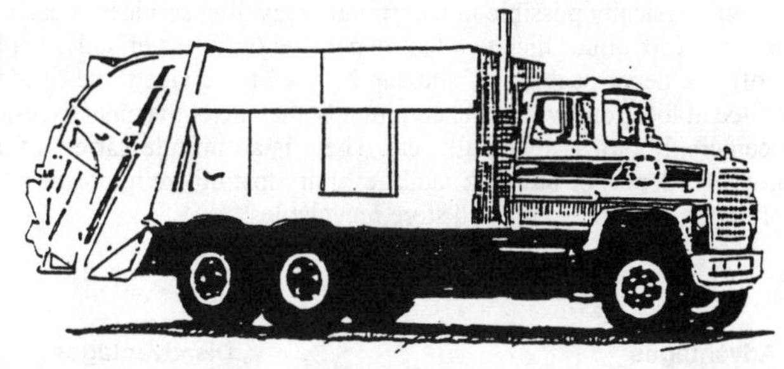

FIGURE 27.6 Rear packer truck.

Advantages

- one-person operation
- "igloo" shape recognized internationally
- materials separated by participants
- number of depot containers at any site can be varied according to demand

Disadvantages

- requires a minimum number of depot locations to be economical
- high capital cost
- not suitable for curbside collection
- depots not integrated with curbside collection

FIGURE 27.7 Igloo and crane truck.

Roll-off containers are also used at multifamily dwellings and at drop-off depots, although their use normally requires residents to bring recyclables to the container. As a result of their long-standing association with trash collection, their use may not be perceived by the residents as a serious attempt to institute a good recycling program. These containers are collected and hauled individually by roll-off trucks (Figure 27.8) or by hook lift trucks (Figure 27.9) that are normally used for trash collection.

More consideration is now being given to incorporating recycling into the design of multifamily structures, rather than trying to add the system after the facilities are designed and constructed. These include multichute systems, carousel units at the bottom of a single dedicated recycling chute, stacked containers on rolling carts located in enlarged garbage/recycling rooms on each floor and the provision of dedicated storage space in each residential unit for segregated recyclables.

When it is not physically possible to incorporate recycling services at each multifamily facility, or when economics dictate a less expensive (and less effective) collection alternative, an off-site depot or drop-off site can be used to service these residents. These are typically sited at locations within the community that are convenient to residents, such as shopping centers, libraries, town halls, etc. There is an infinite variety of customized designs of drop-off (depot) or buyback facilities, but most utilize igloos or compartmentalized roll-off containers to receive and store recyclables

Advantages

- available on a contract basis
- handles a variety of container sizes
- divided roll of boxes can be used as depots

Disadvantages

- not suitable for curbside collection
- not all boxes are compatible with all trucks
- few municipalities handle enough containers to justify owning a truck

FIGURE 27.8 Roll-off truck.

Advantages	Disadvantages
• large variety of containers available	• left hand drive
• costs can be shared with non-recycling operations	• requires a minimum level of containers to be worthwhile
• some containers stack for storage	• depots and curbside operations must be handled independently

FIGURE 27.9 Lift hook truck.

INDUSTRIAL, COMMERCIAL, AND INSTITUTIONAL (ICI) SYSTEMS

There are many point sources of recyclables within a municipality that provide significant quantities of recyclables. These include such facilities as commercial office buildings, hotels, hospitals, schools and universities, and bars and restaurants. As for multifamily dwellings, the design of recycling systems to service these facilities, in most cases, is specific to each location.

Once the internal system is designed to assist in the segregation and storage of recyclables, most of these facilities will utilize roll-out carts, or more frequently, front-end containers for the collection of these materials. The type, size, and number of containers to be used will depend on the number of materials segregated, the recovery rates, the collection vehicle to be used, and the capability of the processing facility or end market to accept commingled or segregated recyclables.

CO-COLLECTION

More and more municipalities throughout North America now have legislated targets for waste diversion. These ambitious targets are requiring a thorough evaluation and implementation of alternative waste management strategies, including the separation, collection, and diversion of more individual residential material streams, such as kitchen organics, leaf and yard waste, household hazardous wastes and garbage at the same time as more traditional recyclables.

European residential collection systems were the first to focus on the separation and collection of a "wet" or primarily organics stream and a "dry" residential stream comprised primarily of recyclables. Programs generally focused on two basic systems: a two stream (clean wet and other, or clean dry and other) system, or a three-stream system (clean wet, clean dry, and garbage). The goal of these programs was to develop a system that accomplished the recovery and quality objectives of each stream, while maximizing collection efficiencies and minimizing collection costs.

As the challenge in North America has been met in recent years to develop a dedicated vehicle to collect curbside recyclables, the current challenge is to design a collection methodology to efficiently and cost-effectively collect all of the other material streams that are now part of a municipality's solid waste management program. The concept of collecting two or more material streams at the same time in the same vehicle is referred to as "co-collection." The rationale behind a co-collection system is that by collecting several waste streams in a single pass, significant reductions may be realized in collection time, and ultimately, collection cost.

Numerous North American truck manufacturers (e.g. Kann, Labrie, McNeilus, Shu-Pak, etc.) have developed multicompartment collection vehicles with compaction capability to meet this need. Recent truck designs have evolved with two- or three-compartment collection vehicles, split either horizontally or vertically. Early versions of vertically split vehicles had more operational difficulties to overcome than vehicles that were split horizontally. For instance, these trucks had to be designed to accommodate the potential of differential weight distribution caused by density differences of the collected material fractions and the weight of compaction equipment. Manufacturers now incorporate both designs in recent co-collection models. Most co-collection vehicles now feature these vehicles with either side-loading (Figure 27.10) or front-loading designs (Figure 27.11) that provide convenient cab step-out and loading.

An additional option for these co-collection vehicles involves cart tippers for use with conventional or divided roll-out carts. This feature, in combination with an alternating-week collection schedule for individual recyclable streams (e.g., fibers and containers), allows the municipal program designer or collection contractor great flexibility in deciding how materials are to be segregated and collected. Within some limitations, the capacity and compaction functions of the individual vehicle compartments can be selected to suit the material mix. An additional advantage of these vehicles is that variable degrees of compaction can be applied to the materials in each compartment as desired to balance capacity concerns (e.g.. high compaction for trash) with efforts to maintain the integrity of

FIGURE 27.10 Side loading co-collection truck.

FIGURE 27.11 Front end co-collection truck.

individual recyclable containers (e.g., low compaction for commingled recyclables to minimize breakage).

Collection of leaf and yard waste as part of a co-collection system has often proven difficult, especially in those geographic areas where seasonal growth results in large volume fluctuations. Most North American co-collection programs now focus on some combination of refuse, kitchen organics, and recyclables. Also, although municipal co-collection programs now operate in medium- to high-density curbside neighborhoods, perhaps their greatest economic advantage is in rural areas (where long travel times between stops for several separate collection vehicles are replaced by a single co-collection vehicle).

Clearly, great strides have been made in the last decade to respond to the needs of municipalities across North America in providing more efficient and cost-effective collection of all waste types. It is expected that these innovations will continue, to the benefit of all concerned.

CHAPTER 28
PROCESSING EQUIPMENT

KENNETH ELY JR.
President
Ely Enterprises Inc.
Cleveland, Ohio

INTRODUCTION

Today's recycling equipment serves diverse situations: factories and distribution centers compress and bundle wastepapers; shopping malls have aluminum collection sites; automobile manufacturers incorporate their own recycling centers. Also, more and more recycling facilities group together various types of equipment. These centers process varieties of materials at once.

Over the years, technical innovations have enhanced processing equipment with higher accuracy, better safety features, a more economical use of energy, and greater speed. This has led to the development of peripheral types of equipment. Shredders shred and pretreat materials prior to baling or compacting. Conveyor systems transport material quickly through all aspects of the recycling process.

In addition, more specific types of technology have developed in response to various material markets. For example, can flatteners evolved because of the need to compress beverage cans for transportation. Also, glass crushers are popular because they produce a form of glass called cullet, which resemble small glass pebbles. (Cullet is the preferred form of glass with glass manufacturers.)

Most recently, recyclers have seen very sophisticated processes developed due to the growth of the recycling industry as a whole. These processes further improve accuracy and speed; especially in the areas of sorting. Eddy current separators pull nonferrous metals from a conveyor line. Similarly, optical glass separation systems sort glass according to color. Optical plastic separation systems also sort plastic according to type. These techniques have come a long way from recycling's humble beginning.

In the twenties and thirties, recycling machinery was very labor-intensive. Balers were often hand-cranked, and it wasn't until the thirties that electrical baling units outsold hand-operated machines. Crude by today's standards, these machines relied on a series of ratchets and chains to bale materials.

Throughout the thirties, forties, and fifties, electromechanical machines were produced that were large and durable. As a testament to this durability, some of these machines are still operating today. At that time, recycling came under the auspices of junk and scrap metal dealers.

Of course, during the sixties, a growing awareness of the need to recycle, coupled with technological advances, produced an array of safer, faster, and more powerful equipment.

The advent of hydraulics meant that material could be easily compressed and reshaped. Today we see the results of these developments in powerful balers, compactors, and densifiers.

In the nineties, recycling equipment is catching up with the computer chip. Manufacturers are producing more and more machinery with programmable intelligence. Such technology yields in-depth information concerning throughput, production time, density, etc. Most types of equipment are available with computerized options. Larger pieces of machinery can even connect to a modem for on-line adjustments by the manufacturer.

Despite this technology, an important concern for potential buyers of recycling equipment is flexibility; are a certain technology and its by-products limited to one specialty market? One should be aware of how widely marketable a certain recycled material will be after processing. Such an awareness can aid a customer in avoiding equipment that will be obsolete a few years after its purchase.

Of course, the truly conscientious customer will make the most successful purchases. Customers should buy from the manufacturers and dealers with the best reputations. Recycling is a field where slipshod equipment and service quickly come to light. The technology is too advanced and the processes too demanding for it to be otherwise.

With today's emphasis on recycling, the multitude of different types of processing equipment is truly vast. This includes convenient appliances such as "reverse" vending machines and under-the-counter can flatteners. For our purposes, however, we will focus on the more commonly used equipment.

BALERS

Since the advent of large-scale recycling in the 1950s and 1960s, the baler has played an increasingly important role in recycling. The reason why is simple: as more varieties of waste products are handled in larger amounts, efficient storage and handling methods become essential. The baler is simply a bundling system, fulfilling a modem, economical need-packaging uniformity. In this case, the "packaging" encompasses recyclable products made from plastics, papers, or metals. The final product is a densely packed cube of sorted or mixed waste products that has been tied with plastic or wire. These waste bales are easily stacked for shipping or storage.

Depending on need, balers are available in a variety of sizes and configurations. These sizes range from smaller, easily transported vertical balers, to larger horizontal balers. All types of balers have similar elements which include: a "feed hopper" area, into which the recyclables are fed; one or more "rams," the flat surfaces which compress the material; and the "baling chamber" or "compression chamber," where bale compression takes place.

Vertical Balers

Vertical balers are the most popular baling units in the marketplace today. Typically, a vertical baler has a ram that moves from an upright position, down through a hopper area, into the baling chamber. This ram is driven from above by a hydraulic cylinder, although some units incorporate two cylinders, which pull the ram down into the compression chamber (Fig. 28.1).

The operator feeds materials into the hopper area through a safety window. This safety window prevents the machine from operating while it is open. Once the hopper is full, the operator simply closes the safety window and pushes a button to begin the compression cycle. Following compression, the ram moves upward again, opening the safety window to accept another charge of materials. This process continues until a bale of a certain density has been formed. Some units today determine density automatically, through use of a pressure relay. Other baling units rely on a visual sighting by the operator to determine the complete formation of a bale.

When a sufficiently dense bale has been formed, the entire mass is bound with wire strapping or plastic. This is accomplished manually, by opening a door to the baling chamber and running precut lengths of wire, or similar material, through a manual tying arrangement. A recycler will typically use 12- to 14-gauge wire, depending on the material being recycled.

Today's vertical balers can produce bales ranging in length from 18 to 72 in. The most common bale today measures 30 by 48 by 60 in, a standard, acceptable mill size for most recycled materials.

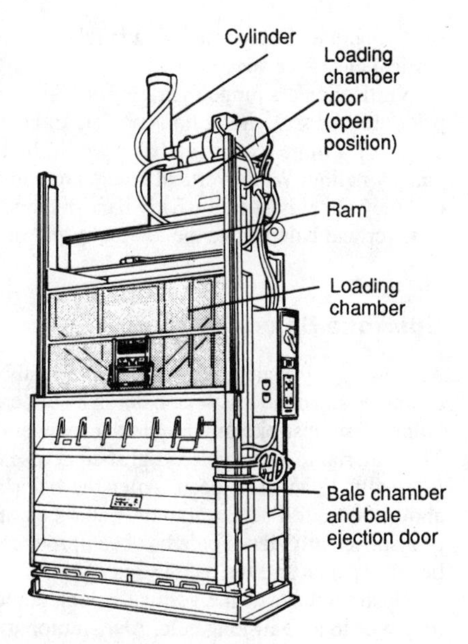

FIGURE 28.1 Vertical baler.

After the bale has been tied, it is ejected from the chamber by means of either a "kick-plate," which ejects the bale from below, or a series of chains connected to the ram which tighten as the ram is raised. Tying time takes between 5 and 10 min, while the overall baling time varies from as little as a half bale an hour to two or three bales an hour. Of course, the type of material being baled determines the bale time.

Vertical balers are very versatile. Recyclers can use the machines to process nearly all grades of waste paper, corrugated cardboard, foam scraps, ferrous and nonferrous metals, steel scrap (provided it is a light gauge), used beverage containers, and plastic bottles. A vertical baler has even been used to compress peppers at a spice plant in Texas for easier transportation across country. There are some technical concerns with certain materials, however. For instance, because newspaper is not very compressible, it requires an even distribution throughout the feed hopper. If this step is not followed properly, bale quality will suffer, and the bale might not hold together. Also, plastic bottles should have all caps removed to ensure that no air is trapped within the bale.

Vertical balers are electrically operated, with motor size ranging from 1 to 30 hp. The power comes from hydraulics, where the motor turns a pump which directs the flow of oil to a hydraulic cylinder. Due to their low horsepower, these machines are very economical to run. Cylinder sizes vary from 2 to 10 in. A typical 60-in baler would incorporate a 6-in cylinder and a 10-hp motor. An important consideration when comparing balers is hydraulic system pressure and cylinder size. Occasionally, a manufacturer will downsize a unit's cylinder size while operating at a higher system pressure. Although this practice keeps the equipment's purchase cost down, it ultimately leads to premature system breakdowns. One would do better to invest in a baler with a large cylinder and a system pressure in the 1800- to 2200-lb/in^2 range.

Vertical balers are usually self-contained and easily transported. Machines are transported on their sides, and after being placed on end and having power connected, they are set to operate. Once installed, a single-cylinder downstroke baler requires about 13 ft of vertical clearance space.

Vertical balers range in price from $6,000 to 40,000. Some of the factors determining price are unit size, motor horsepower, and the size of the hydraulic cylinder. A typical example of a more expensive baler would be a 30-hp unit with a 10-in cylinder for high-density baling. While vertical balers are convenient due to size, one should keep in mind that they are almost exclusively hand-fed and cannot be automated. Of course, because of this, vertical balers have the lowest output of all balers.

Upstroke Balers

Another type of baler, similar to the vertical baler, is the upstroke baler. The most striking characteristic of an upstroke baler is its underground charging chamber (Fig. 28.2). Installation of an upstroke unit requires a foundation and a pit constructed below ground level. The material to be recycled is pushed across the floor and into the baling chamber, which is usually 12 to 18 ft deep. After the baler's doors are closed, the ram moves upward, above the floor line, and into the baler's compression section. The process is repeated until a sufficiently dense bale has been produced, whereupon the doors to the baling chamber are opened and the bale is tied off.

Upstroke balers, like vertical balers, come in different sizes, producing from a 54-in-long bale to a 72-in-long bale. Also, motor sizes vary from 10 to 25 hp. Upstroke balers in the 25-hp range are often called "high-density" balers, because they can produce bales of

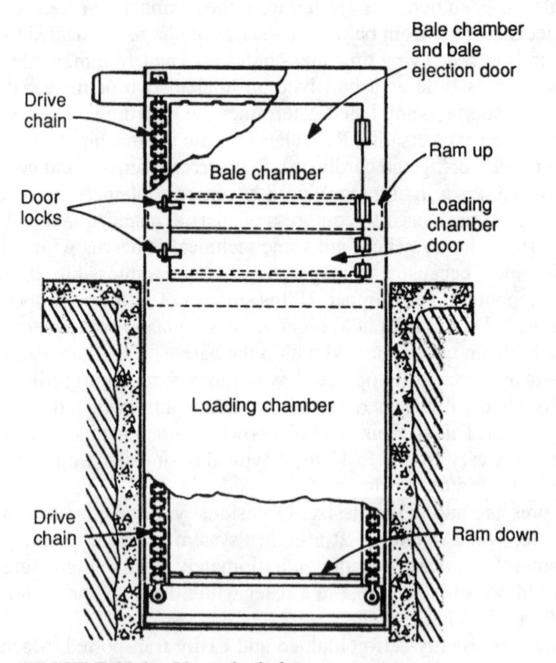

FIGURE 28.2 Upstroke baler.

extra-high density. Unlike the hydraulic rams found in vertical balers, most upstroke baler rams are electromechanical; that is, they are gear-driven and rely on a series of chains to pull the ram up into the baling chamber. There are some hydraulic ram upstroke units on the market, however.

Upstroke balers were in widespread use throughout the fifties and sixties, but have since drastically lost popularity. Nevertheless, some upstroke units are still produced in very small numbers. One can also purchase used units.

The main reason for the upstroke baler's decline is the inherent construction costs. These costs, which include digging a pit, setting a foundation, and installing the unit, are very high. A recycler who installs the unit on leased property faces the prospect of a future change of location. Needless to say, you can't take the foundation and pit with you.

Likewise, resale is difficult, as the new owner must dig out the old unit, transport it to the new location, and then dig a new pit at a new location. Potential customers should beware that spare parts are difficult to find and very expensive. The units themselves are probably already well used. Additionally, the units from the fifties and sixties do not have today's safety features.

All this would lead us to ask why anyone would buy such a unit today. The most obvious reason, besides a possible bargain price, is high-density baling. Accordingly, the majority of upstroke balers sold today are high-density units.

Horizontal Balers

As the name implies, horizontal balers compress materials in a horizontal manner. These balers can process the same materials as vertical balers, although the potential throughput of a horizontal configuration is far greater. Horizontal balers are available with motor sizes ranging from 5 to 150 hp. All horizontal units feature hydraulically driven rams. Higher-end large-volume units also offer such optional features as continuous feed and automatic-tying mechanisms. Of course, these diverse sizes and features also add up to a large spectrum of horizontal unit prices, ranging from $10,000 to $600,000.

Closed-Door Manual-Tie Horizontal Balers. The most basic type of horizontal baler is the closed-door manual-tie horizontal baler, which operates in a fashion similar to vertical balers (Fig. 28.3). The closed-door baler features a hopper area that accepts either conveyor-fed, hand-fed, or cart-dumped material. These units incorporate a photoelectric relay within the hopper area. When enough material has filled the hopper to block this relay, the ram cycles forward, compressing the material from the hopper into the baling area. Because this is a horizontal system, it will accept new charges into its hopper area while the ram is cycling forward. The excess material simply accumulates above the ram and then tumbles into the baling chamber as the ram retracts from the baling chamber.

Like vertical balers, these units compress material in a series of laminations. At a certain point, a pressure setting indicates that enough material has accumulated to form a bale and alerts the machine's operator via a horn or buzzer. The operator must then switch the system from an automatic mode to a manual mode. The ram moves forward one final time to hold the bale in place. The bale is then tied manually.

Although closed-door balers process a wide variety of materials, large-scale baling of certain materials may necessitate the addition of accessory equipment, such as a shredder or a conditioner. These devices prepare material for baling by decreasing its infeed density per cubic foot. This accessory equipment is necessary to avoid jamming or stalling while processing large amounts of wastepaper. Because the unit's photoelectric eye may

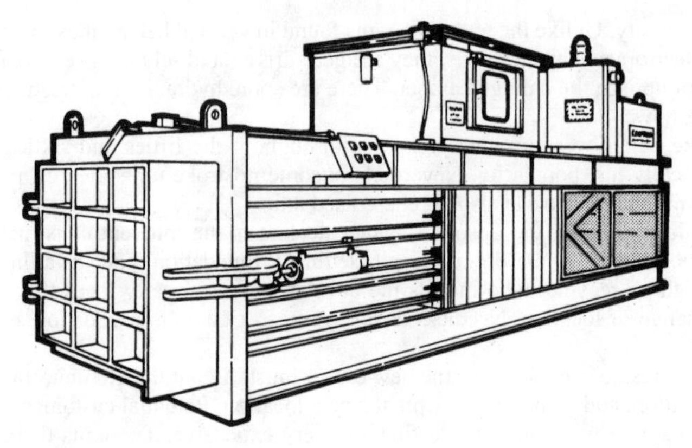

FIGURE 28.3 Closed-door horizontal baler.

not recognize the high density of these papers, too much material may tumble into the hopper area for the ram to compress at one time.

Closed-door horizontal balers are available in different sizes, producing bales ranging from 42 to 72 inches in length, with cross sections ranging from 24 × 24 in to 48 × 48 in. Often, the desired throughput as well as the material to be processed help determine a particular baler's size and features. For example, a machine with a large feed opening would lend itself more to baling bulky cardboard than to baling paper trim.

New closed-door horizontal balers start in price at about $15,000 and can cost as much as $17,000, depending on motor and cylinder size. Motor size varies from 5 to 50 hp, while the cylinder size runs from 4 to 10 in. As with vertical balers, cylinder size and system pressure determine ram pressure, while motor size and pump assembly dictate the cycle time.

Closed-door unit maintenance is relatively easy and inexpensive. Nevertheless, proper use of the machine is essential. Because of the amount of force produced by the machine, misuse will result in substantial damage and very heavy repair costs. Note: operators should pay extra attention to the status of the "wear plates," which are attached to the bottom of the ram and the floor. Since the horizontal ram moves across the floor of the machine, this series of wear plates periodically wear out and must be replaced.

Open-End Automatic-Tie Horizontal Balers. The open-end automatic-tie horizontal baler is similar to the closed-door manual-tie baler, but on a larger scale (Fig. 28.4). The open-end baler operates in the same way as the closed-door baler in regards to infeed and cycling. Instead of compressing the bale against a fixed baling chamber with a door, however, the open-end baler continuously extrudes the material. The open-end baling chamber incorporates tension cylinders, which apply varying degrees of pressure against the baling chamber walls. These tension cylinders are extended during bale compression until the correct bale density is determined by the unit's pressure setting. At this point, the tension cylinders ease their pressure, allowing the compressed formation to escape. To maintain a continuous process, an automatic tying mechanism wraps and ties wire around the emerging bale. Like the closed-door balers, the open-ended versions produce their bales through a series of laminations which push more material into the bale with each ram stroke.

Open-end balers excel in continuous, large-scale applications, such as box plants or

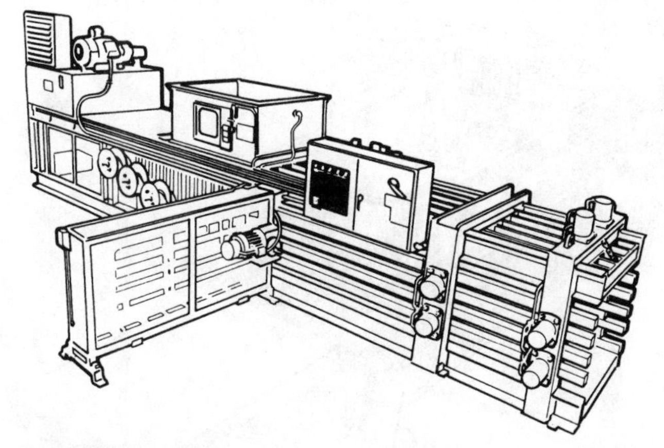

FIGURE 28.4 Open-end autotie horizontal baler.

large wastepaper plants. Like closed-door balers, open-end balers are often accessorized with shredders or conditioners to prepare material for baling.

Open-end units feature motors ranging in size from 20 to up to 150 hp. Cylinder sizes commonly range from 6 to 12 in, while smaller sizes are rare. Open-end unit feed openings differ according to need, from 30 × 30 in to 48 × 72 in. Likewise, baling sizes vary. Of course, since the final bale is extruded continuously, the machine's operator determines the bale length.

Due to larger motor sizes, the operational costs of these machines is somewhat higher than the previously mentioned balers. Also, these units' maintenance procedures and costs reflect their more sophisticated technology (automatic tying system, tension cylinders). Labor costs and processing times are lower, however, due to the automatic tying system.

Two-Ram Horizontal Balers. Two-ram horizontal balers were designed in the late sixties for very large-scale baling without any preparation via a shredder or conditioner. As the name implies, this type of baler incorporates two rams: one to compress the material from the feed opening into the compression chamber in a series of laminations and a second side ram to laterally eject the bale (Fig. 28.5). In addition, these machines feature a shear-knife assembly where the ram and baling chamber are fitted with large blades. Thus, the movement of the ram creates a scissors-like action, which slices excess material protruding from the hopper area into clean, even bales. As with the open-end balers, the bale is then neatly wire-tied via an automatic tying system.

Two-ram balers are ideal for large-scale waste operations, where a wide variety of materials are processed. They will easily bale aluminum cans, cardboard, wastepaper, plastic, and steel scrap.

As with other types of balers, two-ram balers are available in different sizes and with different features. Motor sizes range from 50 to as high as 300 hp. Main ram cylinder sizes typically run from 10 to 18 in, while the ejector ram cylinders are usually between 8 and 14 inches in diameter. Feed openings are large, normally 5 ft wide by 5 to 10 ft long. Depending on the unit, two-ram balers cost between $160,000 and $700,000.

Two-ram balers are versatile machines. Due to their large feed openings and motor sizes, they can handle a formidable variety of recyclables, provided there is an abundant

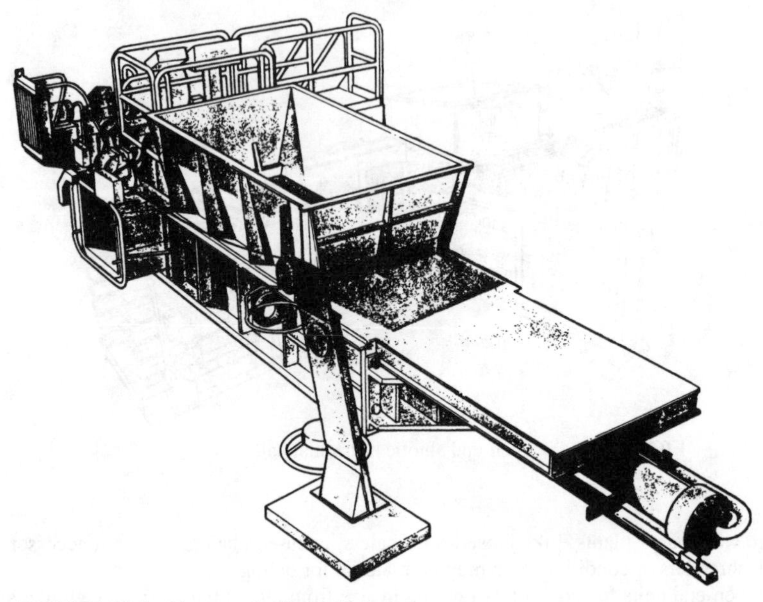

FIGURE 28.5 Two-ram horizontal baler.

supply of material to be processed. It is possible for a two-ram baler to actually process materials faster than an operator can feed it. For this reason, an infeed conveyor is often necessary.

Compared to other balers, two-ram units are expensive to run and maintain. Higher horsepower means higher energy cost. Nevertheless, throughput capacity is also greater. Thus, a two-ram unit's operating cost per ton of material is usually more economical.

Also, these machines are susceptible to a high degree of wear and tear when processing materials on a large-scale. This is illustrated by the periodic replacement of the metal wear plates lining the hopper area, baling chamber, and the ram faces. These metal plates eventually wear down, and new plates must be welded or bolted onto the surfaces.

CONVEYORS

Conveyors set the pace, so to speak, at most processing centers. They move recyclables from one point to another, facilitating the flow of unloading, sorting, processing, bundling, and finally, transportation or storage. Conveyors ensure that this flow is constant.

Different types of conveyors serve many different industries, but for recycling purposes only two types are prominent—the slider-bed conveyor and the direct-drive conveyor. Slider-bed conveyors are usually smaller and less powerful than direct-drive conveyors. Both types have their advantages and disadvantages, and it's important that the right conveyor be used for the right situation.

Slider-Bed Conveyors

A slider-bed conveyor is basically a belt that runs a certain length between two or more pulleys-from a few feet to hundreds of feet (Fig. 28.6). The recyclable material sits on the belt, which moves along as the pulleys turn. A motor turns the pulley via a reducer unit, which is a geared mechanism that applies torque to the pulleys. A frame system holds the motor, reducer, belt, and pulleys secure.

The slider-bed conveyor incorporates a rubber or synthetic belt, stretched taut, running the length of the conveyor. These conveyors rely on friction between the head pulley or driveshaft and the belt to provide movement for the conveyor.

Drive units are located either near the discharge end or in the middle of the conveyor belt. The discharge or "head-shaft" unit, consisting of a motor and reducer unit, powers the conveyor from the point where it expels its material. Thus, the motorized unit pulls the belt upward. The middle location or "center-drive" unit drives the conveyor from beneath the middle of the conveyor's length.

One of the advantages of the center-drive system is that it keeps the drive unit away from contaminants on the surface of the belt. In a recycling operation, materials such as broken glass or liquid can quickly end up in the drive unit. With a drive unit near the conveyor's discharge area, that is more likely to occur. Unfortunately, there isn't always enough room to accommodate a center drive on the underside of a conveyor. Also, a center-drive unit is usually more expensive than a head-shaft unit.

Slider-bed-type conveyors will feed both small and large processing applications, but they are better suited for the smaller. For instance, a slider-bed conveyor with a 12-in-wide belt is ideal for feeding a glass crusher or can flattener. On a larger scale, however, such as a 48-in-wide belt feeding up and into a baler, heavy loads might present difficulties because the system relies on tension to keep the belts moving forward. A heavy load of waste materials might cause too much friction, causing belt slippage or even nonmovement.

An important consideration with slider-belt conveyors is the type of belt and belt

FIGURE 28.6 Slider-bed conveyor.

thickness being used in the system. For some materials, cleated belting may be necessary. Cleats are the upraised backstops on a belt surface that keep material from sliding backward.

Belts should be of a sufficient thickness, especially when processing commingled materials. As with the drive unit, broken glass will get under the belt and slowly wear it down, as well as the steel bed that the belt slides across.

The amount of materials that are allowed to creep under the belt can be greatly diminished by means of "troughing" the conveyor bed. With troughing, the sides of the steel frame bed, and thus the belt running over it, are turned upward, so that the material being transported tumbles toward the center of the belt. This cuts down on the amount of belt surface area that transports material, however. Some units feature side skirting, where the conveyor frame sidewalls overlap the belt, holding it in place and keeping material from getting under the belt. Other units feature corrugated belt sidewalls. Thus, any residue, such as broken glass, is kept from getting under the belt and is expelled with the rest of the material.

Slider-belt conveyors are generally less expensive than direct-drive conveyors. Prices are from $1800 or more, depending on length, motor size, reducer size, and belt width. Nevertheless, a slider-belt conveyor is maintenance-intensive and incurs a lot of wear and tear.

Aside from periodically lubricating the bearings and shaft assemblies, the majority of maintenance involves the belt, which inevitably stretches. Once the belt has stretched, the operator must adjust its tension via the shaft assemblies. This stretching often leads to belt "walking," or shifting heavily to one side or the other. A walking belt quickly results in frayed edges or a torn belt, as the side of the belt rubs against the metal frame. Some innovative features are available which diminish the effects of stretching. One method is the "crowned pulley," where the pulley turning the conveyor belt is tapered slightly lower on its ends than in its center. This serves to add tension to the belt center, so that it doesn't shift easily to either side. Another means of avoiding belt shifting is a guide tab in the bottom of the belt that fits into a notch running the length of the frame bed.

Direct-Drive Conveyors

Direct-drive conveyors excel at moving heavy loads in large quantities; slider-bed conveyors move hundreds of pounds of material an hour, but direct-drive units move thousands of pounds (Fig. 28.7). These qualities are ideal for large recycling facilities and transfer stations, especially for transporting material up and into large compactors and balers. Normally, a direct-drive conveyor will cost more than a similar-sized slider-bed conveyor, and they are often custom built. Usually, a direct-drive unit will be built recessed into the ground, with a long, flat loading section between 10 and 100 ft long. Material is pushed across the floor and onto the belt. The conveyor then carries the material up an incline and toward wherever it is feeding.

These conveyors feature a chain and sprocket system to drive the conveyor. This alleviates the slippage that slider-bed conveyors experience under heavy loads, as neither friction nor belt tautness are involved.

The belts are usually composed of steel segments, connected together in an interlocking "piano hinge" fashion. Variations on the steel belt usually involve rubber, with a steel subframe beneath for support. Steel belts are preferable for more abrasive applications, such as commingled materials, wastepaper, metals, and solid wastes. Both types incorporate chains on both sides of the belt, which mesh with drive and tail sprockets.

The most important belt and chain maintenance concerns are cleaning and lubrication.

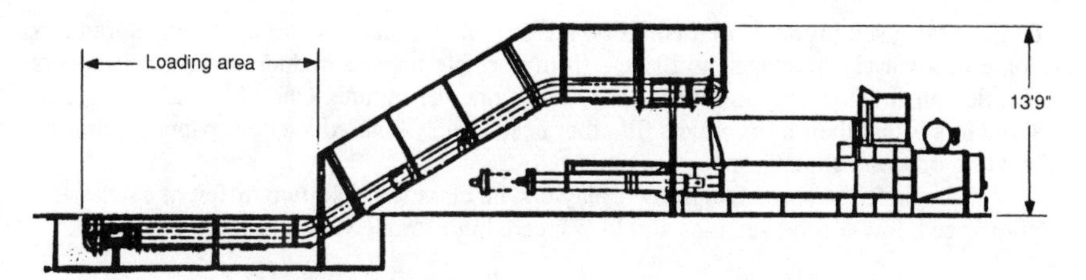

FIGURE 28.7 Direct-drive conveyor.

Chain lubrication and cleaning should take place at least once a week, if not more often. Today, most conveyors can be outfitted with an automatic lubrication system. These systems feature an oil reservoir that lubricates the chain via brushes or valves when the conveyor moves.

Chain design is an important aspect of the conveyor system. Some chain systems incorporate a roller assembly, so that the belt and chain move easily over the frame bed without dragging. Other chain systems do not incorporate rollers, however, and rely on the motor's horsepower to drag the belt over a high-density plastic-lined frame bed. Although chains without rollers are less expensive, they are not conducive to heavy load factors. In a heavy load situation, such as a commingled recycling plant, a chain-driven conveyor system is subject to high wear and tear. In one case at a transfer station, a conveyor's frame bed had to be replaced after only 18 months, because it was not equipped with rollers. As is so often the case with recycling machinery, a bit more money in the initial investment far outweighs potential future repair expenses due to inadequate equipment.

Depending on the application, direct-drive conveyors vary in width from 24 in to 10 ft. In addition, one can choose between different chain types, chain sizes, motor sizes, and reducer sizes, according to the desired carrying capacity. Carrying capacity is an important consideration. At any given time, with the belt loaded with material from base to discharge point, the conveyor should be able to transport its load, starting from a stationary position.

Conveyor purchasers should make sure that a system's frame will adequately support the belt when it is fully loaded. A smart option is a series of "load bars" or "impact bars." These bars run the length of a conveyor and provide support beneath the belt. Load bars keep a belt from deflecting downward while under a heavy load. Usually the underside of the steel belt is outfitted with a small piece of metal called a "wear shoe" so that the belt is not riding directly on the load bar.

Prices for direct-drive conveyors run from $20,000 and up, depending on the conveyor length, width, motor size, and accessory features as described above.

SHREDDERS

Shredders, as their name implies, digest a vast array of objects into much smaller pieces. These objects include scrap metal, plastic, aluminum, and wood. The shredder's versatile "diet" has led to its use in other areas besides recycling. For example, they now serve in construction and demolition for breaking down building materials. On a small scale, this often involves confidential papers or store coupons. On a larger scale, shredders destroy damaged or used products that could be redistributed under a warranty claim. Shredders come in a variety of shapes and sizes, from portable paper shredders to huge shredders that devour flattened automobiles at the rate of one per minute. One of the advantages of shredders is that their by-products fill other needs, such as shredded newspaper for animal bedding or wood chips for a garden or compost.

All the different shredders in use today can be classified into two different categories: high-speed, low-torque versions and low-speed, high-torque versions.

High-Speed, Low-Torque Shredders

The first category, high-speed, low-torque, incorporates a single shaft with either fixed knives or swinging knives and hammers rotating at very high speeds (Fig. 28.8). These speeds vary between 1000 and 3500 r/min, although the latter is somewhat extreme. These knives or hammers work against either a grate through which shredded materials pass or a stationary bed-knife assembly. Shredded materials include aluminum cans, used beverage containers, paper scrap, and as previously mentioned, automobiles.

A high-speed shredder pulverizes whatever it is fed. It relies on brute force and is very noisy. High-horsepower motors are essential, meaning 50 hp and up. Accordingly, electrical costs can be astronomical, and customers are well advised to analyze potential operating costs.

There are possible hazards with high-speed shredders. First of all, because materials are so heavily pulverized, a high-speed unit can produce a lot of dust. This creates a potential fire hazard. Should a knife or hammer hit a metal object, such as a nail, and produce a spark, the dust in the air could ignite immediately. Also, jam-ups could spell disaster. Should a high-speed shredder encounter an object that it cannot cut through, there is no reverse or overload setting. Something has to give way, either the material or the revolving shaft. Preferably, the machine will simply jam up, although shafts have been known to break. Should this occur, one must open the machine and remove the object. Although repair may not be necessary, this removal step consumes time.

Of course, both high- and low-speed shredders are maintenance-intensive due to their violent type of work. With a high-speed shredder, maintenance involves periodically

FIGURE 28.8 High-speed, low-torque shredder: (*a*) with grate; (*b*) without grate.

turning the hammers or knives. Usually, knives are double-sided. Once both sides are worn down, however, one must either replace them or resurface them. Resurfacing entails rewelding a work surface on the knives, followed by resharpening to a cutting edge.

Low-Speed, High-Torque Shredders

Low-speed, high-torque machines utilize two or more shafts with protruding teeth (Fig. 28.9). The shafts rotate counter to each other. These machines require lower horsepower to shred because they apply higher torque, via gear reduction, to the shafts. One could say that high-torque shredders are more subtle than their high-speed counterparts, methodically puncturing and ripping rather than blindly slashing and pulverizing. This sort of design is perfect for shredding tires, due to the tires' high resiliency.

Of course, low-speed shredders require less horsepower than high-speed units. Motor sizes are usually below 100 hp and sometimes even below 50 hp. Additionally, low-speed machines can respond to jam situations. Should someone overload the unit, the motor has the ability to sense, through an overload relay, that it is drawing too much amperage. The unit then reverses itself to clear the jam and again reverses itself in an attempt to recut the material. Should a series of reversals fail to process the overload, the machine will shut itself off and indicate an overload situation.

Another advantage to low-speed shredders is less noise. The machines will run at under 80 db, usually below the industry levels that require earphones. High torque and lower speeds also eliminate huge amounts of dust, greatly reducing the fear of spontaneous

FIGURE 28.9 Low-speed, high-torque shredder.

combustion.

The spectrum of shredder sizes and prices in the market is very broad, ranging from $2500 paper shredders to huge $500,000 scrap metal units. For general recycling facilities, however, a good guideline would be: low-speed shredders cost between $35,000 and $400,000 and high-speed units between $70,000 and $600,000. The main thing is to realize one's precise needs. Needs mean size, type, and amount of material to shred; operating speed; and particle size after shredding. Sometimes this requires a combination of shredders rather than just one unit, where the operator shreds the material and then reshreds the output. The more specific the need, the easier it is to determine the type of shredder. For example, those who are planning on shredding tires should ask questions such as, "Auto or truck tires?" And even then: "Will these be regular or radial tires," and "What is the desired throughput and particle size?"

This is not to say that one shredder cannot process a multitude of different materials. But, it's more prudent to determine as specifically as possible the main use of a shredder. Also, it would be a wise move to study that shredded material's market before making a substantial investment.

COMPACTORS

Compactors became popular in the 1960s, as a reaction to increased hauling and disposal rates. At that time, businesses began searching for a more effective way to dispose of their wastes. The answer was the compactor.

Stationary and Self-Contained Compactors

The first stationary compactors compressed material into roll-off boxes, large metal structures usually measuring 8 × 8 × 22 ft (Fig. 28.10). When enough material was accumulated so that no more could be added, a hauler detached the box and hauled it away to a landfill.

This system was ideal for dry wastes. Some industries, such as restaurants and hospitals, disposed of partial liquid wastes, however. This presented problems. Because the

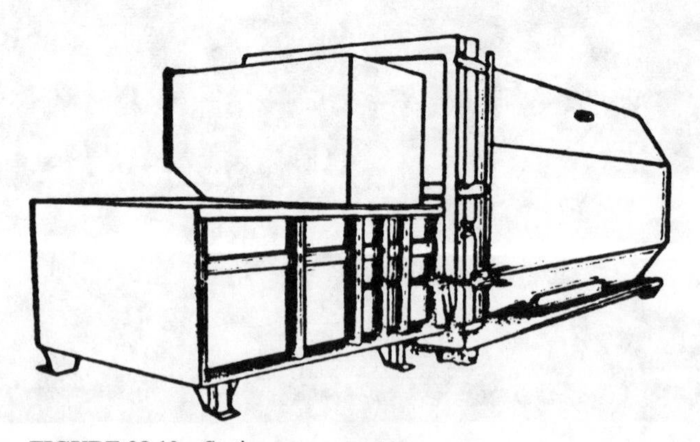

FIGURE 28.10 Stationary compactor.

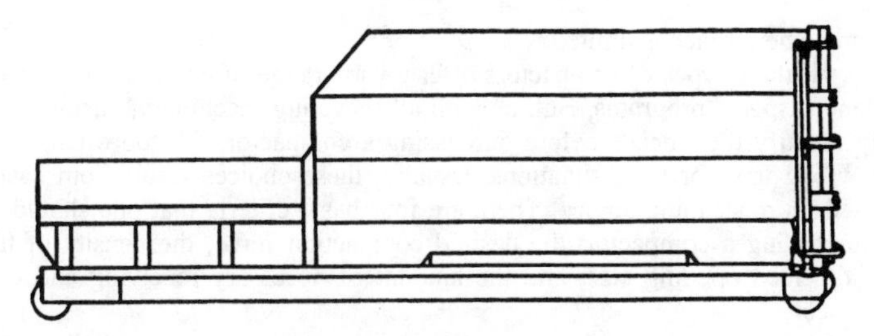

FIGURE 28.11 Self-contained compactor.

compactor was separate from the container, liquid waste resulted in spills and residue. This residue left an odor and often attracted animals and insects. Consequently, the self-contained compactor was introduced (Fig. 28.11).

Self-contained compactors were simply a compactor and a roll-off box housed together on the same platform. For hauling, the electrical power unit was separated from the assembly. These self-contained units included a liquid retention area underneath the compactor to prevent spillage.

Vertical Compactors

Within the last decade, the market has seen the introduction of a third type of compactor, the vertical compactor (Fig. 28.12). This type of compactor is ideal for low-volume appli-

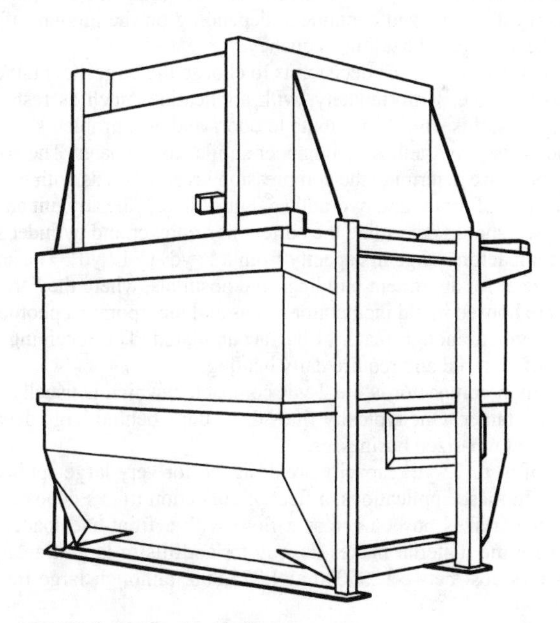

FIGURE 28.12 Vertical compactor.

cations, and where space is limited.

Today, all three types of compactors offer a wide range of choices for individual recycling and disposal programs. But, as with all recycling machinery, customers should precisely identify their needs before purchasing a compactor. All too often, customers buy the wrong unit for their situations. Usually, these choices result from hasty decisions based on misleading advice. There are four basic criteria that one should address before purchasing a compactor: the desired compaction force; the density of the final product; the feed opening size; and the amount of necessary receiving and container storage space.

The first two criteria are synonymous with the terms "compression ratio" and "capacity." Compression ratio measures the before and after effects of a compactor on its material. Thus, a compactor with a 4-to-1 compression ratio will produce material that is four times more compressed than before it was processed. Generally, stationary compactors have a compression ratio of 4 to 1, self-contained units ratios are 3 or 4 to 1, while vertical compactors' ratios are only 2 or 3 to 1. Of course, these are only guidelines. Actual results will depend on the type of material being compacted.

A compactor manufacturer measures its machines' processing capacity in cubic yards. This yardage does not indicate the amount of material processed into the receiving container. Rather, yardage refers to the amount of material that a compactor can receive and process in one stroke. This measurement is not precise, however. For instance, a 2-yd^3 compactor may only actually process 1½ yd^3 of material. That is because this capacity represents an approximation established by the manufacturer. Because manufacturers' ratings are so inconsistent, the National Solid Waste Management Association (NSWMA) established compactor rating guidelines. Therefore, when comparing compactor processing capacities, one should always consider the NSWMA ratings.

Receiving container space is rated by the amount of cubic yards it can contain. One can find nearly any size container for a given situation. One size compactor can be coupled with several different-sized containers, depending on the amount of material to be processed and the anticipated hauling frequency.

One might feel that the easiest decision is to choose the largest container possible and haul as little as possible. Unfortunately, with applications such as restaurants, nursing homes, and hospitals, this would contribute to odors and pest problems.

The compactor uses hydraulics in a manner similar to the baler. The size of the cylinder and system pressure determine the compression level. Also, as with balers, the market offers a large variety of motor and hydraulic cylinder sizes. The amount and type of material to be processed should determine the correct horsepower and cylinder size.

Stationary compactors range in capacity from a ½ yd^3 to 12 yd^3. The smaller capacity compactors are ideal for apartment buildings and hospitals, where they are chute-fed. Often, such units are housed in old incinerator rooms and incorporate a photoelectric relay to start compaction when enough material has accumulated. The receiving container may only hold 2 yd^3 of material and require daily hauling.

Another common compactor is the 2-yd^3 compactor, which is usually coupled with a 40-ft receiving container. One typically finds these units behind large department stores, mall areas, and medium-sized businesses.

Compactors of 6- to 12-yd^3 capacity are designed for very large applications, such as transfer stations. In these applications, a fleet of collection trucks deposits waste in a tipping area. The material is pushed across a floor with a front-end loader into the compactor. From there the material makes its way to landfills in large transfer trailers. Stationary compactors cost between $5000 and $25,000, although large transfer units can cost over $100,000.

Self-contained compactors generally have smaller feed openings and thus smaller capacities. This is the result of both compactor and receiver being contained in one unit. Feed opening sizes are limited to 2 yd^3 and under, while overall unit length doesn't exceed 24 ft. These units' receiving container capacities are usually 36 yd^3. Self-contained units list for between $16,000 and $24,000, depending on size and optional features.

Vertical units feature similar-capacity compactors as self-contained units. Their receiving containers are smaller, from 3 to 8 yd^3. Of course, one doesn't purchase a vertical compactor for its capacity, but rather, for its size.

Vertical compactors are markedly less effective than other compactors. As mentioned earlier, compression ratios do not exceed 2 or 3 to 1. This is because vertical compactors exchange smaller motor and hydraulic features for size convenience. Prices for these units are in the $7,000 to $12,000 range.

All compactors require a firm concrete foundation. Softer foundations, such as asphalt, are simply incompatible. The weight and motion of the compactor will cause the unit to literally sink into the asphalt.

The uses of compactors need not be limited to traditional waste hauling. The machines often prove ideal in recycling facilities for processing commingled residue. After sorting, each material can be loaded into its respective compactor for storage while awaiting processing, or for transportation to another site.

Because compactors are so diverse, they are also ideal for industrial-plant recycling programs. A plant can often substitute a compactor for a baling system. Some large car plants and distribution centers have incorporated two compactors: one to process industrial residues and one to prepare cardboard for recycling. In this situation, the key is to identify a nearby papermill or wastepaper dealer who will accept the compressed cardboard in receiving containers.

GLASS CRUSHERS

Glass crushers pulverize all types of glass, usually containers, into the gravel-size pieces called *cullet*. Cullet is a preferable form of glass for recyclers because it is denser. This fluid form simplifies transportation and is furnace-ready. Thus, glass companies more readily accept cullet because it is one step further along in the meltdown process.

There are several different types of glass crushers on the market today. These vary in complexity from a simple sledgehammer and 55-gal drum device to sophisticated conveyor-fed units. The latter types incorporate conveyors which drop material onto either a set of rotating blades, a rotating hexagonal drum, or a series of rotating chains. There are several brands of conveyor-fed glass crushers offered on the market today, and prices range from $3,000 to 6,500 (Fig. 28.13).

Conveyor-fed glass-crushing units are electrical. The motors are small, usually under 4 hp. Because glass is such an abrasive material, these machines require periodic crushing mechanism replacements. As with shredders, part replacement and careful maintenance come with the territory.

CAN FLATTENERS AND BLOWERS

Can flatteners are a means of flattening beverage cans for ease of handling and transportation (Fig. 28.14). Most machines are geared toward aluminum can flattening because aluminum has a higher value. Can flatteners are coveyor-fed and they use a magnetic charge

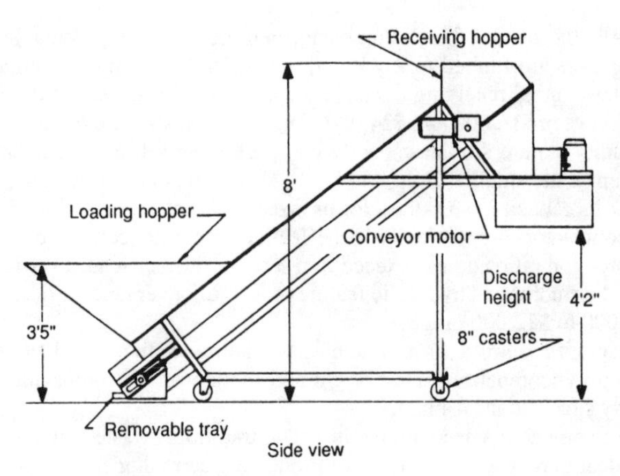

FIGURE 28.13 Glass crusher.

to separate aluminum cans from steel and bimetal cans. The magnet is contained in the head pulley of the conveyor. As the materials pass around the head pulley, ferrous containers remain on the belt. Aluminum containers, which aren't magnetically attracted, simply fall into the flattening device. After the other cans pass around the head pulley, they fall into a reject chute underneath the conveyor.

Once in the flattening area, the aluminum cans are crushed either by a wheel rolling against a stationary plate or two wheels counterrotating against each other. Following this operation, the flattened cans are simply expelled into a blowing tube. This tube transports the cans up into a storage bin via a strong current of air.

As with glass crushers, there are several manufacturers of can flatteners on the market today. The market prices for a can flattener with blower range between $7,000 and $15,000. Most brands feature motor sizes under 10 hp. Since the process is abrasive, identifying which units are most durable is an important concern. Maintenance for these units

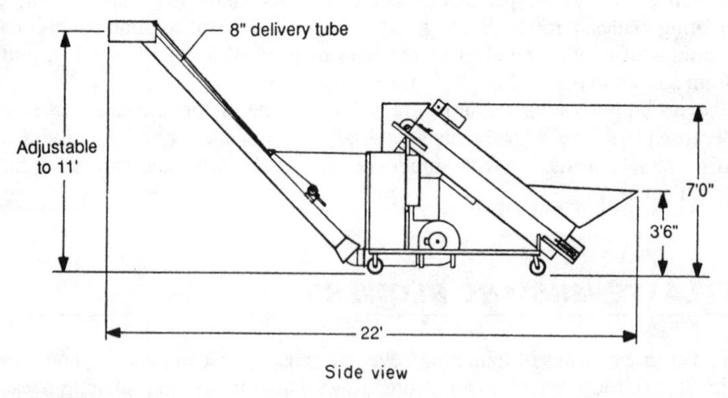

FIGURE 28.14 Can flattener.

consists mainly of lubrication. While the machine may not be as maintenance-intensive as a shredder, for example, one should routinely inspect the unit's condition.

Most can flatteners today are manufactured to process aluminum but not steel, because the head roller is magnetic and difficult to change. Even without the magnet, one encounters difficulties with oversized institutional cans. The problem is that the can will not fit into the crushing mechanism and will simply bounce around above them as they turn. Fortunately, some manufacturers are creating units specifically for steel cans.

CAN DENSIFIERS

For large-scale recycling operations, a more practical method of processing steel and aluminum cans is the can densifier (Fig. 28.15). The can densifier forms the cans into a brick that weighs about 18 lb and measures a cubic foot (14 × 12 × 8 in) in dimension. These bricks are stacked on a skid for transportation to an aluminum smelter or metals-processing facility.

At one time, metal companies paid a premium for the more densely packed material, because it was furnace-ready. Correspondingly, densiflers grew in popularity. Today, the premium is no longer the rule but the exception. The metal companies began to find impurities in the blocks of material. Also, there were sometimes high moisture contents in the materials, due to the cans' previous contents. Flattened cans usually have lower moisture content after being transported through an air blower and stored in trailers. If the materials have too high a moisture content, aluminum companies will deduct a percentage for moisture. Since the metal companies could not perfectly ensure quality control, they discontinued their premiums.

Can densifiers are similar to horizontal balers. Their large cylinders apply tremendous pressure to the cans. The ram faces are specially configured to indent each cube. These indents allow a steel banding to be wrapped around a skid full of cubes. Thus, cubes are uniformly banded and easily stacked.

Another consideration with densifiers is the high volume of cans that are required to form a dense cube. This means that shipments may be somewhat sporadic, affecting an operation's cash flow.

Densifier maintenance is very similar to that of a horizontal baler. Older densifiers

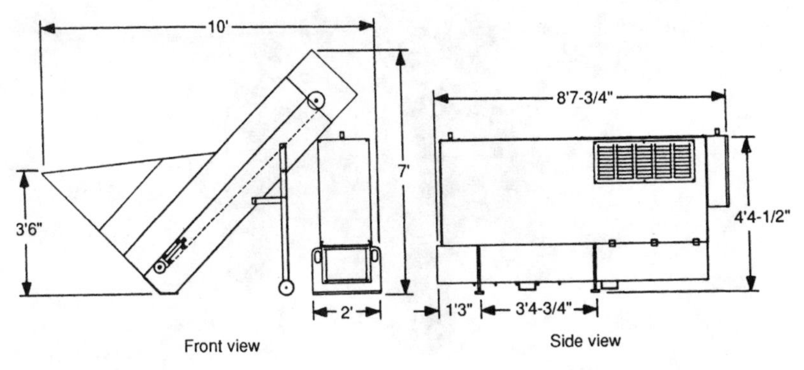

FIGURE 28.15 Can densifier.

have a lot of moving parts and hoppers which require a good deal of maintenance. Today's units, however, are generally very efficient and easy to operate and maintain.

ALLIGATOR SHEARS

An alligator shear is a large machine for cutting metal (Fig. 28.16). A hydraulic cylinder moves the shears up and down when an operator pushes down on a foot pedal. The shear is used to cut and prepare miscellaneous pieces of metal, such as plumbing valves or aluminum siding, for further processing. Most commonly, scrap metal yards rely on these machines for preparing large amounts of material.

Formerly, alligator shears operated continuously through a fly-wheel assembly. Materials were fed through the rapidly moving blades. Unfortunately, the machines had no safety features, and losses of limbs and lives occurred. Today, these shears incorporate a foot pedal and hydraulics in order to ensure safety. These features have also led to higher prices, and today alligator shears start at $12,000 and go up to $40,000 or more, depending on size.

Both blade and cylinder sizes are variable. Blades lengths start at 4 in for small valve-processing applications, and continue on up to 36 in.

HIGH-TECH MACHINERY

The term "high tech" here describes those "cutting-edge" processes that are past the development stage and yet aren't in widespread use. One of the most promising recycling developments today involves the automatic separation of aluminum, plastics, and glass.

FIGURE 28.16 Alligator shear.

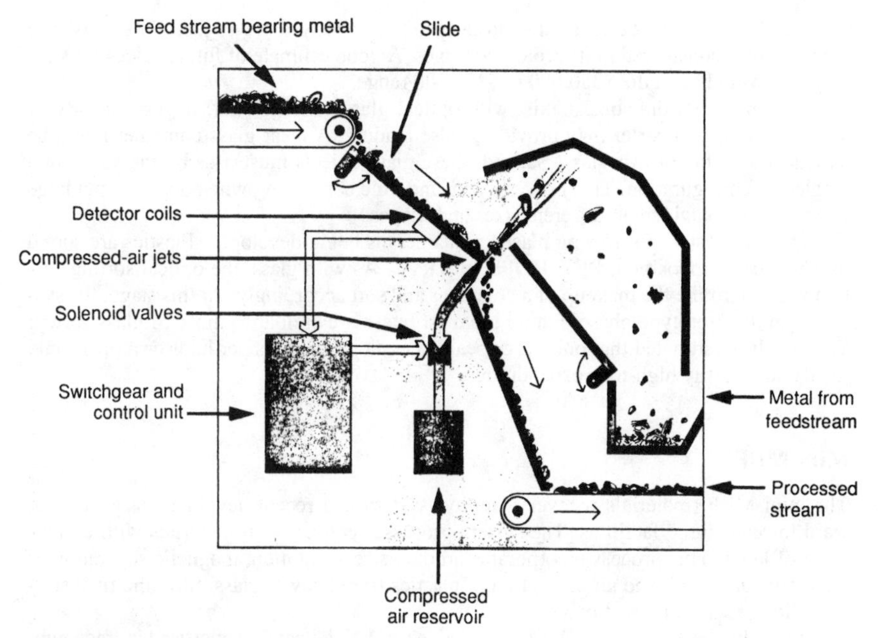

FIGURE 28.17 Eddy current separator.

Eddy Current Separator

An eddy current separator is a device that applies an eddy current, or magnetic field, to commingled recyclables on a conveyor line (Fig. 28.17). This current repels aluminum the way two magnets repel each other. When applied to mixed material, the eddy current separator will quickly and efficiently remove all aluminum from the other materials more efficiently, perhaps, than humans. In the sort term, however, the cost of the eddy current system is considerably higher than labor costs: between $100,000 and $250,000.

On the other hand, potential buyers will need to weigh the economical aspects. Sorting commingled material is an unskilled, low-paying occupation with a high turnover rate. Eddy current separators do not quit their jobs or fail to show up for work. Nor do they file workers' compensation claims or require heavy insurance. And, as this technology progresses, less and less expensive versions of the system should appear on the market.

Eddy current systems will probably play a large role in commingled separation, but the verdict is still out, as the machines haven't existed long enough to precisely establish operating costs or production rates.

Optical Color Screening

Another recently developed recycling technology is the sorting of plastics and glass based on optical screening. In the case of glass, the sorting is by color—green, brown, and clear. Traditionally, this job was carried out by workers on an assembly-like conveyor line. Needless to say, this type of work is also tedious. Reliable workers generally don't stick around a long time. Once again, it's too early to tell if the system will justify its costs. The technology is considered in its prototype stages. A good estimate of future prices for such a system would be in the $250,000 to $300,000 range.

Some potential drawbacks exist with optical glass sorting. First, if glass is dirty or stained, the optical sorter may provide a false reading. A clear glass container might be sorted in with the brown glass. Second, these glass objects must pass by the sorter in a single-file configuration. This proves tricky and time consuming when dealing with huge quantities of containers in different sizes and shapes.

A similar system for sorting plastic containers is being developed. Plastics are sorted by chemical composition: PET, HDPE, and PVC. As with glass, the optical sorting system will determine the makeup of a container and sort accordingly. At this stage, the system is in the prototype phase. Future price estimates are similar to those of glass-sorting systems. It is estimated that only large-scale operations (10 tons per hour or more) could justify such costly high-tech expenditures.

Mini MRF

The mini MRF (materials recovery facility) system is a recent development geared toward lower-volume facilities. The system processes commingled materials with a minimum of labor. This process incorporates product size separation, magnetic separation of aluminum and steel, and air separation of plastics from heavier glass. Still, the final sorting relies on visual inspection.

Currently, these small systems are available and being heavily advertised in trade publications. They cost between $100,000 and $500,000, depending on size and options.

The market for high-tech recycling equipment is still very young. In order for it to "mature," purchasers must possess precise, accurate figures, such as cost and output of the new equipment versus its labor, insurance, and compensation costs. Only then can a buyer justify purchasing such technology.

A final word that cannot be stressed enough: before buying a piece of equipment, know the reputation of its manufacturer and distributor. Resources are available to buyers. They should ask questions such as "Does the manufacturer conform to standards set by the American National Standards Institute, for balers and compactors? Is the manufacturer's equipment approved by Underwriter's Laboratory? Does the manufacturer carry product liability insurance? And, what is the company's history and financial status?"

Because recycling is the upcoming growth industry for the nineties, we are witnessing exciting innovations in processes and technology.

SECTION IV
RECYCLING OPERATIONS

CHAPTER 29
RECYCLING PROGRAM PLANNING AND IMPLEMENTATION

THOMAS A. JONES, JR.
Executive Analyst
R. W. Beck and Associates
Framingham, Massachusetts

KAREN LUKEN
Director of Solid Waste
R. W. Beck and Associates
Cincinnati, Ohio

The planning and implementation of a recycling program, like the proverbial team of people with their eyes closed describing an elephant, can be a complex and frustrating endeavor. Complex because most recycling programs have many parts; frustrating because often the parts do not appear related to one another. Composting yard waste is at least as different from marketing glass as a pachyderm's ear is from its trunk.

Planning and implementation call for expertise in commodity markets, engineering, public finance, environmental law, and many other areas. This chapter focuses on three areas which will be important to a potential recycling program developer: (1) program planning and development, (2) staffing, and (3) financing. Because most multimaterial recycling programs are developed by the public sector, the chapter assumes this perspective.

Virtually every municipality in the United States has considered, and the vast majority have implemented, recycling programs. So program implementation has evolved into program maintenance and refinement. But maintenance and refinement have been difficult for several reasons. First, the markets for recovered materials, which were relatively robust in the early to mid-1990's, collapsed in 1996. For example, the price of newspaper dropped from $150+ per ton to zero. In some cases, cities have had to pay to have their newspaper taken away. As revenues from the sale of recovered materials plummeted, it created serious fiscal binds for recycling programs that had counted on the sale of recovered materials to support their efforts. Second, state and local budget allocations for supporting municipal recycling programs began to shrink as other municipal programs, such as education and police, required more funding. Third, states set ever higher recycling goals—Rhode Island currently has a goal of 70%. So planners were faced with more aggressive goals and smaller budgets to achieve them.

Finally, municipalities no longer have the authority to mandate that all waste go to a specified facility, such as a material recycling facility. This authority is called "flow control," and was ruled unconstitutional in 1994. Although some forms of flow control, such as contractual flow control, have been legally defended, many communities avoid using this authority due to concerns about legal challenges. Because of this, waste streams that used to go to recycling facilities may now go to less expensive disposal facilities.

Despite the shift from program implementation to program maintenance and refinement, this chapter focuses on the implementation process because this provides the most comprehensive view of recycling programs. It is hoped that planners who have already faced the challenges of implementing a program will find the discussion below a useful checklist, as they attempt to keep their program running smoothly, or shift into a higher gear.

PROGRAM PLANNING AND DEVELOPMENT

Too often planning seems to be the refuge of bureaucrats and consultants. Acres of forests have been leveled to provide paper for "planning documents" that are now gathering dust in municipal closets. The first point to be made is that the purpose of planning is not to produce a written document. The effort made to plan must be appropriate to the activity. A walk on the beach does not require a feasibility report. On the other hand, building a material recovery facility requires more than some back-of-the-envelope calculations. Properly done, planning should make an activity easier, more efficient, and even more enjoyable.

One purpose of this section of the chapter is to describe the planning process for a municipal recycling program. It is intended to take the reader from that heady conceptual stage, when the ideas for program activities are developed on paper, through the development of a comprehensive recycling program. It also includes some discussion about how programs are updated over time. Since recycling programs come in all shapes and sizes, this section highlights four critical steps which are common to all programs. A separate subsection is devoted to each step.

The first subsection describes the "idea" stage, where the goals and broad concepts of program options are developed. A community that is contemplating the development of a recycling program faces a wide variety of options, in terms of program components and operations. Although recycling is an ancient and well-established activity, large-scale municipal programs are a relatively recent phenomenon. The chapter outlines the elements which many existing municipal programs now include.

The second subsection focuses on evaluating program options. The Seattle Recycling Potential Assessment (RPA) model, a sophisticated computer model, is used to provide a concrete example of a planning methodology. At this stage of program planning ideas are weighed against each other and the conceptual begins to become real. The third subsection, still using Seattle's RPA model, discusses the overall plan, that is, the choice of options based on overall program goals and available resources. It is here that the dynamic nature of program planning is critical.

The fourth and final subsection describes the development of a schedule. That is, the logically ordered work tasks must be assigned a duration, and be allocated resources. There are a range of approaches to this function, from a loose, plan-as-you-go approach to a more rigorous quantitative approach, such as the Critical Path Method (CPM) or Performance Evaluation and Review Technique (PERT). The rigor a municipality applies to its planning effort will be determined by a number of factors which are also discussed in this final section.

The product of this planning effort should be a program set within the constraints of time and available resources. The plan should provide guidance to the staff, the community leaders, and the citizens.

The first part of the chapter is a general introduction to the planning process and deals broadly with important planning issues. However, it may not provide the necessary amount of detail for a particular community. For this reason, references to specific programs and other sources of information are included throughout the chapter.

Mandatory versus Voluntary Recycling

The question of whether or not a community should require recycling is answered, partially, through state regulation and legislation. According to a 1999 survey by *Biocycle* magazine, every state has a statewide recycling goal, although only seven states have mandated these goals. A few states, such as Massachusetts, cannot mandate recycling, but have enacted waste bans which prohibit certain materials, such as yard waste, cardboard, and newspaper, from landfills and combustion facilities by specified dates. Municipalities in these states must develop recycling programs or face penalties for violating the waste bans.

For communities not affected by state-mandated recycling legislation or waste bans, the issue of whether the recycling program will be mandatory or voluntary should be addressed early in the planning process. Successful programs of either type will have strong public education and promotion. They will also be convenient, offering residents and businesses a relatively easy way to place materials out for collection. No community can expect mandated recycling to replace convenience or public education. They must go hand in hand. The components of a convenient collection program and an effective public education program are described below.

Given a convenient, well-publicized program, voluntary recycling has achieved very high participation rates. Typical mandatory programs may be slightly more successful, but they also incur the cost of enforcement, in the form of recycling inspectors who check to see that trash does not contain recyclable material and lawyers who defend court challenges. Penalties may range from the town's refusal to pick up trash which contains recyclables to a system of graduated fines.

In some instances, communities have passed mandatory recycling legislation but have chosen not to enforce it. This allows the town to show they are serious about recycling without immediately incurring the cost of enforcement. If participation is disappointing in spite of a vigorous public education program, the town always has the option of developing an enforcement mechanism.

As recycling has become institutionalized across the country, the debate over mandatory versus voluntary recycling has quieted. It is no longer a question of whether people should recycle, but how they will recycle.

Program Components

In order to provide some substance to the discussion which follows, the chapter begins with a discussion of some examples of program elements. No effort has been made to present the program of a specific community, or to suggest that the configuration of components presented here is the "best" one. Indeed, if the experience of communities developing recycling programs up to now shows anything, it is that recycling programs are quite site-specific. Factors, including the composition of the waste stream, the availability of secondary material markets, the economy of the region, and the political climate of the community, will influence the final shape of the program.

The discussion of the program components below is not intended to be comprehensive. These components are presented elsewhere in the Handbook in far greater detail. However, a general understanding of these components will be the foundation of the later discussion of activities and sequencing. In some cases, the options available within a component are presented, but in all cases, a general approach is recommended as a part of this typical program.

Recycling is a closed-loop process in which products that are purchased and used are then collected and reused, avoiding the cost and environmental damage associated with waste disposal. This reuse may include the reuse of the product itself, such as a book in a used book exchange, or it may include the reuse of the product material. The making of new glass from recycled cullet is an example of this.

The components of a recycling program should be viewed as links in a chain of activities which make recycling a reality for the community. Each of the links must have its own integrity, but each must also be connected to the other components of the program. The program described below contains five links: *education, collection, processing, marketing,* and *procurement.*

Education and Promotion

The first link in this chain is an educational program. The success of a recycling program will depend to a large degree on the knowledge and enthusiasm of the residents and businesses who are participating in it. From implementation of a new program to the maintenance and development of an ongoing program, an effective program of education and promotion is critical. The program must be well-organized and accurately target its intended audience. For example, programs in Miami, Florida; New York City; and Los Angeles; as well as other large metropolitan areas, translate their written materials and spoken public services announcements into Spanish for Hispanic residents.

For our purposes, the education and promotion can be divided into five elements. First, the citizens and businesses in the service area should be given a general education in recycling. This will not only help any specific program to get under way, but it will prepare the way for additions to the program weeks or months later. Second, the specific program which has been developed must be promoted. People have to know the details. What materials will be recycled? How are the materials to be prepared? How will they be collected? Participants need to know the answer to these questions, among others. Third, there must be a means of keeping people informed about the progress of the program as it develops. Fourth, an ongoing effort at source reduction, eliminating waste altogether, should be a part of the educational component. Finally, school projects should be developed to prepare the next generation of recyclers, and to help reinforce the existing program. Together, these five facets can provide a comprehensive and effective education and promotional program.

General Public Education

For a recycling program to work, the public must grasp a few simple principles about recycling. These principles include an understanding of the "recycling loop," the increased value of materials which have been separated and cleaned, and the importance of markets. The principles can be applied to curbside collection programs, yard waste composting programs, or tire recycling programs.

This kind of education can be accomplished through public service announcements (PSAs) on local radio and television stations, newspaper articles, speakers at service organizations, and promotional activities at community fairs. It is usually helpful if this kind of general education precedes a specific collection program. It can stimulate interest, build enthusiasm, and generate commitment.

The thrust of a general education program should come before specific programs are implemented, but there should be an ongoing effort to educate new residents and reinforce the lessons to long-time citizens.

Our typical program assumes an active general education component, particularly in the early stages of the recycling program.

Program Promotion

Once the policy decisions have been made, the community will be ready to implement a specific recycling program. It may be a curbside collection program; it may be the opening of a drop-off facility; or it may be a multifaceted program reaching a number of different generator types. However, all programs will have certain characteristics that must be communicated to the potential participants.

It will operate within certain geographic boundaries, it will target certain materials for collection, it will require some degree of separation and preparation of materials, it will have some type of schedule, and there may be other aspects to it. For example, if it is a mandatory program, the consequences of nonparticipation should be made clear.

The most effective way of getting the specific details of the program to the potential participants is to contact them directly. This may be done through a mailing which goes to residents along with utility bills, for example, or it can be hand-delivered to homes or apartments. It is also helpful to have a "hotline" to inform residents.

Promotional material should arrive just before the implementation of the program. If it is sent out too far ahead of program implementation, people will forget, lose the information, and grow impatient. If it is sent out after the program has started, people will become confused and frustrated. The timing of this facet of the program is critical.

This is what the City of Denver did. Denver had three pilot collection programs at the same time. For six months, Denver offered a three-bin curbside collection program to some residents. In this program, residents separated their recyclables into three categories, usually paper, glass, and metal. These recyclables were collected in a special compartmentalized truck. The second collection program was a one-bag commingled program; all recyclables were placed in a specially marked bag, which was collected with the rest of the garbage in a rear-loading packer truck. It was then taken to a central facility where it was sorted and prepared for market. The third program involved the sorting of all garbage at a mixed waste processing facility. At the end of six months, each program will be evaluated.

As a result of these pilot studies, the City of Denver decided to use a recycling program that was a hybrid of two pilot programs. Under this program, residents receive two 18 gallon bins. The resident places commingled containers in one and newspapers in the other. The recyclables are then collected in a truck with two compartments, one for mixed containers and one for newspaper, and transported to a central facility.

There is no fee for this service, but residents must subscribe by mailing in a postcard or calling the City. Currently, Denver has a 45–50% subscription rate, with a goal of 60%.

Progress Reports

Once the program has begun, it is a good idea to keep citizens informed about the progress of the program. How much material has been collected? What have the revenues been? How much has been saved through avoided cost? A regular quarterly, semiannual, or annual report is a good way to provide this type of information. It should be set up on a regular schedule and use a format which is easy to understand.

Source Reduction Education

Source reduction refers to programs that reduce the amount of waste which is created at the source that normally generates it. That is, by each individual resident or generator. The idea is to cut the amount of waste that is generated, not just recycle the waste that is produced.

Most of the efforts at source reduction are educational, and they aim at helping waste generators understand how they can reduce their waste. Commercial waste audits are conducted to show how businesses can reduce waste. Environmental shopper programs are developed to show consumers how to cut waste through more intelligent buying habits. Regional waste exchanges can assist industries in finding companies that might be interested in using certain materials that are now being thrown away.

School Programs

The educational programs discussed above focus on the potential participants of a recycling program—residents and businesses. But there are several reasons to incorporate an educational program for children as a part of the overall program. Children are an unusually effective way to reach their parents and other relatives. A child who has been taught the importance of recycling will often convince parents who have otherwise ignored it. In addition, the continuing success of recycling will depend on an educated citizenry and teaching the next generation of citizens the importance of recycling is an excellent investment.

School programs should be an ongoing and regular part of the school's curriculum and can be integrated into a number of different disciplines, such as math, social studies, and economics.

Collection

The second link in the recycling chain is the collection of materials. This can mean collecting the recyclable materials from residents or businesses at the curb, or having the generators drop off the materials at collection centers. Curbside collection can be accomplished in several different ways and a municipality may wish to explore a number of different options simultaneously before settling on one.

However, most municipalities decide on a preferred system before implementing an actual program.

Curbside Collection

For most densely populated towns and cities, the collection of recyclables at the curb is most common. Like the collection of mixed solid waste (MSW) at the curb, the collection of recyclables may be the direct responsibility of the community's public works department, or it may be carried out by private haulers, through contracts, or through an "open market." It is important to remember that the collection of recyclables requires far more continuity than the collection of MSW. First, a specific set of materials is collected (newspaper, clear glass, aluminum cans, etc.). While all families may not place out all these materials, the collection system must be prepared to accept the targeted materials. Second, the materials must be handled in a prescribed way. They may have to be sorted at the curb by the driver, for example. Tops may have to be removed from glass bottles. Each collection vehicle must adhere to the prescribed procedure. Once the materials have been collected, they must be driven to a processing center to be prepared for market.

If the community is collecting recyclables through the department of public works or other municipal agency, it must begin planning well before the public places material on the curb. If the program is to have the materials separated at the curb, either by the resident or the driver, special recycling trucks must be evaluated and selected for purchase. Collection routes must be designed and integrated with the existing MSW collection routes. Drivers must be trained to handle the new equipment and become familiar with the new procedures.

If the collection is to be carried out by a private hauler through a contract, the type of program must be developed, competitive bids solicited and evaluated, a bidder chosen, and a contract negotiated. This process may take a year or more to complete.

Curbside programs are designed for single-family homes, or multifamily homes of three or four. For apartment complexes, another approach must be used. Most apartment recycling takes place at a central facility where residents bring their recyclables. Because the turnover of tenants is usually higher in apartment buildings than in single-family homes, public education programs are particularly important for apartment tenants.

Drop-off Centers

In many small rural communities, recyclables may be collected at a drop-off center. Here again, a certain collection protocol must be observed. Only the targeted materials will be accepted at the drop-off center and they must be prepared and sorted in a specific way. For this type of program, the drop-off center becomes the focus of the program. It must be designed and built to accommodate not only the current needs of the community, but to handle the anticipated growth of the program. This may mean more people using the facility, or more materials accepted at some future date. Once the facility is built, it must be staffed and maintained. Records should be kept of materials collected so that progress can be monitored. Some communities, such as Buffalo, New York, offer residents a choice of either curbside collection or drop-off. This, of course, complicates the job of the planning entity in implementing the program.

Self-haul facilities may also include an area to put yard waste for composting and a "dump-and-pick" operation where recyclables are separated from other disposed trash.

The program outlined here includes both curbside collection and drop-off centers. The two programs need to be coordinated. The drop-off centers must be created and publicized, and a pilot program conducted.

Well-run drop-off programs need not be located in rural areas. One of the oldest and most successful drop-off programs in the country operates in Wellesley, Massachusetts, a suburb about 30 mi from Boston.

Processing

Once the materials have been collected, they must be prepared for market. This is the third link in the chain. Preparation will depend on the type of material collected, the volume of material collected, the demands of the market for each material, and the arrangements for transportation between the processing center and the market. Table 29.1 shows the processing options for several sample materials.

The processing of these materials usually takes place at a dedicated building called a materials recovery facility (MRF) or intermediate processing center (IPC). These types of facilities receive collected materials, sort them, densify them, and ship them to end-use markets.

While there are a variety of possible ownership and operating configurations, the facilities are often owned by the city and operated by a private vendor. Sometimes the city will operate the facility itself, or the facilities can be "merchant" facilities which are owned and operated by a private firm.

The processing facility is a key element of the recycling program. If the facility is to involve a private vendor, the procurement procedure must be set in motion so that the facility is ready once the materials are ready to be collected.

The coordination of the collection system and the processing facility cannot be stressed too much. The facility's receiving area must be compatible with the types of vehicles used for collection. For example, doors must be large enough to accommodate the vehicles. The degree of sorting at the curb must be assumed in the processing facility.

The materials must be prepared to meet the quality standards demanded by the end

TABLE 29.1 Material processing options

Material	Processing options
Newspaper	Baled or loose
Corrugated cardboard	Baled or loose
Metal cans	Crushed, baled, or loose
Glass	Crushed or loose
Plastic containers	Baled, granulated, pelletized
Tires	Shredded or whole

markets. This means that those standards must be well understood and agreed upon, preferably through contracts, before the facility is designed so that the processed materials will be acceptable to the end markets. If they fail to meet those standards, the entire recycling system will break down and the material will have to be landfilled.

Marketing

The fourth link in the recycling chain is marketing. This is the selling of processed materials to an end-use market. Each targeted material will have a separate market with its own price, quality standards, and characteristic fluctuations. For example, newsprint will be sold to a newspaper broker or directly to a mill. The buyer will demand that the paper be relatively free of contaminants and other types of paper when it is delivered. The price will fluctuate depending on a number of factors, including export demand, current supply, and the general strength of the domestic economy.

Every material that the city plans to collect must have a market. In most cases, it will be possible to obtain contracts for the material, although in some areas where markets are soft, strong contracts for some materials will not be available. In these cases, the city must decide whether or not it wishes to accept the risk of collecting and processing these materials in hopes of selling them without a contract.

Procurement

The final link in the recycling chain, and one that is often overlooked, is procurement. The willingness of end-use markets to take material is directly tied to demand for products made from recycled materials. If no one demands newsprint made from recycled fiber, the newsprint mills have little incentive to produce it. On the other hand, if there is a substantial demand, the mills will be actively looking for sources of recycled fiber.

Procurement policies can be developed for two kinds of products. For products that are directly consumed by the city, such as office paper, the city can develop purchasing guidelines which specify that these products must contain a certain amount of recycled material. For example, the city can require that office paper must contain a certain percentage of postconsumer office waste.

The city may develop a closed-loop recycling program where it consumes products it creates. For example, it may use a tub grinder to process construction and demolition debris and then use the resulting aggregate in its own paving and building projects.

The city may also encourage businesses within its jurisdiction to "buy recycled," thereby building up the demand for recycled products. Local newspapers can be urged to use recycled newsprint and plastic fabricators can be urged to use recycled plastic scrap.

Evaluating Program Options: Seattle's Planning Model

It has already been pointed out that the rigor and formality of the planning process varies widely. In Canada's Province of Ontario, the planning process is becoming more rigorous through the efforts of the provincial government, which has been moving to formalize the planning process since the enactment of the Environmental Assessment Act in 1975. It is likely that a number of planning guidelines will be passed down from the province to individual municipalities.

Formalized planning is often found in metropolitan areas such as Seattle, Washington, and San Jose, California, which have large complex programs. Small cities typically use a less formal approach. For example, Loveland, Colorado, a city of 40,000 people north of Denver, has created its program by listening carefully to its own citizens, surveying programs in other parts of the country to see what is working, and "using common sense."

Planning for recycling must involve an examination of all components of the solid waste management system, as well as consideration of broader public-sector issues. Recycling cannot be considered in isolation. Although the way in which such an examination is structured may differ from location to location, the components are quite widely accepted. In general, they include the jurisdiction's waste stream, recycling program options, the effect of recycling programs on disposal, and the solid waste management system costs. The number of ways this information can be arranged is manifold.

One of the most sophisticated planning models now in use is Seattle's Recycling Potential Assessment Model (RPA model). The RPA is a computer model that was developed in 1988 by the Seattle Solid Waste Utility (the "Utility") to "evaluate the feasibility and cost-effectiveness of many recycling options open to the City." The model was used by the Utility in 1997 to estimate program results and costs for solid waste alternatives, which was part of the process to update their solid waste plan, *On the Path to Sustainability*. The following discussion uses the Seattle RPA model as one example of how to evaluate program options.

Waste Stream Forecast

The recycling program should be built on assumptions about the quantity and composition of the waste being generated. Information on the waste stream will be used to size facilities and programs, estimate revenues from secondary material markets, and gauge the success of program goals.

A comprehensive waste stream forecast will provide the answers to three broad questions.

1. *What is the current composition of the waste stream? That is, what materials are available for recycling?* For the sake of program development, the question must be applied to various generator categories-residential, commercial, institutional, and industrial. These categories may be subdivided even further. Residential into single-family and multifamily households; commercial into offices, large retail stores, restaurants, hotels, and so on. These estimates of current waste composition may be based on four-season waste sampling programs that provide a statistically accurate estimate of the composition, or on national estimates which are applied locally. Table 29.2 shows the estimate of waste generated (by weight) developed by the United States Environmental Protection Agency in 1997. An estimate of the composition of the waste generated is one of the crucial inputs to the planning process.

2. *How much waste is currently generated?* The answer to this question is not the total number of tons delivered to the local landfill. Rather, it is the total amount of waste generated, the "gross generation," which includes waste diverted from disposal through

private recycling. That is, rather than discarding corrugated cardboard, some supermarkets may be collecting and baling cardboard on their own, long before a municipal recycling program has begun. The cardboard is waste, part of the gross generation of waste, but the amount of solid waste the supermarket actually sends to the landfill is its "net generation." The City of Seattle, with a strong environmental ethic, estimated that voluntary and private recycling was achieving about a 24% recovery rate from all waste. Knowing how much waste is generated means having an estimate of both gross and net generation. Again, knowing this for various classes of generators will help in targeting recycling programs.

3. *What is going to happen in the future to the quantity and composition of the waste; that is, what social, demographic, and economic forces will affect the generation of waste and will each individual citizen become more or less wasteful?* Over the past 20 years the average amount of waste generated by each American has increased. Will efforts at source reduction curb this trend? Will the population of the area increase, thereby increasing the total amount of waste to be disposed? Will the number of households and the size of households increase? Will the economic climate stimulate new businesses that will generate waste, or will it force bankruptcies and inhibit consumption, thereby decreasing the amount of waste generated? As difficult as these questions may be to answer, some estimate of future generation is necessary. Figure 29.1 shows the projections for waste generation used by Seattle in 1988 as an input for its RPA model.

The RPA model allows for different generation forecasts to be applied to different generator categories, including a forecast of private recycling. The information in the waste stream module is then used to evaluate the effects of various recycling programs.

Recycling Program Options

Program options integrate elements into generator-specific and material-specific approaches. Examples would be backyard composting, curbside yard waste collection, apartment recycling, and drop-off centers. For each option, the RPA model takes participation (how many households actually use the program) and efficiency (how much material is actually diverted) as inputs.

Program costs are also inputs to this module. Costs that are not dependent on program size are fixed costs; those costs that vary with the number of participants or the number of tons collected are variable costs. These costs are developed for each program. Variable costs can be recalculated each time participation or efficiency rates are changed. To show another effect of each program on the solid waste management system, the RPA also calculates the avoided cost for each program. That is, the RPA calculates the cost of disposing of each ton diverted. This is termed the "benefits per ton." Figure 29.2 compares the benefits per ton with costs per ton for the backyard composting program.

The RPA model can combine different sets of programs to form different recycling scenarios. For example, Scenario 1 for RPA includes the following eight programs:

1. Waste reduction
2. Backyard composting
3. Curbside yard waste
4. Self-haul yard waste
5. Curbside recycling
6. Apartment diversion
7. Business and industrial recycling
8. Drop-off recycling

Scenario 2 is the same, except two new programs, apartment recycling and a self-haul dump-and-pick operation, are added. Scenarios can be compared on the basis of costs, benefits per ton, and total tons recycled. Each scenario can also be compared to the city's

TABLE 29.2 Materials generated* in the municipal waste stream, 1960 to 1997 (In thousands of tons and percent of total generation)

	Thousands of Tons							
Materials	1960	1970	1980	1990	1994	1995	1996	1997‡
Paper and Paperboard	29,990	44,310	55,160	72,730	80,840	81,670	79,680	83,840
Glass	6,720	12,740	15,130	13,100	13,350	12,830	12,290	12,010
Metals								
Ferrous	10,300	12,360	12,820	12,640	11,780	11,640	11,830	12,330
Aluminum	340	800	1,730	2,810	3,050	2,960	2,950	3,010
Other Nonferrous	180	670	1,160	1,100	1,350	1,260	1,260	1,270
Total Metals	*10,820*	*13,830*	*15,510*	*16,550*	*16,180*	*15,860*	*16,040*	*16,610*
Plastics	390	2,900	6,830	17,130	19,250	18,900	19,760	21,460
Rubber and Leather	1,840	2,970	4,200	5,790	6,210	6,030	6,200	6,590
Textiles	1,760	2,040	2,530	5,810	7,260	7,400	7,720	8,240
Wood	3,030	3,720	7,010	12,210	11,280	10,440	10,840	11,570
Other†	70	770	2,520	3,190	3,700	3,650	3,690	3,750
Total Materials in Products	54,620	83,280	108,890	146,510	158,080	156,780	156,220	164,080
Other Wastes								
Food Wastes	12,200	12,800	13,000	20,800	21,500	21,740	21,850	21,910
Yard Trimmings	20,000	23,200	27,500	35,000	31,500	29,690	27,920	27,730
Miscellaneous Inorganic Wastes	1,300	1,780	2,250	2,900	3,100	3,150	3,200	3,250
Total Other Wastes	33,500	37,780	42,750	58,700	56,100	54,580	52,970	52,890
Total MSW Generated—Weight	88,120	121,060	151,640	205,210	214,180	211,360	209,190	216,970
	Percent of Total Generation							
Materials	1960	1970	1980	1990	1994	1995	1996	1997
Paper and Paperboard	34.0%	36.6%	36.4%	35.4%	37.7%	38.6%	38.1%	36.6%
Glass	7.6%	10.5%	10.0%	6.4%	6.2%	6.1%	5.9%	5.5%
Metals								
Ferrous	11.7%	10.2%	8.3%	6.2%	5.5%	5.5%	5.7%	5.7%
Aluminum	0.4%	0.7%	1.1%	1.4%	1.4%	1.4%	1.4%	1.4%
Other Nonferrous	180	670	1,160	1,100	1,350	1,260	1,260	1,270
Total Metals	*12.3%*	*11.4%*	*10.2%*	*8.1%*	*7.6%*	*7.5%*	*7.7%*	*7.7%*
Plastics	0.4%	2.4%	4.5%	8.3%	9.0%	8.9%	9.4%	9.9%
Rubber and Leather	2.1%	2.5%	2.8%	2.8%	2.9%	2.9%	3.0%	3.0%
Textiles	2.0%	1.7%	1.7%	2.8%	3.4%	3.5%	3.7%	3.8%
Wood	3.4%	3.1%	4.6%	6.0%	5.3%	4.9%	5.2%	5.3%
Other†	0.1%	0.6%	1.7%	1.6%	1.7%	1.7%	1.8%	1.7%
Total Materials in Products	62.0%	68.8%	71.6%	71.4%	73.8%	74.2%	74.7%	75.6%
Other Wastes								
Food Wastes	13.8%	10.6%	9.6%	10.0%	10.3%	10.4%	10.1%	
Yard Trimmings	22.7%	19.2%	18.1%	17.1%	14.7%	14.0%	13.3%	12.8%
Miscellaneous Inorganic Wastes	1.5%	1.5%	1.5%	1.4%	1.4%	1.5%	1.5%	1.5%
Total Other Wastes	38.0%	31.2%	28.2%	28.6%	26.2%	25.8%	25.3%	24.4%
Total MSW Generated—%	100.0%	100.0%	100.0%	100.0%	100.0%	100.0%	100.0%	100.0%

*Generation before materials recovery or combustion. Does not include construction & demolition debris, industrial process wastes, or certain other wastes.

†Includes electrolytes in batteries and fluff pulp, feces, and urine in disposable diapers.

‡U.S. waste composition in 1997. Does not include construction and demolition debris, industrial process wastes. (*Source:* U.S. EPA *Characterization of Municipal Solid Waste*)

Details may not add to totals due to rounding.

Source: Franklin Associates

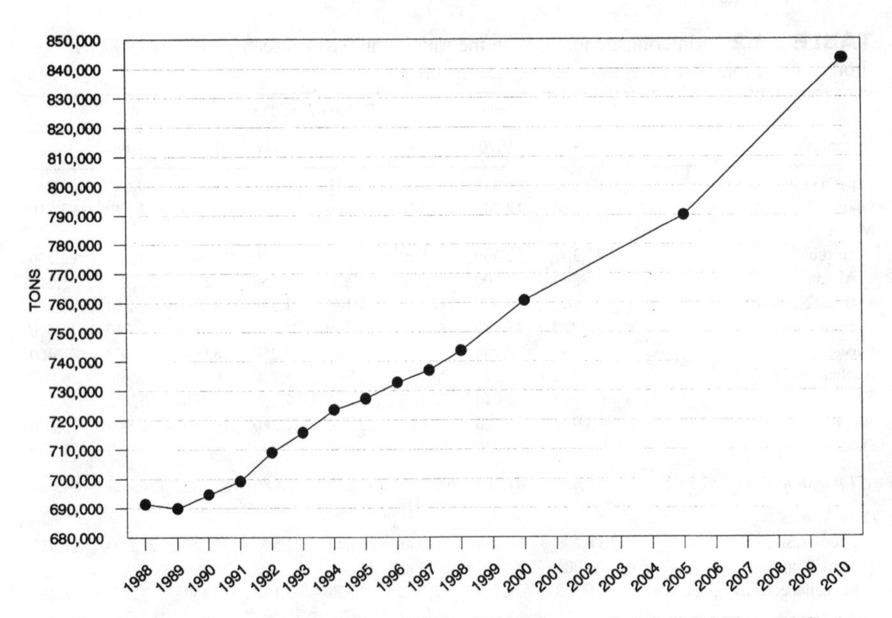

FIGURE 29.1 Seattle waste generation, 1988 to 2010. (*Source: Seattle Solid Waste Utility.*)

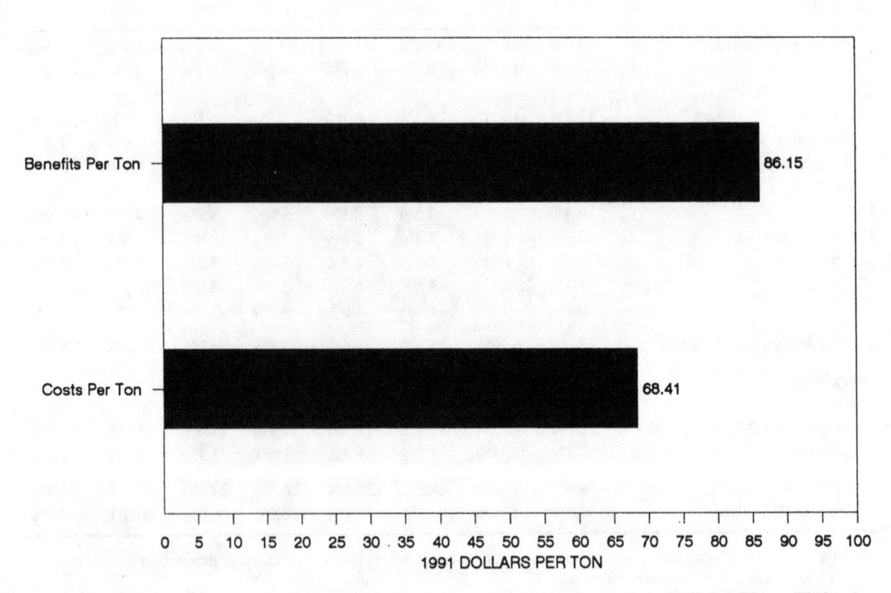

FIGURE 29.2 Cost and benefit of backyard composting. (*Source: Seattle Solid Waste Utility.*)

recycling goal. Because the waste composition estimates, participation and efficiency rates, and costs are projected into the future, the effect of these programs over time can be projected.

The Total System

The RPA model passes the individual program costs onto the system cost module, which aggregates the information about recycling rates and costs from all programs. These aggregated costs are then combined with the costs for disposal of the solid wastes not reduced or recycled. This provides the total system cost for the program. Figure 29.3 projects the percentage of total number of tons recycled for the period from 1988 to 2000 compared to the estimated total waste generated.

Average Rates

Once the total system costs have been calculated in the system cost module, the RPA will calculate the rates needed to recover the revenues needed to operate the programs. These rates are sector-specific; that is, the costs associated with commercial-sector programs are allocated to the commercial revenue requirements. The same process is used for residential programs. The total revenue required for each sector is divided by the number of tons disposed by each sector to determine the average rate.

Because a change in rates may affect the behavior of participants, the RPA model includes a loop which feeds the new rate back into the generation model. If generation and disposal change as a result of the new rates, the program and system costs are recalculated and new revenue requirements and new rates are determined. This iterative process continues until the system is in balance.

Revising the Model

In 1988, the City of Seattle set a recycling goal of 40% by 1990 and the Solid Waste Utility developed its plan with this goal firmly in mind. The recycling plan developed by the Seattle Solid Waste Utility in 1988 was implemented and tracked. Over the next several years, new information on waste composition and generation, private recycling, participa-

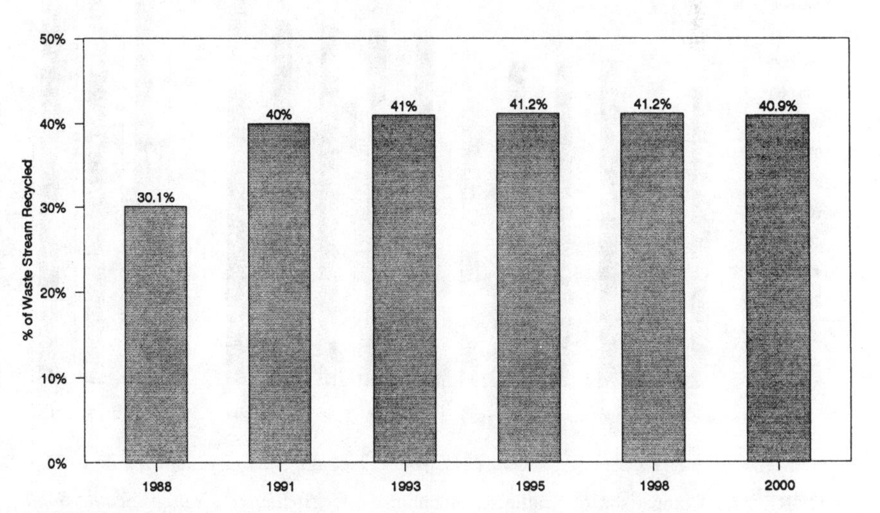

FIGURE 29.3 Seattle waste recycling rate. (*Source: Seattle Solid Waste Utility.*)

tion rates, program efficiencies, and costs became available. This new information was used to update the RPA model in 1991.

According to the RPA model, the city had moved from 30% recycling in 1988 to 40% recycling in 1991. However, more aggressive goals for recycling in future years had been set. The city wished to reach 60% recycling by 1998 through cost-effective programs and the Utility needed to design a set of program options that would achieve this goal. This meant that the RPA model had to be updated, a new set of program options developed, and the model rerun.

Updating the model began with revisiting the waste stream assumptions. The composition of the waste disposed had changed since 1988 as a result of the ongoing recycling programs, as well as economic and social factors. A comparison of the composition of the residential waste in 1988 and 1990 is shown in Fig. 29.4.

In addition to changes in the assumptions about waste composition, two other factors were found to alter the projected amount of recycling in Seattle. First, the method of calculating the overlap between private sector recycling and the Utility-sponsored recycling changed. Second, the three years of operating experience provided more realistic assumptions. These factors resulted in changes which are shown in Fig. 29.5.

Two new and more aggressive recycling scenarios were developed to achieve the 1998 goal of 60% recycling. The new scenarios are outlined in Table 29.3.

The recycling rates of the new scenarios were compared with the city's goals and, as Fig. 29.6 shows, none of the three scenarios met the 1998 goal of 60%. The benefits and costs of the programs were also calculated. Figure 29.7 shows the costs and benefits for Scenario 2, as well as the number of tons recycled by each program in 1998.

Seattle's RPA model allows the Utility to update the model as new information becomes available, revise the programs as goals or other circumstances change, and check

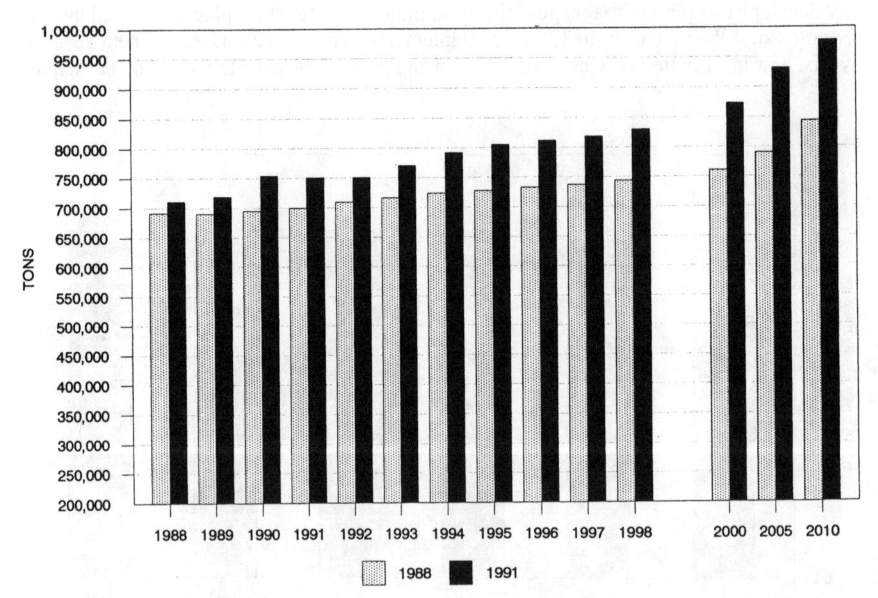

FIGURE 29.4 Comparison of Seattle's residential waste generation. (*Source: Seattle Solid Waste Utility.*)

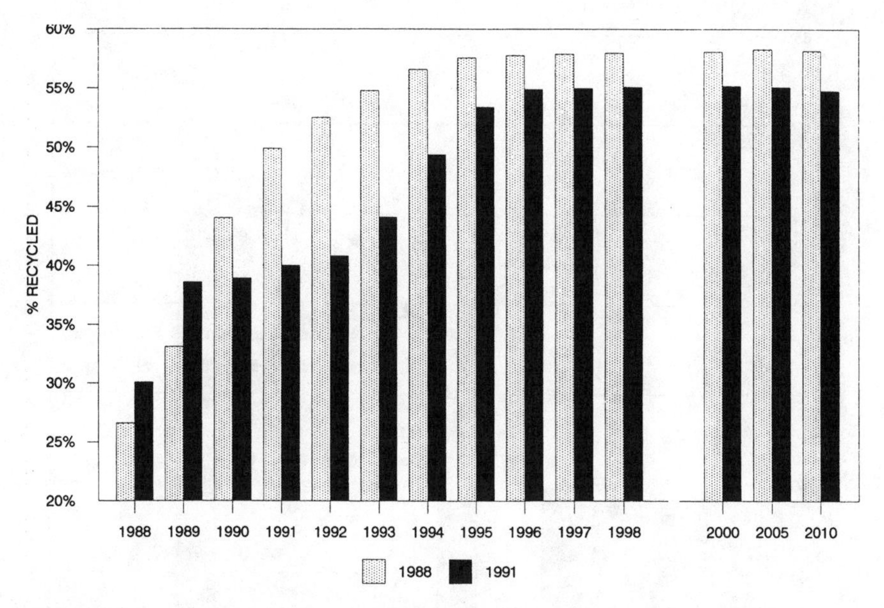

FIGURE 29.5 Comparison of recycling rates. (Includes new factors: three years of operating experience and changed calculation method.) (*Source: Seattle Solid Waste Utility.*)

to see that the whole system is in balance. These are the goals of any planning system, whether the system is a complex computer model like Seattle's or a regular discussion of a rural town's drop-off program.

To move from program design to implementation requires two further steps: definition of the work to be done and scheduling.

TABLE 29.3 Recycling scenarios for Seattle model

	Existing programs	Scenario 1	Scenario 2
Waste reduction	x	x	x
Backyard composting	x		x
Curbside yard waste	x	x	x
Self-haul yard waste	x	x	x
Curbside recycling	x	x	x
Apartment diversion	x	x	x
Business/industry recycling	x	x	x
Drop-off program	x	x	x
Apartment recycling			x
Self-haul dump and pick		x	x
Mandatory commercial paper			x

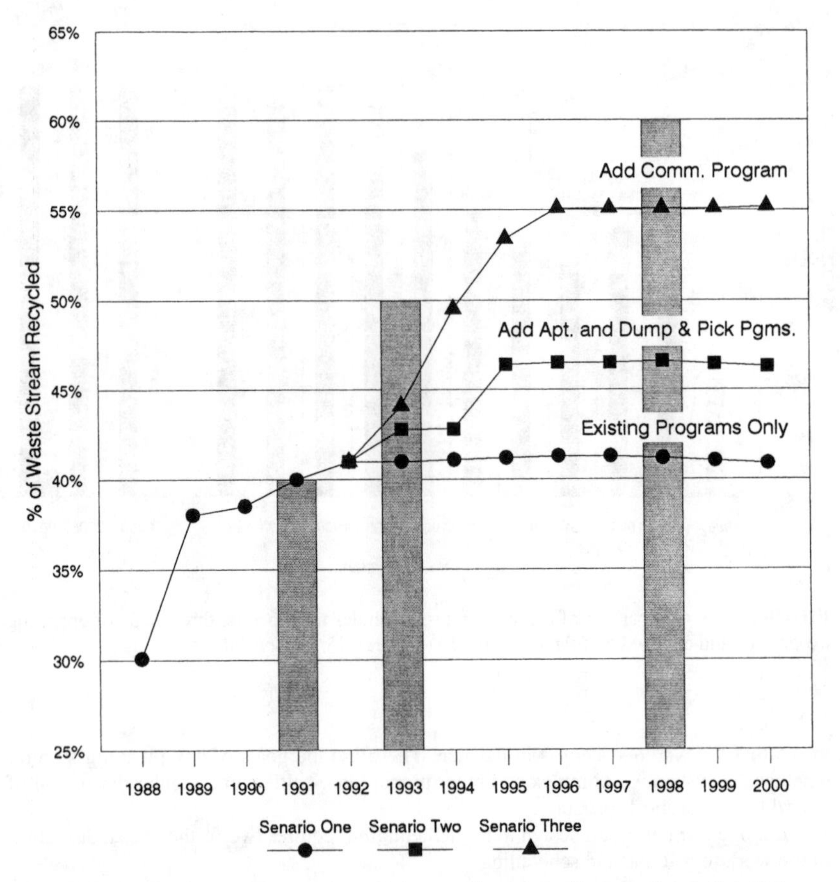

FIGURE 29.6 Recycling rates and Seattle's recycling goals. (*Source: Seattle Solid Waste Utility.*)

Defining the Work

No matter how complex, or simple, the recycling program is, it can be put into action more easily if the activities involved are clearly defined. In practice, the work should be broken down into discreet elements which can be carried out, managed, and monitored with the resources of the organization.

The planner who is defining activities must strike a balance between two extremes. The first is offering very broad rather vague guidelines for staff members. The second is providing meticulously detailed instructions describing every step of every activity. The former leaves staff members wondering what is needed and the latter leaves no room for individual initiative. The best way to strike this balance is to involve those who will be responsible for carrying out the work in defining it.

For example, there is little doubt that developing a school recycling curriculum with the teachers who will be using it in the classroom will result in a better curriculum. The

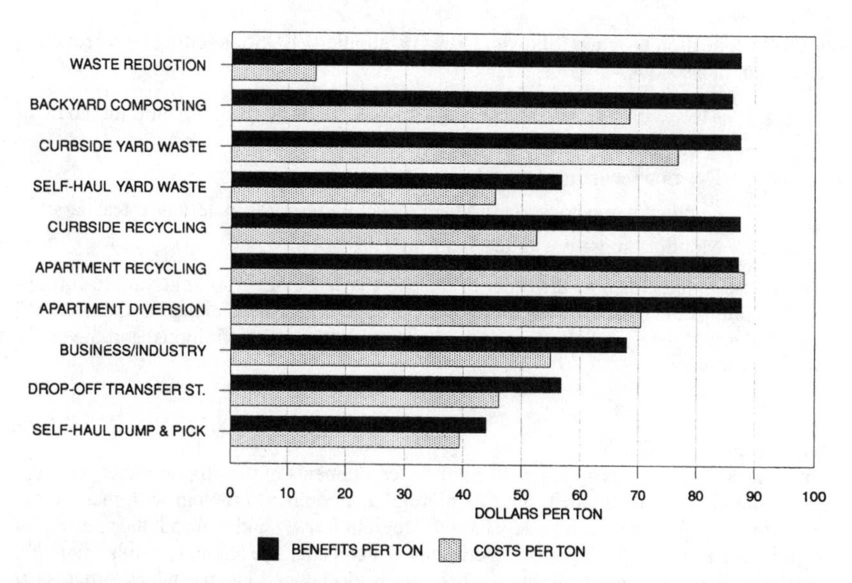

FIGURE 29.7 Benefits and costs of programs in Senario 2. (Added elements: Self-haul dumps and pick, and apartment recycling.) (*Source: Seattle Solid Waste Utility.*).

same thing is true for outlining the steps that will need to be taken to develop a recycling curriculum. The people who do the work are the experts and can help define the jobs that must be done. Furthermore, involving those responsible will generate a sense of commitment that will be essential once the work is started.

But there are other hazards in defining work tasks, even with the help of those doing the work. First, external conditions change over time. Shifts in the economy, movements in the secondary material markets, the promulgation of new solid waste management regulations, and the evolution of processing technologies, for example, often undermine even the most carefully developed instructions. Recycling is a particularly dramatic example of this because it has evolved so rapidly during the past five years. The planner of a recycling program will have to be especially nimble to keep up with the changes in the business. In this sense, a sound plan must be dynamic, adjusting to the shifts in external conditions.

Programs must also be dynamic in terms of growing to meet local demands. That is, recycling programs evolve in scope and sophistication. Almost as soon as a program is defined, there are pressures to enlarge it. For example, Seattle set more aggressive recycling goals even before the previous ones were accomplished. Markets for new materials, such as plastics, may develop. Technology becomes available to process more cost effectively. Residents and businesses want to recycle more kinds of materials. Programs want to keep up with these kinds of demands.

Developing a Work Breakdown Structure (WBDS)

One formal approach to defining the work to be done is called, logically enough, the work breakdown structure (WBDS) of the program. As an example of what a WBDS looks like, consider the school recycling curriculum which a city wishes to develop as a part of its public education program. Shown below is a simple WBDS for setting up a recycling curriculum in a school.

Task 1 Meet with teachers and school officials to discuss and develop the goals of the program.

Task 2 Develop recycling curriculum materials.

Task 3 Conduct workshops to introduce recycling resource materials to teachers.

Task 4 Monitor program with teachers.

In many cases, like this one, the order of the tasks is self-evident. In other, more complicated, situations the most efficient order of the tasks is not clear. In either case, the "logic" of the order will determine the overall program goals, the task definitions, and the availability of resources.

Defining Program Goals

It is illogical to discuss program goals after the components of the program have been determined and the tasks defined, but this highlights a common problem with many recycling programs. While efforts to develop the program move rapidly ahead, the reasons for the program are never clearly articulated. This is a considerable handicap to planning, because it is the program's goals which help set priorities and against which progress is measured.

Although there are a number of different goals which communities cite in justifying their recycling programs, the goals generally fall into one of two broad areas. Although both types of goals generally play a part in most programs, it is instructive, for our purposes, to draw a distinction.

The first type of goal is economic. In areas where tipping fees are high and landfill capacity is limited (or nonexistent), recycling can offer a cost-effective means of managing a significant fraction of the waste stream. Bound by high-costs, diminishing disposal capacity, and an ever stricter regulatory environment, a community in such an area will develop a source-reduction and recycling program to shrink its flow of solid waste, thereby cutting the cost of solid waste disposal. Many communities in the northeastern United States find themselves in this situation.

On the other hand, there are successful recycling programs in communities where the tipping fee is less than $20 per ton and the landfill has a life of 30 years. Many of these communities see recycling as a means of saving energy and natural resources. For example, Boulder, Colorado's, recycling program is not the result of outrageous tipping fees and closing landfills. It is part of a broad range of environmental programs carried out by concerned citizens.

This is not to say that these same environmental concerns are not shared by communities with high tipping fees, or that cost-effective recycling programs are not the goal of communities with low tipping fees. However, it is instructive to note the way in which priorities are ordered by the two types of communities.

A program driven primarily by economics will be evaluated accordingly. Program costs will be developed, evaluated, and justified in terms of the high tipping fees. A recycling program that costs more than the avoided cost of disposal will not be well-received. Financial analysis will play an important role in determining program priorities. In a community without the pressure of high tipping fees, program priorities may be determined by less quantifiable benchmarks, such as minimizing environmental damage or conserving resources.

The point is that a clear articulation of the goals of the program will inform all the decisions that follow. Without such an articulation, confusion, frustration, and wasted resources can result.

Establishing Program Sequence and Timing

The work breakdown structure (WBDS) outlined above presents the work necessary to develop a school recycling curriculum. It should be emphasized again that each individual program will contain different elements and approach the tasks in different ways. Nevertheless, the activities within any program must be carried out in a particular sequence. The timing of tasks and subtasks will be a critical element in the success of the program. Certain tasks must precede others. For example, residents must be given information about the curbside collection of recyclables before the collection trucks arrive so that the targeted materials can be properly prepared. On the other hand, disseminating information too far ahead will raise expectations and enthusiasm prematurely.

It should be noted that the timing of a program can be influenced by many outside factors. For example, as a result of a long period of negative publicity about the closing of the landfill and a period of rapid rate increases, Seattle's Solid Waste Utility felt compelled "do something" positive and do it quickly. The utility implemented a curbside recycling program, although it was not clear how this program would fit with the other recycling programs. Rarely does a city have the luxury of taking as much time to plan as it would like.

The sequencing of activities will be particularly difficult and important at the beginning of the program when many tasks will begin at about the same time. Certain tasks, where the planner has some control, should be given ample time. The time necessary to evaluate proposals, for example, will vary depending on how many proposals are received. Contract negotiations are another area where it is difficult to know ahead of time how much time will be needed. It is wise to allot more than enough time for these types of activities.

A second issue underlying planning efforts of this kind will be. scarcity of resources. Only rarely will a planner have all the resources available to carry out the "ideal" program. In fact, it is likely that the expectations for the program will be very high and the resources available to carry out the program will be very modest. Therefore, priorities must be determined. This brings us back to program goals which, if clearly stated, should help set the task-ordering priorities. All program elements are not equal and a good planner will recognize this early in the planning process.

Sequence Planning with a Gantt Chart

One simple way to present and manipulate the sequence of program activities is with a Gantt chart. This is simply a visual means of showing a number of tasks over a specified period of time. The activities are arranged in sequence and given a visual "weight" according to their duration. Figure 29.8 shows one type of a Gantt chart for the tasks under "School Programs."

Note that each activity is assigned a certain period. It is estimated that the activity can be carried out during this period. The chart indicates the task order, the length of time allotted for each task, and the length of time allotted for the entire group of tasks. Sometimes it is useful to identify milestone events or documents which mark the end of a particular task. This kind of signal is useful to all participants, for it indicates concrete progress toward the overall program goals. For example, the completed curriculum (B) might be a milestone document in the school recycling program activities (Fig. 29.9).

The Gantt chart is particularly useful with a relatively small program. It becomes a less useful tool as the size and complexity of the program grows. It also does not reflect the priorities among many tasks taking place simultaneously. Nevertheless, the exercise

LIST OF TASKS:

TASK 1 MEET WITH TEACHERS AND SCHOOL OFFICIALS TO DISCUSS AND DEVELOP THE GOALS OF THE PROGRAM

TASK 2 DEVELOP RECYCLING CURRICULAR MATERIALS

TASK 3 CONDUCT WORKSHOPS TO INTRIDUCE RECYCLING RESOURCE MATERIALS TO TEACHERS

TASK 4 MONITOR PROGRAM WITH TEACHERS

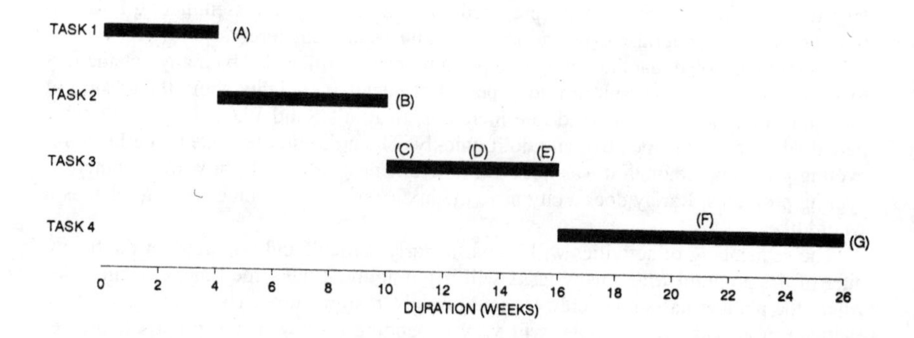

(A) PUBLISH STATEMENT OF GOALS
(B) PUBLISH CURRICULAR MATERIALS
(C) (D) & (E) TEACHER WORKSHOPS
(F) INTERIM PROGRESS REPORT
(G) EVALUATION OF PROGRAM

FIGURE 29.8 Gantt chart for developing recycling curriculum with milestones.

of assigning a time period to each task is a very useful exercise for the planner. It is a necessary first step in developing a program schedule.

Program Scheduling

It was pointed out above that recycling programs may be developed for different reasons, have different goals, and go about accomplishing their goals in different ways. The differences among programs will also be manifest in the approaches to planning. Some programs evolve with little or no planning in the formal sense. Ideas are put forth and argued over, funds sought and won, and programs implemented. The participants understand intuitively what needs to be done and when. Other, larger programs may be more formal, using something akin to a Gantt chart to schedule activities. Programs that are relatively small or that are in the early stages of evolution are often carried out with very little formal planning and this is perfectly appropriate.

LIST OF TASKS:

TASK 1 MEET WITH TEACHERS AND SCHOOL OFFICIALS TO DISCUSS AND DEVELOP THE GOALS OF THE PROGRAM

TASK 2 DEVELOP RECYCLING CURRICULAR MATERIALS

TASK 3 CONDUCT WORKSHOPS TO INTRIDUCE RECYCLING RESOURCE MATERIALS TO TEACHERS

TASK 4 MONITOR PROGRAM WITH TEACHERS

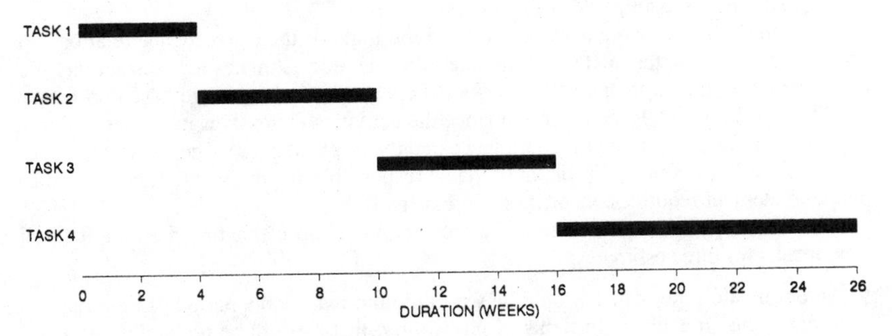

FIGURE 29.9 Gantt chart for developing recycling curriculum for a small program.

However, as programs become larger and sophisticated, the need for more rigorous, formal planning increases. Intuition and simple Gantt charts cannot handle the complexities of the program. For those interested in this more rigorous approach to planning, this section of the chapter concludes with a discussion of program scheduling models.

Program Scheduling Models

The science, and art, of project scheduling has grown rapidly over the past several decades, driven primarily by large contracts for the federal government involving hundreds of tasks, and thousands of people, and millions of dollars. Building hydroelectric dams, Stealth bombers, or space satellites are examples of such projects. These huge projects could not be conducted without some way of ordering the tasks, assigning responsibility, and monitoring progress.

Two examples of scheduling models that are often used for large, multitask programs are the Critical Path method (CPM) and the Program Evaluation and Review Technique (PERT). Both models seek to identify the sequence of tasks and activities which will take the most time. Both models require the construction of a network which articulates the logic of the overall program. Although similar in approach, PERT differs from CPM in terminology and in its ability to incorporate probability into its time estimates. This makes PERT particularly useful in scheduling activities characterized by uncertainty. The PERT model has been used in the example below, although some comments about the CPM model are also provided.

The PERT model will first be considered in terms of time, as a limiting factor, and will then be considered in terms of time, cost, and resources, as limiting factors.

Network Construction—Time Models

Once the separate tasks for the program have been defined, the proper sequence of tasks must be determined. This has already been discussed above. Using this information, a network is built of the tasks. The logic of the network is shown through a series of notations which indicate activities and events. Activities take time and resources; events mark the beginning or completion of activities. For example, negotiating a contract would be an activity. An executed contract is an event. Figure 29.10 shows the notation commonly used to indicate activities and events in the PERT and CPM scheduling models.

The arrows denote activities and the circles or nodes denote events. In developing the network, the precedence relationship of events can be shown by placing the nodes in logical order. The PERT model, and the CPM model, can show a variety of types of precedence relationships, as Fig. 29.11 shows.

After the activities have been defined and the logic of their precedence relationships has been established, the next step is to determine the time estimates for each activity. For CPM, which is the simpler of the two models, a single time estimate based on current conditions is required. Remember that since the activities themselves may change during the course of the project, the time estimate can only be a best guess. The most accurate estimate is likely to come from the individuals responsible for the work. Again, a critical planning element requires consultation with the staff.

The PERT model requires a more rigorous approach to estimating the time for projects. It asks for three estimates:

- An optimistic time, which would be the minimum reasonable period of time required to complete an activity. In terms of probability, there would be probability of 1% or less that the activity would take less time than this.
- A most likely time, which would be the estimate that would be used in the CPM network.
- A pessimistic time, which would be the maximum reasonable period required to complete an activity. In terms of probability, there would be a probability of 1% or less that the activity would take more time than this.

This statistical basis of these estimates will be used to calculate the expected times and the variance of the activity times. Again, whether the CPM model or the PERT model is used, the time estimates should be gathered from those who will be performing the activity.

The PERT model estimates activity times using the three estimates. Consider the tasks identified to develop the collection of recyclables at a drop-off center.

Task 1: Determine location of drop-off centers.

Task 2: Integrate drop-off program with curbside program.

Task 3: Purchase drop-off containers.

Task 4: Arrange for pickup of materials and delivery to processing facility.

FIGURE 29.10 Notation for PERT and CPM models.

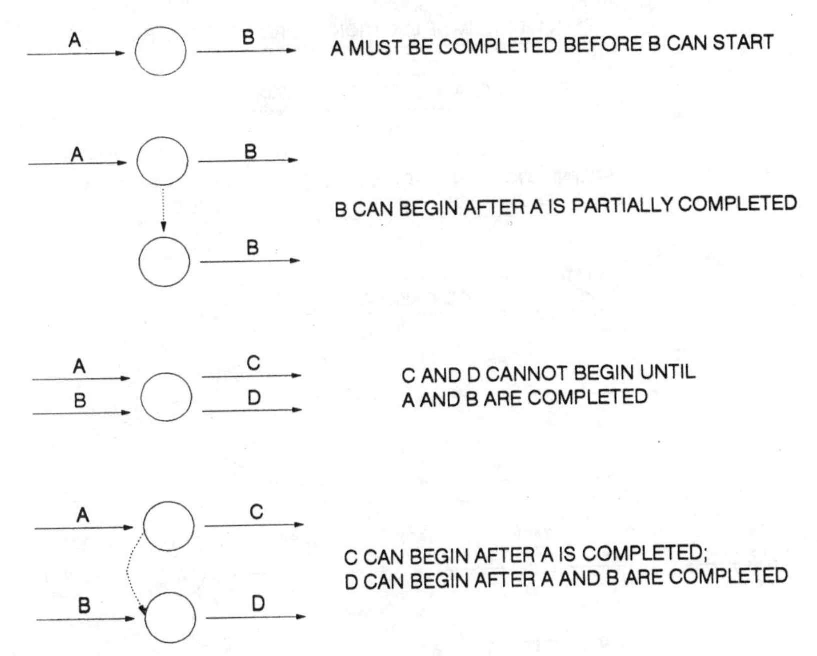
A
B
A MUST BE COMPLETED BEFORE B CAN START
A
B
B
B CAN BEGIN AFTER A IS PARTIALLY COMPLETED
A
B
C
D
C AND D CANNOT BEGIN UNTIL
A AND B ARE COMPLETED
A
C
B
D
C CAN BEGIN AFTER A IS COMPLETED;
D CAN BEGIN AFTER A AND B ARE COMPLETED

FIGURE 29.11 Notation for precedence relationships in PERT and CPM models.

Task 5: Develop promotional materials and signage.

Task 6: Publicize drop-off program.

Task 7: Conduct pilot program.

First, we must designate the precedence relationship among these seven tasks. Using the model notation, we begin the two independent activities (Fig. 29.12*a*).

As the diagram indicates, there are two independent activities in this program, that is, two activities that do not require a preceding activity. The two activities are selecting the drop-off sites and procuring the drop-off containers. It is assumed that the selection of materials to be collected and their preparation has been determined earlier.

Only after the drop-off sites have been selected can the drop-off program and the curbside program be integrated and this is noted as in Fig. 29.12*b*. The arrangements for collecting the materials can only be made when the sites have been selected and the containers procured, which is noted as in Fig. 29.12*c*.

With the information about drop-off sites and collection schedules available, the promotional materials can be developed, the program publicized, and the pilot program begun. Figure 29.12*d* is the notation for the tasks defined under drop-off centers.

Next the estimated time (ET) for each task should be determined. To accomplish the first task, locating the drop-off centers, data on the service area must be gathered and evaluated. Information on population density, traffic patterns, the planned routes for curbside service, and available sites must be assembled. Then a group of potential sites may be chosen and visited to determine a ranking. Finally, the proposed locations should be reviewed by an independent group for reasonableness. The optimistic time (denoted as *a*)

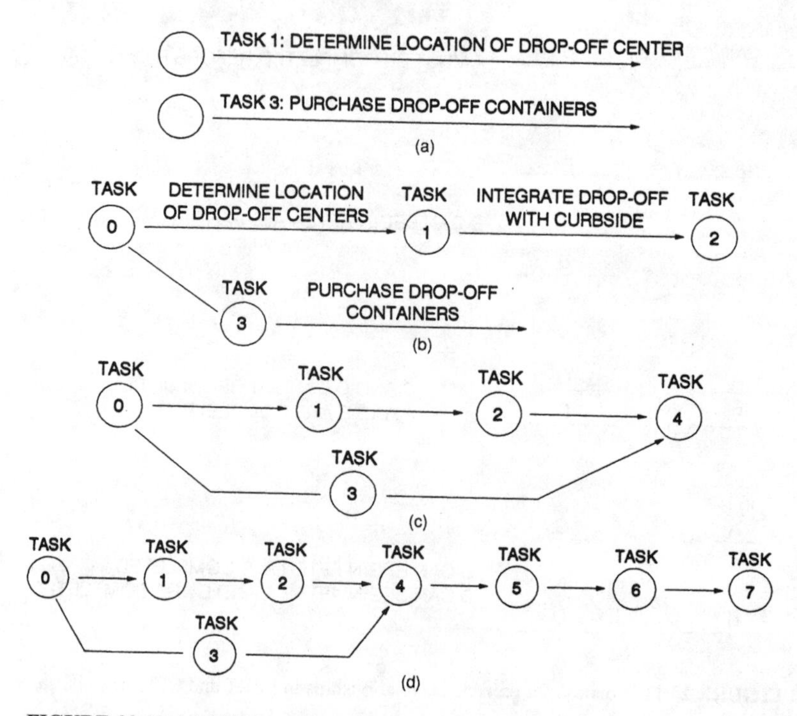

FIGURE 29.12 (*a*) PERT notation for developing drop-off centers—Tasks 1 and 3. (*b*) PERT notation for developing drop-off centers—Tasks 1, 2, and 3. (*c*) PERT notation for developing drop-off centers, Tasks 1 to 4. (*d*) PERT notation for developing drop-off centers, Tasks 1 to 7.

for this task is reported to be 4 weeks by the team who will be responsible for carrying it out. This assumes that all the data are readily available, there will be numerous available sites from which to choose, and the team will work smoothly together.

The most likely time (denoted as b) is 7 weeks. This assumes the data can be obtained after some searching, that sites can be found after some searching, and that the team works together reasonably well. The pessimistic time (denoted as c) is 12 weeks. This assumes that much of the data are not available and must be developed, that the few available sites must be researched carefully, and that there is some friction among team members.

The ET for this task is calculated by using the following formula:

$$\text{ET} = \frac{a + 4b + c}{6} = \frac{4 + 28 + 12}{6} = 7.33 \text{ weeks}$$

This formula is based on a beta distribution which gives the "most likely" time four times the weight of the optimistic and pessimistic distribution.

Next, the variance for each activity is calculated with the formula:

$$\text{Variance} = \left(\frac{b - a}{6}\right)^2 = \left(\frac{12 - 4}{6}\right)^2 = \left(\frac{8}{6}\right)^2 = 1.77$$

TABLE 29.4 Expected times (weeks) and variance for each task

Task	Optimistic time	Most likely time	Pessimistic time	Expected time	Variance
1	4	7	12	7.33	1.77
2	3	6	10	6.17	1.36
3	6	10	15	10.17	2.25
4	2	3	4	3	0.11
5	4	6	12	6.67	1.78
6	2	3	4	3	0.11
7	6	10	20	11	5.44

The greater the difference between the optimistic time and the pessimistic time, the greater the variance. Table 29.4 shows the optimistic, pessimistic, and most likely times for each of the seven tasks, as well as the ET and the variance for each one. The figures used below are hypothetical.

The purpose of all these calculations is to determine the critical path. The path is laid out along the sequence of tasks which must be followed to complete the activity most efficiently. In a program with tens or hundreds of tasks, this is valuable information for the planner. The key to locating the critical path is finding "slack" time. If an activity has slack time, it is not on the critical path. There is no slack for tasks on the critical path. To calculate slack time in the PERT model, we place the ET and variance for each task in the program notation, as shown in Fig. 29.13.

Below the notation, two times are shown for each task. The T(E) of each task is the earliest expected completion time and is calculated by summing all ET's of each task from point zero forward to the given task. Therefore, the T(E) of Task 1 is 7; the T(E) of Task 2 is 7 + 6 or 13; the T(E) for Task 3 is 7 + 6 + 10 or 23, and the T(E) of Task 4 is 7 + 6 + 4 or 17 and so forth. The total T(E) for all seven tasks is 38 weeks.

The T(L) is the latest expected completion time for a task and is found by setting T(L) equal to the T(E) total of 38, and moving backward through the network. Therefore, Task 7 has a T(L) of 38; Task 6 has a T(L) of 27; and so forth.

Note that Task 3 has a T(E) of 10 and a T(L) of 13. The difference of 3 weeks between these two is slack time. That is, Task 3 could be started at week 7 immediately upon completion of task 1, or it could be started at week 12 and be completed at week 16 simultaneous with task 2. The 5 weeks of slack time is between week 7 and week 12. Because Task 3 has slack time, it is not on the critical path. None of the other tasks has slack time. Therefore, the critical path goes through the following tasks –1,2,4,5,6,7, as shown in Fig. 29.14.

It has been pointed out before that the kind of information provided by this model will

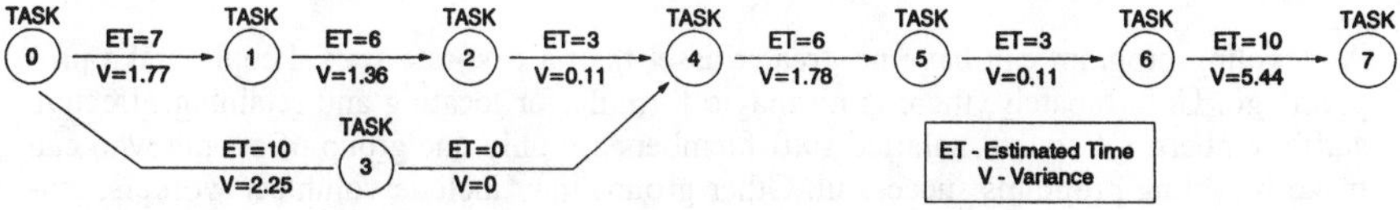

FIGURE 29.13 Notation with estimated time and variance.

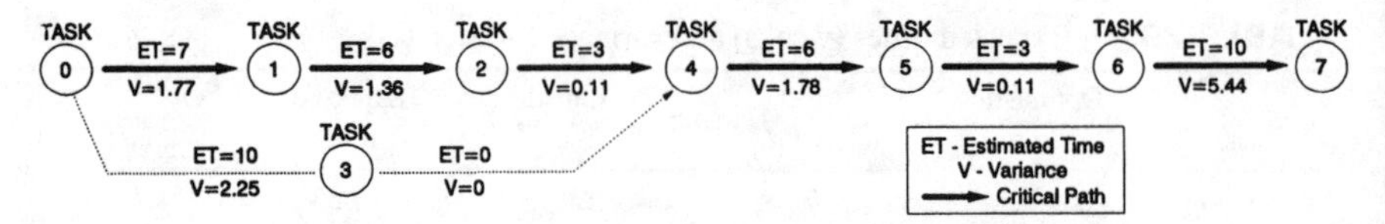

FIGURE 29.14 Notation with estimated time, variance, critical path.

be useful for programs with many components and tasks, with limited resources, and significant investments in staff and other resources. For small programs, it is probably not worth the time and effort to develop this type of network and a simple Gantt chart may be the only scheduling tool necessary.

The PERT model can also determine the probability of completing the project on a given date. For example, let us assume that the City Council has requested that the pilot program for the drop-off centers be completed 38 weeks from today. The PERT model will show the probability of meeting this deadline using the formula:

$$Z = \frac{\text{due date} - \text{earliest ET for last activity}}{\sqrt{\text{sum of variances on the critical path}}}$$

For the drop-off center activities, the values would be

$$Z = \frac{38 \text{ weeks} - 42 \text{ weeks}}{\sqrt{9.61}} = \frac{-4}{3.1} = -1.29$$

Because the PERT model uses a beta distribution, the Z value of 1.29 must be looked up on a table of areas of cumulative standard normal distribution. A value of –1.29 shows a value of 9%. In other words, there is a 9% chance of completing the pilot program within 38 weeks.

The PERT and CPM models can also be used to develop the minimum cost schedules using inputs about the expected cost of activities, given certain time constraints. Often these models require an iterative process which can be accomplished most efficiently with a computer. There is software available for constructing these models.

Once a schedule has been developed, whether it is a Gantt chart for a modest program or a sophisticated computer model capable of spinning out probable completion times, work can begin. The plan should provide direction, set priorities, and indicate activity times. The plan should be revisited regularly to confirm its relevance in the face of changing conditions. It should be a critical tool in building an effective recycling program.

STAFFING

A recycling program can have no greater asset than a first-rate staff. People make programs go. Unfortunately, there is no magic formula for locating and retaining effective staff members. Of course, salaried staff members are only one group of people who can make recycling programs successful. Other groups may include volunteer workers, employees of private sector firms, and consultants.

Recycling is a multidisciplinary activity. A survey of recycling experts was conducted by Rob Grogan, a doctoral student at the Harvard School of Education, to ascertain the skills and knowledge which a recycling coordinator might require. The most important skills identified were, first, managing information and recycling records and, second, understanding the economics of recycling. Other skill areas, in descending order of importance, were publicity and promotion, budgeting, secondary material marketing, local government, vehicle routing and specifications, disposal options: costs and technologies. This is an imposing list.

The first two subsections focus on professional staff members—where to find them and how to recruit and retain them. The third subsection considers the most effective way to supplement the efforts of these recycling professionals.

Locating Potential Staff Members

Large-scale municipal recycling programs have become more prevalent over the last two decades. Due to this, numerous individuals have had the opportunity to work in various areas of recycling, ranging from operating facilities to designing outreach programs.

Unlike a decade ago when programs hired individuals with specialized skills to operate just one component of the program, today's recycling professional is required to understand and execute a multitude of responsibilities. Thus, today's material recovery facility operator may need to be able to write grant applications, generate support for a program among elected officials, and conduct classroom presentations, as well as operate the facility and negotiate contracts with brokers.

It is also essential that today's recycling professional understand the economics and politics of making a recycling program sustainable, as well as its environmental benefits.

The Solid Waste Association of North America offers a Recycling Manger's Certification program and a few colleges and universities have developed, or are developing, specific degree programs in recycling. Many others offer recycling as a part of waste management curriculum which may be part Engineering Department or the Environmental Studies or Natural Sciences Department. The University of Maine, Michigan State University, and the University of Wisconsin have all developed programs in waste management that include recycling. Many graduate schools in public policy use recycling case studies to examine waste management issues. Also, many of the best recruits from colleges are those students who helped to develop or run the college's own recycling program. One clue as to where to find well-trained recyclers is to look for those schools with the best recycling programs.

Often the needs of the program will point toward specialized skills. That is, a program seeking help in secondary markets might look to paper companies or glass manufacturers for candidates. A person who has conducted promotional campaigns for the local public interest research group or environmental organization might be ideally suited to designing a recycling promotional campaign. Other kinds of specialists such as chemists (for insights into plastics recycling), mechanics (for collection vehicle evaluation), or bankers (for help in financing) may also prove valuable. Providing training in recycling to a skilled professional is often the best strategy.

Finally, because recycling is still in a dynamic industry, many highly motivated individuals have been drawn to it. Placing an ad in one of the trade journals can result in responses from people with a broad range of backgrounds. The next step is to convince them to join your staff.

Recruiting and Retaining Staff Members

Often, the private sector has an advantage over the public sector in recruiting and retaining skilled workers. Private firms can usually offer higher salaries, better benefits, and superior working conditions. In fact, some firms look to the public sector for recruits because of the valuable experience a municipal recycling program can provide. Facing this kind of competition, what can the public sector offer?

First, it can appeal to a recruit's sense of social altruism. Public service still attracts many of the country's most able workers, and this is particularly true of those in recycling. There seems to be a strong sense of idealism among those entering the recycling field. Helping to solve a community's waste management problems offers powerful outlet for this idealism.

Second, the job is invariably challenging. The problems are enormous; the resources are limited; and the solutions require a combination of enlightened public policy, appropriate technology, and a balancing of market forces. The attraction of a challenging position that improves community life can be very powerful to a hard-working idealist.

Retaining effective staff members is probably a greater challenge than getting them in the first place. Here, the conditions which keep staff members are no different than those that keep any other kind of worker. Staff members need recognition and positive reinforcement. They should be rewarded for good work with greater responsibility and, hopefully, better pay and benefits. Although a positive working environment—open, collaborative, and responsive—is critical to keeping a talented staff, there are limits to its effectiveness in the face of shrinking real wages and benefits.

Supplementing the Staff

Even the most talented staff may not provide all the expertise and help necessary to run an effective recycling program. There are times when outside help is necessary and it is important to recognize these times and respond appropriately. There are, at lest, two groups of "outsiders" that may effectively supplement the work of the professional staff—volunteers and consultants.

Many recycling staffs use volunteers to provide site-specific assistance in particular programs. For example, curbside collection of recyclables is often supported by volunteer block leader programs. In this type of program, a local resident takes responsibility for encouraging neighbors to recycle. This responsibility can include providing promotional material produced by the city, monitoring the success of recycling in the neighborhood, and meeting with new families to inform them about the program. A similar kind of program is often used in multifamily dwellings. An apartment resident will take charge of the recycling area of the building and encourage new tenants to participate in recycling.

Although volunteer help is not paid, neither is it "free." To be effective, volunteers need to be recruited, kept up-to-date, and thanked (often) for their efforts. This all will take time from the professional staff. Volunteers who are recruited and abandoned will soon disappear. On the other hand, the local assistance that these volunteers provide is almost impossible to duplicate, even if additional staff members could be paid. Volunteers are most valuable in making local contact and providing on-site assistance. They are less valuable as temporary workers in the office, carrying out functions that a paid staff member could just as easily be doing.

Many recycling programs, particularly in small towns or neighborhoods, have begun as volunteer programs. The all-volunteer staff finds a site, arranges for the collection of recyclables, publicizes the program, and markets the material. Countless recycling programs have been started this way. Community leaders too often view these programs as a "free" recycling program which can go on forever. Unfortunately, these volunteer programs seldom become permanent. After a while, sometimes a long while, the volunteers burn out and the recycling program wilts away. Communities with all-volunteer recycling programs should decide if the program is worthwhile. If it is, it deserves public support in the form of professional leadership. Volunteers can still help, but recycling should become a city service with a budget and paid staff. If the program does not warrant this kind of support, the community should not be astonished when it dies a quiet death.

Nonprofit organizations are another kind of volunteer help that can often supplement the work of a professional staff. Environmental groups, such as the Sierra Club and the Audubon Society, often publish helpful technical reports or can be consulted on topics of interest to the program. Legal issues and new solid waste regulations are usually reviewed by organizations such as the Environmental Defense Fund and the Environmental Law Foundation. A familiarity with the work of local and national nonprofit groups can often be useful to a municipal recycling program.

Professional consultants can also supplement the work of a professional staff, if they are used judiciously and well. Consultants can provide a range of services, from very specialized expertise to broad program support. A good consultant should offer the benefits of experience with recycling programs in other areas, specific technical skills, and a high degree of objectivity: Their work usually covers a specified, and limited, period of time. Unfortunately, they are usually quite expensive. However, for a limited project, it may cost less to hire a consultant than to hire a permanent staff member whose salary and benefits must continue after the project is completed.

Consultants are best used when the following three conditions are met. First, the work which the consultant will do is clearly defined and limited. It makes no sense to hire a consultant until the scope of work is clear. Second, the work of the consultant will not duplicate or usurp the work of a staff member. This is not only wasteful of program resources but it can be very dispiriting to the staff member who may feel "replaced" by the consultant. Third, there are adequate resources to monitor the work of the consultant during the project. Turning a consultant loose on a project without adequate monitoring is unfair to the program and the consultant. The costs of hiring a consultant include not only the consultant's fee but the staff time required to work with the consultant during the project. Working together, the staff member and the consultant should develop a product that will serve the program.

FINANCING A RECYCLING PROGRAM

The final section of this chapter addresses the financing of recycling programs. This is a complex and highly technical topic, which is introduced here to provide an overview of financing options. No attempt is made to describe all of the many approaches that specific communities have employed. Rather, this is a menu of potential sources of funds that can be investigated as the program moves toward implementation. All of these sources may not be available to all programs, but at least some of them will be and will deserve a closer look.

In developing a plan to pay for recycling, it is instructive to look at the way in which other large solid waste projects, such as landfills and waste-to-energy facilities, have been financed. Recently there has been a call, particularly by the Environmental Defense Fund, to "level the playing field" for waste management projects. If the recycling program will be responsible for diverting solid waste from disposal, it should be given equal footing with other technologies.

State Grants

In order to encourage recycling, particularly in the light of rapidly diminishing landfill capacity, many states have created grant programs which offer direct funding for the planning and implementation of recycling programs. Northeastern states such as Pennsylvania, New York, Vermont, and Rhode Island have spent millions of dollars to develop recycling programs. State grants may support the writing of feasibility studies for recycling, the implementation of pilot recycling programs, research of secondary materials markets, the purchase of collection vehicles, and many other kinds of programs.

Both Rhode Island and Massachusetts have used state funds to construct material recovery facilities (MRF) and purchase collection vehicles. The MRFs in Johnstown, R.I., and Springfield, Mass., are now being used to process recyclables collected from communities in those states. Unfortunately, as state budgets have begun to show significant deficits, these planning grants and state-supported programs are disappearing. This source of funds will depend, among other things, on the fiscal health of the state.

Revenue from Material Sales

Recycling has the potential to generate revenue through the sale of recycled materials. This potential is limited by the type, volume, and quality of the materials collected, as well as the health of the secondary materials markets. There are two kinds of market risk that have plagued programs recently.

First, price levels for materials can be extremely volatile during a relatively short time period. Due to this, it is almost impossible to develop budgets when revenues are substantially dependent on material sales.

The second risk in relying on revenue from the sale of materials has been the risk of having loads rejected. As more material has entered the market, the level of contamination in some loads has risen, according to the end markets. This, in turn, has led to a tightening of quality standards demanded by the markets. For example, in the past year, some recycling programs have had loads of glass and aluminum rejected because the loads contained too many contaminants. Metal caps and neck rings in the glass, and lead, sand, and water in aluminum cans, led end users to refuse to accept loads. Not only do rejected loads mean a loss of revenue, but they also mean that the cost of collecting and processing the materials has been wasted. Furthermore, the rejected loads must either be landfilled, or reprocessed, incurring additional expenses.

Given these market risks, most programs that look to material sales as a source of funds use very conservative estimates.

Taxes

Many programs seek at least some of their funding through municipal taxes. This is particularly true where recycling is regarded as part of the solid waste management program. That is, the collection and disposal of waste, including recycling, is supported through taxes.

While this would appear to be a ready source of funds, there is fierce competition for tax dollars from other municipal services, such as fire, police, and education. In addition, there are limits to the amount a community can tax its residents and businesses. Raising taxes is generally accompanied by intense public scrutiny, and often by public debate, outcry, and anger. As a result, recycling programs are generally looking for other sources of funds to replace, or at least, supplement, tax revenues.

Tipping Fee Surcharges

Another source of funds for a recycling program is a surcharge. These are the fees charged at landfills and incinerators for disposing or processing of waste, in addition to the regular tipping fee. In many cases, a supplemental amount, as low as $1 or $2, is added to the tipping fee to pay for recycling programs. This has the advantage of assessing those who are generating the waste. One interesting twist to this source of funds is that, as the recycling program becomes more successful, less waste is disposed and the revenue from the tipping fees decreases. When this happens, the answer is to increase the supplement.

User Fees

Many communities have developed municipal services that are supported by the citizens that use the service, rather than by the entire community through the tax base. The "enterprise fund" does not rely on taxes. They have their own set of fees that generate revenues and are budgeted separately.

There is a large and growing variety of user fees for solid waste. The simplest of these is a flat rate for all residents receiving garbage collection service. Another popular option is the "variable can rate" which provide different levels of service for different rates. The city of Renton, Washington, for example, offers two levels of weekly collection, a 30-gal or 15-gal container for a monthly fee. These fees include the cost of collection and disposal of garbage, and the collection, processing, and marketing of recyclables. The variable can rates also provide an economic incentive for recycling and source reduction.

Bonds

Bonds are another source of public sector funds for recycling. Bonds are a common way of raising money for capital-intensive projects such as incinerators and MRFs. Most cities have the ability to incur debt through the issuance of bonds. There are two types of bonds generally used.

General obligation ("GO") bonds are backed by the financial strength of the city or state. However, there are limits on the amount of debt a city or state can incur. If this limit has been reached through the use of GO bonds for other kinds of projects, this option will not be available. This is one limitation in issuing GO bonds. Repayment schedules for GO bonds often require large amounts of money in the first years of operation, which may increase the cost of recycling in the early years when the program is just getting started. Conversely, earlier amortization of the debt will result in lower total interest cost, reducing the cost of recycling during the repayment period of the bonds. It is generally better to have the repayment schedule and the facility revenues closely matched in order to stabilize recycling fees.

The second type of bond is a revenue bond. Unlike GO bonds, which are backed by the financial resources of the city, revenue bonds are backed by revenue generated by the project. The repayment schedule and the revenues can be closely matched. In addition, revenue bonds have the advantage of not affecting the city's debt limit.

A solid waste recycling facility typically has two primary revenue sources: incoming and outgoing recyclables. Revenue from incoming recyclables is generatred through tip fees and revenue from outgoing recyclables is generated from their sale to end users. To receive a revenue bond, recycling facilities depend on flow control to guarantee a certain quantity of incoming recyclables.

However, with the 1994 Supreme Court decision in Carbone vs. Clarkstown, a community's ability to enact flow control was substantially reduced and is now subject to numerous legal challenges. Without flow control, most facilities do not have a guaranteed flow of waste and, consequently, a guaranteed revenue stream. Because of this, bondholders are less likely to issue revenue bonds.

Even if flow control can be achieved, revenues are also dependent on the sale of recyclable materials. The market risk involved in the sale of recyclable materials has already been discussed. Thus, even with flow control, the bondholder may not accept repayment based on material sales because the market risk is too great.

Charging a fee at the gate of the facility will generate revenue. However, if flow control is not an option, a high gate fee will only result in the recyclable going to another facility, loss of fees and lost revenues.

Bonding is a common source of funds. GO bonds will be attractive as long as the city has not reached its debt limit and does not have competing needs for these bonds. If they do, revenue bonds can be used as long as the revenues from the project are certain and will cover the payments to the bondholders.

Private Financing

In cases where public-sector funds are not available to support an entire recycling program, help from the private sector may be sought. Generally, a private firm will risk its own resources, if it believes it has a reasonable opportunity to make a profit.

The collection of recyclables, for example, is often carried out by a private firm. The collection services may be offered through competitive bid by the city that will guarantee the hauler of all the customers in the service area. The City of Seattle has two contract haulers, Waste Management, Inc., and Rabanco, which collect and process recyclables for city residents. If an equitable contract can be worked out, this arrangement has advantages for both the city and the hauler. The hauler is assured of a large group of customers without any competition, and the city receives collection services at a known cost.

A private firm may also build and operate an MRF. Like revenue bondholders, a private firm that puts its capital at risk will want some kind of assurance that all recyclables in the area will be delivered to its facility. A flow-control ordinance or a belief that there will be little or no competition from other facilities can provide this assurance.

A city that relies on a privately owned and operated facility must also realize that it is exposed to risk of abandonment. If the owner of the facility decides to close the doors—because the facility is not making a profit or because a better opportunity has opened up somewhere else—the city will have no recourse but to develop another facility or curtail its recycling program. For this reason, many MRFs are publicly owned, but privately operated.

There are many other financing arrangements such as leasebacks and combinations of the options discussed above. These options offer great flexibility, but they also require careful study. The financing of a recycling program should reflect the local economic and political conditions, as well as maturity of the recycling program and availability of strong secondary material markets, and the number of households and businesses to be served.

CHAPTER 30

PUBLIC AWARENESS PROGRAMS

JOSEPH C. BARBAGALLO, P.E.
Associate
Malcolm Pirnie, Inc.
White Plains, New York

INTRODUCTION

Public Policy and Recycling

Municipalities managing recycling programs face more challenges moving into the next millennium than they did 10 years ago when the programs were being conceived and implemented. Today, municipalities have mandated recycling programs and have implemented public awareness programs that have been quite successful at making the public knowledgable about recycling and its benefits. The challenge to the municipal recycling manager is to keep recycling awareness programs fresh while limiting expenditures to the budgets allocated for these purposes. Indeed, failure to consider continued public education as a technical component, similar to equipment maintenance, and seeking new markets for materials may doom the most organized municipal recycling program.

Of course, the need to "sell" recycling to the public is not an issue. The public's embrace of the recycling ethic is fairly widespread, due in a large measure to the "greening" of our culture. As a matter of fact, according to the USEPA (www.epa.gov/epaoswer/non-hw/muncpl/reduce.htm), "recycling is one of the best environmental success stories of the later 20th century. Recycling, including composting, diverted 57 million tons of material away from landfills and incinerators in 1996, up from 34 million tons in 1990—a 67% increase in just 6 years. By 1996, more than 7,000 curbside collection programs served roughly half of the American population. Curbside programs, along with drop-off and buy-back centers, resulted in the diversion of 27% of the nation's solid waste." Although this diversion rate is an impressive accomplishment, it is far short of the 50% recycling mandate passed by many state legislatures.

Due to the near shortfalls in actual diversion when compared to state mandates, municipal managers of recycling programs are taking a hard second look at the solid waste system to identify other means and methods that can be implemented to improve recycling within their districts. Although it is not the answer to all of the challenges facing the recycling manager, an ongoing public awareness program is still an important part of maintaining and advancing recycling levels.

Public education programs also vary in style and content as much as the recycling projects they promote. Although many similar elements will make up a particular program, the degree to which they are used is often predicated by local conditions. In this sense, no two educational campaigns are alike nor are there necessarily any standard guidelines to be followed during implementation.

This is not to suggest that the forms of communication used to promote participation are simply a matter of random selection. (It should be noted here, the terms *public education, public relations,* and *communications* are interchanged frequently but will generally refer to the promotional aspects of a recycling program.) The packaging of an education program requires careful consideration of each communication process based on frequency, cost, retention, and audience response. Choosing the proper mix of each and acquiring the approval of decision makers is the major challenge facing most practitioners.

Where then to begin? Subsequent sections will examine many of the methods and types of communication used to educate the public on recycling. However, it is worthwhile to develop an understanding of the political, economic, and environmental realities that influence the effectiveness of a recycling program.

The American Experience

There is no doubt that recycling has been a part of American culture for some time. Scrap operations have been around for years, reaching their peaks during the world war periods. Educational efforts of that time called for the support of "our boys overseas" by recycling our scrap into the mechanisms of war. In the late 80's, images of a garbage barge floating along the eastern seaboard of the U.S. sparked renewed enthusiasm for recycling programs. Although these "crises" do not exist today, the call to recycling is no less important as we move into the 21st century. Indeed, the public's collective consciousness is well-versed on the need for and reasons behind recycling as one element of an expanding global environmental perspective.

For this reason, the battle for heightened public awareness is already won. The war still rages, however, as well-intentioned public policy can fail to consider its own impact on local communities. As an example, legislation mandating recycling percentages may fail to consider market-driven forces that can impede success and put undo strain on the entire solid waste system. This, coupled with the strain new laws place on limited municipal resources and the growing presence of the vertically integrated, profit focused private sector waste management companies, puts many recycling programs (which must rely on public support) at a disadvantage.

Communicating

This would suggest that the difficulties with recycling (available markets, ownership of recyclables, equipment procurement, union requirements, etc.) would need to be worked out in advance of local program implementation and any education component. On the other hand, experience shows that many of these details can remain unresolved yet startup may be less than a few weeks away. From a public education posture, this can be disastrous as communicating the specifics of a recycling program may be rushed or haphazardly accomplished.

Flexibility, therefore, is a common feature of the most successful education programs. This is particularly critical to recycling pilot efforts, where full-scale or multimaterial programs may take several years to implement. Clearly, informing the public in an ever-evolving program creates significant communication challenges. Challenges which, if not properly responded to, can severely impact participation rates.

Understanding the communication process, therefore, becomes central to the success of public education programming. The formulation of messages, their delivery using various media, and the follow-up necessary to ensure consistency, requires communications savvy, an element often lost among the more administrative concerns of recycling implementation. Knowing the local community, particularly across socioeconomic lines, is also invaluable to program planning and fundamental to establishing dialogue between local government and the constituencies they will rely on for recycling's success.

The various phases and components used to implement successful public education programs will be examined. Approaches to program management as well as "shopping lists" of activities will be explored. Key to any effort, however, is credibility, both in the messages communicated and, more importantly, in the commitment to the education process. Inasmuch as the public has demanded recycling, municipal officials, business leaders, and the public must be supportive of the recycling education function and recognize its primary role to recycling's success.

STARTING A PUBLIC AWARENESS PROGRAM

"The first steps are often the most difficult." This is certainly a recycling truism. Indeed, gaining cooperation of a varied constituency and building the momentum necessary to move the program through its many cycles can be as important (and risky) as a child's first walk. As with any event, however, planning is fundamental to success and the first step to getting started.

Research

Before any communication can be effective, some level of understanding must be gained regarding the audience to be communicated with, particularly if the communication is attempting to change behavior. This not only requires awareness of various demographic influences, but will rely on the proper interpretation of local attitudes. Research can accomplish this goal, and can be used in many forms. Furthermore, it provides an opportunity to evaluate and project community trends, and ensures that promotional programming remains current and appealing. Research will also determine which resources work best for a particular program, and what funding and media outlets are available. The goal here is to tailor a communications effort to reach a variety of subgroups, with messages that will persuade *most* citizens in the community to recycle.

Audience Identification

There are several means toward gathering audience information, but none are more beneficial than direct interviews with potentially affected constituents. This will require sufficient staff capability, but at a minimum, may involve a recycling coordinator "making the rounds" in an effort to create reliable audience profiles. Door-to-door or telephone surveys, attendance at community meetings, and participation with school and civic events are several ways in which to gather useful information.

In a more formalized method, focus groups can be conducted where community representatives are brought together in a round-table forum to discuss the pros and cons of varying promotional ideas. Local universities may even provide the resources for conducting this type of survey as part of a business class project (e.g., speak with the head of the business department). Additionally, "behind-the-scenes" demographic investigations (reviews of tax roles, school populations, business, etc.) can further refine audience profiles and may uncover additional resources for program implementation (e.g., the city clerk's office can provide direction for acquiring this type of information).

In whatever form, initial research efforts should attempt to answer the following:

- What socioeconomic groups make up the community?
- How aware are citizens of recycling? (What experience have they had?)
- What is the educational level of most constituents?
- Where do the citizens get their information? Radio and TV news programs? Newspapers? Talk shows? Posters at local stores?

As program planning advances, it may be worthwhile to pretest logos or promotional graphics, determining what will serve as the most attractive vehicles for a message. Such pretesting will gather feedback on communication materials before they are printed, recorded, or produced, thereby saving costs. In some programs, public contests are conducted to acquire logos and themes. This public involvement provides excellent publicity to the impending recycling program and is successful at starting up the public education process. Questionnaires can also provide a vehicle for the public to comment and feel as if they are contributing to the program. The level of response to any of these activities also acts as a bellwether for anticipating public participation concerning the recycling issue in their community. Finally, failure to conduct research can be very costly in terms of time and money. Materials that have no audience appeal, are confusing, or at worst, offend, can severely impact recycling participation over the long term.

Identifying Resources

The number of recycling programs that can devote significant resources to the public education function are among the minority. Most recycling programs have spent available capital funds on the more operational requirements of their program (curbside bins, vehicles, salaries). Conversely, more care will need to be given to the education effort, so as to maximize the effectiveness of any resources that are available.

As noted above, all opportunities to involve the community in the planning process as early as possible should be taken advantage of. Citizens groups and local organizations can provide valuable input and assistance in developing a public education plan, as well as the labor for implementation, often on a volunteer basis. As an example, Boy Scouts can assist in the distribution of recycling bins, generating publicity for their local troop and the recycling program as well.

Community Groups and Neighborhood Councils. As a resource, a recycling awareness message can often "piggyback" other communication efforts. For example, the City of San Diego's recycling program spread its message with the help of the *I Love A Clean San Diego* special interest group. Seattle, Washington's, program was assisted by the *Friends of Recycling*. In many instances, "block leader" programs can be initiated where a local resident acts as a resource for a neighborhood's recycling implementation. These individuals can be recruited to communicate face-to-face with resistant residents, or to make initial contact by hand-delivering materials (door to door).

Media. The use of news publications and other forms of the media are a common and economical publicity resource. Media coverage in the form of feature articles and advertisements run in local newspapers, public service announcements issued to local radio and television, as well as community access programming on cable television stations are low-cost ways to reach thousands of community members. Speaking directly with editors to determine how best to use these resources is beneficial in the short term but also goes a long way toward establishing press contacts for longer-term publicity. Steps for properly using the media are outlined in subsequent sections.

The Public School System. Information and recycling materials distributed during school presentations and assemblies do reach children's homes. Teachers and administrators usually welcome the opportunity to host an environmental speaker as well. Many school systems are now developing solid waste curricula in response to the growing public concern (see Fig. 30.1). It is also important to remember that in many households, it is the children who are responsible for taking the trash out. They'll also be given the job to recycle.

Often, solid waste organizations can act as a resource for the public schools to develop a waste management and recycling curriculum. The Rhode Island Solid Waste Management Corporation, for instance, provides educational materials to the public school system for use in recycling experiments and demonstrations. The agency also works closely with a number of schools and universities to promote new environmental policies and programs. The state legislature of Minnesota has gone even further and established the Minnesota Waste Education Program in 1987, in order to develop a unified approach to waste education in schools and communities throughout the state. Among the program's objectives are the development of school programs and curricula on waste topics for grades K to 12, and to advocate and help implement their usage in the public schools.

Utilities. Utilities can often assist with a recycling program, by allowing public education materials to be mailed along with bills and notices as extra inserts. This idea has similar applications for large "in-town" businesses that can distribute information to employees via paychecks, bulletins, and cafeteria displays. Municipal mailings can also incorporate logos and phrases on mailing envelopes to increase awareness (see Fig. 30.2). The use of recycled paper is also helpful.

State and Federal Agencies. Many state departments of environmental protection have created handbooks, promotional materials, and programs for use by local recycling programs. These materials usually have a minimal cost and agency personnel can provide guidance on their use. The U.S. Environmental Protection Agency's Office of Solid Waste in Washington, D.C., can provide resource materials as well (Fig. 30.3).

Environmental and Trade Organizations. There is no shortage in the number of organizations available to provide information on recycling issues from a local, national, and international perspective. The Solid Waste Association of North America (SWANA); American Paper Institute in New York; the National Resource Recovery Association; U.S. Conference of Mayors in Washington, D.C.; Inform, Inc., New York; and the World Watch Institute in Washington, D.C., are but a handful of groups that can provide literature, database information, legal assistance, and public education expertise for a recycling program. Many organizations like the Audubon Society or Nature Conservancy will operate on a local chapter level, providing resources and sometimes staff for even the smallest of programs. As non- or not-for-profit organizations, their assistance can be limited and

Check all the non-paper items thrown away in your classroom and add your own:

1.	**Rubber Bands** ______	11.	______
2.	**Paper Clips** ______	12.	______
3.	**Aluminum Foil** ______	13.	______
4.	**Plastic** ______	14.	______
5.	**Styrofoam** ______	15.	______
6.	**Soda Cans** ______	16.	______
7.	**Glass Jars or Bottles** ______	17.	______
8.	______	18.	______
9.	______	19.	______
10.	______	20.	______

Which items could have been reused at least once?

1.	______	6.	______
2.	______	7.	______
3.	______	8.	______
4.	______	9.	______
5.	______	10.	______

How about several times?

1.	______	6.	______
2.	______	7.	______
3.	______	8.	______
4.	______	9.	______
5.	______	10.	______

FIGURE 30.1 (*a*) Recycling classroom assignments.

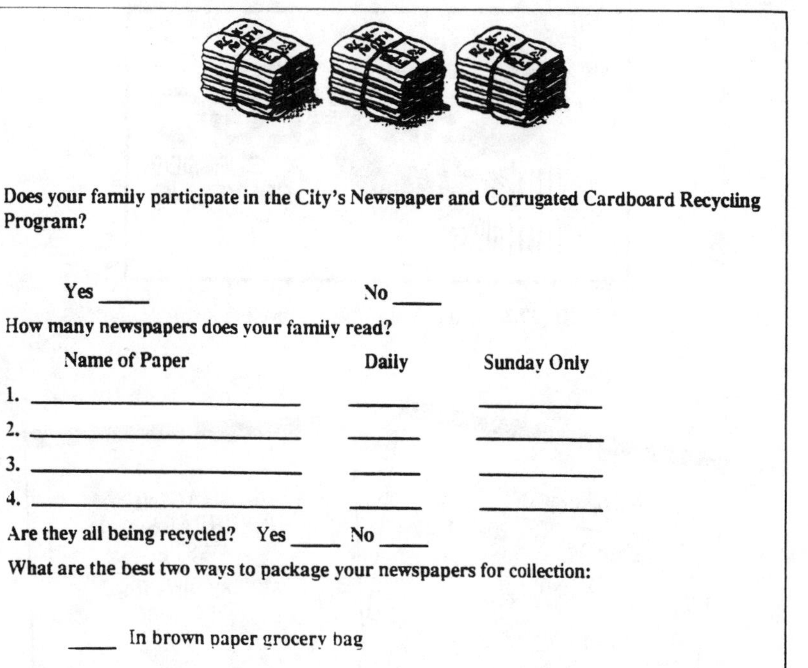

Does your family participate in the City's Newspaper and Corrugated Cardboard Recycling Program?

Yes _____ No _____

How many newspapers does your family read?

	Name of Paper	**Daily**	**Sunday Only**
1.	________________	_____	_______
2.	________________	_____	_______
3.	________________	_____	_______
4.	________________	_____	_______

Are they all being recycled? Yes _____ No _____

What are the best two ways to package your newspapers for collection:

_____ In brown paper grocery bag

_____ In plastic garbage bag

_____ In corrugated box

_____ Bundled with cord

_____ Plastic grocery bag

How many corrugated boxes did your family want to get rid of last month? _____

Have you been flattening your corrugated cardboard boxes and putting them under your newspapers for collection by the City? Yes _____ No _____

FIGURE 30.1 (*Continued*) (*b*) Home assignments. (*Courtesy of Waterbury Regional Resource Recovery Authority.*)

small expenses may be incurred for materials. On the whole, however, these organizations can provide valuable counsel for instituting a public education campaign and supporting it through its various stages.

Community Events. Recycling representatives often get excellent visibility at community fairs and other events, especially when distributing free items such as key chains and buttons. Such events provide another opportunity for face-to-face contact and question

FIGURE 30.2 San Diego curbside recycling logo.

FIGURE 30.3 Learning about Municipal Solid Wastes. (*Courtesy of GRCDA, Silver Spring, Md.*)

answering. As an example, the City of San Diego sent members of its Community Outreach Team to cover a dozen fairs and parties, often accompanied by their recycling mascot, Rascal the Cat (Fig. 30.4). Bloomington, Minn., took advantage of a 1989 city hall rally to kick off curbside pickup, followed by a series of community events.

FIGURE 30.4. Rascal, the Trash Cat.

Creative Design. For writing and designing public education materials, local art or advertising classes may welcome internship programs that provide "real-life" experience. Working with a marketing consultant or firm, perhaps at discounted or pro-bono rates, can also ensure a professional approach to materials creation.

Planning

Planning a public awareness campaign must take into account many factors, including budgetary constraints, the time available until startup, targeted audience, and type of program (i.e., curbside pickup versus dropoff center; single versus multimaterials). Although few recycling efforts will have the time to develop a formalized plan, goals and objectives should be agreed to prior to initiating education efforts so as to minimize confusion during startup. The plan will undoubtedly change as research reveals new information, trouble spots are identified, and the recycling program progresses; however, an initial plan should outline the following.

Audience Identification. As noted above, this is a critical first step that must be planned for appropriately.

Goals. What is to be gained from the education process? To simply state public participation as the campaign goal may be too broad. Certainly, one's aim is to bring the recycling message to individual publics that make up the community. However, very *specific* goals must be set in order to reach the various community segments. Goal setting here refers to determining, planning, and implementing communication strategies and events to penetrate the school system, ethnically diverse areas, low-income neighborhoods, etc. Examples of such goals are "conducting recycling education presentations at all grade schools in the community," "producing and distributing a bi-lingual recycling flyer," and "organizing a series of recycling fairs in low-income apartment complexes."

Public education goals are distinct from collection and capture rate goals, though the two will reflect one another. Several individual education goals successfully met will inevitably increase capture rates.

Available Resources. This includes identifying available money, staff, publicity vehicles, and media outlets. It is important to anticipate how each resource will relate to particular promotional components so as to maximize those resources.

Schedule. Coordinating the public education campaign with the program implementation deadline is critical to campaign success. Heavy publicity too far in ad- vance can become stale, losing audience interest. Too late, and word may not have enough time to get out. Although there are no fast rules for scheduling, several guidelines are offered below. It is also important to take into account local and seasonal community events, which can impact program startup. For example, campaigns that are initiated in early September may have to compete with many "back-to-school" activities and promotions.

Staffing. For many current programs, an official recycling coordinator is not only assigned to conduct the recycling program but must implement the public education component as well. Sometimes an assistant or two, perhaps help from the solid waste management technical staff, can be relied on. It is rare that any given recycling program, regardless of the municipality's size, will have more than a few individuals whose time is purely devoted to the public education function. Often volunteers will play a key role in the communication process, sometimes actually directing the publicity program.

Program staffing decisions depend almost entirely upon the allocated budget, but the norm is for staff members to wear several hats due to limited funds. This may require the public education function to be located in the public works department, mayor's office, or another appropriate area. Access by interested parties, not to mention a "manned" phone are key concerns. (The City of Hollywood, Florida, has a catchy, easy-to-remember hotline number—96-CYCLE.) As a rule of thumb, the more help available, the better chances are for getting out timely messages. For programs that are well-funded, outside public relations consultants and artists or a full-time public relations director can be hired.

However, since funding dictates that staffing traditionally is limited, interagency *resource sharing* is critical to program development. For example, a recycling coordinator can work' with the Department of Transportation to create signs for recycling depots. Statistical information and/or accounting services can be provided by city finance departments. Creative services can be tapped, too. Lacking funds to hire an outside graphic artist, the City of San Diego's recycling program, for example, used the services of the City's own graphic arts department to prepare brochures, flyers, and promotional item artwork.

In all cases, implementing a recycling public education program requires the commitment and cooperation of the entire municipal structure. A meeting with department heads early in program development can help identify contacts and resources available to you within the municipal system.

Recycling Message and Media Strategies

Message and media strategies are planned according to the specific audiences being targeted, and also within budget limitations. Media selected and the messages formulated often need to account for adult and student audiences, low-income areas, and multilingual groups. Due to the higher costs associated with electronic promotional methods, most educational campaigns will rely on printed promotional pieces. Attractive direct mail brochures and flyers are an efficient means of reaching the entire community. They can be mass-mailed to entire Zip Code regions, and at bulk-mail rates if time is not a concern. These pieces are easily saved and displayed for reference in a household, and their design and production can be tailored to the available budget.

Posters, stickers, and cards can provide "point-of-contact" promotions, particularly where recycling activities can most logically take place. For example, a poster placed in an employee lounge adjacent to vending machines can encourage recycling, particularly if receptacles are made available. Similarly, a magnetized stick-on for the family refrigerator acts as a constant reminder in the home.

Bumper stickers, buttons, and other premium items, whether they promote a local politician or global cause, are guaranteed visibility. There is an up-front cost for these items (ranging from a few cents to a dollar or two a piece), and they can be given away or sold for a minimum amount to offset costs. Audience research is critical, because some communities may not be as likely to use a particular type of promotion.

Newspaper advertisements are an excellent means of reaching a large number of households, as a local newspaper is widely read as an information source. Since newspapers are usually among the first materials to be recycled, providing direction on how to recycle "this" paper is an effective promotion. Often these announcements, including pickup schedules, are free of cost as a public service on a space-available basis. Even paid newspaper advertisements can be fairly reasonable, depending on day of publication and frequency of inserts.

Television promotion is far-reaching, and a feature spot will do much to encourage participation and enhance enthusiasm. Of course, paid television advertising is prohibitively expensive and may be too broad for specific audiences being targeted. Television does offer a news vehicle, however, and press releases announcing the startup, outstanding results, or other milestones of a recycling program should be sent in on the chance they might be picked up for local news interest. Visible events (fairs, seminars, contests, fund-raising events, etc.) are also more likely to receive coverage, particularly as many news broadcasts will have "fill time" built into their broadcasts.

By far, the most successful public awareness strategies are those that incorporate as many points of contact with the community as possible. Recycling awareness is accomplished more thoroughly when mailers, newspaper advertisements, public service announcements, buttons, key chains, customized grocery bags, mugs, tee shirts, and special events are used together to penetrate a community, than when only one or two of these are used (Fig. 30.5).

Scheduling

While programs vary, there seems to be a common timetable for scheduling initial public awareness efforts. In scheduling, it is important to keep in mind that if the program's startup is advertised too far in advance, community members may forget about it altogether by the time the date arrives. The most effective startup campaigns usually stagger announcements and information about the campaign, with promotions becoming more frequent as the startup date approaches. A typical schedule is outlined below.

Publicity for a recycling program ideally begins at least several months in advance, beginning with some general recycling articles in local newspapers. Announcements at regularly scheduled meetings of the agency or authority responsible for recycling implementation are most common. Top municipal officials are often the initial spokespersons, with detailed communications being handled by a program coordinator.

Flyers or brochures, on average, are first distributed to recycling program households one month before the program actually begins—at minimum a few weeks ahead of implementation time, announcing what recyclables will be collected and how residents should comply. Often, these promotional materials may be distributed along with curbside bins for programs using this method. Subsequent newspaper advertisements can be run one week before startup. Promotion of the program will generally taper off as recycling is integrated into community life. Many communities have run weekly ads for their program during the first three months, then reduced to monthly ads, using only occasional spots to

(a)

FIGURE 30.5 (a) Curbside recycling doorknob poster.

advertise special events and new recycling tips once the program has taken off or new phases are added to the program.

Public service announcements, poster displays, and special events should attempt to "saturate" targeted audiences during the startup period, hopefully creating a groundswell of interest that will carry the program several weeks into its development. It is not uncommon for participation to taper off, however, as the novelty of the program wears thin. (This may also be due to the up-front collection of recyclables stored in anticipation of program startup and a subsequent drop off due to fewer available recyclables.) For this

(b)

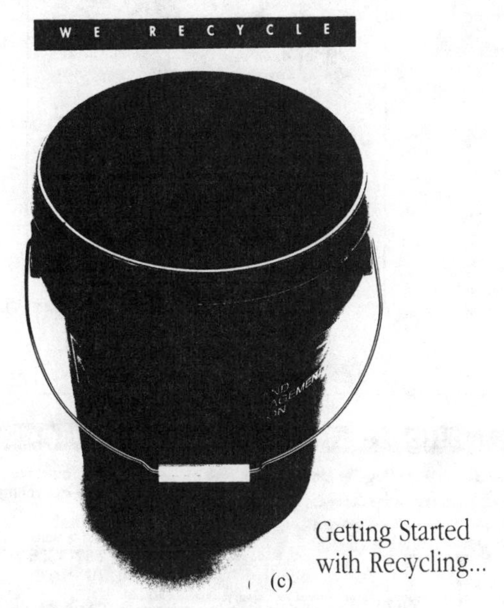

(c)

FIGURE 30.5 (*Continued*) (*b*) Curbside recycling follow-up poster; (*c*) Rhode Island Poster— What Should I Do First?"

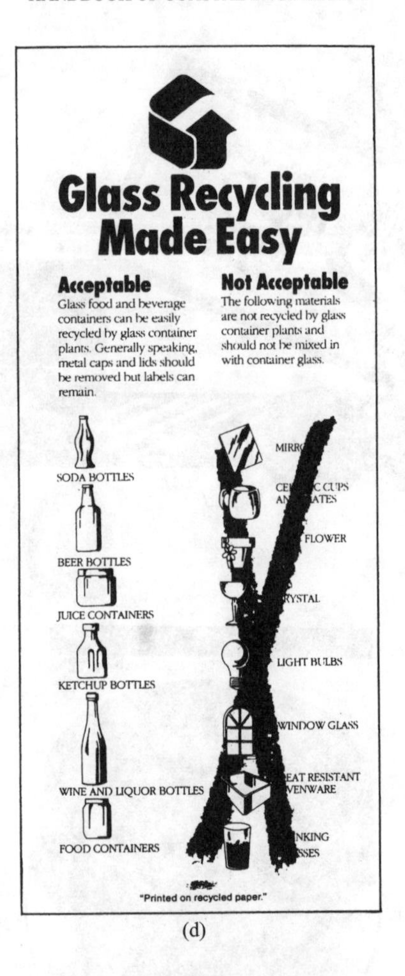

(d)

Bloomington Residents Recycle!

How Much Will I Pay Per Month?

Single Family Homes and Duplexes . . . $1.00

Multiple Family Dwellings, including Apartments, Town Houses, Condos and Mobile Homes . . . . $0.87 per unit

Exemptions:

If you leave for part of the year and shut off your water, you may ask to have the recycling charge discontinued for that period. There is a reconnection charge.

There will not be an exemption for persons who do not choose to use curbside collection service. If you have extraordinary circumstances which you believe exempt you from this charge, you may file a formal appeal within 30 days of the mailing of your bill. Appeal applications may be picked up at the Public Works Building, 9930 Logan Avenue South.

Do You Have Any Recycling Questions?

Call the B.R.R. Hotline
887-9685
TTD: 887-9677

This is printed on recycled paper.

(e)

FIGURE 30.5 (*continued*) (*d*) Glass recycling made easy; (*e*) Bloomington Residents Recycle (B.R.R.).

TABLE 30.1 Planning and Timing Recycling Events

Appropriate time frame	Public education element
1 year to 6 months before startup	Initial plan formulation; research (ongoing)
6 months; ongoing	Goal setting and strategies; phase-in of program
2–3 months before startup	General recycling articles in newspaper
1–2 months before startup	Mailer; backup support of TV and radio public service announcements; general news articles about recycling, condition of landfill, etc.
3–4 weeks before startup	Major newspaper articles and advertisements describing the details of the program
The week of startup	Kickoff event (citywide rally, special day, etc.)
2 weeks to 1 month after startup; ongoing	Evaluation; subsequent reminders as needed

reason, it is critical to maintain the public awareness program on a schedule that fits the allocated budget.

Table 30.1 lists a typical order of events in planning your recycling public awareness program. It is important to remember that flexibility is critical to. the scheduling process. Last-minute program changes, unforeseen difficulties, and even a bout of bad weather can impact the best made plans for any recycling education program. Prudent planners will build options into their educational efforts so that if a particular event fails to materialize, program communications will continue. Of course, this is most important for kickoff events designed to "springboard" a program into the community's mainstream. Having a backup plan not only acts as an insurance policy but may even get implemented, thereby expanding program education efforts.

EFFECTIVE COMMUNICATION STRATEGIES

When designing a recycling public education campaign, practitioners tend to focus initial energies on the creation of tangible promotional items (fact sheets, mailers, etc.). This is usually in response to scheduling pressures that leave little time for planning and preparation. The result is that the public is often besieged with messages on the "whys" of recycling, as these are easier to prepare, rather than the "how to" of a particular program. Often, program specifics are left to a news item in the local paper or a handout placed in a recycling container a week or so in advance. Because of this, key information may not reach an audience and subtle differences in those audiences may not be accounted for, causing low participation rates over the long term.

To avoid this, sufficient audience research should be undertaken as a means toward "packaging" a proper balance of generic and specific communication elements. Although there may be little time for this type of research, this step can be fundamental to program development. Methods for gaining valuable demographic information using limited resources are highlighted in earlier chapters. As a basis for understanding, however, it is necessary to consider the dynamics involved with the communication process and their influence on the messages created for any program.

Persuasion almost always consists of convincing an individual that their participation is beneficial to them. Whether buying a car, selecting an entree, or exercising the right to vote, one's decision to do this or that is largely predicated by some influence. The advertising industry makes its living in this way, creating favorable product images, enticing some kind of consumer reaction, hopefully a positive one. This axiom has a divisive side when taken in the form of a threat, real or perceived, so that participation is coerced. (Often, recycling programs that "mandate" compliance attempt to use this type of strong-arm persuasion.)

Barriers

From any perspective, however, persuasion will require overcoming some barriers. For recycling programs, this usually means recognizing the inconvenience separating materials in the home can cause (and yes, recycling is an inconvenience) and finding an easy, simply communicated means to solve this. Public reaction has generally included complaints that recycling takes too much time in an already time-constrained world; is a dirty process requiring on-going cleaning and storage; or is difficult to comply with, particularly if recyclables must be brought to another location. This perception may be based on personal experience, presuppositions, or collective word of mouth, but is rooted in the nation's "throwaway" disposal habits of years prior, which are not easy to change.

Being Audience-Specific

For these reasons, program messages must be formulated to address the specific perceptions and realities of the target audience, and attempt to minimize any negatives of recycling by appealing to the positive benefits of participation. Socioeconomic factors can often influence these perceptions and be a key to a communication's success. Whereas white-collar, upscale areas may respond to messages promoting the environment and resource savings, urban areas may tend to focus on the economic payback recyclables offer. Benefits therefore should be audience-specific while delivering clearly defined instructions on how the individual is to participate. This can be a tricky process, particularly when limited resources may restrict the frequency of communications developed.

Message Design

Every attempt must be made to avoid complicated, fragmented, or infrequent communications. Inasmuch as the recycling program should be designed to minimize inconvenience, so too, the program's communications must be readily understood and acted upon. Information that is simply stated, upbeat, and repeated often has the greatest chance at success. Step-by-step identification of the who, what, when, and how is imperative. Messages should also be presented through attractive, visually appealing means.

What rules can be applied to the creative development of a recycling communications program? As a first step, program identity should be consistently conveyed and legitimized. Usually this is initiated with the establishment of a program logo and theme. This is important because it will separate the program's messages from the thousands of others that will impact on identified audiences. "Name-brand" recognition, so to speak, is crucial to program participation because once credibility has been established, communications incorporating program logos and themes will have a greater recognition. Similarly, slogans, jingles, or cartoon characters can provide instant recognition so that messages are easily attributed to the program. As examples, Keep America Beautiful, Inc.'s Indian representative remains an instantly recognizable figure for antilitter campaigns, even though this image is decades old. Similarly, Timex's "It Takes a Licking and Keeps on Ticking" has been assimilated into American language as a colloquium for many years now. Closer to the recycling message, San Diego's mascot, Rascal the Cat, and their "We're Not Trash" logo and Rhode Island's Oscar, a seagull-like cartoon mascot of the Ocean State Cleanup and Recycling division of the Department of Environmental Management are assisting each of these diverse recycling programs with their recognition campaigns (see Fig. 30.6).

It is also imperative that messages convey simple ideas. Lengthy descriptions of environmental benefits and "saving the ecosystem" may be interesting, but do little for participation. Participants wish to devote minimal time to recycling. Similarly, the information needed to participate must be easily learned, easily applied, and require little time to assimilate into the daily routine.

This is where proper audience research can be most beneficial. By tailoring messages to identified groups, communications can push the most positive aspects of participation with a given audience. If "saving the environment" is a primary concern among residents in a particular neighborhood, promotional efforts should play to this collective mindset by stressing the benefits recycling will have on the local ecosystem, "Think Globally, Act Locally." Correlations between the amount of newspaper recycled versus the use of virgin trees is an example of this type of promotion. Where economics is an issue, campaigns that stress the savings associated with offsetting landfill disposal costs may be more affective.

Although few programs can afford the slick, first-rate production quality of commercial advertising and promotions, a professional approach to the design of graphics, text, and all promotional materials should be strived for. Inasmuch as tangible communications pieces must deliver information, they also act to represent the organization that has put them out. The goal there is to acquire the highest professional quality so as to improve audience recognition and credibility with the program. Achieving successful results with your printed materials is possible without the use of a professional public relations or graphic design agency, if a few simple guidelines are followed.

Materials should be readable, uncluttered, and simple, with a mixture of text and illustrations that will make the reader want to read the piece. The text should be large enough to read easily, and printed in a standard type style (nothing so fancy or calligraphic that it is pretty but hard to read). Allow for some white space—wide margins, space around pictures, and enough interline spacing on the text will make the piece more attractive and less tedious for the reader (see Fig. 30.7).

The copy should be lively, and be aimed at the average reading level (at approximately the eighth grade level). Try to avoid using technical terms that might be misunderstood or that may "turn off" an audience.

Clean line-art drawings carefully placed on a page can be very effective, so there is no real need for more elaborate drawings and pictures if a budget does not allow them. The lack of real drawing ability can be overcome by using "clip art" packages (inexpensive books of graphics specifically for this purpose), or by tracing and slightly modifying ex-

FIGURE 30.6 (*a*) "Poster—"Hey! We're Not Trash." (*b*) Ocean state cleanup and recycle poster.

Fall Leaves Are Not Garbage This Year!

This Fall the leaves you rake from your yard will not be added to the piles of garbage accumulating in the City's landfill, the North End Disposal Area. Instead, they will be piled at John Coe Park where they will be turned into "black gold", or leaf compost, for use in landscaping and gardening. This is just part of the City's new approach to waste management, and keeps the City ahead of State mandates.

The City encourages residents to compost their leaves on their own property as much as possible. Information on home composting is available from the Recycling Office. However, residents who do not wish to compost their own leaves and want them removed must now put the leaves out for collection in special biodegradable paper bags. The bags will be available to the public in major grocery stores. When full, residents can roll the tops shut on the bags or simply leave them open and standing at the curb. It will be unlawful to dispose of any material other than leaves in these bags, and leaves will not be collected unless they are in the special composting bags.

Collection of the bagged leaves will take place on the last four Wednesdays in November.

However, residents and private haulers can bring leaves (only) directly to the composting site, John Coe Park, starting October 10, 1989. See the back of this page for details and directions to the site, and information regarding City collection.

The composting of Waterbury's leaves is a good idea for many reasons. Most importantly, it saves valuable landfill space for material that cannot be disposed of in other ways. The valuable and useful product which results from this process helps the City save money and resources, and is an excellent example of how a waste material, when looked at in a different way, can be seen as a resource.

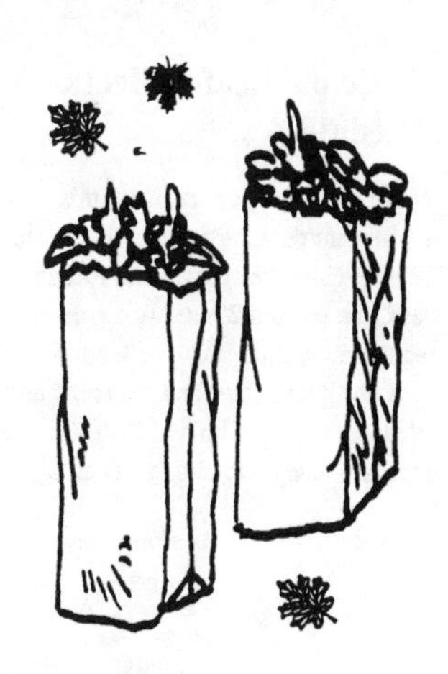

Printed on Recycled Paper

(a)

FIGURE 30.7 (*a*) "Fall Leaves Are Not Garbage This Year."

Let The City Collect Your Leaves, Or......

If you find that your leaves are piling up before the scheduled leaf collection for your neighborhood and want to get rid of them, residents of Waterbury and Wolcott can bring their leaves directly to the composting site at John Coe Park until Saturday, December 16, 1989, loose or in the special bags. Any other containers used to transport the leaves to the site must be taken home after emptying. The site will be open to accept leaves Monday through Saturday, 7:00am to 3:00pm. There will be no charge to bring leaves to the site, however residents should bring proof of residency, such as a driver's license.

John Coe Park is located in northwest Waterbury off of Brookside Road. Please refer to the map at the right for directions.

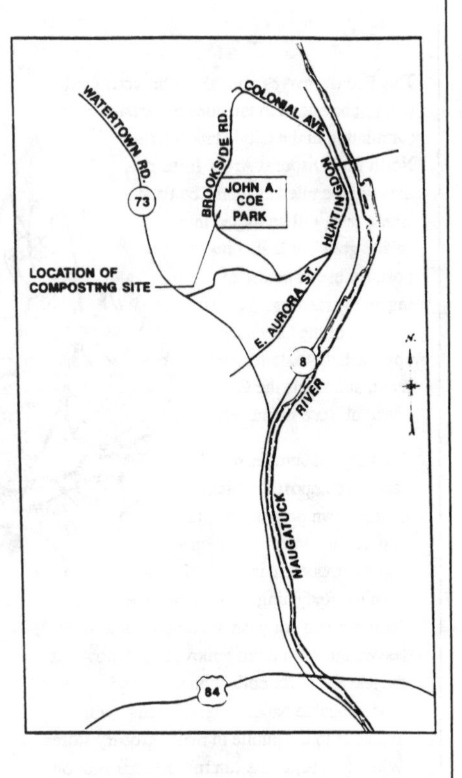

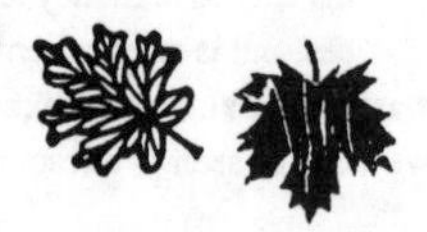

Municipal Leaf Collection Schedule

You'll have two opportunities this November if you want the City to pick up your leaves. These will be the **second** and **fourth** Wednesdays (the 8th and 22nd) for residents with regular municipal waste collection on Mondays and Thursdays, and the **third** and **fifth** Wednesday (the 15th and 29th) for those with regular waste collection on Tuesdays and Fridays.

Information for residents interested in the finished compost will be available from the Refuse Department at 574-6857, or the Recycling Office in the Spring of 1990.

Waterbury Regional Resource Recovery Authority

(b)

FIGURE 30.7 (*Continued*) (*b*) "Let the City Collect Your Leaves Or—."

isting images. Many solid waste and recycling symbols are universal and can be used without restrictions.

The use of color(s) in your pieces is advisable, and will make any material more eye-catching and pleasant to look at. However, the cost of production will rise with each additional color used. In selecting one or two colors to use, coordinate with the city or town's colors, or establish colors that will exclusively identify the recycling program consistently throughout all your printed material. Colors can be selected from a standard color chart obtained from a printer, or incorporate process colors—basic colors that are not custom-mixed.

Make every effort to print materials on recycled materials, indicating this somewhere on the piece with the recycling symbol. It is important that the sponsoring organization practice what it preaches.

RECYCLING PRESS COVERAGE

Media coverage, such as newspaper articles, radio announcements and interviews, and even television spots can be low-cost ways to communicate with hundreds to thousands of community members about a recycling program. By using the media, it is possible to provide the "how to" of participation while at the same time promoting the credibility of the program by way of third-party reference. In order to gain media coverage, it helps to approach the various media proactively, rather than waiting for their call.

Print Media

Local daily and weekly papers are often quite willing to write a feature story regarding a recycling program, just as long as the story has an angle that makes it newsworthy. The kickoff of the program will certainly be news, as might participation rates, outstanding individual or community participants in the program, special collection days, or a program milestone (a goal achieved, the program's anniversary, etc.). Interviews with community recycling leaders or the recycling coordinator also make interesting feature articles.

Press releases sent to the attention of the special features editor, environmental editor, or the general news editor of the newspapers, will help announce events, but a followup phone contact is always helpful to better chances of coverage. Press releases, typed double-spaced on the recycling organization's or municipality's letterhead, should be written in standard press release format: FOR IMMEDIATE RELEASE, or FOR RELEASE ON (date) should be written as the first line. A headline should appear next. The first paragraph begins with a dateline, which includes the date and place of release. Press releases are written in inverted pyramid style, with the most important, newsworthy information at the top, followed by paragraphs of supporting information in order of decreasing importance. A contact person and phone number should appear either at the top or the bottom of the release (see Fig. 30.8).

If a good photo is available, send a black-and-white print along with the release. If the release does not interest the editor, sometimes a captivating photograph and caption will be printed instead. Although a release must compete with many others that often sit piled on a busy editor's desk, one carefully crafted release may be all that's needed to develop a permanent press contact. Of course, when a reporter calls, relevant information should be handy—names, dates, collection figures. If a photograph was not sent with the release, offer to send one or to be available for a photojournalist.

James G. Martin, Governor

James T. Broyhill, Secretary

North Carolina
Department of Economic and Community Development

Release: Immediate Date: October 18, 1989
Contact: Al Ebron (919) 733-2230

STATE-WIDE SERIES OF SOLID WASTE MANAGEMENT SEMINARS SCHEDULED

RALEIGH, NC - The Energy Division, North Carolina Department of Economic and Community Development, is sponsoring the 1989 Seminar on Solid Waste Management in North Carolina, from November 7-16, at six locations throughout the state. Featuring recognized experts in the fields of waste-to-energy, recycling and environmental protection, these full-day seminars are designed to educate public officials, public employees and decision-makers on national, regional and local perspectives of solid waste disposal.

The dates and locations of each seminar include: **Raleigh, NC** (McKimmon Center- NC State University), Tues., November 7; **Washington, NC** (Civic Center), Wed., November 8; **Fayetteville, NC** (Cumberland County Library), Thurs., November 9; **Salisbury, NC** (Holiday Inn), Tues., November 14; **Lenoir, NC** (Holiday Inn), Wed., November 15; **Asheville, NC** (Quality Inn - Biltmore), Thurs., November 16. Further information and a free brochure are available from Mr. Al Ebron at the Energy Division of the NC Department of Economic and Community Development (919-733-2230).

Each seminar will look at solid waste issues from North Carolina's regional perspectives. Issues to be discussed are landfilling, recycling, implementation strategies, economics and ways to integrate methods of reducing and managing waste. Case studies will also be presented.

The State's Regional Councils of Governments and Commissions are contributing professional support to the 1989 seminar series. Speakers include representatives from the North Carolina Recycling Association, Sun Shares, Madison Environmental Alliance, and State and Local Agencies.

FIGURE 30.8 North Carolina Press Release—Seminar.

Press Conferences

Press conferences should be used only when there is *timely, critical, and highly newsworthy* information to convey. Program startups, market identifications, or crisis situations will usually require a formalized meeting between program managers and the press. As an event, press conferences can make or break a recycling program as it attempts to stand up to a naturally critical review by the press.

Like all events, planning is essential. When applicable, a notice sent in advance of the conference is helpful to announce the who, what, when, and where. Although details should be left for the conference, the press will need some indication on what the conference is about prior to devoting a reporter to cover it.

Materials, including a formal press release, should be duplicated for distribution. "Backgrounders," fact sheets that provide information on how the program arrived at this stage in its development, as well as photos, a conference agenda, and names and titles of conference speakers should be available. Much of these materials can be contained in a press kit, a folder in which each of these items can be collected. When budgets allow, these folders can be printed with program logos and designs.

Recognizing that reporters from a variety of media may attend, efforts should be made to accommodate their needs. A conference late in the day may be suitable for TV, but would leave most papers to cover the event in the next day's news. Graphics in a handout may be suitable for reproduction in a newspaper, but TV may respond better to visual displays used during the conference.

For the sake of the program's integrity as well as credibility among the press, it is imperative that conference presenters rehearse statements in preparation. This usually goes without saying. Yet, a harried official, who receives a written statement just prior to walking in front of the lights, is not uncommon. Every effort must be made to prepare spokespersons so as to maintain professionalism and credibility. For those who have never spoken with the press but are required to, it is best to acquire some counsel from a local speaker's group, university, or professional trainer. An early investment in this area can do wonders for a program's communication efforts.

Radio

Radio can be approached in much the same way as print media, but is more selective in audience contact. In addition to on-the-hour news and specials, talk programs may run on a weekly basis providing opportunities for in-depth interviews. Another form of promotion with radio (and TV) is the use of public service announcements (PSAs). PSAs are brief announcements that a radio station will pay or read, free of charge, as a public service. The broadcast of a PSA can reach thousands of people. The only drawback with the PSA is that radio stations tend to play them as "fillers" as time permits, and unlike paid advertising spots, a PSA may not be delivered on a regular schedule or during a prime listening time.

Public Service Announcements

PSAs are usually 15-, 30-. 45-, or 60-second timed spots prerecorded or read live. (Sometimes a local or official or celebrity can be used as the narrator for greater impact.) The shorter the PSA, the more likely it will be played. A 30-second PSA can be fit into many more time slots than a 60-second PSA.

Written PSAs will be read out loud or prerecorded. Language, therefore, should be kept simple, direct, and captivating, with the sentences relatively short, and incorporating action words to avoid creating a boring, stagnant spot (see Fig. 30.9). Repeat important information often (at least twice for phone numbers and pickup days). Most people listen to a radio as they are actively engaged with some other activity. Rarely is a pencil handy, so repetition is necessary. Before submitting a PSA to the radio station, it should be read out loud to detect any trouble spots (tongue twisters, difficult sentences, etc.).

PSAs can be typewritten, double-spaced, with wide margins on letterhead. The timed

"Waterbury Recycles"
Public Service Announcement (PSA)
(15 seconds)

Our garbage is the number one environmental problem facing this City. Unless we find new ways to dispose of our trash, we'll soon have no place to put it—what then?

We know that one part of the solution is recycling, that is, the separation of those materials in our garbage that once again can be put to good use.

Starting ______________, all single-family and some multi-family homes, on city-collected routes only, will be required to separate newspapers from their trash.

Details on how to save your newspapers for special pick-up, at no cost to you, is being sent to you. Please read this material carefully. We need your help so that "Waterbury Recycles!"

FIGURE 30.9 Sample radio public service announcement.

reading length should be indicated on top. A contact name and phone number are also necessary. The radio station will need time to review any PSA, edit, if necessary, and put it through internal production channels. If a PSA is timely, the radio station should be given ample notice to prepare it.

Television

A television features editor can be approached the same as a newspaper features editor—by press release and/or phone call. Television editors work on very tight deadlines and short notice, so in order to have a chance at a television spot, a recycling story must be timely and exceptionally interesting. It also must lend itself to a *visual display*. A story covering a program kickoff will have more visual appeal if, for instance, a colorful mascot is parading around shaking people's hands, or if children are participating in a can-crushing contest.

Opportunities for getting time on television are greater with local, smaller TV stations or cable channels than with larger, national network affiliates, simply because recycling programs are generally locally oriented news and do not lend themselves to standard national broadcasts.

Crisis Communications

When approached by the media regarding a negative aspect of the recycling program (record low participation rates, complaints of missed pickups, program controversy, etc.), it is imperative to maintain openness of communication and a level of cooperation. Never attempt to avoid the media. Better to answer questions directly, while citing past achievements, recognizing current problems, assuring remedial action is under way. Subsequent stories that, at a minimum, provide an official comment will offset criticism that may come from many quarters. Something positive may also be picked up and elaborated on by a third party.

Paid Use of the Media

If budgets allow, paid advertising as a media outlet can be useful. Though relatively expensive, a well-designed, carefully placed advertisement in a newspaper or magazine can provide high visibility. Similarly, a professionally recorded ratio announcement played in the right time slots can be very effective. The frequency of placements will also determine the number of discounts given. Shopping around is always cost-effective, assuming there are several media outlets to choose among. Most TV and radio station and newspaper account managers are also helpful in determining the best use of the advertising dollars and can provide creative assistance with the design of the advertising piece.

Follow-up

Never assume that a reporter's story or even a paid advertisement will get it right. Mistakes can happen, particularly if scheduling is tight. Published articles or news features must be reviewed immediately. If confusion or incorrect information has been propagated by the piece, odds are the phones will soon ring with a confused constituency on the other end. If the wrong date is published in an ad, recyclables could end up on the curb weeks in advance of pickup, immediately throwing a well-planned program into chaos.

When errors are made, the quicker and more comprehensive the response the better. To begin, mistakes should be brought to the attention of the reporter, senior editor, publisher, or owner as the case may warrant. Although a correction would normally be made in follow-up editions or broadcasts, the potential impact of the error may warrant more than a short, easily glanced-over correction notice. Work with editors to approach the issue at the same level of news importance as the original story—not as simply a clerical mistake requiring a quick fix. Also, be cooperative rather than combative, as this will yield beneficial results, particularly if suggestions can be discussed to remedy the situation.

In most cases, however, the information reported will be correct (although not always at the level of importance desired) and useful to the receiving audience. Timing other promotional items (delivery or newsletters, bins, etc.) with an ad or news report helps to create the repetition suggested earlier. Copies of printed articles or ads can also have a second life as reprints for media kits, or handouts to small groups. Be sure to thank editors and reporters as well. They will be needed again as the recycling program develops.

COMMUNITY INVOLVEMENT PROGRAMS

Regardless of the funds available for creating promotional materials, all recycling programs can take advantage of community outreach as the most effective means for generating participation among a desired constituency.

Outreach

Outreach methods used in the public information segments of a recycling campaign most often use group interaction for disseminating specialized information and answering public concerns. Usually aimed at specific target groups, outreach operations, while labor-intensive, are especially useful in harnessing the organized participation of vested interest groups. As a result, these programs require individuals who act in key educational leadership roles. These roles may take the form of speaking to schools and business and citizen groups; directing neighborhood councils that encourage local support of the recycling program; acting as liaisons with teachers and student groups; operating information booths at regional events; leading commercial and residential workshops; or directing or implementing demographic surveys to judge the effectiveness of the program's proposed or actual informational and educational methods. To some extent, specialized recycling guides and monetary incentives such as variable trash rates may also be viewed as part of an outreach effort.

Staff Education

Although each of these outreach programs have singular advantages and disadvantages, individuals must be knowledgeable about the recycling program to be able to communicate effectively on a group or one-on-one basis. Program workers, particularly volunteers, should undergo some type of training regarding the specifics of the recycling program, the concerns of the groups with whom they will be interacting, as well as the municipality where the program will be implemented. To some extent, well-produced videos can also be used effectively in conjunction with trained speakers to supplement presentations.

Programming

One of the more labor-intensive outreach tools, the municipal hot line, is designed to answer public concerns and questions on an individual and immediate basis. The ideal, a 24-hour line, usually supplements telephone workers with an answering machine during off-hours. In addition to directly communicating specific information on an as-needed basis, this tool also has the advantage of using only superficially trained workers, since most questions tend to be general in nature, regarding scheduling, material separation, packaging, etc. Of course, a fact sheet with easily referenced answers to commonly asked questions is helpful and will help maintain consistency of the information relayed.

School programming is an essential method of public education in the truest sense of that term, because practical as well as theoretical recycling information is brought into the learning process. This will not only benefit programs where particular student body may reside, but also fosters life-long habits that will prove beneficial on a global scale. Also, lessons learned are generally incorporated into the home, particularly if they are presented as a family-oriented activity.

In addition to the more typical presentations by a recycling program coordinator, field trips to recycling separation centers or recycling plants answer the question of what happens to the materials after collected. In this way, natural curiosity is replaced with a fundamental understanding of the recycling process, often to level far greater than the parents, who will ultimately have the day's events repeated at the dinner table.

Contests where students create new things out of recyclable materials can be both educational and fun. In the mind of a fourth-grader, a discarded (but clean) plastic milk container takes on many possibilities, particularly if he or she is free to decorate, add to, and generally let his or her imagination produce some tangible new toy or device. This type of program is a favorite with science and art teachers alike, combining resources within a given school. Awards for the top creations can be as simple as the presentation of a certificate by a local official or program mascot. This will increase the publicity capabilities of the event as well. Other contests have included citywide participation in logo creations, slogans, and mascot designs. These events must be coordinated effectively so as not to become overburdened or poorly managed.

Fairs or block parties provide an ideal time to promote a program in particular neighborhoods. A booth, operated by local volunteers, can distribute information and answer questions. Can-crushing contests, guessing the number of recyclables in a container, or art contests are ways to drum up interest and media coverage. Local businesses, including local scrap dealers and paper companies. may, depending on the size of the event, donate materials, food, and beverages (bins for recyclables a must!). Local officials would do well to make an appearance in support of the event.

Cleanup events at local parks and public areas provide an ideal connection with the recycling message. Often, cleanup programs are supported by the local Chamber of Commerce or some other entity that will welcome the support of the recycling program. The twentieth anniversary of Earth Day in 1990 led to many new promotional events and can provide an opportunity for communicating the recycling message.

The establishment of a speakers bureau (or the inclusion of a recycling staff person on an existing bureau) increases the opportunity for communicating with targeted groups. Presentations to civic groups, seniors, or any interested party require little setup and are welcomed by these organizations. Handout materials and the use of a slide or video show will break up the presentation, maintaining interest while saving the energies of a staff person who may have several presentations in a week.

Each of these events can be modified to specific need and according to available resources. It is important, however, to infuse the highest level of quality into all programming that is undertaken. Poorly managed events can result in negative public response. Better to do a few things well then many more which may be poorly received.

MEASURING PROGRAM EFFECTIVENESS

Perhaps the easiest and most common way to assess the effectiveness of a recycling public education program is to simply look at the quantity of recyclables collected and extrapolate a percentage based on prior knowledge of the waste stream. This will certainly provide some concrete indication on whether or not the populace is recycling. By looking at collection data from specific geographical areas or programs, it is also possible to pinpoint how successful individual neighborhoods are. Of course, this does nothing to answer the question "why," and, unless investigated according to other criteria, is of little use as a measurement of promotional efforts for recycling program managers.

Earlier chapters discussed the need to establish public education goals that relate to the communication processes being undertaken for the public education program, as opposed to the quantity of materials collected. This same thinking should apply when judging the effectiveness of an education campaign. Goals judged according to communication principles will yield a beneficial understanding of what is successful and what is not.

Assessments

This will require an organized followup program, incorporating many of the same components used during the initial research phase of the education program. In many respects, many of the interviews and surveys conducted at that time are worth repeating. Questions to be answered can include

- Are audiences clearly defined?
- Do target audiences receive appropriate messages?
- Have communications been frequent enough?
- Can new socioeconomic considerations be identified?
- Are there competing or similar messages?
- Has the proper medium been used for a message?
- What works? What doesn't?

Answers to these questions can be acquired by speaking with many of the individuals and groups originally canvassed during the research phase of the program. It is also helpful to conduct random, spot surveys of residents in neighborhoods where participation is both good and poor. In addition to asking questions on the effectiveness of the communications program, another opportunity has been created to gather information on participant habits, prejudices, and ideas. This "new" information can be valuable for introducing new recyclables into the program or for followup promotional activities. In effect, audience profiles can be refined for further focusing the public education campaign.

Changes

In the event there is a problem, often slight modifications are all that may be necessary to improve a particular form of communication or element of the program. Simply jazzing up promotional items or increasing distribution frequency may have marked results. However, when an action is shown to have minimal or disastrous results, it should be replaced with another, regardless of resources already expended, so as to maintain the integrity of other education elements that are successful.

Measurements

Several quantitative measurements can be borrowed from the marketing and advertising fields, however, to help assess a particular education element. When using a direct-mail piece, for instance, a retention/response rate of only 2 to 5 percent is considered good. Of course for industries who may be mailing to hundreds of thousands of potential customers (as with mail-order firms), the significance of this percentage can be translated in very strong sales figures. When applied to a recycling program, however, a compliance rate of at least 10 to 15 percent should be strived for at the start, gradually increasing based on the extent of followup activities and troubleshooting that is undertaken. For one material (e.g., newspapers), a collection rate over 50 percent is not uncommon at the outset of the program. The amount of returned mail will also give a fairly good indication of the validity of the program mailing list.

For a telephone hot line, several calls an hour just prior to the startup of a program can be encouraging, particularly if the number was publicized through some other mechanism (e.g., brochure mailer, advertisement). Of course, a truer judgment of the overall communications effort can be made based on the type of calls coming in and the degree of confusion being relayed.

Collection personnel can be a leading indicator of the communications success, particularly for residential pickup programs. The condition of recyclables when collected (e.g., washed, bundled, or separated appropriately) or the number of households actually participating on a given street can yield a valuable profile of the community's response. Collection personnel may also have important technical suggestions once a program is under way that could impact future program communications. Many recycling coordinators will also follow collection vehicles during the first weeks of a program to see for themselves the success or pitfalls of program startup.

For drop-off centers, periodic visitations are a must, particularly during various hours of operation. In some cases, poorly marked access or directions will turn potential recyclers away simply due to the frustration of locating the center. Centers should also be maintained for cleanliness and security. Residents who don't feel comfortable when dropping off materials will not participate over the long term.

Summary

Regardless of the changes that the recycling manager has seen over the last decade, public education programming remains an essential component to recycling's success. Not only does the public education function ensure participation, but it can be instrumental in maintaining program integrity over the long term. Flexible in their application, public education programs can be simply designed or full-scale, multifaceted productions requiring any range of costs. In any case, the program must be implemented with consistency and with ongoing attention and maintenance to maximize the benefit.

Staffing needs can vary to include one identified recycling professional or a handful of volunteers combining talents to get the job done. Regardless of resources, however, advance planning is critical, particularly for large-scale, multimaterial programs. This includes the proper research of audiences to be communicated with and the identification of available resources for getting the message out. These messages should incorporate lively text and graphics and can be produced professionally or with nonprofit assistance. Program assessments, based on communications criteria, offer a means to identify needed changes and program successes.

Acknowledgments to Peter Wolf and Lorraine Krupa for their assistance in preparing this chapter.

APPENDIX. RESIDENTIAL RECYCLING QUESTIONNAIRE, CITY OF HOLLYWOOD, FLORIDA

1. City of Hollywood Commissioners plan to implement a voluntary recycling program. Will you participate?

87.6% Yes   6.9% No   5.58% No answer

2. Please check the items you are currently recycling:

36.6% Newsprint   5.7% Plastics

3.3%	Corrugated	9.4%	Used oil
3.7%	Clear glass	10.0%	Batteries
3.8%	Mixed glass	8.1%	Tires
21.3%	Aluminum	42.2%	No answer

3. Would it be more convenient for you if recyclables collection service were provided:

45.9%	Once a week	9.2%	Once a month
20.2%	Twice a week	7.3%	No answer
17.1%	Every other week		

4. Would you prefer that the collection of recyclable materials be made on the first day or second day of your weekly household garbage collection?

32.3%	First day	42.1%	Don't care
19.2%	Second day	6.4%	No answer

5. Benefits identified for recycling programs are listed below. Please indicate the two most important benefits which impact your life style.

(2) 47.9% Clean environment

(4) 33.7% Preserve valuable resources

(5) 22.9% Reduce landfilling and incineration

(3) 37.3% Reduce future costs of your sanitation bill

(1) 50.8% Conservation of the environment 4.6% No answer

6. Do you feel that separating newspapers, glass, and aluminum cans from your garbage and storing them in a separate area until pickup day would be:

25.5% Convenient

62.0% Slightly inconvenient

10.6% Very inconvenient

1.9% No answer

7. Washing some recyclables such as glass and aluminum containers will minimize bug and odor problems. How willing would you be to wash glass and aluminum containers before storage?

25.6%	Very willing	19.1%	Unwilling	3.9%	No answer
46.0%	Willing	5.4%	Don't care		

8. Would you prefer to bring your separated recyclable materials to a convenient drop off location rather than place them at your residence for pickup?

6.1% Yes

79.1% No

10.6% Does not matter

4.2% No answer

9. In terms of YOUR participation in the City's recycling program, how important is it that the program be required by local law?

43.6% Very important

23.0% Somewhat important

26.7% Not important

5.8% No answer

10. Which category describes your residence?

Over 99% Single family home Less than 1% Duplex

Comments:

Favorable 41.1%

Unfavorable 3.7%

No answer 55.2%

CHAPTER 31

TRAINING PERSONNEL AND MANAGERS

BETTY MUISE
Director of Training
Resource Integration Systems, Ltd.
Toronto, Ontario, Canada

INTRODUCTION

The information and skills of a recycling program manager make up a long and diverse list, which new professionals to the field probably find more than a little daunting. In any one day, a recycling program manager may fill in for the forklift operator who did not show up to work; make impromptu repairs to a forklift, baler, or other essential piece of equipment; then don suit and dress shoes and make a convincing presentation to local council members and legislators—all in a day's work! Hopefully, this sort of day will not happen to a recycling program manager too often, but it does illustrate the tremendous range of knowledge and skills that a recycling program manager may need to have.

Based on direct involvement with over 300 recycling coordinators and recycling personnel during the development of training courses in seven states, the following summarizes the primary areas of competencies generally required.

Program Economics and Funding Options

This will vary somewhat depending on future support offered by the state.

- Knowledge of the widest possible range of funding options, including user fees, tax options, revenue sources, and means of financing larger facilities
- Knowledge of the funding options, skill in being able to evaluate the options and make a decision, skill in being able to sell the funding options to public officials and to the public
- Knowledge of the role of full cost accounting

End Markets

- Knowledge of international, national, state, and local market opportunities
- Knowledge of other marketing opportunities such as market cooperatives and market development options

- Knowledge of market specifications (contaminants, shipping requirements, bale specifications, etc.)
- Knowledge of longer-term trends

Recycling Program Management

- Skill in program management such as managing budgets, managing people, and managing equipment (maintenance, proper operation, safety)
- Knowledge of and skill in managing an efficient processing and collection (e.g., routing efficiency and production efficiency) and in communicating with the public and public officials
- Knowledge of full range of equipment options for collection and processing including recently developed or state-of-the-art technologies or approaches including drop-off collection, curbside (dedicated vehicles), curbside (co-collection approaches), multimaterial processing, approaches for integrating organics collection and composting, and opportunities for corresponding savings in garbage collection
- Knowledge and skill in selection of and management of contractors

Promotion

- Knowledge of and ability to coordinate promotion based on the fundamental principles of promotion (consistency, targeting of message, etc.)
- Knowledge of promotion ideas for recycling
- Skill in utilizing community as a resource
- Skill in developing and maintaining strong support of council
- Skill in public and one-on-one presentations
- Skill in dealing with the media

Local Government

- The skill of being able to continually foster local government support (recycling program managers)
- Knowledge of relevant state and local laws

Other Important Topics

- Monitoring
- Design and implementation of recycling programs for multifamily and institutional/commercial and industrial (ICI) sectors
- Composting

In this chapter, an exploration of how training can be used to develop the capability of recycling program staff members and managers will be covered. In the first half of this chapter the principles behind how to define training needs and what the typical training needs in the recycling profession are will be addressed. In the second half of this chapter the considerations necessary in designing a training plan for the recycling program manager and the recycling program staff members and utilizing training practices that accomplish *true* learning—not just the "hear and forget" variety will be addressed.

DEFINING TRAINING NEEDS

Whatever business or profession a program manager is in, and particularly in the recycling field, difficulties arise when trying to squeeze in training opportunities. Generally speaking, it takes time, money, and most importantly, is difficult to find the exact program to meet the needs of each staff member. Frequently, recycling program staff members and managers find themselves wasting time sitting in sessions that are only vaguely useful with their attention slipping back and forth from the session to other concerns at home or at the office. This situation is all too common and truly unfortunate, because it is only by addressing an *individual's own learning needs* that training can be worthwhile and productive.

Learning Needs for Recycling Manager and Staff

When starting to think about potential training initiatives for a recycling manager or for the recycling staff, there can be no better starting point than defining the training needs. In other words, what is the outcome in *performance* terms that the job type requires? Improvement in performance or capability is the bottom line of training and may be accomplished using the following procedures:

1. *Define the full list of performance requirements for the job function* (otherwise known as competencies). For example, two performance requirements for the position of a commercial recycling program coordinator may include:
   a. Ability to understand and apply applicable rules, regulations, policies, and procedures relating to the commercial-sector recycling program.
   b. Ability to communicate effectively both orally and in writing with area business and community leaders and the general public.
2. Evaluate the list generated above and *note whether the competency requires information or knowledge of a particular subject, or whether it requires a proficiency in a certain skill.* For example, the first competency noted above requires broad *knowledge* of the recycling program, whereas the second competency refers to a specific *skill* in being able to deal with the media. See Table 31.1 for examples of required skill competencies and knowledge competencies for various recycling program positions.
3. *Describe the gap between the current and ideal level of proficiency in each competency area for the recycling program manager and for the recycling staff.* For example, using the same example as above, the commercial recycling program coordinator may have an excellent understanding of the planned commercial-sector recycling program, but has limited knowledge of the relationships between various policy-making entities that may impact the successful implementation of the program. Similarly, this person may have an excellent ability to prepare for talks and presentations, but his or her actual delivery and presentation style is weak.

TABLE 31.1 Recycling Job Performance and Competency Requirements

Position	Performance requirement (examples)	Competency requirement	
		Knowledge	Skill
Administrative Recycling Manager	Develop countywide recycling plan consistent with state mandates, Coordinate public and private entities,	Regulations, policies, and procedures relating to the recycling program.	Writing technical reports, implementing programs, communications, and public relations.
Multifamily Recycling Specialist II and Commercial Program Development	Act as a consultant on multifamily dwellings and private haulers, Negotiate contracts,	Rules, policies, and procedures relating to multifamily recycling programs.	Writing and public speaking. Effective public relations abilities.
Recycling Specialist I—Single Family, Institutional Special Wastes	Coordinate the development and maintenance of single-family, institutional, and special waste recycling,	Residential and institutional programs. Resolving hard-to-handle complaints for single-family homes,	Knowledge of word processing and spreadsheet development, typing, and communications skills—oral and written.
Public Information Specialist	Researches and writes publications for the Solid Waste Authority. Prepares annual reports. Develops press kits,	Journalistic principles and practices. Marketing and research techniques. Public relations, typography, and graphic design.	Assemble and write a wide variety of interesting publications. Oral and written communications skills.
Operational Recycling Manager	Responsible for all aspects of the recycling center. Coordinate activities between pilot and contract collection programs and the materials recovery facility.	Principles and practices of recycling programs. Materials, equipment and supplies used in implementing recycling programs.	Oral and written communications skills. Working relationship skills Plan and coordinate projects.
Recycler I	Maintenance of facility. Receiving, processing, and packaging recyclables.	Operate mobile and stationary equipment, i.e., trucks, balers, forklift, crushers, and loaders,	Sufficient physical strength to perform manual work. understand and follow oral and written instructions.
Recycling Collection Supervisor	Supervise and coordinate activities or workers engaged in the curbside collection process.	Of practices, methods, tools, equipment, and materials of the curbside collection program.	Supervise semiskilled workers, operate equipment, public relations, and maintenance of daily logs.
Contract Manager (Recycling)	Management of collection contracts and interlocal agreements with municipal governments.	Contract management techniques and tools. Managerial, accounting, and budget practices of the recycling program.	Computer skills in word processing, spreadsheet, and database. Present clear and concise reports. Establish procedures and contract requirements.

Source: Solid Waste Authority of Palm Beach County, Florida.

This type of analysis results in a very specific description of training needs for the commercial recycling program coordinator that allows the recycling program manager to focus on improving the required performance. The training plan that would be established for this person would not need to address general-level information on recycling program operation, nor provide general training on presentations. Rather, the training plan would be quite specific and would provide:

1. *Training on policy development by entity, responsibility, and contact person:* For example, how are state rules implemented through state regulatory agencies or how do flow-control policies of recovered materials at the state level affect the successful implementation of a commercial-sector recycling program? As much as possible, the trainee would not only observe the process but also communicate directly with decision-making entity's staff persons. Flowcharts and diagrams of the rules, regulations, and the entities developing and implementing them as such may also prove to be useful.
2. *Training and practice on how to deliver presentations:* Several practice sessions could be arranged for the trainee to practice the delivery of presentations. Practice presentations could be made to more experienced colleagues who could offer practical advice and tips. Alternatively, practice sessions could be taped and then self-analyzed. Ideally, mock situations should be created for the trainee to become more comfortable with a range of challenging situations (such as answering difficult questions, dealing with hecklers, etc.). Finally, the trainee could attend any one of the many presentation training seminars that are offered by private training companies.

The example outlined above illustrates the importance of being specific about training needs so that the subsequent training can focus exactly on the performance improvement that is required. By thinking through the training needs of the recycling program manager and the recycling staff in this manner, the first step in ensuring that training efforts are effective both in terms of cost effectiveness and improvements in capability will be achieved.

An inventory of potential competencies is provided for several typical recycling positions in Table 31.1. Every job is different; therefore the inventory will probably require modifications or additions. Use Table 31.1 as a shopping list or prompt to complete a training needs inventory.

ADDRESSING TRAINING NEEDS

Once what the training needs of the recycling program manager and the recycling staff are defined, the next step is to determine how best to address these needs. Not all learning environments that we encounter as adults are ideal. A lot of the learning environments, in fact, are not time efficient and do not allow actual learning to occur. In other words, maybe a recycling staff had fun, or met interesting people, but did they learn? Whether the trainee decides to simply enroll in an existing training course or program or the department decides to offer on-the-job or more formal training, to meet the training needs, it is important to know the ways in which adult learning can be enhanced and ensured.

The Learning Process for Recyclers

Typically at recycling conferences and seminars, attendees are asked to sit back and listen to presentations and watch slides. Then, when coffee time comes, the real learning begins when all of the recyclers form impromptu groups and start sharing recycling success or horror stories and new market contacts and promotion ideas. What we know about how we learn is exemplified by the above description. Recyclers learn when they are actively involved, applying new information to their own situations and drawing upon and using their own experiences or those of their colleagues.

Some general rules of thumb for effective training include

- As much as possible, ensure that the information or skill being addressed in the training is *relevant and practical*. Especially with recyclers, the more practical the better. *On-the-job training is absolutely the best form of training possible.*
- Allow for *practice and application of the new information or skill.* This ensures that the new material will actually be remembered.
- Provide opportunities for participants to *draw upon their own experiences* and to share them and analyze them with others.
- *Treat your trainees with respect and acknowledge that they come into the training with well-developed problem-solving abilities and with experiences that are already relevant to the training.* Trainees are not empty vessels waiting to be filled—they have lived, learned, and have experiences already that you are simply adding to.
- Remember that most people have limited capacity for lecture. After one-half hour, most people begin to lose attention.
- Use a variety of training techniques whenever possible.

Designing a Training Plan

A training plan outlines how training needs will be addressed. Some training needs may be simply met by enrolling in a seminar offered by existing training or recycling organizations. Other learning needs may be met in a more tailored manner such as specially designed in-house training sessions for groups of staff or a mentoring buddy system for a new staff member. In general, a 6-month training plan should be developed (and followed up and revised) for all staff members. Training plans should spell out how each of the training needs that have been identified for each staff member can be met and should draw upon a full variety of training options such as occasional seminars, reading assignments, on-the-job opportunities, and attendance at professional association meetings.

In addition to attending existing training courses, there are a host of possible training techniques or training options that can be used to meet remaining learning needs. Each type of training option is best suited to meet a particular kind of learning need whether it is simple awareness building or skill development. Examples of training options with related accomplishments are shown in Table 31.2.

Several states have developed and delivered special training programs to complement state funding programs or to support the accomplishment of legislated recycling targets or goals. Florida, Maryland, Michigan, Rhode Island, North Carolina, and Ontario, Canada, have or are planning such programs sponsored through their relevant State Department of Environmental Regulations or Department of Natural Resources. Short courses are now offered by several universities (e.g., University of Wisconsin, Rutgers University in New Jersey, Clemson University), trade associations [e.g., Solid Waste Association of North America (SWANA)], and trade journals (e.g., *Waste Age Magazine*).

Managing and Delivering Training

With demanding day-to-day recycling activities and problems, training can get pushed to the back burner. There are three ways to prevent this from happening.

TABLE 31.2 Training Activities and Accomplishments

Training options	Can be used to accomplish:
Individual readings or individual readings with an associated work assignment	General introduction/awareness or preparation for more detailed skill development to follow.
Lecture presentations	Same as above. Use to introduce material that the participant will then be called upon to utilize or apply in some fashion. Keep lecture segments to 1/2 hour and break up the lecture segments with a variety of learning techniques.
Discussions, sharing of experiences, e.g., within staff, or other program operations.	Good way to help broaden participants' exposure to different approaches and programs.
Discussion of issues (e.g., the role of mandatory versus voluntary enforcement of recycling) (with staff, or with others)	When issues are discussed (rather than presented) then a better understanding and empathy for the issue may result. Discussions are a good way to sensitize the participants to key issues or the main themes of the subject matter, before they are covered in more detail.
Role playing and practicing	Most useful for skill development (e.g., developing management skills, communication skills, etc.). Basic principles/characteristics or techniques of the skill should be introduced or the participant should be allowed to "discover" these things as part of the training (e.g., one of the keys to dealing with aggressive journalists is not to be intimidated). Then the training should provide ample opportunity for practicing and in many situations as possible. Trainees should be able to learn a lot from self-criticism, but criticism from peers can also be constructive if done in a nonthreatening manner.
On-the-job training	There is truly no better training opportunity than learning on the job. However, sometimes we neglect these training opportunities. For example, take your new public relations and promotions coordinator to your next public meeting or council meeting, rather than go alone. It is likely to be a terrific training opportunity.

1. Set objectives (in the training plan) and track accomplishment of each objective, A good way to improve the likelihood that the training plan is acted upon is to establish joint accountability between the staff member and manager for the accomplishment of the training objectives.
2. Build training opportunities into the day-to-day job as much as possible. Every day can be a learning experience. Even "disaster days" present learning opportunities, in fact, sometimes the most effective learning occurs when mistakes are made!

TABLE 31.3 Training Plan for a Commercial Recycling Program Coordinator

Performance requirement	Competency requirement		
	Knowledge	Skill	Training option
Supervise the development of commercial/industrial and institutional recycling programs. Interacts with commercial businesses and material markets,	Ability to understand and apply applicable rules, regulations, policies, and procedures relating to the commercial sector recycling program. Degree in public or business administration,	Ability to establish and maintain a credible and effective working relationship with the public, administrative, and governmental officials, news media, and co-workers,	Attendance at state and local government workshops and forums; practice sessions with recycling program manager for public presentations; review of professional market journals; regularly scheduled discussions with market representatives to discuss international, national, state, and local market opportunities.
Integration of these individual projects into the countywide recycling program that includes operation, design, development, and implementation. Ongoing project evaluation and modification.	Two years' experience in business management, program development, or waste management/recycling. Ability to develop and implement commercial programs as directed by the countywide recycling plan.	Ability to understand and prepare technical reports related to commercial recycling.	Technical writing courses; review of existing commercial recycling reports; attendance at trade shows.
Assist with all activities related to multifamily programs.		Ability to communicate effectively both orally and in writing to community leaders and general public,	Practice sessions with colleagues; attendance at training seminars offered by private sector.
Assist the executive director, operations director, and recycling coordinator in development of project procedures, reporting and tracking material recovery, and participation in education programs and fiscal programs.		Ability to speak, understand, and write the English language sufficiently to converse with the general public, peers, and subordinates; respond to inquiries and to make entries on reports and records.	Computer software training (word processing, database management); attendance at state-sponsored recycling coordinator's training course; review and analysis of existing similar programs.
Directs subordinate personnel and perform related work,		Ability to supervise a small group of professional subordinates personally.	Attendance at program management/supervisory training workshops.

Source: Solid Waste Authority of Palm Beach County, Florida.

3. Use other resources as much as possible. For example, if the recycling program is having problems with contaminants (either in collection or processing operation), bring in representatives from markets to talk about the importance of avoiding contaminants or a tour of one of the mills (e.g., a glass plant) to demonstrate the problem with contaminants.

Example Training Plan

An example training plan is provided for a commercial recycling program coordinator in Table 31.3. Again, every job is different and will require: (1) *defining program needs;* (2) *defining job classification* to achieve the program needs; (3) *defining training needs* for the individual selected to fulfill the job; and (4) *designing the training plan* to achieve the training needs objectives.

CHAPTER 32

RECYCLING PROGRAM CONSIDERATIONS, DECISIONS, AND PROCEDURES

BARBARA J. STEVENS, PhD
President
Ecodata, Inc., Westport, Connecticut

INTRODUCTION

This chapter presents various factors for a community to evaluate before deciding on a new or second-generation recycling program. Most communities now have in place a recycling program for single-family dwellings. Many of these communities are deciding how best to modify these existing systems to achieve greater overall efficiencies and high diversions of materials from disposal. These communities may well wish to consider a change in the structure of their recycling program and its interface with solid waste collection programs, as a means to achieve their current program goals. Only a minority of communities have in place effective recycling programs for multifamily dwellings and the commercial sector. While in most cases, these segments of the market may be traditionally reserved for the private sector, there are many options for government involvement in these areas as well as in the traditional government preserve of single-family residential service. The options range from complete reliance upon municipal or public sector workers and equipment to a complete reliance upon the private sector, with various gradations in between.

Wherever possible in this chapter, information is presented in a manner that should be helpful to a recycling planner trying to initiate a new or revise an existing program. The information should provide useful insights into the planning process to other interested individuals and organizations, including advocacy groups, recycling and other solid waste professionals, and interested citizens. The descriptions of the options and the decisions that must be made (see Table 32.1) should help interested parties participate in the processes of option formulation, influence the outcome of the policy selection process, and affect the manner of program implementation.

The next section defines the terms that will be used throughout the chapter. These definitions may not be used identically in all of the literature, but they will be used consistently throughout the remainder of this chapter. This section also contains some examples of communities that employ the various organizational arrangements for their recycling programs. The "Examples of Use" section highlights the factors that a decision maker

TABLE 32.1 Factors to remember when making recycling decisions

Residential sector:
- Programs only affect about half the waste stream
- To recycle 25+% of waste stream, usually need to consider nonresidential waste stream as well
- Apartments and single-family units may have different recycling program requirements

Nonresidential sector:
- Government policy can encourage recycling
- May require different programs than residential sector
- Private sector most often acts as collector/processor in this sector
- Programs affect about half the waste stream

might wish to consider in narrowing the list of options for organizational arrangements for recycling services. Major factors are considered first, such as the degree of recycling that must, by law, be achieved. Recycling services, it is stressed, consist of several program elements, and it is not necessary that each element be provided in the same manner. For example, recycling collection could be handled by a private firm under contract to the community, whereas processing and marketing could be handled by a regional public sector entity. The "Pursuing Each Option" section contains specific suggestions on how to implement the various options for recycling programs. In each section, information is provided for the community wishing to use that option as the means of handling all elements or just one element of the recycling program.

OPTIONS—ORGANIZATIONAL ARRANGEMENTS FOR RECYCLING PROGRAMS

Main Organizational Alternatives

There are four main organizational alternatives for establishing a recycling program or some of the elements of a recycling program. These main organizational alternatives, arrayed in order of the least to the most government involvement, are: private, franchise, contract, and municipal (see Table 32.2). Each is discussed below.

Private is the name given here to the arrangement that involves government the least. Basically, in this arrangement, firms are allowed to provide the services without interference or intervention by the local government. Market forces determine the extent to

TABLE 32.2 Main organizational alternatives

	Percent municipal involvement	
Type of program	0%	100%
Municipal	XXXXXXXXXXXXXXXXXXXXXXXXX	
Contract	XXXXXXX	
Franchise	XXXXX	
Private	XXX	

which the local private sector would initiate a recycling program, the materials that would be recycled, the customers that would be included in the program, and the percentage of the waste stream that would be diverted from the disposal site.

It is possible to have a private sector recycling program with or without government incentives. Private sector recycling with no government incentives generally occurs in areas where economic conditions make such a program profitable. Typically, this condition prevails in areas where there are large generators of high-value recyclable products (such as office buildings generating high-value paper scrap or stores with large corrugated cardboard generation), and/or a high disposal fee. Such conditions typically prevail in the older large urban centers of the Northeastern part of the United States. For example, private sector recycling is common in such cities as New York, Boston, and Philadelphia, to name a few.

Government incentives can be used to encourage private sector recycling in communities where the prevailing economics alone do not justify establishing such a program. Such incentives can be based on microeconomic theory or on a quota system. For example, if the local disposal site is owned by the city, the city could increase the tip fee as a means of encouraging recycling, or the city could pass a regulation that in order to qualify for a license, a waste hauler would have to demonstrate that at least a minimum percentage of materials collected was actually recycled.

The private arrangement is consistent with government licensing of firms involved in the hauling, processing, marketing, and publicizing of recycling programs. Except in the case of a government program requiring haulers to meet a recycling quota, there is usually little enforcement necessary in such a system. However, unless prevailing disposal fees are very high, or unless all waste generators in a community are high-volume generators of high-value recyclables, private sector recycling alone is not likely to serve a majority of waste generators in a community or to result in the highest potential diversion of waste from the disposal site.

The *franchise* arrangement is defined as one in which the local jurisdiction authorizes one firm to provide a given service to a given category of customers in a specified area. The franchise may be exclusive, in which case only one firm is allowed to provide the services specified in the specified area, or nonexclusive, in which case more than one firm is allowed to offer services in the specified area. The holder of the franchise bills its customers directly, and the firm may be required to pay a franchise fee to the community in return for the granting of the franchise. In solid waste collection, franchise fees are often revenue-producers for local governments, with fees often set in the 5–10% of gross revenues range. For recycling, the franchise fee may be negative, providing a subsidy just sufficient to induce the private sector to provide the services.

The *contract* arrangement is very similar to the exclusive franchise arrangement. In this arrangement, a single firm has the government-granted right to provide specific services to a specified category of customers in a specified area. However, unlike the exclusive franchise system, in which the firm receives its revenues via billing customers individually, in the contract arrangement, the firm is paid in one check by the local jurisdiction itself. If there is any billing of the individual customers, it is done by the local government or its agents, but not by the contractor.

Municipal service occurs when employees of the local jurisdiction perform the tasks required to deliver the service, and the equipment required to perform the services is similarly owned or leased by the local jurisdiction. In a municipal arrangement, public sector workers drive the collection vehicles, sort the recyclables, market the commodities derived from the raw recyclables, and enforce participation and containerization requirements. Customers may be billed directly for the services provided, or the services may be financed from general property tax (or other) revenues.

Considerations Regarding the Options

The scale of operations can affect the costs of service to a community. Previous studies have shown that there are significant economies of scale in refuse collection, with per-ton or per-household costs decreasing by about 20% as the scale of operations increases to about five collection vehicles per day. No further economics of scale are available after this point. There is no reason to assume that these results do not pertain to recycling collection as well as refuse collection, as the nature of the routing and collecting operations are very similar. Thus, communities considering more than one arrangement, for example, one area of a city with municipal collection and one area with contract collection, should understand that they may pay extra for diversity, so long as each area is not large enough to require at least five vehicles each day.

Many communities select different recyclables collection methods for different sectors. For example, one arrangement may be used for recyclables service from single and small multifamily buildings, while another arrangement entirely is used for recyclables collection from commercial and large apartment buildings.

Often, communities choose a recycling arrangement to coordinate with the refuse collection alternative that prevails in their community. For much of the country, commercial refuse collection is handled with the private system, and communities with this type of refuse collection are often unwilling to use arrangements for recycling services which will cause local refuse haulers to go out of business. Thus, government incentives in the private system are a popular method of initiating commercial recycling programs in such communities. Often, the residential and small business sectors are combined and have a refuse collection system that differs from that of the large-scale commercial sector. In such communities, the recycling arrangement for the small residential sector will typically also differ from that of the large commercial sector.

Examples of Use

The *private* arrangement is often used for the commercial sector, particularly in areas where high disposal fees favor recycling for economic reasons. Examples of cities where there is a well-developed private sector recycling program include many of the large, older urbanized areas, including New York, Newark (New Jersey), Philadelphia, Boston, and Chicago. Some smaller communities, where the arrangement for refuse collection is the private system, have elected to use the private arrangement for recycling service as well. An example of the latter is Westport, Connecticut, which implemented private recycling when mandated by Connecticut law in 1991.

Private recycling systems can also be implemented with incentives from or ordinances passed by the government. For example, an ordinance could be passed requiring any franchised hauler to offer recycling services (for a fee, where the economics of recycling do not justify the service at no fee) to all customers. Such restrictions are being considered by many cities, including, for example, Babylon, New York. Alternatively, the local government could pass an ordinance requiring property owners in the sector with private recycling service to provide a recycling program on their premises. The State of Connecticut and the City of Portland, Oregon have such requirements.

The *franchise* arrangement for recycling is probably most common in communities granting an exclusive franchise for both refuse collection and recyclables services. In this way, a single fee to the customer can include the refuse and recycling services. Many counties in Florida franchise the collection of refuse and recyclables, incorporating recycling services in the basic solid waste fee. In Illinois, the Villages of Schaumburg and

Streamwood jointly procured the services of refuse collection and recyclables services, awarding the contract in the manner of an exclusive franchise, whereby the selected firm must bill customers for the joint services of refuse and recyclables collection and dispersion. (See Table 32.3 for the outline of the specifications in the franchise agreement for these communities.)

In most areas of the country, the prevailing prices for refuse disposal and recycled commodities do not on their own justify initiation of recycling programs from small generators. Such small generators would include residences and small commercial establishments. For recycling to occur from such establishments, it is typically necessary for the local jurisdiction to subsidize the program in some manner. This is easily accomplished when the organizational arrangement selected for the recycling program is that of a contract. Contracts for recycling services are found in many communities, including Seattle, Washington, San Jose, California, the State of Rhode Island, Norwalk, Connecticut, and Fort Worth, Texas. The contractor can be paid a fixed amount per unit serviced or processed, or a varying amount per unit serviced or processed, with the fee changing according to the actual or expected change in revenues from the sale of commodities.

Cities with their own municipal refuse collection systems very often initiate a municipal recycling program serving the same establishments serviced by the refuse collection system. New York City, Houston, Texas, and Chicago, Illinois are examples of cities that fit this organizational description. In each case, the municipal recycling crews service the residential establishments from which municipal crews collect refuse.

DECISION MATRIX

The Need for More than One Decision Process

A community deciding on the organizational arrangement for recycling services may wish to subdivide the service into two or more groups, perhaps using a different organizational arrangement for the different components. Recycling services, in their entirety, are multifaceted. At a minimum, they include a publicity or education component, an enforcement component, a collection component, a processing component, and a marketing component. The first two components typically involve less monies than the latter three. They are typically handled by specialized departments or private firms, and they are not typically handled by the solid waste professionals who might otherwise collect refuse or collect, process, and market recyclables. Processing and marketing are often treated as a unit, but there is no reason why this must be so. Some firms are able to collect the refuse, but they are unable to process and market the collected materials. A decision maker, then, should apply the decision matrix considerations to each of the components of the recycling service, in the process of deciding how recycling services should be provided in any given community. Table 32.4 shows the options for each program element.

Certain basic considerations affect the level of government involvement that can be expected to be required. For example, in states where recycling goals are mandated by law, the government can expect to be involved in ensuring that the stated goals are actually met. The higher the level of recycling mandated, as a percentage of the waste stream, the more the community is likely to need to make use of programs in addition to market forces to achieve goals. The government involvement would typically occur when programs are mandatory, and the government would take responsibility for enforcing com-

TABLE 32.3 Franchise arrangement for recyclables—list of specification in franchise agreement of Villages of Schaumburg and Streamwood, Illinois

- Definitions
- Service required
  - Residential service
  - Service exclusions
  - Service for certain condominium projects
- Term of contract
  - Initial term
  - Automatic renewal
  - Exclusive contract
  - Renegotiation
- Contractor qualifications
  - Minimum experience
  - Access to landfill
  - Access to materials processing facility
  - Adequate finances
  - Adequate rolling stock
  - Recycle logo
- Contract provisions*
  - Weekly service
  - Resident notification
  - Receptacle location
  - Refuse receptacles
  - Recyclable material receptacles
  - Unlimited number of receptacles
  - Noncontainerized materials
  - Office paper program
  - Additional services
  - Hours of operations
  - Holidays
  - Nonresidential services
  - Septemberfest packer truck
  - Workmanlike performance
  - Emergency provisions
  - Refuse collection vehicles
  - Recyclable material collection vehicles
  - Refuse disposition
  - Recyclable material disposition
  - Right of inspection
  - Monthly report
  - Complaint response
- Compensation
  - Charges for single-family residential service
  - Charges for multifamily residential service
  - Charges for service to village
  - Collection of charges
  - Recyclable material sales revenue

TABLE 32.3 *Continued*

Insurance, bond, and performance provisions
Automobile liability insurance
Liability insurance
Worker's compensation insurance
No limit on insurance amounts
Village culpability
Performance bond
Failure to perform
Adherence to all appropriate legislation

*Although the specifications refer to this as a contract, the selected contractor is not paid by the local governments, but rather "shall look solely to the customer for the payment of the monthly charge." When the firm bills customers directly, the agreement is a franchise, in our terminology.

pliance. Although the government would have to pass the legislation in these cases, the actual day-to-day operations of the various program elements could be conducted according to any of the organizational arrangements listed above.

Factors to Consider in the Decision Making Process

As there are several individual components to the recycling program, it is usually necessary to consider each one separately in the decision process. Then, if the decision is to handle two or more program components with the same organizational arrangement, it may be possible to link the components into the activities of a single governmental entity or private sector procurement. The decision process regarding the best arrangement for each of the various components of the recycling program can be reduced to a consideration of the elements discussed below and listed in Table 32.5.

Costs

Cost of the component under the various arrangements is an important input to the decision making process. Costs can be divided into start-up and ongoing categories. Expenses included in the start-up category are the expenses of personnel recruitment, training, and system planning. Included in system planning, especially for the municipal arrangement, are the costs of specifying and procuring capital equipment or facilities. For the contract

TABLE 32.4 Program element options*

Recycling program element	Who does the work?
Publicity	Public or private
Enforcement	Public or private
Collection	Public or private
Processsing	Public or private
Marketing	Public or private

*Each program element can be considered separately in deciding on the organizational arrangement for a recycling program.

TABLE 32.5 Choosing an organizational arrangement—checklist of factors to consider for each program element

Factor	Municipal	Contract	Franchise	Private
1. Costs				
2. Administrative burden				
3. Experience in recycling				
4. Legal constraints				
5. Customer satisfaction				
6. Source of funds				
7. Community satisfaction				
8. Market risk				
9. Financial risk				
10. Flexibility				
11. Integration with systems				
12. Program effectiveness				

or franchise options, the start-up costs include the costs of preparing the specifications for the procurement and the actual expenses associated with soliciting and reviewing responses, overseeing the selected firm during the start-up period, and planning for ongoing administrative oversight of the contractor.

Estimating ongoing costs varies somewhat for the various arrangements. The different cases are considered below.

Municipal. In order to estimate start-up and ongoing costs, it is necessary to have an estimate of the type and number of personnel and pieces of equipment that will be needed to perform the work. For any of the components, it is necessary to have competent, trained personnel in charge of the work. An important consideration is the quantity of work that will be required. For example, with publicity, if only a part-time level of effort will be required, a community might not wish to hire and train a municipal worker, but might prefer to use the services of a professional advertising agency. However, if the City already has a graphics department, which could supply the services using existing personnel, such an alternative might be attractive.

For the components of the service that are ongoing, including the collection, processing, and marketing of the commodities, the community needs to ask how many people will be required, with what skills, and whether the community has experience in implementing and supervising programs using workers with these skills and responsibilities. Where experienced supervisors are already on board, they can provide estimates of the numbers of individuals needed; consultants may be necessary to form estimates of needs in areas where the community does not have in-house experience.

Similarly, the number and type of pieces of equipment that will be required can be estimated by supervisors already working for the community, or by planners or consultants using information from comparable communities. For processing and collection components, it is necessary to decide how the materials will be sorted by the generators, and how (if at all) the materials will be sorted by the collectors. These decisions affect the costs of the collection vehicles, the work rates of the collectors (and, consequently, the number of collection crews that will be required), and the nature and types of processing lines necessary.

Estimates of the ongoing costs should include some amortization of capital equipment, even though the budgetary process of the community might not include such items in the operating budget. The reason to include amortization of capital equipment in the estimate of the cost of municipal service is to facilitate comparison of the costs of municipal service to the cost of service provided by a private firm. Whereas the city might buy the capital equipment from one budget and cover the costs of personnel from another, the private firm would include the amortized costs of capital equipment and the costs of ongoing personnel in the single price charged to the community. To compare the cost of municipal service to the cost of service from the private sector, the city must follow a similar procedure of aggregating all its costs, using a full cost accounting methodology.

Other ongoing costs that should be considered are the costs of insurance or self-insurance to cover accidents to personnel and property, in the case of jurisdictions that are self-insured. Costs of replacing equipment or containers for recycling need to be included, as do costs of supervision, vehicle repair and maintenance, fuel, other fluids, and the costs of personnel to handle inquiries and complaints.

All of these costs can be aggregated to equal an annual estimate of the costs of providing the service. Table 32.6 provides a checklist of the cost components for municipal service. This estimate can then be divided by the number of service units (households, stops, or tons, depending on how the community thinks of its service requirements) to obtain an estimate of the unit cost of municipal service.

Contract and Franchise Service. The cost of contract and franchise service can be estimated by using data from nearby communities, if nearby communities employ contractors in a similar manner. Alternatively, the relative expected work rates from contractor or franchise crews can be estimated by using ratios observed from refuse collection activities in communities nearby. Finally, a solicitation of proposals or bids could be used as to determine what the private sector would charge to provide the components or components under consideration. Figure 32.1 provides the price bid sheet that private firms were asked to fill in to qualify for award of a contract for refuse collection and recycling service in East Brunswick, NJ. The bid requested was a fee per ton of refuse and per ton of recyclables. East Brunswick saved significantly by procuring refuse and recycling under the same contract, which is referred to here as an integrated procurement.

To the prices quoted by the private sector contractors, the community must add the expected ongoing costs of contract administration in order to obtain the overall expected cost of this alternative. Typically, such costs do not exceed 4% of the total price charged by the contractor. For the contract arrangement, if there is to be billing of individual customers by the government rather than by the private firm, the jurisdiction needs to add the expected cost of such billing to obtain the total cost of the system.

TABLE 32.6 Municipal cost checklist

- Direct labor costs
- Fringes—even if included in the budget of another department
- Capital expenditures—include the amortized costs of these cost components
- Customer inquiry service—labor/fringes/utilities, etc.
- "Hidden" expenses—allowance for self-insurance losses, or workman's compensation claims, when self-insured, etc.
- Supervisory/management costs
- Equipment operation and maintenance costs—even if from another department or no additional workers expected to be hired (if workers have extra time to work on this program, the extra time could also be allotted to another program)
- Facility costs (rent/operation) for dispatch, repair, and management facilities.

TOWNSHIP OF EAST BRUNSWICK
DEPARTMENT OF ADMINISTRATIVE SERVICES
BID SPECIFICATION FOR SOLID WASTE AND RECYCLING SERVICES
JANUARY 1, 1997

TOTAL PRICE BID SHEET, FILL IN UNSHADED BLOCKS. PROPOSER: ______________________
(See 10.01 of contract for listing of programs)

BASE BID	CHARACTERISTICS			1ST Yr. $./TON		Contract Adjustment Factor		Maximum Fee Col. 5a*6a Col. 5b*6
Column #1	SOLID WASTE Freq. Col. #2	RECY-CLABLE Freq. Col. #3	TERM (Years) Col. #4	REFUSE Col. #5a	Recyl. Col. #5b	Col. #6a	Col. #6b	
B1-3	2X/wk	0.5X/dk	3			73010	22190	
B1-5	2X/wk	0.5X/dk	5			138515	40180	
B2-3	2X/1X/wk	0.5X/dk	3			73010	22190	
B2-5	2X/1X/wk	0.5X/dk	5			138515	40180	
B3-3	1X/wk	0.5X/dk	3			73010	22190	
B3-5	1X/wk	0.5X/dk	5			138515	40180	
Bulk 1X/mo C1-3	NO	NO	3	$/ton	Bulk	22317		
Bulk 1X/mo. C1-5	NO	NO	5	$/ton	Bulk	40410		
				$/Year				
C2-3 White gds 1X/	NO	NO	3	$/ton	White Gds.	3.17	0	
C2-5 White gds 1X/	NO	NO	5	$/ton	White Gds.	5.74	0	
				$/Day 2 Crews				
C3-3 Leaf Pickup	NO	NO	3			206.1	0	
C3-5 Leaf Pickup	NO	NO	5			373.1	0	

Note: This procurement asks the bidder to quote a price per ton for pickup of solid waste once or twice a week, and recyclable pickup one every other week, for either a five or three year term. Prices are also requested for bulk item, white good, and leaf pickup. Other communities might request a price per household served, or a lump sum bid.

FIGURE 32.1 Sample price submittal sheet, integrated solid waste and recycling procurement.

Private Service. For service provided under the private alternative, the major start-up cost would be incurred by the private sector firms. Additionally, regulations might be required to provide incentives/requirements to the private sector firms to set up recycling programs. Ongoing costs here are included in the prices charged to customers by the private firms involved in providing the service component(s). A rough estimate of the cost of this alternative can be obtained by considering the prices of contractor service, where similar services are provided pursuant to a government contract, and adding a percentage to reflect the uncertainties of a free market environment as compared to an exclusive territory/market. Alternatively, prices in other communities can be adjusted by relative wage rates prevailing in the areas, to get an estimate of costs in a particular jurisdiction.

All of these cost estimates should be considered in light of cost to the local government, cost to the customers, and overall program cost. Table 32.7 provides a format for this comparison. In a municipal program, the government pays all the program costs, and it may or may not charge customers a user fee that covers these costs. Here, there is the possibility to structure fees so that the public perceives the program's costs as less than they actually are. Similar observations apply to the contract system, where the government pays the private firm the costs for the services it provides, and user fees that equal total program costs may or may not be charged.

The private arrangement, in contrast, is by definition funded by a user fee. Here the local government generally pays for just a small portion of the service costs under this arrangement, and user fees cover the majority of program costs. Thus, in the private arrangement, customers pay full program costs, and they therefore may perceive this service to be more costly than a municipal or contract service where no user fees are charged or where user fees are set at a level well below that required for cost recovery. The selection of an arrangement can affect a community's ability to influence public perception of program costs; choosing an arrangement that requires that customers pay the full cost of recycling may be politically unpopular in some communities. However, this might not be a drawback in areas where use of the facility or service is voluntary.

Administrative Burden

The various arrangements impose different administrative burdens upon the local jurisdiction. In general, the private arrangement requires the least administrative commitment

TABLE 32.7 Comparing public and private sector costs

Arrangement	Cost for city work	Fees to contractor	Overall costs
Municipal—total cost per Figure 32.1			
Contract or franchise			
• Payment to firm			
• Cost of procurement and supervision			
• Billing expense			
• Publicity costs			
• Enforcement costs			
Private			
• Prices charged customers			
• Publicity costs			
• Enforcement costs			

from the local jurisdiction, whereas the municipal arrangement requires the greatest. In between in terms of commitment are the contract and the franchise arrangements. The contract and franchise arrangements require administration, and the contract system requires the community to conduct any individual billing of customers. Often, in recycling, such billing is not an additional cost, as the fee for refuse collection is adjusted to cover the desired portion of the expenses of recycling as well as the costs of refuse collection.

Experience in Providing Recycling Services

The extent to which the local jurisdiction's employees are experienced in providing recycling services, or in providing services similar to those required for recycling, is an important determinant of deciding on whether to create a municipal or other type of service or service component. Many municipalities have experience in collection, but do not have experience in the processing or marketing of the commodities collected. Lack of in-house experience would tend to mitigate against selecting the municipal arrangement, at least for the processing and marketing elements of a recycling program, especially for communities located near large urban areas where private sector firms operate well-established material recovery facilities (MRFs).

Legal Issues

Local ordinances may limit a community's ability to establish certain arrangements. For example, local ordinances may limit the length of time for a contract, or the extent to which a franchise arrangement (which generates funds to the community) may be allowed. Of course, the law can always be changed, or the system modified to conform with the law. For example, San Diego's People's Law of 1919 specifies that there will be no fee charged residents for solid waste services, a provision that appears to eliminate the franchise or private arrangement for residential recycling services, absent a referendum repeal of the law.

Customer Satisfaction

An important issue for local officials is the extent to which citizens approve of services provided by the government. When service is provided by municipal employees, local officials sometimes feel that they have greater control over the quality of service provided than when service is provided by a private firm. However, with properly structured contractual agreements, the private firm can be as fully motivated as an elected official to provide high-quality service. Contracts or franchises can require that firms go back promptly to collect any "misses" and they can even allow for misses that are probably the fault of the resident rather than the fault of the contractor. Similarly, service parameters such as the placement of containers after emptying, the extent of noise, the replacement of covers, and the hours of collection and processing can be specifically stated in contracts, sometimes allowing even more control over performance than when a municipal work force performs the service. There is no evidence that, on average, customer satisfaction differs significantly when service is provided by a public sector entity as compared to a private sector firm.

Sources of Funds

In an economically self-supporting program, funds to provide recycling services come from the generators of the recyclable commodities, the markets buying the processed commodities, and the avoided disposal costs (i.e., the money the community or business saves by not having to pay disposal fees for recycled materials). For large commercial establishments, market forces alone often justify recycling. For the small commercial and residential sectors, market forces alone may not justify setting up curbside (or sometimes even other less-expensive types such as drop-off) recycling programs, so some new source of funds, in addition to those listed above, is generally required. For the small customer, these additional sources are usually general tax revenues or a subsidy from user fees for refuse collection.

In all cases, capital expenditures are usually required before service can begin. With a municipal service, funds are required not only for on-going program support but also for start-up capital expenditures, e.g., to purchase the recycling vehicles and containers and to construct a MRF. Such funds can come from bond revenues, or through lease purchase of equipment. With bond issue, the alternatives include general obligation bonds and special purpose bonds, where the repayment would be secured by the value of the assets purchased with the monies or by a dedicated user fee (e.g., a tip fee at a materials processing facility) revenue source. In the case where borrowed monies are secured by a dedicated revenue stream, it is necessary to be able to guarantee that the materials will flow to the facility charging the fee; for recycling facilities, this means that a commitment to deliver recyclables to a processing facility would have to be obtained, ideally, with the length of the commitment matching the term of the bond issue.

For some communities, limits on the quantity of debt that can be issued have been reached, and alternative sources must be found for capital. The need for capital commitment can be reduced via lease of vehicles and reliance on the private sector to provide processing and marketing of recyclables. Contracting or franchising, or relying on the private arrangement, of course, reduces the need for capital expenditures by the local government. The contract arrangement requires the community to pay the contractor's ongoing expenses; included in these charges is amortization of the capital investments that have been made by the contractor. The contract or franchise alternative allows the community to shift the burden of raising capital from the community to the private sector; the community then guarantees to use the facility for a specified term, paying the contractor's fees from user fees or general tax revenues.

In some areas, user fees for refuse collection are set high enough to pay for recycling services as well. Where significant numbers of recyclables are collected, and where the user charge system is volume- or quantity-based (thereby allowing generators who recycle an opportunity to save money on refuse collection and disposal), such systems can be expected to be very successful in encouraging participation in recycling. This is the model that was used by San Jose, CA; when quantity-based fees were implemented along with an enhanced second-generation recycling program (which included yard waste collection) in the mid 1990s, more than 30% of customers shifted to smaller refuse containers and recycling quantities collected increased by over 50%.

Community Satisfaction, Benefits, and Costs

The very existence of a recycling program provides the opportunity for positive feelings on the part of citizens. Helping to save scarce resources by avoiding disposal and reusing products is basically a desirable environmental objective. However, customer satisfaction can be dramatically affected by program design. Tacoma, WA simplified set-out requirements for its recycling program, added commodities to the list of those accepted for collection, improved publicity, and, between 1995 and 1997, increased by over 300% the tons of materials recycled while increasing the percentage of citizens who described themselves as "completely satisfied" with their recycling service from 67.3% to 92.3%.

Market Risk

In establishing a recycling program, there is the inherent risk of a change in the prevailing prices for processed commodities. To the extent that prices increase, the overall profitability of the program should increase (provided, of course, that materials are not diverted to other programs, such as buy-back centers); however, when prices decrease there is the risk of reduced revenue or loss. As more and more communities begin recycling programs, and as the supply of recovered commodities increases, prices might be expected to drop, at least in the short run until demand adjusts and prices seek new equilibria. One advantage of the arrangements that include the private sector is that the community has a chance to share this market risk with another entity—the involved private firms. With a municipal arrangement, the community alone must bear all the market risk.

Risk of Service Interruption

Communities arranging for local services are often concerned with the possibility of service interruption. For some services, such as refuse collection, interruption in service can have health consequences in addition to nuisance value. Recyclables that are separately collected from refuse pose less of a health hazard if not regularly collected than does regular municipal solid waste. Most of the problems associated with service interruption fall into the category of nuisance rather than health hazard.

Each component of the recycling service can be interrupted. Collection, processing, and marketing can each be halted by a labor dispute. If the component is provided by a contract or franchise, the agreement with the city can exact financial penalties for such service disruptions, and most private firms will obtain other workers to continue to provide service even when faced with a labor action. With municipal or private service, a labor dispute may well result in an interruption of the affected component.

Technical difficulties in equipment deployment or maintenance can also affect service reliability. With experienced providers, however, this is unlikely to occur. Again, contracts can exact financial penalties for failure to process or market commodities in a timely manner. To the author's knowledge, there is no evidence that service disruption is more or less likely to occur with contract or franchise service than with municipal service; instances of such events have not recently occurred. Recent instances of service disruption or threatened service disruption have occurred in areas where a unionized work force provides private refuse collection services to communities (New York and New Jersey areas).

Financial Risks

When any new program is established, estimates must be made concerning the expected costs and revenues. For a recycling program, these estimates are especially volatile, as behavior changes can occur rapidly or slowly, depending upon motivation and other factors. The unit costs and net costs of providing the service depend greatly on the level of participation and the extent to which participants prepare recyclables as directed. Again, the arrangements that involve a private firm provide the opportunity to share these risks with another party, as compared to the municipal delivery option. Some communities feel that they will pay more when there is uncertainty regarding these factors, such as participation rate and diversion quantities, which affect program economics; these communities prefer to shoulder the risk of the program start-up so that they may later provide reliable data on which to solicit a contract. These uncertainties can also be addressed in an initial solicitation with appropriately structured clauses requiring reporting on program achievements.

Flexibility

Many communities perceive that the recycling industry is in a state of rapid technological and institutional change. These communities believe that markets will develop soon for materials that are presently considered nonrecyclable, and that some markets presently available may not continue to prevail. If these beliefs are correct, then it will be desirable to have the type of recycling program that can be easily modified to react to changes in market conditions (Table 32.8).

The municipal arrangement allows easy change in collection, processing, and marketing procedures, without the intervening step of contract renegotiation, which might be re-

TABLE 32.8 Ways to build-in program flexibility

For truckside sorting programs:
- Use equipment with variable numbers of compartments and variable compartment sizes

For commingled collection:
- Use a large enough set-out container to accommodate seasonal variations in recycling quantity

For processing:
- Specify equipment with multiple uses—e.g., loaders, conveyor belts, roll-off containers, balers, etc.
- Aim for modular design to allow expansion without extensive redesign

For marketing:
- Prespecified repayment for additional materials; consider a published price base or indexed value
- Incentive payments for innovative market development

quired were a contract or franchise system selected. Conversely, the private arrangement is driven primarily by market forces, and this arrangement would be expected to react to changing conditions, probably even more swiftly than the municipal system might. Again, it is possible to address some of the flexibility issues through proper structuring of contracts and franchises; perhaps the greatest defense against being "locked out" of a market opportunity due to a restrictive contract or franchise is to avoid unnecessarily long terms for these agreements. In a recycling contract, the term might be considered unnecessarily long if it exceeded the expected useful life of the majority of the capital equipment. As most processing equipment has a shorter life than does a building, even for materials processing facilities, some communities are separating the ownership of the building from the ownership/operation of the facility. The former element of the program might have a term of up to twenty years; the latter would be on the order of about half that, based on expected equipment operating lives.

Whatever the nature of the arrangement, it is important to remember that for program flexibility, it is necessary to involve not only the collector, the processor, and the marketer, but also the generator. For residents and small commercial establishments, initiation of a recycling program often means a change in behavior. Portland, OR created by local ordinance a commercial recycling program that requires multifamily and commercial sector landlords to offer recycling to their tenants, but leaves the choice of which materials to recycle largely to the landlord. Thus, a restaurant can recycle glass and steel cans and plastic bottles, in addition to the required commodities of newspapers and scrap paper. A multifamily complex might select magazines, glass bottles, and steel cans as additional recycling commodities (Figure 32.2).

Integration with Existing Systems

Recycling services are different from refuse collection services, yet each addresses components of the discard stream emanating from residential and commercial establishments in a community. In a new program, particularly, the percentage of the discard stream that will be diverted from the refuse can to the recycling container is unknown. What is known, at least to some degree of accuracy, is the size of the total discard stream. The refuse collection entity, faced with a reduction of unknown magnitude in quantity of waste it must collect is probably unwilling to commit itself in advance to any rerouting or reduction in work force. The entity charged with setting up the recycling program is also uncertain of the quantities of materials that must be collected.

In this situation, often a merging of the two entities can result in greatest efficiencies

MULTIFAMILY RECYCLING PLAN FORM
PORTLAND—1996
(For sites where units are provided with individual bins)

WHAT PROPERTIES ARE AFFECTED?

This recycling requirement affects all complexes of five units and larger, including apartments, condominiums, mobile home parks and moorages, within Portland's Urban Services Boundary. The requirement became effective January 1996. It is from the City's Solid Waste and Recycling Code (Sec. 17.102.180).

WHAT IS REQUIRED

- Each complex is required to have recycling containers for at least five materials.
- **Newspapers** and **scrap paper** are mandatory.
- Each owner may choose another three from:
  - Glass bottles & jars (clear and green)
  - Magazines
  - Corrugated cordboard and kraft paper (brown bags)
  - Plastic bottles (including milk jugs)
  - Steel "tin" cans
- The location(s) of centralized recycling containers must be at least as convenient to residents as the garbage container(s), and must be adequate to hold the amount of materials that accumulates between collections. [For new construction, see also Portland City Code Title 33, Planning and Zoning (Sec. 33.120.260)]
- **Instead of having a centralized container system, an owner may choose to provide each unit with an individual recycling container, so long as collection is weekly, each unit's setout location is convenient, and residents may set out all materials listed above, plus the other materials available at single family residences.**
- The owner/manager must provide written recycling information to all residents at least once a year, and to all new residents when they move in.
- Garbage haulers provide each owner/manager with a copy of this one-page Recycling Plan Form to complete and return.
- The owner/manager is not held legally responsible for whether residents use the recycling system, so long as the above requirements are met.

FOR MORE INFORMATION

Call your hauler or the City of Portland, Recycling Program, (503) 823-7202.

FIGURE 32.2 Government mandated recycling, with program flexibility. Portland, Oregon multifamily and commercial recycling program.

and risk reductions (Figure 32.3). In the short run, assuming no changes in consumption and discard behavior, every ton of material diverted from the refuse stream is a ton of material diverted to the recycling stream. If the same firm or agency collects both the refuse and the recyclables, at least they know the aggregate tonnage that needs to be collected, and any tons lost on the refuse side will be reclaimed on the recycling side. Any workers who can be freed up on the refuse side as a consequence of reduction in waste quantities can be used in the recycling work force.

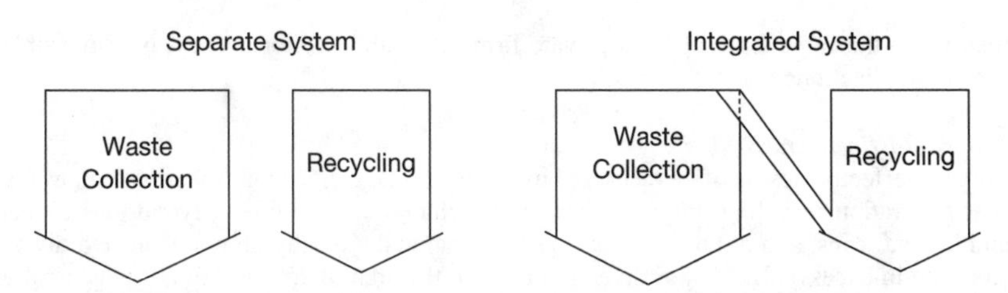

FIGURE 32.3 Integrated waste and recycling programs can save dollars. Usually, integrating the waste collection service with the recycling service allows some savings in waste collection costs. These released resources (manpower, for example) are available for the recycling program.

Generally, a reduction in tonnages can be expected to reduce the costs of refuse collection, but a community will not see this reduction in costs in most refuse contracts or in most franchises as prices are typically quoted in dollars per household, and the actual number of tons of waste will be known with uncertainty except in cases where recycling programs are fully mature. However, if the two services are contracted for together, the aggregate price tends to be lower than when the services are contracted for separately, as the aggregate quantity to collect per household is then a known, whatever the success of the recycling program (see Table 32.9). Many communities, including East Brunswick, NJ, Seattle, WA, and San Jose, CA, contract with a single provider for all solid waste and recycling collection services within a specific contract area when it comes time to contract for recycling services a second time. These integrated contracts remove uncertainty as to the total amount of materials to collect, and they allow the contractor to shift personnel and resources from one collection program to another, as necessary, enhancing overall efficiency.

Another advantage to arranging for refuse and recycling collection to take place in the same manner is that this facilitates coordination of schedules. Most communities have observed that recycling participation increases when recyclables and refuse are collected on the same day. If one private firm is collecting refuse and another private firm is collecting recyclables, arranging for this same-day service is somewhat harder

TABLE 32.9 Cities with integrated systems

These cities with municipal collection have reduced the number of refuse collection crews as their recycling programs have expanded:

- Hollywood, Florida
- Newark, New Jersey
- Norwalk, Connecticut
- Tacoma, Washington

Cities with contract collection have combined their refuse and recycling contracts for an integrated contract, saving significant amounts:

- Cherry Hill, New Jersey—over 27% savings
- Hillsborough County, Florida—over 25%
- San Jose, California—over 10%
- Seattle, Washington—over $2 million per year

then when the same entity, be it a private firm or a public sector agency, is responsible for both collections.

Program Effectiveness

Program effectiveness is often measured by the percentage diversion from disposal. Many first-generation recycling programs have not included commodities beyond glass, steel and bimetal cans, and old newspapers. These programs typically do not achieve a diversion rate in excess of 20%. For diversion rates in the area of 25% to 50%, it is generally necessary to recycle mixed waste paper and yard debris, as well as some plastics. Communities with aging equipment adapted to first-generation programs who wish to step up to the diversion levels that second-generation programs can achieve must analyze the requirements for personnel, financing, and elements of risk as they essentially gear up to initiate a new program.

After considering all the above factors, for each of the components of a recycling program, a community will hopefully be able to determine which organizational arrangement is optimal for its own situation. The decision may be to use one organizational arrangement for one component, and another for the other components of the system. This is a common result of the decision making process. For example, the community could decide to use a regional processing facility (as, for example, is required of communities in New Jersey counties), and to use a municipal work force for residential recyclables collection and to use the private arrangement for commercial recyclables collection. This is the decision adopted by Montclair, New Jersey. San Francisco, California, contracts with a private firm for collection, processing, and marketing of residential recyclables. The decisions of the two communities are consistent with prevailing arrangements for refuse collection; in Montclair, residential refuse collection is via municipal work force, whereas commercial refuse is collected pursuant to the private arrangement; in San Francisco, all refuse is collected by one of two firms with exclusive franchise arrangements.

PURSUING EACH OPTION

Each of the major arrangements is considered below. For each, the roles of the local government are identified and the steps necessary to achieve program implementation are discussed. Throughout, the emphasis is on the steps required of the local government, rather than, say, the steps required of the contractor that might be selected by the local government.

The Private Arrangement

The government interested in encouraging private sector recycling can promote companies that are presently recycling. The government might also deliver recyclables from government buildings to these recyclers. Announcements, education programs, and awards are the major types of government involvement short of providing a direct subsidy or legislating recycling activities.

The government may act directly to effect recycling by modifying local ordinances. For example, the community may require licenses of vehicles used to collect and transport waste materials. The licenses could be granted only to firms that can demonstrate that they actively participate in recycling programs. Perhaps even more strongly, a condition of the license could be demonstrated recycling activities. The demonstration required might be a monthly or annual report indicating the quantity of waste materials collected in the aggregate, and verifiable records, such as weight tickets, indicating that a prespecified level of recycling actually occurred. Such ordinance changes are commonly considered by communities in states (such as Connecticut, Pennsylvania, and California) which have adopted legislation requiring all local jurisdictions to achieve significant levels of recycling of the entire municipal waste stream by specific dates.

It is also possible for the government to subsidize one or more of the elements of a recycling program to encourage private sector recycling via the marketplace. Local tax credits could be offered to businesses that source-separate recyclables or to firms that collect and/or process recyclables. The local government could guarantee to purchase products made of recyclables, thus encouraging entrepreneurs to set up manufacturing facilities. For example, plastics recycling can often be instigated if a secure market for finished products is available. The government could also assist in the siting and permitting of facilities involved in recycling.

Many jurisdictions undertaking such activities, particularly in states that mandate specified levels of recycling, will wish to know how effective their programs actually are. Instituting some type of reporting system will enhance a community's ability to measure program effectiveness. One difficulty in setting up such a program is the likelihood that private arrangement recycling activities will pertain to the discard stream of more than one community. For example, recycling collectors may route their trucks into more than one community, on a daily or weekly basis. Processors may accept wastes from more than one community. Thus, knowing how many tons of recyclables are processed by firms located in a community is not the same as knowing how much private sector recycling is taking place, as a percentage of the community's discard stream. Many states, including Florida, California, New Jersey, and New York, require annual reporting from all local governments responsible for solid waste disposal and recycling; these reports provide aggregate data as to the quantities disposed of and recycled by the public as well as the private sectors. Some cities, such as Portland, OR, require annual reporting of recycling and disposal activities of all refuse and recycling firms licensed by the city.

The Franchise Arrangement

This arrangement applies both to exclusive and nonexclusive arrangements. Each is considered below.

Nonexclusive Franchises

Nonexclusive franchises are very similar to the private arrangement discussed above. The main difference between the private and the nonexclusive franchise arrangement is that in the latter arrangement the community can assess a franchise fee (positive or negative) on the private firm. Typically, franchise fees take the form of a percentage of revenue. Some communities have ordinance provisions requiring that no profits be made from franchise fees. In these communities, the ordinance could be changed, or the fee could be set just to cover the administrative costs of record keeping and franchise monitoring. With the specific reporting requirements of laws such as California's AB 939 and Pennsylvania's Act 101, the record keeping and monitoring functions will not be without significant cost, especially for arrangements, such as the private and nonexclusive franchise, where more than one firm is involved.

To implement this system, the community must draft the franchise agreements, and establish the criteria by which firms will be awarded the franchise. With the nonexclusive system, the franchise is typically awarded to any firm meeting some basic eligibility criteria. Compliance with reporting requirements is a good criterion to include as a condition for continued award of the franchise, as it makes the jurisdiction's reporting job easier.

The franchise agreement can also specify the quality of services that must be performed, and it can specify remedies for failure to achieve specified quality levels. These quality levels could relate to the percentage of materials that are to be recycled, types of materials that will be handled, and to the types of collection equipment that might be required. It is possible to have various levels of franchise fees, with lower fees for firms recycling higher percentages of the discard stream. It might also be possible to structure the franchise payment so that the basic fee would be credited for each ton of materials recycled, up to a total or even a negative overall franchise fee payment due.

In the nonexclusive franchise arrangement, as firms are competing with one another, it is typically not necessary to set rates. In this arrangement, competition among service providers can be relied upon to regulate rates.

It can be difficult to specify exactly which types of services will be provided under this arrangement, however. As the franchise firms bill the customers directly, and pay fees to the local government, the local government does not have the enforcement power of the purse in this arrangement (as it does in the contract arrangement, for example, where the local government pays the private firm). The easiest penalty to enforce—removal of the franchise authorization—may be too harsh for minor service infractions. Less severe penalties, however, may be difficult to enforce. The community could specify in the agreement, for example, that penalty payments would be made by the firm to the city for each occurrence of a specific type of infraction; e.g., if a collector loaded source separated recyclables into a mixed refuse load, the penalty might be $X extra payable to the city.

Exclusive Franchises

The exclusive franchise is the same as a contract, except that billing of customers is done by the firm, so payment is direct from customer to firm instead of from city to firm. The exclusive franchisee is selected in the same way that a contractor is selected, either by competitive procurement or by negotiation.

Because of the exclusive nature of the agreement, it is important that the franchise agreement require the franchisee to provide the specified type of service to the specified category of customer. For recycling services, this means that service must be provided, even if the quantity of recyclables is smaller than the firm might wish. Without competition among service providers, rates are generally established in the franchise agreement. The rates can be set in the procurement itself, with a bidding or price proposing process. In some franchises, the community adopts rate based regulation, auditing the expenses of the firm and setting rates to cover approved expenditures and allow a reasonable rate of return. This method of regulation is common in the Bay Area around San Francisco, and is practiced by communities including San Francisco, Oakland, Hayward, and Fremont. To the extent that the firm providing the services operates in jurisdictions other than the local community, and to the extent that services other than that covered by the franchise are provided, the rate-based rate regulation system presents thorny problems associated with evaluating how to allocate overhead items, how to assess charges by parent and other related companies, and how to evaluate relative efficiency. In most cases, determining the price via a competitive procurement process, with subsequent price adjustments based on predetermined indices over the term of the agreement, is easier and requires less governmental expenditure than does the rate-based rate of return methodology.

The term of the franchise agreement is an important policy variable. Typically, the lowest prices are obtained when the term matches the expected useful life of the major capital equipment to be used in delivering the service. For collection, the major capital expense is the vehicles and recyclables containers, with expected lives in the order of 7 to 10 years; this would be a technologically optimal length of time for the collection agreement. For processing, major equipment can be expected to last from 10 to 15 years, with the building itself having an even longer expected lifetime. Technology here would argue for a long term contract—on the order of 15 to 20 years. However, communities must consider the desire to remain flexible as another determinant of the agreement's term. Desire for flexibility might argue for shorter terms.

Performance guarantees with franchise systems can take the form of a performance bond or irrevocable letter of credit (Figure 32.4). Such an instrument may be called in the instance when the franchisee is determined not to be performing the work. When more minor service difficulties arise, it is desirable to have less dramatic means of exacting desired levels of service. Financial penalties may be difficult to exact, as the flow of funds is

Proposer_General Disposal Corp. Alternative__Base____ Collection Area__All__

Form 2 Surety Intent *(required)*

TO: CITY OF SEATTLE

We have reviewed the Proposal of General Disposal Corp.
(Contractor)
of 1318 North 128th Street Seattle, WA 98133
(Address)

for the following contract:

CITY OF SEATTLE
Residential Solid Waste Services

We understand that Proposals will be received until 2:00 p.m. on January 29, 1999, and wish to advise that should this Proposal be accepted and the Contract awarded to the Contractor listed above, it is our present intention to become surety on the Performance bond required by the Contract.

Any arrangement for the Bonds required by the Contract is a matter between the contractor and ourselves and we assume no liability to the owner or third parties if for any reason we do not execute the requisite bonds.

We are duly licensed to do business in the State of Washington.

Dated: 1/21/99 By: AMERICAN HOME ASSURANCE COMPANY
(Name of Surety)
ANNE E. STRIEBY, ATTORNEY-IN-FACT
(Name of Signatory) (Title)
(Seal) Anne E. Strieby (signature)

FIGURE 32.4 Example of consent of surety for performance bond.

from the private firm to the jurisdiction, rather than the other way around. One way around this difficulty might be to require an escrow account to be established, from which the city might retain the right to withdraw penalty amounts, as per relevant performance provisions of the agreement between the franchisee and the community. Another alternative would be to retain an independent firm to receive the monies (billed by the franchisee) and remit predetermined percentages to the city and the contractor, subject to the city's direction. This alternative, of course, tends to give the franchise the same characteristics as the contract, in terms of performance guarantees; the franchisee, however, would retain the responsiblity of billing customers, a key characteristic of the franchise system.

In procuring the franchise services, it is wise to require respondents to provide only the information that is to be directly useful in evaluating capability and proposed services. If the franchisee is to be allowed to charge the proposed price schedule, and if this is to be determined via a competitive bid process, then it is not necessary to ask the private sector to provide a pro forma worksheet whereby they estimate their expenses and profits in order to arrive at a bid price. Requiring such information (interesting but not necessary to the process) may deter firms from responding and make the procurement less competitive than it otherwise might have been. In general, the simpler the requirements of the respondent, and the more complete the information provided in the procurement (ideally, including the draft contract agreement between the jurisdiction and the to be selected firm), the better the outcome, in terms of lower prices and enhanced competition.

Typically, the franchise is awarded to the financially and technically competent firm that offers to perform the service at the lowest fee to the customers. In some procurements, the jurisdiction may wish to pay a small premium above the lowest price offered in order to award the work to a more experienced or reliable firm. To avoid legal difficulties concerning the evaluation of proposals, it is wise to specify in advance what minimum experience qualifications are required (Table 32.10). For collection of recyclables, there is little justification regarding one type of equipment over another (assuming one is discussing the same level of service—the same type of containers, frequency, etc.). For processing and marketing, some firms may be able to guarantee markets more reliably than others, or may have a more flexible processing system. Such factors can justify the deci-

TABLE 32.10 Minimum experience guidelines

- Experience identical to or related to that required under this procurement
- Experience in which the proposer has used the organizational approach proposed here
- Exerience with the type of equipment to be used here
- Exerience in providing the type of service to be provided here
- Exerience in transition and implementation of services identical to or similar to services required under this procurement
- Exerience in curbside collection of recyclables into a compartmentalized vehicle
- Exerience in processing and marketing the quantities of recylables expected to be collected pursuant to this procurement
- Documented ability to provide necessary equipment as evidenced by vendors' commitments to produce and deliver proposer's selected equipment on proposer's schedule
- Demonstrated ability to locate, prepare, and start the necessary support services for this project
- Demonstrated ability to meet all financial obligations, to perform services even in adverse circumstances, and to obtain necessary financing for performance of the scope of services
- Demonstrated capability to prepare and submit in a timely basis the reports and program documentation required here

sion to award the processing component to a firm that might not be the absolute low price proposer.

The Contract Arrangement

The contract arrangement, which gives the local jurisdiction control over the specification of the services to be provided, as does the exclusive franchise arrangement, also allows the local jurisdiction to control the flow of funds to the firm performing the work. This added feature enhances the community's ability to enforce desired work standards; a community concerned with assuring the quality of work might find a contract arrangement more satisfactory than an exclusive franchise arrangement for this reason.

A contract is typically let pursuant to a private firm's response to a community's issue of a request for proposals or bids (RFP/B). In general, a community must make basic programmatic decisions before the RFP/B can even be issued. These issues are addressed below.

Basic Decisions

The most basic decision any community initiating a recycling program via contract with the private sector must make is the scope of services to be performed by the contractor. A recycling program typically has three main components for each category of generator: collection of recyclable materials; processing of collected recyclable materials (which may include manual and/or mechanical sorting, size reduction, baling, etc.); and marketing of processed recyclable materials. Where an integrated procurement of solid waste and recycling is formulated, basic decisions need to be made regarding refuse services as well as recycling services. Table 32.11 displays the headings of the RFP issued by the City of Seattle for an integrated procurement of single family and multifamily refuse and recycling services.

A community must decide whether it wishes to contract with a single firm for all three components of the recycling program (a "full-service" contract), for just two (typically, collection and processing), or for just collection. If a community decides to contract for anything other than "full-service" recycling, then the community must make additional arrangements for the components of the recycling program not procured under the RFP/B.

Contracting for collection only of recyclable materials is attractive for communities desiring to establish their own municipal processing/marketing program or for communities that are allowed to use a nearby processing plant developed by another community or by the private sector. For example, in New Jersey and Connecticut, communities frequently have access to regional or county processing/marketing plants. Use of such a multicommunity facility is desirable when a community is too small to capture economies of scale in developing its own processing plant, and when the collection method is such that processing is necessary.

Another reason why communities might decide to procure collection services only may be based on the nature of the collection firms in the local market. A community located in a market where major national solid waste management firms operate might reasonably expect a response to a bid requesting any combination of services, ranging from collection only to full service. However, if only small local firms provide refuse collection service in a particular area, or if only one major national firm is present, then the community might wonder how many bids/proposals it might receive in response to a request for full-service recycling. Here, the fear is that the local firms, familiar only with collection, might not respond to a full-service RFP/B, leaving the community to receive but one response.

TABLE 32.11 Integrated solid waste and recycling request for proposals—Seattle, Washington, 1998

TABLE OF CONTENTS

Still another reason why communities might decide to procure collection services only is that recycling services are being procured simultaneously with refuse collection services. In this case, limiting the procurement to collection is often felt to maximize the competition for the contract. Indeed, it is even possible and often desirable to procure recycling services conditionally, at the same time that refuse collection services are being procured. For example, in a 1999 procurement of refuse collection and recycling services, Seattle, WA included the provision that the city would have the right to require the contractor to initiate once a week curbside collection of food waste for a specified fee. This conditional procurement provides the city with an advantageous price for food waste collection, allowing the city to plan for processing and marketing of this commodity.

In sum, to decide the basic structure of the recycling procurement, a city must consider its individual preferences regarding involvement in processing and marketing, and whether processing and marketing alternatives exist and are available to the community. The city must also consider the capability of likely bidders to provide services other than collection, and the city must weigh any expected decrease in numbers of bidders if services other than collection are procured along with collection. The larger the city and the greater the number of large firms operating in the city's local market, the less relevant is this consideration.

Recycling Collection Issues

In preparing a contractual document for recycling collection services (whether a community contracts for full-service recycling or for just recycling collection services alone, these comments will be applicable) a community is wise to specify clearly the services required. A well-written procurement and contract document will specify exactly what is desired, and the document will also allow for future modifications to service specifications. Some of the most important issues to cover in the procurement document and contractual arrangement include the following.

What. The community must specify what materials it wishes the recycling contractor to collect. For curbside programs, these materials are usually those for which there are nearby markets or brokers. If the exact names of the materials cannot be specified, then the number of separate materials should be specified, and limits placed on the relative volume/weight of each of the materials to be collected. This latter comment is most relevant for procurements conducted in advance, such as that of Seattle referred to above.

Who. The request for bids or proposals must specify exactly which households or businesses in the community are to receive the recyclables collection service. Often, it is easiest to specify that the same establishments that receive regular refuse collection service also receive recycling service.

However, the letting of a contract for recycling collection service is a good opportunity to examine the definition of establishments under existing refuse collection arrangements. Many communities providing residential service define residential units as those with fewer than four, three, or two dwelling units per building. Often, when it is time to let a contract, the community finds that there is no ongoing data collection procedure in the jurisdiction that determines exactly how many housing units fit into the specified category. The number of housing units in the specified category is important to the contractor, especially if the contractor is reimbursed on a per-household-served basis. Thus, to avoid onerous and time-consuming surveys of each building in a community to determine the number of dwelling units, it is advisable to establish rules for reimbursement of contractors that are related to regularly collected data in the jurisdiction. The contractor may be required to collect recyclables from all establishments meeting certain criteria; it is not necessary that the contractor's payment be linked to the same definition. For fair treatment of contractors and a minimization of their risk (which can be expected to be translated into a reduced price to the locality), it is desirable if the basis of payment is closely linked to the requirements of service delivery.

When. Here, the community needs to specify how often the recyclable materials will be collected. It has been quite clearly indicated, in cities that have experimented with recyclables collection on the same day as refuse collection (as compared to on a different day), that same-day collection increases participation. This effect may be due to the difficulty of learning new behaviors, peer pressure, or other unknown causes. However, a community that desires to capitalize on this factor must specify not only that recyclable materials will be collected so many times per month, but also that the day of collection will coincide with the scheduled refuse collection for each household serviced. Frequency of collection is very often, in second-generation programs, changed from weekly to biweekly; one-third of communities surveyed in 1997 offered biweekly rather than weekly recycling, and diversion rates were no lower in the group with biweekly service than in the group with weekly service

Why. In order to maximize public participation in recycling programs, it is usually necessary to initiate an education program. (Chapters 9 and 30 of this volume contain more information about this important topic.) This may take many forms, but the direct contact between the collector of recyclable materials and the generator of these materials is usually a key component of any effective program. Local jurisdictions contracting with private firms for collection of recyclables may still want to retain control over the content of the communication between the collector and the customer. This can be accomplished by specifying in the request for proposals that the locality will retain the right to approve materials to be delivered to customers, including door hangers and leaflets and fliers included in bills; the contract must also specify the conditions (e.g., improper segregation or preparation of recyclable materials) under which the collector must deliver one of the items to the household. Table 32.12 highlights some reason for tagging customers' recyclable set-outs. Most localities find that it is wise to require an initial delivery of an informational pamphlet, as well as reinforcing reminders to those households participating incorrectly or not at all.

How. It is important for the community to know how the generator must prepare recyclable materials prior to collection, and in how many compartments it wishes the collector to haul these materials. Clearly, it makes no sense to require more elaborate sorting on the part of the household than on the part of the collector. Imagine the disillusionment of generators who carefully set out separated cans and bottles, then observe a collector loading them willy-nilly into the same compartment. It is possible, however, to require generators to place all recyclables in one container and to require the collector to sort into separate bins in the recycling vehicle prior to delivery to the processing plant. Over the past decade, the number of MRFs has increased from under a hundred to over 500. As more communities add commodities to recycling programs, generator sorting into more than two categories (usually, papers and containers) becomes less and less common.

Where. For residential recycling, materials are picked up either at central drop-off areas or at individual residences. In the latter case, recyclable collection is usually at the

TABLE 32.12 Reasons for tagging recyclables

- Too much contamination (e.g., aluminum foil contaminated with food)
- Nonrecyclables (e.g., porcelain or china mixed with glass jars and bottles)
- Wrong materials (e.g., magazines bundled with newspapers)
- Wrong containerization (e.g., newsprint not properly bundled or tied)

curb. Curbside collection of any material is less costly than backyard or frontyard collection of that material, as curbside collection minimizes walking time to and from the collection vehicle. Curbside collection also allows generators to set their recycling containers out in a visible location, thereby validating their support for recycling. The peer pressure to participate generated by such a process can be important in encouraging households to recycle. A community needs to specify where recyclable materials will be located for pickup. It is also desirable to indicate in the request for bids what containerization requirements will be enforced by the locality.

Specification Preparation

Several general observations concerning the writing of the specifications for work follow. In addition to specifying what must be done, it is important to think ahead to the end of the term, to allow for easy transition, and to assure that no one firm will have an unfair advantage when it becomes time to relet the contract. Important issues in this regard include:

Equipment Ownership. Recycling containers frequently last beyond the initial term of the contract. If the community does not want to give an advantage to the incumbent contractor, it would be wise to specify that containers distributed by the incumbent contractor as a part of establishing the service become the property of the community upon the termination of the contract. Then, the next contractor selected will not have to pay to distribute containers again (this itself is a costly procedure); and the incumbent will not have the competitive advantage of being the only contractor who would not have to do such distribution.

Multiple Service Providers. Many communities, particularly large communities, are concerned that viable competitors will no longer be in business and able to compete, when the term of the first contract is over. To alleviate this worry somewhat, large communities often use a district plan, granting contracts to more than one competitor. Sometimes the competitors include municipal agencies as well as private firms. Cities using this model include Seattle, WA, Boston, MA, Oklahoma City, OK, Phoenix, AZ, Dallas, TX, Fort Worth, TX, and Newark, NJ.

Future Bidding Specifications. It is wise to include clauses in the agreement that would require the contractor to provide information that would assist the city in procuring competitive bids in the future. In franchise agreements, billing files should be turned over from one contractor to another, so that the customer file need not be recreated each time a new agreement is initiated (and so that the incumbent will not have a unique competitive advantage). In contract agreements, it would be helpful to know the quantities of recyclables, ideally by route, and the seasonal patterns, if any, again by route. In San Jose, CA, the recycling contractor collects participation by route and is contractually required to submit these reports to the city. The new refuse and recycling contractors in Seattle will be required to provide Geographic Positioning Satellite (GPS) and radio links between vehicles and customer service agents, allowing real-time communication between drivers, dispatchers, customers, and customer service agents.

Cities need also to consider the "what if" aspects of a contract procurement. Specifying exactly what work is required is important. It is also important to specify what will occur if the required work is not performed as delineated in the contractual documents. The most successful agreements cover the eventualities of a total failure to perform the work as well as less major service irregularities.

Major failures to perform are typically covered by force majeure clauses, defining which causes are deemed not to be the responsibility of the contractor. It is typical not to include any man-made events other than war in this list, thus not including strikes as an incident of force majeure. The jurisdiction is protected against failures caused by the contractor by, typically, a performance bond, which is frequently in the amount of six months' payments to the contractor, and in clauses that allow the community to use the contractor's equipment on an as-necessary basis to provide the service.

Minor failures to deliver service can range from a missed collection (perhaps the responsibility of the generator, who may have placed the container at the curb after the contractor's collection vehicle passed by) to failure to collect on an entire day, to rudeness, failure to monitor recyclables for proper preparation (and appropriate tagging of rejected containers to help educate the public), failure to maintain vehicles properly, failure to provide proper insurance to the jurisdiction, etc. Often, to the extent that these occurrences can be envisioned in advance, they are listed in the contract, and a dollar penalty is associated with each such occurrence. The penalty is set high enough to make it well worth the contractor's management efforts to try to avoid such occurrences, while not so high as to be unreasonable.

Recycling, Processing, and Marketing Issues

The issues that arise in establishing contract parameters for processing are generally similar to those that arise in collection. In processing and marketing, however, there are additional issues relating to the sharing of risk of market price changes for commodities, interaction between the collection contractor and the processing contractor, and control over the processing decisions, regarding which commodities will actually be produced.

The basic issues of delineating what is to be performed, when, where, how, and why apply to processing as to collection. The commodities that are to be received and processed can probably be specified in advance, at least for the start of the contract, but the community may wish to retain some flexibility to change components in mid term in response to changes in market conditions.

To avoid issues of auditing and enforcement, some processing and marketing contracts do not include a provision for a sharing of revenues. In these contracts, either the entire risk of change in market prices is borne by the contractor or the risk is shared in a formula tied to published prices, rather than to revenues received. When the entire risk of processing and marketing is borne by the contractor, the jurisdiction typically pays the contractor a set amount per household or per ton handled. The contractor is then free to keep the profits (if any) from sale of commodities. This is the model adopted by East Brunswick, NJ in its recycling program for household recyclables. Alternatively, the jurisdiction can pay a fee to the contractor, with the magnitude of the fee determined by changes in the market prices for commodities. The City of Philadelphia uses this approach, relying on market prices as published in the *Recycling Times* to compute an index number, which is then applied quarterly to establish the tipping fee for recyclables delivered to the contractor's processing facility.

Another issue arises in setting performance standards for the processor. Typically, communities are interested in recycling as much as possible of the materials delivered to the processor; they would like a minimum of bypass waste sent to ultimate disposal. The processor can guarantee to meet bypass guarantees only if the purity of the incoming feedstock stream can be assured or if the contract allows for loads containing unacceptably high levels of contamination to be rejected. The contract with the collector must be coordinated with that of the processor, to allocate responsibility for rejecting inappropriately prepared materials prior to collection and for rejecting loads that mistakenly included such materials. Collection methods alone can also affect the processing efficiency. For example, breakage of glass in collection reduces the ability to color-sort this commodity;

in areas of the country without markets for mixed glass, breakage in collection may foreclose the possibility of recycling.

Jurisdictions may also wish to assert control over the manner in which commodities are marketed. For example, when yard wastes are composted, the community may wish to develop a brand name and identity, with an ongoing history of product performance. If this is the case, then the contract must provide for community ownership of the brand at the termination of the processing contract.

The Municipal Arrangement

To establish an in-house recycling program, all the planning necessary to set up a contract or franchise arrangement must take place, and more. The jurisdiction must be clear on what materials will be collected, when, from whom, in what type of containers, and delivered to which processing location. Additionally, the jurisdiction must attempt to predict participation and diversion rates, so that volumes and weights of materials to collect can be estimated. The volume and weight estimates affect the selection of recyclables containers for use by households and the specification of recycling vehicles. Although general predictions of volumes of materials will assist in sizing the compartments in the collection vehicles, ideally specified so that all compartments fill up at approximately the same time, it is not necessary to attempt perfection in this specification, as waste stream composition, participation, and even materials eligible for collection can all be expected to change over the useful life of the vehicles.

With work rates and vehicle needs comes the determination of the number of collectors that will be required. In the municipal arrangement, some of these workers may be freed up from refuse collection routes, which can typically be extended as recycling is implemented. The number of workers required is a function of crew size, work rate, distance to the processing facility, type of collection, and personnel practices regarding such matters as vacations, holidays, and length of the work week. When personnel have been recruited, they must be trained in collection methodology, vehicle operation, safety, and routes. Planners and supervisors are typically also required to initiate a municipal program.

Processing facilities require careful specification of equipment components and the physical layout of the equipment and the flow of commodities in the plant. Consideration needs to be given to the type of collection vehicles and the interface between these vehicles and the processing plant. With a well-configured system, it is possible to avoid dumping on a floor, with subsequent loading of recyclables onto conveyors with a bucket or scoop, and instead deposit recyclables directly onto the conveyors, without requiring the additional handling and loading step.

The municipal arrangement theoretically allows the jurisdiction complete flexibility to change systems as desired, but these opportunities must be balanced against the need to change behavior on the part of the public each time such a change is implemented. In addition, once municipal workers, whether unionized or not, become permanently employed in the recycling program, it may be difficult in practice to disrupt established staffing patterns. Thus, even if a new technology becomes available to allow collection or processing with fewer workers, implementation may be resisted by interest groups fearing loss of employment.

The public sector service also leaves the jurisdiction with no private sector partner with whom to share the risk of liability—personal or property, or the risks of changes in the markets for recyclables. An additional risk the municipality shoulders alone is the risk of capital obsolescence. If the jurisdiction finances its own program, then its expenditures on capital are a sunk cost, which may not be fully recoverable if technology changes during the expected useful lifetime of the equipment or plant.

Other liability risks the community bears alone are potential liability for the quality of materials delivered to processors. For example, if contaminated materials are delivered to markets, the impact of these materials may not be observed until after they have been combined with other materials and processed. At this time, it may be discovered that the glass is not the proper color or that the paper has been discolored by laser paper contained in the feedstock or that the cardboard is rendered useless due to odor from a pesticide once contained in the recycled materials. In any of these cases, the typical risk is that the newly produced material cannot be used for its intended purpose, and the community delivering the contaminated material may be liable for the loss on the production load, if it can be proved that contamination originated in the materials delivered by the community. Until legislation is passed that exempts public entities from liability arising from the sale of recycled goods, after the community collects them, this potential exposure will remain. Public entities, with their "deep pocket" resource of taxation, would be especially attractive targets for financial restitution. This exposure, pending legislation, can be lessened through careful disclaimer clauses in purchase and sale contracts.

While there have been no comparative studies of the costs of recycling services provided by the public sector as compared to the private sector, there is no reason to believe that, at least for collection, general results differ from those that have been found for refuse collection. In that area, municipal agencies in large cities with populations over 50,000 deliver service at an average cost significantly in excess of the price similar-sized cities pay to private contractors; there are, however, individual cities that achieve results opposite to this general conclusion. For example, Phoenix, Arizona has a municipal agency that competes against the private sector in that city's procurements of refuse collection services, and the public agency has recently been selected as the low-price proposer for all the refuse collection districts in that city. Thus, cities considering municipal service need to evaluate whether their costs will significantly exceed those of the contract or franchise arrangement.

In any arrangement, there are considerations that affect overall program efficiency and effectiveness. These include allowing adequate time for program initiation, planning for flexibility, changing elements as appropriate in response to changes in the external and the internal environment, and measuring the success of the program on an ongoing basis. These elements are discussed below.

PROGRAM EVALUATION AND CORRECTIVE ACTION

Typical Planning Requirements

A longer time period than might be expected is typically required to implement a recycling program. Consideration of the alternatives and a decision regarding the type of program and the preferred organizational arrangement can occupy periods of several years, especially in large cities with numerous interest groups and affected parties. Even in small communities, it is not unusual for a planning group to spend at least a year deciding upon the basic form of the recycling program.

Once the basic decisions regarding the program have been made, the action steps for implementation must occur. For the private arrangement, necessary legislative changes can sometimes be accomplished quickly, and sometimes they can take years. Franchise and contract arrangements each require somewhat the same procedures: (1) preparation and issue of procurement documents, (2) receipt of and evaluation of proposals or bids, (3) selection and signing of agreement, (4) interval between selection and beginning of work, and (5) beginning of program. Procuring services of the private sector without an adequate time interval for each of these procurement milestones tends to reduce competition among private firms, to decrease their efficiency in preparing the proposal and gearing up for contract initiation, and, consequently, to result in higher prices to the community.

Of particular importance is to allow for the appropriate time interval between award of the contract and beginning of the work. Particularly for a processing facility, where construction and permitting may be required, an interval of between one and two years will probably not be too long, depending upon the local difficulties involved in siting and permitting and construction. Even for programs where the contractor need not site and permit a new facility, such as collection-only contracts, capital equipment must often be procured, and this can take up to six months, in some cases.

In sum, once the planning has been completed for a contract or a franchise arrangement, it typically requires at least a year to procure and achieve start-up of collection services and perhaps another year to procure and achieve start-up of a recyclables processing facility. A similar time frame applies to the municipal arrangement. Whereas in this arrangement it is not necessary to prepare the request for proposals/bids or to evaluate the responses from the private sector, it is necessary to prepare procurement documents for capital equipment, and to comply with governmental requirements for personnel recruitment and training. Typically, with the municipal arrangement, it is also necessary to coordinate the start of the program with the annual budget cycle, especially when new capital equipment is required.

Program Flexibility

Flexibility to add new materials to the list of recyclables or to alter the collection/processing technology in response to new developments is a primary concern of officials of most jurisdictions. Perhaps the primary step a community can take to preserve this option is to enter into contracts and franchises that are not overly long, say about five years. Changing the technology of collection or processing is costly when initiated prior to the normal time for equipment replacement. However, there may be instances when such replacement is required if, for example, technology changed so that there was no market or only a very undesirable market for materials prepared the "old" way. It is certainly possible and probably desirable to include a "contract opener" clause that would allow renegotiation of any or all elements of the contract in the event of major technology change or market shift. The jurisdiction can probably protect itself more easily against market shift that causes the contractor to be unable to market the recyclables by citing such a failure as a cause for contract termination.

Allowing for too much change in the terms of a contract can increase, probably unnecessarily, the price the community must pay. For example, if the contract allows the community to change the commodities to be collected, to change the quantities that must be processed, to change the number of separate commodities that are collected, etc., then the flexibility to the community increases the risks of cost-increasing change to the contractor, and the community can expect to pay more for the flexibility. Ideally, a community interested in a competitive price and flexibility should insert flexibility clauses that do not place all the risk on the contractor and do not cause the community to pay more for an option that is extremely unlikely to occur. For example, it is unlikely that a new collection technology will be developed and implemented within the next few years that would justify changing the collection methods in a contract with just a few years until the expiration of the term.

Program Modification

Communities continue to modify their programs to enhance effectiveness and to reach a wider population. For example, a community might expand a residential recycling program to include small commercial establishments generally located along the regular residential routes. With a private arrangement or a municipal arrangement, these establishments can be incorporated by the existing recycling vehicles, perhaps with some rerouting, depending upon how many such establishments there are. With a contract or franchise arrangement, careful structuring of the procurement and contract documents can also make this type of modification easy. For example, the contract might say that, with six months' notice, the city would have the right to require the contractor to collect recyclables from up to X additional residential or commercial establishments, all to be located within the contract or franchise area, for a unit fee equal to the per-unit fee then prevailing in the contract. Planning ahead for increased service can make implementation in the future easier and less costly. In this example, planning to require the regular recycling contractor to include new categories of generators in the service is more efficient than contracting with a second firm whose collection vehicles would cover the same areas of the community as do those of the regular recycling contractor.

One area that should be highlighted is the relationship between regular refuse collection and recyclables collection. Where the two services are performed by the same firm and covered by the same contract, integration is probably already achieved, so long as the services are provided with a single fee. However, when the community pays one fee for refuse service and another fee for recyclables services, and when each fee is expressed in per-household terms, then better integration can be achieved. When the refuse collection contract comes due, the specification should be changed so that the community gets a rebate, at least equal to the disposal fee per ton times the number of tons of recyclables collected, from the households covered by the contract. When contracts come due at different times, it may be necessary to live with a few years of nonintegration, but the disposal savings from diverting recyclables from the refuse stream is clearly available to the community, by simple respecifying of the collection contract.

Enforcing Program Requirements

Mandatory recycling programs typically involve monitoring for participation. Participation can be enforced by inspection of the residual refuse, to check for presence of materials that should be in the recycling rather than the refuse container. This method is practiced by several of the communities in New Jersey, including Jersey City and Newark, where mandatory recycling programs have the longest track record. The sanitation inspectors generally warn and instruct first-time offenders in proper recycling behavior. Second- and third-time offenders are often fined, and in some communities a court appearance for repeated offenses is a possibility.

Recycling programs can also be enforced by checking on participation in the recycling program itself. Various communities require their collectors to determine participation, often using automatic counters to indicate the number of set-outs on a route. There is just a small leap from this approach to determining which generators on a particular route actually set out recyclables, of which kind, and in what quantities. Several companies offer computerized methods for keeping track of participation rates, quantities of refuse and recyclables, and other customer data. These systems use computers, Geographic Positioning Satellite (GPS) systems, radios, and bar codes for input and transmittal of data between route drivers, customer service agents, dispatchers, billing systems, and customers. Seattle will require that all contractors equip vehicles with the Trakit system, whose performance capabilities are highlighted in Table 32.13. Many companies use hand-held computers to record participation of customers, particularly in the commercial sector.

Voluntary programs, of course, do not enforce participation among customers. However, such programs may employ inspectors to enforce compliance with program regulations and to encourage participation. Economic incentives for participation in recycling, generally taking the form of volume-based fees for solid waste services, have resulted in dramatic increases in recycling quantities and participation in recycling programs in communities throughout the United States.

Measuring Performance

No program can be implemented and maintained at peak efficiency and effectiveness without constant evaluation. For program evaluation, it is necessary to know what is being done, by whom, and at what costs. In a recycling program, desired information includes the following:

- Quantities of recyclable materials collected, by category of generator, by type of material. For example, to know the pounds of newspapers, commingled glass and cans, and yard waste collected from single family homes.

TABLE 32.13 Geographic Positioning Satellite Systems (GPS) and computer recording of route activities

System components:

1. GPS unit reads the truck's location and connects the Barcode Pad & Pen and the Display Unit to the truck's radio
2. Barcod Pad & Pen lets driver send information to base station computer—records no set-outs, service problems, repair needs, etc.
3. Display Unit lets driver receive information such as need to provide a special pickup

System capabilities:

- Automatic monitoring of truck location
- Automatic recording of street location (driver only needs to note number of address where an event occurs)
- Automatic monitoring of truck speed and routing
- Inputs to billing system
- Note if receptacle is wrong size or "extra" materials have been set out
- Provide return trip requests—e.g., if customer notes that recyclables have been tagged, removes the contamination, call the service desk, customer can receive pickup that same day—driver might get a message like:

```
11204 NE 208th PLACE
—CONTAMINATION
REMOVED. PLEASE RETURN.
THANKS!
```

- Percentage diversion of the waste stream, by category of generator. For single-family homes, the aggregate of the refuse collected plus the recyclables collected is the denominator of this percentage, with the quantity of recyclables collected as the numerator. Of course, it is also desirable to compute the overall diversion of the waste stream, using aggregate community waste generation (from all sources) as the denominator.
- Cost of the recycling program. For private sector arrangements, the prices charged by the private firms providing the service may be used as indicative of the cost to the community. For municipal arrangements, it is important to include all the program costs, even those that may not be included in the operating budget, such as amortization of capital equipment, vehicle repair and maintenance, etc.
- Participation and satisfaction with the system. The participation rate can be difficult to measure, as some generators may place recyclables for collection less often than the service is offered. A telephone survey to inquire as to satisfaction can be helpful, and the indicated participation from such a survey typically gives an upper bound to actual participation levels. Address specific recording of containers placed for collection is necessary to compute a totally accurate participation rate. Unless a community has a program whereby user fees are tied to participation in the recycling program, it is probably wiser to spend scarce resources to determine the diversion rate rather than the participation rate.
- Sales of materials and prices at which the materials are sold. It is desirable to obtain this information, even if the payments to the private contractor are not tied to these variables. The information helps the community to develop its economic picture of the recycling program, and the data will be useful in aggregate form for developing additional recycling programs.

While there is no national data base regarding recycling rates and costs, the programs presently being established by the various states are in effect creating the basis for such a system. It is probable that standards for cost accounting and measuring diversion and participation rates will be established in the not too distant future. Any community with a well-thought-out measurement system will be well positioned to adapt to any future record keeping requirements. Chapter 33, "Data Collection and Cost Control," (this volume) addresses these very important topics in greater detail.

CHAPTER 33

DATA COLLECTION, COST CONTROL, AND THE ROLE OF COMPUTERS IN THE RECYCLING INDUSTRY

DANIEL E. STROBRIDGE
Associate
Camp, Dresser & McKee, Inc.
Tampa, Florida

FRANK G. GERLOCK
Senior Solid Waste Planner
Camp, Dresser & McKee, Inc.
Tampa, Florida

RICHARD HLAVKA
President
Green Solutions
South Prairie, Washington

INTRODUCTION

Recycling is an essential component of an integrated waste management system. The success of a recycling program can be measured not only by its ability to reduce the quantity of solid waste being landfilled or incinerated, but also by the costs incurred to operate the program. Economic evaluations are performed to determine which methods of recycling make the most sense economically. For example, is it more cost-effective to operate two-person curbside pickup crews or one-person crews? The collection and analysis of data can help show which are the least-cost methods for program operation.

The capital and operating costs of a comprehensive multimaterial recycling program are significant expenses which will not necessarily be met by material sales revenues. The magnitude of these expenses requires that the system be managed as efficiently as any other public works service. To do this, system data must be collected periodically for a broad number of program parameters. Cost analyses and management decisions can then be based on this information.

In addition to aiding management decisions, the collection of certain data is required by various state and federal regulations. The solid waste regulations in a number of states require documentation of recycling efforts and material quantities recycled. Federal lawmakers are preparing to revise the Resource Conservation and Recovery Act and may include national goals for recycling.

This chapter addresses methods of data collection, suggests which data should be collected to facilitate efficient program management, gives examples of data collection forms and procedures used by existing programs, demonstrates methods of data analysis, and describes techniques for controlling recycling program costs.

DATA COLLECTION

Planning and managing a recycling program requires the collection and analysis of several types of data. A good data collection and management system (DCMS) is a valuable tool that should be used to evaluate the efficiency of a recycling program and to plan program improvements and expansions.

Types of Data

The data needed to plan and manage a recycling program are essentially the same, regardless if the system is operated by a municipality or a private contractor. However, the information required to support a municipally operated program may need to be more comprehensive to ensure proper governmental accountability.

Before data are collected and analyzed, it is important to determine the objectives for use of the data. When these objectives are identified, the DCMS can be designed. For example, data may be used to show:

- Volume of material diverted from the waste stream
- Number of households serviced
- Source of waste (i.e., commercial, residential, multifamily)
- Number of setouts serviced per day
- Participation rate per month
- Participation rate for each material type
- Frequency of participation per month
- Route size
- Time to complete route
- Number of containers per truckload

Data Collection Methods

Information may be collected manually or by electronic automated methods. The choice of which method to use must consider factors such as:

- Available resources (e.g., budget, staff, available technical expertise)
- System size
- Available time
- Data requirements

Until recently, automated systems were principally used for data input, manipulation, and presentation of data; manual methods were used primarily for data collection. However, with the increased development and use of high-technology devices like optical laser disks, geographic information systems, and hand-held computers, it is not possible to implement a completely automated DCMS.

Before data are collected, by manual or automated means, the manner in which they will be analyzed should be determined. Data must be input in a format that will achieve the desired results. Questions to be answered include

- What types of data are being collected?
- What quantity of data is being collected?
- How many runs of data will there be?
- How will the data be organized?

Quantities of Recyclable Materials

Since the major benefits of a recycling program are directly related to the quantities of materials collected, the first task in data collection should be to estimate the amount of material that can be recovered. This can be calculated easily by multiplying the total solid waste generation quantity times the percentage of selected recyclable material in the waste stream times the estimated participation rate. Table 33.1 shows an example of this formula, excluding estimates of participation rates. This initial calculation is extremely helpful in assessing the future accomplishments of a recycling program.

Next, a sensitivity analysis should be performed to determine the yearly tonnages of selected recyclable materials that could be recovered at various recovery rates. This is done by multiplying the selected recyclable material component total annual tonnage by the estimated rate of recovery. Tables 33.2 through 33.5 show the results for rates of recovery at 5, 15, 30, and 50 percent, respectively. This example indicates that a recovery

TABLE 33.1 Indiana Jones County, estimated quantities and composition of recyclable materials in the county's solid waste stream (tons per year, rounded)

		1990	1991	1992	1993	1994	1995	2000
Estimated total solid waste generation		150,123	156,127	162,373	168,868	175,623	181,857	218,228
	Estimated	Recyclable quantities						
Selected recyclable material	percent of waste stream*	1990	1991	1992	1993	1994	1995	2000
Newspaper	12.0	18,015	18,735	19,485	20,264	21,075	21,823	26,187
Corrugated cardboard	10.0	15,012	15,613	16,237	16,887	17,562	18,186	21,823
Office paper	2.5	3,753	3,903	4,059	4,222	4,391	4,546	5,456
Aluminum	1.5	2,252	2,342	2,436	2,533	2,634	2,728	3,273
Plastic (containers)	0.5	751	781	812	844	878	909	1.091
Glass containers	6.0	9,007	9,368	9,742	10,132	10,537	10,911	13,094
Yard waste	15.0	22,519	23,419	24,356	25,330	26,344	27,279	32,734
Bimetallic cans	2.5	3,753	3,903	4,059	4,222	4,391	4,546	5,456
Totals	50.0	75,062	78,064	81,186	84,434	87,812	90,928	109,114

*From national and state waste composition data.

TABLE 33.2 Indiana Jones County estimated recyclable material quantities at 5% recovery rate* (tons/year)

	Total solid waste generated					
	1990	1991	1992	1993	1994	1995
Recyclable material†	150,123	156,127	162,373	168,868	175,623	181,857
Newspaper	901	937	974	1,103	1,054	1,091
Corrugated cardboard	751	781	812	844	878	909
Office paper	188	195	203	211	220	227
Aluminum	113	117	122	127	132	136
Plastic (containers)	38	39	41	42	44	45
Glass	450	468	487	507	527	546
Yard waste	1,126	1,171	1,218	1,267	1,317	1,364
Bimetallic cans	188	195	203	211	220	227
Totals	3,753	3,903	4,059	4,222	4,391	4,546

*Yields a 2.5 percent waste stream reduction. That is, if 5 percent of the total recyclable materials shown in this table were recovered, it would reduce the total solid waste generated in the county by 2.5 percent.

†Recyclable materials shown represent the percentage of the total waste stream composition developed in Table 33.1.

TABLE 33.3 Indiana Jones County estimated recyclable material quantities at 15% percent recovery rate* (tons/year)

	Total solid waste generated					
	1990	1991	1992	1993	1994	1995
Recyclable material†	150,123	156,127	162,373	168,868	175,623	181,857
Newspaper	2,702	2,810	2,923	3,040	3,161	3,273
Corrugated cardboard	2,252	2,342	2,436	2,533	2,634	2,728
Office paper	563	586	609	633	659	682
Aluminum	338	351	365	380	395	409
Plastic (containers)	113	117	122	127	132	136
Glass	1,351	1,405	1,461	1,520	1,581	1,637
Yard waste	3,378	3,513	3,653	3,800	3,952	4,092
Bimetallic cans	563	586	609	633	659	682
Totals	11,259	11,710	12,178	12,665	13,172	13,639

*Yields a 7.5 percent waste stream reduction. That is, if 15 percent of the total recyclable materials shown in this table were recovered, it would reduce the total solid waste generated in the county by 7.5 percent.

†Recyclable materials shown represent the percentage of the total waste stream composition developed In Table 33.1.

TABLE 33.4 Indiana Jones County estimated recyclable material quantities at 30% percent recovery rate* (tons/year)

	Total solid waste generated					
Recyclable	1990	1991	1992	1993	1994	1995
material†	150,123	156,127	162,373	168,868	175,623	181,857
Newspaper	5,404	5,621	5,845	6,079	6,322	6,547
Corrugated cardboard	4,504	4,684	4,871	5,066	5,269	5,456
Office paper	1,126	1,171	1,218	1,267	1,317	1,364
Aluminum	676	703	731	760	790	818
Plastic (containers)	225	234	244	253	263	273
Glass	2,702	2,810	2,923	3,040	3,161	3,273
Yard waste	6,756	7,026	7,307	7,599	7,903	8,184
Bimetallic cans	1,126	1,171	1,218	1,267	1,317	1,364
Totals	22,518	23,421	24,356	25,330	26,343	27,279

*Yields a 15 percent waste stream reduction. That is, if 30 percent of the total recyclable materials shown in this table were recovered, it would reduce the total solid waste generated in the county by 15 percent.

†Recyclable materials shown represent the percentage of the total waste stream composition developed in Table 33.1.

TABLE 33.5 Indiana Jones County estimated recyclable material quantities at 50% percent recovery rate* (tons/year)

	Total solid waste generated					
Recyclable	1990	1991	1992	1993	1994	1995
material†	150,123	156,127	162,373	168,868	175,623	181,857
Newspaper	9,007	9,368	9,742	10,132	10,537	10,911
Corrugated cardboard	7,506	7,807	8,119	8,443	8,781	9,093
Office paper	1,877	1,952	2,030	2,111	2,195	2,273
Aluminum	1,126	1,171	1,218	1,267	1,317	1,364
Plastic (containers)	375	390	406	422	439	455
Glass	4,504	4,684	4,871	5,066	5,269	5,456
Yard waste	11,259	11,710	12,178	12,665	13,172	13,639
Bimetallic cans	1,877	1,952	2,030	2,111	2,195	2,273
Totals	37,531	39,034	40,593	42,217	43,906	45,464

*Yields a 25 percent waste stream reduction. That is, if 50 percent of the total recyclable materials shown in this table were recovered, it would reduce the total solid waste generated in the county by 25 percent.

†Recyclable materials shown represent the percentage of the total waste stream composition developed in Table 33.1.

rate of 50 percent of all recyclable materials would be necessary to achieve a 25 percent reduction in the total waste stream, Of course, the actual waste stream composition and recovery rate for each material will vary.

Materials Market Survey

After recyclable materials have been identified, potential markets must be investigated. A telephone survey will indicate which markets purchase or use the materials targeted for recovery, current market prices and specifications, and willingness of potential markets to execute purchase contracts. Figure 33.1 is a sample questionnaire that may be used for telephone interviews. Table 33.6 shows how the data collected from interviews might be summarized.

The amount of revenue that can be expected from the sale of materials can be estimated using the data collected in the two previous steps. Multiply the estimated quantity of each material by the estimated sale price of that material. Table 33.7 shows a graphic display of projected revenues.

As much information as possible should be gathered on a monthly basis from the materials market contractor. Important considerations are:

- How will diverting the recyclable material contribute to the established recycling goal?
- Does current market value justify the cost of recycling the selected material?
- Do avoided disposal costs make the project cost-effective?

When materials are selected for a recycling program, there are no firm rules that apply universally to all communities. However, diversion goals and marketability typically are the most significant factors.

Route Data

After a recycling program is established, data must be collected and analyzed to determine:

- Level of participation
- Volume of material being collected
- Operational efficiency

More specifically,

- Participation rates for specified time periods
- Participation rates by material type
- Route productivity
- Crew hours worked
- Overall collection rate
- Crew productivity

Data that must be collected include:

- Collection route identification
- Crew identification
- Total number of service addresses
- Total number of setouts

INDIANA JONES COUNTY
MATERIALS MARKET SURVEY QUESTIONNAIRE

COMPANY NAME: ____________________

ADDRESS: ____________________

CONTACT PERSON: ____________________

PHONE NO.: ____________________

Listed below are materials being considered for recovery from the study area's solid waste.

1. Please check the materials your company purchases, and indicate any minimum or maximum quantity you typically purchase and the quantity, if any, you are currently receiving from within Indiana Jones County.

	Minimum Quantity	Maximum Quantity	Quantity from Indiana Jones
____ Newspaper	____	____	____
____ Corrugated Cardboard	____	____	____
____ Other Grades of Paper	____	____	____
____ Light Gauge Ferrous Metals (tin and bi-metal cans)	____	____	____
____ Heavy Gauge Ferrous Metals (appliances and misc.)	____	____	____
____ Other Metals	____	____	____
____ Aluminum Cans	____	____	____
____ Glass	____	____	____
____ Plastic	____	____	____

MCGHOT. 1/9 Page 1

FIGURE 33.1 Materials market survey questionnaire.

INDIANA JONES COUNTY
MATERIALS MARKET SURVEY QUESTIONNAIRE

	Minimum Quantity	Maximum Quantity	Quantity from Indiana Jones
____ Yard Trash	________	________	________
____ Wood Waste	________	________	________
____ Composted Materials	________	________	________
____ Resource Recovery Residue	________	________	________

2. What are your company's specifications (or specific requirements) for the materials you purchase?

____
____
____
____

3. For each material checked in Question No. 1, please indicate the unit price you are presently paying for the material and the basis for your price offering.

Material	Current Unit Purchase Price	Pricing basis
________	________	________
________	________	________
________	________	________
________	________	________
________	________	________
________	________	________

MCGHOT. 1/9 Page 2

FIGURE 33.1 *(continued)*

INDIANA JONES COUNTY
MATERIALS MARKET SURVEY QUESTIONNAIRE

4. Would you be willing to purchase all the material checked in Question No. 1 from the study area?

   ____ Yes ____ No

   If no, please list the materials you would be interested in purchasing.

   ____

   ____

5. Would your company be willing to enter into an intermediate or long-term purchase agreement for materials recovered from the study area's solid waste?

   ____ Yes ____ No ____ Maybe

6. Would your company be willing to offer a floor price (a minimum purchase price regardless of market conditions) for the purchase of these materials?

   ____ Yes ____ No ____ Maybe

7. Would your company be willing to provide containers, processing equipment, or transportation for recyclable materials?

   ____ Yes, Please indicate which ______________________________

   ____

   ____ No

8. Please provide any other information about your company or recycling programs which might be helpful to our project.

   ____

   ____

   ____

   ____

   ____

   ____

MCGIIOT. 1/9 Page 3

FIGURE 33.1 *(continued)*

TABLE 33.6 Indiana Jones County materials market survey

Company & address	Materials purchased	Specifications	Current purchase price	Willing to contract	Offer floor price	Provisions
Tin-Can Man 3021 SW 1st Terr	Aluminum cans	Aluminum only	$0.50-0.70/lb	Yes	Yes	Containers and transportation are highly possible if built into the negotiations.
No Name, OK	Corrugated cardboard	Separate from other materials	$25–40/ton	Yes	Yes	
Contact: Jim McCann 555-555-5555	All other grades of paper, except newsprint		Competitive	Yes	Yes	
Can Reclaim Recovery Corp. Fairview, AL	Aluminum cans	Aluminum only, no impurities (e.g., steel, dirt, plastic)	$0.45–0.47/lb	Yes	Possible	Containers and transportation would be provided.
Contact: Rick Dees 555.555.5555	Glass	Separated by color, no caps or metal rings, labels OK	$50/ton minus delivery and handling costs	Yes	Possible	
	Newspaper	Newspaper only	$20/ton	Yes	Possible	
	Aluminum scrap		Competitive	Yes	Possible	
	Radiators		Competitive	Yes	Possible	
	#1 & #2 copper		Competitive	Yes	Possible	
	Brass		Competitive	Yes	Possible	

TABLE 33.7 Projected revenue from Sale of recyclable material, Indiana Jones County

	15%				30%				50%			
	Volume, tons		Revenue/yr		Volume, tons		Revenue/yr		Volume, tons		Revenue/yr	
Material	1990	1995	1990	1995	1990	1995	1990	1995	1990	1995	1990	1995
Newspaper	2,2702	3,283	81,066	98,190	5,404	5,547	162,120	196,540	9,007	10,911	270,210	327,330
Corrugated cardboard	2,252	2,728	112,600	136,400	4,504	5,456	225,200	272,800	7,506	9,093	375,300	454,650
Bottle plastic	113	136	113,000	136,000	225	273	225,000	273,000	375	455	375,000	455,000
Aluminum cans	378	409	371,800	449,900	676	818	743,600	899,800	1,126	1,364	1,238,600	1,500,400
Glass	1,351	1,637	54,040	65,480	2,702	3273	108,080	130,920	4,504	5,456	180,160	218,240
Ferrous metals	563	682	16,890	20,460	1,126	1,364	33,780	40,920	1,877	2,273	56,310	68,190
Totals	7,319	8,865	749,390	906,430	14,637	17,731	1,447,780	1,808,850	24,395	29,552	2,495,580	3,023,810
Waste stream reduction	4.9%		9.8%	16.4%								

*The prices used are for demonstration purposes only. Actual revenue may vary: newspaper, $30/ton; corrugated cardboard, $50/ton; aluminum, $1000/ton; glass, $40/ton; ferrous metals, $30/ton; PET bottles, $100/ton

- Total number of truckloads
- Truckload weight
- Time on route
- Time between stops
- Time to pick up each setout
- Demography/density
- Miles on route
- Type of service location (single-family, multifamily, commercial)
- Type of materials set out
- Tonnage of materials set out
- Geography
- Weather

The frequency of data collection will vary with the type of data collected. Data such as number of service addresses, demographics, geography, type of service location, and miles on route will be collected only once (unless the route changes).

Information that should be collected monthly would include time between stops and time required to pick up each setout. Data that should be collected daily include collection route and crew identification, number of setouts, number of truckloads, truckload weight, time on route, and type and volume of materials set out.

Route Data Collection Protocol

A sample data collection form is shown in Fig. 33.2. Data collection forms should be simple and easy to complete, with no en route calculations required. The steps for route data collection should be thoroughly rehearsed, with "hands-on" practice by the people who will collect the information. Sample forms should be developed with examples of correct data input, and potential problems should be discussed.

If there are sufficient workers, it is recommended that a trained data collector accompany each driver on the first run. The driver may watch as the data are collected, and questions can be raised and answered. On the second run, the driver should record the data, and the trained data collector should observe and answer questions.

Initially, data logs should be checked daily to ensure that they are completed correctly and generating the appropriate types of information. If adjustments are necessary, it is much easier to make them early in the program.

DATA ANALYSIS

Data analysis must be accurate and timely. The value of careful data analysis cannot be overstated, for it is the analysis that will determine the direction the program should take to assure its continued success.

Methods of Analysis

The data that have been collected must be organized into a format that is convenient for use. Data may be organized manually or with an automated system. If an automated sys-

DATE:	________	No. of Households Served on Route:	________
VEHICLE I.D.:	________	No. of Set-outs:	________
ROUTE I.D.:	________		
OPERATOR:	________		
WEATHER:	________		

Location:	Time	Mileage
Beginning	________	________
Leaving Lot:	________	________
Beginning on Route:	________	________
Leave for Disposal Site:	________	________
Arrival at Disposal Site:	________	________
Leave from Disposal Site:	________	________
Arrival Back on Route:	________	________
Leave for Disposal Site:	________	________
Arrival at Disposal Site:	________	________
Leave from Disposal Site:	________	________
Ending on Lot:	________	________

Disposal Activity		Type of Material	Weight
Load #1	Aluminum, Cans		________
	Newspaper		________
	Glass		________
	Plastic		________
	Yard Waste		________

Prepared By:

________ Operator ________ Date

FIGURE 33.2 Recycling route information.

tem is used, the data use objectives should be considered when a software package is selected. Some elements to consider are:

- The data input screen should imitate the data collection form.
- The data verification process should be incorporated in the program.
- Accounting requirements should be satisfied.

- The system should be flexible enough to allow for change.

When data summaries are required, reports can be generated by Lotus 1-2-3 or dBase III Plus. Reports can be designed to show information and relationships regarding participation, frequency, cost, and many other facets of a recycling program. The following example describes how data may be used:

Data	Use
Tonnage/time period lbs/setout/load	Project total tonnage for collection, transport, processing, and marketing Determine type of pickup and containers
Participation rate	Number and type of trucks
Time/pickup	Participation projections
Pickups/mile	Project staffing
Time/mile	
Volume/time period	
Cost/ton	Capital improvement program
Cost/material	Expand, modify, or delete program
Cost/setout Cost/system component	

Tables 33.8 and 33.9 illustrate typical spreadsheet data.

COST CONTROL

After a system is devised to collect and analyze data, the final step is to use the collected information to control program costs. Program costs can be categorized as

- Collection
- Processing
- Public education
- Marketing, administration/overhead

Costs may be allocated on a full-cost basis (including overhead and indirect costs proportionately assigned to the program), or on a direct-cost basis (including only costs that are distinctly identified with the recycling program, such as labor and equipment). The level of volunteer assistance and contributions received will affect these costs (e.g., donated equipment, donated facilities, public service media time).

TABLE 33.8 Indiana Jones County curbside recycling program contractor monthly statistics, december

	Price	Tin Man, total lbs	Revenue	Cycle One, total lbs	Revenue	Northside, total lbs	Revenue	Central, total lbs	Revenue	County, total lbs	Total revenue
Newspaper	$0.00500	223,000	$1,115.00	227,860	$1,139.30	138,600	$693.00	84,280	$421.40	673,740	$3,368.70
	0.00250	0	0.00	37,135	92.84	0	0.00	0	0.00	37,135	92.84
Glass/clear	0.00500	31,620	158.10	34,920	174.60	21,380	106.90	52,140	260.70	140,060	700.30
Glass/colored	0.00000	25,620	0.00	22,820	0.00	12,900	0.00	0	0.00	61,340	0.00
Glass/both	0.00500	0	0.00	8,219	41.10	0	0.00	0	0.00	8,219	41.10
Aluminum	0.31000	4,840	1,500.40	4,240	1,314.40	3,140	973.40	2,100	651.00	14,320	4,439.20
	0.31000	0	0.00	546	169.26	0	0.00	0	0.00	546	169.26
Plastic	0.00500	0	0.00	0	0.00	0	0.00	8,800	44.00	8,800	44.00
Total		285,080	$2,773.50	335,740	$2,931.49	176,020	$1,773.30	147,320	$1,377.10	944,160	$8,855.39
Total tons		143		168		88		74		472	
Contract homes		11,976		11,987		6,145		8,016		38,124	
Contract rate			$1.595		$1.595		$1.92		$1.89		
Gross cost			19,101.72		19,119.27		11,798.40		15,150.24		65,169.63
Revenue/county			1,386.75		1,465.75		886.65		688.55		4,427.70
Net cost			17,714.97		17,653.52		10,911.75		14,461.69		60,741.93
Cost/home			1.48		1.47		1.78		1.80		6.53
Total drive-bys		48,811		46,794		25,034		33,921		154,560	
Total setouts		17,399		20,829		10,304		12,278		60,810	
Setout rate		36%		45%		41%		36%		39%	
Lbs/setout		16		16		17		12		16	
lbs/home		24		28		29		18		25	
Part/rate											

TABLE 33.9 Indiana Jones County curbside recycling program, 19XX monthly statistics

	January	February	March	April	May	June	July	August	September	October	November	December
Newspaper	171,936	188,580	209,070	204,580	222,250	210,752	191,662	192,110	192,096	220,820	644,887	710,875
Glass	59,723	55,817	58,707	59,933	64,373	60,919	59,719	59,556	56,654	55,901	171,248	209,619
Aluminum	2,964	2,633	2,706	3,809	4,551	4,614	4,608	4,141	4,635	4,579	13,214	14,866
Plastic	—	—	—	—	—	—	—	—	—	—	8,060	8,800
Total lbs	234,623	247,030	270,483	268,322	291,174	276,285	255,989	255,807	253,385	281,300	837,409	944,160
Total tons	117	124	135	134	146	138	128	128	127	141	419	472
Tons to date	117	241	376	510	656	794	922	1,050	1,177	1,317	1,736	2.208
Tin Man homes	5,453	5,979	5,979	5,983	5,984	5,984	5,755	6,000	6,000	6,000	11,976	11,976
Cycle One homes	5,703	5,968	5,973	5,973	5,973	5,991	5,758	5,991	5,758	5,991	11,987	11,987
Northside homes	0	0	0	0	0	0	0	0	0	0	6,145	6,145
Central homes	0	0	0	0	0	0	0	0	0	0	8,016	8,016
Total homes	11,156	11,947	11,952	11,956	11,957	11,975	11,513	11,991	11,758	11,991	38,124	38,124
Total drive-bys	42,672	47,512	48,464	47,691	47,977	47,861	46,538	47,427	46,538	47,947		154,560
Total setouts	13,655	15.204	16,478	15,738	16,792	15,794	15.823	15,651	15,823	16,302	0	60,810
Av. setout rate	32%	32%	34%	33%	35%	33%	34%	33%	34%	34%	ERR	40%
Av. lbs/setout	17	16	16	17	17	17	16	16	16	17	ERR	16
Av. lbs/home/mo.	21	21	23	22	24	23	22	21	22	23	22	25
Revenue total	$3,150.82	$3,033.45	$1,948.61	$2,291.57	$2,288.30	$2,261.82	$2,206.23	$2,061.77	$2,200.36	$2,251.05	$7,822.22	$8,855.39
Revenue/county	$1,575.41	$1,516.73	$974.30	$1,145.79	$1.144.15	$1,130.91	$1,103.12	$1,030.88	$1,100.18	$1,125.52	$3,911.11	$4,427.70
Cost, gross	$19,966.74	$21,385.24	$21,394.14	$21,401.34	$21,403.14	$21,435.18	$20,608.24	$21,463.98	$21,049.24	$21,463.98	$65,169.63	$65,169.63
Cost, net	$18,391.33	$19,868.52	$20,419.84	$20,255.56	$20,258.99	$20,304.27	$19,505.13	$20,433.10	$19,949.06	$20,338.46	$61,258.52	$60,741.94
Cost, home	$1.65	$1.66	$1.71	$1.69	$1.69	$1.70	$1.69	$1.70	$1.70	$1.70	$1.61	$1.59

Collection System Cost

Collection system cost is determined primarily by the method of collection used for a recycling program. However, operation and maintenance (O&M) and capital expenditures must also be considered. Examples of O&M and capital costs are

Capital	O&M
Site purchase	Labor
Construction and improvement	Fuel and maintenance
Collection trucks	Container replacement
Collection containers	Equipment leases

Capital costs incurred for the initial purchase of collection equipment are often the major expense. Of course, these costs will vary in relation to the size of the program. Initial startup costs can be reduced by the purchase of used equipment instead of new equipment. An example of the potential savings is shown below.

Truck Costs—Used Compared to New

	New equipment	Used equipment
Self-unloading trucks	$60,000–110,000	$30,000–45,000
Flatbed with bins	18,000–35,000	6,000–20,000
Self-unloading trailer	7,000–12,000	2,000–10,000
Flatbed trailer	5,000–11,000	2,000–8,000

The cost of labor dominates the ongoing costs of operation and maintenance. Some factors associated with labor costs, such as wage rates, may not be under the control of management. However, overall O&M costs can be controlled by assuring that

- Vehicles and crews are properly sized
- Routes are properly sized and driven
- Participation rates are high
- Fewer material components are sorted into separate compartments

Processing Costs

Some collected materials must be processed before they are sold, and processing may increase the value of some materials. Common examples of processing include:

Newspaper	Sorting, baling
Cans	Separating, crushing, baling
Glass	Sorting, metal removal, crushing

The typical O&M and capital expenditures associated with processing are

Capital	O&M
Site purchase	Labor
Construction improvement	Fuel and maintenance
Processing equipment	Equipment lease

Purchasing used equipment, especially for processing, is an excellent cost control method, since the equipment is rarely used for more than a few hours each day. An example of potential savings is shown below:

Processing Equipment—Used Compared to New

	New equipment	Used equipment
Horizontal baler	$15,000–700,000	$20,000–100,000
Vertical baler	6,000–40,000	2,000–8,000
Glass crusher	3,000–6,500	1,000–4,000
Can separator	3,000–6,000	1,000–2,000
Can flattener	7,000–15,000	1,000–4,000
Can densifier	15,000–32,000	8,000–16,000

Of course, additional processing costs should not exceed additional revenues generated by processed waste.

Public Education Costs

Without public education and promotion of a recycling program, there will be no materials to collect. The cost of public education typically ranges from 1 to 10 percent of the operating budget. Although public education is not a major expense, careful assessment of program operations can ensure that the expenditures are used in the most effective manner.

Marketing, Administration, and Overhead Costs

Marketing costs usually include the costs of locating buyers for the materials and arranging delivery. The labor cost for marketing is generally included under administrative cost. Administration and overhead costs can include items such as

- Program management
- Utilities
- Site lease
- Supplies
- Telephone
- Insurance
- Taxes and permits
- Accounting
- Conferences and training

Cost savings can be realized if in-kind services are used. For example,

- Using existing staff for vehicle maintenance
- Using existing staff for program administration
- Using existing facilities such as offices, phones, and computers

When in-kind services are used, it is difficult to assign an accurate cost-saving value.

SUMMARY

Cost control is an ongoing, dynamic endeavor performed by management. The data collection techniques, as previously outlined, can assist in preparing information for the decision-making process. Current thinking tends to consider recycling an integral part of a comprehensive solid waste system. The costs for a recycling program may range from $30 to $60 per ton of recyclable material collected and transported to market. However, despite the fact that recycling may have a higher cost per ton than another alternative, it is presently at the forefront of public interest. The application of cost control methods will allow a recycling program to be implemented as efficiently as possible.

COMPUTERS IN THE RECYCLING INDUSTRY—INTRODUCTION*

Much has changed since the first edition of the *Handbook* in the area of computers and recycling. Computers and software have made significant inroads into the day-to-day operations of the recycling industry, hence the need for this update.

This section reviews the ways in which computer software can assist the recycling and waste management industry. The recycling industry, like all industries, has a very real need for information gathering and management technologies. The use of computers can provide important, even critical, benefits for creating more efficient operations, tracking financial data, regulatory compliance, and developing business, and can provide other competitive advantages. It is no longer a question of whether a recycling operation needs a computer to be effective, but how much can and should be computerized.

This chapter does not address computer hardware issues or the topic of recycling old computers.

COMPUTER SOFTWARE

Overview

There are several areas that software can potentially assist with, including:

- Management of customers and accounts
- Route management and analysis
- Equipment/fleet maintenance
- Program planning and analysis
- Preparing letters and other documents
- Expense records and other data

The last two items in the above list can be accomplished with the same "off-the-shelf" software programs used by all businesses, such as word processing, spreadsheet, and database software. The first four items, however, may require specialty software, and there are many software programs that have been designed specifically for the recycling

*This section was prepared by Richard Hlavka, President of Green Solutions, South Prairie, Washington.

industry that can assist in these areas. Custom made or modified software is also an option, but the following focuses on commercially available software.

Change is occurring very rapidly in the field of commercial software. Software and hardware are updated frequently, with upgraded software or entirely new programs being released every week. The following is intended to provide a guide for the types of applications that may be of interest to the recycling industry, but potential users will need to conduct their own research to determine the best available software for their needs. One of the factors to consider at that time should be compatibility with other software, including the ability to integrate new software into existing systems as well as anticipated compatibility with future software additions.

Customer Billing and Account Management Software

Software for billing and related purposes has been refined and improved substantially since the DOS-based programs that only allowed bills to be printed out. Many of today's programs stress "customer management " and provide a wide range of options for tracking outstanding amounts and printing a variety of reminder notices; they are compatible with other software programs. Several programs also offer capabilities that could be characterized as route management or that integrate billing with scalehouse records and other functions. Some examples of this type of software follow.

Billing Software

Billing and accounting software takes a variety of forms, from simple database programs that are easy to use to more complex programs with numerous capabilities. Various software programs are available for the full range of operating platforms and for either fixed site or collection operations.

The more complex software programs include functions such as managing the billing process, tracking outstanding bills, handling customer lists, and printing a variety of forms and reports. Some software integrates accounting, payroll, container tracking, and scheduling for recyclable materials, solid waste, and special waste streams.

Other software can track transactions at recycling and waste management facilities. For example, one such program can handle both the financial aspects (such as ticketing, monthly statements, and account status) and management aspects (numerous reports, including amounts and types of waste, average times to unload, and number of customers). This program can also handle multiple sites with data transfer to a central office.

Point of Service

Several types of software collect weight and other data at the point where service is provided to a customer, whether this is at a scale or front office of a recycling facility or en route for a collection operation. For scale records, inputting weight data into a software program provides a reliable, secure and labor-saving data management system that opens up a variety of possibilities for reports and performance measurement tools. Data can be input manually using drop-down menus and other convenient measures, or it can be accomplished automatically using methods such as radio frequency identification (RIFD). For buy-back centers, point-of-purchase software can track customer information as well as address deposit collections and other special needs.

Onboard Scales

Onboard scales and software programs that store and manage the resulting weight data have been significantly improved in the past few years. More work needs to be done to increase the accuracy of on-board scales, but in some cases these are already being used for billing purposes. This is more an issue for waste collection companies, as many garbage haulers have discovered that they are losing money on certain accounts (typically those with extremely dense or heavy garbage), but also an issue for recycling collections, where weight-based charges or credits are used.

Bar Codes

Bar codes on recycling containers are used by a few communities to track participation in curbside recycling programs, usually for the purpose of documenting eligibility for garbage rate discounts. Bar codes and electronic devices on commercial containers are also being used by private collection companies to document the amount of recycling, often in combination with on-board scales.

Route Management Software

Several of the route management software programs defy easy characterization because the programs also perform billing and other functions. Collection companies have sometimes started using a program for its billing capabilities and then later discovered the route management and other functions. The route management and other functions may also become more important if a company adds customers or expands the types of services offered.

The following capabilities are offered by various route management software programs:

Route Design

Several programs exist for designing the most efficient routing of collection stops or optimizing routes. One of the more complex (and expensive) of these uses the Geographic Information System (GIS), customer database, and street database to design, display and print route information.

Tracking Vehicles and Containers

With the growth of Global Positioning System (GPS) technology, an increasing number of companies are finding this useful for tracking the location of vehicles and shipment containers. Among other advantages, tracking collection vehicles while they are en route allows for easier and more efficient additions or deletions to pickup schedules.

A related capability is found in a program that transmits data through a wireless network to handheld computers used by staff working throughout a large recycling facility, enabling them to easily access and report data. This form of two-way communication goes far beyond the radios used at many facilities, as it allows for direct data input or access from the yard or floor of a recycling center.

Tracking Materials

One program is designed to track materials from the point of collection to the point of being shipped out of a facility as a finished product, processed material, or waste. Specific capabilities for this include tracking shipments, scheduling loads, manifesting, tracking inventory, and related functions. Tracking materials internally has a clear benefit for those materials that may be classified as hazardous wastes if not recycled in a timely fashion, but handling systems for all recyclable materials can benefit from the easier operating methods and better productivity reports that this approach allows.

Equipment and Fleet Maintenance Software

As a collection or processing company expands its operation and more equipment is needed, a point may be reached where software will prove invaluable for tracking maintenance and repairs for the equipment. Such software has the following advantages.

Tracking Maintenance Schedules

Software is available that assists in tracking maintenance schedules, allowing this work to be done in a timely fashion and on a schedule that works best for the operation. Conducting maintenance and minor repairs on a regular schedule will help avoid more costly and inconvenient breakdowns.

Replacement Evaluations

Software can evaluate the cost-effectiveness of maintaining a piece of equipment compared to replacing it. To the extent that the software is used to record the amount of downtime for a piece of equipment and the cost of repairs, this information can be used to determine when a piece of equipment should be replaced.

Other Advantages

Many of the software programs that are designed for equipment and fleet maintenance also perform other functions, such as tracking parts inventory, storing warranty information, and generating work orders.

Program Planning and Analysis Software

This type of software includes several programs that may be primarily of interest to planners and consultants. Because of differing designs and capabilities, this group can be further subdivided into the following categories.

General Planning

Several programs are available that can assist with planning for solid waste management systems. These are generally designed to allow the costs and waste diversion rates for several options to be analyzed and compared. These programs may also assist with facility design by determining the capacity needed over the planning period. The area for analysis is generally a municipality, but at least one of these programs is designed to allow options to be considered for a unit as small as a single company.

Recycling Results and Costs

Most of the software in this group consists of spreadsheets that have been produced by various public agencies. These spreadsheets are set up to facilitate analysis of recycling center costs, single-family or apartment recycling costs, or other activities. A few software programs are also available from private sources, and these are generally designed to allow easier comparison of different options for recycling programs. One such program is designed to do this for recycling options at individual businesses.

Other Information

Included in this group are software programs that provide resources for specific types of programs, such as one that is designed as "tool kit" for encouraging waste reduction. This tool kit contains worksheets, templates for letters and other documents, clip art, curricula, tips and strategies for specific businesses and departments, evaluation forms, and more.

In no case can the above software programs completely replace careful consideration of policies and a community's needs, but all of these programs are useful for addressing "what if" scenarios and providing ballpark cost estimates.

Miscellaneous

There is a wide range of other software that is designed to meet specific needs of the recycling and composting industry, including software that

- Controls the composting process and monitors for compliance with regulatory standards
- Is designed to assist with waste composition data collection and analysis, including determining sampling methodologies and addressing the statistical meaningfulness of the results
- Assists in conducting or producing facility audits and safety plans
- Provides reference material such as regulatory information

Finally, if none of the above programs are exactly what is needed, custom-made software is always an option.

CONCLUSION

Every recycling business or agency, large or small, public or private, should examine their operations to determine how computers and software can best benefit them. This examination should begin with an assessment of needs and should also include a review of the technology and software applications that are available. This review should be done on a periodic basis to determine if there are software upgrades or enhancements to existing computer systems that would help recyclers maintain a cost-effective and competitive operation. All of the above software programs are designed with the intent of making recycling operations more efficient and cost-effective, but it is up to the recycler to determine if a trade-off in technology versus labor and other costs is truly advantageous to their bottom line. Potential users should always thoroughly research the software that is available versus their needs before investing time and money in any one program.

CHAPTER 34
QUALITY CONTROL MONITORING FOR RECYCLABLE MATERIALS

STEPHEN A. KATZ
Vice President
and
SCOTT W. SPRING
New England CRINC
North Billerica, Massachusetts

INTRODUCTION

A recycling program's success is largely dependent upon its ability to consistently produce high-quality, marketable end products. Recycling programs must have stable markets for processed materials in order to assure a program's sustainability and cost-effectiveness. Educational, design, and operational measures at all stages of the recycling process—source separation, collection, delivery, presort inspection, and outload—can assure the production of high-quality, salable products and hence program success.

IMPORTANCE OF QUALITY CONTROL

The importance of producing quality recyclable materials is clear: recyclables are commodities and are treated as such by the end markets. The supply and demand for recycled products is constantly in flux, and so are the prices paid to recyclers. In order to reap the highest possible price, recyclers must demonstrate to the end markets that processed material conforms to specifications and is consistently available in sufficient quantities.

Revenues from the sale of recyclables serve to offset the other costs of implementing and sustaining a recycling program. Costs can include

- Feasibility studies, waste audits
- Solid waste management consulting
- Recyclables collection and transfer
- Recycling containers

- Educational materials and publicity
- Recycling facility site and structure
- Processing equipment (conveyors, magnets, crushers, screens)
- Processing costs (labor, supplies, utilities)
- Materials marketing management

Recycling is overwhelmingly a *cost-avoidance measure* for communities and industry. The frequently encountered "cash for trash" mentality ignores the economic realities of establishing a comprehensive municipal or commercial recycling program. While materials do have a certain value in the marketplace, very rarely do the revenues from the sale of recycled materials offset all costs of collecting, processing, and transporting the recyclables. As disposal costs at landfills and incinerators continue to rise, recycling has increasingly become a more attractive waste management tool and cost-avoidance measure.

The importance of processed material quality and high revenue generation for a materials recycling facility is shown in the equation below:

$$\text{Cap costs} + \text{O\&M costs} + \text{Trans. costs} - \text{revs} = \text{net processing costs}$$

where Cap costs = Capital costs
O&M costs = Operations and maintenance costs
Trans costs = Processed materials transportation costs
Revs = Material revenues

The importance of producing high-quality end products is clear. End markets recognize high-quality material and will compensate those recyclers who consistently produce contaminant-free products.

QUALITY CONTROL PROBLEMS

In addition to recyclable materials, processors often receive significant amounts of *rejects*. Rejects may consist of

- Recyclable items not currently accepted (potential recyclables)
- Nonrecyclables
- Hazardous wastes
- Cross-contaminants

A "potential recyclable" might be a piece of scrap metal found in a curbside collection container. Although the recycling program might accept tin cans and other postconsumer metals, there is no guarantee that absolutely all scrap metals are accepted for recycling. "Nonrecyclable" refers to an item typically found in municipal solid waste, such as a coffee filter, mistakenly placed in a recycling container. "Hazardous wastes" includes household hazardous wastes such as paint and automotive products. These potentially harmful wastes often require special handling. Finally, "cross-contamination" can be a problem in multibin systems which require residents to separate various types of recyclables into product-specific bins.

It is ironic that the overzealous recycler is often responsible for rejects in the material stream. Although scrap metal and plastic toys may be *technically* recyclable, they are not always *logistically* recyclable. In order for a municipal or commercial recycling program to succeed, generators need to recycle only designated materials. A high-quality recycling program has the means to detect and remove reject materials.

LINES OF DEFENSE

From the generator to the end market, the recyclables typically pass through many stages. It is possible to monitor materials for quality control throughout the recycling process. An aggressive educational campaign serves to reduce the percentage of contaminants in the material stream. Other lines of defense include the detection and removal of rejects at specific junctures of the recycling process. The multiple lines of defense at a large-scale materials recycling facility (MRF) are described below.

Educating the Generator

Perhaps the single most important and effective quality control measure recyclers can take is the implementation of comprehensive educational campaigns geared toward creating positive recycling habits. Clear and concise educational materials which inform participants about the "dos" and "don'ts" of recycling can have a tremendous impact on the quality of materials received. Explaining the importance of recycling to a specific community or business is also an effective way of raising participation rates and improving material quality.

Educational brochures and newsletters should clearly specify which recyclables are included in the program and what preparation (if any) is required. Graphics can be used effectively to show generators what materials are accepted (see Fig. 34.1). Recyclers should also explicitly list unacceptable (reject) items.

RECYCLE	WE DO ACCEPT	WE DON'T ACCEPT
Glass Jars & Bottles	**All food and beverage glass jars and bottles** • all jars and bottles must be completely empty and rinsed • discard caps and lids	• caps or bottle tops • window glass, dishes, or drinking glasses • ceramics, light bulbs etc. • glass vases or pottery
Newspaper and Corrugated Cardboard	**Newspapers and corrugated cardboard** • both should be clean and dry • bundle or bag your newspaper (paper bags only!) • corrugated cardboard boxes must be cut into individual sheets no larger than 2 1/2 by 3 feet	• newspaper inserts • magazines or phone books • junk mail • cereal, rice or cookie boxes • detergent cartons
Plastic Bottles	**Plastic soda bottles, milk, water and juice jugs and laundry detergent bottles** • all bottles must be completely empty and rinsed •all beverage bottles (soda, milk etc.) must be flattened • discard caps and lids	• cottage cheese or margarine tubs • plastic film or wraps, no plastic bags • styrofoam containers • plastic flower pots • caps or bottle tops
Metal Cans	**All beverage and food cans made of metal** (aluminum and tin cans) • CANS only • all cans must be completely empty • please rinse your cans	• aerosol cans • aluminum foil or pans • aluminum siding or other scrap metal • pie pans or frying pans

FIGURE 34.1 Educational brochure for households.

Common reject items in curbside collection programs include the following:

- Ceramics
- Window glass
- Scrap metal
- Scrap plastics
- Junk mail

Each of these degrade processed material quality and can lower revenues. Some, such as ceramics, can be very damaging to an end-markets' manufacturing equipment. These and other rejects should be explicitly listed as "nonrecyclable."

Information should be distributed at the start of the recycling program, and monthly or quarterly reminders serve to solidify good recycling habits. Educational materials should be tailored to the specific audience that the recycler is trying to reach. Brochures designed for a wealthy, educated community may look very different than those geared toward an inner-city community. Some cities have found it necessary to produce recycling literature in several languages.

Educating the Hauler

Haulers that transport recyclables from curbside collection programs can play an Important role in monitoring and improving material quality. Haulers, in addition to homeowners, need to be educated about the importance of keeping rejects out of the process flow. Training programs for drivers and incentive programs have been successfully used by municipalities across North America.

UNACCEPTABLE FOR RECYCLING DUE TO:

BOTTLES / CANS NOT RINSED CLEAN

METAL / PLASTIC RING NOT REMOVED

CLEAR PLASTIC BOTTLE AND JAR / CAPS NOT REMOVED

LABELS NOT REMOVED FROM CANS

PLASTIC — NOT LABELED 1 or 2

ITEM NOT INCLUDED IN COLLECTION PROGRAM

FIGURE 34.2 Sticker list of unacceptable materials.

Drivers in curbside collection bin programs have the opportunity to inspect commingled materials before they are loaded into the truck. Materials that do not meet delivery standards should be left at the curb so that residents realize what is not acceptable. Some hauling firms have designed special stickers which allow drivers to specify the reason that a material does not meet specifications (Fig. 34.2). This prevents rejects from entering the process flow and simultaneously serves to educate homeowners.

Haulers involved in other types of recycling programs such as drop-offs have fewer opportunities to monitor material quality. Still, procedures may be put in place which hold generators and haulers accountable for the quality of recyclable material.

MATERIALS RECYCLING FACILITIES

Central sorting and processing facilities for postconsumer recyclables are commonly used to consolidate recyclables from a region and prepare them for shipment to market. These facilities are known as materials recycling (or recovery) facilities—MRFs.

Residents separate recyclable materials into one bin and place nonrecyclable refuse in another. The commingled recyclables are delivered to an MRF where they are mechanically sorted into their respective fractions. Recyclable materials in MRF programs typically include the following:

PET plastic containers
HDPE plastic containers
Aluminum food and beverage containers
Aluminum foil and pie tins
Tin-coated steel containers
Glass food and beverage containers
Newspaper
Corrugated cardboard
Magazines
High-grade office paper
Kraft bags

Highly mechanized MRFs allow municipalities to recover 20 to 30 percent of their waste for reuse. Combined with effective composting programs, recovery rates of over 50 percent are possible. With landfill and incineration costs rising exponentially, communities worldwide are turning to recycling as a cost-effective waste management tool.

A well-designed MRF has three main areas of activity:

- Tipping area (recyclables unloading)
- Processing area (sorting, removal of contaminants, baling)
- Shipping area (loading docks, transfer equipment)

Quality control monitoring can take place in all three areas (Fig. 34.3).

Tipping Area Monitoring

Most recycling facilities reserve the right to reject loads which contain over a certain percentage of reject materials. In this way, haulers have an incentive not to deliver nonrecyclables to the MRF. Warnings and fines can prevent haulers from repeating mistakes. Personnel on the tipping floor visually inspect loads of recyclables for large volumes of mistakenly delivered municipal solid waste, yard waste, and household hazardous waste. Often, MRF personnel are equipped with a camera to record the condition of incoming

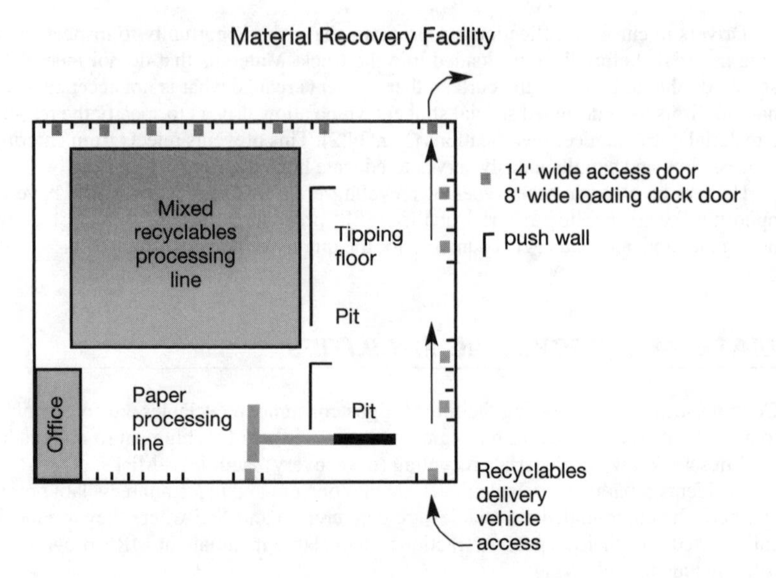

FIGURE 34.3 MRF areas for quality control monitoring.

loads. Tipping floor inspections are especially crucial during the recycling facility startup so that both residents and haulers are informed early on about what is not accepted at the MRF.

Processing Area Monitoring

Regardless of the quality and intensity of education programs, reject materials will continue to arrive at the MRF. Rejects vary in volume depending upon the materials collected and the collection method. In successful, large-scale curbside collection programs, rejects typically equal 3 to 5 percent of the total incoming tonnage.

An inspection, or "presort," station should be a part of every recycling facility design. This is a dedicated station for the removal of nonrecyclable materials and any objects which could injure or damage workers or equipment. In mechanized MRFs, the material is conveyed past the station and rejects are deposited into a chute and onto another conveyor. Reject conveyors from throughout the processing system move material to a central rejects container.

Workers at the presort station are equipped with protective Kevlar gloves and sleeves which eliminate the risk of cuts and scrapes. A thin flow of material at this stage in the processing system is crucial to allow workers to remove reject materials effectively. Well-educated sorters can remove the vast majority of rejects at the inspection station.

Materials Screening

Often reject materials are too small to remove by hand, and hence screening machines are used to remove all small contaminants and fines. Shaker screens remove all materials of

FIGURE 34.4 Automated screens ensure high material quality.

less than a certain diameter, usually about 1 in. Automated screens are a cost-effective way of ensuring high material quality (Fig. 34.4).

Avoiding Cross-Contamination

High-throughput automated facilities must be designed for a specific stream of recyclables in order to avoid cross-contamination. Metals in the plastics stream or glass in the plastics stream are classified by end markets as contaminants. Gravity-based separation of glass and plastics, for example, produces noticeably less cross-contamination than the traditional air classification method. Air classifiers rely on bursts of air to separate lighter materials from heavier glass. However, this system is difficult to regulate and does not handle crushed plastics and aluminum well.

Inclined sorting machines rely on gravity to separate glass from lighter materials. Glass rolls down the belt while lighter-weight plastic and aluminum are carried to the side by rotating chains (Fig. 34.5).

Electromagnets which remove all ferrous containers must be carefully placed in the system to remove only ferrous cans. Glass, paper, and other contaminants may be caught by the moving ferrous containers if the magnet is not at the correct height and orientation (Fig. 34.6).

QUALITY CONTROL CHECK

Once the commingled recyclables have been separated at the MRF, a final quality control check can be performed to verify that all contaminants have been removed. End markets for

FIGURE 34.5 Inclined sorting machine separates glass from lighter materials by gravity.

glass are particularly sensitive to contamination by ceramics, lead glass, light bulbs, etc. In order to maintain glass quality, facility staff can review the crushed or semi-crushed glass during loadout. End markets are very stringent about material conforming to specifications. Higher prices are paid for high-grade materials that meet all requirements.

See Table 34.1 for a list of typical end-market specifications. Final specifications are determined by the individual end users and change as technology advances.

FIGURE 34.6 Electromagnets remove ferrous containers.

TABLE 34.1 Material Specifications

Material	Baling and sorting requirements	Nonconforming materials
Aluminum	Aluminum containers will be baled to industry standards (approximately 52″ × 30″ × 40″—16–20 lb/ft^3). Aluminum will be baled and shipped in trailer loads. Maximum contaminants:	Iron—up to 1.0% per bale Lead—up to 1.0% per bale Moisture—up to 4% per bale
Tin cans	Tin will be baled in approximately 52″ × 30″ × 40″ sized bales weighing 33–40 lb/ft^3, with minimal quantities of food products, labels, etc. remaining. Tin cans will be baled and shipped in trailer loads. Maximum contaminants:	Aluminum—up to 5% per bale Labels remaining—up to 10% per bale Moisture—up to 5% per bale
PET plastic	PET soda bottles are baled in approximately 52″ × 30″ × 40″ sized bales weighing 650–750 lb. PET plastic will be shipped in trailer loads. Color separation is not required.	Total nonconforming materials may not exceed 4% per bale.
HDPE plastic	HDPE plastics bottles will be baled in approximately 52″ × 30″ × 40″ sized bales weighing 650–750 lb. HDPE plastic must be processed so as to be capable of being shipped in trailer loads.	HDPE—not less than 95% per bale Miscellaneous plastic—up to 4% per bale
Glass	Glass will be sorted by color into amber, flint and green subfractions prior to being crushed. Major contaminants are refractory materials, metals, non-container glass and plastics. Paper labels are tolerated. Cullet size will be approximately 1/2″ to 2.″	Flint cullet: not less than 95% flint, up to 2.5% green, up to 2.5% amber Amber cullet: not less than 90% amber, up to 10% flint, up to 10% green Green cullet: not less than 85% green, up to 15% flint, up to 10% amber
Newspaper	Newspaper is baled to a 45″ × 45″ × 50″ sized bale weighing approximately 1175 lb. Material shall consist of sorted fresh newspaper, not sunburned, free from papers other than news, with not more than the normal percentage of rotogravure and colored sections.	Prohibitive materials: None permitted Total nonconforming materials will not exceed 2% per bale
Corrugated	Corrugated containers are baled to 45″ × 45″ × 50″ bales weighing approximately 980 lb. Material shall consist of corrugated containers having liners of test liner, jut or kraft.	Prohibitive materials may not exceed 1% per bale Total nonconforming materials may not exceed 5% per bale

EDUCATION, DESIGN, OPERATIONS

Thorough educational programs, experienced system design, and a well-managed operation are the keys to producing high-quality recyclable materials. Recycling programs that aim to educate participants about the requirements and goals of recycling are more likely to be free of reject materials. Reaching the generator and spreading the word about what rejects are and why they are harmful is the starting point of a successful program.

A carefully engineered processing system improves the detection and recovery of rejects. Cross-contamination will not be a problem if the system and system components are designed specifically for the volume and mix of incoming recyclables.

Recycling facility operations, regardless of size or sophistication, which budget adequate equipment, labor, and resources will be the most effective in minimizing rejects and maximizing material revenues.

CHAPTER 35

RECYCLING AT LARGE COMMERCIAL FACILITIES

CYNDIE ECKMAN
Recycling Program Administrator
Reedy Creek Energy Services
Lake Buena Vista, Florida

IMPORTANT COMMERCIAL RECYCLING FACTORS

Commercial establishments present a potential for a variety of recycling opportunities. Initiation of a recycling program for a large commercial establishment requires a great deal of planning. The size and composition of the establishment's waste stream will indicate potential commodities for recycling. Markets must be identified for the potential recyclables, including packaging specifications of the materials and transportation to recycling markets. Once markets are determined, an implementation plan should be developed to provide methods of collection within the establishment, and packaging of the materials in preparation for delivery to market. In addition to traditional recycling, nontraditional recycling of materials, waste reduction, and purchasing strategies can contribute to an overall increase in recycling performance. As with any recycling program, education and continued evaluation play a critical role. Careful planning will result in a successful commercial recycling program.

It is important to first identify what defines a large commercial establishment. Large commercial establishments typically have waste streams greater than one ton per day. These waste streams include municipal solid waste, hazardous waste, and construction and demolition debris. The establishments themselves may consist of one building occupied by one business, a cluster of buildings occupied by several businesses, or any combination between. Commercial establishments can be grouped in the general categories of entertainment, administrative, industrial, and retail.

Table 35.1 lists examples for each of these categories. Most commercial establishments have routine hours of operation with a constant flow of solid waste. Others, such as sporting events, concerts, and fairs, are random or occasional. The size of a large commercial waste stream can be an economic recycling advantage for the establishment. These establishments can often accumulate large quantities of recyclables in relatively short periods of time, resulting in lower costs for shipping and higher revenues per unit. Industrial facilities routinely have additional alternatives for transportation, such as access to railroads or ports. The economic benefit of size often results in commercial recycling programs that are less costly than landfill disposal and may even result in revenues for the establishment.

TABLE 35.1 Classification of large facilities

Entertainment	Administrative	Industrial	Retail
Entertainment districts	Office buildings	Utilities	Malls/chain stores
Parks/attractions	Schools/universities	Factories	Restaurants
Large public events	Banks/post office	Warehouses	Hotel/motel

CONDUCTING A WASTE COMPOSITION STUDY

As with any recycling program, the first step in initiating a commercial recycling program is to determine what materials are in the waste stream. Existing establishments can accomplish this through a waste composition study. The waste composition will vary greatly between different types of establishments and may include categories of waste not typically found in household waste streams, such as waste tires, concrete, and asphalt. Construction and demolition debris can make up as much as 50% of the total waste stream of some large commercial establishments. Industrial-type facilities may have even higher percentages. Entertainment, administrative, and retail establishments generally have compositions of paper waste equaling or exceeding 25% of their total waste stream. Existing facilities should compare waste composition results against other similar facilities. Planned establishments can compare the waste compositions of similar establishments to predict their waste stream composition. The following example illustrates the use of a waste composition study.

Example 1. The ABC Hotel and Convention Center is planning to open in six months. The new hotel will have 600 rooms, a convention center, a restaurant, and a gift shop. The county in which the hotel is located has a 40% mandatory recycling goal for all commercial establishments. ABC does not know what materials they will be able to recycle, but they have three other hotels in different states. The three existing hotels each have 500–700 rooms, a convention center, one or more restaurants, and one or more gift shops. ABC conducts a waste composition study on the three existing hotels. Table 35.2 shows the percentage of eight different material categories sorted for in the study.

Since the three existing hotels have a similar number of rooms, a convention center, and similar retail establishments, an average from the three studies can be used to predict

TABLE 35.2 Waste composition analysis

Material	ABC Hotel 1	ABC Hotel 2	ABC Hotel 3	New ABC Hotel
Cardboard	8.7%	6.3%	7.7%	7.6%
Office paper	2.2%	1.2%	2.4%	1.9%
Newspaper	4.2%	6.3%	5.9%	5.5%
Metal cans	1.1%	0.8%	2.0%	1.3%
Glass bottles	6.9%	4.2%	6.3%	5.8%
Plastic bottles	0.8%	1.1%	1.9%	1.3%
Food waste	34.6%	36.8%	29.9%	33.8%
Other	41.4%	43.2%	44.1%	42.9%

the waste composition of the new hotel. The last column in Table 35.2 shows the projected waste composition. If the new ABC Hotel initiates a recycling program for the first eight materials that comprises 57.1% of the total projected waste stream, and is able to recycle 70% of those available materials, they will able to meet the 40% recycling goal.

CONTRACT AGREEMENT FACTORS

Once the waste composition is determined, local markets must be investigated to determine the availability of recycling programs in the area. Many solid waste collection companies offer commercial recycling collection. Typically, large collection containers such as 8-yard boxes or large rolling carts are provided by the hauler and placed in the service area of the commercial establishment. Collection may be commingled, allowing cardboard and office paper to be mixed, or source-separated, requiring that cardboard is placed in a collection container separate from the collection container for office paper. Collection companies will require a contract for services. Variations of a collection contract should be considered carefully before entering into an agreement. Collection usually involves a standard fee each time a container is emptied. There may be additional fees for the weight of the material in the container. There may also be penalty charges incurred if the container is blocked, contaminated, damaged, or overloaded. Revenues for recyclables vary greatly depending upon market conditions. Some collection contracts include provisions for the sharing of revenues when markets are high and the sharing of costs when markets are low. Others factor revenues into the collection fee by basing revenues on market averages. A careful comparison of the establishment's waste composition against average markets is very important when determining contract specifications.

Example 2. Food Supermarkets is considering a contractual agreement with a local recycling collection company for the collection of cardboard. The supermarket generates an average of 15 tons of cardboard each month. The collection company will provide an 8-yard collection container for the cardboard. If the boxes are flattened prior to placing them into the 8-yard box, about 1000 pounds of cardboard can be placed in the box. The average market value for cardboard is $30/ton and it costs the collection company $20/ton for collection and processing. Food Supermarkets has the following choices for the contractual agreement:

1. Pay a flat rate of $12 each time the 8-yard box is collected
2. Pay a rate of $20 each time the 8-yard box is collected and split the revenues in half with the collection company
3. Pay a rate of $25/ton of cardboard collected and no fee for collection of the box

Which rate would be most beneficial to Food Supermarkets? With a generation of 15 tons/month, the 8-yard box will be collected about 30 times per month. The following is the average monthly cost to Food Supermarkets for recycling cardboard.

1. $12/pickup × 30 pickups = $360
2. $20/pickup × 30 pickups = $600, $15/ton revenue share × 15 tons = $225, $600 collection fee – $225 revenues = $375
3. $25/ton × 15 tons = $375

In this example, the first choice would be the most efficient on the average, even though there may be months when markets are high for cardboard and revenues may be received.

LANDFILL AND MRF OPTIONS

Recycling options may be available at local landfills, directly with recyclers, and at composting operations. Many landfills accept materials for recycling such as landscape debris, construction debris, scrap metal, waste tires, and many hazardous wastes. Landfills charge a tipping fee for materials delivered, which may be agreed upon through a contract with the landfill. Landfills may discount the tipping fee for source-separated materials, such as a collection box containing only clean wood. Some landfills have sorting operations to remove metals and other recyclables. Commercial establishments that have the ability to transport materials, such as warehouses, may find it economical to deliver materials directly to a recycler. Many local material recovery facilities (MRFs) will charge a tipping fee to accept materials. Materials delivered may be required to meet certain commingled or source-separation criteria prior to tipping on their floor. Other commodity-specific recyclers, such as scrap metal dealers, will accept only source-separated materials, but will usually pay fair market value for them. There are a variety of composting facilities that will accept various materials. Many use untreated wood and landscape debris as amendments for the composting process. Some compost organics such as food waste. Similar to MRFs, composting facilities will typically charge a tipping fee to accept materials. Depending on the waste composition of an establishment, one of the above methods for recycling or a combination of the above methods may be needed to accomplish the desired rate of recycling.

COMMERCIAL CHAIN STORES

Commercial entities that have a chain of stores or multiple facilities have additional options for recycling. Commercial chains with central warehousing can backhaul recyclables from deliveries. The backhauled recyclables can be accumulated for truckload volumes. Many times recyclers will offer higher revenues for such bulk quantities of recyclables. If a high volume of cardboard is generated from the individual chain locations, vertical balers may be a consideration. Vertical balers allow locations to bale the cardboard on-site, which is more space-efficient and efficient for transport. Cardboard recyclers will typically offer higher revenue for baled cardboard. Balers can be used for other materials such as stretch wrap, metal cans, or plastic bottles, if those materials are generated in high volumes.

Evaluating Recycling Feasibility for Each Material

Once recyclable materials have been identified, markets have been located, and transportation has been investigated, the feasibility of recycling should be determined. It is important that feasibility is determined prior to implementing any commercial recycling program. Each commodity should be evaluated separately, including all costs incurred and costs avoided as part of the evaluation. The following example demonstrates the evaluation of two commodities for large-scale commercial recycling.

Example 3. Leisure Time Golf Club is considering the addition of an aluminum can recycling program. The club consists of two 18-hole golf courses, a putting green, a clubhouse with a retail store, and a restaurant with a bar. A local recycler has offered to place an 8-yard box for collection of the aluminum cans at a charge of $15 per pickup and a share in revenues of $0.30/lb. The cost for collection of solid waste is $40/ton. Based on a waste composition study, Leisure Time is estimated to generate 0.50 tons of aluminum cans per month. It's custodial crew is projected to spend three man-hours per month at a rate of $16/hour for additional labor related to the recycling of aluminum cans. The cost for inside collection of containers in the clubhouse and on the golf course is expected to be $700. Is this a feasible recycling program?

First, compare apples to apples by converting all of the costs to $/month.

- At two pickups per month, collection costs will be $30/month
- Revenues = ($.30/lb × 2000 lbs/ton) × 0.50 tons/month = $300/month
- Cost avoided for landfill disposal = $40/ton × 0.50 tons = $20/month
- Labor cost = 3 hours/month × $16/hour = $48/month
- New container cost spread over five years = $11.67/month (these containers are in addition to those supplied by the hauler)

Revenues and cost avoided for landfill disposal amount to a gain of $320/month. Collection costs, labor costs, and container costs amount to an expense of $89.67/month. The net result is a gain of $230.33/month. In this particular example, the recycling of aluminum cans would be a feasible program. At larger entertainment facilities this net gain can be substantial. Not all recycling programs will result in revenue. Other factors such as recycling goals, public image, and environmental concern have nonmonetary value that may contribute to an establishment's decision to recycle.

PLANNING AT A SPORTS STADIUM

Planning for the initial setup of a large-scale commercial recycling program can be critical to the success of the overall program. The first step is to determine the points of generation for each recyclable to be captured. Many large commercial establishments, such as entertainment districts, contain a mixture of smaller commercial entities, such as gift shops, restaurants, and nightclubs. Each smaller entity will have a unique waste stream and will be able to recycle different materials. When determining generation points, knowledge of the packaging that purchased items will be delivered in and what type of containers will be disposed of is important. Table 35.3 demonstrates possible points of generation for a typical baseball stadium.

Though food may be generated at the food stand locations, capturing uncontaminated food from customers may be difficult, therefore this option was left blank. Notice that glass was not chosen as an option for entrance and exit areas. Glass is typically not given

TABLE 35.3 Locating generation points at a baseball stadium

Location	Cardboard	Glass	Cans	Food	Office Paper
Entrances/exits			✓		
Food stands			✓		
Food back area	✓	✓	✓	✓	
Merchandise stands	✓				✓
Administrative offices	✓		✓		✓
Ticket booth					✓

to customers in this type of environment, for safety reasons, so this option may not be feasible. In this example, it is also assumed that beverages will be available in cans. Monitoring the purchasing habits of customers will provide information on the volume of recyclables that are available for collection at any given time. Once generation points are decided upon, physical collection points should be determined. Space must be available for containers, and containers should be visible.

SETTING UP A TEST PHASE

In cases where the feasibility of recycling is questionable or the volume available is undetermined, setting up a test phase will provide a method of determining these values in a real environment without investing in a full-scale recycling program. A test phase should have a set begin date, a set end date, and a set location that is representative of the whole location. A shopping mall may choose to test the collection of aluminum cans by placing a few collection containers near an entrance that is in close proximity to food vendors serving beverages in cans. A university may choose one building containing a vending machine area as a test location. A plan for measuring results must be established. This may be done by weighing the amount of cans collected in small time intervals, such as daily, to see if the capture rate improves over time. Measuring contamination levels is also important during testing. If more trash is found than aluminum cans in the recycling bins, changes may need to be made. Testing various placements of the containers and signage on the containers also plays a part in the testing phase. Once testing is complete, results can be evaluated and determinations made on continuing or expanding the program.

COLLECTION CONTAINER CONSIDERATIONS

Collection containers have an important function in commercial recycling. In large public areas, such as parks or fairs, containers should be large and obvious. The placement and visual effect of the containers is critical to effectively capture large quantities of uncontaminated recyclables. Large public events may involve various nationalities; therefore, it is important to use graphical labeling on the containers. The chasing arrows recycling symbol is universal and effective. Pictures of the intended materials are also helpful. Special lids with holes or slots sized for the intended type of recyclable may help to discourage other materials from being placed in the receptacle. Recycling containers should be placed near trash containers to reduce contamination. Containers should be emptied frequently during crowded events. If a trash container is full, visitors will tend to use the closest container available for trash, even if it is labeled for recycling. In industrial areas special containers may need to be established to accumulate certain recyclables. A properly labeled 55-gallon drum may be the collection container for aerosol cans or batteries. Areas may be designated for the accumulation of broken pallets to be ground and recycled into mulch.

STAFFING CONSIDERATIONS

The addition of recycling to any establishment requires staff to implement and maintain the program. Depending on the extent of the recycling program, additional custodial labor may be needed, solely for the purpose of recycling. In areas, such as parks, where the public has access to recycling, contamination will occur on some level, and custodial staff may need to perform some sorting to eliminate contaminants. They may need to prepare the recyclables for transport to the recyclers. This may involve operation of equipment such as balers or stacking materials on pallets. Maintenance staff will be responsible for the maintenance of any equipment and also for the upkeep of the recycling containers. Administrative staff will be necessary to provide education, oversee the program as a whole, monitor the program progress and feasibility, evaluate new programs, and provide feedback.

COMMUNICATING TO THE PUBLIC

Education for large-scale commercial recycling can be divided into two areas; public education and staff education. Public education is necessary for any programs that the public has access to. The key points to communicate to the public while they are at the commercial establishment are the fact that recycling is available, what can be recycled, and where it can be recycled. This can be accomplished in different ways, depending on the nature of the establishment. In any event, the containers should be clearly labeled with pictures to communicate to all nationalities and the universal recycling symbol should be visible. Events that have a brochure or schedule of activities should include a statement that recycling is available. Stadium-type events can include an announcement to advise the public that recycling is available and where. Chain stores, such as supermarkets, that provide recycling of grocery bags should place collection containers near entranceways with obvious labels. These chain stores could advertise recycling on sale flyers, bags, or cash register receipts. Establishments that practice recycling that is not visible to the public can advertise their accomplishments as well. Restaurant chains can place a recycling statement on their menu. Merchandise shops can hang signs in visible areas to display their recycling achievements.

STAFF EDUCATION

Education for staff involves teaching them about general recycling and about what programs are available at their working location. This can be accomplished though training classes, company newsletters, intranet sites, staff meeting, and bulletin boards. Program feedback to staff is very important to keep them involved in the program. Contests with prizes can be used to create competition, keeping staff interested in the program. New staff members should be educated about recycling during initial training. Education will increase participation in recycling and help to reduce contamination.

RECYCLING AT ANNUAL EVENTS

Not all recycling programs are continuous. Recycling at special events requires not only the planning stages that have been discussed thus far, but a breakdown period as well. A temporary event, such as an annual fair, usually involves a group that oversees the operation of the fair. This group coordinates with the various vendors to ensure that each exhibit is set up on time, managed properly during the event, and broken down properly once the event is over. This same group would most likely oversee a recycling program in place during such an event. The following example shows one possibility for the steps this group might take to set up a recycling program at such an event.

Example 4. Jane is responsible for setting up a recycling program during the South County Annual Fair in May. She is to ensure that all vendors participate, that containers are available for the public, and that the program is properly communicated to the public. These are the steps that Jane followed:

- Research past fairs at any location that practiced recycling. Gather information on materials recovered, placement of containers, costs, lessons learned, and recycling participation
- Establish a budget for recycling
- Investigate available recycling markets and hauling services for potential commodities that will be generated
- Determine feasibility of each commodity and share with management team to determine which commodities will be recycled
- Notify vendors of recycling requirement and solicit feedback on individual needs for recycling
- Enter agreement or contract with hauler(s) and custodial services to provide recycling service. Include feedback from the hauler on weights and types of materials collected for measurement purposes in the contract
- Rent or order necessary containers for collection of materials. Some organizations may loan or donate containers for publicity
- Include recycling statements on flyers, maps, posters, and other fair media. Plan for announcements to be made at various time intervals to remind visitors about recycling
- Ensure that containers are in place prior to start of the fair
- Monitor the progress of the program routinely during the fair, fine-tuning as needed to improve capture rates and limit contamination
- Ensure that recycling continues during breakdown of the event
- Gather and prepare feedback to and from the hauler(s), vendors, visitors, custodial staff, and management

One-time recycling can include other events such as area clean-ups, phone book drives, construction projects, and Earth Day events. In each case, the general steps of Example 4 can be followed.

IMPORTANCE OF RECYCLING FEEDBACK AND MEASURING PERFORMANCE

Measurements and recycling feedback can be used to monitor the success of the program and share that success with employees or the public. In administrative and industrial areas where the source of recyclables is primarily generated and handled by the employees, feedback is important to maintain interest in the program. By soliciting feedback from solid waste and recycling haulers, a recycling rate can be calculated. The recycling rates in Figure 35.1 were calculated by dividing the total amount of solid waste generated by the amount of solid waste recycled.

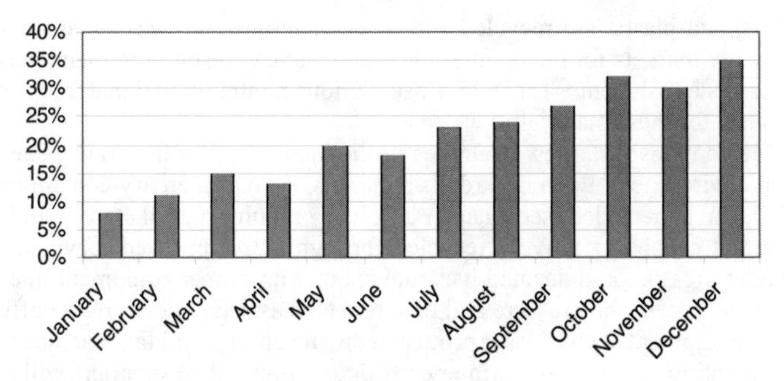

FIGURE 35.1 Monthly recycling rates.

Another way to depict recycling performance may be by showing the tonnage of each material recycled for a given period of time, as in Figure 35.2.

KEEPING UP WITH IMPROVED TECHNOLOGY

Reeducation is another important step for a successful commercial recycling program. Reminders in company newsletters, on bulletin boards, on Internet sites, and other visible areas help get the point across. New recycling programs should be evaluated on a regular basis. Improved technologies or new recycling industries in the area might present new opportunities to recycle materials that did not previously have a recycling market. Local demand for recyclables may increase the market value of materials, resulting in a more feasible material to recycle.

MARKETS FOR SPECIAL MATERIALS

Large commercial establishments frequently produce waste streams that contain materials outside of the normal recycling markets. Many times, traditional materials such as papers,

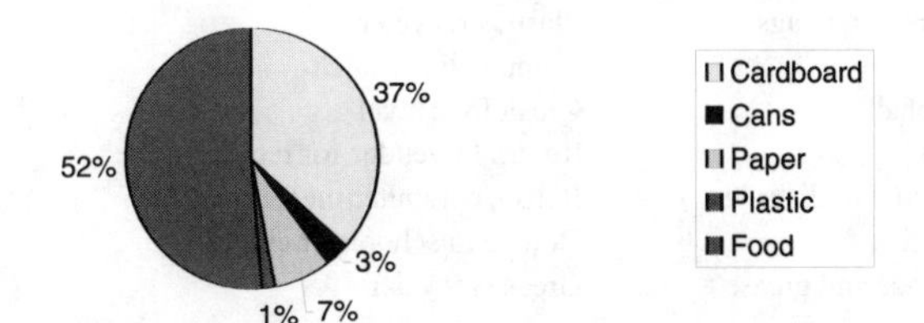

FIGURE 35.2 2000 Recycling materials summary (by weight).

metals, glass, and plastic are recycled and other materials are overlooked for recycling value. There are markets for many other materials that can result in tremendous cost savings for these establishments. Table 35.4 lists various nontraditional materials and examples of markets for these materials.

This table contains just a few examples of the many uses for materials generated in a commercial environment. Even hazardous materials, such as mercury-containing devices and used oil, can be recycled (see Chapter 21). They require special storage and handling procedures, but can be effectively recycled through EPA approved recyclers. Aerosol cans are under pressure and deemed hazardous, but with proper equipment, the cans can be punctured to release the pressure and then recycled as scrap metal. Large office buildings that use equipment such as laser printers, copy machines, and fax machines will generate large quantities of toner and printer cartridges. Some office suppliers will offer a rebate for returned used cartridges or at least offer to collect and recycle the used cartridges. Many used materials are still very reusable and should not be discarded. Local charities and organizations rely on donations such as building supplies, equipment, clothing, and food. Warehouses that contain food or other perishable items should regularly monitor expiration dates. Any items nearing expiration that are in danger of not being used should be donated quickly to ensure their usage and avoid disposal in the solid waste stream. Contracts with vendors should include recycling.

TABLE 35.4 Recycling of nontraditional materials in large facilities

Commodity	Use/market
Textiles	Rags, donation (shelters, charity, etc.)
Batteries	EPA approved recycler
Polystyrene foam	Reuse, give to shipping companies, polystyrene recycler
Carpet	Donate if usable, carpet recycler
Food waste	Pig farmers, composting, feed pelletizer, food bank
Usable plywood	Donate to affordable housing, sell
Non-pressure treated wood	Composting, mulch
Usable appliances	Donation, sell
Bubble wrap	Reuse, give to shipping companies
Wood pallets	Reuse, refurbish broken pallets, return to vendor, composting, mulch
Electronics	Sell parts, donate, electronics recycler
Plastic stretch wrap and bags	Plastics recycler
Landscape debris	Composting, mulch
Concrete and asphalt	Crush for gravel
Printer cartridges	Return to vendor for recycling
Computer media (CDs, disks, etc.)	Reuse, computer media recycler
Used office supplies	Donate to schools or charity
Used kitchen grease and grease trap grease	Grease recycler
Waste tires	Waste tire recycler

THINK REUSE

Vendors can reuse shipping materials such as pallets, bubble wrap, and other filler materials. This will not only save disposal costs, but will save the vendor money for the purchase of the shipping materials. Watching the materials that are placed into the solid waste dumpsters will identify what materials are being disposed of in large quantities. These materials can then be targeted for locating a market. Large commercial establishments tend to have many departments with differing functions, and should look for ways to reuse items between departments. For example, a water park may have a camera shop that loads film for the customer and accumulates empty film canisters. The same water park may have a lab that tests water samples and purchases small containers to hold the samples. The camera shop could give the lab the film canisters for their water samples. This would reduce the waste of the canisters and save the purchasing cost of the sample containers. Reusable materials should always be chosen over disposable items unless there is a particular need for the disposable item. Office complexes should be encouraged to use refillable pens and pencils. Industrial workers should use durable gloves that can be refurbished and reused instead of disposable gloves. Employees can be issued reusable coffee mugs instead of providing polystyrene cups. Vendors can be required to reuse packaging containers and pallets. An effective reuse policy can minimize the size of a large commercial facility's waste stream.

MANAGING PROCUREMENT

One of the best ways for large commercial establishments to increase recycling is through careful procurement. The procurement services department of any commercial establishment is typically the department that evaluates and contracts for routine purchasing of materials. This department should look for materials packaged in recyclable containers made with recycled-content durable, reusable products. When purchasing materials packaged in recyclable containers there may be more than one option. Weigh the cost of purchase against the market value of the container to determine which purchasing choice is more efficient.

Example 5. A large office complex is negotiating a contract with a beverage supplier to supply soft drinks that will be used in conference rooms and other areas throughout the complex. They have a choice between soft drinks in aluminum cans and in plastic bottles. A case of aluminum cans can be purchased at $2.00/case and a case of plastic bottles at $1.85/case. Aluminum cans have a recycling value of $0.45/lb and plastic bottles have a recycling value of $0.10/lb. Which is the better buy?

This choice can be made with little calculation. Assume that the weight of one empty case of aluminum cans and one empty case of plastic bottles are equal to one pound each. The cost of the aluminum cans less the recycling value is equal to $1.55. For the plastic bottles the cost less the recycling value is equal to $1.75. Though the initial cost for purchasing cans is higher in this example, the end result is actually more cost-effective.

Recycled-content products should be purchased when feasible. Many times, recycled-content products cost the same or are less expensive than products made with virgin materials. Durable products may cost more initially, but will last longer and reduce waste. Industrial businesses tend to use large-scale equipment, making durability an important factor. The procurement services department should include clauses in contracts requiring the use of recycled content in specially made items, requiring the use of minimal packaging, or requiring that the vendor reuse packaging. Effective purchasing can result in increased recycling.

In general, cost-effective recycling programs can be initiated at large commercial establishments with successful results. By knowing the waste that is produced, the potential market, and the alternative purchasing choices, many items can be diverted from landfill disposal. Careful planning prior to initiating any recycling program will result in a successful system for recovering materials. Cost savings and landfill avoidance can be found by evaluating materials that are not traditionally recycled. Many items that are still usable can be donated or sold. By practicing a well-rounded reduction, reuse, and recycling program, commercial establishments can have a big impact on the volume of material disposed of in landfills.

SECTION V

CASE HISTORIES

CHAPTER 36
CASE HISTORIES

LISA WAGNER HALEY
Environmental Scientist
Haley Environmental Consulting and Engineering
Kirkland, Washington

KEVIN McCARTHY
Recycling Specialist, CH_2M Hill
Sacramento, California

DAVID C. STURTEVANT
Solid Waste Project Manager, CH_2M Hill
Bellevue, Washington

*A SMALL RURAL COMMUNITY: THE ARCATA COMMUNITY RECYCLING CENTER'S PROGRAM**

Program Overview

An excellent example of a comprehensive, well-established recycling system is that managed by the Arcata Community Recycling Center (Fig. 36.1). This private nonprofit recycling center, located in northern California, has been in operation for 20 years. Unlike many programs that have been established in response to recent attention to the solid waste crisis, this program was borne out of a general concern for the environment, particularly for resource conservation. The program truly is comprehensive—incorporating education on conservation and waste minimization with a system for exchange of reusable items with the collection, processing, and marketing of many different recyclable materials.

Demographics

The center is located in Arcata, California (population 15,300), on the northern California coast in Humboldt County. The town is approximately 300 miles north of San Francisco and 95 miles south of the Oregon border. The county is rural in nature in terms of population densities, distance from metropolitan areas, and its resource-based economy. Several

*Case history prepared by Lisa Wagner Haley with the support of Kate Krebs, Executive Director, Arcata Community Recycling Center, Inc.

FIGURE 36.1

small communities are located within the 3600 mi^2 area of the county. Many are isolated by narrow roads and rugged terrain. Arcata itself is a typical small university town, located in an area of higher population density.

Most of the economy historically has been based on natural resource industries. Within the town is Humboldt State University, which is also a major employer of local residents. In recent years, the area has become increasingly popular to tourists. Many residents, therefore, are employed in the retail or service sector.

Solid Waste Collection Services

Arcata's residential and commercial waste is handled by a private hauler. The city council, which reserves the city's right to approve the rates, holds a franchise agreement with the private company. Solid waste is collected on a weekly basis. Customers pay a flat fee for one-can service and additional charges are added if the household places more than one 30 gal can at the curb. Commercial establishments arrange billing based on the frequency and volume of service. The waste is taken to a transfer station in a neighboring city, where it is compacted and hauled to the nearby landfill. The site owner estimates the site has at least 10 more years of operating capacity.

Program History

The original recycling operations began in 1971, as a project of Northcoast Environmental Center (NEC), a nonprofit consortium of environmental groups from the Arcata area. A recycling drop-off site was established, which was staffed completely by volunteers. The majority of organization of the program was carried out by three people, who researched the feasibility of the project, identified markets for materials collected, and rallied volunteers to staff the site and donate the use of trucks for transporting collected materials to market. The drop-off site was established on a vacant lot in Arcata and was open to the public on Saturdays. Volunteers received, sorted, and processed the materials donated. Materials accepted included newspaper, aluminum, and glass. A local scrap dealer accepted the aluminum, and the other materials were shipped by rail to markets in central California. One hundred tons of recyclables were collected the first year.

The program soon came to be known as the Arcata Community Recycling Center, and its operations continued to be cosponsored by the NEC. The program received funds from the county's federal revenue share in 1973, which were used to purchase the program's first collection vehicle. This also allowed them to move their operations to the present site at 9th and N Streets. Here, they increased the hours they remained open to the public and increased business collection service, they received additional funds to purchase a baler, forklift, and a 2 ton truck. The county also provided assistance by providing labor through the VISTA, CETA, JIPTA, and other job-training programs. In the seventies, programs such as these were viewed as part of environmental and resource conservation, not as a part of solid waste management, and this was the motivation behind the county support.

Financial assistance from the center's founding organization, the NEC, provided a means for hiring a project director. Wesley Chesbro was hired to fill this role. He now serves as a Humboldt County Commissioner.

The ACRC negotiated an agreement with the Arcata Garbage Company to collect recyclable materials from their residential and commercial customers. The city of Arcata allocated $600 to expand these collection services. High-grade paper and cardboard were collected from area businesses and newspaper, glass, and aluminum were collected from the residences. Also the ACRC set up multimaterial drop-off sites in five of the area's outlying towns. The center began to buy back newspaper from community groups who collected them as a fund-raising activity.

In 1979, the ACRC incorporated separately as a nonprofit organization from the Northcoast Environmental Center. The recycling center received a $62,500 grant from the California Waste Management Board to design and construct a mobile collection vehicle to be used for commercial collection routes.

In 1981, the center designed and implemented "Project Recycle" at Humboldt State University, the city of Arcata and Humboldt County offices to collect high-grade ledger paper.

When the secondary markets' crash devastated many recycling programs, the center survived by laying off staff and cutting back services. This included ceasing pickup of materials from outlying rural towns. Luckily, though, community groups continued the pickups. Other local recycling programs closed.

Understanding the importance of regional communication and cooperative efforts, the center obtained funding from Apple computers to establish the Western Rural Microcomputer Network. The network linked five rural collection centers (Bend, Oregon; Grants Pass, Oregon; Arcata, California; Chico and Visalia, California) to allow them to share information on markets and transportation.

ACRC encouraged community groups to establish recycling as part of their fund-raising activities by designing a "community buy-back service" in 1983. In 1985, the program was awarded the "Best Recycling Center in California" award by the California Resource Recovery Association.

From 1985 to 1986, a capital fund drive was conducted to raise $46,000 of the necessary $90,000 to purchase a processing and warehouse site. A pilot program was implemented in 1985 called the Neighborhood Recycling Network. Sets of three canisters for glass, aluminum cans, and newspaper were placed in the two trial neighborhoods, and block leaders were recruited to help spread the word about the program. This project was jointly funded by the Department of Conservation, Division of Recycling, city of Arcata, and Humboldt Area Foundation. The total cost of the pilot project was $12,500.

ACRC was selected as one of seven models used by the Ford Foundation funded report "Case Studies in Rural Solid Waste Recycling" in 1987. This program was the only one west of the Mississippi chosen for study.

The program received the "Best Integrated into the Community Recycling Center" from the California Integrated Waste Management Board and the "Special Achievement in California Recycling" award from the California Resource Recovery Association.

In order to conduct research on the feasibility of developing local markets for secondary materials, funding was obtained from the state Rural Renaissance and the Ford, Shalan, and Irving foundations. The research findings were disseminated locally, statewide, and nationally.

The Environmental Protection Agency Region IX awarded a contract to ACRC to develop a small-scale secondary glass remanufacturing facility on the north coast, the commercial collection pickup has become more sophisticated. The truck used for collection had been equipped with a mobile cardboard baler to allow for baling the cardboard while en route. Cardboard and office paper are currently collected from approximately 250 establishments and comprises almost 35% of all materials collected by the ACRC. Some businesses have agreed to pay the ACRC an amount equal to their monthly garbage bill savings for this service.

A Rural Recycling Network was designed for collection service based on the Neighborhood Recycling Network model developed earlier for two outlying communities. Within a year a Raise the Roof fund drive collected over $30,000 to pay for construction of a roof over the buy-back receiving yard. The level of materials processed rose from 100 tons in the first year of operation to over 500 tons currently.

The administrators of this program have obviously made significant efforts to obtain funding from private foundations and government agencies. These sources have proven critical for many aspects of operations. However, the sale of recyclables provided the majority and the most consistent portion of ACRC revenues.

Funding sources from 1971 to 1991 area as follows:

Private foundations and corporate donors

Humboldt Area Foundation

Shalan Foundation

James Irvine Foundation

Tides Foundation

Apple Computers

Owens Illinois Brockway

Alcoa Recycling

Northwest Paper Fibres

Weyerhauser West Coast

Simpson Paper Company

Government agencies

California Waste Management Board

California Department of Conservation, Division of Recycling

U.S. Environmental Protection Agency, Region IX

Office of Appropriate Technology

County of Humboldt, Private Industry Council

County of Humboldt, Redwood Region Economic Development

County of Humboldt, Youth Employment Training

City of Arcata

The ACRC Program Today

The Arcata Community Recycling Center manages a complete program that includes a donation drop-off and buy-back site, a community buy-back program, commercial collection, Humboldt State University campus collection, a self-service oil change station, a reusables depot, a neighborhood recycling network, a community recycling education program, and more.

Current operations for the ACRC have been in the same location since 1974. The recycling center accepts aluminum cans, newspaper, glass bottles and jars (clear, green, and brown), plastic PET bottles, office paper (sorted white paper, colored paper, computer printout paper), cardboard and brown paper bags. Over 50% of these materials are received as drop-off donations from area residents. Operations at the facility are managed by five core staff, including a full-time executive director, three part-time operations staff, and a part-time reusables depot manager. The center relies on about five part-time individuals from county work programs for materials processing.

Permanent staff and labor from the county's general relief program are responsible for processing the materials to prepare them for market. They mechanically flatten and blow the cans into an Alcoa semi-truck; sort and pack unbroken wine bottles by size and shape, and crush the glass by color; bale the cardboard; and accumulate newspaper and high-grade paper in gaylords. Materials are stored in and around the ACRC building until marketed.

The ACRC currently has two flat-bed trucks (1½ and 2½ ton capacities) for collection of recyclables. Other equipment includes two magnetic metal separators; platform scale; large and small baler; four bottle crushers; forklift; pallet jack; bins, barrels, and gaylords for storage; metal container bins for recycling drop-off; aluminum can flattener and blower; banding gun; hand tools; and a small paper shredder.

Public education efforts include fall ads in the local papers, door hangers, and brochures. Presentations are given to community groups; "block captains" and other volunteers spread the word. The ACRC keeps funding low and time as a higher priority in advertising the program. Surveys are conducted to gauge the success of these efforts. A 1984 survey revealed that 75% of Arcata's residents knew about the program and 60% made use of it at least once a month.

The center is open Wednesday through Saturday from 9:00 A.M. to 5:00 P.M.; 24 hour recycling bins are located at the back gate of the center. Earnings from recyclables can be donated to one of over 300 groups on a posted Community Buyback Program list, or donated to the center.

In addition to accepting donations, California Redemption Value carbonated beverage containers can be redeemed for marked prices. These containers must be kept separate from other containers. Scrap value is offered for aluminum cans, glass bottles and jars, and newspapers.

The center is organized into stations that are clearly marked with instructions on how the materials are to be accepted. (See Fig. 36.2.)

Commercial Collection Routes

In Arcata and neighboring communities, the ACRC offers a free collection service for pickup of office paper and cardboard boxes. Participants in this program include approximately 250 schools, offices, and businesses, both large and small. A copy of the "Office Paper Recycling Guide" is displayed in Fig. 36.3.

Neighborhood Recycling Network

As many program managers are learning, an effective way of advertising a recycling program is by making use of community resources. The ACRC has established a neighborhood recycling network, where trained volunteer block leaders distribute pamphlets and

HOW TO RECYCLE

IT'S EASY!

Separate and sort your recyclable materials at home immediately after you have used them. Call us for tips on how to set up a recycling area in your home or office.

OPEN FOR BUYBACK & DONATIONS

Corner of 9th and N Streets
Arcata

HOURS:

Wednesday through Saturday,
9 a.m. to 5 p.m.

- emptied & rinsed
- flattening not required
- New: all aluminum cat food cans
- CA REDEMPTION VALUE*
- NO "tin" cans (soup, beans, etc.)

- bundled in paper bags or with twine
- folded flat and stacked (must be dry)
- NO magazines, office paper or envelopes, phone books, or junk mail

- emptied & rinsed
- lids & labels are O.K.
- sort by color: clear, green, brown
- CA REDEMPTION VALUE*
- NO mirrors, ceramics or windows

- remove caps
- CA REDEMPTION VALUE*
- NO milk jugs or juice containers

- sort white paper, colored paper, computer print-out paper (xerox copies are O.K.)
- NO envelopes
- commercial collection available—see SERVICES

- flatten boxes and fold bags (must be dry)
- brown corrugated cardboard only
- NO six-pack cartons or waxed boxes
- commercial collection available—see SERVICES

FIGURE 36.2

information on details of the ACRC recycling program. Displayed in Fig. 36.4 is an example of a flyer that briefly discusses the "ethic" of recycling and provides instructions for preparing materials.

Community Recycling Education

To educate the community about recycling, speakers have been trained, slide shows have been developed, and tours have been established to promote awareness and participation

OFFICE PAPER RECYCLING GUIDE

COMPUTER PAPER

Continuous computer paper with perforated strips may be recycled as is. Here are the two recyclable types:

- **white with green stripes computer paper**
- **white with blue stripes computer paper**

Plain white computer paper may be put in the white paper recycling bin.

WHITE PAPER

It's a high grade of paper, and therefore valuable. White paper consists of the following:

- **white letterhead stationery**
- **plain white bond copying paper (NO off-white or natural)**
- **white typing and writing paper**
- **white forms**
- **white carbonless forms**
- **white manila tab cards (index cards)**
- **white cover stock**
- **adding machine tape**
- **white lined composition paper**
- **other dull-finish white paper**

COLORED PAPER

Keep colored paper separate from white and computer paper. Colored paper consists of the following:

- **colored letterhead stationery**
- **colored bond copying paper, including off-white and natural (NO goldenrod or florescents)**
- **colored typing and writing paper**
- **colored forms**
- **colored carbonless forms**
- **colored manila tab cards (index cards)**
- **colored cover stock**
- **other dull-finish colored papers**

Printed on Recycled Paper, Of Course!

FIGURE 36.3

UNACCEPTABLE MATERIALS

Be careful not to "contaminate" your paper recycling containers with unacceptable materials. If you're recycling an old report, for example, you will have to take it apart and separate into white paper, colored paper, and throw away plastic covers or other binding. If you're recycling a white paper form, be sure to throw away the carbon. WHEN IT COMES TO OFFICE PAPER RECYCLING, IF IN DOUBT, THROW IT OUT! Here's a handy list of what NOT to put in your office paper recycling containers:

- magazines and catalogs
- junk mail
- **ENVELOPES**
- tissue paper
- FAX paper
- coated (glossy) paper
- goldenrod or florescent copy papers
- ditto masters
- blueprint papers
- photographs
- phone books (no part may be recycled)
- carbon paper
- adhesive stickers, labels or tape
- waxed paper
- waxed paper delivery boxes
- waxed 6-pack cartons
- paper plates, cups, food wrappers or containers
- styrofoam
- plastic
- vellum sheet protectors
- rubber bands
- paper clips
- metal fasteners (STAPLES ARE O.K.)
- bindings
- construction paper or art paper (NO crayon, oil or acrylic paint)
- cellophane
- garbage

NOTE: NEWSPAPERS, CARDBOARD AND BROWN PAPER BAGS are recyclable! Keep them separate from office paper, but do recycle them. MANILA FILE FOLDERS may be reused or brought to the recycling center to be reused. Bring in your LASER-PRINTER PAPER to have it tested for recyclability.

ARCATA COMMUNITY RECYCLING CENTER is open for office paper recycling Wednesday through Saturday, 9 a.m. to 5 p.m. at the corner of 9th and N Streets. Drop off your office paper in the warehouse in the clearly marked recycling boxes. Call 822-4542 if you have any questions...and thanks for recycling!

FIGURE 36.3 (*Continued*)

of the community. Area schools can participate in recycling projects, receive in-class presentations, and learn why recycling is so important for our future.

Humboldt State University Recycling Program

In cooperation with the Arcata Community Recycling Center, recycling containers are located throughout the Humboldt State University campus. The HSU Associated Students office provides information for students interested in participating in the program.

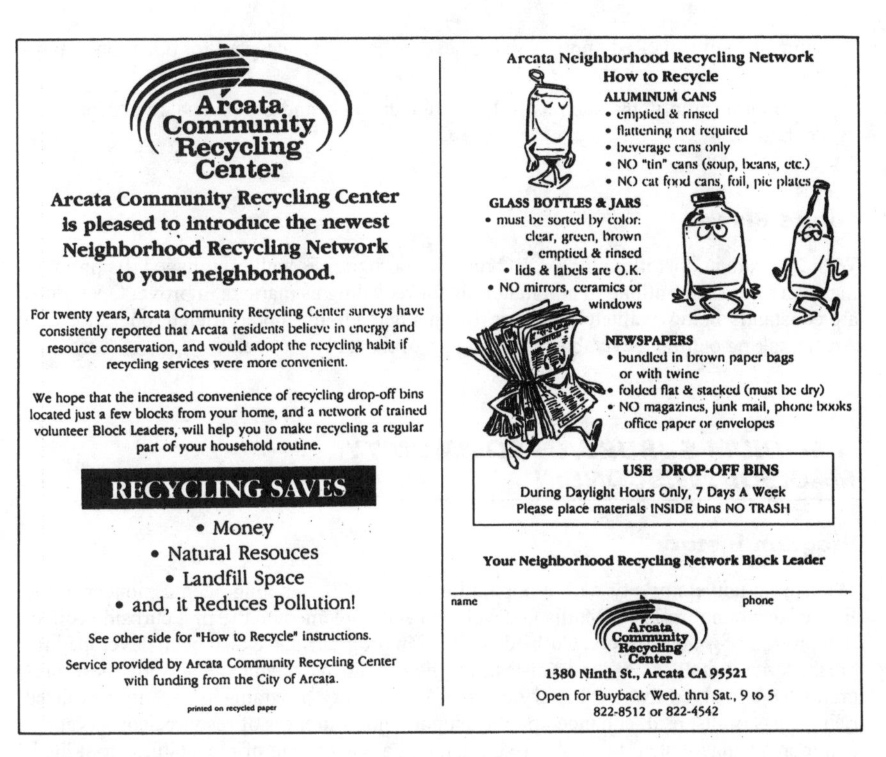

FIGURE 36.4

Self-Service Oil Change Station

A garage is maintained for those interested in a do-it-yourself oil change. The garage is kept clean and dry and tools are available for loan. Citizens can change their own motor oil, leave it in the ACRC tank, and it will be recycled. A $1.00 donation is requested for this service.

Reusables Depot

Donations are accepted of clothing, books, and household items that are in good, reusable condition. On approval, they also accept furniture, building materials, reusable packaging materials, and other items for resale. Typical items available for purchase include canning jars, homebrew bottles, and gallon jugs as well as kitchen goods. Items can be purchased at bargain prices or buy-back earnings can be swapped for depot items.

Other Services

Reusable Styrofoam peanuts can be donated or taken out.

Reusable paper egg cartons can be brought in or taken out.

Bimetal California Redemption Value beverage cans are accepted for donation or buy-back.

Aluminum scrap items such as foil, TV dinner trays, and pie plates are accepted for donation only.

Future Plans

With 20 years in operation, the ACRC recycling program is well established. Its program managers hope to continue to add materials for recycling as markets improve. Operations are constantly being evaluated and improvements made for more efficient operations. In Arcata, taking out the recyclables is just a part of the daily routine.

*A MEDIUM SUBURBAN COMMUNITY: MADISON, WISCONSIN**

Program History

Madison, a university city with a population of 190,000, has long been a pioneer in the field of recycling. In 1968, Madison developed and implemented the first curbside collection program for newspapers. Curbside collection programs have also been developed for brush, large metal items, yard wastes, and household recyclables. Many of these programs have evolved from voluntary efforts to mandatory programs as the city has faced rising costs of disposal, enhanced public support and awareness of resource conservation, and more stringent state mandates (e.g., banning a wide range of recyclables from landfills).

Program Description

Madison has been actively establishing, implementing, and fine-tuning a number of recycling programs targeting the following materials:

- Newspapers
- Brush, tree trimmings, and logs
- Large metal items (i.e., white goods)
- Yard wastes (i.e., leaves, grass clippings, and garden wastes)
- Waste oil
- Office paper
- Scrap metals
- Asphalt and concrete
- Lead acid batteries
- Paint products
- Telephone books

*Case history prepared by Kevin McCarthy and David C. Sturtevant.

The programs described below, will achieve an estimated 1991 recycling rate of 41.5%.

Madison's curbside collection of newspaper was begun in 1968 as a voluntary, pilot program and later expanded citywide in 1970. On the same day as their refuse collection, residents place bundled newspapers curbside, and the material is collected and loaded into special racks attached to refuse packer trucks (Fig. 36.5). The collected material is ultimately transferred to semi-trailers and shipped to market in Alsip, Illinois. A total of 3420 tons of newspaper were collected for recycling.

Beginning in 1972 for tree cuttings (i.e., logs) from city forestry operations and 1976 for residential brush and tree trimmings, Madison has chipped these materials and transferred the chips to local composting facilities. Chips are also made available to the general public. Residents are offered the following services:

- Separate monthly curbside collection of brush and tree trimmings from April through October on a scheduled basis.
- Separate curbside collection of Christmas trees in January.

The collected materials are chipped at the curb or at a centralized processing facility. A total of 16,269 tons of brush and tree trimmings were collected.

Since 1976, Madison has offered separate weekly curbside collection of white goods (i.e., stoves, refrigerators, water heaters, downspouts, gutters, and miscellaneous recre-

FIGURE 36.5 Special rack for newspapers attached to refuse packer.

ational equipment). When providing residents with large or bulky item collection, city crews set aside metal items for separate collection. These items are later collected with a dump truck, transferred for storage in 30-yd3 roll-off bins, and the bins are serviced by a local scrap dealer. Over 1000 tons were recovered for recycling.

In 1978, the city initiated a waste oil recycling program by establishing three drop-off sites. Two additional sites were added in 1979. Each site consists of a 275-gal tank mounted on a 3 ft × 5 ft × 6-in concrete pad. Each tank is equipped with a special pouring spout. Metal containers are also available at each site so residents can discard containers used to transport the oil. A total of 42,866 gal of waste oil was reclaimed.

Madison began recycling leaves in 1980. Nine years later, mandatory recycling of all yard wastes was implemented. The city provides separate scheduled curbside collection of yard wastes with two collections in April and three in the fall. Residents place the materials in loose piles at the curb and city crews use a "dust pan" approach to collect; rear-loader packer trucks are equipped with a dust pan attachment into which city crews sweep the material (Fig. 36.6). The collected material is transported to area farms for direct soil applications or to compost facilities operated by Dane County. From May through September, Madison does not offer curbside collection services for yard waste but operates three drop-off centers for these materials. In addition, the city encourages residents to reduce their yard waste generation by leaving grass clippings on lawns and/or practicing backyard composting. Nearly 20,000 tons of yard wastes are collected for recycling in 1990.

Madison's interest in household recyclables dates back to the 1987 startup of two drop-off centers for glass containers. In 1989, this program was expanded to 13 drop-off centers, all of which accept glass, aluminum cans, tin cans, and PET and HDPE plastic containers. Over 720 tons of these materials were collected for recycling when the program started.

Curbside recycling of household recyclables was initiated in 1989 through the efforts of four neighborhood groups conducting biweekly curbside collection of commingled bagged recyclables. With the February 11, 1991, kickoff of the city's mandatory weekly curbside collection of household recyclables, including cardboard, these groups discontinued their efforts. The city's program serves approximately 58,000 residential housing units and consists of residents purchasing specially marked clear plastic bags from locally designated retail outlets, placing bottles and cans into the bags, and setting the bags curbside on their regular refuse collection day. Residents are also asked to flatten and bundle cardboard containers for curbside placement alongside the bagged recyclables. City crews pick up the recyclables using a recycling collection truck and transport the material to a MRF. The MRF is responsible for processing and marketing the collected recyclables.

The city's park division has also developed a program for collecting recyclable containers. Recycling containers are strategically located throughout city parks and are also provided for special events such as summer concerts.

As for other city department recycling efforts, since 1975 the city has required all employees to recycle office paper. This resulted in the recovery of 199 tons of paper. The city's public works department is separating out and recovering scrap metals. In addition, Madison has been recycling both asphalt from city streets and concrete from sidewalks since 1975. Asphalt is either reprocessed by city crews or sold to an asphalt manufacturer. Concrete is crushed into aggregate by a sand and gravel operation. A total of 19,634 tons of asphalt and 3000 tons of concrete were recycled in a year.

Within the past two years, Madison has implemented recovery programs for paint products and telephone books. Telephone books can be placed curbside and picked up by city crews. Telephone books can also be returned to drop-off sites provided by a local department store chain. As for paint products, Madison and Dane counties offer a program

FIGURE 36.6 Dust pan attachment to sweep yard waste into rear-end refuse truck.

whereby residents may drop off selected items, which are later reused, recycled, or disposed at a hazardous material facility.

Finally, the remaining materials in Madison's waste stream are sent to a "reduction plant" that processes combustible materials into refuse-derived fuel (RDF). The RDF is shipped to a local utility where the material is burned to generate steam and electrical power. Noncombustible materials, except ferrous metals, are separated and transferred for disposal at a landfill. Ferrous metals are magnetically separated out from the noncombustible materials and recycled. RDF accounts for 21.9% of Madison's waste stream and landfilling is used for 36.3%.

Public Education and Information

Madison's significant educational activities have focused on the implementation of yard waste recycling and the recently implemented curbside recycling program for household recyclables. The city's most comprehensive efforts to date have focused on the curbside recycling program. The city, with the assistance of a public relations firm, has been conducting a multifaceted outreach program consisting of

- Neighborhood and community group outreach through meetings and a newsletter
- Use of direct mail for a brochure entitled the "Recyclopedia"
- Development and distribution of posters
- Development and distribution of TV, radio, and transit PSAs using local and national celebrities
- Press conferences and other media events
- Special mailings to businesses and landlords

Program Funding

Currently, Madison's recycling programs are financed primarily through property taxes and with revenue from the sale of recyclables. However, based on the provisions of a recently adopted statewide recycling law, the state will be reimbursing some portion of the city's net recycling program costs. The city is also considering moving away from a property tax-based funding system to a volume-based refuse pricing system for its residential customers. See the 1990 Summary Sheet (Table 36.1).

Future Program Activities

Madison has a busy agenda for future program activities, such as

- Expanding curbside collection to include magazines, polystyrene, and other plastic containers specified by the state
- Providing assistance to businesses and institutions in setting up office paper recycling programs
- Developing recycling opportunities for bulk wood wastes
- Expanding existing household hazardous waste recycling programs
- Integrating existing and new curbside collection systems so as to minimize costs and lessen energy impacts
- Encouraging source-reduction efforts

TABLE 36.1 Summary sheet for City of Madison, Wisconsin, solid waste collection and disposal costs

Total annual costs	$7,567,210
Revenues	203,458
Net annual costs	$7,363,752
Total tonnage	182,280
Net cost per ton	$ 40.39
Net annual cost per capita (Population, 191,262)	$ 38.50 ($3.21/mo.)
Net annual cost per residential unit (Residential units, 52,985)	$138.98 ($11.58/mo.)
1990 Curbside recycling start-up costs	
Labor	$12,592.30
Benefits	3,833.29
Mileage	1,064.48
Special services	1,844.15
Special supplies	100.00
Office equipment	222.85
Advertising	88,600.00
	$108,257.07

Lessons for Other Programs

With its experience in implementing a diverse set of recycling programs, Madison has learned some important lessons including

- Focusing program education from the perspective of letting customers know "why, how, when, where, and what to recycle;" this will reduce possibilities for consumer confusion.
- Recognizing some of the hidden costs (e.g., site maintenance, disposal of materials not accepted, etc.) associated with properly running drop-off centers.
- Developing contingency plans for a program starting off too successfully (e.g., higher recovery of materials than expected).
- Planning any program changes far in advance.

A MEDIUM URBAN COMMUNITY: SANTA FE, NEW MEXICO*

Program History

Given several inherent obstacles to recycling in New Mexico, including low waste disposal fees and limited proximity to recycling markets, recycling in New Mexico made great strides with the kickoff of Santa Fe's curbside recycling program on July 2, 1990;

*Case study prepared by Kevin McCarthy and David C. Sturtevant.

New Mexico's first curbside collection program. The City of Santa Fe (population 60,000) and the local Keep America Beautiful affiliate, Santa Fe Beautiful, have implemented other programs including

- Telephone book recycling
- Christmas tree recycling

Program Description

Santa Fe's voluntary curbside recycling program provides 19,000 residences (i.e., housing units in 16 plexes or below) with weekly curbside collection of commingled glass bottles and jars, aluminum cans, tin cans, PET soda bottles, clear HDPE milk jugs, and newspapers. Residents place commingled recyclables into an 18 gal container and set the container out curbside the same day as their refuse collection. Materials are collected and sorted into bins on a recycling collection truck operated by a private company under contract to the city. The materials are later processed and sold to markets in the southwest and midwest. The city's program has proven popular with citizens as reflected in a monthly participation rate of 80%. The curbside program diverted 1012 tons of glass, 270 tons of metals, 44 tons of plastic, and 1126 tons of newspapers.

As for telephone book recycling, in cooperation with a local telephone book distributor, drop-off sites have been established. In addition, Santa Fe Beautiful has worked with the local telephone company to promote their collection efforts. Collected telephone books are marketed to a firm that shreds the books for use as insulation material.

Christmas tree recycling has been a regular event during the holiday season for the past six to seven years. Christmas trees are collected curbside and chipped and used for mulch. The city also chips tree trimmings and shrubbery found on street medians.

While encouraging Christmas tree recycling, Santa Fe Beautiful promotes the purchase of living trees instead of cut trees through a "Giving Tree Project." Local nurseries reported a tripling of living tree sales.

Public Education and Information

Santa Fe has carried out a variety of community education efforts. For its curbside recycling program, an extensive campaign was launched complete with "Carlos Coyote," the program mascot, brochures printed in English and Spanish, public service announcements (PSAs), media events, and a personal letter from the mayor to all curbside recipients. Ongoing public education includes weekly Carlos Coyote cartoons in the Sunday newspaper, attendance by Carlos Coyote at parades and other local events, PSAs, and a volunteer block leader program (Figs. 36.7–36.11). Other recycling educational activities include an annual school awareness program for K-6 grades, bumper sticker giveaways, distribution of brochures, public presentations, and Santa Fe Beautiful Litter Lympics events.

Program Funding

So far, the city's recycling activities have been funded through surplus capital improvement funds and the general fund. The city's curbside recycling program was financed through $350,000 from a surplus capital improvement fund. Revenues from the sale of re-

FIGURE 36.7

cyclables are returned to the city's general fund to help offset program costs. That year, the city allocated $425,000 from the general fund to pay for the program. These funds are paid out to a private contractor that provides the collection, processing, and marketing services for the program under a four-year contract with the city.

Educational programs sponsored by Santa Fe Beautiful are funded through a combination of city monies, grants, in-kind services and donations, and membership dues.

SCRUFFY — DID YOU EVER STOP TO THINK ABOUT THE REASONS WHY PEOPLE LITTER?
NO PAL — LET'S HAVE THE LECTURE!
© 1989 Ken Peterson
CARLOS COYOTE SAYS —
PEOPLE LITTER WHERE THEY FEEL NO OWNERSHIP. PLEASE BE RESPONSIBLE FOR THE PROPER DISPOSAL OF YOUR WASTE!
SANTA FE BEAUTIFUL
SANTA FE COUNTY
THE NEW MEXICAN

FIGURE 36.8

IT MEANS A PERSON WHO KNOWS WE NEED PLACES TO DUMP STUFF BUT SAYS — "NOT IN MY BACKYARD!"
WHAT'S A "NIMBY" CARLOS?
© 1989 Ken Peterson
CARLOS COYOTE SAYS —
EVERYONE KNOWS WE NEED A PLACE TO DISPOSE OF OUR SOLID WASTE — BUT NO ONE WANTS IT LOCATED CLOSE TO HIS OR HER PROPERTY. THE CITY OF SANTA FE REMINDS YOU . . . RECYCLING WILL EXTEND THE LIFE OF OUR LANDFILLS.
473-7220
SANTA FE BEAUTIFUL
SANTA FE COUNTY
THE NEW MEXICAN

FIGURE 36.9

YOU HAVE TO GUARD AGAINST THESE SANTA FE WINDS!
© 1990 Ken Peterson
NEWSPAPERS MUST BE SECURE WHEN PLACED IN CONTAINER • IN A PAPER SACK • TIED WITH STRING ON TOP OF THE CONTAINER • OR PLACED IN THE SIDE OF THE CONTAINER, SO WINDS WON'T BLOW THEM AWAY.
FOR INFORMATION OR SUGGESTIONS ON THE CITY CURBSIDE RECYCLING PROGRAM PLEASE CALL 473-7892.
473-7220
SANTA FE BEAUTIFUL
SANTA FE COUNTY
THE NEW MEXICAN

FIGURE 36.10

NO-WE CAN'T USE ANYTHING MADE OUT OF STYROFOAM
HOW ABOUT STYROFOAM PLATES?
ENVIRONMENTAL CONTROL INC
© 1990 Ken Peterson
USE PRODUCTS PACKAGED IN CONTAINERS USING RECYCLED MATERIALS.
WHEN SHOPPING, LOOK FOR BOXES THAT ARE GREY IN COLOR AND/OR HAVE THIS SYMBOL ON THE SIDE WHICH MEANS THEY CONTAIN RECYCLED FIBERS.
BUY PRODUCTS IN CONTAINERS THAT ARE RECYCLABLE IN SANTA FE.
REMEMBER NOT ALL PRODUCTS ARE RECYCLABLE HERE IN SANTA FE SO WHEN YOU HAVE A CHOICE, BUY THOSE THAT ARE.
473-7220
SANTA FE BEAUTIFUL
SANTA FE COUNTY
THE NEW MEXICAN

FIGURE 36.11

Future Program Activities

With recent passage of a comprehensive recycling law in New Mexico, Santa Fe and other cities will be under increasing pressure to expand existing programs and start up new recycling efforts. Currently, Santa Fe is focusing its efforts on initiating commercial sector recycling activities.

Lessons for Other Programs

The early success of Santa Fe's recycling efforts is attributed to a multisector participatory approach coupled with strong support from elected officials. In addition, a particularly effective approach has been to stress the win-win community benefits of recycling. Finally, since the city relies on general fund revenues to finance their programs, a strong focus should be receiving a stable revenue stream from recycled materials sales by developing long-range pricing arrangements.

*A LARGE URBAN COMMUNITY: THE AUSTIN, TEXAS, RECYCLING PROGRAM**

Program Overview

Recycling Week, "Cash for Trash," and a fleet of Recycling Block Leaders are the components of success for the City of Austin's recycling program (Fig. 36.12). Voluntary curbside pickup is offered to households for commingled recyclables in 5 gal plastic buckets or 14 gal recycling containers. The program began with a pilot project in 1982 and completed its expansion phase in 1989, bringing weekly collection to 110,000 households. Commercial collection of office paper is offered to public schools and businesses.

Demographics

Austin, the capital of Texas, is populated by 481,000 people and is growing at a rate just under 1% per annum. The city limits encompass 185 mi^2 with a population density of four persons per acre (Fig. 36.13).

Presently, unemployment is 4.4% and is gradually decreasing. Major public-sector employers include the University of Texas, Bergstrom Air Force Base, and various agencies of federal, state, and local government. High-technology systems and services dominate Austin's corporate employment market.

Solid Waste System

Refuse is accepted at three different landfills, two private and one public (operated by the city of Austin). The city landfill tracks municipal collections, which include 110,000 single-family residential accounts that produce approximately 160,000 tons of solid waste per year. Using data obtained from the City Planning Department and the Department of

*Case study prepared by Lisa Wagner Haley with the support of the Austin Recycling Program staff.

FIGURE 36.12

Health, the average statewide per capita generation rate of solid waste is calculated to be 5 lb per person per day.

Presently, Texas has one-sixth of the nation's operating landfills, and tipping fees average approximately $15 per ton. Therefore, Austin has been able to implement waste-reduction programs as a preventive measure, rather than a reaction to a disposal crisis. They are making the fullest use of existing resources and carefully weighing future benefits against immediate program costs.

The city provides solid waste collection service to the area's single-family dwellings, while private haulers collect from multifamily dwellings and businesses.

Program History

During the 1970s, a growing number of citizens began recycling residential waste materials at drop-off centers established and maintained by Ecology Action (EA), a nonprofit volunteer organization. By 1980, EA was operating nine multimaterial centers, servicing newspaper recycling bins on the University of Texas campus, and initiating an office paper recycling program.

In 1981, when the market for old newspaper crashed, the organization was forced to cut back on the scale of its operations. Citizen recyclers appealed to the city council for a municipal curbside program. A citizens' task force was appointed and a consultant was commissioned to develop a solid waste study and a 20-year plan.

In 1981, the city adopted the Solid Waste Management Plan, which called for an integrated program, in-

FIGURE 36.13

cluding a pilot curbside recycling program, technical assistance for backyard or other decentralized composting, a waste-to-energy facility, transfer stations, and permitting of a new landfill.

In 1982, a pilot curbside recycling program was established in two neighborhoods to service 3000 homes. In 1983, the service area was expanded to include 12,000 homes. The final expansion phase of the program was completed, bringing weekly curbside collection to 110,000 households. In 1983, the city abandoned its attempt to permit a new landfill in the face of organized opposition on the part of neighboring residents.

To promote and advertise the recycling program, block leaders were recruited in 1983. Five gallon buckets were made available from the local fire stations (Fig. 36.14).

In 1984, a referendum was approved to construct a 600-ton/day waste-to-energy plant. In 1984, the project was canceled due to revised economic projections, persistent environmental concerns, and the ubiquitous NIMBY (not-in-my-backyard) syndrome.

During September, a week was designated "Recycling Week." During this week, new developments in the program are publicized, new block leaders are actively recruited, and businesses, schools, and citizens groups are recognized for their achievements and participation in the recycling and waste-reduction program. Also, a privately sponsored program was implemented, called "Cash for Trash"; $100 in cash prizes are offered to randomly selected curbside recyclers.

In September 1989, city staff conducted an informal survey of local recycling businesses, government agencies, and nonprofit groups to arrive at an educated estimate of the total amount of recycling activity in Austin. All contacts were asked to estimate the volume of materials received from outside Austin and deduct them from their totals. The figures in Table 36.2 are based on responses to this survey, with the exception of City of Austin figures, which are based on recovery rates.

Figures were also calculated for items such as auto bodies, car batteries, clothing, household items, small appliances, books, waste oil, and paint. These items were calculat-

BLOCK LEADERS NEEDED!

We're looking for ~~a few~~ lots of good people!

FIGURE 36.14

TABLE 36.2 Amount of material recycled in Austin survey

Materials	City of Austin, tons	Commercial, tons	Nonprofit, tons
Metal	500	65,060	62
Glass	2,800	2,750	300
Paper	6,700	35,000	1,780
Yard waste	2,790	4,800	5,628
Subtotals	12,790	107,610	7,770
Total: 128,170 tons/year = approximately 23.5% recycling rate			

ed separately because they are not normally found in the waste stream, are only temporarily kept out of the waste stream by repair and/or reuse, or because they are not considered solid waste (Fig. 36.15).

Program Description

The program is a voluntary system offering weekly curbside collection of materials. Newspapers, corrugated boxboard, and kraft paper grocery bags are collected together;

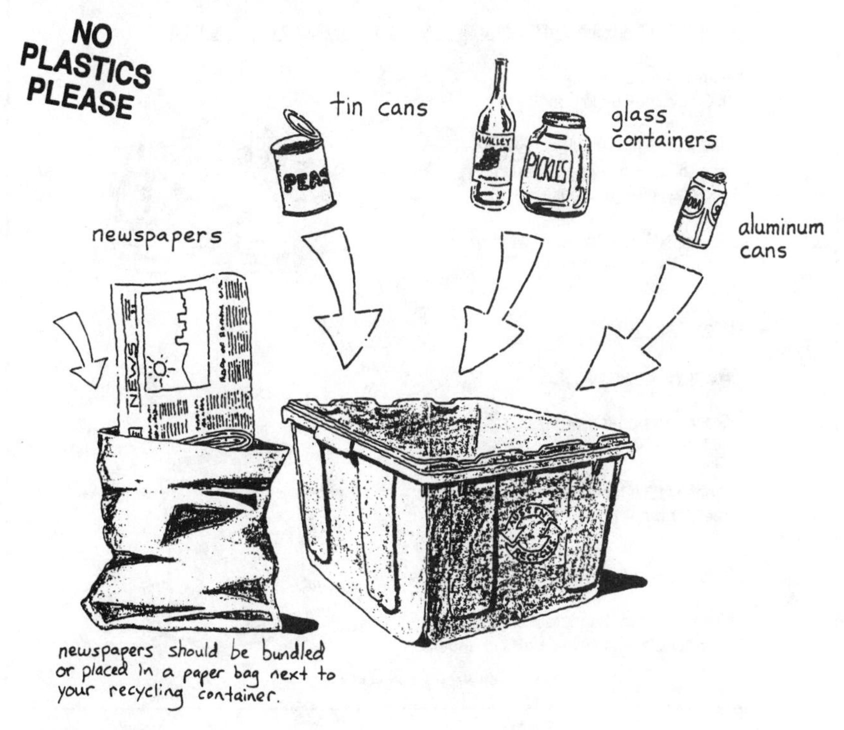

FIGURE 36.15

glass containers and cans (both steel and aluminum) may be commingled and set out in a separate container. In approximately 40% of the service area, a single 14 gal container is used for all materials (this was started in August 1990).

Block Leaders

To promote the program, Austin has enlisted over 1050 block leaders. The block leaders make home visits to their neighbors, distributing pamphlets (Fig. 36.16) on recycling and

HOW TO RECYCLE WITH THE CITY OF AUSTIN RECYCLING PROGRAM

Recycling Hotline
479-6753

Environmental and Conservation Services

* Place the materials listed below at the curb (separate from garbage) in containers by <u>8:00 a.m.</u> on your recycling day (see map).

<u>GLASS BOTTLES & JARS, ALUMINUM & STEEL FOOD CANS & FOIL</u>

* Remove food residue; lids & labels may be left on
* Place together in a paper grocery bag, box, or any open container (no plastic bags, please)
* Please <u>NO</u> aerosol cans, window glass, Pyrex, or ceramics

<u>NEWSPAPERS</u>

* Remove advertising inserts
* Stack in a paper grocery bag, or bundle with string
* Please <u>NO</u> telephone books, mail, or magazines

<u>CORRUGATED BOXES</u>

* Flatten to less than 3 feet on any side
* Bag/bundle with newspapers or separately

Printed on Recycled Paper

FIGURE 36.16

composting, yard signs, and bumper stickers. They provide explanations of how materials should be prepared for collection.

Recycling Week

Recycling Week was implemented to increase awareness of the city's recycling program by highlighting events and new developments in the program. Local newspapers, radio stations, and television networks contribute to the effort by covering the events and circulating press releases prepared by the city's public information office.

Tonnage of materials collected, calls to the recycling hot line and the number of volunteer block leaders recruited all increased by significant amounts during and after this public awareness and education activity.

Keep Austin Beautiful (a local chapter of the association Keep America Beautiful) promotes their commercial recycling programs during Recycling Week. The program promotes recycling in public schools and honors the school with the highest per capita recycling totals at the end of the school year. The Keep America Beautiful Clean Recycler Program promotes recycling and responsible waste management practices in Austin businesses.

Drop-off Recycling

In addition to curbside collection, drop-off boxes are located around the city. Residents who do not have curbside pickup use these boxes to deposit their newspaper, and at some locations, their glass and metal as well. The sites are managed by private recyclers.

The city contracts with a private recycler for the operation of a recycling station at the landfill. The station receives 400 to 500 tons of material each year, primarily large appliances and bulky scrap metal. Landfill customers are assessed a $10 surcharge for the disposal of major appliances, which provides an incentive for them to return them to the drop-off station instead.

Several buy-back operations also exist in the area. These businesses buy newspapers, cardboard, glass, aluminum cans, and bulky metals. Some also accept used clothes, appliances, furniture, and building materials, which are often repaired and resold.

Yard Waste Recycling

A pilot leaf collection program diverted more than 1300 tons of yard waste during the peak leaf-fall periods in late autumn and early spring utilizing regular garbage collection vehicles and crews (Fig. 36.17). Collection personnel identify and pick up leaves as they are normally set out, so that no special procedures are required by residents. The leaves are taken to a wastewater treatment facility, where inmates of a local correctional complex remove plastic bags and other extraneous materials. The leaves are then mixed into windrows of treated sewage sludge along with wood chips provided by the city's tree service contractor. The composting process, which takes 10 to 12 weeks to complete, yields a finished product that is marketed commercially through local garden shops and landscape businesses. Forty-five percent of Austin's sewage sludge is treated in this manner. The remainder of the treated sludge is applied to agricultural land leased by the city and an area farmer, with profits from the annual hay crop shared by both parties.

The city council appropriated funds to purchase a brush shredder for the diversion of wood waste at the landfill. The material generated by this operation will be used in the composting operation described above, made available for on-site composting, sold as a mulch, an industrial fuel supplement, or used as cover material for the landfill.

Christmas Tree Recycling

For the past six years, the Austin Parks and Recreation Department has collected Christmas trees for recycling on the two weekends following Christmas. Each year, over 45,000

COMPOSTING

A Guide To
Organic Recycling

Environmental and Conservation
Service Department

FIGURE 36.17

trees were collected by over 200 volunteers at 13 drop-off locations, with roll-off containers provided by local waste haulers. Tree donors receive pine seedlings in return for their participation in the event. The trees are shredded with chippers donated by local tree service companies and applied as mulch to city park lands.

Home Chemical Collection
In the spring, the annual Home Chemical Collection Day provides city residents the opportunity to safely dispose of hazardous material that would otherwise go to the landfill.

Nearly 4000 gal of used motor oil and 398 auto batteries were recycled as part of the event; 1200 gal of usable paint was reclaimed for use in local housing rehabilitation projects.

Procurement Policies

In an effort to set an example of waste reduction and recycling in municipal facilities, the city council established a purchasing policy, which created a 10% price preference for the purchase of recycled paper products by city agencies. Prior to the city council's Comprehensive Recycling Resolution of January 1990, which redefined recycled paper to conform with the latest EPA procurement guidelines, the city's purchasing department experienced some difficulty in receiving bids for fine paper with a certified percentage of post-consumer fiber. There has been no such problem with coarse paper products.

Markets

Periodically, recycling brokers are invited to bid on all curbside materials, and given the opportunity, through their bids, to suggest additional recyclables that could be collected by the city (Fig. 36.18). Local delivery is specified, forcing out-of-town businesses to work with local brokers or establish a local recycling operation. In the most recent bid invitation, bidders were allowed to bid on both separated and commingled materials. Both a floor price (not to be lowered during the term of the contract) and an escalator price (started as a percentage of a particular market quotation) was solicited for each material.

A three-year contract for all materials was awarded to ACCO Waste Paper, allowing the city the economies of commingled collections and a single point of delivery. To process commingled containers, ACCO has established a separate mechanical and manual sorting line, employing 14 mentally retarded adults through a local social service agency to staff the new facility.

The final market for Austin's newspaper is a deinking mill in Mexico City that produces recycled newsprint. Glass container cullet is sold to the Owens-Brockway plant in Waco, Texas, where it is recycled into new containers. Aluminum cans are sent to Alcoa plants in Arkansas and Tennessee, and tin cans are processed into structural reinforcing rods at the Structural Metals minimill in Seguin, Texas.

FIGURE 36.18

*A MAJOR URBAN COMMUNITY: LOS ANGELES, CALIFORNIA**

Program History

Moving beyond a divisive battle over development of a waste-to-energy project, the Los Angeles City Council in May 1987 called for a mandatory, citywide recycling program. In June 1988, Mayor Tom Bradley established a goal of reducing and recycling 50% of the waste stream collected by the city; the city's Bureau of Sanitation collects trash from all single-family residential units and small apartment buildings (i.e., four units or less). A year later, the city issued a Recycling Implementation Plan (RIP) that outlined a series of recommendations for a comprehensive recycling program. During the summer of 1989, the city moved forward with the RIP and created a new Recycling and Waste Reduction Division (RWRD) within the Bureau of Sanitation to implement residential recycling programs and citywide education and information activities. To address the remaining and largest portion of the waste stream, the city created an Integrated Solid Waste Management Office (ISWMO) directly under the Board of Public Works to help foster, initiate, and plan for commercial-sector recycling.

In 1989, the State of California passed Assembly Bill 939, which charged municipalities with reducing by 50% the amount of waste going into landfills by the year 2000. In response, the City prompted its Bureau of Sanitation to implement a residential recycling program.

In 1990, that program—the LA Resource Program—operating under a directive by then Mayor Tom Bradley was implemented citywide as the official waste reduction recycling effort for the City.

Some two years prior to the deadline of AB 939 of the City of Los Angeles met the 50% reduction requirement.

Revamped throughout the years, and now simply referred to as the Automated Recycling program, the 16 gallon yellow bin is a thing of the past. City residents now use a single-stream commingled automated recycling method. The 90 gallon container is used to recycle all clean paper, including unwanted mail, magazines and newspapers; glass bottles and jars; plastic bottles; and metal and aluminum cans. The Automated Recycling Program consists of three automated containers—Blue for recyclables, green for composting of yard trimmings, and black for refuse. Since 1990, the City has sponsored a Christmas Tree Recycling campaign. Residents are encouraged to recycle the trees curbside, or at any of the 15 drop-off locations throughout the city.

Program Description

The City of Los Angeles has begun to aggressively implement an integrated set of recycling programs addressing both residential and commercial waste generators. In September 1990, the city started the first phase of its mandatory curbside recycling program with subsequent and final rollout over a three-year period for collection of household recyclables and a five-year period for yard debris. The program phase-in is necessary due to the large number of residential units to be served [i.e., nearly 800,000 homes, concurrent phase-in of automated waste collection, and need for sufficient lead time for development of material recovery facilities (MRFs)] and a yard-waste composting facility. The city is

*Case history prepared by Kevin McCarthy and David C. Sturtevant.

in the process of developing mandatory enforcement procedures and an antiscavenging ordinance.

The actual program consists of weekly curbside collection, on the same day as refuse collection, of commingled glass bottles and jars, aluminum cans, tin cans, PET soda bottles, and clear HDPE containers placed by residents in 14 gal plasbins (Fig. 36.19) and also bagged or bundled newspapers placed on top of the bins. Recycling collection trucks transfer the material to a MRF and the material is separated and marketed for sale. In addition, an automated refuse collection truck picks up yard debris placed curbside in 60 gal containers (Fig. 36.20). The collected yard debris is transferred to a yard debris processing facility. Since 1989, the city has conducted a Christmas tree recycling program. Beginning this year, the program will consist of pilot curbside collection and seven drop-off centers. Residents bringing their tree to a drop-off center will be offered free compost.

As for commercial-sector recycling, the ISWMO has initiated a generator-based approach that focuses on identifying and targeting those sectors generating the most waste. ISWMO is currently developing a waste characterization database for targeted generators and later will develop specific waste reduction and recycling strategies for these generators.

In the interim, ISWMO has focused its energies on the following activities:

- Creating trade association (e.g., grocers, restaurants, hospitals, and landscapers) recycling task forces

Reciclar. Ahora es tan fácil como sacar la basura.

Recycling. Now it's as easy as taking out the trash.

FIGURE 36.19 Multilanguage recycling posters.

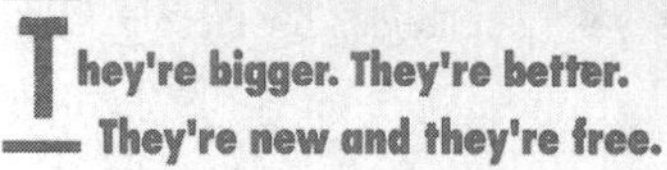

They're bigger. They're better.
They're new and they're free.

Son más grandes. Son mejores.
Son nuevos y son gratis.

FIGURE 36.20 Multilanguage recycling posters.

- Developing and sponsoring a series of workshops (e.g., office recycling, hospital recycling, hotel recycling, and composting) outlining successful programs, marketing/procurement issues, and waste audit procedures
- Making available videotapes on workshops held to date
- Developing recycling resource guides (e.g., on recycled paper, recycled paint, thrift shop directory, recycled toner cartridges, and "grasscycling")
- Producing a newsletter highlighting ISWMO activities and noteworthy recycling efforts in the community

Solid Resources Citywide Recycling Division (SRCRD)

This division of the Bureau of Sanitation addresses nonresidential recycling and reuse issues. SRCRD publishes a directory called "Put it to Good Re-Use LA." The directory includes an index of materials that can be donated, including appliances, building materials, computers, toys, watches, even eyeglasses, and a list of organizations that will accept these donations. Donated materials are diverted from landfills, helping the City meet legislated waste diversion goals. Copies may be obtained by calling (213) 847-144, faxing requests to (213) 847-3054, or via e-mail at SRCRD @san.ci.la.ca.us.

On the waste reduction front, the city is implementing several programs including:

- A "junk-mail" campaign
- Pilot backyard composting program servicing approximately 30,000 participants

- A “Don’t-Bag-It” outreach effort encouraging residents to leave grass clippings on their lawns

Public Education and Information

In the spring of 1991, the city hired three public relations contractors to assist with public education and information activities in three areas: curbside recycling program, citywide recycling awareness, and school-age recycling curriculum. Since 1996, the Board of Public Works, Public Affairs Office, working closely with the Bureau of Sanitation is charged with publicizing the Automated Recycling Program. Public education and outreach efforts include radio, television, and print campaigns, school outreach, a speaker’s bureau, and other marketing elements. Curbside recycling education is under way in the city and consists of an overall multilingual (i.e., English, Spanish, Korean, Mandarin, and Cantonese) campaign of community meetings, door-to-door canvassing, door hangars, and display booths exhibited throughout the city (Figs. 26.21 and 26.22). Citywide recycling awareness will be kicked off this fall with an ad campaign for television and radio, newspapers, and billboards. Finally, a school-age curriculum was begun citywide in September.

Program Funding

The City’s recycling programs are financed using general fund revenues. This is accomplished through two departments: (1) the ISWMO with a 1991 budget of $4 million to plan and facilitate recycling in the private commercial sector, and (2) the Bureau of Sanitation, Division of Recycling, for the operation of municipal residential recycling activities. Some proceeds come from the Automated Recycling Progam.

Future Program Activities

The city will continue to roll out its curbside recycling program to all city residents over the next three to five years. Additionally, the city recently completed a 12-week pilot program for collection of mixed paper and will be determining when to add this material to its curbside program. Other materials under consideration include corrugated containers, colored HDPE containers, and mixed plastics.

As noted above, the ISWMO will be developing generator-specific waste reduction and recycling strategies for those generators producing the greatest amount of trash in the city. Such a focus on these generators will give the city the best return on its investment in term of monies spent to increase the recovery of recyclables. In addition, the city will then be able to identify other generators that will need more technical and/or financial assistance to reduce or recover material from their waste streams.

Lessons for Other Programs

The unique challenges facing the city’s recycling efforts are primarily a result of its enormous and culturally diverse population, currently at nearly 3.5 million residents. Particular challenges include program funding and education. With the predominant use of general fund revenues to fund a city’s program, as with the City of Los Angeles, strong

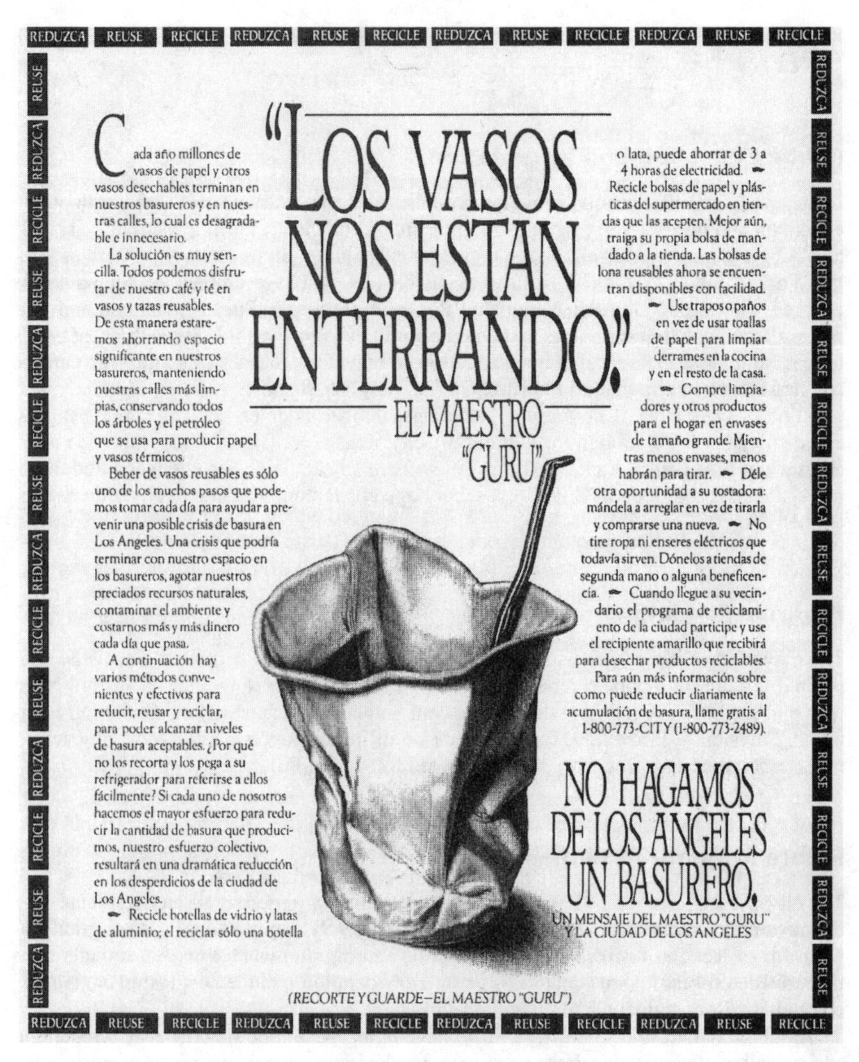

FIGURE 36.21 Spanish language recycling poster.

emphasis should be given to including avoided disposal cost savings as well as material sales revenues in the program balance sheet; this allows for a more accurate assessment of program costs and benefits. While seemingly obvious, this is especially important given the fiscal constraints and realities facing most local governments.

As for recycling education, the city emphasizes citizen involvement through providing them with opportunities to participate and incentives to do so. These messages must be tailored to the particular audience. However, given the recent deluge of "green marketing" messages (e.g., diapers are compostable or polystyrene is recyclable), it is particularly crucial that citizens clearly understand the recycling opportunities (e.g, types and ways to recycle materials in their area) offered to them.

FIGURE 36.22 Spanish language recycling poster.

*A SMALL RURAL COUNTY: AUBURN, WASHINGTON'S RECYCLING PROGRAM**

Program Overview

The City of Auburn is located in King County, Washington, approximately 20 miles from the city of Seattle (Fig. 36.23). The county's comprehensive solid waste management

*Case history prepared by Lisa Wagner Haley. The City of Auburn, Solid Waste Division is acknowledged for their support for the preparation of this case history.

FIGURE 36.23

plan includes an aggressive recycling goal that requires that communities reduce their waste by 50% by the year 1995 and 65% by the year 2000. In order to meet this goal, the City of Auburn has implemented a voluntary waste reduction and recycling program. Residents take their recyclables to multimaterial drop sites and have the option of signing up for the curbside collection of yard waste.

Demographics

The City of Auburn is a 20.7 mi^2 incorporated municipality located in the Green River Valley of King County Washington, positioned between Seattle and Tacoma. It is estimated that the population included 34,150 citizens. (Fig. 36.24).

The city's origin is that of an agricultural community, and now lures individuals from many different backgrounds. The areas largest employer is nearby Boeing, Inc. The largest sector of employment within the city is in manufacturing, second is in trade.

The city boasts easy access to state highways and close proximity to both the port of Seattle (the closest deep-water port to the far east) and the port of Tacoma.

Program Description

The program for waste reduction and recycling consists of several components, integrating technical assistance, increased public educational campaigns, and increased level of services.

The waste reduction element includes

- A school education program
- The King County Master Composter program
- Promoting reuse through an annual swap meet, a newspaper column, and dissemination of informational material
- Source reduction and procurement workshops for business and industry

The residential recycling element includes

FIGURE 36.24

- An expanded drop-off program of 20 to 30 multimaterial sites for collecting materials from single- and multifamily residents
- Promotion through mailers, instructional brochures, and a media campaign through the local news

The commercial/industrial sector element includes

- A technical assistance program designed to help businesses identify waste reduction opportunities and set up their own recycling systems

- Collection services offered by private contractors
- A privately owned commercial buy-back center
- Promotional and informational materials including a sign-up card, a self-audit form, fact sheets, a cost/savings worksheet, and a news media campaign

The yard waste management element includes

- An education campaign for home composting, including a demonstration site and volunteer outreach
- Curbside yard waste collection
- Program promotion through a mailer and instructional, brochure

The rates incentive program element is structured to encourage

- Reduction
- Participation in the recycling program

The legislative element of the program

- Encourages allocating space for recycling through land-use codes
- Discourages scavenging and theft through ordinances
- Establishes a procurement policy that reduces waste and mandates the purchase of products made from recycled materials

The monitoring component of the program will track

- Residential and commercial solid waste tonnage
- Recycling rates: participation and material amounts by program

After only two months of implementation of these new program components, the tonnages of materials collected in some instances more than doubled! (Fig. 36.25).

The program, which is offered to citizens, consists of conveniently located multimaterial drop sites and optional curbside collection of yard waste. The rate incentive structure was implemented so that the more residents recycle and reduce waste, the less they pay for garbage service.

The reason for using a drop-off system instead of implementing curbside pickup is that a private contractor has already been capturing an estimated 3.3% of the total waste stream through a program consisting of 75 drop boxes and because the service can be provided to residents at no cost to them, where a curbside program costs an additional $2 to $5 per month. The contractor is willing to expand this system by establishing new drop-off sites and adding to the types of materials accepted for recycling.

Currently, the only method of disposal available to Auburn is to take solid waste to King County's Cedar Hills landfill. At the county's rate of disposal, the landfill will reach capacity in 20 years. By reducing the volume of waste in King County going to the landfill, the county hopes to add an additional 20 years to the life of the facility. The tipping fee at the county landfill increased from $47 to $66 per ton in one year.

The current rate structure for collection of solid waste, including taxes, is as shown in the Table 36.3.

Residents have the option of signing up for the service level they feel would best suit their needs. These program changes were implemented May 1, 1991. Customers of the garbage service were all provided a "Residential Recycling Guide," which describes the entire program, discusses incentives for recycling (not only the reduction of disposal costs, but saving landfill space, conserving energy, reducing reliance on natural re-

FIGURE 36.25 Recycling poster, King County, Washington.

sources, and keeping the city clean) (Fig. 36.26). They are provided the choice of phoning in their service level or returning a mail-in card printed in the handbook. If they do not change their service level by the start of their billing date, they are automatically billed at the one-can rate. They are reminded that they will be charged at the additional can rate for each extra can they set out.

The instructions in the handbook are simple and basic, as is the program. Instructions are provided on how to set up a home recycling center, using cardboard boxes and paper bags as receptacles for the materials. Materials collected include newspaper, mixed paper,

TABLE 36.3 Rate structure for collection of solid waste

Garbage service	Rate	90-gal yard waste toter
10 gal can*	$ 3.00	$4.50
20 gal can	6.00	4.50
1 can†	7.00†	4.50
2 cans‡	14.50	4.50
60 gal toter	14.50	4.50
90 gal toter	22.00	4.50

*To qualify for the 10 or 20 gal service, customers must sign up for the yard waste toter or show proof that they are composting.

†Can is a standard 30 gal size.

‡Each additional can is $7.50 per can.

Auburn
Residential
Recycling
Guide

FIGURE 36.26 King County, Washington, recycling guide.

corrugated, three colors of separated glass, aluminum, two types of plastics, and tin. An estimate is given on the space required for accumulating a month's worth of material as 3 ft by 3 ft. Drop boxes are situated in convenient locations, such as school, church and store parking lots. We all accept that you leave the store carrying bags of groceries. This method promotes the idea that you also arrive with bags of recyclables and leave them in the bin before you do your shopping. Some neighborhoods pool their recyclables and take turns taking them to a drop site.

A key factor pointed out in the booklet is to find a method that is convenient and stick with it. Several places in the book list the city Recycling Coordinators telephone number, which can be called for assistance.

Specific instructions for material preparation are shown in the pamphlet reprinted in Fig. 36.27. The pamphlet concludes with a listing of all area drop-off centers, categorized by the material each accepts. It also provides a listing of resources, sign-up instructions, and a mail-in card.

Future Plans

Since mailing of the booklets, plastics (PET and HDPE) were added to the materials accepted at the drop sites.

The city's contractor plans to construct a materials processing facility in the near future. This will allow for the collection of even more types of materials and will also allow the residents one more choice in how they like to recycle.

The city staff is investigating the feasibility of a used-oil collection program.

*A MEDIUM SUBURBAN COUNTY: CHAMPAIGN COUNTY, ILLINOIS**

Program History

Founded in 1978, the Community Recycling Center (CRC) is a not-for-profit multimaterial processing center with a goal of maximizing recovery of recyclables from the Champaign County, Illinois, waste stream (population 120,000). Over the past 10 years, CRC has been actively involved with implementing and carrying out a variety of recycling activities, including

- Curbside recycling
- Used motor oil collections
- Household hazardous waste collection events
- Buy-back and drop-off centers
- Multimaterial commercial collection programs
- University recycling programs
- Cooperative purchasing of recycled paper
- Community outreach and recycling education

Because of the success of many of these programs, CRC was selected "The Best Recycling Center in the Country" in 1990 by the National Recycling Coalition.

Program Description

Currently, CRC is providing the collection, processing, and marketing of recyclables from the following sources:

- A buy-back center and drop-off center located at its processing facility in Champaign
- Nine rural village drop-off sites throughout Champaign County

*Case history prepared by Ken McCarthy and David C. Sturtevant.

PAPER

Paper is classified into various grades; NEWSPRINT, CARDBOARD (known as "corrugated"), HIGH-GRADE or LEDGER PAPER, and MIXED PAPER (advertising mail, magazines, and non-corrugated cardboard).

All of these paper materials are accepted at Auburn's drop sites.

Collect clean, dry paper material in a box at home and then empty into the appropriate bin at the drop site.

Please do not recycle paper towels, disposable diapers, paper plates, tissues, greasy, waxy or plastic coated paper or candy wrappers.

When paper is recycled, it is made into all kinds of things like cereal boxes, new paper and even building materials.

Remember, for every ton of newspaper recycled, 17 trees are saved.

GLASS

There are many ways to re-use your glass jars. Fill them with nuts and bolts, use them for canning, storing dry goods, or use them to make terrariums and other crafts.

Even if you re-use your glass bottles and jars, you will still have a lot left over.

RECYCLE THEM! Wash them and remove the lids and rings. Be careful not to break them. Take all of your glass bottles and jars to a conveniently located drop site. The bins are marked "Brown", "Green", and "Clear", so you can separate by color.

The glass you recycle will be melted down and re-made into new glass containers.

Drinking glasses, window glass, light bulbs or mirrors are <u>not</u> acceptable at this time.

FIGURE 36.27 Recycling instructions—King County, Washington.

HOUSEHOLD METALS

If you look around your house and garage, you will find many recyclable metal items. While all of the ones listed may not be collected at the drop sites, they all can be recycled or re-used. For ideas of where to take those odd metals, call 1-800-RECYCLE or Auburn's Recycling Coordinator at 931-3047.

Some examples of metals you may find around the home:
*ALUMINUM FOIL, PIE PLATES, TV DINNER TRAYS
*LAWN MOWERS
*BATTERIES AND OTHER CAR PARTS
*ALUMINUM LAWN CHAIRS, LADDERS, SCREEN DOORS
*COPPER WIRE AND PIPE
*SHEET METAL, APPLIANCES

You may take Aluminum Cans, Foil, Pie Plates and TV Dinner Trays to the drop sites. Be sure to rinse off all food material.

Tin Cans are also acceptable at the drop site. To prepare the cans for recycling, wash them, remove the labels, remove both ends and flatten.

Aluminum and Tin are processed and recycled differently so you need to collect them separately. If you have trouble telling them apart, tin cans have a side seam and are attracted by magnets.

"Americans throw away enough aluminum every three months to rebuild our entire commercial air fleet."

FIGURE 36.27 (*Continued*)

YARD WASTE

In Auburn, yard waste is 17% of the total waste stream. That means you can drastically reduce your waste and the amount of waste going to the landfill by either composting your yard trimmings at home, or by participating in the curbside collection of yard waste.

When you sign up for this service, the contracted hauler will supply you with a 90 gallon wheeled toter. In this toter you may put lawn trimmings, weeds, and twigs up to 3" in diameter. Yard waste in plastic bags will not be accepted.

Animal and Food Waste is not acceptable in your yard waste container.

Your yard trimmings will then be taken to a composting facility where it will be composted into a nutrient rich soil conditioner. Many of the composting facilities bag and re-sell this compost in local stores.

If you would like more information on home composting, you may call Auburn's Recycling Coordinator at 931-3047, or you may call King County Composting Information at 296-4466.

FIGURE 36.27 (*Continued*)

- 300 Champaign-Urbana businesses (e.g., bars, restaurants, offices, and schools)
- Administrative and academic buildings at the University of Illinois and Parkland College

CRC's buy-back center and drop-off center accept materials 5 days a week. The materials taken include glass bottles and jars, aluminum cans, tin cans, bimetal cans, PET soda bottles, clear HDPE milk jugs, newspaper, magazines, cardboard, and high-grade paper. Over 1000 tons of materials were collected and CRC paid out over $241,000 to recyclers using the buy-back center.

The nine rural village drop-off centers make up what is known as the Hometown Recycling Program. Established in 1988, this was Illinois's first countywide rural recycling program and is operated by CRC under a full-service contract with Champaign County (Fig. 36.28). Drop-off sites are available 24 hours a day and accept glass bottles and jars, aluminum cans, tin cans, bimetal cans, clear HDPE milk jugs, and newspapers. Nearly 600 tons of material were collected.

CRC also operates a multimaterial commercial collection program in the cities of Champaign and Urbana. CRC provides businesses with 55 gal drums for the collection of some or all of the following materials: glass bottles and jars, aluminum cans, tin cans, bimetal cans, newspaper, cardboard, and high-grade paper (Fig. 36.29). CRC empties the bins and spots new bins for each business participating in the program. In addition, CRC can also provide businesses with in-house collection containers. CRC charges $6 per pickup for the collection service. Approximately 1500 tons were collected.

Initiated in 1981, CRC handles the collection, processing, and marketing of glass bottles and jars, aluminum cans, tin cans, bimetal cans, clear HDPE containers, cardboard, and high-grade paper from administrative and academic buildings on the University of Illinois and Parkland College campuses.

In five years CRC handled all facets of the curbside recycling programs for the cities of Champaign and Urbana. However, due to an inability of CRC and the cities to reach

FIGURE 36.28 Rural village drop-off center, Champaign County, Illinois.

FIGURE 36.29 Removing drop-off center section, Champaign County.

agreement on a new service contract, another contractor was selected to run the programs. CRC continues to collect materials from a curbside recycling program serving county residents. In addition, negotiations are under way regarding CRC providing education and outreach for Champaign, Urbana, and the county's curbside recycling programs.

Along with CRC's recycling efforts in Champaign County, other efforts are spearheaded by the Intergovernmental Solid Waste Disposal Association (ISWDA). This agency was formed in July 1986 by the cities of Champaign and Urbana and Champaign County. ISWDA has been involved with preparing a solid waste management plan for the county, sponsoring household hazardous waste collection events, and development of a material recovery facility (MRF)/ transfer facility. Most noteworthy, ISWDA has pursued development of a MRF and selected a vendor to design, build, and operate the facility. As currently envisioned, the MRF will serve dual functions of processing mixed residential and commercial, construction and demolition waste, and source-separated recyclables. Given the source-separation approach of past and existing programs implemented by CRC, the relative priority and sizing of the facility to meet these dual functions has raised considerable debate in the Champaign County area.

Public Education and Information

With its comprehensive and diverse recycling programs in place, CRC has long focused on public education and information. In 1980, CRC started the first "Recycling Week" in Illinois, and in 1983 was the first recycling center to hire an education coordinator. Educational activities have included producing numerous booklets, flyers, and fact packs covering composting, sources of environmentally friendly home products, sources of commercial collection containers, recycling facts, and other areas. Videos and books are also available at a CRC library. CRC has conducted active community outreach with direct educational services provided to over 4000 schoolchildren and adults.

Program Funding

CRC's starting operating budget of $1.1 million was derived primarily from the sale of recyclable materials and from service contracts with Champaign County, commercial collection accounts, and other programs. A small portion of the budget also comes from individual and corporate donations, memberships, and state grants.

Future Program Activities

Building on its existing programs, CRC plans the following initiatives:

- Find private sponsors for drop-off centers in Champaign and Urbana.
- Establish recycling programs for areas not currently served within Champaign and Urbana.
- Establish recycling programs in large apartment buildings (i.e., five or more units).
- Expand commercial-sector recycling.
- Market products made from recycled materials.
- Expand recycling education programs.

Regarding the first initiative, CRC previously operated eight drop-off centers in the Champaign and Urbana area, but discontinued operations due to financial difficulties. Thus, they are seeking to reestablish drop-off centers with private-sector assistance (e.g., from grocery stores).

Lessons for Other Programs

Given the not-for-profit status of CRC, strong consideration must be given to how such a status or role integrates with the efforts of local governments. In the case of CRC and ISWDA, while pursuing seemingly complementary courses, disagreements have arisen over their respective approaches to recovering recyclables from the waste stream. A current dispute over the relative merits of and/or levels of source separation versus mixed-waste processing will be a reoccurring debate across the country as the public, elected officials, private industry, and environmentalists grapple with how best to maximize recycling.

*A LARGE URBAN COUNTY: MECKLENBURG COUNTY'S RECYCLING PROGRAM**

Program Overview

Mecklenburg County's (North Carolina) recycling program has three objectives: to meet the 1993 state recycling goal of 25%; to economically produce high-quality recyclables; and to encourage within the community and region recycling and conservation (Fig. 36.30). To meet these goals, the county has implemented a multi-faceted recycling pro-

*Case history prepared by Lisa Wagner Haley with the support of the Mecklenburg County Solid Waste staff.

gram, with three distinct areas of concentration: residential, industrial/commercial, and wood and yard waste composting.

Charlotte, the county's principal city, boasts one of the largest publicly run curbside recycling programs in the country. The program began with pilot curbside collection, which provided the county the necessary data for sizing the now operational materials recovery facility and allowed program managers to evaluate options for the permanent collection program.

FIGURE 36.30

Other components of the program include a series of drop-off recycling centers (including one at the landfill), yard waste collection, and an aggressive business recycling program.

Demographics

Recent population estimates show Mecklenburg County has approximately 511,433 citizens. The county's principal city is Charlotte. The area is referred to as the "trucking capital of the United States" and many feel it is becoming the financial capital of the southeast. Major employers in the area include IBM, Duke Power Company, First Union Bank, and NCNB (Fig. 36.31).

Solid Waste Management

Solid waste from Mecklenburg County is currently accepted at three facilities (Fig. 36.32). The county's Harrisburg Road landfill, which opened in 1974, has almost reached the end of its 17-year life. Waste is also taken to a private landfill in an adjacent county, and to the 85,000 ton/year capacity mass burn waste to energy plant. The location of a new county landfill is pending site approval from the state.

Program History

In 1975, the first request was made by Mecklenburg County citizens that they be provided drop-off centers to bring in materials for recycling. In 1982, funds were first appropriated to begin a promotional strategy under the guidance of a public relations and marketing firm, to inform the public about the concept and need for recycling.

Beginning in 1983, voters began showing their support for making waste to energy a part of their waste disposal by voting to set aside general obligation bonds approved to fund the construction of the facility. Also in 1983, the county started its wood waste recycling program with the purchase of a tub grinder that was used to shred wood scraps, pallets, and tree limbs.

In 1984, the Mecklenburg County Board of Commissioners began developing a comprehensive plan for waste disposal. Phase I of the voluntary pilot curbside recyclables collection program was implemented in February 1987. The program expanded to include 16,000 homes by August 1989.

FIGURE 36.31

FIGURE 36.32

During the pilot stage of the curbside collection program, three types of round and square containers were tested. Participating residents were polled and it was found that the 14 gal square container was the favorite. The color "warm red" was chosen to symbolize the urgency of the need to recycle. In 1986, a temporary materials processing facility was purchased.

In June 1988, the North Mecklenburg Household Waste and Recyclables center opened in the northern part of the county. Prior to this, north county residents who did not receive garbage and curbside recycling collection services had no choice but to travel over 30 mi to the county-operated landfill. A pilot yard waste program was established in 1988, which produced 3500 tons of mulch in its mulch first year of operation.

A pilot program was initiated in 1987. Recyclables were taken to a temporary 12 ton/day materials recovery facility where materials were prepared for market. Plastic bottles were granulated and stored in large gaylord boxes awaiting shipment to Southeast Container in Asheville, N.C. A manual sorting line was used where employees separated cans from bottles, and glass was separated by color. A magnetic separator was added later as a more efficient method of separating cans. Aluminum was purchased by Republic Alloys Inc. and glass was purchased by Owens Brockway.

In searching for a location to site a new materials recovery processing facility, the county encountered large-scale public opposition, similar to that experienced by those opposing the siting of landfills and incinerators. A neighborhood group even sued the county for fear that the facility would bring noise, odor, and unwanted traffic to their neighborhood.

The county selected Fairfield County Redemption (FCR) of Stratford, Connecticut, to locate, construct, and operate its material recovery facility. They flew citizens to their Connecticut facility to show them first hand how little the recycling center impacted the surrounding community.

FCR secured a five year contract with the county to operate the new facility. For the first year, 75% of profits exceeding $41,000 were shared with the county. The contract has since been renegotiated. FCR is responsible for marketing the materials.

Unique features of the building include a gift shop where visitors can purchase goods made of recyclable material, an exhibit area where larger items can be displayed, and a conference area where lectures can be presented.

Program Description

Currently, 109,000, or 85% of the 121,000 single-family homes in Mecklenburg County, are served by curbside recycling. A single 14 gal container is provided to residents in which plastic (HDPE and PET), glass, steel, bimetal, and aluminum beverage and food containers are commingled along with newspaper, which is stacked on top or underneath the container (Fig. 36.33). The theme RECYCLE NOW and the warm red color of the boxes were chosen to emphasize the urgent need to recycle.

Collection of recyclables occurs weekly on the same day as curbside trash collection. Recycling trucks with a single operator/collector transport the commingled materials (paper separated) to the material recovery facility. Participation rates for curbside collection exceed 80% on a monthly basis, i.e., 80% of the homes that have a red box fill it with recyclables and place it at the curb at least once a month. On an average, each home that recycles is contributing 425 lb of recyclables a year to the program and collectively diverting 19,500 tons of resources from burial in the local landfills.

The percentage of materials collected each week by weight are

Newspaper, 73%

Glass, 16%

Plastics, 4%

Aluminum, 1.5%

Residue, 5.5%

The residential recycling program also includes a network of drop centers for recyclables not included in the curbside recycling program. There are 14 drop centers with 8 more in the planning stage. In addition to the same materials collected at curbside, some of the drop centers accept corrugated containers, used motor oil, and lead-acid batteries (Fig. 36.34).

The North Mecklenburg Household Waste and Recyclables Drop Center, which opened in June 1988, provides residents in the northern part of the county with a conveniently located facility to drop household waste and recyclables. To deposit household waste at the center, residents must bring in the following

FIGURE 36.33

Questions or Concerns
About Drop Center Recycling?
Call 704-336-6087
Monday-Friday 9 a.m.-5 p.m.

FIGURE 36.34

amounts of recyclables: passenger car, 3 bags; single-axle trailer, 6 bags; van or pickup truck, 6 bags. The center is staffed by one person and is opened three days a week. Approximately 135 tons of recyclables were received at the center during its first year of operation.

The most extensive material recovery effort is located at the Harrisburg Road Landfill Recycling Drop Center, where in addition to the materials mentioned, a wide variety of scrap metals are accepted from incoming loads of waste.

Residential recycling provided 77 tons per day of highly marketable materials to the county's privately owned material recovery facility. This figure increased to 82 tons/day. The facility was designed to handle 65 to 70 tons/day, with the flexibility for expansion. The facility employs a unique system, combining the processing center with an amphitheater, a specially designed observation gallery, computers that allow users to interactively learn about recycling, and finally, exhibits and a gift shop featuring products made with recycled materials.

In the beginning, a tipping fee of $7.50 per ton was paid to FCR by the county for every ton of recyclables delivered to the material recovery facility through municipally operated collection programs. The county operated the 14 drop-off centers and encouraged home owners who are not part of the curbside collection program to bring their recyclables, free of charge, to any of the drop-off centers (Fig. 36.35).

During the first six months of operation, January to June 1990, the facility processed 7975 tons of newspaper, glass jars and bottles, PET and HDPE plastic containers, and aluminum and steel cans. Revenues of $315,912 were generated from the sale of materials. From July 1990 to June 1991, 19,955.2 tons of newspaper, glass, plastic, steel and aluminum were delivered to the MRF (Fig. 36.36).

The estimated recovery rate of recyclables from the solid waste stream is 25%. On an average week, Charlotte residents recycle 384 tons of material.

Commercial and industrial sector recycling activities are maintained through a Business Waste Recycling program. An easy-to-follow guide is provided to interested businesses (Fig. 36.37). It explains the possibilities, practicalities, and benefits of starting a recycling program at the workplace. Examples are provided of typical large-quantity items found at restaurants and bars, grocery stores, manufacturing firms, and offices. The guide covers all aspects of a program-from designating a recycling coordinator to designing the program, securing markets and monitoring success.

FIGURE 36.35

A special program was created by the county staff called the PAPER CHASE that focuses on recovery of high-grade (white ledger and computer) paper in office buildings. PAPER CHASE has been implemented in most county and municipal buildings and is recovering 70 tons per year of office paper (Fig. 36.38).

Yard waste is not accepted at the drop centers, but it is collected at curbside on a weekly basis. Residents are encouraged to place the yard waste in 40 gal fiber drums or trash containers for curbside pickup in order to prevent unnecessary operational problems due to plastic bags.

A compost processing facility operations on a 15-acre site will target 30,000 tons of material per year.

Before, yard waste was processed at an 8-acre mulch/compost facility, located in the northern part of the county. The site adjoins the North Mecklenburg Household Waste and Recyclables Drop Center. Local county residents may brought their clean yard waste free of charge to the facility from 8:00 A.M. to 3:30 P.M., Monday through Saturday.

Yard waste is debagged and processed through a slow-speed shredder and a tub grinder for initial processing before being placed into windrows for biological treatment. Detailed testing and cultivation, in addition to proper windrow turning, ensure desired characteristics. The final process involves separating the product into two different sizes. The product larger than ⅜ in is sold as mulch, and the particles smaller than ⅜ in are sold as compost. Mulch and compost are purchased from the facility Monday through Saturday, 8:00 A.M. to 3:30 P.M.

FIGURE 36.36

The program managers have found that to ensure high-quality, contaminant-free

FIGURE 36.37 Business Waste Recycling Guide, Mecklenburg County, N.C.

materials for all aspects of the program, they need to devote a heavy emphasis to promotion, education, and convenience for the recycler (Fig. 36.39).

A public relations and marketing firm was hired to assist in promoting the Mecklenburg County program. Numerous press releases were issued to television and radio stations, periodic media briefings were conducted, and special news opportunities were created. In addition to news coverage, most of the local newspapers and several of the local television and radio stations ran public service announcements and editorials supporting recycling. Additionally, the community-owned public television station produced two 30-mm documentaries on the solid waste crisis and the steps Mecklenburg County was taking to address the problem.

FIGURE 36.38

FIGURE 36.39

The county sponsored recycling conferences, which were designed to attract business people and news coverage. Annually, the county presents awards to outstanding recyclers, both individuals and organizations. Specially designed materials are sent to targeted audiences, such as those living in curbside collection areas, selected businesses, and potential conference attendees.

Future Plans

The program recently added collection of household hazardous wastes in 1992. The materials collected if not recycled, is otherwise properly disposed of by a private vendor. Recommendations for an ongoing program are being developed.

*A LARGE RESORT TYPE COMMUNITY, CITY OF VIRGINIA BEACH, VIRGINIA**

With a population reaching 440,000 and a customized residential recycling program, the City of Virginia Beach, Virginia, with a public–private partnership with Tidewater Fibre Corp. (TFC), has reached an amazing 76% recycling rate with its automated, commingled curbside program. This complements the city's existing automated waste collection service by diverting approximately 40,000 tons of material, or 20% per year, from the residential waste stream.

The curbside program, originally designed by a team of Virginia Beach employees,

*Prepared by Tina DiSalvo Fries, public information officer, City of Virginia Beach, Virginia. Reprinted with permission from *Waste Age* magazine, November 1999.

called for the investment by TFC, located in Chesapeake, Virginia, of $8 million to build a single stream, automated materials recovery facility (MRF) with a capacity of approximately 60,000 tons per year. The new 35,000 square-foot facility included. an automated sorting and processing system, a fleet of twenty trucks, Heil 7000 automated bodies, and 113,000 95 gallon automated containers for residents.

Background

In 1988, to reduce the amount of landfilled waste, Virginia Beach began collecting newspapers at drop-off centers located at public schools and participated in a regional curbside recycling service. This service, however, required residents to sort the materials at the curb. By 1996, Virginia Beach began searching for a more user-friendly program. During this time, the city established 50 drop-off centers to handle recyclables.

The Virginia Beach Department of Public Works Waste Management Division worked to create a service that would meet the expectations of both residents and city leadership. A Request for Proposals (RFPO was issued, and out of five private contractors, TFC was selected to provide the service. In March 1997, TFC and the City of Virginia Beach signed a five-year renewable contract. A key TFC strength is its ability to market recyclables. The company has a number of long-standing contacts with purchasing agents, buyers, and sellers. Equipment can be bought and other aspects of a program can be learned, but if you do not have a market for your recyclables the program can collapse. TFC markets products, the largest volume being paper, primarily in the southeastern United States, but also targets international markets

Task Areas

To help ensure the new program's success, a team of city staff and TFC representatives outlined 16 major tasks and 129 individual tasks. The task areas included

- Internal training
- Risk management
- Route development
- Container distribution
- Required resources
- Route map production
- Townhouse collection
- Billing procedures
- Recycling database development
- Media/public relations plan development
- Education plan development
- Communication between the city and TFC
- Drop-off recycling centers
- Communications with the public
- Temporary container storage areas
- Implementation time line

By November 1997, with the help of TFC, Virginia Beach delivered containers to approximately 113,000 homes over a three-month period. When the project was fully implemented, curbside recycling service was provided to residents every other week on their normal waste collection day. Single-family homes received 95 gallon automated containers and townhomes received 18 gallon bins.

The Bottom Line

The cost of Virginia Beach's curbside recycling service and collection at 20 drop-off sites, down from 50 sites, is approximately $4 million annually. Additionally, by diverting more than 40,000 tons of waste to the new recycling system, the city is saving $2 million annually in avoided collection and disposal costs. This, coupled with the city's waste management division reducing its operating costs by $800,000, has brought the net cost for recycling services down to $1 per household per month. More noteworthy, however, is that the new system generates income of which Virginia Beach receives 15% net revenue share, but is not responsible for financial losses. Currently, TFC pays Virginia Beach between $3000 to $4000 per month in revenue.

The numbers tell the story (see Table 36.4). After introducing the new automated program, Virginia Beach's recycling participation rate, previously 50% with the manual bin program, increased to 76%. Likewise, the tonnages of recyclables collected increased from 10,000 to 40,000 tons annually. Overall, this collection program has a much higher participation rate than other programs because the bins are bigger and are automated, Because of the new recycling program, trucks are picking up about 25% less garbage than before. Previously, there were a lot of calls from customers who needed a special garbage pick-up in addition to their regular collection day. Now, thanks to recycling, there is more room in their trash containers and consequently, there has been a reduction in those calls But the benefit of the commingled program obviously is volume. These large containers collect five times as much as the small bins and they have a lid that helps keep the materials dry and prevents them from blowing around.

Educating the Public

Residents are given many gentle reminders to recycle. For example, the Virginia Beach's Public Works Management Office greets visitors with a sign that says "This porch was

TABLE 36.4 Virginia Beach recycling program statistics and results

Customers	115,000
Participation rate	76%
Program type	Automated, commingled, 95 gallon cart
Yearly tonnage recycled	34,575
Cost per household/month*	$1
Cost per ton*	$39.46
Annual revenue	$25,269
Waste stream reduced	20%
City's population	439,889
Number of square land miles	258

*Includes avoided disposal costs, avoided collection costs, and revenue.

built with lumber made from plastic milk jugs collected through our city recycling service." An extensive marketing plan is used to interest the public in the new program and to let them know how easy the commingled program would make recycling for them.

The newly distributed carts also included a note. "Please give this new curbside recycling service a 30-day trial. If it doesn't meet your expectations, please call 430-2450. Let's make this recycling program work, together." As a result, 99% of residents accepted the carts. Those who refused the carts could still use drop-off centers to dispose of their recyclables. Other marketing components included brochures that coaxed residents, and posters hung in local recreation centers, libraries, and other public meeting sites. As a result, the recycling participation rate soared, doubling and tripling that of other municipalities in the area. More recently, education efforts have shifted from selling recycling residents to coaching them on what materials are acceptable.

Boosting Quality Control

TEC truck drivers play an important role in quality control. While on their routes, drivers will place "tag" notices on carts that contain inappropriate materials. A list of first-time tagged residents is turned over to the City managers who then send letters thanking them for recycling but listing unacceptable items seen in their containers. Usually, one letter is enough to get these tagged residents to comply.

Each quarter, the City runs an ad in the local newspaper about its recycling progress. Usually, the message reads, "We've collected XXXX number of cans and bottles and we thank you. But don't forget to pay attention to the products you're putting in the cart."

SECTION VI

RECYCLING IN OTHER COUNTRIES

CHAPTER 37
RECYCLING IN OTHER COUNTRIES

HERBERT F. LUND, P.E.
Editor-in-Chief

Although there are many areas around the world where recycling has not yet materialized, there are many other countries that are just as active recyclers as the United States and a few, like Germany, The Netherlands, and Japan, that are more aggressive. Because of this diversity of world-wide recycling activity, we have prepared this new section on recycling in other countries.

Our coverage includes the European Union (EU) countries, parts of Asia and the Far East, Canada, Scandinavia, Central and South America, and some Caribbean nations. For the convenience of professionals seeking new recycling information about foreign countries, this special new section has been divided into two main subsections: (1) Legislation, Rules, and Directives and (2) New Technology Developments.

LEGISLATION, RULES, AND DIRECTIVES

The European Union (EU) Umbrella

First formed with the west European nations (except Norway), the EU is now, besides reducing trade barriers, heavily involved in legislation, rules, initiatives and directives concerning the environment, solid and hazardous waste disposal, and recycling.

Proposed EU Law on Waste Electrical and Electronic Equipment Challenged

A proposed EU ban on the use of recycled hazardous wastes in electrical and electronic equipment by the European Commission's Environmental Directorate (DGXI) may break several international trade rules and provoke trade battles on foreign-made products containing such materials as lead and mercury. According to the Brussels-based law firm Hunton & Williams, who are advisors to the U.S. electronics industry, the proposed ban on the use of certain hazardous substances would be an illegal trade barrier because it would effectively ban imports of products. foreign-made products. If the hazardous waste were properly pretreated before disposal or recycling, the substances would not harm the environment, according to Hunton & Williams. Further, the firm believes the World Trade Organization would judge such a ban unnecessary because alternate policies enforcing incineration and landfill policies would prevent environmental damage.

If, on the other hand, the European Commission could prove that the recycled substances were harmful to the environment, then the proposed ban should not be limited to electrical products but include all other products.

Landfill Directive Imposes Strict New Rules

From the United Kingdom's (UK) Brunel University, Center for Environmental Research, Hillary Stone, Associate Lecturer in Environmental Law, evaluates the Directive aimed at reducing the amount of waste going to landfills as well as promoting recycling and recovery. Currently, the volume of biodegradable waste landfilled is scheduled to be reduced to 25% of the 1995 levels by the year 2016. Further, after landfill closure, the landfill operator would still be responsible for managing a site for a minimum of thirty years. EU member states have until 2001 to implement the Directive into each nation's law.

Three types of landfill sites are defined; hazardous waste, nonhazardous waste, and inert waste. Codisposal of hazardous and domestic wastes will be banned. The following materials will no longer be landfilled; explosives, corrosive or inflammable wastes, liquid wastes, infectious hospital wastes, and other wastes not meeting certain landfill criteria. As this Directive results in higher full costs to manage landfill sites, governments in Europe are beginning to recognize the benefits of recycling. In order to increase recycling rates, they must take action on recycling collection, reprocessing, and marketing. There is a need for a precise definition of biodegradable waste for all countries to adopt. The 35% original target from 1995 levels is called for by the year 2016. Countries like the U.K. that landfill 85% of their municipal waste will be granted an additional four years to reach this target. Stone feels the 35% level should be reduced to 25%.

Countries Face Court Action Over Waste Laws

A number of EU member countries are being threatened by the European Commission for failure to comply with the 1993 waste shipments directive and the 1991 hazardous waste directive. In Italy, some companies failed to obtain prior authorization on hazardous wastes. The government exempted companies from prior authorization without meeting Directive conditions. Also the Commission alleges that Germany is "unduly restricting" waste shipments to other EU countries for recovery in cement kilns by setting restrictive conditions, such as minimum calorific value. If this value is not met, the shipments are targeted for disposal rather than recovery. If the Commission allowed Germany to continue this restrictive procedure, it would be legal to curtail exports under the "self-sufficiency" principle. Germany cannot append "additional conditions" to the basic concept of recovery.

European Association Criticizes EU Packaging Law

Late in 1999, the European Recovery and Recycling Association (ERRA), whose members include such companies as Coca Cola, Heineken, Nestles, and Procter & Gamble, want the current packaging and packaging waste legislation changed. Rather than using current recycling targets, the association suggests replacing these targets with an absolute limit on the amount of packaging going into landfills. Their reasoning is that landfill diversion targets would make a greater contribution to sustaining the packaging chain than would higher recycling targets. It is the ERRA's contention that the EU Commission is placing too much emphasis on maximizing recycling and not enough on reducing environmental problems. A recent Swedish study found only a 5% environmental benefit from money spent on compliance with packaging recycling targets.

British Judge Rules Against EU Definition of "Waste"

In a case initiated by the British Metals Federation challenging the UK Government's interpretation of the European Union's definition of "waste," Judge J. Carnwath ruled that ferrous and nonferrous scrap requiring no further processing are not considered "waste," but are now legally classified as "raw material." The Federation legal challenge against the UK Environmental Agency originally argued that all recyclable metals should be removed from the "waste" category. However, the judge did not comply with this view completely.

Impact of "Extended Producer Responsibility" (EPR) or "Takeback" Laws

Rather than government collecting recyclables and being fully responsible for marketing the materials, the "extended producer responsibility" (EPR) concept puts the "end-of-life" waste management responsibility back on the companies that made the product. The idea is politically attractive because it usually alleviates the need for local governments to raise money to pay for recycling costs.

The EPR concept started in Europe but Taiwan actually had a "takeback" law on its books late in the 1980s. The earliest manifestation of "EPR" was the German Packaging Ordinance enacted in 1991. This law holds producers and sellers responsible for the takeback and recycling of packaging at the end of its useful life and precludes the use of public money for this purpose. The catalyst for this ordinance was a looming shortage of landfill capacity, the "not in my back yard" syndrome for incinerators, and the fact that when the Berlin Wall went down, West Germany could no longer ship wastes in East Germany.

Before Germany enacted their packaging ordinance, landfill tip fees ranged from $200 to $300 per ton. The German governments goal was to force industry to consider disposal of its packaging waste as a part of a products life cycle.

Germany's law was aggressive. For example, it had 80% recovery mandates. During the first two years, unforeseen problems arose. For instance, the costs of implementation exceeded expectations. Additionally, the program produced an overabundance of secondary materials that were leaving Germany and upsetting the materials markets of their European neighbors.

The German system remains rather unique. It requires industry to set up a "dual system," using separate trucks collecting the packaging waste. Because the law only allowed this new "DSD" (Duales System Deutschland) 18 months to set up for 80 million people, the waste haulers jacked up their prices about 30%. As a result, DSD struggles with a $2 billion system, the most expensive in the world. The system could not fail, because if it did, retailers would have to take back directly, so the retailers propped up DSD.

Not deterred by these difficulties, Austria followed with a similar law. Other countries enacted their own version of "EPR" for packaging.

In 1994, the European Union enacted its Directive on Packaging and Packaging waste to harmonize package reduction and recovery requirements across member countries and to alleviate the influx of secondary materials resulting from Germany's mandates. Most member countries have passed laws to comply with the directive, but these laws are taking different forms in different countries.

As of 2000, twenty-nine countries have takeback laws for packaging. Only Germany and Austria require a separate private system, however. In most countries, government continues to collect the materials but industry must take over with its "producer responsibility organizations" or "PRO's" to ensure recovery of materials.

But the issue of EPR did not stop with packaging. In the United States and 15 other countries, laws have been enacted to require takeback of certain types of batteries, and as of 2000, nine countries had takeback laws or authority on a wide range of electronic products as well.

A few highlights of recycling policies around the globe follow.

European Union

The European Union (EU) is moving on a number of takeback-related Directives. They include electronics waste, end-of-life vehicles, and a proposed Directive on household hazardous waste, The 1991 Directive on battery takeback is slated to be strengthened in the near future. In the meantime, seven European countries have already enacted different takeback laws for electronics, not waiting for the EU to act. The European Union also enacted a landfill directive that will require a 65% reduction in degradable waste going to landfills by 2016.

Packaging Directive

The EU Packaging Directive requires, for example, that by 2001, countries "recover" a minimum of 50% of their used packaging, with material recycling at 25% minimum, and no material recycled at less than 15%. To address the initial problems posed by Germany's stringent packaging law, the Directive also says that countries mandating more than 65% recovery or 45% recycling must prove they have adequate internal capacity to handle the material. Greece, Portugal, and Ireland need only meet a 25% recycling standard. The European Commission is proposing to expand the targets even more in 2000, as required in the original directive. There have been many delays due to severe industry opposition and debate over how well the current targets are being met,

Growth of "Green Dot"

As a result of the Directive, not 15 but more than 18 European countries now have takeback laws for packaging, including Switzerland, Norway (not in the EU), and Eastern European countries that are trying to gain entrance to the EU. The system is very fragmented, with each country having different requirements for fees and reports from manufacturers and importers. It is estimated that the system is costing companies about $10 billion per year in fees and calculations alone in Europe.

The "Green Dot" symbol is owned by Germany's Duales System Deutschland (DSD), which owns the copyright throughout the world. It has licensed the symbol in eight countries, with plans to expand to five more by 2002. It formed an umbrella group, PRO EUROPE, to oversee licensing arid attempt to coordinate the activities of its nations.

Germany

Under Germany's system, packaging waste is collected separately in yellow bins and producers pay fees to DSD to cover the cost of collecting and recycling this material. Municipal governments no longer collect or recycle packaging waste. However, glass goes to retailers and bottle banks, and paper goes to drop-off bins. Some electronics waste is recycled, but a law to require takeback is on hold. Batteries must also be recovered. Unique to Germany there is a refillables quota of 72% on most drinks except milk. The quota, which is highly controversial in Europe, was not met in 1998. The government is mulling over whether to amend the law or enforce a mandated deposit on the noncompliant containers Another law—which may conflict with the dual system—will require pretreatment of most wastes before they enter landfill by 2016. This is based on a EU Directive.

Car Manufacturers Balk at EU Directive to Take Back End-of-Life-Vehicles (ELVs). With Volkswagen and DaimlerChrysler being the most outspoken opponents, lawyers for the German Association of Car Manufacturers contend that the proposed European Union's ELV Directive is inconsistent with European law, since manufacturers were not properly informed. The European Association of Car Manufacturers (ACEA) has said it was considering bringing the case before the European Court of Justice. According to a report in the UK *Financial Times,* the takeback directive on ELVs would cost car manufacturers about ten billion Euros.

Sweden

New Proposed Swedish Ordinance on Electronic Waste. In 1998, the Swedish government sent an ordinance proposal to the European Commission (EC) dealing with extended producer responsibility (EPR) for electric and electronic equipment. Since there was no comment by the end of 1999 from the Commission or other EU members, the Swedish government feels it can now bring the ordinance into force.

The proposed ordinance defines the producer as a manufacturer, importer, or distributor of electric and electronic equipment. It directs the producer to:

- Take back free of charge any old equipment when the customer buys new equipment
- Inform householders and others about the take-back obligation
- Explain and discuss the specific take-back scheme on request from a municipality
- Treat the waste from electric and electronic products in an environmentally sound manner
- Inform recyclers about product content in order to facilitate recycling
- Provide the Swedish Environmental Protection Agency (SEPA) with any facts necessary to ensure compliance with the ordinance

An August 1999 draft includes a highly controversial provision making the manufacturers responsible for collecting and recycling "historical" equipment produced before the ordinance becomes effective at the product's end-of-life. Orgalime, the industry association, estimates that the proposed new ordinance would cost the industry Euro 15 billion annually. The association says the industry could not have anticipated retroactive legislation for products whose average lifespan is between 8 and 20 years.

Denmark

There are no specific obligations on industry to meet recovery and recycling targets in Denmark. The Danes believe that their existing voluntary agreements will enable the targets to be met. There are just voluntary agreements—an agreement of August 1994 on the recycling and reuse of plastic and board transport packaging, and an agreement made in December 1992 on glass recycling.

However, there are certain restrictions on packaging that predate the EU Directive. Several laws ban the use of cans for beer and carbonated soft drinks. Nonrefillable glass and plastic containers for these products are permitted for imports, but only on condition that a deposit, return and recycling system is set up which is equivalent in its effects to the refill system Danish producers must use. These restrictions are currently being challenged by the EU Commission.

A voluntary agreement from 1991 to reduce the use of polyvinyl chloride (PVC) in Denmark has just been cancelled by the Environment Minister. Further measures against PVC are expected.

There have been heavy taxes on beverage containers for many years. but a new packaging tax came into force on January 1, 1999 that covers all types of packaging. Whereas packaging manufacturers paid the old beverage container taxes (which still remain in force), the new taxes are paid by packer/fillers and importers. The government expects the costs to be passed on to the consumer. Flexible plastics packaging is subject to a particularly high rate of tax, and this may lead to exporters of goods being asked to change their packaging or else contribute to the taxes payable. There are voluntary agreements for take-back of batteries and various electronics.

Austria

Basic obligations to take back used sales, secondary and transport packaging and ensure it is reused, recovered, or recycled were established by the Packaging Ordinance of 1992. Importers can fulfill these obligations by joining the Green Dot recovery organization, ARA. Fees are paid to ARA both for consumer and transport packaging by importers or by packer/fillers. Packaging recovered by ARA must be printed with the Green Dot. Austria also has takeback obligations for batteries, deposits on refrigerators, flourescent tubes, and separate collection of household hazardous waste. The batteries law covers all batteries for recycling.

Belgium

The Inter-regional Co-operation Agreement of 1997 requires importers first placing packaged products on the Belgian market to take back as much used packaging as is necessary to meet the very high recovery and recycling targets laid down. Companies have the option of joining an authorized recovery organization. The recovery organization FOST Plus uses the Green Dot, so packaging exported to Belgium must be marked accordingly. A new organization, Val-I-Pac, has been set up to handle transport packaging. Belgium also has an eco-tax on various packaging, publications, disposable cameras, pens, tableware, and other items. Two of the regions are moving on electronics takeback, and all batteries must be recycled.

France

France has a "shared responsibility" law, in which industry takes over once collected material has been sorted The national recycling program for packaging, Eco-Emballages, which currently charges manufacturers a flat fee "per package" to help fund collection and recycling of packaging, will move to a sliding-scale fee based on volume, weight, packaging material and recyclability The new fees structure is intended to reward packagers that reduce packaging volume and use easily recyclable materials. Under this approach, the charges for some types of plastic packaging go up; other packaging materials, such as paper, will cost relatively less. Landfilling of recyclable wastes will be prohibited by 2002, and there are landfill taxes. Takeback for batteries is mandatory and there is a voluntary agreement on autos.

Italy

A 1997 Legislative Decree implements the Packaging and Packaging Waste Directive and other EU waste measures. It imposes requirements on the whole packaging chain to meet the recovery and recycling targets. Producers can do this by either joining the National Packaging Consortium (CONAI), establishing a deposit system or establishing their own collection and recycling system.

Importers and distributors must submit annual data on the amount of packaging material placed on the market and that which is reused or recycled. CONAI covers transport packaging as well as sales packaging. Takeback is required for electronics items, as there is authorization to require deposits on such items.

Spain

A 1998 law implements the Directive on packaging. Spanish distributors must either join the Ecoembes recovery organization or set up their own deposit system to ensure the return of packaging for recycling or recovery. Packaging that is not part of a recovery scheme must be marked with an appropriate symbol to be determined by the Environment Ministry. Packaging that is covered by Ecoembes must carry the Green Dot symbol. The amount of packaging entering the waste stream must fall by 10% between 1997 and 2001. Packer/fillers and importers likely to generate more packaging waste than the thresholds laid down (material by weight) must prepare waste prevention plans every three years. In practice, these may involve asking foreign suppliers to change their packaging. Spain has authority on electronics takeback, and voluntary initiatives are being started in 2000.

The Spanish iron and steel and metal recycling industries have come to an agreement with Spain's government regulatory bodies on a protocol establishing measures to detect and control radioactive materials in scrap metal. This agreement follows a 1998 accident in which a foundry smelting radioactive scrap metal spread a plume of radioactive pollution across Europe.

The Netherlands

The Environmental Advisory Council, in a report called "The Netherlands and the European Environment," stresses vigorous implementation of the European Union's environmental, legislation at the national level. According to the Council, there is no need to pursue a specific and stringent national environmental policy within the European Union framework. Further, the report urges stricter maintenance of the EU legislation in member states and more stringent application of environmental conditions to restrict further diminish the effectiveness of policies on world trade.

Dutch Government Aggressive on Car "Take-Back" Recycling. Because The Netherlands has a very high population density, higher than Japan, and there is very little space for landfill waste disposal, the government has taken aggressive steps to encourage alternatives to waste disposal by imposing taxes on the dumping of waste and by implementing a policy of Extended Producer Responsibility (EPR), also called a "take-back" system. A waste tax was adopted to make landfilling less attractive than incineration. With the EPR system, the government shifted the environmental impact of a product back to the producer. Table 37.1 lists main waste streams and EPR measures developed voluntarily, mandated by legislation, or a combination of the two.

A prime example of applying the EPR system is car recycling. As of 1998, the amount of end-of-life vehicles recycled reached 86% by weight, up from 75% in 1995. Producer responsibility was introduced for car recycling as a voluntary measure by the industry itself. Critical to achieving this increase was the formation of Auto Recycling Nederland (ARN) in 1993 to implement the EPR system for vehicles. Dismantler companies certified by ARN receive a fee for removing specific materials (see Figure 37.1 for functions of car dismantlers) such as rubber, foam from from seats, liquids, and certain plastics. Funding of the ARN operation is by a mandatory fee of 150 Guilders for

TABLE 37.1 Voluntary and mandatory applications of extended producer responsibility (EPR) in The Netherlands*

Waste stream	Voluntary	Legislation
Car wrecks	X	
Car tires		X
Electrical and electronic household appliances		X
Batteries		X
Packaging waste	X	
Paper/board	X	
Agricultural foils	X	
PVC cladding units	X	
PVC pipes	X	
Photographic hazardous waste	X	

*Waste streams for which EPR is operational in The Netherlands [3].

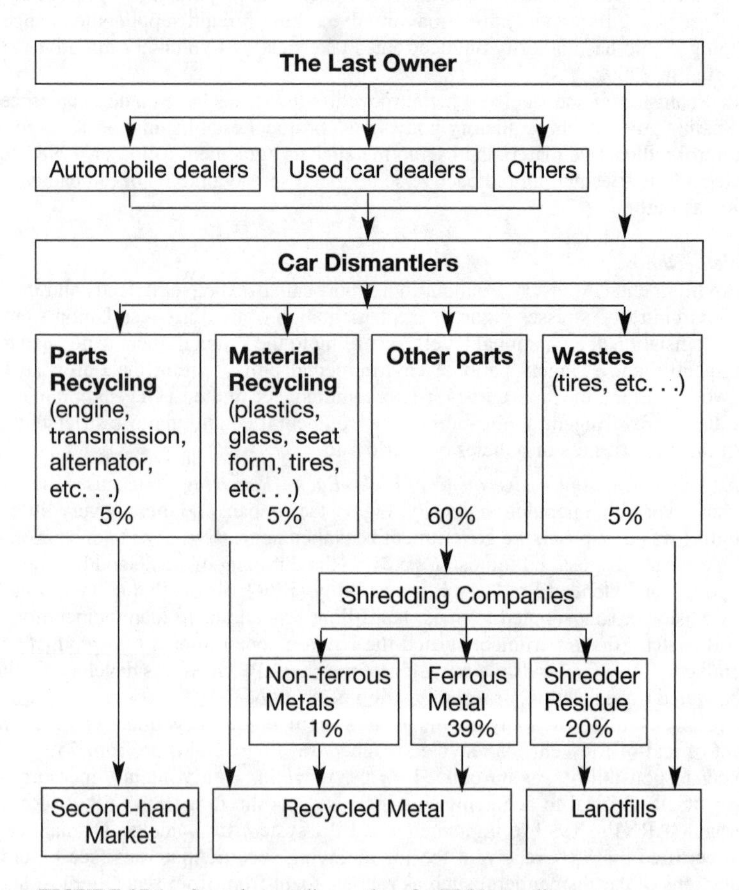

FIGURE 37.1 Steps by car dismantlers for ELV recycling.

waste disposal effective in January 1998. ARN processes all end-of-life vehicles without any charge to the last owner. Listed below is the average composition of shredded vehicle residue.

Plastics	35%	Ferrous metals	8%
Urethane	16%	Nonferrous metals	4%
Fiber	13%	Glass	7%
Rubber	7%	Wire harness	5%
Wood	3%	Paper	2%

Source: Recycling International Magazine, Arnhem, The Netherlands. Hylke Jan Glass, Department of Applied Earth Sciences, Delft University of Technology, The Netherlands.

Norway

Government Report Targets Waste Reduction

In its first annual "Government Environmental Policy and State of the Nation" report, Norway has set a range of targets for reducing waste and links them to the government's annual draft budget. Specific targets include recycling (for re-use or for conversion to energy) of 75% of all waste materials by 2010, and reduction of waste generated to a rate below that of economic growth.

Japan

Government Tightens Waste Disposal Laws on Illegal Dumping

After a shipment of recovered paper bound for the Philippines was found to contain considerable nonpaper wastes, including medical wastes, the entire shipment was rejected to conform with international Basel Convention. When the shipment was returned to Japan, Nisso, the waste processing company, had gone bankrupt, the owner disappeared and the managing director committed suicide. The entire load was incinerated at a cost of $5 million to be born by the Japanese taxpayer.

Because of this unfortunate incident, the Japanese government is revising its waste disposal laws to strengthen the responsibility of the waste producer. They will be forced to restore any site where illegal dumping has taken place to its original condition. The revision will impose a responsibility on the landfill landowner who made his site available to the illegal dumper and who will also have to restore the site to its original condition.

Furthermore, Japanese legislators also are preparing various amendments to existing site disposal laws. Examples are restrictions on the transfer of landfill ownership to thwart land brokers from making unreasonable profits at a time when landfill space has become tight. Presently, city authorities have handled municipal wastes but not industrial wastes. As cases of illegal dumping the government increase, the government may consider operating landfills to accept industrial wastes. (*Source:* Hideo Itoh, Yokohama, *Recycling International.*)

Japan's Approach to End-of-Life Vehicle (ELV) Recycling

Recycling targets for Japan's annual 5 million ELVs have been developed by the Ministry of International Trade and Industry (MITI). As part of this program, MITI launched recycling initiatives (Table 37.2) that stimulate voluntary measures and fee competition among recycling companies. The Waste Disposal Law became legally binding in April

TABLE 37.2 Recycling initiative for end-of-life vehicles adopted by Japan Ministry of International Trade and Industry (MITI)

Problems associated with the disposal of ELVs	Recycle Initiative
Shortage of disposal sites Excessive amount of waste Hazardous materials in waste	Set target values (*)
(Fundamental problems)	Draw up schedules for voluntary activities by related industries)
Improper dumping Illegal dumping	Introduce manifest system (prepare manifest system specifically designed for automobiles)
Technical development	
Passing on expenses incurred in processing Regional problems (e.g., slow concession procedure at municipalities)	Set up frameworks to facilitate communication among parties concerned (businesses, researchers, governments, municipalities, etc.) by email, fax, etc. (e.g., the establishment of the Information Center for Motor Vehicle Recycling.)

1996; it specifies that shredded residue is a waste that requires specially controlled landfills. Table 37.3 indicates the value of ELVs in Japan at various recycling stages. Because the last car owner pays for the cost of dumping, the car recycling business in Japan remains profitable.

The absence of direct manufacturing responsibility for processing ELVs has encouraged manufacturers to study designs for recycling and new technologies for ELV recycling. Table 37.4 shows the voluntary measures taken by Toyota for research and development of new technologies. In response to the MITI Recycling Initiative, Japanese car manufacturers established a new corporate research center for car recycling to collect information from manufacturers on recycling technologies. Recycling companies are provided with manuals on efficient dismantling and recycling of ELVs, including names of hazardous materials.

Under Japanese ELV recycling policy, an infrastructure separates the recycling business from manufacturing, thereby leaving the recycling business open to free market competition. Presently, most efforts in Japan focus on developing cars that are more easi-

TABLE 37.3 Trade values of ELVs during various recycling stages

	Last owner to dealer	Dealer to dismantler	Dismantler to shredder
Positive value	35.3%	50.8% (–0.2%)*	48.1%
Zero value	28.2%	19.8% (–8.0%)*	28.8%
Negative value	34.1%	22.2% (+8.2%)*	19.2%

*Indicates the change from the data of 1995

TABLE 37.4 Voluntary measures adopted by Toyota for recycling ELVs

Subject	Topic	Actions
Design for recycling	Attaining 90% or more of recyclable rate for new models by 2000	Using fewer varieties of plastics Design for easy dismantling and separation Using more recycled parts
	Diminishing negative impact to the environment	Less lead to be used (compared to 1996, one half by 2000 and one third by 20005) Design of air bag for disposal (change of substances)
Researches on recycling technologies	Researches on recycling technologies and proper processing	Development of better ways to process shredder residue and attainment of minimizing the amount of shredder residue to be landfilled to one fifth (opened a full-scale recycling plant with a capacity of more than 100 metric tons of scrap a day in 1998) Development on the ways to recycle Nickel Hydrogen batteries
Creating better recycling system	Creating the infrastructure for recycling and proper processing	Efficient bumper recycling through auto dealers Creating the infrastructure for collecting and destroying CFC 12 through auto dealers Creating the infrastructure for Nickel Hydrogen battery recycling

ly recycled. (*Source: Recycling International Magazine*, Arnhem, The Netherlands. Toshio Kyosal, Department of Applied Earth Sciences Delft University of Technology, The Netherlands.)

Most plastic packaging discarded by households in Japan is incinerated. A package recycling law of 1997 required mandatory collection of most containers starting in 2000. Local governments must collect the discarded packaging; business is responsible for recycling and reusing the materials into saleable products. A PRO collects fees on materials that do not have a street value, and the fees are based on actual capacity to recover the material. There is also a law to require recovery of TVs, appliances, and air conditioners. However, a new umbrella law expected to pass in 2000, will supercede some of the other laws, and it is expected that takeback will also be required for more electronics, as well as construction and demolition debris.

Taiwan

Taiwan has had takeback laws on its books since 1989. However the government passed a law in 1998 that switched the entire system from the private sector to the government sector. Recycling is required for a wide range of items, including certain packaging, electronics, all batteries, used oil, autos, and other items. Companies must pay fees to government-run consortiums that are supposed to ensure full recycling. A proposed revision to the waste law would include tax breaks fox recycling equipment. The law requires a special four-in-one symbol on affected items, though the government may abandon this requirement. (*Source: Recycle Laws International*, Raymond Communications Inc. College Park, Maryland.)

China

China to Tighten Import Controls on Recycled Equipment and Materials

An informed source has stated that China plans to increase controls on imports of cables, used electronic equipment, and nonrecyclable materials. In order to curtail illegal import of reusable electronic equipment, China may refuse containers carrying entire keyboards, mouses, printed circuit boards, and computer processing units. By the year 2000, imports of old copper cable and electronic equipment may be reduced by 30%. Furthermore, China would not allow imports of materials or nonrecyclables containing more that 4% impurities. For instance, some Chinese ports consider rubber as nonrecyclable.

Canada

Canadians continue to embrace recycling at the curb and elsewhere but Canada posts lower recycling rates than the United States. The average Canadian recycling rate is 25.6%. This does not include the far-northern rural areas. Canada is definitely wedded to the container deposit system. All ten provinces have an expanded deposit system. Even the Yukon collects some aseptic containers. Both Alberta and Saskatchewan have voluntary industry-paid levies on milk jug take-back schemes to help ensure recovery. A number of provinces have initiated various take-back programs for special wastes. Leading the pack is British Columbia, with programs for paint, solvents, pesticides, gasoline, used oil, and pharmaceuticals.

Interestingly, Canadians have gone straight to producer responsibility. For example, only a few provinces have any purchasing preferences for recycled products or grants/loans to businesses for recycling. None have tax credits for recycling businesses. There is a National Packaging Protocol, which calls for a 50% reduction in packaging by the year 2000. Industry claims it has already met this goal, largely through weight reduction, switching to plastics, reusing pallets, and recovering old corrugated containers. By 2000, a number of provinces should pass legislation on issues ranging from paint take-back programs to landfill bans. Toronto continues its battle over Blue Box funding and deposits, and it fights with Michigan over trash exports (see Table 37.5). (*Source: State Recycling Laws Update*, Raymond Communications, Inc. College Park, Maryland.)

Brazil

Brazil enacted recent laws that require recovery of tires and batteries. Its legislature and another regulatory body are moving on new laws that will require takeback of a wide range of packaging. The federal bill targets plastic packaging only. Meanwhile, Rio de Janiero enacted a plastic packaging takeback law in 2000.

Caribbean States

Overview of Waste Recycling

Of the Organization of Eastern Caribbean States (OECS), only two had recycling initiatives of significance: Barbados and Trinidad and Tobago.

TABLE 37.5 Sample recycling regulations from five Canadian provinces (as of May 1999)

Province	Extended product responsibility (EPR)	Environmental taxes/fees	Container deposits	Rate of recycling reduction
British Columbia	Used oil, waste paint, solvents, pesticides, pharmaceuticals	Deposits, EPR only	10 to 30 cents, All Beverage Containers except milk, milk substitutes, infant formulas, and meal replacements	26% Diversion per capita to 32.5%
New Brunswick	Starting used oil	Half of deposits produce $14 million for Environmental Fund	10 cents, 500 ml 20 cents, 500 ml All beverage containers except milk	19%
Nova Scotia	Used tires, more expected voluntarily	Voluntary contributions from industry of C$1 to 2 million EPR fees on tires Stewardship on materials expected	10 cents, 500 ml 20 cents, 500 ml All except milk	35% Diversion rate
Ontario	None. Demanded by municipal governments	10 cent tax on nonrefillable beer cans	10 to 40 cents Beer and soft drinks	32%
Quebec	Recycled old newspaper content proposed	Packaging tax proposed	Beer and soft drinks 5 cents nonrefillable. 10 cents refillable	28%

Source: Raymond Communications, College Park, Maryland.

Barbados. The recycling initiatives in Barbados are by the private sector (8 firms or more) for such recyclable commodities as paper, glass, ferrous and nonferrous metals, PET bottles, and lead–acid batteries. Markets are generally overseas (e.g., Trinidad, Miami, etc.). Very little assistance is received from Government; but this is expected to change as a result of recommendations by consultants for a more in-depth approach to waste reduction and recycling.

Trinidad and Tobago. The recycling initiatives in Trinidad and Tobago are by the private sector, although the central coordinating body is a state-owned company, Solid Waste Management Company, Limited. Recyclables are typical (e.g., paper, plastic. etc.) with two exceptions—oily waste to the local oil company and bottle waste to the local glassworks. Three layers of recycling activity are recognized:

1. Bottom Layer: Junkmen and salvagers work at the household gate and at the main landfill site, and an estimated 38 tons are salvaged each week.

2. Middle Layer: Represented by the small salvage dealers who are the middlemen between salvagers/haulers and the users/dealers/processors.
3. Top Layer: These are the secondary materials processors/brokers who are independent contractors for the recycling process, except for the glass and sugar industries, which enjoy a certain measure of in-house recycling.

Other Countries

The existing recycling initiatives in other Caribbean countries are basically at-home and disposal-site salvaging by some members of the public (See Table 37.6).

TABLE 37.6 Existing initiatives in waste recycling

Caribbean country/territory	Public Sector		Private sector		General public
	Program	Incentives	Companies recyclables	Tons/ year	
Barbados	Proposed	Proposed	8 companies or more	13,600	Salvagers at landfill
Belize	None	None	—	—	Salvagers at landfill
Jamaica	Proposed	Proposed	—	—	Salvagers at landfill
Antigua and Barbuda	None	None	1 company	—	Salvagers Yachtees
Dominica	None	None	—	—	Salvagers at landfill
Grenada	None	None	—	—	Salvagers at landfill
Montserrat	None	None	—	—	Salvagers at landfill
St Kitts and Nevis	None	None	—	—	Salvagers at landfill
St. Lucia	None	None	—	—	Salvagers at landfill
St. Vincent	None	None	—	—	Salvagers at landfill
Trinidad and Tobago	Partial	None	7 recyclable materials	20,000	Organized salvaging at landfills

Acknowledgment is made to the following sources: Manfred Beck, Publisher and Editor, *Recycling International* Magazine, Arnhem, The Netherlands; Randy Brown, Executive Director, Clean Islands International Pasadena, Maryland; Michele Raymond, Publisher, *Recycling Laws International*, Raymond Communications, Inc., College Park, Maryland; *Business News Americas*, Santiago, Chile.

In the absence of a national waste reduction/recycling program and the accompanying polices, legislation and procedures, a number of incentives are being considered for application in Barbados and some other countries.

- Increased tipping fees at landfill sites, with lesser fees for recyclers
- Businesses engaged in recycling activities to receive the same concessions (e.g., duty-free) as local manufacturers
- Introduction of an Environmental Levy on specific items for disposal
- Establishment of a collection system for recyclables, including drop-off depots
- Development of a public education/information program
- Additional incentives, financial and otherwise, being discussed are: mandatory/legislated separation of solid waste, tax holiday, foreign exchange considerations, concessionary interest rates on loans, preferential access to donor funds, etc

In the meantime, there are-virtually no incentives for waste reduction/recycling in Caribbean countries.

NEW TECHNOLOGY DEVELOPMENTS

Austria

New PET Recycling Process Costs Less Than Virgin Plastic
A new system has been developed by an Austrian company, Enema, to process PET plastic recyclables more profitably. With this process, PET bottle flakes are converted into flat sheet film that eliminates the need for pelletization and the external conditioning steps of crystallization and predrying. Thus, the process reduces the cost of storage and transportation of materials. By reducing the number of times the material is exposed to heat, the quality of the reclaimed material is enhanced. Costs are between 50 to 60% less than that of raw materials.

Canada

Robots Sort Plastic Recyclables for Improved Quality and Efficiency
BrainTech, Inc. of North Vancouver, British Columbia, a company that specializes in robotics, has developed a prototype sorting system to separate plastic recyclables by robot eye. The project, called the Sorting Precision Quality Recycling System, uses robotic arms and sensors that read the infrared spectra of the recyclables. Using robotics to sort plastics is not the same as using the technology to assemble automobiles. Plastic bottles have different shapes and different sizes. There is the need to recognize these shapes and sizes on the line Existing technology can deal with a row of plastic bottles and knock down certain kinds, such as PET, to separate them. But the robotic and infrared technology identifies plastic at a more accurate level. [For details, call (800) 825-0644.]

Used Clothing to be Sorted by Automatic Computerized System
A revolutionary sorting system for used clothing has been developed by Fastrac Integrated Systems, Inc., Ontario, Canada. Advantages claimed are: reduced labor (between 33 and 50% compared to manual sorting), less space used, better accuracy, less worker fa-

tigue, and general, valuable information on operator output and quality of mixed rags. The system does not move material but instead efficiently sorts used clothing.

Czech Republic

Ozone Used to Recycle Tires into Granules

At an estimated cost of $0.70 per kg of rubber granulate, a Czech company, Pneu Demont spol.s.r.o of Pilsen, has patented a recycling process that converts waste tires into granulate within an ozone-rich atmosphere. Untreated waste tires are continuously introduced into a sealed tunnel through a trap and transported and stretched by a system of drums to ensure that ozone enters the pores in the rubber. At room temperature, depending on the ozone concentration and mechanical loads involved, the tire granulation process takes between six to ten minutes and retains the original physical and chemical properties of the rubber. Montana Steel, whose owner is one of the six patent holders, started a pilot facility that ran for a year; the first production plant started up in 2000. The production capacity of the plant will be 700 kg of rubber granulate per hour and will produce five different sizes of granules, up to 40 mm. Unshredded steel cords and textile fibers are exposed and separated without mechanical or magnetic systems.

France

New Oxyreducer Recycling Process for Heavy Metals and Battery Wastes

In Le Havre, France, Swiss recycling company Citron built a new recycling plant processing 23,000 metric tons of heavy metals and battery wastes to yield secondary ferrous and nonferrous metals. One of the main advantages is the processing of waste batteries. Compared to most other processes, this recycling method can accept unsorted input batteries directly from collection areas. Organic fractions or poisonous substances are completely destroyed at a reaction temperature of 1350 °C. One of the main innovations of this 35–40 minute process is a sophisticated burner technology that leads to forming vertical layers in the atmosphere inside the furnace. The packed beds always remain in the reduction

FIGURE 37.2 The new Citron recycling plant in the port area of Le Havre, France. (Courtesy of *Recycling International* Magazine, Arnhem, The Netherlands.)

zone. But in the higher layers, the gases formed in the piling through pyrolysis or evaporation reach an oxidizing zone where organic substances containing zinc, cadmium, and lead, are oxidized. Mercury is not oxidized but remains in the gaseous state This oxidizing stage allows complete recovery of the energy potential of the organic materials. As a result of the high temperatures, other unwanted substances evaporate and exit the furnace as waste gas.

The newly developed process can handle a very broad range of possible waste materials allied to high tolerance for unwanted substances in the input materials. Since there is no laborious sorting, an important process cost is eliminated. Other economies are achieved by direct input from collection, which eliminates costly middle steps involving sorting plants, recycling facilities, transportation, and storage.

Germany

New Online Environmental Protection Internet Market Survey

What is called "the largest on-line internet market survey for environmental protection" has been undertaken by the German International Transfer Centre for Environmental Technology (ITUT). The database lists details of 8600 German companies that are active in the environmental field, including manufacturers of machinery and equipment and providers of services for the recycling industry. A company listing is available free of charge. The website can be viewed at www.Umfis.de.

Japan

Two New Processes Convert Steel Mill Wastes into High Quality Metal

Kobe Steel Ltd. and Midrex Direct Reduction Corporation, a U.S.-based international ironmaking technology company, have developed processes called Fastmet and Fastmelt to recycle steel mill wastes and recover iron and zinc. Motivation for this process was steelmakers' concerns for disposal of iron-bearing wastes, threat of closure of on-site landfills, recovery of valuable ferrous metals, controlling steelmaking raw material costs, conserving capital, and coke oven and sintering plant environmental problems.

To address these concerns, the Midrex Direct Reduction Process was selected as the basis for the new coal-based Fastmet and Fastmelt processes. The Fastmet process converts steel mill wastes into metallic iron in a rotary hearth furnace using carbon as the reducing agent (see Figure 37.3). In Japan at Kobe Steel's Kakogawa Steel Works, a commercial Fastmet prototype plant has been in operation since December 1995. The plants' successes include: stable operation producing a highly metallized product, optimization of the pelletizing process, high productivity, minimum raw material consumption, use of steel mill wastes as raw materials, and briquette and product melting tests.

The Fastmelt process (see Figure 37.4) uses the Fastmet product to produce ingots or briquettes, which are then fed into a specially designed electric melter to produce a high-quality liquid iron known as Fastiron. For projects to produce blast-furnace-grade liquid iron from waste steel materials, Kobe Steel and Midrex will work with National Recovery Systems for the agglomeration step and EMC International for the smelter.

Fastmet and Fastmelt plants can be designed for 150,000 to more than a million tons per year. The benefits of this system include generation of minimal fines, use of inexpensive binders and little or minimum postagglomeration drying, as well as reduced complexity of the plant and improved productivity. The product has a typical metallization level of 85–92% and a carbon content of 2–4%.

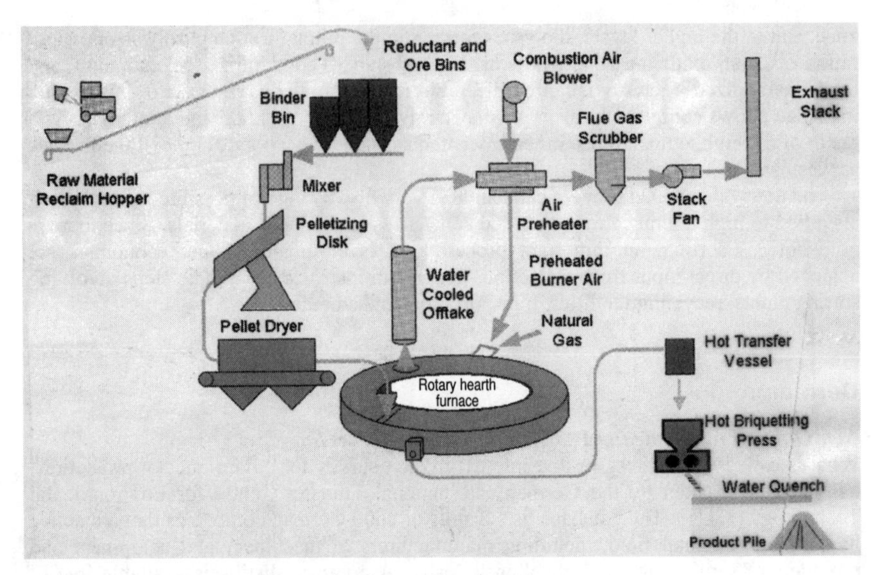

FIGURE 37.3 Fastmet process simulator and melter.

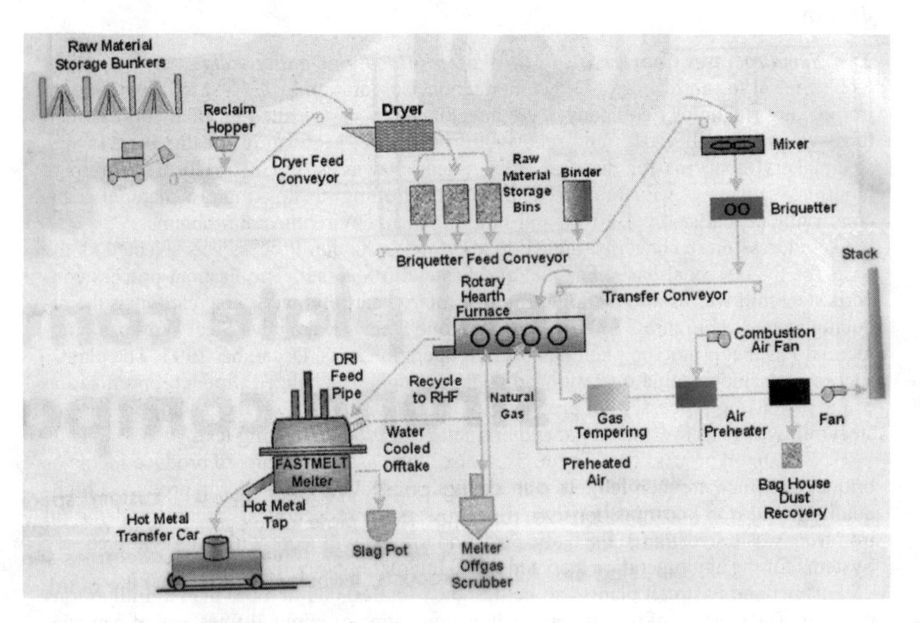

FIGURE 37.4 Fastmelt process flowsheet.

Coconut Shell Powder Helps Plastic Degradation

A Japanese scientist and inventor, Ichiro Sugimoto, may have come up with a solution to degrade plastic materials in landfills much faster by use of an additive made from coconut shell powder. The president and founder of Sanco-Wise Co., Tokyo, has spent the last seven years developing a plastic that biodegrades in a backyard garden within a week under the right conditions. His efforts showed how microrganisms fed on the coconut shell powder mixed in with the plastic resin, causing the plastic material to break down in the soil.

Kitchen waste bags made from his polymer can be buried in a backyard garden. During the June through September period when the average rainfall in Tokyo is 6½ inches, the bags can decompose within one week. Another application is seedling pots made with the degradable powder. By placing the pots in the ground with the seedling and soil, the pot will decompose within four months, permitting the root system to spread and the plant to grow strong.

But the most creative application can be in replacing "white goods" body frames with a polymer containing the coconut shell powder. Since Japan has a new law that mandates that states that it shall be the original manufacturer's responsibility to pay for the disposal costs of used equipment, appliance metal housing had to be collected and either crushed or shredded for recycling, this new biodegradable polymer could be used to fabricate housings that will decompose quickly in the landfills. Manufacturers could save money and landfills could have extended lives.

The Netherlands

The Dutch Car Recycling Association and the Technical University of Delft have founded a new "Centre of Excellence" to carry out research on new recycling techniques for cars aimed at recovering more materials End-of-Life Vehicles (ELVs). Controlling the costs of car recycling is another research goal, which is to start within the next two years. The Dutch expect to recycle 95% of ELVs by 2015. The research projects will focus primarily on separation of plastics, processing shredded wastes, and determining how many car parts can be recycled. With the active support of the Dutch Car Recycling Association, cooperation should be extensive from the recycling industry.

Modern Glass Recycling Attacks Contaminants for Higher Quality

Looking at the history of glass recycling, the advent of automated sorting for color and contaminants has prompted a major increase in glass recycling processing plants. No longer does expensive and inadequate hand sorting of cullet constitute a bottleneck. Now modern plants can process up to 250,000 tons per year without the use of more labor, since equipment in a modern processing plant is monitored by computer.

A continuing contaminant problem involves papers, which, in the presence of liquids, degrade to a sticky powder that covers the cullet. This can lead to clogging of screen openings or the gradual decrease in screen aperture. Further, the papers can camouflage cullet or form paper balls, both of which can disrupt the readings of the sensor-based sorting machines.

By increasing the size of the samples, sensors can detect contaminants and also determine the nature of the contaminants (shape, size, and color). Since these features often characterize a certain type of contaminant, the sensing process generates an estimate for the concentration of each contaminant. The standard spacing of current sensors is 4 mm. Recently, miniaturization has led to the introduction of spacings of only 2 mm between the sensors. More high-tech developments are currently under investigation at the Delft

University of Technology, Department of Raw Materials, Mijnbouwstr, 120,2628 RX, Delft.

Recycling Car Plastic Fuel Tanks

In cooperation with European manufacturers, the Dutch Car Recycling Association, ARN, and the plastic recycling company, Leto Recycling, Almelo, plans are underway to start up a new fuel tank recycling plant at the end of 2000.

The association, ARN, expects to recycle 95% of Dutch end-of-life vehicles by means of high-grade handling and reduction of shredded residue. Recycling fuel tanks can make a significant contribution towards this goal. Furthermore, all of the plastic fuel tanks are made of the same plastic (HDPE), making dismantling and processing relatively easy.

Major Breakthrough in Eddy Current Metals Separation

Until recently, separation of metals with particle sizes below 5 mm was not considered possible by means of eddy current technology. In the course of experiments with a combination of eddy current technologies and wet processing, a new principle was discovered for the separation of 0.5 mm to 10 mm metal particles from waste streams. The new process, called "Magnus separation" or "Wet Eddy Current Separation," is based on a hydrodynamic effect that makes a spinning particle drift away horizontally as it settles in water (see Figure 37.5).

In understandable terms, the Magnus force alters the path of a spinning tennis or golf ball from its normal ballistic trajectory. However, the technological application of this force emerges only when spinning objects reach their terminal equilibrium velocity in a fluid. A simple experiment, shown in Figure 37.5, shows marbles rolling down an inclined surface into a glass container filled with water and demonstrates the principle. Particle size in water should typically be restricted to below 20 mm.

Separation based on the Magnus effect requires rotation of those particles destined for

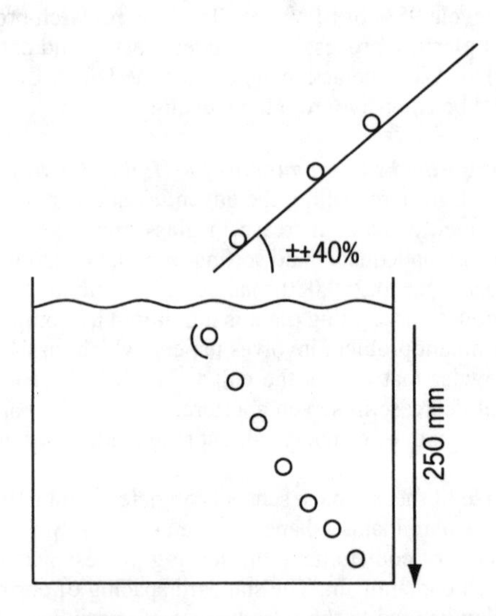

FIGURE 37.5 Experimental demonstration of the Magnus effect.

removal. A rotating magnet induces selective spinning in well-conducting particles, usually nonferrous metals that are present in the waste mixture. Using this separation principle, metals can therefore be separated from nonmetals or from poor conductors. Figure 37.6 illustrates two design arrangements for the Magnus Separator.

This new technology for separating metal particles of 0.5 to 10 mm can be applied to processing residue from wire chopping, electronic scraps, and fines from car scrap. Other potential uses are foundry sand and contaminated soils. Fine fractions of incinerator bottom ash could be another target for the Magnus Separator. This currently untreated material contains substantial amounts of aluminum, brass, copper, and zinc that is lost in land-

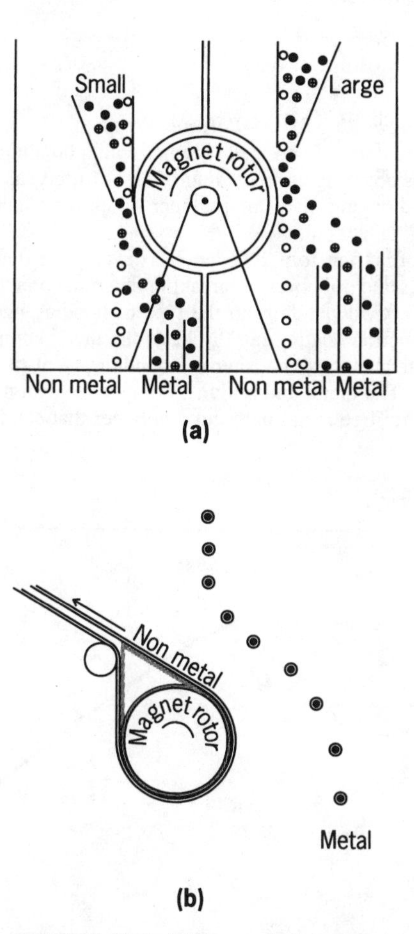

FIGURE 37.6 Different design options for the Magnus separator for processing coarse and fine particles together at enhanced capacity (a) and for separation of course feed containing magnetic material at enhanced efficiency (b).

fills. Recovery of these metals improves the environmental and engineering qualities of bottom ash as a secondary building material. Metal particles in bottom ash start around one mm with the Magnus separator taking out the 1–2 mm fraction. A 2–15 mm sample from a Dutch incinerator passed through the machine at 8.5 tons per hour; it contained 2.4% aluminum and 2.4% of copper, brass and zinc.

High throughput is a major advantage. Even with relatively small particle sizes, feeding the material as a slurry blanket rather than a monolayer results in relatively high throughput. Materials can be fed into both aides of the rotor, thus doubling the throughput. Actually, one side of the rotor is better suited for larger materials, whereas the other side should be used for sizes less the 6 mm. Possible applications include separation of lead and copper from cable scrap.

Sweden

New Technology for Debonding Metal–Polymer Scrap

Metal–polymer parts are normally bonded together with a bonding agent. New European Union (EU) regulations covering car recycling of end-of life vehicles (ELVs), as well as regulations governing dumping of metal–polymer scrap, have increased the demand for new economical recycling technologies.

Linlan Debonding of Staffanstorp, Sweden, has developed a new technique to separate and recover both metals and polymers by breaking down the bonding agent with heat between 200 °C and 300 °C, depending on the type of bonding agent. A combination of magnetic field and induction heating rapidly heats the metal component of a compound product to a uniform temperature, causing the bonding agent to break down while the polymer remains solid. The metal and polymer components can then be easily separated to leave a clean, polymer-free metal part and a polymer that can be reused or burned as fuel.

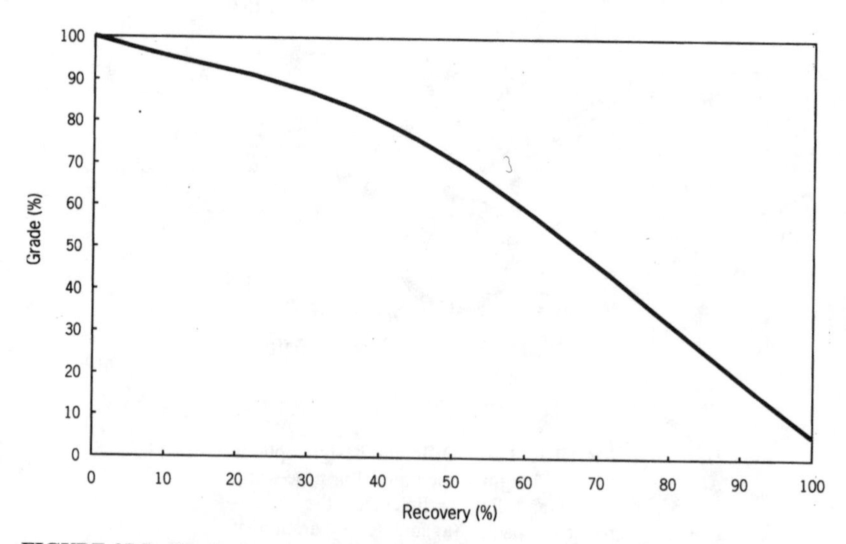

FIGURE 37.7 Grade-recovery curve for the nonferrous concentrate from 2–15 mm bottom ash. (*Source: Recycling International* Magazine, Arnhem, The Netherlands.

The value of the reclaimed parts of these products is high and is comparable to new parts rather than scrap value. The metal–polymer parts in this process are manufacturing rejects and the metal components can be used with new polymer coatings. Debonding equipment for recycling will have a capacity of 2000 kg per hour for metal–polymer scrap products such as truck wheels, pump housings, flanges, truck vibration mounts, and rollers with a diameter up to 400 mm. Other applications include process tubes with an outer diameter of 400 mm, which are covered with polymer on the inside or outside, and tanks up to three meters in diameter with polymer either on the inside or outside. Still another potential application is the separation of aluminum–polymer products used in the building industry.

United Kingdom

Built-in Electronic Fraud Tracking System Watches over Scrap Yards

Early in the 90's, Rob Horan, Manager of Information Technology of Mayer Parry Recycling quickly found there was no computer software suitable for scrap yard operations. It was then that Horan decided to develop a completely new information software that was tailor-made for a metals recycling company.

One of the many side benefits of this new software was a fraud tracking operation. The software logs everything that is weighed and therefore can check for any irregularities. For instance, if a truck weighs much more or less than its almost identical predecessor, the computer will automatically flag a warning. This fraud tracking function also allows the operator to identify legitimate reasons why an action occurs, weeding out coincidences and leaving the oddities open for review and inspection. The software is a protection for the yard operators because everything must be recorded. The fraud tracking system is unique in the metals scrap industry and Horan reports uncovering actual frauds within the company.

The software can record the intake of materials and subsequently identify separated grades of metals. It also shows delay and run time information, working hours, plant downtime, delay hours and no-scrap hours. An additional element identifies approximate shredder yield to keep track of input–output availability through a batch control option; it is possible to analyze each batch by its particular number and thus detect the volume of each individual metal separated from the batch of nonferrous metals and the volume of contaminating dirt. With this software, the company only pays a supplier for the metallic content of the load and not for any old tires buried in it.

European Union Directive Precipitates Aluminum "Dross" Recycling

Dross, a scum on the surface of molten aluminum similar to the foamy head in a just poured bottle of beer, had been considered an annoying waste product until a legislative directive from the European Union forced corrective recycling action with the development of thermal aluminum recovery. This process resulted in the creation of a new machine by J. McIntyre Machinery Co., Nottingham, England. It is called Tardis.

Not only does the machine recover more than 20% of aluminum from the dross, but harmful emission are 99% removed, nasty odors eliminated from the community, and unsightly waste piles removed. Equipment cost payback of less than twelve months is reported. In Brazil, two Tardis units costing $400,000 had payback times of two months after aluminum recovery.

Tardis is a dross pressing and cooling system consisting of twin hydraulic rams with specially shaped heads that fit into a similarly sized pot. Dross is raked into the pot while it is still hot and then simultaneously pressed and cooled. Fine particles of aluminum within the dross coagulate into large pieces while free liquid aluminum drains through a hole in the bottom of the pot into a series of collecting trays. The Tartis head is air-cooled, promoting rapid cooling of the dross and thereby preventing the materials from thermiting. This reduces oxidation and conserves metal material. The final result of the process is as aluminum ingot, an aluminum cone the shape of the pot, and cooled dross. The ingots and cones are melted on-site and the enriched cold dross can be recycled on or off the site. (*Source:* Manfred Beck, *Recycling International.*)

The following sources for this section are acknowledged: *Recycling International* Magazine, Arnhem, The Netherlands; *Waste News*, a Crain Publication, Akron, Ohio.

APPENDIX A

RECYCLING INFORMATION AND SOURCES

TERRY GROGAN
Chief, Municipal Waste Reduction Branch
U.S. Environmental Protection Agency
Washington, DC

HERBERT F. LUND, P.E.
Editor-in-Chief
Coconut Creek, Florida

FEDERAL AGENCIES IN THE UNITED STATES

Environmental Protection Agency

EPA ADMINISTRATOR
U.S. Environmental Protection Agency
12th and Pennsylvania Avenue, NW
Washington, DC 20460
(202) 564-4700

MUNICIPAL AND INDUSTRIAL SOLID WASTE DIVISION (Recycling)
U.S. Environmental Protection Agency
2800 Crystal Drive, 8th Floor
Arlington, VA 22202
(703) 308-8254

OFFICE OF SOLID WASTE
U.S. Environmental Protection Agency
401 "M" Street, SW
Washington, D.C. 20460
(202) 260-4610

DIRECTOR OF SOLID WASTE PROGRAMS
U.S. Environmental Protection Agency
2800 Crystal Drive, 8th Floor
Arlington, Virginia 22202
(703) 308-8895

EPA Regional Offices—Solid Waste Contacts

Region 1 (Connecticut, Vermont Massachusetts, Rhode Island, New Hampshire, Maine)
1 Congress St., Suite 1100 SPP
Boston, MA 02114-2023
(617) 918-1813 / FAX (617) 918-1810

Region 2 (New Jersey, New York, Puerto Rico, Virgin Islands)
290 Broadway, New York, NY
10007-1866, Mail Code 2DEPP-RPB
(212) 637-4125/FAX (212) 637-4437

Region 3 (Delaware, Maryland, West Virginia, Pennsylvania, Virginia, District of Columbia)
1650 Arch St., Philadelphia, PA 19103-2029, Mail Code 3WC21
(215) 814-3298/FAX (215)814-3163

Region 4 (Alabama, Florida, Georgia, Kentucky, Mississippi, North and South Carolina, Tennessee)
61 Forsyth St. Atlanta, GA 30303-3104, Mail Code 4WD-RPB/RSS
(404) 562-8449/FAX (404) 562-8439

Region 5 (Illinois, Indiana, Ohio, Minnesota, Michigan, Wisconsin)
77 West Jackson Blvd., Chicago, IL 60604-3590,Mail Code DW-8J
(312) 886-0976/FAX (312) 353-4788

Region 6 (Arkansas, Louisiana, Texas, New Mexico, Oklahoma)
1445 Ross Ave., Dallas, TX 75202-2733
Mail Code 6PD-U
(214) 665-6760/FAX (214) 665-7263

Region 7 (Iowa, Kansas, Missouri, Nebraska)
901 North 5th Street, Kansas City, KS 66101 ARTD/SWPP
(913) 551-7523/FAX (913) 551-7947

Region 8 (Colorado, Montana, Utah, North and South Dakota, Wyoming)
999 18th Street, Suite 500
Denver, CO 80202-2466,MC 8P-P3T
(303) 312-6286/FAX (303) 312-6044

Region 9 (Arizona, California, Guam, Nevada, Hawaii, Republic of Palau, Commonwealth of the Northern Mariana Islands, American Samoa, Federated States of Micronesia, The Republic of the Marshall Islands)
75 Hawthorne St. Mail Code WST-7
San Francisco, CA 94105
(415) 744-1284/FAX (415) 744-1044

Region 10 (Alaska, Idaho, Oregon, Washington)
1200 6th Avenue, Seattle, WA 98101
Mail Code WCM-128
(206) 553-6517/FAX (206) 553-8509

STATE AGENCIES

ALABAMA
Dept. of Economic and Community Affairs
P.O. Box 5690
Montgomery, AL 36 103-5690
(334) 242-5336

ALASKA
Dept. of Environmental Conservation
Division of Environmental Health
410 Willoughby Ave., Ste. 105
Juneau, AK 99801-1795
(907) 465-5162

ARIZONA
Dept. of Environmental Quality
Waste Programs Division
3033 North Central Ave
Phoenix, AZ 85012
(602) 207-4208

ARKANSAS
Dept. of Pollution Control and Ecology
Solid Waste Management Division
8001 National Drive
Little Rock, AR 72209
(501) 682-0600

CALIFORNIA
Integrated Waste Management Board
8800 Cal Center Drive
Sacramento, CA 95828
(916) 255-2200

COLORADO
Dept. of Public Health and Environment
Hazardous Materials and Waste Management Division
HMWMD-B2
4300 Cherry Creek Dr. South
Denver, CO 80246-1530
(303) 692-3320

DELAWARE
Dept. of Natural Resources and
Environmental Control
Solid Waste Management Branch
89 Kings Highway
Dover, DE 19901
(301) 739-3820

DISTRICT OF COLUMBIA
Department of Public Works
65 K Street NE, Lower Level
Washington, DC 20002
(202) 939-7192

FLORIDA
Dept. of Environmental Protection
Division of Waste Management
3900 Commonwealth Blvd. M.S. 10
Tallahassee, FL 32399-3000
(850) 487-3299

GEORGIA
Department of Natural Resources
Environmental Protection Division
205 Butler Street, SE, Room 1154
Atlanta, GA 30334
(404) 362-2692

HAWAII
Department of Health
Environmental Management Division
919 Ala Moana Blvd, Ste. 212
Honolulu, HI 96814
(808) 586-4226

IDAHO
Division of Environmental Quality
1410 North Hilton Street
Boise, ID 83706
(208) 373-0417

ILLINOIS
Environmental Protection Agency
2200 Churchill Road
1021 N. Grand Avenue East
Springfield, IL 62702
(217) 785-2800

INDIANA
Department of Environmental Management
Office of Solid and Hazardous Waste Mgmt
P.O. Box 6015
Indianapolis, IN 46206
(317) 233-3656

IOWA
Department of Natural Resources
Waste Management Assistance
Division
502 E. 9th Street
Des Moines, IA 50319
(515) 281-4867

KANSAS
Department of Health and Environment
Division of Environment
Forbes Field, Building 740
Topeka, KS 66620
(785) 296-1600

KENTUCKY
Department for Environmental
Protection
Division of Waste Management
14 Reilly Road
Frankfort, KE 40601
(502) 564-6716

LOUISIANA
Department of Environmental Protection
Office of Waste Services
P.O. Box 82178
Baton Rouge, LA 70884
(225) 765-0355

MAINE
Department of Environmental Protection
Bureau of Solid Waste Management
State House Station 17
Augusta, ME 04333
(207) 582-8740

MARYLAND
Department of the Environment
Hazardous and Solid Waste Management
Administration
2500 C Highway
Baltimore, MD 21224
(301) 631-3386

MASSACHUSETTS
Department of Environmental
Protection
Bureau of Waste Prevention
One Winter Street
Boston, MA 02108
(617) 292-5961

MICHIGAN
Department of Environmental Quality
Waste Management Division
P.O. Box 30473
Lansing, MI 48933
(517) 373-2730

MINNESOTA
Minnesota Pollution Control Agency
520 Lafayette Road
St. Paul, MN 55155
(612) 296-6300

MISSISSIPPI
Department of Environmental Quality
Environmental Resource Center
P.O. Box 20305
Jackson, Mississippi 39289
(601) 961-5666

MISSOURI
Department of Natural Resources
Environmental Quality Division
P.O. Box 176
Jefferson City, MS 65102
(800) 334-6946

MONTANA
Department of Environmental Quality
Pollution Prevention Bureau
P.O. Box 200901
Helena, MT 59620
(406) 444-6697

NEBRASKA
Department of Environmental Quality
Air/Waste Management Division
P.O. Box 98922
Lincoln, NE 68509
(402) 471-2186

NEVADA
Division of Environmental Protection
Bureau of Waste Management
333 West Nye Lane
Carson City, NV 89706
(702) 687-4670

NEW HAMPSHIRE
Department of Environmental Services
Waste Management Division
6 Hazen Drive
Concord, NH 03301
(603) 271-2900

NEW JERSEY
Department of Environmental Protection
P.O. Box 414
Trenton, NJ 08625
(609) 530-8208

NEW MEXICO
Environmental Protection Division
Solid Waste Bureau
Harold Runnels Building
1190 St. Francis Drive
Santa Fe, NM 87505
(505) 827-2775

NEW YORK
Department of Environmental Conservation
Division of Solid and Hazardous Materials
50 Wolf Road
Albany, NY 12233
(518) 457-6934

NORTH CAROLINA
Department of Environment and Natural Resources
Division of Waste Management
P.O. Box 29569
Raleigh, NC 27626
(919) 715-6500

NORTH DAKOTA
Department of Health
Division of Waste Management
P.O. Box 5520
Bismarck, ND 58506
(701) 328-5166

OHIO
Environmental Protection Agency
Division of Solid and Infectious Waste Mgmt
Lazarus Government Center
P.O. Box 1049
Columbus, OH 43216
(614) 644-2621

OKLAHOMA
Department of Environmental Quality
Waste Management Division
P.O. Box 1677
Oklahoma City, OK 73101
(405) 702-5100

OREGON
Department of Environmental Quality
Waste Management Division
811 SW Sixth Avenue
Portland, OR 97204
(503) 229-5913

PENNSYLVANIA
Department of Environmental Protection
Bureau of Waste Management
Rachel Carson State Office Building
400 Market Street
Harrisburg, PA 17105
(717) 787-7382

RHODE ISLAND
Department of Environmental Management
Office of Waste Management
235 Promenade Street
Providence, RI 02908
(401) 222-2797

SOUTH CAROLINA
Department of Health and Environmental Control
Bureau of Solid and Hazardous Waste Management
2600 Bull Street
Columbia, SC 29201
(803) 896-4000

SOUTH DAKOTA
Department of Environment and Natural Resources
Waste Management Program
Foss Building
523 East Capitol
Pierre, SD 57501
(605) 773-3153

TENNESSEE
Department of Environment and Conservation
Division of Solid Waste Management
L&C Tower, 5th Floor
401 Church Street
Nashville, TN 37243
(615) 532-0780

TEXAS
Texas Natural Resource Conservation
Commission-Office of Waste Management
P.O. Box 13087
Austin, TX 78753
(512) 239-2104

UTAH
Department of Environmental Quality
Division of Solid and Hazardous Waste
P.O. Box 144880
Salt Lake City, UT 84114
(801) 538-6170

VERMONT
Department of Environmental Conservation
Agency of Natural Resources
103 Main Street
Waterbury, VT 05671
(802) 241-3888

VIRGINIA
Department of Environmental Quality
101 North 14th Street
James Monroe Building, 11th Floor
Richmond, VA 23219
(804) 225-2667

WASHINGTON
Department of Ecology
Solid Waste and Financial Assistance Program
P.O. Box 47600
Olympia, WA 98504
(360) 407-7455

WEST VIRGINIA
Division of Environmental Protect.
Office of Waste Management
1356 Hansford Street
Charleston, WV 25301
(304) 558-5929

WISCONSIN
Department of Natural Resources
Bureau of Waste Management
P.O. Box 7921
Madison, WI 53702
(608) 264-6032

WYOMING
Department of Environmental Quality
Division of Solid Waste
122 West 25th Street
Cheyenne, WY 82002
(307) 777-7752

SOLID WASTE INFORMATION SOURCES SPONSORED BY THE FEDERAL GOVERNMENT

Center for Environmental Research Information: Central point of distribution for EPA research results and reports; (513) 569-7562

Asbestos Ombudsman: Responds to questions and concerns about asbestos in schools issues; (800) 368-5888; (202) 557-1938

Emergency Planning and Community Right-to-Know Information Hot Line: Provides communities and individuals with help in preparing for accidental releases of toxic chemicals. This hot line, which complements the RCRA/Superfund Hot line, is maintained as an information resource rather than an emergency number; (800) 535-0202; (202) 479-2449

National Appropriate Technology Assistance Service: Provides information on waste and materials management in reference to energy issues. Operated by the U.S. Department of Energy; (800) 428-2525

National Pesticides Telecommunications Network Hot Line: Provides information on pesticide-related health, toxicity, and minor cleanup to physicians, veterinarians, fire departments, government agencies, and the general public. Also provides impartial information on pesticide products, basic safety practices, health and environmental effects, and cleanup and disposal procedures; (800) 858-7378; (806) 743-3091 (in Texas)

New England Solid Waste Research Library: Research library for 650 subject headings (updated quarterly). Most information focuses on New England, although information is available on other U.S. regions and Europe; (617) 573-9687

RCRA/CERCLA Superfund Hot Line: Responds to questions from the public and regulated community on the Resource Conservation and Recovery Act, and the Comprehensive Environmental Response, Compensation and Liability Act (Superfund). Responds for requests for RCRA and Superfund documents; (800) 424-9346; (202) 382-3000

Rural Information Center: Library of information pertaining to rural issues sponsored by the U.S. Department of Agriculture; (301) 344-2547

SWICH (Solid Waste Information Clearinghouse): USEPA/Solid Waste Association of North America-sponsored national information center on solid waste information. Information is available on-line as a bulletin board and a 6200-volume library and through fax and mail requests; modem: (301) 585-0204; Fax: (301) 585-0297

Toxic Substances Control Act (TSCA) Assistance Information Service: Provides information and publications about toxic substances, including asbestos; (202) 554-1404

USEPA Procurement Hot Line: Distribute copies of various federal procurement guidelines and provides lists of manufacturers and distributors of products meeting these guidelines. (703) 941-4452

MSW Factbook: A paperless, electronic reference manual containing more than 200 "screens" of useful facts, figures, tables, and information about MSW. Can be viewed on EPA's Web site or installed on any computer equipped with a mouse and Microsoft Windows.

Characterization of Municipal Solid Waste in the United States 1997 Update: Describes the national MSW stream. Includes information on MSW generation, recovery, and discard quantities; per capita generation and discard rates; residential and commercial portions of MSW generation; the role of source reduction and other trends in MSW management; and projections for MSW generation and management through 2010.

Decision-Maker's Guide to Solid Waste Management, Volume II: Contains technical and economic information to assist solid waste management practitioners in planning, managing, and operating MSW programs and facilities. Includes suggestions for best practices when planning or evaluating waste and recycling collection systems, source reduction and composting programs, public education, and landfill and combustion issues.

For additional information on solid waste issues, call the EPA RCRA Hotline at (800) 424-9346

REGIONAL WASTE EXCHANGES

EPA Region 1 (Connecticut, Massachusetts, New Hampshire, Rhode Island, Vermont)

New Hampshire Material Exchange (NHME)
WasteCap of New Hampshire
122 North Main Street
Concord, NH 03301
(603)224-5388

Material Exchange
1037 State Street
Bridgeport, CT 06605
(203) 335-3452

Maine Materials Exchange
93 Maquoit Drive
Freeport, ME 04032
(207) 865-6621

Berkshires Materials Exchange
Center for Ecological Technology
112 Elm Street
Pittsfield MA 01201
(413) 445-4556

Vermont Business Materials Exchange
P.O. Box 935
Brattleboro, VT 05302
(802) 257-7505

EPA Region 2 (New Jersey, New York, Puerto Rico, Virgin Islands)

Hudson Valley Materials Exchange (HVME)
207 Milton Turnpike
Milton, NY 12547
(914) 795-5507

New Jersey Industrial Waste Info. Exchange
50 West State Street, Ste 1310
Trenton, NJ 08608
(609) 989-7888

EPA Region 3 (District of Columbia, Delaware, Maryland, Pennsylvania, Virginia, West Virginia)

Northeast Industrial Waste Exchange (NIWE)
P.O. Box 2171
Annapolis, MD 21404
(410) 280-2080

EPA Region 4 (Alabama, Georgia, Florida, Kentucky, Mississippi, North Carolina, South Carolina, Tennessee)

EnviroShare Materials Exchange
Hall County Resource Recovery
P.O. Drawer 1435
Gainesville, GA 30503
(770) 535-8284

Florida Recycling Material System (FRMS)
2207 NW 13th Street, Ste D
Gainesville, FL 32609
(352) 846-0183

Kentucky Industrial Materials Exchange
Kentucky Pollution Prevention Center
420 Academic Building
University of Louisville
Louisville, KY 40292

Mississippi Technical Assistance Program
P.O. Box 9595
Mississippi State, MS 39762
(601) 325-8454

Southeast Waste Exchange
Urban Institute University of North Carolina
9201 University City Blvd.
Charlotte, NC 28223
(704) 547-4289

Southern Waste Information Exchange
P.O. Box 960
Tallahassee, FL 32302
(904) 386-6280 or (800) 441-7949

Tennessee Materials Exchange
226 Capitol Blvd. Building, Ste 605
Nashville, TN 37219
(615) 532-8881

EPA Region 5 (Illinois, Indiana, Michigan, Ohio, Wisconsin)

Illinois Industrial Materials Exchange
Illinois EPA
P.O. Box 19276
Springfield, IL 62794
(217) 782-0450

Indiana Materials Exchange
133 West Market Street
Box 263
Indianapolis, IN 46204
(800) 968-8764

Minnesota Technical Assistance Program
Materials Exchange
1313 Fifth Street SE, Ste 207
Minneapolis, MN 55414
(612) 627-4646 or (800) 247-0015

Northeast Minnesota Waste Exchange
Western Lake Superior Sanitary District
2626 Courtland Street
Duluth, MN 55806
(218) 722-3336

Southeast Minnesota Materials Exchange
171 West Third Street
Winona, MN 55987
(507) 457-6464

Southwest Minnesota Materials Exchange
Nobels County/Environmental Services
P.O. Box 757
Worthington, MN 56187
(507) 372-8227

EPA Region 6 (Arkansas, Louisiana, New Mexico, Oklahoma, Texas)

Arkansas Industrial Development Commission
One Capitol Mall, Room 4B2 15
Little Rock, AR 72201
(501) 682-7325

Oklahoma Waste Exchange Program
Department of Environmental Quality
Public Information and Education Office
1000 NE 10th Street
Oklahoma City, OK 73117
(405) 271-7353

Resource Exchange Network for Eliminating
Waste (RENEW)
Texas Natural Resource Conservation
Commission (MC-1 12)
Austin, TX 78711
(512) 239-3171

Transcontinental Materials Exchange
Department of Civil Engineering
Louisiana State University
Baton Rouge, LA 70803
(504) 388-4594

EPA Region 7 (Iowa, Kansas, Missouri, Nebraska)

Iowa Waste Reduction Center
By-Product and Waste Search Center
University of Northern Iowa
75 BRC
Cedar Falls, IA 50614
(319) 273-2079

Kansas Materials Exchange
P.O. Box 152
Hutchinson, KS 67504
(316) 662-0551

Missouri Waste Exchange Service
325 Jefferson Street
Box 744
Jefferson City, MO 65102
(573) 751-4919

EPA Region 8 (Colorado, Montana, North Dakota, South Dakota, Utah, Wyoming)

Montana Materials Exchange
MSU Extension Service
Taylor Hall
Bozeman, MT 59717
(406) 994-1748

Colorado Materials Exchange
University of Colorado Recycling Services
University Memorial Center, Room 331
Campus Box 207
Boulder, CO 80309
(303) 492-8307

Arizona Waste Exchange
4725 East Sunrise Drive, Ste 215
Tucson, AZ
(602) 299-7716

California Materials Exchange (CALMAX)
8800 Cal Center Drive
Sacramento, CA 95826
(916) 255-2369

California Waste Exchange (CWE)
P.O. Box 806
Sacramento, CA 95812
(916) 322-4742

Hawaii Materials Exchange
P.O. Box 121
Wailuku, HI 96793
(808) 667-7744

EPA Region 10 (Alaska, Idaho, Oregon, Washington)

Alaska Materials Exchange
Alaska Dept of Environmental Conservation
555 Cordova Street
Anchorage, AK 99501
(907) 269-7586

Industrial Materials Exchange
506 Second Avenue, Room 201
Seattle, WA 98104
(206) 296-4899

National Materials Exchanges
Chicago Board of Trade Recyclables Exchange
141 West Jackson Blvd
Chicago, IL 60604
(312) 341-7955

Other Waste Exchanges

Alberta Waste Materials Exchange
William Kay
Industrial Development Department
Alberta Research Council
PO Box 8330, Postal Station F
Edmonton, Alberta
Canada T6H 5X2
(403) 450-5408

Canadian Waste Materials Exchange
Dr. Robert Laughlin
Ortech International
Sheridan Park Research Community
Mississauga, Ontario
Canada L5K 1B3

Industrial Material Exchange Service
Diane Shockey
PO Box 19276
2200 Churchill Rd., #31
Springfield, IL 62794-9276
(217) 782-0450

Montana Industrial Waste Exchange
Sharon Miller
PO Box 1730
Helena, MT 59624
(406) 442-2405

Pacific Materials Exchange
Bob Since
South 3707 Godfrey Blvd.
Spokane, WA 99204
(509) 623-4244

Resource Exchange & News
Kay Ostorwski
400 Ann Street, NW
Suite 301A
Grand Rapids, MI 49505
(616) 363-3262

Texas Water Commission
Hope Castillo
P0 Box 13087
Austin, TX 787113087
(512) 463-7773

NATIONAL PERIODICALS

AMERICAN CITY & COUNTY
Communications Channel, Inc.
255 Barfield Road
Atlanta, GA 30328
(770) 955-9970

BIOCYCLE Magazine
The J.G. Press
419 State St., 2nd Floor
Emmaus, PA 18049
(610) 967-4135

BROWN FIELDS REPORT
King Publishing Group, Inc.
627 National Press Building;
Washington, DC 20045
(202) 638-4260/FAX (202) 662-9719

C & D RECYCLER
P.O. Box 22123
Cleveland, OH 44101-9901
(800) 456-0707

COMPOSTING NEWS
McEntee Media Corporation
13727 Holland Road
Cleveland, OH 44142-3920
(216) 362-7979/FAX (216) 362-6553

CONTAINER RECYCLING REPORT
P.O. Box 42270
Portland, OR 97242-0270
(503) 233-1305/FAX (503) 1356

ENVIRONMENTAL INDUSTRIES NEWS
Environmental Industries Association
4301 Connecticut Ave., N.W.
Washington, DC 20008,Suite 300
(202) 244-4700/FAX (202) 966-4818

GREENWIRE
National Journals Environmental News Service
3129 Mount Vernon Road
Alexandria, VA 22305
(703) 518-4600/FAX (703) 518-8702

HAZARDOUS WASTE NEWS
Business Publishers, Inc.
8737 Colesville Road
Silver Spring, MD 20910-3928
(301) 587-6300/FAX (301) 587-1O81

MSM MANAGEMENT
Forester Communications, Inc.
5638 Hollister St., Suite 301
Santa Barbara, CA 93117
(805) 681-1300/FAX (805) 681-1212

THE N R C CONNECTION
National Recycling Coalition, Inc.
1727 King Street, Suite 105
Alexandria, VA 22314
(703) 683-9025/FAX (703) 683-7026

THE PAPER STOCK REPORT
McEntee Media Corporation
13727 Holland Road
Cleveland,OH 44142-3920
(216) 362-7979/FAX (216) 362-6553

PLASTIC RECYCLING UPDATE
P.O. Box 42270
Portland, OR 97242-0270
(503) 233-1305/FAX (503) 233-1356

POLLUTION PREVENTION INFORMATION CLEARING HOUSE
U.S. Environmental Protection Agency
Office of Pollution Prevention
401 "M" St., S.W. MS 7409
Washington, DC 20460
(202) 260-1023

RAYMOND COMMUNICATIONS, INC.
Publisher of
State Recycling Laws Update
5111 Berwyn Road - Suite 115
College Park, MD 20740
(301) 345-4237/Fax: (301) 345-4768
www.Raymond.com

THE RECYCLING MAGNET
Steel Recycling Institute
Foster Plaza 10, 680 Andersen Drive
Pittsburgh, PA 15220
(800) 876-7274/Fax: (412) 922-3213

RECYCLED PAPER NEWS
McEntee Media Corporation
13727 Holland Road
Cleveland, OH 44142-3920
(216) 362-7979/Fax: (216) 362-6553

RECYCLING TODAY
G I E Media, Inc.
4012 Bridge Ave.
Cleveland,OH 44113-3320
(800) 456-0707/Fax: (216) 961-0364

RESOURCE RECOVERY REPORT
P.O. Box 3356
Warrenton, VA 20188
(540) 347-4500/Fax: (540) 349-4540

RESOURCE RECYCLING
P.0. Box 42270
Portland OR 97242-0270
(503) 233-1305/Fax: (503) 233-1356

RESOURCES
The Environmental Resources Management Group, Inc.
855 Springdale Drive
Exton, PA 19341
(800) 544-3117

REUSABLE NEWS
U.S. Environmental Protection Agency
Office of Solid Waste and Emergency Resources
Aril Rius Bldg., 1200 Pennsylvania Ave, NW
Washington, DC 20460
RCRA Hot Line (800) 424-9346

SCRAP
Institute of Scrap Recycling Industries
1325 G Street, N.W.
Washington, DC, 20005-3104
(202) 662-8527, -8525, -8594

SCRAP TIRE NEWS
Recycling Research Institute
P.O. Box 714, 133 Mountain Road
Suffield, CT 06078
(860) 668-5422/FAX (860) 668-5651
(Also annual State Scrap Tire Legislation and Directory of Scrap Tire and Rubber Users)

THE SOFT DRINK RECYCLER
National Soft Drink Association
Environmental Affairs Division
1101 16th Street, N.W.
Washington, DC 20220-0806
(202) 463-6700

SOLID WASTE DIGEST-NATIONAL EDITION
Chartwell Information Publishers
805 Cameron Street
Alexandria, VA 22314
(800) 234-8692/FAX (703) 519-7881

SOLID WASTE REPORT
Business Publishers, Inc.
87237 Colesville Road
Silver Spring, MD 20910-3928
(30l) 587-6300/Fax(301) 587-1081

STATE RECYCLING LAWS UPDATE
Raymond Communications, Inc.
5111 Berwyn Road, Suite 115
College park, MD 20740
(301) 345-4237/Fax: (301) 345-4768

WASTE AGE
6151 Powers Ferry Road, N,W.
Atlanta, GA 33039
(770) 618-0112

WASTE NEWS
Cram Communications, Inc.
1725 Merriman Road
Akron, OH 44313-5251
(330) 836-9180/Fax: (330) 836-1692

WASTE REDUCTION TIPS
McEntee Media Corporation
13727 Holland Road
Cleveland, OH 44142-3920
(216) 362-7979/Fax: (216) 362-6553

INTERNATIONAL PERIODICALS

BUSINESS NEWS AMERICAS
Carmencita 106
Las Condes
Santiago, Chile
011-562-233-0302/FAX 011-562-232-9376
(Covering Central & South Americas and Mexico)

ISWA TIMES
International Solid Waste Association
ISWA General Secretariat
Vester Farminagsgade 29
DK-l780 Copenhagen V, Denmark
Phone 011 +45 33 1.5 65 65

RECYCLING INTERNATIONAL MAGAZINE
P.O. Box 2098
6802 CP Arnhem, The Netherlands
Phone 011 +31 26.3120.994
FAX 011 +31 26.3120.630

RECYCLINC LAWS INTERNATIONAL
Raymond Communications
5111 Berwyn Road, Suite 115
College Park, MD 20740
(301) 345-4237/FAX (301) 345-4768

WARMER BULLETIN
The Journal of the World Resources Foundation
Bridge House, High Street
Tunbridge, Kent TN9 lDP, England
011 44+0+ 1732-368333

ASSOCIATION RESOURCES

AIR AND WASTE MANAGEMENT ASSOCIATION
MUNICIPAL WASTE COMMITTEE
Fort Duquesne Boulevard
P.O. Box 2861
Pittsburgh, PA 15230
(412) 232-3444; Fax: (412) 232-3450

ALUMINUM ASSOCIATION
900 19th Street, NW
Suite 300
Washington, DC 20006
(202) 862-5100: Fax: (202) 862-5164

AMERICAN FOREST & PAPER ASSOCIATION
1111 19th Street, NW
Washington, DC 20036
(202) 463-2700
E-mail: info@afandpa.ccmail.compuserve.com

AMERICAN PLASTICS COUNCIL
(Society of the Plastics Industry)
1275 K Street, NW
Suite 400
Washington, DC 20005
(202) 371-5319; Fax: (202) 371-5679

AMERICAN SOCIETY OF MECHANICAL ENGINEERS SOLID WASTE PROCESSING DIVISION
345 East 47th Street
New York, NY 10017-2392
(212) 705-7722; Fax: (212) 705-7674
(800) 843-2763

ASEPTIC PACKAGING COUNCIL
1225 "I" (Eye) Street, NW
Suite 500
Washington, DC 20005
(800) 277-8088; Fax: (202) 333-5987

ASPHALT RECYCLING & RECLAIMING ASSOCIATION
3 Church Circle
Suite 250
Annapolis, MD 21401
(410) 267-0023

ASSOCIATION OF BATTERY RECYCLERS
1901 North 66th Street
Tampa, FL 33619
(813) 626-6151; Fax: (813) 622-8388

ASSOCIATION OF STATE & TERRITORIAL SOLID WASTE MANAGEMENT OFFICIALS
444 North Capitol Street, NW
Suite 315
Washington, DC 20001
(202) 624-5828; Fax: (202) 624-7875

AMERICAN SOCIETY FOR TESTING & MATERIALS COMMITTEE D-34 ON WASTE MANAGEMENT
Attn: Gilbert Bourcier
c/o Old Dominion Engineering Services Company
13900 Elmstead Road
Midlothian, VA 23113
(804) 794-6437; Fax: (804)794-5160

CAN MANUFACTURERS INSTITUTE
1625 Massachusetts Avenue, NW
Washington, DC 20036
(202) 232-4677; Fax: (202) 232-5756

COMMITTEE FOR ENVIRONMENTALLY EFFECTIVE PACKAGING
601 13th Street, NW
Suite 900, South
Washington, DC 20005
(202) 783-5594; Fax: (202) 783-5595

COMPOSTING COUNCIL
114 South Pitt Street
Alexandria, VA 22314-3112
(703) 739-2401; Fax: (703) 739-2407

CONTAINER RECYCLING INSTITUTE
1400 16th Street, NW
Washington, DC 20036-2217
(202) 797-6839; Fax: (202) 797-5437

CORNELL WASTE MANAGEMENT INSTITUTE
Center for the Environment
100 Rice Hall
Cornell University
Ithaca, NY 14853-5601
(607) 255-1187; Fax: (607) 255-8207

COUNCIL OF STATE GOVERNMENTS
Center for the Environment
P.O. Box 11910
Lexington, KY 40578-19 10
(606) 244-8000; Fax: (606) 244-8001

DIRECT MARKETING ASSOCIATION
120 Avenue of the Americas
New York, NY 10036-8096
(212) 768-7277; Fax: (212) 768-4546

ENVIRONMENTAL DEFENSE FUND
257 Park Avenue, South
New York, NY 10010
(212) 505-2100; Fax: (212) 505-2375

ENVIRONMENTAL INDUSTRIES ASSOCIATION
(formerly the National Solid Waste Management Association)
4301 Connecticut Ave., N.W Suite 300
Washington, DC 20008
(202) 244-4700; Fax: (202) 966-4818

ENVIRONMENTAL INFORMATION
(Formerly: National SWM Assoc.)
4915 Auburn Avenue
Suite 303
Bethesda, MD 20814
(301) 961-4999; Fax: (301) 961-3094

ENVIRONMENT CANADA
Prevention & Treatment Division
351 Saint Joseph Boulevard
12th Floor
Hull, Quebec K1A 0H3 CANADA
(819) 953-0616; Fax: (819) 953-6881

FOOD MARKETING INSTITUTE
800 Connecticut Avenue, NW
Washington, DC 20006
(202) 452-8444; Fax: (202) 429-4529

FOODSERVICE & PACKAGING INSTITUTE
1901 North Moore Street
Suite 1111
Arlington, VA 22209
(703) 527-7505; Fax: (703) 527-7512

GLASS PACKAGING INSTITUTE
1627 K Street, NW
Suite 800
Washington, DC 20006
(202) 887-4850; Fax: (202) 785-5377

GROCERY MANUFACTURERS OF AMERICA
1010 Wisconsin Avenue. NW
Suite 900
Washington, DC 20007-3694
(202) 337-9400; Fax: (202) 337-4508

INSTITUTE FOR LOCAL SELF-RELIANCE
2425 18th Street, NW
Washington, DC 20009
(202) 232-4108; Fax: (202) 332-0463

INSTITUTE OF PACKAGING PROFESSIONALS
481 Carlisle Drive
Herndon, VA 22070-4823
(703) 318-8970; Fax: (703) 318-0310

INSTITUTE OF SCRAP RECYCLING INDUSTRIES
1325 G Street, NW
Suite 1000
Washington, DC 20005
(202) 737-1770; Fax: (202) 626-0900

INTEGRATED WASTE SERVICES ASSOCIATION
1401 H Street, NW, Suite 220
Washington, DC 20005
(202) 467-6240; Fax: (202) 467-6225

INTERNATIONAL CITY/COUNTY MANAGEMENT ASSOCIATION
777 North Capitol Street, NE, Suite 500
Washington, DC 20036-2217
(202) 797-6839; Fax: (202) 797-5437

NATIONAL CONFERENCE OF STATE LEGISLATURES
1560 Broadway
Suite 700
Denver, CO 80202
(303) 830-2200; Fax: (303) 863-8003

NATIONAL GOVERNORS ASSOCIATION
444 North Capitol Street, NW
Suite 267
Washington, DC 20001-1512
(202) 624-5300; Fax: (202) 624-5313

NATIONAL LEAGUE OF CITIES
1301 Pennsylvania Avenue, NW
6th Floor
Washington, DC 20004
(202) 626-3 000; Fax: (202) 626-3043

NATIONAL RECYCLING COALITION
1727 King Street
Suite 105
Alexandria, VA 22314
(703) 683-9025; Fax: (703) 683-9026

NATIONAL SOFT DRINK ASSOCIATION
Solid Waste Management Department
1101 16th Street, NW
Washington, DC 20036-4877
(202) 463-6732; Fax: (202) 463-8178

NATIONAL SOLID WASTE MANAGEMENT ASSOCIATION
(see: Environmental Industries Association)

NATIONAL TIRE DEALERS & RETREADERS
ASSOCIATION
1250 "I" (Eye) Street, NW
Suite 400
Washington, DC 20005-3989
(202) 789-2300; Fax: (202) 682-3999

NATURAL RESOURCES DEFENSE COUNCIL
40 West 20th Street
New York, NY 10011
(212) 727-2700; Fax: (212) 727-1773
E-mail: nrdcinfo@nrdc.org

NORTH AMERICAN HAZARDOUS MATERIALS MANAGEMENT ASSOCIATION
Washington, DC 20002-4201
(202) 289-4262; Fax: (202) 962-3500

INTERNATIONAL SOLID WASTE ASSOCIATION
Bremerholm 1
DK-1069 Copenhagen K
DENMARK
011 +45+33+914491; Fax: 011+45+33+919188

KEEP AMERICA BEAUTIFUL, INCORPORATED
1010 Washington Boulevard
Stamford, CT 06901
(203) 323 8987, Fax: (203) 325-9199

LEAGUE OF WOMEN VOTERS
1730 M Street, NW
Washington, DC 20036
(202) 429-1965; Fax: (202) 429-0854

MUNICIPAL WASTE MANAGEMENT ASSOCIATION
(See: U.S. Conference of Mayors)

NATIONAL ASSOCIATION FOR PLASTIC CONTAINER RECOVERY
3770 Nations Bank Corporate Center
100 North Tryon Street
Charlotte, NC 28202
(704) 358-8882; Fax: (704) 358-8769

NATIONAL ASSOCIATION OF COUNTIES
440 First Street, NW
Washington, DC 20001
(202) 393-6226; Fax: (202) 393-2630

NATIONAL ASSOCIATION OF REGIONAL COUNCILS
1700 K Street, NW
Suite 1300
Washington, DC 20006
(202) 457-0710; Fax: (202) 296-9352

NATIONAL ASSOCIATION OF TOWNS & TOWNSHIPS
1522 K Street, NW
Suite 1010
Washington, DC 20005-1202
(202) 624-3550; Fax: (202) 289-7996

SCRAP TIRE MANAGEMENT COUNCIL
1400 K Street, NW
Suite 900
Washington, DC 20005
(202) 682- 4880. Fax: (202) 682-4854

SOLID WASTE ASSOCIATION OF NORTH AMERICA (SWANA)
1100 Wayne Avenue
Suite 700, P.O. Box 7219
Silver Spring, MD 20907-72 19
(301) 585-2898; Fax: (301) 589-7068

STEEL RECYCLING INSTITUTE
Foster Plaza 10
680 Anderson Drive
Pittsburgh, PA 15220-3213
(800) 876-7274; Fax: (412) 922-3213
(412) 922-2772

U.S. CONFERENCE OF MAYORS MUNICIPAL WASTE MANAGEMENT ASSOCIATION
1620 "I" (Eye) Street, NW
6th Floor
Washington, DC 20006
(202) 293-7330; Fax: (202) 293-2352

THE WASTE WATCH CENTER
16 Haverill Street
Andover, MA 01810
(508) 470-3044; Fax: (508) 470-3384

WORLD RESOURCE FOUNDATION
Bridge House, High Street,
Tunbridge, Kent TN9 1DP, UK
01 1+44+0+1732+368333; Fax:01 1+44+0+1732±368337

WORLD RESOURCES INSTITUTE
1709 New York Avenue, NW
Washington, DC 20006
(202) 638-6300; Fax: (202) 638-0036

CHECKLIST: EVALUATING RECYCLING COMPANY SERVICES

After determining what materials are available for recycling, the coordinator should contact reliable recycling companies that can provide full service to your business. The coordinator will benefit from using the services of the recycling company, especially from sharing ideas on the best type of recycling program for your individual company. When interviewing recycling companies ask the following questions:

1. Will the recycling company help you organize and promote your program?
2. Has the recycling company done this type of business program in the past? Will they give references?

3. Is the recycling company willing to sign a long-term contract?
4. Will the recycling company provide only "scheduled" or "on call" pickups? How much notice is required?
5. Can the recycling company provide the equipment you need to ensure the success of your program? Your needs may include baling equipment, containers for workstation separation, central collection and dock containers, or a storage trailer.
6. Will the recycling company assure payment for your recyclables? What are the terms?
7. Will the recycling company pick up your paper in the event that the value of your recyclables decline?
8. Will the recycling company mandate a minimum pickup weight requirement? Is there a pickup charge?
9. Will the recycling company also shred your confidential documents? Will they securely store them until their destruction date?

TABLE A.1 Conversion figures

	Weight, lb
Glass (average weights per unit)	
10 oz single-serving juice container	0.19
12 oz container	0.23
20 oz container	0.38
75-gal caddy filled with glass (uncrushed)	174
46-gal bin filled with glass (uncrushed)	107
40-yd roll-off filled with glass	24,000
1 yd^3 glass	600
Aluminum (average weights per unit)	
10.oz single-serving container	0.06
90-gal caddy filled with aluminum cans	23
46.gal bin filled with aluminum cans	12
40-yd roll-off filled with steel cans	8,000
1 yd^3 aluminum	60
Tinplate steel (average weights per unit)	
10 oz food or beverage can	0.20
12.oz food or beverage can	0.07
18 oz food or beverage can	0.11
75-gal caddy filled with steel cans	54
46-gal caddy filled with steel cans	33
40 yd roll-off filled with steel cans	8,000
1 yd^3 tinplate steel	200
Fine paper, newspaper, old corrugated (OCC) (average weights per unit)	
75 gal caddy filled with fine paper or newspaper	212
1 school newspaper	0.05
1 daily newspaper	0.7
1 yd^3 fine paper	500
40.yd container filled with flattened OCC	6,000
1 yd^3 flattened OCC	150
1 yd^3 newspaper	500

TABLE A.2 Conversion factors for recyclables

Material	Volume	Weight, lb
Newsprint, loose	1 yd^3	360–800
Newsprint, compacted	1 yd^3	720–1000
Newsprint	12-in stack	35
Corrugated cardboard, loose	1 yd^3	300
Corrugated cardboard, baled	1 yd^3	1000–1200
Glass, whole bottles	1 yd^3	600–1000
Glass, semi crushed	1 yd^3	1000–1800
Glass, crushed (mechanically)	1 yd^3	800–2700
Glass, whole bottles	One full grocery bag	16
Glass, uncrushed to manually broken	55 gal drum	125–500
PET soda bottles, whole, loose	1 yd^3	30–40
PET soda bottles, whole, loose	Gaylord*	40–53
PET soda bottles, baled	30″ × 48″ × 60″	500
PET soda bottles, granulated	Gaylord*	700–750
PET soda bottles, granulated	Semi.load	30,000
Film, baled	30″ × 42″ × 48″	1100
Film, baled	Semi-load	44,000
HPDE (dairy only), whole, loose	1 yd^3	24
HPDE (dairy only), baled	30″ × 48″ × 60″	500–800
HPDE (mixed), baled	30″ × 48″ × 60″	600–900
HPDE (mixed), granulated	Gaylord~	800–1000
HPDE (mixed), granulated	Semi-load	42,000
Mixed PET & dairy, whole, loose	1 yd^3	Average 32
Mixed PET, dairy and other rigid, whole, loose	1 yd^3	Average 38
Mixed rigid, no film or dairy, whole, loose	1 yd^3	Average 49
Mixed rigid, no film, granulated	Gaylord*	500–1000
Mixed rigid & film, densified by mixed plastic mold technology	1 yd^3	Average 60
Aluminum cans, whole	1 yd^3	50-74
Aluminum cans, whole	1 full kraft paper grocery bag	Average 1.5
Aluminum cans	55 gal plastic bag	13–20
Ferrous cans, whole	1 yd^3	150
Ferrous cans, flattened	1 yd^3	850
Leaves, uncompacted	1 yd^3	250–500
Leaves, compacted	1 yd^3	320–450
Leaves, vacuumed	1 yd^3	350
Wood chips	1 yd^3	500
Mulch	1 yd^3	200–300
Grass clippings	1 yd^3	400–1500
Used motor oil	1 gal	7
Tire, passenger car	1	12
Tire, truck	1	60
Food waste, solid and liquid fats	55 gal drum	412

*Gaylord size most commonly used 40″ × 48″ × 36″
Source: National Recycling Coalition, Washington, DC

TABLE A.3 Recycling audit worksheet

All waste bins inside and outside the building should be checked and the relevant data and comments recorded in the space provided below.

Bin #	Composition by material, %	Is it a typical composition? (Y/N)	Contamination* H,M,L†	source	Collection frequency	Appearance or size constraints for additional bins	Comments
	Glass ______ Metal ______ Organics ______ Fine paper ______ OCC ______ Plastics ______ Other ______						
	Glass ______ Metal ______ Organics ______ Fine paper ______ OCC ______ Plastics ______ Other ______						
	Glass ______ Metal ______ Organics ______ Fine paper ______ OCC ______ Plastics ______ Other ______						

*Contamination refers to the contamination of potential recyclables with other wastes that deem the product unmarketable, e.g., food grease on OCC.

†H = High; M = Medium; L = Low.

TABLE A.4 Recovery estimates worksheet

Company/Institution: ____

Contact: ______________________ Phone: ______________________________

Address: ____

Target material: ____

Recovery estimates can be obtained by using consumption data or waste generation data. The actual recovery will be below calculated consumption and generation levels. Losses of material due to contamination, discarding of recyclable materials off-site, etc., must be incorporated into the final recovery estimate.

Consumption Data

Amount consumed (provided by contact): ________________ lb/wk.

If units come in a variety of sizes, calculate the average size before using conversion figures.

____

____

If the data available are units/wk.; convert to lb/wk.

__________ units/wk. x __________ lb/unit* = __________ lb/wk.

**Use Conversion Figures Sheet, Table A.1.*

Estimated Recovery:

consumption (lb/wk) – _________ material loss recovery (lb/wk.) due to off-site disposal, staff participation level, contamination, etc. = _________ potential recovery (lb/wk.)

GeneratIon Data

Information gathered in waste audit can be used to calculate generation rates:

_________ refuse container volume in yd^3 × _________ % target material in container

_________ # of full containers/wk. = _________ yd^3 generated/wk.

Estimated Recovery

_________ consumption (lb/wk) – _________ material loss recovery (lb/wk.) due to off-site disposal, staff participation level, contamination, etc. = _________ potential recovery (lb/wk.)

CONVERSIONS (lb to yd^3)

Volume estimates may be required to help determine how quickly your truck will fill and what volume of storage containers the site will need.

Use Table A.1.

___________ estimated recovery (lb/wk.)/weight of 1 yd^3 of material (lb) = _______________ estimated recovery (yd^3/wk.)

TABLE A.5 Storage container data and evaluation

Container type	Volume capacity, yd^3	Material capacity, lb						Advantage/disadvantage
				Cans				
		News	Glass	Steel	Alum.	PET	Commingled	
32 gal garbage can Standard can, available from hardware stores Suitable for any manual collection system	0.18	N/A	108	29	11	5.4	49	A Easy to get Inexpensive D Must be manually unloaded Holds small amount
55 gal drum Used drums Suitable for any manual collection system	0.3	N/A	180	48	18	9	82	A Low or no cost Can be located inside or outside D Not mobile Must be manually unloaded
Semi-automated collection cart 90 gal wheeled plastic carts Serviced by rear-loading packer or recycling truck fitted wit hydraulic lift	0.5	250	300	80	30	15	138	A Holds large volume yet has narrow design to fit inside buildings Highly mobile Can be mechanically loaded D Somewhat expensive
Bulk lift container (steel) 2 or 3 yd^3 on casters Modify lids to only accept recyclables Serviced by front- or rear-end loader or Easy Mobile Container System	2	1,000	1,200	320	120	60	552	A Large capacity Portable Can be mechanically serviced D Somewhat expensive Must usually be located outside building
Igloo (fiberglass) 1.5 to 4 yd^3 Serviced by roll-off truck with compartments, fitted with hydraulic crane	1.5–4	1.5 yd^3 = 750 4 yd^3 = 2000	1.5 yd^3 = 900 4 yd^3 = 2,400	1.5 yd^3 = 240 4 yd^3 = 640	1.5 yd^3 = 90 4 yd^3 = 240	1.5 yd^3 = 45 4 yd^3 = 120	414 1,104	A Attractive Can be mechanically serviced D Expensive Small openings Locate outside, 36 ft^2 per igloo
Roll-off container Can divide into sections Need to cover open tops	10–60 common size is 40 yd^3	40 yd^3 = 20,000	40 yd^3 = 24.000	40 yd^3 = 6,400	40 yd^3 = 2,400	40 yd^3 = 1,200	11,040	A Very large capacity Use for storage and transport D Expensive Locate outside, about 100 ft^2

TABLE A.6 Monthly total weight of glass recyclables worksheet

	Package size	Single container weight, oz	Number of cases per month		Average case weight, lb		Weight
Beer							
Beer	12 oz	7		×	10.5	=	
Heavy bottle	12 oz	11		×	17.1	=	______
							Subtotal
Liquor	1.75	34		×	12.75	=	
Liter	18			×	13.5	=	
	750 ml	15.5		×	11.6	=	______
							Subtotal
House wine	4 liter	41		×	10.25	=	
	3 liter	36		×	9.0	=	
	1.5 liter	28		×	10.5	=	______
							Subtotal
Varietal wine	1.5 liter	28		×	10.5	=	
	750 ml	15.5		×	11.6	=	
	375 ml	13		×	19.5	=	
	187 ml	7.5		×	11.25	=	______
							Subtotal
Champagne	1.5 liter	40		×	15	=	
	1.0 liter	32		×	24	=	
	750 ml	23		×	17.25	=	
	187 ml	13		×	19.5	=	______
							Subtotal
Other							
Wine cooler	12 oz	7		×	10.5	=	
Bar mixes	10 oz	5.6		×	8.4	=	
Mineral water	11 oz	9		×	13.5	=	
Perrier water	6.5 oz	5.3		×	7.9	=	______
							Subtotal
Total weight of monthly glass containers used							Total

TABLE A.7 Weekly waste generation by occupied square foot

Generator segment	Waste production, lb per occupied ft^2 per week	lb/year
Office	0.05	2.6
Industrial	0.06	3.12
Transportation, communication, and utilities	0.10	5.2
Retail	0.22	11.44
Wholesale/warehouse and distribution (WWA)	0.06	3.12
Public and institutional (Public)	0.04	2.08

Source: Westchester County's Solid waste Management Plan, Malcolm Pirnie, Inc., White Plains, N.Y.

TABLE A.8 Commercial waste quantity and composition

Generator	Paper	Cardboard	Plastic	Metals	Other
Office	65%	15%	6%	2%	12%
Industrial	35%	20%	25%	6%	14%
Retail	35%	40%	8%	1%	16%
Transportation, communication, utilities	20%	15%	15%	5%	45%
Wholesale/warehouse and distribution	25%	32%	25%	7%	11%
Public	45%	10%	5%	6%	34%

Source: Westchester County's Solid Waste Management Plan. Malcolm Pirnie, Inc., White Plains, N.Y.

APPENDIX B
GLOSSARY

HERBERT F. LUND
Editor in Chief

BELLE LUND
Coconut Creek, Florida

A

abatement The reduction in landfill pollution by source reduction and waste recycling.

acid gas scrubber A device that removes particulate and gaseous impurities from a gas stream. This generally involves the spraying of an alkaline solid or liquid, and sometimes the use of condensation or absorbent particles.

acrylonitrile butadiene styrene A high-durability plastic-rubber blend. The acronym ABS is commonly used.

acute exposure Receipt of a large dose of a hazardous substance over a short period of time.

ADF Advanced disposal fee; a fee for disposing of a product which is included in the product's price.

aeration The process of exposing bulk material, such as compost, to air. *Forced aeration* refers to the use of blowers in compost piles.

aerobic A biochemical process or condition occurring in the presence of oxygen.

aerobic digestion The utilization of organic waste as a substrate for the growth of bacteria which function in the presence of oxygen to stabilize the waste and reduce its volume. The products of this decomposition are carbon dioxide, water, and a remainder consisting of inorganic compounds, undigested organic material, and water.

agricultural wastes Solid wastes of plant and animal origin, which result from the production and processing of farm or agricultural products, including manures, orchard and vineyard prunings, and crop residues, which are removed from the site of generation for solid waste management. Agricultural refers to SIC Codes 011 through 0291.

air emissions Solid particulates (such as unburned carbon) and gaseous pollutants (such as oxides of nitrogen or sulfur) or odors. These can result from a broad variety of activities including exhaust from vehicles, combustion devices, landfills, compost piles, street sweepings, excavation, demolition, etc.

air knife A blower device that employs an airstream to push selected material(s) off a conveyor.

air pollution The presence of unwanted material in the air in excess of standards. The term "unwanted material" here refers to material in sufficient concentrations, present for a sufficient time and under circumstances to interfere significantly with health, comfort, or welfare of persons, or with the full use and enjoyment of property.

air classification A process in which a stream of air is used to separate mixed material according to the size, density, and aerodynamic drag of the pieces.

algal bloom Population explosion of algae (simple one-celled or many-celled, usually aquatic, plants) in surface waters. Algal blooms are associated with nutrient-rich runoff from composting facilities or landfills.

aluminum can or container Any food or beverage container that is composed of at least 94 percent aluminum.

alternatives Other possible ways of dealing with, treating, or disposing of wastes.

amber cullet Broken brown glass containers.

anaerobic A biochemical process or condition occurring in the absence of oxygen.

anaerobic digestion The utilization of organic waste as a substrate for the growth of bacteria that function in the absence of oxygen to reduce the volume of waste. The bacteria consume the carbon in the waste as their energy source and convert it to gaseous products. Properly controlled, anaerobic digestion will produce a mixture of methane and carbon dioxide, with a sludge remainder consisting of inorganic compounds, undigested organic material and water.

animal bedding An agricultural product, occasionally made from waste paper, for use in livestock quarters.

animal and food processing wastes Waste materials generated in canneries, slaughterhouses, packing plants, or similar industries.

antiscavenge ordinance A governmental regulation prohibiting the unauthorized collection of secondary materials set out for pickup by a designated collector.

aquifer A geologic formation, group of formations, or part of a formation capable of yielding a significant amount of groundwater to wells, springs, or surface water.

asbestos Fibrous forms of various hydrated minerals, including chrysotile (fibrous serpentine), crocidolite (fibrous reinbecktite), amosite (fibrous cumingtonite-grunerite), fibrous tremolite, fibrous actinolite, and fibrous anthophyllite.

ash The residue that remains after a fuel or solid waste has been burned, consisting primarily of noncombustible materials. (See also bottom ash and fly ash.)

ash pit A pit or hopper located below or near a furnace where residue is accumulated and from which it is removed.

ASTM American Society for Testing and Materials.

avoided costs Solid waste management cost savings resulting from a recycling program. One cost saving can be avoided disposal fees. Another avoided cost can be the saving in garbage collection costs through rerouting and extended truck life.

auto-tie A mechanical device that automatically wraps a bale with wire. "away" An unknown place where people "throw" things and expect never to deal with them again; in reality there is no such place.

B

back-end materials recovery Secondary materials recovery from incinerated municipal solid waste.

backyard composting The controlled biodegradation of leaves, grass clippings, and/or other yard wastes on the site where they were generated.

baghouse A municipal waste combustion facility air emission control device consisting of a series of fabric filters through which MWC (municipal waste combustion) flue gases are passed to remove particulates prior to atmospheric dispersion.

bale A densified and bound cube of recyclable material, such as waste paper, scrap metal, or rags.

baler A machine used to compress recyclables into bundles to reduce volume. Balers are often used on newspaper, plastics, and corrugated cardboard.

ballistic separator A device used in some composting operations that separates inorganic materials from organic matter.

base load A continuous stream or electrical output over a given period of time at design conditions.

baseline recycling systems Systems that identify whether a new recycling collection program is diverting material that otherwise would have been disposed or whether that material is being diverted from another recycling collection system. Example: Municipal source separation and curbside collection programs are to be introduced in an area with active private sector recycling.

beneficiation In recycling, the mechanical process of removing contaminants and cleaning scrap glass containers. Originally a mining industry term for the treatment of a material to improve its form or properties, such as the crushing of ore to remove impurities.

beverage industry recycling program A state coalition of beverage producers, packagers, wholesalers, and retailers that undertake activities in support of recycling, particularly buy-back centers.

biodegradable A substance or material that can be broken down into simpler compounds by microorganisms and other decomposers such as fungi.

biodegradable material Waste material that is capable of being broken down by microorganisms into simple, stable compounds such as carbon dioxide and water. Most organic wastes, such as food wastes and paper, are biodegradable.

bioreactor system An engineered system designed to rapidly stabilize the decomposable organic waste constituents in a municipal solid waste landfill by controlling the biological processes for maximum waste reduction. The ultimate advantage of the bioreactor system is extending the useful life of the landfill.

biosolids Municipal sewage sludge that is a primarily organic semisolid product resulting from the wastewater treatment process that can be beneficially recycled.

block-leader promotion The use of volunteers to promote recycling collection service in a specific block or neighborhood.

bimetal can or container Any metal container composed of at least two different types of metals, such as a steel container with an aluminum top.

bogus corrugating medium The fluted middle of corrugated containers made entirely from waste paper.

biomass Any organic (wood, agricultural, or vegetative) matter; key components are carbon and oxygen.

bottle bank A mobile, divided bin used for receiving, storing, and transporting glass containers for recycling.

bottle bill Legislation requiring deposits on beverage containers; appropriately called Beverage Container Deposit Law (BCDL).

bottom ash The nonairborne combustion residue from burning fuel in a boiler. The material falls to the bottom of the boiler and is removed mechanically. Bottom ash constitutes the major portion (about 90 percent) of the total ash created by the combustion of solid waste.

broker An individual or group of individuals that act as an agent or intermediary between the sellers and buyers of recyclable materials.

brownfield Property that is no longer used for its original purpose and may be contaminated. Examples would be old gas stations or abandoned factories.

Btu (British thermal unit) Unit of measure for the amount of energy a given material contains (e.g., energy released as heat during combustion is measured in Btus). Technically, 1 Btu is the quantity of heat required to raise the temperature of one pound of water one degree Fahrenheit.

buffer zone Neutral area that acts as a protective barrier separating two conflicting forces. An area that acts to minimize the impact of pollutants on the environment or public welfare. For example, a buffer zone is established between a composting facility and neighboring residents to minimize odor problems.

bulk-cullet box A pallet-sized reusable corrugated container used to ship cullet. See also gaylord container.

bulking agent A material used to add volume to another material to make it more porous to airflow. For example, municipal solid waste may act as a bulking agent when mixed with water treatment sludge.

bulky waste Large items of refuse including, but not limited to, appliances, furniture, large auto parts, nonhazardous construction and demolition materials, trees, branches, and stumps that cannot be handled by normal solid waste processing, collection, and disposal methods.

buy-back recycling center A facility that pays a fee for the delivery and transfer of ownership to the facility of source-separated materials for the purpose of recycling or composting.

bypass waste Solid waste that has been contractually committed to a facility, but which is diverted when the facility is unavailable or is not able to process due to size, etc.

C

capital costs Those direct costs incurred in order to acquire real property assets such as land, buildings, and building additions; site improvements; machinery; and equipment.

capture rate A standard reporting practice of all materials collected or captured as related to the total available designated materials. The rate should include collected deposit containers if part of the recycling program. Capture rate equals designated materials recovered divided by the total designated materials available.

carcass The foundation structure of a tire, including sidewalls, bead, and cord.

CERCLA Comprehensive Environmental Response, Compensation, and Liability Act (Superfund Act), 1980. This act provides funds for emergency cleanup of spills and cleanup of abandoned or inactive hazardous waste sites.

centralized yard waste composting System utilizing a central facility within a politically defined area with the purpose of composting yard wastes.

chain-flail crusher A simple, low-volume glass container crusher using a motor-driven chain.

charcoal A dark or black porous carbon prepared from vegetable or animal substances (as from wood by charring in a kiln from which air is excluded).

chronic exposure Receipt of a small dose of a hazardous substance over a long period of time.

claw truck A specially designed attachment to a front-end loader used to pick up loose yard waste at the curb.

Clean Air Act Act passed by Congress to have the air "safe enough to protect the publics health" by May 31, 1975. Required the setting of National Ambient Air Quality Standards (NAAQS) for major primary air pollutants.

Clean Water Act Act passed by Congress to protect the nation's water resources. Requires EPA to establish a system of national effluent standards for major water pollutants, requires all municipalities to use secondary sewage treatment by 1988, sets interim goals of making all U.S. waters safe for fishing and swimming, allows point-source discharges of pollutants into waterways only with a permit from EPA, requires all industries to use the best practicable technology (BPT) for control of conventional and nonconventional pollutants and to use the best available technology (BAT) that is reasonable or affordable.

closed-loop recycling The process in which an item is recycled back into the same product (as old aluminum cans are made into new cans).

co-collection The collection of ordinary household garbage in combination with special bags of source-separated recyclables.

co-composting Simultaneous composting of two or more diverse waste streams.

coding In the context of solid waste, coding refers to a system to identify recyclable materials. The coding system for plastic packaging utilizes a three-sided arrow with a number in the center and letters underneath. The number and letters indicate the resin from which each container is made: 1 = PETE (polyethylene terephthalate), 2 = HDPE (high density polyethylene), 3 = V (vinyl), 4 = LDPE (low-density polyethylene), 5 = PP (polypropylene), 6 = PS (polystyrene), and 7 = other/mixed plastics. Noncoded containers are recycled through mixed plastics processes. To help recycling sorters, the code is molded into the bottom of bottles with a capacity of 16 oz or more and other containers with a capacity of 8 oz or more.

codisposal Burning of municipal solid waste with other material, particularly dewatered sewage sludge.

cofiring/burning Municipal solid waste in a combustion unit along with other fuel, especially coal.

cogeneration Production of two forms of energy from one source.

collection The act of picking up and moving solid waste from its location of generation to a disposal area, such as a transfer station, resource recovery facility, or landfill.

combustible Various materials in the waste stream that are burnable, such as paper, plastic, law clippings, leaves, and other organic materials.

combustion is a municipal solid waste (MSW) practice that helps reduce the amount of needed landfill space by reducing the waste volume.

commercial sector One of the four sectors of the community that generates garbage. Designed for profit.

commercial solid waste Solid waste originating from stores; business offices; commercial warehouses; hospitals, educational, health care, military, and correctional institutions, nonprofit research organizations; and government offices. Commercial solid waste refers to SIC Codes 401 through 4939, 4961, and 4971 (transportation, communications, and *certain* utilities), 501 through 5999 (wholesale and retail trade), 601 through 6799 (finance, insurance, and real estate), 701 through 8748 (public and private service industries such as hospitals and hotels), and 911 through 9721 (public administration). *Commercial solid wastes do not include construction and demolition waste.*

commercial unit A site zoned for a commercial business that generates commercial solid wastes.

commercial waste Waste materials originating in wholesale, retail, institutional, or service establishments such as office buildings, stores, markets, theaters, hotels, and warehouses.

commingled recyclables A mixture of several recyclable materials into one container.

compacting drop box A roll-off box attached to a compacting device for receiving and compressing a secondary material, such as old corrugated containers.

compactor Power-driven device used to compress materials to a smaller volume.

composite liner A liner composed of both a plastic and soil component.

composition A set of identified solid waste materials, categorized into waste categories and waste types.

compost A humuslike relatively stable material resulting from the biological decomposition or breakdown of organic materials.

compost substrate Organic biodegradable material that can be used as a feedstock for a composting process.

composting The controlled biological decomposition of organic solid waste under aerobic conditions. Organic wastes such as food scraps and yard trimmings interact with microorganisms (mainly bacteria and fungi) to produce a humus-like substance.

composting facility A permitted solid waste facility at which composting is conducted that produces a product meeting the definition of "compost."

concentration The amount of one substance contained in a unit of another substance.

conservation The planned management of a natural resource to prevent exploitation, destruction, or neglect.

construction and demolition (C&D) waste Solid wastes, such as building materials and packaging and rubble resulting from construction, remodeling, repair, and demolition operations on pavements, houses, commercial buildings, and other structures. Construction refers to SIC Codes 152 through 1794, 1796, and 1799. Demolition refers to SIC Code 1795.

consumption The amount of any resource (material or energy) used in a given time.

container A receptacle or a flexible covering for the shipment of goods.

container cullet Broken scrap glass bottles and jars. See also *cullet.*

container deposit legislation Laws that require monetary deposits to be levied on beverage containers. The money is returned to the consumer when the containers are returned to the retailer. Also called "bottle bills."

contaminant Anything that becomes mixed with a recyclable commodity that prevents the end user from using the commodity. Examples would be food waste in a glass jar or phone books mixed in with newspapers.

corrosive Defined for regulatory purposes as a substance having a pH level below 2 or above 12.5, or a substance capable of dissolving or breaking down other substances, particularly metals, or causing skin burns.

corrugated container According to SIC Code 2653, a paperboard container fabricated from two layers of kraft linerboard sandwiched around a corrugating medium. Kraft linerboard means paperboard made from wood pulp produced by a modified sulfate pulping process, with basis weight ranging from 18 to 200 lb, manufactured for use as facing material for corrugated or solid fiber containers. Linerboard also may mean that material that is made from reclaimed paper stock.

corrugating medium Fluted paperboard used in making corrugated boxes. Paperboard made from chemical or semichemical wood pulps, straw, or reclaimed paper stock, and folded to form permanent corrugations.

corrugated paper Paper or cardboard manufactured in a series of wrinkles or folds, or into alternating ridges and grooves.

cost-effective A measurement of cost compared to an unvalued output (e.g., the cost per ton of solid waste collected) such that the lower the cost, the more cost-effective the action.

counts Population and household see population and household counts.

crusher A mechanical device used to break secondary materials such as glass bottles into smaller pieces.

cryogenic processing The freezing and cracking of secondary materials to assist in separation.

cullet Broken or waste glass used in the manufacture of new glass.

cultivation In composting, accelerating the decomposition of biodegradable wastes by turning, watering, aerating, loosening, and/or inoculating the waste with microorganisms or fertilizer to lower the carbon/nitrogen ratio.

curbside collection Collection of recyclable materials at the curb, often from special containers, to be brought to various processing facilities.

curbside recycling program Refers to a program that sponsors scheduled pick-up of recyclable items from household curbs.

cycle A periodically repeated sequence of events.

curbside-separate To separate commingled recyclables prior to placement in individual compartments in a truck providing curbside collection service; this task is performed by the collector.

D

decompose To separate into constituent parts or elements or into simpler compounds; to undergo chemical breakdown; to decay or rot as a result of microbial and fungal action.

decomposition The act of undergoing breakdown into constituent parts. Breaking down into component parts or basic elements. degradable Capable of being broken down into smaller components by chemical, physical, or biological means.

degradability Ability of materials to break down, by bacterial (biodegradable) or ultraviolet (photodegradable) action.

degradable plastics Plastics specifically developed for special products that are formulated to break down after exposure to sunlight or microbes. By law, six-pack rings are degradable; however, they gradually degrade, causing litter and posing a hazard to birds and marine animals.

degradation (Also biodegradation) A natural process that involves assimilation or consumption of a material by living organisms.

deink The removal of ink, filler, and other nonfibrous material from printed waste paper.

delacquer Process used to remove lacquer from scrap metals, such as aluminum cans.

demurrage (1) The detention of a truck or railroad car for the loading or unloading of secondary materials. (2) The compensation paid to the shipper for detaining the truck or railroad car.

densified refuse-derived fuel (d-RDF) Refuse-derived fuel that has been compressed or compacted through such processes as pelletizing, briquetting or extruding, causing improvements in certain handling or burning characteristics.

densifier A machine developed in the 1980s to compress used aluminum cans into a small, dense brick.

Department of Environmental Regulation (DER) In some states DER is the agency charged with the enforcement of environmental and recycling laws.

deposit Matter deposited by a natural process; a natural accumulation of iron ore, coal; money paid as security.

designated materials Each material designated for collection and described in the same manner that the materials are specified for separation. For example, when citing glass, specify whether it is color-separated or mixed; for plastics, whether soda bottles, or PET, or whether yard wastes are collected separately.

detinner A company that buys steel cans and each tin mill products, and removes the tin through any of several processes, selling the detinned steel to steel mills and foundries, and the recovered tin to its appropriate markets.

detinning Removing tin from "tin" cans by a chemical process to make both the tin and steel more easily recycled.

devulcanization The processing of scrap tires by use of a thermochemical reaction.

discard rate A numerical figure describing the average pounds of solid waste generated per person each day. The average discard rate during the late 1990's reported by the U.S. Environmental Protection Agency was over 3 pounds.

dioxin The generic name for a group of organic chemical compounds formally known as polychlorinated dibenzo-p-dioxins. Heterocyclic hydrocarbons that occur as toxic impurities, especially in herbicides.

direct energy Vigorous exertion of power from the source without interruption.

"dirty" MRF See **full MRF, waste recovery facility (WRF), mixed-waste processing facility.**

disposable Something that is designed to be used once and then thrown away.

disposable The process of solid waste management through landfilling or transformation at permitted solid waste facilities, or other repository intended for permanent containment of waste.

disposal capacity The capacity, expressed in either weight in tons or its volumetric equivalent in cubic yards, which is either currently available at a permitted solid waste landfill or will be needed for the disposal of solid waste generated within the jurisdiction over a specified period of time.

disposal cost savings Savings of reduced waste-hauling requirements, avoided tipping fees, and other operational cost savings related to waste disposal because of the operation of a recycling program.

disposal facility A collection of equipment and associated land area that serves to receive waste and dispose of it. The facility may incorporate one or disposal methods.

disposal index Materials disposed per capita at time of measurement divided by materials disposed per capita at reference time. A per capita disposal index measures the difference in quantities of materials disposed with reference to a base period.

disposal obligation The obligation of the county to provide for the disposal of all solid waste generated in each contract community and in the unincorporated county and delivered to a resource recovery system disposal facility or transfer station designated pursuant to the plan of operations.

disposal surcharge A special fee levied against waste disposal volumes.

diversion alternative Any activity, existing or occurring in the future, which has been, is, or will be implemented by a jurisdiction that could result in or promote the diversion of solid waste through source reduction, recycling, or composting, from solid waste landfills and transformation facilities.

diversion rate The amount of material recovered divided by the amount of material recovered and material disposed. A measure of the amount of waste material being diverted for recycling compared with the total amount that was previously thrown away. It describes quantities diverted from land filling, incineration, or by exporting to another disposal site.

door hanger A printed card, distributed to households and hung on a door, promoting a recycling service.

DOT Department of Transportation.

Downstroke baler A baling device in which the compression ram and platten move down vertically in the chamber.

drained whole batteries A scrap metal grade consisting of lead-acid batteries free of liquid and extraneous materials.

drop-off center A method of collecting recyclable or compostable materials in which the materials are taken by individuals to collection sites, or centers, and deposited into designated containers.

dump A site where mixed wastes are indiscriminately deposited without controls or regard to the protection of the environment: now illegal.

durability The ability of a product to be used without significant deterioration for its intended purpose for a period greater than the mean useful product lifespan of similar products.

E

ecosystem A system made up of a community of living things and the physical and chemical environment with which they interact.

eddy-current separation An electromagnetic technique for separating aluminum from a mixture of materials.

effluent The liquid leaving wastewater treatment systems.

embedded energy The sum of all the energy involved in product development, transportation, use, and disposal.

electrostatic precipitator (ESP) A gas-cleaning device that collects entrained particulates by placing an electrical charge on them and attracting them onto opposite-charged collecting electrodes. They are installed in the back end of the incineration process to reduce air emissions.

emission Discharge of a gas into atmospheric circulation.

emission standard A rule or measurement established to regulate or control the amount of a given pollutant that may be discharged into the atmosphere.

emissions The solid, liquid, and gaseous substances exhausted to the environment.

endanger To expose to danger or put in a position of peril.

end market or end use The use or uses of a diverted material or product that has been returned to the economic mainstream, whether or not this return is through sale of the material or product. The material or product can have a value that is less than the solid waste disposal cost.

energy Ability to do work by moving matter or by causing a transfer of heat between two objects at different temperatures.

energy recovery The conversion of solid waste into energy or a marketable fuel. A form of resource recovery in which the organic fraction of waste is converted to some form of usable energy, such as burning processed or raw refuse, to produce steam.

enterprise fund A fund for a specific purpose that is self-supporting from the revenue it generates.

entanglement Process whereby animals get caught and die in nets or other materials that have been discarded.

environment The external conditions of an organism or population; the term "the environment" generally refers to the sum total of conditions—physical and biological—in which organisms live.

environmental impact statement (EIS) A document prepared by EPA or under EPA guidance (generally a consultant hired by the applicant and supervised by EPA) that identifies and analyzes in detail the environmental impacts of a proposed action. Individual states also may prepare and issue an ETS as regulated by state law. Such state documents may be called environmental impact reports (EIR).

environmental quality The overall health of an environment determined by comparison to a set of standards.

enviroshopping The act of purchasing merchandise with consideration for the environment. An awareness of how products impact the Earth's environment and natural resources and how it changes a consumer's choice in purchasing.

EPA U.S. Environmental Protection Agency; the federal agency charged with the enforcement of all federal regulations having to do with air and water pollution, radiation and pesticide hazard, ecological research, and solid waste disposal.

export density The preferred density of a processed secondary material destined for shipment to another country.

external costs Of, relating to, or connected with outside expenses.

F

facility operator Full-service contractors or other operators of a part of a resource recovery system.

feasible A specified program, method, or other activity can, on the basis of cost, technical requirements and time frame for accomplishment, be undertaken to achieve the objectives and tasks identified by a jurisdiction in a countywide integrated waste management plan.

feasibility analysis A detailed investigation and report to determine whether a particular project is suitable, reasonable to pursue, and capable of being successfully completed.

fee Dollar amount charged by a community to pay for services; see **tipping fee**.

ferrous Pertaining to, or derived from, iron. (In resource recovery, often used to refer to materials that can be removed from the waste stream by magnetic separation.)

ferrous metals Any iron or steel scrap that has an iron content sufficient for magnetic separation.

ferrous scrap dealer A business that acquires used commercial and consumer steel products and processes them to be recycled in steel mills.

fiber The threadlike particles that make up a sheet of paper. The sources of paper fiber are diverse and include recycled paper, trees, rice, cotton, and linen.

fine A penalty in a dollar amount finite having limits or being limited; not endless in quantity or duration.

flint glass Clear or uncolored glass.

fluffer A device used to fluff waste paper in order to improve baling effectiveness.

fly ash All solids including ash, charred papers, cinders, dusty soot, or other matter that rise with the hot gases from combustion rather than falling with the bottom ash. Fly ash is a minor portion (about 10 percent) of the total ash produced from combustion of solid waste, is suspended in the flue gas after combustion, and is removed by pollution control equipment.

flow control A legal or economic means by which waste is directed to particular destinations. For example, an ordinance requiring that certain wastes be sent to a combustion facility is waste flow control.

food waste All animal and vegetable solid wastes generated by food facilities, as defined in California Health and Safety Code section 27521, or from residences, that result from the storage, preparation, cooking, or handling of food.

forced deposits A term for container deposit legislation used by opponents of such measures. See *container deposit legislations*.

foreign cullet A glass industry term for cullet supplied to a glass producer from an outside source.

front-end loader (1) A solid waste collection truck that has a power-driven loading mechanism at the front; (2) a vehicle with a power-driven scoop or bucket at the front, used to load secondary materials into processing equipment or shipping containers.

front-end recovery The salvage of reusable materials, most often the inorganic fraction of solid waste, prior to the processing or combusting of the organic fraction. Some processes for front-end recovery are grinding, shredding, magnetic separation, screening, and hand sorting.

full material recovery facility (MRF) A process for removing recyclables and creating a compostlike product from the total of full mixed municipal solid waste (MSW) stream. Differs from a "clean" MRF that processes only commingled recyclables (see *WRF, "dirty" MRF*).

furnace An enclosed refractory or waterwall structure where the preheating, drying, igniting, and burning take place.

G

garbage Solid waste consisting of putrescible animal and vegetable waste materials resulting from the handling, preparation, cooking, and consumption of food, including waste materials from markets, storage facilities, handling and sale of produce, and other food products. Generally defined as wet food waste but not synonymous with "trash," "refuse," "rubbish," or solid waste.

gas control system A system at a landfill designed to prevent explosion and fires due to the accumulation of methane concentrations and damage to vegetation on final cover of closed portions of a landfill or vegetation beyond the perimeter of the property on which the landfill is located and to prevent objectionable odors off-site.

gaseous emissions Waste gases released into the atmosphere as a by-product of combustion.

gas scrubber A device where a caustic solution is contacted with exhaust gases to neutralize certain combustion products, primarily sulfur oxides (SO_x) and secondary chlorine (Cl).

gaylord container A large reusable corrugated container used for shipping materials (dimensions approximately 40 by 48 by 37 in).

generation rate Total tons diverted, recovered, and disposed within reference time divided by the population. The annual per capita generation rate is the total tons generated in 1 year divided by the population of residents.

generator Any person, by site or location, whose act or process produces a solid waste; the initial discarder of a material.

glass An inorganic substance consisting of a mixture of silicates.

glassmaking The process of making glass from raw materials (lime, sand, soda, and cullet).

Government Refuse Collection and Disposal Association A Silver Spring, Maryland, organization representing municipalities that collect and/or dispose of solid waste. Also known as GRCDA. (Name changed to SWANA)

government sector One of the four sectors of the community that generates garbage. The administration of the public policy and affairs of an area.

grab sample A single sample of a secondary material taken at no set time for evaluation or testing.

grade A classification of recycled products that separates them by composition, previous use, or source. There are four main grades of recovered paper: corrugated/Kraft paper, newspaper, high-grade papers, and mixed papers.

grains per cubic foot A measure of airborne particulates or dust expressed in weight (grains) per unit of gas (1 ft^3). One pound equals 7000 grains.

granulator A mechanical device that produces small plastic particles.

grapple A type of crane bucket having more than two teeth.

grate A device used to support the solid fuel or solid waste in a furnace during drying, ignition, or combustion. Openings are provided for passage of combustion air.

gravel Loose rounded fragments of rock.

gravity separation The separation of mixed materials based on the differences of material size and specific gravity.

green waste Usually refers to a combination of large brush, stumps, and yard waste; in some cases, may include some food waste.

gross national product (GNP) The total market value of all the goods and services produced by a nation during a specified time period.

groundcover Material used to cover the soil surface to control erosion and leaching, shade the ground, and offer protection from excessive heaving and freezing. Some ground covers are produced from yard waste compost.

groundwater Water beneath the earth's surface that fills underground pockets (known as aquifers) and moves between soil particles and rock, supplying wells and springs.

growth rate Estimation of progressive development; the rate at which a population or anything else grows.

H

habitat Place or type of place where an organism or community of organisms lives and thrives; contains food, water, shelter, and space.

hammermill shredder A broad group of machines that crush, chip, or grind materials. Hammermill shredders typically employ high-speed rotating equipment with fixed or pivoting hammers on a horizontal or vertical shaft.

hammermill A type of crusher used to break up waste materials into smaller pieces or particles, which operates by using rotating and flailing heavy hammers.

haul distance The distance a collection vehicle travels from its last pickup stop to the solid waste transfer station, processing facility, or sanitary landfill.

haulers Those persons, firms, or corporations or governmental agencies responsible (under either oral or written contract, or otherwise) for the collection of solid waste within the geographic boundaries of the contract community(ies) or the unincorporated county and the transportation and delivery of such solid waste to the resource recovery system as directed in the plan of operations.

hazard Having one or more of the characteristics that cause a substance or combination of substances to qualify as a hazardous material, as defined by section 66084 of Title 22 of the California Code of Regulations.

hazardous material Chemical or product that poses a significant threat to human health and/or the environment while being transported.

hazardous substance Chemical that is dangerous to human health and/or the environment while being stored or used.

hazardous waste Waste that because of its quantity, concentration, or physical, chemical, or infectious characteristics may pose a substantial present or potential hazard to human health or the environment when improperly treated, stored, transported, disposed of, or otherwise managed.

HDPE (high-density polyethylene) A recyclable plastic, used for items such as milk containers, detergent containers, and base cups of plastic soft drink bottles.

heavy-media separation The use of a fluid medium to separate materials. The fluid's density lies between the heavy and light fractions being separated.

heavy metals Hazardous elements including cadmium, mercury, and lead that may be found in the waste stream as part of discarded items such as batteries, lighting fixtures, colorants, and inks.

high-grade paper Relatively valuable types of paper such as computer printout, white ledger, and tab cards. Also used to refer to industrial trimmings at paper mills that are recycled.

high-grade waste paper Waste paper with the most value, consisting of the pulp substitute and deinking high-grade categories.

home sector One of the four sectors of the community that generates garbage. As it refers to environment; a place of origin.

horizontal baler A baling device in which the ram and platten move horizontally in the chamber.

household hazardous waste Those wastes resulting from products purchased by the general public for household use that, because of their quantity, concentration, or physical, chemical, or infectious characteristics, may pose a substantial known or potential hazard to human health or the environment when improperly treated, disposed, or otherwise managed. (See Chapter 21.)

household hazardous waste collection A program activity in which household hazardous wastes are brought to a designated collection point where the household hazardous wastes are separated for temporary storage and ultimate recycling, treatment, or disposal.

humus The organic portion of soil providing nutrition for plant life: a dark substance resulting from the partial decay of plant and/or animal matter.

Hydrapulper Trade name for a pulp mill machine that uses a rotor and blades to mix dry fibers, such as waste paper, and water to produce a pulp slurry.

hydrogeology The study of surface and subsurface water.

hydro-seeding A process in which grass seeds are mixed with fertilizer, green dye, a mulch product, and water, then sprayed onto a bare dirt area. The mulch will hold the seeds in place until they sprout and helps the soil hold the water, while the fertilizer helps the seeds grow. The mulch portion of such a mixture is usually made from some type of recycled paper.

I

igloo A half-sphere container used at drop-off centers for the receipt and storage of residential recyclable materials, such as glass and metal containers.

ignitable A substance that is capable of burning rapidly and has a flash point less than 140°F.

illegal dumping Disposing of waste in an improper manner and/or location and in violation of waste disposal laws.

impact An effect on the environment or on living things.

impermeable Restricts the movement of products through the surface.

implementation The accomplishment of the program tasks as identified in each component.

incidental catch Accidental capture and drowning of nontarget animals, such as porpoises, in fishing nets.

incineration An engineered process involving burning or combustion to thermally degrade waste materials. Incinerators must meet clean air standards. This process is used particularly for organic wastes. The wastes are reduced by oxidation and will normally sustain combustion without the use of additional fuel.

incinerator A facility designed for the controlled burning of waste; reduces waste volume by converting waste into gases and relatively small amounts of ash; may offer potential for energy recovery.

incinerator ash The remnants of solid waste after combustion, including noncombustibles (e.g., metals) and soot.

industrial solid waste Solid waste originating from mechanized manufacturing facilities, factories, refineries, construction and demolition projects, and publicly operated treatment works, and/or solid wastes placed in debris boxes.

industrial unit A site zoned for an industrial business that generates industrial solid wastes.

industrial waste Materials discarded from industrial operations or derived from industrial operations or manufacturing processes, all nonhazardous solid wastes other than residential, commercial, and institutional. May also include small quantities of waste generated from cafeterias, offices, or retail sales departments on same premises. Industrial waste includes all wastes generated by activities such as demolition and construction, manufacturing, agricultural operations, wholesale trade, and mining.

inert solids or inert waste A nonliquid solid waste including, but not limited to, soil and concrete, that does not contain hazardous waste or soluble pollutants at concentrations in excess of water-quality objectives established by a regional water board pursuant to Division 7 (commencing with section 13000) of the California Water Code and does not contain significant quantities of decomposable solid waste.

infectious waste Waste containing pathogens or biologically active material that because of its type, concentration, or quantity is capable of transmitting disease to persons exposed to the waste.

informed decision A conclusion or course of action based on facts.

infrastructure A substructure or underlying foundation: those facilities upon which a system or society depends; for example: roads, schools, power plants, communication networks, and transportation systems.

ingestion Eating; swallowing.

inorganic Not composed of once-living material (e.g., minerals); generally, composed of chemical compounds not principally based on the element carbon.

inorganic refuse Noncombustible waste material made from substances composed of matter other than plant, animal, or certain chemical compounds of carbon. Examples are metals and glass.

inorganic waste Waste composed of matter other than plant or animal (i.e., contains no carbon).

in-plant waste Waste generated in manufacturing processes. Such might be recovered through internal recycling, energy recovery, and/or through a salvage dealer. Also referred to as preconsumer waste.

institutional waste Waste materials originating in schools, jails, hospitals, nursing homes, research institutions, and public buildings. The materials include packaging materials, food wastes, and disposable products.

integrated solid waste management A practice of using several alternative waste management techniques to manage and dispose of specific components of the municipal solid waste stream. Waste management alternatives include source reduction, recycling, composting, energy recovery, and landfilling.

integrated waste management A solid waste management strategy that ranks the preferred alternatives in the following order: source reduction and reuse, recycling, resource recovery, and landfill disposal.

intensive recycling A concept promoted by opponents of waste-to-energy systems, whereby municipal recycling efforts target all recyclables in the waste stream.

intermediate processing center (IPC) Usually refers to a facility that processes residentially collected mixed recyclables into new products for market; often used interchangeably with materials recovery facility (MRF). A facility where recyclables that have been separated from the rest of the waste are brought to be separated and prepared for market (crushed, baled, etc.). An IPC can be designed to handle commingled or separated recyclables or both.

intermodal shipping The linking of two forms of transportation, such as trucks and railroads, to ship materials. For example, one might use intermodal shipping by loading secondary materials in a truck trailer, having it trucked to a railroad yard, putting the trailer on a railcar, moving the trailer by rail and unloading it for truck delivery to the receiving mill.

internal costs Expenses of, relating to, or occurring within the confines of an organized structure.

in-vessel composting A composting method in which the compost is continuously and mechanically mixed and aerated in a large, contained area.

investment tax credit A reduction in taxes permitted for the purchase and installation of specific types of equipment and other investments.

J

jurisdiction The city or county responsible for preparing any one or all of the following: the countywide integrated waste management plan, or the countywide siting element.

L

landfill A large, outdoor area for waste disposal; in sanitary landfills, waste is layered and covered with soil. Landfills usually have liner systems and other safeguards to prevent groundwater contamination.

landfill by-products Chemicals and gases that result from the biodegradation of waste in a landfill or interaction with rain and environmental conditions. Two byproducts that must be monitored are leachates and methane gas.

landfill liner Impermeable layers of heavy plastic, clay, and gravel that protect against groundwater contamination. Most sanitary landfills have at least two plastic liners or layers of plastic and clay.

large-quantity generator Sources such as industries and agriculture that generate more than 1000 kg of hazardous waste per month.

leachate Liquid that has percolated through solid waste or another medium and has extracted, dissolved, or suspended materials from it, which may include potentially harmful materials. Leachate collection and treatment is of primary concern at municipal waste landfills.

limited Restricted in number or supply, such as limited natural resources.

limited supply Restricted amount of a product or resource available at a given time.

liner A continuous layer of low-permeability natural or synthetic materials beneath or on the sides of a landfill or landfill trench that controls the downward or lateral escape of leachate.

liner A layer of natural clay or manufactured material (various plastics) that serves as a barrier to prevent leachate from reaching or mixing with groundwater in landfills, lagoons, etc.

litter Highly visible solid waste discarded outside the established collection disposal system. (Solid waste properly placed in containers is often referred to as trash and garbage; uncontainerized, it is referred to as litter.) Litter accounts for about 2 percent of municipal solid waste.

logger A mechanical device used to flatten scrap metal such as white goods. Many loggers are mobile and are taken periodically to disposal sites to process collected scrap metal.

long-term impact Future effect of an action, such as an oil spill.

low-grade paper Less valuable types of paper such as mixed office paper, corrugated paperboard, and newspaper.

LULU Locally unwanted land use: for example, jails, airports, and landfills.

M

magnet A body having the property of attracting iron and producing a magnetic field external to itself.

magnetic separator Equipment usually consisting of a belt, drum, or pulley with a permanent or electromagnet and used to attract and remove magnetic materials from other materials.

magnet separation A system to remove ferrous metals from other materials in a mixed municipal waste stream. Magnets are used to attract the ferrous metals.

mandatory recycling Programs that by law require consumers to separate trash so that some or all recyclable materials are not burned or dumped in landfills.

manmade Made by people rather than that which occurs naturally.

manual separation The separation of recyclable or compostible materials from waste by hand sorting.

manufactured (materials) Substances no longer in their natural or original state; products of a manufacturing process.

manufacturing sector One of the four sectors of the community that generates garbage. Responsible for making, developing consumer goods.

marine wastes Solid wastes generated from marine vessels and ocean work platforms, solid wastes washed onto ocean beaches, and litter discarded on ocean beaches.

market development A method of increasing the demand for recovered materials so that end markets for the materials are established, improved, or stabilized and thereby become more reliable.

MARPOL Annex V An international agreement that bans the dumping of plastic trash at sea.

mass-burn facility A type of incinerator that burns solid waste without any attempt to separate recyclables or process waste before burning.

mass combustion The burning of as-received, unprocessed refuse in furnaces designed exclusively for solid waste disposal/energy recovery.

mass burn A municipal waste combustion technology in which solid waste is burned in a controlled system without prior sorting or processing.

materials market The combined commercial interests that buy recyclable materials and process them for reuse. The demand for goods made of recycled materials determines the economic feasibility of recycling.

materials recovery Extraction of materials from the waste stream for reuse or recycling. Examples include source separation, front-end recovery, in-plant recycling, post-combustion recovery, leaf composting, etc.

materials recovery facility (MRF) A permitted solid waste facility where solid wastes or *recyclable* materials are sorted or separated, by hand or by use of machinery, for the purposes of recycling or compacting. Same as an IPC. Sometimes, the term "MRF" is used to refer to a mixed-waste processing facility; in which case, it is sometimes called a "dirty" or "full MRF." (Also see *WRF, Waste Recovery Facility*) When MRF is used synonymously with IPC, it is sometimes called a "clean MRF."

mechanical pulp Pulp produced by grinding wood into fibers.

mechanical separation The separation of waste into various components using mechanical means, such as cyclones, trommels, and screens.

medium-term planning period A period beginning in the year 1996 and ending in the year 2000.

metal A mineral source that is a good conductor of electricity and heat, and yield basic oxides and hydroxides. One of the hidden treasures in garbage.

methane An odorless, colorless, flammable, and explosive gas produced by municipal solid waste undergoing anaerobic decomposition. Methane is emitted from municipal solid waste landfills, can be used as fuel.

microorganisms Microscopically small living organisms that digest decomposable materials through metabolic activity. Microorganisms are active in the composting process.

midnight dumper An idiomatic term for an individual or business that disposes of waste in an illegal, stealthy manner.

mineral A naturally occurring substance of inorganic origin and internal crystalline structure; for example, metals are obtained from mineral resources.

mixed paper A waste type that is a mixture, unsegregated by color or quality, of at least two of the following paper wastes: newspaper, corrugated cardboard, office paper, computer paper, white paper, coated paper stock, or other paper wastes.

mixed refuse Garbage or refuse that is in a fully commingled state at the point of generation.

mixed-waste processing facility A facility that processes mixed refuse to remove recyclables and, sometimes, refuse-derived fuel and/or a compost substrate.

model Graphic, mathematical, verbal, or physical representation of a process or phenomenon.

modular combustion unit A self-contained, typically shop-assembled, incinerator designed to handle small quantities of solid waste. Several "modules" or units may be combined in a plant, as needed, depending on the quantity of waste to be processed.

modular incinerator Smaller-scale waste combustion units prefabricated at a manufacturing facility and transported to the MWC facility site.

monitoring well A well created to check the quality of the substance within, usually water.

mulch Ground or mixed yard wastes placed around plants to prevent evaporation of moisture and freezing of roots and to nourish the soil.

mulch mowing The practice of leaving grass clippings on the lawn after mowing rather than bagging them for curbside collection or using them as compost or for mulch.

multi-material A collection or processing system handling more than one secondary material. For example, a drop-off center accepting newspaper and aluminum cans is considered a multimaterial operation.

municipal solid waste (MSW) Includes nonhazardous waste generated in households and commercial and business establishments and institutions; excludes industrial process wastes, demolition wastes, agricultural wastes, mining wastes, abandoned automobiles, ashes, street sweepings, and sewage sludge.

municipal solid waste composting The controlled degradation of municipal solid waste including after some form of preprocessing to remove noncompostible inorganic materials.

municipal wastewater The combined residential, commercial, institutional, and industrial wastewater generated in a given municipal area.

N

National Ambient Air Quality Standards Federal standards that limit the concentration of particulates, sulfur dioxide, nitrogen dioxide, ozone, carbon monoxide, and lead in the atmosphere.

natural Determined by nature, occurring in conformity with the ordinary course of nature; a state of nature untouched by civilization.

natural resource Material or energy obtained from the environment that is used to meet human needs; material or energy resources not made by humans.

net diversion rate The fraction of total refuse not disposed as a result of recycling.

NIMBY (not in my back yard) Refers to the fact that people want the convenience of products and proper disposal of the waste generated by their use of products, provided the disposal area is not located near them.

nitrogen A tasteless, odorless gas that constitutes 78 percent of the atmosphere by volume. One of the essential ingredients of composting.

nonbiodegradable A substance that will not decompose under normal atmospheric conditions.

nonferrous metals Any metal scraps that have value and that are derived from metals other than iron and its alloys in steel, such as aluminum, copper, brass, bronze, lead, zinc, and other metals, and to which a magnet will not adhere.

nonpoint source Undefined wastewater discharges such as runoff from urban, agricultural, or strip-mined areas that do not originate from a specific point.

nonrecyclable Not capable of being recycled or used again.

nonrecyclable paper Discarded paper that has no market value because of its physical or chemical or biological characteristics or properties.

nonrenewable (resource) Not capable of being naturally restored or replenished; resources available in a fixed amount (stock) in the earth's crust; they can be exhausted either because they are not replaced by natural processes (copper) or because they are replaced more slowly than they are used (oil and coal).

normally disposed of Those waste categories and waste types that (1) have been demonstrated by the Solid Waste Generation Study to be in a solid waste stream attributed to the jurisdiction as of January 1, 1990; (2) that are deposited at permitted solid waste landfills or transformation facilities subsequent to any recycling or composting activities at those solid waste facilities; and (3) that are allowed to be considered in the establishment of the base amount of solid waste from which source reduction, recycling, and composting levels shall be calculated.

O

old corrugated containers As a paper-stock grade, baled corrugated containers having liners of test liner, jute, or kraft. The boxes are generated in retail stores, factories, and homes when merchandise is removed from them. A common acronym is 0CC.

old newspaper (ONP) Any newsprint that is separated from other types of solid waste or collected separately from other types of solid waste and made available for reuse and that may be used as a raw material in the manufacture of a new paper product.

open dump A site where solid waste is illegally discarded in an uncontrolled area.

operational costs Those direct costs incurred in maintaining the ongoing operation of a program or facility. Operational costs do not include capital costs.

organic Composed of living or once-living matter; more broadly, composed of chemical compounds principally based on the element carbon, excluding carbon dioxide.

organic waste Solid wastes originating from living organisms and their metabolic waste products, and those made from petroleum, that contain naturally produced organic compounds and are biologically decomposable by microbial and fungal action into the constituent compounds of water, carbon dioxide, and other simpler organic compounds.

other plastics All waste plastics except polyethylene terephthalate (PET) containers, film plastics, and high-density polyethylene (HDPE) containers.

outthrow Waste paper so manufactured or treated or in such a form as to be unsuitable at another grade. For instance, old newspapers are an outthrow when selling old corrugated containers but not an outthrow of mixed waste paper.

overissue newspaper Printed newspapers that were not circulated and are available for recycling.

over-the-scale trade The volume of business at a dealer or processor generated from purchasing materials delivered by independent scavengers, peddlers, individuals, and others. Also called door trade.

P

packaging Any of a variety of plastics, papers, cardboard, metals, ceramics, glass, wood, and paperboard used to make containers for foods, household and industrial products.

packer (1) A processing operation where waste paper is converted into paper stock and baled for shipment to consumers. (2) A solid waste collection vehicle employing a compaction mechanism.

paper Made from the pulp of trees. Paper is digested in a sulfurous solution, bleached and rolled into long sheets. Acid rain and dioxin are standard byproducts in this manufacturing process.

paperboard A type of matted or sheeted fibrous product. In common terms, paperboard is distinguished from paper by being heavier, thicker, and more rigid. See also specialty products.

partially allocated costs The costs of adding a recycling program to an existing operation such as a waste hauling company or public works department. Also known as incremental costs.

participant Any household that contributes any materials at least once during a specified tracking period.

participation rate A measure of the number of people participating in a recycling program compared to the total number that could be participating. Participation rate is calculated by dividing the number of households source-separating by the total number of households served. In a setting where there are single-family detached homes, it is possible to calculate the participation rate by keeping a set-out log, by address, for each and every household. Twelve weeks is a reasonable tracking period. In assessing true participation some households may separate materials for recycling, but sell or donate them elsewhere, rather than set them out for pickup.

participation/set-out ratio A multiplier used to estimate participation where set-outs are easily counted but it is not possible to conduct a full participation survey. As a short cut, the participation rate can be extrapolated from the set-out rate and the participation rate obtained from a sample area.

$$\text{Participation/set-out ratio} = \frac{\text{participation rate}}{\text{set-out rate}}$$

As a hypothetical example, a participation/set-out ratio of 2.5 might be derived from a program with a 30 percent weekly set-out rate and 75 percent participation rate that was documented from a representative portion of the route:

$$\frac{\text{75 percent participation rate}}{\text{30 percent set-out rate}} = 2.5$$

If another similar weekly program were found to have a 25 percent set-out rate, the 2.5 participation/set-out ratio could be applied to estimate the participation rate:

25.0	set-out rate
× 2.5	participation/set-out ratio
62.5	participation

The multipliers found to be accurate and useful are 2 to 2.5 for weekly programs, 1.5 for biweekly programs, and 1 for monthly programs. An interesting and potentially useful finding would be the point in a program's history that participation rates begin to "decay," and require a renewed educational and publicity efforts. One reason that participation drops is that one in five households changes its address each year. Thus approximately 20 percent of households would require more than an annual educational message.

personification Attribution of personal qualities; representation of a thing or abstraction as a person or by the human form.

PET (polyethylene terephthalate) A plastic resin used to make packaging, particularly soft drink bottles.

petroleum A mineral resource that is a complex mixture of hydrocarbons, an oily, flammable bituminous liquid, occurring in many places in the upper strata of the earth.

photodegradable Refers to plastics that will decompose if left exposed to light.

plan of operations The plan for the operation of a resource recovery system.

planned obsolescence The practice of producing goods that have a very short life so that more goods will have to be produced.

plastic resins Chemical components of plastics.

plastics Synthetic materials consisting of large molecules called polymers derived from petrochemicals (compared to natural polymers such as cellulose, starch, and natural rubbers).

particulate matter (PM) Tiny pieces of matter resulting from the combustion process that can have harmful health effects on those who breathe them. Pollution control at MWC facilities is designed to limit particulate emissions.

particulates Suspended small particles of ash, charred paper, dust, soot, or other partially incinerated matter carried in the flue gas.

passbys The total number of potential participants on a residential recycling collection route.

pathogen An organism capable of causing disease.

pelletizer A machine that produces chips or granules. Pelletizers are commonly used in plastics processing.

percolate To ooze or trickle through a permeable substance. Groundwater may percolate into the bottom of an unlined landfill.

permeable Having pores or openings that permit liquids or gases to pass through.

permits The official approval and permission to proceed with an activity controlled by the permitting authority. Several permits from different authorities may be required for a single operation.

permitted capacity That volume in cubic yards or weight in tons that a solid waste facility is allowed to receive, on a periodic basis, under the terms and conditions of that solid waste facility's current Solid Waste Facilities Permit issued by the local enforcement agency.

permitted landfill A solid waste landfill for which there exists a current Solid Waste Facilities Permit issued by the local enforcement agency.

permitted solid waste facility A solid waste facility for which there exists a Solid Waste Facilities Permit issued by the local enforcement agency.

platten The rectangular face of a baling ram. The platten pushes or compresses the secondary material into the baling chamber.

point of generation The physical location where the generator discards material (mixed refuse and/or separated recyclables).

point source Specific, identifiable end-of-pipe discharges of wastes into receiving bodies of water; for example, municipal sewage treatment plants, industrial wastewater treatment systems, and animal feedlots.

pollutants Any solid, liquid, or gaseous matter that is in excess of natural levels or established standards.

pollute To contaminate; to make impure.

pollution Harmful substances deposited in the environment by the discharge of waste, leading to the contamination of soil, water, or the atmosphere.

polyethylenes A group of resins created by polymerizing ethylene gas. The two major categories are high-density polyethylene and low-density polyethylene.

polyethylene terephthalate A lightweight, transparent, rigid polymer resistant to chemical and moisture, and with good insulating properties. polymer A large molecule containing a chain of chemically linked subunits (monomers).

polyolefins A plastics subgroup including polyethylene and polypropylene.

polystyrene A hard, dimensionally stable thermoplastic that is easily molded. PS is a common acronym.

polyvinyl chloride A plastic made by polymerization of vinyl chloride with peroxide catalysts. A common acronym is PVC.

population and household counts Include a count or an estimate of the number of persons the household serves and the housing densities within the program area. Because many resort communities experience seasonal fluctuations in population figures should indicate whether it is based on year-round residents, peak population, or a calculated average.

population, housing, and land use descriptions These descriptions should include median age, income levels, and education, type of housing, density of development, and proportions of single- or multifamily homes: Inclusion of a description of housing density is a good indicator of urbanization and thus good background information for comparison of programs. Type and extent of commercial development. Gross leasable area (GLA) can serve as a rough indicator; it can be derived from a census of retail trade and may be available from county economic development offices. Mention should be made of any unusual factors that would cause the waste stream to differ from the community it is being compared to or from national averages. These factors would include climate, generation of special industrial or vegetative wastes, unusually large business concerns, government agencies or institutions, transient resort populations, large retirement population, a large student population associated with colleges and universities, or vacation area.

postconsumer recycling The reuse of materials generated from residential and commercial waste, excluding recycling of material from industrial processes that has not reached the consumer, such as glass broken in the manufacturing process.

preconsumer Recyclable items that are collected from mills and manufacturing plants as by-products of their manufacturing process. These include over-runs on printing jobs, trimmings from envelopes, glass bottles that break prior to shipping, and excess plastic materials from molds.

precycling Activities such as source and size reduction, material selection when shopping, and reducing toxicity of products in manufacturing prior to recycling that helps reduce the amounts of municipal solid wastes generated. A term coined in Berkeley, California, which involves a commitment to improve the environment through conscious decision making when shopping; for example, replacing plastic coffee cups with porcelain mugs or cardboard egg crates replacing plastic containers.

private collection The collecting of solid wastes for which citizens or firms, individually or in limited groups, pay collectors or private operating agencies. Also known as private disposal.

privatization The assumption of responsibility for a public service by the private sector, under contract to local government or directly to the receivers of the service.

processable waste That portion of the solid waste stream that is capable of being processed in a mass burn resource recovery facility, including all forms of household and other garbage, trash, rubbish, refuse, combustible agricultural, commercial and light industrial waste, commercial waste, leaves and brush, paper and cardboard, plastics, wood and lumber, rags, carpeting, occasional tires, wood furniture, mattresses, stumps, wood pallets, timber, tree limbs, ties, and logs, not separated and recycled at the source of generation, but excluding unacceptable waste and unprocessible waste.

processed Treated, or made by a special process or treatment, especially when involving synthesis or artificial modification.

processing The procedures used to prepare, refine, preserve, or otherwise change the initial form of materials or products.

product An outcome or an object; the amount, quantity, or total produced.

program The full range of source reduction, recycling, composting, special waste, or household hazardous waste activities undertaken by or in the jurisdiction or relating to management of the jurisdiction's waste stream to achieve the objectives identified in the source reduction, recycling, composting, special waste, and household hazardous waste components, respectively.

pulp A moist mixture of fibers from which paper is made.

pulp substitutes Unprinted, clean waste paper that can be used directly in papermaking.

purchase preference A preference provided to a wholesale or retail commodity dealer that is based upon the percentage amount that the costs of products made from recycled materials may exceed that of similar nonrecycled products and still be deemed the lowest bid.

PURPA The Public Utilities Regulatory Policies Act of 1978. A federal law whose key provision mandates private utilities to buy power commissions and equal to the "avoided cost" of power production to the utility. The act is intended to guarantee a market for small producers of electricity at rates equal or close to the utilities' marginal production costs.

putrescible waste Solid wastes that are capable of being decomposed by microorganisms with sufficient rapidity to cause nuisances from odors or gases and capable of providing food for, or attracting, birds and disease vectors.

PVC plastic (polyvinyl chloride) A typically insoluble plastic used in packaging, pipes, detergent bottles, wraps, etc.

pyrolysis The process of chemically decomposing an organic substance by heating it in an oxygen-deficient atmosphere. High temperatures and closed chambers are used. The major products from pyrolysis of solid waste are water, carbon monoxide, and hydrogen. Some processes produce an oillike liquid of undetermined chemical composition. The gas may contain hydrocarbons, and frequently there is process residue of ash and a carbon char.

R

rack collection The collection of old newspapers at the same time as residential waste collection. The waste paper is placed in a side or front rack attached to the waste collection truck.

radioactive A substance capable of giving off high-energy particles or rays as a result of spontaneous disintegration of atomic nuclei.

RAO Responsibility assumption overload; a phenomenon whereby individuals feel powerless regarding a problem.

rate structure That set of prices established by a jurisdiction, special district (as defined in Government Code section 56036), or other rate-setting authority to compensate the jurisdiction, special district, or rate-setting authority for the partial or full costs of the collection, processing, recycling, composting, and/or transformation or landfill disposal of solid wastes.

raw materials Substances still in their natural or original state, before processing or manufacturing; or the starting materials for a manufacturing process.

RCRA Resource Conservation and Recovery Act of 1976; requires states to develop solid waste management plans and prohibits open dumps; identifies lists of hazardous wastes and sets the standards for their disposal.

reactive For regulatory purposes, defined as a substance that tends to react spontaneously with air or water, to explode when dropped, or to give off toxic gases.

recover To reclaim a resource embedded in waste.

recovered materials Those materials that have known recycling potential, can be feasibly recycled, and have been diverted or removed from the solid waste stream for sale.

recovered material Material that has been retrieved or diverted from disposal or transformation for the purpose of recycling, reuse, or composting; does not include those materials generated from and reused on site for manufacturing purposes.

recovery rate All discarded materials that have been recovered through various recovery strategies including yard waste, composting, and reuse. Designated materials recovered plus returned via deposit divided by the total designated materials available.

recyclables Materials that still have useful physical or chemical properties after serving their original purpose and that can, therefore, be reused or remanufactured into additional products. Waste materials that are collected, separated, and used as raw material.

recycle To separate a given material from waste and process it so that it can be used again in a form similar to its original use; for example, newspapers recycled into newspapers or cardboard.

recycled Composed of materials that have been processed and used again.

recycled content That portion of a product that is made from recycled materials. This percentage may include both pre- and postconsumer materials. Buying products with recycled content helps support recycling.

recycling The act of extracting materials from the waste stream and reusing them.

Recycling generally includes collection, separation, processing, marketing, and the creation of a new product or material from used products or materials. In general usage, recycling refers to the separation of recyclable materials such as newspaper, aluminum, other metals or glass from the waste. This includes recycling of materials from municipal waste, often done through separation by individuals or specially designed materials recovery facilities; industrial in-plant recycling; and recycling by commercial establishments.

a. Recycling, *primary* is remaking the recyclable material into the same material in a process that can be separated a number of times (e.g., newspapers into newspapers, glass containers into glass containers)
b. Recycling, *secondary* is remaking the recyclable material into a material that has the potential to be recycled again (e.g., newspaper into recycled paperboard)
c. Recycling, *tertiary* is remaking the recyclable material into a product that is unlikely to be recycled again (e.g., glass into asphalt, paper into tissue paper)

Recycling prevents the emission of many greenhouse gases, saves energy, supplies valuable raw materials to industry, and reduces the need for new landfills and combustors. In the late 1990's, recycling in the United States prevented the release of 33 million tons of carbon into the air, roughly the amount emitted by 25 million cars.

recycling bin A container in which to place recyclables.

recycling center A place where recyclable items are taken for processing.

recycling loop A process through which materials that might otherwise be wasted are collected and processed for conversion into new products that otherwise would have been discarded.

recycling processor A generic term for businesses and operations that prepare secondary materials for sale to end users. Waste paper dealers, scrap metal yards, drop-off centers, and buy-back centers are examples of recycling processors.

recycling program Should include the following: types of collection equipment used, collection schedule, route configuration, frequency of collection per household, whether curbside set-out containers are provided by the program, publicity and educational activities and budget, financial evaluation (costs, revenues, and savings), processing and handling procedures, market prices, ordinances and enforcement activities.

recycling rate A percentage ratio of the weight of solid waste collected for recycling to the total solid waste weight collected for disposal in landfills and waste incinerators. In several areas the recycling rate may include waste materials such as scrap metal from cars and salvaged steel from old ships and other materials collected by private buy-back centers. A true recycling rate should be based on tonnages of municipal and commercial solid waste normally disposed of in landfills. Scrap metals from private buy-back centers never enter landfills. The basic purpose of determining the recycling rate is monitoring recycling progress for diverting solid wastes from the landfills.

reference waste Varies by different classification schemes to describe waste types. For example, when a community reports on its "waste," it may refer to all of the wastes delivered to a disposal site, all of the wastes collected by the public works department, or all of the wastes generated by residences, that may or may not include multifamily residences. Wastes are defined by management strategy, collection sector, and type of generator, respectively.

reference waste classification Uses two major categories: municipal solid waste (MSW) and industrial waste. MSW further consists of four subdivisions: residential single-family, residential multifamily, institutional, and commercial.

refractory A material that can withstand dramatic heat variations. Used to construct conventional combustion chambers in incinerators. Currently, waterwall systems are becoming more common.

refuse-derived fuel (RDF) A solid fuel obtained from municipal solid waste as a result of mechanical process or sequence of operations, that improves the physical, mechanical, or combustion characteristics compared to the original unsegregated feed product or unprocessed solid waste. Usually, noncombustibles and recyclable materials are removed. The fuel may be sized for the specific requirements of the furnace where it will be burned, producing a "fluff' or shredded RDF. In some processes, RDF may be compressed into pellets or cubes, producing a densified RDF (d-RDF).

refuse-derived fuel facility A type of incinerator that separates recyclables from solid waste before it is burned.

region The combined geographic area of two or more incorporated areas; two or more unincorporated areas; or any combination of incorporated and unincorporated areas.

regrind Ground-up recyclable plastics.

reject One of the 4 R's of solid waste management and part of the precycling process, whereby a person consciously choses not to purchase or consume an item or energy source.

rejects Material rejected at the beginning of processing.

renewable (resource) Capable of being naturally restored or replenished; a resource capable of being replaced by natural ecological cycles or sound management practices.

renewable resources A naturally occurring raw material or form of energy, such as the sun, wind, falling water, biofuels, fish, and trees, derived from an endless or cyclical source, where, through management of natural means, replacement roughly equals consumption ("sustained yield").

request for bid A mechanism for seeking bidders to supply recycling goods and services or to purchase secondary materials. An acronym is RFB.

request for proposal A mechanism for seeking qualified firms or individuals to supply recycling goods or services. A common acronym is RFP.

request for qualifications A mechanism for determining the experience, skills, financial resources, or expertise of a potential bidder or proposer. Commonly abbreviated as REQ.

repairability The ability of a product or package to be restored to a working or usable state at a cost that is less than the replacement cost of the product or package.

re-refining The use of petroleum-refining techniques on used motor oil to produce lubrication stocks.

residential solid waste Solid waste originating from private single-family or multiple-family dwellings.

residential unit A site occupied by a building that is zoned for residential occupation and whose occupants generate residential solid wastes.

residential waste Waste materials generated in houses and apartments. The materials include paper, cardboard beverage and food cans, plastics, food wastes, glass containers, old clothes, garden wastes, etc.

residue Materials remaining after processing, incineration, composting, or recycling have been completed. Residues are usually disposed of in landfills.

resource A natural or synthetic material that can be used to make something else; for example, wood resources are made into paper and old bottles can be made into new ones.

Resource Conservation and Recovery Act of 1976 (RCRA) This law amends the Solid Waste Disposal Act of 1965 and expands on the Resource Recovery Act of 1970 to provide a program to regulate hazardous waste; to eliminate open dumping; to promote solid waste management programs through financial and technical assistance; to further solid waste management options in rural communities through government grants; and to conduct research, development and demonstrate programs for the betterment of solid waste management, resource conservation and recovery practices.

resource recovery A term describing the extraction and utilization of materials and energy from the waste stream. Materials recovered, for example, would include paper, metals, and glass that can be used as "raw materials" in the manufacture of new products. Energy is recovered by utilizing components of waste as a fuel or feedstock for chemical or biological conversion to some form of fuel or steam. An integrated resource recovery program may include recycling, waste-to-energy, composting, and/or other components, with a landfill for residue disposal.

retention basin An area designed to retain runoff and prevent erosion and pollution.

returnable Can be returned for deposit and/or reuse.

reusability The ability of a product or package to be used more than once in its same form.

reuse The use of a product more than once in its same form for the same purpose; e.g., a soft-drink bottle is reused when it is refined to the bottling company for refilling; finding new functions for objects and materials that have outgrown their original use; to use again.

reverse vending machine A machine that accepts empty beverage containers (or other items) and rewards the donor with a cash refund.

roll-off container A large waste container that fits onto a tractor trailer that can be dropped off and picked up hydraulically.

rotary kiln An enclosed waterwall, cylindrical barrel-shaped device that is utilized for the combustion of materials at high temperatures. Agitation of the material is accomplished through slow rotation of the barrel.

rubber An amorphous polymer of isoprene derived from natural latex of certain tropical plants or from petroleum.

rubber-asphalt A product that combines ground-up scrap tires and asphalt. It is primarily used in highway, runway, and street projects as a stress-absorbing membrane interlayer.

rubbish Nonputrescible solid waste (excluding ashes), consisting of both combustible and noncombustible waste materials.

runoff Water (originating as precipitation) that flows across the surface of the ground—rather than soaking into it—eventually entering bodies of water; may pick up and carry with it a variety of suspended or dissolved substances. In many cases, the runoff is a leachate composed of toxic compounds.

S

sanitary landfill A method of disposing of refuse on land without creating nuisances or hazards to public health or safety. Careful preparation of the fill area and control of water drainage are required to assure proper landfilling. To confine the refuse to the smallest practical area and reduce it to the smallest practical volume, heavy tractor-like equipment is used to spread, compact, and usually cover the waste daily with at least six inches of compacted soil. Modern, properly engineered sanitary landfills are lined with compacted clay or an artificial (plastic) liner; have leachate collection systems to remove the leachate for treatment and disposal; and have systems to collect and remove methane gas generated in the landfill.

salvage The controlled removal of solid waste materials at a permitted solid waste facility for recycling, reuse, composting, or transformation.

scavenger One who illegally removes materials at any point in the solid waste management system.

scavenging The uncontrolled and unauthorized removal of materials at any point in the solid waste management system.

scrap Products that have completed their useful life, such as appliances, cars, construction materials, ships, and postconsumer steel cans; also includes new scrap materials that result as by-products when metals are processed and products are manufactured. Steel scrap is recycled in steel mills to make new steel products.

scrubber A device for removing unwanted dust particles, liquids, or gaseous substances from an airstream by spraying the airstream with a liquid (usually water or a caustic solution) or forcing the air through a series of baths; common antipollution device that uses a liquid or slurry spray to remove acid gases and particulates from municipal waste combustion facility flue gases.

seasonal Those periods of time during the calendar year that are identifiable by distinct cyclical patterns of local climate, demography, trade, or commerce.

secondary ingestion Process whereby plastics or other materials eaten by animals at low levels of the food chain show up in animals at higher levels of the food chain.

secondary material A material that is used in place of a primary or raw material in manufacturing a product.

secure landfill A landfill designed to prevent the entry of water and the escape of leachate by the use of impermeable liners.

separate collection A system in which specific portions of the waste stream are collected separately from the rest to facilitate recycling or otherwise improve solid waste management.

set-out A quantity of material placed for collection. Usually a set-out denotes one household's entire collection of recyclable materials, but in urban areas, where housing density makes it difficult to identify ownership of materials, each separate container or bundle is counted as a set-out. A single household, for example, may have three set-outs: commingled glass, metals, and newspapers.

set-out rate An empirical measure obtained by counting the number of households that set out materials on their assigned collection day and the number of households in the service area. The set-out rate is not a measurement of true participation, as participants may choose to set out materials less frequently than the service is provided. Whenever a set-out rate is cited, it is desirable to also provide information about the geographic boundaries, population, and type of households served (single-family, or mixed), the period of time when counts are made, and whether it is per route or all routes together. The time of day that observations are made is important to note, if not done simultaneously with pickup. An accurate set-out count could only be made at the actual time of collection.

$$\text{Set-out rate} = \frac{\text{number of individual set-outs on collection day}}{\text{total number of households served}}$$

short-term impact Immediate effect of an action, such as an oil spill.

short-term planning period A period beginning in the year 1991 and ending in the year 1995.

shredder A mechanical device used to break up waste materials into smaller pieces by tearing, shearing, cutting, and impact action.

shrinkage The difference in the purchase weight of a secondary material and the actual weight of the material when consumed.

SIC code The standards published in the U.S. Standard Industrial Classification Manual (1987).

sludge Solid matter that settles to the bottom of septic tanks or wastewater treatment plant sedimentation tanks; must be processed by bacterial digestion or other methods, or pumped out for land disposal, incineration, or composting.

small-quantity generator Sources such as small businesses and institutions that generate less than 1000 kg of hazardous waste per month.

soil The loose top layer of the earth's surface in which plant life can grow.

soil amendment Any material, such as yard waste compost, added to the soil to improve soil chemistry.

soil conditioner Any material, such as yard waste compost, added to the soil to improve the physical soil structure.

soil liner Landfill liner composed of compacted soil used for the containment of leachate.

solid waste Garbage, refuse, sludges, and other discarded solid materials, including those from industrial, commercial, and agricultural operations, and from community activities; does not include solids or dissolved materials in domestic sewage or other significant pollutants in water resources, such as silt, dissolved or suspended solids in industrial waste water effluents, dissolved materials in irrigation return flow or other common pollutants; any nonliquid, nongaseous waste.

solid waste disposal facility Any solid waste management facility that is the final resting place for solid waste, including landfills and incineration facilities that produce ash from the process of incinerating municipal solid waste.

solid waste generation study The study undertaken by a jurisdiction to characterize its solid waste stream.

solid waste management The systematic administration of activities that provide for the collection, source separation, storage, transportation, transfer, processing, treatment, and disposal of solid waste.

sort The process of separating materials into specific categories.

sorted color ledger As a paper stock grade, consists of printed or unprinted sheets, shavings and cuttings of colored or white ledger, bond, writing and other papers. This grade must be free of treated, coated, padded, or heavily printed stock.

sorted white ledger As a paper stock grade, consists of printed or unprinted sheets, shavings, books and cuttings of white ledger, bond, writing and other papers. This grade must be free of treated, coated, padded, or heavily printed stock.

source reduction The design, manufacture, acquisition, and reuse of materials so as to minimize the quantity and/or toxicity of waste produced. Source reduction prevents waste either by redesigning products or by otherwise changing societal patterns of consumption, use, and waste generation. Any action that avoids the creation of waste by reducing waste at the source including redesigning products or packaging so that less material is used. Source reduction prevents emissions of many greenhouse gases, thus reducing the need for new landfills and combustors.

source separation The segregation of various specific materials from the waste stream at the point of generation. For example, households separating paper, metals, and glass from the rest of their wastes. Source separation makes recycling simpler and easier. Residences source-separate recyclables as part of a curbside recycling program.

source separation legislation Legislation that is intended to facilitate collection of designated materials for recycling, composting, or reuse by specifying how such materials are to be segregated and set out for collection. It usually prohibits mixing of such designated materials with wastes that are to be disposed. This is often referred to as *mandatory recycling legislation*, and discourages disposal and encourages recycling of selected items within municipally collected waste stream.

special news deink quality As a paper stock grade, consists of baled, sorted, fresh dry newspapers, free from magazines and containing no more than the normal percentage of colored sections.

special wastes Refers to items that require special, separate handling, such as furniture, mattresses, tree trunks, white goods, concrete, and asphalt, tires, hazardous wastes, bulky wastes, and used oil.

specialty products A category of fiber products including insulation board, roofing felt, cellulose insulation, animal bedding, hydromulch, and molded pulp. The other two principal forms of fiber products are paper and paperboard.

spreader stoker A horizontal moving grate, perforated to permit feeding underfire air for combustion, typically used in waterwall furnaces designed to burn refuse-derived fuel. Fuel is introduced by pneumatic or mechanical means over the grate in a manner that spreads the fuel over the stoker (grate); 30 to 50 percent burns in suspension, with the remainder burning on the grate.

stack (chimney, flue) A vertical passage for conducting products of combustion to the atmosphere.

stack emissions Air emissions from combustion facility stacks.

static pile system A composting method in which air ducts are generally installed under or in the base of compost piles, or windrows, so that air can be circulated through the pile.

statistically representative Those representative and random samples of units that are taken from a population sample. For the purpose of this definition, population sample includes, but is not limited to, a sample from a population of solid waste generation sites, solid waste facilities and recycling facilities, or a population of items of materials and solid wastes in a refuse load of solid waste.

steel A malleable alloy of iron and carbon, which is 100 percent recyclable, and also has recycled content. Steel is used to make a variety of products, such as cans, car parts, appliances, construction materials, tools, toys, and hundreds of other products for consumer and commercial use.

steel can A rigid container made exclusively or primarily of steel. Used to store food, beverages, paint, and a variety of other household and consumer products, all of which are 100 percent recyclable.

stewardship Taking responsibility; caring for and protecting an entity such as resources and the planet.

stoichiometric air The amount of air theoretically required to provide the exact amount of oxygen for total combustion of a fuel. Municipal solid waste incineration technologies make use of both sub-stoichiometric and excess air processes.

styrofoam Also known as polystyrene, a synthetic material consisting of large molecules called polymers derived from petrochemicals. Experts agree that styrofoam will never decompose.

substitute To put or use in the place of another.

Subtitle C The hazardous waste section of the Resource Conservation and Recovery Act (RCRA).

Subtitle D The solid, nonhazardous waste section of the Resource Conservation and Recovery Act (RCRA).

Subtitle F Section of the Resource Conservation and Recovery Act (RCRA) requiring the federal government to actively participate in procurement programs fostering the recovery and use of recycled materials and energy.

sulfate pulp Kraft pulp produced by chemical methods using an alkaline solution of caustic soda and sodium sulfite. Sulfate pulp is used primarily in paperboard and coarse paper grades.

sulfite pulp Acid pulp produced by chemically cooking wood using sulfurous acid. Sulfite pulp is used for most printing and tissue grades of paper.

Superfund Common name for the Comprehensive Environmental Response, Compensation and Liability Act (CERCLA) to clean up abandoned or inactive hazardous waste dump sites.

surface water management system Systems designed, constructed, operated, and maintained to prevent surface water flow onto waste-filled areas.

sustainable The ability to support, endure, or keep up. Meeting present needs without compromising future resources. In order for recycling to be sustainable, it is essential for the public to buy products made with recycled materials.

SWMA The 1988 Florida Solid Waste Management Act.

T

tare The weight of extraneous material, such as pallets, strapping, bulkhead, and sideboards, that is deducted from the gross weight of a secondary material shipment to obtain net weight.

thermomechanical pulp Pulp produced by heating wood, then subjecting it to repeated compressions and stress relaxations between opposite bars and grooves to break the wood into fibers.

thermoplastics Plastic material that can be melted to a liquid or semifluid state, which then rehardens when cooled.

thermosets Plastic material set to permanent shapes when heat and pressure are applied during forming; they cannot be softened again when reheated.

throwaway lifestyle A way of living characterized by a high level of product consumption and discarding, especially if the products are meant for one-time usage.

throwaway society A society characterized by a throwaway lifestyle (see above).

time period As a specific time period and a particular start date for a recycling program measurement.

tin A natural element used as a coating on steel cans to stabilize the flavors of the contents. The use of tin as a coating material goes back in history to about 300 B.C.

tin can A term sometimes used to describe a steel food can—foods were first canned in the early 1800s. Technological developments have allowed for the tin coating on a can to become progressively thinner, to the point that tin now represents less than one-third of 1 percent of the weight of a steel can.

tire-derived fuel (TDF) A form of fuel consisting of scrap tires shredded into chips (TDF).

top dressing A covering material, such as yard waste compost, spread on soil without being plowed under.

tipping fee A fee, usually dollars per ton, for the unloading or dumping of waste at a landfill, transfer station, recycling center, or waste-to-energy facility, usually stated in dollars per ton; also called a disposal or service fee.

tipping floor Unloading area for vehicles that are delivering municipal solid waste to a transfer station or municipal waste combustion facility.

ton A unit of weight in the U.S. Customary System of Measurement, an avoirdupois unit equal to 2000 pounds. Also called short ton or net ton; equals 0.907 metric tons.

toxic Defined for regulatory purposes as a substance containing poison and posing a substantial threat to human health and/or the environment.

trade waste A European term for recyclable materials such as envelope cuttings and boxboard cuttings generated by manufacturers.

transfer station A place or facility where waste materials are taken from smaller collection vehicles (e.g., compactor trucks) and placed in larger transportation units (e.g., over-the-road tractor trailers or barges) for movement to disposal areas, usually landfills. In some transfer operations, compaction or separation may be done at the station.

transformation facility A facility whose principal function is to convert, combust, or otherwise process solid waste by incineration, pyrolysis, destructive distillation, or gasification, or to chemically or biologically process solid wastes, for the purpose of volume reduction, synthetic fuel production, or energy recovery. Transformation facility does not include a composting facility.

trash A term used for wastes that usually do not include food wastes and ashes but may include other organic materials, such as plant trimmings, or material considered worthless, unnecessary, or offensive that is usually thrown away. Generally defined as dry waste material, but in common usage it is a synonym for rubbish or refuse.

trommel A perforated, rotating, slightly declined cylinder that may be used in resource recovery facilities to break open trash bags, remove glass in large enough pieces for recovery and remove small abrasive items such as stones and dirt. Trommels have been used to remove steel cans from incinerator residue.

tub grinder Machine to grind or chip wood wastes for mulching, composting, or size reduction.

turbidity Cloudiness of a liquid.

U

unacceptable waste Motor vehicles, trailers, comparable bulky items of machinery or equipment, highly inflammable substances, hazardous waste, sludges, pathological and biological wastes, liquid wastes, sewage, manure, explosives and ordinance materials, and radioactive materials. Also includes any other material not permitted by law or regulation to be disposed of at a landfill unless such landfill is specifically designed, constructed, and licensed or permitted to receive such material. None of such material shall constitute either processable waste or unprocessible waste.

unprocessible waste That portion of the solid waste stream that is predominantly noncombustible and therefore should not be processed in a mass burn re- source recovery system; includes, but is not limited to, metal furniture and appliances; concrete rubble; mixed roofing materials; noncombustible building debris; rock, gravel, and other earthen materials; equipment; wire and cable; and any item of solid waste exceeding 6 feet in any one of its dimensions or being in whole or in part of a solid mass, the solid mass portion of which has dimensions such that a sphere with a diameter of 8 inches could be contained within such solid mass portion, and processable waste (to the extent that it is contained in the normal unprocessible waste stream); excludes unacceptable waste.

upstroke baler A baling device in which the compression ram and platten move upward into the chamber. Pit balers are a type of upstroke baler.

U.S. Environmental Protection Agency (EPA) The federal agency created in 1970 and charged with the enforcement of all federal regulations having to do with environmental pollutants.

used beverage cans Cans generated from the consumption of beer, soft drinks, juice, and other beverages. The reference is typically to used aluminum cans. Also known as UBC.

used brown kraft As a paper stock grade, consists of baled brown kraft bags free of objectionable contents.

used oil Oil that has been utilized for a purpose and is ready to be discarded or recycled.

V

variable container rate A charge for solid waste services based on the volume of waste generated measured by the number of containers set out for collection.

variable can rate A charge for solid waste services based on the volume of waste generated, measured by the number of containers set out for collection.

vector An agent, such as an insect, snake, rodent, or animal capable of mechanically or biologically transferring a pathogen from one organism to another.

vegetative waste Waste materials from farms, plant nurseries, and greenhouses that are produced from the raising of plants. This waste includes such residues as plant stalks, hulls, leaves, and tree wastes processed through a wood chipper.

vermicomposting The use of worms to digest raw or stabilized organic waste.

vertical baler Downstroke or upstroke baler.

vibrating screen A mechanical device that sorts material according to size.

vinyl A polymer of a vinyl compound, derived from ethylene.

virgin (materials) Term describing raw materials as yet unused; for example, virgin aluminum has not yet been fabricated into cans; compare recycled aluminum.

volume A three-dimensional measurement of the capacity of a region of space or a container. Volume is commonly expressed in terms of cubic yards or cubic meters. Volume is not expressed in terms of mass or weight.

volume-based rates A system of charging for garbage pickup that charges the waste generator rates based on the volume of waste collected, so that the greater the volume of waste collected, the higher the charge. "Pay-by-the-bag" systems and variable can rates are types of volume-based rates.

volume reduction The processing of waste materials so as to decrease the amount of space the materials occupy. Reduction is presently accomplished by three major processes: (1) mechanical, which uses compaction techniques (baling, sanitary landfills, etc.) and shredding; (2) thermal, which is achieved by heat (incineration) and can reduce volume by 80 to 90 percent; and (3) biological, in which the organic waste fraction is degraded by bacterial action (composting, etc.).

volumetric net diversion rate Same as net diversion rate, except the fraction is measured as a volume instead of a weight.

voluntary separation The participation in waste recycling willingly, as opposed to mandatory recycling.

W

waste Anything that is discarded, useless, or unwanted; opposite of conserve, as in "to waste."

waste categories The grouping of solid wastes with similar properties into major solid waste classes, such as grouping together office, corrugated, and newspaper as a paper waste category, as identified by a solid waste classification system, except where a component-specific requirement provides alternative means of classification.

waste composition The relative amount of various types of materials in a specific waste stream.

waste diversion To divert solid waste, in accordance with all applicable federal, state, and local requirements, from disposal at solid waste landfills or transformation facilities through source reduction, recycling, or composting.

waste diversion credit A financial incentive provided to municipalities or private recycling operations based on the tonnage diverted from the waste stream.

waste exchange A system that allows the waste from one activity to be used as a resource in another activity.

waste exchange A computer and catalog network that redirects waste materials back into the manufacturing or reuse process by matching companies generating specific wastes with companies that use those wastes as manufacturing inputs.

waste generator Any person whose act or process produces solid waste, or whose act first causes solid waste to become subject to regulation.

waste management (Also integrated solid waste management) A practice of using alternative techniques to manage and dispose of specific components of the municipal solid waste stream. Waste management alternatives include source reduction, recycling, composting, energy recovery, and landfilling.

waste minimization An action leading to the reduction of waste generation, particularly by industrial firms.

waste paper Recyclable paper and paperboard.

waste paper hydromulch A growing medium produced from waste paper.

waste recovery facility (WRF) A process for separation of recyclables and creating a compostlike material from the total of full mixed municipal solid waste stream. Differs from a "clean" MRF that processes only commingled recyclables (See *"full"* or *"Dirty" MRFs*).

waste reduction Reducing the amount or type of waste generated. Sometimes used synonymously with source reduction, decreasing the quality of materials and/or products that must be disposed. Simple volume reduction as crushing, baling, yard waste chipping.

waste stream A term describing the total flow of solid waste from homes, businesses, institutions, and manufacturing plants that must be recycled, burned, or disposed of in landfills; or any segment thereof, such as the "residential waste stream" or the "recyclable waste stream." The total waste produced by a community or society, as it moves from origin to disposal.

waste type Identified wastes having the characteristics of a group or class of wastes that are distinguishable from any other waste type, except where a component-specific requirement provides alternative means of classification.

water table Level below the earth's surface at which the ground becomes saturated with water. Landfills and composting facilities facilities are designed with respect to the water table in order to minimize potential contamination.

water treatment plant A facility designed to improve water quality; removes impurities from drinking water or cleans sewage prior to final discharge.

waterwall furnace Furnace constructed with walls of welded steel tubes through water is circulated to absorb the heat of combustion. These furnaces can be used as incinerators. The steam or hot water thus generated may be put to a useful purpose or simply used to carry the heat away to the outside environment.

waterwall incinerator Waste combustion facility utilizing lined steel tubes filled with circulating water to cool the combustion chamber. Heat from the combustion gases is transferred to the water. The resultant steam is sold or used to generate electricity.

weight-based rates A system of charging for garbage pickup that charges based on the weight of garbage collected, so that the greater the weight collected, the higher the charge. The logistics of implementing this system are currently being experimented with.

wetland Area that is regularly wet or flooded and has a water table that stands at or above the land surface for at least part of the year. Coastal wetlands extend back from estuaries and include salt marshes, tidal basins, marshes, and mangrove swamps. Inland freshwater wetlands consist of swamps, marshes, and bogs. Federal regulations apply to landfills sited at or near wetlands.

wet scrubber Antipollution device in which a lime slurry (dry lime mixed with water) is injected into the flue gas stream to remove acid gases and particulates.

white goods A term used to describe large appliances such as refrigerators, washers, and dryers. The terminology was derived from the standard white color of these appliances that existed until recent years.

windrow A large, elongated pile of composting material.

windrow system A composting system in which waste is placed in windrows to compost and either aerated (in static pile system) or turned periodically.

windrowing The placement and management of compostable material in piled rows, where microorganisms break down organic material into a finished compost product.

wood waste Solid waste consisting of wood pieces or particles that are generated from the manufacturing or product of wood products, harvesting, processing, or storage of raw wood materials, or construction and demolition activities.

Y

yard trash Vegetative matter resulting from landscaping maintenance or land clearing operations and includes materials such as tree and shrub trimmings, grass clippings, palm fronds, trees, and tree stumps.

yard waste or yard debris Leaves, grass clippings, prunings, and other natural organic matter discarded from yards and gardens. Yard wastes may also include stumps and brush, but these materials are not normally handled at composting facilities.

REFERENCES

1. National Resource Recovery Association, The United States Conference of Mayors, "A Solid Waste Management Glossary," Washington, D.C.
2. Florida Department of Education, Tallahassee, Florida.
3. Steel Recycling Institute, Pittsburgh, Pennsylvania.
4. The National Recycling Coalition," Measurement Standards and Reporting Guidelines." Washington, D.C.
5. California Integrated Waste Management Board-Title 14, Chapter 9, Article 3-Definitions.
6. Resource Recycling, Inc., "Glossary of Recycling Terms and Acronyms."

APPENDIX C
ABBREVIATIONS

ABM Abrasive Blast Media
ADF Advanced disposal fee
APC Air Pollution Control
ANSI American National Standards Institute
APWA American Public Works Association
ARMI Asphalt Rubber Membrane Interlayer
BAN Bond anticipation note
BMP Best Management Practice
BRBA Buy Recycled Business Alliance
Btu British thermal unit
C&D Construction and Demolition Debris
CEM Continuous Emissions Monitoring
CERCLA Comprehensive Environmental Response, Compensation and Liability Act
CFR Code of Federal Regulations
CHEMTREC Emergency Information Service of the Chemical Industry
CI Activated Carbon Injection System
CPRR Center for Plastic Recycling Research
CSS Curbside sort
CSWS Council for Solid Waste Solution
CSWMP County Solid Waste Management Plan
CY Calendar Year
DEP Department of Environmental Protection
DER Department of Environmental Regulation
DOH Department of Health
DOT Department of Transportation
DRC Database of Recycled Commodities
EIS Environmental impact statement
EPA Environmental Protection Agency
EPtox Extraction procedure toxicity test
ESP Electrostatic precipitator

FCM Fully commingled
FF Filter Fabric
FMC Filter Manufacturers Council
GO bond General obligation bond
HDPE High-density polyethylene #2
HID High Intensity Discharge
HHW Household hazardous waste
HSWA Hazardous and Solid Waste Act of 1984
IPC Intermediate processing center
IRB Industrial revenue bonds
JTR Jobs Through Recycling
KABS Keep America Beautiful Systems
kw Kilowatt
LAER Lowest achievable emission rate
LDPE Low-density polyethylene
MACT Maximum Achievable Control Technology
MCD Mercury-Containing Device
MCL Mercury-Containing Lamp
MEMA Motor and Equipment Manufacturers Association
MR Mixed degradables
MRF Materials recovery facility
MSW Municipal solid waste
MSW-RDF Municipal solid waste and refuse-derived fuel processing facility
MWC Municipal waste combustor
MW Megawatt
NAAOS National Ambient Air Quality Standards
NAHMMA North American Hazardous Materials Management Association
NAPCOR National Association of Plastic Container Recovery
NESHAP National Emission Standards for Hazardous Air Pollutants
Ni-Cd Nickel–Cadmium
NIMBY Not in my backyard
NIOSH National Institute for Occupational Safety and Health
NOX Nitrogen Oxides
NPDES National Pollutant Discharge Elimination System
NRC National Recycling Coalition
NSPS New Source Performance Standards
NSWMA National Solid Wastes Management Association
OCC Old corrugated cardboard
ONP Old newspaper

OSHA Occupational Safety and Health Act
P2 (PP) Pollution Prevention
PCB Polychlorinated biphenyl
PCDD Polychlorinated dibenzodioxin
PCDF Polychlorinated dibenzofuran
PCRB Pollution Control Review Board
PET Polyethylene terephthalate
PP Polypropylene
PPB Parts per billion
PPIS Pollution Prevention Incentives for States
PPM Parts per million
PRBA Portable Rechargable Battery Association
PRCNJ Plastic Recycling Corporation of New Jersey
PSA Public service announcement
PSD Prevention of significant deterioration
PUOCC Public Used Oil Collection Center
PVC Polyvinyl Chloride
QA/QC Quality Assurance/Quality Control
RAN Revenue anticipation note
RBAC Recycling Business Assistance Center
RCRA Resource Conservation and Recovery Act
RDF Refuse-derived fuel
RFQ Request for qualifications
RFP Request for proposals
RMPF Recovered Materials Processing Facility
RSM Recovered Screened Material
SARA Superfund Amendment and Recovery Act
SIC Standard Industrial Classification
SIP State implementation plan
SLA Sealed Lead–Acid
SOP Standard Operating Procedures
SQG Small-quantity generator
SWANA Solid Waste Association of North America
SWDA Solid Waste Disposal Act
SWIX Southern Waste Information eXchange
SWMA Solid Waste Management Act
TAN Tax anticipation note
TC Toxicity Characteristic
TCDD Tetrachloro-p-dibenzodioxin

TPD Tons per day

TSS Truckside sort

UBC Used beverage containers

UOF Used Oil Filter

US EPA United States Environmental Protection Agency

UWR Universal Waste Rule

VOC Volatile organic compound

WCS Waste characterization study

WPF Waste processing facility

WRF Waste recovery facility

WTE Waste-to-Energy

REFERENCES

1. The Solid Waste Association of North America (SWANA), Silver Spring, Maryland.
2. National Recycling Coalition Measurement Standards and Reporting Guidelines, Washington, D.C.
3. Westchester County, New York Solid Waste Management Plan.
4. Solid Waste Management in Florida, July 1999.

INDEX

About the Editor

Herbert F. Lund is the former recycling manager for the city of Hollywood, Florida, a professional engineer, and an independent recycling consultant. He is a major contributor to professional journals, the recipient of the Jesse H. Neal Editorial Achievement Award for best single article in a trade magazine, a fellow of the American Society of Mechanical Engineers, and editor in chief of *The McGraw-Hill Industrial Pollution Control Handbook* as well as the first edition of this book. He chairs the Recycling Seminar in cooperation with the Florida Bureau of Solid and Hazardous Waste, held annually in Key West, Florida. Mr. Lund resides in Coconut Creek, Florida.